Yohsuke Kamide
Abraham C.-L. Chian
(Eds.)

Handbook of the Solar-Terrestrial Environment

Yohsuke Kamide
Abraham C.-L. Chian
(Eds.)

Handbook of the Solar-Terrestrial Environment

With 255 figures, including 63 color figures

Springer

Phys
po## 615-836228

Yohsuke Kamide

Research Institute for Sustainable Humanosphere
Kyoto University
Uji 611-0011
Japan
e-mail: kamide@rish.kyoto-u.ac.jp

Abraham C.-L. Chian

National Institute for Space Research (INPE)
and World Institute for Space Environment Research (WISER)
P. O. Box 515
São José dos Campos-SP 12227-010
Brazil
e-mail: achian@dge.inpe.br

Library of Congress Control Number: 2007922927

ISBN 978-3-540-46314-6
Springer Berlin Heidelberg New York

Book DOI: 10.1007/b104478

Springer is a part of Springer Science+Business Media
springer.com

© Springer-Verlag Berlin Heidelberg 2007

Typesetting and Production:
LE-TeX Jelonek, Schmidt & Vöckler GbR, Leipzig, Germany
Cover: WMXDesign GmbH, Heidelberg, Germany

SPIN 11367758   55/3180/YL - 5 4 3 2 1 0   Printed on acid-free paper

Cover illustration: An example of a dramatic auroral display in the break-up region of an auroral substorm. What is not seen in the picture is the rapid variations in time and space that characterize auroral break-ups (courtesy of Torbjörn Lövgren, Swedish Institute of Space Physics).

DB
11/20/07

D.L.

Preface

As a star in the universe, the Sun is constantly releasing energy into space, as much as 3.9×10^{33} erg/s. This energy emission basically consists of three modes. The first mode of solar energy is the so-called blackbody radiation, commonly known as sunlight, and the second mode of solar electromagnetic emission, such as X rays and UV radiation, is mostly absorbed above the Earth's stratosphere. The third mode of solar energy emission is in the form of particles having a wide range of energies from less than 1 keV to more than 1 GeV. It is convenient to group these particles into lower-energy particles and higher-energy particles, which are referred to as the solar wind and solar cosmic rays, respectively.

Ever since the solar system was formed about 4.6 billion years ago, the Sun has continuously irradiated Earth, making life on Earth possible. Gradually, an environment conducive to human life emerged. The space in the universe that surrounds the Earth is called the solar-terrestrial environment, or the geospace. The study of the solar-terrestrial environment tries not only to unveil complex Sun-Earth relationships in terms of various physical processes that occur between the Sun to the Earth, but also to better understand our position and role in the universe. This area of research has recently become increasingly important as mankind begins to use the near-Earth space as part of our domain through space communications and the space station.

The study of the solar-terrestrial environment endeavors to understand quantitatively the conditions of the Earth's magnetosphere and its upper atmosphere, including the ionosphere and the thermosphere, influenced by the activity of the solar atmosphere and the solar wind that travels in the interplanetary space. Typical signatures of this chain of processes observable in the geospace are represented by geomagnetic storms and magnetospheric substorms, during which auroras in the polar sky become very active. These processes cover a wide range of time and spatial scales, making observations in the solar-terrestrial environment complicated and the understanding of processes difficult. In the early days, the phenomena in each plasma region were studied separately, but with the progress of research, we realized the importance of treating the whole chain of processes as an entity because of strong interactions between various regions within the solar-terrestrial system. On the basis of extensive satellite observations and computer simulations over the past two decades, it has become possible to analyze specifically the close coupling of different regions in the solar-terrestrial environment.

This handbook presents our current knowledge of the basic processes in the solar-terrestrial environment. The order of the twenty chapters in this handbook is such that the readers can first comprehend the energy flow process from the Sun to the Earth, followed by chapters discussing the fundamental physical principles or concepts that enable the readers to understand the essence of the processes, such as waves and instabilities in space plasmas, magnetic field reconnection, and nonlinear plasma processes, that are needed to interpret the dynamic phenomena occurring in the geospace system. Next, important signatures for the variability in the solar-terrestrial environment such as aurora, substorms, geomagnetic storms, and magnetic micropulsations are discussed. We also feel that it is quite timely to launch this book because the effects of space weather and space climate, that are applications of the solar-terrestrial research, have recently become an important societal concern. The final chapters of this book deal with the planets and comets, which must be undergoing plasma processes similar to those in the solar-terrestrial environment.

Our aim is for this handbook to serve as a reference for researchers working in the area of space science and geophysics. However, it is also written in a style accessible to graduate students majoring in those fields. In fact most of the chapters, after having been prepared for this book, were presented as tutorial lectures at the Advanced School on Space Environment: Solar Terrestrial Physics (ASSE 2006), organized jointly by the World Institute for Space Environment Research (WISER) and the International School of Space Science (ISSS), from 10 to 16 September 2006, in L'Aquila, Italy. Furthermore the authors and the editors also have the intention to provide undergraduate students in science and engineering with the opportunity to discover that there is a fascinating field of research in which they can interpret exciting phenomena in terms of the basic physical laws they learned in their classrooms. The references in each chapter are limited to books, review papers, and the most important seminal papers.

Each chapter has been reviewed by two referees. The editors wish to thank the following reviewers who have kindly participated in the evaluation of the chapters: J.H. Allen, T. Aso, F. Bagenal, W. Baumjohann, G. Brodin, M. Chen, C.R. Clauer, I. Daglis, Y. Feldstein, B. Fraser, T. Fuller-Rowell, K.-H. Glassmeier, T.I. Gombosi, N. Gopalswamy, J. Haigh, R. Harrison, A. Hood, M. Hoshino, W.-H. Ip, A. Klimas, L. Lanzerotti, A.T.Y. Lui, A.A. Mamun, D. Moss, T. Mukai, M. Neugebauer, M. Ossendrijver, G. Parks, A.D. Richmond, G. Rostoker, M. Ruohoniemi, M. Rycroft, M. Schulz, K. Shibata, J. Slavin, J.-P. St Maurice, R.A. Treumann, and D. Webb.

Kyoto, São José dos Campos
June 2007

Y. Kamide
A.C.-L. Chian

Contents

Part 2 The Earth

Part 3 Space Plasmas

1 An Overview of the Solar–Terrestrial Environment

Abraham C.-L. Chian and Yohsuke Kamide

An overview of the solar–terrestrial environment is presented. First, we review the early historical development of solar–terrestrial science, and introduce our current view of the Sun, solar wind, magnetosphere-ionosphere-thermosphere, geomagnetism and geomagnetic storms/substorms, aurora, planets and comets, and cosmic rays. The Sun-Earth relation is discussed. In addition to solar influence, the Earth's environment is impacted by cosmic rays of galactic origin and other cosmic sources such as sporadic gamma-ray bursts from magnetars. The solar–terrestrial environment is highly dynamic, dominated by a wealth of complex phenomena involving waves, instabilities, and turbulence. The study of the solar–terrestrial environment is essential to improve our ability to monitor and forecast space weather and space climate, and contributes significantly to the development of plasma astrophysics and controlled thermonuclear fusion.

Contents

Abraham C.-L. Chian and Yohsuke Kamide, An Overview of the Solar–Terrestrial Environment.
In: Y. Kamide/A. Chian, Handbook of the Solar-Terrestrial Environment. pp. 1–23 (2007)
DOI: 10.1007/11367758_1

1.1 Introduction

The term solar–terrestrial environment refers to the regions of the Earth's atmosphere, ionosphere and magnetosphere, which are influenced by the physical conditions in the solar interior, solar atmosphere, and solar wind, as well as in galactic cosmic rays. An artistic illustration of the solar–terrestrial environment is given in Fig. 1.1.

The study of the solar–terrestrial environment is both, old and at the same time new. It is old in the sense that the early studies of the influence of solar activity such as sunspots on the magnetic disturbances and auroras on the Earth's surface contributed to the discovery of the electromagnetic laws in the last centuries which we rely on today. It also covers the subject of the relationship between the Sun and Earth's atmosphere, and the possible effect of solar activity on the climate which was known for long time. On the other hand, the progress of the solar–terrestrial environment research has accelerated dramatically in recent years since the availability of satellite observations of space, in particular in the near-Earth environment. Interplanetary spacecraft have also observed the inner regions and outer regions of the solar system.

The aim of this chapter is to introduce the fundamental concepts of the solar–terrestrial environment, which will motivate the readers to deepen their understanding of this fascinating field of research through reading other chapters. The plan of this chapter is as follows. In Sect. 1.2, we present an overview of our current knowledge of the solar–terrestrial environment by separating the discussion under seven headings: Sun, solar wind, magnetosphere-ionosphere-thermosphere, geomagnetism and geomagnetic storms/substorms, aurora, planets and comets, and cosmic rays. Following the overview of each subject, we will give a short account of the early historical development of each subject. In Sect. 1.3, we will explore the complex nature of the solar–terrestrial environment, covering the basic concepts of linear waves, instabilities, nonlinear waves, nonlinear wave–wave and wave–particle interactions, and turbulence; only qualitative discussions will be given in this section. Applications of the solar–terrestrial environment research in space weather and space climate, plasma astrophysics, and controlled thermonuclear fusion are discussed in Sect. 1.4. Concluding remarks are given in Sect. 1.5.

1.2 Overview and History of Solar-Terrestrial Environment Research

In this section, we give an overview of the solar–terrestrial environment research and its early historical development. Each sub-section will begin with a short description of our present knowledge of each subject, followed by a brief account of the early history. Further information on the physics of solar–terrestrial environment, known as space physics, can be found in other chapters of this book as well as in the books by Nishida (1982), Priest (1984), Melrose (1991), Parks (1991), Hargreaves (1992), Kivelson and Russell (1995), Baumjohann and Treumann (1996), Gombosi (1998), and Prölss (2004). Details on the historical and recent developments of the solar–terrestrial environment research can be obtained from the review papers by Rishbeth (2001) and Stern (2002), the special volume of Journal of Geophysical Research edited by Gombosi, Hultqvist, and Kamide (1994), and the books edited by Chian and the WISER Team (2003), and Jatenco-Pereira et al. (2005). References will not be cited in the remainder of this section since they are readily available in the aforementioned publications.

1.2.1 Sun

Overview

The Sun is an ordinary star of spectral type G2V with magnitude of 4.8. The Sun's age is about 4.5 billion years; the solar mass is 330 thousands times that of the Earth; and the solar radius is 109 times that of the Earth. The distance between the Sun and the Earth is 1.4959787×10^{11} m, which is defined as one astronomical unit (AU); sunlight takes about 8 minutes to reach the earth. The Sun is a giant ball of high-temperature ionized gas (plasma) held together by its own gravitational force. It consists mainly of hydrogen (90%) and helium (10%).

The solar interior is composed of a rigidly rotating *core* at a temperature of 15 million K, producing energy through nuclear fusion of hydrogen into helium. A by-product of this thermonuclear reaction are solar neutrinos. The core is surrounded by the *radiative zone* and the *convection zone*. The convection zone is rotating differentially with a period of 26 days near the equator and 37 days near the poles. The Sun is oscillating globally;

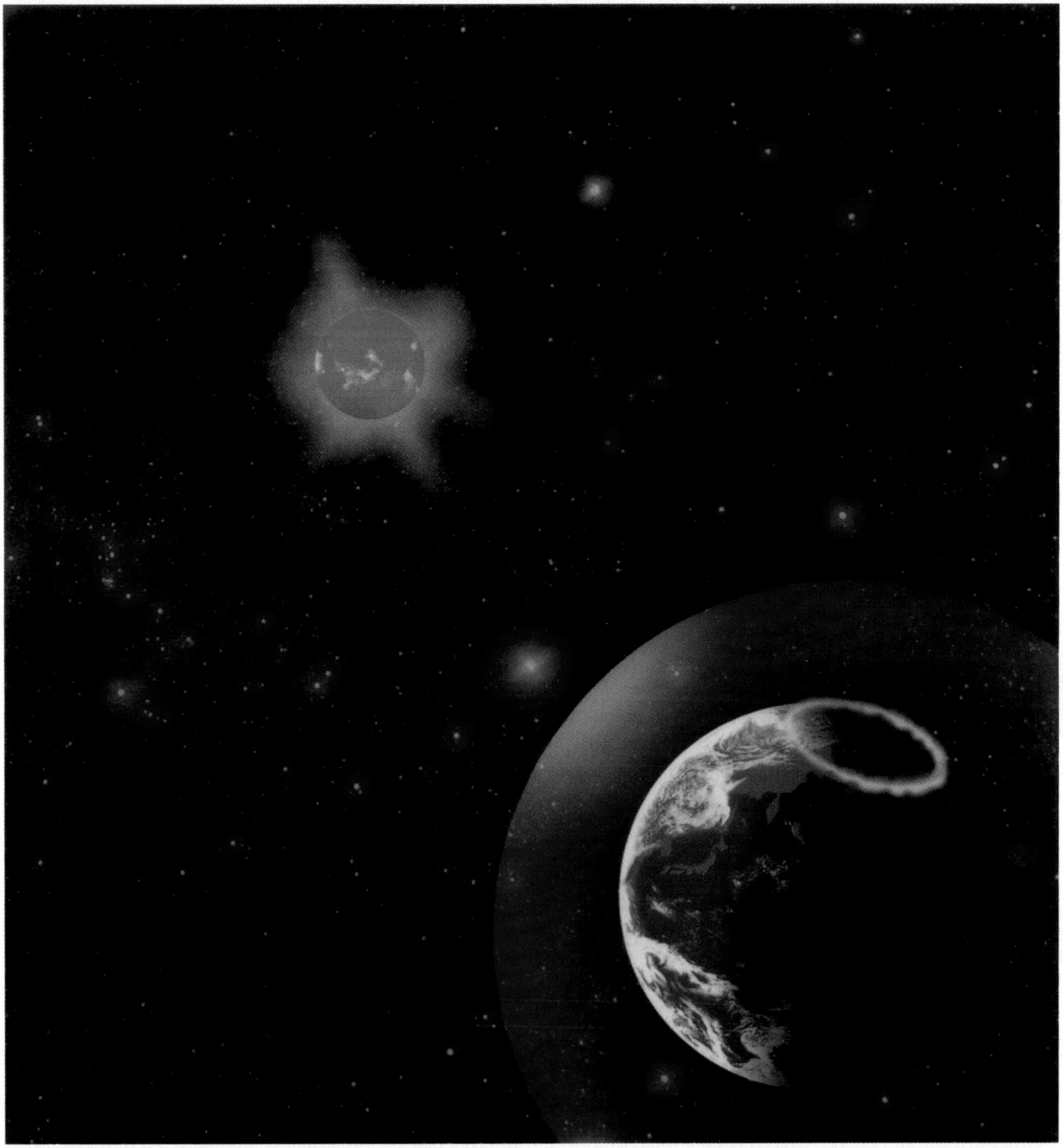

Fig. 1.1. An artistic illustration of the solar–terrestrial environment. [Adapted from the Solar-Terrestrial Environment Laboratory]

these oscillations are sound waves which can be used to probe the solar interior based on the *helioseismology* network around the world. Solar activity is controlled by the magnetic fields generated by the combined action of convection and differential rotation of a nonlinear dynamo in the solar interior. The variation of solar activity produces the so-called *solar cycles*.

The solar atmosphere consists of three layers. The lowest layer is the *photosphere* that represents the top of the convection zone and is covered with a granular pattern outlining the turbulent convection cells; most of the sunlight is emitted from the photosphere. A white-light picture of the Sun occasionally displays dark spots known as *sunspots*, which are surrounded by brighter areas known as *active regions*. The magnetic field of the photosphere is on the order of 1 gauss or less outside and 3000 to 4000 gauss inside sunspots. xSolar active regions may sometimes brighten abruptly, giving rise to *solar flares*. In addition, thin dark structures known as *filaments* or *prominences* are seen in the active regions. The *chromosphere* and *corona* lie above the photosphere. The minimum temperature in the photosphere is 4200 K, which increases gradually, reaching 2 million K in the corona.

The solar atmosphere is highly structured and dynamic. Soft X-ray images of the Sun show *coronal loops* and *bright points*, where magnetic fields and plasmas are interacting; disturbances in geomagnetic activities are caused by these interactions. Figure 1.2 shows a soft X-ray image of the Sun taken by the Yohkoh satellite. Dark regions known as *coronal holes* are seen in the image of Fig. 1.2, where the magnetic field is open and through which the solar wind is streaming outward. Huge erupting bubbles known as *coronal mass ejections* (CMEs) are observed ahead of erupting prominences.

History

In the third century BC, Aristarchus of Samos proposed a heliocentric model of the solar system, which considers the Earth as one of several planets that revolve around the Sun, and the earth rotates about its own axis. However, this heliocentric model of the solar system was neglected for a long time. Up to the sixteenth century AD, the geocentric model of the universe formulated in the book *Almagest* by king C. Ptolemy of Alexandria in 140 AD was universally accepted, with the Sun and all the celestial bodies revolving around the Earth. In 1543, N. Copernicus published the book *De Revolutionibus Orbium Coelestium* (Latin for "On the Revolutions of the Heavenly Spheres"), which revived the heliocentric model of Aristarchus and gave the correct order for the known planets in the solar system. This heliocentric model was confirmed by the observations and calculations of planetary motions around the Sun by T. Brahe, J. Kepler and Galileo Galilei in the sixteenth and seventeenth centuries.

In 1609, Galileo Galilei was one of the first to use the newly invented telescope to observe sunspots. In 1687, I. Newton published his *Principia* on the laws of motion that govern the dynamics of the objects under the influence of a force, including the motion of a planet around the Sun. In 1843, the 11-year sunspot cycle was found by H. Schwabe. In 1848, R. Wolf collected earlier observations to trace sunspot cycles before the published data of H. Schwabe and introduced the "Zurich sunspot number", an empirical criterion for the number of sunspots, taking into account the fact that they usually appear in tight groups; he discovered that the length of the sunspot cycle varies with an average value close to 11 years. In 1852, E. Sabine observed an association between the sunspot cycle and the occurrence of large geomagnetic disturbances.

R. Carrington studied the rotation of sunspots around the Sun and found that their period and other properties vary with solar latitude; on 1 September 1859 he discovered a white light solar flare which was followed a day later by an intense geomagnetic storm, and suggested a causal link. In 1863, R. Carrington discovered the differential solar rotation. G. Hale (1868–1938) used the spectroheliograph which he developed to establish that solar flares are associated with the brightening in the H-alpha light and big flares often precede geomagnetic storms. He showed that sunspots are strongly magnetized with a typical field of 1500 gauss. Sunspots usually are seen in pairs of opposite polarity, suggesting that the magnetic field lines emerge from the Sun in one of the spots belonging to a pair and reenter at the other conjugate spot. The sunspot cycle is a magnetic phenomenon with an average period of 22 cycles involving a reversal of polarity in each 11-year solar cycle. In 1893, E. Maunder

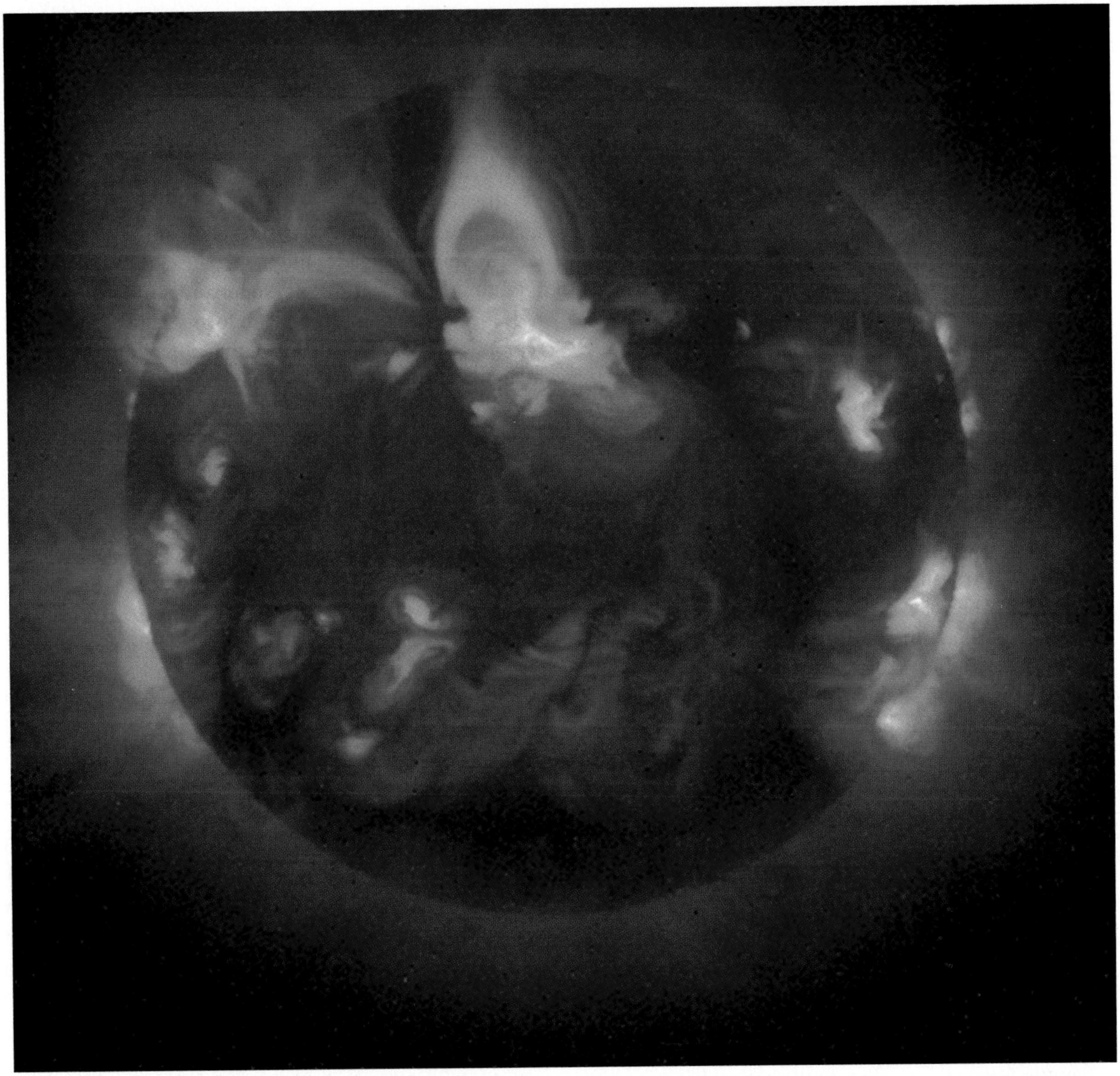

Fig. 1.2. X-ray image of the Sun taken by the Yohkho satellite. [Courtesy of JAXA/ISAS]

identified a period of nearly absence of sunspots in 1645–1715, now known as the *Maunder Minimum*. In 1949, H. Friedman and his colleagues recorded in a rocket experiment X-rays emitted from solar active regions. H. Babcock in 1960 used a solar magnetograph he built to observe the general dipole field of the Sun and determined its strength to be of the order of 5 gauss.

1.2.2 Solar Wind

Overview

The solar wind results from the expansion of the solar atmosphere, forming a *supersonic flow* of ionized plasma and magnetic field that permeates the interplanetary medium. This stream is the consequence of the pressure difference between the solar corona and interplanetary space which drives the solar plasma radially outward, escaping from the influence of solar gravity. The solar wind consists mainly of protons and electrons, with a small admixture of ionized helium and heavy ions. A weak magnetic field of a few nano Teslas near the earth is embedded in the solar wind plasma, oriented in a direction nearly parallel to the *ecliptic plane* which is defined by the plane of the Earth's orbit around the Sun. At 1 AU the solar wind magnetic field makes an angle of roughly 45 degrees to a line from the Sun to the observer. Within less than 2 solar radii away from the photosphere the complex structure of the solar magnetic field reduces to this simple, radially directed structure. Due to the high conductivity of the solar wind, the solar wind magnetic field is frozen into the solar wind plasma and is convectively transported outward into the interplanetary medium.

The solar wind is strongly affected by changes in solar activity. It transmits the effects of solar variability to the planets. The solar rotation winds up the solar wind magnetic field lines into an Archimedean spiral. Hence, with increasing radial distance from the Sun the initially radial magnetic field turns gradually into a toroidal direction. Due to the solar rotation with a period of 27 days, the interplanetary magnetic pattern shows a recurrent behavior of 27 days. In addition to this recurrent variation in the properties of the solar wind, there are sporadic interplanetary disturbances caused by solar activities such as shock waves, coronal mass ejections, and flares.

Two types of solar wind are observed. The fast solar wind originates from coronal holes where the magnetic field is open. It has a velocity between 400 km/s and 800 km/s. During solar minimum the slow solar wind, with a velocity between 250 km/s and 400 km/s, originates from regions close to the heliospheric current sheet at the heliomagnetic equator, during solar maximum it originates above the coronal Helmet streamers in active regions where the magnetic field lines are closed. At a certain distance from the Sun the slow and fast winds collide and an interaction region develops between fast and slow streams. Since these structures rotate with the Sun, they are known as *corotating interaction regions* or CIRs.

In the outer region of the solar system where the solar wind is slowed down by the supersonic interstellar wind, a shock is formed. This shock is known as the *termination shock* of the solar wind. Between the termination shock and the outer boundary of the heliosphere, the *heliopause*, lies the *heliosheath*. Here the plasma is subsonic. The heliosphere is the cavity in the local interstellar medium which is structured by the solar wind and the frozen-in solar magnetic field. The heliopause separates it from the interstellar medium. Outside the heliopause the interstellar medium is braked in its supersonic motion by the presence of the heliosphere and forms a bow shock. The region between this heliospheric bow shock and the heliopause contains dense interstellar plasma called the *ion wall*.

History

In 1897, the electron was discovered by J. Thomson in his experiment with beams of negatively charged particles using an evacuated glass bulb which had been patented by T. Edison in 1883. J. Thomson also studied positive ions contained in the conducting gas of his experiment. The discovery of the electron opened the way to the physical explanation of ionization and conductivity in an ionized gas. K. Birkeland in 1916 and F. Lindemann in 1919 predicted that the solar wind should be composed of both, negative electrons and positive ions. In 1923, I. Langmuir coined the term *plasma* to describe the ensemble of electrons and ions in a gas discharge.

A streaming plasma of electrons and ions was introduced by S. Chapman and V. Ferraro in 1930 in their model of geomagnetic storms. In 1943, C. Hoffmeister observed that a comet tail is not exactly radial but lags behind the comet's radial direction by about 5 degrees, and revived the idea of solar corpuscular radiation. In 1951, L. Biermann suggested that the observation that comet tails always are pointing away from the Sun is due to the existence of corpuscular radiation continuously blowing away from the Sun, which predicted the continuous presence of the solar wind. In 1957, H. Alfvén developed a model for the formation of a cometary tail, proposing that the solar wind magnetic field is draped over the comet due to the solar wind motion. E. Parker in 1958 showed that the solar atmosphere cannot be in static equilibrium but is continually expanding. This effect should give rise to a solar wind. In 1959, the Luna 1 spacecraft measured the solar wind for the first time.

In 1961, H. Bridge and B. Rossi detected the solar wind with the instrument aboard the Explorer 10 spacecraft.

In 1960, P. Coleman, L. Davis, Jr. and C. Sonnett reported the first in situ measurement of the interplanetary magnetic field using the magnetometer of the Pioneer 5 spacecraft. In 1962, the Mariner 2 spacecraft provided a detailed study of the solar wind plasma, including the observation of an interplanetary shock by C. Sonnett on 7 October 1962 which was associated with the sudden commencement of a geomagnetic storm. In 1962, I. Axford and P. Kellogg independently predicted the existence of a planetary bow shock in front of the Earth. This prediction was confirmed by the IMP-1 spacecraft one year later. In 1961, the Explorer 10 spacecraft detected the crossing of the magnetopause for the first time. In 1964, the OGO 1 spacecraft obtained an accurate detection of the bow shock. In 1968, J. Bell and A. Hewish discovered pulsars while observing interplanetary scintillations using radio techniques; this technique is still in use for monitoring the propagation of solar disturbances in the solar wind. In 1974, J. Gosling developed the concept of Coronal Mass Ejections (CMEs) based on in situ and coronograph images of solar mass ejecta and related them to the geomagnetic activity.

In 1955, L. Davis, Jr. suggested the existence of the heliosphere, the region of space surrounding the Sun formed by the interaction of the local interstellar medium with the solar wind. The Pioneer 10 and 11, and Voyager 1 and 2 deep-space spacecraft launched in the 1970s, after exploring the outer planets are heading towards the heliopause and the interstellar medium. In 1993, D. Gurnett reported the first evidence of the heliopause based on the 2 – 3 kHz radio emissions coming from the heliopause and detected by the Voyager 1 and 2 spacecraft.

1.2.3 Magnetosphere–Ionosphere–Thermosphere

Overview

When the solar wind impinges on the Earth's dipolar magnetic field, it is slowed down and deflected around it, forming a cavity called the *magnetosphere*. Since the solar wind moves at supersonic speed, a *bow shock* is formed in front of the magnetosphere. There the solar wind plasma is decelerated, and a substantial fraction of its kinetic energy is converted into thermal energy. The region of thermalized subsonic plasma flow behind the bow shock is known as the *magnetosheath*. The boundary between the magnetosheath and the magnetosphere is the *magnetopause*. The solar wind kinetic pressure modifies the outer configuration of the Earth's dipolar magnetic field; in the dayside magnetosphere the field is compressed, whereas in the nightside magnetosphere the field is stretched out into a *magnetotail*.

The magnetosphere is filled with magnetospheric plasma composed mostly of electrons and protons. The sources of these particles are both the solar wind and the ionosphere. Most of the magnetotail plasma is concentrated in the *plasma sheet* around the tail midplane. Near the Earth, the plasma sheet reaches the high-latitude ionosphere. The outer region of the magnetotail is called the magnetotail *lobe*, which contains a highly rarefied plasma. The inner magnetosphere contains two populations of confined plasma: a low-energy or cold relatively dense plasma component on the closed nearly dipolar geomagnetic field lines at low- and mid-latitudes filling the *plasmasphere*, and a rather energetic but very dilute plasma component that occupies the same field lines and has become famous under the name of the Van Allen *radiation belts*. Inner and outer radiation belts consist of trapped energetic electrons and protons on nearly stable orbits, bouncing along the magnetic field lines between the northern and southern hemispheres and at the same time drifting azimuthally and encircling the Earth.

Solar radiation in the ultraviolet (UV) and extreme ultraviolet (EUV) wavelengths is responsible for the photoionization of the Earth's upper atmosphere, known as the *ionosphere*, which acts as a transition region from the fully ionized magnetospheric plasma to the neutral atmosphere. Energetic particles from the galaxy, Sun, magnetosphere, and ionosphere are also sources of the ionosphere ionization via impact ionization. The ionospheric plasma is partially ionized, consisting of a mixture of charged and neutral particles. This implies that both Coulomb collisions and neutral collisions contribute to its electrical conductivity. The ionosphere forms at altitudes above 80 km and extends to higher altitudes, merging into the plasmasphere. The dense and cold plasmaspheric plasma is of ionospheric origin and coexists spatially with the radiation belts, extending to about 3 to 5 Earth radii. Being frozen into the closed geomagnetic field lines the plasmasphere corotates with the Earth. It has the shape of a torus; the outer boundary of the plasmasphere is called the *plasmapause*.

The *lower* ionosphere below an altitude of about 90 km is called the D-region, which is very weakly ionized and highly collisional. The *upper* ionosphere consists of the E-region which has its ionization peak at about 110 km, and the F-region. The F-region itself consists of two layers, the F1-layer at an altitude around 200 km and the F2-layer at an altitude around 300 km.

The regions of the atmosphere are primarily classified by their temperature. The temperature decreases with increasing altitude in the lower region of atmosphere known as the *troposphere*, which is bounded by the *tropopause* at a height of 10 to 12 km. The *stratosphere* above that altitude is a region of increasing temperature. A maximum due to ozone absorption appears at around 50 km altitude where the *stratopause* is situated. The temperature decreases again in the *mesosphere* (or middle atmosphere) reaching a minimum at the *mesopause* at 80 to 85 km altitude, above which lies the thermosphere, where the temperature increases again dramatically with altitude. The thermosphere is the upper region of the atmosphere extending up to approximately 500 km. It experiences dynamical changes being controlled by the variable output from the Sun.

History

The famous German mathematician and physicist C. Gauss (1777–1855) in Göttingen, when analyzing the magnetic field on the Earth' surface applying his newly developed potential theory, was the first to postulate that in addition to its sources inside the body of the Earth the geomagnetic field should have external sources. He left unidentified but ingeniously attributed to currents flowing somewhere at a few hundred km above the Earth's surface. In 1882, B. Stewart proposed that the electrically conducting region of the upper atmosphere is the most likely location of the electric currents that are responsible for the solar-modulated variations in the Earth's surface magnetic field. S. Arrhénius suggested in 1888 that the conducting layer in the upper atmosphere could be produced by solar radiation. In 1901, G. Marconi carried out the first experiment on transatlantic radio transmissions between England and Canada, which introduced the radio technique to study the solar–terrestrial environment. In 1902, A. Kennely and O. Heaviside independently postulated the existence of a conducting layer in the upper atmosphere, known initially as the Kennely–Heaviside layer, later coined *ionosphere* by R. Watson-Watt and E. Appleton in 1926, to explain Marconi's radio experiment. The existence and altitude of the ionosphere were verified in 1924 by E. Appleton using radio waves. G. Breit and M. Tuve confirmed the experiment of E. Appleton in 1925. The radio sounding method of the ionosphere was based on the ionosonde which is still in use today. It was developed by G. Breit and M. Tuve, which sends short pulses of radio signals at vertical incidence and measures the timing of the reflected signal in order to infer the altitude and the local plasma density of the reflecting layer. In 1927, E. Appleton discovered two distinct layers in the ionosphere, the E and F regions. In the same year, P. Pedersen developed a theory of radio wave propagation in the ionosphere and derived the vertical structure of the ionosphere. One year later, in 1928 L. Austin showed that the strength of the wireless transatlantic radio signal exhibits a modulation with solar cycle. Subsequently, in 1931 S. Chapman developed a theory of upper atmospheric ionization that explains the structuring of the ionosphere into layers. Later in 1948 F. Hoyle and D. Bates proposed that the ionospheric E-layer is produced by solar X-rays. J. Van Allen and his team in the late 1950s developed rocket experiments in the ionosphere which detected energetic electrons in the polar regions and observed the radiation associated with the precipitation of these electrons.

S. Chapman and V. Ferraro proposed in 1930/1931 that a corpuscular stream emitted occasionally by the Sun would compress the geomagnetic field and cause a transient increase of the magnetic field measured on the Earth's surface. In this way they explained the sudden commencement observed at the start of a geomagnetic storm. In this context Chapman and Ferraro were also the first to propose the formation of a temporary diamagnetic ring current flowing during the interaction of the corpuscular stream around the earth in the equatorial region in order to account for the magnetic field decrease in the storm main phase following the sudden commencement. The formation of a geomagnetic cavity, now known as the *magnetosphere*, a term coined by T. Gold in 1959, carved out of the solar wind plasma by the geomagnetic field as the solar wind approaches the Earth was first calculated based on the balance between solar wind pressure and magnetic energy density of the geomagnetic field by V. Zhigulev and E. Romishevskii in 1959. The history of these events can be found in Hultqvist (2001). The International Geophysical Year (IGY) was organized in 1957 to promote geophysical studies worldwide. These studies led to a rapid growth of the research in the ionosphere and magnetosphere. The first artificial satellite Sputnik I was launched on 4 October 1957 by the Soviet Union. The United States followed in early 1958, launching the Explorer 1 satellite, which immediately led to the discovery of a belt of intense trapped energetic protons above the magnetic equator of the magnetosphere by J. Van Allen. The outer radiation belt was observed in 1958 by Pioneer 3. In 1959, K. Gringauz used ion traps aboard the LUNIK 2 spacecraft to discover the plasmasphere. The magnetopause, the boundary between magnetosphere and solar wind, was observed by Explorer 12 in 1962. In 1963, N. Ness discovered the magnetotail, and the IMP 1 satellite made the first mapping of the magnetotail. In 1966, C. Kennel and H. Petscheck formulated the stable trapping theory of charged particles in the Earth's magnetosphere which provided the key step for understanding the presence and stability of the Van Allen radiation belts. In 1968, the plasma sheet was detected by the OGO 1 and OGO 3 satellites. In 1970, the Vela satellites observed the thinning/thickening of the plasma sheet associated with substorms. In 1983, ISEE-3 explored the distant magnetotail, before heading for comet Giacobini-Zinner.

In 1886, a natural radio emission in the audio-frequencies, now known as *whistlers*, was detected in a telephone line in Austria. In 1919, H. Barkhausen reported the observation of these radio signals during World War I and suggested that they were due to meteorological influences. T. Eckersley confirmed this phenomenon in 1925 and ascribed it to the dispersion of an electrical impulse in a medium loaded with free ions. In 1935, he concluded that *whistlers* are the result of the dispersion of a burst of electromagnetic noise propagating through the ionosphere. In 1953, L. Storey, analyzing the radio dispersion induced by lightning, discovered the existence of plasma above the ionosphere and found that *whistlers* travel back and forth along the geomagnetic field lines of the dipole mirror field. He also discovered other types of VLF radio emissions not related to lightning which are now known to be generated in the magnetosphere. H. Alfvén, in 1942, developed the theory of low-frequency magnetic waves in an electrically charged fluid subject to the Lorentz force. These are the famous *Alfvén waves* named after him. This theory was the discovery of *magnetohydrodynamics* which describes the behavior of magnetized plasmas as conducting fluids. The existence of *Alfvén waves* in space plasmas was predicted by H. Alfvén. In 1959, V. Troitskaya proposed ULF wave pulsations as a diagnostic tool of magnetospheric processes, and in 1974 D. Gurnett discovered that the auroral plasma is an intense source of radio emissions known as *auroral kilometric radiation* (AKR).

1.2.4 Geomagnetism and Geomagnetic Storms/Substorms

Overview

Variations in the Earth's magnetic field are generated by a number of different sources. At the Earth's surface, such variations are produced in part by currents in the Earth's interior, i.e., currents flowing in the liquid transition region between the solid Earth's core and the Earth's mantle by a process known as the Earth's *dynamo*. The dynamics of ionosphere and magnetosphere is dominated by various current systems. The distortion of the Earth's dipole magnetic field by the solar wind is accompanied by electric currents. The compression of the dayside magnetic field is associated with the magnetopause current on the magnetopause surface. The tail-like field of the nightside magnetosphere is accompanied by the tail current flowing on the tail surface and the neutral sheet current in the central plasma sheet, separating the northern and southern lobes. The ring current flows around the Earth in the westward direction at radial distances of several (5 – 8) Earth radii and is carried primarily by 20 – 200 keV protons.

A number of current systems flow in the ionosphere at altitudes of about 100 to 150 km, including the equatorial electroject near the magnetic equator, the Sq currents in the dayside mid-latitude ionosphere, and the auroral electrojects in the auroral oval. In addition to these "perpendicular" currents, field-aligned currents flow parallel to the earth's magnetic field lines, which connect the magnetospheric currents to the polar ionospheric currents. Most of the field-aligned currents are carried by electrons. Part of the ring current is also involved in the magnetosphere-ionosphere coupling. It flows near dusk in the equatorial magnetosphere, is connected to the ionosphere via the field-aligned currents and closed in the ionosphere via the ionospheric current system. Similarly, part of the tail current is diverted into the ionosphere by field-aligned currents, giving rise to the substorm wedge current system.

All of the currents above the ionosphere are controlled by the dynamic pressure of the solar wind (which depends on the velocity and density of the solar wind) and the dawn-dusk component of the interplanetary electric field (which depends on magnitude of the southward component of the interplanetary magnetic field). The variations in any of these parameters are responsible for the corresponding variations in the magnetospheric-ionospheric current system. These changes are the origin of the variation of the Earth's magnetic field, known as *geomagnetic activity*. Geomagnetic indices such as *Dst, Kp, aa, ap, AL, AU, AE* and *PC* have been developed to monitor the geomagnetic activity on a global scale, which is related to solar activity.

The basic type of geomagnetic activity is the polar magnetic substorm. One of the proposed time sequences of the events in a substorm is as follows. During the substorm growth phase the reconnection processes at the dayside magnetopause provide free energy which is stored in the magnetotail, which evolves towards an unstable state. During the growth phase, the plasma sheet shrinks in its width and thins out. The onset of substorm expansion is associated with the formation of a neutral line in the near-Earth plasma sheet. In between the new and old neutral lines plasmoids are formed which flow out along the magnetotail under the action of the field line tension caused by the reconnection process. As the substorm progresses the plasma sheet expands and the neutral line displaces towards the distant magnetotail in order to recover its equilibrium position. An indication of substorms is the appearance of discrete aurorae. Geomagnetic disturbances accompany the auroral activity, which is evidenced by an increase of current perturbations in the ionosphere. Typical auroral disturbances produce perturbations of the auroral electrojects with amplitudes in the range of 200 to 2000 nT and with durations of 1 to 3 hours.

When the coupling of the solar wind to the magnetosphere becomes strong and extends over long times, it leads to a worldwide geomagnetic storm. The development of such a geomagnetic storm can be defined by the development of a ring current flowing in the inner magnetosphere, which is monitored by the *Dst* index. The *Dst* index is derived from the instantaneous longitudinal average of the mid-latitude magnetic disturbances. A great magnetic storm usually starts with a sudden increase in the geomagnetic field at the earth's surface, the sudden commencement. The enhancement of the magnetic field may last for several hours. This initial phase is followed by a decrease in *Dst*, which is known as the storm main phase. It is followed by a rapid first recovery phase followed by a long, slow second recovery phase. A typical geomagnetic storm lasts between 1 to 5 days. It is empirically known that the main phase of geomagnetic storms is characterized by the frequent occurrence of intense substorms.

History

Ancient Chinese scientists had studied and learned about magnetism in nature and had even invented the compass. The first navigational compasses were widely used on Chinese ships by the eleventh century AD, and during the naval expeditions of Cheng Ho to India, Sri Lanka, the Arab Peninsula, and Egypt in the fourteenth and fifteenth centuries AD. The discovery of the compass spread from China to Europe, and the compass was used by European navigators in their great sea voyages in the fifteenth century AD. In 1600 W. Gilbert published his famous book *De Magnete* (Latin for "On the Magnet") in which he proposed that the Earth would be a giant magnet, thus explaining the unique directivity of the compass needle. E. Halley published two books on geomagnetism in 1683 and 1692, and two geomagnetic charts in 1701 and 1702, based on his two scientific expeditions to the North and South Atlantic Oceans.

G. Graham used the sensitive compass which he had developed to observe in 1722 that the geomagnetic field undergoes delicate and rapid fluctuations. His observations were first confirmed in 1740 by A. Celcius and verified further by O. Hiorter who discovered the diurnal *Sq* variation of the geomagnetic field. On 5 April 1741, O. Hiorter discovered that the geomagnetic and auroral activities were correlated; simultaneous observations by G. Graham confirmed the occurrence of strong geomagnetic activity on that day.

The development of geomagnetism was supported by the experimental and theoretical works on magnetism and electromagnetism by H. Oersted (1777–1851) who discovered the link between electricity and magnetism by noting that an electric current produces a magnetic field and by the investigations of A. Ampère (1777–1836) who provided the physical explanation for Oersted's discovery by developing Ampère's law which relates the magnetic field to currents. Later M. Faraday (1791–1867) demonstrated that an electric current induces a magnetic field, which is known as Faraday's law, and introduced the concept of the lines of force (i.e., field lines) in order to describe the pattern and strength of the magnetic force. J. Maxwell (1831–1879) formulated Maxwell's equations that govern the electromagnetic field. He concluded from the equations that an electromagnetic field can propagate as a wave at the velocity of light through free space, a fundamental conclusion which was brilliantly confirmed experimentally by H. Hertz in 1886. H. Lorentz (1853–1928) formulated the Lorentz force equation which describes the effect of an electromagnetic field on the motion of charged particles. He constructed a theory that the atoms consist of charged particles and that the oscillations of these charged particles are the source of light.

The term magnetic storm, meaning a worldwide disturbance of geomagnetic field, was coined by A. von Humboldt (1769–1859) based on his geomagnetic data collected during his expedition to Latin America; he also proposed the first worldwide network of magnetometers to study geomagnetism. C. Gauss (1777–1855) developed the mathematical description of the geomagnetic field and set up a network of widely spaced magnetometers to make simultaneous measurements of the geomagnetic field. His mathematical analysis of these data enabled the separation of the contributions originated by the upper atmosphere from those originated from the Earth's interior. A global network of four geomagnetic observatories was established by British scientists by 1839. The data from this network was used by E. Sabine to show in 1852 that geomagnetic variations are a worldwide phenomenon and the variation of the intensity of geomagnetic disturbances is correlated with the sunspot cycle. In 1852, R. Wolf independently proposed the relation between solar cycles and geomagnetic activities shortly after Sabine. The solar flare first observed by R. Carrington on 1 September 1859 was followed a day later by one of the strongest geomagnetic storms ever recorded, with auroras seen as far south as Panama. In 1892, E. Maunder noted that large sunspot groups were associated with large geomagnetic storms, and in 1904 he found 27-day recurrence of geomagnetic activity related to the rotation rate of the Sun.

The year 1900 was a landmark for solar–terrestrial physics thanks to the theoretical ideas put forward by O. Lodge and G. Fitzgerald that geomagnetic storms are caused by the passage near the earth of the magnetic disturbances emanated from the sunspot regions. T. Gold, following the original idea of Chapman and Ferraro, suggested in 1955 that interplanetary shocks are the reason of magnetospheric compressions in the initial phase of geomagnetic storms, and S. Singer in 1957 attributed the main phase of a geomagnetic storm to the enhanced circular motion of trapped energetic particles around the Earth forming a diamagnetic ion ring current. Increases in geomagnetic activity associated with increases of the southward component of the interplanetary magnetic field were observed.

A. Dessler and E. Parker in 1958 developed a hydromagnetic theory of magnetic storms to explain the magnetic depression observed on the Earth as being due to the ring current. In 1961, I. Axford, in cooperation with C. Hines, and J. Dungey developed complementary models of the magnetosphere. Axford and Hines' model of a closed magnetosphere assumed a viscous interaction at the magnetopause to drive magnetospheric convection, while Dungey's model predicted an open magnetosphere which incorporated dayside and nightside reconnection of the magnetic field in order to account for the electromagnetic solar-wind magnetosphere coupling.

1.2.5 Aurora

Overview

The aurora is a phenomenon that is typically observed at high latitudes. It is a visible manifestation of the solar–terrestrial connection. The aurora borealis appears in the high northern latitudes, and the aurora australis appears in the high southern latitudes. Energetic electrons accelerated by field-aligned electric fields in the polar ionosphere precipitate into the upper atmosphere. The aurora is a source of X-ray, UV, infrared, radio, and optical radiations. The optical aurora is generated when these electrons hit the neutral atoms or molecules in the atmosphere, emitting electromagnetic line radiations in the range from ultraviolet (UV) to infrared (IR) due to excitation and ionization of neutrals. The green color commonly seen in the aurora is related to the atomic oxygen line at 557.7 nm which occurs typically at altitudes from 100 to 200 km. At higher altitudes, the auroral red line of atomic oxygen at 630.0 nm is emitted. Violet or blue auroral line emissions of molecular nitrogen are radiated at 391.4, 427.0 and 470.0 nm. Red aurora is produced also by energetic protons.

The aurora is usually visible at latitudes of about 65 – 70 degrees inside a region called the auroral oval. It is organized into distinct structures or auroral forms, and often appears as thin band-like structures known as *auroral arcs* which are aligned with the east–west direction. The precipitating auroral electrons can also emit bremsstrahlung X-ray radiation. Energetic electrons in the auroral region of the ionosphere-magnetosphere generate various types of non-thermal radio emissions such as the *auroral kilometric radiation* (AKR). Some of these auroral radio emissions are detected on the ground.

The aurora is intensified in the substorm process. At the substorm expansion onset, quiet auroral arcs abruptly explode into bright arcs, becoming highly dynamical and structured. Both are accompanied by intensifications of auroral UV, X-ray and radio-wave emissions. In 1964 S. Akasofu found that the evolution of large-scale auroras follows a fixed pattern. He also attributed the auroral morphological changes to geomagnetic disturbances and introduced the concept of polar magnetic substorms. Substorms in the magnetosphere were detected in 1968 by the OGO 5 satellite. In the late 1970s, the ISSE 1 and 2 satellites obtained clear evidence of the accelerated plasma flows at the magnetopause and in the magnetotail associated with magnetic field reconnection. A large-scale aurora usually begins with quiet arcs of fairly low intensity, elongated roughly in the geomagnetic east-west direction. After some time the aurora starts to move equatorward, increases in intensity and may develop ray structures that take the form of less regular bands. Afterwards, the sky explodes all of a sudden and the aurora spreads over the entire sky. At the same time, the aurora expands rapidly, primarily poleward, changing in form and intensity. This unique behavior is called an *auroral break-up*, signalling the expansion onset of a magnetospheric substorm. With progressing time, the aurora becomes weaker and diffuse. It is the beginning of the recovery phase.

Auroral activity is a global phenomenon, which is closely related to solar wind changes and geomagnetic activity. The energy which has been dynamically transferred into the magnetosphere from the solar wind is dissipated. The global pattern of the auroral current system, i.e., the behavior of the *auroral electrojet*, reflects the auroral response to changes of the interplanetary magnetic field.

History

In 1716, E. Halley noted that the orientation of the auroral curtains are aligned with the projections of the Earth's magnetic field into the upper atmosphere. In 1770, J. Wilcke reported that auroral rays extend upward along the direction of the geomagnetic field. In 1790, H. Cavendish used triangulation to estimate the height of auroras. In 1859, E. Loomis mapped the occurrence and location of aurora and plotted it to obtain an oval-shaped belt around the geomagnetic pole, known as the *auroral oval*, which is displaced by about 20 to 25 degrees from the pole. In 1878, H. Becquerel proposed that particles such as protons are ejected from the Sun and guided by the Earth's magnetic field to auroral latitudes. Using the extensive data on the geomagnetic disturbances related to auroras from his 1902–3 expedition to northern Norway, K. Birkeland suggested the existence of field-aligned electric currents associated with auroras. Inspired by his terrella laboratory experiment showing that electrons incident on a magnetic dipole inside a model Earth generates radiation patterns similar to the auroras, Birkeland suggested that the electrons that excite auroras must have come from the Sun. Birkeland's work led C. Störmer in 1907 to develop a theory for the motion of charged particles in the Earth's dipole magnetic field. Further, on the basis of triangulation technique, he used a camera to accurately determine the height of the aurora. Störmer's calculations showed that the charged-particle orbit spirals around the magnetic field and bounces back and forth along it. This motion was confirmed later by the discovery of the radiation belts. Indeed, most auroras are due to electrons, even though these electrons are not of solar but magnetospheric origin. Some auroras are related to protons; the first observation of the proton aurora was reported by L. Vegard in 1939.

In the early 1970s, C. Anger obtained the first global space-based auroral image by a scanning photometer onboard the ISIS-2 spacecraft. Subsequently, auroral imagery from a series of the MDSP satellites became available, identifying substorm changes of detailed forms of auroral structures. The launch of more advanced spacecraft, such as Dynamics Explorer 1, Viking, Polar, and Image made it possible to obtain the global auroral distribution that can be used not only to serve as the substorm reference in timing but also to estimate instantaneous distributions of the electric conductivity in the polar ionosphere.

1.2.6 Planets and Comets

Overview

The interaction of the solar wind with other planets and comets produces planetary and cometary magnetospheres/ionospheres. The study of other planets and comets provides valuable insights into the physical processes in the solar–terrestrial environment.

The solar system planets are divided into two groups: the inner planets are Earth-like planets Mercury, Venus, Earth, Mars, and the outer planets Jupiter, Saturn, Uranus and Neptune are giant gaseous planets. The Earth-like planets have comparable sizes and atmospheres, although with distinct densities, temperatures and compositions. The outer giant gaseous planets consist mainly of hydrogen and helium. They have moons and rings that affect the dynamics of their magnetospheres. The outermost tiny dwarf planet Pluto seems to be a solid body again. Pluto was probably not formed together with the solar system, but is a captured asteroid.

The properties of planetary magnetospheres depend on the magnetic moments, rotation periods, the inclinations of the dipole axes with respect to the rotation axes and the ecliptic plane, plasma sources and sinks of the planets, and the local conditions of the solar wind. In front of all planetary magnetospheres a bow shock is formed where the supersonic solar wind is decelerated to sub-magnetosonic speeds. The structures behind the bow shocks are different. Since the magnetic fields of Mars and Venus are weak, no well-structured magnetospheres develop. Instead, the interaction of the solar wind with the ionosphere of Venus produces an ionopause/magnetopause that separates the solar wind and planetary plasmas.

Earth-like magnetospheres are found at Mercury, Saturn, Uranus and Neptune, but the dynamics, structures and sizes of these magnetospheres are very different. Jupiter's magnetosphere is much more complex than the Earth-like ones because the fast rotation of Jupiter generates centrifugal forces that stretch the magnetosphere outward into the equatorial plane, resulting in a rather flat magnetosphere with a plasma sheet in the dayside equatorial plane. In addition to the solar wind, the moons and rings also contribute plasma sources to the magnetospheres of the outer planets. Radiation belts are observed in the magnetospheres of the outer planets.

When a comet approaches the Sun, its atmosphere becomes huge although its icy nucleus is small with only a few kilometers in diameter. The solar wind-comet interaction produces a gravitationally unbound, sublimated cometary neutral atmosphere flowing outward from the nucleus at speeds of about 1 km/s. The resulting ionosphere adopts the velocity of the expanding neutrals and creates a planet-sized cavity. The boundary of this cavity is known as the contact surface, and the boundary above the contact surface where the composition transition occurs is known as the cometopause. A weak bow shock develops in the mass-loaded plasma where the solar wind flow is slowed down. In comets, mass loading implies that the flowing background plasma becomes loaded with heavy ions of atmospheric origin and as a result is slowed down.

History

In the early 1960s, the Mariner 2, 4, and 5 spacecraft studied Venus and Mars. Venera 9 and 10 were launched in 1975, the Pioneer Venus spacecraft was launched in 1978, and the Phobos spacecraft was launched in 1989. They provided additional information on the interaction of the solar wind with Venus and Mars, and their atmospheres. In 1974–5, the Mariner 10 spacecraft observed the magnetosphere of Mercury. In 1976, the Viking spacecraft executed a soft-landing on Mars. In the 1970s and 1980s, the Pioneer 11 and 12, and Voyager 1 and 2 studied the magnetospheres of Jupiter, Saturn, Uranus and Neptune.

T. Brahe in 1577 observed a comet and used the method of parallaxes to show that the comet was located outside the Earth's atmosphere. In 1951 C. Hoffmeister observed that a comet tail lags behind the comet's radial direction by about 5 degrees. L. Biermann in 1951 suggested the existence of solar wind based on observations of cometary tails. H. Alfvén in 1957 developed a model of the comet tail. In 1985, the ICE spacecraft studied the comet Giacobini-Zinner. In 1986, the VEGA 1 and 2, Giotto, Suisei, and Sakigake spacecraft studied the comet Halley and its magnetosphere. These spacecraft revealed that the solar wind interaction with comets is radically different from that with a magnetized planet. Analysis of particle and magnetic field data from the encounter of Giotto with the comet Halley show the complex plasma structure of the cometary bow shock, dominated by heavy mass-loading ions. In addition to bow shock, other boundaries were identified in the solar wind-comet interaction region including the contact surface, the ion pileup boundary, the cometopause, and the magnetic pileup boundaries.

1.2.7 Cosmic Rays

Overview

Cosmic rays of galactic origin are continually and isotropically incident on the solar system. They consist mainly of hydrogen and helium nuclei, electrons, and heavier ions such as C, O, and Fe. The galactic cosmic ray intensity measured on the ground and by the spacecraft is anti-correlated with the sunspot number, because the intensity of galactic cosmic rays below a few giga-electron-volts is modulated by the solar activity, with a minimum at the solar maximum. This decrease in the intensity of galactic cosmic rays is known as the *Forbush decrease*. To reach the Earth from the interstellar medium, galactic cosmic rays have to propagate from the heliopause to the inner heliosphere. During their propagation in the heliosphere, galactic cosmic rays experience the Lorentz force and are scattered by irregularities such as the Alfvén and magnetohydrodynamic turbulence in the interplanetary medium. In addition, they are blocked and scattered by transient inhomogeneities such as magnetic clouds, traveling interplanetary shocks and shocks at the corotating interaction regions. With increasing solar activity the fluctuations of the irregularities and transient inhomogeneities in the heliosphere increase, thus fewer galactic cosmic rays manage to penetrate the inner heliosphere and the Earth.

History

In 1912, V. Hess used the ground-based and balloon-borne detectors to discover that cosmic rays are incident upon the Earth from space. In 1938, S. Forbush found that the variations of the cosmic ray intensity measured on the Earth are correlated with changes in the geomagnetic field intensity and noted that a worldwide change in the Earth's magnetic field intensity during geomagnetic storms was correlated with a rapid decrease of cosmic ray intensity, which became known as the *Forbush decrease*. P. Blackett (1897–1974) was successful in generating cosmic ray showers in cloud chambers. E. Fermi developed in 1949 a theory of cosmic rays acceleration between two moving magnetic fields. In 1954, Forbush showed that his ion chamber measurements between 1937 and 1952 were negatively correlated with the sunspot number over a solar cycle. In 1955, J. Simpson interpreted this negative correlation in terms of the modulation of galactic cosmic rays by the solar cycle, and suggested that the interplanetary magnetic fields may prevent galactic cosmic rays from entering the solar system near the Earth's orbit. Simpson developed the neutron monitor to measure cosmic rays both on the ground and in space, and in 1960 Simpson and his group used the cosmic ray dectector aboard the first deep space spacecraft Pioneer 5 to prove that an interplanetary shock had caused a Forbush decrease.

1.3 Nature of the Solar-Terrestrial Environment

The solar–terrestrial environment consists of fully ionized plasmas (Sun, solar wind, magnetosphere), partially ionized plasmas (the ionosphere), and neutral fluids (the Earth's atmosphere). The Sun-Earth relation depends on a chain of coupling processes involving the solar interior-solar atmosphere-solar wind-magnetosphere-ionosphere-atmosphere interactions. Space plasmas, cometary and planetary atmospheres are complex by nature dominated by waves, instabilities and turbulence. The nature of solar–terrestrial environment can be studied by theoretical analysis, laboratory, ground, and space observations, as well as computer simulations. In this section, we provide an introduction to the basic concepts of waves, instabilities and turbulence in the solar–terrestrial environment. We will adopt a qualitative approach without going into quantitative details. Further information can be obtained in other chapters of this book. Basic plasma physics is discussed in the books by Chen (1984), Clemmow and Dougherty (1989), Stix (1992), Bittencourt (2004), Tajima (2004), and Gurnett and Bhattacharjee (2005). Basic space physics is discussed in the books by Priest (1984), Melrose (1991), Gary (1993), Baumjohann and Treumann (1996), Krishan (1999), Priest and Forbes (2000), Kallenrode (2001), Gurnett and Bhattacharjee (2005), and Jatenco-Pereira et al. (2005). Nonlinear plasma physics is discussed in the books by Sagdeev and Galeev (1969), Tsytovich (1970), Hasegawa (1975), Burlaga (1995), Treumann and Baumjohann (1997), and Biskamp (2003). Recent advances on nonlinear processes in the solar–terrestrial environment can be found in Chian and the WISER Team (2003), Büchner et al. (2005), Chian et al. (2005), Jatenco-Pereira et al. (2005), Lui et al. (2005), and Sulem et al. (2005).

1.3.1 Linear Waves

Disturbances in fluids and plasmas propagate as waves which transport the energy of perturbations from one region to another. A wave is characterized by its amplitude and phase (frequency and wavenumber). If the amplitude of the disturbances is small, linear waves can be represented as a superposition of plane waves using Fourier analysis. In a dispersive medium such as a plasma, wave propagation follows a dispersion relation between wave frequency and wavenumber. The velocity of wave phase propagation is called *phase velocity*; the velocity of wave energy flow is called *group velocity*. In general, the phase and group velocities of a wave are different.

Two types of waves can propagate in a plasma. The first type is electromagnetic waves which reduce to light waves in a vacuum; examples of electromagnetic waves in magnetized plasmas are whistler and Alfvén waves. Whistlers are high-frequency right-hand circularly polarized electromagnetic waves which can be excited by a lightning in one hemisphere and travel to the other hemisphere along the Earth's magnetic field lines through the magnetosphere. Due to the wave dispersion of whistler waves, higher-frequency waves have higher group and phase velocities, thus whistlers will be detected in a frequency-time sonogram as a falling tone. Whistlers are observed in the magnetosphere during substorms as well. They can also be induced by atmospheric lightnings. Shear Alfvén waves are low-frequency electromagnetic waves propagating parallel to the ambient magnetic field that represent string-like oscillations of the ambient magnetic field. Alfvén waves are observed in the solar atmosphere, solar wind, and planetary magnetospheres.

The second type is plasma waves which are internal plasma oscillations; examples of plasma waves are Langmuir waves and ion-acoustic waves. Langmuir waves are high-frequency electron plasma waves with frequency close to the plasma frequency (which is a function of the plasma density). They are related to the oscillatory motion of the electrons driven by the electrostatic force that restores charge neutrality in the plasma. Langmuir waves are observed in the solar wind in connection with type-III solar radio emissions, in the auroral plasmas, and upstream of planetary bow shocks, interplanetary shocks and the heliosphere's termination shock. Ion-acoustic waves are low-frequency ion plasma oscillations related to the oscillatory motion of the ions, which have similar properties as sound waves in a gaseous medium. Ion-acoustic waves are observed in the solar wind, upstream and downstream of bow shocks and interplanetary shocks, and in the auroral ionosphere.

A cut-off of a wave occurs in a plasma when the wavenumber vanishes, leading to wave reflection. In an unmagnetized plasma, the cut-off frequency of an electromagnetic wave is the plasma frequency; hence, an electromagnetic wave can only propagate above the plasma frequency. The ionosonde technique used for determining the density profile of the ionosphere is based on the reflection of a radio wave at a critical layer of the ionosphere due to the wave cut-off. Resonance of a wave occurs when the wavenumber becomes infinite, leading to wave absorption. For an electromagnetic wave propagating along the ambient magnetic field, a resonance occurs at the electron cyclotron frequency for a right-hand circularly polarized electron cyclotron wave, and a resonance occurs at the ion cyclotron frequency for a left-hand circularly polarized ion cyclotron wave. These resonances lead to wave-particle interactions and wave damping or wave growth. Whistlers can accelerate electrons in the magnetosphere to relativistic energies via the electron cyclotron resonance, sometimes producing so-called *killer electrons* which can damage the artificial satellites.

Wave damping can be caused by collisional or collisionless processes. Collisional wave dissipation is due to binary collisions between either charged or neutral particles. Examples of collisionless wave dissipation are Landau damping and cyclotron damping associated with wave-particle interactions. Landau damping occurs when the particle velocities are close to the phase velocity of the plasma wave. Cyclotron damping occurs when the particle velocities parallel to the ambient magnetic field are close to the wave phase velocities parallel to the ambient magnetic field, Doppler-shifted by the cyclotron frequencies of either electrons or ions, depending on the polarization of the wave. Landau-damping and cyclotron damping are important mechanisms for waves to heat and accelerate space plasmas.

1.3.2 Instabilities

In general, the solar–terrestrial environment is in a non-equilibrium state due to a variety of instabilities that occur in the plasma and fluid systems. Instabilities in space plasmas and the earth's atmosphere are driven by a multitude of free energy sources such as velocity shear, gravity, temperature anisotropy, electron and ion beams, and currents. An unstable wave is characterized by a complex wave frequency, whose real part describes the rate of wave oscillations, and the imaginary part describes the growth rate of the instability. The growth rate can be obtained by seeking the complex solution of a plasma dispersion relation. Plasma instabilities can be classified into macroinstabilities and microinstabilities; the former occur on scales comparable to the bulk scales of the plasma, the latter occur on scales comparable to the particle motion.

Macroinstabilities are fluid in nature and can be studied by fluid and MHD (magnetohydrodynamic) equations. They are instabilities in configuration space, thus a macroinstability lowers the energy state of a system by distorting its configuration. Examples of macroinstabilities are the Kelvin–Helmholtz instability and the Rayleigh–Taylor instability. The Kelvin–Helmholtz instability is produced by velocity shear flows in fluids and plasmas, such as the transition region between the magnetosheath and the magnetosphere. Since the magnetosheath plasma is flowing along the magnetopause around the magnetosphere which has its own flow velocity, the coupling between the magnetosheath and magnetospheric flows causes ripples to grow at the interface by the Kelvin–Helmholtz instability. This velocity shear-driven macroinstability contributes to plasma and momentum transport from the magnetosheath across the magnetopause to the magnetosphere by mixing the two regions.

The Rayleigh–Taylor instability, also known as the interchange instability, is a macroinstability at a fluid or plasma boundary under the influence of a gravitational field. At the boundary, the gravitational field causes ripples to grow at the interface leading to the formation of density bubbles. The growth rate of the Rayleigh–Taylor instability is a function of the gravitational acceleration. This macroinstability occurs frequently at the equatorial ionosphere, where collisions with neutrals modify considerably the instability leading to the formation of plasma density bubbles known as *equatorial spread-F*. These plasma bubbles are the origin of ionospheric scintillations of the GPS signals.

Microinstabilities are kinetic in nature and can be studied on the basis of the Boltzmann–Vlasov equations. Microinstabilities are velocity space instabilities related to the distribution function of the plasma particles, and appear when the distribution function of the plasma particles departs from a Maxwellian distribution. There are two types of microinstabilities: electromagnetic and electrostatic instabilities. An electromagnetic microinstability results from the growth of electromagnetic waves due to growing current densities in the plasma. An example of electromagnetic microinstabilities is the anisotropy-driven instability responsible for the growth of whistler waves in the magnetosphere. Whistler instability may be excited by the free energy stored in the temperature anisotropy of hot electrons in the magnetospheric plasma, whose growth rate and instability threshold are functions of the temperature anisotropy. During substorms the enhanced temperature anisotropies are induced by plasma convection from the magnetotail into the inner magnetosphere, resulting in the excitation of broadband whistler waves. These whistler waves can interact strongly with hot electrons of the radiation belts and plasma sheet via electron cyclotron resonance, causing the enhanced precipitation of energetic electrons in the auroral regions and in the region of South Atlantic Magnetic Anomaly.

An electrostatic microinstability results from the growth of electrostatic plasma waves due to growing charge densities in the plasma. Examples of electrostatic instabilities are the bump-in-tail instability and the ion-acoustic instability. When an electron beam interacts with a background plasma, a bump-in-tail configuration appears on the distribution function of the plasma electrons. The free energy of the electron beam can produce a beam-plasma instability, leading to the growth of Langmuir waves with frequencies near the background plasma frequency. The growth rate of this instability depends on the electron beam velocity and the instability can be excited if the electron beam velocity exceeds a threshold related to the thermal velocity of the background plasma. Langmuir waves driven by the interaction of the energetic electron beams emanating from the solar active regions with the solar atmosphere and solar wind plasmas, via the bump-in-tail instability, are responsible for the generation of type-III solar radio emissions by solar flares.

Ion-acoustic waves can be produced by an electrostatic microinstability due to an electron current flowing in the plasma. The combined distribution function consisting of hot drifting electrons and cold immobile ions indicates that an ion-acoustic instability can develop in the region of the electron distribution function where its slope is positive. The growth rate for this current-driven instability depends on the electron drift velocity. It is excited provided the electron drift velocity exceeds the ion acoustic velocity and the electron temperature is much larger than the ion temperature. The unstable ion-acoustic waves excited by the electron current propagate in the direction of the current, at a speed slightly lower than the speed of the current. An ion-acoustic instability can be driven by the field-aligned currents in the auroral ionosphere, as evidenced by the density irregularities observed along the auroral field lines. Other types of electrostatic and electromagnetic waves can also be induced by currents in space plasmas.

1.3.3 Nonlinear Waves

The solar–terrestrial environment is an intrinsically nonlinear system. Nevertheless the linear description is often very good even though it is strictly valid only at the initial stage of a growing instability when the amplitude of the unstable waves is still infinitesimally small. When the instability grows, the disturbances reach finite-amplitudes and nonlinear effects begin to affect the system behavior. In the saturated state of an instability, the system dynamics is governed by nonlinear effects. In addition to the finite-amplitude effect of disturbances, a wealth of other nonlinear effects may appear in fluids and plasmas. These effects occur on the kinetic level and include the distortion of the undisturbed particle orbits, the interaction between the unstably excited particles and the waves, and the interactions between the waves themselves in which particles are included only to higher order.

There are many types of nonlinear waves in the solar–terrestrial environment both on the macroscopic and on the microscopic scales. Examples of nonlinear waves on the microscopic scales are solitons and double layers. Well-known examples of macroscopic-scale nonlinear waves is the collisionless shock waves. An ion-acoustic soliton results from a balance between wave steepening and wave dispersion. Without wave dispersion, the natural tendency of a wave to steepen, which happens due to the nonlinearity of the medium, will lead to wave breaking. The ion-acoustic soliton has a bell-shaped wave form which propagates at a constant velocity across a uniform plasma. It is stable and travels a long distance without changing its shape. There is an inverse relation between the amplitude of a soliton and its width, i.e., the larger the amplitude of the soliton, the narrower its width. Solitons preserve their shapes and velocities after a collision with another soliton. Ion-acoustic solitons can also evolve into double layers due to the reflection and transmission of plasma particles in the solitons themselves. In contrast to solitons which do not contain a net potential drop across the soliton, double layers are nonlinear structures containing net potential drops. Hence, a series of double layers if aligned along the magnetic field with correct polarity will add up to produce large electric potential drops along a magnetic field line. Ion-acoustic solitons, ion-acoustic shocks and double layers with large electric fields have been observed by satellites in the regions of auroral plasmas in conjunction with field-aligned currents and ion beams. Contrary to the balance between steepening and dispersion, a balance between wave steepening and dissipation leads to the formation of a shock wave which has a ramp-shape curve. The thickness of the shock ramp is related to the dissipation and the shock velocity. In the presence of dissipation in addition to dispersion, an ion-acoustic soliton evolves into an oscillatory shock structure. The oscillations occur either upstream or downstream of the shock depending on the external conditions.

A shock is a discontinuity which divides a continuous medium into two different regimes: the regions upstream and downstream of a shock. In a gas-dynamic shock, the physical process is dominated by binary collisions between the molecules. Since the space plasma density is low, in space plasmas such as the solar corona and the solar wind, particle collisions are rare. Hence, shocks in space plasma are *collisionless shocks*. Examples of collisionless shocks are the Earth's bow shock and interplanetary shocks. A bow shock is produced by the slowing down of the supersonic solar wind by the Earth's magnetosphere, forming a standing shock wave in front of the dayside magnetosphere. Interplanetary shocks are the result of mass ejected from solar active regions. These masses travel from the solar atmosphere across the interplanetary medium. Apparently all interplanetary shocks are driven by such CMEs, but only a portion of the CMEs drive an interplanetary shock. Shocks are stable for a long time, for example, interplanetary shocks can reach the outer boundary of the heliosphere. Three types of collisionless shocks are found in the solar–terrestrial environment: fast, intermediate, and slow shocks. Planetary bow shocks and most interplanetary shocks are fast MHD shocks. Only a few slow MHD shocks have been identified in the solar wind. However, intermediate and slow MHD shocks may be more common in the solar corona and have been proposed to exist in relation to magnetic field reconnections.

1.3.4 Turbulence

Turbulence is a nonlinear phenomenon where stochastic multiscale processes and deterministic chaotic processes coexist, characterized by the presence of incoherent as well as coherent spatio-temporal fluctuations. Space plasmas and atmospheric fluids are dynamically evolving turbulent systems whose behavior is governed by nonlinear wave-wave interactions and nonlinear wave-particle interactions.

Nonlinear wave-wave interactions occur if the waves are resonant or phase-synchronized, described by the phase-matching conditions (i.e., the resonant relations of the wave frequencies and the wave vectors) which represent physically the conservation of wave energy and momentum. Wave coupling can be treated, approximately, either as coherent interactions (e.g., parametric interactions) where the wave phases are fixed, or incoherent interactions (e.g., the random-phase approximation) where the wave phases are random. Examples of nonlinear wave-wave interactions are three-wave processes and four-wave processes. For example, two oppositely propagating Langmuir waves can interact to generate a radio wave at the second harmonic plasma frequency, which can explain the origin of nonthermal radio emissions generated by either electron beams accelerated by solar flares or interplanetary shocks in the solar corona and solar wind. Nonlinear wave-wave interactions involving Langmuir waves and ion-acoustic waves have been observed in the solar wind in connection with type-III solar radio emissions.

Wave-particle interactions occur if the waves are resonant with particles. In addition to the linear wave-particle interactions such as Landau damping and electron- and ion-cyclotron resonances, various types of nonlinear wave-particle interactions can take place. An important effect of large-amplitude waves is particle trapping, where particles can become trapped in a wave potential trough if the particle kinetic energy in the wave frame is less than the potential energy of the wave. During the evolution of an instability, as the amplitude of wave perturbations increases the particle trapping can readily occur. Both trapped and untrapped plasma particles contribute to the nonlinear evolution of waves and instabilities. Particle trapping removes part of the resonant particles from the particle distribution function which inhibits the ability of these particles to inject energy to the instability, hence leading to the saturation of the instability. For example, the pitch-angle diffusion resulting from the wave-particle interactions involving large-amplitude whistler waves and radiation-belt electrons in the plasmasphere can deplete the resonant electrons and scatter them into the loss cone, leading to enhanced particle precipitation into the auroral atmospheres. Bernstein–Greene–Kruskal waves, known as *BGK waves*, are nonlinear plasma waves resulting from particle trapping and untrapping. For example, the Polar and Cluster satellites near the Earth's magnetosphere, magnetopause and bow shock observed large-amplitude solitary waves called *electron holes*, which are related to BGK waves.

A typical power density spectrum of turbulence in the solar–terrestrial environment shows power-law in frequency and wavenumber, which is an indication of energy cascade and multiscale interactions. Energy transfer in turbulence can occur via either direct cascade or inverse cascade mechanisms. In the direct cascade mechanism, the energy is transferred from large scales to small scales, whereas in an indirect cascade mechanism the energy is transferred from small scales to large scales. For example, in the nonlinear evolution stage of the collisional Rayleigh–Taylor instability in the equatorial ionosphere the plasma bubbles can evolve from long wavelengths (of the order of kilometers) to shorter wavelengths (of the order of meters) via the direct cascade mechanism, producing a broadband power spectrum.

The turbulence in the solar–terrestrial environment is *intermittent*, exhibiting spatio-temporal variations that switch randomly between bursting periods of large-amplitude fluctuations and quiescent periods of low-amplitude fluctuations. Such intermittent behavior becomes more pronounced in small scales. The statistical approach to turbulence shows that the probability distribution functions of fluctuations are of nearly Gaussian shape at large scales, but become non-Gaussian with sharper peaks and fatter tails as the scales get smaller. This implies that extreme events, i.e., large-amplitude fluctuations, have a higher probability of occurrence than if they are normally distributed. The intermittent coherent (non-Gaussian) structures are localized regions of turbulence where finite phase correlation exists, and they have a typical lifetime longer than the background of stochastic fluctuations.

Turbulence can exhibit chaotic behavior as well. *Chaos* in atmospheres was discovered by E. Lorenz in 1963 and has contributed significantly to the study of nonlinear wave-wave and wave-particle interactions as well as turbulence in fluids and plasmas. A chaotic system shows sensitive dependence on the system's initial conditions so that nearby orbits will diverge exponentially in time and space. A chaotic system demonstrates also sensitive dependence on small variations of the system parameters. Order and chaos can coexist in a nonlinear dynamical system. The ordered state is described by a *stable periodic orbit* and the chaotic state (i.e., a *chaotic attractor*) is described by an infinite set of *unstable periodic orbits*. The dynamical systems approach to turbulence can elucidate the nonlinear dynamics and structures of the solar–terrestrial environment, for example, the Alfvén intermittent turbulence in the solar wind.

In 2D and 3D, turbulence consists of two components: an incoherent component of background flow and a coherent component related to vortices. The coherent component is a collection of nonlinear coherent structures of multi-scales, associated with localized regions of concentrated vorticity such as vortices in a turbulent shear flow. Vortices are fairly stable and can persist for a large number of vortex rotation periods. Due to their long lifetimes, vortices play a major role in the transport of mass and momentum in plasmas and fluids.

The study of phase synchronization in a system of coupled oscillators has improved our understanding of a variety of nonlinear phenomena in physical, chemical, and biological systems. Synchronization may explain the formation of nonlinear coherent structures such as solitons and vortices in turbulence. The concept of synchronization of coupled periodic oscillators has been generalized to coupled chaotic oscillators. For example, the imperfect phase synchronization of the fundamental spectral components with distinct scales in fluids and plasmas can be the origin of the intermittent bursts of wave energy in the turbulence. There is a finite phase coherence in wave interactions in turbulence, as evidenced in the MHD turbulence observed upstream of the Earth's bow shock.

1.4 Applications

1.4.1 Space Weather and Space Climate

Sun and Earth are related not only through the impact of solar radiation on the Earth's weather and climate, but also through the impact of solar wind on the Earth's space weather and space climate. Space weather (or space climate) is the short-term (or long-term) variabilities in the plasma environment of the Earth and other planets. The study of solar–terrestrial environment involves the investigation of short-term as well as long-term evolution of solar and geomagnetic activities, which has applications for the forecasting of space weather and space climate.

Space weather is linked to the conditions on the Sun and in the solar wind, magnetosphere, ionosphere, and thermosphere, that can influence the performance and reliability of space-borne and ground-based technological systems and can endanger human life or health. The dynamic solar wind-magnetosphere coupling gives rise to dynamic changes in geomagnetic activities such as geomagnetic storms and substorms, with typical durations of days and hours, respectively. These space weather phenomena can affect anthropogenic systems such as satellites, navigations, telecommunications, power transmission lines, gas pipelines, and the safety of astronauts and airline passengers.

On the other hand, space climate is related to the long-term trends of solar variability, with time scales of the order of tens and hundreds of years or more, as well as the long-term dynamics of geomagnetic field including its polarity reversal that occurs about every 100 thousand years. The study of space climate requires the understanding of the complex coupled Sun-solar wind-magnetosphere-ionosphere-atmosphere-ocean system. Solar wind magnetic fields and geomagnetic fields play crucial roles in space climate by modulating the precipitation of high-energy galactic cosmic rays and solar particles into the Earth's atmosphere. One example of large social disruption of space climate is the 70 years-long mini ice age during the Maunder Minimum of 1645–1715, during which the solar magnetic activity was greatly reduced as indicated by the low number of sunspots. Space climate can be studied using past records of important physical parameters. Historical data and proxy archives, such as the time series of tree rings, coral band densities, and ice cores, provide the means to reconstruct the past history of space climate, contributing to the improvement of space climate forecasting.

1.4.2 Plasma Astrophysics

Most of the visible matter in the universe is in the plasma state. We discuss in this section a few examples of astrophysical plasmas. The Sun is an ordinary main sequence star (or dwarf star) of spectral type G2, with magnetic field in the photosphere on the order of 1 gauss or less outside and 3 to 4 kgauss inside sunspots. It is expected that many other stars in our galaxy have properties similar to the Sun. These solar-like stars have a magnetic field strong enough to control the dynamics and structure of its atmosphere, similar to the solar atmosphere. Optical, radio, ultraviolet, and X-ray observations have shown that solar-like activity (stellar spots, chromospheres, transition regions, coronae, and stellar winds) is seen in dwarf stars of spectral type G-M, dwarf stars of spectral type A7-F7, and T Tauri stars among others.

Global magnetic fields are observed in variable Ap stars, which are peculiar stars with enhanced lines of Fe-peak elements and of the rare earth elements. The surface magnetic fields are predominantly dipolar, ranging from a few hundred gauss to 34 kgauss. The measured global magnetic fields vary in phase with the spectrum and light variations. The periodic variations in spectrum, light, and magnetic field are due to the stellar rotation with the period of the observed variations equal to the period of rotation.

Strong magnetic fields have been detected in compact stellar objects such as white dwarfs, neutron stars, and magnetars. White dwarfs have magnetic fields of around 1×10^6 to 5×10^8 gauss. Neutron stars are associated with pulsars that emit periodic beams of coherent radio waves at their rotation period. The magnetic field of pulsars is in the range of 10^{11} to 10^{13} gauss. Pulsar radio waves are generated in the pulsar magnetosphere made up of strongly magnetized electron-positron plasma resulting from pair-production. Magnetars are highly magnetized neutron stars formed in a supernova explosion, with magnetic fields of around 10^{15} gauss.

The magnetic fields also play important roles in other astrophysical plasmas such as in the star formation and evolution, exoplanets, accretion disks, stellar and extragalactic jets and outflows, interstellar and intergalactic media, galactic center, and in the primordial universe. The study of physical processes in the solar–terrestrial environment has relevant applications to plasma astrophysics. For example, the investigation of magnetic field reconnection, collisionless shocks, particle acceleration, and plasma heating in space plasmas can improve our understanding of similar processes in astrophysical plasmas.

1.4.3 Controlled Thermonuclear Fusion

The worldwide energy consumption has been growing rapidly in recent years, which increases the demand for new energy sources since the world reserve of fossil fuels such as petroleum and coal are being depleted in a fast rate. A promising energy source is controlled thermonuclear fusion which derives usable energy from the fusion of light nuclides such as deuterium (D), tritium (T), helium-3, and lithium. Deuterium exists abundantly in nature, for example, it is readily available in sea water. The fusion process itself does not leave long-lived radioactive products, hence the problem of radioactive-waste disposal is much less serious than nuclear fission reactors. The success of nuclear fusion would provide virtually limitless energy supply.

The plasma for nuclear fusion reactors requires very high temperatures, involving the creation in the nuclear reactors of plasma conditions similar to the interior of the Sun. Such high temperatures would allow the ions to reach high enough velocities to overcome their mutual Coulomb repulsions, so that collision and fusion can occur. To maintain the plasma temperature, the power produced by the thermalnuclear fusion reactions must exceed the energy loss due to the bremsstrahlung radiation emitted by plasma electrons. This radiation is emitted when electrons are accelerated due to elastic collisions with ions. An ignition temperature can be determined by equating the power produced by a given nuclear reaction to the power emitted by the bremsstrahlung radiation. For the D-T reaction, the ignition temperature is about 4 keV. The aim of nuclear fusion research is to produce energy output larger than the energy input used to heat the plasma. The breakeven is given by the Lawson criterion related to the product of the plasma density and the confinement time. For D-T reaction, the Lawson criterion is about 10^{14} $cm^{-3}s$.

Two of the problems facing the thermonuclear fusion development are plasma confinement and plasma heating. Two types of plasma confinement are being developed: (1) magnetic confinement by a tokamak which has a strong toroidal magnetic field supplemented by a poloidal magnetic field produced by a large current in the plasma; and (2) inertial confinement by using high-power lasers to ignite nuclear fusion reaction in a pellet. A variety of plasma instabilities appear in magnetic and inertial confinements. For example, resistive tearing mode instability, current-driven instabilities, and drift instabilities are found in tokamaks; Raleigh–Taylor instabilities and parametric instabilities such as stimulated Raman scattering and stimulated Brillouin scattering are found in laser-fusion. One of the main challenges of nuclear fusion research is to control and suppress these plasma instabilities. Plasma heating in tokamaks and laser-fusion are closely related to wave processes and nonlinear processes in plasmas. For example, tokamaks can be heated by radio-frequency waves using Alfvén waves, electron or ion cyclotron waves, and lower-hybrid waves. To heat a tokamak efficiently, it requires an optimization of wave excitation, wave propagation, wave absorption and thermalization.

Solar-terrestrial environment provides an ideal natural laboratory for studying these fundamental physical processes in plasmas. Most waves and instabilities that appear in tokamak and laser-fusion plasmas are found in space plasmas. Many problems facing the nuclear fusion research, such as plasma heating by radio-frequency waves, are being studied in solar–terrestrial physics. Therefore, advances in solar–terrestrial environment research have contributed and will continue to contribute to the progress of controlled thermonuclear fusion in the years to come (Tajima, 2004; Gurnett and Bhattacharjee, 2005).

1.5 Concluding Remarks

Man landed on the moon for the first time on 20 July 1969. This opens the door for man's migration to the outer space, which may extend eventually to other planets and moons, such as Mars. Many nations are taking part in the activities in space stations and the development of manned space vehicles. As space travel and space exploration become a reality, we need to improve our capability to monitor and forecast space weather in the space environment, which impact on the health and safety of astronauts and space travelers.

On 16 December 2004, the Voyager 1 spacecraft crossed the termination shock of the heliosphere, and will continue its journey to the heliosheath, to the bow shock of the solar system, and into the interstellar wind. The information of the physical processes in the outer heliosphere and the interstellar space will help us to understand the impact of the local interstellar medium on the solar system.

In addition to solar activities and galactic cosmic rays, other cosmic sources may impact on the solar–terrestrial environment. For example, on 27 December 2004, powerful gamma-ray bursts emitted by a magnetar in our galaxy arrives on the Earth and produced significant impact on the Earth's magnetosphere and ionosphere. This indicates the need to deepen our understanding of the relation between the cosmos and the solar–terrestrial environment.

Acknowledgement. This work was supported in part by the Grant-in-Aid for Creative Scientific Research "The Basic Study of Space Weather Prediction" (grant 17GS0208) of the Ministry of Education, Culture, Sports, Science and Technology (MEXT) of Japan, and CNPq of Brazil. The authors wish to thank Dr. R. Treumann and Dr. A. Klimas for their valuable comments. A.C.-L. Chian gratefully acknowledges the award of a visiting professor fellowship by Nagoya University and the kind hospitality of the Solar-Terrestrial Environment Laboratory.

References

Baumjohann, W. and R.A. Treumann, Basic Space Plasma Physics. Imperial College Press, London, 1996.
Biskamp, D., Magnetohydrodynamic Turbulence. Cambridge University Press, Cambridge, 2003.
Bittencourt, J.A., Fundamentals of Plasma Physics. Springer, Berlin, 2004.
Büchner, J., S. Chapman, A.C.-L. Chian, A.S. Sharma, D. Vassiliadis, and N. Watkins (Eds.), Nonlinear and Multiscale Phenomena in Space Plasmas. Proceedings of the Nonlinear Geophysics Session of the AGU Fall Meeting in 2004. Nonlinear Processes in Geophysics, special issue, 2005. (http://www.copernicus.org/EGU/npg/special_issues.html)
Burlaga, L.F., Interplanetary Magnetohydrodynamics. Oxford University Press, New York, 1995.
Chen, F.F., Introduction to Plasma Physics and Controlled Fusion, 2nd Ed. Plenum Press, New York, 1984.
Chian, A.C.-L., and the WISER Team (Eds.), Advances in Space Environment Research, Vol. 1, Kluwer Academic Publishers, Dordrecht, 2003; also in Space Sci. Rev., 107, Nos. 1–2, pp. 1–540, 2003.
Chian, A.C.-L., J. Büchner, P. Chu, and P. Watkins (Eds.), Nonlinear Processes in Solar-Terrestrial Physics and Dynamics of Earth-Ocean-Space System. Proceedings of the WISER Workshop on Earth-Oceans-Space (EOS 2004). Nonlinear Processes in Geophysics, special issue, 2005. (http://www.copernicus.org/EGU/npg/special_issues.html)
Clemmow, P.C., and Dougherty, J.P., Electrodynamics of Particles and Plasmas. Perseus Books, New York, 1989.
Gary, S.P., Theory of Space Plasma Microinstabilities. Cambridge University Press, Cambridge, 1993.
Gombosi, T.I., Physics of the Space Environment. Cambridge University Press, Cambridge, 1998.
Gombosi, T.I., B. Hultqvist, and Y. Kamide (Eds.), J. Geophys. Res., 99, No. A10, pp. 19099–19212, 1994.
Gurnett, D.A., and A. Bhattacharjee, Introduction to Plasma Physics: With Space and Laboratory Applications. Cambridge University Press, Cambridge, 2005.
Hargreaves, J.K., The Solar-Terrestrial Environment. Cambridge University Press, Cambridge, 1992.
Hasegawa, A., Plasma Instabilities and Nonlinear Effects. Springer Verlag, Heidelberg, 1975.
Hultqvist, B., Earth's Magnetosphere, in J.A.M. Bleeker, J. Geiss, and M.C.E. Huber (Eds.), The Century of Space Science, Vol. 2, pp. 1529–1557. Kluwer, Dordrecht, 2001.
Jatenco-Pereira, V., A.C.-L. Chian, J.F. Valdes-Galicia, and M.A. Shea (Eds.), Fundamentals of Space Environment Science, Elsevier, Amsterdam, 2005; also in Adv. Space Res., 35, No. 5, pp. 705–973, 2005.
Kallenrode, M.-B., Space Physics: An Introduction to Plasmas and Particles in the Heliosphere and Magnetospheres. Springer, Berlin, 2001.
Kivelson, M.G., and C.T. Russell (Eds.), Introduction to Space Physics. Cambridge University Press, Cambridge, 1995.
Krishan, V., Astrophysical Plasmas and Fluids. Kluwer Academic Publishers, Dordrecht, 1999.
Lui, A.T.Y., Y. Kamide, and G. Consolini (Eds.), Multiscale Coupling of Sun-Earth Processes. Elsevier, Amsterdam, 2005.
Melrose, D.B., Instabilities in Space and Laboratory Plasmas. Cambridge University Press, Cambridge, 1991.
Nishida, A. (Ed.), Magnetospheric Plasma Physics. Center for Academic Publications Japan, Tokyo, 1982.
Parks, G.K., Physics of Space Plasmas: An Introduction. Addison-Wesley, Redwood City, 1991.
Priest, E., Solar Magnetohydrodynamics. Reidel, Dordrecht, 1984.
Priest, E., and T. Forbes, Magnetic Reconnection: MHD Theory and Applications. Cambridge University Press, Cambridge, 2000.
Prölss, G.W., Physics of the Earth's Space Environment. Springer-Verlag, Berlin, 2004.
Rishbeth, H., The centenary of solar–terrestrial physics, J. Atmos. Solar-Terr. Phys., 63, 1883, 2001.
Sagdeev, R.Z., and Galeev, A.A., Nonlinear Plasma Theory. W.A. Benjamin, New York, 1969.
Stern, D.P., A millennium of geomagnetism, Rev. Geophysics, 40 (3), 1-1, 2002.
Stix, T.H., Waves in Plasmas. AIP, New York, 1992.

Sulem, P.-L., T. Passot, A.C.-L. Chian, and J. Büchner (Eds.), Advances in Space Environment Turbulence. Proceedings of WISER Workshop on Space Environment Turbulence (ALFVEN 2004). Nonlinear Processes in Geophysics, special issue, 2005. (http://www.copernicus.org/EGU/npg/special_issues.html)

Tajima, T., Computational Plasma Physics: With Applications to Fusion and Astrophysics. Westview, Oxford, 2004.

Treumann, R.A., and W. Baumjohann, Advanced Space Plasma Physics. Imperial College Press, London, 1997.

Tsytovich, V.N., Nonlinear Effects in Plasmas. Plenum Press, New York, 1970.

Part 1

The Sun

2 The Solar Interior – Radial Structure, Rotation, Solar Activity Cycle

Axel Brandenburg

Some basic properties of the solar convection zone are considered and the use of helioseismology as an observational tool to determine its depth and internal angular velocity is discussed. Aspects of solar magnetism are described and explained in the framework of dynamo theory. The main focus is on mean field theories for the Sun's magnetic field and its differential rotation.

Contents

Axel Brandenburg, The Solar Interior – Radial Structure, Rotation, Solar Activity Cycle.
In: Y. Kamide/A. Chian, Handbook of the Solar-Terrestrial Environment. pp. 27–54 (2007)
DOI: 10.1007/11367758_2

2.1 Introduction

The purpose of this chapter is to discuss the conditions that lead to the magnetic activity observed at the surface of the Sun. Only to a first approximation is the Sun steady and spherically symmetric. A more detailed inspection reveals fully three-dimensional small scale turbulent motions and magnetic fields together with larger scale flows and magnetic fields that lack any symmetry. The cause of the large scale and small scale magnetic fields, as well as large scale circulation and differential rotation, is believed to be the turbulent convection which, in turn, is caused by the increased radiative diffusivity turning much of the radiative energy flux into convective energy flux.

The magnetic field is driven by a self-excited dynamo mechanism, which converts part of the kinetic energy into magnetic energy. As in technical dynamos the term 'self-excited' refers to the fact that part of the electric power generated by induction is also used to sustain the ambient magnetic field around the moving conductors. How the conversion of kinetic energy into magnetic energy works will be discussed in some detail in this chapter. The kinetic energy responsible for this process can be divided into (i) small scale irregular turbulent motions (convection) and (ii) large scale differential rotation and meridional circulation. It is the anisotropy of the small scale motions that is responsible for making the rotation nonuniform. Furthermore, lack of mirror symmetry of the small scale motions is responsible for producing large scale magnetic fields. This process is explained in many text books, e.g. Moffatt (1978), Parker (1979), Krause & Rädler (1980), Stix (2002), or Rüdiger & Hollerbach (2004).

The magnetic field is also responsible for linking solar variability to natural climate variations on Earth. Changes in the Sun's magnetic activity affect the solar irradiance by only 0.1%, which is generally regarded as being too small to affect the climate. However, the UV radiation is more strongly modulated and may affect the climate. According to an alternative proposal, the Sun's magnetic field shields the galactic cosmic radiation, which may affect the production of nucleation sites for cloud formation that in turn affects the climate. Thus, an increase in the solar field strength increases the shielding, decreases the cosmic ray flux on Earth, decreases the cloud cover, and hence increases the temperature. This chain of events is rather simplified, and there can be drastically different effects from high or low clouds, for example. For a recent review of this rapidly developing field see Marsh & Svensmark (2000).

We begin by discussing the theoretical foundations governing the properties of turbulent convection zones, and discuss then helioseismology as an observational tool to determine, for example, the location of the bottom of the convection zone as well as the internal angular velocity. We turn then attention to the properties of the Sun's magnetic field and discuss dynamo theory as its theoretical basis. Magnetic field generation is caused both by the turbulent convection and by the large scale differential rotation, which itself is a consequence of turbulent convection, as will be discussed in the last section of this chapter. Only a bare minimum of references can be given here, and we have to restrict ourselves mostly to reviews which give an exhaustive overview of the original literature. Original papers are here quoted mainly in connection with figures used in the present text.

2.2 Radial Structure

In order to determine the depth of the convection zone in the Sun and the approximate convective velocities it is necessary to solve the equations governing the radial structure of a star. For this purpose the Sun can be regarded as spherically symmetric. The equations governing the radial structure of the Sun (or a star) are quite plausible and easily derived. They can be written as a set of four ordinary differential equation, namely the

- equation for the Sun's gravitational field (Poisson equation),
- hydrostatic equilibrium (momentum equation),
- thermal equilibrium (energy equation),
- radiative equilibrium (radiation transport equation, convection).

These are given in all standard text books on stellar structure (e.g. Kippenhahn & Weigert 1990). In the following we discuss only a subset of these equations in order to describe some essential properties of the solar convection zone.

2.2.1 Global Aspects

The rate of energy production of the Sun, i.e. its luminosity, is $L_\odot = 4 \times 10^{26}$ W or 4×10^{33} erg s^{-1}. The total intercepted by the Earth is only a small fraction,

$$\frac{\pi R_{\rm E}^2}{4\pi R_{\rm E\odot}^2} = 4 \times 10^{-10} , \tag{2.1}$$

where $R_{\rm E}$ is the radius of the Earth (6400 km) and $R_{\rm E\odot}$ is the distance between the Earth and the Sun (= 1 AU = 1.5×10^{11} m). Thus, the total power reaching the projected surface of the Earth is $4 \times 10^{-10} \times 4 \times 10^{26}$ W = 1.6×10^{17} W. This is still a lot compared with the total global energy consumption, which was 1.4×10^{13} W in the year 2001.

The total thermal energy content of the Sun can be approximated by half its potential energy (Virial theorem), i.e.

$$E_{\rm th} \approx \frac{GM_\odot^2}{2R_\odot} = 2 \times 10^{41}\,{\rm J} , \tag{2.2}$$

where $G \approx 7 \times 10^{-11}\,{\rm m^3\,kg^{-1}\,s^{-2}}$ is Newton's constant, $M_\odot \approx 2 \times 10^{30}$ kg is the mass of the Sun, and $R_\odot \approx 7 \times 10^8$ m is its radius. The time it would take to use up all this energy to sustain the observed luminosity is the Kelvin–Helmholtz time,

$$\tau_{\rm KH} = E_{\rm th}/L_\odot \approx 10^7\,{\rm yr} , \tag{2.3}$$

which is long compared with time scales we could observe directly, but short compared with the age of the Sun and the solar system (5×10^9 yr). Therefore, gravitational energy (which is extremely efficient in powering quasars!) cannot be the mechanism powering the Sun. This motivated the search for an alternative explanation, which led eventually to the discovery of the nuclear energy source of stars.

The similarity between gravitational and thermal energies can be used to estimate the central temperature of a star by equating $GM/R = \mathcal{R}T_{\rm c}/\mu$. For the Sun this gives

$$T_{\rm c} \sim \frac{\mu}{\mathcal{R}}\frac{GM}{R} = 1.5 \times 10^7\,{\rm K} \tag{2.4}$$

for its central temperature. Here, $\mathcal{R} \approx 8300\,{\rm m^2 s^{-2}\,K^{-1}}$ is the universal gas constant and $\mu \approx 0.6$ is the non-dimensional mean molecular weight for a typical mixture of hydrogen and helium. The estimate (2.4) happens to be surprisingly accurate. This relation also tells us that the central temperature of the Sun is only determined by its mass and radius, and not, as one might have expected, by the luminosity or the effectiveness of the nuclear reactions taking place in center of the Sun.

2.2.2 Thermal and Hydrostatic Equilibrium

The condition of hydrostatic equilibrium can be written in the form

$$0 = -\frac{1}{\rho}\nabla p + \boldsymbol{g} , \tag{2.5}$$

where ρ is the density, p is the pressure, and $\boldsymbol{g}$ is the gravitational acceleration. In the spherically symmetric case we have $\boldsymbol{g} = -(GM_{\rm r}/r^2, 0, 0)$ in spherical polar coordinates, where $M_{\rm r}$ is the mass inside a sphere of radius r. Equation (2.5) is readily solved in the special case where the radial dependence of the density is polytropic, i.e. $\rho(r) \sim T(r)^m$, where m is the polytropic index. This yields

$$T(r) = T_{\rm c} - \frac{1}{1+m}\frac{\mu}{\mathcal{R}}\int_0^r \frac{GM_{\rm r}}{r^2}\,{\rm d}r . \tag{2.6}$$

So, in the outer parts of the Sun, where $M_{\rm r} \approx$ const, (2.6) can be integrated, which shows that the temperature has a term that is proportional to $1/r$.

Significant amounts of energy can only be produced in the inner parts of the Sun where the temperatures are high enough for nuclear reactions to take place. The central temperature is characterized by the condition of thermal energy equilibrium, which quantifies the rate of change of the local luminosity, $L_{\rm r}$, with radius. Outside the core, nuclear reactions no longer take place, so $L_{\rm r}$ can be considered constant. The radiative flux is given by $F = L_{\rm r}/(4\pi r^2)$, which thus decreases like $1/r^2$ in the outer parts.

In the bulk of the Sun, energy is transported by photon diffusion: the optical mean-free path is short compared with other relevant length scales (e.g. pressure scale height), so we are in the optically thick limit and can use the diffusion approximation for photons. The radiative flux, F, is therefore in the negative direction of and proportional to the gradient of the radiative energy density, aT^4, where $a = 7.57 \times 10^{-15}\,{\rm erg\,cm^{-3}K^{-4}}$ is the radiation-density constant. The connection between fluxes and concentration gradients is generally referred to as Fickian diffusion. As in kinetic gas theory, the diffusion coefficient is 1/3 times the typical particle velocity (= speed of light c) and the mean free path ℓ of the photons, so

$$F = -\frac{1}{3}c\ell\frac{\rm d}{{\rm d}r}(aT^4) = -\frac{3}{4}ac\ell T^3\frac{{\rm d}T}{{\rm d}r} \approx -K\frac{{\rm d}T}{{\rm d}r} , \tag{2.7}$$

which is basically the condition of radiative equilibrium. Here we have introduced the radiative conductivity K. The photon mean free path is usually expressed in terms of the opacity κ, which is the effective cross-section per unit mass, so $\ell = (\rho\kappa)^{-1}$. Expressing a in terms of the Stefan–Boltzmann constant, $\sigma_{\rm SB} = ac/4$, we have

$$K = \frac{16\sigma_{\rm SB} T^3}{3\kappa\rho} . \tag{2.8}$$

An approximation for the opacity κ that is commonly used for analytic considerations is Kramer's formula

$$\kappa = \kappa_0 \rho T^{-7/2} \quad \text{(Kramer's opacity)} , \tag{2.9}$$

where $\kappa_0 = 6.6 \times 10^{18}\,\mathrm{m^5\,K^{7/2}\,kg^{-2}}$ for so-called free–free transitions where two charged particles form a system which can absorb and emit radiation. This value may well be up to 30 times larger if the gas is rich in heavier elements, so it is a good electron supplier and bound-free processes (ionization of neutral hydrogen by a photon) become important as well. In practice, a good value is $\kappa_0 \approx 10^{20}\,\mathrm{m^5\,K^{7/2}\,kg^{-2}}$ (corresponding to $10^{24}\,\mathrm{cm^5\,K^{7/2}\,g^{-2}}$). With Kramer's formula, the conductivity is

$$K = \frac{16\sigma_{\rm SB} T^{13/2}}{3\kappa_0\rho^2} . \tag{2.10}$$

For a polytropic stratification, i.e. when the density is given by a power law of the temperature, $\rho \sim T^m$, we have

$$K \sim T^{13/2-2m} , \tag{2.11}$$

which is constant for an effective polytropic index $m = 13/4 = 3.25$. This gives indeed a reasonable representation of the stratification of stars in convectively stable regions throughout the inner parts of the Sun. At the bottom of the solar convection zone the density is about $200\,\mathrm{kg\,m^{-3}}$ and the temperature is about $2 \times 10^6\,\mathrm{K}$. This gives $K = (3\ldots100) \times 10^9\,\mathrm{kg\,m\,s^{-3}K^{-1}}$. In order to carry the solar flux the average temperature gradient has to be around $0.01\,\mathrm{K/m}$.

2.2.3 Transition to Adiabatic Stratification

In reality K does change slowly with height. Therefore the polytropic index effectively changes with height. If $m < 13/4$, then K decreases with decreasing T. However, in order to transport the required energy flux, the temperature gradient has to increase, so the polytropic index decreases further, until it reaches a critical value where the specific entropy gradient reverses sign. This leads to the onset of Rayleigh–Benard convection.

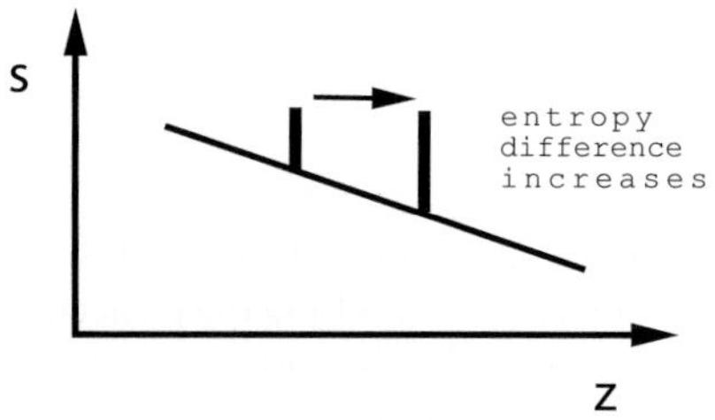

Fig. 2.1. Specific entropy profile for an unstable atmosphere. The difference in specific entropy between the blob and the surroundings increases as the blob ascends. Gravity points in the negative z direction, so $\boldsymbol{g} \cdot \boldsymbol{\nabla} s > 0$ in this case

Specific entropy is an important quantity, because it does not change in the absence of local heating or cooling processes. For a perfect gas, and ignoring partial ionization effects, the specific entropy can be defined, up to a constant s_0, as

$$s = c_{\rm v} \ln p - c_{\rm p} \ln \rho + s_0 , \tag{2.12}$$

where $c_{\rm p}$ and $c_{\rm v}$ are the specific heats at constant pressure and constant volume, respectively. Their ratio is $\gamma = c_{\rm p}/c_{\rm v}$, which is 5/3 for a monatomic gas, and their difference is $c_{\rm p} - c_{\rm v} = \mathcal{R}/\mu$.

If the specific entropy of the environment decreases in the upward direction, an upward moving blob of gas will develop excess entropy; see Fig. 2.1. Assuming pressure equilibrium across the blob, (2.12) shows that a positive entropy excess δs corresponds to a density deficit, $-c_{\rm p}\delta \ln \rho$. Thus, the blob will be lighter than its surroundings and will therefore be buoyant, which drives the convection. Likewise, a downward moving blob will become heavier and fall even faster.

Using an equation of state for a perfect gas, i.e. $p = (\mathcal{R}/\mu)\rho T$ we have $\mathrm{d}\ln p = \mathrm{d}\ln\rho + \mathrm{d}\ln T$, and therefore (2.12) gives

$$\frac{\mu}{\mathcal{R}} \frac{\mathrm{d}s}{\mathrm{d}\ln T} = \frac{1}{\gamma - 1} - m . \tag{2.13}$$

This shows that, once m drops below $1/(\gamma - 1) = 1.5$, specific entropy decreases in the upward direction, i.e. in the direction of decreasing temperature. As a result, convection sets in which rapidly mixes the gas and causes the specific entropy to be nearly constant, keeping the effective value of m always close to the critical value of 1.5.

In order to calculate the actual stratification, we need to solve (2.5) and (2.7) together with the equations describing the increase of M_r and L_r with radius. Assuming that M_r and L_r are constant (valid far enough away from the core), we are left with two equations, which we express in terms of $\ln p$ and $\ln T$, so

$$\frac{d\ln p}{dr} = -\frac{\mu}{\mathcal{R}T}\frac{GM_r}{r^2}, \tag{2.14}$$

$$\frac{d\ln T}{dr} = -\frac{1}{KT}\frac{L_r}{4\pi r^2}. \tag{2.15}$$

It is convenient to integrate these equations in the form

$$\frac{d\ln p}{dr} = -\frac{1}{H_p} \quad \text{and} \quad \frac{d\ln T}{dr} = -\frac{\nabla}{H_p}, \tag{2.16}$$

where the symbol ∇ is commonly used in astrophysics for the local value of $d\ln T/d\ln p$, and $H_p = \mathcal{R}T/(\mu g)$ is the local pressure scale height. In the convectively stable regions, i.e. where $m > 3/2$ (corresponding to $\nabla < \nabla_{ad} = 2/5$, and neglecting partial ionization effects), we have $\nabla = \nabla_{rad}$, where ∇_{rad} can be found by dividing (2.15) by (2.14), so

$$\nabla_{rad} = \frac{1}{K}\frac{\mathcal{R}}{\mu}\frac{L_r}{4\pi GM_r}. \tag{2.17}$$

Inside convection zones, on the other hand, ∇ is replaced by ∇_{ad}, so in general we can write $\nabla = \min(\nabla_{rad}, \nabla_{ad})$. In Fig. 2.2 we show solutions obtained by integrating from $r = 500$ Mm (1 Mm = 1000 km) upward using $T = 4\times10^6$ K as starting value with ρ chosen such that the resulting value of m is either just below or just above $13/4 = 3.25$.

The considerations above have demonstrated that m must indeed be quite close to $13/4 = 3.25$ in the radiative interior, but that its value decreases over a depth of about 50 Mm to the adiabatic value of 1.5 just below the bottom of the convection zone. The precise location of the bottom of the convection zone depends on the value of specific entropy in the bulk of the convection zone; see the middle panel of Fig. 2.2. This value depends on the detailed surface physics and in particular the value of the opacity at the top of the convection zone. Here the Kramers opacity is no longer appropriate and the opacity from producing a negative hydrogen ion by polarizing a neutral hydrogen atom through a nearby charge becomes extremely important.

2.2.4 Mixing Length Theory and Convection Simulations

The approximation of setting $\nabla = 2/5$ in the unstable region becomes poor near the surface layers where density is small and energy transport by turbulent elements less efficient. In fact, if the specific entropy were completely constant throughout the convection zone, there would be no net exchange of entropy by the turbulent elements. The definition for the convective flux is

$$\boldsymbol{F}_{conv} = \overline{(\rho \boldsymbol{u})' c_p T'}, \tag{2.18}$$

where overbars denote horizontal averages and primes denote fluctuations about these averages. A mean field calculation shows that $\boldsymbol{F}_{conv}$ is proportional to the negative entropy gradient (see the monograph by Rüdiger 1989),

$$\boldsymbol{F}_{conv} = -\chi_t\overline{\rho}\,\overline{T}\nabla\overline{s} \quad (\text{if } \boldsymbol{g}\cdot\nabla\overline{s} > 0), \tag{2.19}$$

where χ_t is a turbulent diffusion coefficient. In the following we omit the overbars for simplicity. Note that, by comparison with (2.7), in a turbulent environment Fickian diffusion down the *temperature* gradient gets effectively replaced by a similar diffusion down the *entropy* gradient. As with all other types of diffusion coefficients, the diffusion coefficient is proportional to the speed of the fluid parcels accomplishing the diffusion, and the length over which such parcels stay coherent (i.e. the mean free path which is commonly also denoted as the mixing length). Thus, we have

$$\chi_t = \frac{1}{3}u_{rms}\ell. \tag{2.20}$$

The subscript t indicates that this coefficient applies to turbulent transport of averaged fields. Given that the total flux is known, and also the fractional contribution from the radiative flux, we know also the convective flux. Thus, (2.19) can be used to determine the radial entropy gradient, provided we know χ_t, and hence u_{rms} and ℓ.

A natural length scale in the problem is the scale height, so we assume that the mixing length is some frac-

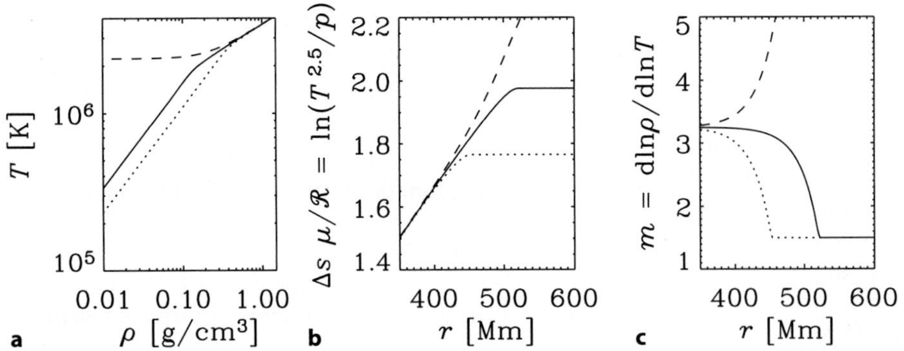

Fig. 2.2a–c. Solutions of (2.16) and (2.17), starting the integration at $r = 350$ Mm with $T = 4 \times 10^6$ K and three different values of the density: $\rho = 1.340$ (*dashed line*), $\rho = 1.345$ (*solid line*), and $\rho = 1.350$ (*dotted line*). The *left hand panel* shows temperature versus density. In this panel the integration goes from *right* to *left* (i.e. in the upward direction toward lower density). The *middle panel* shows the radial specific entropy profile; note that for the two cases with $\rho \le 1.345$ (*solid* and *dotted lines*) a convection zone develops at $r \approx 500$ and 420 Mm, respectively. These two cases correspond to cases where $m < 13/4$ at the lower boundary, as seen from the *right hand panel*. Note the positive entropy gradient indicating stability. In the last two panels the integration goes from *left* to *right*

tion $\alpha_{\rm mix}$ of the local vertical pressure scale height, i.e.

$$\ell = \alpha_{\rm mix} H_{\rm p} \,. \tag{2.21}$$

The scaling of the rms velocity is constrained by (2.18). Assuming that temperature and velocities are well correlated (warm always up, cool always down), we can also write

$$F_{\rm conv} \approx \rho u_{\rm rms} c_{\rm p} \delta T \,, \tag{2.22}$$

where $\delta T = (\overline{T'^2})^{1/2}$ is the rms temperature fluctuation. The relative proportion, with which convection produces velocity and temperature fluctuations, can be estimated by balancing the buoyancy force of a blob against its drag force, so $F_{\rm buoy} = F_{\rm D}^{\rm (turb)}$ and therefore $\delta\rho\, gV = C_{\rm D} \rho u_{\rm rms}^2 S$, where $C_{\rm D}$ is the drag coefficient, V is the volume of the blob, and S its cross-sectional area. We parameterize the ratio $V/(C_{\rm D}S) = \alpha_{\rm vol} H_{\rm p}$, where $\alpha_{\rm vol}$ is a nondimensional factor of order unity characterizing the blob's volume to surface ratio. Assuming pressure equilibrium we have furthermore $|\delta\rho/\rho| = |\delta T/T|$, and

$$u_{\rm rms}^2 = \alpha_{\rm vol}\, g H_{\rm p} \frac{\delta T}{T} \,. \tag{2.23}$$

Thus, $u_{\rm rms}$ is proportional to $(\delta T/T)^{1/2}$, so $F_{\rm conv}$ is proportional to $(\delta T/T)^{3/2}$, and therefore

$$\delta T/T \sim F_{\rm conv}^{2/3} \quad \text{and} \quad u_{\rm rms}/c_{\rm s} \sim F_{\rm conv}^{1/3} \,. \tag{2.24}$$

These scaling relations hold also locally at each depth; see Fig. 2.3, where we show that in simulations of Rayleigh–Benard convection; the vertical profiles of the normalized mean squared vertical velocity, $\langle u_z^2 \rangle / c_{\rm s}^2$, and of the relative temperature variance, $\delta T/T$, are indeed locally proportional to $[F_{\rm conv}/(\rho c_{\rm s}^3)]^{2/3}$. In Fig. 2.3 the nondimensional coefficients are $k_T \approx 1.1$ and $k_u \approx 0.4$, which implies

$$F_{\rm conv} \approx k_u^{-3/2} \rho u_{\rm rms}^3 \,, \tag{2.25}$$

where $k^{-3/2} \approx 4$ and $\langle u_z^2 \rangle = u_{\rm rms}^2$ has been used. Using $F_{\rm conv} = 7 \times 10^7\ {\rm W\,m^{-2}}$ and $\rho = 10\ {\rm kg\,m^{-3}}$ at a depth of about 40 Mm this equation implies $u_{\rm rms} = 120\ {\rm m\,s^{-1}}$.

Using (2.19) and the fact that $\chi_{\rm t} \propto u_{\rm rms} \propto F_{\rm conv}^{1/3}$ we have $F_{\rm conv} \propto F_{\rm conv}^{1/3} |{\rm d}s/{\rm d}z|$, or[1]

$$|{\rm d}s/{\rm d}z| \propto F_{\rm conv}^{2/3} \,. \tag{2.26}$$

[1] A more rigorous calculation using the equations above shows that

$$\frac{{\rm d}s/c_{\rm p}}{{\rm d}z} = -\frac{k}{H_{\rm p}} \left(\frac{F_{\rm tot}}{\rho c_{\rm s}^3} \right)^{2/3} ,$$

$$\text{where} \quad k = 3 \frac{\gamma - 1}{\alpha_{\rm mix}} \left[\alpha_{\rm vol} \left(1 - \frac{1}{\gamma} \right) \right]^{-1/3} ,$$

and $k \approx 1$ for $\alpha_{\rm mix} = \alpha_{\rm vol} = 2$.

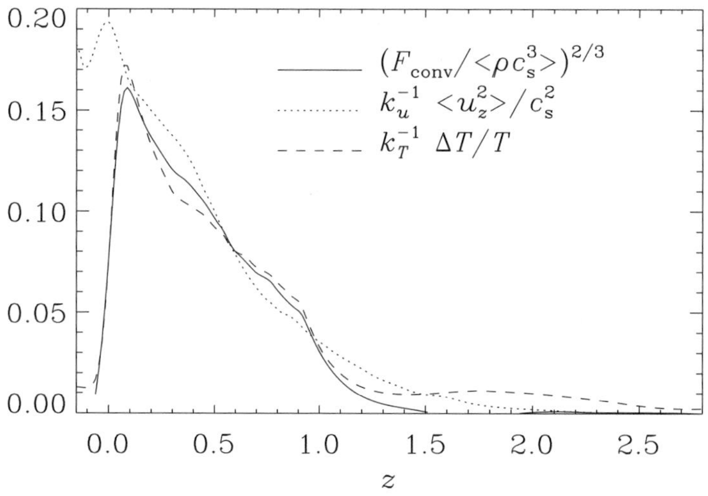

Fig. 2.3. Vertical profiles of the normalized mean squared vertical velocity fluctuations and temperature fluctuations, compared with the normalized convective flux raised to the power 2/3. Note the good agreement between the three curves within the convection zone proper. In this plot, z denotes depth. The positions $z = 0$ and 1 corresponds to the top and bottom of the convection zone, respectively. There is a lower overshoot layer for $z > 1$ and an upper overshoot layer for $z < 0$. [Adapted from Brandenburg et al. (2005)]

In calculating the specific entropy gradient, we can, as a first approximation, assume that $F_{\rm conv}$ is approximately the total flux. However, it would not be difficult to calculate the entropy gradient self-consistently by solving a cubic equation. We also note that the entropy gradient is related to ∇ by

$$\frac{\mathrm{d}s/c_{\rm p}}{\mathrm{d}\ln p} = \nabla - \nabla_{\rm ad}\,, \quad \text{where} \quad \nabla_{\rm ad} = 1 - \frac{1}{\gamma}\,. \qquad (2.27)$$

A solution of the full system of equations, which include more realistic physics than what has been described here, has been given by Spruit (1974); see Table 2.1. The rms velocities are about half as big as expected from (2.25).

Near and beyond the upper and lower boundaries of the convection zones the approximation (2.23) becomes bad, because it ignores the fact that convective elements have inertia and can therefore overshoot a significant distance into the stably stratified regions. In those layers where the entropy gradient has reversed, a downward moving fluid parcel becomes hotter than its surroundings. Thus, in these layers the convection carries convective flux downward, so its sign is reversed. Simulations have clearly demonstrated that, owing to strong stratification, convection will be highly inhomogeneous, with narrow downdrafts and broad upwellings. This leads to a characteristic (but irregular) pattern of convection; see, e.g., the text book by Stix (2002).

Table 2.1. The solar mixing length model of Spruit (1974)

z [Mm]	T [K]	ρ [g cm^{-3}]	$H_{\rm p}$ [Mm]	$u_{\rm rms}$ [m/s]	τ [d]	$\nu_{\rm t}$ [cm^2/s]	$\Omega_{0\tau}$
24	1.8×10^5	0.004	8	70	1.3	1.5×10^{12}	0.6
39	3.0×10^5	0.010	13	56	2.8	2.0×10^{12}	1.3
155	1.6×10^6	0.12	48	25	22	3.2×10^{12}	10
198	2.2×10^6	0.20	56	4	157	0.6×10^{12}	70

The precise location of the bottom of the convection zone in now fairly well determined from detailed models of stellar structure, where the full evolution from a zero-age main sequence star to a chemically evolved star where some of the hydrogen has been burnt into helium and other elements, has been taken into account. An even more accurate and quite independent determination of the bottom of the convection zone and the overall stratification is possible through helioseismology. This will be discussed in the next section.

2.3 Helioseismology

The Sun exhibits so-called five-minute oscillations that are best seen in spectral line shifts. These oscillations were first thought to be the oscillatory response of the atmosphere to convection granules pushing upwards into stably stratified layers. This idea turned out to be wrong, because the oscillations are actually *global* oscillations penetrating deep layers of the Sun. In fact, they are just sound waves that are *trapped* in a cavity formed by reflection at the top and refraction in deeper layers. At the top, sound waves cannot penetrate if their wave length exceeds the scale on which density changes. The refraction in deeper layers is caused by the higher wave speed of the wave front in those parts that are deeper in the Sun. This makes the wave front bend back up again.

The decisive observation came when a wavenumber–frequency (or $k-\omega$) diagram was produced that showed that these modes have long term and large scale spatio-temporal coherence with wavenumbers corresponding to 20 – 60 Mm; see Fig. 2.4. By now the determination of $k-\omega$ diagrams has grown to a mature and standard tool in solar physics.

2.3.1 Qualitative Description

Since the beginning of the eighties, standing acoustic waves in the Sun have been studied in great detail. It has become possible to measure directly (i.e. without the use of a solar model):

(i) the radial dependence of the sound speed, $c_s(r)$, which is proportional to the temperature. Note that $c_s^2 = \gamma p/\rho = \gamma \mathcal{R} T/\mu$, but the mean molecular weight increases near the core due to the nuclear reaction products.
(ii) the radial and latitudinal dependence of the internal angular velocity, $\Omega = \Omega(r,\theta)$, throughout the Sun.

This technique is called *helioseismology*, because it is mathematically similar to the techniques used in seismology of the Earth's interior. Qualitatively, the radial dependence of the sound speed can be measured, because standing sound waves of different horizontal wave number penetrate to different depths. Therefore, the frequencies of those different waves depend on how exactly the sound speed changes with depth. Since the Sun rotates, the waves that travel in the direction of rotation (i.e. toward us) are blue-shifted, and those that travel against the direction of rotation (i.e. away from us) are red-shifted. Therefore, the frequencies are split, depending on the amount of rotation in different layers. There are many reviews on the subject (e.g., Demarque & Guenther 1999). Here we follow the text book by Stix (2002).

Acoustic waves are possible, because they are constantly being excited by the "noise" generated in the convection zone via stochastic excitation. The random fluctuations in the convection are turbulent and contain power at all frequencies. Now the Sun is a harmonic oscillator for sound waves and the different sound modes can be excited stochastically.

Helioseismology has now grown to be immensely sophisticated and more accurate data have emerged from observations with the Michelson Doppler Imager aboard the SOHO spacecraft, located at the inner Lagrange point between Sun and Earth, and also the GONG project (GONG = Global Oscillation Network Group). The latter involves six stations around the globe to eliminate nightly gaps in the data.

2.3.2 Inverting the Frequency Spectrum

As with a violin string, the acoustic frequency of the wave increases as the wavelength decreases. More precisely, the frequency is given by $\nu = c_s/\lambda$, where λ is the wavelength and c_s is the sound speed. We will also use the circular frequency $\omega = 2\pi\nu$ with $\omega = c_s k$, where $k = 2\pi/\lambda$ is the wavenumber. If sound waves travel an oblique path then we can express the wavenumber in terms of its horizontal and vertical wavenumbers, k_h and k_v, respectively. We do this because only the horizontal wavenumber can be observed. This corresponds to the horizontal pattern in Fig. 2.4. Thus, we have

$$k^2 = k_h^2 + k_v^2 \,. \tag{2.28}$$

The number of radial nodes of the wave is given by the number of waves that fit into that part of the Sun where the corresponding wave can travel. This part of the Sun is referred to as the cavity. The larger the cavity, the more nodes there are for a given wavelength. The number of modes n is given by

$$n = 2\Delta r/\lambda = 2\Delta r \frac{k_v}{2\pi} = \Delta r k_v/\pi \,, \tag{2.29}$$

where Δr is the depth of the cavity. If the sound speed and hence k_v depend on radius, this formula must be generalized to

$$n = \frac{1}{\pi} \int_{r_{\min}}^{R_\odot} k_v \, \mathrm{d}r \,, \tag{2.30}$$

supposing the cavity to be the spherical shell $r_{\min} < r < R_\odot$.

The horizontal pattern of the proper oscillation is described by spherical harmonics with indices l and m,

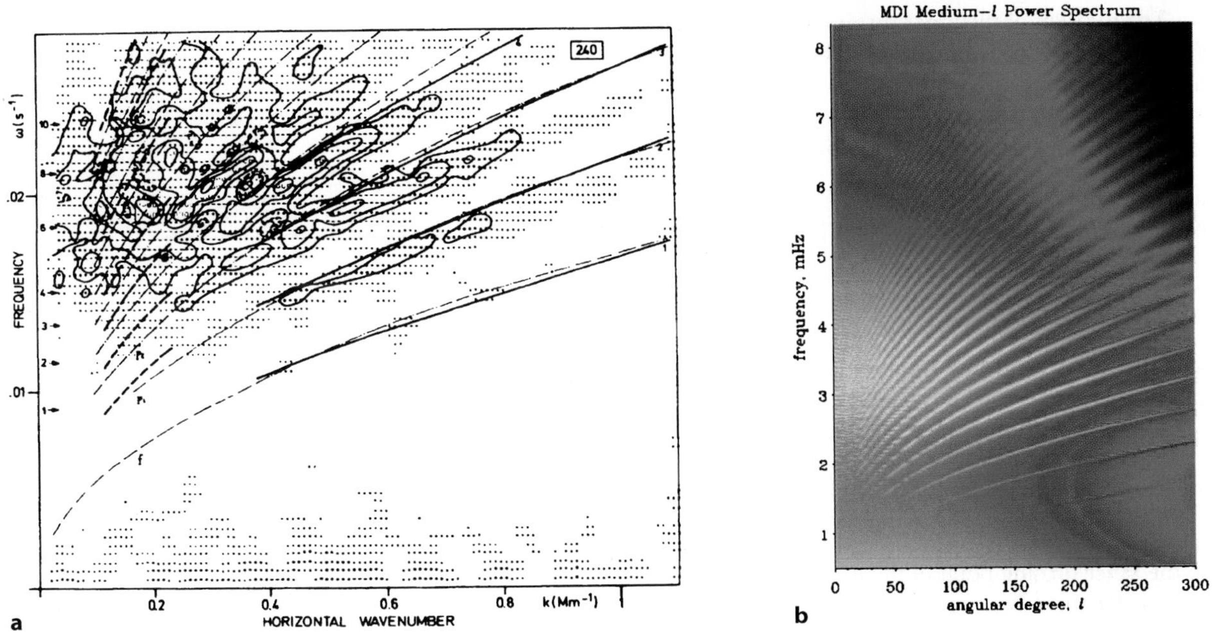

Fig. 2.4a,b. Comparison between the $k_h - \omega$ (or $l - \nu$) diagrams obtained by Deubner in 1975 (**a**), and by the SOHO/MDI team in 2000 (**b**). The figure by Deubner, where he compares observations with the predictions of Ulrich (1970), proved that the 5-min oscillations were global modes. Courtesy F.-L. Deubner (**a**) and P. Scherrer (**b**)

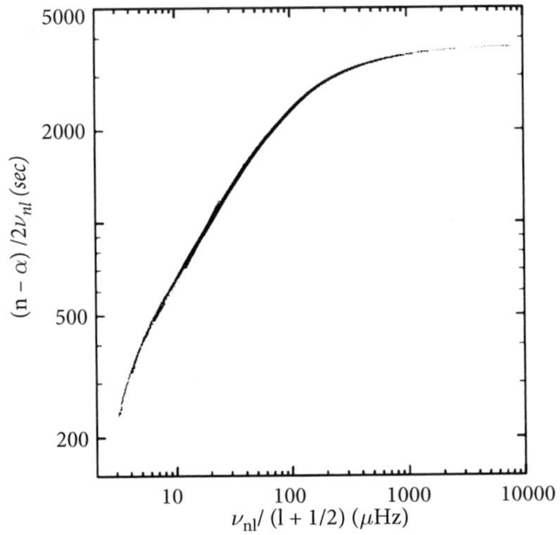

Fig. 2.5. The Duvall law. The *vertical axis* (ordinate) corresponds to F in (2.37) and the *horizontal axis* (abscissa) is basically u^{-1}. He found this law well before its significance was understood in terms of one of the functions in Abel's integral transformation. [Courtesy J. Christensen-Dalsgaard et al. (1985)]

hence the horizontal wavenumber is

$$k_\mathrm{h}^2 = \frac{\ell(\ell+1)}{r^2}\,, \tag{2.31}$$

and we can write

$$k_\mathrm{v} = \sqrt{\frac{\omega_{nl}^2}{c_\mathrm{s}^2} - \frac{\ell(\ell+1)}{r^2}} = \frac{\omega_{nl}}{r}\sqrt{\frac{r^2}{c_\mathrm{s}^2} - \frac{\ell(\ell+1)}{\omega_{nl}^2}}\,. \tag{2.32}$$

where the subscripts of ω_{nl} denote the radial order n and the spherical harmonic degree l of the modes. Therefore, the number n of radial nodes is given by

$$\frac{\pi(n+\alpha)}{\omega_{nl}} = \int_{r_\mathrm{min}}^{R_\odot} \sqrt{\frac{r^2}{c_\mathrm{s}^2} - \frac{\ell(\ell+1)}{\omega_{nl}^2}}\,\frac{\mathrm{d}r}{r}\,, \tag{2.33}$$

where an empirically (or otherwise) determined phase shift $\alpha \approx 1.5$ accounts for the fact that the standing waves are confined by barriers that are "soft" and extended, rather than rigid and fixed.

The location of the inner turning radius is given by the point where the wavevector has become completely horizontal. Using

$$\omega_{nl}^2/c_s^2 = k^2 = k_h^2 + k_v^2 , \tag{2.34}$$

together with $k_v = 0$ at $r = r_{min}$ and $k_h^2 = \ell(\ell+1)/r^2$, we have $(r_{min}/c_s)^2 = \ell(\ell+1)/\omega_{nl}^2$. This implies that

$$r_{min} = \frac{c_s}{\omega_{nl}}\sqrt{\ell(\ell+1)} , \tag{2.35}$$

so only modes with low ℓ values have turning points close to the center and can be used to examine the Sun's core. We now introduce new variables

$$\xi = \frac{r^2}{c_s^2}, \qquad u = \frac{\ell(\ell+1)}{\omega_{nl}^2} , \tag{2.36}$$

so the inner turning point of the modes corresponds to $\xi = u$. Furthermore, we denote the left hand side of (2.33) by $F(u)$, so we can write

$$F(u) = \int_u^{\xi_\odot} \sqrt{\xi - u}\frac{\mathrm{d}\ln r}{\mathrm{d}\xi}\mathrm{d}\xi , \tag{2.37}$$

where the location of the inner refraction point corresponds to $u = \xi$. The function $F(u)$ was obtained from observations by Duvall (1982) on the grounds that this combination of data makes the different branches collapse onto one (see Fig. 2.5). He discovered this well before its significance was understood by Gough (1985) several years later.

Since we know $F(u)$ from observations and are interested in the connection between r and ξ (i.e. r and c_s), we interpret (2.37) as an integral equation for the unknown function $r(\xi)$. Most integral equations cannot be solved in closed form, but this one can. Gough (1985) realized that it can be cast in the form of Abel's integral equation. The pair of complementary equations (primes denote derivatives) is

$$F(u) = \int_u^{\xi_\odot} \sqrt{\xi - u}G'(\xi)\mathrm{d}\xi , \tag{2.38}$$

$$G(\xi) = \frac{2}{\pi}\int_\xi^{\xi_\odot} \frac{1}{\sqrt{\xi - u}}F'(u)\mathrm{d}u . \tag{2.39}$$

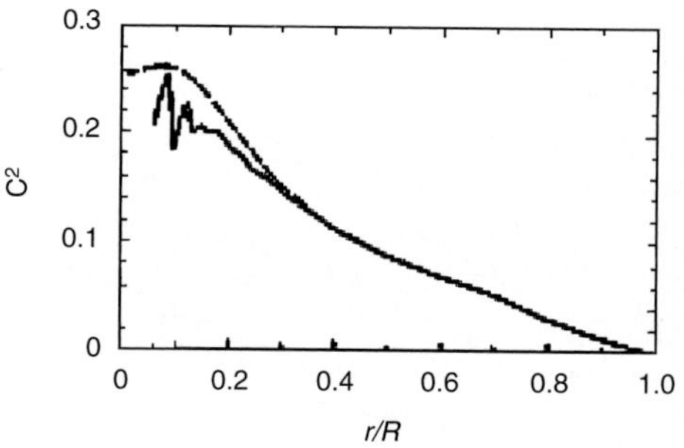

Fig. 2.6. Radial dependence of the sound speed on radius in the Sun. Note the change in slope near a radius of 0.7 solar radii. The oscillations near the center are not physical. The theoretical model (*dotted line*) is in fair agreement with the direct measurements. The sound speed has its maximum not in the center, because the mean molecular weight μ increases towards the center, which causes c_s to decrease. (We recall that $c_s^2(r) = \gamma\mathcal{R}T/\mu$.) [Adapted from Stix (2002)]

Inserting the definitions for ξ and u into (2.39), we obtain

$$\begin{aligned}\int_\xi^{\xi_\odot} \frac{F'(u)}{\sqrt{u-\xi}}\mathrm{d}u &= -\frac{\pi}{2}\int_\xi^{\xi_\odot} \frac{\mathrm{d}\ln r}{\mathrm{d}\xi'}\mathrm{d}\xi' \\ &= -\frac{\pi}{2}\ln r\Big|_{\xi'=\xi}^{\xi_\odot} = \frac{\pi}{2}\ln\frac{r(\xi)}{R_\odot} .\end{aligned} \tag{2.40}$$

This equation can be solved for $r = r(\xi)$:

$$r(\xi) = R_\odot \exp\left(\frac{2}{\pi}\int_\xi^{\xi_\odot} \frac{F'(u)}{\sqrt{u-\xi}}\mathrm{d}u\right) . \tag{2.41}$$

This is the final result of inverting the integral equation (2.37). It establishes the link between the observable function $F(u)$ and the function $r(\xi)$, from which the radial profile of the sound velocity c_s can be obtained. Figure 2.6 gives the result of an inversion procedure that computes the radial dependence of the sound speed on depth, using the detailed frequency spectrum as input.

It should be noted, however, that this approach is usually not practical when input data are noisy. Instead, a minimization procedure is often used where the resulting function is by construction smooth. This procedure falls under the general name of *inverse theory* and is frequently used in various branches of astrophysics.

Historically, the model independent determination of the sound speed and thereby the temperature in the center of the Sun has been important in connection with understanding the origin of the solar neutrino problem. In fact, the solar neutrino flux was measured to be only one third of that originally expected. A lower core temperature could have resolved this mismatch, but this possibility was then ruled out by helioseismology. Now we know that there are neutrino oscillations leading to a continuous interchange between the three different neutrino species, which explains the observed neutrino flux of just one species.

2.3.3 The Solar Abundance Problem

Opacities depend largely on the abundance of heavier elements. The solar models calculated with the old tables agreed quite well using the conventional abundance ratio of heavier elements to hydrogen, $Z/X = 0.023$. However, the abundancies were based on fits of observed spectra to synthetic line spectra calculated from model atmospheres. These models parameterize the three-dimensional convection only rather crudely. New synthetic line spectra calculated from three-dimensional time-dependent hydrodynamical models of the solar atmosphere give a lower value of the solar oxygen abundance. With the new values ($Z/X = 0.017$) it became difficult to reconcile the previously good agreement between stellar models and helioseismology. The solution to this problem is still unclear, but there is now evidence that the solar neon abundance may have been underestimated. A neon abundance enhanced by about 2.5 is sufficient to restore the good agreement found previously.

The detailed stratification depends quite sensitively on the equation of state, $p = p(\rho, T)$. However, the uncertainties in the theoretically determined equation of state are now quite small and cannot be held responsible for reconciling the helioseismic mismatch after adopting the revised solar abundancies.

2.3.4 Internal Solar Rotation Rate

Another important problem is to calculate the internal rotation rate of the Sun (Fig. 2.7). This has already been possible for the past 20 years, but the accuracy has been ever improving. We will not discuss here the mathematics in any further detail, but refer instead to the review by Thompson et al. (2003). The basic technique involves the prior calculation of kernel functions, $K_{nlm}(r, \theta)$, that are independent of Ω, such that the rotational frequency splitting can be expressed as

$$\omega_{nlm} - \omega_{nl0} = m \int_0^R \int_0^\pi K_{nlm}(r, \theta)\Omega(r, \theta) r \, \mathrm{d}r \, \mathrm{d}\theta \tag{2.42}$$

Several robust features that have emerged from the work of several groups include.

- The contours of constant angular velocity do not show a tendency of alignment with the axis of rotation, as one would have expected, and as many theoretical models still show.
- The angular velocity in the radiative interior is nearly constant, so there is no rapidly rotating core, as has sometimes been speculated.
- There is a narrow transition layer at the bottom of the convection zone, where the latitudinal differential rotation goes over into rigid rotation (i.e. the tachocline). Below 30° latitude the radial angular velocity gradient is here positive, i.e. $\partial\Omega/\partial r > 0$, in contrast that what is demanded by conventional dynamo theories.
- Near the top layers (outer 5%) the angular velocity gradient is negative and quite sharp.

A completely model-independent knowledge about the internal rotation rate of the Sun has proved to be invaluable for the theory of the magnetic field in the Sun, for its rotation history, and for solar dynamo theory. Prior to the advent of helioseismology some 25 years ago, the idea of a rapidly rotating core was quite plausible, because at birth the Sun is believed to have spun at least 50 times faster than now, and because in the Sun the viscous time scale exceeds the age of the Sun. The fact that also the core has spun down means that there must be some efficient torques accomplishing the angular momentum transport inside the Sun. A likely candidate is the magnetic field. It it indeed well known that only a weak poloidal field is needed to brake the rotation of the radiative interior.

Helioseismology has indicated that the transition from latitudinal differential rotation in the bulk of the convection zone to nearly rigid rotation in the radiative

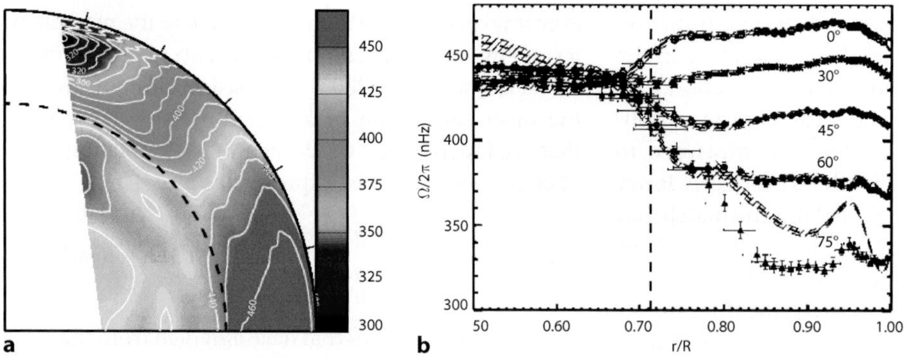

Fig. 2.7a,b. Angular velocity profile in the solar interior inferred from helioseismology (after Thompson et al. 2003). In panel (**a**) a two-dimensional (*latitude-radius*) rotational inversion is shown based on the Subtractive Optimally Localized Averaging (SOLA) technique. In panel (**b**) the angular velocity is plotted as a function of radius for several selected latitudes, based on both SOLA (*symbols*, with 1σ error bars) and regularized least squares (RLS; *dashed lines*) inversion techniques. *Dashed lines* indicate the base of the convection zone. All inversions are based on data from the Michelson Doppler Imager aboard the SOHO spacecraft, averaged over 144 days. Inversions become unreliable close to the rotation axis, represented by *white areas* in panel (**a**). Note also that global modes are only sensitive to the rotation component which is symmetric about the equator (courtesy M.J. Thompson et al. 2003)

interior is relatively sharp. This transition region is called the tachocline. The idea of a sharp transition region has problems of its own, because viscous spreading would tend to smooth the transition with time. The solution to this problem was thought to be related to the effect of a mostly horizontal turbulence. However, it can be argued that the rigidity of the radiative interior is constantly maintained by the presence of a weak magnetic field of about 1 G; see Rüdiger & Hollerbach (2004) for a recent monograph covering also this aspect.

2.3.5 Local Helioseismology

At larger values of ℓ the coherence time of the waves becomes rather short and the modes are no longer global and take on a more local character. There are various techniques that use these modes to extract information about local variations of sound speed and local flows. The most popular method is the ring diagram technique. For a detailed review see Gizon & Birch (2005). Among other things this method has demonstrated the presence of converging flows around sunspots and a rather shallow temperature subsurface structure. However, a serious shortcoming of the present approach is the neglect of magnetosonic and Alfvén waves.

2.4 Solar Activity Cycle

In the following we discuss some basic properties of the solar magnetic field. Its main feature is the 11 year cycle, as manifested in the (approximately) eleven year variation of the sunspot number. Sunspots are associated with sites of a strong magnetic field of about 2 – 3 kG peak field strength. Sunspots appear typically at about $\pm 30°$ latitude at the beginning of each cycle, i.e. when the sunspot number begins to rise again. During the course of the cycle, spots appear at progressively lower latitudes. At the end of the cycle, sunspots appear at low latitudes of about $\pm 4°$. Again, detailed references cannot be given here, but we refer to the paper by Solanki et al. (2006) for a recent review.

2.4.1 The Butterfly Diagram

Although the detailed mechanism of their formation is still uncertain, it seems that sunspots form when a certain threshold field is exceeded, so they occur usually only below $\pm 30°$ latitude. However, magnetic fields can still be detected at higher latitudes all the way up to the poles using the Zeeman effect. Figure 2.8 shows, as a function of latitude and time, the normal component of the azimuthally averaged surface field, $\overline{\boldsymbol{B}}(R_\odot, \theta, t)$, where

$$\overline{\boldsymbol{B}}(r,\theta,t) = \int_0^{2\pi} \overline{B}\frac{\mathrm{d}\phi}{2\pi} . \tag{2.43}$$

Such diagrams, which can also be produced for the mean number of sunspots as function of time and latitude, are generally referred to as butterfly diagrams.

Although the field strength in sunspots is about 2 kG, when the field is averaged in longitude only a small net field of about ±20 G remains. Near the poles the magnetic field is more clearly defined because it fluctuates less strongly in time near the poles than at lower latitudes. A characteristic feature is that the polar field changes sign shortly after each sunspot maximum.

At intermediate latitudes $|\cos\theta| = 0.5\ldots0.7$, corresponding to a latitude, $90° - \theta$, of $\pm(30°\ldots45°)$, there are characteristic streaks of magnetic activity that seem to move poleward over a short time ($\sim 1\ldots2$ yr). These streaks are rather suggestive of systematic advection by poleward meridional circulation near the surface. This indicates that the streaks are really just a consequence of the remaining flux of decaying active regions being advected poleward from lower latitudes. Looking at a plot of the magnetic field at poorer resolution would show what is known as the polar branch, whose presence has been found previously through various other proxies (e.g. through the migration of the line where prominences occur). This has been reviewed in detail by Stix (1974).

In summary, the cyclic variation of the field together with its latitudinal migration, and the alternating orientation of bipolar magnetic regions are the main systematic properties of the solar magnetic field. In Sect. 2.5 we discuss theoretical approaches to the present understanding of this phenomenon.

2.4.2 Cyclic Activity on Other Solar-Like Stars

It should be noted that magnetic activity and activity cycles are not unique to the Sun. In fact, many stars with outer convection zones display magnetic activity, as is evidenced by proxies such as the H and K line emission within the Calcium absorption line. This H and K line emission is caused by hot plasma that is confined in the magnetic flux tubes in the coronae of these stars. Among the solar-like stars of spectral type G and solar-like rotation, many have cyclic magnetic activity while others show time-independent magnetic activity that is believed to be associated with the possibility that these stars are in a grand minimum, such as the famous Maunder minimum.

2.4.3 Grand Minima

Grand minima are recurrent states of global magnetic inactivity of a star. This behavior may be associated with the chaotic nature of the underlying dynamo process. For the Sun this behavior is evidenced through the record of the Carbon 14 isotope concentrations in tree rings as well as through the Beryllium 10 isotope concentrations of ice core drillings from Greenland. It is interesting to note that during the Maunder minimum between 1645 and 1700 the magnetic activity was not completely suppressed; ^{10}Be still show cyclic variability, albeit with a somewhat longer period of about 15 years. Shortly after the Sun emerged from the Maunder minimum the sunspot activity was confined only to the northern hemisphere. This type of latitudinal asymmetry has been seen in some dynamo models that display sporadically a mixture of modes that are symmetric and antisymmetric about the equator. For the Sun, some of the earlier grand minima have specific names such as the Spörer minimum (1420–1530), the Wolf Minimum (1280–1340), and the Oort minimum (1010–1050).

Grand minima can be important for the Earth's climate. For example the Maunder minimum is associated with the 'Little Ice Age' that occurred from 1560 to 1850. During the 500 years before that the Sun was particularly active as is evidenced by the high levels of ^{14}C production: this was the period when wine was made from grapes grown in England and when the Vikings colonized Greenland.

By combining different proxies of solar activity, several typical time scales can be identified, the Schwabe 11-year cycle, the 88-year Gleissberg cycle, the 205-year De Vries cycle, and the 2100 or 2300 year Hallstatt cycle.

2.4.4 Active Regions and Active Longitudes

Active regions are complexes of magnetic activity out of which sunspots, flares, coronal mass ejections, and several other phenomena emerge with some preference over other regions. These regions tend to be bipolar, i.e. they come in pairs of opposite polarity and are roughly aligned with the east–west direction.

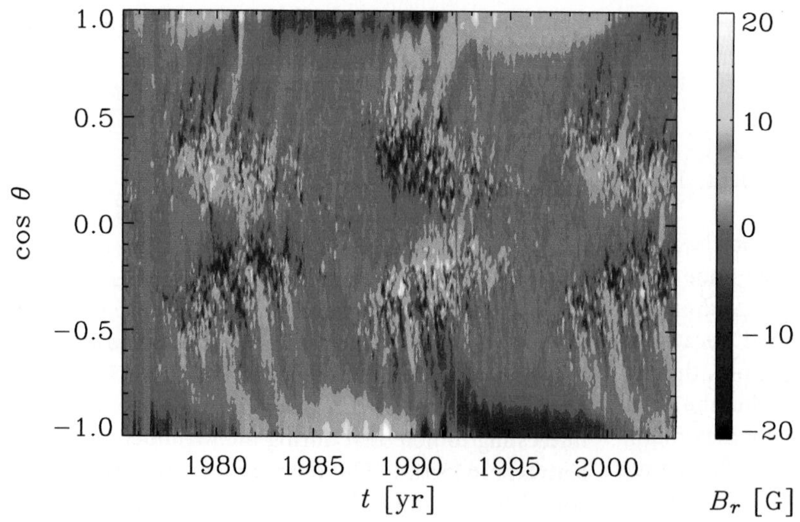

Fig. 2.8. Longitudinally averaged radial component of the observed solar magnetic field as a function of cos(colatitude) and time. *Dark shades* denote negative values and *light shades* denote positive values. Note the sign changes both in time and across the equator (courtesy of R. Knaack)

Over periods of up to half a year active regions appear preferentially at the same longitude and follow a latitude-dependent rotation law. An analysis of solar magnetograms show that at the beginning of each cycle, when most of the activity occurs at about ±30° latitude, the rotation rate of the active longitudes is less than at the end of each cycle, when the typical latitude is only ±4° latitude. There are various reports that these longitudes might be stable over longer periods of time (so-called active longitudes), but this is still very much a matter of debate.

The notion of field line anchoring is occasionally used in connection with sunspot proper motions. Long before the internal angular velocity was determined via helioseismology, it was known that sunspots rotate faster than the surface plasma. Moreover, young sunspots rotate faster than old sunspots. A common interpretation is that young sunspots are still anchored at a greater depth than older ones, and that therefore the internal angular velocity must decrease with height. This provided also the basis for the classical mean field dynamo theory of the solar cycle according to which the radial angular velocity gradient has to be negative. This will be discussed in more detail in Sect. 2.5.

With the advance of helioseismology, it has become clear that at low latitudes the angular velocity decreases with radius throughout the bulk of the convection zone. A negative radial gradient exists only in the upper 30 Mm (sometimes referred to as the supergranulation layer). Indeed, the very youngest sunspots have a rotation rate that is comparable to or even slightly in excess of the fastest angular velocity seen with helioseismology anywhere in the Sun (i.e. at $r/R_\odot \approx 0.95$).

2.4.5 Torsional Oscillations

At the solar surface the angular velocity varies with the 11 year cycle. In other words, $\overline{\Omega}$ at the surface (at $r = R_\odot$) is not only a function of colatitude θ, but also of time. The pattern of $\overline{\Omega}\ (R_\odot, \theta, t)$ shows an equatorward migration, similar to the butterfly diagram of the mean poloidal magnetic field in Fig. 2.8. Helioseismology has now established that this pattern extends at least half way into the convection zone. At the bottom of the convection zone the 11 year variation is not (yet?) observed, but there is possibly a 1.3 year modulation of the local angular velocity, although this is still unclear and debated (see the review by Thompson et al. 2003). In recent years this 1.3 year modulation has gone away, but it has been speculated that the presence of a modulation may depend on the phase in the cycle.

The 11 year cyclic modulation is known as torsional oscillation, but model calculations demonstrate that these oscillations can be understood as a direct response to the varying magnetic field. The amplitude of the torsional oscillations is about 8%, suggesting that magnetic effects must be moderate and the fields of sub-equipartition strength.

2.5 Dynamo Theory

Given that the magnetic decay times in astrophysical plasmas are generally very long, there have been a number of attempts in the literature to explain the Sun's magnetic field in terms of a primordial, frozen-in field. Such approaches tend to be rather sketchy when it comes to predicting any quantitative details that can be tested. Dynamo theory, on the other hand, provides a self-consistent framework of magnetic field generation in general that can be tested against direct simulations. Owing to the turbulent nature of the flows, such dynamos are generally referred to as "turbulent dynamos". Unfortunately, early simulations did not reproduce the solar behavior very well. The reason for this may simply be that, for example, the resolution was insufficient to capture important details. The failure to explain the observations has led to a number of ad hoc assumptions and modifications that are not satisfactory. At the same time, dynamo theory itself has experienced some important extensions that followed from trying to explain a long standing mismatch between simulations and theory, even under rather idealized conditions such as forced turbulence in a periodic domain. In this section we can only outline the basic aspects of dynamo theory. For a more extensive review, especially of the recent developments, see Ossendrijver (2003) and Brandenburg & Subramanian (2005).

2.5.1 The Induction Equation

At the heart of dynamo theory is the induction equation, which is just the Faraday equation together with Ohm's law, i.e.

$$\frac{\partial \boldsymbol{B}}{\partial t} = -\nabla \times \boldsymbol{E} \quad \text{and} \quad \boldsymbol{J} = \sigma\left(\boldsymbol{E} + \boldsymbol{U} \times \boldsymbol{B}\right) , \qquad (2.44)$$

respectively. The initial conditions furthermore must obey $\nabla \cdot \boldsymbol{B} = 0$. Eliminating $\boldsymbol{E}$ yields

$$\frac{\partial \boldsymbol{B}}{\partial t} = \nabla \times \left(\boldsymbol{U} \times \boldsymbol{B} - \boldsymbol{J}/\sigma\right) . \qquad (2.45)$$

Then, using Ampere's law (ignoring the Faraday displacement current), $\boldsymbol{J} = \nabla \times \boldsymbol{B}/\mu_0$, where μ_0 is the vacuum permeability, one obtains the induction equation in a form that reveals the diffusive nature of the last term as $\ldots + \eta\nabla^2\boldsymbol{B}$, where $\eta = (\sigma\mu_0)^{-1}$.

A complete theory of magnetic field evolution must include also the momentum equation, because the magnetic field will react back on the velocity field through the Lorentz force, $\boldsymbol{J} \times \boldsymbol{B}$, so

$$\rho\frac{\mathrm{d}\boldsymbol{U}}{\mathrm{d}t} = -\nabla p + \boldsymbol{J} \times \boldsymbol{B} + \boldsymbol{F} , \qquad (2.46)$$

together with the continuity equation, $\partial\rho/\partial t = -\nabla \cdot (\rho\boldsymbol{U})$. In (2.46), $\boldsymbol{F}$ subsumes a range of possible additional forces such as viscous and gravitational forces, as well as possibly Coriolis and centrifugal forces.

To study the dynamo problem, the complete set of equations is often solved using fully three-dimensional simulations both in Cartesian and in spherical geometries. Especially in early papers, the continuity equation has been replaced by the incompressibility condition, $\nabla \cdot \boldsymbol{U} = 0$, or by the anelastic approximation, $\nabla \cdot (\rho\boldsymbol{U}) = 0$. In both cases, ρ no longer obeys an explicitly time-dependent equation, and yet ρ can of course change via the equation of state (pressure and temperature are still changing). These approximations are technically similar to that of neglecting the Faraday displacement current.

As long as the magnetic field is weak, i.e. $\boldsymbol{B}^2/\mu_0 \ll \rho\boldsymbol{U}^2$ at all scales and all locations, it may be permissible to assume $\boldsymbol{U}$ as given and to solve only the induction equation for $\boldsymbol{B}$. This is called the kinematic dynamo problem.

Meanwhile some types of dynamos have been verified in experiments. One is the Ponomarenko-like dynamo that consists of a swirling flow surrounded by a nonrotating counterflow (Gailitis et al. 2001). The flow is driven by propellers and leads to self-excited dynamo action when the propellers exceed about 1800 revolutions per minute (30 Hz), producing peak fields of up to 1 kG. Another experiment consists of an array of 52 connected tubes with an internal winding structure through which liquid sodium is pumped, making the flow strongly helical with nearly uniform kinetic helicity density within the dynamo module containing the pipes (Stieglitz & Müller 2001). Such a flow is particularly interesting because it allows meaningful averages to be taken, making this problem amenable to a mean field treatment. The mean field approach is important in solar physics and will be discussed in Sect. 2.5.3. First, however, we discuss the case where no mean field is produced and only a small scale field may be generated.

2.5.2 Small Scale Dynamo Action

There is an important distinction between small scale and large scale turbulent dynamos. This is mainly a distinction by the typical scale of the field. Both types of dynamos have in general a turbulent component, but large scale dynamos have an additional component on a scale larger than the typical scale of the turbulence. Physically, this can be caused by the effects of anisotropies, helicity, and/or shear. These large scale dynamos are amenable to mean field modeling (see below). On the other hand, small scale dynamo action is possible under fully isotropic conditions. This process has been studied both analytically and numerically; see Brandenburg & Subramanian (2005) for a review. Indeed, small scale dynamos tend to be quite prominent in simulations, perhaps more so than what is realistic. This may be a consequence of having used unrealistically large values of the magnetic Prandtl number, as will be discussed in the following.

The strength of the small scale dynamo depends significantly on the value of the magnetic Prandtl number $\mathrm{Pr_M} \equiv \nu/\eta$, i.e. the ratio of the viscosity, ν, to the magnetic diffusivity η. In the Sun, $\mathrm{Pr_M}$ varies between 10^{-7} and 10^{-4} between the top and the bottom of the convection zone, but it is always well below unity. In this case the Kolmogorov cutoff scale of the kinetic energy spectrum of the turbulence is much smaller than the resistive cutoff scale of the magnetic energy spectrum. Therefore, at the resistive scale where the small scale dynamo would operate fastest, the velocity is still in its inertial range where the spatial variation of the velocity is much more pronounced than it would be near the Kolmogorov scale, relevant for a magnetic Prandtl number of order unity. This tends to inhibit small scale dynamo action. In many simulations $\mathrm{Pr_M}$ is close to unity, because otherwise the magnetic Reynolds number would be too small for the dynamo to be excited. As a consequence, the production of small scale field may be exaggerated in simulations. It is therefore possible that in the Sun small scale dynamo action is less important, and that large scale dynamo action is by comparison much more prominent, than found in simulations. An example may be the simulations of Brun et al. (2004), which are currently the highest resolution turbulence simulations of solar-like convection in spherical shell geometry. Here the magnetic field is indeed mostly of small scale.

In mean field models only the large scale field is modeled. This large scale field is governed both by turbulent magnetic diffusion as well as non-diffusive contributions such as the famous α

effect. As will be explained in the next section, this means that the mean electromotive force has a component parallel to the mean field, so it has a term of the form $\alpha\overline{\boldsymbol{B}}$; see Brandenburg & Subramanian (2005) for a recent review. However, once a large scale field is present, the turbulent motions (which are always present) will wind up and mix the large scale field and will hence also produce a small scale field. This does *not* represent small scale dynamo action, even though there is a small scale field; if the large scale field is absent, the small scale field disappears.

Let us emphasize again that the Sun does possess a large scale field, with spatio-temporal order, as is evidenced by Fig. 2.8. This automatically implies a small scale field. In addition, there may be small scale dynamo action occurring locally in the near-surface layers where the Coriolis force is comparatively weak, but this depends on whether or not small scale dynamo action is inhibited by a small value of the magnetic Prandtl number.

2.5.3 Mean Field Theory

The mean field approach allows the complicated three-dimensional dynamics to be treated in a statistical manner. The averaged equations are then only two-dimensional. In some cases, e.g. in Cartesian geometry, it can be useful to define two-dimensional averages, so that the resulting mean field equations are only one-dimensional. In the following we describe the essential features of this approach. By averaging the induction equation (2.45), e.g. according to the toroidal averaging procedure, we obtain

$$\frac{\partial \overline{\boldsymbol{B}}}{\partial t} = \nabla \times \left(\overline{\boldsymbol{U}} \times \overline{\boldsymbol{B}} + \overline{\boldsymbol{\mathcal{E}}} - \eta\mu_0 \overline{\boldsymbol{J}} \right) , \qquad (2.47)$$

where $\overline{\boldsymbol{\mathcal{E}}} = \overline{\boldsymbol{u} \times \boldsymbol{b}}$ is the mean electromotive force from the small scale magnetic and velocity fields, with $\boldsymbol{u} = \boldsymbol{U} - \overline{\boldsymbol{U}}$ and $\boldsymbol{b} = \boldsymbol{B} - \overline{\boldsymbol{B}}$ being the fluctuations, i.e. the deviations from the corresponding averages.

There are two quite different approaches to calculating $\overline{\boldsymbol{\mathcal{E}}}$ and its dependence on $\overline{\boldsymbol{B}}$. The first order

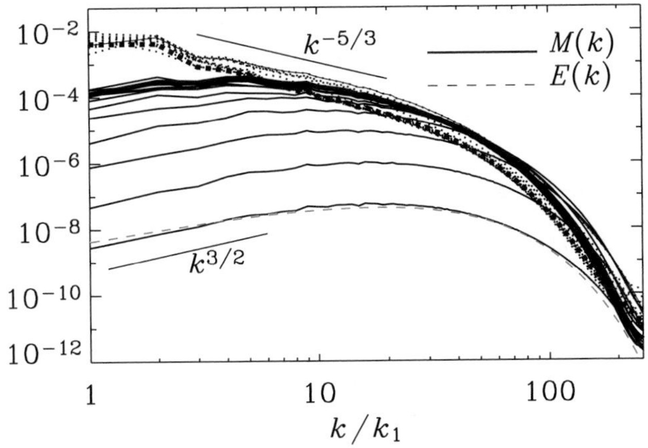

Fig. 2.9. Magnetic and kinetic energy spectra from a non-helical turbulence simulation with $P_m = 1$. The kinetic energy is indicated as a *dashed line* (except for the first time displayed where it is shown as a *thin solid line*). At early times the magnetic energy spectrum follows the $k^{3/2}$ Kazantsev (1968) law (the *dashed line* gives the fit to the analytic spectrum), while the kinetic energy shows a short $k^{-5/3}$ range. The Reynolds number is $u_{\rm rms}/(\nu k_{\rm f}) \approx 600$ and 512^3 meshpoints were used. The time difference between the spectra is about 14 $(k_{\rm f}u_{\rm rms})^{-1}$. [Adapted from Brandenburg & Subramanian (2005)]

smoothing approximation uses just the linearized evolution equation for $\boldsymbol{b}$, while the tau approximation uses also the linearized momentum equation together with a closure hypothesis for the higher order triple correlation terms. For references and historical aspects we refer to the review by Brandenburg & Subramanian (2005). Both approaches predict the presence of terms of the form

$$\overline{E}_i = \alpha_{ip}\overline{B}_p + \eta_{ipl}\overline{B}_{p,l} \,, \tag{2.48}$$

where a comma denotes partial differentiation. The tau approximation gives

$$\alpha_{ip} = -\tau\epsilon_{ijk}\overline{u_k u_{j,p}} + \tau\epsilon_{ijk}\overline{b_k b_{j,p}}/\rho_0 \,, \tag{2.49}$$

where τ is the correlation time. However, within the first order smoothing approximation the magnetic term in α_{ip} is absent. In order to illuminate the meaning of these tensors, it is useful to make the assumption of isotropy, $\tilde{\alpha}_{ip} = \tilde{\alpha}\delta_{ip}$ and $\tilde{\eta}_{ipl} = \tilde{\eta}_{\rm t}\epsilon_{ipl}$. This yields

$$\tilde{\alpha} = -\frac{1}{3}\left(\overline{\omega\cdot\boldsymbol{u}} - \overline{\boldsymbol{j}\cdot\boldsymbol{b}}/\rho_0\right) \,, \quad \tilde{\eta}_{\rm t} = \frac{1}{3}\overline{\boldsymbol{u}^2} \,, \tag{2.50}$$

where $\omega = \nabla\times\boldsymbol{u}$ is the small scale vorticity and $\boldsymbol{j} = \nabla\times\boldsymbol{b}/\mu_0$ is the small scale current density. Thus, $\tilde{\alpha}$ is proportional to the residual helicity, i.e. the difference between kinetic and current helicities, and $\eta_{\rm t}$ is proportional to the mean square velocity.

Using a closure assumption for the triple correlations we have, under the assumption of isotropy, the important result

$$\alpha = -\frac{1}{3}\tau\left(\overline{\omega\cdot\boldsymbol{u}} - \overline{\boldsymbol{j}\cdot\boldsymbol{b}}/\rho_0\right) \,, \quad \eta_{\rm t} = \frac{1}{3}\tau\overline{\boldsymbol{u}^2} \,. \tag{2.51}$$

The electromotive force takes then the form

$$\overline{\boldsymbol{\mathcal{E}}} = \alpha\overline{\boldsymbol{B}} - \eta_{\rm t}\mu_0\overline{\boldsymbol{J}} \,. \tag{2.52}$$

This equation shows that the electromotive force does indeed have a component in the direction of the mean field (with coefficient α). The $\eta_{\rm t}$ term corresponds to a contribution of the electromotive force that is formally similar to the microscopic diffusion term, $\eta\mu_0\overline{\boldsymbol{J}}$, in (2.47). Therefore one speaks also of the total magnetic diffusivity, $\eta_{\rm T} = \eta + \eta_{\rm t}$. The presence of the α term, on the other hand, has no correspondence to the non-turbulent case, and it is this term that invalidates Cowling's anti-dynamo theorem for mean fields. Indeed, there are simple self-excited (exponentially growing) solutions already in a one-dimensional model (see below).

Equation 2.51 shows that the presence of an α effect is closely linked to the presence of kinetic and/or current helicity, while turbulent magnetic diffusion is always present when there is a small scale turbulent velocity field. This shows immediately that just increasing the turbulence (without also increasing the helicity) tends to *diminish* turbulent mean field dynamo action, rather than enhancing it, as one might have thought.

The formalism discussed above does not address the production of kinetic helicity in the Sun. This can be calculated perturbatively by considering the effects of vertical density and turbulent intensity stratification and rotation. At lowest order one finds

$$\alpha_{\phi\phi} = -\frac{16}{15}\tau^2 u_{\rm rms}^2 \Omega\cdot\nabla\ln(\rho u_{\rm rms}) + \ldots \tag{2.53}$$

for the first term in (2.49). For details we refer to the reviews by Rüdiger & Hollerbach (2004) and Branden-

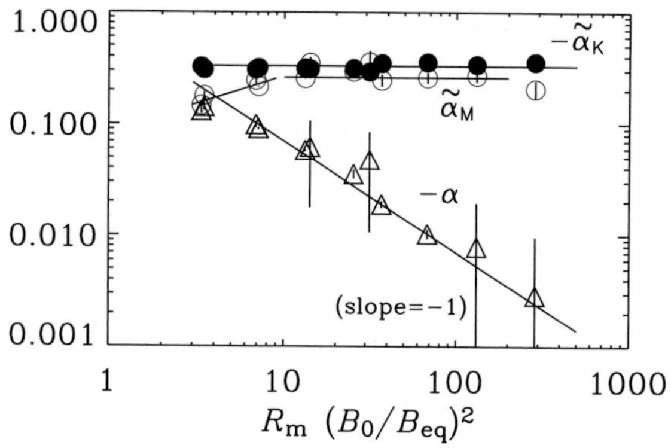

Fig. 2.10. R_m dependence of the normalized α compared with dependence of $\tilde{\alpha}_K^{(k)}$ for B_0/B_{eq} around unity. *Vertical bars* give error estimates. The *vertical bars on the data points* give estimates of the error (see text). [Adapted from Brandenburg & Subramanian (2005)]

burg & Subramanian (2005). The magnetic contribution to the α effect proportional to $\overline{\boldsymbol{j}\cdot\boldsymbol{b}}$ (in the isotropic case) still needs to be added to the right hand side of (2.53). This $\overline{\boldsymbol{j}\cdot\boldsymbol{b}}$ contribution is mainly the result of the dynamo itself, which tends to built up small scale current helicity along with the large scale magnetic field. Thus, the value of $\overline{\boldsymbol{j}\cdot\boldsymbol{b}}$ cannot be obtained independently of the actual solution to the dynamo problem.

2.5.4 Numerical Determination of α

A simple way of determining α numerically is by imposing a constant field of strength $\boldsymbol{B}_0$ over a domain of simulated turbulence. Since the mean field is constant, i.e. $\overline{\boldsymbol{B}} = \boldsymbol{B}_0$, the mean current density vanishes in (2.52), so α can directly be determined by measuring the electromotive force, $\overline{\boldsymbol{u}\times\boldsymbol{b}}$, in the direction of the imposed field, and dividing one by the other. In other words, $\alpha = \overline{\boldsymbol{u}\times\boldsymbol{b}}\cdot\boldsymbol{B}_0/B_0^2$. The values of α collapse onto a single line. Looking at (2.51), such a decline of α can only come about if either τ or $\overline{\omega\cdot\boldsymbol{u}}$ decrease with B_0, or, alternatively, if $\overline{\omega\cdot\boldsymbol{u}}$ and $\overline{\boldsymbol{j}\cdot\boldsymbol{b}}$ approach each other. It is quite clear from the data that neither $\overline{\omega\cdot\boldsymbol{u}}$ nor τ decrease and that instead there is, at least for small values of $R_m(B_0/B_{eq})^2$, a tendency for $\overline{\omega\cdot\boldsymbol{u}}$ and $\overline{\boldsymbol{j}\cdot\boldsymbol{b}}$ to approach one another.

The "catastrophic" decrease of α with decreasing η is directly a consequence of magnetic helicity conservation in a closed or periodic domain, but this can be alleviated in the presence of helicity fluxes out of the domain. We return to this discussion in Sect. 2.6.3 when we consider the consequences for the nonlinear saturation of the dynamo effect.

2.5.5 Other Effects

There are a number of other effects that contribute to the algebraic relationship between the electromotive force and the mean field. One is a pumping effect associated with the antisymmetric components of the α tensor,

$$\alpha_{ij}^{(A)} = \frac{1}{2}(\alpha_{ij} + \alpha_{ji}) \equiv -\frac{1}{2}\epsilon_{ijk}\gamma_k \quad \text{(pumping)}\,, \quad (2.54)$$

where γ_k is the pumping velocity. This name is motivated by the fact that the term $\alpha_{ij}^{(A)}\overline{B}_j$ can also be written as $(\boldsymbol{\gamma}\times\overline{\boldsymbol{B}})_i$. This shows that the vector $\boldsymbol{\gamma}$ plays the role of an effective advection velocity.

The pumping effect is sometimes called turbulent diamagnetism. This has to do with a remarkable relation between pumping velocity and turbulent magnetic diffusion,

$$\boldsymbol{\gamma} = -\frac{1}{2}\nabla\eta_t\,. \quad (2.55)$$

Calculating the contribution to the electromotive force from this term together with the turbulent diffusion term gives

$$\overline{\boldsymbol{\mathcal{E}}} = \ldots - \frac{1}{2}\nabla\eta_t\times\overline{\boldsymbol{B}} - \eta_t\nabla\times\overline{\boldsymbol{B}} = \ldots - \frac{1}{\sigma_t}\nabla\times\left(\overline{\boldsymbol{B}}/\mu_t\right)\,, \quad (2.56)$$

where

$$\sigma_t = \sigma(\eta_t/\eta)^{-1/2} \quad \text{and} \quad \mu_t = \mu_0(\eta_t/\eta)^{-1/2} \quad (2.57)$$

are turbulent conductivity and turbulent permeability, respectively. (The normalization with the microscopic values of σ and μ_0 is done in order for the turbulent values of σ_t and μ_t to have correct dimensions.)

Another potentially important term is an effect of the form $\boldsymbol{\delta}\times\overline{\boldsymbol{J}}$, which has long been known to be able to produce dynamo action if its components are of the appropriate sign relative to the orientation of shear. It is clear that $\boldsymbol{\delta}$ must be an axial vector, and both the local angular velocity, $\boldsymbol{\Omega}$, as well as the vorticity of the mean flow, $\overline{\boldsymbol{W}} = \nabla\times\overline{\boldsymbol{U}}$ are known to contribute. Dynamo action is only possible when $\boldsymbol{\delta}$ and $\overline{\boldsymbol{W}}$ are antiparallel. It is still not quite clear from turbulence calculations whether the orientation of the vector $\boldsymbol{\delta}$ relative to the shear is appropriate for dynamo action in the convection zone.

2.6 Models of the Solar Cycle

2.6.1 One-Dimensional Models

It has long been known that an α effect combined with differential rotation can cause oscillatory propagating solutions. In order to appreciate the possibility of oscillatory self-excited solutions, let us consider one-dimensional solutions, allowing for variations only in the z direction, but field components still pointing in the two directions. Applied to the Sun, we may think of the z direction being latitude (= negative colatitude, $-\theta$), x being radius, and y being longitude, so $(x, y, z) \longrightarrow (r, \phi, -\theta)$. Let us consider a mean flow of the form $\overline{\boldsymbol{U}} = (0, Sx, 0)$, i.e. the flow has only a y component that varies linearly in the x direction. We write the field in the form $\overline{\boldsymbol{B}}(z,t) = (-\overline{A}_y', \overline{B}_y, 0)$, where a prime denotes a z derivative. The corresponding dynamo equation can then be written as

$$\dot{\overline{A}}_y = \alpha \overline{B}_y + (\eta + \eta_t)\overline{A}_y'' , \tag{2.58}$$

$$\dot{\overline{B}}_y = S\overline{B}_x + (\eta + \eta_t)\overline{B}_y'' , \tag{2.59}$$

where we have neglected a term $(\alpha\overline{B}_x)'$ in comparison with $S\overline{B}_x$ in the second equation. (Here $\overline{B}_x = -\overline{A}_y'$ is the radial field.)

Solutions to these equations are frequently discussed in the literature (e.g. Moffatt 1978, Brandenburg & Subramanian 2005). It is instructive to consider first solutions in an unbounded domain, e.g. $0 < z < L_z$, so the solutions are of the form

$$\overline{\boldsymbol{B}}(z,t) = \mathrm{Re}\left[\hat{\boldsymbol{B}}_k \exp\left(\mathrm{i}kz + \lambda t\right)\right] . \tag{2.60}$$

There are two physically meaningful solutions. Both have an oscillatory component, but one of them can also have an exponentially growing component such that real and imaginary parts of λ are given by

$$\mathrm{Re}\lambda = -\eta_\mathrm{T} k^2 + \left|\frac{1}{2}\alpha S k\right|^{1/2} , \tag{2.61}$$

$$\mathrm{Im}\lambda \equiv -\omega_\mathrm{cyc} = \left|\frac{1}{2}\alpha S k\right|^{1/2} . \tag{2.62}$$

The solutions are oscillatory with the cycle period ω_cyc. This shows that, in the approximation where the $(\alpha\overline{B}_x)'$ term is neglected (valid when $Sk \gg \alpha$), the mean field dynamo is excited when the dynamo number,

$$D = \left|\frac{1}{2}\alpha S k\right|^{1/2} / (\eta_\mathrm{T} k^2) , \tag{2.63}$$

exceeds a critical value that is in this simple model $D_\mathrm{crit} = 1$.

A number of important conclusions can be drawn based on this simple model. (i) The cycle frequency is proportional to $\sqrt{\alpha S}$, but becomes equal to $\eta_\mathrm{T} k^2$ in the marginal or nonlinearly saturated cases. (ii) There are dynamo waves with a pattern speed proportional to $\eta_\mathrm{T} k$ propagating along contours of constant shear. For example, for radial angular velocity contours with angular velocity decreasing outwards, and for a positive α in the northern hemisphere, the propagation is equatorward. If the sign of either S or α is reversed, the propagation direction is reversed too.

For more realistic applications to the Sun one must solve the mean field dynamo equations in at least two dimensions over a spherical domain with appropriate profiles for α, η_T, and Ω. In the following we discuss four different dynamo scenarios that have been studied over the years.

If the flow is assumed given, no feedback via the Lorentz force is allowed, so the dynamo equations are linear and the magnetic energy would eventually grow beyond all bounds. In reality, the magnetic field will affect the flow and hence $\overline{\boldsymbol{U}}$, as well as α, η_t, and other turbulent transport coefficients will be affected. We will postpone the discussion of the nonlinear behavior to Sect. 2.6.3.

2.6.2 Different Solar Dynamo Scenarios

A traditional and also quite natural approach is to calculate the profiles for α and η_t using the results from mean field theory such as (2.53) and to take the profiles for the rms velocity and the correlation time, $\tau = \ell/u_\mathrm{rms}$, from stellar mixing length models using $\ell = \alpha_\mathrm{mix} H_\mathrm{p}$ for the mixing length, where H_p is the pressure scale height. For $\Omega(r,\theta)$ one often uses results from helioseismology. In Fig. 2.11 we reproduce the results of an early paper where the $\Omega(r,\theta)$ profile was synthesized from a collection of different helioseismology results then available. The α and η_t profiles, as well as profiles describing some other effects (such as pumping and $\Omega \times \overline{J}$ effects) where taken from a solar mixing length model. In this model an equatorward migration is achieved in a limited range in radius where $\partial\Omega/\partial r < 0$. In this model this is around $r = 0.8R_\odot$. Note also that in this case $\overline{B}_r$ and $\overline{B}_\phi$ are approximately in antiphase, as is also seen in the Sun.

Distributed dynamos have been criticized on the grounds that magnetic buoyancy will rapidly remove the magnetic field from the convection zone. Since then, helioseismology has shown that the radial Ω gradient is virtually zero in the bulk of the convection zone and only at the bottom is there a finite gradient, but it is positive at latitudes below $\pm 30°$. This may still yield an equatorward migration in the butterfly diagram, because (2.53) would predict that at the bottom of the convection zone, where the magnitude of the positive $\nabla_r \ln u_{rms}$ gradient exceeds that of the negative $\nabla_r \ln \rho$ gradient. This changes the sign of α, and makes it negative near the bottom of the convection zone in the northern hemisphere. This led to the idea of the overshoot dynamo that is believed to operate only in a thin layer at or just below the convection zone proper. Such dynamos have been considered by a number of different groups.

In Fig. 2.12 we show the result of an overshoot dynamo calculation. An important problem that emerges from such an approach is that when the dynamo layer is too thin, the toroidal flux belts are too close to each other in latitude. This leads to the conclusion that the thickness of the dynamo region should not be less than 35 Mm. At the bottom of the convection zone this corresponds to half a pressure scale height. However, this value is already rather large and no longer supported by helioseismology, which predicts the thickness of the overshoot layer to be about 7 Mm or less.

Another variant of this approach is the interface dynamo. The main difference here is that α is assumed to operate in the bulk of the convection zone, but it is still taken to be negative, so as to give equatorward migration. Also important is the sharp jump in η_t at the bottom of the convection zone. However, when the latitudinal variation of the angular velocity is included, no satisfactory butterfly diagram is obtained.

A completely different class of dynamos are the flux transport dynamos that are governed by the effect of meridional circulation transporting surface flux to the poles and flux along the tachocline toward the equator. The α effect is now assumed positive, so in the absence of meridional circulation the dynamo wave would propagate poleward. However, under certain conditions, meridional circulation can actually reverse the direction of propagation of the dynamo wave. A calculation with a realistic solar angular velocity profile has been presented by Dikpati & Charbonneau (1999); see Fig. 2.13. They establish a detailed scaling law for the dependence of the cycle period on the circulation speed, the α effect (or source term), and the turbulent magnetic diffusivity. To a good approximation they find the cycle period to be inversely proportional to the circulation speed.

With such a variety of different models and assumptions (most of them ignoring what was previously derived for $\alpha(r,\theta)$, $\eta_t(r,\theta)$, and other transport effects), dynamo theory has been perceived as rather arbitrary. One reason for this level of arbitrariness that developed in modeling the solar dynamo is that the effects of nonlinearity are not well understood. This might affect the properties of the dynamo coefficients in the saturated state making them quite different from those obtained in linear theory. In the following we sketch briefly the tremendous developments on nonlinear saturation that have occurred in the past few years.

2.6.3 Nonlinear Saturation

The effects of nonlinearity can be divided into macroscopic and microscopic effects. The former is simply the result of $\overline{B}$ on $\overline{U}$, as described by the Lorentz force, $\overline{J} \times \overline{B}$, in the mean field momentum equation. This effect is sometimes also referred to as the Malkus–Proctor effect and has been incorporated to various degree of sophistication in a number of models starting with incompressible models in the context of the geodynamo and the solar dynamo.

The microscopic feedback can be subdivided into two different contributions. The effect of $\overline{B}$ on the turbulent velocity (conventional α quenching), and the more direct effect of the small scale current helicity, $\overline{j \cdot b}$ (or $\epsilon_{ijk}\overline{b_k b_{j,p}}$ in the anisotropic case), on the α effect or, more precisely, on the electromotive force. The latter can, under some conditions, lead to catastrophic α quenching; see Brandenburg & Subramanian (2005) for a review of this vast field of recent research.

The $\overline{j \cdot b}$ term cannot be implemented directly, because it is necessary to have a theory for how $\overline{j \cdot b}$ de-

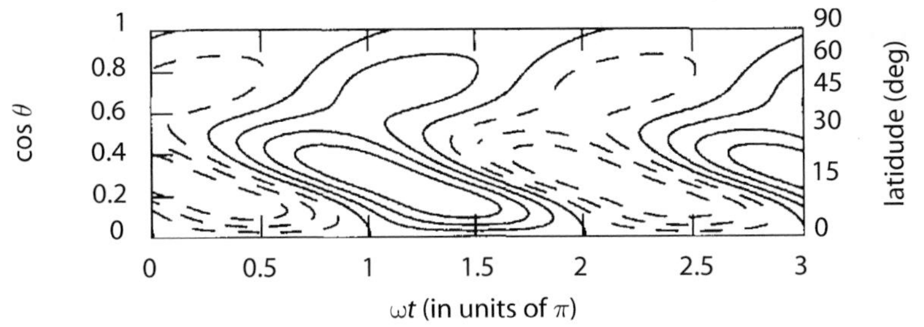

Fig. 2.11. Butterfly diagram of B_ϕ taken at reference depth $r = 0.85R_\odot$. [Adapted from Brandenburg & Tuominen (1988)]

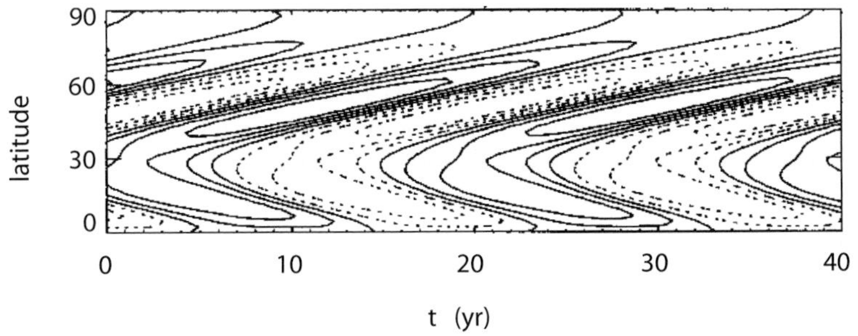

Fig. 2.12. Butterfly diagram of B_ϕ evaluated at the bottom of the convection zone at $r = 0.7R_\odot$. [Adapted from Rüdiger & Brandenburg (1995)]

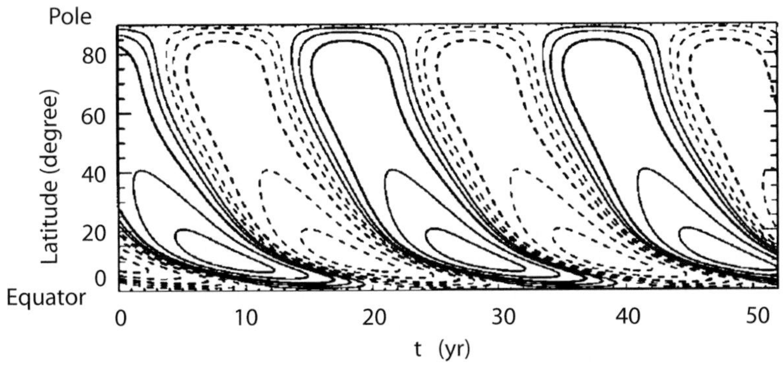

Fig. 2.13. Butterfly diagram of B_ϕ at $r = 0.7R_\odot$. The maximum circulation speed at the surface is $15\,\mathrm{ms}^{-1}$ and the turbulent magnetic diffusivity is assumed to be $\eta_t = 3\times10^{11}\,\mathrm{cm}^2\mathrm{s}^{-1}$. [Courtesy of Dikpati & Charbonneau (1999)]

pends on the mean field. Under some idealized conditions (steady state, triply-periodic boundary conditions) the answer can be obtained from the general evolution equation for magnetic helicity, which reads

$$\frac{\partial}{\partial t}(\boldsymbol{A}\cdot\boldsymbol{B}) + \nabla\cdot\boldsymbol{F}_{\mathrm{H}} = -2\eta\mu_0(\boldsymbol{J}\cdot\boldsymbol{B})\ . \tag{2.64}$$

Here, $\boldsymbol{A}$ is the magnetic vector potential with $\boldsymbol{B} = \nabla\times\boldsymbol{A}$, while $\boldsymbol{A}\cdot\boldsymbol{B}$ is the magnetic helicity density, and $\boldsymbol{F}_{\mathrm{H}}$ is its flux. Magnetic helicity and its flux are gauge-dependent, i.e. they are not invariant under the transformation $\boldsymbol{A}\to\boldsymbol{A}' = \boldsymbol{A} + \nabla\Lambda$. However, when averaging over a triply-periodic volume this ambiguity disappears and $\langle\boldsymbol{A}\cdot\boldsymbol{B}\rangle$ is gauge-invariant and obeys

$$\frac{\mathrm{d}}{\mathrm{d}t}\langle\boldsymbol{A}\cdot\boldsymbol{B}\rangle = -2\eta\mu_0\langle\boldsymbol{J}\cdot\boldsymbol{B}\rangle\ . \tag{2.65}$$

(The spatial average of a divergence also vanishes for triply periodic domains.) We see that in the steady state, $\mathrm{d}/\mathrm{d}t = 0$, so $\langle\boldsymbol{J}\cdot\boldsymbol{B}\rangle = 0$. Splitting into large scale and small scale contributions, we have $\langle\boldsymbol{j}\cdot\boldsymbol{b}\rangle = -\langle\overline{\boldsymbol{J}}\cdot\overline{\boldsymbol{B}}\rangle$. This connects the small scale current helicity explicitly with the properties of the large scale field.

The same procedure can still be applied in the unsteady case by considering magnetic helicity evolution for the large scale and small scale components, i.e. for $\langle\overline{\boldsymbol{A}}\cdot\overline{\boldsymbol{B}}\rangle$ and $\langle\boldsymbol{a}\cdot\boldsymbol{b}\rangle$. The evolution of $\langle\overline{\boldsymbol{A}}\cdot\overline{\boldsymbol{B}}\rangle$ follows straightforwardly from the mean field equations, which shows that there is continuous production of large scale magnetic helicity given by $2\overline{\boldsymbol{\mathcal{E}}}\cdot\overline{\boldsymbol{B}}$. In order not to produce any net magnetic helicity, as required by (2.65), the evolution of $\langle\boldsymbol{a}\cdot\boldsymbol{b}\rangle$ has the same term but with the opposite sign. Furthermore, under isotropic conditions, $\langle\boldsymbol{a}\cdot\boldsymbol{b}\rangle$ is proportional to $\langle \boldsymbol{j} \cdot \boldsymbol{b} \rangle$, which in turn is proportional to the magnetic contribution to the α effect. Finally, the restriction to triply periodic boundary conditions can be relaxed (and hence a flux divergence can be permitted) if there is sufficient scale separation, i.e. if the energy carrying scale of the turbulence is clearly smaller than the domain size.

This then leads to an explicit evolution equation for the magnetic α effect,

$$\frac{\partial \alpha_{\mathrm{M}}}{\partial t} + \nabla \cdot \overline{\boldsymbol{F}}_{\alpha} = -2\eta_{\mathrm{t}} k_{\mathrm{f}}^2 \left(\frac{\overline{\boldsymbol{\mathcal{E}}} \cdot \overline{\boldsymbol{B}}}{B_{\mathrm{eq}}^2} + \frac{\alpha_{\mathrm{M}}}{R_{\mathrm{m}}} \right) , \qquad (2.66)$$

where $\alpha_{\mathrm{M}} = \frac{1}{3}\tau \overline{\boldsymbol{j} \cdot \boldsymbol{b}}$ is the magnetic α effect and $\overline{\boldsymbol{F}}_{\alpha} = \frac{1}{3}\tau \overline{\boldsymbol{F}}_{\mathrm{C}}$, where $F_{\mathrm{C}} \approx k_{\mathrm{f}}^2$ is the current helicity flux. This so-called dynamical α quenching equation is able to reproduce the resistively slow saturation behavior, found in simulations of helically driven turbulence. In the steady state, this equation predicts for $\alpha = \alpha_{\mathrm{K}} + \alpha_{\mathrm{M}}$

$$\alpha = \frac{\alpha_{\mathrm{K}} + R_{\mathrm{m}} \left(\eta_{\mathrm{t}} \overline{\boldsymbol{J}} \cdot \overline{\boldsymbol{B}} + \nabla \cdot \overline{\boldsymbol{F}}_{\alpha} \right)}{1 + R_{\mathrm{m}} \overline{\boldsymbol{B}}^2 / B_{\mathrm{eq}}^2} . \qquad (2.67)$$

Note that in the special case of periodic domains, used in some simulations where $\overline{\boldsymbol{J}} = 0$ and $\nabla \cdot \overline{\boldsymbol{F}}_{\alpha} = 0$, this equation predicts catastrophic quenching, i.e. $\alpha = \alpha_{\mathrm{K}} / (1 + R_{\mathrm{m}} \overline{\boldsymbol{B}}^2 / B_{\mathrm{eq}}^2)$, so α is suppressed relative to its kinematic value α_{K} in a strongly Reynolds number-dependent fashion – as seen in Fig. 2.10.

In the case of open boundaries, there is a flux of magnetic helicity. Under the two-scale hypothesis this can be defined in a gauge-invariant manner. Magnetic helicity fluxes provide a way to escape the otherwise resistively limited saturation and catastrophic quenching. Several simulations and mean field models have confirmed this. Although dynamical quenching has already been applied to solar dynamo models it remains to be seen to what extent the previously discussed conclusions about distributed versus overshoot layer dynamos are affected, and what the role of meridional circulation is in such a model.

2.6.4 Location of the Dynamo

It is generally believed that the magnetic field emergence in the form of sunspots is deeply rooted and associated with strong toroidal flux tubes of strength up to 100 kG, as predicted by the so-called thin flux tube models. When parts of the flux tube become destabilized due to magnetic buoyancy, it rises to the surface to form a sunspot pair. However, there are some open questions: how are such coherent tubes generated and what prevents them from breaking up during the ascent over 20 pressure scale heights? Alternatively, the usual mean field dynamo would actually predict magnetic field generation distributed over the entire convection zone. Sunspot formation would mainly be associated with local flux concentration within regions of enhanced net flux. This picture is appealing in many ways and has been discussed in more detail in Brandenburg (2005). However, although both pictures (deep rooted versus distributed dynamo) have received some support from mean field modeling, there is still no global turbulence simulations that reproduces the solar activity cycle without questionable assumptions.

2.7 Differential Rotation

It became clear from the discussion in Sect. 2.6 that differential rotation plays an important role in producing a large scale magnetic field in the Sun. It may also be important for the dynamo in disposing of its excess small scale current helicity, as discussed in the previous section. In this section we discuss the theoretical basis for explaining the origin and properties of solar and stellar differential rotation.

2.7.1 Mean Field Theory of Differential Rotation

The origin of differential rotation has long been understood to be a consequence of the anisotropy of convection. It has long been clear that the vertical exchange of momentum by convection should lead to a tendency toward constant angular momentum in the radial direction, i.e. $\overline{\Omega}\varpi^2 = \mathrm{const}$, and hence the mean angular velocity scales with radius like $\overline{\Omega}(r) \sim r^{-2}$. Here, $\varpi = r \sin\theta$ denotes the cylindrical radius (i.e. the distance from the rotation axis).

The $r\phi$ component of the viscous stress tensor contributes to the angular momentum equation,

$$\frac{\partial}{\partial t}\left(\rho \varpi^2 \overline{\Omega}\right) + \nabla \cdot \left[\rho \varpi \left(\overline{\boldsymbol{U}} \overline{U}_{\phi} + \overline{\boldsymbol{u} u_{\phi}}\right)\right] = 0 , \qquad (2.68)$$

where $\overline{u_i u_j} = Q_{ij}$ are the components of the Reynolds tensor. In spherical coordinates the full mean velocity vector is written as $\overline{\boldsymbol{U}} = (\overline{U}_\varpi, \varpi\overline{\Omega}, \overline{U}_z)$.

The early treatment in terms of an anisotropic viscosity tensor was purely phenomenological. A rigorous calculation of the Reynolds stresses shows that the mean Reynolds stress tensor is described not only by diffusive components that are proportional to the components of the rate of strain tensor of the mean flow, but that there are also non-diffusive components that are directly proportional to the local angular velocity. In particular the $r\phi$ and $\theta\phi$ components of the Reynolds tensor are of interest for driving r and θ gradients of $\overline{U}_\phi \equiv \varpi\overline{\Omega}$. Thus, for ordinary isotropic turbulent viscosity one has, using Cartesian index notation,

$$Q_{ij} = -\nu_t\left(\overline{U}_{i,j} + \overline{U}_{j,i}\right) - \zeta_t\delta_{ij}\overline{U}_{k,k}\,, \qquad (2.69)$$

where ζ_t is a turbulent bulk viscosity, and commas denote partial differentiation. This expression implies in particular that

$$Q_{\theta\phi} = -\nu_t \sin\theta\frac{\partial\overline{\Omega}}{\partial\theta}\,. \qquad (2.70)$$

Note that for the Sun, where $\partial\overline{\Omega}/\partial\theta > 0$ in the northern hemisphere, this formula would predict that $Q_{\theta\phi}$ is negative in the northern hemisphere. However, it was noted long ago from correlation measurements of sunspot proper motions that $Q_{\theta\phi}$ is in fact *positive* in the northern hemisphere. The observed profile of $Q_{\theta\phi}$ is also known as the Ward profile. The observed positive sign was used to motivate that there must be an additional term in the expression for Q_{ij}. Using a closure approach, such as the first order smoothing approximation that is often used to calculate the α effect in dynamo theory, one can find the coefficients in the expansion

$$Q_{ij} = \Lambda_{ijk}\overline{\Omega}_k - \mathcal{N}_{ijkl}\overline{U}_{k,l}\,, \qquad (2.71)$$

where Λ_{ijk} describes the so-called Λ effect and $\mathcal{N}_{ijkl}$ is the turbulent viscosity tensor. The viscosity tensor $\mathcal{N}_{ijkl}$ must in general be anisotropic. When anisotropies are included, $\mathcal{N}_{ijkl}$ gets modified (but it retains its overall diffusive properties), and Λ_{ijk} takes the form

$$\Lambda_{ijk}\overline{\Omega}_k = \begin{pmatrix} 0 & 0 & V\sin\theta \\ 0 & 0 & H\cos\theta \\ V\sin\theta & H\cos\theta & 0 \end{pmatrix} \overline{\Omega}\,, \qquad (2.72)$$

where V and H are still functions of radius, latitude, and time; V is thought to be responsible for driving vertical differential rotation ($\partial\overline{\Omega}/\partial r \neq 0$) while H is responsible for latitudinal differential rotation ($\partial\overline{\Omega}/\partial\theta \neq 0$).

The first order smoothing approximation predicts the following useful approximations for V and H:

$$V \approx \tau\left(\overline{u_\phi^2} - \overline{u_r^2}\right)\,, \qquad (2.73)$$

$$H \approx \tau\left(\overline{u_\phi^2} - \overline{u_\theta^2}\right)\,. \qquad (2.74)$$

These expressions show that when the rms velocity in the radial direction is larger than in the azimuthal direction we must expect $V < 0$ and hence $\partial\overline{\Omega}/\partial r < 0$. In the Sun, this effect is responsible for the negative radial shear near the surface where strong downdrafts may be responsible for a comparatively large value of $\overline{u_r^2}$. Likewise, when the rms velocity in the latitudinal direction is larger than in the azimuthal direction we expect $H < 0$ and hence $\partial\overline{\Omega}/\partial\theta < 0$, so the equator would spin slower than the poles. This does not apply to the Sun, but it may be the case in some stars, especially when the flows are dominated by large scale meridional circulation.

2.7.2 The Λ Effect from Turbulence Simulations

Several of the relationships described above have been tested using convection simulations, both in local Cartesian boxes located at different latitudes as well as in global spherical shells. Generally, the various simulations agree in that the sign of the horizontal Reynolds stress is positive in the northern hemisphere and negative in the southern, reproducing thus the Ward profile. The simulations also show that the off-diagonal components of the turbulent heat transport tensor are mostly positive in the northern hemisphere, and negative in the southern hemisphere. This agrees with the sign required if the baroclinic term is to produce a tendency toward spoke-like angular velocity contours. Simulations also reproduce the sudden drop of angular velocity at the top of the convection zone. This agrees with a predominantly negative sign of the vertical Reynolds stress at a similar depth. Furthermore, some of the more recent simulations show an unexpectedly sharp increase of the horizontal Reynolds stress just near the equator (at around ±5° latitude), before changing sign right at the equator. The significance of this result for the solar differential rotation pattern is still unclear.

2.7.3 Meridional Flow and the Baroclinic Term

According to the formalism described in the previous section, a finite differential rotation can be obtained by ignoring meridional flows and solving (2.68) in isolation. However, this would only be a poor approximation that becomes quickly invalid when the angular velocity becomes large compared with the turbulent viscous decay rate. This is quantified by the Taylor number

$$\mathrm{Ta} = \left(2\overline{\Omega}_0 R^2/\nu_\mathrm{t}\right)^2 . \tag{2.75}$$

Using the first order smoothing expression from Rüdiger (1989), $\nu_\mathrm{t} = (2/15)\,\tau u_\mathrm{rms}^2$, we have for values typical for the Sun (see Table 2.1), i.e. $\nu_\mathrm{t} \approx 10^{12}\ \mathrm{cm}^2/\mathrm{s}$, $\mathrm{Ta} \approx 10^9$. This value of Ta is rather large so that nonlinearities produce strong deviations from linear theory.

As the value of Ta is increased, the Coriolis force increases, which then drives a meridional flow. This meridional flow first increases with increasing values of Ta, but then it reaches a maximum at $\mathrm{Ta} \approx 3 \times 10^5$, and later declines with increasing values of Ta. (The solar value is $\mathrm{Ta} \approx 3 \times 10^7$.) This decline is because eventually the Coriolis force can no longer be balanced against advection or diffusion terms. This can best be seen by considering the curl of the momentum equation,

$$\frac{\partial W_\phi}{\partial t} + \varpi \overline{\boldsymbol{U}}\cdot\nabla\left(\frac{W_\phi}{\varpi}\right) - \nu_\mathrm{t} \mathrm{D}^2 W_\phi = \varpi \frac{\partial \overline{\Omega}^2}{\partial z} + \hat{\phi}\cdot\overline{\nabla T \times \nabla S} . \tag{2.76}$$

We recall that we consider here a nonrotating frame of reference, so there is no Coriolis force. Nevertheless, part of the inertial term takes a form that is quite similar to the Coriolis term, but here $\overline{\Omega}$ is a function of position, while in the Coriolis term the angular velocity would normally be a constant.

In the barotropic case one has $\nabla T \parallel \nabla S$ so there is no baroclinic term, i.e. $\hat{\phi}\cdot(\nabla T \times \nabla S) = 0$. So, if viscous and inertial terms are small, which is indeed the case for rapid rotation, then $\partial\overline{\Omega}^2/\partial z$ has to vanish, so $\overline{\Omega}$ would be constant along cylinders; see Fig. 2.14. It is generally believed that the main reason for $\overline{\Omega}$ not having cylindrical contours in the Sun is connected with the presence of the baroclinic term. The presence of magnetic forces may also play a role, but unlike the baroclinic term, magnetic forces tend to produce a rather variable $\overline{\Omega}$ patterns, often connected with rapid motions near the poles where the inertia is lower.

Currently the highest resolution simulations of global convection in spherical shells are those by Miesch et al. (2000). These simulations show a great amount of detail and reproduce some basic features of the Sun's differential rotation such as the more rapidly spinning equator. However, in low latitudes they show strongly cylindrical $\overline{\Omega}$ contours that deviate markedly from the more spoke-like contours inferred for the Sun using helioseismology. These simulations also do not show the near-surface shear layer where the rotation rate drops by over 20 nHz over the last 30 Mm below the surface.

Mean field simulations using the Λ effect show surprisingly good agreement with the helioseismologically inferred $\overline{\Omega}$ pattern, and they are also beginning to address the problem of the near-surface shear layer. In these simulations it is indeed the baroclinic term that is responsible for causing the departure from cylindrical contours. This, in turn, is caused by an anisotropy of the turbulent heat conductivity which causes a slight enhancement in temperature and entropy at the poles. In the bulk of the convection zone the entropy is nearly constant, so the radial entropy variation is smallest compared with the radial temperature variation. It is therefore primarily the latitudinal entropy variation that determines the baroclinic term, with

$$\varpi \frac{\partial \overline{\Omega}^2}{\partial z} \approx -\hat{\phi}\cdot\overline{\nabla T \times \nabla S} \approx -\frac{1}{r}\frac{\partial \overline{T}}{\partial r}\frac{\partial \overline{S}}{\partial \theta} < 0 . \tag{2.77}$$

The inequality shows that negative values of $\partial\overline{\Omega}^2/\partial z$ require that the pole is slightly warmer than the equator ($\partial\overline{S}/\partial\theta < 0$). However, this effect is so weak that it cannot at present be observed. Allowing for these conditions in a simulation may require particular care in the treatment of the outer boundary conditions. In Fig. 2.15

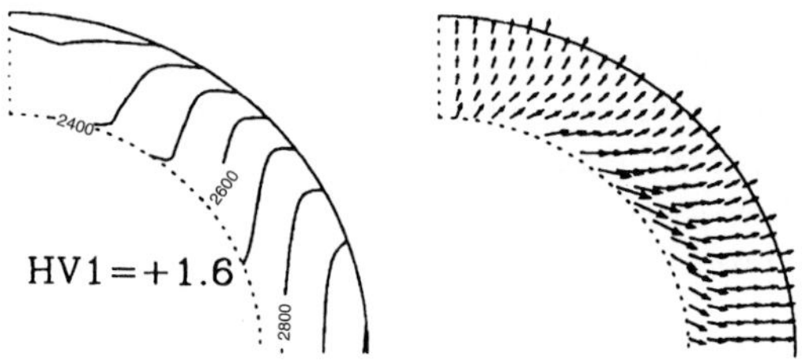

Fig. 2.15. Contours of angular velocity (*left*) and turbulent convective energy flux (*right*) for a model with anisotropic heat transfer tensor. [Adapted from Brandenburg et al. (1992)]

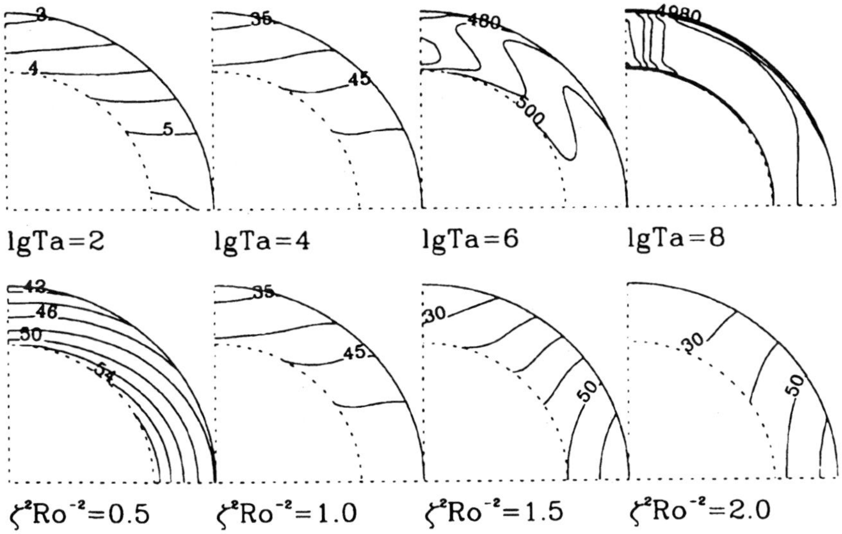

Fig. 2.14. Contours of constant $\overline{\Omega}$ for different values of Taylor number (*upper panel*) and different values of the inverse Rossby number, affecting the relative importance of H over V (*lower panel*). [Adapted from Brandenburg et al. (1990)]

we show the plots of angular velocity contours and convective energy transport in a model with anisotropic turbulent conductivity tensor, χ_{ij}. Given that the flux, $\boldsymbol{F}$, is proportional to $-\chi_{ij}\nabla_j\overline{S}$, a negative $\partial\overline{S}/\partial\theta$ can be produced from a positive F_r with a positive value of $\chi_{r\theta}$.

In the discussion above we ignored in the last step a possible correlation between entropy and temperature fluctuations, i.e. a contribution from the term $\overline{\nabla T' \times \nabla S'}$ where primes denote fluctuations. Such correlations, if of suitable sign, might provide yet a further explanation for a non-zero value of $\partial\overline{\Omega}^2/\partial z$.

2.7.4 Near-Surface Shear Layer

The first results of helioseismology indicated significantly higher angular velocities in the sub-surface than what is seen at the surface using Doppler measurements. This apparent conflict is now resolved in that helioseismological inversions of the data from the SOHO spacecraft show a sharp negative gradient, connecting the observed surface values smoothly with the local maximum of the angular velocity at about 35 Mm depth; see Fig. 2.16.

The theory of this negative near-surface shear layer is still a matter of ongoing research, but it is clear that negative shear would generally be the result of predominantly vertical turbulent velocities such as strong downdrafts near the radiating surface. However, such a layer that is dominated by strong downdrafts was only thought to be several megameters deep, and not several tens of megameters. With an improved theory for the anisotropy of the turbulence especially near the surface layers, one obtains a clear radial decline of the local angular velocity near the surface, although still not quite as much as is observed; see Fig. 2.17. In any case, these results do at least reproduce the near-surface shear layer qualitatively correctly. A proper understanding of this layer is now quite timely in view of the fact that near-surface shear is likely to contribute to the production of strong toroidal fields.

2.7.5 Magnetic Effects

In Sect. 2.4.5 we mentioned the torsional oscillations, which is a cyclic modulation of the latitudinal profile of the angular velocity at the surface of the sun. Model calculations suggest that these oscillations can well be modeled by restoring the Lorentz force by adding a term $-\varpi\overline{\boldsymbol{B}B_\phi}$ under the divergence in (2.68). Unfortunately, given that there is no definitive solar dynamo model, models for the Sun's torsional oscillations are equally preliminary and still a matter ongoing research.

In this connection it may be worth noting that there are also magnetic effects on other properties of the sun, most notably luminosity variations (by about 0.1%) and changes of the Sun's quadrupole moment. The latter does not really seem to be important for the Sun, but in close binaries this effect leads to measurable changes in the orbital period.

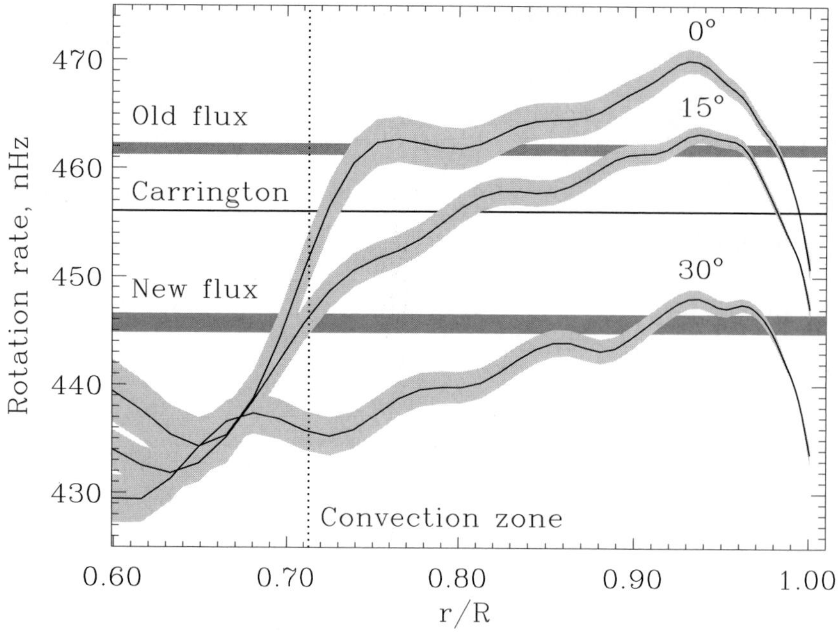

Fig. 2.16. Radial profiles of the internal solar rotation rate, as inferred from helioseismology (sidereal, i.e. in a fixed frame). The rotation rate of active zones at the beginning of the cycle (at ≈30° latitude) and near the end (at ≈4°) is indicated by horizontal bars, which intersect the profiles of rotation rate at $r/R_{\odot} \approx 0.97$. For orientation, the conventionally defined Carrington rotation period of 27.3 days (synodic value, corresponding to 424 nHz) has been translated to the sidereal value of 456 nHz. Courtesy of Benevolenskaya et al. (1999)

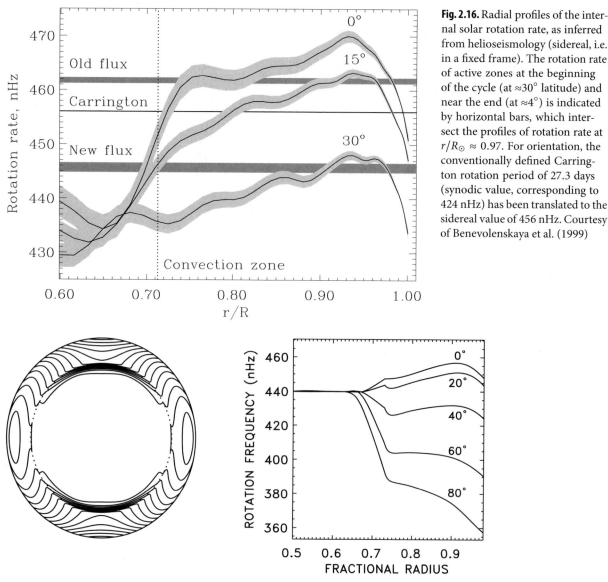

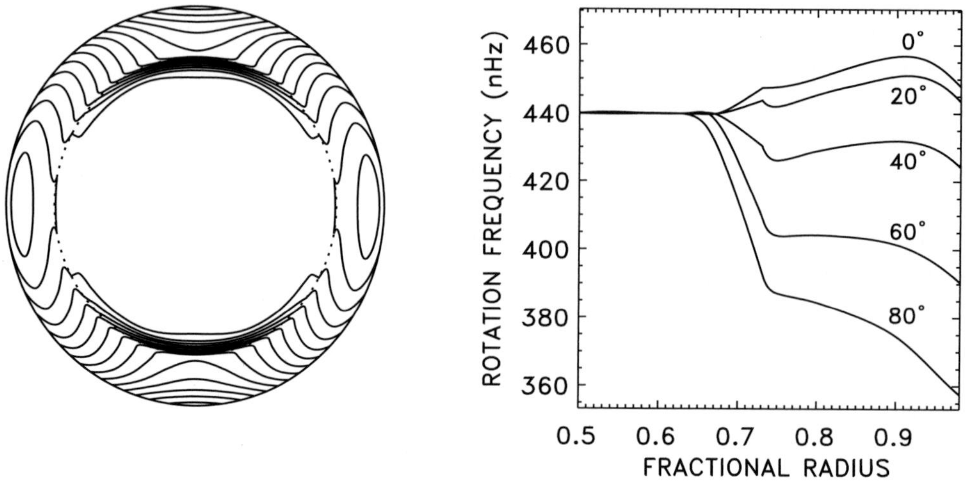

Fig. 2.17. Rotation law obtained by Kitchatinov & Rüdiger (2005) taking the anisotropy of the turbulence near the surface into account. [Courtesy Kitchatinov & Rüdiger (2005)]

2.8 Conclusions

In the past few decades there have been significant developments in understanding the physics of the Sun. Even regarding the radial structure of the Sun, which was thought to be qualitatively well understood, major revisions have emerged just recently with the refinement of three-dimensional simulations of solar granulation. Such simulations have led to new spectral line fits that imply a drastically reduced abundance of the heavier elements. This has consequences for the opacities that affect the deep parts of the Sun's interior.

There are many aspects of solar physics where a detailed understanding of the three-dimensional flow pattern of the Sun is crucial. It is not surprising that effects involving details of the turbulent flow field in the solar convection zone, such as the theory of differential rotation and magnetic field generation, provide other examples where the three-dimensional dynamics is important. Fully three-dimensional simulations of solar convection with magnetic fields produce flow and magnetic field structures in great detail, but at present they deviate in some important aspects from the Sun (e.g. the fraction of small scale to large scale field is rather large; and the angular velocity contours are still too strongly aligned with the rotation axis). Some tentative explanations are available (magnetic Prandtl number not small enough in the simulations to reduce or even suppress small scale dynamo action, and surface conditions not realistic enough to allow for sufficiently large a baroclinic term). Future advances in computer technology will bring a steady increase in numerical resolution. However, increase of spatial resolution by a factor of two will always be very difficult when close to the machine capacity. Substantial progress may rather hinge on new insights that may emerge from a closer interrelation between local simulations where turbulence is well resolved and mean field calculations that benefit from input and calibration of detailed simulations.

Acknowledgement. I thank David Moss and an anonymous referee for making useful suggestions that have helped improving this chapter, and Dr. Y. Kamide for his patience. I apologize again for not having been able to acknowledge, through appropriate references, the many achievements that have been reported here. I have therefore mainly been referencing review material that in turn quote the relevant original work.

References

Bahcall, J. N., Basu, S., & Serenelli, A. M., "What Is the Neon Abundance of the Sun?," *Astrophys. J.* **631**, 1281–1285 (2005)

Benevolenskaya, E. E., Hoeksema, J. T., Kosovichev, A. G., & Scherrer, P. H., "The interaction of new and old magnetic fluxes at the beginning of solar cycle 23," *Astrophys. J.* **517**, L163–L166 (1999)

Brandenburg, A., , "The case for a distributed solar dynamo shaped by near-surface shear," *Astrophys. J.* **625**, 539–547 (2005)

Brandenburg, A., & Tuominen, I., "Variation of magnetic fields and flows during the solar cycle," *Adv. Space Sci.* **8**, 185–189 (1988)

Brandenburg, A., & Subramanian, K., "Astrophysical magnetic fields and nonlinear dynamo theory," *Phys. Rep.* **417**, 1–209 (2005)

Brandenburg, A., Moss, D., Rüdiger, G., & Tuominen, I., "The nonlinear solar dynamo and differential rotation: A Taylor number puzzle?" *Solar Phys.* **128**, 243–251 (1990)

Brandenburg, A., Moss, D., & Tuominen, I., "Stratification and thermodynamics in mean-field dynamos," *Astron. Astrophys.* **265**, 328–344 (1992)

Brandenburg, A., Chan, K. L., Nordlund, Å., & Stein, R. F., "Effect of the radiative background flux in convection," *Astron. Nachr.* **326**, 681–692 (2005)

Brun, A. S., Miesch, M. S. & Toomre, J., "Global-scale turbulent convection and magnetic dynamo action in the solar envelope," *Astrophys. J.* **614**, 1073–1098 (2004)

Christensen-Dalsgaard, J., Duvall, T. L., Jr., Gough, D. O., Harvey, J. W., & Rhodes, E. J., Jr., "Speed of sound in the solar interior," *Nature* **315**, 378–382 (1985)

Demarque, P., & Guenther, D. B., "Helioseismology: Probing the interior of a star," *Proc. Nat. Acad. Sci.* **96**, 5356–5359 (1999)

Deubner, F.-L., , "Observations of low wavenumber nonradial eigenmodes of Sun," *Astron. Astrophys.* **44**, 371–379 (1975)

Dikpati, M., & Charbonneau, P., "A Babcock-Leighton flux transport dynamo with solar-like differential rotation," *Astrophys. J.* **518**, 508–520 (1999)

Duvall, T. L., Jr., , "A dispersion law for solar oscillations," *Nature* **300**, 242–243 (1982)

Gailitis, A., Lielausis, O., Platacis, E., et al. , "Magnetic field saturation in the Riga dynamo experiment," *Phys. Rev. Letters* **86**, 3024–3027 (2001)

Gizon, L., & Birch, A. C., "Local Helioseismology," *Living Rev. Solar Phys.* **2** (2005), http://www.livingreviews.org/lrsp-2005-6

Gough, D., "Inverting helioseismic data," *Solar Phys.* **100**, 65–99 (1985)

Kazantsev, A. P., "Enhancement of a magnetic field by a conducting fluid," *Sov. Phys. JETP* **26**, 1031–1034 (1968)

Kippenhahn, R. & Weigert, A. *Stellar structure and evolution.* Springer: Berlin (1990)

Kitchatinov, L. L. & Rüdiger, G., "Differential rotation and meridional flow in the solar convection zone and beneath," *Astron. Nachr.* **326**, 379–385 (2005)

Krause, F., & Rädler, K.-H., *Mean-Field Magnetohydrodynamics and Dynamo Theory.* Pergamon Press, Oxford (1980)

Miesch, M. S., Elliott, J. R., Toomre, J. et al., "Three-dimensional spherical simulations of solar convection. I. Differential rotation and pattern evolution achieved with laminar and turbulent states," *Astrophys. J.* **532**, 593–615 (2000)

Marsh, N., & Svensmark, H., "Cosmic rays, clouds, and climate," *Spa. Sci. Rev.* **94**, 215–230 (2000)

Mihalas, D., *Stellar Atmospheres*. W. H. Freeman: San Francisco (1978)
Moffatt, H. K., *Magnetic field generation in electrically conducting fluids*. Cambridge University Press, Cambridge (1978)
Ossendrijver, M., "The solar dynamo," *Astron. Astrophys. Rev.* **11**, 287–367 (2003)
Parker, E. N., *Cosmical Magnetic Fields*. Clarendon Press, Oxford (1979)
Rüdiger, G., *Differential rotation and stellar convection: Sun and solar-type stars*. Gordon & Breach, New York (1989)
Rüdiger, G. & Brandenburg, A., "A solar dynamo in the overshoot layer: cycle period and butterfly diagram," *Astron. Astrophys.* **296**, 557–566 (1995)
Rüdiger, G., & Hollerbach, R., *The magnetic universe*. Wiley-VCH, Weinheim (2004)
Solanki, S. K., Inhester, B., Schüssler, M., "The solar magnetic field," *Rep. Prog. Phys.* **69**, 563–668 (2006)
Spruit, H. C., "A model of the solar convection zone," *Solar Phys.* **34**, 277–290 (1974)
Stieglitz, R., & Müller, U., "Experimental demonstration of a homogeneous two-scale dynamo," *Phys. Fluids* **13**, 561–564 (2001)
Stix, M., "Comments on the solar dynamo," *Astron. Astrophys.* **37**, 121–133 (1974)
Stix, M., *The Sun: an introduction*. Springer-Verlag, Berlin (2002)
Thompson, M. J., Christensen-Dalsgaard, J., Miesch, M. S., & Toomre, J., "The internal rotation of the Sun," *Ann. Rev. Astron. Astrophys.* **41**, 599–643 (2003)
Ulrich, R. K., "The five-minute oscillations on the solar surface," *Astrophys. J.* **162**, 993–1002 (1970)

3 Solar Atmosphere

Eric R. Priest

There has been a revolution in understanding the nature of the Sun's atmosphere that has been stimulated by high-resolution satellite observations together with the realisation that the atmosphere is a magnetic world in which plasma interacts with magnetic fields in complex and highly nonlinear ways. We describe here the recent surprises that have arisen about the photosphere and corona especially from the SoHO satellite. Then we describe the role of the magnetic field and summarise the main properties of magnetic waves and magnetic reconnection. Finally, we discuss in some detail solar flares and the possible ways in which the solar corona is heated.

Contents

Eric R. Priest, Solar Atmosphere.
In: Y. Kamide/A. Chian, Handbook of the Solar-Terrestrial Environment. pp. 55–93 (2007)
DOI: 10.1007/11367758_3

3.1 Introduction

The Sun's atmosphere consists of three regions. The *photosphere* emits most of the Sun's light and consists of a thin layer only $\frac{1}{2}$ Mm thick with a density of 10^{23} particles per cubic metre (one hundredth of that in the Earth's atmosphere). The *chromosphere* is rarer and more transparent with a density of 10^{17} m^{-3}. The *corona* is even less dense with a density of 10^{15} m^{-3}: it extends the Earth's orbit (where the density is 10^7 m^{-3}) and beyond.

At first sight one may expect the temperature to decrease as we go away from the solar surface, and at first it does do so to a minimum value of 4200 K. However, beyond that it rises slowly through the chromosphere and then rapidly in a narrow transition region to a few million degrees in the corona. It was only in 1940 that it was realised the corona is so hot.

The photosphere is the top of the convection zone and is a seething mass of continuously changing material. A granular structure (*granulation*) covers the Sun at this level, with each cell having a diameter of about 1 Mm and a lifetime of typically 5 min. Also a larger pattern (*supergranulation*) is present with a scale of 15 Mm: material rises in the centre of a cell, moves outward at about 500 m s^{-1} and moves down at the edges. The whole Sun rotates with a period of 25 days near the equator, but 31 days near the poles.

A white-light picture shows up dark spots – the sunspots – in two bands, one north of the equator and the other to the south. However, by observing the Sun at different wavelengths, pictures of the atmosphere at different altitudes can be obtained. For example a so-called $H\alpha$ filter reveals the chromosphere (Fig. 3.1) with a great deal of structure. The areas around sunspots are brighter than normal and are known as *active regions*. Occasionally, such a region may brighten very rapidly to give a *solar flare*. Also, there are thin dark structures known as *filaments* or *prominences*.

The corona is observed at eclipses as a faint halo of very low density and high temperature (Fig. 3.2). Low down the magnetic field tends to be closed, further out it is radial, pulled out by the solar wind. On average, there is only one eclipse per year, lasting two minutes, so artificial eclipses have been created by a coronograph, a telescope with a disc that blots out the glare of the photosphere: this is difficult because the corona is only one-millionth of the normal brightness of the Sun, about as bright as the full moon, so normally the corona cannot be seen through the dazzling light of the photosphere.

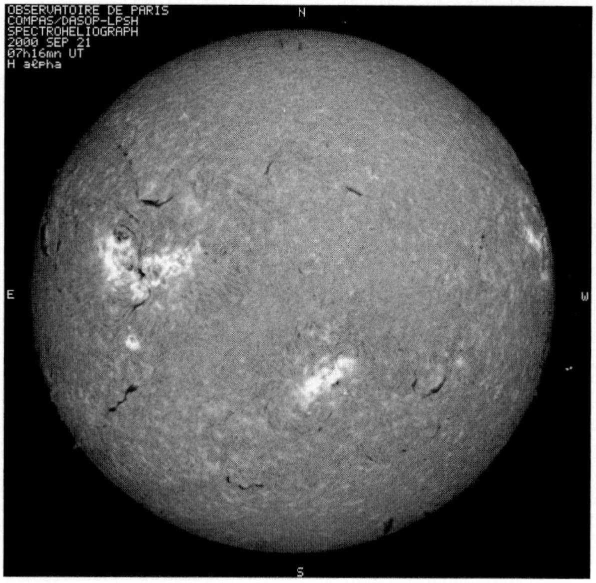

Fig. 3.1. Chromosphere in $H\alpha$, showing active regions (*bright*) and filaments or prominences (*thin dark ribbons* on the disk). (Courtesy B. Schmieder, Meudon Observatory)

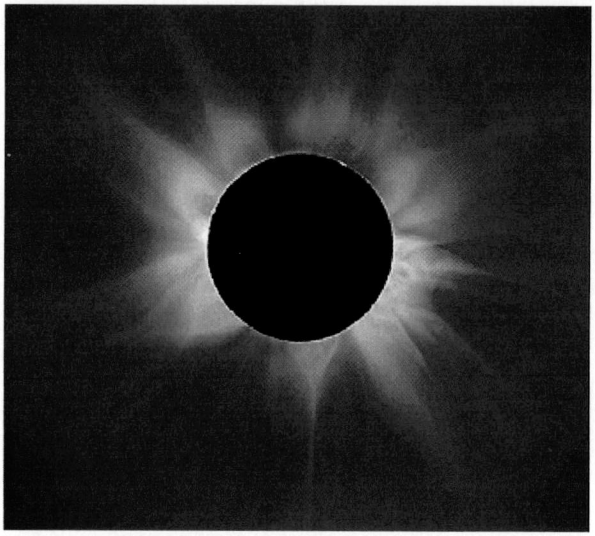

Fig. 3.2. Corona in white light during an eclipse (Courtesy High Altitude Observatory)

3.1.1 Solar Activity

Several types of transient activity are observed. *Sunspots* are dark, cool areas in the photosphere with extremely strong magnetic fields, up to 3000 G (Fig. 3.3). They often occur in pairs where a large flux tube pokes through the solar surface. They occur in two zones either side of the equator and the number of spots varies with an eleven-year period, the sunspot cycle. A sunspot group is surrounded by a region of moderate field strength (about 100 G), an *active region*, which is hotter and brighter than its surroundings.

The existence of sunspots has been known since at least the 4th century B.C. They can be as large as 20 Mm in diameter, each consisting of a central dark umbra at a temperature of 4100 K, surrounded by a penumbra of light and dark radial filaments. The field is almost vertical in the umbra, and more horizontal in the penumbra, and magnetic models of its structure have been constructed. There is a radial Evershed outflow in the penumbra of 6 km s^{-1}.

Some sunspots are unipolar, some bipolar, and others more complex. They can last for up to 100 days or so, and they can occur in two zones on either side of the equator. The number of spots varies, with an 11-year period, but there were very few during the Maunder minimum (1645–1715), when the Earth's climate was cooler than normal. The sunspot zones start at high latitudes and move toward the equator as the cycle progresses. Sunspot pairs exhibit polarity rules, such that the leading spot of a pair in one hemisphere tends to show the same polarity. At the start of a new cycle, the polarity of new spots changes.

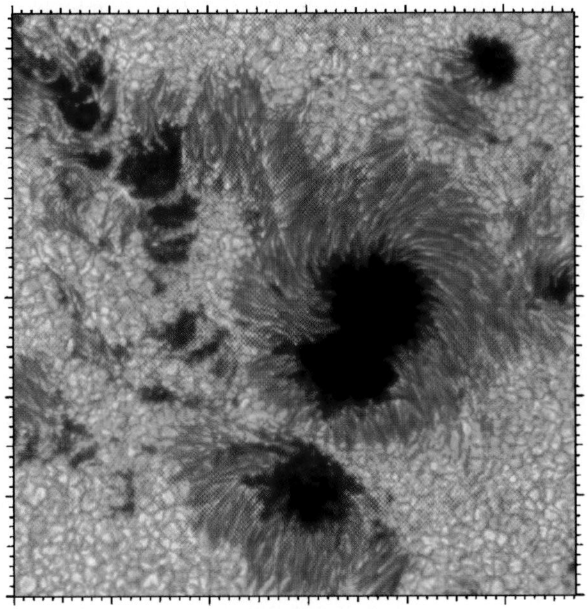

Fig. 3.3. A close-up of a sunspot (Swedish Solar Telescope, G. Scharmer)

A sunspot is dark because it is cooler than the surrounding photosphere. Cooling occurs locally, because the magnetic field inhibits convection and thus allows the spot temperature to become lower. In the normal photosphere, convection mixes the surface and the hotter subsurface layers and thus makes the surface hotter than it would otherwise be. A magnetic flux tube below the surface tends to rise by the process of *magnetic buoyancy*. Lateral total (plasma plus magnetic) pressure balance between the flux tube and its surrounding field-free region (denoted by subscript zero) implies

$$p + \frac{B^2}{2\mu} = p_0 \tag{3.1}$$

and so

$$p < p_0 \tag{3.2}$$

If the temperature difference is not too great, this in turn implies

$$\rho < \rho_0 \tag{3.3}$$

so that the tube is less dense than its surroundings and experiences an upward buoyancy force. When the tube rises and breaks through the solar surface, it can then create a pair of sunspots of opposite polarity, as often observed.

Prominences are vertical sheets of very dense cool material in the corona, observed in projection on the disc as thin, dark filaments (Fig. 3.4). Whereas spots usually fade after a few weeks, prominences keep growing for months up to 1000 Mm in length. They are the most stable of all surface features and can endure for nine months. Sometimes a prominence becomes unstable and erupts outwards. *Solar flares* are rapid brightenings in the chromosphere and corona near sunspots.

3.1.2 The Solar Revolution

The traditional view of the Sun was of a well understood object with a spherically symmetric atmosphere

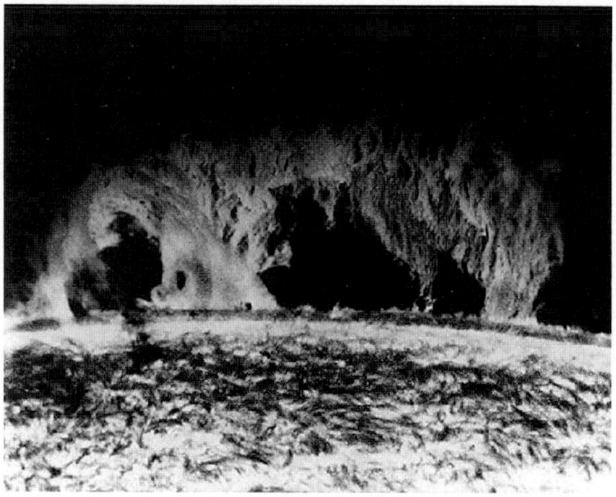

Fig. 3.4. A prominence (Big Bear Solar Observatory)

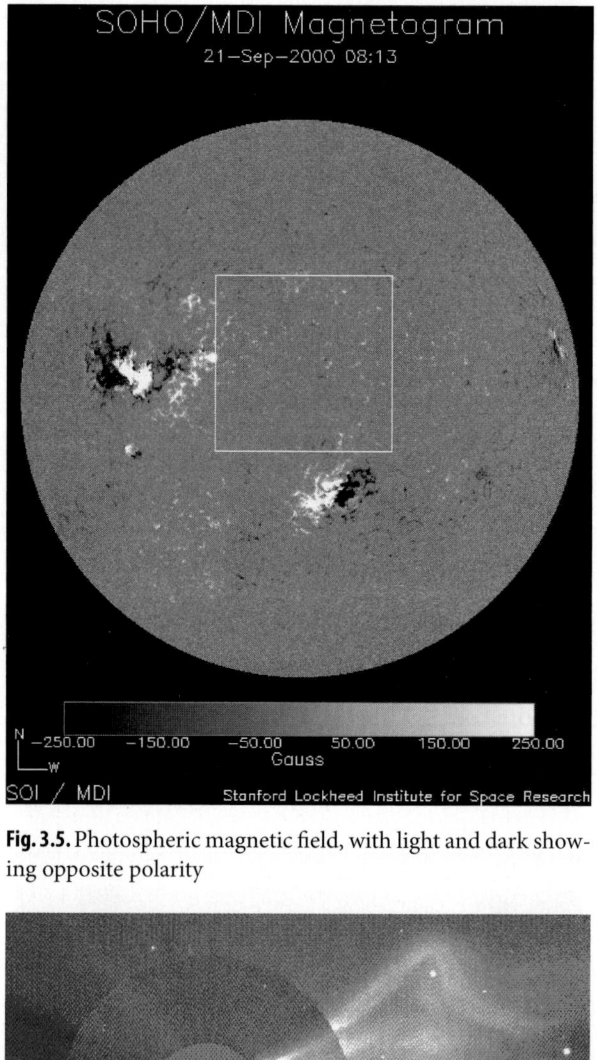

Fig. 3.5. Photospheric magnetic field, with light and dark showing opposite polarity

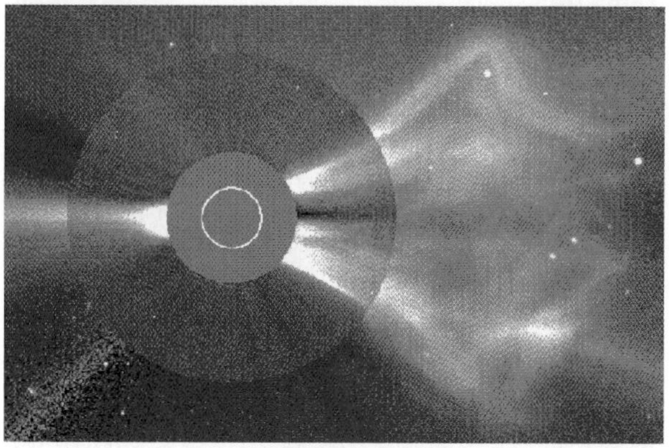

Fig. 3.6. A coronal mass ejection and an erupting prominence (Lasco Coronograph on SOHO)

and a magnetic field that is negligible ($\approx 1\,$G) except in sunspots; the atmosphere was heated by sound waves and an excess pressure drove a spherically symmetric expansion, the solar wind.

Many features of this old view have been completely transformed because of high-resolution observations from the ground of the photosphere and corona and X-ray observations from satellites of the corona. We now realise that the plasma atmosphere of the Sun is highly structured and dynamic and that most of what we see is caused by the magnetic field.

There is a similar change of thinking in astrophysics, where now the magnetic field is realised to be crucial in e.g., star formation, stellar activity (cycles, spots, coronae, flares), magnetospheres of compact objects (white dwarfs, neutron stars, and black holes), jets and accretion discs. However, the Sun continues to be a Rosetta stone for astronomy because here we can study many of the basic physical questions in depth.

In what ways has the traditional picture of the Sun changed? Many key topics are not well understood at all, such as the detailed internal structure, coronal heating, the origin of the solar wind, and the causes of eruptions and of solar flares. Also, there have recently been many dramatic discoveries:

i The Sun is oscillating globally. Many different normal modes of vibration have been detected and are being used to probe the interior and deduce its structure, just as seismology is used to infer the Earth's internal structure (see Sect. 3.2).

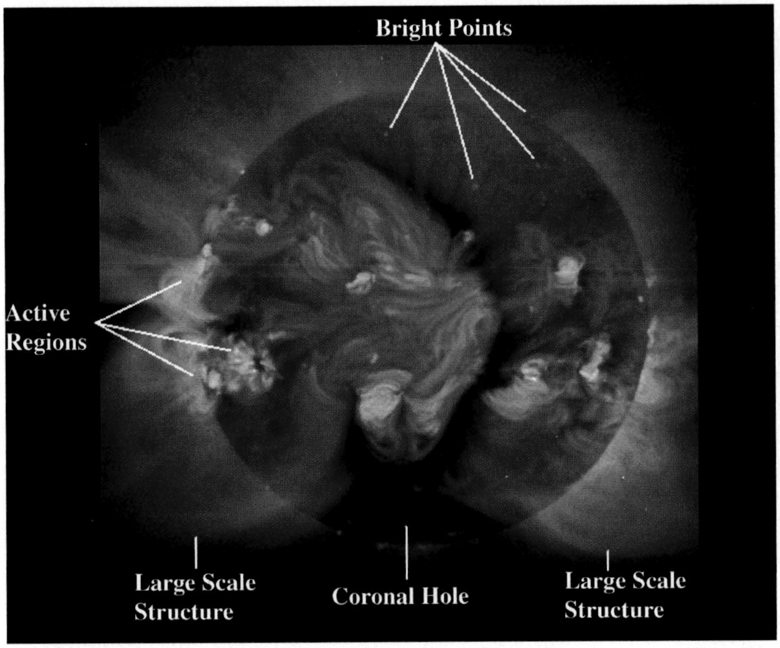

Fig. 3.7. A soft X-ray picture of the corona from the Yohkoh satellite

ii The magnetic field outside active regions is concentrated to form intense flux tubes of a thousand gauss at the edges of supergranule cells, whence they are carried by the supergranule flow (Fig. 3.5). Another earlier surprise was that active regions form a global pattern, with black polarity to the left in the northern hemisphere in Fig. 3.5 and to the right in the southern. This pattern reverses with the start of a new sunspot cycle. It occurs because differential rotation in the interior shears up poloidal flux and creates toroidal flux of one polarity in the north and the opposite polarity in the south. It is when such toroidal flux rises by magnetic buoyancy that it creates a pair of sunspots where it breaks through the surface.

iii Many new details of solar flares have been revealed by the RHESSI satellite, the most important being the detections of X-rays, the imaging of hard X-rays and information about the particle acceleration process.

iv With coronographs hugh erupting bubbles, called coronal mass ejections, have been seen propagating ahead of erupting prominences (Fig. 3.6).

v The corona has been revealed directly by soft X-ray telescopes (Fig. 3.7). It is an intriguing new world with myriads of loops and possesses a three-fold structure of coronal holes (magnetically open regions from which the fast solar wind is escaping), coronal loops and X-ray bright points (where magnetic fields are interacting).

3.1.3 Recent Surprises

Photosphere

The photosphere is covered with turbulent convection cells, namely, granulation (with typical diameters of 1 Mm) and supergranulation (with diameters of 15 Mm).

Amazing images in white light at 0.1 arcsec from the Swedish Solar Telescope reveal incredible detail in and between the granulation (Scharmer et al., 2002; van der Voort et al., 2004). Tiny bright points are probable locations of intense magnetic flux tubes at the edges of supergranules, where it was thought until this year that 90% of the quiet-Sun flux is located. However, close-ups of a few granules (Fig. 3.8) reveal for the first time bright points, "flowers" and ribbons in the intergranular lanes around granules and suggest the presence of many more intense magnetic tubes throughout the centres of supergranules at the granule boundaries. Indeed, Trujillo-Bueno et al. (2004) have suggested five or six times more magnetic flux resides there than we thought previously.

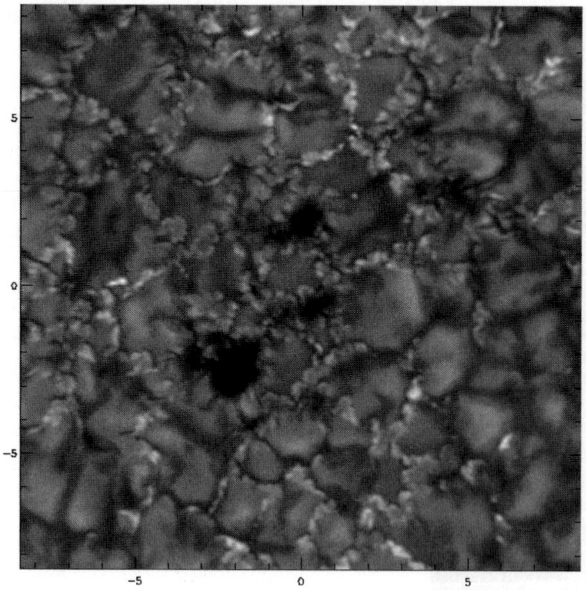

Fig. 3.8. A close up of a few granule cells in the photosphere (Swedish Solar Telescope, La Palma, G Scharmer)

Sunspots too are revealing new surprises. In the dark umbra of a spot the magnetic field is close to vertical. The striated penumbra surrounds the umbra and possesses a spreading magnetic field that is far from vertical. However, the penumbra is certainly not simple: recent observations from the Swedish Solar Telescope (Scharmer et al., 2002) have revealed bright flows moving both inward and outward, together with strange dark cores. Also, the bright penumbral filaments are thought to be at intermediate angles to the vertical and to represent magnetic field lines that go far from the sunspot, whereas the dark filaments are lower-lying and so return to the solar surface close to the spot, probably held down by granular pumping (Thomas et al., 2002), as sketched in Fig. 3.9.

The Corona

The corona can be viewed during a solar eclipse, and an early surprise was the discovery by Grotrian and Edlen (1940) that the coronal temperature is a million degrees or so. In the corona, the magnetic field dominates the plasma, both heating it somehow and creating its beautiful structure. The corona can also be observed direct with an X-ray or euv telescope, and indeed Yohkoh has revealed it to be a magnetic world with an amazingly rich variety of MHD phenomena.

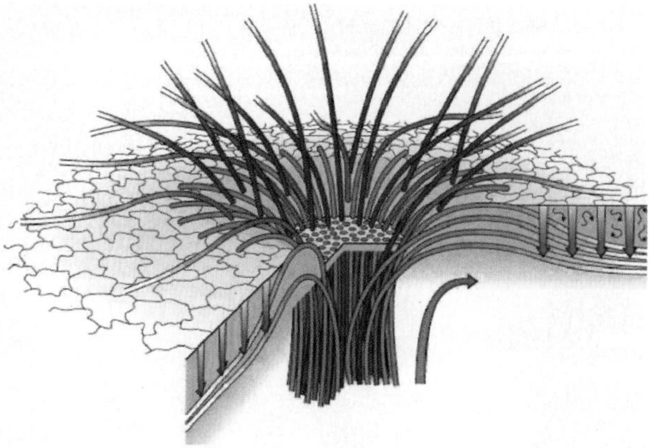

Fig. 3.9. Model of sunspot structure (Thomas et al., 2002)

Earlier rocket images and images from Skylab showed that the corona consists of dots called X-ray bright points, together with coronal holes (dark regions from which the fast solar wind escapes), coronal loops and active regions which are rather fuzzy. However, the TRACE mission has shown active regions in incredible detail and that the corona is made up of intricate loops of plasma aligned along the magnetic field (Fig. 3.10).

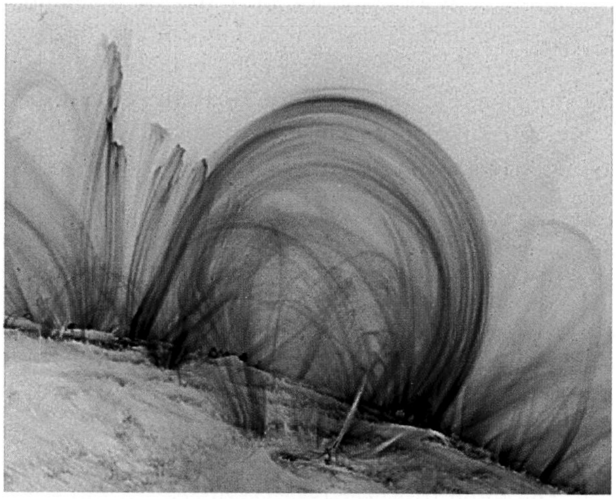

Fig. 3.10. Coronal loops imaged by TRACE

The Solar and Heliospheric Observatory (SoHO)

SoHO was launched in 1995 and is orbiting the Sun at the L1 point in phase with the Earth. A joint ESA-NASA mission, it is observing the Sun continuously for the first time and has transformed our understanding of the Sun. It has produced many surprises, notably in the solar interior and corona.

Solar Flares and Coronal Mass Ejections

An important question is: how do eruptive solar flares and coronal mass ejections occur? The LASCO instrument on SoHO is a coronograph which has observed the huge ejections of mass called *Coronal Mass Ejections* (*CME's*), which can sometimes reach the Earth and disrupt communications and space satellites.

On October 28, 2004, an incredibly large and complicated group of sunspots was crossing the solar disc and spawned the 3rd largest solar flare ever recorded. It produced a halo CME, namely, one that produces a halo round the Sun (Fig. 3.11) since in general it is either coming right towards the Earth or is moving away from it. This particular one was travelling towards the Earth at 2000 km s^{-1}, five times faster than normal.

High-energy particles taking only an hour to reach SoHO (by comparison with the CME itself, which takes a couple of days) produced "snow" as they bombarded the CCD detector plates (see Fig. 3.11), and when the CME did reach Earth it produced beautiful aurora that we viewed eagerly in St Andrews for a couple of nights. One week later, when the sunspot group had reached the limb of the Sun, the fireworks continued as it gave birth to the largest solar flare ever recorded.

The overall picture of what happens in an eruptive flare is that a sheared and twisted magnetic tube with an overlying arcade either loses equilibrium (Priest and Forbes, 1990; Forbes and Priest, 1995) or goes kink unstable or breaks out (Antiochus et al., 1999; Maclean et al., 2005). As the tube erupts, it drives reconnection in the arcade under the erupting tube. The reconnection heats a loop to high temperatures, which then cools down and drains as new loops are heated and the reconnection location rises. The result is the appearance of a rising arcade of hot loops with cool loops beneath them. A particularly fine example was caught by the TRACE satellite and the RHESSI flare satellite on April 21, 2002 (Fig. 3.12). The RHESSI contours of hard X-ray flux at 2 – 25 keV show emission from the reconnecting current sheet above the 1.5 MK TRACE loops, while the

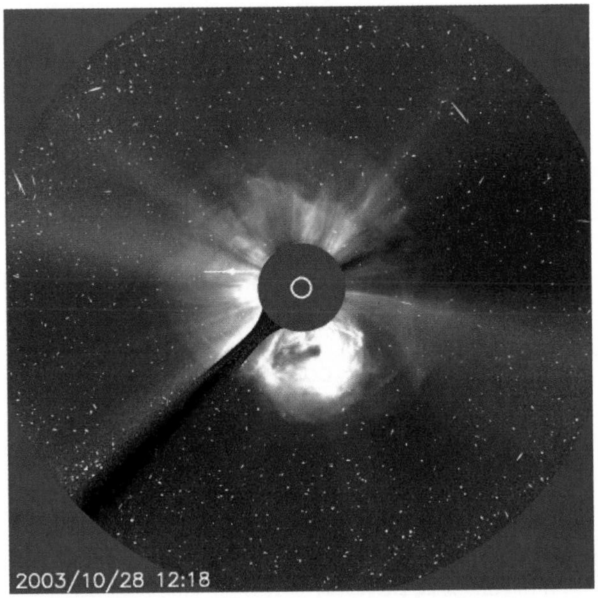

Fig. 3.11. CME on October 28, 2003, viewed by LASCO

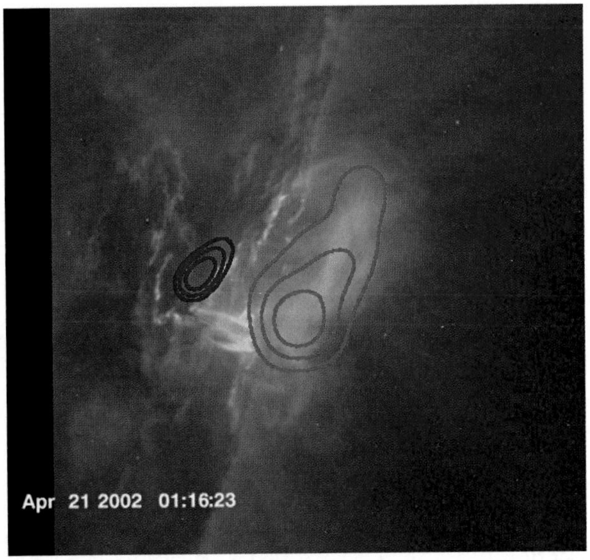

Fig. 3.12. Overlay of RHESSI contours of hard X-ray flux and TRACE image in 195 A during a solar flare (courtesy P. Gallagher)

50 – 100 keV contours show emission from high-energy electrons accelerated in the reconnection and having travelled down to the feet of the loops. Particle acceleration is thought to be partly by DC acceleration in the current sheet and partly by Fermi and betatron acceleration in the field lines that are springing downwards after reconnection.

Coronal Heating

Another question is: how is the solar corona heated to several million degrees by comparison with the photospheric temperature of only 6000 K? We know that the magnetic field is responsible and the mechanism is, in my view, likely to be magnetic reconnection, but the exact process is still uncertain. A key discovery from SoHO is, however, the existence of the *magnetic carpet* (Schrijver et al., 1997), the fact that the photospheric sources of the coronal magnetic field are highly fragmentary and concentrated into intense flux tubes threading the solar surface. These sources are also highly dynamic, magnetic flux emerging continually in the quiet Sun and then undergoing processes of fragmentation, merging and cancellation, in such a way that the quiet Sun flux is reprocessed very quickly, in only 14 h (Hagenaar, 2001).

Recently, using observed quiet-Sun magnetograms from the MDI instrument on SoHO, Close et al. (2004) have constructed the coronal field lines and studied their statistical properties. For the region they considered, 50% of the flux closed down within 2.5 Mm of the photosphere and 95% within 25 Mm, the remaining 5% extending to larger distances or being open (Fig. 3.13). They then tracked the motion of individual magnetic fragments in the magnetogram and recalculated the coronal field lines and their connectivity. In doing so, they discovered the startling fact that the time for all the field lines in the quiet Sun to change their connections is only 1.5 h. In other words, an incredible amount of reconnection is continually taking place – indeed, enough to provide the required heating of the corona.

Furthermore, a *Coronal Tectonics Model* has been proposed (Priest et al., 2002), which seeks to determine the effect of the magnetic carpet on coronal heating (see Sect. 3.5.4 (p. 88)). Each observed coronal loop reaches down to the surface in many sources, so the flux from each of these tiny sources is separated by separatrix surfaces (*separatrices*). As the sources move around, they generate current sheets on the separatrices and separators, where reconnection and heating takes place. In other words, the idea is that the corona is filled with myriads of separatrix and separator current sheets continually heating impulsively.

3.2 The Role of the Magnetic Field

3.2.1 Basic Equations

The Sun's magnetic field has several effects on the plasma. Some are passive: it may channel particles,

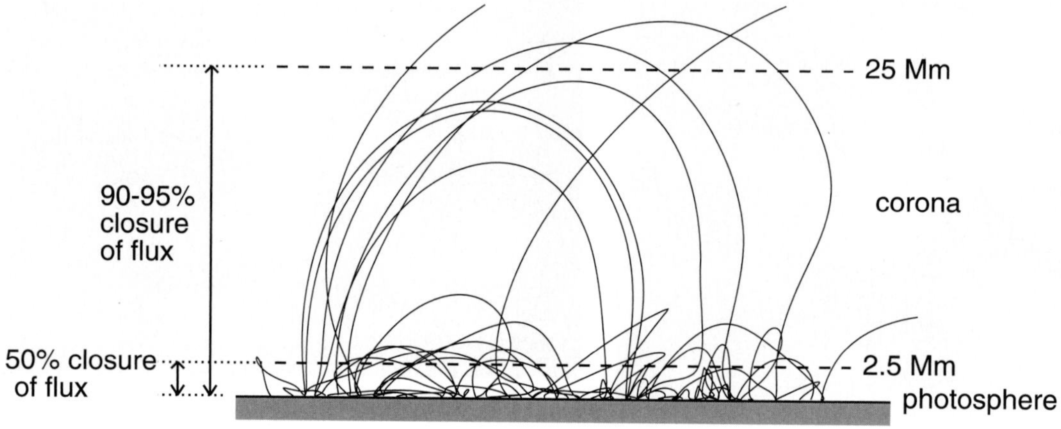

Fig. 3.13. Magnetic field lines in the quiet Sun (Close et al., 2004)

plasma and heat and may thermally insulate one part of the plasma from a neighbouring part. But some effects are active: the magnetic field may exert a force on the plasma and thus create structure or accelerate the plasma; it may store energy for a while and then suddenly release it; it may support waves or drive instabilities. The interaction between a plasma and a magnetic field can be modelled according to the principles of magnetohydrodynamics (MHD), in which the plasma can be treated as a continuous medium.

The equations of MHD unify the equations of slow electromagnetism and fluid mechanics. Maxwell's equations comprise Ampere's law, the vanishing of the divergence of the magnetic field, Faraday's law and Poisson's equation. These are accompanied by Ohm's law, the continuity equation, the momentum equation, and the ideal-gas law. In addition, an energy equation is required in order to determine the temperature (T) and thus close the system. In MHD, the displacement current $\partial \boldsymbol{D}/\partial t$ is neglected, which is valid when the plasma speed $\boldsymbol{v}$ is much slower than the speed of light.

The resulting equations may seem at first to be rather complicated, but they reduce to two main equations, one for the plasma velocity $\boldsymbol{v}$ and one for the magnetic field $\boldsymbol{B}$. Ampere's law becomes

$$\boldsymbol{j} = \nabla \times \frac{\boldsymbol{B}}{\mu_0} , \tag{3.4}$$

which determines the current density $\boldsymbol{j}$ once $\boldsymbol{B}$ is known. In order of magnitude,

$$j \sim \frac{B}{\mu_0 L} , \tag{3.5}$$

where L is the scale length for magnetic variations. Ohm's law is

$$\boldsymbol{E} = -\boldsymbol{v} \times \boldsymbol{B} + \frac{\boldsymbol{j}}{\sigma} , \tag{3.6}$$

which determines the electric field in terms of $\boldsymbol{v}$ and $\boldsymbol{B}$. Taking the curl and using Faraday's law will give the first of our two main equations, namely, the *induction equation*:

$$\frac{\partial \boldsymbol{B}}{\partial t} = \nabla \times (\boldsymbol{v} \times \boldsymbol{B}) + \eta \nabla^2 \boldsymbol{B} , \tag{3.7}$$

where $\eta = 1/(\mu_0 \sigma)$ is the *magnetic diffusivity* and here is assumed uniform.

The ratio of the first term to the second term on the right of (3.7) is the *magnetic Reynolds number*:

$$R_{\mathrm{m}} = \frac{vL}{\eta} = \mu_0 \sigma v L , \tag{3.8}$$

which for most solar phenomena on global length scales (~ 1 Mm, say) is enormous ($10^6 - 10^{12}$). Here L is a characteristic scale length for changes of the field and the flow. Thus the magnetic field is frozen to the plasma, and the electric field does not drive the current but is simply $\boldsymbol{E} = -\boldsymbol{v} \times \boldsymbol{B}$. The exception is in intense current concentrations or sheets, where L is so small that $R_{\mathrm{m}} \approx 1$. If the first term on the left of (3.7) is negligible, we have a simple diffusion equation:

$$\frac{\partial \boldsymbol{B}}{\partial t} = \eta \nabla^2 \boldsymbol{B} , \tag{3.9}$$

which implies that irregularities diffuse away on a time scale

$$\tau_{\mathrm{d}} = \frac{L^2}{\eta} , \tag{3.10}$$

known as the *diffusion time*, and with a (diffusion) speed

$$v_{\mathrm{d}} = \frac{\eta}{L} , \tag{3.11}$$

where τ_{d} is the time scale for magnetic energy conversion into heat by ohmic dissipation and is normally very long. For example, a length scale L of 10^7 m and a temperature of 10^6 K give a diffusion time of 10^{14} s! Thus it is only in regions of intense magnetic gradient (and therefore enormous current density) that L is small enough to produce time scales of interest in, for example, coronal heating or solar flares. Such scales may be present in shock waves or equilibria with current sheets or in reconnecting configurations.

The second main equation of MHD is the momentum equation:

$$\rho \frac{\mathrm{d}\boldsymbol{v}}{\mathrm{d}t} = -\nabla p + \boldsymbol{j} \times \boldsymbol{B} + \rho \boldsymbol{g} . \tag{3.12}$$

On the right-hand side, the first two terms represent the effects of thermal pressure and curvature. When the plasma beta,

$$\beta = \frac{2\mu p}{B^2} , \tag{3.13}$$

is small, the magnetic forces usually dominate the thermal pressure forces. This occurs, for example, in active regions. Equating the left-hand side of (3.12) to the magnetic force in order of magnitude gives a speed of

$$v = \frac{B}{(\mu_0 \rho)^{\frac{1}{2}}} \equiv v_A \tag{3.14}$$

which is the *Alfvén speed* and is the typical speed to which magnetic forces can accelerate plasma. Equating the sizes of the first and third terms on the right of (3.12) (with $p = R\rho T$) gives a length-scale of

$$L = \frac{RT}{g} \equiv H\,, \tag{3.15}$$

which is known as the *scale height* for the fall-off of the pressure with height. For example, it is about 500 km in the chromosphere, and 50,000 km in the corona. It explains why the pressure decreases with height so rapidly in the photosphere, and much more slowly in the corona. For a simple one-dimensional atmosphere with $p = p(z)$, such a hydrostatic balance gives

$$\frac{\mathrm{d}p}{\mathrm{d}z} + \rho g = 0\,, \tag{3.16}$$

where $\rho = p/(RT)$, and so, if the temperature is locally uniform, the solution is

$$p = p_0\,\mathrm{e}^{-z/H} \tag{3.17}$$

which exhibits the exponential pressure fall-off explicitly.

The magnetic force can be decomposed by writing

$$\boldsymbol{j}\times\boldsymbol{B} = (\boldsymbol{\nabla}\times\boldsymbol{B})\times\frac{\boldsymbol{B}}{\mu_0} = -\boldsymbol{\nabla}\left(\frac{B^2}{2\mu_0}\right) + \frac{(\boldsymbol{B}\cdot\boldsymbol{\nabla})}{\mu_0}\boldsymbol{B}\,, \tag{3.18}$$

in which the first term on the right represents the effect of a magnetic-pressure force acting from regions of high to low magnetic pressure $[B^2/(2\mu_0)]$. This has the same form as the normal plasma pressure gradient ∇p, in which, for example, a pressure $p(x)$ that is increasing with x produces a pressure force $-\mathrm{d}p/\mathrm{d}x$ in the negative x-direction. The second term in (3.18) represents the effect of a magnetic tension effect that gives a force when the field lines are curved and thus tends to shorten them. By putting $\boldsymbol{B} = B\boldsymbol{b}$, where $\boldsymbol{b}$ is a unit vector along the field, it can be written

$$(\boldsymbol{B}\cdot\boldsymbol{\nabla})\frac{\boldsymbol{B}}{\mu_0} = B\frac{\mathrm{d}}{\mathrm{d}s}\left(\frac{B\boldsymbol{b}}{\mu_0}\right) = \frac{B^2}{\mu_0}\frac{\mathrm{d}\boldsymbol{b}}{\mathrm{d}s} + \frac{B}{\mu_0}\frac{\mathrm{d}B}{\mathrm{d}s}\boldsymbol{b} \tag{3.19}$$

in which the first term on the right is

$$-\frac{B^2\boldsymbol{n}}{\mu_0 R} \tag{3.20}$$

in terms of the unit vector $\boldsymbol{n}$ along the principal normal and the radius of curvature R. The second term on the right cancels the gradient in pressure along the magnetic field in (3.10), so that there is (obviously) no $\boldsymbol{j}\times\boldsymbol{B}$ force along the magnetic field.

The two equations (3.7) and (3.12) are supplemented by the continuity equation

$$\frac{\mathrm{d}\rho}{\mathrm{d}t} + \rho\boldsymbol{\nabla}\cdot\boldsymbol{v} = 0 \tag{3.21}$$

and an energy equation for the upper solar atmosphere

$$\frac{\rho^\gamma}{\gamma-1}\frac{\mathrm{d}}{\mathrm{d}t}\left(\frac{p}{\rho^\gamma}\right) = -\boldsymbol{\nabla}\cdot(\kappa\boldsymbol{\nabla}T) - \rho^2 Q(T) + \frac{j^2}{\sigma}\,, \tag{3.22}$$

which describes how the entropy of a moving element of plasma changes because of three effects on the right-hand side: the conduction of heat, which tends to equalise temperatures along the magnetic field; the optically thin radiation, with a temperature dependence $Q(T)$ and ohmic heating. κ is the thermal conductivity. For temperatures between 10^5 K and a few times 10^6 K, Q decreases with temperature, and this tends to drive a *radiative instability*, because, if the temperature falls, the radiation will increase, and so the plasma will tend to cool further. In the lower solar atmosphere, optically thick radiative transfer effects come into play.

Equilibria of sunspots, prominences, coronal loops and other solar structures are described by the force balance

$$\boldsymbol{j}\times\boldsymbol{B} - \boldsymbol{\nabla}p + \rho\boldsymbol{g} = 0\,. \tag{3.23}$$

Along the magnetic field there is no contribution from the magnetic force, and so we have a hydrostatic balance between pressure gradients and gravity. In places such as active regions, where the magnetic field dominates, (3.23) reduces to the disarmingly simple form

$$\boldsymbol{j}\times\boldsymbol{B} = 0 \tag{3.24}$$

and the fields are said to be *force-free*, where

$$\boldsymbol{j} = \nabla \times \frac{\boldsymbol{B}}{\mu} \tag{3.25}$$

and

$$\nabla \cdot \boldsymbol{B} = 0\,. \tag{3.26}$$

Thus the electric current is parallel to the magnetic field, and so

$$\nabla \times \boldsymbol{B} = \alpha \boldsymbol{B} \tag{3.27}$$

where α is a scalar function of position. Taking the divergence of (3.27) gives

$$\boldsymbol{B} \cdot \nabla \alpha = 0 \tag{3.28}$$

so that α is constant along a field line. If α is uniform, the curl of (3.27) yields

$$(\nabla^2 + \alpha^2)\boldsymbol{B} = 0\,. \tag{3.29}$$

Solutions to this are known as linear or constant-α fields and are well understood. The particular case $\alpha = 0$ gives potential fields with zero current. Of all the fields in a finite volume with a given value for the normal component on the boundary, the field that has the smallest magnetic energy is the potential field.

There are many different kinds of MHD instabilities which are relevant to the Sun, as described by Bateman (1978). They include the following: interchange modes, in which field lines are wrapped around plasma in a concave manner; Rayleigh–Taylor instability, in which plasma is supported by a field against gravity, which may create structure in prominences; sausage and kink modes of a flux tube; Kelvin–Helmholtz instability, in which plasma flows over a magnetic surface; resistive modes of a sheared magnetic field, which drive reconnection; convective instability when a temperature gradient is too large, which can concentrate flux tubes in the photosphere; radiative instability, which creates cool loops and prominences up in the corona; and magnetic buoyancy instability of a magnetic field that decreases with height, which causes flux tubes to rise through the convection zone. In each case, the question of nonlinear development and saturation is important.

Further details of the MHD equations and their use in modelling solar phenomena can be found elsewhere e.g., Priest (1982).

3.2.2 Magnetic Waves

In a gas there are sound waves which propagate equally in all directions at the sound speed. In a plasma there are also waves, but they are of several types. Waves are very important in the solar atmosphere and throughout the cosmos. For example, they may be seen propagating out of sunspots or away from large solar flares. They are also a prime candidate for heating the solar atmosphere.

Alfvén and Compressional Alfvén Waves

On disturbing a uniform magnetic field, one would expect, by analogy with an elastic band, the magnetic tension to make a wave propagate along the field lines with speed

$$v_{\mathrm{A}} = \sqrt{\left(\frac{\text{tension}}{\rho}\right)} = \frac{B}{\sqrt{(\mu\rho)}}\,, \tag{3.30}$$

known as the *Alfvén speed*, since the tension is B^2/μ. Consider, therefore, an ideal plasma (with negligible dissipation), initially at rest with a uniform field $\boldsymbol{B}_0 = B_0\hat{\boldsymbol{z}}$ and density ρ_0. The effect of a disturbance is to introduce a velocity $\boldsymbol{v}'$ and to make the other variables $\boldsymbol{B}_0+\boldsymbol{B}'$, $\rho_0+\rho'$. Suppose it is so small that squares and products of v', B', ρ' can be neglected. Then, the pressureless MHD equations become

$$\frac{\partial \boldsymbol{B}'}{\partial t} = \nabla \times (\boldsymbol{v}' \times \boldsymbol{B}_0)\,, \tag{3.31}$$

$$\rho_0 \frac{\partial \boldsymbol{v}'}{\partial t} = (\nabla \times \boldsymbol{B}') \times \frac{\boldsymbol{B}_0}{\mu}\,, \tag{3.32}$$

where $\nabla \cdot B' = 0$ and ρ' is given by

$$\frac{\partial \rho'}{\partial t} + \nabla \cdot (\rho_0 \boldsymbol{v}') = 0\,, \tag{3.33}$$

and p' by $p' = c_s^2 \rho'$.

Look for wave-like solutions by supposing that the perturbation quantities behave like $\exp[\mathrm{i}(\boldsymbol{k}\cdot\boldsymbol{r} - \omega t)]$ so that (3.31) and (3.32) reduce to

$$-\omega \boldsymbol{B}' = \boldsymbol{k} \times (\boldsymbol{v}' \times \boldsymbol{B}_0) = (\boldsymbol{B}_0 \cdot \boldsymbol{k})\boldsymbol{v}' - \boldsymbol{B}_0(\boldsymbol{k}\cdot\boldsymbol{v}')\,, \tag{3.34}$$

$$-\mu\rho_0\omega \boldsymbol{v}' = (\boldsymbol{k} \times \boldsymbol{B}') \times \boldsymbol{B}_0 = \boldsymbol{B}'(\boldsymbol{B}_0 \cdot \boldsymbol{k}) - \boldsymbol{k}(\boldsymbol{B}' \cdot \boldsymbol{B}_0)\,, \tag{3.35}$$

where $\boldsymbol{k}\cdot\boldsymbol{B}'=0$, $p'=c_s^2\rho'$ and

$$-\omega\rho' + \rho_0\boldsymbol{k}\cdot\boldsymbol{v}' = 0\,. \tag{3.36}$$

For waves propagating at some angle θ to $\boldsymbol{B}_0$, we assume $\boldsymbol{k}\cdot\boldsymbol{v}'=0$. Then (3.34) and (3.35) become

$$-\omega\boldsymbol{B}' = (\boldsymbol{B}_0\cdot\boldsymbol{k})\boldsymbol{v}'\,,$$
$$-\mu\rho\omega\boldsymbol{v}' = (\boldsymbol{B}_0\cdot\boldsymbol{k})\boldsymbol{B}' - \boldsymbol{k}(\boldsymbol{B}'\cdot\boldsymbol{B}_0)\,,$$

where again $\boldsymbol{B}'\cdot\boldsymbol{B}_0=0$ (which may be seen most easily by taking $\boldsymbol{k}\cdot$ the second equation and remembering that $\boldsymbol{k}\cdot\boldsymbol{v}'=\boldsymbol{k}\cdot\boldsymbol{B}'=0$). These two equations imply the dispersion relation for waves, namely

$$\omega^2 = k^2 v_A^2\cos^2\theta\,, \tag{3.37}$$

since $\boldsymbol{B}_0\cdot\boldsymbol{k}=B_0 k\cos\theta$ and $v_A^2=B_0^2/(\mu\rho_0)$.

Alfvén waves are transverse in the sense that $\boldsymbol{v}'$ is perpendicular to the direction $\boldsymbol{k}$ of propagation and so by (3.36) ρ' (and therefore also p') vanishes. The phase velocity is $\omega/k=\pm v_A\cos\theta$ in the direction of $\boldsymbol{k}$, so that the speed of propagation depends on the direction and there is no propagation perpendicular to $\boldsymbol{B}_0$ ($\theta=\frac{1}{2}\pi$).

Consider now the case when $\boldsymbol{k}\cdot\boldsymbol{v}'\neq 0$. Substitute for $\boldsymbol{B}'$ from (3.34) in (3.35), so that

$$\mu\rho_0\omega^2\boldsymbol{v}' = \left[(\boldsymbol{B}_0\cdot\boldsymbol{k})\boldsymbol{v}' - \boldsymbol{B}_0(\boldsymbol{k}\cdot\boldsymbol{v}')\right] \times (\boldsymbol{B}_0\cdot\boldsymbol{k}) - \boldsymbol{k}\left[(\boldsymbol{B}_0\cdot\boldsymbol{k})(\boldsymbol{v}'\cdot\boldsymbol{B}_0) - B_0^2(\boldsymbol{k}\cdot\boldsymbol{v}')\right]\,. \tag{3.38}$$

This represents three linear homogeneous equations for three unknowns (v_x', v_y', v_z') and so in principle the determinant of coefficients would give a relation between the coefficients, namely the dispersion relation. But since $\boldsymbol{v}'$ only appears in the forms $\boldsymbol{v}'$, $\boldsymbol{k}\cdot\boldsymbol{v}'$ and $\boldsymbol{B}_0\cdot\boldsymbol{v}'$, we make take in turn $\boldsymbol{B}_0\cdot$ and $\boldsymbol{k}\cdot$ this equation to obtain two equations for $\boldsymbol{k}\cdot\boldsymbol{v}'$ and $\boldsymbol{B}_0\cdot\boldsymbol{v}'$, namely

$$\mu\rho_0\omega^2(\boldsymbol{B}_0\cdot\boldsymbol{v}') = 0\,, \tag{3.39}$$

and

$$\mu\rho_0\omega^2(\boldsymbol{k}\cdot\boldsymbol{v}') = k^2B_0^2(\boldsymbol{k}\cdot\boldsymbol{v}')\,. \tag{3.40}$$

Thus, from (3.40), either $\boldsymbol{k}\cdot\boldsymbol{v}'=0$ or

$$\omega^2 = k^2 v_A^2\,, \tag{3.41}$$

which is the dispersion relation for compressional waves.

These waves propagate equally in all directions, like sound waves, and, since $\boldsymbol{k}\cdot\boldsymbol{v}'\neq 0$, (3.36) implies that ρ' and p' are in general non-zero. For propagation across the field ($\boldsymbol{k}\cdot\boldsymbol{B}_0=0$) it can easily be seen from (3.38) that $\boldsymbol{v}'$ is parallel to $\boldsymbol{k}$ and therefore the mode is longitudinal.

Magnetoacoustic Waves

There are two waves when the pressure vanishes, namely the and compressional waves, and one wave when the magnetic field vanishes, namely the sound wave. If pressure fluctuations are included in the MHD equations by adding a term $-\boldsymbol{\nabla}p'$ to the right of (3.32), the effect is to add a term $-\mu\boldsymbol{k}p'$ to the right of (3.35) and $\boldsymbol{k}c_s^2\mu\rho_0(\boldsymbol{k}\cdot\boldsymbol{v}')$ to the right of (3.38). The waves (for which $\boldsymbol{k}\cdot\boldsymbol{v}'$ vanishes) are unaltered since ρ' and p' vanish. However, the sound and compressional waves are coupled together to give two *magneto-acoustic waves*.

Equations (3.39) and (3.40) become

$$-\omega^2(\boldsymbol{B}_0\cdot\boldsymbol{v}') = -(\boldsymbol{B}_0\cdot\boldsymbol{k})c_s^2(\boldsymbol{k}\cdot\boldsymbol{v}')\,, \tag{3.42}$$

$$\left(\mu\rho_0\omega^2 - k^2c_s^2\mu\rho_0 - k^2B_0^2\right)(\boldsymbol{k}\cdot\boldsymbol{v}') = -k^2(\boldsymbol{B}_0\cdot\boldsymbol{k})(\boldsymbol{B}_0\cdot\boldsymbol{v}')\,. \tag{3.43}$$

Thus, either $\boldsymbol{k}\cdot\boldsymbol{v}'$ and $\boldsymbol{B}_0\cdot\boldsymbol{v}'$ both vanish, when (3.38) gives the dispersion relation (37) for waves; or the above two equations imply

$$\omega^4 - \omega^2k^2\left(c_s^2+v_A^2\right) + c_s^2v_A^2k^4\cos^2\theta = 0\,. \tag{3.44}$$

This is the dispersion relation for slow and fast magnetoacoustic waves. The smallest root for ω^2/k^2 gives the slow mode and the largest the fast mode. The particular cases $p_0=0$ (i.e. $c_s^2=0$) and $B_0=0$ (i.e. $v_A^2=0$) reduce to the dispersion relations ($\omega^2=k^2v_A^2$ and $\omega^2=k^2c_s^2$) for compressional and sound waves, respectively, as expected.

3.2.3 Magnetic Reconnection

The induction equation

$$\frac{\partial\boldsymbol{B}}{\partial t} = \nabla\times(\boldsymbol{v}\times\boldsymbol{B}) + \eta\nabla^2\boldsymbol{B} \tag{3.45}$$

shows that the magnetic field changes due to advection and diffusion. The time-scale for diffusion is $\tau_d = L^2/\eta$, which, as we have seen, is very long for typical global length-scales, and the speed of diffusion is $v_d = \eta/L$.

For example, a current sheet diffuses away and converts magnetic energy into heat ohmically. The field lines diffuse in through the plasma and cancel, so that the region of diffused field spreads out at v_d. Therefore a steady state may be produced if magnetic flux (and plasma) are carried in at the same rate as it is trying to diffuse. However, in order to do so, we need to create an extremely small length-scale L (and therefore large magnetic gradient $\nabla \boldsymbol{B}$ and current $\boldsymbol{j}$). Furthermore, although the magnetic field may be destroyed by cancellation as it comes in, the plasma itself cannot be destroyed and needs to flow out sideways, as illustrated in the following model.

In what follows we describe the basic processes of diffusion, annihilation in two dimensions. The extension to three dimensions is a matter for current research, as reviewed in Priest and Forbes (2000), Chap. 8.

Magnetic Annihilation

Suppose we have a steady state flow

$$v_x = -\frac{V_0 x}{a}, \quad v_y = \frac{V_0 y}{a}, \tag{3.46}$$

so that the streamlines are the rectangular hyperbolae (xy = constant). A property of Eq. (3.46) is that $\nabla \cdot \boldsymbol{v} = 0$, so that the steady-state continuity equation reduces to $(\boldsymbol{v} \cdot \nabla)\rho = 0$ which implies that the density (ρ) is uniform (if it is constant at the edge). The flow vanishes at the origin and therefore represents an incompressible, *stagnation-point flow*.

Suppose now that the magnetic field lines are straight with $\boldsymbol{B} = B(x)\hat{\boldsymbol{y}}$ and that they reverse sign at $x = 0$. Then in Ohm's Law,

$$\boldsymbol{E} + \boldsymbol{v} \times \boldsymbol{B} = \eta \nabla \times \boldsymbol{B}, \tag{3.47}$$

both $\boldsymbol{v} \times \boldsymbol{B}$, $\nabla \times \boldsymbol{B}$ and therefore $\boldsymbol{E}$ are directly purely in the z-direction. Thus for a steady state with $\boldsymbol{E} = E(x, y)\hat{\boldsymbol{z}}$, the equation $\nabla \times \boldsymbol{E} = 0$ implies that $\partial E/\partial y = \partial E/\partial x = 0$, so that

$$E = \text{constant}. \tag{3.48}$$

In the present case Ohm's Law reduces to

$$E - \frac{V_0 x}{a} B = \eta \frac{\mathrm{d}B}{\mathrm{d}x}. \tag{3.49}$$

Now, when x is sufficiently large, the right-hand side of (3.49) is negligible and $B \approx E/(V_0 x)$, whereas when x is very small the second term is negligible and $B \approx Ex/\eta$. These approximate solutions are indicated by dashed curves in Fig. 3.14b. When x is large the magnetic field lines are are frozen to the plasma and are carried inwards, whereas when x is small the magnetic field diffuses through the plasma. The division between these two extremes, i.e. the half-width of the resulting current sheet, occurs (by equating the two approximations for B) when $x = (a\eta/V_0)^{\frac{1}{2}}$. The steady-state equation of motion, however, is also satisfied and so the above solution represents an exact solution of the nonlinear MHD equations – one of the very few that exists.

Sweet Parker

The Sweet–Parker model consists of a simple diffusion region of length $2L$ and width $2l$ between oppositely directed fields, for which an order-of-magnitude analysis is conducted. First of all, suppose the input flow speed and magnetic field are v_i, B_i, respectively, and ask: what is the outflow speed?

The electric current in order of magnitude is $j \approx B_i/(\mu l)$ and so the Lorentz force along the sheet is $(\boldsymbol{j} \times \boldsymbol{B})_x \approx jB_0 = B_i B_0/(\mu l)$. This force accelerates the plasma from the rest at the neutral point to v_0 over a distance L and so, by equating the magnitude of $\rho(\boldsymbol{v} \cdot \boldsymbol{n})v_x$ to the above Lorentz force, we have

$$\rho \frac{v_0^2}{L} \approx \frac{B_i B_0}{\mu l}. \tag{3.50}$$

However, from $\nabla \cdot \boldsymbol{B} = 0$,

$$\frac{B_0}{l} \approx \frac{B_i}{L}, \tag{3.51}$$

and so the right-hand side of (3.50) may be rewritten as $B_i^2/(\mu L)$ and we have

$$v_0^2 = \frac{B_i^2}{\mu \rho} \equiv v_{Ai}^2, \tag{3.52}$$

where v_{Ai} is the Alfvén speed at the inflow. Not surprisingly, therefore, the magnetic force accelerates the plasma to the Alfvén speed.

The next question is: how fast can field lines and plasma enter the diffusion region (at v_i)? First of all, note

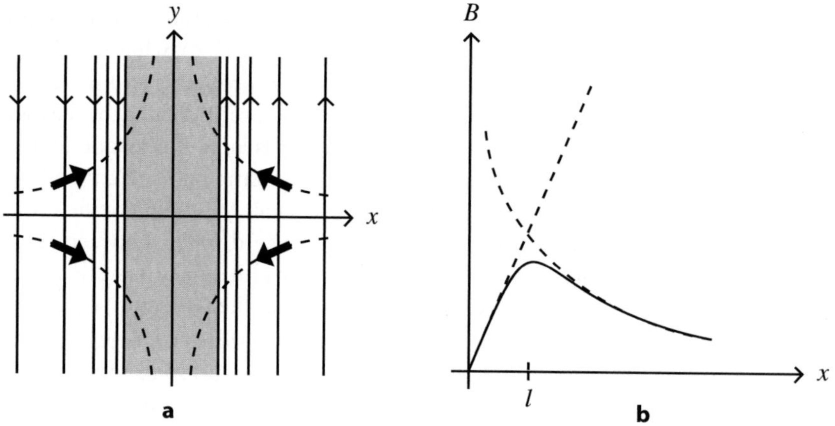

Fig. 3.14. (**a**) Stagnation point flow creating a shaded current sheet. (**b**) Magnetic field profile with small-x and large-x approximations shown as *dashed curves*

that for a steady state the plasma must carry the field lines in at the same speed that they are trying to diffuse outward, so that

$$v_i = \frac{\eta}{l} . \tag{3.53}$$

Also conservation of mass implies that the rate $(4\rho L v_i)$ at which mass is entering the sheet must equal the rate $(4\rho l v_0)$ at which it is leaving, so that

$$L v_i = l v_{Ai} . \tag{3.54}$$

The width l may be eliminated between these two equations to give $v_i^2 = \eta v_{Ai}/L$, or in dimensionless form

$$M_i = \frac{1}{R_{mi}^{\frac{1}{2}}} \tag{3.55}$$

in terms of the Alfvén Mach number

$$M = \frac{v}{v_A} \tag{3.56}$$

and the magnetic Reynolds number

$$R_m = \frac{L v_A}{\eta} \tag{3.57}$$

based on the Alfvén speed.

The fast regimes of reconnection that we shall consider next contain a tiny Sweet–Parker diffusion region around the X-point, and the flow speed and magnetic field at large distances L_e from the X-point are denoted by v_e and B_e. The properties of reconnection models depend on two dimensionless parameters, namely the reconnection rate $(M_e = v_e/v_{Ae})$ and the global magnetic Reynolds number $(R_{me} = L_e v_{Ae}/\eta)$.

Reconnection is said to be "fast" when the reconnection rate (M_e) is much larger than the Sweet–Parker rate (3.55). Properties at the inflow to the diffusion region (denoted by subscript "i") may be related to the "external" values at large distances (denoted by subscript "e"). Thus flux conservation $(v_i B_i = v_e B_e)$ may be written in dimensionless form as

$$\frac{M_i}{M_e} = \frac{B_e^2}{B_i^2} . \tag{3.58}$$

Thus, once B_i/B_e has been determined from a model of the external region outside the diffused region, (3.58) determines M_i/M_e and the dimensions of the diffusion region follow in terms of M_e and R_{me} from (3.53) and (3.54).

Petschek Model (1964)

Petschek's regime is "almost uniform" in the sense that the field in the inflow region is a small perturbation to a uniform field B_e. It is also potential in the nsense that there is no current in the inflow region. Most of the energy conversion takes place at standing slow-mode shocks, which are almost switch-off in nature (Fig. 3.15). These shock waves accelerate and heat the plasma, with $\frac{2}{5}$ of the inflowing magnetic energy being changed to heat and $\frac{3}{5}$ to kinetic energy.

The Petschek analysis is straightforward. The inflow region consists of slightly curved field lines and the magnetic field is a uniform horizontal field $B_e\hat{\boldsymbol{x}}$, plus a solution of Laplace's equation which vanishes at large distances and which has a normal component of, say, B_N

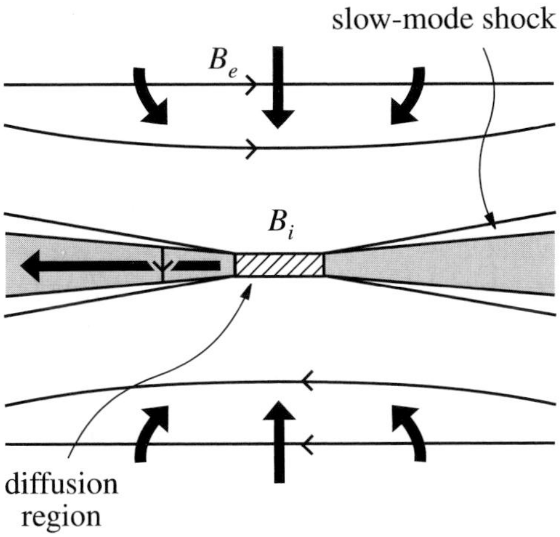

Fig. 3.15. Petschek's model

at the shock waves and zero at the diffusion region. To lowest order, the inclination of the shocks may be neglected, and so the problem is to find a solution in the upper half-plane which vanishes at infinity and which equals $2B_N$ between L and L_e on the x-axis and, by symmetry $-2B_N$ between $-L_e$ and $-L$. Now, we may regard the normal component on the x-axis as being produced by a continuous series of poles. If each pole produces a field m/r at a distance r, then the flux produced in the upper half-plane by that pole will be πm: but, if the pole occupies a distance dx of the x-axis, the flux is also $2B_N\,dx$, so that $m = 2B_N/\pi$ and integrating along the x-axis gives the field at the origin produced by the poles as

$$\frac{1}{\pi}\int_{-L_e}^{L}\frac{2B_N}{x}\,dx - \frac{1}{\pi}\int_{L}^{L_e}\frac{2B_N}{x}\,dx\,. \qquad (3.59)$$

Adding this to the field (B_e) at infinity gives

$$B_i = B_e - \frac{4B_N}{\pi}\log\frac{L_e}{L}\,. \qquad (3.60)$$

But at the shock waves, remembering that slow shock travel at the Alfvén speed based on the normal field, $B_N/\sqrt{(\mu\rho)} = v_e$, so that (3.60) becomes

$$B_i = B_e\left(1 - \frac{4M_e}{\pi}\log\frac{L_e}{L}\right), \qquad (3.61)$$

which is the expression for B_i that we have been seeking.

Since $M_e \ll 1$ and $B_i \approx B_e$, the scalings for the diffusion region dimensions become

$$\frac{L}{L_e} \approx \frac{1}{R_{me}M_e^2}\,, \quad \frac{l}{L_e} \approx \frac{1}{R_{me}M_e}\,, \qquad (3.62)$$

which shows that the dimensions of the central region decrease as the magnetic Reynolds number (R_{me}) or reconnection rate (M_e) increase. Petschek suggested that the mechanism chokes itself off when B_i becomes too small, and so he estimated a *maximum reconnection rate* M_e^* by putting $B_i = \frac{1}{2}B_e$ in (3.61) to give

$$M_e^* \approx \frac{\pi}{8\log R_{me}}\,. \qquad (3.63)$$

In value this is typically 0.01 since $\log R_{me}$ is slowly varying, and so we see that for typical R_{me} values this is much faster than the Sweet–Parker rate.

Thus, for twenty years the problem of fast reconnection was thought to have been solved completely (by Petschek), until, in 1986 a new generation of reconnection models was proposed (with Petschek's mechanism as a special case). Also, high-resolution numerical experiments were undertaken, which have demonstrated that these regimes of fast reconnection can indeed occur, provided the resistivity is (as expected in practice) either localised (Biskamp, 1986; Yan et al., 1992; Ugai and Tsuda, 1977; Ugai, 1995) or quasi-uniform (Baty et al., 2006).

Unified Theory of Fast Almost-Uniform Reconnection (Priest and Forbes, 1986)

I was puzzled to find that often numerical experiments have quite different properties from Petschek's theory, such as diverging rather than slightly converging inflow and different scalings of the diffusion region dimensions with reconnection rate. The object is to analyse the inflow region and find B_i/B_e and therefore how M_i varies with the reconnection rate (M_e). The analysis is to take the steady ideal, 2D, MHD equations

$$\rho(\boldsymbol{v}\cdot\nabla)\boldsymbol{v} = -\nabla p + \boldsymbol{j}\times\boldsymbol{B}\,, \qquad (3.64a)$$

$$\boldsymbol{E} + \boldsymbol{v}\times\boldsymbol{B} = 0 \qquad (3.64b)$$

where $\nabla\cdot\boldsymbol{v} = 0$, $\nabla\cdot\boldsymbol{B} = 0$, $\boldsymbol{j} = \nabla\times\boldsymbol{B}/\mu$ and $\boldsymbol{E}$ is constant, and then to seek solutions

$$\boldsymbol{B} = \boldsymbol{B}_0 + M_e\boldsymbol{B}_1 + \ldots\,, \quad \boldsymbol{v} = M_e\boldsymbol{v}_1 + \ldots \qquad (3.65)$$

that are a small perturbation about a uniform field $\boldsymbol{B}_0 = B_e\hat{\boldsymbol{x}}$. In these expansions $M_e \ll 1$ is the expansion parameter.

Neglecting the pressure gradient, to lowest order (3.64a) becomes

$$j_1 B_0 = 0 \tag{3.66}$$

or, if $(B_{1x}, B_{1y}) = (\partial A_1/\partial y, -\partial A_1/\partial x)$,

$$\nabla^2 A_1 = 0 \, . \tag{3.67}$$

The boundary conditions are to impose $B_{1x} = 0$ on the top boundary $\partial B_{1y}/\partial x = 0$ on the side boundaries and

$$B_{1y} = f(x) = \begin{cases} 2B_N, \; L < x < L_e \, , \\ 2B_N x/L, \; -L < x < L \, , \\ -2B_N \; - L_e < x < -L \, . \end{cases} \tag{3.68}$$

The resulting separable solutions are

$$\begin{aligned} B_{1x} &= -\sum_0^\infty a_n \cos\left[\left(n + \frac{1}{2}\right)\pi x\right] \\ &\quad \times \sinh\left[\left(n + \frac{1}{2}\right)\pi (1 - y)\right], \\ B_{1y} &= \sum_0^\infty a_n \sin\left[\left(n + \frac{1}{2}\right)\pi x\right] \\ &\quad \times \cosh\left[\left(n + \frac{1}{2}\right)\pi (1 - y)\right], \end{aligned} \tag{3.69}$$

where

$$a_n = \frac{4B_N \sinh\left[\left(n + \frac{1}{2}\right)\pi L(L_e)\right]}{L/L_e \left(n + \frac{1}{2}\right)^2 \pi^2 \cosh\left[\left(n + \frac{1}{2}\right)\pi\right]} \tag{3.70}$$

From these expressions we can calculate B_i/B_e and substitute in (3.58) in place of (3.61). The resulting graphs of M_e against M_i for given R_{me} show that for a given R_{me} there is indeed a maximum reconnection rate (M_e^*), as Petschek had surmised. Furthermore, the variation of M_e^* with R_{me} is very close to Petschek's estimate.

But now, how can we generalise this analysis by relaxing one of the assumptions? What we decided to do is include pressure gradients, so that terms $\mathrm{d}p_1/\mathrm{d}y$ and $-(\mu/B_e)\mathrm{d}p_1/\mathrm{d}y$ are added to the right hand sides of (3.66) and (3.67). But the effect on the solution (3.69) is simply to add a constant $a_n b$ to each term in the sum for B_{1x}.

The new parameter b produces a whole range of different regimes: $b = 0$ gives Petschek's regime (a weak fast-mode expansion); $b = 1$ gives a Sonnerup–like regime (a weak slow-mode expansion); $b < 0$ gives a family of slow-mode compressions; $0 < b < 1$ gives a hybrid family of slow- and fast- mode expansions; $b > 1$ gives a flux pile-up family of slow-mode expansions. For a *compression* the pressure increases as the plasma moves in towards the diffusion region, whereas for an *expansion* the pressure decreases. *Slow-mode* behaviour has the pressure and magnetic field behaving differently, so that for example a slow-mode compression makes the pressure increase and the magnetic field decrease. *Fast-mode* behaviour has the plasma and magnetic field both increasing or decreasing together.

For $b = 1$, M_e increases linearly with M_i; for $b = 0$ the Petschek maximum is found; all other regimes with $b < 1$ also possess a maximum reconnection rate, although when $0 < b < 1$ the maximum rate is faster than Petschek's when $b > 1$ there is no maximum rate, within the limitations of the theory.

Thus the solutions depend on the parameter b, which is in turn determined by the nature of the flow on the inflow boundary since v_x at the corner ($x = L_e$, $y = L_e$) is proportional to $b - 2/\pi$. When $b < 0$ the streamlines near the axis $x = 0$ are converging and so tend to compress the plasma and hence produce slow-mode compressions. When $b > 1$ the streamlines diverge and so tend to expand the plasma, producing slow-mode expansions: we refer to this as a "flux pile–up regime", since the magnetic field lines come closer together and the field strength increases as they are carried in. Another feature is that the central diffusion regions are much larger for the flux pile-up regime than the Petschek regime. These solutions are confirmed by numerical experiments when the magnetic diffusivity is enhanced in the diffusion region (Biskamp, 1994).

3.3 Prominences

Prominences appear as thin dark filaments on the disk in $H\alpha$ pictures, but in reality they are huge vertical sheets of plasma a hundred times cooler and denser than the surrounding corona. Densities typically are $10^{16} - 10^{17}\ \mathrm{m}^{-3}$, and temperatures are 5000 – 8000 K; the dimensions are

200 Mm long, 50 Mm high, and 6 Mm wide. They remain stable for months and are supported against gravity by a magnetic field of strength 0.5 – 1 mT (5 – 10 G) and inclined at a small angle (15 degrees) to the prominence axis. There is much fine-scale structure in the form of thin threads of width 300 km, although their cause is unknown. Plage (or active-region) prominences are smaller and lower than their large quiescent cousins, with densities greater than 10^{17} m^{-3} and fields of 2 – 20 mT (20 – 200 G).

Prominences lie above a reversal in the line-of-sight photospheric magnetic field, but the direction in which the field passes through the prominence may be normal (the same as one would expect from a simple arcade above the photospheric polarity) or inverse (in the opposite direction), as exemplified in the models of Kippenhahn and Schluter (1957) and Kuperus and Raadu (1974), respectively. Leroy (1989) found that one-third of his sample of prominences were of normal polarity, and two-thirds were of inverse polarity. There are also complex flow patterns in prominences, namely upflows of 3 km s^{-1} in $H\alpha$ when viewed in the disk and upflows on either side of a prominence of 6 – 10 km s^{-1} at 10^5 K.

Prominences probably form because of radiative instability, which can be demonstrated most simply as follows: consider a uniform hot equilibrium in an arcade of density ρ_0 and temperature T_0 between mechanical heating $(h_0\rho_0)$ and radiation simply proportional to density squared

$$0 = h_0\rho_0 - Q_0\rho_0^2 = 0 \tag{3.71}$$

and perturb this at constant pressure

$$\rho c_p \frac{\partial T}{\partial t} = h_0\rho - Q_0\rho_0^2 + \kappa_\parallel \frac{\partial^2 T}{\partial s^2} \tag{3.72}$$

where the latter term represents thermal conduction in a direction s along the magnetic field. Writing the perturbed temperature as

$$T = T_0 + T_1 \exp\left(\omega t + \frac{2\pi i s}{L}\right) \tag{3.73}$$

then gives the growth rate as

$$\omega = \frac{Q_0\rho_0}{c_p T_0} - \frac{\kappa_\parallel 4\pi^2}{\rho_0 L^2} \,. \tag{3.74}$$

When the loop length L is small, ω is negative, and conduction damps away the perturbation, but when L exceeds a critical value, radiative instability occurs. The classic model for support of a prominence is due to Kippenhahn and Schluter (1957), who assume a uniform temperature (T) and horizontal field (B_x), but a vertical field $[B_z(x)]$ and pressure $[p(x)]$ that vary with horizontal position (x). The horizontal and vertical components of force balance are then

$$p + \frac{B^2}{2\mu} = \text{constant} \tag{3.75}$$

and

$$\rho g = \frac{\mathrm{d}B_z}{\mathrm{d}x}\frac{B_x}{\mu} \tag{3.76}$$

so that the magnetic field both supports and compresses the plasma. The solution of these two equations, with $\rho = p/(RT)$, is

$$B_z = B_0 \tanh\frac{x}{l}\,, \quad p = \frac{B_0^2}{2\mu}\operatorname{sech}^2\frac{x}{l}\,, \tag{3.77}$$

where the prominence width l is

$$l = \frac{B_x H}{B_0} \tag{3.78}$$

and the vertical field tends to $= \pm B_0$ as x approaches $= \pm\infty$. This solution has been extended to include temperature variations and to allow variations with height (Ballester and Priest, 1987).

Later, twisted-flux-tube models for prominences were proposed (Priest et al., 1989), that agree much better with observations than do the classic models. The basic geometry is a large-scale flux tube that is slowly twisted up either by Coriolis forces or by flux cancellation (Martin, 1986; van Ballegooijen and Martens, 1989). Eventually, a dip with upward curvature near the summit is formed and at that point the prominence begins to form by radiative instability. As the twist or flux cancellation continues, the prominence grows in length, until the twist or length becomes too great.

3.4 Solar Flares

3.4.1 Introduction

The solar flare is a beautiful, awe-inspiring event which involves many branches of solar physics. Here, however, I shall discuss only the magnetohydrodynamics of the flare in order to determine the background environment or mould within which particle acceleration and emissions (from radio through the visible to X-rays and γ-rays) are occurring. After a summary of the observations, I shall describe the theory for the energy conversion process and the magnetic eruption. There is a subtle interaction between the MHD and microscopic plasma physics of the flare. The MHD provides the current sheets, shock waves and fluid turbulence, while the microscopic theory can determine in principle the plasma turbulent transport coefficients and the particle acceleration.

The amount of magnetic energy contained in coronal structures is certainly sufficient for a flare. For example, an arcade of field strength 500 gauss, radius 20 Mm, length 100 Mm and shear angle 45° contains an excess energy of 6×10^{25} j (6×10^{32} erg), sufficient for a large flare, whereas a loop of field strength 500 G, radius 5 Mm, length 100 Mm and twist 2π contains an excess energy of 7×10^{23} j, sufficient for a small flare. Also, there is enough time to store the energy in excess of potential by slow photospheric motions, injecting the required Poynting flux. For example, footpoints moving at 1 km s^{-1} cover a distance of 100 Mm a day.

However, the time-scale for energy release due to magnetic diffusion (ohmic dissipation) over a length-scale l is $\tau_d = l^2/\eta$ when $\eta = (\mu\sigma)^{-1}$ is the magnetic diffusivity and σ is the electrical conductivity; and, for a global length-scale $l \sim 10$ Mm, this time is of order 10^{11} seconds, much too long for a flare. The result is that one needs to create extremely small structures with length-scales of 1 km or less in order to release the energy fast enough. The main questions we therefore need to answer from a magnetohydrodynamic viewpoint are: how does the magnetic structure go unstable and how in detail is the energy released?

A typical large solar flare has three phases. During the *preflare phase* one sees the slow rise of a prominence (for typically half an hour), together with a soft X-ray brightening and the initiation of a coronal mass ejection. At the *rise phase* (for a few minutes) the prominence erupts much more quickly and there is a steep rise in $H\alpha$ and soft X-ray emission; also hard X-ray spikes are present together with impulsive EUV and microwave bursts. Type II and III radio bursts may be present and two bright ribbons form in the chromosphere. Throughout the *main phase* for many hours the intensity declines slowly and the ribbons separate, being joined by a rising arcade of loops, with hot X-ray loops located above cool $H\alpha$ loops.

Several new observational features were emphasized during the 1980's and 90's. Firstly, the density in the corona is found to increase due to evaporation from the chromosphere. Secondly, the high-temperature part of the flare takes the form of a loop or an arcade. Thirdly, often, but not always, emerging flux is observed close to the prominence before it erupts (Heyvaerts et al., 1977). Fourthly, two main types of flare are observed, namely eruptive flares (accompanied by a coronal mass ejection) and confined flares, although they have many features in common (Priest, 1981; Shibata, 1999).

The effects on the Earth can be considerable. In March 1989 there was a historic flare which ejected a great plasmoid that reached the Earth after a couple of days. It compressed the Earth's magnetic field, which drove by induction an electric current through the national electricity grid in Canada. This tripped the safety mechanisms and caused a power cut in the whole of Quebec. Also a great aurora was produced which was seen as far south as Italy and Jamaica. Short-wave radio and tv communications around the globe were disrupted because of the disturbance to the ionosphere, and compass readings were distorted by 10 degrees posing a threat to aeroplane and boat navigation. The SMM satellite was slowed down, its altitude falling by 1 km. Also a Norwegian oil company abandoned surveying because of the effects on the delicate magnetic sensors used by oil explorers to steer the drill heads.

The overall picture of a large solar flare (Fig. 3.16) is that during the *preflare phase* an active-region prominence and its overlying arcade rises slowly due to some kind of weak eruptive instability or nonequilibrium. At the flare *onset* the field lines that have been stretched out start to break and reconnect, which releases energy impulsively and causes the prominence suddenly to erupt much more rapidly. During the *main phase*, reconnection continues and creates hot X-ray loops and $H\alpha$ loops with $H\alpha$ ribbons at their footpoints. The increase in altitude of the reconnection point causes the locations of the hot loops to rise and of the chromospheric ribbons to move apart.

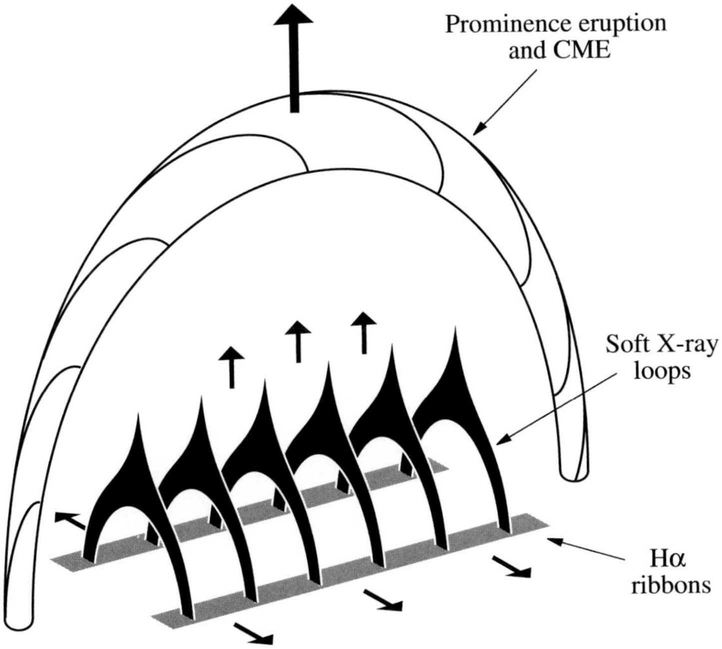

Fig. 3.16. The overall scenario for a solar flare

3.4.2 Energy Release by Magnetic Reconnection

Role of Emerging Flux and Moving Satellite Spots

New magnetic flux emerging vertically through the photosphere or satellite sunspots moving rapidly horizontally are often observed before flares. They may have two roles. The first is the creation of small (simple-loop) flares. As new flux emerges or satellites move they create a current sheet at some height h between the new or satellite flux and the ambient field. A small flare may then begin when h reaches a critical value for the current density to exceed the threshold for strong micro-turbulence. The critical height has been estimated by solving the energy balance within the sheet (Heyvaerts et al., 1977). The second role is the triggering of large flares when the overlying magnetic configuration contains a lot of stored magnetic energy in excess of potential. The new flux or satellite spots may push up a prominence until the critical height (h) for instability is exceeded. They may also rip away by reconnection some of the field lines lying over the prominence and helping to keep it down.

The role of reconnection in large flares is to release the magnetic energy, both in the impulsive and main phase, and sometimes to trigger this release (Shibata, 1999). The main questions that MHD addresses are the cause of the eruption and the details of the energy conversion process. Numerical experiments have focussed on two problems: first, the details of the closure process, whereby the stretched-out field lines reconnect and close back down; secondly, the global eruption in response to footpoint motions.

Building on earlier ideas by Carmichael, Sturrock and Hirayama, Kopp and Pneuman (1976) built a theoretical model for the creation by reconnection of flare loops in a two-ribbon flare, with the loops rising and the ribbons separating as the reconnection point rises. Then Cargill and Priest (1982) showed that the magnetic shocks that propagate from the reconnection site can heat the plasma to the observed temperatures of sometimes 10^7 K. Subsequently, the plasma cools and falls to give the classical $H\alpha$ loops with plasma draining down. Detailed modelling of the positions of the loops has been undertaken by Poletto and Kopp (1988) and a kinematic analysis to deduce the resulting electric fields was presented by Forbes and Priest (1982a).

Forbes and Priest (1982b) and Forbes and Priest (1984) set up a numerical experiment with initially vertical field lines in stretched-out equilibrium and

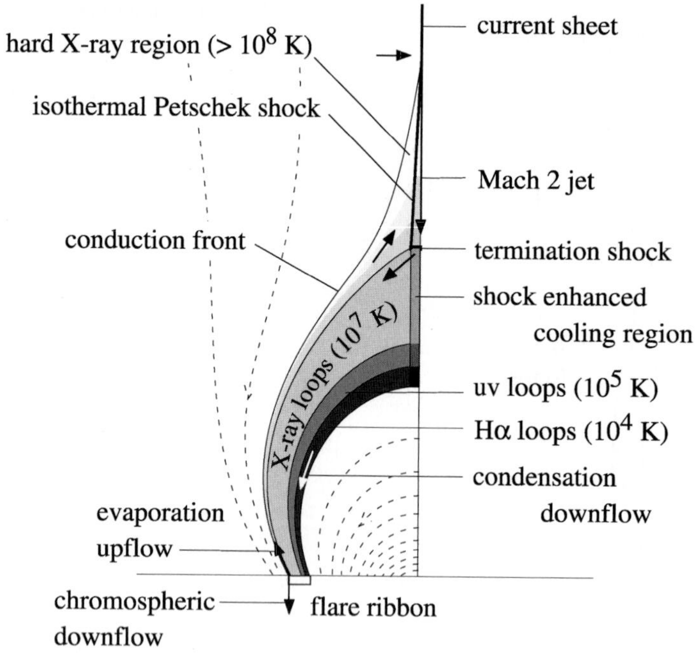

Fig. 3.17. Schematic creation of flare loops and ribbons by reconnection (courtesy T. Forbes)

anchored at their lower ends to the base of the numerical box due to photospheric line-tying. The experiment showed how the field lines begin to reconnect by the tearing-mode instability and then close down, creating closed loops in a quasi-steady manner. The current density contours reveal the location of the diffusion region, and the slow-mode shock waves.

The basic picture of the closure process has been refined considerably (Fig. 3.17), and numerical experiments have revealed new features of relevance to the observations (Forbes et al., 1989) as follows:

1. The quasi-steady reconnection may be modulated in a time-dependent manner as the reconnection enters an impulsive bursty regime in which the central diffusion region is so long that it tears, with neutral point pairs being slowly created and then rapidly annihilated. This process may explain the sudden jumps in loop height that are observed, and the impulsive nature of hard X-ray emission.
2. A fast-mode shock stands in the flow below the reconnection region and slows down the supermagnetosonic stream of plasma as it encounters the obstacle of closed field lines below it (Fig. 3.17). At the same time it compresses the plasma. The increase in density drastically reduces the radiative time-scale and triggers a thermal condensation which creates cool $H\alpha$ loops below the hot X-ray loops.
3. A reversed deflection current deflects the flow around the stagnation region.
4. The slow-mode shock wave in the presence of thermal conduction splits up into a conduction front (across which the temperature rises), together with an isothermal shock wave (across which the density increases).
5. Evaporation is driven from the chromosphere, both by the conduction front and by high-energy particles which travel from the reconnection site along the separatrix ahead of the conduction front. This greatly enhances the density in the hot X-ray loops, in agreement with observations.
6. When the magnetic field is smaller than 10 G and the field component out of the plane is large enough, a different regime of reconnection is found with submagnetosonic streams of plasma ejected down the reconnection region. The result is that no fast-shock or rapid condensation is created. Such a regime is appropriate to the eruption of quiescent prominences outside active regions.

3.4.3 Conditions for Flare Occurrence

Observational Pointers

In a large flare the overall scenario (Priest, 1981) is that during the preflare phase a prominence and its highly sheared field rise slowly. Then at flare onset we have the impulsive phase and a rapid eruption of the prominence, which probably occur because of the start of reconnection below the rising prominence. But why does the eruption occur? Why does the flare start?

For large flares several necessary conditions have been proposed. The obvious one is storage of free energy in a nonpotential field, since the photospheric changes during a flare are small. Also flaring has been associated with strong magnetic gradients. Two important necessary conditions are "Shear in the Corona" and "Complexity". Shear before flares is often observed in the chromosphere, as shown in $H\alpha$ fibrils and also in the photosphere, as shown in magnetograms near the polarity inversion line. These are suggestive of shear in the overlying corona.

Most major flares have a *filament activation and eruption*. Since we know that the magnetic field in a prominence has a shear angle of at least 75°, the main role of prominences is likely to be as an indicator of high shear in the corona.

The Emerging Flux Model (Heyvaerts et al., 1977) was developed primarily to explain complexity (in the form of emerging flux) and the presence of widely separated kernels. It suggested that, after new flux emerges, a current sheet becomes turbulent when it reaches a critical height. Also, the type depends on the magnetic environment, so that, if the overlying field contains no excess stored energy, only a small flare occurs, whereas, if a lot of stored energy is present, the emerging flux may trigger its release in a larger volume. Indeed, flares often begin at remote footpoints in a way that is naturally explained by emerging flux. Interacting flux is in general a regular part of active region evolution, as revealed in impacting polarities or parasitic polarities or delta spots. Often the basic structure of a flare is now believed to be an interaction of several bipoles or multiple loops, as seen in $H\alpha$ in X-rays and in radio images.

The role of rapid spot motions or rotations, as seen for instance in the relative motion of new and old spots, is to increaseshear and sometimes to increase complexity. Complexity is likely to be necessary because it allows reconnection between separate structures to release excess energy that is previously stored in one or other structure. In addition, when spots move and create a force-free field, new equilibria with different connections and lower energy may become available and the flare then represents a violent transfer to a new equilibrium.

Furthermore, the Yohkoh mission has lead to many advances in understanding of flares, the most notable being the paper by Masuda et al. (1994), which revealed the properties of hard X-ray impulsive sources located at the tops of loops.

Theoretical Ideas

Theory has suggested several ways in which a flare may start. In principle it seems that one just needs to solve the MHD equations $\boldsymbol{j} \times \boldsymbol{B} = 0$ and $\partial \boldsymbol{B}/\partial t = \nabla \times (\boldsymbol{v} \times \boldsymbol{B})$ for an evolution through a series of force-free equilibria due to footpoint motions. There are several possibilities for the evolution of the magnetic energy $W(t)$ when a critical point is reached (Fig. 3.18). First of all, a transcritical or pitchfork bifurcation may occur in which the equilibrium becomes linearly unstable but nonlinearly stable due to the presence of a new nearby stable equilibrium along which the evolution proceeds with no energy release. Secondly, a subcritical bifurcation may take place due to nonlinear instability or metastability, so that the system jumps down to a new equilibrium with energy release. Thirdly, a state of nonequilibrium may appear with no neighbouring equilibrium, so that a catastrophe occurs with energy release. It is hoped that such behaviour will show up in numerical computations.

Mikic et al (1988) modelled the global eruption of a coronal arcade numerically. They have a periodic set of arcades and impose a shearing motion of amplitude $0.01v_{\mathrm{a}}$ at the base. With 100×256 mesh points their magnetic Reynolds number is 10^4. The arcade evolves through a series of equilibria, and then at some point reconnects and forms a plasmoid which is ejected out of the top of the numerical box.

So why does eruption occur? One possibility is that it results from the interaction of two separate regions, as in the Emerging Flux Model (Heyvaerts et al 1977) and as Mikic et al (1988) and Biskamp and Welter (1989) model numerically. This appears to be the case in some, but not all, flares. More recent suggestions are as follows.

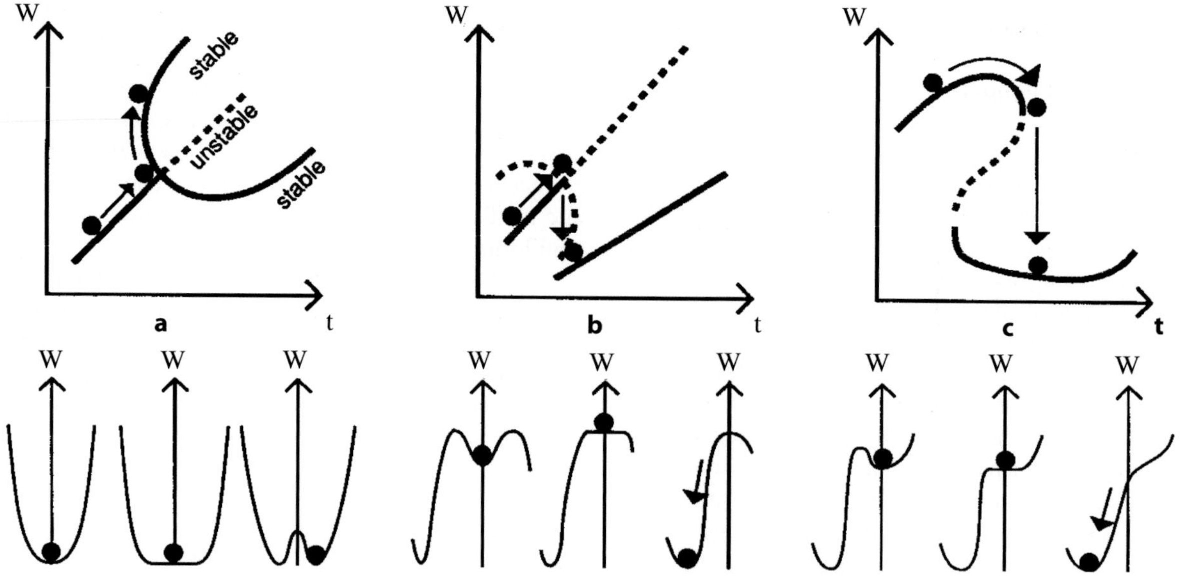

Fig. 3.18. (**a**) Pitch-fork bifurcation, (**b**) subcritical bifurcation (**c**) nonequilibrium

3.4.4 Catastrophe and Instability Models for Eruption

A linear instability analysis has the limitation of saying nothing about the nonlinear development, and so more recently the possibility of magnetic non-equilibrium or catastrophe has been considered.

Priest and Forbes (1990) set up a model for equilibrium and eruption by regarding the prominence as an electric current filament situated at height h in a background active-region modelled by a magnetic line dipole situated below the photosphere. As the filament current or twist increases so the prominence rises slowly through a series of equilibria.

But when a critical twist or current is exceeded there is no neighbouring equilibrium and a magnetic non-equilibrium or catastrophe takes place, with the unbalanced forces causing the prominence to erupt (Fig. 3.19). The velocity $(\mathrm{d}h/\mathrm{d}t)$ of the prominence as a function of time or height may be obtained by solving its equation of motion. $(m\,\mathrm{d}^2h/\mathrm{d}t^2 = IB^*)$, where m is the prominence mass, I its current and B^* the field at the prominence due to the background field.

When no reconnection is allowed one finds that beyond the non-equilibrium point the magnetic energy declines and the filament speed increases with height. Thus, in theory, the prominence erupts even if no reconnection is allowed. When reconnection is allowed, however, it is then driven by the eruption. The magnetic work declines more rapidly and the prominence erupts faster, with an energy release eight times faster. The resulting large electric field in the reconnection region may well be important for accelerating fast particles. The creation of the reconnecting current sheet below the erupting prominence is followed numerically (Forbes, 1991), and plots of density and current density contours reveal the presence of a fast-mode shock travelling ahead of the prominence together with slow-mode shocks near the reconnection site.

3.5 Coronal Heating

3.5.1 Introduction

Understanding how the solar corona is heated to a few million degrees by comparison with the photospheric temperature of only 6000 K is one of the major challenges in astronomy or plasma physics. Until the space age the only way of glimpsing the Sun's outer atmosphere was during a solar eclipse (Fig. 3.2), when the Moon cuts out the glare of the photosphere in white light and the much fainter corona comes into view with beautiful structures that are dominated for the most part by the magnetic field.

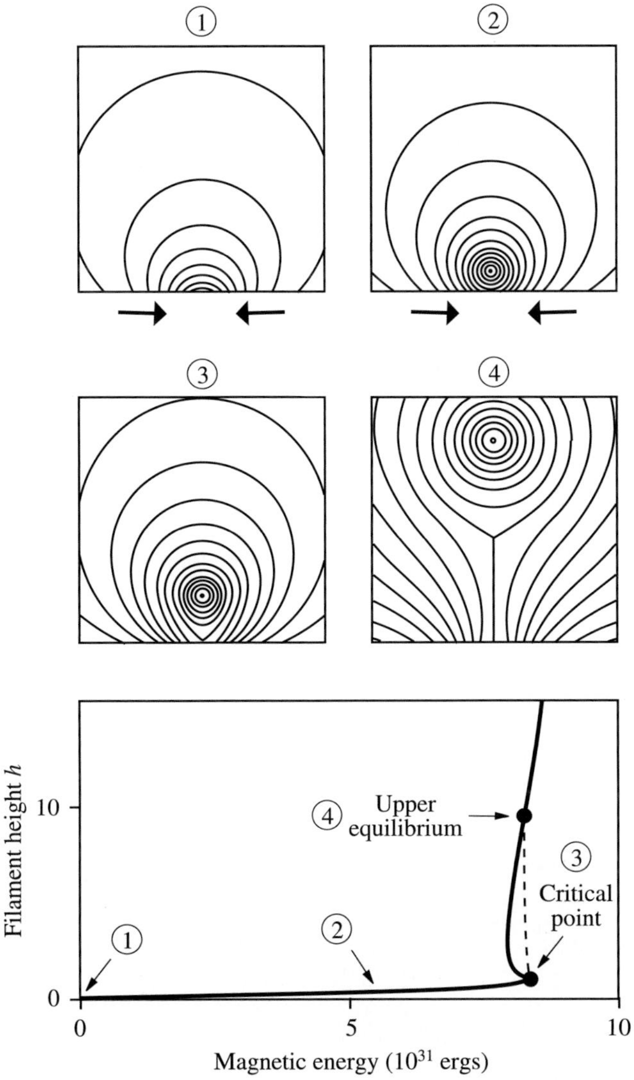

Fig. 3.19. The catastrophe model for flare eruption of Forbes and Priest

The energy required to heat the corona is typically 300 W m^{-2} (3×10^5 erg cm s^{-1}) in a quiet region or coronal hole and 5000 Wm^{-2} in an active region (Withbroe and Noyes, 1977). The energy flux from the solar surface due to photospheric motions moving the footpoints of coronal magnetic fields is plentiful: since $\boldsymbol{E} = -\boldsymbol{v} \times \boldsymbol{B}$, the Poynting flux is

$$\frac{\boldsymbol{E} \times \boldsymbol{B}}{\mu} \approx \frac{v_h B_h B_v}{\mu} , \qquad (3.79)$$

where v_h is the horizontal velocity and B_h, B_v are the horizontal and vertical components of the magnetic field. Thus, in order of magnitude, a typical v_h of 0.1 km s^{-1}, B_v of 200 G and B_h of 100 G would give a Poynting flux of 10^4 W m^{-2}. However, although B_v is measured well, the value of B_h is highly uncertain and depends on the nature of the coronal interactions, and therefore on the heating mechanism. Also, the details of how the energy flux is converted into heat and the efficiency of the various proposed heating mechanisms have not yet been determined.

The coronal magnetic field is incredibly complex and such complexity may be described in terms of the *magnetic skeleton*, which consists of a series of null points, *separatrices* (surfaces of field lines that generally originate from the fans of nulls and separate topologically distinct regions of space) and *separators* (field lines which join one 3D null point to another and represent the intersection of two separatrices). Current sheets tend to form and dissipate at separatrices and separators, where the magnetic connectivity of coronal footpoints is discontinuous, but they can also do so at quasi-separatrices where the magnetic connectivity has steep gradients (Priest and Démoulin, 1985; Titov et al., 2002).

It is possible that the different coronal structures are all heated by the same mechanism, but it is also possible that they are heated by different processes. Two general classes of model have been proposed for heating the corona. The first is MHD waves (Sect. 3.5.3), which may dissipate either by phase mixing or by resonant absorption or by shock dissipation (Kudoh and Shibata, 1999). The second class is magnetic reconnection (Sect. 3.5.4), either at null points or in the absence of null points. Furthermore, reconnection itself can heat the plasma either directly by ohmic heating or indirectly in a variety of ways, since it can generate waves or jets which subsequently dissipate ohmically or viscously.

One part of the coronal heating problem appears to have been solved, since it has been shown convincingly that X-ray bright points are probably heated (according to the Converging Flux Model) by magnetic reconnection driven in the corona by footpoint motions (Pages 83). However, the heating mechanisms of the other structures are at present unknown. The most likely mechanism for heating coronal holes is probably magnetic waves: a particularly attractive option is by high-frequency waves between 1 Hz and 1 kHz, generated by rapid, tiny reconnection events in supergranule boundaries (Axford and McKenzie, 1996; McKenzie et al., 1997; McKenzie and Sukhorakova, 1998), especially since the resulting ion-cyclotron or kinetic Alfvén waves may also be driving the fast solar wind and explaining the huge line-broadening that is seen with the UVCS instrument on SOHO (Kohl et al., 1997).

In most of the corona reconnection is now widely regarded as the most likely mechanism for coronal heating, especially since Yohkoh and SOHO observations have given a wide range of evidence in favour of reconnection at work in the corona (Pages 81–83). Several ways have been proposed in which the corona may be heated by magnetic reconnection, namely: driven reconnection (Page 83), turbulent relaxation (Page 83–86) and coronal tectonics (Pages 88–90), which grew out of the earlier braiding ideas (Pages 86–88) and incorporates binary reconnection, separator heating and separatrix heating.

Until 1998, it had been the general belief that the observational errors are so great that nothing could be inferred from measured loop properties about the form of the heating term in the energy balance equation (Chiuderi, 1981). However, Priest et al. (1998) suggested that in principle one should be able to deduce form of the heating if the temperature and density along a loop are known. They applied this new philosophy in a preliminary manner to a large loop system in soft X-ray images from the Japanese satellite Yohkoh. The large-scale corona consists of large magnetic loop systems that dominate the corona at solar minimum (e.g., Figs. 3.7 and 3.20), and are also present outside active regions and coronal holes at solar maximum when the global X-ray intensity is an order of magnitude higher (Acton, 1996). How is this large-scale corona heated?

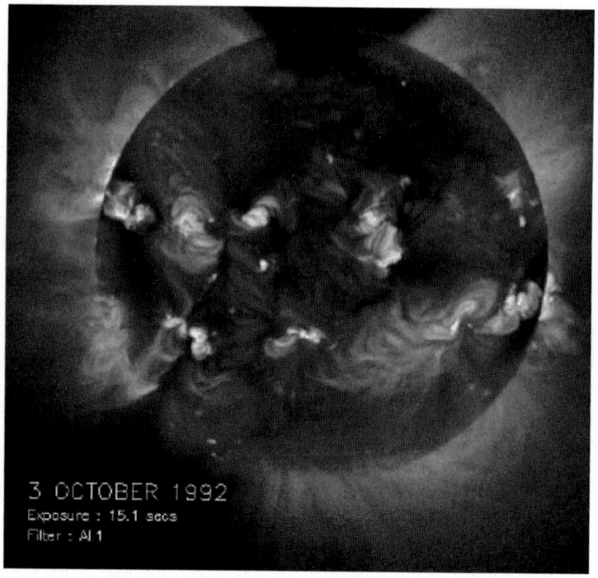

Fig. 3.20. Global image of the Sun in soft X-rays from the Japanese satellite Yohkoh (Courtesy Len Culhane)

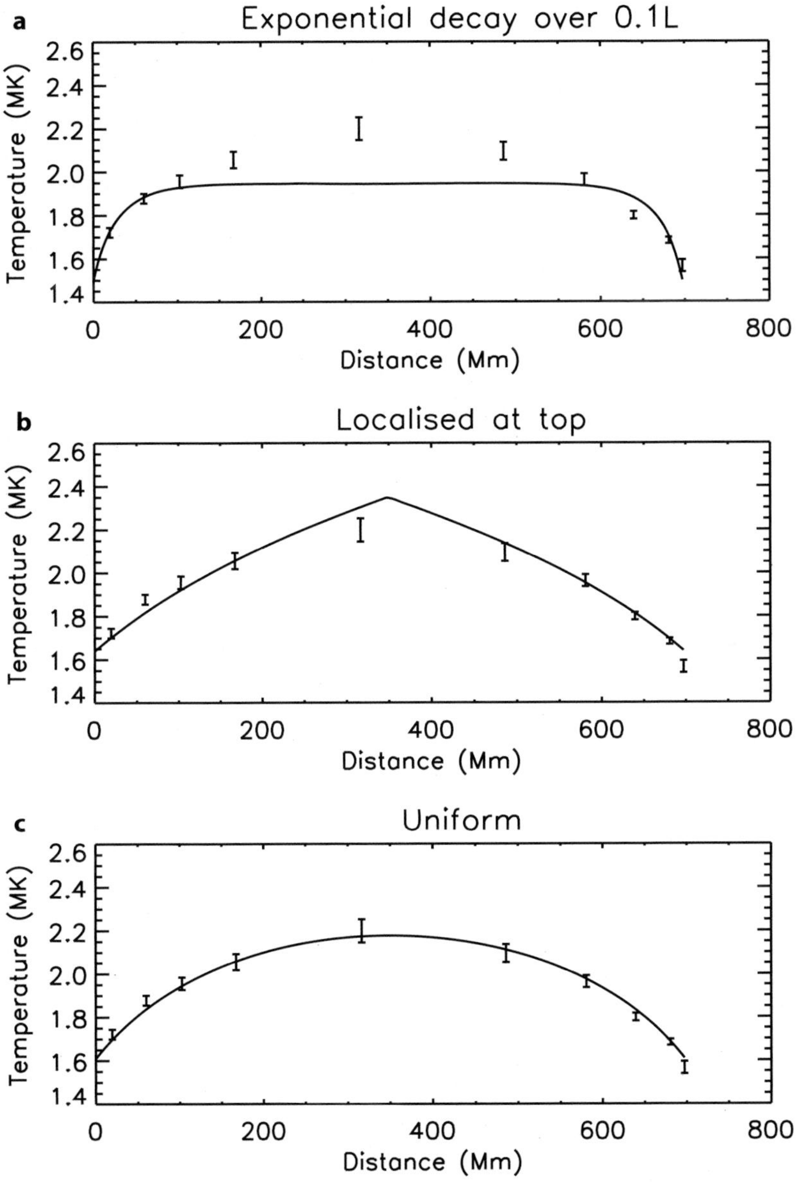

Fig. 3.21a–c. A comparison of the observed loop temperature profile with three models, in which heat is (**a**) dumped near the feet, (**b**) localised at the summit, and (**c**) deposited uniformly through the loop

If, on the one hand, the large-scale coronal loops were heated by turbulent reconnection in many small current sheets due to Coronal Tectonics (Pages 88–90), the heat would tend to be deposited fairly uniformly through the high coronal part of the loop. If, on the other hand, the heating were by long-wavelength Alfvén waves standing in the loop and possessing a maximum amplitude at the summit, it would tend to be dumped near the summit. Furthermore, if the heating were by XBP's or other reconnection processes near the solar surface, then the heat would be liberated mainly near the loop feet. A steady state thermal balance between such heating and thermal conduction would then produce a temperature profile along the loop from one coronal footpoint to another that has the variable $T^{\frac{7}{2}}$ being a quadratic function in the first case, or a pointed function in the second case, or having a steep footpoint rise and a flat summit profile in the third case.

Realising that the temperature profile is, therefore, highly sensitive to the nature of the heating mechanism, Priest et al. (1998) used the Yohkoh Soft X-ray Telescope to compare the temperature along a large loop (Fig. 3.20) with a series of models in order to deduce the likely form of heating. The most likely values of the parameters of the models were found by minimising χ^2.

The observed temperatures rise from about 1.4 MK near the observed feet to about 2.2 MK at the summit, and it can be seen in Fig. 3.21a that the model with the heat concentrated near 1.4 MK gives a very poor fit, whereas heat focused at the summit (Fig. 3.21b) produces a better fit. However, uniform heating between 1.4 MK and 2.2 MK (Fig. 3.21c) fits best of all and therefore provides preliminary evidence that the heating mechanism deposits the energy fairly uniformly along the observed length of the loop, at least for this example and for this temperature range.

Of the existing models, the one which can most easily explain the uniform heating in the high-temperature part of the loop is the Coronal Tectonics Model. (However, nonlinear Alfvén wave dissipation by coupling to magnetoacoustic modes that form shocks can also give uniform heating, Moriyasu et al. (2004).) Moreover, the tectonics model would suggest that, lower down at the feet of the loop below the measured temperatures, the heating should be much greater due to the carpet dissipation there. However, the importance of this work was not so much in the tentative conclusions of the particular loop system (which may well change when better observations and models are used in future) as in the suggestion that the observed temperature can indeed be used to deduce the form of the heating and therefore to put limitations on the likely heating mechanism.

3.5.2 Numerical Experiment on Global Active Region Heating

The response of the corona to photospheric footpoint motions is in most models (such as the coronal tectonics model), likely to be a localised heating in many small regions that is highly intermittent and impulsive in space and time. Gudiksen and Nordlund (2005a,b) have conducted a remarkable 3D MHD computational experiment that demonstrates this well using as realistic physics as is possible at present. They started with an initial stratified atmosphere and an initial magnetic field that is a potential extrapolation of an MDI magnetogram (but scaled down by a factor 4). In the lower chromosphere and photosphere an artificial cooling keeps the temperature close to its initial value. At the photospheric base they then imposed a simulated random granular pattern with a maximum amplitude of $300\,\mathrm{m\,s^{-1}}$ and three scales of 1.3, 2.5 and 5.1 Mm (although nothing corresponding to supergranulation). A grid of 150^3 was used in a computational box of $60 \times 60 \times 37$ Mm. Horizontal periodicity was assumed with vanishing vertical velocity and vertical temperature gradient on the upper boundary.

In the numerical experiment they found that the Poynting flux through the lower boundary did indeed dissipate ohmically in the atmosphere, and maintained a temperature of 10^6 K as expected. The ohmic heating decreased with height through the photosphere and chromosphere by a factor of 10^4. Most of the coronal heating was from intermittent short-period reconnection events representing about 8% of the Poynting flux or about $2 \times 10^6\,\mathrm{erg\,cm^{-2}\,s^{-1}}$. It was a factor of 4 larger than the radiation and was proportional to the square of the magnetic field. The transition region where the temperature reached 10^5 K was highly intermittent in space and time, ranging between 2.7 Mm and 12.3 Mm in height with an average of 5 Mm. At a height of 25 Mm, the plasma density and temperature varied continually in space and time by a factor 100 from 10^8 to $10^{10}\,\mathrm{cm^{-3}}$ and 10^4 to 10^6 K, respectively. The average mean density in the corona was roughly constant with height due to flows as high as $400\,\mathrm{km\,s^{-1}}$ (average $20\,\mathrm{km\,s^{-1}}$) as plasma was continually heated and cooled. The resulting configuration was approximately a nonlinear force-free field but close to potential. Simulated images in the TRACE 171 Å and 195 Å bands were remarkably realistic in a qualitative sense (Fig. 3.22), although the TRACE 171 intensity was a factor 10 too high, and there was no high-temperature (4 MK) present.

For the assumed parameters, the above experiment is able to maintain a model corona at 1 MK. The authors hope that at more realistic solar parameters, where, for example, the magnetic Reynolds number (R_m) is at least a factor of a million higher, the energy would somehow cascade down to the appropriate scales and somehow dissipate in a way that is independent of resolution and R_m. Whether or not that is true, one cannot at present tell: for example, the braiding experiments of

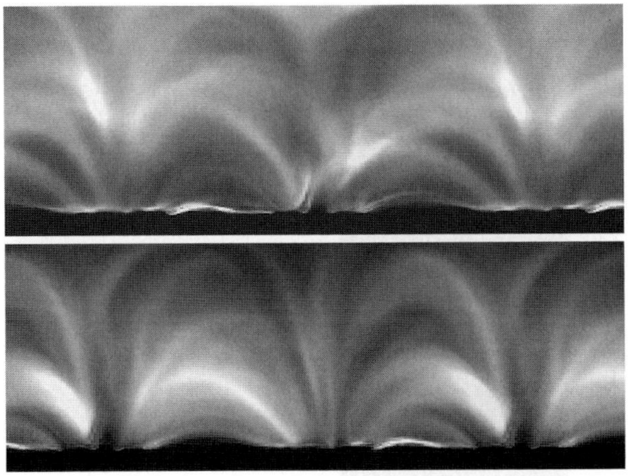

Fig. 3.22. Emulated TRACE 171 (*top*) and 195 (*bottom*) images from the computational experiment of Gudiksen and Nordlund (2005)

Galsgaard and Nordlund (1996) give a heating rate that is far from constant since it increases by a factor of between 1.3 and 2.5 as the spacial resolution is increased by a factor 2, depending on the experiment. Therefore, alternative scenarios and models that consider such dissipation at much higher R_m values (e.g., Sect. 3.5.4) are of great complementary value.

3.5.3 Heating by MHD Waves

Biermann (1946) and Schwarzchild (1948) suggested the heating of the upper atmosphere by sound waves that are generated by turbulence in the convection zone and then steepen to form shock waves as they propagate upwards. Indeed, the effect of photospheric oscillations on an overlying nonmagnetic atmosphere has been graphically demonstrated by Carlsson and Stein (2002) in models of a dynamic atmosphere.

However, it is now thought that all the corona and most of the atmosphere is dominated by the magnetic field and so heating by acoustic shocks is only relevant in nonmagnetic parts of the photosphere and low chromosphere. Although MHD waves are not thought to be the dominant form of coronal heating in closed regions, understanding how they dissipate in complex magnetic configurations by phase mixing or resonant absorption is important in view of the new field of coronal seismology, which is a potential means of determining many physical parameters in the corona. Furthermore, they remain a natural mechanism for heating in coronal holes, where reconnection is much less effective, and one could argue that, if waves are the heating mechanism in magnetically open regions, why should they not also be operating in coronal loops? Of interest here are the MHD simulations of nonlinear low-frequency waves in an open tube by Suzuki and Inutsuka (2005), which extend the previous studies of Kudoh and Shibata (1999) and Moriyasu et al. (2004).

3.5.4 Heating by Magnetic Reconnection

In view of the amazing complexity of the coronal magnetic field and the fragmentation and restless motion of its sources in the photosphere, coronal structures are continually changing and interacting with one another, so that magnetic reconnection is a natural way of heating the corona. Recent space satellites such as Yohkoh, SOHO and TRACE have provided much evidence of reconnection at work (see below), although a definitive observational test of exactly how reconnection is operating will await the next generation of satellites (such as Solar-B, SDO and Solar Orbiter).

There are several viable theories for heating coronal loops, one of which is to describe reconnection in many small regions in terms of MHD turbulence ideas (Pages 83–86). Another has been to develop Parker's (1972) earlier concept of nanoflares and braiding (Pages 86–88) to take account of the *magnetic carpet* and therefore greatly enhance the effectiveness of heating according to the *Coronal Tectonics Model* (Pages 88–90). The idea here is that the presence of many small magnetic sources in the photosphere creates a highly complex coronal topology containing myriads of separatrix surfaces where current sheets form and dissipate in response to photospheric motions. Several aspects of the coronal tectonics ideas are described here, including the basic model, binary reconnection, and a comparison of separator and separatrix heating.

Evidence from Yohkoh and SOHO for Reconnection

Yohkoh and SOHO have given important clues about the nature of coronal heating. When large-scale fields close down after eruptions, Yohkoh discovered that they do so in characteristic cusp-shaped structures (Tsuneta et al., 1992). In addition, observations of the temperature in active regions show that all the hottest loops are either cusps or pairs of apparently interacting structures, which is highly suggestive of reconnection (Yoshida and Tsuneta, 1996). In addition, many X-ray jets have been discovered by Shibata et al. (1996) – another clear signature of reconnection at work. These jets can extend for more than half a solar radius with a flow speed in excess of 200 km s^{-1}.

SOHO has three instruments observing the low corona (EIT, SUMER, and CDS) and two observing the outer corona (UVCS and LASCO), all of which have spotted the results of reconnection. Indeed, SOHO has demonstrated that reconnection gives an elegant explanation for many diverse phenomena, such as explosive events, blinkers, possibly tornadoes, X-ray bright points, the variation of the magnetic carpet, and the existence of the large-scale corona.

You can, of course, never prove a theory with observations – you can only disprove it, which is rather sad for theorists! But now there has been a real paradigm shift. Whereas previously reconnection was a fascinating concept which exercised the imagination of theorists like myself, now there are so many SOHO and Yohkoh observations that fit beautifully into place with the eyes of reconnection that it has become the natural explanation for many coronal heating phenomena. Let us then say a little about each of the above manifestations of energy release by magnetic reconnection.

Many *explosive events* have been observed with SUMER by Innes et al. (1997). They have presented an example in Si IV with a step-size of 1 arc sec, which reveals bidirectional jets that have been interpreted as jets accelerated in opposite directions from a reconnection site (Fig. 3.23). Furthermore, Chae et al. (1998) have compared an explosive event with Big Bear magnetograms and found that it is located over a site where magnetic fragments are approaching one another and, presumably, driving reconnection in the overlying atmosphere.

With the CDS instrument Harrison (1997) has discovered brightenings in the transition region which he has christened *blinkers*. They are located in the network and last typically 10 min. Berghmans and Clette (1998) have considered brightenings with EIT in He II at 80,000 K. They are similar to blinkers but have a wide variety of sizes and time-scales. From observations of 10,000 such events they have determined their statistical properties and found clear correlations of duration, intensity, and area with energy. Furthermore, Krucker et al. (1998) have observed microflares with EIT. They have energy of $10^{18} - 10^{19}$ J ($10^{25} - 10^{26}$ erg), an energy spectrum of $W^{-2.6}$, where W is the energy of an event, and they contribute 20% of the quiet-Sun heating. Even smaller-scale events have been studied with the Transition Region and Coronal Explorer (TRACE) data by Parnell and Jupp (2000), who find that the spectrum continues down to at least 10^{24} erg at a slope between W^{-2} and $W^{-2.6}$. If it continued further to 4×10^{22} erg at a slope of $W^{-2.6}$, nanoflaring events of such an energy could provide all the heating of the quiet corona.

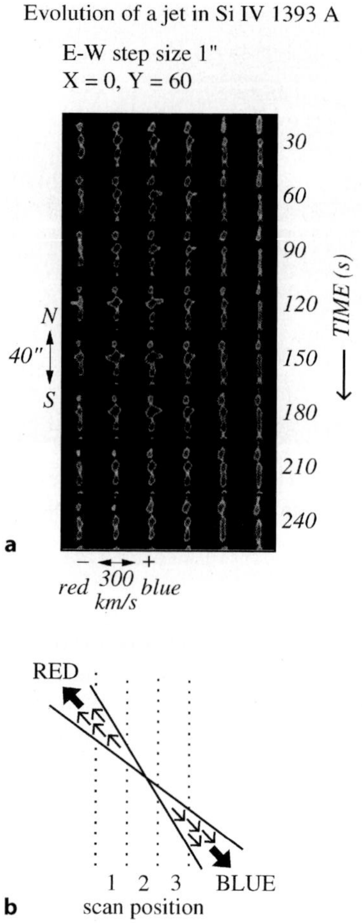

Fig. 3.23a,b. SUMER spectra of an explosive event in Si IV with a step-size from left to right in each row of 1 arc sec (courtesy D. Innes)

Pike and Mason (1998) have observed macrospicules in coronal holes with the CDS instrument on SOHO. The macrospicules are found to be rotating at typically $20\,\mathrm{km\,s^{-1}}$ but occasionally at $150\,\mathrm{km\,s^{-1}}$, and they christened such structures *tornadoes*. These could well be a consequence of 3D reconnection, since twisted structures are naturally produced by the conservation of global magnetic helicity. Furthermore, do ordinary spicules also possess such twist – and is the same true for even finer-scale structures that could be driving ion-cyclotron waves and possibly accelerating the fast solar wind?

X-ray Bright Points: the Converging Flux Model

All over the surface of the Sun one finds very small regions (≈ 3 arc sec) which are bright in X-rays and have a duration of a few hours or less. Because these regions appeared point-like in early X-ray telescopes, they are called X-ray bright points or XBPs for short. Some XBPs are located above emerging flux and are explained by the Emerging Flux Model (Heyvaerts et al., 1977), but most are situated in the corona above pairs of opposite polarity magnetic fragments that are approaching one another (Harvey, 1988). As they collide, these fragments appear to annihilate one another in a process which observers refer to as *cancellation* (Martin et al., 1985). Although it has been suggested that cancellation might simply be the result of the submergence of a simple loop, this is thought to be rather unlikely (Priest, 1987), because there is no obvious way that submergence would produce an overlying coronal brightening. It is much more likely that reconnection is taking place above the cancellation site.

The Converging Flux Model shown in Figs. 3.24 and 3.25 explains how cancellation can lead to the appearance of an XBP (Priest et al., 1994). Because of the overlying field in the cancellation region, a null-point does not form until the opposite polarities are sufficiently close. The null point first appears at the surface and then moves upwards as the polarities approach. However, continued motion eventually causes the null-point to reverse direction and sink back into the photosphere. In most cases magnetic flux emerges in a supergranule cell and then moves to the boundary, so that one polarity tends to accumulate while the other reconnects with opposite-polarity network and forms a bright point (Fig. 3.25). This structure matches well the predictions made by a three-dimensional version of the model (Parnell et al., 1994).

Falconer et al. (1999) hunted for X-ray bright points with the EIT (Extreme ultra-violet Imaging Telescope) instrument on SOHO in a large square region of side $0.6R_\odot$. They applied a filter to remove the background haze, and this showed up many smaller bright points than normal, which they called micro-bright points. Comparison with a Kitt Peak magnetogram showed that the normal bright points lie over large magnetic fragments of mixed polarity, which are close to each other and are presumably driving reconnection in the overlying corona. The *micro-bright points* all lie in the network and most of them also lie over mixed polarity, so this is consistent with bright points being the large-scale end of a much larger spectrum of reconnection events heating the corona by the Converging Flux Model mechanism.

Relaxation by MHD Turbulence

At the same time as coronal structures are trying to evolve through nonlinear force-free equilibria in response to footpoint notions, they also tend to relax by 3D reconnection towards linear force-free states that conserve global magnetic helicity. Turbulent relaxation tends not to destroy magnetic helicity when the magnetic Reynolds number is very large, but it can convert it from one kind to another, such as from mutual to self helicity. Heyvaerts and Priest (1984) suggested that the corona may be heated by relaxation as it evolves through a series of linear force-free states, satisfying $\nabla \times B = \alpha \boldsymbol{B}$, with the footpoint connections not preserved but the force-free constant (α) determined from the evolution of relative magnetic helicity (Berger and Field, 1984)

$$H_\mathrm{m} = \int (\boldsymbol{A} + \boldsymbol{A}_0)(\boldsymbol{B} - \boldsymbol{B}_0)\,\mathrm{d}V\,, \qquad (3.80)$$

where $\boldsymbol{A}$ is the vector potential and $\boldsymbol{A}_0$, $\boldsymbol{B}_0$ refer to corresponding potential values with the same normal field at the boundary. Boundary motions cause the magnetic helicity to change in time according to

$$\frac{\mathrm{d}H_\mathrm{m}}{\mathrm{d}t} = 2\int (\boldsymbol{B}\cdot\boldsymbol{A}_0)(\boldsymbol{v}\cdot\boldsymbol{n}) - (\boldsymbol{v}\cdot\boldsymbol{A}_0)(\boldsymbol{B}\cdot\boldsymbol{n})\,\mathrm{d}S\,. \qquad (3.81)$$

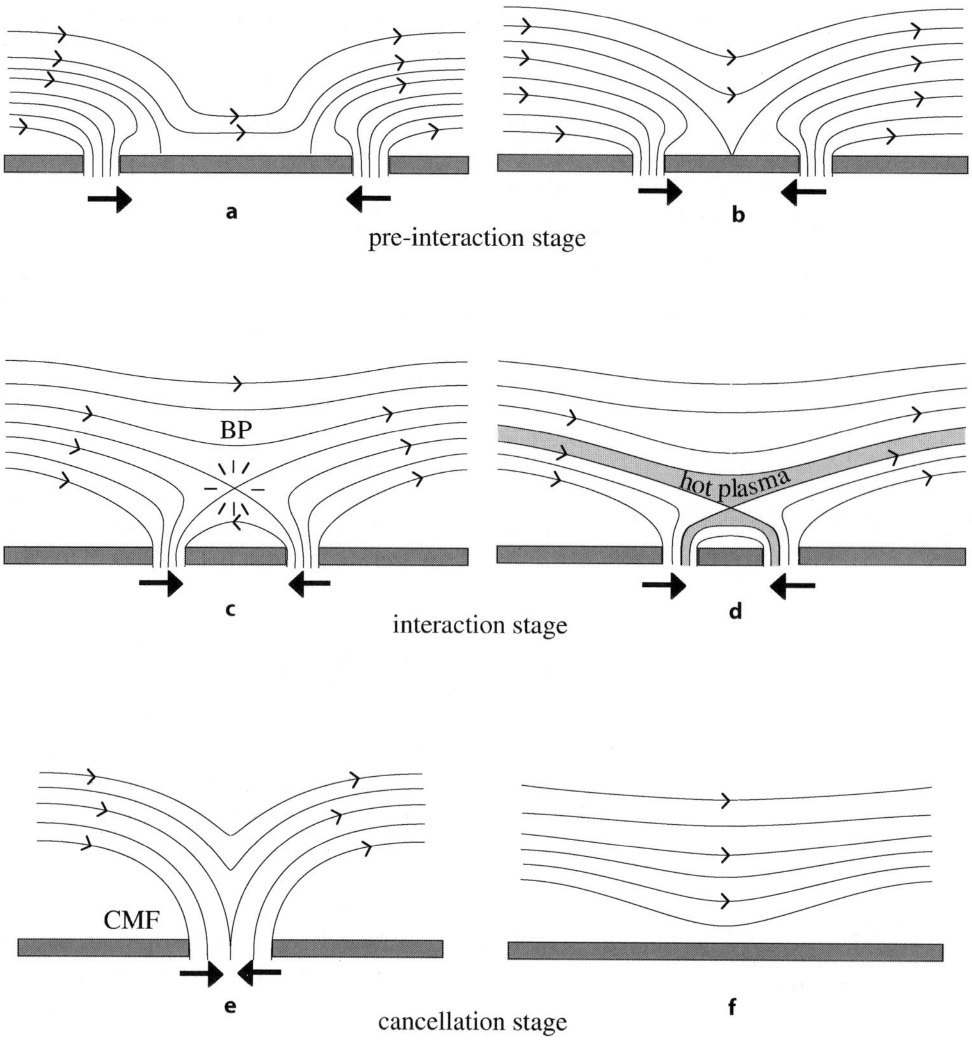

Fig. 3.24a–f. The Converging Flux Model for an X-ray bright point, showing the stages in the approach and interaction of two equal and opposite magnetic fragments. (**a**) In the pre-interaction stage an X-point does not form because the fragments are too far apart. (**b**) Once they are close enough, an X-point appears at the base. (**c**) As the fragments move yet closer, the X-point initially rises upward into the corona to create an X-ray bright point (BP) with filamentary extensions, but then (**d**) the X-point starts to move downwards again. (**e**) Finally, when the fragments meet in the photosphere as a cancelling magnetic feature (CMF), the coronal X-point disappears (**f**) and so eventually do the fragments

Conceptually, photospheric motions tend to build up energy in a nonlinear force-free field by reconnection. The resulting heating flux is of the form

$$F_{\mathrm{H}} = \frac{B^2 v_0}{\mu}\frac{\tau_{\mathrm{d}}}{\tau_0}, \tag{3.82}$$

where, as before, τ_{d} is the dissipation time and τ_0 the time-scale for footpoint motions. This is the same as Parker's (1979) result when τ_{d} is replaced by the reconnection time (d/v_{R}) and τ_0 is replaced by the convection time (L/v_0).

Several extensions of the basic theory have been constructed. Vekstein et al. (1991) suggested *intermediate*

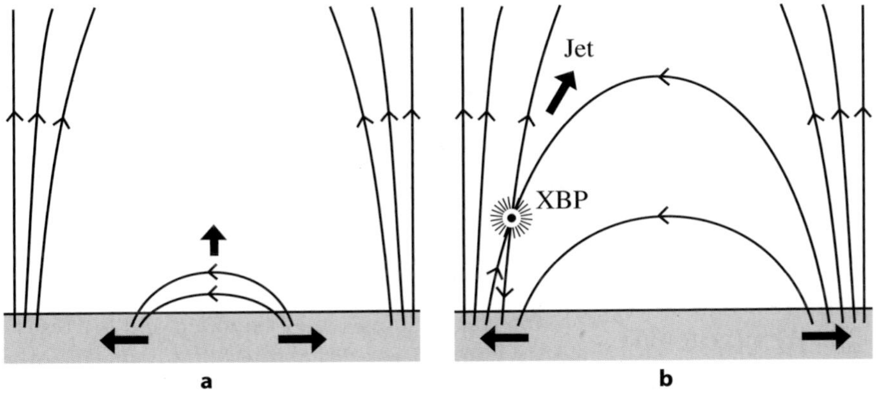

Fig. 3.25a,b. The creation of an X-ray bright point at the edge of a supergranule cell, according to the Converging Flux Model

relaxation to a state between the nonlinear and linear fields, while Vekstein et al. (1993) suggested the corona is in a state of *partial relaxation* with the closed fields being relaxed linear force-free states and the open fields being potential. The basic analysis has been applied by Browning and Priest (1986) to a set of closely packed flux tubes and by Dixon et al. (1988) to an axisymmetric flux tube.

Important physical quantities are the global ideal invariants, which are conserved in the absence of dissipation. In 2D MHD the global invariants (Montgomery, 1983; Frisch et al., 1975) are the energy, correlation (or mean-square vector potential), and cross helicity

$$W_2 = \int \frac{\rho v^2}{2} + \frac{B^2}{2\mu}\,\mathrm{d}S\,, \quad a = \int \frac{A^2}{2}\,\mathrm{d}S\,, \quad H_2 = \int \boldsymbol{v}\cdot\boldsymbol{B}\,\mathrm{d}S\,, \tag{3.83}$$

while in 3D we have the energy, magnetic helicity, and cross helicity

$$W_3 = \int \frac{\rho v^2}{2} + \frac{B^2}{2\mu}\,\mathrm{d}V\,, \quad H = \int \boldsymbol{A}\cdot\boldsymbol{B}\,\mathrm{d}V\,, \quad H_3 = \int \boldsymbol{v}\cdot\boldsymbol{B}\,\mathrm{d}V\,, \tag{3.84}$$

where $\boldsymbol{B} = \nabla \times \boldsymbol{A}$.

In Fourier space these global invariants undergo *cascades*, which are direct if the transfer is from large to small wavelengths and indirect (or inverse) if it is in the other direction. The energy has either a Kolmogorov spectrum ($\sim k^{-\frac{5}{3}}$) or a Kraichnan spectrum ($\sim k^{-\frac{3}{2}}$) and has a direct cascade towards small wavelengths, whereas the correlation A ($\sim k^{-\frac{1}{3}}$) in 2D and magnetic helicity H ($\sim k^{-2}$) have indirect cascades towards large wavelengths.

Ting et al. (1986) have conducted a series of experiments on two-dimensional MHD in which they find *selective decay* when the initial kinetic energy is much smaller than the magnetic energy and the cross helicity (normalised) is less than the energy. Here the magnetic energy decays faster than the magnetic helicity and so has a direct cascade towards small wavelengths, since nonlinear interactions tend to replenish it. At the same time the magnetic helicity has an indirect (or inverse) cascade, and the magnetic field tends towards a force-free state. In contrast, when the kinetic and magnetic energies are similar and the cross-helicity is of a similar size to the energy, Ting et al. find a process of dynamic alignment with a tendency towards Alfvénic states having $\boldsymbol{v} = \pm\boldsymbol{B}/\sqrt{\mu\rho}$ (see also Biskamp (1994)).

Applications of the theory to coronal heating have been made by several authors. Sturrock and Uchida (1981) calculated the rate of increase of stored energy due to random twisting of a flux tube through force-free states as

$$\frac{B^2}{\mu L}\left\langle v_\mathrm{p}^2\right\rangle \tau_\mathrm{p}\,, \tag{3.85}$$

where v_p is the photospheric velocity, τ_p is the correlation time of the footpoint motions, and the angular brackets indicate a mean value. They assume that the dissipation time $\tau_\mathrm{d} > \tau_\mathrm{p}$, so that the free energy can continue to be stored. In contrast, Heyvaerts and Priest (1984) included a dissipation mechanism and assumed $\tau_\mathrm{d} < \tau_\mathrm{p}$. They obtained a heating flux of

$$F_\mathrm{H} = \left(\frac{B^2}{\mu L} v_\mathrm{p}^2 \tau_\mathrm{p}\right)\left(\frac{L}{L + l_\mathrm{p}}\right)^2 \frac{\tau_\mathrm{d}}{\tau_\mathrm{p}}\,, \tag{3.86}$$

where L is the loop half-length and l_p is the length scale for photospheric motions. The last two factors in the above expression are less than unity and show how the efficiency of the process is limited.

Van Ballegooijen (1985) discussed the initial stages of the cascade of energy due to random footpoint motions. He assumed that the states are force-free and the length scale (l_p) for photospheric motions is much smaller than the loop length. He found that the mean-square current density (j^2) increases exponentially [$\sim \exp(t/\tau_p)$]. Gomez and Ferro Fontan (1988) have applied two-dimensional MHD turbulence theory to twisted coronal loops and have suggested the injection of energy at a specific wave number (k_p), followed by a cascade of energy like $k^{-\frac{5}{3}}$ to a dissipation wave number k_d, together with an inverse cascade of mean-square potential like $k^{-\frac{1}{3}}$. Also, Gomez et al. (1993) have measured fine-scale structure in NIXT (Normal Incidence X-ray Telescope) images and find the intensity has a k^{-3} spectrum, although only over one order of magnitude.

Many coronal heating mechanisms, such as braiding and current-sheet formation or resistive instabilities or waves, all lead to a state of MHD turbulence, so how can we analyse such a state? Heyvaerts and Priest (1984) made a start by adapting Taylor's relaxation theory to the coronal environment, in which the field lines thread the boundary rather than being parallel to it.

Although many mechanisms produce a turbulent state, they are incomplete in the sense that there is a free parameter present, such as τ_d in the above equation, or a correlation or a relaxation time. In other words, the mechanisms do not determine the heating flux (F_H) in terms of photospheric motions alone. Heyvaerts and Priest (1992) therefore began a new approach in which they assume photospheric motions inject energy into the corona and maintain it in a turbulent state with an unknown turbulent magnetic diffusivity (η^*) and viscosity (ν^*) that are different from the classical values. There are two parts to their theory. First of all, they calculate the global MHD state driven by boundary motions, which gives F_H in terms of ν^*. Secondly, they invoke cascade theories of MHD turbulence to determine the ν^* and the η^* that result from F_H. In other words, the circle is completed and F_H is is determined independently of ν^* and η^*. They applied their general philosophy to a simple example of one-dimensional random photospheric motions producing a two-dimensional coronal magnetic field.

Suppose the dimensionless boundary motions are $\pm V(x)\hat{\mathbf{y}}$ (with Fourier coefficients V_n) at $z = \pm L$ and produce motions $v(x,z)\hat{\mathbf{y}}$ and field $B_0\hat{\mathbf{z}} + B_y(x,z)\hat{\mathbf{y}}$ within the volume between $z = -L$ and $z = L$. Then the steady MHD equations of motion and induction reduce simply to

$$0 = \frac{B_0}{\mu}\frac{\partial B_y}{\partial z} + \rho\nu\nabla^2 v_y\,, \quad 0 = B_0\frac{\partial v_y}{\partial z} + \eta\nabla^2 B_y\,. \quad (3.87)$$

The solutions may easily be found and the resulting Poynting energy flux through the boundary over a width h is

$$F_H = \frac{B_0^2 v_{A0}}{\mu}\Sigma\frac{V_n^2 H^*}{\eta/(L v_{A0})}\left(1 + 2\lambda_n^2/\sqrt{1+4\lambda_n^2}\right) \times \frac{\sin h\left(\sqrt{1+4\lambda_n^2}/H^*\right) + \sin h(1/H^*)}{\cos h\left(\sqrt{1+4\lambda_n^2}/H^*\right) - \cos h(1/H^*)}\,, \quad (3.88)$$

where $H^* = \sqrt{\eta^* \nu^*}/(L v_{A0})$ is the inverse Hartmann number and $\lambda_n = H^* n\pi L/h$. For the second step, invoking Pouquet et al. (1976) theory gives $\nu^* = \eta^*$ and, if a is the half-width of the loop,

$$F_H = \frac{27[\nu^{*2}/(L v_{A0})]\pi^3}{2a^3/L^3}\frac{B_0^2 v_{A0}}{\mu}\,, \quad (3.89)$$

so that equating the two above expressions for F_H gives a single equation for ν^*. They found typically for a quiet region loop that a density of $2 \times 10^{16}\ \mathrm{m^{-3}}$ and a magnetic field of $3\times10^{-3} - 5\times10^{-3}$ tesla (30 – 50 G) produces a heating F_H of $2.4 - 5.5\times10^2\ \mathrm{W\,m^{-2}}$ and a turbulent velocity of 24 – 33 km s^{-1} whereas values of $5 \times 10^{16}\ \mathrm{m^{-3}}$ and 10^{-2} tesla (100 G) for an active region loop give $2 \times 10^3\ \mathrm{W\,m^{-2}}$ for the heating and 40 km s^{-1} for the turbulent velocities. Given the limitations of the model, these reasonable values are very encouraging. Inverarity and Priest (1995 a,b) then went on to apply the theory to a twisted flux tube and to turbulent heating by waves due to more rapid footpoint motions.

Magnetic Dissipation, Braiding and Nanoflares

When photospheric motions are *sufficiently slow* and the wavelength *sufficiently long*, a wave description ceases to be helpful. Instead the coronal magnetic configuration *evolves passively through a series of equilibria*, which store energy in excess of potential. This energy has come originally from the photospheric motion. The electric currents associated with such large-scale equilibria produce negligible ohmic heating. The only way that magnetic field (i.e., ohmic) dissipation can produce the necessary coronal heating is for the magnetic field changes and accompanying electric currents to be concentrated in extremely intense *current sheets*, *current sheaths* (around flux tubes) or *current filaments*. If the current density is so strong that the width of such a current concentration is less than (typically) a few metres, the dissipation may be considerably enhanced by the presence of plasma turbulence.

Provided current sheets, sheaths or filaments can be formed, they produce a rapid conversion of magnetic energy into heat (by ohmic dissipation), bulk kinetic energy and fast-particle energy, in a manner that has been studied extensively in connection with the more violent heating of a solar flare. This suggests that, especially in the strong magnetic field of an active region, the corona is in a state of *ceaseless activity* and is being heated by many tiny micro-flarings (10^{26} erg) or nanoflarings (10^{23} erg) that are continually generated by the photospheric motion below.

The features of heating by *magnetic* (or *current*) *dissipation* that needs to be understood concerns the way in which current sheets, sheaths or filaments are formed, are maintained (if necessary) and decay. The order-of-magnitude estimates of Tucker (1973) and others are described below, to determine how thin the resistive regions need to be to provide the necessary heating. Current sheets have received the most attention; they made be formed either by pushing topologically distinct regions against one another or by magnetic non-equilibrium. In the former case, they are maintained for as long as the external footpoint motion continues. Current filaments may be created as a result of tearing-mode instability or thermal instability.

Order of Magnitude

Tucker (1973) and Levine (1974) were among the first to suggest coronal heating by the dissipation of non-potential magnetic fields. They considered neutral current sheets dispersed throughout active regions, and they established qualitatively that *current dissipation* could provide enough heat for the corona. Tucker supposes that magnetic energy is being stored at a rate

$$\frac{\mathrm{d}W_{\mathrm{m}}}{\mathrm{d}t} \sim \frac{vB^2}{2\mu}L \tag{3.90}$$

by photospheric motions (v) that twist a magnetic field of strength B over an area L^2. The energy is at the same time being dissipated ohmically, at a rate

$$D \sim \frac{j^2}{\sigma}L^3 \tag{3.91}$$

for currents (j) distributed uniformly through the volume (L^3) of the active region. If the magnetic field is being twisted up faster than it is relaxing ohmically, the excess energy will be stored until it is released as, for instance, a *solar flare*. But, if the two rates (3.90) and (3.91) are equal, the active region will maintain a steady state. The effective twisting speed v that is needed to provide a heat input of, say, $3000\,\mathrm{W\,m^{-2}}$ to the corona can be found from Eq. (3.90) as $v \sim 100\,\mathrm{m\,s^{-1}}$ for a photospheric field strength of 100 G. Furthermore, uniform dissipation throughout the active region with a classical Coulomb electrical conductivity (σ) requires a current density that can be estimated by equating DL^{-3} from Eq. (3.91) to $3000\,\mathrm{W\,m^{-2}}$. With $L \sim 1000$ km, Tucker finds $j \sim 30\,\mathrm{A\,m^{-2}}$. Since this corresponds to the rather large magnetic field gradient of $0.4\,\mathrm{G\,m^{-1}}$, he suggests that the dissipation is concentrated at thin current sheets rather than distributed uniformly. The ohmic dissipation inside sheets may be greatly enhanced above normal because of the much larger electric currents and the possibility of plasma turbulence, but the sheets occupy only a small fraction of the active-region volume. For each sheet of thickness l^* and area L^{*2} with an electric current $j^* \sim B/(\mu l^*)$ and a turbulent electrical conductivity (σ^*), the rate of heat generation is

$$D^* \sim \frac{j^{*2}}{\sigma^*}L^{*2}l^* \quad \text{or} \quad D^* \sim \frac{B^2}{\mu^2\sigma^*}\frac{L^{*2}}{l^*}\,. \tag{3.92}$$

Tucker adopts a turbulent conductivity that is about a million times smaller than the classical value and assumes a sheet width of 10 m, consistent with the critical current for turbulent onset. He finds that only a few current sheets of length $L^* = 1000$ km are necessary to generate the heat that is required for an active region. Levine (1974) suggested that the tangled nature of coronal magnetic fields produces many small current sheets that are collapsing. During the collapse, particles are accelerated and then thermalised by Coulomb collisions in the surrounding region.

Current Sheets and Braiding

Current sheets may be formed in several ways. One is by the interaction of topologically separate parts of the magnetic configuration of, say, and active region. High-resolution observations of the *photospheric* magnetic field exhibit a highly complex magnetic pattern with frequent changes of polarity. The *coronal* field is also complex, with many distinct magnetic flux tubes shown up by X-ray and EUV pictures. As the photospheric footpoints of coronal waves move, so the neighbouring coronal flux-tubes will respond and interact with one another, either moving further apart or coming closer together. At the interface between the two tubes, a *current sheet* is formed, the magnetic field reconnects, and magnetic energy is released in the process. Such magnetic dissipation takes place not only when neighbouring magnetic field lines are oppositely directed, but also when the field lines are inclined at a non-zero angle.

The formation of current sheets when new magnetic flux is emerging from below the photosphere has been studied in connection with solar flares, but the same calculations are applicable when magnetic flux is evolving rather than emerging. In particular, it must be stressed that the current sheet is a response to the applied photospheric motions. If the neighbouring footpoints move relative to one another at a certain speed, then the corona will just respond by creating a current sheet and allowing magnetic reconnection at that speed. Furthermore, the reconnection and associated dissipation is maintained as long as the relative footpoint-motion continues, with the dimensions of the current sheet depending on the magnetic field strength and photospheric speed. Conditions inside the sheet will only be turbulent if the resulting sheet width is small enough. It should also be noted that slow magnetoacoustic shock waves radiate from the ends of the current sheet and that fine jets of plasma are emitted between pairs of shocks. As plasma comes in slowly from the sides, the bulk of the heat is released as these shock waves rather in the central current itself.

Current sheets may also develop when magnetoacoustic equilibrium becomes *unstable* or even ceases to exist, a situation known as *non-equilibrium*. In a simple bipolar magnetic field when the photospheric footpoints move slowly, the low $-\beta$ corona responds by establishing a series of force-free configurations. In general, however, the coronal magnetic field is much more complex than this, and it contains topologically distinct flux systems. Current sheets may be formed according to Parker (1972) by the braiding of magnetic fields around one another and their dissipation leads to *nanoflare* heating in many small locations. A three-dimensional resistive MHD numerical experiment by Galsgaard and Nordlund (1996) has shown that the resulting current sheets are highly complex, that the braiding is roughly by one turn before reconnection sets in, and that the resulting heating along a loop is rather uniform Galsgaard et al. (1999), as shown in Fig. 3.26.

Coronal Tectonics

The surface of the Sun is covered with a multitude of magnetic sources which are continually moving around and which produce a highly complex magnetic field in the overlying corona, known as the *magnetic carpet* (Schrijver et al., 1997). A key question which the Coronal Tectonics model seeks to address is: what is the effect of the relative motions of photospheric sources in driving reconnection and, therefore heating, in the overlying corona? One possibility is *separator heating* due to the high-order interactions of several sources, but a more fundamental process is the binary interaction due to pairs of sources (see below).

Basic model

In determining the effect of the magnetic carpet on coronal heating, three factors are important: the concentration of flux in the photosphere into discrete intense flux tubes; their continual motion; and the fact that the global topography of the complex coronal field consists of a collection of topologically separate volumes divided from one another by separatrix surfaces.

A *Coronal Tectonics Model* for coronal heating (Priest et al., 2002) takes account of these three factors. Each coronal loop has a magnetic field that links the solar surface in many sources. The flux from each source is topologically distinct and is separated from each other by separatrix surfaces (Fig. 3.27). As the sources move, the coronal magnetic field slips and

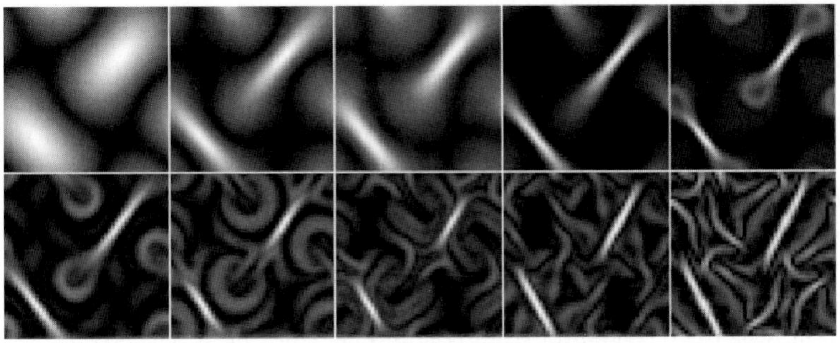

Fig. 3.26. The current filamentation due to braiding in a section at several times. (Galsgaard and Nordlund, 1996)

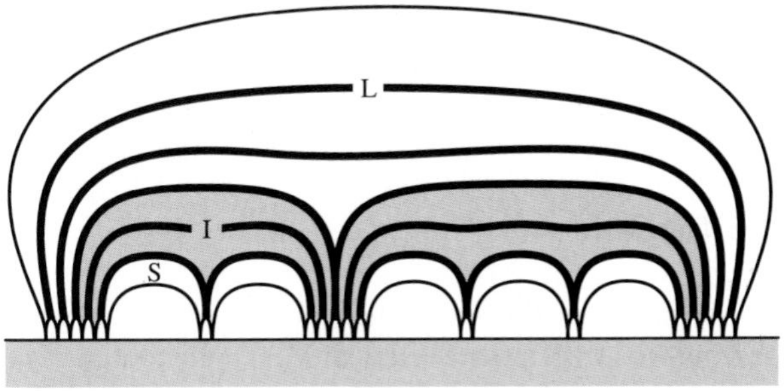

Fig. 3.27. A schematic of a coronal loop consisting of many sub-volumes, each linked to a separate source and divided from one another by *separatrix* surfaces

forms current sheets along the separatrices, which then reconnect and heat. Thus, in our view, the corona is filled with myriads of separatrix current sheets continually forming and dissipating.

But the fundamental flux units in the photosphere are likely to be intense flux tubes with fields of 1200 G, diameters of 100 km (or less) and fluxes of 3×10^{17} Mx (or less). A simple X-ray bright point thus links to a hundred sources and each TRACE loop probably consists of at least 10 finer, as yet unresolved, loops.

Whereas (Parker, 1972)'s braiding model assumes complex footpoint motions acting on a uniform field, Priest et al. (2002) consider the effect of simple motions on an array of flux tubes that is anchored in small discrete sources. For a simple model consisting of an array of flux tubes anchored in two parallel planes, they have demonstrated the formation of current sheets and estimated the heating. A more realistic model would have the sources asymmetrically placed so as to create many more separatrices, or, more realistic still, it would place all the sources on one plane and have mixed polarity. The basic principles would, however, be unchanged.

The results give a uniform heating along each separatrix, so that each (sub-telescopic) coronal flux tube would be heated uniformly. But at least 50% of the photospheric flux closes low down in the magnetic carpet (Close et al., 2004) so the reamaining 50% forms large-scale connections. Thus, the magnetic carpet would be heated more effectively than the large-scale corona. Unresolved observations of coronal loops would give enhanced heating near the loop feet in the carpet, while the upper parts of coronal loops would be heated uniformly but less strongly.

Coronal Recycling Time

Photospheric sources of the coronal magnetic field are highly fragmentary and concentrated into intense flux tubes threading the solar surface. They are also highly dynamic, with magnetic flux emerging continually in the quiet Sun and then undergoing processes of fragmentation, merging and cancellation, in such a way that the quiet Sun photospheric flux is reprocessed very quickly, in only 14 h (Hagenaar, 2001).

Close et al. (2004) wondered what the corresponding coronal reprocessing time is, and have used observed quiet-Sun magnetograms from the MDI instrument on SOHO to construct the coronal magnetic field lines and study their statistical properties.

For the region they considered, 50% of the flux closed down within 2.5 Mm of the photosphere and 95% within 25 Mm, the remaining 5% extending to larger distances or being open (Fig. 3.13). They then traced the motion of individual magnetic fragments in the magnetogram and recalculated the coronal field lines and their connectivity. In so doing, they discovered the startling fact the time for all the field lines in the quiet Sun to change their connections is only 1.5 h. In other words, an incredible amount of reconnection is continually taking place – indeed, enough to provide the required heating of the corona.

Binary Reconnection

The skeleton of the field due to two unbalanced sources (stars) in the photosphere includes a null point closer to the smaller source which possesses a spine that joins the null point to the weaker source and to infinity. It also possesses a fan surface of field lines that arch over the weaker source in the form of a dome which intersects the photosphere in a dashed curve. Part of the magnetic flux from the stronger source lies below the separatrix dome, while the remaining flux lies above it and links out to distant sources.

A new suggestion is that the fundamental heating mechanism is one of so-called "binary reconnection" due to the motion of a given magnetic source relative to its nearest neighbour. The heating is due to several effects: (1) the 3D reconnection of field lines that start up joining the sources and end up joining the largest source to other more distant sources; (2) the viscous or resistive damping of the waves that are emitted by the sources as their relative orientation rotates; and (3) the relaxation of the nonlinear force-free fields that join the two sources and that are built up by the relative motion of the two sources. For details see Priest et al. (2003).

Separator and Separatrix Reconnection

Several distinct types of reconnection are associated with null points, namely, spine, fan, and separator reconnection (Priest and Titov, 1996) and in particular separator reconnection is a prime candidate for coronal heating. Numerical experiments have been conducted on this possibility (Galsgaard and Nordlund, 1997; Parnell and Galsgaard, 2004) and the way in which it operates has been studied in detail by Longcope and coworkers.

Having shown how a current sheet may form along a separator (Longcope, 1996), a stick-slip model for reconnection was developed together with the concept of a "Minimum-Current Corona" (Longcope, 1996). The assumption is that, after slow motions of the photospheric footpoints, the corona relaxes to a flux-constrained equilibrium in which the magnetic fluxes within each domain are conserved but the field lines within each domain can slip through the plasma, or move their footpoints (Longcope, 2001). Such equilibria have potential magnetic fields in each domain and current sheets along the separators. The theory has been applied to X-ray bright points (Longcope, 1998).

Priest et al. (2005) stress that in general the effect of slow photospheric motions on complex coronal magnetic configurations will be to generate three forms of current, namely, distributed currents throughout the volume, current sheets on separators and current sheets on separatrices (Fig. 3.28). They compare energy storage and heating at separators and separatrices by using reduced MHD to model coronal loops that are much longer than they are wide.

3.6 Conclusion

As we have seen, the Sun is intrinsically an object of great fascination, with a rich variety of MHD phenomena that are not yet well understood, but are being modelled by analytical and numerical computations. There are two theoretical possibilities for heating the solar corona, namely, magnetic dissipation either in Alfvén waves or in current sheets. Prominences are created by radiative instability and are supported in large flux tubes. Solar flares are due to an eruptive MHD instability or catastrophe, followed by reconnection as the magnetic field closes down. Many of these basic processes on the Sun occur elsewhere in the solar system and indeed in other astronomical objects. So, in the future we hope that there will be more cross-fertilization between solar physicists and magnetosphericists and that together we may understand and appreciate the beautiful universe in which we live.

There have been many developments in the basic MHD theory for the solar atmosphere during the past few years. A combination of numerical experimentation and analytical theory has greatly increased our understanding of the basic process of magnetic reconnection for solar flares. Numerical experiments of reconnection have refined and developed the basic picture of field line clusure that creates flare loops and have explained many observaitonal features. Also, the eruption of a prominence at the start of such a flare may well be caused by a process of magnetic non-equilibrium when the length or height of the prominence is too great. In future, many developments are expected from more

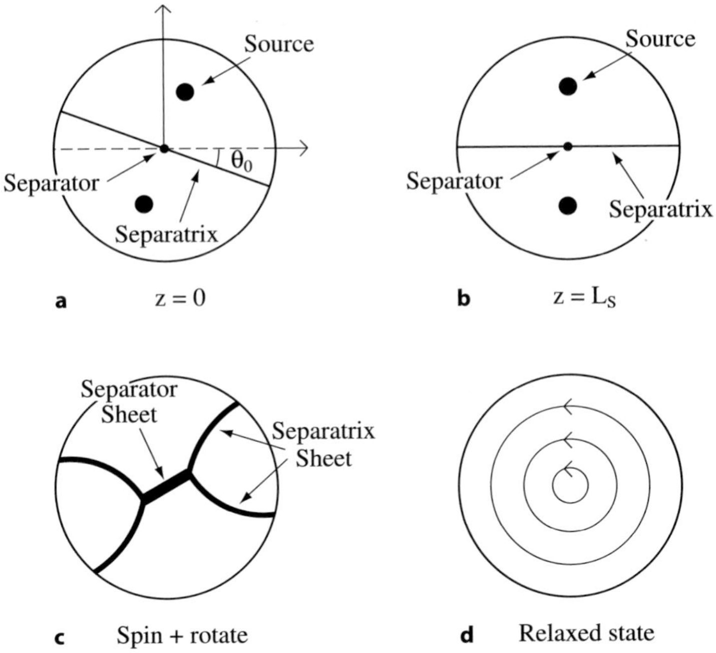

Fig. 3.28a–d. Skeleton of the field due to two unbalanced sources (stars)

sophisticated numerical experiments and from studies of the three-dimensional aspects. Also, one hopes that the microscopic plasma physicists can make use of the MHD studies, so that further progress can be made on the microscopic processes such as particle acceleration, that together we may understand this beautiful phenomenon.

References

Acton, L., 1996, Cool Stars, Stellar Systems and the Sun 109, 45

Antiochos, S., Devore, C., and Klimchuk, J., 1999, Astrophys. J. 510, 485

Axford, W. and McKenzie, J., 1996, Astrophys. Space Sci. 243, 1

Ballester, J. and Priest, E., 1987, Solar Phys. 109, 335

Bateman, G., 1978, Instabilities, MIT Press, Cambridge, UK

Baty, H., Priest, E.R., and Forbes, T., 2006, Phys. Plasmas

Berger, M. and Field, G., 1984, J. Fluid Mech. 147, 133

Berghmans, D. and Clette, F., 1998, A Cross-roads for European Solar and heliospheric Physics, eds E.R. Priest, F. Moreno-Insertis, and R.A. Harris ESA SP-417, Noordwijk, 229

Biermann, L., 1946, Naturwissenschaften 33, 118

Biskamp, D., 1986, Phys. Fluids 29, 1520

Biskamp, D., 1994, Physics Reports 237(4), 179

Browning, P. and Priest, E., 1986, Astron. Astrophys. 159, 129

Cargill, P. and Priest, E., 1982, Solar Phys. 76, 357

Carlsson, M. and Stein, R., 2002, Astrophys. J. 572, 626

Chae, J., Schuhle, U., and Lemaire, P., 1998, Astrophys. J. 505, 957

Chiuderi, C., 1981, Solar Phenonema in Stars and Stellar Systems, eds R.M. Bonnet and A.K. Dupree, 269

Close, R., Parnell, C., and Priest, E., 2004, Astrophys. J. 612, LSI

Dixon, A., Browning, P., and Priest, E., 1988, Geophys. Astrophys. Fluid Dyn. 40, 294

Falconer, D., Moore, R., and Porter, J., 1999, Astron. Soc. Pacific, Provo Utah

Forbes, T., Malherbe, J., and Priest, E.R., 1989, Solar Phys. 120, 285

Forbes, T. and Priest, E., 1982a, Solar Phys. 81, 303

Forbes, T. and Priest, E., 1982b, Planet. Space Sci. 30, 1183

Forbes, T. and Priest, E., 1984, Solar Phys. 94, 315

Forbes, T. and Priest, E., 1995, Astrophys. J. 446, 377

Frisch, U., Pouquet, A., Leorat, J., and Mazure, A., 1975, J. Fluid Mech. 68, 769

Galsgaard, K., Mackay, D., Priest, E.R., and Nordlund, Å., 1999, Solar Phys. 189, 95

Galsgaard, K. and Nordlund, Å., 1997, J. Geophys. Res. 102, 231

Galsgaard, K. and Nordlund, Å., 1996, J. Geophys. Res. 101, 13445

Gomez, D. and Ferro Fontan, C., 1988, Solar Phys. 116, 33

Gomez, D., Marten, P., and Golub, L., 1993, Astrophys. J. 405, 767
Gudiksen, B. and Nordlund, Å., 2005a, Astrophys. J. 618, 1031
Gudiksen, B. and Nordlund, Å., 2005b, Astrophys. J. 618, 1020
Hagenaar, H., 2001, Astrophys. J. 555, 448
Harrison, R., 1997, Solar Phys. 175, 467
Heyvaerts, J. and Priest, E., 1984, Astron. Astrophys. 137, 63
Heyvaerts, J. and Priest, E., 1992, Astrophys. J. 390, 297
Heyvaerts, J., Priest, E., and Rust, D., 1977, Astrophys. J. 216, 123
Innes, D., Inhester, B., Axford, W., and Wilhelm, K., 1997, Nature 386, 811
Kippenhahn, R. and Schluter, A., 1957, Zs. Ap. 43, 36
Kohl, J., Noci, G., Antonucci, E., Tondello, G., Huber, M., Gardner, L., Nicolosi, P., Strachan, L., Fineschi, S., Raymond, J., Ramoli, M., Spadara, D., Panasyuk, A., Siegmund, O., Benna, C., Garavella, A., Cranmer, S., Giordano, S., Karovska, M., Martin, R., Michels, J., Modigliani, A., Naletto, G., Pernechele, C., Poletto, G., and Smith, P., 1997, Solar Phys. 175, 613
Kopp, R. and Pneuman, G., 1976, Solar Phys. 50, 85
Krucker, S., Benz, A., Bastian, T., and Acton, L., 1998, Astrophys. J. 499
Kudoh, T. and Shibata, K., 1999, Astrophys. J. 514, 493
Kuperus, M. and Raadu, M., 1974, Astron. Astrophys. 31, 189
Leroy, J., 1989, in Dynamics and structure of quiescent solar prominences; Proceedings of the Workshop, Palma de Mallorca, pp 77–113
Levine, R., 1974, Astrophys. J. 190, 457
Longcope, D., 1996, Solar Phys. 169, 91
Longcope, D., 1998, Astrophys. J. 507, 433
Longcope, D.W., 2001, Phys. Plasmas 8, 5277
Maclean, R., Beveridge, C., Longcope, D.W., Brown, D.S., and Priest, E.R., 2005, Proc. Roy. Soc. A, 461, 2099–2120
Martin, S., 1986, in Coronal and Prominence Plasmas, pp 73–80
Martin, S., Livi, S., and Wang, J., 1985, Astrophys. J. 38, 929
Masuda, S., Kosugi, T., Hara, H., Tsuneta, S., and Ogawara, Y., 1994, Nature 371(6497), 495
McKenzie, J., Axford, W., and Banaskiewicz, M., 1997, Geophys. Res. Lett. 24, 2877
McKenzie, J. and Sukhorakova, G., 1998, Astron. Astrophys. 330, 1145
Montgomery, D., 1983, in Solar Wind Five, pp 107–130
Moriyasu, S., Kudoh, T., Yokoyama, T., and Shibata, K., 2004, Astrophys. J. Letts. 601, L107
Parker, E., 1972, Astrophys. J. 174, 499
Parnell, C. and Galsgaard, K., 2004, Astron. Astrophys. 428, 595
Parnell, C. and Jupp, P., 2000, Astrophys. J. 529, 554
Parnell, C., Priest, E., and Golub, L., 1994, Solar Phys. 151, 57
Pike, C. and Mason, H., 1998, Solar Phys. 182, 333
Poletto, G. and Kopp, R., 1988, Solar Phys. 116, 163
Pouquet, A., Frisch, U., and Leorat, J., 1976, J. Fluid Mech. 77, 321
Priest, E., 1981, Solar Flare Magnetohydrodynamics, Gordon and Breach Science Publishers
Priest, E., 1987, in A.W.E. Schroter, M. Vazquez (ed.), The Role of Fine-Scale Magnetic Fields on the Structure of the Solar Atmosphere, pp 297–316, Camb. Univ. Press.
Priest, E. and Démoulin, P., 1995, J. Geophys. Res. 100, 23, 443
Priest, E. and Forbes, T., 1990, Solar Phys. 126, 319
Priest, E., Heyvaerts, J., and Title, A., 2002, Astrophys. J. 576, 533
Priest, E., Hood, A., and Anzer, U., 1989, Astrophys. J. 344, 1010
Priest, E., Longcope, D., and Heyvaerts, J., 2005, Astrophys. J. 624, 1057
Priest, E., Parnell, C., and Martin, S., 1994, Astrophys. J. 427, 459
Priest, E. and Titov, V., 1996, Phil. Trans. Roy. Soc. Lond. 355, 2951
Priest, E.R., Foley, C., Heyvaerts, J., Arber, T., Culhane, J., and Acton, L., 1998, Nature 393, 545
Priest, E.R. and Forbes, T.G., 2000, Magnetic Reconnection: MUD Theory and Applications, Cambridge University Press, Cambridge, UK
Priest, E.R., Longcope, D., and Titov, V., 2003, Astrophys. J. 598, 667
Scharmer, G., Gudiksen, B., Kiselman, D., Lofdahl, M., and van der Voort, L., 2002, Nature 420, 151
Schrijver, C., Title, A., Ballegooijen, A., Hagenaar, H., and Shine, R., 1997, Astrophys. J. 487, 424
Schwarzchild, M., 1948, Astrophys. J. 107, 1
Shibata, K., 1999, Astrophys. and Space Sci. 264, 129
Shibata, K., Shimojo, M., Yohoyama, T., and Ohyama, M., 1996, Magnetic Reconnection in the Solar Atmosphere eds R.D. Bently and J.T. Mariska, 29
Sturrock, R. and Uchida, Y., 1981, Astrophys. J. 246, 331
Suzuki, T. and Inutsuka, S., 2005, Astrophys. J. Letts. 632, L49
Thomas, J., Weiss, N., Tobias, S., and Brummel, N., 2002, Nature 420, 390
Ting, A., Mattaeus, W.H., and Montgomery, D., 1986, Phys. Fluids 29, 3261
Titov, V., Hornig, G., and Demoulin, P., 2002, J. Geophys. Res. 7, doi:10.1029/2001JA000278
Trujillo-Bueno, J., Shchukina, N., and Asensio Ramos, A., 2004, Nature 430, 326
Tsuneta, S., Takahashi, T., Acton, L., Bruner, M., Harvey, K., and Ogawara, Y., 1992, Publ. Astron. Soc. Japan 44, L211
Tucker, W., 1973, Astrophys. J. 186, 285
Ugai, M., 1995, Phys. Plasmas 2, 338
Ugai, M. and Tsuda, T., 1977, J. Plasma Phys. 17, 337

van Ballegooijen, A., 1985, Astrophys. J. 298, 421
van Ballegooijen, A. and Martens, P., 1989, Astrophys. J. 343, 971
van der Voort, L., Lofdahl, M., Kiselman, D., and Scharmer, G., 2004, Astron. Astrophys. 414, 717
Vekstein, G., Priest, E., and Steele, C., 1991, Solar Phys. 131, 297
Vekstein, G., Priest, E., and Steele, C., 1993, Astrophys. J. 417, 781
Withbroe, G. and Noyes, R., 1977, Ann. Rev. Astron. Astrophys. 15, 363
Yan, M., Lee, L., and Priest, E., 1992, J. Geophys. Res. pp 8277–8293
Yoshida, T. and Tsuneta, S., 1996, Astrophys. J. 459, 342

4 Solar Wind

Eugene N. Parker

The circumstances that create the solar wind were established from ground based observations, and the essential physics is illustrated directly by those early observations. For instance, the continual emission of solar corpuscular radiation in all directions from the Sun was indicated by the anti-solar acceleration of the gaseous comet tails. The million degree temperature of the corona of the Sun was recognized, and it was appreciated that the corona is strongly bound by the gravitational field of the Sun. As a consequence of the thermal conductivity and the extended active heating of the corona, the million degree temperature extends far out into space, to where the gas density is very small, and the gravitational binding energy falls below the enthalpy of the gas. The outer corona expands away to infinity, reaching supersonic speed. This is, of course, the solar wind, or solar corpuscular radiation, responsible for the anti-solar comet tails.

Contents

Eugene N. Parker, Solar Wind.
In: Y. Kamide/A. Chian, Handbook of the Solar-Terrestrial Environment. pp. 95–116 (2007)
DOI: 10.1007/11367758_4

4.1 Introduction

A hundred years ago one could not have guessed that the corona of the Sun, observed for millennia during total eclipses, controls space throughout the solar system, all the way out through the Kuiper belt to more than 100 AU. The first evidence of the dynamical out reach of the Sun was recognized about a hundred years ago, when it was discovered that a large flare on the Sun is often followed a day or two later by a jiggling of the magnetic field of Earth. The transit time of one or two days indicated Sun-Earth transit velocities of the order of 10^3 km/s, suggesting particles emitted from the Sun. By 1919 it was recognized that the particle emission, called *solar corpuscular radiation*, must consist of equal numbers of electrons and protons. For if the numbers were not equal, the Sun would quickly charge to enormous electrostatic potential, selectively controlling the relative numbers of escaping electrons and protons. Note that one mole of electrons raises the potential of the Sun to 1.3×10^6 V, while the present rate of emission of both electrons and ions is approximately 10^{12} mol/s.

Apart from the occasional bursts of solar corpuscular radiation, space was considered to be a hard vacuum, capable of sustaining strong electrostatic fields. The mechanism for accelerating the solar corpuscular radiation to 10^3 km/s was not known, but vaguely associated with the magnetic active regions on the Sun wherein flares occur (cf. Parker, 2001 for more detail).

The purpose of this chapter is to expound the basic physics of the solar corona and its gradual outward acceleration to form the supersonic 300 – 1000 km/s solar wind. Today the solar wind is studied quantitatively with instruments carried on spacecraft as far away as Ulysses, passing over the poles of the Sun at 2 AU, and Voyager I, now beyond the termination shock. The more detailed investigation of the time variations, composition, and magnetic structures in the wind throughout the inner solar system has shown the hundred and one quirks and idiosyncrasies of the coronal expansion and its magnetic fields in the escaping solar wind. The Space Age brought direct detection and measurement of the basic solar wind velocity and density, and then went on to study the remarkable behavior of the ion and electron temperatures, the peculiar variations of the ion pressures parallel and perpendicular to the magnetic field, and the vigorous small-scale magnetic fluctuations superposed on the mean large-scale magnetic field. So the details of the wind are a fascinating study and help to point the way to the coronal heating that creates the wind in the first place. However, we are looking here only at the basic physical necessity for the wind, rather than the diverse array of associated phenomena. So the theoretical development is confined to steady radial expansion. The actual situation is more complicated, of course, with time dependent variations of coronal conditions around the Sun.

The theoretical considerations and the essential ground based observations that led to recognition of the solar wind and its origin in coronal expansion are simple and predate the space age. For the fact is that the solar wind is responsible for geomagnetic fluctuations, the aurora, cosmic ray variations, and comet tail dynamics, all observable from the surface of Earth. The problem, of course, was to understand that these diverse phenomena collectively implied the existence of coronal expansion and the basic solar wind phenomenon. So we present here a brief account of the ground based observations and theoretical interpretations that indicate the existence of the solar wind and then go into the theoretical dynamical properties of the solar corona in the basic case of steady radial outflow.

4.2 The Corona

The essential physics of the solar wind had its beginning around 1940 when the work of Grotrian (1939), Lyot (1939), and Edlen (1942) established direct spectroscopic confirmation of the 10^6 K degree electron temperature of the solar corona, already suggested – but not taken seriously – by the scale height of the corona observed during eclipses of the Sun. It followed that the corona is principally fully ionized hydrogen, with about ten percent fully ionized helium and lesser amounts of the heavier elements. Billings (1966) subsequently observed the widths of several coronal emission lines and showed that the emitting ions also have temperatures of the order of 10^6 K. In fact the ion temperatures tend to be higher than the electron temperature (cf. Aschwanden, 2004).

Chapman (1954) worked out the kinetic properties of ionized hydrogen, showing that the thermal conductivity has the large value

$$K(T) = 6 \times 10^{-7} T^{5/2} \,\mathrm{erg/cm\,s\,K} \qquad (4.1)$$

as a consequence of the high thermal velocity of the electrons. He applied this result to the outward extension of the temperature of the corona, supposing that the corona is heated only at its base, $r = a$. For steady conditions of radial heat flow the thermal conduction equation,

$$\frac{1}{r^2}\frac{\mathrm{d}}{\mathrm{d}r}\left[r^2 K(T)\frac{\mathrm{d}T}{\mathrm{d}r}\right] = 0\,, \tag{4.2}$$

becomes

$$\frac{\mathrm{d}}{\mathrm{d}r}\left(r^2 T^{5/2}\frac{\mathrm{d}T}{\mathrm{d}r}\right) = 0\,, \tag{4.3}$$

which has the solution

$$T(r) = T(a)\left(\frac{a}{r}\right)^{2/7} \tag{4.4}$$

if we require that $T(r)$ falls to zero at $r = \infty$ (Chapman, 1959). With $T(a) = 10^6$ K, the outward heat flux is then $2K(T)T/7a = 2.5 \times 10^3$ erg/cm^2 s, for a total of 1.5×10^{26} erg/s over the entire corona ($a = 7 \times 10^{10}$ cm). Thermal electromagnetic emission is small enough to be neglected in the gross heat budget of the corona. So this simple conduction model gives a reasonable picture of the outward heat flow and the extension of the coronal temperature into interplanetary space.

For the record, note that the thermal emission from the corona is described by the energy ε radiated per unit volume and time (Weyman, 1960) with

$$\varepsilon = 1 \times 10^{-23} N^2\ \mathrm{erg/cm^3\,s}\,,$$

and varying but little with temperature over the range $0.5 - 3 \times 10^6$ K. The number density of the coronal gas is denoted by N. Most of the emission is from the few electrons remaining bound to the heavier ions, so the expected increase of emission with rising temperature is largely compensated by the diminishing number of bound electrons.

For the typical coronal density $N = 10^8\ /\mathrm{cm}^3$ it follows that $\varepsilon = 10^{-7}$ erg/cm^3 s. The corona is optically thin, and the characteristic coronal thickness is given by the pressure scale height $\Lambda = 2kT(a)/Mg$ for ionized hydrogen. Here g is the gravitational acceleration $GM_{\odot}/a^2 = 2.7 \times 10^4$ cm/s^2 in terms of the mass $M_{\odot}$ (2×10^{33} gm) of the Sun. The effective molecular mass M is taken to be 2×10^{-24} gm, instead of 1.66×10^{-24} gm, to compensate crudely for the helium in the corona. For $T = 10^6$ K it follows that $\Lambda = 5 \times 10^9$ cm and $\varepsilon\Lambda = 5 \times 10^2$ erg/cm^2 s. This is to be compared with the outward heat flow of 25×10^2 erg/cm^2 s. So the thermal emission is a modest heat loss at the base of the corona, declining outward in proportion to N^2, and rapidly becoming negligible. There is, of course, a larger downward heat flow from the base of the corona into the chromosphere, but that does not concern the temperature distribution outward from $r = a$.

4.3 Outward Decline of Density and Pressure

The essential feature of Chapman's calculation of coronal temperature distribution is the slow outward decline of the temperature, described by (4.4). For instance, with a representing the radius of the Sun, the orbit of Earth, at 1 AU, lies at $r = 220a$, where the temperature has fallen by the factor $220^{2/7} = 4.7$. That is to say, at 1 AU the temperature is $T = 0.21T(a)$, or 2.1×10^5 K for $T(a) = 10^6$ K back at the Sun.

Barometric equilibrium of the pressure $2NkT$ of a static corona with electron and ion temperature $T(r)$ in the gravitational field of the Sun is described by

$$\frac{\mathrm{d}}{\mathrm{d}r}2NkT = -\frac{GM_{\odot}}{r^2}NM\,,$$

where $M_{\odot}$ is again the mass (2×10^{33} gm) of the Sun, M is the mass of a hydrogen atom, and N is the ion number density. It follows that the pressure $p = 2NkT$ is given by

$$p(r) = p(a)\exp\left[-\int_a^r \mathrm{d}r\frac{GM_{\odot}M}{2r^2kT(r)}\right], \tag{4.5}$$

$$= p(a)\exp\left\{Q\left[1-\left(\frac{a}{r}\right)^{5/7}\right]\right\}, \tag{4.6}$$

and

$$N(r) = N(a)\left(\frac{r}{a}\right)^{2/7}\exp\left\{-Q\left[1-\left(\frac{a}{r}\right)^{5/7}\right]\right\}, \tag{4.7}$$

where, for convenience, we have written

$$Q = \frac{7GM_{\odot}M}{10akT(a)} = \frac{1.9 \times 10^7}{T(a)}\,. \tag{4.8}$$

Thus for $T(a) = 10^6$ K, Q has the value 19.0, using the round number $M = 2 \times 10^{-24}$ gm again. It is instructive to examine this static coronal model in some detail, because it carries over into the basis for coronal expansion.

As already noted, the pressure scale height $2kT/Mg$ at the base of the corona is 5×10^9 cm, where $g = 2.7 \times 10^4$ cm/s^2 and $T = 10^6$ K. The pressure falls to the value

$$p(\infty) = p(a)\exp(-Q) \tag{4.9}$$

at infinity, rather than to zero. Note further that, with the assumption that T falls to zero at infinity, the density increases without bound, in proportion to $r^{2/7}$, in order that the pressure approach the finite limit given by Eq. (4.9). The minimum density $N_{\min}$ lies at $r/a = (5Q/2)^{7/5}$ and is given by

$$N_{\min} = N(a)\left(\frac{5Q}{2}\right)^{2/5} \exp\left(\frac{2}{5} - Q\right).$$

Had we supposed that T approached some finite limit at large r, the density would remain finite, of course.

Now $Q = 19$ for $T(a) = 10^6$ K, from which it follows that $p(\infty) = 5.4 \times 10^{-9} p(a)$, while $N_{\min} = 4 \times 10^{-6} N(a)$ at about $r = 220a = 1$ AU with $N_{\min} = 400\,/\text{cm}^3$ for $N(a) = 10^8\,/\text{cm}^3$. The pressure at the Sun is $p(a) = 2.8 \times 10^{-2}$ dyn/cm^2. Far from the Sun, $p(\infty) = 1.5 \times 10^{-10}$ dyn/cm^2.

Chapman (1959) pointed out that Earth orbits through the outer corona of the Sun. That is to say, the corona fills the inner solar system. He was particularly interested in how the corona might conduct heat into the tenuous outer atmosphere of Earth.

4.4 Comets and Solar Corpuscular Radiation

At about this same time Biermann (1948, 1951, 1957) studied the phenomenon of the anti-solar acceleration of comet tails. The gaseous tails of comets always point away from the Sun, regardless of the direction of motion of the comet. The standard explanation for the anti-solar orientation of comet tails was the radiation pressure of sunlight, particularly the UV that is strongly absorbed by the ions in the comet tail. Observations of the vigorous anti-solar acceleration of the small inhomogeneities in comet tails sometimes showed outward accelerations a hundred times larger than the inward acceleration of solar gravity following an outburst of flaring at the Sun. Biermann pointed out that the absorption cross sections for the ions making up the gaseous comet tails could be closely estimated from theory, and turned out to be far too small for radiation pressure to accomplish the strong anti-solar acceleration. Then Biermann made the fundamental point that, if solar electromagnetic radiation is inadequate, there remains only the possibility that solar corpuscular radiation is responsible.

Charge exchange between the solar corpuscular radiation and the atoms and ions of the comet tail is the strongest interparticle interaction, from which Biermann concluded that the solar corpuscular radiation has a number density of the order of $10^3\,/\text{cm}^3$ at the orbit of Earth, presumably with outward velocities of the order of 10^3 km/s from active events on the Sun. This density estimate was not far from the estimated 500 electrons/cm^3 inferred from the brightness of the strongly polarized zodiacal light, assuming that the zodiacal light represents Thomson scattering of sunlight from interplanetary electrons (Elssaser, 1954; Blackwell, 1955, 1956).

4.5 Cosmic Ray Variations

Another approach was begun by Simpson, who invented the cosmic ray neutron monitor to study the curious variations of the cosmic ray intensity. Forbush (1937, 1954) discovered that the cosmic ray intensity declines abruptly by a few percent at the time when the solar corpuscular radiation from a flare event on the Sun arrived at Earth. He used ion chambers, responding mainly to μ mesons, created by the impact of cosmic ray protons of 10 – 30 GeV with the upper atmosphere of Earth. Simpson (1951, Simpson, Fonger, and Tremaine, 1953) developed the neutron monitor to detect the neutrons, produced mainly by the more numerous lower energy cosmic ray protons of 1 – 10 GeV, which would be more sensitive to what ever was producing the variations in the cosmic ray intensity. The Forbush-type decreases, as well as the 11-year cycle of cosmic ray variation with the general level of solar magnetic activity, showed up with amplitudes of 10 – 20% percent. Simpson constructed a series of neutron monitor stations from the geomagnetic equator (Huancayo, Peru), where the geomagnetic field excludes cosmic rays below about 15 GeV, to Chicago, where cosmic rays are freely admitted above about 2 GeV. Comparing the responses of the individual neutron monitors gave the energy spectrum of the cosmic ray variations. On that basis he was able to exclude the idea of large-scale electrostatic effects in space. Electrostatics was a popular idea at the time based on the prevailing concept that space is absolutely free of charged particles, except for the few cosmic ray particles themselves.

It is obvious that the electrostatic enthusiasts were not paying attention to either the phenomenon of solar corpuscular radiation or zodiacal light, implying interplanetary electron densities (of the order of $10^3/\mathrm{cm}^3$) that would prohibit the assumed electrostatic potentials of 10^9 V, or more, to modulate the cosmic ray intensity above 1 GeV per particle. Such large potential differences would convert any free interplanetary electrons or solar corpuscular radiation into relativistic particles, greatly enhancing the cosmic ray bombardment of Earth at those times when the potential was switched on. In any case, producing a Forbush decrease with an adverse potential would have the effect of diminishing the energy of all cosmic rays by about the same amount. Simpson's energy spectrum showed instead that the cosmic ray particles were simply increasingly diminished toward lower energies rather than shifted either up or down the energy spectrum. That looked more like the work of a magnetic field that deflected and turned back cosmic ray particles without much change of particle energies.

Then came the giant cosmic ray flare of 23 February 1956. The prompt arrival at Earth of protons (up to about 30 GeV) coming from the direction of the explosive flare on the Sun indicated that any magnetic fields between Earth and Sun must be more or less radial, i.e. a direct magnetic connection. Then within about 10 minutes of the first arriving particles the space around Earth filled with particles moving in all directions, and then those particles leaked away over the next hour or two. It was clear that not far beyond Earth there was a magnetic barrier that impeded further escape of the relativistic protons from the flare on the Sun. Thus the cosmic ray flare produced a flash picture of the general layout of the magnetic fields in interplanetary space (Meyer, Parker, and Simpson, 1956).

4.6 Plasma in Interplanetary Space

So there were magnetic fields in space, and that implied space filled with plasma in order to retain the fields. For without a plasma a magnetic field is free to escape at the speed of light. For plasma one turned to Chapman's extended corona and to Biermann's solar corpuscular radiation.

Now there was a feature of Biermann's comet tail analysis that escaped general attention. We must understand that comets swing by the Sun at random times, irrespective of the cycle of magnetic sunspot and flare activity. Most comets come by at low heliographic latitudes, but occasionally a comet passes by at high heliocentric latitude, more or less over the poles of the Sun. The essential point is that in no case was the anti-solar acceleration of the gaseous tail found to be absent. It followed that the Sun emits solar corpuscular radiation in all directions at all times. Hence, the origin of the solar corpuscular radiation does not depend on magnetic active regions on the Sun, which are missing at sunspot minimum and do not ever appear at high latitudes. The origin must be some common ongoing effect that is always present all around the Sun. The production of solar corpuscular radiation is more vigorous at the time of a large flare, but never absent. And that was the fundamental clue to the existence of what came to be called the *solar wind*.

However, the idea ran into immediate difficulty, because the solar corpuscular radiation, consisting of equal number of electrons and ions, i.e. is really a plasma – a vigorous outward streaming of ionized gas from the Sun at 10^3 km/s. On the other hand, space is filled with Chapman's extended solar corona. By 1958 it was known that two tenuous collisionless plasmas do not freely interpenetrate, because their relative bulk velocities excite electron and ion plasma oscillations which lock the two together, the kinetic energy of the initial relative motion converting into the vigorous oscillations (Pierce, 1949; Parker, 1958a, Petschek, 1958). So the ideas put forth by Chapman and Biermann on an extended static corona and an outward streaming corpuscular radiation seemed to be mutually exclusive. Yet neither concept could be dismissed. The resolution of the contradiction lay in recognizing that Chapman's static coronal model near the Sun must in some way become Biermann's solar corpuscular radiation at larger distance from the Sun. The corona must be a dynamical entity rather than a purely static atmosphere.

Earlier researches (Parker, 1957) had made it clear that the large-scale bulk motion of a plasma is described by the equations of hydrodynamics and magnetohydrodynamics regardless of whether the plasma is collisionless or collision dominated. So the next step was to examine the radial component of the hydrodynamic momentum equation for an atmosphere held captive to the Sun by gravity.

4.7 The State of the Corona

It is essential at this point to have clearly in mind the forces that might be involved in coronal expansion. So we take a close look at the relative values of the gas pressure, gravitational potential energy, thermal velocities, etc. to construct a simple physical picture of the corona. At 10^6 K the coronal gas is fully ionized except for a few electrons attached to the occasional heavier ion (Ca, Fe, Si, etc.), so for the present discussion it is sufficient to approximate the gas as a mixture of hydrogen and a little helium, with a mean ion mass of approximately $M = 2 \times 10^{-24}$ gm, while approximating the pressure $p(r)$ as $2N(r)kT(r)$ for ionized hydrogen. As already noted, the plasma pressure $p(a)$ at the base of the corona is 2.8×10^{-2} dyn/cm^2 for $N(a) = 10^8$ ions/cm^2. The rms ion thermal velocity is $[3kT(a)/M]^{1/2} = 1.45 \times 10^7$ cm/s, while the rms electron thermal velocity is 6.2×10^8 cm/s. The thermal energy $3kT(a)$ for each ion–electron pair is 4×10^{-10} erg, or about 250 eV, i.e. 125 eV per particle at 10^6 K. The gravitational binding energy $GM_{\odot}M/a$ is 3.8×10^{-9} erg, or about nine times the thermal energy of the ion electron pair. The essential point is that the corona is strongly bound by the gravitational field of the Sun. Without strong binding, there could be no quasi-steady supersonic solar wind, as we shall soon see.

Consider, then, Chapman's static barometric equilibrium model described by Eq. (4.5). As already noted, the pressure diminishes outward to a nonvanishing limit at infinity. For $T(a) = 10^6$ K we find that $p(\infty) \approx 1.5 \times 10^{-10}$ dyn/cm^2. It is this residual pressure at large r that is the origin of the solar wind. The existing inward pressure at infinity is the pressure of the interstellar gas, magnetic field, and cosmic rays, totaling about 10^{-12} dyn/cm^2, or less than $10^{-2}\, p(\infty)$.

Note from Eq. (4.5) that the pressure of the corona falls to zero at infinity if, and only if, the integral in the exponential becomes large without limit with increasing r. That requires a $T(r)$ declining asymptotically faster than $1/r$. That is to say, only if the thermal energy per electron ion pair declines faster that the gravitational potential energy $1/r$ is the corona gravitationally bound at large r. Thus, if $T(r)$ declines as $1/r^q$ with $q < 1$, there is a distance beyond which the thermal energy kT exceeds the binding energy, and the gas is free to escape, even though nearer the Sun the gas is strongly bound.

In the Chapman conduction model of the corona, we have $q = 2/7$ with the gravitational energy larger than the thermal energy by a factor $n = 9$ at the base. Thus the thermal energy of an electron–ion pair becomes equal to the gravitational binding energy where

$$\left(\frac{a}{r}\right)^{2/7} = n\left(\frac{a}{r}\right),$$

or

$$r/a = n^{7/5}.$$

With $n = 9$, this gives $r = 22a$. So beyond $22a$ the thermal energy exceeds the gravitational binding energy, and the gas streams away to infinity. The gravitationally bound gas in $r < 22a$ expands upward to replace the escaping gas in $r > 22a$, and the outward streaming continues.

In fact it is really the enthalpy rather than the thermal energy that should be considered here because the gas from below does work as it presses up behind the escaping gas. The enthalpy density is larger than the thermal energy density by the factor γ, where γ is the ratio of specific heats, equal to $5/3$ for a monatomic gas. The corresponding value of n is $9 \times 3/5 = 5.4$. The enthalpy exceeds the gravitational potential energy everywhere beyond $r = 10.6a$, and escape prevails. We might define the *top* of the corona to lie at $10.6a$ where gravitational binding becomes smaller than the enthalpy. Below the top at $10.6a$ the corona is approximately in hydrostatic equilibrium, and beyond the top the corona expands outward into space. So hydrodynamics enters the picture at this point.

4.8 Theoretical Foundations of Hydrodynamics and Magnetohydrodynamics

For the past 50 years, or more, there has existed a subculture in the space science and astrophysics community centered around the abiding faith that the familiar hydrodynamic equations

$$\frac{\partial N}{\partial t} + \nabla \cdot (N\boldsymbol{v}) = 0$$

$$NM\left[\frac{\partial \boldsymbol{v}}{\partial t} + (\boldsymbol{v} \cdot \nabla)\,\boldsymbol{v}\right] = -\nabla p + \boldsymbol{F}$$

for a fluid with velocity $\boldsymbol{v}$ in the presence of an applied force $\boldsymbol{F}$ (dyn/cm^3), apply only to liquids and to collision-dominated gases. The assertion is that the equations cannot apply to a collisionless plasma because without collisions the pressure is undefined, the thermal velocity distribution is not Maxwellian, etc. The subculture goes on to declare that magnetohydrodynamics (MHD) does not apply when a magnetic field is present in a collisionless plasma because the scalar Ohm's law is not applicable and because Ampere's law may not be satisfied, etc. The list of nonreasons is extensive. The subculture declares that it is the electric current $\boldsymbol{j}$, driven by the electric field $\boldsymbol{E}$, that is the cause of the magnetic field $\boldsymbol{B}$. Hence $\boldsymbol{j}$ and $\boldsymbol{E}$ are the fundamental physical variables. They turn to the generalized Ohm's law to relate $\boldsymbol{j}$ and $\boldsymbol{B}$ and then to Ampere's law to compute $\boldsymbol{B}$ from $\boldsymbol{j}$, but Newton's equations of motion and Maxwell's equations of electrodynamics do not play a central role. In contrast, if one starts with Newton and Maxwell, the result is the well known partial differential equations of MHD in terms of the bulk plasma velocity $\boldsymbol{v}$ and the magnetic field $\boldsymbol{B}$ carried in the plasma. Now one can, of course, use the electric drift velocity to replace $\boldsymbol{v}$ by $\boldsymbol{E}$ and $\boldsymbol{B}$ in the MHD equations and then use the Biot–Savart integral to replace $\boldsymbol{B}$ by $\boldsymbol{j}$. However, the resulting equations in $\boldsymbol{j}$ and $\boldsymbol{E}$ are intractable global nonlinear integro-differential equations. So the subculture shuns that approach and turns to a variety of declarations. For instance, it is asserted that the electric field $\boldsymbol{E} = -\boldsymbol{v} \times \boldsymbol{B}/c$ in the solar wind, as observed in the frame of reference of Earth, penetrates into the geomagnetic field and drives the convection of the field. It is declared that the familiar laws of electric circuits in the laboratory are applicable to the dynamical electric currents associated, through Ampere's law, with the magnetic field $\boldsymbol{B}$. Thus it is declared that the dynamics can be understood in terms of the inductance of the magnetic field and its currents. Hence, for instance, blocking the flow of $\boldsymbol{j}$ by shutting off the conductivity of the plasma would produce an enormous *emf*. Unfortunately Newton, Maxwell, and Lorentz know nothing of these effects.

Normally we would not feel it worthwhile to comment on the personal faith of others. However in recent years this fundamentalism has become increasingly aggressive, insisting that their view is the only one, when in fact it is not one at all. The interested reader is referred to the recent review by Parks (2004; see also Melrose, 1995; Spicer, 1982; Alfven and Carlquist, 1967) in which Parks asserts that MHD cannot apply to the ideal infinitely conducting fluid, because with infinite conductivity the electric field $\boldsymbol{E}'$ in the frame of the moving fluid is identically zero. Hence there can be no electric field to create the time dependent electric currents implied by the MHD induction equation for the time varying magnetic field,

$$\frac{\partial \boldsymbol{B}}{\partial t} = \nabla \times (\boldsymbol{v} \times \boldsymbol{B}) \ . \tag{4.10}$$

So MHD is alleged to be unworkable. Curiously, Parks fails to note that if one considers an arbitrarily large but finite conductivity, there is no problem with MHD.

The subculture does not respond to numerous challenges (cf. Parker, 1996a,b, 2000; Vasyliunas, 2001, 2005), evidently feeling no obligation to work from Newton and Maxwell. So it is useful here to outline for the reader the theoretical basis for hydrodynamics and MHD, which are unavoidable if one believes in conservation of mass, momentum, and energy, in the Faraday induction equation, and in Lorentz transformations between moving reference frames.

The basic requirement for the application of the familiar hydrodynamic equations to the large-scale bulk motion of a tenuous, and perhaps collisionless, gas or plasma is that the number density N be large enough to be locally statistically well defined. Consider a bulk fluid motion with characteristic scale Λ. It is necessary to have a well defined fluid density on some much smaller scale λ so that the spatial derivatives of density, bulk velocity, pressure, etc. are well defined. This requires that $N\lambda^3 \gg 1$. The required statistical precision is determined by the purpose of the dynamical calculations of the bulk velocity, of course, but the criterion is adequately met in every realistic large-scale case of which we are aware, even including the relativistic galactic cosmic ray gas, for which $N = 10^{-10}/\mathrm{cm}^3$.

The solar wind with a density $N = 5$ ions/cm^3 at the orbit of Earth serves as an illustration. The bulk flow of the wind around the magnetosphere of Earth involves scales as small as $\Lambda \approx 10^8$ cm, for which the microscale λ might be chosen as small as 10^5 cm. Then $N\lambda^3 = 5 \times 10^{15}$ ions, and N is defined over the dimension λ to within a statistical uncertainty of one part in 0.7×10^8. That seems quite adequate for most purposes, where uncertainties in the actual conditions are vastly larger. Note, of course, that the bow shock in the solar wind involves structure, comparable to the ion cyclotron radius, that is too small for hydrodynamics, so that its treatment involves plasma kinetics.

Now, if the number density N is well defined, then so is the mean bulk velocity, the mean random thermal energy, the pressure etc. Denote the velocity of the individual particle by u_i, equal to the sum of the well defined local mean bulk velocity v_i and the thermal velocity w_i of the particle relative to the mean velocity,

$$u_i = v_i + w_i \, .$$

where the index i refers to the vector component of the velocity, with $i = 1, 2, 3$. Then with $N\lambda^3 \gg 1$, the sum Σ of the momenta of all the particles in the volume $V = \lambda^3$ is

$$\Sigma M u_i = \Sigma M v_i + \Sigma M w_i \tag{4.11}$$

in the simple case that all the particles have the same mass M. In particular, $\Sigma M w_i$ is equal to zero. Thus

$$N M v_i = \frac{1}{V} \Sigma M u_i \tag{4.12}$$

where NMv_i is the mean momentum density.

The particles are conserved, of course, with or without collisions, and so also is the bulk momentum. Now the time rate of change of the density of any conserved quantity is equal to the negative divergence of the flux of that density. Thus the time rate of change of the number density is equal to the negative divergence of the mean particle flux,

$$\frac{\partial N}{\partial t} = -\frac{\partial}{\partial x_j} N v_j \, . \tag{4.13}$$

Similarly, the time rate of change of the momentum density NMv_i is equal to the negative divergence of the flux of momentum density. The jth component of the flux of Mu_i is $\Sigma M u_i u_j$. So writing out u_i in terms of $v_i + w_i$, we have

$$\frac{\partial}{\partial t} N M v_i = -\frac{\partial}{\partial x_j} \frac{1}{V} \Sigma M \left(v_i v_j + w_i v_j + v_i w_j + w_i w_j \right) .$$

The terms first order in w_k vanish, leaving

$$\frac{\partial}{\partial t} N M v_i = -\frac{\partial}{\partial x_j} N M v_i v_j - \frac{\partial p_{ij}}{\partial x_j} \tag{4.14}$$

where the pressure tensor p_{ij} is defined as the thermal flux in the j-direction of the momentum density of the thermal motions in the i-direction,

$$p_{ij} = \frac{1}{V} \Sigma M w_i w_j \, . \tag{4.15}$$

Thus p_{ij} represents the stress, i.e. the flux of momentum density, transported by the thermal motions. The Reynolds stress tensor R_{ij}, representing the mean momentum density NMv_i transported by the mean velocity v_j in the j-direction, is

$$R_{ij} = N M v_i v_j \, . \tag{4.16}$$

Thus

$$\frac{\partial}{\partial t} N M v_i = -\frac{\partial R_{ij}}{\partial x_j} - \frac{\partial p_{ij}}{\partial x_j} \, . \tag{4.17}$$

Multiply Eq. (4.13) by v_i and subtract from Eq. (4.14), producing the hydrodynamic momentum equation in the familiar form

$$N M \left(\frac{\partial v_i}{\partial t} + v_j \frac{\partial v_i}{\partial x_j} \right) = -\frac{\partial p_{ij}}{\partial x_j} \, .$$

If there is an applied force F_i per unit volume, e.g. gravity, then, obviously, the time rate of change of the momentum is described by

$$N M \left(\frac{\partial v_i}{\partial t} + v_j \frac{\partial v_j}{\partial x_j} \right) = -\frac{\partial p_{ij}}{\partial x_j} + F_i \, . \tag{4.18}$$

The essential point is that the hydrodynamic equations (4.13) and (4.18) are nothing more than the statement of conservation of matter and Newton's requirement that the time rate of change of the mean momentum is equal to the applied force and to the applied stress transmitted by the thermal motions. The two equations are inescapable, with or without collisions and with or without a Maxwellian velocity distribution.

The question of interparticle collisions appears only when we come to compute the time variation of the pressure tensor p_{ij}. In a collisionless plasma without magnetic field the kinetic energy of the individual particle motions in each of the three directions is conserved. The diagonal terms of p_{ij} represent those three kinetic energies, which follows from the tensor equation for the time rate of change of p_{ij} in terms of the negative divergence of the flux of kinetic energy density of the particle motions. Thus, we begin with

$$\frac{\partial}{\partial t}\Sigma M u_i u_j = -\frac{\partial}{\partial x_k}\Sigma M u_i u_j u_k \,,$$

from which we obtain the familiar result

$$\frac{\partial p_{ij}}{\partial t} + v_k\frac{\partial p_{ij}}{\partial x_k} = -p_{ij}\frac{\partial v_k}{\partial x_k} - p_{ik}\frac{\partial v_j}{\partial x_k} - p_{kj}\frac{\partial v_i}{\partial x_k} - \frac{\partial T_{ijk}}{\partial x_k} \tag{4.19}$$

where the flux of thermal energy is given by the heat flow tensor

$$T_{ijk} = \frac{1}{V}\Sigma M w_i w_j w_k \,. \tag{4.20}$$

Equation (4.19) is easily extended to the collisional case by introducing linear scattering (or other) terms between the components of p_{ij}. In the presence of a magnetic field the components of p_{ij} parallel and perpendicular to the magnetic field provide the usual MHD motion perpendicular to $\boldsymbol{B}$. For a more detailed discussion, see (Parker, 1957, 2007).

4.9 Kinetic Conditions in the Corona

It is evident from the fact that the solar corona is heated to 10^6 K that something, presumably plasma waves and MHD waves, interacts with the thermal motions of the ions and electrons, somehow accelerating them to thermal velocities of the order of $1-2\times10^7$ cm/s and almost 10^9 cm/s, respectively. The scattering and acceleration of the solar wind ions is evidently a powerful effect far out into space. For instance, spacecraft plasma detectors show that at a distance of 1 AU the ion temperature may fall to little more than 10^4 K, as one might expect from adiabatic expansion. At times, however, the ions show temperatures up toward 10^6 K, indicating strong heating well out toward 1 AU. Indeed, the ion thermal motions in the radial direction and the two perpendicular directions are often comparable, even though the expansion of the wind is primarily in the two directions perpendicular to the radial direction. Presumably at such times the ions are heated in interplanetary space by Alfven waves, whistlers, and ion cyclotron waves (Busnardo-Neto, et al., 1976; Li, et al., 1999a,b) presumably originating in the microflares in the coronal holes back at the Sun (Martin, 1984; Porter and Moore, 1988). In contrast, the electron temperatures are relatively well behaved, controlled mostly by thermal conduction from the 10^6 K corona near the Sun. They are affected but little by waves with frequencies below the electron cyclotron frequency.

Finally, note that any strong thermal anisotropy in the solar wind is unstable to the growth of plasma waves whose creation feeds on the anisotropy. In summary, then, it is clear that Coulomb collisions are by no means the only mechanism for pushing the thermal motions toward isotropy. The creation, absorption, and scattering of plasma waves plays a strong role where interparticle collisions are not important.

The role of interparticle Coulomb collisions may be seen from the approximate expression for the ion mean free path λ appropriate for the temperatures and densities of the solar corona and wind,

$$\lambda \approx 1.4\times10^{-13}w^4/N\,\text{cm}\,.$$

Thus, at the base of the corona, where $T = 10^6$ K and $N = 10^8/\text{cm}^3$, the rms thermal velocity w is typically 1.4×10^7 cm/s, and the mean free path turns out to be 6×10^9 cm, to be compared with the pressure scale height of 5×10^9 cm. The plasma is collision dominated. At 1 AU, where N might be 5 ions/cm^3 and the ion temperature 2×10^5 K, we have $w = 0.7\times10^7$ cm/s and a mean free path of about 3×10^{13} cm, whereas the characteristic scale of the density gradient (N proportional to $1/r^2$) is 0.5 AU or 0.75×10^{13} cm. So the solar wind plasma is relatively free of interparticle Coulomb collisions. On the other hand, at temperatures of 10^5 K and below Coulomb collisions play a role.

We conclude from these facts that, whether collisional or not, the expanding corona and solar wind are a hydrodynamic "meteorological" phenomenon, depending on the little known nonuniform heat input distribution starting at the Sun. So in studying the dynamics of the expanding corona and wind, we introduce temperature distributions in an ad hoc manner, striving to understand the general physical consequences of a diversity of possible temperature profiles, rather than producing a coronal model based on specific assumptions.

4.10 Magnetohydrodynamics

A word on magnetohydrodynamics is in order before turning to the dynamics of the expanding corona. The introduction of a magnetic field into a plasma converts the hydrodynamics of the large-scale bulk motion into magnetohydrodynamics, in which the magnetic field is carried bodily with the bulk motion of the plasma and the magnetic stresses contribute to the force F_i applied to the plasma in Eq. (4.18). The magnetic stress is described by the Maxwell stress tensor

$$M_{ij} = -\delta_{ij}\frac{B^2}{8\pi} + \frac{B_i B_j}{4\pi} , \qquad (4.21)$$

where the first term on the right hand side represents the isotropic magnetic pressure $B^2/8\pi$, and the second term represents the tension $B^2/4\pi$ in the direction of the field. This magnetic stress system may be in equilibrium on its own, with $\partial M_{ij}/\partial x_j = 0$, exerting no force on the plasma. Or the magnetic field may transmit some of its stress to the plasma, with $F_i = \partial M_{ij}/\partial x_j$ in the momentum equation (4.18).

Consider, then, the Faraday induction equation for the time variation of the magnetic field B_i, written conveniently as

$$\frac{\partial \boldsymbol{B}}{\partial t} = -c\nabla \times \boldsymbol{E} . \qquad (4.22)$$

The essential point is that a plasma, or even a partially ionized gas, with enough free electrons and ions to form a good conductor of electricity, cannot support an electric field $\boldsymbol{E}'$ in its own moving frame of reference. That is to say, such a conducting gas has no significant electrical insulating ability. To see what this implies, suppose that the plasma is collision dominated, so that the scalar Ohm's law applies, with

$$\boldsymbol{j} = \sigma \boldsymbol{E}' \qquad (4.23)$$

Ampere's law requires a current density $\boldsymbol{j}$, given by

$$4\pi \boldsymbol{j} = c\nabla \times \boldsymbol{B} \qquad (4.24)$$

as the field is carried and deformed in the swirling plasma. It follows that

$$\boldsymbol{E}' = \frac{c}{4\pi\sigma}\nabla \times \boldsymbol{B} . \qquad (4.25)$$

Thus, for a given $\nabla \times \boldsymbol{B}$, $\boldsymbol{E}'$ is small in the limit of large σ. As an example, the conductivity, $\sigma = 2\times10^7\,T^{3/2}/\mathrm{s}$, is typically 2×10^{16} /s in the corona. The characteristic electric field driving the current in a magnetic field B with characteristic scale l is then of the order of

$$E' = cB/4\pi l\sigma .$$

Hence the characteristic potential difference $E'l$ for a field of 10 gauss in the corona is of the order of 10^{-6} statvolts, or 3×10^{-4} V across whatever scale l one might imagine.

In the opposite extreme of a collisionless plasma the ions and electrons are accelerated in the electric field $\boldsymbol{E}$ applied in the laboratory frame of reference, so that they both take up the electric drift velocity

$$\boldsymbol{v}_\mathrm{d} = c\frac{\boldsymbol{E}\times\boldsymbol{B}}{B^2} . \qquad (4.26)$$

In that drifting frame of reference there is no electric field, $\boldsymbol{E}' = 0$, so there is no further acceleration, unless the applied electric field changes. So with, or without collisions, there is no significant electric field $\boldsymbol{E}'$ in the frame of reference of the moving plasma.

Given the electric field $\boldsymbol{E}$ and magnetic field $\boldsymbol{B}$ in the laboratory reference frame, the nonrelativistic Lorentz transformation gives the field $\boldsymbol{E}'$ and $\boldsymbol{B}'$ in the frame moving with the bulk plasma velocity $\boldsymbol{v}$ relative to the laboratory, where

$$\boldsymbol{E}' = \boldsymbol{E} + \boldsymbol{v}\times\boldsymbol{B}/c , \quad \boldsymbol{B}' = \boldsymbol{B} , \qquad (4.27)$$

neglecting terms $O(v^2/c^2)$ compared to one. It follows that

$$\boldsymbol{E} = -\boldsymbol{v}\times\boldsymbol{B}/c + \boldsymbol{E}' . \qquad (4.28)$$

Substituting this into Eq. (4.22) yields

$$\frac{\partial \boldsymbol{B}}{\partial t} = \nabla \times (\boldsymbol{v}\times\boldsymbol{B}) - c\nabla\times\boldsymbol{E}' .$$

With $\boldsymbol{E}'$ very small, if not identically zero, this reduces to the magnetohydrodynamic induction equation

$$\frac{\partial \boldsymbol{B}}{\partial t} = \nabla \times (\boldsymbol{v}\times\boldsymbol{B}) , \qquad (4.29)$$

which states that the magnetic field is transported bodily with the bulk velocity $\boldsymbol{v}$ of the plasma (Parker, 1957). Thus, besides $\boldsymbol{B}$, there is an electric field

$$\boldsymbol{E} = -\boldsymbol{v} \times \boldsymbol{B}/c \tag{4.30}$$

in the laboratory because there is no electric field in the frame of reference moving with the plasma. The magnetic field cannot slip relative to the plasma without inducing an electric field $\boldsymbol{E}' \neq 0$ in the plasma, which is not allowed by the high electrical conductivity of the plasma. Substituting the electric field of Eq. (4.30) into the Faraday induction equation leads to Eq. (4.29) and the phenomenon of magnetohydrodynamics.

One can see the reality of the bulk transport of the magnetic field by computing the Poynting vector

$$\boldsymbol{P} = c\frac{\boldsymbol{E} \times \boldsymbol{B}}{4\pi} ,$$

representing the flux of electromagnetic energy. With the electric field given by Eq. (4.30), it follows that

$$\boldsymbol{P} = \boldsymbol{v}_{\perp}\frac{B^2}{4\pi} \tag{4.31}$$

where $\boldsymbol{v}_{\perp}$ is the component of $\boldsymbol{v}$ perpendicular to $\boldsymbol{B}$, and $B^2/4\pi$ is the magnetic enthalpy density. So the magnetic enthalpy moves exactly with the fluid motion perpendicular to $\boldsymbol{B}$. There is, of course, no way to define locally the fluid velocity parallel to $\boldsymbol{B}$, nor any inductive effects from such motion.

Magnetohydrodynamics enters the solar wind picture when we consider the interaction of the solar wind with the magnetic fields of the Sun. So with this sketch of the theoretical foundations of hydrodynamics and magnetohydrodynamics, consider the problem of the dynamics of the corona of the Sun, tightly bound by gravity but with the kinetic temperature extending far out into space.

4.11 Hydrodynamic Expansion of the Solar Corona

It is sufficient for illustrating the general phenomenon of coronal expansion to treat the simple case of steady hydrodynamic radial outflow of the coronal gas. Thus conservation of matter requires that

$$N(r)v(r)r^2 = N(a)v(a)a^2 \tag{4.32}$$

for $r > a$. The momentum equation is

$$N(r)\, Mv(r)\frac{\mathrm{d}v}{\mathrm{d}r} = -\frac{\mathrm{d}}{\mathrm{d}r}2N(r)\, kT(r) - \frac{GM_{\odot}MN(r)}{r^2} . \tag{4.33}$$

It is convenient to introduce the characteristic thermal velocity $U(r) = [2kT(r)/M]^{1/2}$, so that

$$v\frac{\mathrm{d}v}{\mathrm{d}r} = -U^2\frac{1}{N}\frac{\mathrm{d}N}{\mathrm{d}r} - \frac{\mathrm{d}U^2}{\mathrm{d}r} - \frac{GM_{\odot}}{r^2} . \tag{4.34}$$

Then since

$$\frac{1}{N}\frac{\mathrm{d}N}{\mathrm{d}r} = -\frac{2}{r} - \frac{1}{v}\frac{\mathrm{d}v}{\mathrm{d}r} , \tag{4.35}$$

the density N can be eliminated from Eq. (4.34), yielding

$$\frac{\mathrm{d}v}{\mathrm{d}r}\left(v - \frac{U^2}{v}\right) = \frac{2U^2}{r} - \frac{GM_{\odot}}{r^2} - \frac{\mathrm{d}U^2}{\mathrm{d}r} . \tag{4.36}$$

We are interested in a solution to Eq. (4.36) that (a) begins with $v \ll U$ ($U \approx 1.2 \times 10^7$ cm/s for $T = 10^6$ K) at the base of the corona, $r = a$, simulating Chapman's static corona, and (b) increasing to $v \gg U$ for $r \gg a$, simulating Biermann's solar corpuscular radiation. So we are interested in a solution of Eq. (4.36) for which $\mathrm{d}v/\mathrm{d}r > 0$. Near the Sun, with $v^2 < U^2$, it is evident that the factor $v - U^2/v$ on the left hand side of the equation is negative. Inspection of the right hand side, neglecting the relatively small term $\mathrm{d}U^2/\mathrm{d}r$, indicates that it is also negative as a consequence of the strong gravitational binding, $2U^2 < GM_{\odot}/r$. Hence there is a solution with $\mathrm{d}v/\mathrm{d}r > 0$.

Far from the Sun the solution for $v > U$ makes the factor $v - U^2/v$ positive, while the term $GM_{\odot}/r^2$ becomes less than $2U^2/r$ because of the more rapid decline of $1/r^2$. So again we have $\mathrm{d}v/\mathrm{d}r > 0$. It is clear, however, that $v - U^2/v$ must pass through zero at the same point $r = r_c$ where the right hand side vanishes if the solution $v(r)$ is to be well behaved. That is to say, we require that $v(r_c) = U(r_c)$ at the critical point r_c where the right hand side vanishes, i.e. where

$$r^2\frac{\mathrm{d}}{\mathrm{d}r}\frac{U^2}{r^2} + \frac{GM_{\odot}}{r^2} = 0 . \tag{4.37}$$

For the simple case of uniform coronal temperature, this gives

$$r_c = \frac{GM_{\odot}}{2U^2} . \tag{4.38}$$

So the expansion velocity v crosses over the ion thermal velocity U at the point where the expansion velocity and the thermal velocity are each equal to half the gravitational escape velocity $(2GM_{\odot}/r_c)^{1/2}$.

It is instructive to examine the solution $v(r)$ in the neighborhood of the critical point, writing

$$r = r_c + \xi \,, \; v = U + u \,. \tag{4.39}$$

Then to first order in ξ and u, Eq. (4.36) reduces to

$$u \frac{\mathrm{d}u}{\mathrm{d}\xi} = U^2 \frac{\xi}{r_c^2} \,.$$

It follows that

$$u^2 = U^2 \left(\frac{\xi}{r_c} \right)^2 + K \,, \tag{4.40}$$

where K is the arbitrary integration constant. The solutions crossing the critical point ($K = 0$) are $u/U = \pm\xi/r_c$. For outward increasing $v(r)$ we have

$$\frac{u}{U} = +\frac{\xi}{r_c} \,. \tag{4.41}$$

In fact, for the case at hand, $\mathrm{d}T/\mathrm{d}r = 0$, the momentum equation (4.36) is easily integrated to

$$\frac{v^2}{U^2} - \ln \frac{v^2}{U^2} = 4 \ln \frac{r}{r_c} + 4\frac{r_c}{r} - 3 + C \tag{4.42}$$

where C is the integration constant. The form of the family of solutions on the (v, r) plane is sketched in Fig. 4.1. The solution $C = 0$, passing across the critical point from small v near the Sun to large v at great distance, is shown by the heavy line across the critical point, satisfying the boundary conditions (a) $v = 0$ at $r = 0$, and (b) $v > U$ at $r = \infty$, so that the density, and therefore the pressure, fall to zero with increasing r. No other member of the family of solutions has this property.

Let us see to what extent we can understand the basic physics of this steady expansion of the corona to $v(\infty) > U$. Recall from the discussion of the static equilibrium of the corona that the outer regions of the strongly bound corona are free to expand away to infinity. Then note that Eq. (4.42) can be written

$$\frac{v^2(r)}{U^2} = \left(\frac{r_c}{r} \right)^4 \exp\left[3 - \frac{4r_c}{r} + \frac{v^2(r)}{U^2} \right] .$$

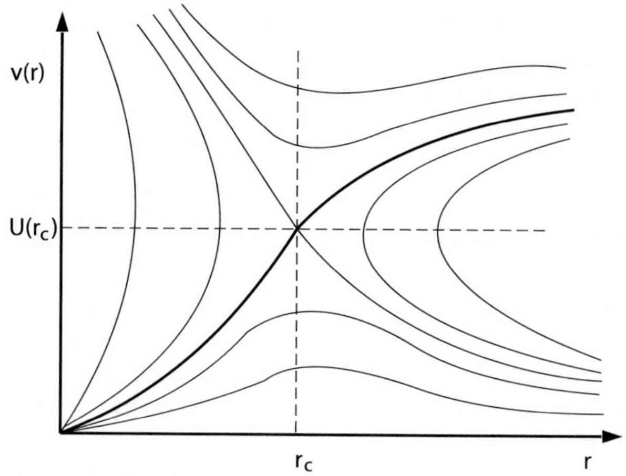

Fig. 4.1. A sketch of the family of solutions to the momentum equation (4.36), given by Eq. (4.42) for an isothermal corona. The *heavy line* indicates the solution for $C = 0$, passing from small velocity $v(r)$ near the Sun to nonvanishing $v(r)$, and, hence, vanishing pressure, at infinity

For $r \ll r_c$ and $v^2(r) \ll U^2$ it follows that

$$\frac{v^2(r)}{U^2} \cong \left(\frac{r_c}{r} \right)^4 \left[1 + \frac{v^2(r)}{U^2} + \ldots \right] \exp\left(3 - \frac{4r_c}{r} \right) . \tag{4.43}$$

On the other hand, for $r \gg r_c$ and $v(r) \gg U$, Eq. (4.42) can be written as

$$\begin{aligned} \frac{v^2(r)}{U^2} &= 4 \ln \frac{r}{r_c} - 3 + \frac{4r_c}{r} + \ln \frac{v^2(r)}{U^2} \\ &\cong 4 \ln \frac{r}{r_c} - 3 + \frac{4r_c}{r} \\ &\quad + \ln \left\{ 4 \ln \frac{r}{r_c} - 3 + \frac{4r_c}{r} \right. \\ &\quad \left. + \ln \left[4 \ln \frac{r}{r_c} - 3 + \frac{4r_c}{r} + \ldots \right] \right\} \end{aligned} \tag{4.44}$$

The convergence of this continued logarithmic expression is poor, but it is obvious that the velocity $v(r)$ increases asymptotically without limit with increasing r/r_c, in proportion to $2\left[\ln(r/r_c)\right]^{1/2}$. The energy for this is the continual input of thermal energy to maintain the temperature T of the expanding gas. The extended coronal heating implied by the isothermal coronal model represents the interplanetary "afterburner" along the same lines as the afterburner in an aircraft jet engine, adding and burning fuel as the exhaust gas expands out through the exhaust channel.

Cranmer (2004) has pointed out that the transcendental relation (4.42) between $v(r)/U$ and r/r_c has a direct solution in terms of the Lambert W function, representing the analytic solution to that transcendental form. The Lambert W function $x = W(y)$ is the inverse of the function $y = W \exp W$. Then

$$\frac{v^2(r)}{U^2} = -W\left[-\left(\frac{r_c}{c}\right)^4 \exp\left(3 - \frac{4r_c}{r}\right)\right]$$

The Lambert W function is multivalued, and one must choose the correct branch to work with the expanding corona (see also, Velli, 2001).

Now an obvious question is how does the isothermal corona expand when the temperature T is reduced to successively lower values. Note, then, that there is no minimum temperature T for reaching the asymptotic supersonic velocity. The effect of reducing T, or U, is to move the critical point farther out, as is obvious from Eq. (4.38). The expansion velocity continues to increase without limit with increasing r, so a specified velocity is achieved only very much farther out. The essential physical point is that the density falls to negligible values. It follows from Eq. (4.31) and Eq. (4.43) that the density at the critical radius r_c is

$$N(r_c) = N(a)\frac{a^2 v(a)}{r_c^2 v(r_c)} \qquad (4.45)$$

$$\cong N(a)\exp\left(\frac{3}{2} - \frac{GM_\odot}{aU^2}\right), \qquad (4.46)$$

diminishing very rapidly with declining U^2 when $U^2 \ll GM_\odot/a$. So there is a theoretical supersonic expansion for any isothermal corona, but the expanding corona quickly becomes too tenuous to impose itself on the surrounding interstellar space. For instance, a coronal temperature of 0.75×10^6 K gives $Q = 25.4$ in Eq. (4.9) for the pressure of a static corona at large distance. It follows that $p(\infty) = 0.9 \times 10^{-11} p(a)$. Thus, for $p(a) = 2.8 \times 10^{-2}$ dyn/cm^2 and $N(a) = 10^8$ /cm^3 the result is $p(\infty) = 2.5 \times 10^{-13}$ dyn/cm^2, somewhat less than the interstellar gas and field pressures, of the order of 1×10^{-12} dyn/cm^2. It is evident that a corona at 0.75×10^6 K could be confined by the interstellar pressure to static equilibrium with an outer boundary at some finite distance from the Sun.

Consider what happens, then, if the temperature is increased above 10^6 K. The critical point, given by Eq. (4.38) moves inward, reaching the base of the corona at $r = a$ when T is about 7×10^6 K. There is no subsonic expansion region, and the gas at the base of the corona explodes outward into space with an initial speed U. The heat input required to maintain 7×10^6 K or more in the face of this massive outflow would become a significant fraction of the total solar luminosity. The situation would be quite different from the strongly throttled outflow of the tenuous outer reaches of the corona of 1×10^6 K.

4.12 Sufficient Conditions on Coronal Temperature

The isothermal corona is a serious idealization when carried to arbitrarily large r, where, surely, the temperature of the gas must decline. The obvious question is the limiting rate of decline sufficient to produce a supersonic expansion. A convenient formal device for investigating this question is a polytropic relation of temperature and pressure with the density $N(r)$. Write

$$T(r) = T(a)\left[\frac{N(r)}{N(a)}\right]^{\alpha-1}, \qquad (4.47)$$

so that the temperature declines with declining $N(r)$ for $\alpha > 1$. Then

$$p(r) = p(a)\left[\frac{N(r)}{N(a)}\right]^{\alpha}, \qquad (4.48)$$

and $\alpha = 1$ represents the isothermal case already discussed, while $\alpha = 5/3$ represents adiabatic expansion, i.e. no heat input in $r > a$. The degree to which α is less than 5/3 is a measure of the heat input to the expanding coronal gas.

The momentum equation (4.32) can be written

$$v\frac{dv}{dr} + \frac{\alpha}{\alpha-1}U^2(a)\frac{d}{dr}\left(\frac{N}{N(a)}\right)^{\alpha-1} + \frac{GM_\odot}{r^2} = 0, \qquad (4.49)$$

where $U^2(a) = 2kT(a)/M$. This equation can be integrated to

$$\frac{v^2(r)}{2U^2(a)} + \frac{\alpha}{\alpha-1}\left[\frac{N(r)}{N(a)}\right]^{\alpha-1} = \frac{GM_\odot}{r} + S, \qquad (4.50)$$

where S is the constant of integration, chosen so that the solution passes across the critical point. A detailed analysis can be found in Parker (1960a, 1963). The essential point is that expansion at large r with nonvanishing velocity $v(\infty)$ provides a density declining as $1/r^2$ (see Eq. (4.31)) so that the temperature declines as $1/r^{2(\alpha-1)}$. It follows that the thermal energy declines less rapidly than the gravitational potential energy if, and only if, $\alpha < 3/2$. In that case there is some distance r_T beyond which the thermal energy exceeds the gravitational binding energy, so that expansion is inevitable, presumably with a velocity comparable to the thermal velocity at r_T. The temperature falls to zero with increasing r so that the expansion is supersonic. So there is no minimum temperature $T(a)$ or $T(r_c)$ for expansion when $\alpha < 3/2$, but, as with the isothermal corona, the expanding gas becomes too tenuous to assert itself, and the corona is effectively confined to static equilibrium by the interstellar pressure.

Then it is not without interest to consider the circumstances for uniform expansion velocity, $\mathrm{d}v/\mathrm{d}r = 0$. It is readily shown from Eq. (4.36) that

$$U^2(r) = \frac{GM_\odot}{3r} + Cr^2 ,$$

where C is an arbitrary constant. If the temperature has a minimum at $r = r_m$, then

$$U^2(r) = \frac{GM_\odot}{3r}\left(1 + \frac{r^3}{2r_m^3}\right) ,$$

with the minimum value

$$U^2(r_m) = \frac{GM_\odot}{2r_m} .$$

This special temperature distribution would require coronal heating beyond r_m.

Finally note that the *local* acceleration of the wind beyond r_c is *enhanced* by the local rate of *decline* of the temperature. A decline in temperature is conducive to local acceleration for the simple reason that the pressure behind a given element of gas accelerates the gas only when the pressure ahead is diminished. Equation (4.36) can be written

$$\frac{\mathrm{d}v}{\mathrm{d}r}\left(v - \frac{U^2}{v}\right) = -r^2\frac{\mathrm{d}}{\mathrm{d}r}\left(\frac{U^2}{r^2}\right) - \frac{GM_\odot}{r^2} .$$

Beyond the critical point the factor $v - U^2/v$ is positive, and the first term on the right hand side represents the local contribution of the temperature to the acceleration. Thus for instance, an isothermal corona, $\mathrm{d}U^2/\mathrm{d}r = 0$, maintained by local heating, provides only the local acceleration $2U^2/r$, neglecting the diminished gravitational attraction of the Sun. On the other hand, switching off the heat supply, so that the temperature declines adiabatically, the acceleration is substantially enhanced. It follows from Eqs. (4.35) and (4.47) that

$$\frac{1}{U^2}\frac{\mathrm{d}U^2}{\mathrm{d}r} \approx -(\alpha-1)\left(\frac{2}{r} + \frac{1}{v}\frac{\mathrm{d}v}{\mathrm{d}r}\right)$$

Then far beyond the critical point r_c we have

$$\begin{aligned}\frac{1}{v}\frac{\mathrm{d}v}{\mathrm{d}r}\left(v^2 - U^2\right) &\approx \frac{2U^2}{r} - \frac{\mathrm{d}U^2}{\mathrm{d}r} ,\\ &\approx \frac{\alpha\left(v^2 - U^2\right)}{v^2 - \alpha U^2}\frac{2U^2}{r} ,\\ &\approx \alpha\left[1 + (\alpha-1)\frac{U^2}{v^2} + \alpha(\alpha-1)\frac{U^4}{v^4} + \ldots\right]\frac{2U^2}{r} .\end{aligned}$$

The local acceleration is enhanced by a factor in excess of $\alpha = 5/3$. Needless to say, while the local acceleration is enhanced, the final asymptotic velocity is better served by a temperature that does not diminish so much.

4.13 Analogy with Expansion Through a Laval Nozzle

It was pointed out many years ago (Clauser, 1960) that the expansion of the solar corona to supersonic velocity is closely analogous to the expansion of a gas through a Laval nozzle to supersonic velocity in a vacuum. The Laval nozzle consists of a constriction in the flow channel, followed by a broad flaring out of the channel into a vast region of reduced gas pressure. The gas is pumped steadily under pressure through the channel. The flow speed increases as the gas flows into the constriction or throat, reaching the speed of sound in the throat, beyond which the gas expands freely down the flaring channel into the vacuum beyond. Denote the cross sectional area of the channel by $A(s)$, the steady flow velocity by $v(s)$, and the number density of the gas by $N(s)$, where s denotes distance measured along the channel. Conservation of matter requires

$$N(s)\,v(s)\,A(s) = N(0)\,v(0)\,A(0) \tag{4.51}$$

so that

$$\frac{1}{N}\frac{\mathrm{d}N}{\mathrm{d}s} = -\frac{1}{A}\frac{\mathrm{d}A}{\mathrm{d}s} - \frac{1}{v}\frac{\mathrm{d}v}{\mathrm{d}s} \tag{4.52}$$

in place of Eqs. (4.32) and (4.35). The momentum equation

$$N(s)\,Mv(s)\,\frac{\mathrm{d}v}{\mathrm{d}s} = -\frac{\mathrm{d}}{\mathrm{d}s}2N(s)\,kT(s) \qquad (4.53)$$

takes the form

$$\frac{\mathrm{d}v}{\mathrm{d}s}\left(v - \frac{U^2}{v}\right) = \frac{U^2}{A}\frac{\mathrm{d}A}{\mathrm{d}s} - \frac{\mathrm{d}U^2}{\mathrm{d}s}\,, \qquad (4.54)$$

where again $U^2(s) = 2kT(s)/M$. The critical point $s = s_c$ is located where the right hand side vanishes,

$$\frac{\mathrm{d}}{\mathrm{d}s}\left(\frac{U^2}{A}\right) = 0\,, \qquad (4.55)$$

and $v(s_c)$ is equal to $U(s_c)$. For uniform temperature the critical point would lie at the throat of the nozzle, where A passes through the minimum value A_c, where

$$\frac{\mathrm{d}A}{\mathrm{d}s} = 0\,. \qquad (4.56)$$

Beyond the constricted throat at s_c, A increases without bound, and the gas expands without limit to zero pressure and a finite supersonic flow velocity.

For the isothermal case Eq. (4.54) integrates to

$$\frac{v^2}{U^2} - \ln\frac{v^2}{U^2} = 2\ln\frac{A}{A_c} + 1 \qquad (4.57)$$

for the solution passing across the critical point from subsonic to supersonic flow (Parker, 1958b). In the limit of large A/A_c the velocity has the asymptotic form

$$\frac{v}{U} \approx \left(2\ln\frac{A}{A_c}\right)^{1/2} + O(1)\,. \qquad (4.58)$$

Forcing the gas into the narrow throat accelerates the initial slow subsonic flow to the speed of sound without much change in density, in the manner described by Eq. (4.51). Beyond the throat the gas expands into the vacuum beyond, accelerating beyond the speed of sound at the same time that the temperature and the speed of sound decline asymptotically to zero. The analogy with the gravitationally confined corona of the Sun is obvious. The corona is in quasi-equilibrium inside the critical point, with rapid upward decline of $N(r)$, so any outward flow velocity increases rapidly with height, in the manner described by Eq. (4.32), reaching the thermal velocity U at the critical radius r_c. Beyond r_c the dynamics goes over into expansion into the relative vacuum at infinity. Thus, in the absence of a gravitational field, the corona would simply explode away into space whereas the gravitational field throttles back this explosion to a fast tenuous flow out through the "top" of the atmosphere at r_c. It is this gravitational throttling of the density that causes the expansion velocity to increase to U.

4.14 Gravitational Throttling of Coronal Expansion

We can use the concept of gravitationally throttled expansion to construct a formal mathematical scheme that provides an iterative solution to the hydrodynamic equations (4.32) and (4.33) for general temperature $T(r)$ in terms of a succession of quadratures. We illustrate the method here for the simple isothermal case, where the exact solution is available for comparison. In place of Eq. (4.32) write

$$v(r) = U\frac{N(r_c)}{N(r)}\left(\frac{r_c}{r}\right)^2 \qquad (4.59)$$

for the solution crossing the critical point. Now for $r < r_c$, the density $N(r)$ is not far from the hydrostatic equilibrium form given by Eq. (4.5). That is to say, neglecting the inertial term $v\,\mathrm{d}v/\mathrm{d}r$ the density $N(r)$ follows upon integration of Eq. (4.33). The result is conveniently written

$$N(r) = N(r_c)\exp 2\left(\frac{r_c}{r} - 1\right) \qquad (4.60)$$

for the isothermal corona, T = constant. The expansion velocity $v(r)$ follows from Eq. (4.59) as

$$v(r) \cong U\left(\frac{r_c}{r}\right)^2 \exp\left[-2\left(\frac{r_c}{r} - 1\right)\right]. \qquad (4.61)$$

Hence, at the base of the corona ($r = a$)

$$v(a) \cong U\left(\frac{r_c}{a}\right)^2 \exp\left[-2\left(\frac{r_c}{a} - 1\right)\right]. \qquad (4.62)$$

This is to be compared with the approximate form (4.43)

$$v(a) \cong U\left(\frac{r_c}{a}\right)^2 \exp\left[-2\left(\frac{r_c}{a} - \frac{3}{4}\right)\right] \qquad (4.63)$$

obtained from the exact solution (4.42) for $v(a) \ll U$. It is apparent that the result given by Eq. (4.62) is too large at $r = a$ by the factor $\exp(0.5) = 1.649$, as a consequence of the breakdown of the approximation $v^2 \ll U^2$ approaching $r = r_c$.

The next iteration involves using Eq. (4.61) to approximate the inertial term $v\,dv/dr$ neglected in computing Eq. (4.60), thereby obtaining a more accurate expression for $N(r)$ and $v(r)$ in terms of $T(r)$ and $N(a)$. The details of successive iterations can be found in Parker (1964a,b, 1965), where it is shown that the process converges fairly rapidly. The essential point is that this reduction to successive quadratures works for arbitrary $T(r)$ and provides the rate of outflow of mass in terms of $N(a)$ for a given $T(r)$. The convergence to the exact solution depends on the corona being strongly bound by gravity so that the outflow is strongly throttled and the corona is close to static equilibrium almost all the way out to the critical point.

Beyond the critical point the gas expands into a vacuum and a different iterative scheme becomes possible. Integrate Eq. (4.34) outward from the critical point r_c, with the result that

$$\begin{aligned}\frac{1}{2}v^2(r) + U^2(r) - \frac{GM_\odot}{r} &= \frac{3}{2}U^2(r_c) - \frac{GM_\odot}{r_c} + 2\int_{r_c}^{r} d\lambda \frac{U^2(\lambda)}{\lambda} \\ &+ \frac{1}{2}\int_{r_c}^{r} d\lambda \frac{U^2(\lambda)}{v^2(\lambda)}\frac{dv^2}{d\lambda}\end{aligned}$$

for the solution passing across the critical point. The second integral becomes much smaller than the first as $U(r)$ declines and $v(r)$ increases, so it is neglected in the first iteration. Thus for a specified $T(r)$, i.e. $U^2(r)$, the expansion velocity follows. That velocity is then used to provided an approximate value for the second integral in the next iteration, etc. The convergence of the iteration scheme is based on $U^2/v^2 < 1$ in $r > r_c$.

4.15 Wind Density and Solar Mass Loss

It is evident from the foregoing discussion that the density of the solar wind at 1 AU is determined by the density of the corona at the critical point, which is determined by the temperature within the corona, $a < r < r_c$. Thus Eq. (4.60) provides the approximate relation

$$N(r_c) = N(a)\exp\left[-2\left(\frac{r_c}{a} - 1\right)\right] \quad (4.64)$$

for an isothermal atmosphere. The mass loss over a solid angle Ω is, then, $\Omega r_c^2 U N(r_c)$. This is equal to $\Omega a^2 v(a) N(a)$, of course, with $v(a)$ given by Eq. (4.62). The essential point is that the density and mass loss depend very sensitively on $T(r)$ throughout the corona, so they cannot be deduced from theory unless $T(r)$ is known with precision.

This lack of predictability played a curious role in the early development of the theory of coronal expansion and the solar wind before the Space Age got underway (Parker, 1958). Recall from section IV that the density of the solar corpuscular radiation was estimated from observations to be 500 – 1000 electrons and ions/cm^3 at 1 AU. The estimate of the electron density was based on the assumption that the strongly polarized zodiacal light represents Thomson scattering of sunlight from electrons in interplanetary space. The ion density was based on the assumption that the particles of the solar corpuscular radiation interact directly with the cometary ions through charge exchange. However, with the advent of direct measurements in space (Snyder and Neugebauer, 1964; Neugebauer and Snyder, 1964, 1966, 1967) the density turned out to be of the order of 5 ions/cm^3, one hundred times smaller than the previous estimates. The hydrodynamic theory of coronal expansion easily accommodated either the estimated 500 /cm^3 or the 5 /cm^3 given the modest uncertainty in $T(r)$ near the Sun.

The initial gross overestimate of the interplanetary electron and ion density arose from the fact that the zodiacal light actually represents the scattering of sunlight from interplanetary dust grains in gravitational orbit around the Sun, rather than from free electrons. The light scattered from dust grains is strongly polarized, just as from free electrons. We note that the individual orbits of the dust grains gradually spirally inward toward the Sun as a consequence of the Poynting–Robertson effect at the same time that the grains are perturbed by the magnetic field carried in the solar wind (Parker, 1964c). Then it turns out that it is the large-scale magnetic field carried in the solar wind that interacts with the ionized comet tail, rather than charge exchange between solar wind and cometary ions. So none of the pre-Space Age density estimates inferred from observations was correct.

To understand the extreme density–temperature sensitivity, note from Eq. (4.38) that a temperature of 1×10^6 K gives $r_c/a = 6.8$, while 2×10^6 K gives half that, 3.4. Thus from Eq. (4.64) it follows that $N(r_c)/N(a)$ is equal to $\exp(-11.6)$ at 1×10^6 K and $\exp(-4.8)$ at 2×10^6 K, differing by a factor of $\exp 6.8 = 900$. Recalling that r_c at 2×10^6 K is half what it is at 1×10^6 K, the density at 1 AU is diminished by a factor of $(r_c/a)^2$ at the higher temperature for the same velocity ratio v/U, thereby reducing 900 to a final factor of 225. The theoretical uncertainty in any estimate of the density of the wind is obvious. This should be kept in mind, particularly when contemplating the mass loss represented by the stellar winds of other stars where no direct spectroscopic observations are available (cf. Lamers and Cassinelli, 1999).

In contrast with the density of the solar wind, the wind speed varies more or less in proportion to $T(r)^{1/2}$ and with the geometrical divergence of the expansion. The two outstanding velocity states of the solar wind are the slow dense wind, with velocities in the range 300 – 400 km/s and densities fluctuating about 5 – 10 /cm^3, and the fast tenuous wind, with velocities in the range 500 – 800 km/s and densities of 1 – 3 /cm^3 (Cranmer, 2002). Note that the ion flux does not change much, lying in the general vicinity of $2 - 3 \times 10^8$ ions/cm^3. Theory tells us that the density of the wind is increased when the temperature of the corona, $a < r < r_c$, is increased slightly and diminished when the temperature is decreased slightly. Then the velocity of the wind is controlled in large degree by the temperature in the "afterburner" region, $r > r_c$.

It should also be kept in mind that the pressure of the waves that heat the corona beyond the critical point also contribute momentum to the wind (cf. Belcher, 1971; Alazraki and Clouturier, 1971; Hundhausen, 1972), although it is still not clear to what degree. Then it goes without saying that the precise radial profile of the heat input to the corona is not known, and can be inferred only indirectly from what is known of the ion and electron thermal velocity distribution, the degree of ionization of the heavier atoms in the wind, and the bulk flow velocity $v(r)$.

It also must be appreciated that the radial divergence of the coronal expansion beyond the critical point influences that velocity for a given temperature distribution (Parker, 1958b). The more rapid the divergence, the greater the local acceleration of the gas.

So, in summary, there are many parameters that can be adjusted in constructing a specific hydrodynamic model of coronal expansion and the solar wind. It will require substantially more observational data over an extended range of radius toward the Sun to provide a truly unique model with all the parameters fixed by the observational data The basic difficulty seems to be the complicated temperature structure of the solar corona, with the kinetic temperature for the electrons and the ions arising in different ways. The heavier ions are generally hotter than the hydrogen ions, presumably as a consequence of wave heating. Unfortunately we have no direct quantitative knowledge of the heat input to the corona. As already noted, Martin (1984) and Porter and Moore (1988) have suggested that the expanding portions of the corona, particularly the coronal holes, are heated by the dissipation of plasma waves etc. produced by the microflaring in the region. The effect of wave heating far out into interplanetary space can be seen in the varying anisotropy and near isotropy in the wind at 1 AU, already noted in Sect. 4.9.

The gross magnitude of the heating is readily estimated from the observed mass loss. A unit of mass of coronal matter requires an amount of energy $GM_\odot/a$ to lift it from $r = a$ to $r = \infty$ and an additional energy $v^2/2$ to accelerate it to the solar wind velocity v. Note that $GM_\odot/a = 2 \times 10^{15}$ erg/gm for $a = 7 \times 10^{10}$ cm, corresponding to a gravitational escape velocity of 630 km/s. So in the fast wind the final kinetic energy is about the same as the work done in lifting the gas away from the Sun, for a total of 4×10^{15} erg/gm. The solar mass loss of 1×10^{12} gm/s requires an energy input of 4×10^{27} erg/s. Spreading this over the entire surface of the Sun, $4\pi a^2 = 6 \times 10^{22}$ cm^2, requires 0.7×10^5 erg/cm^2 s. Recall that the outward conductive flow of heat from $r = a$ was of the order of 3×10^3 erg/cm^2 s for the simple case of a static corona at 10^6 K, and three times as much at 1.5×10^6 K. The heat conduction downward into the chromosphere is much larger because the temperature gradient is so much steeper. With a characteristic scale of 2×10^8 cm, one obtains 3×10^6 erg/cm^2 s. However it has been pointed out that the downward heat flow is channeled along the magnetic fields that converge downward into tiny intense magnetic fibrils at the visible surface, typically reducing the cross section by a factor of 10^2 or more and stifling the downward heat conduction by a similar factor. So we can guess that the total heat input to the expanding corona must be at least as large as 10^5 erg/cm^2 s (cf. Withbroe and Noyes, 1977; Withbroe, 1988).

4.16 Magnetic Fields and Streams in the Solar Wind

The magnetic fields in the solar wind, typically $4-6 \times 10^{-5}$ gauss at 1 AU (Ness, Scearce, and Seek, 1964), represent magnetic field stretched out from the regions of expanding corona back at the Sun. These are the weak field regions of the quiet corona, with mean values typically 10 gauss or less. In contrast, the strong bipolar magnetic fields ($\approx$ 100 gauss) of the active corona constrain the corona, preventing expansion. So, in fact, the solar wind comes from only those portions of the solar corona with magnetic fields not in excess of something of the order of 10 gauss. The fast solar wind is a product of the coronal holes, while the slow dense phase of the wind appears to issue from the periphery of active regions of the corona (Wilcox and Ness, 1965; Ness, Hundhausen, and Bame, 1967).

The question now is what magnetic fields we should expect to see in interplanetary space (Parker, 1958b). Magnetohydrodynamics decrees that the expanding corona carries with it whatever magnetic fields are embedded in it, stretching the field out into space as the plasma departs from the Sun. If the Sun did not rotate, the field lines would be radial, along with the radial outflow. The field intensity would decline outward as $1/r^2$, and the field at a large distance r at the angular position (θ, ϕ) would be related to the field back at some radial distance $r = b > a$ near the Sun by

$$B_r(r, \theta, \phi) = B_r(b, \theta, \phi) \left(\frac{b}{r}\right)^2 .$$

However, given the 25 day rotation period of the low latitude portion of the Sun (angular velocity $\Omega \cong 3\times10^{-6}$ /s) the otherwise radial field is twisted into a spiral. Starting at $r = b$ with azimuthal velocity Ωb, conservation of angular momentum provides the azimuthal velocity

$$v_\phi = \Omega \frac{b^2}{r}$$

in interplanetary space. If $b = 3a/2$, with $a = 7 \times 10^{10}$ cm for the radius of the Sun, then $\Omega b \cong 3$ km/s. At 1 AU, where $r \cong 140b \cong 220a$, the result is $v_\phi \cong 0.02$ km/s. This is negligible compared to the wind velocity of 300 km/s or more, so each element of solar wind plasma at 1 AU moves essentially radially in this idealized picture of a steady uniform wind.

Consider two elements of solar wind plasma departing from a point on the Sun ($r = b$) at the polar angle θ_0 and azimuthal angle ϕ_0 at the successive times t and $t+\Delta t$ with fixed radial velocity v. The two points are separated in radial distance by $v\Delta t$ and in azimuth by the angle $\Omega\Delta t$, representing a distance $r\Omega\Delta t \sin\theta$. The polar angle θ, measured from the spin axis of the Sun, remains constant for such radial outflow, so $\theta = \theta_0$. Thus the ratio of the radial separation to the azimuthal separation is

$$\frac{dr}{r \sin\theta \, d\phi} = \frac{v}{\Omega r \sin\theta} .$$

Integrating this relation yields the Archimedean spiral trajectory

$$r - b = \frac{v(\phi - \phi_0)}{\Omega}, \quad \theta = \theta_0$$

for a stream of plasma from the fixed point (b, θ_O, ϕ_0) back near the rotating Sun. The magnetic field entrained in this stream follows the same spiral trajectory, and the radial component of the field at (r, θ, ϕ) out in space connects to the point $(b, \theta, \phi + \Omega r/v)$ back near the Sun. The radial field component is, therefore,

$$B_r(r, \theta, \phi) \cong B_r(b, \theta, \phi + \Omega r/v) \left(\frac{b}{r}\right)^2 .$$

The azimuthal component of the field is $B_\phi = B_r r \times \sin\theta \, d\phi/dr$, so that (Parker, 1958b, 1963, Chap. X)

$$B_\phi(r, \theta, \phi) = B_r(r, \theta, \phi) \frac{\Omega r \sin\theta}{v}$$
$$= B_r(b, \theta, \phi + \Omega r/v) \left(\frac{b\Omega \sin\theta}{v}\right) \frac{b}{r} .$$

Note that the azimuthal field component declines as $1/r$, so that in the asteroid belt and beyond the field is essentially azimuthal. Direct observation in space (Ness, Scearce, and seek, 1964; Burlaga, 1997) confirmed this general spiral pattern, on which there is superimposed strong small-scale fluctuations associated with turbulence in the wind.

This is an appropriate point to note again that the wind velocity, assumed to have a uniform value v in the theoretical discussion, actually varies substantially around the Sun as a consequence of the variation of coronal conditions. Thus, for instance, near the equatorial plane of the Sun, a slow wind with velocity v_1 issuing from $\phi = \phi_1$ may be accompanied by a fast wind with the larger velocity v_2 from $\phi = \phi_2$. The spiral paths are

$$r - b = \frac{v_1}{\Omega}(\phi - \phi_1)$$

and

$$r - b = \frac{v_2}{\Omega}(\phi - \phi_2) .$$

The paths collide (intersect) where

$$r - b = \frac{v_1 v_2 (\phi_2 - \phi_1)}{\Omega (v_2 - v_1)} .$$

Consider the example where $v_1 = 400\,\mathrm{km/s}$ and $v_2 = 600\,\mathrm{km/s}$, with $\phi_2 - \phi_1 = \pi/2$.

The two streams collide at $r - b = 6 \times 10^{13}$ cm = 4 AU (cf. Parker, 1963, p. 152) where the spiral is more nearly azimuthal than radial. The fast stream runs into the rear (sunward) side of the slow stream with a relative velocity of 200 km/s. The magnetic fields in both streams are compressed and there is a forward and backward shock propagating away from the surface of contact. The shocks are sometimes strong enough to accelerate a few of the ions to energies of an MeV or more. These spiral *interplanetary interaction regions* are the principal variation in the distant solar wind (Burlaga, 1997).

4.17 Discussion

An obvious question is the place of origin of the magnetic fields in the solar wind among the diverse patches of field in the quiet and in the active regions back at the Sun. We would also like to understand the cause of the fast and slow phases of the solar wind. Neugebauer, et al. (1998, 2002) note that active regions produce slow wind and, remarkably, substreams of fairly fast wind, whereas the coronal holes produce broad streams of fast wind. More recently Schrijver and Derosa (2003) have carried out an extensive mapping back to the Sun from the magnetic fields observed by SOHO and TRACE near 1 AU. They find that the total magnetic unsigned magnetic flux, i.e. the total number of lines of force, regardless of the direction of the field, extending out through the heliosphere is 4×10^{22} Maxwells. This is about 5 percent of the total unsigned flux through the surface of the Sun during times of maximum magnetic activity, and about 20% during minimum activity. In particular, they find that during solar minimum the magnetic flux in the heliosphere can be traced mostly to the extended polar coronal holes, whereas during solar maximum the field originates from a dozen different "disjoint" regions, including plages and even sunspots.

Then Gloeckler, Zurbuchen, and Geiss (2003) have established the remarkable anti-correlation between the speed of the wind and the temperature of the corona where the wind originated back at the Sun. This puzzling observational result motivated Fisk (2003) to examine the magnetic reconnection between the open magnetic flux, along which the expanding corona escapes from the Sun, and the closed loops of strong bipolar magnetic fields of active regions. He showed that substantial reconnection is expected, leading to the escape of some of the very hot gas from the coronal loops into the surrounding open field, and, he estimates, providing a wind for which v^2 is approximately proportional to $1/T$. This is in agreement with the observed anti-correlation between v and T. Thus we may at last have an understanding of the slow wind associated with magnetic regions (Gloeckler, Zurbuchen, and Geiss, 2003). Fisk's suggestion makes it clear that both slow and fast winds may originate from the magnetically active regions, in agreement with the earlier observational analysis of Neugebauer, et al. (2002).

Finally, looking to the outer reaches of the heliosphere, there is a standing shock in the solar wind far out where the impact pressure of the wind is opposed by the interstellar pressure. Theoretical estimates put the shock somewhere in the vicinity of 100 – 120 AU, and the Voyager 1 spacecraft, crossed the shock in December 2005 at 94 AU. The magnetic field immediately upstream from the shock is of the order of 10^{-2} of the 2×10^{-5} gauss observed at 1 AU, and the density of the wind is of the order of 10^{-4} of the mean value of about 5 ions/cm^3 at 1 AU. These numbers are small compared to the interstellar magnetic field, estimated to be $3 - 5 \times 10^{-6}$ gauss, and local interstellar gas density of about 0.1 ions/cm^3

with an equal number of neutral hydrogen atoms, respectively. It is particle acceleration at the termination shock that is believed to be responsible for the so called *anomalous cosmic rays* observed here in the inner solar system.

The solar wind, then, sweeps out an enormous volume of space, called the *heliosphere* because the region is dominated by the Sun through the dynamical solar wind. The galactic cosmic rays are continually swept back by the magnetic fields carried outward in the solar wind, leading to a variable reduced density deep within the heliosphere (Parker, 1963), as first studied by Forbush, Simpson, and others. The dynamics of the interaction of the heliosphere with the local interstellar wind is an ongoing subject of study (Parker, 1960b, 1963 Chap. XI; Zank and Frisch, 1999; Zank, 1999; Müller, Frisch, and Zank, 2006), with the hope that one day there will be direct observational study with fast moving space craft. For it must be appreciated that the light transit time to 100 AU is 800 minutes, i.e. about 13 h, and Voyager 1 has already been voyaging for more than a quarter of a century. Ordinary chemical rocket power is able to get there but only in times comparable to the length of the entire active career of a professional scientist or interested citizen.

References

Alazraki, G. and P. Couturier 1971, Solar wind acceleration caused by the gradient of Alfven wave pressure, *Astron. Astrophys.* **13**, 380–389

Alfven, H. and P. Carlquist 1967, Currents in the solar atmosphere and a theory of solar flares, *Solar Phys.* **1**, 220–228

Aschwanden, M. 2004, *Physics of the Solar Corona*, Springer, Praxis Publishing, Chichester, UK

Belcher, J.W. 1971, Alfvenic wave pressure and the solar wind, *Astrophys. J.* **168**, 509–524

Biermann, L. 1948, Über die Ursache der chromosphärischen Turbulenz und des UV-Exzesses der Sonnenstrahlung, *Zeit. f. Astrophys.* **25**, 161–169

Biermann, L. 1951, Kometenschweife und solare Korpuskular Strahlung, *Zeit. f. Astrophys.* **29**, 274–286

Biermann, L. 1957, Solar corpuscular radiation and the interplanetary gas, *Observatory* **77**, 109–110

Billings, D.E. 1966, *A Guide to the Solar Corona*, Academic Press, New York

Blackwell, D.E. 1955, A study of the outer corona from a high altitude aircraft of the eclipse of 1954 June 30. I Observational data, *Mon. Not. Roy. Astron. Soc.* **115**, 629–649

Blackwell, D.E. 1956, A study of the outer corona from a high altitude aircraft of the eclipse of 1954 June 30. II Electron densities in the outer corona and zodiacal light region, *Mon. Not. Roy. Astron. Soc.* **116**, 50–68

Burlaga, L.F. 1997, *Interplanetary Magnetohydrodynamics*, Oxford University Press, New York

Busnardo-Neto, J., J. Dawson, T. Kaminura, and A.T. Lin 1976, Ion-cyclotron resonance heating of plasmas and associated longitudinal heating, *Phys. Rev. Lett.* **36**, 28–31

Chapman, S. 1954, The viscosity and thermal conductivity of a completely ionized gas, *Astrophys. J.* **120**, 151–155

Chapman, S. 1959, Interplanetary space and Earth's outermost atmosphere, *Proc. Roy. Soc. London A*, **253**, 462–481

Clauser, F. 1960, in 4th *Symposium on Cosmical Gas Dynamics*, Varenna, Italy

Cranmer, S.R. 2002, Coronal holes and the high-speed solar wind, *Space Sci. Rev.* **101**, 229–294

Cranmer, S.R. 2004, New views of the solar wind with the Lambert *W* function 2004, *American J. Phys.* **72**, 1397–1403

Edlen, J.A. 1942, Die Deuting der Emissionlinien im Spectrum der Sonnen Korona, *Zeit. f. Astrophys.* **22**, 30–64

Elsässer, H. 1954, Die raumliche Verteiling der Zodiak allicht Materie, *Zeit. f. Astrophys.* **33**, 274–290

Fisk, L.A. 2003, Acceleration of the solar wind as a result of reconnection of open magnetic flux with coronal loops, *J. Geophys. Res.* **108**, A(4) 1157

Forbush, S.E. 1937, On the diurnal variation of the cosmic ray intensity, *Terrestrial Magnetism and Atmos. Electricity* **42**, 1–16

Forbush, S.E. 1954, Worldwide cosmic ray variations, 1937–1952, *J. Geophys. Res.* **59**, 525–542

Gloeckler, G., T.H. Zurbuchen, and J. Geiss 2003, Implications of the observed anti-correlation between solar wind speed and coronal electron temperatures, *J. Geophys. Res.* **108** A(4), 1158

Grotrian, W. 1939, Sonnen und Ionosphäre, *Naturwiss.* **27**, 555–563, 569–577

Hundhausen, A. 1972, Coronal Expansion and the Solar Wind, Springer-Verlag, pp. 61–67

Lamers, H.J.G.L.M. and J.P. Cassinelli 1999, *Introduction to Stellar Winds*, Cambridge University Press, Cambridge, UK

Li, X., S.R. Habbal, J.V. Hollweg, and R. Esser, 1999a, Heating and cooling of protons by turbulence-driven ion cyclotron waves in the solar wind, *J. Geophys. Res.* **104**, 2521–2535

Li, X., S.R. Habbal, S.V. Hollweg, and R. Esser, 1999b, Proton temperature anisotropy in the fast solar wind: Turbulence driven dispersive ion cyclotron waves, in *Solar Wind Nine*,

American Institute of Physics, AIP Conference Proceedings 471, ed. S.R. Habbal, R. Esser, J.V. Hollweg, and P.A. Isenberg. pp. 531–534

Lyot, B. 1939, A study of the solar corona and prominences without an eclipse, *Mon. Not. Roy. Astron. Soc.* **99**, 580–594

Martin, S. 1984, in *Dynamical Processes in Quiet Stellar Atmospheres*, ed. S.L. Keil, National Solar Observatory, Sunspot, New Mexico, p. 30

Melrose, D.B. 1995, Current paths in the corona and energy release in solar flares, *Astrophys. J.* **451**, 391–401

Meyer, P., E.N. Parker, and J.A. Simpson 1956, Solar cosmic rays of February 1956 and their propagation through interplanetary space, *Phys. Rev.* **104**, 768–783

Müller, H.R., P.C. Frisch, and G.P. Zank, 2006, Heliospheric response to different possible interstellar environments, *Astrophys. J.* **647**, 1491–1505

Ness, N.F., A.J. Hundhausen, and S.J. Bame 1967, Observations of the interplanetary medium: Vela 3 and Imp 3, *J. Geophys. Res.* **76**, 6643–6665

Ness, N.F., C.S. Scearce, and J.B. Seek 1964, Initial results of the IMP 1 magnetic field experiment, *J. Geophys. Res.* **69**, 3531–3569

Neugebauer, M. and C.W. Snyder 1964, Mariner 2 measurements of the solar wind, in *The Solar Wind*, ed. R.J. Mackin and M. Neugebauer, Pergamon Press, Oxford, pp. 3–21

Neugebauer, M. and C.W. Snyder 1966, Mariner 2 observations of the solar wind: I Average properties, *J. Geophys. Res.* **71**, 4469–4484

Neugebauer, M. and C.W. Snyder 1967, Mariner 2 observations of the solar wind: II Relation of plasma properties to the magnetic field, *J. Geophys. Res.* **72**, 1823–1828

Neugebauer, M., et al. 1998, Spatial structure of the solar wind and comparisons with solar data and models, *J. Geophys. Res.* **103**, 14587–14599

Neugebauer, M., P.C. Liewer, E.J. Smith, R.M. Skoug, and T.H. Zurbuchen 2002, Sources of the solar wind at solar activity maximum, *J. Geophys. Res.* **107**, A(12), 1488

Parker, E.N. 1957, Newtonian development of the dynamical properties of ionized gases of low density, *Phys. Rev.* **107**, 924–933

Parker, E.N. 1958a, Dynamical instability in an anisotropic ionized gas of low density, *Phys. Rev.* **109**, 1874–1876

Parker, E.N. 1958b, Dynamics of the interplanetary gas and magnetic field, *Astrophys. J.* **128**, 644–670

Parker, E.N. 1960a, The hydrodynamic theory of coronal corpuscular radiation and stellar winds, *Astrophys. J.* **132**, 821–866

Parker, E.N. 1960b, The stellar wind regions, *Astrophys. J.* **134**, 20–27

Parker, E.N. 1963, *Interplanetary Dynamical Processes*, Wiley, Interscience, New York

Parker, E.N. 1964a, Dynamical properties of stellar coronas and stellar winds. I. Integration of the momentum equation, *Astrophys. J.* **139**, 72–92

Parker, E.N. 1964b, Dynamical properties of stellar coronas and stellar winds. II Integration of the heat flow equation, *Astrophys. J.* **139**, 93–121

Parker, E.N. 1964c, The perturbation of interplanetary dust grains by the solar wind, *Astrophys. J.* **139**, 951–958

Parker, E.N. 1965, Dynamical theory of the solar wind, *Space Sci. Rev.* **4**, 666–708

Parker, E.N. 1996a, The alternative paradigm for magnetospheric physics, *J. Geophys. Res.* **101**, 10587–10625

Parker, E.N. 1996b, Comment on "Current paths in the corona and energy release in solar flares", *Astrophys. J.* **471**, 489–496

Parker, E.N. 2000, Newton, Maxwell and magnetospheric physics, in *Geophysical Monograph 18*, American Geophysical Union, pp. 1–10

Parker, E.N. 2001, A history of the solar wind concept, in *The Century of Space Science*, Kluwer Academic Publishers, the Netherlands

Parker, E.N. 2007, Conversaations on electric and magnetic fields in the cosmos, Princeton University Press, Princeton

Parks, G.K. 2004, Why space physics needs to go beyond the MHD box, *Space Sci. Rev.* **113**, 97–125

Petschek, H.E. 1958, Aerodynamic dissipation, *Rev. Mod. Phys.* **30**, 966–972

Pierce, J.R. 1949, Possible fluctuations in electron streams due to ions, *J. Applied Phys.* **19**, 231–236

Porter, J.G. and R.L. Moore, 1988, in Proceedings of the Sacramento Peak Summer Symposium, ed. R.C. Altrock, National Solar Observatory, Sunspot, New Mexico, p. 30

Schrijver, C.J., and M.L. Derosa 2003, Photospheric and heliospheric magnetic fields, *Solar Phys.* **212**, 165–200

Simpson, J.A. 1951, Neutrons produced in the atmosphere by cosmic radiations, *Phys. Rev.* **83**, 1175–1188

Simpson, J.A., W. Fonger, and S. Treiman 1953, Cosmic radiation intensity-time variations and their origins. I Neutron intensity variation method and meteorological factors, *Phys. Rev.* **90**, 934–950

Snyder, C.W. and M. Neugebauer 1964, Interplanetary solar wind measurements by Mariner II, *Space Res.* **4**, 89–113

Spicer, D.S. 1982, Magnetic energy storage and conversion in the solar atmosphere, *Space Sci. Rev.* **31**, 351–435

Vasyliunas, V.M. 2001, Electric field and plasma flows, What drives what? *Geophys. Res. Lett.* **28**, 2177–2180

Vasyliunas, V.M. 2005, Time evolution of electric fields and currents and the generalized Ohm's law, *Ann. Geophys.* **23**, 1347–1354

Velli, M. 2001, Hydrodynamics of solar wind expansion, *Astrophys. Space Sci.* **277**, 157–167

Weyman, R. 1960, Coronal evaporation as a possible mechanism for mass loss from red giants, *Astrophys. J.* **132**, 380–403

Wilcox, J.M. and N.F. Ness 1965, Quasi-stationary corotating structure in the interplanetary medium, *J. Geophys. Res.* **70**, 5793–5805

Withbroe, G.L. 1988, The temperature structure, mass, and energy flow in the corona and inner solar wind, *Astrophys. J.* **325**, 442–467

Withbroe, G.L. and R.W. Noyes 1977, Mass and energy flow in the solar chromosphere and corona, *Annual Rev, Astron. Astrophys.* **15**, 363–387

Zank, G.P. 1999, Interaction of the solar wind with the local interstellar medium: A theoretical perspective, *Space Sci. Rev.* **89**, 413–688

Zank, G.P. and P.C. Frisch 1999, Consequences of a change in the galactic environment of the Sun, *Astrophys. J.* **518**, 965–973

5 Coronal Mass Ejection

Peter J. Cargill and Louise K. Harra

This chapter reviews the properties of Coronal Mass Ejections (CMEs) at the Sun and in the interplanetary medium. CMEs can eject between 10^{12} and 10^{13} kg of plasma from the Sun at speeds up to 2500 km/s. They are driven outwards by magnetic forces associated with the coronal magnetic field. In the interplanetary medium, they can have an organised magnetic structure and fast CMEs are often accompanied by strong shock waves. In particular, periods of southward interplanetary magnetic field in excess of a few hours occur in CMEs and lead to major geomagnetic disturbances.

Contents

Peter J. Cargill and Louise K. Harra, Coronal Mass Ejection.
In: Y. Kamide/A. Chian, Handbook of the Solar-Terrestrial Environment. pp. 117–132 (2007)
DOI: 10.1007/11367758_5

5.1 Introduction

Coronal Mass Ejections (hereafter CMEs) are a major form of solar activity. They involve the expulsion of large amounts of plasma and magnetic flux at high speeds from the solar corona into the solar wind, are believed to be responsible for the acceleration of coronal ions to high energies, and their manifestation in the solar wind, the Interplanetary Coronal Mass Ejection (hereafter ICME), is responsible for many major disturbances to the Earth's space environment. CMEs are thus not only important from the viewpoint of solar physics, but are a very important aspect in the study of solar terrestrial relations.

A CME can be observed in the outer solar atmosphere using a coronagraph, and in the interplanetary medium by spacecraft equipped to measure the in situ magnetic fields and plasmas. Global interplanetary measurements are also becoming possible using all-sky cameras and emission at radio wavelength. It is useful to state the definition of a CME due to Hundhausen (1993): "an observable change in coronal structure that (1) occurs on a timescale between a few minutes and several hours and (2) involves the appearance of a new discrete, bright white-light feature in the coronagraph field of view". An example is shown in Fig. 5.1. This is one of many observed by the Large Angle Spectrometric Coronagraph (LASCO) on the Solar and Heliospheric Observatory (SOHO) spacecraft.

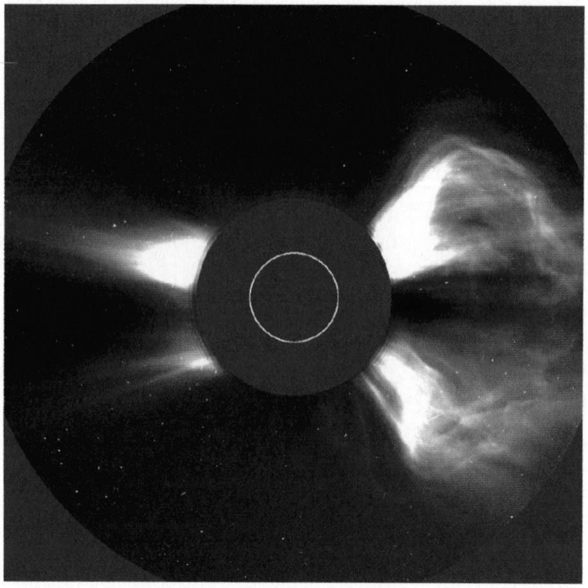

Fig. 5.1. A CME seen by the LASCO coronagraph on the SOHO spacecraft. The *central circle* marks the location of the solar disk, and the *outer circle* is the extent of the coronagraph's occulting disk. The CME is on the *right* of the picture, and is moving away from the Sun with a significant component of motion perpendicular to the Sun–Earth line. [Figure courtesy of the LASCO consortium]

The above definition is fairly rigorous, but does not address some important points. The "appearance of … features" can be taken to imply the motion of coronal plasma outwards: in other words an ejection of mass. Indeed up to 10^{13} kg can be ejected at speeds up to 2000 km/s. The role of the magnetic field in CMEs observed at the Sun is difficult to quantify. An outward force must act on the CME to enable it to escape from the Sun's gravity, and this is almost certainly the Lorentz force associated with coronal currents. But Zeeman splitting is practically undetectable in the corona due to thermal line broadening, so measurements of the field are difficult.

In the interplanetary medium, ICMEs are often (but not always) recognisable by the presence of a coherent magnetic structure lasting up to a day at 1 AU, and a preceding shock wave. An example is shown in Fig. 5.2. It is important to have observations both of the onset of a CME and its manifestation in the interplanetary medium. As we will see, such joint observations have proved to be of great importance.

Thirty five years after the first detection, one might imagine that CMEs and ICMEs are well understood. As we will see, there exists a vast range of measured parameters: speed, occurrence rate, latitudinal distribution, ICME speed, magnetic field strength etc. But major questions associated with CMEs persist. One can summarise them as follows:

- What is the pre-eruption configuration of the coronal magnetic field?
- What causes the CME to begin and how is the CME accelerated?
- What is the relationship (if any) between CMEs and solar flares?
- How does the CME interact with the solar wind to give the observed ICME and how does this depend on the CME properties at the Sun?

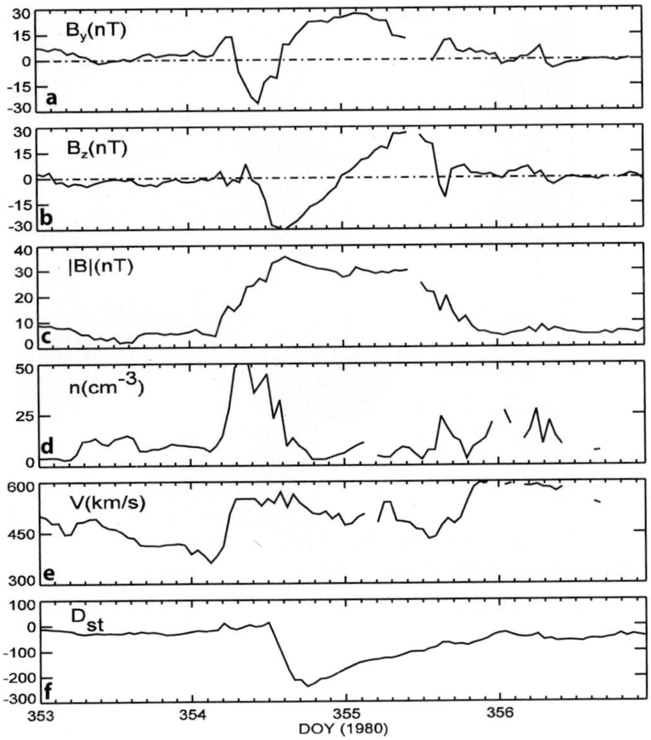

Fig. 5.2a–f. An ICME measured by the ISEE-3 spacecraft. The *panels* show: the *y*-, *z*-magnetic field components in Geocentric Solar Ecliptic (GSE) coordinates, the magnetic field magnitude, the plasma density and velocity, and the geomagnetic Dst index. [Data courtesy of OMNIweb]

In the final section of the chapter we discuss how future observations and theoretical modelling may lead to their resolution. Recommended review papers are Hundhausen (1999) and Gopalswamy (2005).

5.2 CMEs at the Sun

At the present time, the importance of CMEs is taken for granted, yet this is a recent state of affairs. While there is speculative evidence for the ejection of material from the corona during a 19th century eclipse, and the phenomenon of solar prominence eruption (closely related to CMEs) has been known since the early 20th century, serious CME research dates back to 1973 and observations made from the Skylab observatory. In the interplanetary medium ICMEs have been detected since the start of the space age, albeit without on occasions it being realised what they were.

Detection of CMEs in white light require a coronagraph. Such an instrument positions an occulting disk across the solar disk, so that very faint Thomson-scattered light from the outer corona is measurable. Observations from space are desirable so as to extend the field of view to large distances from the Sun, as well as to ensure that simultaneous measurements of the associated high-temperature corona are made. Other methods of CME detection are described in Hudson and Cliver (2001).

The first space-based observation of a CME was made in 1971 and Skylab provided the first opportunity for an extended survey of the outer corona. Four space-based coronagraphs have operated between 1973 and the present time. Skylab made observations for a total of 6 months in 1973 and 1974 near the minimum of the solar cycle. Its coronagraph had a field of view between 1.5 and $5R_S$. The Solwind coronagraph on the P78-1 spacecraft operated between 1978 and 1985 (the maximum and declining phase of the cycle) and had a field of view from 2.5 to $10R_S$. The Solar Maximum Mission (SMM) operated in 1980 (the maximum) and 1984–1989 (the minimum and ascending phase of the cycle) and its coronagraph had a field of view from $1.6 - 6R_S$. The three coronagraphs that comprise the Large Angle Spectoscopic Coronagraph (LASCO C1, C2 and C3) on the Solar and Heliospheric Observatory

(SOHO) spacecraft have operated since 1996 (almost a complete solar cycle) and have fields of view from 1.1 – 3, 1.5 – 6 and 4 – 30R_s respectively. Ground-based facilities such as those on Mauna Loa have also been of importance.

5.2.1 Properties

Morphology

As seen by coronagraphs, CMEs come in all shapes and sizes, and often resist a clear categorisation. Historically, a different paradigm of a CME often arose from each instrument, presumably as a result of different sensitivity, field of view etc. Building on the results from Skylab and Solwind, anaylsis of data from instruments on SMM and SOHO have begun to provide a basic picture of CMEs.

Probably the most important morphological paradigm to emerge has been the "three part CME structure" (e.g. Low, 1996; Hundhausen, 1999). This was noted in some Skylab observations, but was fully appreciated in the CMEs of August 5 and 18 1980, and the latter is shown in Fig. 5.3. The pre-eruption configuration consisted of a prominence, above which was a dark region, known as the prominence cavity, with both being located at the base of a large helmet streamer. As the eruption proceeded, a three-part structure corresponding to these regions was seen moving away from the Sun. Strong evidence for this configuration has been found in numerous other SMM and SOHO observations.

It was suggested by Low (1996) that the prominence cavity was a large magnetic flux rope, with the prominence being suspended at the bottom. Other than the prominence, the flux rope contains only low-density plasma, and so appears as a dark region (cavity) in coronagraph images. While the cause of these eruptions is unknown, Hundhausen (1993) noted that the pre-eruption streamers undergo an inflation that terminates in an eruption. Of course it should also be pointed out that a very significant number of CMEs do not have a 3 part structure. The literature of Skylab, P78-1 and SMM results contains numerous references to "loops, clouds, bubbles, halos and spikes", and the line of sight along which one views the CME also is of importance in ascertaining its three-dimensional geometry.

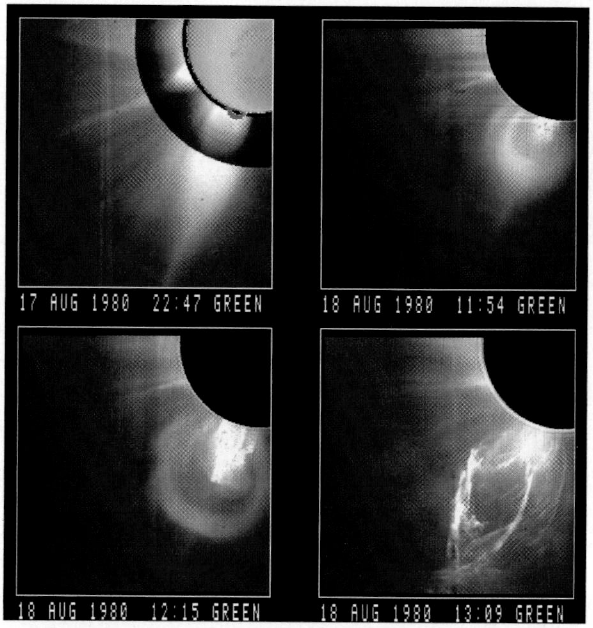

Fig. 5.3. The CME of August 18, 1980. The *four panels* show the pre-eruption helmet streamer, its initial eruption with a trailing cavity, the full three-part structure with the leading bright edge, the cavity, and the prominence, and finally the post-eruption prominence material. [Figure courtesy of HAO]

Rate, Location and Size

The rate of CMEs depends on the phase of the solar cycle. Events occur at a rate of roughly 0.5 per day at solar minimum and between 2 and 5 a day at maximum (Hundhausen, 1993; St Cyr et al., 2000; Yashiro et al., 2004).

CMEs typically come from near the equator at solar minimum, and occur over a wider range of latitudes at solar maximum (Hundhausen, 1993; St Cyr et al., 2000; Yashiro et al., 2004), on occasions occurring north or south of 60°. Hundhausen (1993) noted that the distribution of latitudes does not correspond to those of sunspots or flares, but does correspond to larger structures such as prominences and streamers.

CMEs are vast structures. Hundhausen (1993) noted that the distribution of angular widths peaks at around 40°, with a tail extending up to 100°. St Cyr et al. (2000) and Yashiro et al. (2004) found a similar result, but with (a) a slight increase near the maximum of the solar cycle and (b) a more extensive tail, with events having widths between 100 and 360°. It is almost certain that these wide CMEs are either Earthward or anti-Earthward-directed events filling a large part of the sky. These are the halo CMEs that have become a familiar aspect of the SOHO results.

Mass, Velocity, Acceleration and Energy

Estimates of the average mass and energy of a CME has varied little between the various data sets. Hundhausen (1999) quotes 4×10^{12} kg for Solwind and 2.5×10^{12} kg for SMM events. A survey of LASCO events by Vourlidas et al. (2000) suggests a similar number. If we recall that the total solar wind mass loss is about 10^{14} kg per day, CMEs are thus a small contributor to the total solar mass (and angular momentum) loss.

CME velocities span a wide range of values, and are usually defined as the velocity in the plane of the sky. SMM results showed a peak at about 350 km/s, but with a distribution extending below 100 km/s and up to 2000 km/s (Hundhausen, 1999). Extensive studies of CME speeds have been carried out using the LASCO database. St Cyr et al. (2000) examined the speed of 640 CMEs occurring between 1996 and 1998 and found a similar distribution of speeds. However, Sheeley et al. (1999) claimed that CMEs fell into two different types: gradual CMEs with a speed in the range 400 – 600 km/s and apparently associated with prominence eruptions and fast CMEs with speeds in excess of 750 km/s, and associated with flares.

While a combination of ground-based and SMM observations showed outward acceleration in the inner corona (St Cyr et al., 1999), it is only the high sensitivity and large field of view of the SOHO coronagraphs that has made possible the study of CME acceleration to large radial distances. However, St Cyr et al. (2000) were able to ascertain acceleration reliably in only 17% of CMEs. When averaged over the field of view of the outer two LASCO coronagraphs, they found accelerations in the range 1.4 – 49 m/s^2. They found no cases of deceleration. Sheeley et al. (1999) found accelerations of similar magnitudes to St Cyr et al., but also found cases where deceleration occurred. These latter events were mostly associated with fast CMEs moving perpendicular to the plane of the sky, whereas fast CMEs in the plane of the sky travelled at roughly constant velocity. Sheeley et al. suggested that CMEs moving perpendicular to the plane of the sky were in fact being decelerated at much larger radial distances ($> 50R_s$), which would not be seen in the field of view when the CME is in the plane of the sky. Yashiro et al. (2004) note that slow CMEs tend to be accelerated and fast ones decelerated while Jing et al. (2005) have suggested that the acceleration occurs close to the Sun, inside $1.5R_s$.

Finally, in order to assess the processes responsible for the CME motion, estimates of the energies are needed. The kinetic and potential energies are straightforward to obtain from coronagraph data. The three data sets mentioned above give an average kinetic energy of close to 3×10^{23} J: SMM gave an average potential energy of 5.4×10^{23} J, and LASCO somewhat less (Vourladis et al., 2000). However, in view of the fact that CMEs are almost certainly driven by forces associated with the magnetic field, it is the magnetic energy that is of major interest. As noted above, direct measurements of coronal magnetic fields are difficult.

5.2.2 What Causes CMEs: Observational Evidence

When discussing the solar origin of CMEs, it is essential to address their relationship with solar flares. A close association cannot be debated: as long ago as 1859 Carrington noted that a large solar flare (the first ever observed) was associated with a major geomagnetic storm, presumably due to a CME interacting with the terrestrial magnetosphere. In the next 120 years, the relative ease of making observations of solar flares, which were known to be associated with significant energy release at the Sun, led to a picture of solar flares as the main cause of geomagnetic activity.

This picture was challenged in important papers by Harrison (1986) and Gosling (1993), the latter viewing this explanation of geomagnetic activity as the "Solar Flare Myth", and argued that CMEs were the most important influence. The debate over Gosling's contention has continued, but a present-day consensus is that CMEs, flares, filament (or prominence) eruptions, and other forms of energy release are closely related.

Of great importance in the last 15 years has been a series of spacecraft with instruments observing the Sun over a range of wavelengths between the optical and γ-ray. An overview of many of these observations is the aim of the following subsections.

Eruptive Flares

Eruptive solar flares are a logical place to begin this discussion since historically they represent one of the earliest efforts to address solar eruptions (e.g. Kopp and Pneuman, 1976; Priest and Forbes, 1999). These authors proposed that a simple, bipolar magnetic configuration is opened up through an eruption of the pre-flare corona. The oppositely-directed open field lines then return to a closed state through the magnetic reconnection process (Fig. 5.4), with associated plasma heating leading to the formation of post-flare loops which can be observed for many hours.

In this scenario the transient referred to in Fig. 5.4 which opens up the magnetic field is most likely to be a filament eruption. [Filaments and their eruption are discussed more fully in Sect. 5.2.2 (p. 123)]. If we identify the filament eruption as being the CME, then the CME precedes (and indeed causes) the flare.

In reality things are more complex. While there are strong links between such long duration flares and CMEs, finding the source of a CME in such events is difficult. In addition, it has been observed that the association between flares and CMEs was not limited to long duration events, with short duration and low intensity flares (level C or less) also showing an association. Indeed, over half of CME-associated flares have durations of <2 h, and the flare and CME onsets can occur within 10 minutes of each other. However, if a long duration flare has occurred, one can be reasonably confident that its location is also the source region of the CME (e.g. Kahler, 1992).

It is clear that magnetic reconnection plays an important role in the above CME scenario. However, other CME onset scenarios invoke coronal reconnection as a *cause* of the filament eruption. The search for observational evidence of reconnection has thus been an important goal of the Yohkoh, SOHO, Transition Region and Coronal Explorer (TRACE) and Ramaty High Energy Solar Spectroscopy Imager (RHESSI) missions. Evidence is sought from both the form of the coronal topology, as outlined by the observed plasma structures, and the mass motions expected during reconnection. Hard X-ray images of a sharply-pointed cusp-like structure have been seen in the loop-top regions of both eruptive and compact (confined) flares (e.g. Masuda et al., 1995). This geometry is anticipated in many theories of reconnection, for example as shown in the right hand panel of Fig. 5.4.

Jets (or plasmoids) have been observed over a range of scales in the corona with velocities between 40 and 400 km/s (Ohyama and Shibata, 1998). Moving dark voids have also been observed in the late phases of a large flare, moving towards the solar surface with velocities of between 45 and 500 km/s (McKenzie 2000). Although in both cases the velocities are lower than the typical expected coronal Alfvén speed of at least 1000 km/s, the jets and voids are characteristic of the outflow regions from reconnection sites. However, evidence of the required

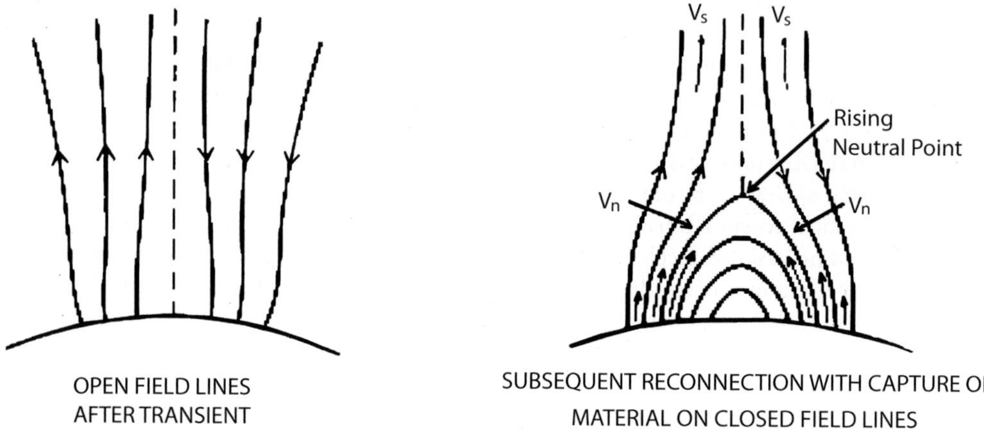

Fig. 5.4. The model of Kopp and Pneuman (1976). A transient opens up the magnetic field lines. They reconnect releasing energy, which produces a flare. The reconnection point is seen to 'rise' with time

reconnection-associated inflows has proved difficult to pin down.

Not all eruptive flares show features consistent with the "classical" picture of Fig. 5.4. Some are largely confined by the coronal magnetic field with no outflows. If no CME was observed then the flare was also without an ejection. Although very high intensity flares are often related to CMEs, this correlation can fail for some magnetic configurations. Green et al. (2002) analyzed an X1.2 flare caused by the interaction between two pre-existing loops in the corona. During the flare outward motion was seen, but was brought to a halt by the overlying magnetic field. Andrews (2003) found that 40% of M-class flares did not have an associated CME.

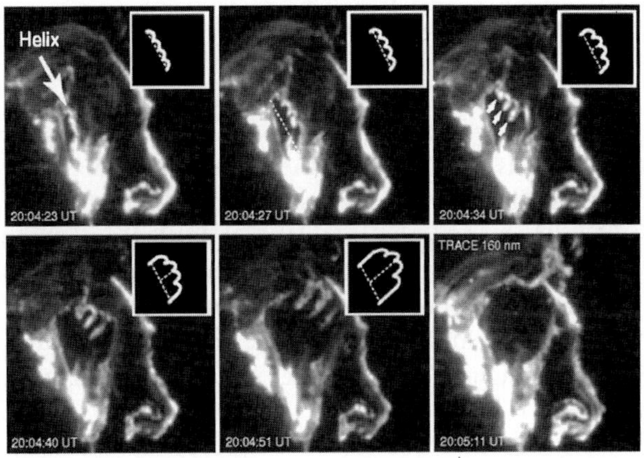

Fig. 5.5. The helical structure of an eruption seen in a flare on the 15th July 2002. The observations were made in the 160 nm band from TRACE. Gary and Moore (2004: Fig. 1)

Filament Eruptions

We argued above that filament eruptions were a plausible cause of the eruptive flare, and so filaments are of considerable importance for understanding the solar origin of CMEs. [Filaments and prominences are the same phenomena. When seen on the disk they appear as dark features as they are cooler than the surrounding plasma and are known as filaments. When they are seen in profile on the limb they are known are prominences.]

Filaments are structures that can suspend cool ($< 10^4$ K) plasma above the surface of the Sun by magnetic forces. They tend to have a complex and twisted magnetic field, and so can store considerable magnetic energy. This twisted complexity can be inferred by the direction of plasma flows when a filament erupts: Evidence of twist can be seen in blue and then red shifts as a filament lifts off from the Sun. Evidence for twist is sometimes seen in images: Fig. 5.5 (Gary and Moore, 2004) shows helical structure (implying twist) in an eruption. A helical structure such as this is what forms the central core of the three-part CME structure discussed earlier.

The scale of filaments encompasses a wide range between structures extending a considerable fraction of a solar radius, to miniature ones. They are seen at all latitudes – even in the polar regions, and all can erupt. Eruptions can be divided into those that occur in the quiet Sun or in active regions. Those in active regions tend to be fast, reaching speeds of hundreds of km/s. The eruptions in the quiet Sun tend to be a lot slower with speeds reaching only tens of km/s or even slower for a filament eruption in a north polar crown region. However, quiet and active region eruptions have similar physical processes but are occurring on difference scales. The magnetic field strength is one of the biggest differences.

For a filament eruption to occur, there must be some process that destabilizes it. In a study of over 100 filament eruptions, Jing et al. (2004) showed that in 68% of their cases destabilization of the filament was strongly associated with new magnetic flux emerging through the photosphere. This is consistent with the earlier result that eruption is more likely if the filament is located near emerging magnetic field (e.g. Feynman and Martin 1995). Alternatively, flare-like brightenings have been found to occur underneath a rising filament, perhaps consistent with reconnection there. This could be a trigger for the eruption itself or just a secondary effect.

However, erupting filaments do not always fit the standard model described earlier, with differences in both structure and timing. Gary and Moore (2004) described a case when quadrapolar reconnection occurred initially, removing the overlying magnetic field, and hence allowing the filament to erupt. Reconnection thus occurs first, as proposed in the "breakout" model of Antiochos et al. (1999). There is also substantial evidence that filaments begin to show dynamical changes at least 2 h before an eruption occurs.

The Importance of the Magnetic Configuration

We have noted that filaments are magnetically complex structures, so the next issue that needs to be examined is the nature of the pre-eruption magnetic field. An important clue may lie in observations of S-shaped structures (sigmoids) that are believed to be indicative of twist in a magnetic structure. Twisted magnetic fields contain electromagnetic energy associated with field-aligned currents, and such structures may be magnetically unstable, and so susceptible to eruption.

Sterling and Hudson (1997: Fig. 5.6) showed an excellent example of a sigmoid which erupts and subsequently relaxed to a cusp-shape following a flare and CME. Sigmoids were first investigated by Rust and Kumar (1996) who found that many large transient X-ray brightenings associated with CMEs were S-shaped. This was interpreted as being due to a MHD helical kink instability that arises due to excessive twist in a magnetic field. It has been found that S-shaped structures are more likely to erupt. However predicting when exactly this will occur, and knowing how much twist is required before an eruption is still not known. It is also difficult to determine due to line of sight effects whether a region is really twisted.

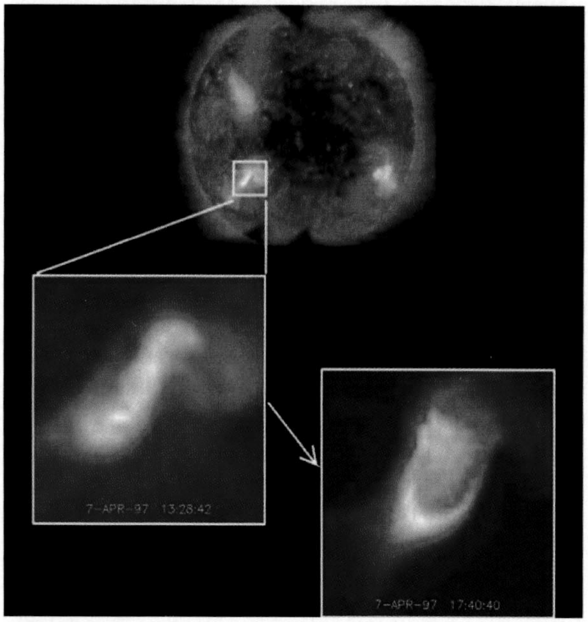

Fig. 5.6. A full Sun Yohkoh X-ray image showing the location of an S-shaped loop system. A close-up view before the eruption is shown *on the left* and a post-eruption image *on the right*. The S-shaped, twisted loop erupts and returns to a cusp shaped feature indicative of a relaxed, potential state. From Sterling and Hudson (1997)

An important conserved quantity associated with twisted magnetic fields is magnetic helicity, which also describes magnetic linkage and knottedness. Helicity can be generated in the corona by differential rotation, although case studies have suggested that this was inadequate. In particular, Green et al. (2002) studied a CME-prolific active region for 5 solar rotations and showed that differential rotation could not provide enough helicity in this case to balance that measured in the CMEs. Instead, it was postulated that the required helicity is provided by twisting fields in the sub-photospheric part of the magnetic flux system forming the active region.

CMEs are thought to be a means for the removal of magnetic helicity from the Sun into interplanetary space (e.g. Rust and Kumar, 1996), so playing an important part in both the evolution of active regions as well as the reversal of the polarity of the magnetic field over a solar cycle. Démoulin et al. (2002) studied an active region over six rotations and showed that flares and CMEs occurred during the early stages of the active region evolution due to the interaction of the emerging flux with the pre-existing flux, while CMEs continued after flaring dies down in the 3rd rotation. The late CMEs may, for this relatively magnetically simple active region, occur when magnetic shear accumulates due to differential rotation, with the CME releasing excess shear and helicity.

The importance of helicity was further emphasized by the analysis of over 100 large intensity flares by Nindos and Andrews (2004). They showed that active region helicity tends to be smaller when CMEs are not produced, although Wang et al. (2004) showed that emerging flux with the opposite sense of helicity is also important in initiating CMEs.

The sign of the magnetic helicity also provides clues to the orientation of the magnetic field in a CME, as well as about the form of the CME at 1 AU. A recent study of Fazakerley et al. (2005) found that the orientation and direction of active region loops involved in a CME was consistent with that found by the time the CME reached the Earth as measured by Cluster spacecraft.

Other Coronal Signatures of CMEs

Coronal Dimming

One of the best signatures for CME onset is that of coronal dimming – that is a loss of coronal plasma, whose signatures can persist for many hours. It is a common method of determining the CME launch location. Although dimming can be seen in some events by eye, it is normally observed by taking difference images. Observations made in soft X-rays by the Yohkoh spacecraft showed dimming related to the presence of long duration flares at the limb (e.g. Hiei et al., 1993). Dimming can also be observed on the disk and a good example associated with a halo CME is described by Sterling and Hudson (1997) who estimated the mass loss as 10^{11} kg. This is less than the typical CME mass inferred from coronagraph observations, suggesting that at least part of the mass loss can be detected by this method. Observations of dimming often demonstrate that plasma removal is on a global scale. In many cases the dimming is transequatorial, and is observed at large distances from the eruption site.

Although dimming is commonly used to determine the source region of CMEs, some caution must be exercised, especially when using narrow-band instruments. This is because a reduction in coronal plasma seen in, for example, the 171 Å images from EIT or TRACE could also be due to a change in temperature. The method of generating difference images is also important, with running and fixed difference images producing very different results. However, observations of dimming using spectroscopic techniques can confirm a density reduction and outflowing plasma (e.g. Harra and Sterling, 2001).

Coronal Waves

The EUV imaging Telescope (EIT) on SOHO has observed a phenomenon that appears to be a wave traveling across the solar disk with speeds of a few hundreds of km/s comprised of a bright front and trailing dimming region and a CME is nearly always observed at the same time (e.g. Biesecker et al., 2002). One explanation was that these are a coronal manifestation of the familiar Moreton waves, seen in H_α, and believed to be due to a shock wave emanating from a flare site. While the speed of the coronal waves are slow, (a few hundred km/s), this is not out of the range of velocities observed in H_α. An alternative explanation put forward by Delaneé (2000) was that the coronal wave was actually a CME lifting off the disk with the wave being the stretching or opening of closed magnetic field lines in response to an erupting filament. Indeed the first coronal wave observed by TRACE was consistent with this scenario (Harra and Sterling, 2003: Fig. 5.7). One of the main difficulties in understanding 'coronal waves' is the poor time cadence of the EIT instrument. Sometimes the wave front is seen in only one image. What we do know is that they are closely associated with CMEs and hence are a phenomena that should be concentrated on in the future with missions such as STEREO, Solar Dynamics Observatory (SDO) and Solar-B.

Figure 5.7 shows the difference images of this event, with the impact of the disturbance on the loops in the northern active region being apparent. A filament eruption took place which appeared to have a similar velocity as the main wave front.

Large-Scale Interconnecting Loops

CMEs are large-scale structures, and it is important that when we try to understand them we look towards the global response of the corona to an eruption of plasma. A significant fraction of all active regions possess trans-equatorial loops. An investigation by Tsuneta (1996) found that there is evidence for magnetic reconnection between the two active regions, and the overall structure of the large trans-equatorial loop changes. Further evidence of reconnection in these large structures with rising loops, a cusp shaped structure and cool loops lying underneath hot loops has also been observed. In addition a transequatorial loop has been observed to disappear and forms part of a CME (Khan and Hudson, 2000). It was determined that a flare shock wave caused the destabilisation of this large-scale loop.

5.2.3 Theoretical Ideas

A successful theoretical model for a CME must explain how a pre-existing coronal structure gains enough kinetic energy to lift its mass out of the Sun's gravitational field and into interplanetary space. In addition, since some CMEs are "fast": i.e. kinetic energy ≫ gravitational

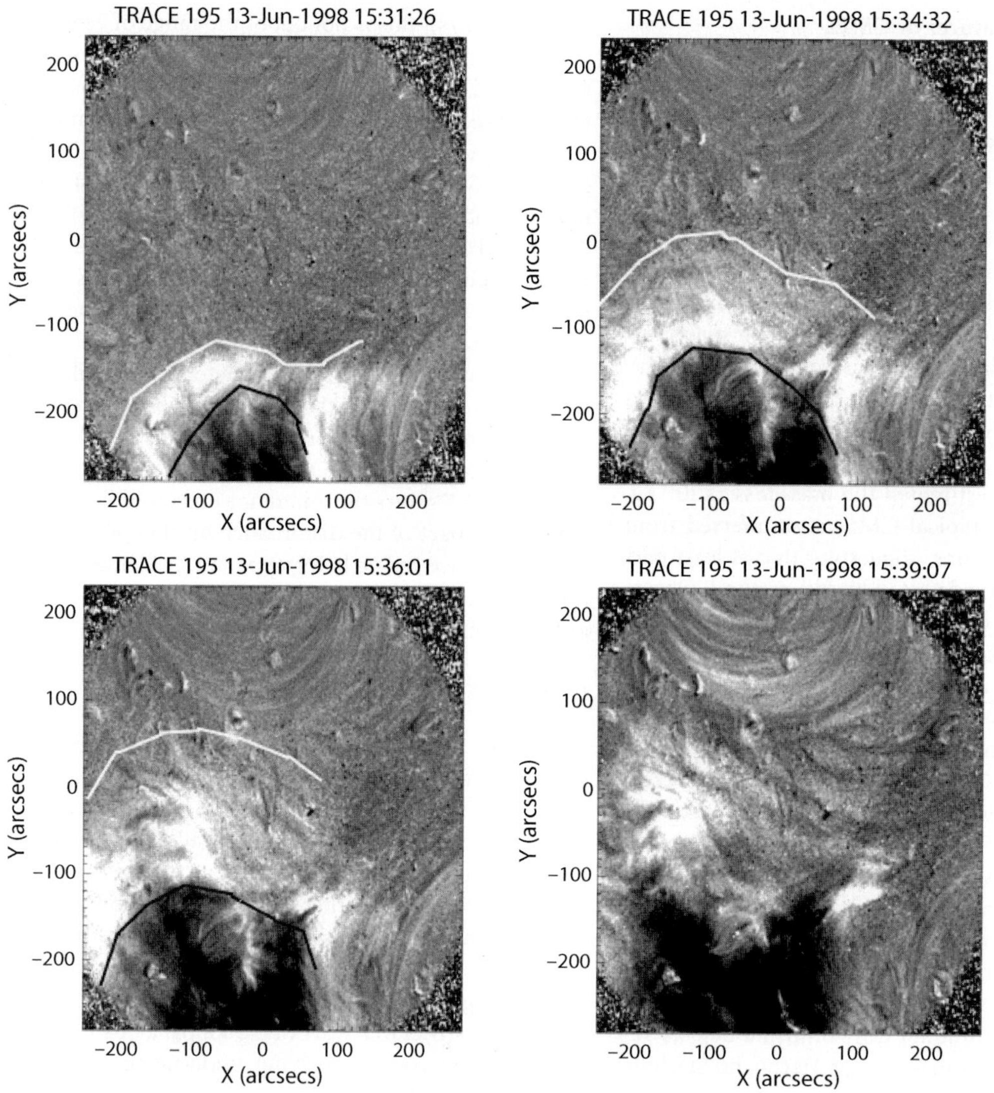

Fig. 5.7. This sequence of images shows percentage difference images highlighting the bright wave front of a coronal wave. The *outer white line* shows the approximate location of the outer edge of the coronal wave. From Harra and Sterling (2003)

potential energy, it is not enough to just account for escape from the Sun: one needs to account for "escape with style".

Although ideas were developed in the 1970s that purported to account for CMEs as the consequence of a coronal point explosion (e.g. a flare), this scenario does not involve the expulsion of plasma and magnetic field from the Sun, and so is not consistent with current knowledge. Instead, the major emphasis is on the development of models involving the eruption of large coronal structure such as filaments and overlying helmet streamers (see Zhang and Low (2005) for a review).

As discussed in Sect. 5.2.2 (p. 122), the generic ideas being discussed today date back to the 1960s when models of eruptive (or two-ribbon) flares involving the "opening up" of the coronal field to interplanetary space were first proposed. The idea is that, given a pre-existing coronal structure, slow motions of the photosphere (under 1 km/s) gradually inject energy into the corona, leading to a magnetically stressed state. Since the coronal Alfvén speed is likely to be well in excess of 1000 km/s, the corona evolves through a series of equilibrium in response to these photospheric motions. However, there comes a time when equilibrium is no longer possible, and an eruption occurs. This idea received a major setback in the work of begun by Aly (1984) who showed that such an open field was in fact in a higher energy state than the original magnetic configuration, and so it was difficult to see energetically how such an eruption could happen.

While Aly's model is quite simple, it has proved to be robust. However, observations from SMM suggested that CMEs were often associated with the eruption of streamers, indicating that the above generic ideas may be correct, so efforts have been directed to producing models that do not rely on Aly's assumptions.

Broadly speaking, recent developments have focussed on the role that magnetic flux ropes play in CME onset. The genesis of this idea can be found in the reviews of Low (1996) and Hundhausen (1999). Observations revealed that in many helmet streamers there is an embedded filament that is located in a relatively tenuous region, defined as a magnetic cavity. The cavity was postulated to be a magnetic flux rope. Subsequent eruptions of such structures showed a clear "three-part" structure moving away from the Sun: first the streamer as a bright leading edge, then a dark region, the cavity, and finally the dense prominence material.

Three-dimensional computational models discussed by Amari et al. (2000) first introduce a large amount of energy into the corona by shearing motion of the photosphere, and subsequently create a coronal flux rope due to low-altitude reconnection due to the introduction of new photospheric magnetic flux of opposite polarity to the original. This leads to a situation in which eruption is energetically favourable, and an eruption occurs.

An alternative approach that also introduces a flux rope is discussed in the 2.5 dimensional models of Antiochos et al. (1999). In this case the flux rope *formation* is an essential part of the CME eruption. This is a feature in common with many models developed in the past two decades (e.g. Priest and Forbes, 1999), but the difference here is that a more complex initial coronal magnetic environment than a simple streamer is introduced. It is argued that the erupting streamer is enveloped by a large-scale dipole field with neighbouring octopole contributions. On the application of footpoint shear the streamer rises, and reconnects with the overlying field. Relieved of the magnetic tension forces confining it to the Sun, the streamer magnetic flux moves outward, and at the same time initiates reconnection behind it, both forming a flux rope and initiating rapid outward motion.

A third class of flux rope model has been proposed by Chen (1996). In distinction from the above two examples, photospheric motions are not invoked to inject a Poynting flux into the corona. Instead, it is argued that sub-photospheric processes can lead to a generation of an electric current parallel to the flux rope axis. This forces a curved flux rope to rise. By varying the injection profile, a wide range of outward velocities can be generated. However, these ideas, while developed within the framework of magnetohydrodynamics, have never been addressed by computational tools, in part because the model is inherently three-dimensional.

All of these models are conceptual, and criticisms of each can, and are, made. But it should be noted that in each case they are the result of many years of development during which numerous other scenarios have been ruled out. To see this evolution, it is instructive to read any review of the theory of CME onset dating from 1990 or before.

How does one verify these scenarios and what are the major obstacles? Increasingly advanced computational models have a major role to play, but on their own will not solve anything. Progress will come from the experimental data obtained from spacecraft-based instrumentation. We discuss some future opportunities in the final section, but one point needs to be made here and that concerns the coronal magnetic field. It has been clear for 30 years that forces associated with the coronal field must be responsible for the acceleration of the CME. Nothing else has any chance of working. Yet the field vector in the corona cannot be measured at this time. Preliminary measurements using IR coronal emission lines are now becoming available and this is an area of solar physics that needs to be given the highest priority.

5.3 Interplanetary Coronal Mass Ejections

To observe CMEs in the interplanetary medium, one generally relies on in situ sampling of particles and magnetic fields by spacecraft. The optimal way to do this is to monitor the solar wind continually at 1 AU. This can be achieved by placing a spacecraft in orbit around the L1 point, although spacecraft in Earth orbit can also act as monitors provided their apogee is sufficiently Sunward and they spend most of their time in the solar wind. The International Sun–Earth Explorer (ISEE-3) spacecraft was at L1 between 1978 and 1982, and since 1997, the NASA Advanced Composition Explorer (ACE) spacecraft have been there. In addition, the Interplanetary Monitor (IMP) series of spacecraft and the WIND spacecraft have spent much of their time in the solar wind.

When making measurements of ICMEs at 1 AU, one is trying to achieve two things: (1) characterise the plasma and magnetic field properties of the ICME there, and (2) relate these measurements back to an event (hopefully a CME) at the Sun. The former is straightforward, but in fact has limitations that we discuss below. Regarding the latter goal, it has only been really achievable since 1996 when the SOHO and Yohkoh spacecraft began observing the Sun jointly, with the WIND spacecraft monitoring the solar wind. Previous studies, such as those involving data from ISEE-3, often relied on ground-based solar measurements such as prominence eruptions to establish a relationship between solar and interplanetary measurements. Gosling (1996) provides a good overview of ICMEs.

5.3.1 Properties at 1 AU

The most important aspects of the ICME are (a) its velocity with respect to the solar wind, (b) its local magnetic structure and (c) its global topology. Observations at 1 AU can shed light on all of these.

ICME Velocity

One needs to define what is meant by ICME velocity. Many ICMEs are preceded by a shock wave that accelerates the solar wind, forming a high-speed turbulent sheath region between the ICME proper and the ambient solar wind. Once the ejecta proper is entered, the velocity often decreases over 10 – 20 h to something approaching the ambient solar wind value. Defining the ICME velocity as being that in the ejecta, values at 1 AU range between 350 and 750 km/s, with a few stragglers at the high-velocity end (e.g. Gopalswamy et al., 2001). This is usually faster than the solar wind, but on occasions is slower. It is important to realise that the velocity differential between the ICME and solar wind is considerably less at 1 AU than near the Sun where very large CME speeds are detected. Thus the ICME has undergone a significant interaction with the solar wind between the Sun and 1 AU that for fast CMEs leads to a loss of kinetic energy (Cargill, 2004). Another important difference between magnetic clouds and the solar wind lies in the composition states (e.g. Lynch et al., 2003), which can provide important clues about their origin and relation to flares.

Magnetic Structure

An important property of ICMEs is that a significant fraction have smooth magnetic field profiles that changed on timescales of hours as well as lower than usual plasma temperatures. Magnetic field strengths of 20 – 30 nT are observed, and occasionally the fields are as strong as 60 nT. Such ICMEs are named magnetic clouds. Roughly 30 – 40% of ICMEs are magnetic clouds. Their long duration implies that they are vast structures, often being 0.25 AU in diameter, and taking a day to pass the Earth.

Inferences About Global Structure

The principal difficulty with in situ ICME measurements is that one samples a time-series that corresponds to a one dimensional "cut" through a three-dimensional structure. To obtain anything more requires either additional spacecraft, and/or some modelling that may rely on questionable assumptions. As we discuss below, there are a limited number of multi-spacecraft measurements of ICME, but with spacecraft separations usually too large to provide serious constraints on the 3-D structure of the ICME. Nonetheless one can infer some important constraints from single spacecraft data.

Magnetic flux ropes are believed to be common structures in the solar and space environment. They are approximately cylindrical regions of plasma where the magnetic field lines form a helical structure due to a field-aligned electric current. The magnetic field profile observed at a magnetic cloud is very reminiscent of what one would expect to see on passing through the centre of a magnetic flux rope (e.g. Lepping et al., 1990).

Another issue is whether the field lines in this flux rope are rooted at the Sun. Evidence for the attachment of both ends to the Sun partly comes from the high energy electrons that carry the solar wind heat flux. In the ambient solar wind, one expects to see only a heat flux along the magnetic field in one direction, away from the Sun, since the other end of any field line goes somewhere in the cool interplanetary medium. On the other hand ICMEs very often have a bi-directional heat flux electrons, indicating that both ends of the field lines were rooted in the solar corona. This is the expected result for a flux rope ICME of solar origin, although there are exceptions. This leads to a common sketch of an ICME shown in Fig. 5.8.

Models of magnetic flux ropes have also been used to infer the local 3-D structure (e.g. Lepping et al., 1990). These depend on assumptions of cylindrical symmetry, as well as assuming energy states that may not be realistic and certainly cannot account for the field enhancements seen in many ICMEs. More realistic modelling suggests that these are not sensible assumptions (e.g. Cargill and Schmidt, 2002). Alternative topologies for magnetic clouds include closed plasmoids, although the expected magnetic field structure of a plasmoid is not observed, and this also creates difficulties in accounting for the bi-directional heat flux electrons.

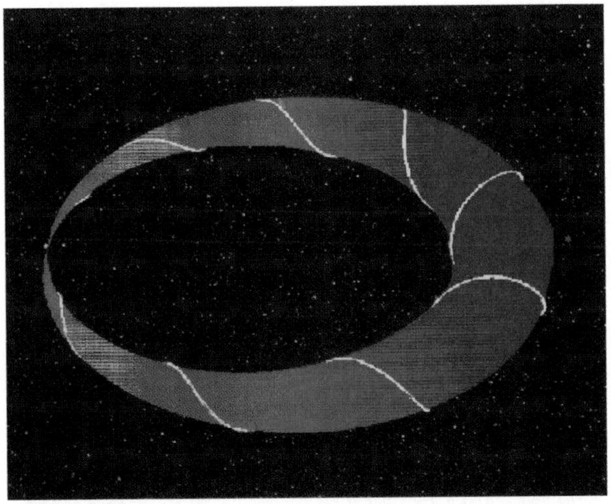

Fig. 5.8. A sketch of an ICME modelled as a large, curved magnetic flux rope with both ends rooted at the Sun. [Figure courtesy of J. Chen]

There have been very few multi-spacecraft observations of ICMEs, and those that have been made were serendipitous. [It should be noted here that although the four spacecraft Cluster mission has encountered CMEs when it has been outside the bow shock, the spacecraft separation is so small ($< 10^4$ km) that little information can be obtained about the ICME itself, although one can infer leading shock geometries to some accuracy (e.g. Fazakerley et al., 2005).] Observations by spacecraft separated by > 1 AU suggest large differences in ICME structure depending on the solar wind properties, while smaller separations suggest coherence only on scales much less than 1 AU (e.g. Mulligan et al., 1999). It is expected that both remote sensing and in situ observations from the forthcoming STEREO mission will make major advances in understanding 3-D ICME structure.

Properties of ICMEs Beyond 1 AU

With the exception of some Voyager ICME encounters, the propagation of ICMEs beyond 1 AU was not studied in any detail until the mid-1990s. The Ulysses mission with its initial phase towards Jupiter and the subsequent out-of-the-ecliptic orbit has observed numerous ICMEs at both large heliocentric distances and at high latitudes.

One would expect that continual interaction of the ICME with the solar wind would lead to a gradual merging of the two, albeit with magnetic cloud structures persisting to large distances. Magnetic clouds are indeed evident at all locations, but something interesting was found in regions of high-speed wind. Ulysses observed ICMEs that were characterised by a very low plasma density, and a relatively strong forward and reverse shock pair. These CMEs are "over-expanding", having begun life with a large excess of plasma pressure, which had expended its energy into creating the shock pairs. This scenario has been confirmed by hydrodynamic and multi-dimensional MHD simulations (Cargill et al., 2000) who also demonstrated that the over-expansion was essential for the maintenance of the flux rope geometry at large distances. The relationship between these and ICMEs seen at 1 AU is unclear.

5.3.2 Putting the Solar and Interplanetary Parts Together

The combination of good observations of CMEs and ICMEs would suggest that it should be possible to form a unified picture of a CME from onset to arrival at the Earth. There are difficulties with this. One is the lack of information of the coronal magnetic field vector, which is associated with the outward acceleration. A second is the difficulty with measuring CMEs directed Earthward, which are visible only as "halos". The best quality images are of CMEs moving at 90° to the Sun–Earth line when one can also measure an accurate velocity in the plane of the sky. Earthward-directed CMEs are visible as a projection on the plane perpendicular to the Sun–Earth line, so that evaluating their velocity Earthward is tricky.

A major unifying theme between CMEs and ICMEs are magnetic flux ropes. Magnetic clouds are a well-established subset of ICMEs, and since it is unlikely that such an organised structure could form spontaneously from a turbulent solar wind, its origin must be solar. The most attractive picture is that the magnetic cloud is the residue of a large solar loop-like structure, and is in fact the prominence cavity discussed in the three-part CME model in Sect. 5.2. The first statistical association that backed up this conjecture was presented by Wilson and Hildner (1986), and coronagraph results are also suggestive of a flux rope (Chen et al., 1997).

Further evidence for this picture comes from the sense of field rotation in a magnetic cloud at 1 AU. Bothmer and Schwenn (1998) noted that in the 1974–1982 timeframe, magnetic clouds tended to arrive at the Earth with a southward IMF leading, but towards the end of this period (i.e. around the time when the global field of the Sun reversed), the opposite was seen. They further argued that these senses of rotation were consistent with the magnetic fields in the associated prominences.

5.4 Conclusions and Future Prospects

Huge progress has been made in the past decade in understanding the cause of coronal mass ejections, and their behaviour as they travel through the interplanetary medium towards the Earth. However significant gaps in our knowledge have also become apparent. Over the next decade it is anticipated that a number of space missions will go a long way to providing detailed answers to the following questions:

1. *What is the origin of the Sun's twisted magnetic fields?* There are suggestions that a combination of differential rotation and sub-surface twisting are what provides the magnetic field with the level of non-potential magnetic energy required to produce a CME. Exploration of the link between the sub-surface regions and outer atmosphere will be possible with the launch of the Solar Dynamics Observatory in 2008. This mission's goal is to determine the flows and interaction below the surface and link these to the outer atmosphere.
2. *What causes the ejection of plasma and magnetic fields?* We know the answer to this lies in the magnetic field, and in particular the twist and shear in flux ropes. The Solar-B mission to be launched in 2006 has a vector magnetograph on board which will permit continuous accurate measurements of surface magnetic field strength, flows and helicity, as well as linking these to the response seen in the outer atmosphere. In the absence of a coronagraph on Solar-B, joint observations with SOHO and STEREO will be important
3. *How does an ejection leave the Sun?* By their nature, single spacecraft observations cannot infer the exact motion of a CME as it leaves the Sun, although with halo events one can often infer the region of origin. The NASA STEREO mission will be launched in 2006 with two identical spacecraft observing at different locations. This will provide the first 3-D view of CMEs, including those moving along the Sun–Earth line, with the all-important information on direction and structure.
4. *How does a CME travel through the solar system?* The link between remote sensing observations on the Sun and in-situ measurements is very difficult. The ESA Solar Orbiter mission is designed to provide such links. The spacecraft will be placed in an orbit around the Sun and its apogee will be 0.22 AU. The combination of remote sensing and in-situ measurements close in to the Sun will allow us to see a dynamical response in the atmosphere and track it

as ejections of mass travel away from the Sun. Also, the two-point NASA STEREO mission will permit in-situ observations of the multi-dimensional structure of ICMEs near 1 AU using both in situ instruments as well as remote detection provided by an all-sky camera and radio instruments.

5. *Can models accurately track a CME from eruption to 1 AU and beyond?* A major development in past years has been the generation of a new range of models that track CMEs from the Sun to interplanetary space (e.g. Riley et al., 2003). Multi-point measurements are essential to establish their validity.

The combination of these exciting missions will probe the origin and progression of CMEs in the solar system in a way we cannot currently achieve. This will be a fascinating time, and new and unexpected results are guaranteed.

Acknowledgement. PC acknowledges the support of a PPARC senior research fellowship. We are grateful to the many people/institutions/organisations for their figures and to Richard Harrison and an anonymous reviewer for comments on this chapter.

References

Aly, J.J., On some properties of force-free magnetic fields in infinite regions of space, ApJ., 283, 349, 1984.

Amari, T., J.F. Luciani, Z. Mikic, and J. Linker, A twisted flux rope model for coronal mass ejections and two-ribbon flares, ApJ. Lett., 529, L49, 2000.

Andrews, M.D., A search for CMEs associated with big flares, Solar Phys., 218, 261, 2003.

Antiochos, S.K., C.R. DeVore and J.A. Klimchuk, A model for solar coronal mass ejections, ApJ., 510, 485, 1999.

Biesecker, D.A., D.C. Myers, B.J. Thompson, D.M. Hammer and A. Vourlidas, Solar phenomena associated with "EIT waves", ApJ., 569, 1009, 2002.

Bothmer, V. and R. Schwenn, The structure and origin of magnetic clouds in the solar wind, Annales Geophys., 16, 1, 1998.

Cargill, P.J., J. Schmidt, D.S. Spicer, and S.T. Zalesak, The magnetic structure of over-expanding CMEs, JGR, 105, 7509, 2000.

Cargill, P.J. and J.M. Schmidt, Modeling interplanetary CMEs using magnetohydrodynamic simulations, Ann. Geophys., 20, 879, 2002.

Cargill, P.J., On the aerodynamic drag force acting on interplanetary coronal mass ejections, Solar Phys., 221, 135, 2004.

Chen, J., Theory of prominence eruption and propagation: interplanetary consequences, JGR, 101, 27, 499, 1996.

Chen, J. et al., Evidence of an Erupting Magnetic Flux Rope: LASCO Coronal Mass Ejection of 1997 April 13, ApJ., 490, L191, 1997.

Delannée, C., Another view of the EIT wave phenomenon, ApJ., 545, 512, 2000.

Démoulin, P., C.H. Mandrini, L. van Driel-Gesztelyi, B.J. Thompson, S. Plunkett, Zs. Kovári, G. Aulanier and A. Young, What is the source of the magnetic helicity shed by CMEs? The long-term helicity budget of AR 7978, A&A, 382, 650, 2002.

Fazakerley, A et al., Relating near-Earth observations of an interplanetary coronal mass ejection to the conditions at its site of origin in the solar corona, GRL, 32, 10.1029/2005GL022842, 2005.

Feynman, J. and S.F. Martin, The initiation of coronal mass ejections by newly emerging magnetic flux, JGR, 100, 3355, 1995.

Gary, G.A. and R.L. Moore, Eruption of a multiple-turn helical magnetic flux tube in a large flare: Evidence for external and internal reconnection that fits the breakout model ofsSolar magnetic eruptions, ApJ., 611, 545, 2004.

Gopalswamy, N., A global picture of CMEs in the inner heliosphere, in The Sun and the Heliosphere as an integrated system, Chapter 8, Springer, 2005.

Gopalswamy, N., A. Lara, R.P. Lepping, M.L. Kaiser, D. Berdichevsky and O.C. St Cyr, Predicting the 1 AU arrival times of coronal mass ejections, JGR, 106, 29207, 2001.

Gosling, J.T., Co-rotating and transient solar wind flows in three dimensions, Ann. Rev. Astron, Astrophys., 34, 35, 1996.

Gosling, J.T., The solar flare myth, JGR, 98, 18, 937, 1993.

Green, L.M., M.C. López Fuentes, C.H. Mandrini, P. Démoulin, L. van Driel-Gesztelyi and J.L. Culhane, The magnetic helicity budget of a CME-prolific active region, Solar Phys., 208, 43, 2002.

Harra, L.K. and A.C. Sterling, Material outflow from coronal intensity dimming regions during coronal mass ejection onset, ApJ., 561, L215, 2001.

Harra, L.K. and A.C. Sterling, Imaging and spectroscopic investigations of a solar coronal wave: Properties of the wave front and associated erupting material, ApJ., 587, 429, 2003.

Harrison, R.A., Solar coronal mass ejections and flares, A&A, 162, 283, 1986.

Hiei, E., A.J. Hundhausen and D.G. Sime, Reformation of a coronal helmet streamer by magnetic reconnection after a coronal mass ejection, GRL, 20, 2785, 1993.

Hudson, H.S. and E.W. Cliver, Observing coronal mass ejections without coronagraphs, JGR, 106, 25199, 2001.

Hundhausen, A.J., Sizes and locations of coronal mass ejections: SMM observations from 1980 and 1984–1989, JGR., 98, 13, 177, 1993.

Hundhausen, A.J., Coronal mass ejections, in K.T. Strong et al. (Eds), *The many faces of the Sun*, Springer, p 143, 1999.

Jing, J., V. B. Yurchyshyn, G. Yang, Y. Xu and H. Wang, On the relationship between filament eruptions, flares and coronal mass ejections, ApJ., 614, 1054, 2004.

Jing, J., J. Qiu, J. Lin, M. Qu, Y. Xu, and H. Wang, Magnetic reconnection rate and the flux-rope acceleration of two-ribbon flares, ApJ., 620, 1085, 2005.

Kahler, S.W., Solar flares and coronal mass ejections, ARA&A, 30, 113, 1992.

Khan, J.I. and H.S. Hudson, Homologous sudden disappearances of trans-equatorial interconnecting loops in the solar corona, GRL, 27, 1083, 2000.

Kopp, R.A. and G.W. Pneuman, Magnetic reconnection in the corona and the loop prominence phenomena, Solar Phys., 50, 85, 1976.

Lepping, R.P., J.A. Jones, and L.F. Burlaga, Magnetic field structure of interplanetary magnetic clouds at 1 AU, JGR, 95, 11, 957, 1990.

Low, B.C., Solar activity and the corona, Solar Phys., 167, 217, 1996.

Lynch, B.J., T.H. Zurbuchen, L.A. Fisk and S.K. Antiochos, Internal structure of magnetic clouds: plasma and composition, JGR 108, 10.1029/2002JA009591, 2003.

McKenzie, D.E., Supra-arcade downflows in long duration flare events, Solar Phys., 195, 318, 2000.

Masuda, S., T. Kosugi, H. Hiro, T. Sakao, K. Shibata, and S. Tsuneta, Hard X-ray source and the primary energy release site in solar flares, PASJ, 47, 677, 1995.

Mulligan, T. et al., Inter-comparison of NEAR and Wind interplanetary coronal mass ejection observations, J. Geophys. Res., 104, 28, 217, 1999.

Nindos, A., and M.D. Andrews, The association of big flares and coronal mass ejections: what is the role of magnetic helicity?, ApJ., 616, L175, 2004.

Ohyama, M. and K. Shibata, X-Ray Plasma ejection associated with an impulsive flare on 1992 October 5: Physical conditions of X-Ray plasma ejection, ApJ., 499, 934, 1998.

Priest, E.R. and T.G. Forbes, Magnetic Reconnection, CUP, 1999.

Riley, P., J.A. Linker, Z. Mikić, D. Odstrcil, T.H. Zurbuchen, D. Lario, and R.P. Lepping, Using an MHD simulation to interpret the global context of a coronal mass ejection observed by two spacecraft, JGR, 108, 10.1029/2002JA009760, 2003.

Rust, D.M. and A. Kumar, Evidence for helically kinked magnetic flux ropes in solar eruptions, ApJ Lett., 464, L199, 1996.

St Cyr, O.C., J.T. Burkepile, A.J. Hundhausen, A.R. Lecinski, A comparison of ground-based and spacecraft observations of coronal mass ejections from 1980–1989, J. Geophys. Res., 104, 12493, 1999.

St Cyr, O.C. et al., Properties of coronal mass ejections: SOHO LASCO observations from January 1996 to June 1998, JGR, 105, 18169, 2000.

Sheeley, N.R., J.H. Walters, Y.M. Wang and R.A. Howard, Continuous tracking of coronal outflows: two kinds of coronal mass ejections, JGR, 104, 24739, 1999.

Sterling, A.C. and H.S. Hudson, Yohkoh SXT Observations of X-Ray "Dimming" Associated with a Halo Coronal Mass Ejection, ApJ., 491, L55, 1997.

Tsuneta, S., Interacting Active Regions in the Solar Corona, ApJ., 456, L63, 1996.

Vourladis, A., P. Subramanian, K.P. Dere and R. Howard, LASCO measurements of the energetics of coronal mass ejections, ApJ., 534, 456, 2000.

Wang, J., G. Zhou and J. Zhang, Helicity patterns of coronal mass ejections-associated active regions, ApJ., 615, 1021, 2004.

Wilson, R.M. and E. Hildner, On the association of magnetic clouds with disappearing filaments, JGR, 91, 5867, 1986.

Yashiro, S. et al., A catalogue of white light CMEs observed by the SOHO spacecraft, JGR, 109, doi:10.0129/2003JA010282, 2004.

Zhang, M. and B.C. Low, The hydromagnetic nature of solar coronal mass ejections, Ann. Rev. of Astron. and Astrophys., 43, 103, 2005.

6 Solar Radio Emissions

Jean-Louis Bougeret and Monique Pick

Solar radio emissions are produced across the entire radio band of the electromagnetic spectrum, encompassing 10 orders of magnitude of wavelengths from sub-millimeter to kilometer (or from THz frequencies to a few kHz). Radio emission from the thermal Sun (the corona at one million degree) can be observed, but also – and this is one of the richnesses of the radio diagnostics –, a variety of intense, sporadic events: radio bursts. Solar radio bursts are produced through non-thermal radiation mechanisms and trace energetic phenomena in the solar corona and in the interplanetary medium. Sources can be observed from chromospheric levels to the limits of the heliophere. Hence, radio observations provide a powerful diagnostics on the solar atmosphere and on a large variety of dynamical phenomena occurring in the heliosphere.

This chapter is an introduction to solar radio physics. Observations of radio bursts reveal various forms of activity and acceleration processes which are associated to large scale eruptive phenomena including flares, filament eruptions, CMEs and shocks. CMEs and the interplanetary shock waves are the most spectacular large-scale manifestations in the solar corona and interplanetary medium. Radio imaging and spectral observations provide signatures on the initial steps, the development of CMEs and their progression through the interplanetary medium and provides unique diagnostics on the associated shock wave and energetic electron events. Radio observations offer very significant contributions to understand solar activity, how solar activity affects the interplanetary medium and the consequence on the solar-terrestrial environment.

Contents

Jean-Louis Bougeret and Monique Pick, Solar Radio Emissions.
In: Y. Kamide/A. Chian, Handbook of the Solar-Terrestrial Environment. pp. 133–151 (2007)
DOI: 10.1007/11367758_6

6.1 Introduction

This chapter of the Handbook of Solar-Terrestrial Environment gives a brief introduction to solar radio astronomy. It is by no means a review of the present status of solar research using radio techniques but rather a presentation of a number of clues to help the reader to understand radio diagnostics, using heuristic examples. Indeed, contrary to all other chapters in this handbook, the object is a given spectral domain ranging from sub-millimeter to kilometer wavelengths (or from THz to kHz radio frequencies) covering ten decades of wavelengths (or frequencies). This overall radio band actually covers a very broad range of solar structures and phenomena. We will show in this chapter that radio observations often give unique information on the diagnostics and interpretation of a large variety of physical phenomena. Figure 6.1 summarizes the different wavelength and frequency ranges in the radio band and shows the "radio window" that can be observed from the ground. The limits of the shortest and longest wavelengths that can be observed from the ground are determined, respectively, by the transparency of the earth's atmosphere and by the frequency cut-off of the ionosphere. Satellite observations are needed to extend the observable radio spectrum to the full radio range of the electromagnetic spectrum.

Radio measurements require a broad range of techniques, using different types of instruments and sensors or antennae. Above all, they cover a broad range of solar phenomena, from micro-flares and millisecond events to coronal mass ejections and interplanetary shock waves, from quiescent structures in the corona to long lasting structures in the interplanetary medium. Indeed the Sun is a radio emitter in the entire radio spectral domain, sometimes a very strong one (bursts of radiation can be 40 dB or more above the background). Radio radiations can be produced basically at any distance from the Sun, from chromospheric levels to several astronomical units (AU).

Radio physics encompasses three techniques: Radar astronomy, interplanetary (and interstellar) scintillation, and the observation of radio sources, corresponding, respectively, to backscatter, forward scatter and emission of radio waves. In this chapter, we will consider only the latter: Radio astronomy.

On the one hand, radio astronomy, particularly at wavelengths longer than a few cm, suffers from the difficulty of obtaining spatial resolution; for instance, kilometer size arrays are needed at meter wavelengths. Indeed the resolution of an instrument is proportional to the operating wavelength divided by the instrument aperture (size of dish, antenna or array). On the other hand, extremely high time resolutions can be attained: Sporadic events, some of them lasting a fraction of a sec-

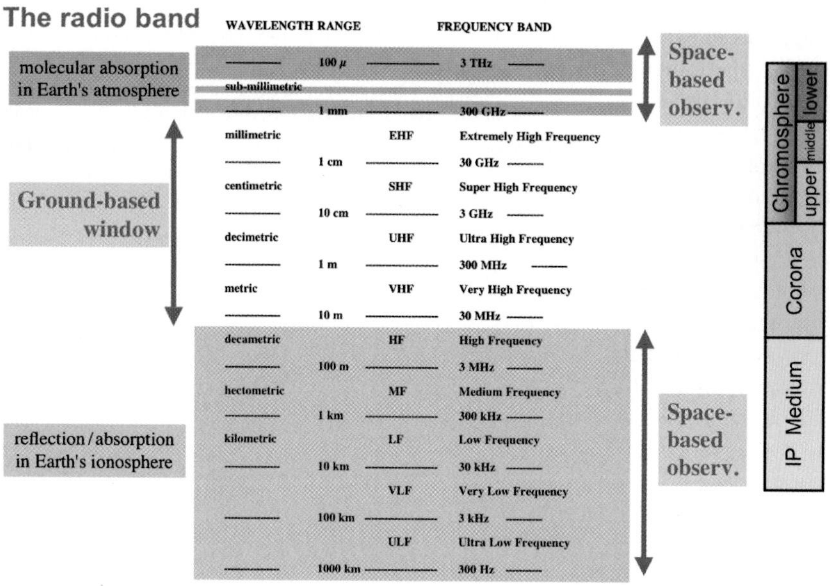

Fig. 6.1. The radio range. The shorter wavelengths are absorbed in the Earth's atmosphere and the longer wavelengths are blocked by the terrestrial ionosphere. The *bar on the right* indicates regions of the solar environment that can be studied (see Fig. 6.2)

ond, can be very accurately timed and compared with observations at other wavelengths (X-rays, gamma-rays). This opens the favored connection between particle measurements and radio observations. Indeed, radio traces energetic electrons and brings a unique, supplementary diagnostic based on remote sensing.

The field of solar radio astronomy is very rapidly evolving. If the development of new generation instruments, in particular radio heliographs, can be credited by a wealth of new observations and diagnostics, the complementarity of radio diagnostics to other diagnostics in order to study a large variety of phenomena is more and more obvious. This will appear clearly in the present chapter.

In general, observations at different wavelengths sample different heights in the solar atmosphere, with longer wavelengths referring to higher heights above the photosphere. This is essentially because radiation mechanisms are mostly related to plasma mechanisms in which the radiated frequency is of the order of the local plasma frequency; at the level of the radiation source, this plasma frequency is proportional to the square root of the local electron density. This is clearly shown in Fig. 6.2. Therefore a given event and its related effects can in principle be probed from the bottom of the corona to a large distance in the interplanetary medium, beyond the earth orbit.

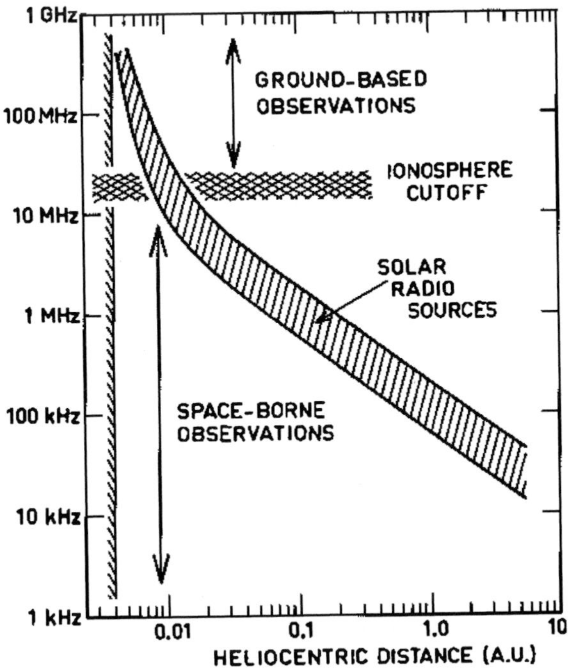

Fig. 6.2. Distribution of the observed frequency of radio sources versus altitude in the corona and the interplanetary medium

Global flux measurements are made by radiometers and spectrographs. Spatially resolved observations are obtained by radio interferometers. Radio heliographs are solar dedicated instruments that provide full images of the Sun. At longer wavelengths, the terrestrial ionosphere blocks the cosmic radio waves and the instrument has to be embarked on an earth satellite or a space probe, in order to track the development of solar events at larger distances from the Sun.

Excellent reviews have been written on solar radio astronomy. A few of them are mentioned below: Kundu, 1965; Zheleznyakov, 1977; Kruger, 1979; Solar Radio Physics, 1985; Melrose and McPhédran, 1991; Benz, 1993; Bastian et al., 1998; Cairns, Robinson, and Zank, 2000; Aschwanden, 2002; Pick, 2006. The most extensive and up-to-date reviews can be found in Solar and Space Weather Radiophysics: Current Status and Future Developments, a set of 17 chapters on solar radio physics including radar astronomy and scintillations (Gary and Keller, 2004).

6.2 Radio Wave Propagation

6.2.1 Basics

Radio waves can only propagate in a medium where the refractive index is real. The refractive index n for a plasma, neglecting the magnetic field, is $n^2 = 1 - f_p^2/f^2$ where f_p and f are, respectively, the plasma and observing frequencies. The electron density N_e, thus the plasma frequency $f_p \sim N_e^2$ of the solar atmosphere decreases monotonically with increasing altitude; therefore, for each frequency, there is an altitude below which propagation does not take place and the range of heights above the photosphere that is observable by a radio telescope is determined by its frequency window. This is illustrated in Figs. 6.1 and 6.2. At 1AU, the plasma frequency is 30 kHz.

In an isotropic medium, the radiation path is determined by the Snell–Descartes law:

$$n \sin \alpha = \text{constant} \ ,$$

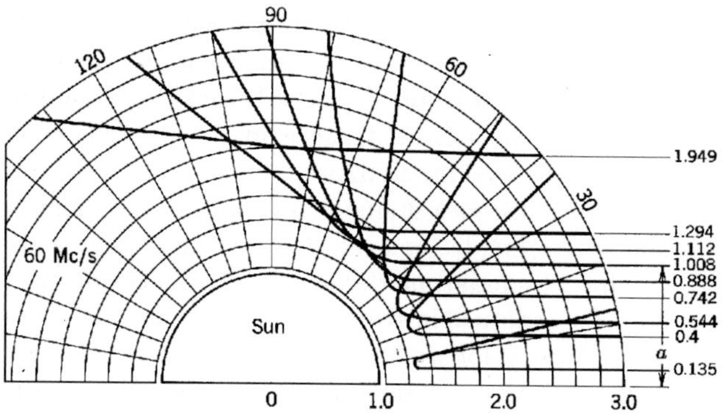

Fig. 6.3. Ray trajectories in the solar atmosphere for 60 MHz. The radial scale is in units of 705,000 km (= photospheric radius + 7650 km). The numbers on the *curves* give the dimension α in the same unit (from Jaeger and Westfold, 1949)

where α is the angle of propagation relative to the gradient of the refractive index. For increasing index of refraction, as is the case for radiation propagating from the Sun to the Earth, the radiation is refracted toward the direction of the gradient, as illustrated in Fig. 6.3.

Moreover, fluctuations in the refractive index, due to coronal inhomogeneities, cause multipath propagation between the source and the observer. This effect modifies the observed properties of the radio sources, introduces an angular broadening and therefore prevents very high angular observations of radio emissions. Finally, the limits of the shortest and longest wavelengths that can be observed from ground are fixed, respectively, by the transparency of the Earth's atmosphere and by the frequency cut-off of the ionosphere; indeed, below approximately 10 MHz, which is the plasma frequency of the ionosphere, the radio waves cannot propagate down to the ground. Satellite observations extend the observable radio spectrum.

Radio emissions result from thermal and non-thermal mechanisms. For thermal emissions, the "source" function B_f is given by the Planck function (Rayleigh–Jeans limit):

$$B_f = k_b T f^2/c^2 ,$$

where k_b is the Boltzman constant, T is the temperature, f the frequency and c the speed of light.

For non-thermal emissions, an effective temperature T_{eff} can be similarly defined:

$$B_f = k_b T_{eff} f^2/c^2 .$$

Spatially resolved observations are limited in angular resolution to a solid angle Ω. The intensity I_f is defined as the received power emitted per unit area, unit frequency and unit solid angle:

$$I_f = k_b T_b f^2/c^2 ,$$

where T_b is called the brightness temperature.

T_b and T_{eff} (or T) are related by the radiative transfer equation:

$$dT_b/d\tau = -T_b + T_{eff} ,$$

where, $d\tau = k\,dr$ is the elemental optical depth, k is the absorption coefficient of electromagnetic waves, and r the distance along the line of sight.

If T (or T_{eff}) is constant (for example, for the quiet corona)

$$T_b = T(1 - (\exp -\tau)) .$$

For an optically thick source, $T_b = T$; for an optically thin source $\tau \ll 1$, $T_b \sim \tau T$ (for example, emission of some coronal structures).

The flux of a radio source is the power received per unit surface and unit frequency. It is expressed in solar flux unit ($sfu = 10^{-22}\,\mathrm{Wm^{-2}Hz^{-1}} = 10^4$ Jansky).

In the presence of a magnetic field and in the case of the "cold plasma approximation", two modes of propagation exist, the extraordinary, x, and the ordinary, o, modes. The polarization characteristics of the observed radio waves are determined by the emission mechanism and/or by the propagation conditions. In general, the two electromagnetic modes propagate independently (weak mode coupling) and, if the radiation travels through a region in which the longitudinal component of the magnetic field changes sign, so does the orientation of the circular polarization. Note that this is not the case under strong mode coupling conditions.

6.2.2 Scattering of Radio Waves

Several analyses, based on the ISEE-3 observations (Steinberg et al., 1984; 1985) showed that the angular size of the primary source is at least doubled by scattering of the radiation from the interplanetary density inhomogeneities. Moreover, type III trajectories determined by ISEE-3 tend to show anomalous behaviors at distances beyond 0.5 AU. In order to reduce those limitations, Reiner and Stone (1988; 1989) proposed a model interpretation of type III radio bursts characteristics taking into account two essential factors: The known finite extent of the radio source and the directivity of the radio emission. Their results are based on an in-depth analysis of the ISEE-3 data and on extensive running of computer simulation codes. They provide new insights into the physical characteristics of the radio source region and the dynamics of the underlying electron event. Reiner and Stone demonstrated that the finite extent of the primary radio source and the beaming of the emitted radiation alone were sufficient to account for the observed spatial aspects of the radio azimuths and relative burst intensities. This comprehensive model will certainly give clues to the new situations of the STEREO mission.

6.3 Thermal Radiation from the Sun

Radio observations of the Sun probes its atmosphere at different altitudes; thermal emissions originate from regions with distinct physical parameters whose intensity varies in time. The origin of this slowly varying component (SVC), which is superimposed over the emission of the steady Sun, depends on the observation frequency.

6.3.1 Microwave Domain

In the microwave domain, low temperature plasmas in the corona, such as dark filaments, are detected as absorbing objects on the disk and as emitting ones above the limb. Above the active regions, the emission is produced by thermal bremsstrahlung (or free–free emission), resulting from the Coulomb interaction between the plasma electrons and the corresponding ions, and by gyroresonance emission of the electrons in the presence of a magnetic field. Thermal gyroemission resonance (Zheleznyakov, 1962; Kakinuma and Swarup, 1962) results from thermal electrons spiralling along coronal magnetic field lines and is emitted at low harmonics of the local electron gyrofrequency ($f = sf_b$) where $s = 1,2,3$ and $f_b = 2.8\,10^6 B$. Above the active regions, the enhanced magnetic field strength increases the gyrofrequency and the gyroresonance emission can be the dominant thermal radiation process between approximately 3 and 15 GHz. The magnetic field does not exceed a few hundred gauss in the faculae but, it can reach values up to few thousand gauss in the Sunspot regions. Above 15 GHz, free–free emission will dominate for all regions with magnetic fields below 2000 Gauss (this value corresponds to the third harmonic of the gyro frequency). An image of the Sun at 17 GHz, seen by the Nobeyama Radio Heliograph (NoRH, Nakajima et al., 1994), is displayed in Fig. 6.4.

Note that at 2.8 GHz (10 cm) (flux used as an index of solar activity) the emission above active regions is a combination of the "faculae" (bremsstrahlung) and of the "Sunspot" (gyroresonance) components. Note also that an occasional contribution of non-thermal emission in some limited areas of the active regions cannot be excluded (Drago-Chiuderi et al., 1987).

It is important to point out that spatially resolved observations of the polarization provide local coronal magnetic field measurements. Quantitative and accurate measurements of this magnetic field bring strong constraints on the calculation of magnetic fields extrapolated from measured photospheric fields.

6.3.2 Decimeter–Meter Domain

Below 0.5 GHz, at metric wavelengths, the thermal emission originates from higher in the corona, where the magnetic field does not exceed a few gauss. At these wavelengths, the SVC is sensitive to large-scale structures of the corona, including coronal holes, arches and streamer belt. These emissions are unpolarized. Radio imaging allows one, for example, to follow the evolution of coronal holes that are the source of high velocity solar wind streams detected in the interplanetary medium. Figure 6.5 shows four radio images of the corona, seen by the Nançay Radio Heliograph with the corresponding EUV image (NRH, Kerdraon and Delouis, 1997).

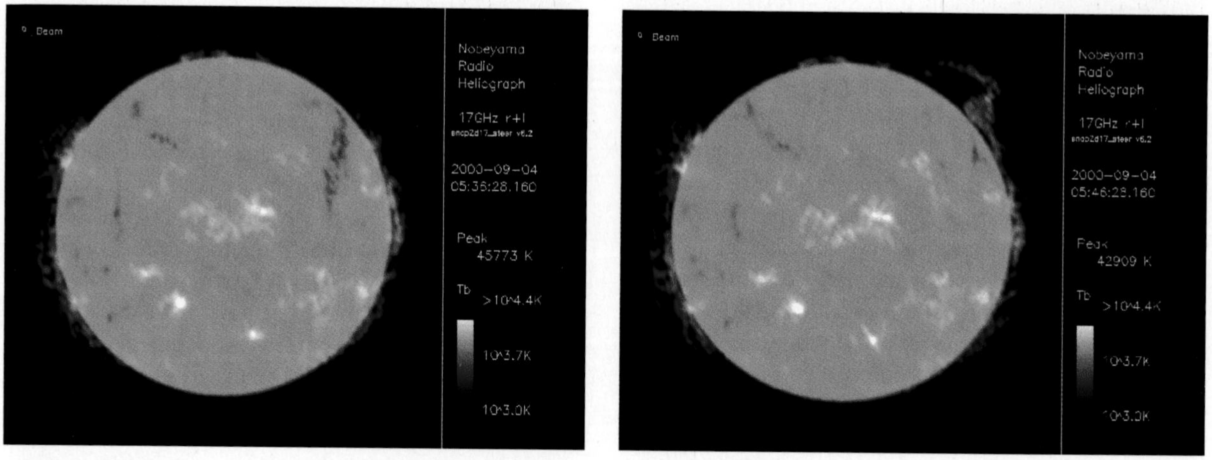

Fig. 6.4. The Sun seen at 17 GHz by the Nobeyama Radioheliograph. Note the filament eruption (upper-right). The time interval between the two images is ten minutes. Images courtesy of the Nobeyama Radio Observatory, NAOJ, http://www.nro.nao.ac.jp

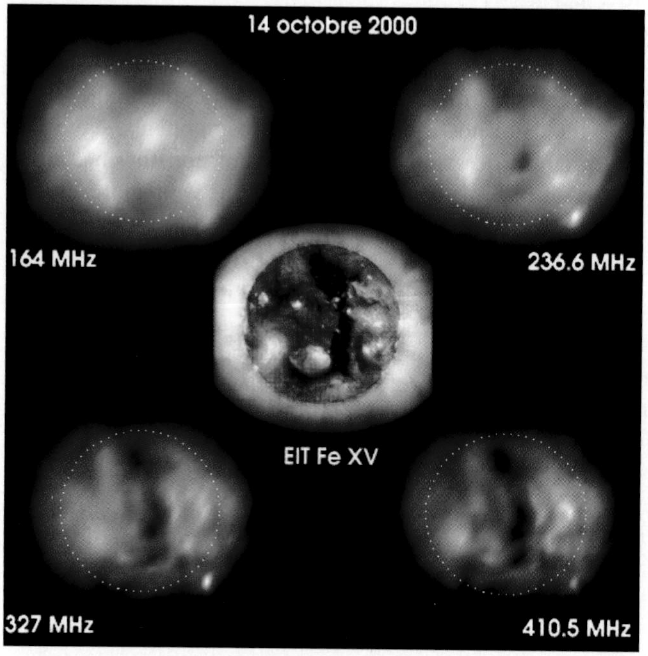

Fig. 6.5. The corona seen at four different radio frequencies on 14 October 2000 by the Nançay Radioheliograph. The coronal hole seen by the EIT instrument on SOHO (center image) is clearly seen at 327 and 410.5 MHz, corresponding to the lower corona

6.4 Solar Radio Bursts

A large variety of radio bursts have been reported in the literature; they present distinct spectral characteristics and can last from a fraction of a second to several hours. They have been historically classified into several morphological types (Wild et al., 1963). Radio emission offers many diagnostic tools for addressing questions such as energy release, particle acceleration and energy transport. Figure 6.6 shows a "dynamic spectrum" from the Hiraiso radio spectrograph. The radio intensity is coded with a gray-scale and displayed as a function of time and frequency. Time is generally displayed in abscissa and frequency in ordinates, with two possible presentations: Frequency increasing with increasing ordinate (from 25 MHz to 2500 MHz in Fig. 6.6); or

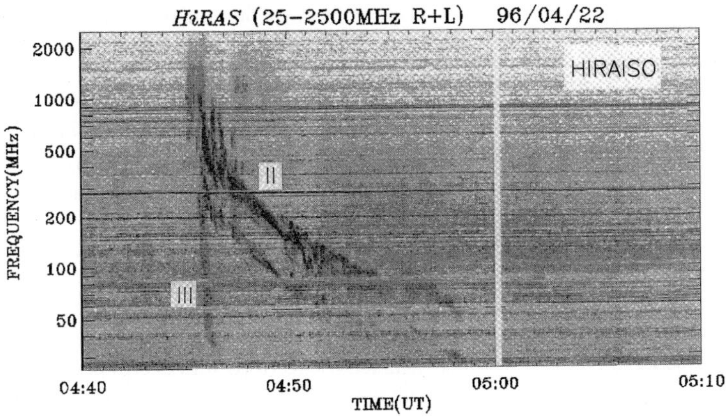

Fig. 6.6. Dynamic spectrum of 22 April 1996 from Hiraiso Terrestrial Research Center: Type III and type II bursts emitted in the corona (after Gopalswamy et al., 1998)

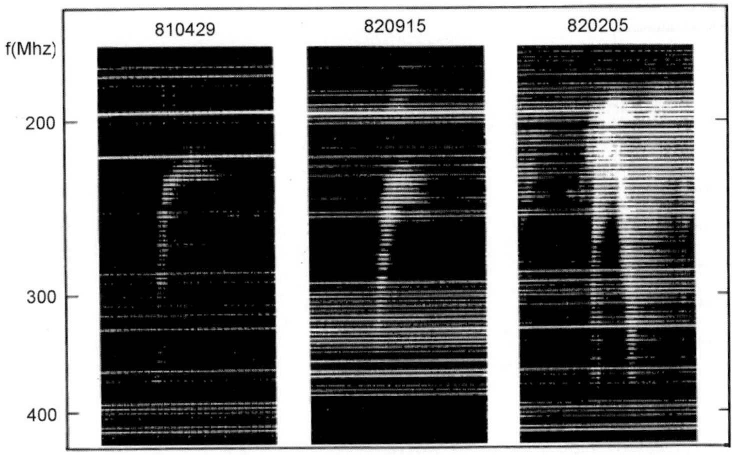

Fig. 6.7. Dynamic spectrum displayed on film for two type U bursts and a "type N" burst (Poquérusse et al., 1984; Caroubalos et al., 1987)

frequency decreasing with increasing ordinate (as in Figs. 6.7, 6.9, 6.10, 6.11, 6.12 and 6.19). The latter case (reversed frequency scale) suggests that the ordinate can be interpreted as distance from the Sun or height above the photosphere, as clearly shown in Fig. 6.12.

6.4.1 Emission Mechanisms

Radio emission produced by non-thermal electrons is attributed to incoherent gyro synchrotron emission and to coherent plasma emission. Their relative contribution depends upon the observing frequency. Gyrosynchrotron emission, which is proportional to the number of radiating electrons, usually dominates at frequency above about 3 GHz. This emission provides diagnostics on the ambient plasma, on the energy range of radiating electrons and on their evolution. Coherent plasma emission (electrons radiating in-phase most often dominates at frequencies below 3 GHz and is caused by plasma instabilities driving various modes that produce observable electromagnetic radio waves. Classification of these radio bursts and their relation to emission mechanisms remain arbitrary in many respects. One can, however, identify the radio bursts due to: (i) the travelling of electron beams or shocks through the corona, then through the interplanetary medium; (ii) trapped electrons in coronal loops; and (iii) regions of coronal reconnection or restructuring.

6.4.2 Electron Beams

Particularly important are the radio bursts produced by electron beams propagating along the magnetic field lines of force; these bursts contain many pieces of information on the electron acceleration, injection and prop-

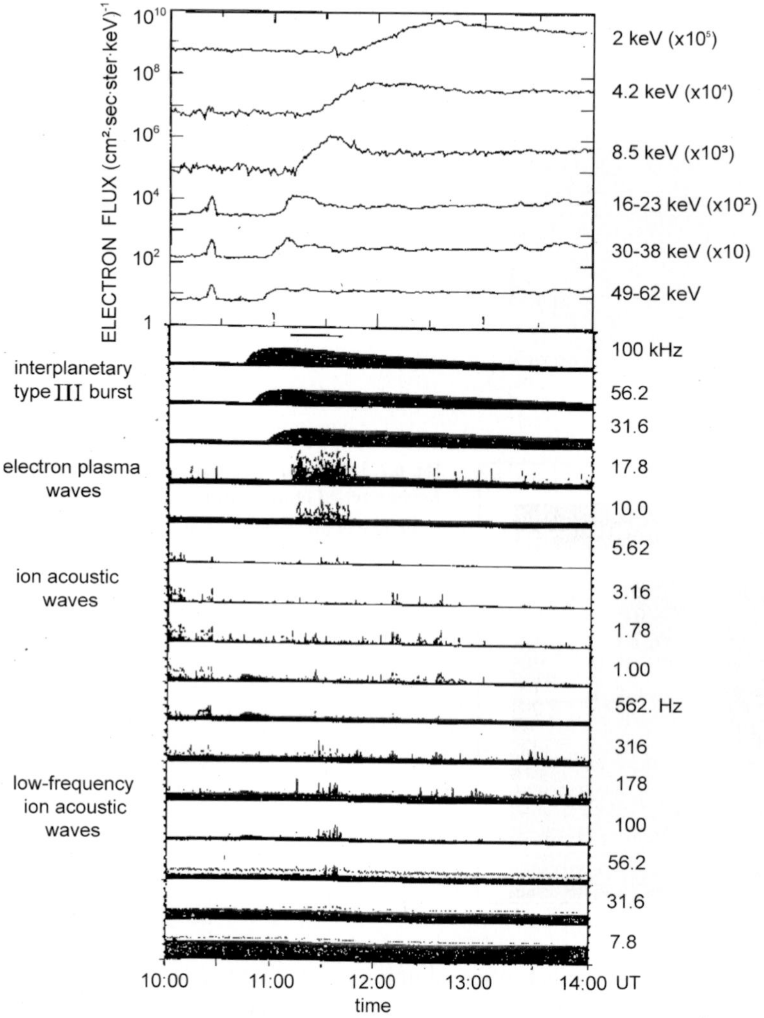

Fig. 6.8. Electron flux and waves observed by the ISEE-3 satellite. The *top panel* shows non-thermal electrons in different energy channels. The slower electrons arrive later. The four upper channels in *the bottom panel* show an interplanetary type III burst and its drift in frequency. Langmuir electron plasma waves are detected exactly when the type III burst is detected at the local plasma frequency, near the 17.8 kHz channel. A comparison between *top* and *bottom panels* shows that these waves are excited by electrons with energy of approximately 8.5 keV. Ion acoustic waves are also detected (Lin et al., 1986)

agation mechanisms, on the beam characteristics and on the medium itself.

Type III bursts are due to electron beams propagating along open magnetic field lines and exciting Langmuir waves at the local coronal plasma frequency which, in turn, produce a radio emission at the fundamental or at the second harmonic of the plasma frequency. These beams travel through the corona at speeds of a fraction of the light velocity. Because the density (thus the plasma frequency) decreases outward from the Sun, the emission drifts rapidly from high to lower frequencies as shown in Fig. 6.6. When streams propagate along a coronal loop, a U or J shape is observed on spectrograms as illustrated in Fig. 6.7. Such spectrograms display the gray or color scale as a function of frequency and time. In some cases, the descending beam has been observed to be reflected at the foot of the arch and ascending again (magnetic mirror). These events were dubbed "type N" by Caroubalos et al. (1987), the letter N being suggestive of the shape of the burst, as shown in the third panel of Fig. 6.7.

Electron streams travel through the interplanetary medium; interplanetary (IP) hectometric–kilometric type III bursts are detected by instruments aboard space missions. Figure 6.8 illustrates the progression of an IP type III burst, the in-situ detection of the Langmuir waves and of the electron beams. Combining radio remote sensing tracking of the electron beams with in situ measurements of the same beam has led to major progress in the theory of radiation from suprathermal electrons in a plasma.

Stochastic growth theory (SGT) can explain in detail the Langmuir waves and electron beams of type III bursts and Earth's foreshock (Robinson 1992, 1995, Robinson, Cairns and Gurnett 1993, Cairns and Robinson 1997, 1998, 1999) and is a natural theory to apply to type II bursts.

Directional measurements, based on the analysis of the spin modulation of the receiver signal, are used to find the source location of IP type III bursts. The source locations projected on the ecliptic plane follow an Archimedean spiral (Fainberg et al., 1972; Baumback et al., 1976).

Radio observations, in the dm–m wavelength range, also provide pieces of information on the acceleration region of the electron beams. Type III bursts above 1 GHz generally have a downward motion in the corona. At lower frequencies, typically 0.4 – 1 GHz, pairs of type III bursts drifting in opposite directions are observed. An example is shown in Fig. 6.9. The frequency location of the change from upward to downward moving beams corresponds to the region of electron acceleration (near 550 MHz in the present example). Another information is given by narrow frequency spikes that are observed just above the starting frequency of type III bursts; these spikes are closely related to the acceleration region (Benz et al., 1996). Radio bursts produced by electron beams often occur during the impulsive phase of solar flares. It will be seen in Sect. 6.4.4 how the observation of radio emission produced by electron beams has contributed to the understanding and development of flare models.

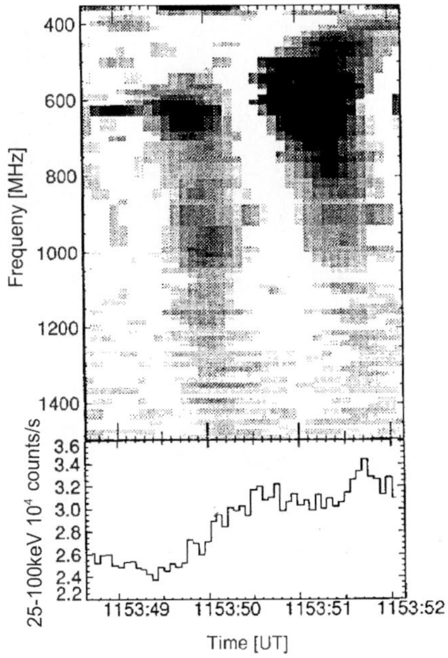

Fig. 6.9. *Top*: Radio flux density of solar type III emission observed by the Zurich spectrometer. Two reverse slope bursts mark two downgoing beams. The second is accompanied by a simultaneous upgoing beam ($f \leq 550$ MHz). *Bottom*: Hard X-ray counting rate measured by BATSE on the GRO satellite (from Aschwanden et al., 1993)

6.4.3 Remote Tracking of Collisionless Shock Waves

Type II bursts are caused by electrons accelerated upstream of coronal or interplanetary shocks. Radio radiation is observed at the local plasma frequency and/or at its harmonics, resulting from a beam-plasma instability similar to that producing the type III radiation. Two types of shocks are considered to produce the type II bursts: The coronal shocks which generate metric type II bursts (Fig. 6.6) and the coronal mass ejection (CME) driven shocks that generate IP type II bursts (Wagner and MacQueen, 1983). Figure 6.10 shows an IP type II burst measured by the WAVES radiospectrograph aboard the WIND spacecraft (Bougeret et al., 1995). Metric type II bursts are associated with flares and most often vanish before reaching the high corona. The general finding is that they are produced by blast coronal waves. Indeed, they are associated with Hα Moreton waves (Moreton and Ramsey, 1960). Moreton waves travel away from the flare region with speeds ranging from a few hundred to a few thousand km/s. They are interpreted as the chromospheric trace of MHD coronal waves responsible for the type II burst.

IP type II bursts are excited by CME-driven shocks formed, most often, at an altitude above 1 solar radius. Figure 6.11 shows that IP type II bursts are confined in an instantaneous narrow frequency band. Metric and IP type II bursts, when observed together, have distinct drift rates and distinct locations. It cannot be excluded,

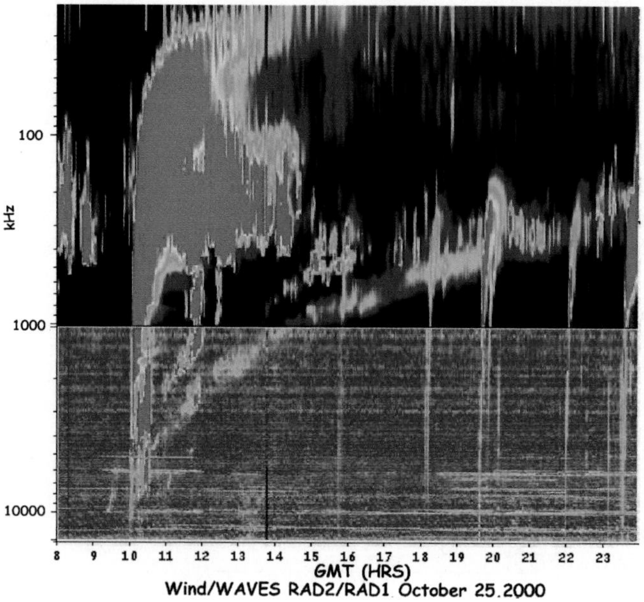

Fig. 6.10. Dynamic spectrum from the WAVES experiment on the Wind spacecraft. 16 hours of data are displayed. Frequency is decreasing upward, corresponding to distances from the Sun ranging from about 2 solar radii at the bottom to 150 at the top. The limit between the two analyzers, seen at 1000 kHz, correspond to about 10 solar radii. The red, fast drifting burst is the IP type III burst. It is saturated on the chosen color scale in order to reveal the fainter, slow drifting IP type II burst. The fundamental emission appears as a few bright points (in particular near 14–16 UT), while the harmonic is more regular and can be tracked for a period of over 14 hours

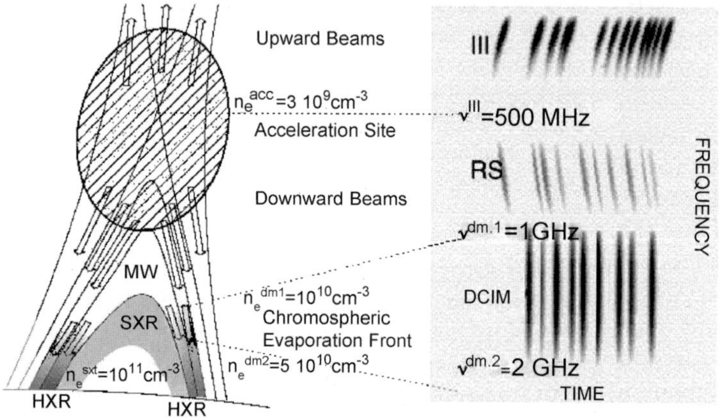

Fig. 6.11. *Left panel*: Diagram of a flare model envisioning magnetic reconnection and chromosphere evaporation processes. *Right panel*: A dynamic radio spectrum with radio bursts indicated in a frequency versus time diagram. The acceleration site is located at the cusp from where electron beams are accelerated in upward (type III) and downward directions (reverse bursts). Downward-precipitating electron beams can be traced as decimetric bursts with almost infinite rate in the 1 – 2 GHz range (from Aschwanden and Benz, 1997)

however, that metric and IP type II bursts are produced by the same shock, but from different sections of the shock front, for instance the metric component from the flanks and the IP bursts from the nose.

Gopalswamy et al. (2001a) showed that the average speed of 101 CMEs associated with a radio decameter–hectometer type II burst is more than two times larger than that of all CMEs (~450 km/s). CMEs associated with a radio type II are also much wider. Hence, radio traces preferentially fast and broad CMEs, those which are more likely to have an impact on the Earth's environment. The same authors also showed that most CMEs with associated radio type II signatures were also associated to Solar Energetic Particle (SEP) events, another indication to the relevance of radio observations to space weather.

6.4.4 Radio Emission Following Flares and Large-Scale Disturbances

During flares, a large part of the energy release is initially converted into energetic electrons and ions. The plasma is heated and particles are accelerated to relativistic energies on a short time scale. A large flare may

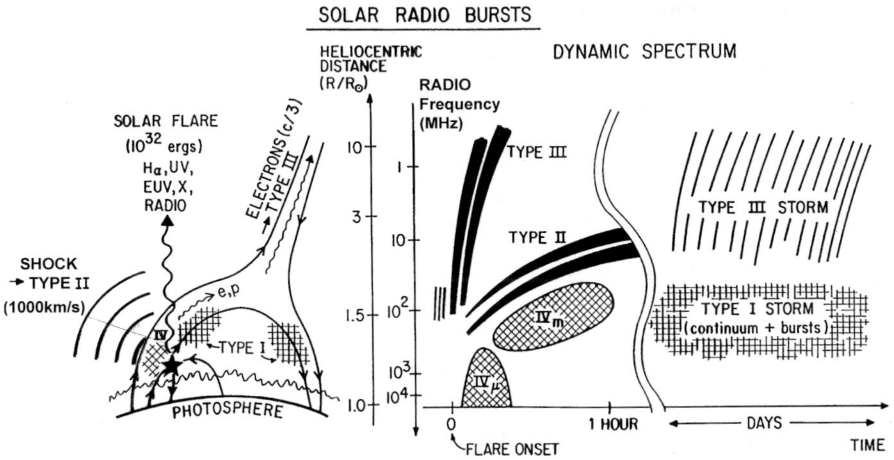

Fig. 6.12. Sketch of a complete radio event. A typical helmet streamer structure is shown on the *left*. The corresponding radio events are shown in the dynamic spectrum on the *right*. The two ordinate scales in the *middle* show the approximate correspondence between frequency and distance

require the acceleration of 10^{37} electron s^{-1} to energies >20 keV within tens of seconds. Synchrotron or gyrosynchrotron emission of electrons is observed even in the smallest flares. Synchrotron emission produced by MeV electrons dominates at frequencies > 3 GHz. Microwaves, soft (SXR) and hard (HXR) radiations are spatially and temporally correlated: They are produced by the same population of electrons accelerated and partly trapped in loop systems. SXR emission (heated plasma) is produced in coronal magnetic loops. The dominant sources of HXR emission are located in conjugate magnetic foot points in the low corona. Microwave emission is produced in the entire volume accessible to nonthermal electrons (e.g. Bastian et al., 1998). Figure 6.11 shows a schematic representation of a flare model derived from radio observations of electron beams and from X-ray observations.

Radio emissions associated with flares at metric and decimetric wavelengths correspond to a wide variety of bursts. Broadband incoherent gyrosynchrotron radiation and coherent plasma radiation are both commonly observed in this frequency range. Radio observations, and also occasionally, gamma ray observations (e.g. Kanbach et al., 1993) revealed that this sequence of flare acceleration in the low corona can be followed by time extended acceleration, lasting from tens of minutes to hours. Figure 6.12 gives a simplified overview on the radio bursts associated with flares and of the subsequent activity. After the flare episode has taken place, energetic electrons produce type IV radio bursts that are broadband continua. Moving type IV bursts are produced by electrons radiating in ascending arches, whereas stationary type IV bursts are observed for several hours and, then, evolve gradually into noise storms that may last from days to weeks, and sometimes reappear at the following solar rotation, just as the associated sunspot group does.

Noise storms are the most common form of activity at meter and decameter wavelengths; narrowband, spiky type I bursts (type III at longer wavelengths) are superimposed on the continuum. Noise storms are produced by suprathermal electrons accelerated continuously over time scales of hours or days.

Figure 6.13 (Bougeret et al., 1984a) shows the history of solar radio emissions at a variety of frequencies, averaged over one day, for a period of almost 1.5 years, as observed by the radio investigation on the ISEE-3 spacecraft. The successive peaks of the upper plots describe interplanetary radio storms that may last up to ten days or more. They correspond to an almost continuous radiation emitted from streams of supra-thermal electrons injected near active regions. The radio structures assume an Archimedean spiral shape (Bougeret et al., 1984b), due to the combination of the solar wind expansion with the solar rotation. Such structures can be seen at radio frequencies rotating from east to west, as the Sun rotates. The analysis of these structures yield information on the solar wind density and velocity above active regions (where they are observed). The derived velocities tend to be smaller than the average solar wind velocity, which raises the question of the acceleration of the slow solar wind above solar active regions.

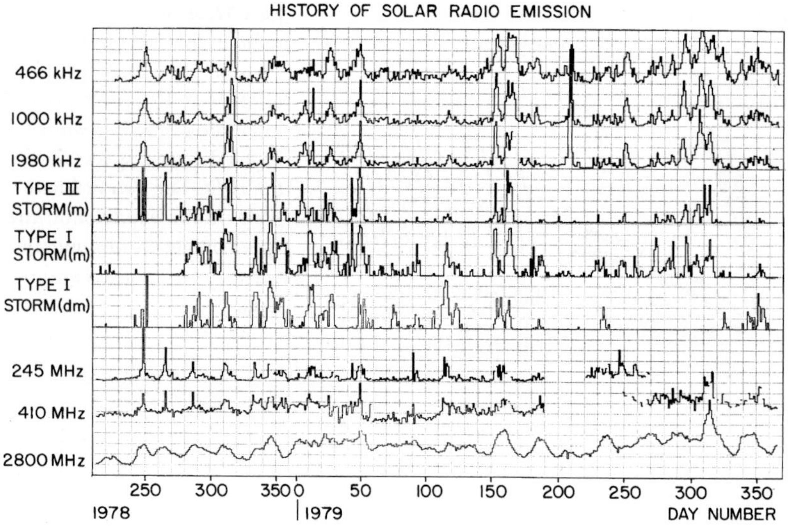

Fig. 6.13. Time history of radio emissions observed by ISEE-3 and storms indices (coronal type I and type III storms) computed from ground-based observations. This demonstrate the extension of active regions through IP space (from Bougeret et al., 1984)

Finally, the current understanding of this variety of radio signatures is that they are associated with large-scale eruptive phenomena including flares, filament eruptions, CMEs and shocks. These phenomena often occur simultaneously and are due to a fast release of magnetic energy. This coincidence emphasizes the difficulty of understanding the link between solar processes and solar energetic particle events (SEP) measured in the interplanetary medium. Radio observations, however, help remove the uncertainty, because they provide independent information often corresponding to a different scale of local events involving micro-physics and wave-particle plasma processes.

6.5 In Situ Wave and Particle Measurements

In situ measurements are part of radio and plasma wave investigations at both the observation stage (they are observed in space by the same radio instruments and displayed on the same dynamic spectrum) and the interpretation stage (a good description of in situ waves is essential to the understanding of the radio wave production mechanisms).

As has already been mentioned, the theoretical progresses on the radiation from beams of suprathermal electrons in the solar atmosphere were mostly based on comprehensive observations from space including both radio and in situ observations (Lin et al., 1986; Robinson et al., 1993; Cairns and Robinson 1997, 1998, 1999).

Bale et al. (1999) provided the first direct observations of energetic electrons and bursty Langmuir

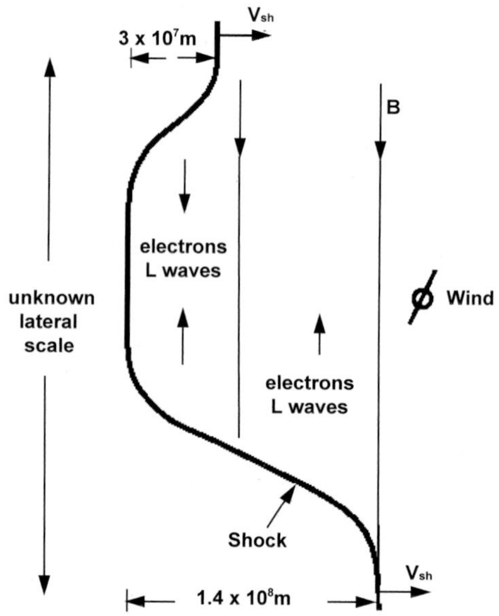

Fig. 6.14. Schematic showing the detailed shock geometry inferred by Bale et al. (1999) using in situ measurements in an active source region of an interplanetary type II burst observed by the WIND spacecraft

waves in an active type II source region upstream of a CME-driven shock (Fig. 6.14). These observations specifically show the production of radiation in a source region moving toward the spacecraft; the arrival of energetic electrons first anti-parallel to B and then parallel to B, which drove high levels of Langmuir waves, and the disappearance of the streaming electrons and Langmuir waves when the shock wave passed over the spacecraft. This shows strong similarities to the Earth's foreshock.

One of the most conspicuous features displayed on the dynamic spectra at low frequency (~30 kHz) is the quasi-thermal noise plasma line, which is almost always present near the local plasma frequency. The shape of this line can be accurately modeled and parameters of the electron distribution function can be derived, such as the electron density and temperature (Meyer-Vernet and Perche, 1989).

6.6 Radio Signatures of Coronal and Interplanetary Coronal Mass Ejections

Coronal mass ejections (CMEs) are the most spectacular large-scale manifestations of activity occurring at the Sun. They correspond to the destabilization of a large portion of the corona. 10^{14} – 10^{16} g of matter are ejected into the heliosphere with speeds ranging from 100 to more than 2000 km/s (Gopalswamy, 2004). Typical CMEs seen by coronagraphs span about 50° in angular extent with a few outstanding events reaching angular extents greater than 100°. CMEs are often associated with eruptive prominences (EP) or disappearing filaments on the solar disk. In that case, the CMEs contain three distinct regions: A bright compression front that surrounds a dark cavity and a bright core inside. The bright core is formed by the material ejected from the cool and dense prominence. CMEs have frequently a much more complex structure than the one just described: They involve multiple magnetic flux systems and neutral lines. As already mentioned, CMEs and flares can be produced jointly but it is now well established that there is no causal relationship between these two phenomena. CMEs can also drive interplanetary shocks (type II bursts) that together with Earth-directed CMEs can produce large transient interplanetary and geomagnetic disturbances (Sect.16.3). Therefore, their study is essential for understanding and predicting space weather conditions. CMEs were most often observed by white-light orbiting coronagraphs which occult the solar disk and which cannot observe their initiation and early stages of development. Since 1996, the LASCO instrument (large angular spectrometric coronagraph) aboard the SOHO mission has provided for the first time a large field of view from 1.1 to approximately 30 solar radii (Brueckner et al., 1995). Combined radio/EUV/LASCO observations have led to new insights on the physics of CMEs.

Radio telescopes can observe both the solar disk and the corona out to a few solar radii with a very high cadence (<0.1 s), while the cadence of CME observations made by space-borne instruments is severely limited.

Figure 6.15 shows the evolution of a prominence observed in radio, at microwave lengths, which was associated with a CME. Prominences have a low temperature (8000°), a high density (~10^{10} – 10^{11} cm^{-3}) and are optically thick at microwave frequencies. The prominence first exhibited a progressive evolution then, during the eruption, it heats up and expands rapidly. Prominences are likely to become CME cores. At decimeter wavelengths, observations of filament eruptions showed that both the filament and the precursor to the white light CME cavity can be detected. There is a clear continuity between the radio thermal depression observed in the low corona, which is identified to the filament cavity, and the corresponding CME when they are observed together (Marqué et al., 2002).

Solar activity in the metric range is often observed in association with CMEs and in the absence of flares; a close spatial and temporal relationship exists between noise storm enhancements and white light transient activity such as CMEs or additional material in the corona at the vicinity of the radio source (Kerdraon et al., 1983). The origin of this non-thermal activity is due to emerging magnetic loops interacting with overlying loops and leading to large scale coronal magnetic reconfiguration. In the presence of flares, strong radio bursts are observed over a broad frequency range. This is discussed in the next section.

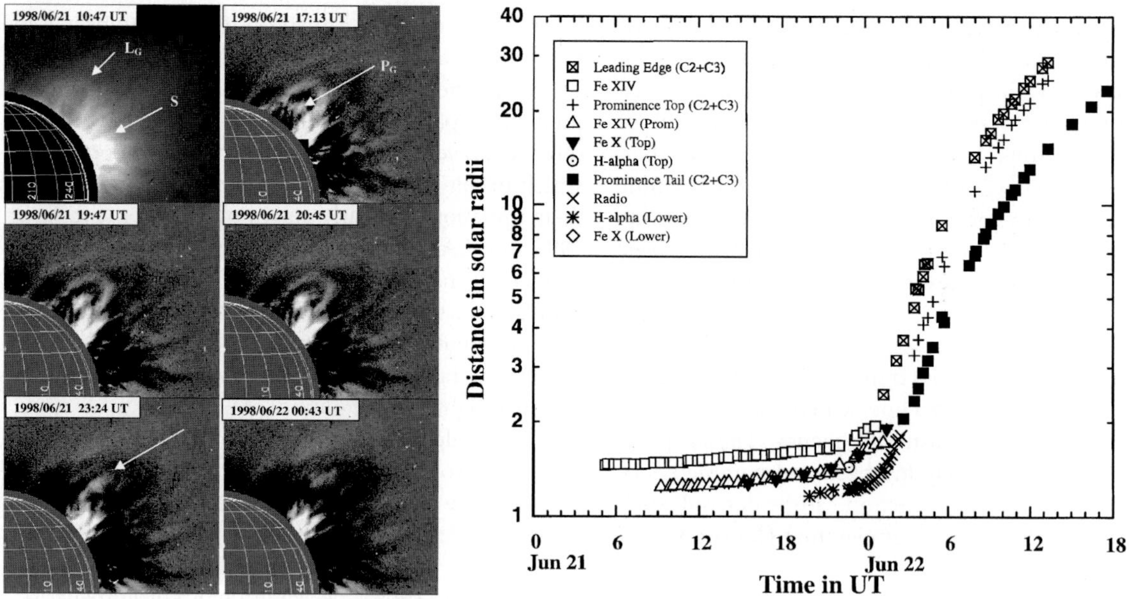

Fig. 6.15. *Left panel*: Time-lapse images taken by LASCO-C1 coronagraph in Fe XIV emission line. The field of view is 1.1 – 3 Rs. All the images shown here have been subtracted from a reference image taken before the occurrence of the CME, except the one at 10:47 UT. The image at 10:47 UT is an on-line image with a nearby continuum and is given in order to show the bright streamer adjacent to the CME. *Right panel*: Plot of height versus time (on log scale) for different features of the CME, viz, the leading edge, the prominence top and the tail, as measured from the images obtained at different wavelengths by various instruments (from Srivastava et al., 2000)

6.6.1 Flare/CME Events: Lift-Off and Angular Spread in the Corona

Radio imaging of fast flare CME events shows that many CMEs originate from a rather small coronal region in the vicinity of the flare site and expand by successive magnetic interactions at larger and larger distances from the flare site; signatures of these interactions are detected by bursts of non-thermal origin, in the dm–m wavelength domain. CMEs reach their full extent in the low corona (below 2Rs) on a time scale of a few minutes. This time scale corresponds to disturbances with speeds of 1000 km/s that were identified with coronal type II bursts by on-the-disk observations. Hence, tracing the space time evolution of the metric emission, with a temporal cadence of typically 10 s, easily allows one to follow the opening of CMEs. This is illustrated, for both a limb and an on-disk event, in Fig. 6.16 for a limb event (Maia et al. 1999) and in Fig. 6.16 for an on-disk event (Pohjolainen et al., 2001). The bottom panel in Fig. 6.17 displays two plots showing the space time evolution of the limb event observed by the NRH at 164 MHz: By integrating the NRH solar images in the north–south and east–west directions respectively. This figure exhibits a clear radio signature of the progression in latitude of the CME event displayed in the upper panel and the corresponding speed of the disturbance can be easily estimated (Pick et al., 2003). These events suggest that large interconnecting loops are ejected together. The triggering of these events could be caused by coronal reconnection at a magnetic null point leading to the destabilization of the system and the ejection of large interconnecting loops, as suggested by Maia et al. (2003).

6.6.2 Direct Radio CME Imaging

For events associated with flaring regions located behind the limb, faint emissions can be detected over

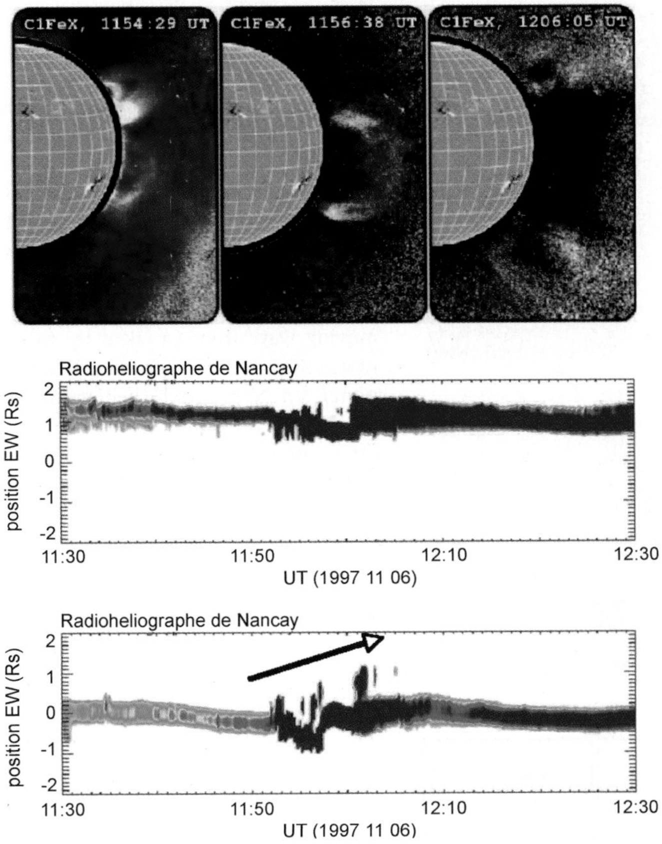

Fig. 6.16. CME event on 6 November 1997. *Bottom panel*: Two one-dimensional plots along east–west *top* and north–south *bottom* directions showing the space-time evolution of the event at 164 MHz. The radio source displacement traces the CME opening seen in the *upper panel* (LASCO coronagraph) (Maia et al., 1999; Pick et al., 2003)

the solar limb and directly compared to white light coronagraph emissions. Radio loops can be detected behind the CME front. The emission is non-thermal gyrosynchrotron from electrons with energy of a few MeV. These observations, presently rather rare, are important for obtaining information on the CME structures, such as constraints on the thermal plasma density (Bastian et al., 2001). An example is displayed in Fig. 6.18.

6.6.3 Interplanetary Coronal Mass Ejections

It is well established that large, non-recurrent geomagnetic storms are caused by interplanetary shocks and interplanetary coronal mass ejections (ICMEs). ICMEs, observed in the solar wind, most often include an IP shock preceding a turbulent plasma sheath and an ejecta. The term "magnetic cloud" refers to ICMEs for which ejecta are characterized by a smooth and continuous rotation of the magnetic field (flux rope) (Burlaga et al., 1981). Many CMEs do not produce large disturbances in the solar wind. Moreover, it is often difficult to make a unique association between ICMEs and CMEs. One underlines the importance of on-disk solar observations to identify the solar events and their solar origin which will produce the solar wind disturbances propagating out and reaching the Earth after a transit time of 2 – 4 days. It appears that propagating interplanetary disturbances are distorted by the slow and fast solar winds and by their interaction with the heliospheric current sheet. The radio diagnostics is helping trace such interactions.

Interaction between fast and slow CMEs have been identified by long-wavelength radio and white

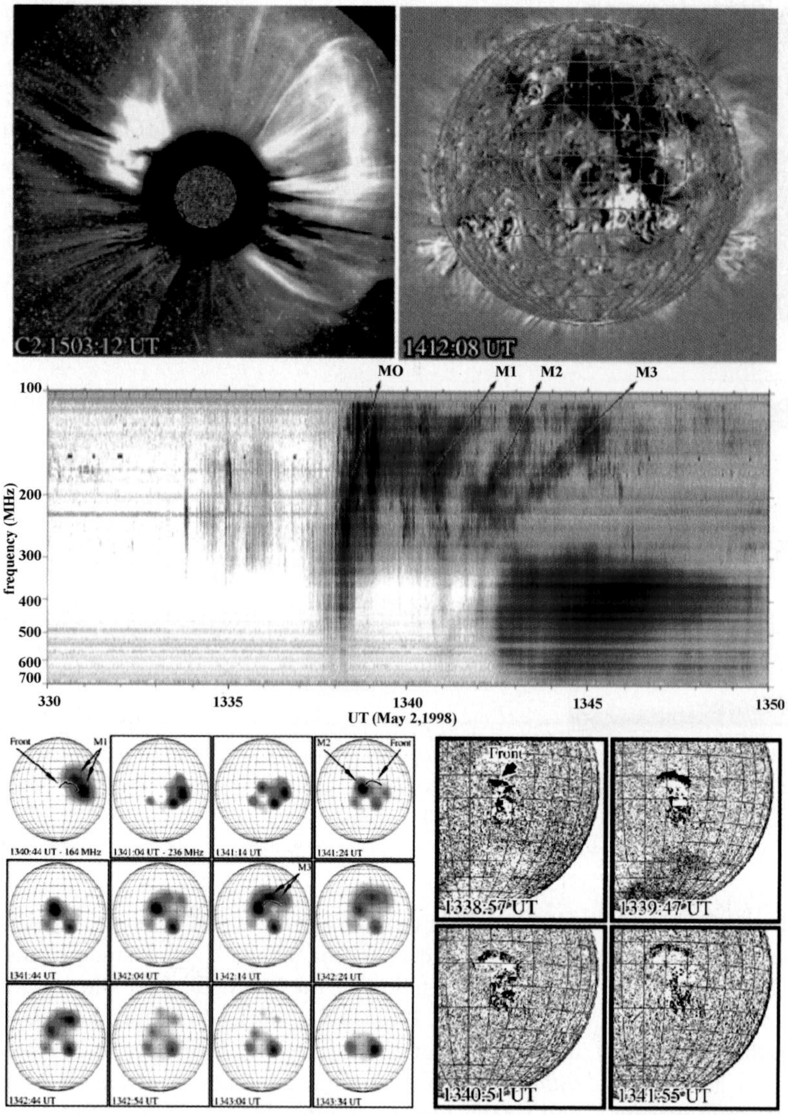

Fig. 6.17. CME event on 2 May 1998. *Upper panel, left*: SOHO LASCO coronagraph image showing the halo CME. *Upper panel right*; SOHO EIT at 195Å showing the EIT dimming region. *Middle panel*: Artemis IV radio spectra. The spectral drifting sources are labelled M0 for the one in the flash phase and M1, M2, M3 for the type II-like emission. *Lower panel, right*: Running difference of Kanzelhohe Hα images showing the moving wave front (marked by an arrow in the first difference image). *Lower panel, left*: Nancay radio heliograph images at 164 MHz and 236 MHz showing the location of the sources labelled M1, M2, M3 at selected times (adapted from Pohjolainen et al., 2001)

light observations; during these interactions, intense, localized radio emissions are detected, as shown in Fig. 6.19. Many such cases have been studied by Gopalswamy et al. (2001b; 2002a). Gopalswamy et al. (2002b) inferred that the efficiency of the CME-driven shocks is enhanced as they propagate through the preceding CMEs and that they accelerate SEPs from the material of the preceding CMEs rather than from the quiet solar wind (Gopalswamy et al., 2004).

6.7 Conclusions: The Relevance of Radio Observations to the Understanding of the Solar-Terrestrial Environment

In this chapter, we have presented several key areas for which radio observations offer significant insights in the understanding of solar activity and its link with the interplanetary medium. Radio emissions can address extremely broad science topics, including the nature and

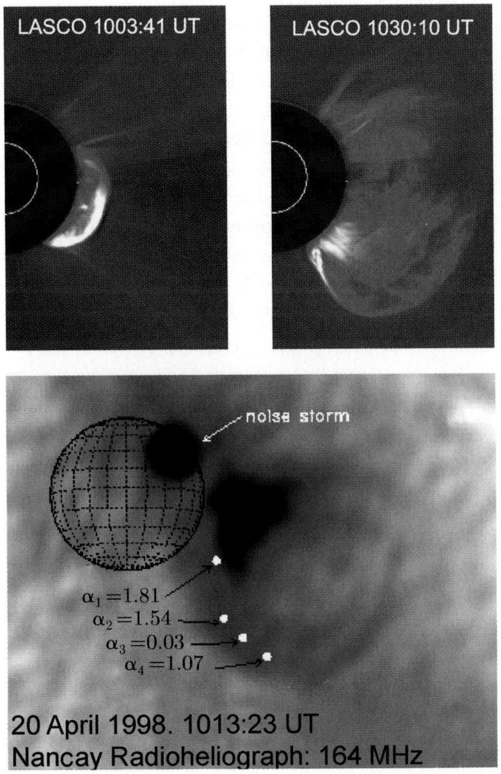

Fig. 6.18. CME event on 20 April 1998: *Upper panel*: Composite LASCO SOHO coronagraph images (C1 and C2) showing the images of a CME. *Lower panel*: Snapshot map of the corresponding radio CME at 164 MHz at 10:13 UT; the brightness of the radio image is saturated in the low corona in order to reveal the faint radio arcade; the emission is gyrosynchrotron. Numbers on the same figure correspond to spectral indices measured at four locations and not discussed h ere (from Bastian et al., 2001)

evolution of coronal and interplanetary magnetic fields, the physics of solar flares, the quiet Sun and quiescent structures, the drivers of space weather (particles and shocks, the acceleration of energetic electrons).

We have shown that radio emissions can bring information that bridges macro-scales and micro-physics by remotely detecting either structures that would not be otherwise visible or sites of particle acceleration where energy transfers occur. Indeed, radio emissions can detect very localized signatures, with a very high signal-to-noise ratio. The timing can be extremely accurate. Non-thermal sources can exhibit significant evolution over 1 second timescales and the instruments – even the imaging radioheliographs – can provide a time resolution much better than a second (25 images per second for the Nançay radioheliograph).

Major progresses are expected from ground-based solar dedicated radio telescopes designed and optimized to produce high resolution, and high-dynamic-range images over a broad range of radio frequencies. Such is the frequency agile solar radiotelescope (FASR) that is to be operated in the range ~0.05 – 24 GHz enabling a wide variety of radio-diagnostic tools to be exploited in order to study the Sun from the mid-chromosphere to coronal heights (Gary and Keller, 2004). The power of radio observations lies in their flexibility, for instance, in accessing the many facets of CMEs (prominence eruption, bulk material, shocks, waves, particle acceleration).

The radio diagnostics needs to be improved, however. Long wavelength observations lack spatial resolution. The size of an IP radio source can be huge (a typical size can be 1/2 of the distance of the source to the Sun). Still, the direction of source centroids can be measured within a few degrees in the IP medium, even from a single spacecraft, allowing us to track radio events (beams of electrons, shock waves) through IP space. The STEREO mission will carry a radio and plasma wave instrument covering the frequency range from 2.5 kHz to 16 MHz, providing a global survey of radio emissions from about 3 Rs from the center of the Sun to the orbit of the earth. The two identical spacecraft will provide simultaneous measurements from two directions. This will allow us to determine the source locations more accurately, thus providing a better knowledge of directional effects and propagation effects. A similar radio instrument on the Wind spacecraft will strongly enhance this stereo technique.

Coordinated observations with ground-based and space radio instruments and with observations in other parts of the electromagnetic spectrum is certainly the clue to progress in the field of solar-terrestrial physics. The radio domain, accessible from both ground and space, has an important potential that remains to be exploited using improved instruments and techniques.

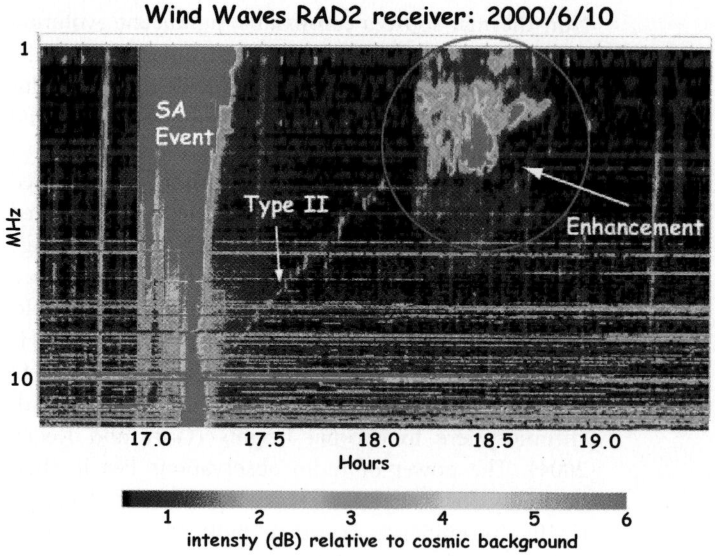

Fig. 6.19. Example of an hectometric interplanetary type II burst observed by the Wind/WAVES spectrograph in the 1 – 14 MHz range. The thin faint drifting line is a type II burst. A bright emission is detected between 18.12 and 18.48 UT, corresponding to a fast CME overtaking a slower one. The red vertical burst is a "shock-associated" event, proposed to have been accelerated as the shock was progressing in the corona at ~1 solar radius above the photosphere. (from Gopalswamy et al., 2001b)

References

Aschwanden, M.J., Benz, A.O., and Schwartz, R.A. (1993): Astrophys. J., **417**, 790.

Aschwanden, M.J., and Benz, A.O. (1997): Astrophys. J., **480**, 825–39.

Aschwanden, M.J. (2002): Space Sci. Rev., **101**, 1–2.

Bastian, T.S., Pick, M., Kerdraon, D., Maia, D. and Vourlidas, A. (2001): Astrophys. J., **20**, L 65–69.

Bastian, T.S., Benz, A.O., and Gary, D.E. (1998): Ann. Rev. Astron. Astrophys., **36**, 131—88.

Baumback, M.M., Kurth, W.S., and Gurnett, D.A. (1976): Solar Phys., **48**, 361–380.

Benz, A.O. (1993): in Plasma Astrophysics, Kluwer Academic Publishers, Dordrecht, the Netherlands, **184**.

Benz, A.O., Csillaghy, A., and Aschwanden, M.J. (1996): Astron. Astrophys., **309**, 291–300.

Bothmer, V., and Rust, D.M. (1999): in Coronal Mass Ejections, Geophysical Monograph **99**, 139–146.

Bougeret, J.L., Fainberg, J., and Stone, R.G. (1984a): Astron. Astrophys., **136**, 255–262.

Bougeret, J.L., Fainberg, J., and Stone, R.G. (1984b): Astron. Astrophys., **141**, 17–24.

Bougeret, J.L., Kaiser, M.L., Kellog, P.J. et al. (1995): Space Sci. Rev, **71**, 231–263.

Bougeret, J.L., Zarka, P., Caroubalos, C., Karlický, M., Leblanc, Y., Maroulis, D., Hillaris, A., Moussas, X., Alissandrakis, C. E., Dumas, G. and Perche, C. (1998), GRL, **25**, 2513

Burlaga, L., Sittler, E., Mariani, F., and Schwenn, R.: (1981): J. Geophys. Res, **86**, 6673–6684.

Brueckner, G.E., and 14 coauthors (1995): Solar Phys., **162**, 357–402.

Cairns, I.H., and Robinson, P. (1997): Geophys. Res. Lett., **24**, 369.

Cairns, I.H., and Robinson, P. (1998): Astrophys. J., **509**, 471.

Cairns, I.H., and Robinson, P. (1999): Phys. Rev. Lett., **82**, 3069.

Cairns, I.H., Robinson, P., and Zank, G.P. (2000): PASA, **17**, 22.

Cane, H.V., Sheeley Jr., N.R., and Howard, R.A. (1987): J. Geophys. Res., **92**, 9869–9874.

Caroubalos, C., Poquérusse, M, Bougeret, J.-L., and Crépel, R. (1987): Astrophys. J., **319**, 503–513.

Drago-Chiuderi, F., Alissandrakis, C., and Hagyard, M. (1987): Solar Phys, **112**, 89.

Fainberg, J., Evans, L.G, and Stone, R.G. (1972): Science, **178**, 743.

Gary, D.E., and Keller, C.U. (2004): Solar and Space Weather Radiophysics – Current Status and Future Developments. Edited by Dale, E. Gary, Center for Solar-Terrestrial Research, New Jersey Institute of Technology, Newark, N.J., U.S.A.; Christoph, U. Keller, National Solar Observatory, Tucson, AZ, U.S.A. Astrophysics and Space Science Library, Volume 314, Kluwer Academic Publishers, Dordrecht.

Gopalswamy, N. and 8 coauthors (1998): J. Geophys. Res., **105**, A1, 307–316.

Gopalswamy, N., Yashiro, S., Kaiser, M.L., Howard, R.A., and Bougeret, J.L. (2001a): J. Geophys. Res., **106**, 29219.

Gopalswamy, N., Yashiro, S., Kaiser, M.L., Howard, R.A., and Bougeret, J.L. (2001b): Astrophys. J. Lett., **548**, L 91–94.

Gopalswamy, N., Yashiro, S., Kaiser, M.L., Howard, R.A., and Bougeret, J.-L. (2002a): GRL, **29**, 106.

Gopalswamy, N., Yashiro, S., Michalek, G., Kaiser, M.L., Howard, R.A., Reames, D.V., Leske, R., and von Rosenvinge, T. (2002b): Astrophys. J. Lett., **572**, L 103–107.

Gopalswamy, N. (2004): in The Sun and the Heliosphere as an integrated system, eds. Poletto and Suess, Kluwer, Boston, 201.

Gopalswamy, N., et al. (2004): J. Geophys. Res., **109**, 12105.

Jaeger, J.C., and Westfold, K.C. (1949): Australian, J. Sci. Res., **A2**, 322.

Kakinuma, T., and Swarup, G. (1962): Astrophys. J., **39**, 5.

Kanbach, G. and 9 coauthors (1983): Astron. Astrophys., **97**, 349–353.

Kerdraon, A., Pick, M., and Trottet, G. (1983): Astrophys. J. Lett., **265**, L 19.

Kerdraon, A., and Delouis, J.M. (1997): in Coronal Physics from Radio and Space Observations, ed. G. Trottet, Springer, Berlin Heidelberg New York, **483**, 192–201.

Kundu, M.R. (1965): Solar Radioastronomy, Interscience Publishers, a division of John Wiley and Sons, New York, USA.

Kruger, A. (1979): Introduction to Solar Radio Astronomy and Radio Physics, D. Reidel, Dordrecht, the Netherlands.

Lin, R.P., Levedahl, W.K., Lotko, W., Gurnett, D.A., and Scarf, F.L. (1986): ApJ, **308**, 954.

Maia, D., Vourlidas, A., Pick, M., Howard, R., Schwenn, R., and Magalhaes, A. (1999): J. Geophys. Res., **104**, 12, 57.

Maia, D., Aulanier, G., Wang, S.J., Pick, M., Malherbe, J.M., and Delaboudinière, J.P. (2003): Astron. Astrophys. **405**, 467–472.

Marqué, C., Lantos, P., and Delaboudiniere, J.P. (2002): Astron. Astrophys., **387**, 317–325.

Melrose, D.B. and McPhedran, R.C. (1991): Electromagnetic Processes in Dispersive Media, Cambridge Unversity Press, Cambridge, UK.

Meyer-Vernet, N., and Perche, C. (1989): J. Geophys. Res., **94**, 2405–2415.

Moreton, G.E., and Ramsey, H.E. (1960): PASP, **72**, 428, 357.

Nakajima, H. and 26 coauthors (1994): Proc. IEEE, **82**, 705.

Pick, M., Maia, D., Marqué. (2003): Advances in Space Res., in press.

Pick, M., (2006): Lecture Notes in Physics **699**, Springer, Berlin Heidelberg New York, 119–142.

Pohjolainen, S.D., and 9 coauthors. (2001): ApJ, **556**, 421–431.

Poquérusse, M., Bougeret, J.L., and Caroubalos, C. (1984): Astron. and Astrophys., **136**, 10–16.

Reiner, M.J., and Stone, R.G. (1988): Astron. Astrophys., **206**, 316–335.

Reiner, M.J., and Stone, R.G. (1989): Astron. Astrophys., **217**, 251–269.

Robinson, P.A. (1992): Solar Phys., **139**, 147.

Robinson, P.A. (1995): Phys. Plasmas, **2**, 1466.

Robinson, P.A., Cairns, I.H., and Gurnett, D.A. (1993): Astrophys. J., **407**, 790.

Solar Radio Physics (1985): Cambridge University Press, eds D. J. McLean and N. R. Labrum.

Srivastava, N., Schwenn, R., Inhester, B., Martin, S.F., and Hanaoka, Y. (2000): Astrophys. J., **534**, 468–481.

Steinberg, J.-L., Dulk, G.A., Hoang, S., Lecacheux, A., and Aubier, M.G. (1984): Astron. Astrophys., **140**, 39–48.

Steinberg, J.-L., Hoang, S., and Dulk, G.A. (1985): Astron. Astrophys., **150**, 205–216.

Wagner, W.J., and MacQueen, R.M. (1983): Astron. Astrophys., **120**, 136.

Wild, J.P., Smerd, S.F., and Weiss, A.A. (1963): Ann. Rev. Astron. Astrophys., **1**, 291.

Zheleznyakov, V.V. (1962): Soviet Astron., **6**, 3.

Zheleznyakov, V.V. (1977): Radio Emission of the Sun and Planets, Pergamon Press, Inc.

Part 2

The Earth

7 Magnetosphere

Michael Schulz

Earth's magnetosphere is a fascinating repository of cold plasma, hot plasma, and energetic charged particles, all acting under the kinematical and dynamical influence of electric and magnetic fields, as well as under the influence of waves and instabilities, which the plasmas themselves control in turn. This chapter constitutes a largely theoretical overview of Earth's magnetosphere and the physical processes that govern it.

Contents

Michael Schulz, Magnetosphere.
In: Y. Kamide/A. Chian, Handbook of the Solar-Terrestrial Environment. pp. 155–188 (2007)
DOI: 10.1007/11367758_7

7.1 Introduction

The magnetosphere (see Fig. 7.1) occupies a region of space within which the geomagnetic field is (in first approximation) confined by the solar wind. It is bounded by a thin current layer called the magnetopause, which is shaped somewhat like a windsock, and preceded upstream by a hyperboloidal bow shock through which the solar wind makes a transition from super-magnetosonic to sub-magnetosonic flow velocity. A gap $\sim 2\text{–}3\,R_E$ (Earth radii, $1\,R_E = 6371.2$ km by convention) separates the bow shock from the magnetopause along the Earth-Sun line because the magnetopause itself presents a blunt obstacle to the flowing solar wind. The size and shape of the magnetosphere are largely determined by pressure balance between the shocked solar wind and the compressed geomagnetic field. The nose of the magnetosphere is located (under nominal solar-wind conditions) about $10\,R_E$ upstream from the best-fitting point dipole.

Orientation of the magnetosphere is controlled by the direction of the solar wind, which is a highly ionized plasma that flows outward from the Sun at velocities $u \sim 300 - 800$ km/s, with a mean value $\langle u \rangle \sim 400$ km/s at low (near-ecliptic) latitudes in the heliosphere. The solar-wind plasma is mostly hydrogenic ($\sim 95\%$ by number density) but also includes some helium, oxygen, and other ions in various charge states. The total electron density N_e at $r = 1$ AU ($\approx 215 r_\odot$) in the heliosphere varies considerably about its mean value $\langle N_e \rangle \sim 8\,\text{cm}^{-3}$. (For reference: 1 AU $\approx$ 149.6 Gm, astronomical unit; $r_\odot \approx 696$ Mm, solar radius.) The radial component of solar-wind velocity tends to increase with heliospheric latitude (measured relative to the heliospheric current sheet) and thus shows both 27-day and semi-annual variations at Earth. (The heliospheric current sheet makes 14.5 rotations per year relative to inertial space, but only 13.5 per year from Earth's perspective.) Sudden variations in solar-wind speed and direction are associated with disturbances such as coronal mass ejections and interplanetary shocks, which can directly impact the magnetosphere.

Magnetospheric dynamical processes are influenced not only by the solar wind itself, but also by the interplanetary magnetic field (IMF), which emanates from the Sun and has its direction in the heliosphere largely controlled by the solar wind. The solar-wind speed u at Earth's orbit is ~ 8 times the local Alfvén speed c_A, which means that the solar wind has an energy density ~ 64 times the local magnetic energy density $B^2/2\mu_0$. (This chapter is written mostly in SI units, système internationale, also called mks, in which $\mu_0 \equiv 4\pi \times 10^{-7}$ henry/meter denotes the permeability of free space. The energy density in cgs units would be written as $B^2/8\pi$.)

The strong inequality $u \gg c_A$ in most of the heliosphere implies that plasma flow determines the magnetic-field configuration rather than the other way around. In order for a radially moving plasma element to remain "forever" on the same field line as the Sun rotates, the field line itself must describe a large-scale Archimedes spiral (Parker, 1963, pp. 138–139) in the heliosphere, being wrapped once around the Sun in the radial distance Δr (≈ 5.1 AU for $u \approx 350$ km/s) that such a plasma element travels during each solar rotation (≈ 25.2 days for global-scale magnetic features). This model for the IMF is known as the *Parker spiral*, and these particular parameters correspond to a local ratio $B_\varphi/B_r \approx 0.393\pi$ between the azimuthal and radial components of $\boldsymbol{B}_{\text{IMF}}$ at Earth's orbit (hence to an angle $\approx 51°$ between $\boldsymbol{B}_{\text{IMF}}$ and the radial direction).

The angle between $\boldsymbol{B}_{\text{IMF}}$ and $\boldsymbol{u}$ implies the presence of an electric field $\boldsymbol{E} = -\boldsymbol{u} \times \boldsymbol{B}_{\text{IMF}}$ upstream from the bow shock. Both the solar-wind plasma and the IMF change direction and undergo compression upon crossing the bow shock and thus entering the magnetosheath. The interplanetary electric field maps into the magnetosheath as well. Through a process called *magnetic reconnection*, particles and fields (both electric and magnetic) from the magnetosheath can penetrate through the magnetopause and thereby enter the magnetosphere. Reconnection at the magnetopause is favored in regions where the magnetosheath $\boldsymbol{B}$ field makes an angle $> 90°$ with the magnetospheric $\boldsymbol{B}$ field, and especially so at places where the magnetosheath $\boldsymbol{B}$ field is anti-parallel to the magnetospheric $\boldsymbol{B}$ field. Such penetration is of major importance for magnetospheric dynamics.

The present chapter offers a largely theoretical overview of magnetospheric structure and dynamics. It provides examples of modeling techniques that aim for realistic descriptions of the magnetospheric configuration, as well as modeling techniques that lend themselves to analytical calculations. Finally, this

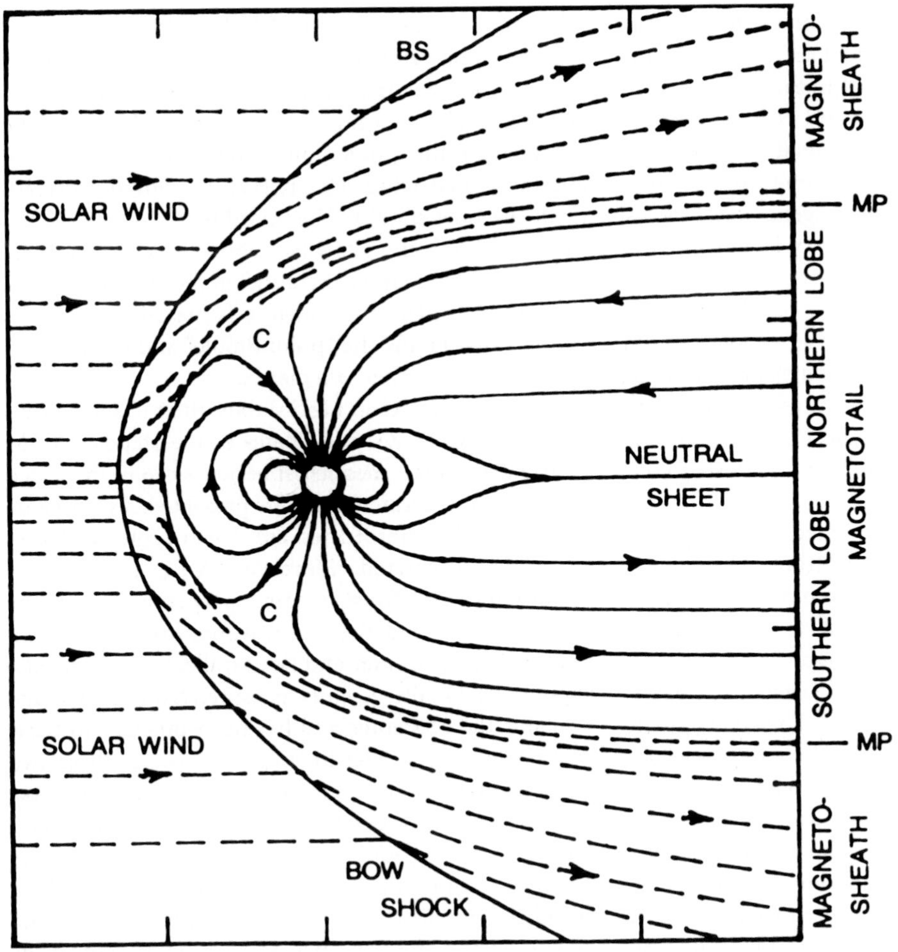

Fig. 7.1. Schematic illustration of solar-wind streamlines (*dashed curves*), magnetic field lines (*solid curves*) in the magnetosphere, important plasma boundary surfaces (magnetopause, bow shock, neutral sheet), and important regions of space (magnetotail lobes, cleft region, magnetosheath) in and around the magnetosphere. The plane of this figure is the noon-midnight meridional plane, which in this context contains the magnetic dipole axis (treated here as normal to the solar-wind velocity $\boldsymbol{u}$, hence for "magnetic equinox" conditions) and lies parallel to $\boldsymbol{u}$. Arrows on dashed curves indicate direction of $\boldsymbol{u}$; arrows on solid curves indicate direction of $\boldsymbol{B}$. The magnetopause (MP, marked in right margin) corresponds to the innermost dashed curve, except that this becomes indistinguishable from the outermost dayside field line shown (*solid curve*) at latitudes $\lesssim 30°$ in the noon meridian. The solid curve (not a field line) upstream from the magnetopause represents the bow shock (BS). The region between the bow shock and the magnetopause is known as the magnetosheath (identified in right margin). Magnetotail lobes (also identified in right margin) belong to the magnetosphere and are separated (as shown) by the nightside neutral sheet, which provides pathways for magnetosheath plasma entry from the flanks. Dayside cleft regions (C, also know as "cusp" regions) belong to the magneto-sphere but provide additional pathways for access of magnetosheath plasma and energetic particles to the magnetosphere

chapter provides an exposition of dynamical processes that govern charged-particle populations in the magnetosphere and lead to formation and evolution of (for example) the plasmasphere, the radiation belts, the ring current, and the aurora.

7.2 Magnetic Configuration

The magnetosphere is anchored in space by the best-fitting point dipole, which is displaced from the geographic center by about 550 km toward a point 2.6°

south of Iwo Jima. The geomagnetic dipole moment $\boldsymbol{\mu}_{\mathrm{E}}$ (≈ 0.300 G-R_{E}^3 at present) is tilted 10.3° relative to the Earth's rotation axis; it lies parallel to the plane defined by geographic meridians 108.2° E and 71.8° W. Because of the offset, however, the geomagnetic dipole axis pierces the Earth's surface about 7° and 14°, respectively, from the northern and southern geographic poles. The corresponding magnetic geometry is illustrated in Fig. 7.2.

It is customary in magnetospheric physics to measure magnetic colatitude θ from the offset-dipole axis and an azimuthal coordinate φ (called *magnetic local time*, MLT) from the plane that contains the offset-dipole axis and lies parallel to the Sun-Earth line. Magnetic-latitude contours in Fig. 7.2 are separated by 10°, and MLT contours are separated by 15°(= 1 h).

The magnetosphere itself is oriented by the apparent direction of the solar wind in Earth's frame of reference. This means that account must be taken of Earth's orbital velocity ($\approx$ 30 km/s) around the Sun, as well as the radial and any non-radial component of solar-wind velocity in the heliosphere, in order to account well for magnetosphere's orientation in space. The geomagnetic tail's orientation is thus essentially (but not perfectly) anti-sunward. The effect of Earth's orbital velocity on magnetospheric orientation is known as *aberration*. There is no particular name for the magnetospheric consequences of an azimuthal component of solar-wind velocity in the heliosphere. Earth's orbital velocity ($\approx$30 km/s) leads to an aberration ($\approx$4.9° for solar-wind velocity $u \approx 350$ km/s) in the apparent mean direction of the solar wind relative to the Sun-Earth line, but this effect is partially offset by the azimuthal component ($\sim$10 km/s) of the solar-wind velocity at $r = 1$ AU in the heliosphere.

An angle of major significance for magnetospheric geometry is the angle ψ between the dipole moment $\boldsymbol{\mu}_{\mathrm{E}}$ and the solar-wind velocity $\boldsymbol{u}$ in Earth's reference frame. This angle varies systematically (with season of year and hour of day) between 56.3° and 123.7°, as well as sporadically in response to variations in solar-wind velocity. The extrema in ψ correspond to "magnetic-solstice" conditions, whereas $\psi = 90°$ corresponds to "magnetic equinox." The plane that contains the offset-dipole axis and lies parallel to the aberrated solar-wind velocity thus bisects the magnetosphere between AM and PM halves that are (in first approximation) mirror images of each other.

Figure 7.3 (Schulz and McNab, 1996) illustrates an idealized magnetosphere at magnetic equinox (upper panels) and near magnetic solstice (lower panels), respectively. The coordinate ξ is measured from a plane that contains the "point dipole" and lies transverse to the aberrated solar-wind velocity (i.e., as realized upstream from the bow shock). The coordinate η in Fig. 7.3 is measured from the plane (see above) that contains the offset-dipole axis and lies parallel to the aberrated solar-wind velocity. The coordinate ζ is measured from the plane that contains the "point dipole" and lies perpendicular to these other two planes.

The coordinates in Fig. 7.3 thus differ in subtle ways from the (X, Y, Z) coordinates of the standard GSM (geocentric solar magnetospheric) system in which data from spacecraft are usually presented. As the name implies, GSM coordinates are Earth-centered rather than dipole-centered. Moreover, X_{GSM} is measured toward the Sun from a plane perpendicular to the Earth-Sun line, and Y_{GSM} is measured from a plane that contains the geocenter and lies parallel to the magnetic dipole axis. Finally, Z_{GSM} is measured from a plane that contains the geocenter and lies perpendicular to these other two planes. The differences between these two right-handed systems, (ξ, η, ζ) and GSM, are unimportant for general orientation, except that ξ corresponds roughly to $-X_{\mathrm{GSM}}$ and η corresponds roughly to $-Y_{\mathrm{GSM}}$. However, some care must be taken when making detailed comparisons between theoretical models and *in situ* data, to be sure that both are expressed in the same coordinate system for this purpose.

The model $\boldsymbol{B}$ field illustrated in Fig. 7.3 is derived from a scalar potential expanded in spherical harmonics:

$$\begin{aligned}\boldsymbol{B}(r,\theta,\varphi) &= \\ &- g_1^0 \nabla[(a^3/r^2)\cos\theta] \\ &- (a^3/b^2)\nabla \sum_{n=1}^{\bar{N}} \sum_{m=0}^{n} \bar{g}_n^m (r/b)^n P_n^m(\theta)\cos m\varphi\,, \quad (7.1)\end{aligned}$$

where $a = 1 R_{\mathrm{E}}$ and b is the distance from the point dipole to the "nose" of the magnetopause. The colatitude θ in (7.1) is measured from the magnetic dipole axis, and the MLT coordinate φ is measured from the plane that contains the dipole axis and lies parallel to the aberrated solar-wind velocity $\boldsymbol{u}$. The expansion coefficients $\{\bar{g}_n^m\}$ in (7.1) are chosen (Schulz and McNab, 1996) so as to

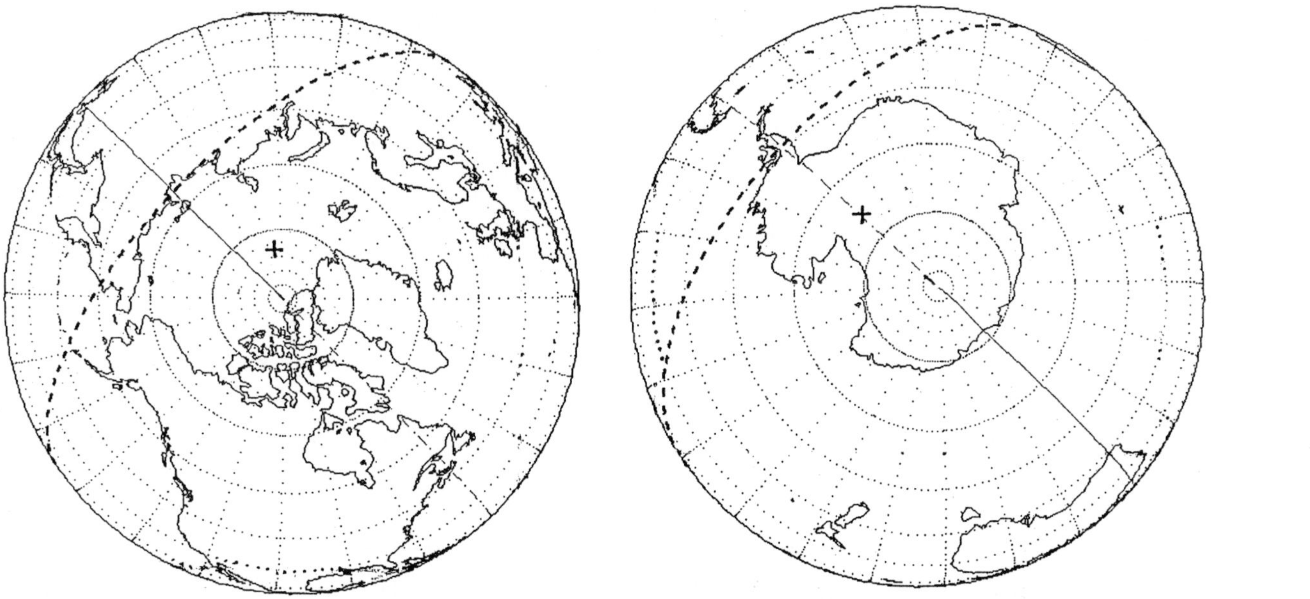

Fig. 7.2. Contours (*dotted*) of constant magnetic (offset-dipole) latitude and constant magnetic local time (MLT) on Earth's surface in "snapshots" taken 71.6 h apart (southern at 04:28.5 UT, 30 Jan 2002; northern at 04:04.5 UT, 2 Feb 2002). Magnetic-latitude contours are spaced by 10°, MLT contours by 15°, dots on MLT contours by 5° latitude. Solid (*lightly dashed*) meridian denotes noon (midnight) MLT. Plus signs (+) denote geographic poles. Heavy dashed curve denotes day-night terminator. (Specific times and dates UT are unimportant for illustrating the magnetic geometry but do control terminator locations and MLT labels of magnetic meridians.) This map was generated from software written by S.M. Petrinec for <http://pixie.spasci.com>, the website about Polar's Ionospheric X-ray Imaging Experiment (PIXIE)

minimize the variational quantity

$$\sigma \equiv \alpha \int_{\mathrm{mp}} (\hat{\boldsymbol{n}} \cdot \boldsymbol{B})^2 \, \mathrm{d}A_{\mathrm{mp}} + (1-\alpha) \int_{\mathrm{xt}} (\hat{\boldsymbol{n}} \times \boldsymbol{B})^2 \, \mathrm{d}A_{\mathrm{xt}} , \tag{7.2}$$

where $\hat{\boldsymbol{n}}$ is the local unit vector directed normally outward from the source surface and α is a parameter that assigns relative weights to equal elements of area on the magnetopausal (concave-tailward) and cross-tail (slightly concave-sunward) portions of the *source surface* in this model. For $\alpha = 0.5$ (as in Fig. 7.3) the weights assigned to equal areas are equal, but the area A_{mp} of the magnetopausal portion is about twice the area A_{xt} of the cross-tail portion.

Each tail field line beyond the source surface in Fig. 7.3 is specified by an equation of the form

$$(\rho/\rho_\infty)^{2.280} = \tanh[(0.61338/b)(\xi + b)] \tag{7.3}$$

proposed by Schulz and McNab (1996), where $\rho^2 = \eta^2 + \zeta^2$ and ρ_∞ is the asymptotic distance of the field line from the ξ axis as $\xi \to +\infty$. The magnetopause itself corresponds to $\rho_\infty = \rho_\infty^* = 1.7696b$ in (7.3). This expression is a parametrized representation of the contour along which the magnetopause modeled numerically by Mead and Beard (1964) intersects the equatorial plane. Surfaces transverse to tail field lines satisfy equations of the form

$$(0.61338\rho/b)^2 = 2.280\{\sinh^2[(0.61338/b)(\xi_0 + b)] - \sinh^2[(0.61338/b)(\xi + b)]\} , \tag{7.4}$$

where ξ_0 denotes the value of ξ at $\rho = 0$. The innermost such cross-tail surface bears the label $\xi_0 = \xi_0^*$ (= b in Fig. 7.3) and constitutes the cross-tail portion of the source surface.

Minimization of σ in (7.2) for $\alpha = 0.5$ yields the values of $\bar{g}_n^m/g_1^0$ shown in Table 7.1 (Schulz and McNab, 1996) for $\psi = 0°$ and $\psi = 90°$. Values of $\bar{g}_n^m/g_1^0$ for intermediate ψ are given by superposition of the results for $\psi = 0°$ (weighted by cos ψ) and the results for $\psi = 90°$ (weighted by sin ψ). As is conventional in geomagnetism

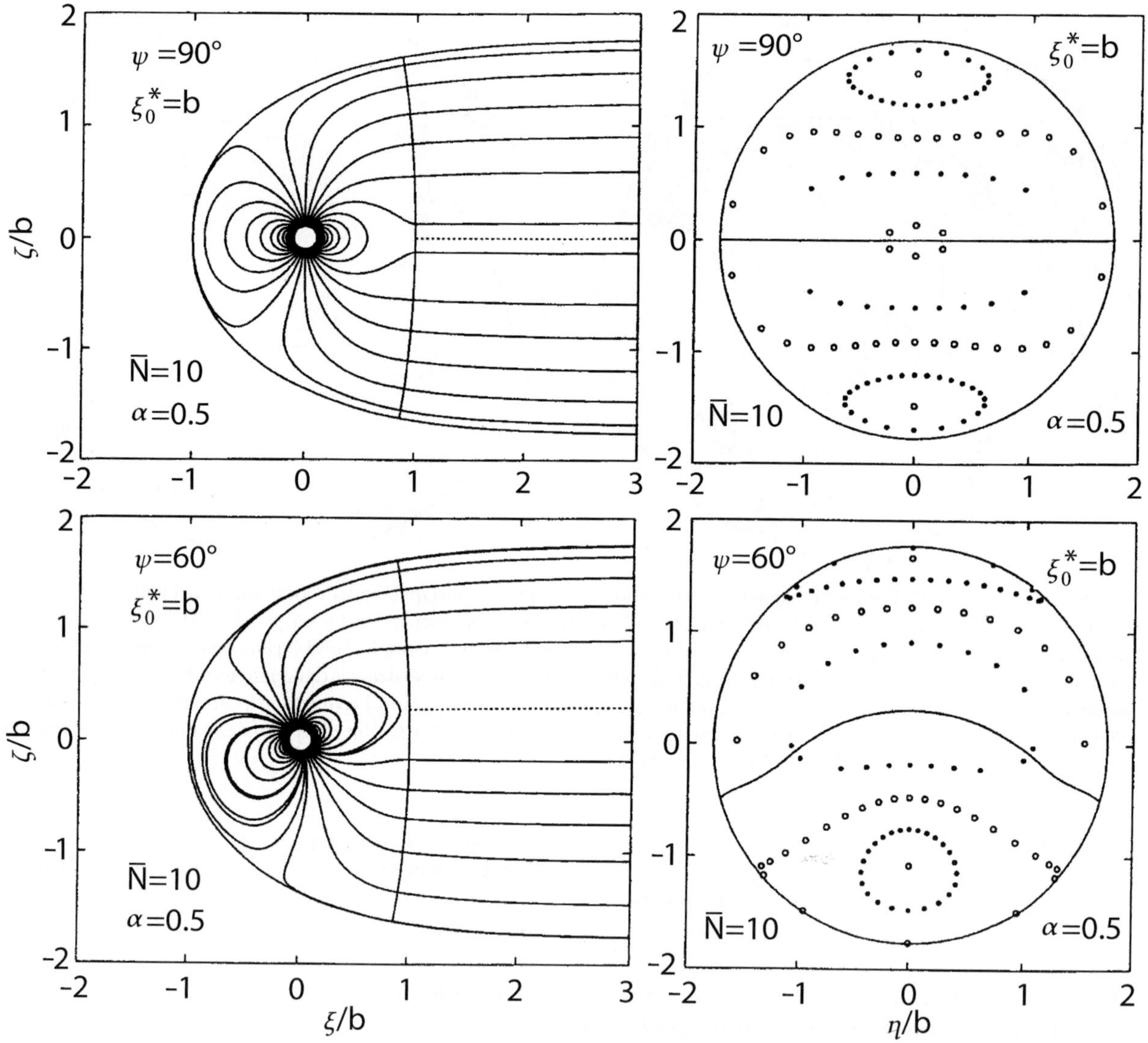

Fig. 7.3. Left-hand panels show representative field lines in "noon-midnight" meridional plane, which contains the tail axis ($\rho = 0$) and the magnetic dipole axis. Coordinate ζ is measured from the plane that perpendicularly bisects this meridional plane along the tail axis. Selected field lines emanate from the "planetary surface" (dipole-centered sphere of radius $r = a = b/10$) at 5° intervals of magnetic latitude Λ (namely ±90°, ±85°, ±80°, ..., 0°). Right-hand panels show intersections of representative field lines (emanating from "planetary surface" at 5° intervals of magnetic latitude Λ and 15° intervals of MLT) for corresponding values of ψ (angle between geomagnetic dipole moment and aberrated solar-wind velocity). Open (filled) circles correspond to Λ = odd (even) multiples of 5°

and in space research, the coefficients $\bar{g}_n^m$ in Table 7.1 bear Schmidt (1935) normalization, which means that the associated Legendre functions in (7.1) are given by

$$P_n^m(\theta) \equiv \left[\frac{2(n-m)!}{(n+m)!}\right]^{1/2} \frac{\sin^m\theta}{2^n n!} \frac{d^{n+m}(-\sin^2\theta)^n}{d(\cos\theta)^{n+m}} \tag{7.5}$$

for $m > 0$ but by $(1/2)^{1/2}$ of this for $m = 0$. Usually omitted for all values of m in mathematics books and from library routines for computers, the square-bracketed factor in (7.5) makes the contribution of degree n to the mean value of B^2, produced over a sphere of radius $r < \min(b, \xi_0^*)$ by magnetospheric currents, proportional to the unweighted sum (over m) of the squares of the expansion coefficients $\bar{g}_n^m$. In other words, the relative influence of various terms (with different values of m) in the expansion for $\boldsymbol{B}$ is accurately reflected by the relative magnitudes of their Schmidt-normalized expansion coefficients.

The $\boldsymbol{B}$ field illustrated in Fig. 7.3 and the considerations behind it represent only one of many approaches to magnetospheric modeling. A different school of thought, exemplified by the work of Tsyganenko (1995), calls for the magnetospheric $\boldsymbol{B}$ field to be represented by globally fitting empirical data on $\boldsymbol{B}$ to a series of analytical functions corresponding to various magnetospheric current systems. Representative results shown

Table 7.1. Optimized values of $\bar{g}_n^m/g_1^0$ in (7.1) for $\alpha = 1/2$ and $\bar{N} = 10$

n	m	$\psi = 90°$	n	m	$\psi = 90°$
1	0	+ 0.4132050264	6	3	+ 0.0091354162
2	1	− 0.5157444974	6	5	− 0.0078231031
3	0	+ 0.0109183660	7	0	+ 0.0013895776
3	2	− 0.0140955499	7	2	− 0.0019297482
4	1	+ 0.0388763351	7	4	+ 0.0018102643
4	3	− 0.0342857048	7	6	− 0.0015384288
5	0	− 0.0127488209	8	1	+ 0.0027969966
5	2	+ 0.0174182053	8	3	− 0.0027159069
5	4	− 0.0150846083	8	5	− 0.0027159069
6	1	− 0.0096295742	8	7	− 0.0021368641
n	m	$\psi = 90°$	n	m	$\psi = 0°$
9	0	+ 0.0001260966	1	0	+ 0.8299144795
9	2	− 0.0001763350	2	0	+ 0.5094588076
9	4	+ 0.0001699207	3	0	− 0.0279829513
9	6	− 0.0001573160	4	0	+ 0.0724585944
9	8	+ 0.0001324011	5	0	− 0.0425294886
10	1	− 0.0003848127	6	0	+ 0.0211107989
10	3	+ 0.0003776193	7	0	− 0.0058803973
10	5	− 0.0003618782	8	0	+ 0.0070342413
10	7	+ 0.0003336352	9	0	+ 0.0004304780
10	9	− 0.0002798767	10	0	+ 0.0011620026

in Fig. 7.4 are based on a non-linear least-squares fit to measurements of the unit vector $\hat{\boldsymbol{B}}$ in a magnetospheric database. The model magnetopause (dashed curve) in Fig. 7.4 is an ellipsoidally capped cylinder (Sibeck et al., 1991) satisfying the equation

$$\rho^2 + 0.14(\xi - \xi_c)^2 = 808.7(P_0/P)^{1/3}a^2 \qquad (7.6)$$

at $\xi \le \xi_c = 65(P_0/P)^{1/6}a$ and $\rho^2 = 808.7(P_0/P)^{1/3}a^2$ at $\xi \ge \xi_c$, where P is the dynamic pressure of the solar wind at 1 AU and $P_0 = 2.04$ nPa (1 pascal $\equiv$ 1 newton/m^2) is a nominal reference value for P. The focal points of the ellipsoid in this model are located on the ξ axis at $\xi \approx (65 \pm 70.482)(P_0/P)^{1/6}a$, the nose of the magnetopause is located at $\xi = -b \approx -11.003(P_0/P)^{1/6}a$, and the asymptotic tail radius is $\rho_\infty^* \approx 2.5846b$.

For simplified modeling of magnetospheric processes, it is often convenient to use one of the magnetic-field representations illustrated in Fig. 7.5 (dipole field plus uniform northward or southward $\Delta\boldsymbol{B}$, anti-parallel or parallel to the dipole axis). These are both axisymmetric models for $\boldsymbol{B}$, but in some ways they resemble the day and night sides (respectively) of a real magnetosphere. The polar *neutral points* (at which $B = 0$) on the boundary sphere of radius $r = b = (2\mu_E/|\Delta\boldsymbol{B}|)^{1/3}$ in Fig. 7.5a (northward $\Delta\boldsymbol{B}$) correspond conceptually to the mid-latitude dayside neutral points (field-bifurcation sites surrounded by "cleft" regions) in Figs. 7.3 and 7.4, whereas the circular equatorial neutral line at $r = b = (\mu_E/|\Delta\boldsymbol{B}|)^{1/3}$ in Fig. 7.5b (southward $\Delta\boldsymbol{B}$) corresponds to the nightside neutral sheet in Fig. 7.3 and nightside current sheet in Fig. 7.4. Indeed, the southward $\Delta\boldsymbol{B}$ itself in Fig. 7.5b corresponds topologically to the tail field in Fig. 7.3.

The advantage of using either field model illustrated in Fig. 7.5 stems from analytical simplicity. For example, the model magnetic field $\boldsymbol{B}(r,\theta)$ corresponding to Fig. 7.5a is given by

$$\boldsymbol{B}(r,\theta) = \nabla[(\mu_E/r^2)\cos\theta + 2(\mu_E/b^3)r\cos\theta], \qquad (7.7a)$$

whereas that corresponding to Fig. 7.5b is given by

$$\boldsymbol{B}(r,\theta) = \nabla[(\mu_E/r^2)\cos\theta - (\mu_E/b^3)r\cos\theta]. \qquad (7.7b)$$

Moreover, the equation of a magnetic field line is

$$r = La[1 - (r/b)^3]\sin^2\theta \qquad (7.8a)$$

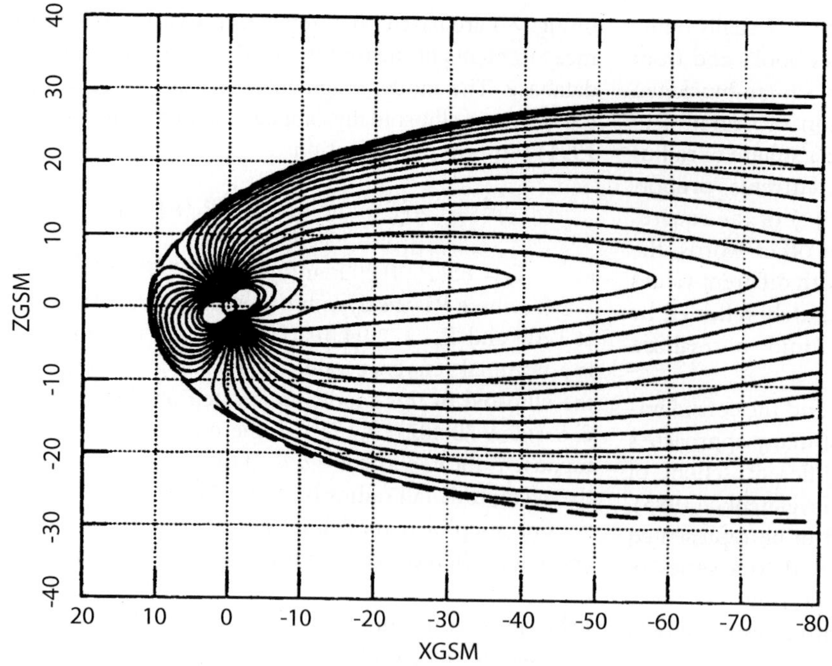

Fig. 7.4. Representative magnetic field lines in noon-midnight meridional plane, according to data-based model of Tsyganenko (1995) including ring-current and tail fields confined for $\psi = 70°$ by ellipsoidal magnetopause specified by (7.6)

in Fig. 7.5a and

$$r = La[1 + (1/2)(r/b)^3]\sin^2\theta \qquad (7.8b)$$

in Fig. 7.5b, where L is a label inversely proportional to the amount of magnetic flux enclosed by the corresponding magnetic shell. In particular, the field line from (7.8b) that intersects the equator ($\sin\theta = 1$) at the neutral line ($r = b$) in (7.7b) bears the label $L^* = 2b/3a$. The magnetic shell $L = L^*$ constitutes the boundary surface between closed and open field lines in (7.8b).

It is often convenient to represent $\boldsymbol{B}(r, \theta, \varphi)$ as the curl of a vector potential $\boldsymbol{A}(r, \theta, \varphi) = \alpha\nabla\beta$, where $\alpha(r, \theta, \varphi)$ is a measure of magnetic flux and $\beta(r, \theta, \varphi)$ is a measure of magnetic azimuth (both being constant along field lines). Such a representation of $\boldsymbol{B} = \nabla\alpha \times \nabla\beta$ is known as an *Euler-potential* representation. For example, the Euler potential functions $\alpha(r, \theta) = (\mu_E/La)$ and $\beta = \varphi$ would generate the axisymmetric model magnetic fields specified by (7.7a) and (7.7b). These two model magnetic fields (illustrated in Fig. 7.5) are also current-free except on their boundary surfaces. Indeed, they have been represented in (7.7a) and (7.7b), respectively, as gradients of scalar potential functions (enclosed in square brackets) whose equipotential surfaces are everywhere orthogonal to $\boldsymbol{B}$. These scalar potential functions might also serve as useful coordinates for specifying positions along a field line in Fig. 7.5a and 7.5b, respectively.

The boundary surface between closed and open magnetic field lines in Fig. 7.5b maps to the ionosphere at $r = r^* = 6481.2$ km (i.e., at altitude 110 km) along a contour of constant colatitude $\theta = \theta^*$, given by

$$\sin^2\theta^* = [1 + (1/2)(r^*/b)^3]^{-1}(3r^*/2b)\,. \qquad (7.9)$$

This corresponds roughly to the poleward boundary of the auroral oval (e.g., $\theta^* \approx 20.18°$ for $b \approx 12.82a$, chosen so that this last closed field line intersects the Earth's surface at $\theta = 20°$). It follows from (7.8b) that the label L^* of this last closed field line (which maps from the circular equatorial neutral line at $r = b$) is given by $L^* = 2b/3a$ (≈ 8.5466 in this example, hence $|\Delta\boldsymbol{B}| = 14.24$ nT).

The ionospheric boundary between closed and open field lines in a realistic magnetosphere has its centroid offset by several degrees of latitude along the midnight ($\varphi = 0°$) meridian rather than coinciding with the magnetic pole. Thus, for example, boundaries between closed and open field lines in Fig. 7.6 apply on a sphere of radius $b/10$ (i.e., on the Earth's surface) for the field models illustrated in Fig. 7.3. There is an asymmetry between "winter" and "summer" magnetic polar caps

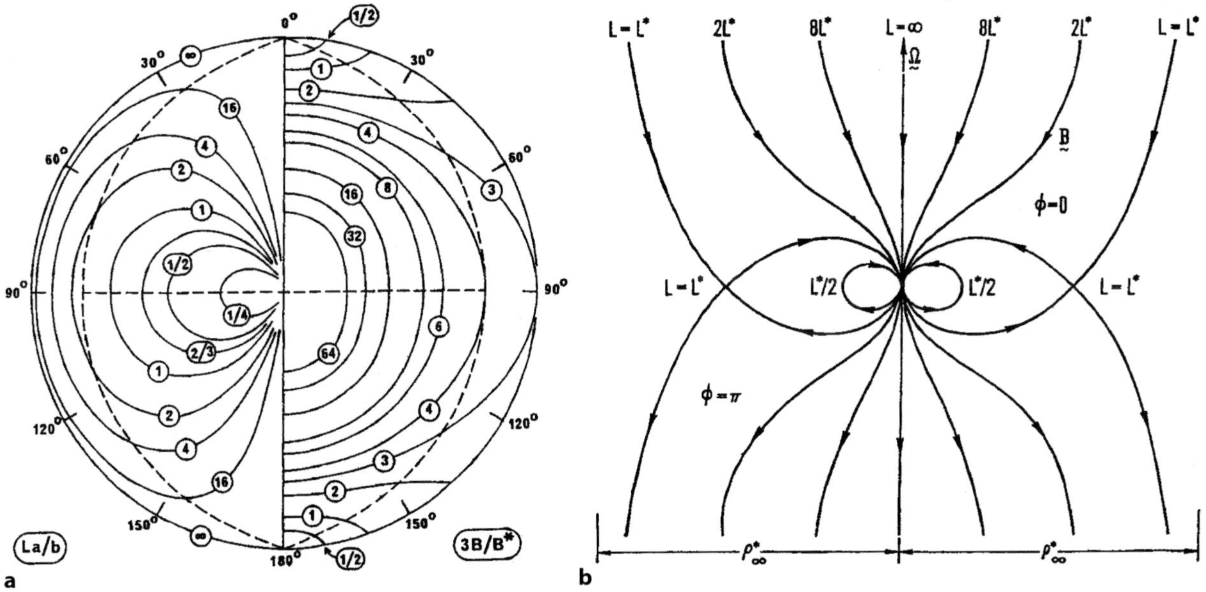

Fig. 7.5a,b. *Left-hand panel* (**a**) shows idealized axisymmetric model magnetospheric specified by (7.7a). The left side of this diagram shows selected field lines identified by La/b according to (7.8a). The *right side of the left panel* shows isogauss contours identified by $3B/B^*$, where $B^* = \mu_E/b^3$ is the value of B obtained from (7.7a) at $(r, \theta) = (b, \pi/2)$. Dashed line and curves show the locus of minima in B along field lines. *Right-hand panel* (**b**) shows selected field lines, as given by (7.8b) in idealized axisymmetric model magnetosphere specified by (7.7b). The equatorial plane in this model contains a circular neutral line ($B = 0$ contour) at $r = b$ on the magnetic shell $L = L^*$

for $\psi \neq 90°$, although they must contain equal amounts of magnetic flux. Simulation results obtained from field models like that shown in Fig. 7.5b should be interpreted relative to coordinates that conform to a more realistic boundary between closed and open field lines, rather than literally in terms of axisymmetric θ and φ labels. Because of current systems not included in (7.1), the magnetic polar caps become larger in diameter and more nearly centered on their respective poles with increasing levels of geomagnetic activity (e.g., Feldstein, 1963).

Another unrealistic feature of the model underlying Fig. 7.5 is that all its field lines lie in meridional planes. Except for the noon-midnight plane of symmetry and perhaps for the tail region, field lines with $L \gtrsim 5$ in a more realistic magnetosphere must lie instead on curved surfaces so as to fit inside the magnetosphere. Figure 7.7 shows projections of such field lines onto the equatorial plane for $\psi = 90°$. The $\boldsymbol{B}$-field model for Fig. 7.7 is that specified by (7.1). Since most field lines in this model lie on surfaces of non-constant φ, there is a further question as to how simulation results obtained from axisymmetric models like that shown in Fig. 7.5b should be interpreted. Experience has shown that a good φ coordinate for the ionospheric footpoint of a field line serves as a good φ coordinate for the same field line throughout the magnetosphere. The same is true of a good L coordinate.

Finally, Fig. 7.8 deals with the magnetic configuration of the *magnetosheath*, which resides between the bow shock and the magnetopause. This is a turbulent region of shocked solar wind, within which the time-averaged magnetic field can be modeled with some confidence. The field lines in Figs. 7.8a and 7.8b have been traced from results of a hydrodynamic simulation by Spreiter et al. (1966) for components of the interplanetary magnetic field (IMF) respectively parallel and transverse to the upstream solar-wind velocity. Superposition of the corresponding $\boldsymbol{B}$ fields for arbitrary IMF direction should hold at this level of approximation. This was, in other words, a fluid model but not MHD. As such, it tended to overestimate the

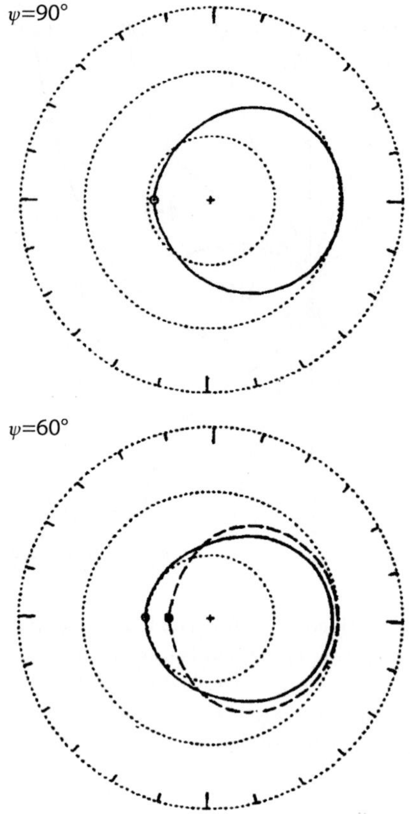

Fig. 7.6. Boundaries (on Earth's surface, $r = a = b/10$) between closed and open magnetic field lines, as obtained from field model of Schulz and McNab (1996) for same selected values of ψ as in Fig. 7.3. Solid (dashed) boundary contours pertain to winter (summer) hemisphere for $\psi = 60°$. Sun symbols ($\odot$) denote mappings of dayside neutral points on noon meridian (Sun is at the left.). Dotted circles correspond to magnetic latitudes 60°, 70°, and 80° on Earth's surface. Plus signs (+) denote magnetic poles

transverse component of $\boldsymbol{B}$ just upstream from the nose of the magnetopause. Development of fully nonlinear MHD models for magnetosheath flows has since been described by Stahara (2002). The magnetosheath is important for magnetospheric dynamics, both because it processes the solar-wind plasma (some of which enters the magnetosphere to become part of the nightside plasma sheet) and because it processes the IMF (part of which enters the magnetosphere via *reconnection* and thereby transmits externally imposed electric fields to the magnetospheric interior).

7.3 Magnetospheric Electric Fields

Electric fields in the magnetosphere can arise from several distinct physical effects, and it is usual to model these various electric fields separately. For example, rigid rotation of the magnetosphere with the Earth at angular velocity $\boldsymbol{\Omega}$ in Fig. 7.5 leads to a meridional electrostatic field given (in SI units) by

$$\begin{aligned} \boldsymbol{E} &= -(\boldsymbol{\Omega} \times \boldsymbol{r}) \times \boldsymbol{B} \\ &= (\boldsymbol{\Omega} \times \boldsymbol{r}) \times [(r \sin\theta)^{-1} \hat{\boldsymbol{\varphi}} \times \nabla(\mu_{\mathrm{E}}/La)] \\ &= -\Omega \nabla(\mu_{\mathrm{E}}/La) . \end{aligned} \tag{7.10}$$

in a non-rotating frame of reference (such that the Sun remains at fixed $\varphi = 180°$, as in Fig. 7.7). This is for an axisymmetric $\boldsymbol{B}$-field model, in which the magnetic dipole moment $\boldsymbol{\mu}_{\mathrm{E}}$ coincides with the rotation axis. If the magnetic field slips relative to the ionosphere, so that the rotation rate Ω in (7.10) varies with L, then the electric field

$$\boldsymbol{E} = -(\mu_{\mathrm{E}}/a) \nabla \int_0^{1/L} \Omega(L') \, \mathrm{d}(1/L') \tag{7.11}$$

in an inertial (fixed-Sun) frame of reference remains derivable from the scalar potential specified in (7.11).

Another important magnetospheric electric field is that associated with *magnetospheric convection*, which can be understood to result from a partial penetration of the interplanetary electric field into the magnetosphere. The interplanetary electric field $\boldsymbol{E}_{\mathrm{i}} = -\boldsymbol{u} \times \boldsymbol{B}_{\mathrm{i}}$ arises because the interplanetary magnetic field $\boldsymbol{B}_{\mathrm{i}}$ typically has a component transverse to the solar-wind velocity $\boldsymbol{u}$ upstream from the Earth's bow shock. The local E_{i} ($\sim 1\,\mathrm{V/km}$ at $R = 1\,\mathrm{AU}$) is at least 10^7 times stronger than the ambipolar electric field in the local solar wind. The direction of $\boldsymbol{E}_{\mathrm{i}}$ in heliospheric coordinates $(\hat{\boldsymbol{R}}, \hat{\boldsymbol{T}}, \hat{\boldsymbol{N}})$ is mostly meridional $(\hat{\boldsymbol{N}})$ to the extent that $\boldsymbol{B}_{\mathrm{i}}$ follows the Parker spiral, but $\boldsymbol{E}_{\mathrm{i}}$ can acquire a significant azimuthal $(\hat{\boldsymbol{T}})$ component (relative to the Sun's rotation axis) because of interplanetary disturbances that turn $\boldsymbol{B}_{\mathrm{i}}$ sharply northward or southward. Moreover, some such changes in $\boldsymbol{B}_{\mathrm{i}}$ are magnified in the magnetosheath because the flow of solar wind (which controls the direction of $\boldsymbol{B}_{\mathrm{i}}$) is largely diverted around the magnetopause there.

Partial penetration of the interplanetary electric field $\boldsymbol{E}_{\mathrm{i}}$ into the tail of the magnetosphere is nicely illustrated with the aid of Fig. 7.5b, in which the asymptote of the tail magnetopause is a cylinder of radius $\rho_\infty^* = 3^{1/2}b$, where b is the radius of the neutral line in (7.7b). More generally, the cylindrical coordinate $\rho \equiv r\sin\theta$ in (7.8b) attains an asymptotic value $\rho_\infty = (2b^3/La)^{1/2} \equiv (3L^*/L)^{1/2}b$ along any tail field line with label $L \geq L^* \equiv 2b/3a$. This geometry suggests a model in which $\boldsymbol{E}_\mathrm{i}$ is asymptotically uniform for $\rho_\infty \gg \rho_\infty^*$ and the cross-tail field $\boldsymbol{E}_\mathrm{t}$ at $\rho_\infty < \rho_\infty^*$ is asymptotically uniform at $|z| \gg b$:

$$\boldsymbol{E} = \nabla\{[E_\mathrm{i} - (\rho_\infty^*/\rho_\infty)^2(E_\mathrm{i} - E_\mathrm{t})]\rho_\infty \sin\varphi\}\,, \quad \rho_\infty > \rho_\infty^* \tag{7.12a}$$

$$\boldsymbol{E} = \nabla\{E_\mathrm{t}\rho_\infty \sin\varphi\} = \nabla\{(3L^*/L)^{1/2}E_\mathrm{t} b\sin\varphi\}\,, \quad \rho_\infty < \rho_\infty^* \tag{7.12b}$$

The quantities in curly brackets define the scalar potential from which the interplanetary and cross-tail electric fields follow. This potential is continuous at $\rho_\infty = \rho_\infty^*$, but its normal derivative is discontinuous there. The condition $\rho_\infty < \rho_\infty^*$ in (7.12b) is equivalent to $L > L^*$ in the tail of the model magnetosphere specified by (7.7b). The electric field at $L < L^*$ is typically modeled as

$$\boldsymbol{E} = \nabla\{3^{1/2}(L/L^*)^{n+1}E_\mathrm{t} b\sin\varphi\},\ L < L^*\,, \tag{7.12c}$$

where n is an adjustable parameter. The models of Brice (1967) and Nishida (1966) are usually understood to stipulate $n = 0$. The models of Stern (1974, 1975) and Volland (1973, 1975) essentially specify $n = 1$, so as to provide for partial "shielding" of the convection electric field in the inner magnetosphere. Either way, the electrostatic scalar potential remains continuous at $L = L^*$, while its normal derivative is discontinuous there.

Figure 7.9 shows equipotential contours corresponding to (7.12b) and (7.12c) on the ionosphere, idealized as in (7.9) as a sphere of radius $r = r^* = 6481.2$ km. The relationship between L in (7.12) and sin θ in Fig. 7.9 is specified by (7.8b) for this purpose. The contour labels in Fig. 7.9 represent values of the convection electrostatic potential for $n = 0$ and $n = 1$, normalized (divided) by $3^{1/2}E_\mathrm{t}b$, which is half the full potential drop across the tail (or polar cap), for $b = 12.82a$. For example, an asymptotically uniform electric field $E_\mathrm{t} = 0.1$ V/km across the tail would correspond to a cross-tail potential drop ≈ 28.3 kV and a normalizing factor ≈ 14.15 kV in Fig. 7.9. This would correspond to very quiet geomagnetic conditions, as measured by the index known as *Kp* (*planetarische Kennziffer*).

The *Kp* index is a quasi-logarithmic measure of *magnetic variability* in the two horizontal components of $\boldsymbol{B}$ at 13 mid-latitude observatories widely distributed in geographic longitude on the Earth's surface (Mayaud, 1980, p. 42). Each observatory provides an integer value (0–9) eight times per day for the local K index, which describes the measured peak-to-peak magnetic variability over each 3-h interval of UT at that station. For example, bin boundaries (which can be regarded as amounting to $K = 0.5, 1.5, 2.5, \ldots, 8.5$) for the historically important station at Niemegk (33 km west of Potsdam, geomagnetic latitude = 51.9° as of year 2000) correspond to magnetic variability of 5, 10, 20, 40, 70, 120, 200, 330, and 500 nT. Bin boundaries are scaled empirically for other stations, so as to factor-out "typical" variations with (for example) magnetic latitude and ground conductivity, before averaging the corresponding values of K from participating observatories to obtain *Kp*. The global average $\langle K\rangle$ over stations widely distributed in longitude serves to suppress effects of day-night and other azimuthal asymmetries. Finally, the resulting $\langle K\rangle$ is rounded to the nearest third of an integer so as to classify *Kp* into one of 28 bins (called 0o, 0+, 1−, 1o, 1+, 2−, 2o, 2+, 3−, …, 9o) for the eight time intervals (UT =00–03, 03–06, 06–09, …, 21–24) into which each day is partitioned. For example, the suffix o (some-times written as a small zero) on a *Kp* bin label denotes a value of $\langle K\rangle$ within ±1/6 of the corresponding integer, but values of $\langle K\rangle < 0$ and $\langle K\rangle > 9$ are not possible under the standard convention for specifying K.

In some respects it is interesting to regard *Kp* instead as a continuous variable that can be manipulated analytically. For example, Kivelson (1976) has estimated from semi-empirical considerations that an equatorially uniform convection electric field $\boldsymbol{E}_\mathrm{c}$ in the inner magnetosphere would have an equatorial strength $E_\mathrm{c} = 0.44(1 - 0.097Kp)^{-2}$ V/km for $Kp \leq 8$. Applied to the model specified by (7.12c) for $n = 0$, this would imply

$$E_\mathrm{t} = 0.17(1 - 0.097Kp)^{-2}\ \mathrm{V/km}\,, \tag{7.13}$$

hence $E_\mathrm{t} = 3.38$ V/km for $Kp = 8$ and $E_\mathrm{t} = 10.5$ V/km if extrapolated to $Kp = 9$. (Equations in this paragraph express semi-empirical rather than causal relationships between *Kp* and $\boldsymbol{E}_\mathrm{c}$. It is clear from a dynamical standpoint that increases in *Kp* are caused by increases in E_t rather than the other way around.)

Similarly, Snyder et al. (1963) demonstrated a now-famous correlation between *Kp* and solar-wind velocity,

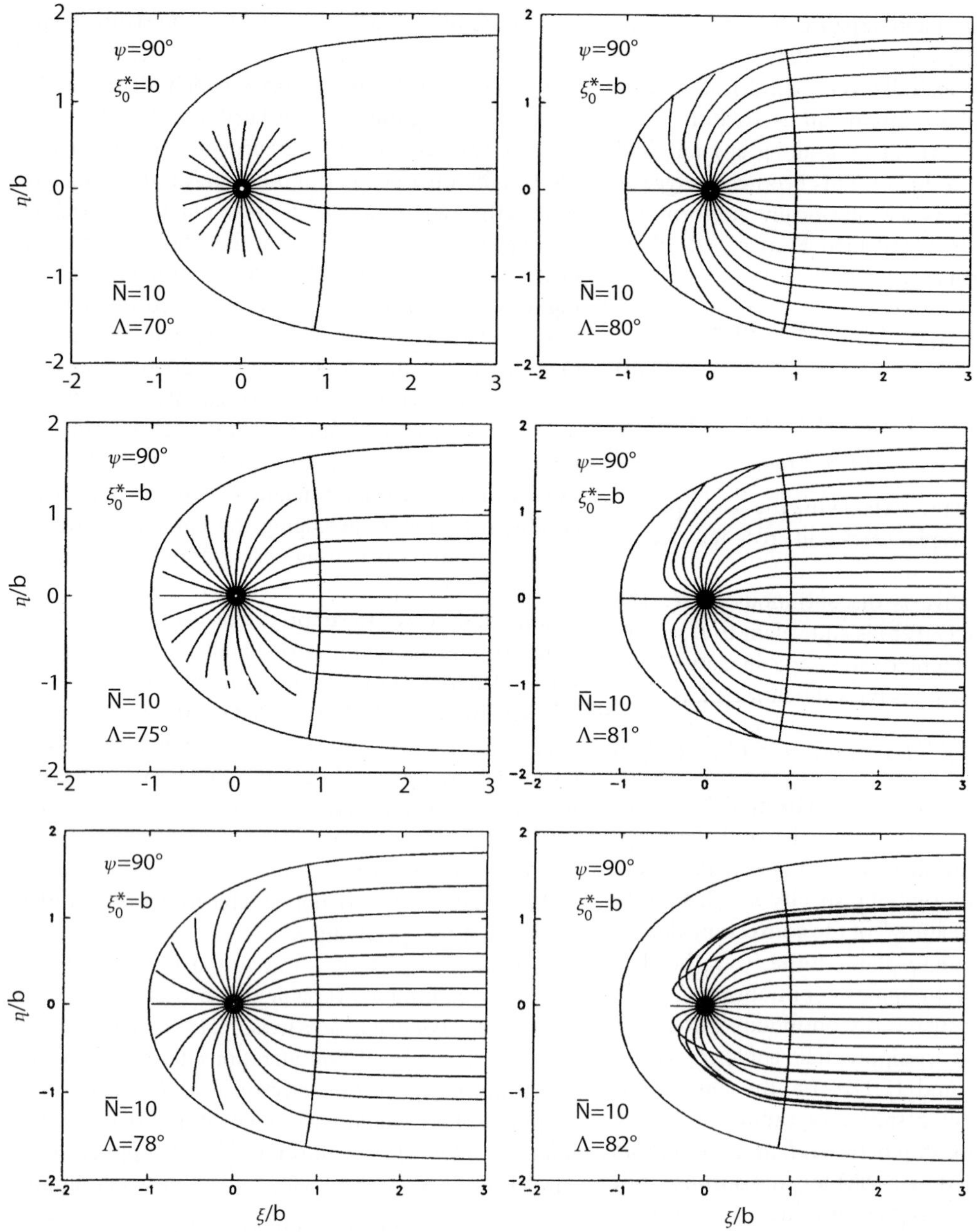

Fig. 7.7. Projections of selected field lines from model of Schulz and McNab (1996) onto equatorial plane for $\psi = 90^\circ$. Field lines in any panel intersect Earth's surface ($r = a = b/10$) at the same "invariant" magnetic latitude Λ but at MLT values (φ) separated by 1 h (i.e., by 15° of magnetic longitude). Field lines not lying in the noon-midnight meridional plane are "bent backward" (away from the Sun) by an amount that increases with Λ for closed field lines. At least some field lines extend into the geomagnetic tail for $\Lambda > 69.5^\circ$, and all do so for $\Lambda > 81.03^\circ$ in this model

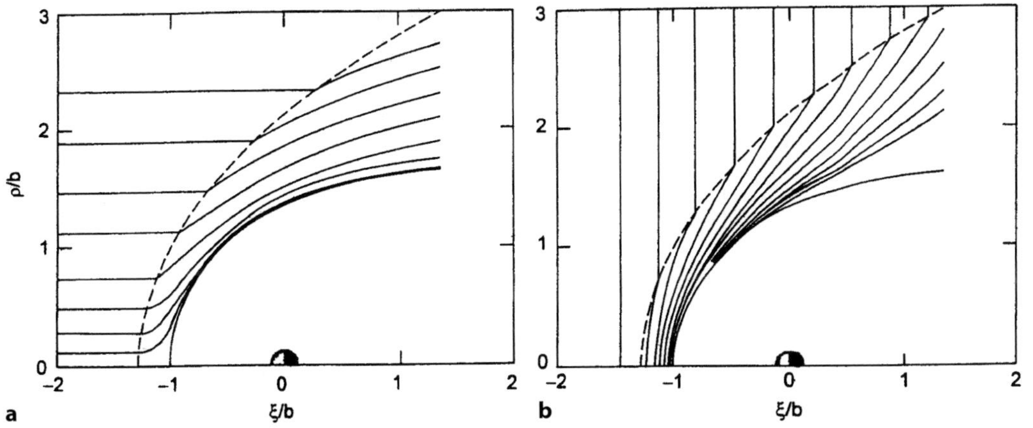

Fig. 7.8a,b. Representative flow streamlines (**a**), which are also magnetic field lines for $\boldsymbol{B}_{\mathrm{IMF}}$ parallel to $\boldsymbol{u}$, and field lines (**b**) for $\boldsymbol{B}_{\mathrm{IMF}}$ transverse to $\boldsymbol{u}$, in gas-dynamic model of Spreiter et al. (1966) with solar-wind Mach number $M = 8$

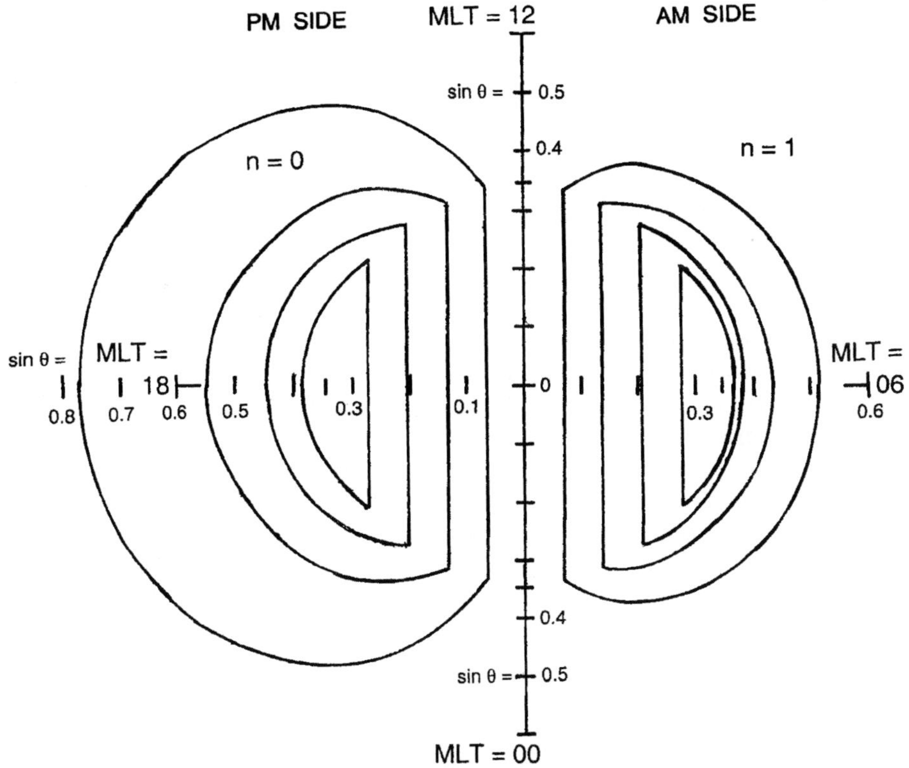

Fig. 7.9. Equipotential contours as projected onto a sphere at altitude 110 km above Earth's surface (seen from high above the north magnetic pole) in two versions of the convection electric field model specified by (7.12). Values of electric potential are spaced by $0.1\Delta V$, where ΔV is the total potential drop across the polar cap ($\sin\theta \leq 0.345$ at altitude $h = 110$ km). The noon-midnight meridian is the contour of zero potential in this model. Polar-cap equipotentials from (7.12b) are straight lines (independent of n) in this projection, and equipotential contours are kinked in this model at the boundary between closed and open magnetic field lines (i.e., at $\sin\theta = 0.345$). The PM (AM) side of this diagram shows equipotentials for $n = 0$ ($n = 1$) in (7.12c) in the region of closed magnetic field lines (i.e., for values of "radial" coordinate $\sin\theta \geq 0.345$). The association of PM (AM) with $n = 0$ ($n = 1$) is just a plotting stratagem to save space, since the model is not really asymmetric between AM and PM

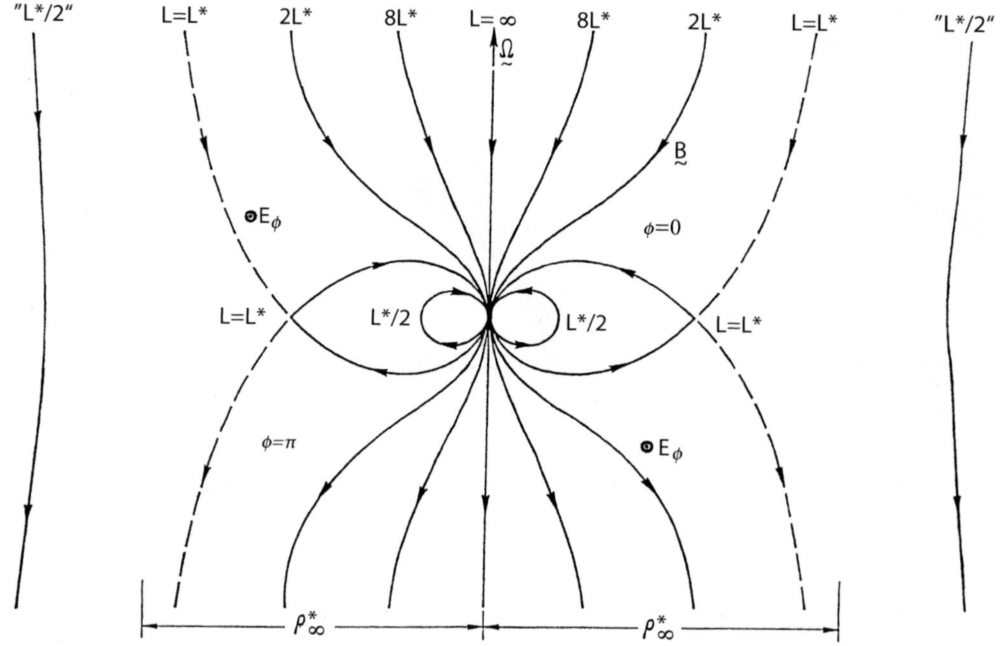

Fig. 7.10. Idealized model magnetosphere based on (7.8b), including "interplanetary" field lines with $L < L^*$. The field specified by (7.7b) can be modified by an arbitrary factor (either positive or negative) outside the "magnetopause" (surface of revolution corresponding to $L = L^*$, *dashed curve*), since azimuthal currents at $L = L^*$ can support an arbitrary inequality in magnitude and/or sign between the interplanetary and tail fields in this axisymmetric model

such that $u \approx (330 + 68Kp)$km/s, where u is clearly the cause and Kp is clearly the result. Part of the reported increase in Kp with u may be related to the enhancement of southward $\boldsymbol{B}_\mathrm{i}$ that can accompany disturbances in the solar wind, which tend to increase in strength with the value of u. (Values of northward $\boldsymbol{B}_\mathrm{i}$ can be similarly enhanced in solar-wind disturbances, but these seem to have a smaller effect on geomagnetic activity.)

The physical process that allows a fraction ε of the interplanetary electric field E_i to penetrate the tail of the magnetosphere is known as *magnetic reconnection*, and the ratio $\varepsilon \equiv E_\mathrm{t}/E_\mathrm{i}$ is known as the *reconnection efficiency*. Reconnection is much more efficient when the $\hat{z}$ component of $\boldsymbol{B}_\mathrm{i}$ is southward rather than northward. This is usually attributed to the fact that the magnetospheric $\boldsymbol{B}$ field is northward near the nose of the magnetopause in Fig. 7.3, and that the clash in directions of $\boldsymbol{B}$ there should lead to plasma instabilities (e.g., tearing mode) that disrupt the current sheet that constitutes the magnetopause. In any case, it is a southward component of $\boldsymbol{B}_\mathrm{i}$ that would generate an east-west component of $\boldsymbol{E}_\mathrm{i}$ with the sign (dawn toward dusk, AM toward PM, opposite to Earth's orbital motion around the Sun) that is normally measured (or inferred) for $\boldsymbol{E}_\mathrm{t}$ inside the magnetosphere.

The open-ended (interplanetary) magnetic field lines labeled "$L^*/2$" in Fig. 7.10 illustrate this $\boldsymbol{B}$-field configuration in the context of (7.8b). In general, open-ended field lines with $L < L^*$ carry the electric potential function specified in (7.12a). However, since the strength of $\boldsymbol{B}$ need not be continuous across the tail magnetopause in Fig. 7.10, the $\boldsymbol{B}$ field outside this model magnetosphere can be regarded as any multiple $B_\mathrm{i}/B_\mathrm{t}$ of the magnetic field specified by (7.7b). Any resulting discontinuity in $\boldsymbol{B}$ (even in the sign of $\boldsymbol{B}$) across the model magnetopause ($L = L^*$) could be supported by an appropriately prescribed azimuthal surface current there.

The simple magnetic field model specified by (7.7b) is useful for thinking concretely about magnetospheric processes and even for performing simple calculations, but the real magnetosphere has a more complicated ge-

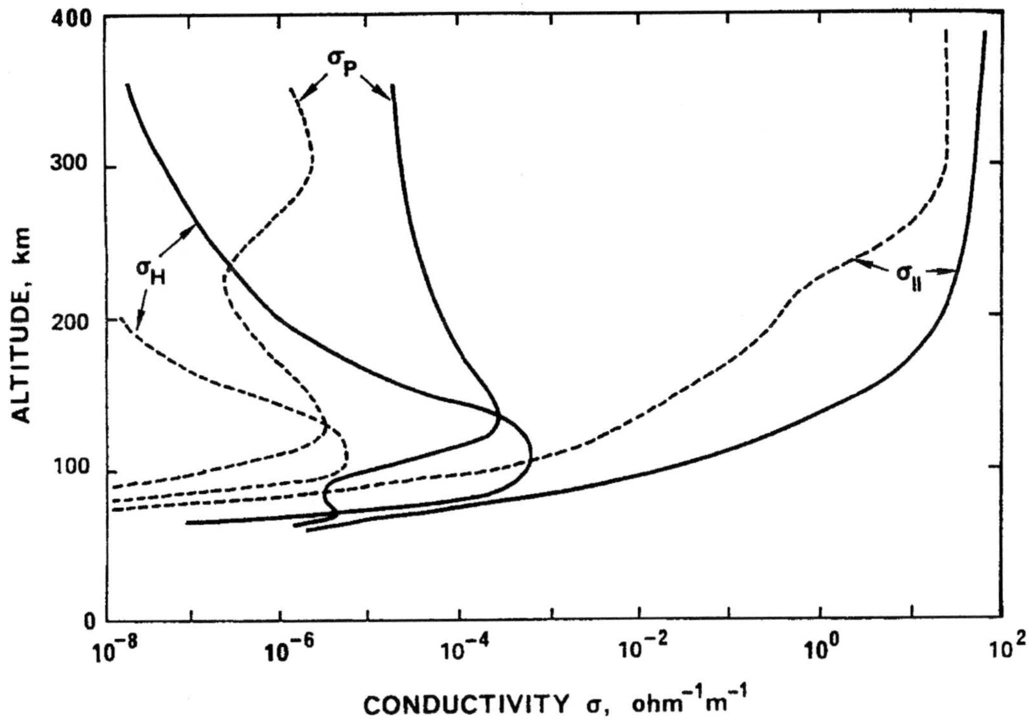

Fig. 7.11. Daytime (*solid curves*) and nighttime (*dashed curves*) electrical conductivity profiles (Hall, Pedersen, and Ohmic) in Earth's ionosphere (based on Giraud and Petit, 1978, p. 182)

ometry than this. Moreover, the pattern of reconnection on the magnetopause can easily deviate from the MLT symmetry that would correspond to (7.12). The convection electric field is influenced also by magnetospheric currents that connect through the ionosphere, where Pedersen and Hall conductivities are distributed somewhat differently in altitude (see Fig. 7.11) and in MLT from each other. The net result of ionospheric Hall conductance is that the pattern of equipotential contours is somewhat rotated in MLT, relative to what is seen in Fig. 7.9. For example, Yasuhara and Akasofu (1977) took account of both Hall and Pedersen conductances in a more sophisticated theoretical calculation, from which they obtained the results shown in Fig. 7.12. For various reasons, including those mentioned in this paragraph, the convection electric field in a real magnetosphere has a more complicated spatial structure than (7.12b) and (7.12c) would suggest. Several empirical models have been devised to capture this complexity by assimilating data of various types into a theoretically reasonable framework.

One such model (Richmond and Kamide, 1988) is known as AMIE (Assimilative Mapping of Ionospheric Electrodynamics). The AMIE procedure involves a formal expansion of the electric scalar potential $V(\theta, \varphi; t)$ at altitude $h = 110$ km in terms of a particular set of orthonormal and continuously differentiable basis functions, designed to provide improved spatial resolution at magnetic latitudes $\gtrsim 50°$ at the cost of diminished spatial resolution at lower latitudes. More precisely, the basis functions involve associated Legendre functions $P_\nu^m(\theta)$ of non-integer degree ν and integer order m at the higher latitudes and trigonometric functions $\tan^m(\theta/2) + \text{ctn}^m(\theta/2)$ at the lower latitudes, with admissible values of ν determined by matching logarithmic derivatives of the corresponding basis functions at (for example) $\theta \approx 40°$ and thus also at $\theta \approx 140°$. Azimuthal variation of $V(\theta, \varphi; t)$ is expressed as usual in terms of $\sin m\varphi$ and $\cos m\varphi$ for integer values of m. Representative families of equipotential contours obtained from real data by the AMIE procedure (Richmond and Kamide, 1988) are shown in Fig. 7.13.

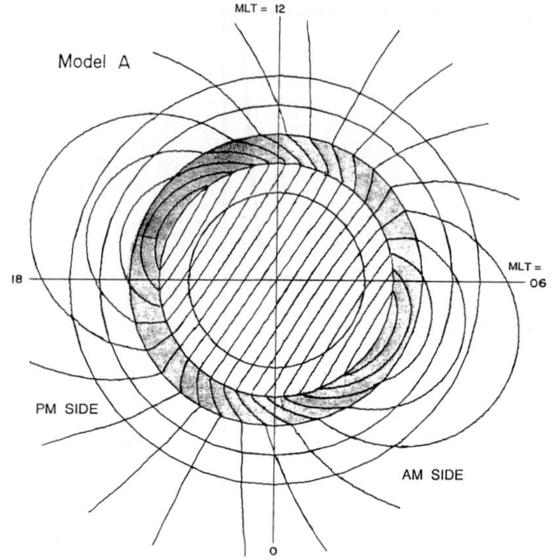

Fig. 7.12. Electric equipotential contours in the ionosphere, as computed from a more sophisticated model (Yasuhara and Akasofu, 1977), taking account of Hall and Pedersen conductances

A somewhat different empirical procedure for deriving $V(\theta, \varphi; t)$ on the ionosphere from real data has been developed by Weimer (1995 et seq.), who employed associated Legendre functions $P_n^m(\theta')$ with integer values of n but achieved improved spatial resolution at the higher latitudes by linearly "shrinking" the latitude coordinate so that (for example) $\theta' = 90°$ as the argument of P_n^m corresponds to a (possibly φ-dependent) magnetic colatitude $\theta \le 45°$ on the ionosphere. Weimer's models benefit from continual improvement, but those mentioned so far share (with AMIE) the disadvantage of being constructed from continuously differentiable basis functions. His model electric fields $\boldsymbol{E}$ thus also lack a discontinuity in the meridional component of $\boldsymbol{E}$ across the boundary between closed and open magnetic field lines, a discontinuity that could lead (see below) to the formation of auroral arcs there. (Recent work by Weimer (2005) shows promise of overcoming this deficiency.)

Figure 7.14, based on (7.12b) and (7.12c), is purely theoretical but shows the desired discontinuities in E_r and E_θ at the boundary between closed and open magnetic field lines. Indeed, the meridional component of $\boldsymbol{E}$ points toward the discontinuity on the PM side of the magnetosphere and away from the discontinuity on the AM side. The resulting ionospheric Pedersen current would likewise be discontinuous, and so Ampère's Law would require magnetospheric currents along $\boldsymbol{B}$ on the boundary surface between closed and open magnetic field lines. These correspond to what are known in the

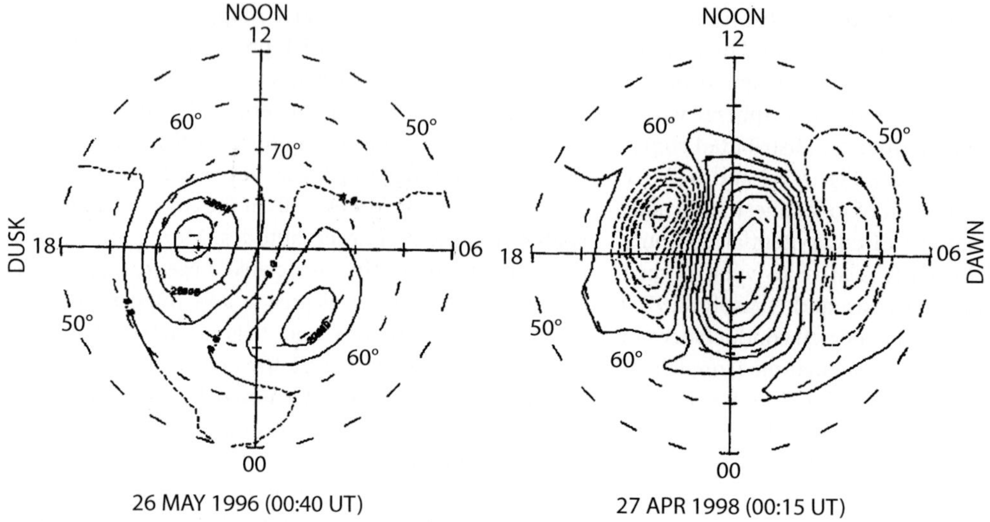

a $\mathbf{B}_{IMF}$ = (+4.0, -3.0, +0.1 nT ΔV = 59 ± 3.8 kV **b** $\mathbf{B}_{IMF}$ = (+1.9, -2.7, -3.7) n.T ΔV = 66 kV

Fig. 7.13a,b. Examples of equipotentials obtained from geophysical data by method of Richmond and Kamide (1988) and made available on AMIE website (http://web.hao.ucar.edu/public/research/ tiso/amie/AMIE_head.html)

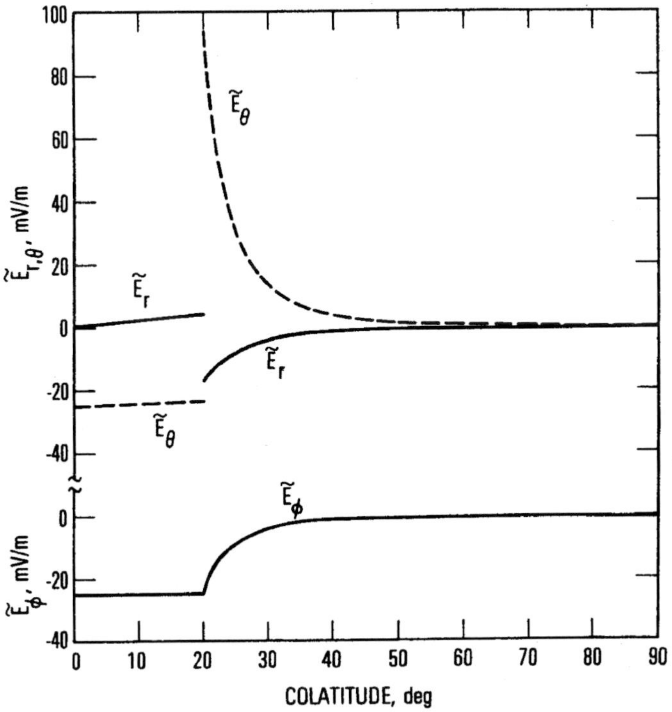

Fig. 7.14. Upper curves show meridional (r, θ) electric field (with discontinuity at $\theta = 20°$, corresponding to $L = L^* = 8.5466$) in dawn-dusk meridional plane at $r = a$. Lower curve shows azimuthal (ϕ) electric field perpendicular to noon-midnight meridional plane. This model $\boldsymbol{E}$ field, based on (7.12) with $n = 1$ and $E_t = 0.3849$ V/km, was used by Straus and Schulz (1976) in a study of ionospheric dynamics

real magnetosphere as *Region-I Birkeland currents*. Such currents need to flow upward along $\boldsymbol{B}$ on the PM side of the idealized magnetosphere and downward along $\boldsymbol{B}$ on the AM side, in order to conserve current at $L = L^*$. Region-I Birkeland currents form thin sheets because they reside (ideally) at $L = L^*$ in this model. They bifurcate at the magnetospheric equator upon meeting the inner edge of the cross-tail current.

Magnetospheric plasma densities are typically inadequate to maintain Region-I Birkeland currents of the required strength from electron and ion motions at thermal velocities alone, and magnetic mirror forces inhibit the precipitation of charged particles from the hot plasma ($\kappa T_e \sim 1-2$ keV for electrons, $\kappa T_p \sim 4-8$ keV for protons) that typically populates the equatorial magnetosphere at $L \approx L^*$. Substantial electric fields parallel or anti-parallel to $\boldsymbol{B}$ (upward on the PM side, downward on the AM side) thus tend to develop in the thin boundary layer at $L \approx L^*$, in order to sustain the requisite Region-I Birkeland current there. The corresponding electric potential drop between the ionosphere and equatorial magnetosphere (but mostly at altitudes $< 10^4$ km) at $L \approx L^*$ can amount to ~ 1 kV under geomagnetically quiet conditions and $\sim 5-10$ kV during major geomagnetic storms. This is one of several mechanisms commonly invoked to explain the appearance of *auroral arcs* (caused by precipitating energetic electrons) and *auroral ion beams* (traveling upward along $\boldsymbol{B}$) on the PM side of the *auroral oval* (located near the boundary between closed and open magnetic field lines). Spatial discontinuities or other sharp gradients in ionospheric electrical conductance can yield similar effects, as can any other causes of spatial discontinuities in ionospheric electric fields.

Conversely, the ionospheric Pedersen current driven by E_θ at $L < L^*$ in Fig. 7.14 increases smoothly with latitude in a way that requires a latitudinally distributed downward current along $\boldsymbol{B}$ on the PM side and a latitudinally distributed upward current along $\boldsymbol{B}$ on the AM side, in order to satisfy Ampère's Law there. These latitudinally distributed currents at $L < L^*$ in Fig. 7.14 correspond to what are known in the real magnetosphere as *Region-II Birkeland currents*. These can amount to 80% or more of the adjacent (and oppositely directed) Region-I currents, but they are distributed more widely in L and thus correspond to much smaller current densities and much smaller parallel (to $\boldsymbol{B}$) electric fields than are found in Region I. This last statement reflects the smoothly monotonic variation of E_θ with latitude in Fig. 7.14 for $L < L^*$, in contrast to the discontinuity in E_θ encountered at $L = L^*$. Ionospheric Hall currents and their azimuthal variations, as well as azimuthal variations of ionospheric Pedersen currents, also contribute to the eventual distribution of magnetospheric Birkeland currents in L and φ, but intrinsic latitudinal variations of ionospheric Pedersen currents account for the essential features noted above. In fact, the global distribution of Birkeland currents is (for various reasons) more complicated than the present discussion might suggest. A famous empirical study of Birkeland currents (deduced from satellite-magnetometer data) revealed the global pattern shown in Fig. 7.15. The anomalously large width of Region I in Fig. 7.15 can be attributed in part to data-binning, as well as to temporal variations in the latitudes at which Region-I currents enter (AM) or leave (PM) the ionosphere.

Finally, there is a need to consider *induced electric fields* associated with temporal variations of $\boldsymbol{B}$. A simple example is the electric field induced by allowing μ_E and/or b in (7.7a) or (7.7b) to vary with time. In these (axisymmetric) cases the induced electric field would be given by

$$\boldsymbol{E} = -\hat{\boldsymbol{\varphi}}(\dot{\mu}_E/r^2)[1 + 2(r/b)^3]\sin\theta - 3\hat{\boldsymbol{\varphi}}\mu_E(r/b^4)\dot{b}\sin\theta \tag{7.14a}$$

or by

$$\boldsymbol{E} = -\hat{\boldsymbol{\varphi}}(\dot{\mu}_E/r^2)[1 - (r/b)^3]\sin\theta + 3\hat{\boldsymbol{\varphi}}\mu_E(r/2b^4)\dot{b}\sin\theta\,, \tag{7.14b}$$

respectively. The secular decrease (~6.5% per century) of the geomagnetic dipole moment μ_E is formally included in (7.14) but is unimportant for most calculations of magnetospheric particle dynamics. The usual sense in which μ_E varies with time is that temporal variations of b (or of equivalent magnetospheric parameters) induce azimuthal currents in the ionosphere and/or in the Earth itself, so as to exclude (in first approximation) any externally imposed magnetic perturbation from an Earth-centered sphere of radius r_c. Such an exclusion would require

$$\dot{\mu}_E/\mu_E = -3[1 + 2(r_c/b)^3]^{-1}(r_c/b)^3(\dot{b}/b) \tag{7.15a}$$

in (7.14a) and

$$\dot{\mu}_E/\mu_E = -[1 - (r_c/b)^3]^{-1}(r_c/b)^3(3\dot{b}/2b) \tag{7.15b}$$

in (7.14b), respectively. The corresponding azimuthal surface currents at $r = r_c$ would thus induce a dipole moment $\Delta\mu_E$ that may either add to or detract from the dipole moment associated with Earth's main field.

The induced electric fields $\boldsymbol{E}$ specified by (7.14a) and (7.14b) are (of course) respectively perpendicular to $\boldsymbol{B}$, as specified by (7.7a) and (7.7b). It is considered somewhat of a paradigm in magnetospheric physics that electric fields (except in special cases such as auroral arcs) can be regarded as perpendicular to $\boldsymbol{B}$ for the purpose of calculating their spatial distributions. This is especially so for induced electric fields. In effect, any component of $\boldsymbol{E}$ parallel to $\boldsymbol{B}$ is regarded as arising from a separate physical process. Using this paradigm, Fälthammar (1968) deduced the lowest-order generalization of (7.14) to models such as (7.1), in which the magnetic field $\boldsymbol{B}$ is not axisymmetric. The result (valid up to second order in r/b) is given by

$$\begin{aligned}\boldsymbol{E} = &-\hat{\boldsymbol{\varphi}}(3r/2b)\dot{b}(a/b)^3\bar{g}_1^0\sin\theta\\ &-\hat{\boldsymbol{\varphi}}(8r^2/21b^2)\dot{b}(a/b)^3\bar{g}_2^1(7\sin^2\theta - 3)3^{1/2}\cos\varphi\\ &-(\hat{\boldsymbol{r}}\sin\theta - 2\hat{\boldsymbol{\theta}}\cos\theta)\dot{b}(4r^2/7b^2)3^{1/2}(a/b)^3\bar{g}_2^1\sin\varphi\end{aligned} \tag{7.16}$$

if Earth-induction currents are neglected. Schulz and Eviatar (1969) showed how to extend the calculation of induced electric fields term-by-term to ever higher order in r/b.

There is a philosophical question as to whether such calculations of induced electric fields are truly unique, since (it is argued) one can always add an electric field (perpendicular to $\boldsymbol{B}$) that is the gradient of a scalar potential dependent only on magnetic field-line labels (so that magnetic field lines are electric equipotentials). Since the curl of a gradient is always zero, it is argued that the resulting superposition still satisfies Maxwell's equations and (in particular) Faraday's Law of electromagnetic induction. The answer seems to be that (7.14) has a certain "irreducible simplicity" that would be contaminated by arbitrarily adding anything to it. As long as additional terms are generated out of necessity and by the same algorithm, to maintain $\boldsymbol{E}\cdot\boldsymbol{B} = 0$ and satisfy Maxwell's equations to ever higher

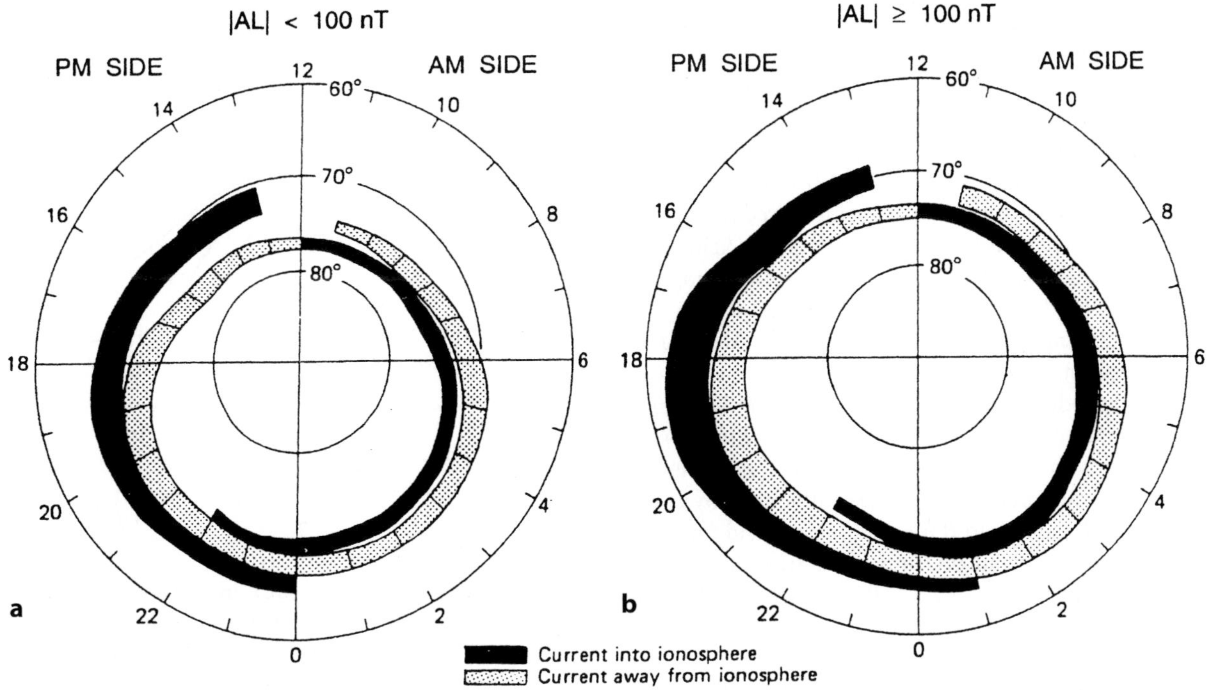

Fig. 7.15a,b. Polar projection showing average locations in which Birkeland currents were inferred to enter the ionosphere (*black shading*) or exit the ionosphere (*stippled shading*) during intervals of larger (> 100 nT) and smaller (< 100 nT) values of the auroral electrojet index *AL* (Iijima and Potemra, 1978; Mayaud, 1980, pp. 96–99). Region I (Region II) corresponds to the higher-latitude (lower-latitude) portion of the Birkeland current system

order in r/b when $\boldsymbol{B}$ is not axisymmetric, it seems that this construction of the induced $\boldsymbol{E}$ is indeed unique. Birmingham and Jones (1968) have reached a similar conclusion on somewhat different grounds. Of course, axisymmetric expressions such as (7.10) or (7.11) can be added to (7.14) if they describe physical processes of interest (e.g., corotation in the example cited), but these would not logically be mistaken for part of an induced electric field, such as would arise from an offset dipole or from a component of $\boldsymbol{\mu}_{\mathrm{E}}$ perpendicular to Earth's rotation axis.

For most practical purposes, it is considered unnecessary to actually calculate the induced electric field, since the effect of such an $\boldsymbol{E}$ field is usually just to impel a charged particle to remain attached to its "original" field line as that field line changes configuration as a consequence of having $\partial\boldsymbol{B}/\partial t \neq \boldsymbol{0}$. Interesting particle-transport effects then result formally from gradient-curvature drifts relative to such magnetic field lines.

7.4 Magnetospheric Charged Particles

Charged particles in the magnetosphere are usually classified as belonging to the cold plasma ($\lesssim 10$ eV), or to radiation belts ($\gtrsim 200$ keV), or to a population spanning intermediate energies ($\sim 1-200$ keV) characteristic of the ring current and nightside plasma sheet. The historical reason for this subdivision is that instruments of different kinds are traditionally used for making the requisite measurements. From a theoretical standpoint, however, different mathematical techniques are used for modeling particle behavior in different regions of phase space. In this context the above demarcations are only roughly applicable.

From a conceptual standpoint, *cold plasma* consists of ions and electrons that drift essentially at velocity $\boldsymbol{u}_{\mathrm{E}} = (1/B^2)(\boldsymbol{E}\times\boldsymbol{B})$ across the ambient magnetic field. By way of contrast, *radiation-belt particles* drift across $\boldsymbol{B}$ at energy-dependent rates determined largely by the sum of $\boldsymbol{u}_{\mathrm{g}} = (p_\perp^2/2mqB^3)(\boldsymbol{B}\times\nabla B)$ and $\boldsymbol{u}_{\mathrm{c}} = (p_\parallel^2/mqB^2)\boldsymbol{B}\times$

$(\partial\hat{\boldsymbol{B}}/\partial s)$, known as the gradient-curvature drift velocity, to which $\boldsymbol{u}_E$ is an almost negligible addendum. (The quantities $p_\perp$ and $p_{||}$ in $\boldsymbol{u}_g$ and $\boldsymbol{u}_c$ denote components of particle momentum $\boldsymbol{p}$ perpendicular and parallel to $\boldsymbol{B}$. The mass m denotes relativistic mass if applicable, and the direction of drift is determined by the sign of the charge q. The coordinate s, usually measured from minima in B, denotes distance along a magnetic field line.) The *plasma sheet* and *ring current* constitute a population of ions and electrons for which $\boldsymbol{u}_E$ and $\boldsymbol{u}_g + \boldsymbol{u}_c$ are comparably important from a kinematical and dynamical standpoint.

Ions and electrons can also move along $\boldsymbol{B}$. Such motions are influenced by gravity and by centrifugal forces associated with curvature of $\boldsymbol{E} \times \boldsymbol{B}$ drift trajectories in the case of cold plasma. They are influenced by components of $\boldsymbol{E}$ parallel to $\boldsymbol{B}$ for cold plasma and for plasmasheet particles in auroral arcs. They are influenced by magnetic-mirror forces $\boldsymbol{F}_\mu = -\hat{\boldsymbol{B}}(p_\perp^2/2mB)(\hat{\boldsymbol{B}} \cdot \nabla B)$ in the case of plasmasheet, ring-current, and radiation-belt particles.

The resulting motions, along and across $\boldsymbol{B}$, can be bounded or not bounded within the magnetosphere. Particles for which the motion is (at least temporarily) bounded within the magnetosphere are said to be *trapped*. Those that escape into the ionosphere are said to *precipitate* there. Some particles can escape across the magnetopause or into the tail, either because of drifts or because they are individually too energetic for the magnetosphere to contain them. Galactic cosmic rays and solar energetic particles typically belong to this last category.

Figure 7.16 illustrates the drift of cold plasma in the equatorial magnetosphere when the magnetic field $\boldsymbol{B}(r,\theta)$ specified by (7.7b), the convection electric field specified by (7.12c) for $n = 1$, and the corotational electric field specified by (7.10) are regarded as time-independent. Cold plasma in this approximation drifts along equipotentials of the combined convection and corotation electric fields. Equipotential contours in Fig. 7.16 bear labels normalized by the (negative) value of the corotation electric potential $\Omega\mu_E/La$ at $L = L^*$, which in this case amounts to 1/8 of the full cross-magnetospheric potential drop $\Delta V^* = 2bE_t3^{1/2}$ associated with magnetospheric convection. The total electric field (convection plus corotation) in this case vanishes at $L = L^*/2$ on the dusk meridian, and

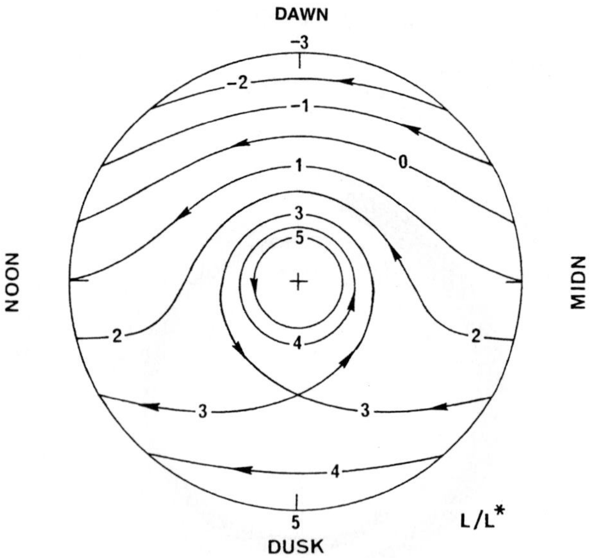

Fig. 7.16. Polar plot, L/L^* vs MLT in equatorial plane, of equipotential contours for combined (convection plus corotation) electric field based on (7.12c) with $n = 1$ and (7.10). Contour labels indicate ratio of total electric potential to that produced at $L = L^*$ by corotation alone (Schulz, 1991, p. 141)

thus on a magnetic field line with $V = 3\Omega\mu_E/L^*a$, which is negative for $\mu_E = -0.3\,\text{G-}R_E^3$. Earth's convection electric field points across $\boldsymbol{B}$ from dawn toward dusk, and the corotation electric field points inward across L.

A more general result within the framework of this model is that the combined electrostatic potential $V(L,\varphi)$ is given in normalized form by

$$(L^*a/\Omega\mu_E)V(L,\varphi) = (L^*/L) - 3^{1/2}(L^*a/\Omega\mu_E)(L/L^*)^{n+1}E_tb\sin\varphi\,, \tag{7.17}$$

which constitutes a quadratic (cubic) equation for L/L^* as a function of $\sin\varphi$ if $n = 0$ (if $n = 1$) for each value of $(L^*a/\Omega\mu_E)V$ at $L < L^*$. The x-type stagnation point in the equatorial drift pattern (cf. Fig. 7.16) corresponds to $L = L_x$, where

$$(L_x/L^*)^{n+2} = (n+1)^{-1}(2\Omega\mu_E/L^*a)(\Delta V^*)^{-1} \tag{7.18}$$

on the dusk meridian ($\sin\varphi = -1$) if (7.18) yields $L_x/L^* \le 1$. The last closed equipotential drift path then corresponds to the value of $(L^* a/\Omega\mu_E)V$ specified by (7.17) at $\sin\varphi = -1$ for this value of L/L^*. In case (7.18) yields $L_x/L^* > 1$ (as might happen during geomagnetically very quiet time intervals) there is no x-type stagnation point, and the last closed equipotential drift path just grazes the equatorial neutral line ($r = b$) at $\sin\varphi = -1$. The corresponding value of $(L^* a/\Omega\mu_E)V$ in this case is found by specifying $L/L^* = 1$.

The equatorial cold-plasma density N should vary as B^2 along any equipotential drift path in Fig. 7.16. The rationale for this expectation follows from a simple calculation of $\nabla \cdot \boldsymbol{u}_E$:

$$\begin{aligned}\nabla \cdot \boldsymbol{u}_E &= \nabla \cdot [B^{-2}(\boldsymbol{E} \times \boldsymbol{B})] \\ &= -B^{-2}[\boldsymbol{B} \cdot (\nabla \times \boldsymbol{E}) - \boldsymbol{E} \cdot (\nabla \times \boldsymbol{B})] \\ &\quad - B^{-4}(\boldsymbol{E} \times \boldsymbol{B}) \cdot \nabla(B^2) \\ &= -\partial(\ln B)/\partial t - (\mu_0/B^2)(\boldsymbol{E} \cdot \boldsymbol{J}) \\ &\quad - (\varepsilon_0\mu_0/2B^2)\partial(E^2)/\partial t - \boldsymbol{u}_E \cdot \nabla(\ln B^2) .\end{aligned} \quad (7.19)$$

If the magnetic field $\boldsymbol{B}$ is time-independent (such that $\partial\boldsymbol{B}/\partial t = \boldsymbol{0}$) and locally current-free (such that $\boldsymbol{J} = \boldsymbol{0}$), and if the displacement current is neglected as usual, then $\nabla \times \boldsymbol{E} = \nabla \times \boldsymbol{B} = \boldsymbol{0}$. In this case it follows from the continuity equation $(\partial N/\partial t) + \nabla \cdot (\boldsymbol{u}_E N) = 0$ that

$$\mathrm{d}N/\mathrm{d}t = (\partial N/\partial t) + (\boldsymbol{u}_E \cdot \nabla N) = -N(\nabla \cdot \boldsymbol{u}_E) \quad (7.20a)$$

and thus from (7.19) that

$$\mathrm{d}(\ln N)/\mathrm{d}t - \mathrm{d}(\ln B^2)/\mathrm{d}t = \mathrm{d}\ln(N/B^2)/\mathrm{d}t = 0 . \quad (7.20b)$$

It follows that (except for *in situ* production or loss, and except for any transport of cold plasma along $\boldsymbol{B}$) the ratio N/B^2 should remain constant along any $\boldsymbol{E} \times \boldsymbol{B}$ drift trajectory. Except perhaps for seasonal or dipole-tilt effects, the component of cold-plasma velocity along $\boldsymbol{B}$ and its first derivative with respect to s should vanish at or near the magnetic equator, thereby removing the reservation about plasma transport along $\boldsymbol{B}$.

Figure 7.17 shows the radial profiles of equatorial cold-plasma density that result at dawn ($\varphi = \pi/2$) and at dusk ($\varphi = 3\pi/2$) when the above scaling ($N \propto B^2$) is applied to a hypothetical noon-midnight density profile specified at $R \le 4$ by

$$\begin{aligned}N(R) &= 991 \exp[0.95(4 - R)]\,\mathrm{cm}^{-3} \\ &\quad + 10^6 \exp[54(1 - R)]\,\mathrm{cm}^{-3} ,\end{aligned} \quad (7.21)$$

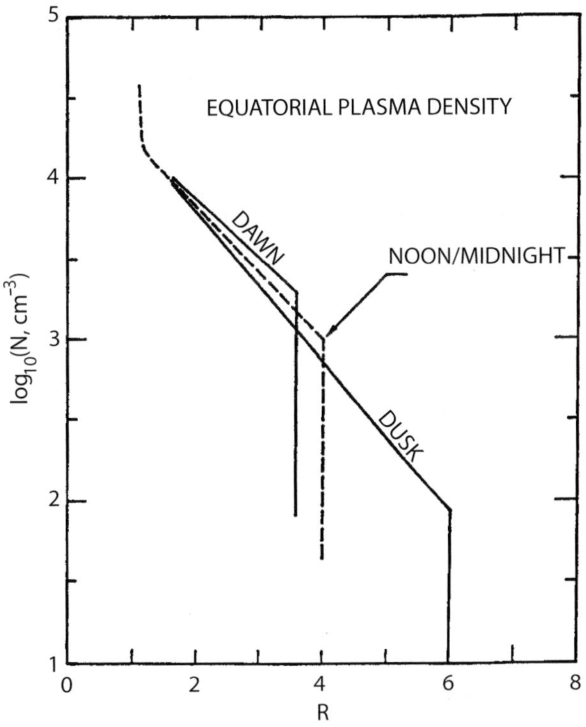

Fig. 7.17. Equatorial plasmaspheric ion or electron density profiles, as mapped from (7.21) in noon-midnight meridional plane to the dawn and dusk meridians, holding N/B^2 = constant along $\boldsymbol{E} \times \boldsymbol{B}$ drift trajectories as prescribed by (7.20b)

where R denotes (equatorial) geocentric distance in units of Earth radii (R_E). The first term of (7.21) corresponds to protons (H^+ ions), whereas the second term corresponds mostly to oxygen ions (O^+).

The reason for truncating (7.21) at some particular value of R is that the cold-plasma density profile in the real magnetosphere shows such a discontinuity, known as the *plasmapause* (Carpenter, 1966). This corresponds in Fig. 7.16 to the boundary between closed and open equipotential drift paths. Cold plasma on open drift trajectories escapes quickly to the dayside magnetopause, while cold plasma on closed drift trajectories continues to circulate as an upward extension of the ionosphere. It follows from (7.17) that the equation of the plasmapause in this time-independent model is

$$(n+2)(L/L_x) = [n + 1 - (L/L_x)^{n+2} \sin\varphi] \quad (7.22)$$

if (7.18) yields $L_x/L^* \le 1$. Thus, the plasmapause is encountered at $L/L_x = (n+1)/(n+2)$ for $\varphi = \pi$ (noon) and for $\varphi = 0$ (midnight), and so truncation of the noon-midnight density profile at $R = 4$ in Fig. 7.17 corresponds to truncation of the dusk ($\varphi = 3\pi/2$) density profile at $R \approx 6$. The plasmapause maps to higher latitudes along magnetic field lines, as specified by (7.8b) in this model. The region of higher cold-plasma density enclosed by the plasmapause is known as the *plasmasphere*. The cold-plasma density N typically increases with latitude (i.e., with decreasing r) along any field line inside the plasmasphere, so as to connect smoothly with the corresponding topside ionospheric plasma-density profile at an altitude ~ 400 km.

The foregoing results pertain to a time-independent model of the plasmasphere, such as might be encountered under geomagnetically quiet conditions. The same principles apply if the convection electric field varies significantly with time, but in this case the discontinuity in cold-plasma density drifts (along with representative particles on it) at instantaneous velocity $\boldsymbol{u}_E = (1/B^2)(\boldsymbol{E} \times \boldsymbol{B})$ rather than following any particular electric equipotential contour. An example is shown in Fig. 7.18. Here it has been assumed (Gorney and Evans, 1987) that the plasmapause in (a), established under time-independent conditions for $t < 0$, is suddenly subjected to a quintupling of the convection electric field specified by (7.12c) for $n = 1$. The transition to a new steady state, with its plasmapause shown in (f) for $t = \infty$, is illustrated in (b)–(e). The electric field $\boldsymbol{E}$ is time-independent throughout this sequence, but the shape of the plasmasphere continues to evolve as shown. The new plasmasphere, reduced in diameter by a factor of $5^{1/3}$, is established not by contraction but by erosion of the old plasmasphere, as the excess plasma (outside the new "last closed equipotential contour") drifts mostly sunward to the dayside magnetopause.

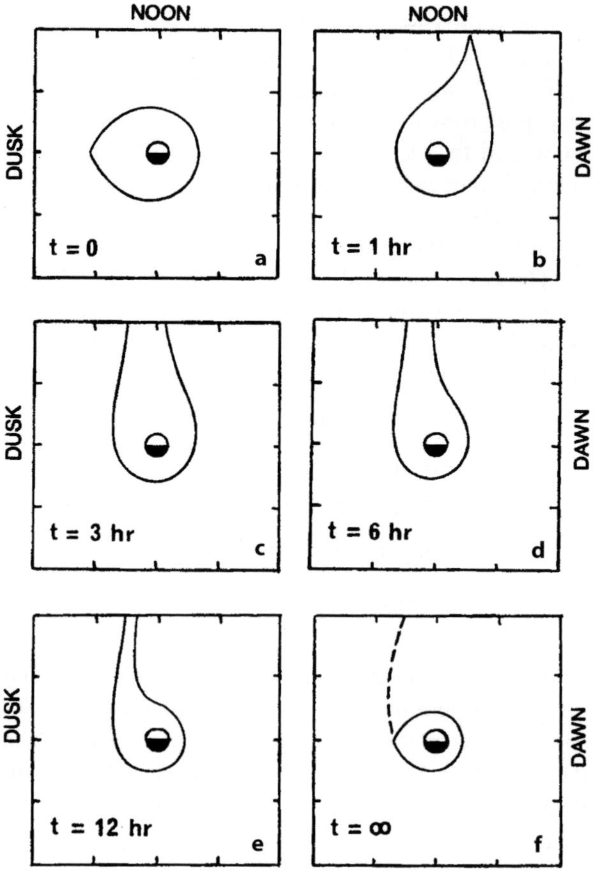

Fig. 7.18a–f. Evolution of plasmapause location after a sudden quintupling (at time $t = 0$) of E_t in (7.12c) with $n = 1$ (Gorney and Evans, 1987). Equatorial area illustrated is $|L \cos\varphi| \le 10$ by $|L \sin\varphi| \le 10$ in each panel. Plasma appendage (*dashed curve*) remains attached to plasmasphere but becomes very narrow as $t \to \infty$. Moreover, the plasma density N along it decreases as B^2 as the associated plasma drifts sunward (toward top of figure)

The original plasmasphere in Fig. 7.18a would eventually be restored if the convection electric field were reduced by a factor of 5, back to its initial value. After the "last closed equipotential contour" had regained its original dimensions, refilling would be achieved by plasma upflow from the ionosphere. This is believed to be a supersonic upflow similar to the *polar wind* of Banks and Holzer (1969). As plasma streams from the northern and southern foot points of a field line interpenetrate, small-angle scattering near the magnetic equator is believed to trap some of the constituent ions between magnetic mirror points (see below) and thus allow build-up of plasma density along the field line. This scenario seems to require a two-step process, in that Coulomb (proton-proton) collisions would provide the requisite small-angle scattering only after an equatorial plasma density ~ 20 cm^{-3} had been established as a consequence of proton scattering by waves generated through a two-stream plasma instability (Schulz and Koons, 1972).

Substantially different kinematical considerations apply to *radiation belts*, which consist of ions and electrons with kinetic energies $\gtrsim 200$ keV. For such particles the main drifts are $\boldsymbol{u}_g = (p_\perp^2/2mqB^3)(\boldsymbol{B} \times \nabla B)$ and $\boldsymbol{u}_c = (p_\parallel^2/mqB^2)\boldsymbol{B} \times (\partial\hat{\boldsymbol{B}}/\partial s)$, known as gradient and curvature drifts, rather than $\boldsymbol{u}_E = (1/B^2)(\boldsymbol{E} \times \boldsymbol{B})$. The gradient-curvature drift in the geomagnetic field is eastward for electrons (with charge $q < 0$) and westward for positive ions ($q > 0$). The logic of gradient drift is that particles of magnetic moment $\mu = p_\perp^2/2mB$ experience a net force $\boldsymbol{F}_g = -\mu\nabla B$ as they gyrate in an inhomogeneous $\boldsymbol{B}$ field, much as they experience an electric force $q\boldsymbol{E}$ in the presence of an electric field. The resulting drift velocity $\boldsymbol{u}_g = (\mu/qB^2)(\boldsymbol{B} \times \nabla B)$ transverse to $\boldsymbol{B}$ is thus found by substituting $\boldsymbol{E} \to -(\mu/q)\nabla B$ in the expression for $\boldsymbol{u}_E = (1/B^2)(\boldsymbol{E} \times \boldsymbol{B})$. Similarly, a charged particle with velocity component $p_\parallel/m$ along $\boldsymbol{B}$, as its center of gyration "tries" to follow the magnetic field line, must experience a "centrifugal" force $\boldsymbol{F}_c = -(p_\parallel^2/m)(\partial\hat{\boldsymbol{B}}/\partial s)$ in consequence of the field line's local curvature $\partial\hat{\boldsymbol{B}}/\partial s$. The resulting drift velocity $\boldsymbol{u}_c = (p_\parallel^2/mqB^2)\boldsymbol{B} \times (\partial\hat{\boldsymbol{B}}/\partial s)$ is found by substituting $\boldsymbol{E} \to -(p_\parallel^2/mq)(\partial\hat{\boldsymbol{B}}/\partial s)$ in the above expression for $\boldsymbol{u}_E$.

A particle's magnetic moment $\mu = p_\perp^2/2\gamma m_0 B$ is closely related to a quantity $M = p_\perp^2/2m_0 B$ known as its *first adiabatic invariant*, where $\gamma \equiv m/m_0$ is the ratio of relativistic mass (m) to rest mass (m_0). The first invariant M is a conserved quantity except for processes that interfere with particle gyration (at frequency $\Omega_1/2\pi = qB/2\pi m$) about the local magnetic field line, essentially because M is proportional to the canonical action integral (see below) associated with particle gyration. It follows that a particle's value of $p_\perp^2$ $(= 2m_0 MB)$ varies in proportion to the value of B at its center of gyration under these conditions. Since electric fields parallel to $\boldsymbol{B}$ are considered negligible for particles with radiation-belt energies, a particle's component of momentum parallel to $\boldsymbol{B}$ is given by

$$p_\parallel \equiv (p^2 - p_\perp^2)^{1/2} = (p^2 - 2m_0 MB)^{1/2} = p[1 - (B/B_m)]^{1/2}, \quad (7.23)$$

where B_m $(= p^2/2m_0 M)$ is known as the *mirror-point field* because it is the value of B required to make $p_\parallel = 0$.

A charged particle is thus (in principle) trapped between points at which $B = B_m$ along a magnetic field line, and the motion of its center of gyration between such mirror points is known as *bounce motion*. The time required for a particle's *guiding center* (as the center of gyration is known because it tends to follow the local magnetic field line) to travel from either mirror point to the other and back is known as the particle's *bounce period*. If the particle's mirror points are sufficiently near the *magnetic equator* (locus of minima in B along field lines), then the bounce motion is like that of a simple harmonic oscillator, and the *bounce frequency* is given by $\Omega_2/2\pi = (1/2\pi)(M/\gamma m)^{1/2}(\partial^2 B/\partial s^2)_0^{1/2}$, where the subscript 0 denotes evaluation at the magnetic equator (i.e., at $s = 0$). The gyrofrequency $\Omega_1/2\pi$ in this limit is given by $\Omega_1/2\pi = qB_0/2\pi m$, where B_0 is the equatorial (i.e., minimum) value of B along the field line of interest, and (if the magnetic-field model is axisymmetric) the drift frequency $\Omega_3/2\pi$ in this limit is given by $\Omega_3/2\pi = u_g/2\pi r_0$, where r_0 is the radius of the drift orbit (guiding-center trajectory in the limit $B_m \to B_0$ = constant). Figure 7.19 shows gyration, bounce, and drift frequencies for such *equatorially mirroring* protons and electrons (with electric fields neglected) over a wide range of particle kinetic energies and L values in a dipolar $\boldsymbol{B}$ field. Figure 7.20 is a schematic illustration of proton gyration and drift in this limit, showing that gradient drift is essentially a gyration-overshoot effect of the radial variation of B_0 in dipolar field geometry.

It is useful to define the *local pitch angle* α between $\boldsymbol{p}$ and $\boldsymbol{B}$, as well as the *equatorial pitch angle* $\alpha_0 \equiv \sin^{-1}[(B_0/B)^{1/2}\sin\alpha]$ that the same particle would attain at $B = B_0$ if M were conserved along the magnetic field line of interest. Thus, the mirror-point field B_m is given by $B_m = B_0\csc^2\alpha_0$. If no such B_m can be found at altitude $h \gtrsim h_c \approx 110$ km, then the particle is said to be in the *loss cone*. For example, the equatorial loss-cone aperture $\alpha_c = \sin^{-1}[(B_0/B_c)^{1/2}]$ for a mirror ratio $B_c/B_0 = 100$ (corresponding to $L \approx 4$) would be about 5.74°. Particles with equatorial pitch angles $\alpha_0 > \alpha_c$ are (at least temporarily) trapped, and particles that mirror at the magnetic equator have $\alpha_0 = 90°$.

The *adiabatic theory of charged-particle motion* is based on the premise that gyration, bounce, and drift are well separated in frequency ($\Omega_3 \ll \Omega_2 \ll \Omega_1$), so that a canonical action integral

$$J_i = \oint_i (\boldsymbol{p} + q\boldsymbol{A}) \cdot d\boldsymbol{l} \quad (7.24)$$

can be associated (in the Hamilton-Jacobi sense) with each element ($i = 1, 2, 3$) of the motion. The *canonical momentum* $\boldsymbol{p} + q\boldsymbol{A}$ in such integrals consists of a par-

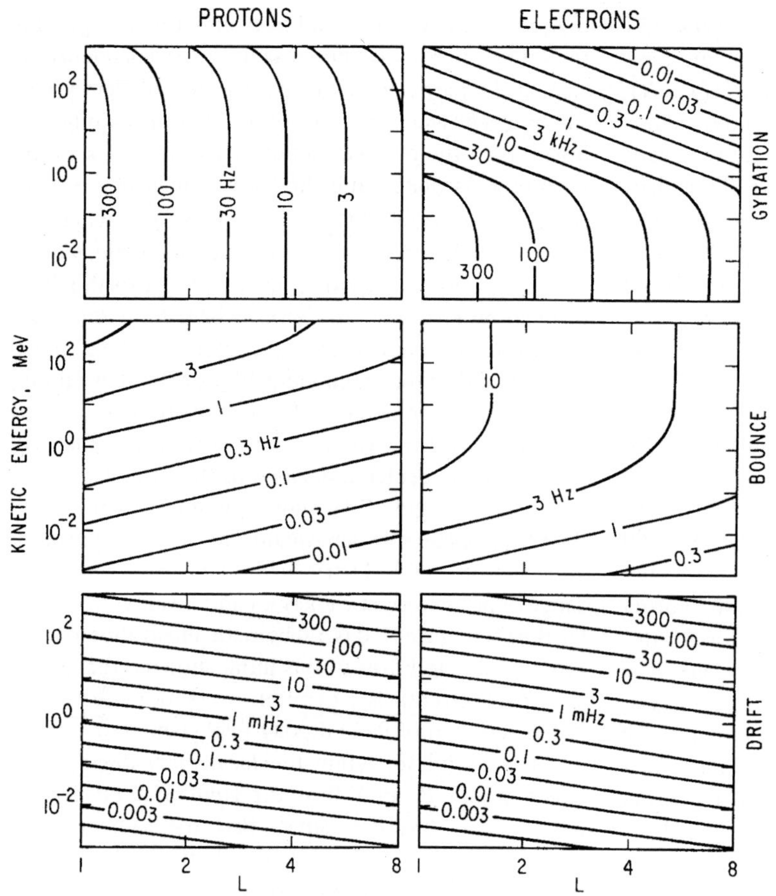

Fig. 7.19. Contours (Schulz and Lanzerotti, 1974, p. 13) of constant adiabatic gyration ($i = 1$), bounce ($i = 2$), and drift ($i = 3$) frequency (Ω_i/π) for equatorially mirroring particles in the geomagnetic dipole field. The inequality $\Omega_1 \gg \Omega_2 \gg \Omega_3$ required for adiabatic motion fails to hold only in the upper-right corners (kinetic energy ~1 GeV, L ~8) on these panels

ticle term $\boldsymbol{p}$ and a vector potential term $q\boldsymbol{A}$. The arc-length element $\mathrm{d}\boldsymbol{l}$ in (7.24) corresponds to the gyration, bounce, or drift trajectory ($i = 1, 2, 3$, respectively). The first action integral works out to $J_1 = \pi p_\perp^2/|q|B$ when due account is taken of both terms (e.g., Schulz and Lanzerotti, 1974, p. 11). The *first adiabatic invariant M* is defined by convention so that $M \equiv |q|J_1/2\pi m_0 = p_\perp^2/2m_0B$. Thus, the magnetic moment μ is equal to M/γ. The second action integral J_2 (also known as the *second adiabatic invariant J*) is just the integral of $p_{||}$ along the field line from one mirror point to the other and back. The third action integral J_3 works out to $J_3 = q\Phi$, where Φ (known as the *third adiabatic invariant*) is the amount of magnetic flux enclosed by the drift path (or by the corresponding *drift shell* in case the particle mirrors off-equator), since the contribution from the $\boldsymbol{p}$ term is negligible compared to that from the $q\boldsymbol{A}$ term if the inequality $\Omega_3 << \Omega_2 << \Omega_1$ is well satisfied. Otherwise, as in the upper-right corners of panels in Fig. 7.19, particle motion can become chaotic and is better simulated by directly applying the Lorentz force equation $\boldsymbol{F} = q\boldsymbol{E} + (q/m)(\boldsymbol{p} \times \boldsymbol{B})$.

Radiation belts are toroidal regions of enhanced trapped-particle intensity, enhanced relative to surrounding regions of space. A representative example is shown in Fig. 7.21. The adiabatic theory of charged-particle motion may seem to provide an explanation for why radiation belts (once formed) might persist. However, the dynamical processes that lead to actual formation and eventual decay of radiation-belt particle intensities entail a (usually weak) violation of one or more adiabatic invariants, and the current state of any particle-radiation environment is the result of past and present competition between such dynamical processes. Radiation-belt dynamics is thus usually formulated in terms of diffusion (or other transport) in phase

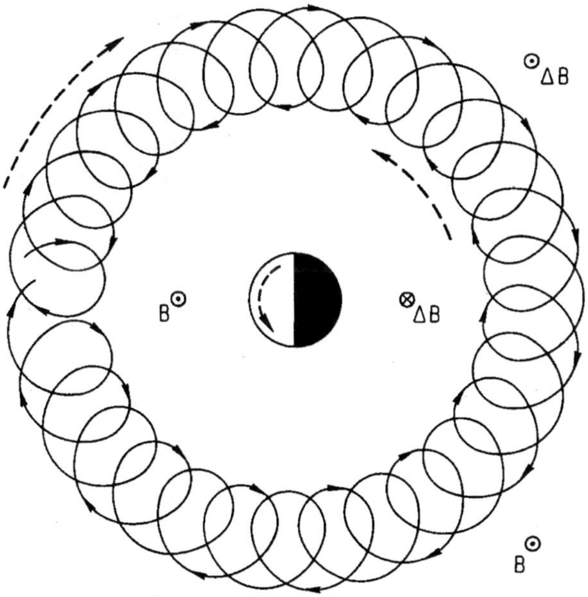

Fig. 7.20. Schematic representation (Schulz and Lanzerotti, 1974, p. 26) of the gyration and azimuthal drift (*solid curve*) of an equatorially mirroring proton, with associated current patterns (*dashed curves*) and the magnetic field perturbations ($\Delta\boldsymbol{B}$) that they produce

space, relative to the adiabatic invariants as canonical coordinates. Formally speaking, the basic equation for this is

$$\frac{\partial \bar{f}}{\partial t} = \sum_{i=1}^{3}\sum_{j=1}^{3}\frac{\partial}{\partial J_i}\left[D_{ij}\frac{\partial \bar{f}}{\partial J_j}\right] - \sum_{i=1}^{3}\frac{\partial}{\partial J_i}\left[\left\langle\frac{\mathrm{d}J_i}{\mathrm{d}t}\right\rangle_{\mathrm{v}}\bar{f}\right] - \frac{\bar{f}}{\tau_q} + \bar{S}, \tag{7.25}$$

where $\bar{f}$ denotes the phase-space density (averaged over gyration, bounce, and drift) and the D_{ij} denote components of the (similarly averaged) diffusion tensor. The need for (7.25) to describe the evolution of $\bar{f}$ (rather than of the local phase-space density f) comes from the fact that the physical processes leading to D_{ij} are typically described by Hamiltonian mechanics, according to which f itself would remain constant along any dynamical trajectory (Liouville's Theorem).

The local phase-space density f is equal to $1/p^2$ times the differential (in energy) unidirectional (in pitch angle) particle flux, which is typically reported in units of $\mathrm{cm^{-2}s^{-1}ster^{-1}keV^{-1}}$. In practice, the diffusion tensor often partitions itself into (a) a term that describes *radial diffusion* (stochastic violation of Φ) at constant M and J and (b) a block that describes *pitch-angle diffusion* (stochastic violation of M and J) at almost constant particle energy. Transformation to the

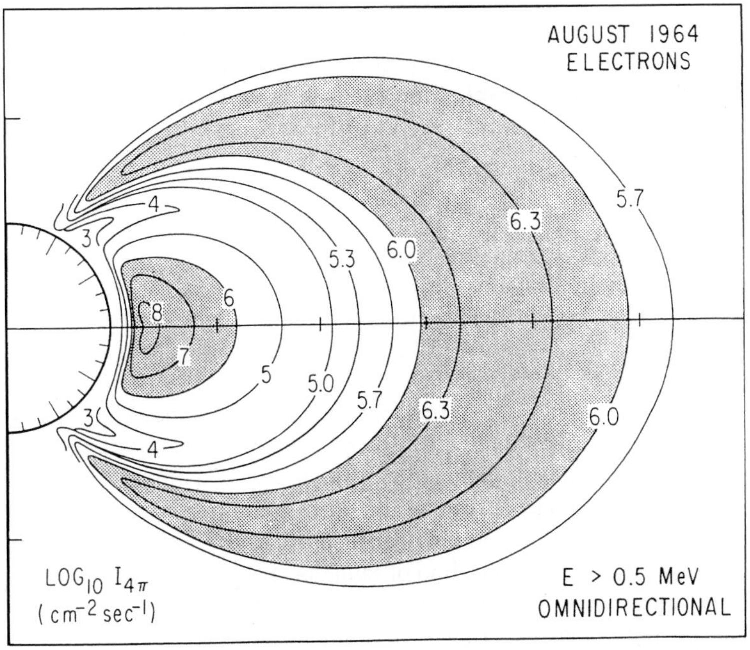

Fig. 7.21. Contours of constant integral omnidirectional electron flux $I_{4\pi}$ in Earth's radiation environment (Vette et al., 1966, p. 20). Shaded regions (Schulz and Lanzerotti, 1974, p. 1) correspond to inner and outer radiation belts ($I_{4\pi} > 10^6\ \mathrm{cm^{-2}s^{-1}}$). As A.L. Vampola (personal communication, ca. 1980) has noted, the inner belt still contained significant fluxes of artificially injected electrons for several years after the "Starfish" nuclear event (0900 UT, 9 July 1962)

more convenient variables $L \equiv 2\pi\mu_E/a\Phi$ and $x \equiv \cos\alpha_0$ then yields

$$\sum_{i=1}^{3}\sum_{j=1}^{3}\frac{\partial}{\partial J_i}\left[D_{ij}\frac{\partial \bar{f}}{\partial J_j}\right] = L^2\frac{\partial}{\partial L}\left[\frac{D_{LL}}{L^2}\frac{\partial \bar{f}}{\partial L}\right]_{M,J} + \frac{\Omega_2}{x}\frac{\partial}{\partial x}\left[\frac{xD_{xx}}{\Omega_2}\frac{\partial \bar{f}}{\partial x}\right]_{p,L} \quad (7.26)$$

after due account is taken of the requisite Jacobian factors. It has been common in various applications of radiation-belt physics to regard pitch-angle diffusion either as negligible, or as sufficient to impel attainment of the lowest eigenmode $g_0(x)$ of the pitch-angle diffusion operator in (7.26), or as strong enough to randomize the equatorial pitch angle α_0 on the half-bounce time scale (π/Ω_2), so as to make the pitch-angle distribution fully isotropic at scalar-momentum (p) values of interest. Under approximations such as these it is permissible to replace the pitch-angle diffusion operator in (7.26) by an L-dependent factor $-\lambda$, and thus the corresponding term in (7.26) by $-\lambda\bar{f}$, with $\lambda = 0$ or λ_0 (lowest eigenvalue of the pitch-angle diffusion operator) or Ω_2/π (for a particle in the loss cone), whichever is appropriate to the underlying assumption about D_{xx}. Even for a particle in the loss cone, the effective loss rate is limited to twice the bounce frequency because a particle requires a quarter bounce period to reach the atmosphere from the equator, and the equatorial pitch-angle distribution then requires another quarter bounce period to recognize that the particle has been lost (e.g., Lyons, 1973).

Lyons and Thorne (1973) showed how (7.25) can lead to a non-monotonic electron radiation intensity profile with two relative maxima in L at any energy $E \gtrsim 100\,\text{keV}$, even though the solution for $\bar{f}$at any value of M is a monotonically increasing function of L. Their study involved construction of a sophisticated theoretical model for the loss rate λ as a function of energy and L value, from a superposition of pitch-angle scattering by Coulomb collisions and whistler-mode (electromagnetic-cyclotron) waves. Results for $\bar{f}$ at selected values of M (with $J = 0$) and for the corresponding differential unidirectional flux $(= p^2\bar{f})$ at selected energies are plotted against L in Fig. 7.22. Normalized radial profiles such as those shown in the left panel of Fig. 7.22 were weighted by the measured spectrum ($\bar{f}$ vs M) at $L = 5.5$ in order to obtain values for mapping to the right panel. For each energy and L value in the right panel, it was necessary to select the weighted value of $\bar{f}$ for the corresponding value of $M = p^2/2m_0B_m$ (with $B_m = B_0$ for $J = 0$) from the left panel (in which only a representative few of the radial profiles for $\bar{f}$ actually used are shown). A more recent analytic model (Schulz, 1991, p. 232; Chen and Schulz, 2001) for the loss rate of radiation-belt electrons is $\lambda = \lambda_0^{\text{coll}} + \lambda_0^{\text{wpi}}$, with

$$\lambda_0^{\text{coll}} \approx 1.9 \times 10^{-4}\gamma(\gamma^2 - 1)^{-3/2}(L - 1.15)^{-2.36}\text{day}^{-1} \quad (7.27\text{a})$$

and

$$\lambda_0^{\text{wpi}} \approx \min[0.08(E,\text{MeV})^{-1.32},\ 0.4(E,\text{MeV})^{1.33}\exp(4.6L - 13.8)]\text{day}^{-1}. \quad (7.27\text{b})$$

This is based on a combination of theoretical (Albert, 1994) and semi-empirical considerations, including actual measurements of electron lifetimes in radiation belts.

The radial diffusion coefficient D_{LL} in (7.26) is usually regarded as a consequence of temporal variations in large-scale magnetospheric electric fields, both induced (Kellogg, 1959; Parker, 1960) and electrostatic (Fälthammar, 1968). The corresponding contributions to D_{LL} are readily calculable (under model-dependent assumptions) in terms of global field-fluctuation spectra that are (however) quite difficult to measure accurately. An alternative semi-empirical fit (Croley et al., 1976) based on long-term inner-zone proton data yielded

$$D_{LL} = 7 \times 10^{-9}L^{10}\text{day}^{-1} + 1 \times 10^{-10}L^{10}(\gamma Z M_0/M)^2\text{day}^{-1} \quad (7.28)$$

for equatorially mirroring particles of various species, for which Z denotes the charge state, $\gamma = m/m_0$, and $M_0 \equiv 1\,\text{GeV/G}$. However, the numerical magnitudes in (7.28) can easily vary by factors $\sim 10^3$ with geomagnetic activity (as measured by Kp or by some other appropriate index). Meanwhile, attention has been directed toward geomagnetic pulsations as a further cause of radial diffusion and energization (Elkington et al., 1999) of radiation-belt particles, as well as toward *in situ* energization of radiation-belt electrons by whistler-mode waves, given that this interaction is significantly inelastic under certain circumstances (e.g., for oblique

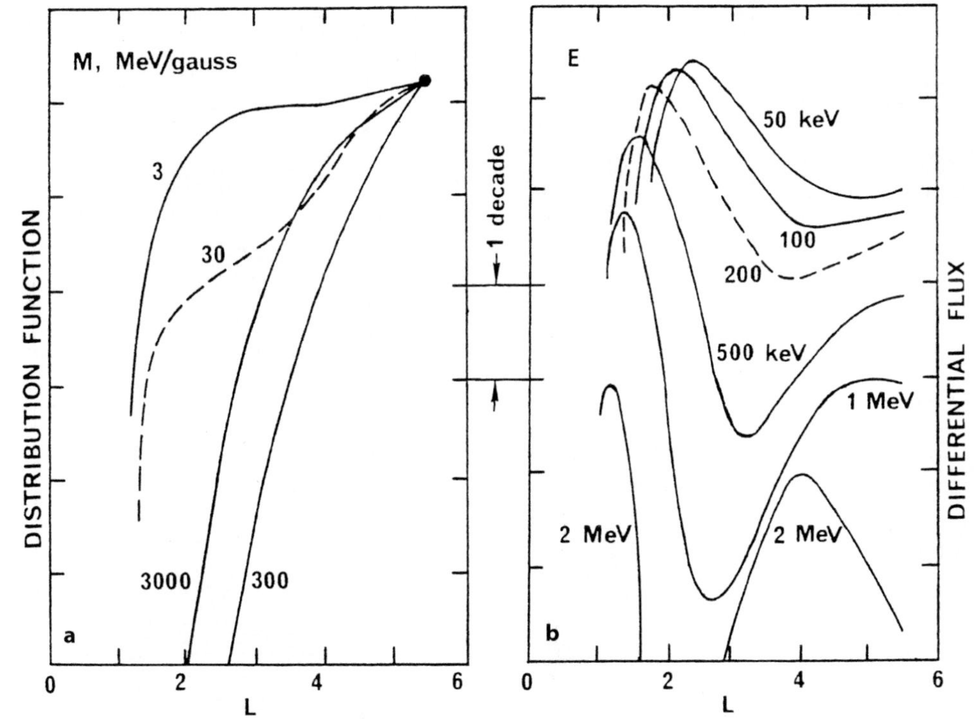

Fig. 7.22a,b. Predicted steady-state profiles (**a**) of phase-space density $\bar{f}$ at constant M, obtained (Lyons and Thorne, 1973; Schulz, 1975, p. 500) from a model similar to (7.25)–(7.28) and normalized to a common value at $L = 5.5$; and predicted differential particle flux (**b**) at specified kinetic energies, normalized by prescribing the measured energy spectrum at $L = 5.5$ (Lyons and Thorne, 1973)

wave propagation, and where the wave frequency $\omega/2\pi$ matches the *relativistic* electron gyrofrequency).

Additional terms on the RHS of (7.25) describe frictional forces (denoted by subscript ν) such as Coulomb drag, *in situ* loss processes such as charge exchange (characterized by lifetime τ_q), and distributed sources (characterized by $\bar{S}$) associated (for example) with beta decay of neutrons emitted from the upper atmosphere because of cosmic-ray bombardment. Charge exchange can also provide a distributed source for ions of a particular charge state (e.g., for He^+) at the expense of an adjacent charge state (e.g., He^{++}) or through the ionization of fast energetic neutral atoms.

The radial diffusion coefficient specified by (7.28) cannot be extrapolated indefinitely toward small values of M/M_0, since a factor $[1 + (\Omega_3\tau)^{-2}]^{-1}$ not shown in (7.28) imposes an upper bound $\sim 1 \times 10^{-4} L^6 \text{day}^{-1}$ on the second term. The parameter τ (~ 1200 s) in this factor represents the characteristic time for exponential decay of impulses in the convection electric field, on which this part of the model for radial diffusion is based. The result quoted in (7.28) corresponds to $\Omega_3\tau \gg 1$, but values of M for which (7.28) thereby fails ($M \lesssim L^2$MeV/G) correspond to kinetic energies ($\lesssim 300L^{-1}$keV) characteristic of the ring current and plasma sheet (i.e., of magnetospheric *hot plasma*) rather than of the radiation belts. This is a régime in which the underlying adiabatic drift trajectories are seriously complicated by effects of the cross-magnetospheric convection electric field, to the extent that the diffusion model underlying (7.28) no longer applies anyway. Examples of such trajectories are shown in Fig. 7.23.

Diffusion relative to closed trajectories in Fig. 7.23 can be formulated in principle, but any boundary conditions on drift-averaged phase-space density $\bar{f}$ in such a formulation should be imposed at the boundary between closed and open drift paths. For modeling stormtime access of hot plasma to the inner magnetosphere

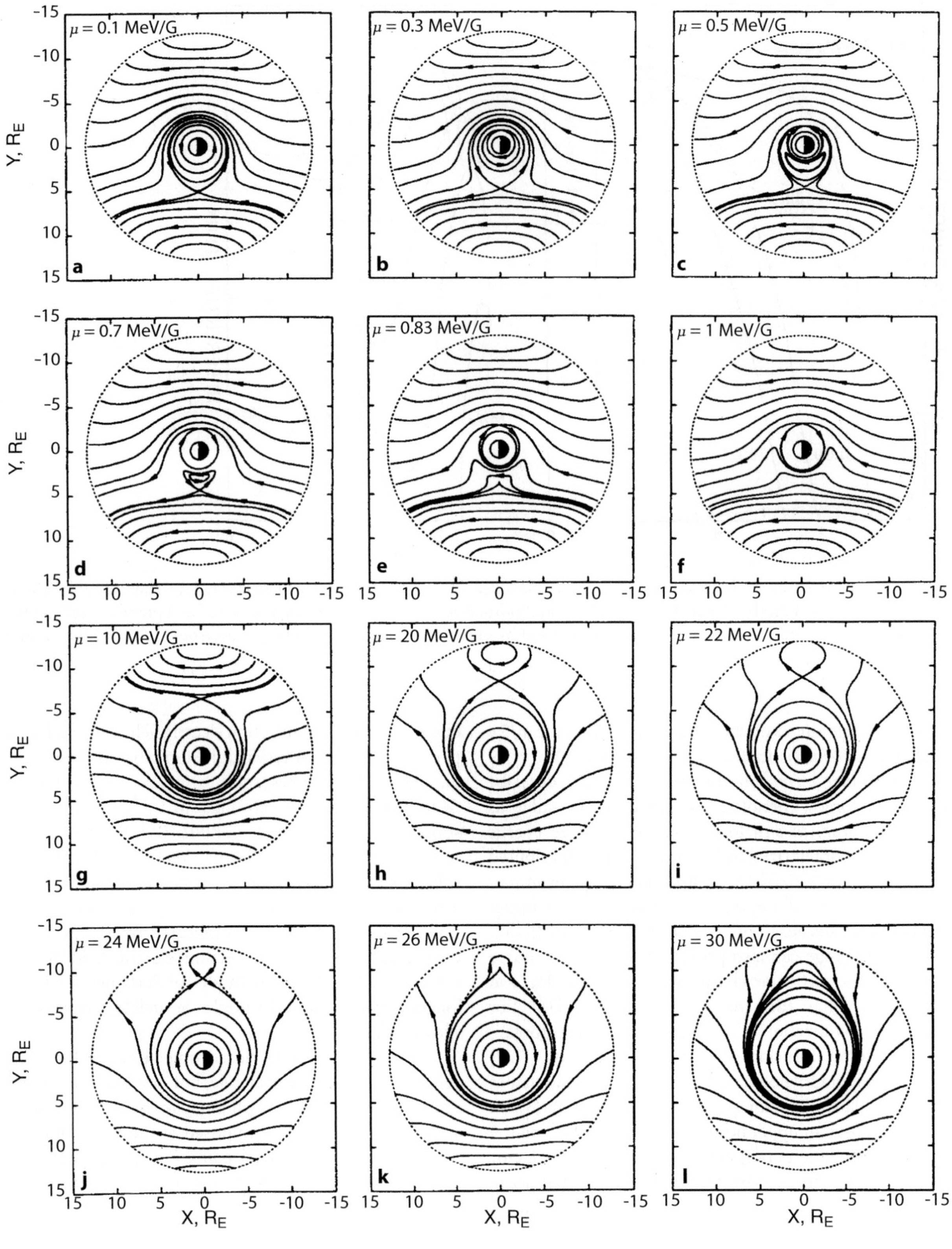
μ = 0.1 MeV/G
μ = 0.3 MeV/G
μ = 0.5 MeV/G
μ = 0.7 MeV/G
μ = 0.83 MeV/G
μ = 1 MeV/G
μ = 10 MeV/G
μ = 20 MeV/G
μ = 22 MeV/G
μ = 24 MeV/G
μ = 26 MeV/G
μ = 30 MeV/G
a
b
c
d
e
f
g
h
i
j
k
l
Y, R_E
X, R_E

Fig. 7.23a–l. Quiet-time drift trajectories of equatorially mirroring singly charged nonrelativistic ions having specified values of first adiabatic invariant μ (Chen et al., 1994)

from $L = L^*$, it is usual to work with the phase-space density f itself before performing a drift average. This requires explicit (rather than merely spectral) specification of temporal variations in the convection electric field. The model for one such study (Chen et al., 1994) is shown in Fig. 7.24. Here the onset times and amplitudes of impulses in the convection electric field were chosen by a random-number generator, but a 600-s "dead time" was imposed after each impulse onset before another could occur. Impulse onsets were also suppressed after 3 h, after 6 h, and after 12 h in order to simulate storm main phases of these various durations. The result was a convective/diffusive transport of magnetospheric hot plasma, so as to form a stormtime ring current that increased in intensity with increasing main-phase duration. Normalized radial profiles of the corresponding equatorial $\Delta \boldsymbol{B}$ are shown in Fig. 7.25. The stormtime ring current is believed to make a major contribution to the geomagnetic index *Dst*, which is measured by deviations of the horizontal component of $\boldsymbol{B}$ from its average value at several low-latitude stations around the Earth. Such values of $\Delta \boldsymbol{B}$ are typically magnified up to 50% by Earth induction, as in (7.15). The value of $\Delta \boldsymbol{B}$ found (by extrapolation) at $R = 0$ is proportional to the total energy content of particles trapped in the geomagnetic field.

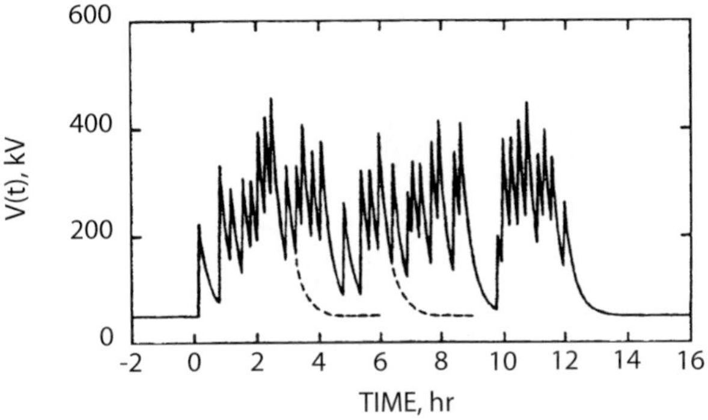

Fig. 7.24. Cross-tail potential drop $V(t)$ for the model storm treated by Chen et al. (1994). This consisted of a 50-kV quiescent value plus a superposition of exponentially decaying impulses with amplitudes and onset times determined by a random-number generator, but with 600 s of dead time after each actual impulse

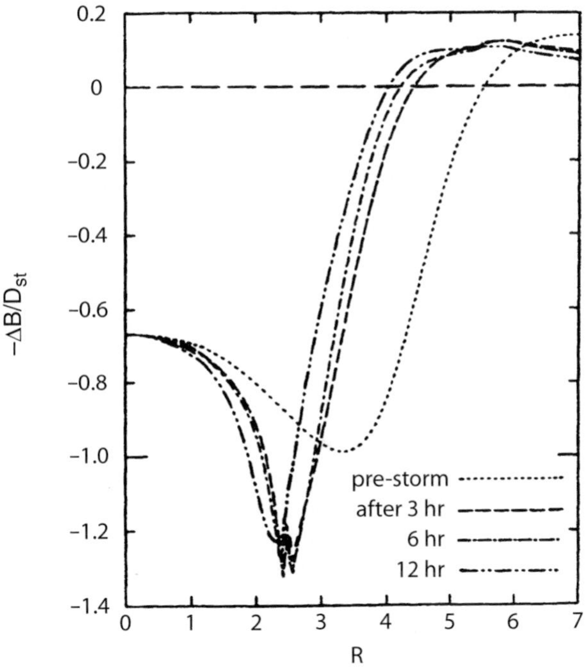

Fig. 7.25. Radial profiles of *Dst*-normalized magnetic-field perturbations produced by the transport model of Chen et al. (1994), before storm onset and after model main phases of selected durations (3 h, 6 h, 12 h)

Figure 7.26 shows the temporal variation of *Dst* over seven days spanning the especially large and well studied geomagnetic storm of 15–16 July 2000. The storm began (as do many large storms) with a *sudden commencement* (sc, marked by arrow), which corresponded to a compression of the magnetosphere by plasma from a (solar-) coronal mass ejection (CME) that had occurred on 14 July 2000. The sc was followed by a gradual decrease in *Dst* over several hours and then by a sharp and deep decrease in *Dst* to about −300 nT, corresponding largely to the build-up of ring-current intensity. (Part of the measured stormtime decrease in *Dst* can be attributed to build-up of the tail field B_t and to decompression of the magnetosphere back to its normal size as the solar-wind pressure pulse associated with the CME moves downstream.) The decrease of *Dst* toward strongly negative values, which can span 3 – 12 h or more, corresponds to

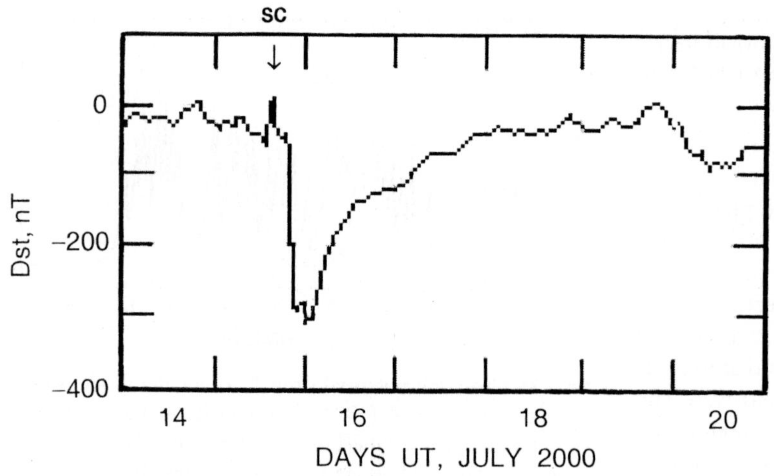

Fig. 7.26. Seven-day plot of final *Dst* index from WDC-C2 in Kyoto (http://swdcdb.kugi.kyoto-u.ac.jp/dstdir/), associated with major geomagnetic storm of 15–16 July 2000. Arrow denotes time of the storm's sudden commencement (sc)

the *main phase* of the storm. This is followed by the *recovery phase*, during which *Dst* increases (over several days) back toward zero.

Recovery phase itself often seems to consist of two stages, an early rapid stage during which the enhanced hot plasma escapes along open drift paths (as in Fig. 7.23) across the dayside magnetopause and a later (more gradual) stage in which the ring current decays via charge exchange of its constituent ions with the neutral (atomic hydrogen) exosphere. However, the ring current's lifetime against charge exchange itself grows longer with time because charge-exchange cross sections vary with energy and ion species, so that the surviving hot plasma increasingly consists (in ion composition and spectral content) of ions with ever-longer lifetimes. Moreover, the decay of the ring current itself induces an azimuthal electric field that serves to energize the surviving ions and electrons, thereby prolonging a storm's recovery phase relative to a model in which $\boldsymbol{B}(r,\theta)$ is specified naively by (7.7b).

Ions (especially H^+ and O^+) contribute more than electrons to the total ring current, mainly because plasmasheet ions typically have higher temperatures than plasmasheet electrons. The plasma sheet is populated in part from the shocked solar wind via the magnetosheath and in part from the ionosphere via auroral ion beams and related plasma flows along $\boldsymbol{B}$ (see Fig. 7.27). Plasmasheet ions and electrons in the nightside current-sheet region (see Figs. 7.3–7.4) are further energized by the cross-tail extension of the convection electric field (e.g., Speiser and Lyons, 1984) before they embark on trajectories of the sort illustrated (for ions) in Fig. 7.23, where they experience further energization through gradient-curvature drift in the direction of $q\boldsymbol{E}$. Particles at $L \gtrsim 0.8L^*$ in Fig. 7.23 often undergo strong pitch-angle diffusion concurrently, in which case they are better approximated as conserving a fourth adiabatic invariant

$$\Lambda \equiv p^3\Psi \equiv p^3 \oint \frac{ds}{B} \tag{7.29}$$

instead of M and J. This Λ is essentially a phase-space volume: the product of a momentum-space volume $(4\pi p^3/3)$ and a flux-tube volume Ψ (per unit magnetic flux), with the factor $4\pi/3$ dropped for algebraic convenience. An analytical approximation (Schulz, 1998) for Ψ in the $\boldsymbol{B}$-field model specified by (7.7b) and (7.8b), accurate within 0.2% for all $L < L^*$, is

$$\Psi \approx (L^4a^4/\mu_E)\left\{(32/35) - \left[2.045 + 1.045(r_0/b)^3 + 0.095(r_0/b)^6 + 0.075(r_0/b)^9\right]\ln\left[1-(r_0/b)^3\right]\right\}, \tag{7.30}$$

where r_0 (regarded as a function of L) denotes the (equatorial) value of r corresponding to $\sin\theta = 1$ in (7.8b). Thus, in the limit that the equatorial pitch angle α_0 is fully randomized over the unit sphere on half the bounce time scale, the relativistic Hamiltonian

$$H = \left[(\Lambda/\Psi)^{2/3}c^2 + m_0^2c^4\right]^{1/2} - m_0c^2 + qV(L,\varphi) \tag{7.31}$$

can be expressed as an analytical function of L and φ, and the resulting bounce-averaged drifts can be calcu-

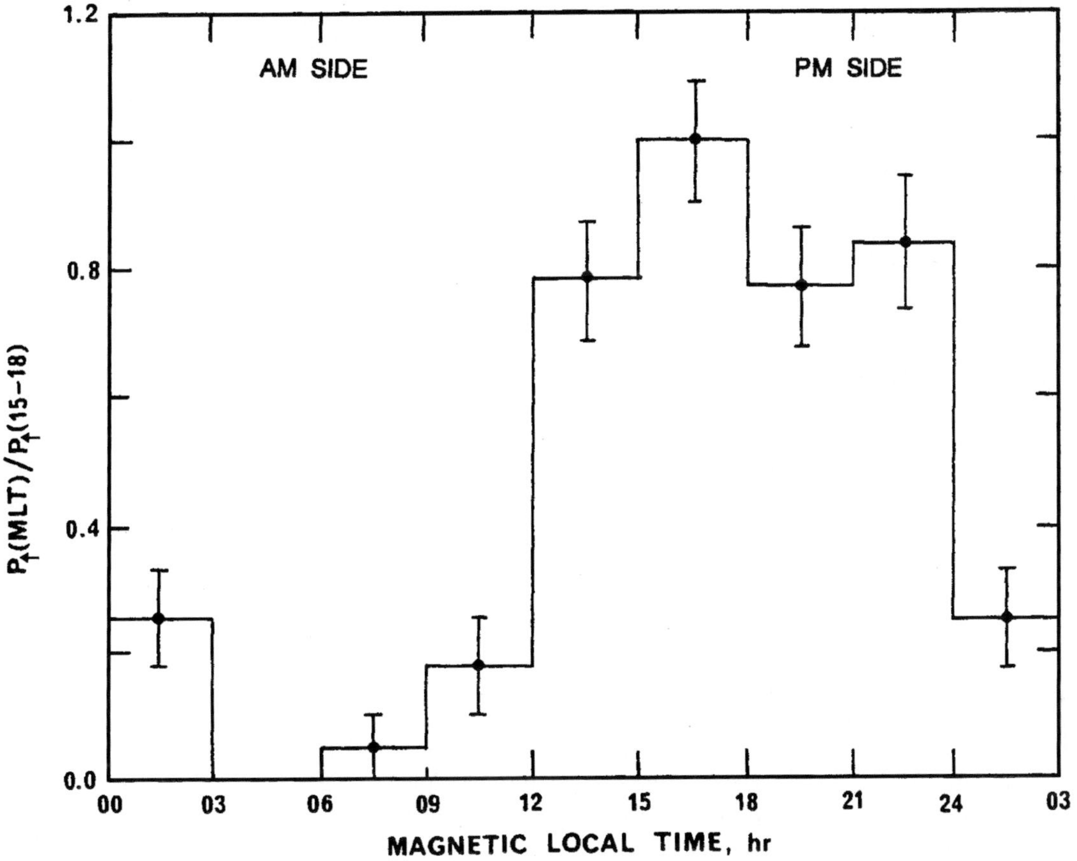

Fig. 7.27. Relative probability-of-occurrence (normalized to 1.0 for the MLT interval 15 – 18 h) of upward-directed ion beams in altitude range 6000 – 8000 km above the auroral oval, according to a statistical study by Ghielmetti et al. (1978)

lated from the equations of Hamilton-Jacobi mechanics:

$$\begin{aligned} \mathrm{d}J_3/\mathrm{d}t &= -(\partial H/\partial\varphi)_\Lambda \\ \Rightarrow \mathrm{d}L/\mathrm{d}t &= +(c/q)(L^2 a/\mu_\mathrm{E})(\partial H/\partial\varphi)_\Lambda \qquad (7.32a) \\ \mathrm{d}\varphi/\mathrm{d}t &= +(\partial H/\partial J_3)_\Lambda \\ \Rightarrow \mathrm{d}\varphi/\mathrm{d}t &= -(c/q)(L^2 a/\mu_\mathrm{E})(\partial H/\partial L)_\Lambda , \qquad (7.32b) \end{aligned}$$

since J_3 is inversely proportional to L. [This answer is right, but the formal derivation is more subtle, since J_3 is technically the integral of $q\mathbf{A}$ around a *closed* path. However, the drift model can be made temporarily axisymmetric (i.e., independent of φ) in this case by temporarily removing the convection electric field and calculating $\mathrm{d}\varphi/\mathrm{d}t$ from $(\partial H/\partial L)_\Lambda$ as in (7.32b). Then the convective $\mathbf{E} \times \mathbf{B}$ drifts in L and φ, which indeed satisfy (7.32) locally, can be added back in.] The flux-tube volume Ψ specified by (7.30) also enters the expression

$$\lambda_0^\mathrm{str} \approx (B_\mathrm{n} + B_\mathrm{s})(1-\eta)(m/p)(4\Psi B_\mathrm{n} B_\mathrm{s})^{-1} \qquad (7.33)$$

for the particle loss rate against strong pitch-angle diffusion. The parameter η in (7.33) is a *backscatter coefficient* not previously mentioned in this chapter. The values of B at the northern (n) and southern (s) ionospheric foot points of the field line of interest enter because they are inversely proportional to the cross-sectional areas of the corresponding unit magnetic flux tube (of volume Ψ) there. These areas need not be equal, since the point dipole from which r is measured in (7.7b) and (7.8b) need not be centered geographically within the Earth (cf. Fig. 7.2).

It follows from (7.30)–(7.31) that charged particles undergoing strong pitch-angle diffusion gain kinetic

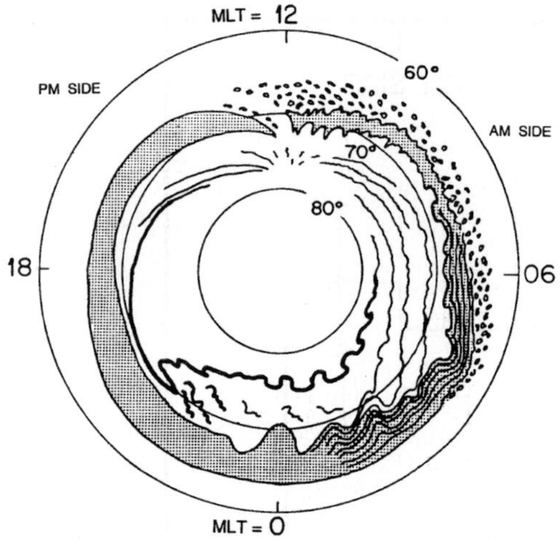

Fig. 7.28. Schematic representation (Akasofu, 1978, p. 76) of geomagnetic polar region, showing major characteristics of the visible aurora. Shaded areas correspond to diffuse aurora, solid features to discrete arcs

energy from transport toward lower L values, but somewhat less strongly than equatorially mirroring particles do. Precipitation of such particles (mainly electrons) into the atmosphere from the plasma sheet at a rate indicated by (7.33) generates the *diffuse aurora*. Electron precipitation from the plasma sheet at somewhat higher latitudes (nearer to $L = L^*$, the boundary between closed and open field lines) is often assisted by parallel (to $\boldsymbol{B}$) electric fields associated with Region-I Birkeland currents (cf. Fig. 7.15) and auroral ion beams (cf. Fig. 7.27), especially on the PM side (MLT = 12 – 24 h) of the auroral oval. Auroras generated this way are known as *discrete arcs*. These often appear as folded curtains of illumination extending across the sky. The overall pattern of auroral-electron precipitation and emitted luminosity, as if observed from above, is illustrated in Fig. 7.28. Auroral particle precipitation in turn influences electrical conductivities in the ionosphere (cf. Fig. 7.11), mainly through energy deposition that tends to ionize neutral atoms.

Enhanced auroral activity is especially associated with a recurrent phenomenon known as the magnetospheric *substorm*. This nomenclature came from a belief that substorms were essential components of geomagnetic storms, but in fact "isolated" substorms not associated with storms can occur also. Thus, the association between storms and substorms may be mostly statistical rather than dynamical. The substorm process begins with a growth in the diameter of the auroral oval, and this *growth phase* can be simulated by allowing the parameter b in (7.7b) and (7.8b) to decrease gradually with time from a nominal quiescent value $\approx 20a$, which corresponds to a polar-cap diameter $\approx 32°$ at $r = a$ (cf. Fig. 7.6). For example, a decrease in b from $20a$ to $10a$ would enlarge the polar-cap diameter to about 46°. The growth phase of a substorm is followed by a sudden decrease in polar-cap diameter (e.g., back to its original 32° as $b \to 20a$), during which the entire band of latitudes between the original auroral oval and its enlarged counterpart is filled with auroral illumination. This is known as substorm *expansion phase*, since it features a poleward "expansion" of the illuminated region. The beginning of the expansion phase is known as *substorm onset*. The region of illumination during expansion phase is much thicker in latitude on the night side than on the day side because the polar cap itself is asymmetric between day and night (cf. Fig. 7.6). The precipitating electrons responsible for this enhanced auroral emission have been energized by the induced electric field, or (equivalently) by the reduction of Ψ (at fixed L) associated in (7.30) with the sudden increase in b and corresponding decrease in r_0. Finally, the auroral electron precipitation and associated light emission from this special band of latitudes gradually decays away as the precipitation described by (7.33) depletes the associated magnetic flux tubes of their hot plasma during substorm *recovery phase*.

7.5 Summary

This chapter constitutes an abbreviated overview of the magnetosphere from a largely theoretical perspective. The unifying theme here is that magnetic fields, electric fields, and charged particles are not easy to treat separately, since they continually interact with each other as constituent parts of the global magnetospheric system. Additional topics in magnetospheric physics, as well as further details on some of the topics treated here, are covered at length elsewhere [e.g., Schulz and Lanzerotti, 1974; Schulz, 1975, 1991; references cited therein] and by many other investigators. Magnetospheric physics is a fascinating field of research, encompassing an enormous variety of natural phenomena that will surely challenge our individual and collective imaginations for many decades into the future.

Acknowledgement. The author thanks N. F. Ness for the nomination and Y. Kamide for the invitation to prepare this chapter. The author also thanks S. M. Petrinec, M. C. McNab, M. A. Rinaldi, and M. W. Chen for collaborations that generated Figs. 7.2, 7.6–7.8, 7.17, and 7.23–7.25, respectively. This work was supported in part by NASA contract NAS5-30372 for PIXIE data interpretation, in part by NSF grant ATM-0201989 for Geospace Environment Modeling (GEM), and in part by NSF grant ATM-0548915 for self-consistent ring-current modeling.

The author has benefited from reading colleagues' articles too numerous to cite in the subject-area of this chapter. However, this is not a review article, and the reference list is admittedly unbalanced. The goal has been to provide access to further details and further sources of information on topics treated here, or in some cases to acknowledge use of graphics first published elsewhere. In some instances the first report of a particular discovery has been left out in favor of a later but seemingly better treatment of the same topic.

References

Akasofu, S.-I. (1977) Physics of Magnetospheric Substorms, 599 pp., Reidel, Dordrecht

Albert, J.M. (1994) Quasi-linear pitch angle diffusion coefficients: Retaining high harmonics, J. Geophys. Res., 99, 23741–23745

Banks, P.M., and Holzer, T.E. (1969) High-latitude plasma transport: The polar wind, J. Geophys. Res., 74, 6317–6332

Birmingham, T.J., and Jones, F.C. (1968) Identification of moving magnetic field lines, J. Geophys. Res., 73, 5505–5510

Brice, N.M. (1967) Bulk motion of the magnetosphere, J. Geophys. Res., 72, 5193–5211

Carpenter, D.L. (1966) Whistler studies of the plasmapause in the magnetosphere: 1. Temporal variations in the position of the knee and some evidence on plasma motions near the knee, J. Geophys. Res., 71, 693–709

Chen, M.W., Lyons, L.R., and Schulz, M. (1994) Simulations of phase space distributions of stormtime proton ring current protons, J. Geophys. Res., 99, 5745–5759

Croley, D.R., Jr., Schulz, M., and Blake, J.B. (1976) Radial diffusion of inner-zone protons: Observations and variational analysis, J. Geophys. Res., 81, 585–594

Elkington, S.R., Hudson, M.K., and Chan, A.A. (1999) Acceleration of relativistic electrons via drift-resonant interaction with toroidal-mode Pc-5 ULF oscillations, Geophys. Res. Lett., 26, 3273–3276

Fälthammar, C.-G. (1968) Radial diffusion by violation of the third adiabatic invariant, in Earth's Particles and Fields (edited by McCormac, B.M.), pp. 157–169, Reinhold, New York and London

Feldstein, Y.I. (1963) Some problems concerning the morphology of auroras and magnetic disturbances at high latitudes, Geomagn. i. Aeronom., 3, 227–239; English transl.: Geomagn. Aeron., 3, 183–192

Ghielmetti, A.G., Johnson, R.G., Sharp, R.D., and Shelley, E.G. (1978) The latitudinal, diurnal, and altitudinal distributions of upward flowing ions of ionospheric origin, Geophys. Res. Lett., 5, 59–62

Giraud, A., and Petit, M. (1978) Ionospheric Techniques and Phenomena, 264 pp., Reidel, Dordrecht

Gorney, D.J., and Evans, D.S. (1987) The low-latitude auroral boundary: Steady state and time-dependent representations, J. Geophys. Res., 92, 13537–13545

Iijima, T., and Potemra, T.A. (1976) Field-aligned currents in the dayside cusp observed by Triad, J. Geophys. Res., 81, 5971–5979

Iijima, T., and Potemra, T.A. (1978) Large-scale characteristics of field-aligned currents associated with substorms, J. Geophys. Res., 83, 599–615

Kivelson, M.G. (1976) Magnetospheric electric fields and their variation with geomagnetic activity, Rev. Geophys. Space Phys., 14, 189–197

Lyons, L.R. (1973) Comments on pitch-angle diffusion in the radiation belts, J. Geophys. Res., 78, 6793–6797

Lyons, L.R., and Thorne, R.M. (1973) Equilibrium structure of radiation belt electrons, J. Geophys. Res., 78, 2142–2149

Mayaud, P.N. (1980) Derivation, Meaning, and Use of Geomagnetic Indices, Geophys. Monogr. 22, Am. Geophys. Union, Washington, DC

Mead, G.D., and Beard, D.B. (1964) Shape of the geomagnetic field solar wind boundary, J. Geophys. Res., 69, 1169–1179

Nishida, A. (1966) Formation of a plasmapause, or magnetospheric plasma knee, by combined action of magnetospheric convection and plasma escape from the tail, J. Geophys. Res., 71, 5669–5679

Parker, E.N. (1963) Interplanetary Dynamical Processes, Wiley/Interscience, New York

Richmond, A.D., and Kamide, Y. (1988) Mapping of electrodynamic features of the high-latitude ionosphere from localized observations: Techniques, J. Geophys. Res., 93, 5741–5759

Schmidt, A. (1935) Tafeln der normierten Kugelfunktionen, sowie Formeln zur Entwicklung, Engelhard-Reyer, Gotha

Schulz, M. (1975) Geomagnetically trapped radiation, Space Sci. Rev., 17, 481–536

Schulz, M. (1991) The magnetosphere, in Geomagnetism, vol. 4, edited by Jacobs, J.A., pp. 87–293, Academic Press, London and San Diego

Schulz, M. (1998) Particle drift and loss rates under strong pitch-angle diffusion in Dungey's model magnetosphere, J. Geophys. Res., 103, 61–67 and 15013

Schulz, M., and Eviatar, A. (1969) Diffusion of equatorial particles in the outer radiation zone, J. Geophys. Res., 74, 2182–2192

Schulz, M., and Koons, H.C. (1972) Thermalization of colliding ions streams beyond the plasmapause, J. Geophys. Res, 77, 248–254

Schulz, M., and Lanzerotti, L.J. (1974) Particle Diffusion in the Radiation Belts, 215 pp., Springer, Heidelberg

Schulz, M., and McNab, M.C. (1996) Source-surface modeling of planetary magnetospheres, J. Geophys. Res., 101, 5095–5118

Sibeck, D.G., Lopez, R.E., and Roelof, E.C. (1991) Solar wind control of the magnetopause shape, location, and motion, J. Geophys. Res., 96, 5489–5495

Snyder, C.W., Neugebauer, M., and Rao, U.R. (1963) The solar wind velocity and its correlation with cosmic-ray variations and with solar and geomagnetic activity, J. Geophys. Res., 68, 6361–6370

Speiser, T.W., and Lyons, L.R. (1984) Comparison of an analytical approximation for particle motion in a current sheet with precise numerical calculations, J. Geophys. Res., 89, 147–158

Spreiter, J.R., Summers, A.L., and Alksne, A.Y. (1966) Hydromagnetic flow around the magnetosphere, Planet. Space Sci., 14, 223–253

Stahara, S.S. (2002) Adventures in the magnetosheath: Two decades of modeling and planetary applications of the Spreiter magnetosheath model, Planet. Space Sci., 50, 421–442

Stern, D. (1974) Models of the Earth's electric field, Rept. X-602-74-159, NASA Goddard Space Flight Center, Greenbelt, MD

Stern, D.P. (1975) Quantitative models of magnetic and electric fields in the magnetosphere, Rept. X-602-75-90, NASA Goddard Space Flight Center, Greenbelt, MD

Straus, J.M., and Schulz, M. (1976) Magnetospheric convection and upper-atmospheric dynamics, J. Geophys. Res., 81, 5822–5832

Tsyganenko, N.A. (1995) Modeling the Earth's magnetospheric magnetic field confined within a realistic magnetopause, J. Geophys. Res., 100, 5599–5612

Vette, J.I., Lucero, A.B., and Wright, J.A. (1966) Models of the Trapped Radiation Environment, Volume II: Inner and Outer Zone Electrons, NASA SP-3024, Washington, DC

Volland, H. (1973) A semi-empirical model of large-scale magnetospheric electric fields, J. Geophys. Res., 78, 171–180

Volland, H. (1975) Models of global electric fields within the magnetosphere, Ann. Géophys., 31, 154–173

Weimer, D.R. (1995) Models of high-latitude electric potentials derived with a least error fit of spherical harmonic coefficients, J. Geophys. Res., 100, 19595–19607

Weimer D.R. (2005) Improved ionospheric electrodynamic models and application to calculating Joule heating rates, J. Geophys. Res., 110, A05306, doi:10.1029/2004JA010884

Yasuhara, F., and Akasofu, S.-I. (1977) Field-aligned currents and ionospheric electric fields, J. Geophys. Res., 82, 1279–1284

8 Ionosphere

Pierre-Louis Blelly and Denis Alcaydé

The Earth's ionosphere (at an altitude range of approximately 60–2000 km) is historically the region of the atmosphere that affects the propagation of radio waves. It is strongly related to the atmosphere; its reservoir of charged particles is created by ionization of atmospheric neutral gaseous compounds. The electrical properties of the ionospheric plasma cause major electrodynamical couplings between the magnetosphere and the atmosphere. Hence, it plays an important role in the particle, momentum and energy transfer processes between the Earth's magnetosphere and atmosphere. It is therefore a key region in the Sun–Earth connection system.

This chapter presents a brief introduction to the physics of the Earth's ionosphere. After a basic description of the processes leading to the creation of ions and electrons in the ionosphere, the main mechanisms involving the ionosphere in the dynamics of the Earth's magnetosphere are discussed, with a particular emphasis on the energization and transport of ionospheric particles. At the end of the chapter some ground-based powerful tools that are routinely used to sound the ionosphere are briefly introduced.

Contents

Pierre-Louis Blelly and Denis Alcaydé, Ionosphere.
In: Y. Kamide/A. Chian, Handbook of the Solar-Terrestrial Environment. pp. 189–220 (2007)
DOI: 10.1007/11367758_8

The diurnal variations of the Earth's magnetic field were known already in the XVIIth century, but in 1839 Gauss was the first to affirm that those variations were due to external sources and to speculate about the existence of a conductive layer in the atmosphere. Well before the discovery of the electron by Thomson in 1897, Stewart expanded on this idea and in 1882 he proposed the theory of the terrestrial dynamo, which gave an explanation of those variations. The end of the XIXth century was also marked by the development of the theory of electromagnetism and the first attempts at radio communication. In 1901 Marconi succeeded in the first radio transmission across the Atlantic. Because of the curvature of the Earth, this connection was made possible only by a deflection of the waves in the atmosphere. However, the importance of the deflection was explained by the diffraction of radio waves in the atmosphere. Based on Stewart's idea, in 1902 Kennely and Heaviside suggested the existence of a conducting layer, made of free electrical charges, which was able to mirror the electromagnetic waves. Then, the idea emerged that because this layer may result from solar ultraviolet radiation, this implies that the Sun exerts control on the deflecting capacity of this layer. This theory of free electrical charges remained controversial for about 20 years. Appleton and Barnett (1925) proved the existence of this layer by studying the reflection of electromagnetic waves of different frequencies using interferometry techniques. This discovery was soon confirmed by Breit and Tuve (1925) who carried out a survey by sending radio pulses. Both experiments provided estimations of the height at which the layer is located, and thus really marked the beginnings of ionospheric physics. Watson-Watt (1929) suggested calling this layer *ionosphere* and Hartree (1931) proposed the first magneto-ionic theory for the propagation of electromagnetic waves in plasmas.

Hence, the ionosphere is historically the region of the atmosphere that can affect the propagation of radio waves. It can also be seen as the region where charged particles are created, since the ionosphere emerges from the ionization of the neutral gaseous compounds of the atmosphere. One may also define the ionosphere as the atmospheric layer where the charged particles are created and have a typical energy lower than 1 eV, more likely close to the neutral thermal energy (a few tenths of an eV).

The ionospheric layer is approximately located between 60 and 2000 km altitude, but the most important contribution lies in the 90–1000 km region. The region above approximately 2000 km can be considered as being in the magnetosphere, which is the subject of Chap. 7.

The ionosphere has a peculiar position in the Sun–Earth system. It is strongly related to the atmosphere, which constitutes its reservoir of charged particles, and the electrical properties of the ionospheric plasma are at the origin of major electrodynamical couplings in the magnetosphere and the solar wind.

This chapter presents the main mechanisms involving the ionosphere in the dynamics of the Earth's magnetospheric system. After a basic description of the processes leading to the creation of ions and electrons in the ionosphere, we discuss the mechanisms of energization and transport of ionospheric particles. The end of the chapter is devoted to some ground-based instruments used to routinely sound the ionosphere; these are powerful tools to access the different scales of the dynamics of the ionosphere. This chapter is intended to give a general overview of the ionosphere and should not be considered as an exhaustive presentation of ionospheric physics. A more complete description is given in Schunk and Nagy (2000).

8.1 Production and Structure

8.1.1 Ionization Processes

Basically, the ionosphere is a by-product of the interaction between the Sun and the Earth environment. As a matter of fact, the ionosphere emerges from the shielding effect exerted by the neutral atmosphere against the penetration of particles (in a wide sense) coming from outside. The energy of these incoming particles is absorbed by collisions with the atmosphere. If their energy is high enough, the collisions can result in the ionization of neutral atoms or molecules. These particles can be either photons coming from the Sun (see Chap. 3), cosmic rays coming from the interplanetary medium, or particles coming from the solar wind (see Chap. 4) or the magnetosphere (see Chap. 7).

As discussed in Chap. 9, the dominant process in the thermosphere is the molecular diffusion in the gravity field, which acts to separate the various atmospheric components with respect to their mass. The lighter species are transported at higher altitudes and each species is in hydrostatic equilibrium with its own scale height. As the temperature gradients in the atmosphere are much lower than the density gradients, we can neglect the dependence on the altitude of the neutral temperature T_n. Moreover, if we assume that the gravity field g is almost uniform all over the altitude range of interest, the concentration $n_n(z)$ at altitude z of the neutral species n of mass m_n given by the barometric law is

$$n_n(z) = n_n^o e^{-\frac{m_n g}{k_b T_n}(z-z_o)} = n_n^o e^{-\frac{z-z_o}{H_n}}, \qquad (8.1)$$

where n_n^o is the concentration of the species n at the reference level z_o, k_b is the Boltzmann constant and $H_n = k_b T_n / m_n g$ is the scale height of species n.

The gravity acts as a mass filter, concentrating the heavier species such as N_2 and O_2 in the lower part of the thermosphere, where they dominate below 200 km. The less heavy species O typically becomes the major species between 200 and 600 km. Finally, the light species H becomes the preponderant species in the exosphere, above 600 km.

The two major sources for the thermal plasma that we consider are the photoionization and the ionizing collision impact on the neutral atmosphere of protons or electrons. The photons are clearly identified as coming from the Sun, while the protons and electrons can come either from the solar wind or the magnetosphere. In this section, we focus on the basic principles of the creation of the thermal plasma from these two ionization processes.

Solar EUV Flux

The first ionization thresholds are 12.1 eV for O_2, 13.6 eV for H and O, and 15.6 eV for N_2. Then, for the main neutrals, only radiations with wavelengths lower than 100 nm (~ 12.4 eV) can produce ions. These wavelengths correspond to the extreme ultra-violet (EUV) band of the solar radiation spectrum, which originates in the chromosphere. It is composed of a continuum and strong emission lines contributing equivalently to the ionization of the neutral atmosphere (Richards et al. 1994, 2006).

The flux coming in at the top of the atmosphere is absorbed by the neutrals as it penetrates within the atmosphere. Due to a wavelength dependency of the neutrals absorption cross section, this absorption is not uniform over the entire EUV spectrum. For simplification, we assume that the atmosphere is made of only one constituent (species n), distributed radially according to the barometric law (8.1). At these wavelengths, the only alteration of the solar flux is an attenuation of the intensity, which is well described by the Beer–Lambert law. Then, the solar flux penetrates within the atmosphere along a straight line (the line of sight) until a point where it is completely absorbed. If s is the distance covered from the top of the atmosphere along the line of sight, the variation $dI(s, \lambda)$ of the intensity $I(s, \lambda)$ of the solar flux at wavelength λ, is given by

$$dI(s, \lambda) = -\sigma_a(\lambda) n_n(z) I(s, \lambda)\, ds, \qquad (8.2)$$

where $\sigma_a(\lambda)$ is the absorption cross section at the wavelength λ. In a multi-component atmosphere, the total variation is the sum of the contribution of each neutral constituent.

Equation (8.2) can be integrated providing a relationship $\frac{ds}{dz}$ between the distance s along the line of sight and the given location (determined by the altitude z). This can be achieved by introducing the solar zenith angle χ, which is the angle between the local vertical and the direction of the Sun (the Sun at zenith: $\chi = 0°$, nightside: $\chi > 90°$).

Integrating (8.2) over the altitude z gives

$$I(z, \lambda) = I_\infty(\lambda)\, e^{-\tau}, \qquad (8.3)$$

where $I_\infty(\lambda)$ is the intensity flux at the topside of the atmosphere. In this equation, we have introduced the optical depth τ

$$\tau \equiv \int_\infty^z \sigma_a(\lambda) n_n(z) \frac{ds}{dz}\, dz = \sigma_a\, n_n(z)\, \mathrm{Chap}(z, \chi, H_n), \qquad (8.4)$$

where $\mathrm{Chap}(z, \chi, H_n)$ is the Chapman function. This function represents the equivalent height of the atmospheric column density along the line of sight, and $n_n(z)\,\mathrm{Chap}(z, \chi, H_n)$ is thereby the number of particles present in the column density crossed by the solar radiation. Its expression is complex for high values of χ, corresponding either to sunset or sunrise, or the high latitude atmosphere, because the computation of the curvilinear abscissa along the line of sight is not straightforward in spherical geometry. However, for low values of χ (typically below 60°), we can obtain simplified expressions by assuming that the Earth is flat. In such a case we have

$$ds = -\frac{1}{\cos(\chi)}dz\,, \tag{8.5}$$

where dz is the vertical displacement associated with the displacement ds along the line of sight length. The Chapman function then writes as

$$\mathrm{Chap}(z, \chi, H_n) = \frac{H_n}{\cos(\chi)} \tag{8.6}$$

and the optical depth τ becomes

$$\tau = \sigma_a n_n(z)\frac{H_n}{\cos(\chi)}\,. \tag{8.7}$$

This parameter is important because it characterizes the power of a solar flux radiation to produce ionization at a given altitude. It varies exponentially with altitude since it is proportional to the atmospheric concentration, and thus τ sharply increases above 1. We may consider that the solar radiation is completely absorbed at the altitude where $\tau = 1$.

Figure 8.1 shows this altitude where optical depth is equal to 1 for solar flux between 5 and 100 nm and for three different values of the solar zenith angle χ. Independently of χ, the ionizing part of the solar flux is stopped at altitudes higher than 100 km. The range 40–80 nm is stopped at the higher altitudes because of the stronger absorption cross section for all the main neutral species (N_2, O_2 and O) in that wavelength range. At a given wavelength, the penetration depth is roughly a linear function of $1/\cos(\chi)$ as indicated by (8.7, but because the slope is proportional to σ_a, the variation in the range 40–80 nm is larger than for other wavelengths.

The ion emerges from the ionization of its neutral parent. If $\sigma_i(\lambda)$ is the ionization cross section for the wavelength λ, then the production $P_i(z, \lambda)$ of the ion i is given by

$$P_i(z, \lambda) = \sigma_i(\lambda)n_n(z)I(z, \lambda)\,. \tag{8.8}$$

Obviously, we have $\sigma_i < \sigma_a$ since σ_a is the total absorption and thereby comprises the part of the flux absorbed in the ionization. If we define z_o the altitude where $\tau = 1$, when the sun is at zenith (i.e. $\chi = 0°$), then we have

$$\sigma_a\, n_n(z_o)H_n = 1 \tag{8.9}$$

and the production is given by

$$P_i(z, \lambda) = P_o \exp\left\{-\frac{z - z_o}{H_n} + 1 - \frac{1}{\cos(\chi)}e^{-\frac{z-z_o}{H_n}}\right\}, \tag{8.10}$$

where $P_o = \sigma_i(\lambda)n_n^o I_\infty(\lambda)/e$ is the ion production at the reference altitude z_0, for zenith conditions. $P_i(z, \lambda)$ is called the Chapman production function. Its variation comprises two contributions in the exponential function. The first one is a linear function of the altitude and corresponds to the barometric decrease of the neutral constituent. The second term is due to the absorption of the flux, through τ. The absorption is very low at a high altitude, where τ is small, and thus the production varies with the altitude in the same way as the concentration. The optical depth τ sharply increases after it reaches a value close to 1 because of the exponential variation, and there the production no longer has the same variation as at a high altitude. The dependence on χ is such that when χ increases, the altitude where the maximum is reached is translated to an upper altitude, and the amplitude of the production is decreased.

The real process is obtained by first summing the basic mechanisms over all the neutral species and then summing over all the wavelengths. This leads to ion production profiles spread over a wide range of altitudes, with the introduction of ionization layer characteristics of the neutrals present in the atmosphere. In Sect. 8.1.2, we present a complete view of the production of ions with the typical ion photoproduction profiles shown in Fig. 8.3.

Precipitation

In theory, the ionization process associated with the precipitation is intimately related to the transport of energetic particles in the atmosphere and requires a complete approach. However, the basic principles can be presented without such a complex approach and in this section, we describe a simple mechanism to evaluate ion production. The main difference with photoionization is that no Beer–Lambert law applies. However, we can develop a scheme that results in a similar approach (Rees, 1989). First of all, due to the interaction mechanism involved in the ionizing impact collision, some energy is lost in the creation of an ion-electron pair. The amount of energy lost depends on the energy of the incident particle and on the nature of the target particle, but we

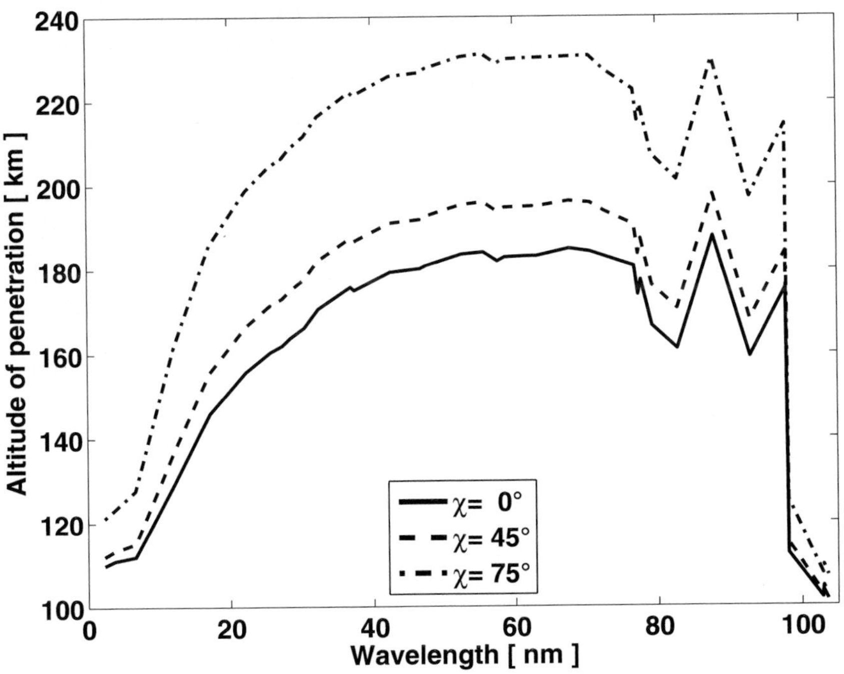

Fig. 8.1. Altitude of solar flux penetration between 5 and 100 nm for three different solar zenith angles χ: $0°, 45°, 75°$. At these altitudes the optical depth is equal to 1 and the intensity flux is thus divided by a factor e

can use a mean value $\Delta\varepsilon$ for all ionizing collisions with a specific incident particle. From experimental measurements it has been determined that $\Delta\varepsilon = 35$ eV is a good compromise for electron precipitation (Rees, 1989). For proton precipitation, Galand et al. (1999) have computed that $\Delta\varepsilon$ varies with the incident energy from 20 eV (proton incident energy of 1 keV) to 30 eV (proton incident energy of 20 keV).

Thus, a particle of initial energy E is likely to produce $E/\Delta\varepsilon$ ions and electrons before it is stopped (i.e. before its energy is lower than $\Delta\varepsilon$). In the case of photons, we consider the attenuation of the intensity flux, while in the case of particle precipitation, we consider the decrease in energy. The distance that a particle covers before it is stopped depends on the number of collisions. For practical reasons associated with the experimental devices used to measure this parameter, experimenters define the penetration depth $R(E)$, which represents the total mass encountered since the entry point. This parameter is preferred to the distance covered, but both are characterizations of the same property, since the collision cross section increases roughly with the mass.

Different expressions have been fitted from complete calculations. For electron precipitation, the following expression can be used (Rees, 1989):

$$R(E) = 4.3 \times 10^{-6} + 5.36 \times 10^{-5} E^{1.67} . \quad (8.11)$$

$R(E)$ is expressed in kg m^{-2} and E is in keV. This expression is valid between 200 and 50 keV. For proton precipitation, Rees (1989) gives a similar expression for incident energy between 1 keV and 100 keV

$$R(E) = 5.05 \times 10^{-5} E^{0.75} . \quad (8.12)$$

Recently, Galand et al. (1999) gave a more complete but also more complex expression for incident proton energy between 1 keV and 20 keV.

In practice, the absorption of the energy is not uniform along the particle trajectory. To account for this phenomenon, we introduce a normalized energy deposition function Λ, which represents the distribution of the ionization along the trajectory. Different shapes have been determined, depending on the characteristics of the particle beam (Rees, 1989). The path along the trajectory is determined by the normalized scattering depth $r(z)$

$$r(z) = \frac{1}{R(E)} \int_z^\infty \rho(z')\,\mathrm{d}z' , \quad (8.13)$$

where $\rho(z) = m_n n_n(z)$ is the mass density at altitude z.

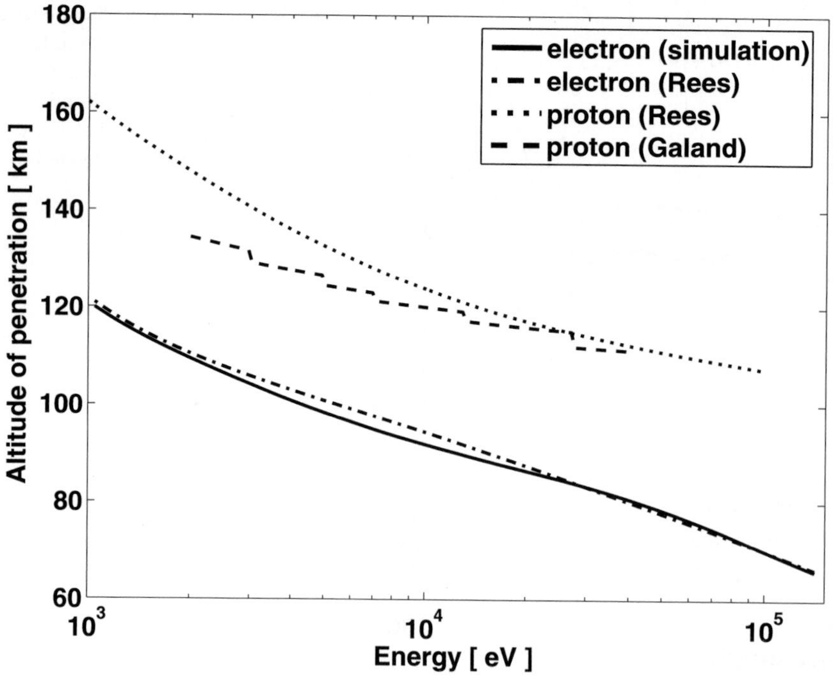

Fig. 8.2. Altitude of penetration of precipitating particles as a function of the energy, i.e. the altitude at which $r(z)$ (see (8.13)) is equal to 1. The curves are plotted for electrons for a complete calculation with a numerical model (*full line*), Rees, (1989) expression ((8.11), *dashed-dotted line*) and for protons: Rees, (1989) expression ((8.12), *dotted line*) and Galand et al. (1999) expression (*dashed line*)

This parameter defines the penetration of the particle from the entry point ($r(z) = 0$) to the altitude where the particle is stopped ($r(z) = 1$). Λ is expressed as a function of this parameter and $\Lambda(r(z))\rho(z)/R(E)$ thus represents the proportion of collisions occurring at altitude z.

If we consider a monoenergetic particle beam of initial energy E (in eV) and flux F (in $\mathrm{m}^{-2}\,\mathrm{s}^{-1}$), $FE/\Delta\varepsilon$ is the total ionization rate issued from the beam. The ion production (in $\mathrm{m}^{-3}\,\mathrm{s}^{-1}$) can then be written as

$$P(z,E) = F\frac{E}{\Delta\varepsilon}\Lambda(r(z))\frac{\rho(z)}{R(E)} \qquad (8.14)$$

and the rate of deposition of energy per unit volume at altitude z is roughly given by

$$\varepsilon(z,E) = \Delta\varepsilon\, P(z,E)\,. \qquad (8.15)$$

If we have a spectral energy distribution instead of a monoenergetic flux, the production is obtained by summing over the energy. This simplified approach is made for a mono-constituent atmosphere, and similarly to what is done for the photons, it can be extended to a multi-constituent atmosphere. We can compute the altitude of penetration of particles, which correspond to the altitude where $r(z) = 1$. Figure 8.2 shows the altitude where electrons or protons are stopped as a function of their initial energy. For a same energy, protons are stopped at higher altitudes (about 20–40 km) than electrons, which is consistent with the interaction cross sections of these particles, which are higher for protons for than for electrons. We see that the expression given by Rees (1989) compares very well with a complete calculation obtained by solving transport for suprathermal electrons.

8.1.2 Primary Ionospheric Outputs

Although simple, the two mechanisms presented above explain the primary production of ions and electrons rather well. However, the ionization does not stop there and a cascade can be observed, which leads to further ion production.

Electrons

As we have mentioned, each ionization process creates an ion-electron pair and absorbs a certain amount of energy. Due to the mass ratio, the emerging electron gets the most important part of the energy above the ionization threshold. Then almost all the electrons are created with energy well above the thermal energy. These electrons are called suprathermal electrons; they may have enough energy to be able to produce another ionization by a collision impact. In that case, they behave like the precipitating electrons discussed in the previous section. They cross a certain distance before they are stopped and are the source of a secondary production. This secondary production may be important; for instance in the case of photoelectrons, the secondary production can be about 1/3 of the primary production, above the E region. This production contributes to increasing the overall production. The cascade operates until the energy of these electrons is much lower than the energy required for ionization. When the energy of these suprathermal electrons reach low values (below 1 eV), they become thermal electrons.

To describe the mechanism of secondary production correctly, a complete calculation of the transport of the suprathermal electrons is necessary. The transport equation is basically given by the Boltzmann equation, which in the case of the Earth with its magnetic field can be simplified (Oran and Strickland, 1978; Lilenstein and Blelly, 2002).

The degradation of the energy of these suprathermal electrons is such that the concentration ratio between suprathermal and thermal electrons is less than 10^{-4} in the ionosphere and thus does not contribute significantly to the total electron pressure. However, this suprathermal electron population is the major source of energy for the thermal electron population. Indeed, the collisions of the suprathermal electrons on the thermal electrons result in the energization of the latter. The energy transfer is modeled well by a continuous loss approximation, with a drag force $\boldsymbol{F}_{\mathrm{drag}}$, acting on every suprathermal, given by

$$\boldsymbol{F}_{\mathrm{drag}} = -n_{\mathrm{e}} L(E) \frac{\boldsymbol{v}}{v} , \tag{8.16}$$

where $\boldsymbol{v}$ is the velocity of the suprathermal electron of energy E and $L(E)$ the energy loss function given by (Swartz and Nisbet, 1972)

$$L(E) = \frac{5.39 \times 10^{-3}}{E^{0.94} n_{\mathrm{e}}^{0.03}} \left(\frac{E - E_{\mathrm{th}}}{E - 0.53 E_{\mathrm{th}}} \right)^{2.36} . \tag{8.17}$$

In this expression, $E_{\mathrm{th}} = 8.61 \times 10^{-5}\, T_{\mathrm{e}}$ is the energy of the thermal electrons (in eV), with T_{e} the electron temperature (in Kelvin) and $n_{\mathrm{e}} L(E)$ in eV m^{-1}.

As discussed in Sect. 8.2.2, this energy transfer is important for the thermal electron energy balance and maximizes where the suprathermals are the most important, that is, almost in the region of maximum of production, around 150–160 km.

To summarize, the electrons are separated into two populations, which interact differently with the atmospheric and ionospheric compounds. They are treated separately, but their strong coupling is accounted for by the inclusion of a mutual interaction, modeled by the drag force $\boldsymbol{F}_{\mathrm{drag}}$, and characterized by the energy loss $L(E)$.

Primary Ions

Due to the mechanisms involved in the ionization, the emerging ions are strongly related to the neutrals present in the atmosphere. The mass separation is a strong parameter for the ion distribution observed in the ionosphere. At low altitude, the production of O_2^+ and N_2^+ is the most important. Above 200 km, the main ion production concerns O^+. At high altitude, above 500 km, H^+ production is the most important. Because the Lyman α line emission is very intense and is not stopped above 90 km, there is a significant production of NO^+ below 90 km, although NO is a very minor constituent of the atmosphere. N^+ is an exception to the standard rule, which says that an ion results from the ionization of its neutral parent: this ion is produced at an important level that is not at all compatible with the concentration of its neutral parent. In fact, this ion is a by-product of the ionization of N_2, for which about 20% corresponds to a dissociative ionization. Figure 8.3 shows the primary and secondary ion photoproduction for the main species in the ionosphere (O_2^+, N_2^+, O^+, H^+ and N^+). The total ion production spreads over a wide region between 100 km and 200 km (400 km as far as H^+ are concerned), and the produced major ion changes according to the atmospheric mass separation. The secondary production of N_2^+ ion is more than one order of magnitude higher around 100 km than the primary production, showing that the secondary production is not a negligible feature.

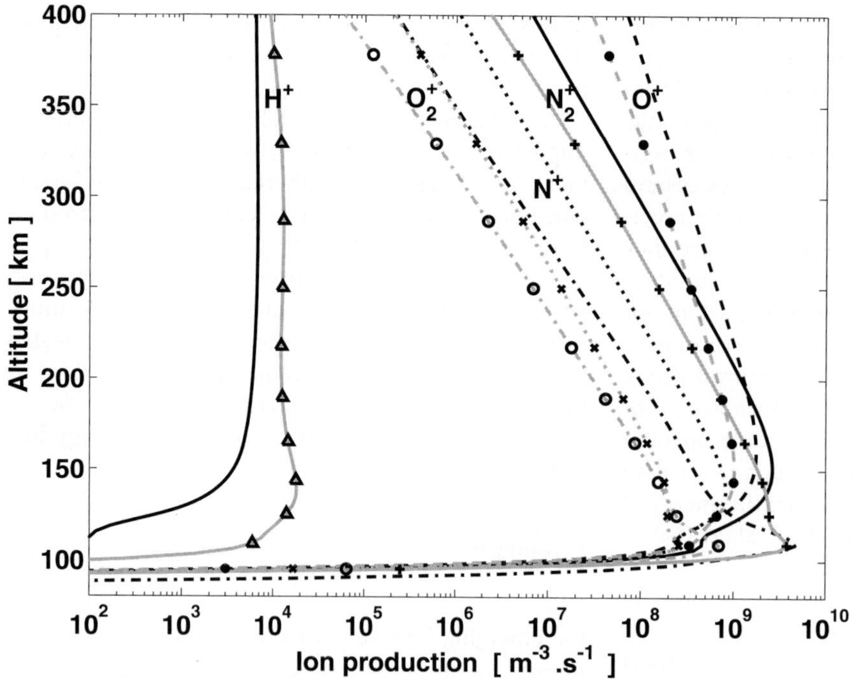

Fig. 8.3. Primary and secondary photoproduction of N_2^+ (primary: *full line*, secondary: *full line with* +), O_2^+ (primary: *dot-dashed line*, secondary: *dot-dashed line with* ○), O^+ (primary: *dashed line*, secondary: *dashed line with* ∗), H^+ (primary: full line, secondary: full line with △), and N^+ (primary: *dotted line*, secondary: *dotted line with* ×) for a χ angle of 24°. The primary production profiles are in *black*, while the secondary production profiles are in *gray*

8.1.3 Ionospheric Structure

The atmosphere is very active from a chemical point of view. In its lower part, where the neutrals are more dense, the chemical rates can be such that the primary ions are quite instantaneously transformed into other ionized species. At an upper altitude, these rates may be in competition with transport processes. The result is a highly inhomogeneous stratified structure, which was first discovered by radio experiments. They were initially referenced alphabetically, but with time the nomenclature concentrated on the main ionospheric layers, the *D*, *E* and *F* regions, the latter region being further divided into F_1 and F_2 regions, to account for its complex structure. The evolution of the concentration n_s of a species s is given by the continuity equation

$$\frac{\partial n_s}{\partial t} + \nabla \cdot (n_s \boldsymbol{u}_s) = P_s - L_s n_s \,, \tag{8.18}$$

where $\boldsymbol{u}_s$ is the velocity of the species *s*. The chemistry is present on the right hand side through the production term P_s, which corresponds to the sum of the production rates, including the processes mentioned in the previous section and the production resulting from the chemistry, and the loss term $L_s n_s$, which corresponds to the loss rate of the species *s*, exclusively by chemical reactions. Section 8.2 is devoted to ionospheric dynamics and therein we discuss the influence of the transport on the concentration, which is important for the upper ionosphere. In this section, we concentrate on the chemistry, which is the dominant process in the lower ionosphere and thereby we assume a steady state solution characterized by the chemical equilibrium

$$P_s = L_s n_s \,. \tag{8.19}$$

Ion Chemistry

As described in previous section, the ionosphere is primarily composed of positive ions and electrons. The chemistry of these ions is a preponderant process up to about 200 km and it is dominated by three main kinds of reactions:

Charge exchange

$$A^+ + B \rightarrow A + B^+ \,,$$

ion charge exchange

$$A^+ + BC \rightarrow AB^+ + C$$
$$AB^+ + C \rightarrow A + BC^+ \,,$$

dissociative electron recombination

$$AB^+ + e^- \rightarrow A + B \,.$$

A charge exchange reaction mostly occurs between an ion and its neutral parent and thus is a reversible process. However, due to the close value of the ionization potential of neutral atoms O and H, the charge exchange reaction between O^+ and H^+ is an important feature of the upper ionosphere. The electron recombination mechanism concerns the molecular ions and controls the lower ionosphere.

Figure 8.4 presents a synopsis of the chemical scheme in the ionosphere above 90 km. Since not all reactions are of equal importance at the same altitude, the ionsphere is stratified vertically depending on the dominant chemical process. Moreover, since the most important reactions are energy-dependent, the ionospheric structure is very sensitive to the coupling between the ionosphere and the atmosphere or magnetosphere.

Cluster Ions and Negative Ions: The *D* Region

The simple mechanism presented in the previous section applies well to the thermosphere, where the concentrations are low enough to restrain the reactive collisions to binary collisions and where a few species can play a role in the chemistry.

Below 90 km, the atmosphere is dense enough to allow for ternary reactions and the hydrate molecules (OH, H_2O, ...) have concentration levels that make them a significant contributor to the chemistry of this region.

This region is called the *D* region and extends approximately from 60 to 80 km. There are two mechanisms at the origin of the *D* region chemistry.

First, the Lyman α emission line (121.5 nm) is not energetic enough to ionize the main neutral of the thermosphere and thus is not stopped above 90 km and penetrates the region below. The nitric oxide NO is present below 90 km and has an ionization potential of 9.6 eV, which allows the molecule to be ionized by Lyman α. Although NO is a very minor constituent of the atmosphere, the intensity of this line is such that a significant amount of NO^+ ion is created. Besides this, O_2^+ ion is still created at a comparable level.

The molecular oxygen O_2 also has a strong electron affinity, which may result in an electron attachment through the ternary reaction

$$O_2 + e^- + M \rightarrow O_2^- + M \,, \tag{8.20}$$

where M stands for a molecule (N_2 or O_2). This mechanism also applies to atomic oxygen O. These processes lead to the creation of negative ions O_2^- and O^-.

All these ions are embedded in chains of chemical reactions that remove them. The most simple mechanism to remove O_2^- ion is the electron detachment reaction

$$O_2^- + O \rightarrow O_3 + e^- \,. \tag{8.21}$$

Other complex chemical schemes, involving CO_2 and NO, transform these primary negative ions into negative cluster ions such as NO_2^-, NO_3^-, CO_3^- or CO_4^- and hydrated negative ions (see Fig. 3 and Table 2 in Turunen et al., 1996). The positive ions are more involved in chemistry with hydrates (see Fig. 2 and Table 2 in Turunen et al., 1996) and lead to the creation of hydrated protons $H^+(H_2O)_n$.

In the end, the *D* region is a region where positive ions (primary and cluster ions), negative ions (primary and cluster ions) and electrons coexist at concentration levels that ensure the quasi-neutrality. In some conditions, essentially at night, the electrons may disappear and the quasi-neutrality results from the charge balance between positive and negative ions. During the day, the negative ions are all removed by the chemistry and electrons ensure the quasi-neutrality. Simple models, based on four species (electrons, positive ions, negative ions and positive hydrated ions) have been developed using overall reactions, which mask the complexity of the chemical scheme but allow for calculations of the positive and negative charges in the medium (Rodriguez and Inan, 1994).

While the *D* region is principally controlled by sunlight, essentially X-rays and the Lyman α emission line, in the case of solar or geomagnetic perturbations, high energy particles may penetrate deep into the atmosphere (see Fig. 8.2) and contribute significantly to its enhancement. In normal conditions, the concentration in the *D* region remains at low values (below 10^{10} m^{-3}) and the dynamics there is controlled by the neutral atmosphere.

Molecular Ions: The *E* and F_1 Regions

Below 200 km, the ionosphere is dominated by molecular ions O_2^+ and NO^+. The production of O_2^+ results

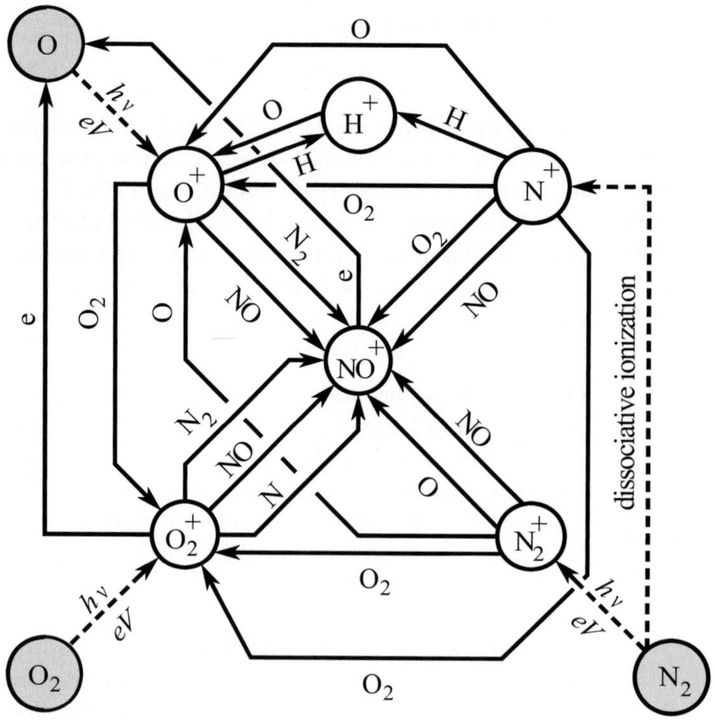

Fig. 8.4. Synopsis of the ionospheric chemical scheme above 90 km. The neutral reactants are shown on the arrows and the initial ionizing processes are labeled $h\nu$ for photoionization and eV for precipitation. The outside neutrals are those involved in the primary ionization processes. NO^+ is in the center because of its importance in the ionospheric chemistry

mainly from the ionization of its neutral parent O_2 and the charge exchange reaction with N_2^+. NO^+ is a by-product of the chemistry and is principally produced by ion charge exchange reaction with N_2^+ and O^+

$$N_2^+ + O \rightarrow NO^+ + N\,, \tag{8.22}$$

$$O^+ + N_2 \rightarrow NO^+ + O\,. \tag{8.23}$$

The reactions involving N_2^+ strongly affect its concentration and although its neutral parent N_2 is the major neutral, N_2^+ is a minor species of the ionosphere. These two reactions are very sensitive to the temperature of the ion (St.-Maurice and Torr, 1978, Sheen and St.-Maurice, 2004) and can be responsible for a strong alteration of the ion composition (see Fig. 8.5).

The production of molecular ions is almost essentially balanced by dissociative electron recombination

$$O_2^+ + e^- \rightarrow O + O \tag{8.24}$$

and

$$NO^+ + e^- \rightarrow N + O\,. \tag{8.25}$$

These reactions have recombination rates that depend on the electron thermal energy (temperature). The rate decreases when temperature increases (Rees, 1989). Consequently, in the case of electron heating (see Sect. 8.2.2) by either plasma instability or strong currents, the electron concentration is likely to increase. Since this concentration is a key parameter in the calculation of the conductivities (see Sect. 8.2.1), such an increase implies a change in the conductivities and thereby in the electrodynamics, which in turn affects the ionosphere–magnetosphere coupling.

However, in a standard situation, the mean recombination rate is almost constant and a simple model can be derived for this region. If we combine O_2^+ and NO^+ in a single molecular ion called M^+, then we can write $n_M = n_e$. Moreover, since all the primary ions resulting from the photoionization are quite immediately eliminated to contribute to the production of the molecular ion M^+, the production rate of the mean ion is equal to the production rate of the electrons P_e. This molecular species is in chemical equilibrium between its production by ionization and its loss by electron recombination, the rate of which is then proportional to n_e^2. Thus, the concentration n_M of this mean ion can easily be derived from the chemical balance

$$P_e = \alpha n_e^2 \tag{8.26}$$

with $\alpha \approx 1.5 \times 10^{-13}\,m^3\,s^{-1}$.

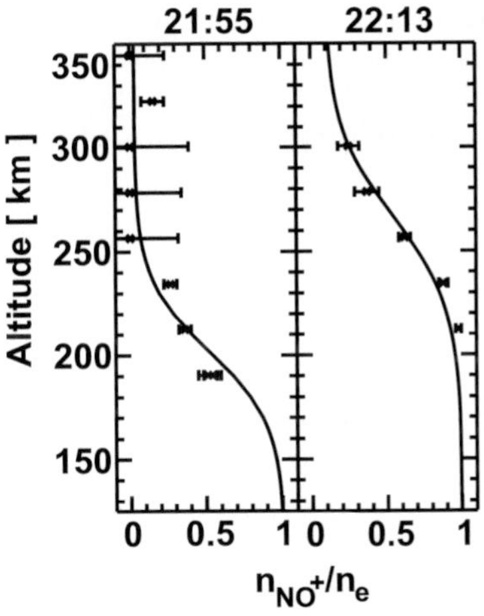

Fig. 8.5. Temporal evolution of the molecular composition ratio (n_{NO^+}/n_e) profiles above EISCAT in the case of a strong convection event (Lathuillère et al., 1997). Crosses are the result of non-Maxwellian analysis and the full lines are the result of a numerical modeling of the event. At 21:55 UT, the soft electron precipitation is such that the transition from molecular to atomic ions is around 200 km. At 22:13 UT, a convection field of 65 mV m^{-1} heats the ions and the transition altitude increases up to 255 km

As a consequence, the electron density has a maximum located at the same altitude as the maximum of the electron production rate. This defines the so-called α–Chapman layer.

Figure 8.3 shows that the electron production rate has two maxima below 200 km. The first maximum occurs around 120 km and corresponds mainly to the production of the O_2^+ ion. This corresponds to a first region called the E region, extending from 90 to approximately 140 km. A second maximum occurs around 150 km and is mainly due to the production of N_2^+ and O^+ ions, which are sources for the NO^+ ion. This corresponds to a second region called the F_1 region, which extends from about 140 km, up to approximately 180–200 km. This region is dominated exclusively by the NO^+ ion. However, the atomic ions start becoming significant species of the ionosphere and become the dominant species above 200 km. Then, the peak around 150 km is not really marked and is masked by the contribution of the atomic species.

Above 180 km, the chemistry is no longer dominated by electron recombination and corresponds to a transition from molecular to atomic ions. We can define the transition altitude between molecular and atomic ions as the altitude where molecular ion concentration is equal to atomic ion concentration. This transition altitude depends on solar illumination: depending on the solar zenith angle, this altitude normally varies from 180 km (low values of χ) to 200 km (high values of χ).

However, the transition altitude is also very sensitive to geomagnetic activity. As a matter of fact, the reaction (8.22) increases with the internal energy, defined as the mean thermal energy in the ion-neutral mass center frame. When ions are strongly heated by friction on neutrals, as is the case for convection (see Sect. 8.2.2), the chemical rate significantly increases and the chemical equilibrium is shifted in favor of NO^+ ions (Sheehan and St.-Maurice, 2004). Figure 8.5 shows an example of such a sharp variation of the transition altitude (Lathuillère et al., 1997). This figure represents the composition ratio (n_{NO^+}/n_e) profiles above EISCAT radar 20 min before a strong convection electric field event (left panel) and just after the event (right panel): the transition altitudes changes from 200 km at 21:55 UT to 255 km at 22:13 UT, when the convection electric field reaches 65 mV m^{-1}. The full line represents the result of a numerical modeling of the event and the crosses are the result of a special analysis of the EISCAT incoherent scatter measurement (see Sect. 8.3.1 and Lathuillère et al., 1997). Such strong variations are typical features of the auroral region and are characteristics of the couplings existing in the magnetospheric system.

Atomic Ions: The F_2 Region and the Upper Ionosphere

Above 180–200 km, O^+ becomes the dominant ion. At these altitudes, the N_2 concentration is so low that the reaction (8.23) does not balance the production of the O^+ ion. The concentration of O^+ thus increases. The characteristic growth time is about 20 min, which means that in absence of other reactions balancing the production, concentration of O^+ would be multiplied by $e^3 \approx 20$ every hour, during the day. With a characteristic time of about 20 min above 300 km, diffusion (see Sect. 8.2.3) is the only process able to balance the ion production. These processes define the F_2 ionospheric layer, which is thus determined by the competition between the production of O^+ and the diffusion of this ion through the neutral atmosphere. This layer has a maximum that is reached between 250 and 400 km. Although

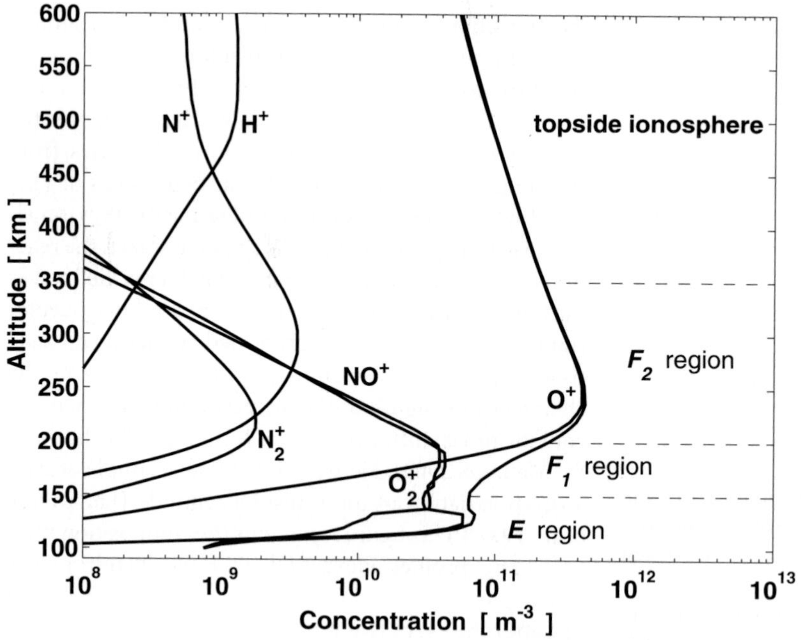

Fig. 8.6. Standard vertical profiles of the main ionospheric ions above 90 km: O_2^+, N_2^+, O^+, H^+ and N^+. The different ionospheric regions discussed in the text are shown

the concentration profile in the F_2 region presents a similar structure that in the E and F_1 regions, it has a different origin since it is not a result of a chemical balance. Above the F_2 peak, the structure is controlled by transport processes discussed in Sect. 8.2.3.

As a result of the balance between the different mechanisms, the ionosphere has a vertical structure that is highly variable, with the heavier ions in the lower part and the lighter ions in the upper part. Figure 8.6 shows the different ionospheric regions and the vertical structure of the ionosphere above 90 km for a typical ionosphere, as discussed in the previous sections.

8.2 Dynamics and Couplings

As mentioned in the Introduction, the characteristic energy of the ionospheric plasma is well below 1 eV. However, the different couplings either with the atmosphere or with the magnetosphere, contribute to energizing this plasma. Depending on the mechanism, the energy gained may result in heating (thermal energy gain) or acceleration (kinetic energy gain). Moreover, the energy gain may vary from a few tenths of eV to a few tens of eV. In the transition region between the ionosphere and the magnetosphere, the processes can then be relayed by other energization mechanisms that are responsible for strong energy gain (up to a few keV). In this section, we discuss the standard processes constituting the engine of the particle transfer between the ionosphere and the magnetosphere.

8.2.1 Electrodynamics

The dynamics of a species s is described by the momentum equation driving the temporal evolution of its velocity $\boldsymbol{u}_s$

$$n_s m_s \frac{\partial \boldsymbol{u}_s}{\partial t} + n_s m_s (\boldsymbol{u}_s \cdot \nabla) \boldsymbol{u}_s + \nabla p_s - n_s [m_s \boldsymbol{g} + e_s (\boldsymbol{E} + \boldsymbol{u}_s \times \boldsymbol{B})] = n_s m_s \sum_t \nu_{st} \left[\boldsymbol{u}_t - \boldsymbol{u}_s + z_{st} \frac{m_s m_t}{k_b (m_s T_t + m_t T_s)} \left(\frac{\boldsymbol{q}_s}{n_s m_s} - \frac{\boldsymbol{q}_t}{n_t m_t} \right) \right] . \quad (8.27)$$

In this equation, $\partial/\partial t$ is the partial derivative with respect to time t (not to be confused with species t), $\boldsymbol{q}_s$ is the heat flow of the species s, and $p_s = n_s k_b T_s$ is its kinetic pressure, $\boldsymbol{g}$ is the acceleration due to gravity, $\boldsymbol{E}$ and $\boldsymbol{B}$ are the electric and magnetic fields, and

e_s is the electric charge of species s. The momentum collision frequency ν_{st} between species s and t verifies $n_s m_s \nu_{st} = n_t m_t \nu_{ts}$. The nature of interaction between the species s and t is accounted for through the geometric factor z_{st} (Schunk, 1977). For a Coulomb interaction, we have $z_{st} = \frac{3}{5}$, while we have $z_{st} = 0$ for Maxwell molecule interaction (collision frequency independent of the energy) used for non-resonant ion-neutral collisions. The heat flow contribution in the momentum equation accounts for thermal diffusion effects. In the case of charged particles, it reduces to the electron heat flow contribution and corresponds to the Seebeck effect. It may become significant above 200 km where electron-neutrals collisions are weak, but for the sake of argument we will neglect it in the rest of this section.

In the plasma, the combined effects of the collisions and the magnetization means that ions and electrons cannot move freely and the overall dynamics is controlled by the electrodynamics through the electric field $\boldsymbol{E}$ and the current $\boldsymbol{J}$. The problem is thus to establish the generalized Ohm's law connecting $\boldsymbol{E}$ and $\boldsymbol{J}$, which is the base of ionospheric electrodynamics and thereby of plasma dynamics. At macroscopic scales, the plasma must ensure the quasi-neutrality condition (for singly charged particles)

$$n_e = \sum_i n_i \,. \tag{8.28}$$

Assuming that ions are positive and single-charged, the current density $\boldsymbol{J}$ is then given by

$$\boldsymbol{J} = e\left(\sum_i n_i \boldsymbol{u}_i - n_e \boldsymbol{u}_e\right) = e\sum_i n_i\left(\boldsymbol{u}_i - \boldsymbol{u}_e\right) . \tag{8.29}$$

In the following, we assume for the sake of simplification that the atmosphere has only one component (the major species) and that the ionospheric plasma is made of a major ion (labeled i) and electrons, with $n_i = n_e$.

The Generalized Ohm's Law

In the lower ionosphere, collisions are strong enough to play a significant role in the momentum equations. Assuming steady state balance and small flows, we can simplify them and write for the ions

$$\begin{aligned} &\nabla p_i - n_i m_i \boldsymbol{g} - n_i e\left(\boldsymbol{E} + \boldsymbol{u}_i \times \boldsymbol{B}\right) \\ &= n_i m_i \nu_{in}\left(\boldsymbol{u}_n - \boldsymbol{u}_i\right) + n_i m_i \nu_{ie}\left(\boldsymbol{u}_e - \boldsymbol{u}_i\right) \end{aligned} \tag{8.30}$$

and for the electrons

$$\begin{aligned} &\nabla p_e + n_e e\left(\boldsymbol{E} + \boldsymbol{u}_e \times \boldsymbol{B}\right) \\ &= n_e m_e \nu_{en}\left(\boldsymbol{u}_n - \boldsymbol{u}_e\right) + n_e m_e \nu_{ei}\left(\boldsymbol{u}_i - \boldsymbol{u}_e\right) . \end{aligned} \tag{8.31}$$

The presence of the neutral wind $\boldsymbol{u}_n$ is meaningful, since the neutral dynamics can significantly affect the electrodynamics, essentially in the equatorial region. We introduce the velocities of the particles $\boldsymbol{u}_i{}' = \boldsymbol{u}_i - \boldsymbol{u}_n$, $\boldsymbol{u}_e{}' = \boldsymbol{u}_e - \boldsymbol{u}_n$, and the electric field $\boldsymbol{E}' = \boldsymbol{E} + \boldsymbol{u}_n \times \boldsymbol{B} = \boldsymbol{E}_\parallel{}' + \boldsymbol{E}_\perp{}'$, in the frame of the neutrals, where $\boldsymbol{E}_\parallel{}'$ and $\boldsymbol{E}_\perp{}'$ are the electric fields parallel and perpendicular to $\boldsymbol{B}$, respectively. We can write

$$\begin{aligned} &m_i\left(\nu_{in} n_i \boldsymbol{u}_i' + \frac{e\boldsymbol{B}}{m_i} \times n_i \boldsymbol{u}_i'\right) \\ &= e n_e \boldsymbol{E}' - \nabla p_i + n_e m_i \boldsymbol{g} - \frac{m_e \nu_{ei}}{e}\boldsymbol{J} , \end{aligned} \tag{8.32}$$

$$\begin{aligned} &m_e\left(\nu_{en} n_e \boldsymbol{u}_e' - \frac{e\boldsymbol{B}}{m_e} \times n_e \boldsymbol{u}_e'\right) \\ &= -\, e n_e \boldsymbol{E}' - \nabla p_e + \frac{m_e \nu_{ei}}{e}\boldsymbol{J} , \end{aligned} \tag{8.33}$$

where we have used $n_i m_i \nu_{ie} = n_e m_e \nu_{ei}$. To derive the relationship between $\boldsymbol{J}$ and $\boldsymbol{E}'$, we introduce the mobility tensors $\overline{\overline{\mu}}_i$ and $\overline{\overline{\mu}}_e$ of the ions and electrons, defined by

$$\phi_s = n_s \boldsymbol{u}_s = \overline{\overline{\mu}}_s \cdot \boldsymbol{F}_s , \tag{8.34}$$

where ϕ_s is the flux of species s and $\boldsymbol{F}_s$ is the sum of the terms acting on species s (i.e. body forces and gradients). We obtain them by inversion of the left hand terms in (8.32) and (8.33). In a coordinate system where $\boldsymbol{E}_\perp'$ is aligned with the first vector and the magnetic field $\boldsymbol{B}$ is aligned with the third vector, we write

$$\overline{\overline{\mu}}_i = \frac{1}{m_i}\begin{pmatrix} \frac{\nu_{in}}{\nu_{in}^2+\Omega_i^2} & \frac{\Omega_i}{\nu_{in}^2+\Omega_i^2} & 0 \\ -\frac{\Omega_i}{\nu_{in}^2+\Omega_i^2} & \frac{\nu_{in}}{\nu_{in}^2+\Omega_i^2} & 0 \\ 0 & 0 & \frac{1}{\nu_{in}} \end{pmatrix} \tag{8.35}$$

and

$$\overline{\overline{\mu}}_e = \frac{1}{m_e}\begin{pmatrix} \frac{\nu_{en}}{\nu_{en}^2+\Omega_e^2} & -\frac{\Omega_e}{\nu_{en}^2+\Omega_e^2} & 0 \\ \frac{\Omega_e}{\nu_{en}^2+\Omega_e^2} & \frac{\nu_{en}}{\nu_{en}^2+\Omega_e^2} & 0 \\ 0 & 0 & \frac{1}{\nu_{en}} \end{pmatrix} , \tag{8.36}$$

where $\Omega_i = eB/m_e$ and $\Omega_e = eB/m_e$ are the gyrofrequencies of the ions and electrons, respectively. B is the coordinate of the magnetic field in the system mentioned above and can be either positive or negative.

By combining (8.30) and (8.31), we can express the current and after lengthy calculations, derive the generalized Ohm's law

$$\boldsymbol{J} = \overline{\overline{\sigma}} \cdot \boldsymbol{E}\,, \tag{8.37}$$

where the conductivity tensor $\overline{\overline{\sigma}}$ is given by

$$\overline{\overline{\sigma}} = \sigma \overline{\overline{I}} - \sigma \left[\overline{\overline{I}} + m_e \nu_{ei} \left(\overline{\overline{\mu}}_i + \overline{\overline{\mu}}_e\right)\right]^{-1} \tag{8.38}$$

with $\overline{\overline{I}}$ being the identity dyadic tensor. In this equation, $\sigma = \frac{e^2 n_e}{m_e \nu_{ei}}$ is the fully ionized plasma conductivity.

Since ν_{ei} is proportional to n_i ($\approx n_e$), σ is independent of the plasma concentration. This characterizes the fact that the plasma is a neutral medium at large scales.

In the coordinate system introduced above, this tensor can be written as

$$\overline{\overline{\sigma}} = \begin{pmatrix} \sigma_P & \sigma_H & 0 \\ -\sigma_H & \sigma_P & 0 \\ 0 & 0 & \sigma_\parallel \end{pmatrix}, \tag{8.39}$$

where

$$\begin{cases} \sigma_P = \left[1 - \dfrac{1}{m_e \nu_{ei}} \dfrac{\alpha}{\alpha^2 + \beta^2}\right] \sigma\,, \\[2ex] \sigma_H = \dfrac{1}{m_e \nu_{ei}} \dfrac{\beta}{\alpha^2 + \beta^2} \sigma\,, \\[2ex] \sigma_\parallel = \left[\dfrac{\dfrac{1}{m_e \nu_{en}} + \dfrac{1}{m_i \nu_{in}}}{\dfrac{1}{m_e \nu_{ei}} + \dfrac{1}{m_e \nu_{en}} + \dfrac{1}{m_i \nu_{in}}}\right] \sigma \end{cases} \tag{8.40}$$

and

$$\begin{cases} \alpha = \dfrac{1}{m_e \nu_{ei}} + \dfrac{1}{m_e} \dfrac{\nu_{en}}{\nu_{en}^2 + \Omega_e^2} + \dfrac{1}{m_i} \dfrac{\nu_{in}}{\nu_{in}^2 + \Omega_i^2}\,, \\[2ex] \beta = \dfrac{1}{m_e} \dfrac{\Omega_e}{\nu_{en}^2 + \Omega_e^2} - \dfrac{1}{m_i} \dfrac{\Omega_i}{\nu_{in}^2 + \Omega_i^2}\,. \end{cases} \tag{8.41}$$

σ_P is called the Pedersen conductivity and is the plasma conductivity in the direction of $\boldsymbol{E}'_\perp$, the electric field perpendicular to the magnetic field. σ_H is called the Hall conductivity and is the conductivity of the plasma in the direction perpendicular to both the electric and magnetic fields. $\sigma_\parallel$ is called the parallel conductivity and is the conductivity of the plasma along the magnetic field line; it is independent of $\boldsymbol{B}$.

These expressions can easily be extended to a multi-component ionosphere by replacing m_i in the equation by $\frac{n_i}{n_e} m_i$ and summing over the ions. In such an extension, ν_{ei} should be replaced by $\nu_e^i = \sum_i \nu_{ei}$, the total electron–ion collision frequency.

Figure 8.7 shows typical vertical profiles of the three conductivities, compared to the vertical profile of σ, which varies with altitude because of its dependence on T_e. The first remark is that $\sigma_\parallel$ is always higher that σ_H and σ_P (about two orders of magnitude higher around 120–150 km) and thus the field-aligned direction does not significantly contribute to the electrodynamics above 250 km: the high conductivity means that the field line does not experience any significant potential drop between its lower boundary (E region) and its upper boundary (in the magnetosphere). Below 150 km, the field-aligned direction may contribute to the overall electrodynamics, especially in the vicinity of auroral arcs where a highly variable ionospheric structure is observed.

The expressions of the conductivities mask the fact that they all depend on the electron concentration n_e, except σ, and consequently they are close to zero in regions where the concentration is low (in the lower ionosphere). Three regions emerge from the equations, which are related to the control exerted by the magnetic field on the conductivities and thereby on the ionospheric electrodynamics:

- Below 80 km: $\nu_{in} \gg \Omega_i$ and $\nu_{en} \gg \Omega_e$.

The collisions with neutrals dominate and the particles do not feel the magnetic field; the ionosphere can be considered as unmagnetized. The conductivities are very low (because of the low value of n_e), σ_H is negligible and σ_P is very close to $\sigma_\parallel$. The conductivity tensor reduces to a scalar.

- Between 90 and 120 km: $\nu_{in} \gg \Omega_i$ and $\nu_{en} \ll \Omega_e$.

The electrons feel the magnetic field. They are trapped in the magnetic field lines and they cannot move freely across them. The ions are still dominated by collisions

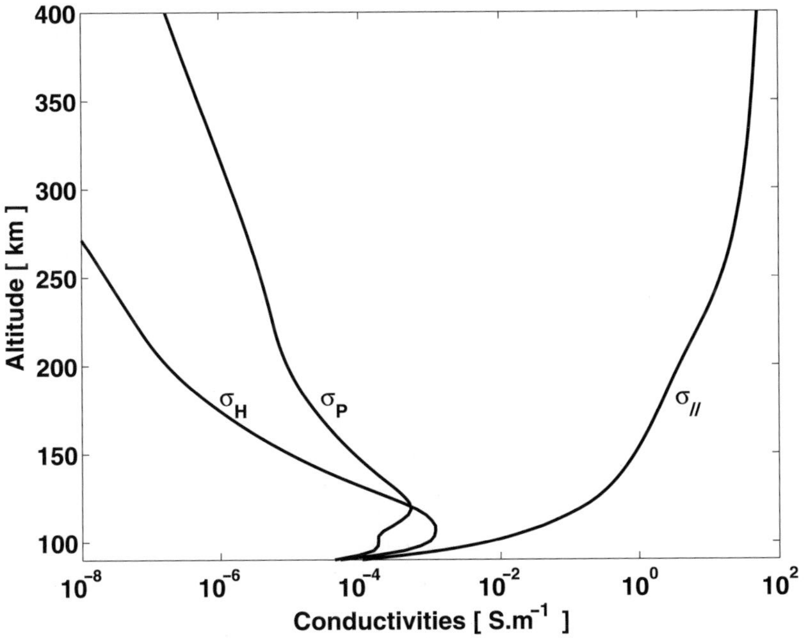

Fig. 8.7. Typical vertical profiles of the Pedersen (σ_P), Hall (σ_H) and parallel ($\sigma_\parallel$) conductivities, compared to the fully ionized plasma conductivity σ. When $\sigma_\parallel$ tends to σ, both σ_P and σ_H tend to zero and the conductivity tensor $\overline{\overline{\sigma}}$ reduces to the only parallel component

with neutrals. This difference between ions and electrons is such that currents perpendicular to $\boldsymbol{B}$ are most likely to develop in this region. The perpendicular conductivities reach a maximum around 110 km, where the perpendicular current is maximal. This is the region of the closure of the ionospheric currents.

- Above 150 km: $\nu_{\text{in}} \ll \Omega_{\text{i}}$ and $\nu_{\text{en}} \ll \Omega_{\text{e}}$.

All the charged species feel the magnetic field and are trapped in the magnetic field lines. They cannot move freely in any direction perpendicular to the magnetic field. Perpendicular conductivities become negligible while parallel conductivity increases and reaches a constant value above 500 km, where the collisions with neutrals no longer influence the plasma dynamics.

The electrons contribute to the Hall current (associated with the Hall conductivity), but not to the Pedersen current (associated with the Pedersen conductivity), which is dominated by the ions.

Polarization Electric Field

Above the electrodynamics region, perpendicular conductivities become weak. Then, no perpendicular current exists and due to the conservation of the current ($\nabla \cdot \boldsymbol{J} = 0$), the field aligned current amplitude varies similarly to the magnetic field strength. At higher altitudes, the collisions become weak and the projection of (8.33) along the magnetic field line can be written as

$$eE_\parallel = -\frac{1}{n_\text{e}} \nabla_\parallel p_\text{e} \tag{8.42}$$

by dropping all unimportant terms. The right hand term is the electrostatic contribution to the electric field, which arises from the charge separation in the plasma; the electron pressure gradient transports the electrons away from the ions and an electric field develops from the constrained charge separation in order to maintain the electrons close to the ions, so that the quasi-neutrality remains valid at macroscopic scales. Assuming that the magnetic field is vertical and the electron temperature is constant, this equation reduces to

$$eE = -\frac{k_\text{b} T_\text{e}}{n_\text{e}} \frac{\mathrm{d}n_\text{e}}{\mathrm{d}z} \,. \tag{8.43}$$

Equation (8.43) expresses the fact that the ionospheric electrons are in thermodynamical equilibrium in the potential field ϕ, from which the polarization electric field derives. Their distribution is given by the Boltzmann factor $e^{-\frac{e\phi}{k_b T_e}}$, similarly to what happens for the neutral atmosphere distribution in the gravity field.

8.2.2 Energetics

The energetics of species s is driven by the two coupled transport equations involving the pressure p_s and the heat flow $\boldsymbol{q}_s$ (see Schunk, 1977; Schunk and Nagy, 2000 for further details) as follows:

– Energy

$$\frac{\partial}{\partial t}\left(\frac{3}{2}p_s\right) + \boldsymbol{u}_s \cdot \nabla\left(\frac{3}{2}p_s\right) + \frac{5}{2}p_s\left(\nabla \cdot \boldsymbol{u}_s\right) + \nabla \cdot \boldsymbol{q}_s = \frac{\delta E_s}{\delta t} . \quad (8.44)$$

– Heat flow

$$\frac{\partial \boldsymbol{q}_s}{\partial t} + \left(\boldsymbol{u}_s \cdot \nabla\right)\boldsymbol{q}_s + \frac{7}{5}\left(\boldsymbol{q}_s \cdot \nabla\right)\boldsymbol{u}_s + \frac{7}{5}\boldsymbol{q}_s\left(\nabla \cdot \boldsymbol{u}_s\right) + \frac{2}{5}\left(\nabla \boldsymbol{u}_s\right) \cdot \boldsymbol{q}_s + \frac{5}{2}\frac{k_b p_s}{m_s}\nabla T_s - \frac{e_s}{m_s}\boldsymbol{q}_s \times \boldsymbol{B} = \frac{\delta \boldsymbol{q}_s}{\delta t} . \quad (8.45)$$

In these two equations, the transport terms are on the left hand side, while the local energy exchange terms driven by the collisions or other local processes are on the right hand side, gathered in the generic terms $\delta E_s/\delta t$ and $\delta \boldsymbol{q}_s/\delta t$. To a first approximation, these terms are expressed as

$$\frac{\delta E_s}{\delta t} = \sum_t n_s \frac{m_s}{m_s + m_t}\nu_{st}\left[3k_b\left(T_t - T_s\right) + m_t\left(\boldsymbol{u}_s - \boldsymbol{u}_t\right)^2\right] + \Theta_s - \Lambda_s , \quad (8.46)$$

where Θ_s and Λ_s stand for heating and cooling processes, respectively, which are not related to energy transfer by elastic processes. These terms are essentially meant for the electrons and correspond to heating by the suprathermals (see Sect. 8.1.2 or cooling by inelastic collisions with neutrals (excitation of internal modes of the neutrals). These terms may also correspond to plasma instabilities, which result in heating of the different species.

$$\frac{\delta \boldsymbol{q}_s}{\delta t} = \sum_t \nu_{st}\left\{\frac{n_s m_s}{n_t m_t}\left(D_{st}^{(4)} + \frac{5}{2}z_{st}\frac{m_t T_s}{m_s T_t + m_t T_s}\right)\boldsymbol{q}_t - \left(D_{st}^{(1)} + \frac{5}{2}z_{st}\frac{m_t T_s}{m_s T_t + m_t T_s}\right)\boldsymbol{q}_s - \frac{5}{2}\frac{m_t z_{st}}{m_s + m_t}p_s\left(\boldsymbol{u}_t - \boldsymbol{u}_s\right)\right\} - \frac{2}{5}z''_{ss}\nu_{ss}\boldsymbol{q}_s , \quad (8.47)$$

where $D_{st}^{(1)}$, $D_{st}^{(4)}$ and z''_{st} are geometric factors depending on the type of interaction (Schunk, 1977).

Due to their mass, electrons and ions behave differently and the dominant terms in these equations differ. Moreover, depending on the region in the ionosphere, the dominant terms may change. Schunk and Nagy (2000) give a complete description of these equations for the ionospheric species. In this chapter, we just provide the general trends.

The ionosphere is an open, multi-species medium where exchanges are not always sufficient to ensure the homogeneity of the temperatures. The situation is such that the energy balance in the ionosphere results in different temperatures for the different species. Schematically, the temperature of the different species in the upper atmosphere are ordered as follows:

$$T_n \leq T_i \leq T_e . \quad (8.48)$$

This hierarchy is the general trend and may be altered under special conditions, such as strong convection events. In the following, we describe the energy balance for the electrons and the ions and more precisely the competition between local collisional processes and transport mechanisms, which leads to the hierarchy given by (8.48). To facilitate the discussion, in the following we consider a steady state balance.

Electron Energy Balance

General Case

Basically, heat transfer from electrons to neutrals and ions is slow because of the small electron mass. This has consequences for the electron energetics. First, concerning the energy balance, the transfer of energy is ensured by inelastic collisions with neutrals because the classical elastic collisions are not strong enough.

To leading order, the electron energy equations can be written as follows:

- Energy

$$\frac{2}{3k_b n_e}\nabla\cdot\boldsymbol{q}_e+\sum_t \frac{m_e}{m_e+m_t}\nu_{et}T_e$$
$$=\frac{2}{3k_b n_e}\Theta_e+\sum_i \frac{m_e}{m_e+m_i}\nu_{ei}T_i$$
$$-\frac{2}{3k_b n_e}\Lambda_e+\sum_n \frac{m_e}{m_e+m_n}\nu_{en}T_n\,. \quad (8.49)$$

- Heat flow

$$\frac{5k_b p_e}{2m_e}\nabla T_e=-\left(\nu_e^n+\frac{13}{10}\nu_e^i+\frac{4}{5}\nu_{ee}\right)q_e-\frac{3k_b T_e}{2e}\nu_e^i\boldsymbol{J}\,. \quad (8.50)$$

In (8.49), Θ_e is the heating source of the thermal electrons coming from the suprathermal electrons (Schunk and Nagy, 1978) (see Sect. 8.1.2). It is derived from the computation of the cooling down of the suprathermal electrons and depends on the energy distribution of this population; it is related to the drag force of the suprathermal (8.16) and depends on $L(E)$ (8.17). Schunk and Nagy (1978) and Schunk and Nagy (2000) give a complete description of this term. Λ_e is the sum of all the inelastic collisions of the thermal electrons with the neutrals that lead to the cooling of the thermal electrons. This concerns the vibrational and rotational excitation of the two homonuclear molecules N_2 and O_2 and the excitation of the fine structure of atomic oxygen O. We do not give the expression for these energy transfers, since the complete calculations are rather fastidious and are developed by Schunk and Nagy (2000) and by Pavlov (1998a, 1998b, 1999).

These terms dominate the energy equation below 200 km, where the neutral atmosphere is dense enough. Following the mass stratification, the contribution of the neutral molecules is preponderant below 150–180 km, while the contribution of atomic oxygen is important around 200 km. Above this, in the F_2 region and the upper ionosphere, the heat transfer due to elastic collisions of the electrons on the ions is the most important contribution to the collisional heat transfer. The departure of the electron temperature from the neutral ones results from two contributions. The first one is a local energy deposition Θ_e that comes mainly from the thermalization of the suprathermal electrons (see Sect. 8.1.2). It maximizes around 150 km and decreases more slowly than the cooling terms. The second contribution is a transport process and starts to become significant above about 300 km; this is the heat transfer due to the electron heat flow. Above 400 km, the evolution of the electron temperature is driven by the heat flow and the energy loss of the electrons on the heavy species (ions and neutrals) becomes negligible.

In (8.50), ν_e^i is the total momentum collision frequency of the electrons with the ions (see Sect. 8.2.1) and $\nu_e^n=\sum_n \nu_{en}$ is the total momentum collision frequency of the electrons with neutrals. The contribution of the current to the heat transport corresponds to the thermoelectric effect; it can be outlined by rewriting (8.50) to express the electron heat flow

$$\boldsymbol{q}_e=-\underbrace{\frac{5}{4}\frac{5k_b p_e}{m_e\nu_{ee}\left(2+\frac{13}{4}\sqrt{2}+\frac{5}{2}\frac{\nu_e^n}{\nu_{ee}}\right)}}_{\kappa_e}\nabla T_e$$
$$-\frac{5}{4}\frac{3\sqrt{2}}{\left(2+\frac{13}{4}\sqrt{2}+\frac{5}{2}\frac{\nu_e^n}{\nu_{ee}}\right)}\frac{k_b T_e}{e}\boldsymbol{J}\,. \quad (8.51)$$

Equation (8.51) shows that there are two contributions to the electron heat flow. The first one corresponds to the classical Fourier's law and is related to the thermal conduction of the electron gas which allows energy transport when temperature gradient appears. The expression for the thermal conductivity κ_e is given by (8.51), and since both p_e and ν_{ee} are proportional to n_e, the thermal conductivity does not depend on the electron concentration at high altitudes, where the collisions with neutrals are negligible compared to the collisions with ions. The plasma can then be considered as fully ionized. In this limit, we have $\kappa_e=\alpha T_e^{5/2}=5.14\times10^{-12}\,T_e^{5/2}$, with T_e in K. The second contribution is the thermoelectric effect; it is proportional to the current density and to the electron temperature.

In the Earth ionosphere, field aligned transport is much more efficient to ensure the energy transfer than transport in any other direction. Then the equations above can be reduced to the 1D projection along the field aligned direction: at mid and high latitudes, this corresponds almost to the vertical direction, while in the equatorial region, it corresponds to the south-north horizontal direction. The field aligned current is thus responsible for the thermoelectric effect. It is almost constant with altitude above the F_2 region, and thus in the case of an upward current, this effect leads to a positive feedback: any increase of the current implies an enhancement of the downward heat flow, which in turn will imply an enhancement of the temperature.

Above the F_2 region and in the case of no parallel current, because the electron has a weak mass, we can approximate the electron energy (8.44) by

$$\nabla \cdot \boldsymbol{q}_{\mathrm{e}} = 0 \,, \tag{8.52}$$

which expresses the fact that the heat transport is conservative and thus that the electrons do not exchange heat with the other populations.

We can project the equation along the magnetic field line, using a dipolar coordinate system. This coordinate system is based on the dipolar magnetic field configuration and is characterized by s the curvilinear abscissa along the magnetic field line, and A the cross section of a magnetic flux tube, which verifies $\mathrm{d}(A.B)/\mathrm{d}s = 0$ (B is the amplitude of the magnetic field). The projection of (8.52) along the magnetic field line is straightforward because the divergence operator $\nabla\cdot$ reduces to $1/A\partial(A.)/\partial s$. Finally, we can integrate (8.52), with the expression of $\boldsymbol{q}_{\mathrm{e}}$ given by (8.51) and we come up with the following expression of the electron temperature profile along the magnetic field line between the lower abscissa s_{o} and the upper abscissa s_∞:

$$T_{\mathrm{e}}(s) = \left(\frac{7}{4\alpha} q_{\mathrm{e}}^{\infty} s_\infty^3 \left(\frac{1}{s^2} - \frac{1}{s_{\mathrm{o}}^2} \right) + T_{\mathrm{e}}^{o\ 7/2} \right)^{2/7} , \tag{8.53}$$

where α comes from $\kappa_{\mathrm{e}} = \alpha T_{\mathrm{e}}^{5/2}$ (see 8.51 and hereafter), $T_{\mathrm{e}}^{\mathrm{o}}$ is the temperature at s_{o} and q_{e}^{∞} is the heat flow at s_∞. This expression can been used to infer the energy input at the top of the field line defined by q_{e}^{∞}, which thereby is an important parameter to characterize the couplings between the ionosphere and the magnetosphere (Blelly and Alcaydé, 1994).

Instabilities

The heating by the suprathermal electrons is not the only source of energy for the thermal electrons. As a matter of fact, the electrodynamics controls the motion of the electrons above 90 km (see the previous section), and up to about 130 km electrons and ions can move in different directions. This resulting differential drift may be strong enough to trigger a two-stream plasma instability below 120 km, known as the Farley–Buneman instability (Farley, 1963; Buneman, 1963). This instability is likely to grow in the E region around 110 km. Theoretical work has been done to understand the way this instability develops in the plasma and to evaluate the amount of energy it brings to the thermal electrons (St.-Maurice and Laher, 1985). The heating is not negligible for the plasma, since it may lead to an enhancement of the electron temperatures reaching 1000 K with an ambient convection electric field of 60 mV m^{-1} in a region where the electrons are normally thermalized. This may have some consequences on the electrodynamics, because as the electron recombination decreases when the electron temperature increases, the high temperatures enhance the electron concentration. Therefore, this alters the conductivities around 110 km in the region where electrodynamics is important.

Ion Heating and Cooling

As is the case for diffusion, the behavior of major ions and minor ions may differ, essentially because in the low regions of the ionosphere the minor ion species are the lighter species, like H^+. However, the energy balance is much easier to establish for the ions than for the electrons. As a matter of fact, due to their mass, transport processes are negligible compared to local collisional processes in the ionosphere (except for H^+). Since only elastic collisions are to be considered, energy equation (8.44) can be written in the very simple form

$$\frac{\delta E_{\mathrm{i}}}{\delta t} = 0 \,. \tag{8.54}$$

We are discussing the case of the dominant ion below 600 km and thus we exclude H^+. In this equation, two sources of heating are clearly identified for the ions. The most important source is the Coulomb collisions with the electrons. The other source is the frictional heating of the ions against the dominant species. The main source of cooling is the collisions with neutrals. Thus, we can write the balance equation in the following form:

$$\begin{aligned} &\nu_{\mathrm{in}} \frac{m_{\mathrm{i}}}{m_{\mathrm{i}} + m_{\mathrm{n}}} \left(T_{\mathrm{i}} - T_{\mathrm{n}} \right) \\ &= \nu_{\mathrm{ie}} \left(T_{\mathrm{e}} - T_{\mathrm{i}} \right) + \frac{\nu_{\mathrm{in}}}{3k_{\mathrm{b}}} \frac{m_{\mathrm{i}} m_{\mathrm{n}}}{m_{\mathrm{i}} + m_{\mathrm{n}}} \left(\boldsymbol{u}_{\mathrm{i}} - \boldsymbol{u}_{n} \right)^2 , \end{aligned} \tag{8.55}$$

where the cooling term is on the left hand side and the heating term is on the right hand side. We can then determine the ion temperature, which writes as

$$T_i = \frac{m_i \nu_{in}}{(m_i + m_n)\nu_{ie} + m_i \nu_{in}} [T_n + \underbrace{\frac{m_n}{3k_b} (\boldsymbol{u}_i - \boldsymbol{u}_n)^2}_{\delta T_n}] + \frac{(m_i + m_n)\nu_{ie}}{(m_i + m_n)\nu_{ie} + m_i \nu_{in}} T_e \tag{8.56}$$

This expression is valid in the collision dominated regime below about 500 km. The frictional heating can be seen as a departure δT_n from the neutral temperature. If this heating is too important, which is the case for a strong convection electric field, the ion distribution function is no longer isotropic (St.-Maurice and Schunk, 1976). In such a case, the former equation is still valid for the mean temperature, but we need to account for two other temperatures: that parallel to the magnetic field and that perpendicular to the magnetic field. We do not discuss this point here and we refer the interested reader to Blelly and Schunk (1993), Schunk and Nagy (2000) for a complete description of the equations in such a case.

At low altitudes, ν_{in}/ν_{ie} is high and T_i is close to T_n. As the altitude increases, this ratio decreases and the ion temperature comes close to T_e. Such a property can be used to infer some neutral atmosphere parameters such as the atomic oxygen concentration or the neutral temperature (Bauer et al., 1970).

Figure 8.8 illustrates the different features of the energetics in the ionosphere discussed above. This figure shows the field-aligned profiles of ion (dot-dashed lines) and electron (full lines) temperatures at 10:20 TU (circles) and 10:25 TU (crosses) on 29 October 2003, as measured by EISCAT UHF radar (see Sect. 8.3.1). This period is taken within the so-called "2003 Halloween event", a major solar event which strongly disturbed the magnetospheric system, particularly the ionosphere. Within the five minute period separating the two measurements, a strong and short-time ionosphere-magnetosphere coupling event starts: the first measurement is made before the event, while the second measurement is made during the event. Before the event, the electron temperature is relatively large compared to standard situation and reaches about 6000 K at 500 km, while the ion temperature reaches 2000 K. No strong convection electric field is observed and the temperature below 300 km indicates that the neutral temperature is around 1000 K, both ion and electron temperatures being equal below 150 km. Five minutes later, the situation is completely different. The ion temperature around 150 km becomes higher than the electron temperature. This indicates the presence of a convection electric field heating the ions. The apparent decrease of the temperature above 150 km is due to a wrong molecular ion composition ratio used in the analysis of the incoherent spectra (see Sect. 8.3.1). As a matter of fact, due to the presence of the electric field, the molecular ions are more important than in a normal situation (see Fig. 8.5) and thus the ion temperature is underestimated. Around 110 km, there is a strong local enhancement of the electron temperature, which is a characteristic of the Farley–Buneman instability induced by the convection electric field (see Sect. 8.2.2). At a higher altitude, the electron temperature reaches about 10,000 K, because of a strong field-aligned current associated to particle precipitation occurring at that time. The ions are then heated by the electrons and the ion temperature reaches 3000 K at 500 km.

8.2.3 Field Aligned Transport

Diffusion

In a collisional medium, the diffusive process is a major source of particle transport. It is initiated by the presence of inhomogeneities that create concentration gradients in the medium. In the case of the ionosphere, such a mechanism exists, but it is altered by the fact that the particles are charged. In the atmosphere, diffusion is likely to develop preferentially along the vertical direction. However, to account for the control by the magnetic field, we discuss diffusion along the magnetic field line and thereby we focus on the mid-latitude or high-latitude ionosphere so that magnetic field lines do not deviate too much from the vertical. Therefore, we will assume that the magnetic field line is vertical. Since in the rest of this section we are dealing with major and minor ions, the ion 1 will refer to the major ion and the ion 2 will refer to the minor species. Furthermore, for simplification, we assume that the neutral atmosphere is composed of only one component labeled n, which is the neutral parent of the major ion (i.e. same mass).

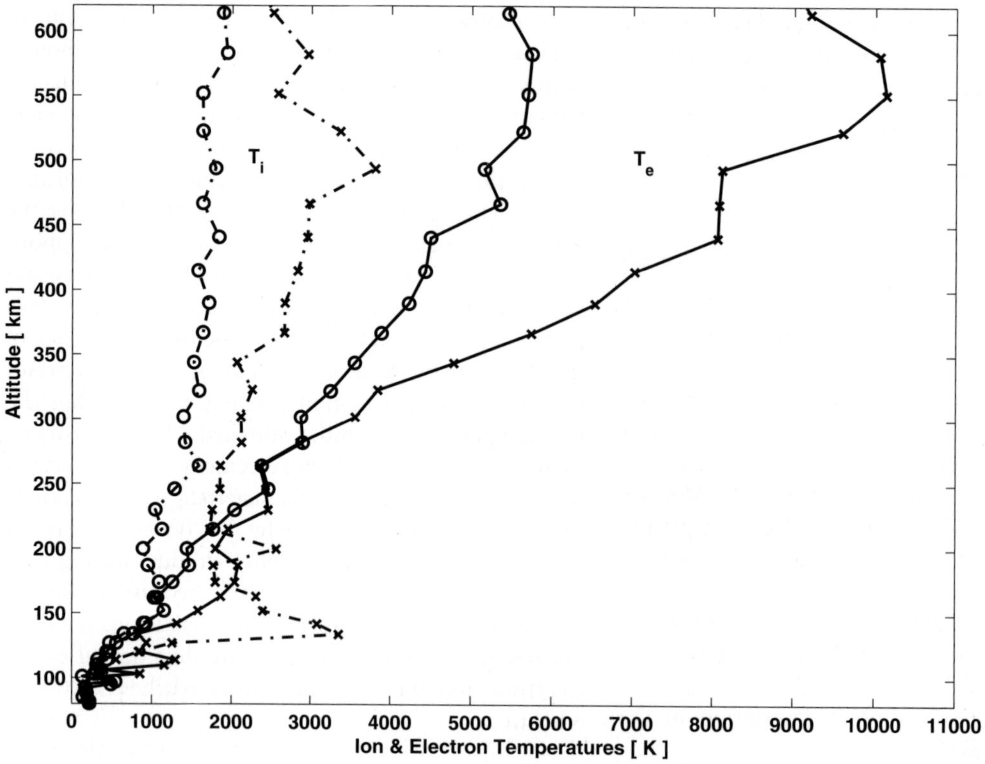

Fig. 8.8. Field aligned profiles of ion (*dot-dashed lines*) and electron (*full lines*) temperatures at 10:20 TU (*circles*) and 10:25 TU on 29 October 2003, as measured by EISCAT UHF radar. The effects of the convection and field-aligned currents discussed in this section are quite visible around 110 km (electron Farley–Buneman instability), around 150-200 km (ion Joule heating) and in the upper ionosphere (electron temperature enhancement due to the thermoelectric effect)

Ambipolar Diffusion

Ambipolar diffusion develops in a region where collisions with neutral species are frequent. It concerns the dominant ion, essentially O^+ in the F_2 region and above. Keeping the important terms in the momentum equation (8.32) and using the expression for the polarization electric field given in (8.43), the projection along the vertical axis gives

$$n_1(u_1 - u_n) = -\frac{k_b T_1}{m_1 \nu_{1n}} \frac{dn_1}{dz} - \frac{k_b T_e}{m_1 \nu_{1n}} \frac{n_1}{n_e} \frac{dn_e}{dz} - \frac{n_1}{\nu_{1n}} g . \quad (8.57)$$

Using the approximation $n_1 \approx n_e$ and assuming that the atmosphere is at rest, the ion flux can be expressed as

$$n_1 u_1 = -D_a \frac{dn_1}{dz} - \frac{n_1}{\nu_{1n}} g . \quad (8.58)$$

In (8.58), we have introduced the ambipolar diffusion coefficient $D_a = \frac{k_b T_p}{m_1 \nu_{1n}}$, where $T_p = T_1 + T_e$ is the plasma temperature. This coefficient characterizes the simultaneous diffusion of the ions and the electrons within the atmosphere and thereby is characteristic of the plasma. The polarization electric field maintains the quasi-neutrality at a macroscopic level and forces the electrons to move with the ions; hence this diffusion of the plasma does not induce any current. This coefficient differs from the molecular diffusion coefficient $D_1 = \frac{k_b T_1}{m_1 \nu_{1n}}$, which characterizes the motion of the ions without the polarization field. Introducing the plasma scale height $H_p = k_b(T_1 + T_e)/m_1 g$, then we can write (8.58) as

$$n_1 u_1 = -D_a \left(\frac{dn_1}{dz} + \frac{n_1}{H_p} \right) . \quad (8.59)$$

Since ν_{in} is proportional to the concentration of the neutral species n, D_a has the same scale height H_n as the neutral. The expression of the flux can then be introduced in the continuity equation (8.18), projected along the vertical axis. We obtain the following differential equation:

$$\frac{d^2 n_1}{dz^2} + \left(\frac{1}{H_n} + \frac{1}{H_p}\right)\frac{dn_1}{dz} + \left(\frac{1}{H_n H_p} - \frac{L_1}{D_a}\right) n_1 = -\frac{P_1}{D_a} . \tag{8.60}$$

We are looking for a solution in a region where chemistry is negligible (P_1 and L_1 are set to 0). If n_1^o, ϕ_1^o and D_a^0 are the concentration, the flux and ambipolar diffusion coefficient of species 1 at a reference altitude z_o, respectively, the concentration $n_1(z)$ at altitude z is given by

$$n_1(z) = \underbrace{\left(n_1^o - \frac{\phi_1^o}{D_a^o}\frac{H_n H_p}{H_p - H_n}\right) e^{-\frac{z-z_o}{H_p}}}_{N_1} + \underbrace{\frac{\phi_1^o}{D_a^o}\frac{H_n H_p}{H_p - H_n} e^{-\frac{z-z_o}{H_n}}}_{N_2} . \tag{8.61}$$

The concentration decreases with altitude. Since the concentration increases in the chemistry region, there is an altitude for which the concentration reaches a maximum, which corresponds to the F_2 peak. This peak is typically located around 250 km and we can set z_o just above this altitude.

The solution is the sum of two terms: N_1 decreasing with the plasma scale height and N_2 decreasing with the scale height of the neutral atmosphere. N_1 corresponds to the hydrostatic solution for the plasma if ϕ_1^o is set to 0. The presence of the electric field affects the ion scale height which would be the one of the hydrostatic equilibrium if the ions were alone in the medium. This change in the scale height makes it possible to maintain the neutrality of plasma at any altitude. N_2 characterizes the diffusion of the ion in the atmosphere. In the case of a positive flux ϕ_1^o, there is an upper limit ϕ_1^{lim}, which guarantees the positivity of both solutions N_1 and N_2

$$\phi_1^o \le \phi_1^{\mathrm{lim}} = D_a^o n_1^o \left(\frac{1}{H_n} - \frac{1}{H_p}\right) . \tag{8.62}$$

The flux ϕ_1^{lim} corresponds to the maximum possible flux for the ions that is compatible with the diffusion mechanism; the ion is then in limiting flux with the neutral scale height. N_1 is then set to 0.

Since the plasma temperature is higher than the neutral temperature, the plasma scale height is larger than the neutral scale height and the solution N_1 prevails at high altitude, as long as the ion remains the major species. In this limit ($n_1 \simeq N_1$), we can give a new expression for the polarization electric field (8.43)

$$eE = m_1 \frac{T_e}{T_1 + T_e} g . \tag{8.63}$$

Replacing the ion temperature T_1 by the plasma temperature T_p, we have included the effect of the polarization force in the ion pressure gradient. Similarly, we could integrate this term in the gravity force by keeping the ion temperature and replacing the mass m_1 by the effective mass $m_1' = m_1 \frac{T_1}{T_1+T_e}$. Thus, the effect of the polarization field can be interpreted as a reduction of the effective mass of the ion; since $T_e \ge T_1$, the effective ion mass is less than half the mass. This notion of effective mass can be applied to the other species, in particular to the light species.

Light Species Diffusion

The treatment of the diffusion of a minor species in the ionosphere is similar to what is presented above. However, in case of a minor light species like H^+ or He^+, Coulomb collisions with the major ion, namely O^+, are much more important than collisions with neutrals above 250 km. Thus, only the major ion is retained in the transport equation. In such conditions, the minor ion flux can be written as

$$n_2(u_2 - u_1) = -D_2 \frac{dn_2}{dz} - \frac{m_2 - \frac{T_e}{T_1+T_e} m_1}{m_2} \frac{n_2}{\nu_{21}} g , \tag{8.64}$$

where $D_2 = \frac{k_b T_1}{m_2 \nu_{21}}$ is the molecular diffusion coefficient of ion 2 through ion 1. The polarization field acts to reduce the effective mass, which becomes $m_2' = m_2 - \frac{T_e}{T_1+T_e} m_1$. Depending on the mass ratio between the two ions, m_2' can be either positive or negative. In the case of O^+ and H^+, the effective mass of the protons is less than −7 amu, which means that H^+ has a negative effective mass. The consequence is that the ion is accelerated upwards, with an acceleration larger than 7 g. The continuity equation applied to the minor species gives the following differential equation

$$\frac{d^2 n_2}{dz^2} + \left(\frac{1}{H_p} + \frac{1}{H_2'}\right)\frac{dn_2}{dz} + \frac{1}{H_p H_2'} n_2 = 0 , \tag{8.65}$$

where we have introduced the effective ion scale height H_2'

$$\frac{1}{H_2'} = \frac{\left(m_2 - m_1 \frac{T_e}{T_i + T_e}\right) g}{k_b (T_i + T_e)} . \tag{8.66}$$

The solution is then

$$n_2(z) = \underbrace{\left(n_2^o - \frac{\phi_2^o}{D_2^o} \frac{H_n H_2'}{H_2' - H_n}\right) e^{-\frac{z-z_0}{H_2'}}}_{N_3} + \underbrace{\frac{\phi_2^o}{D_2^o} \frac{H_2' H_p}{H_2' - H_n} e^{-\frac{z-z_0}{H_p}}}_{N_4} . \tag{8.67}$$

The first solution N_3 corresponds to the equilibrium with the effective scale height H_2' and is obtained in the absence of flux ($\phi_2^o = 0$). In the case of H^+, since H_2' is negative, the concentration thus increases with altitude; it is the hydrostatic equilibrium for the ion H^+ with the effective negative mass m_2'.

The second solution N_4 corresponds to the diffusive equilibrium with the scale height of the plasma. Assuming that the only source of the H^+ ion is the region of production by chemistry, which is located at a lower altitude, we must impose an upward H^+ ion flux ($\phi_2^o > 0$) and thus the ion escapes from the ionosphere. This flux has an upper limiting value $\phi_2^{\lim}$ defined by

$$\phi_2^{lim} = D_2^o n_2^o \left(\frac{1}{H_p} - \frac{1}{H_2'}\right) , \tag{8.68}$$

$\phi_2^{\lim}$ is the maximum flux the ion can reach by diffusion through the major ion. This diffusion flux is the mechanism initiating the polar wind, which is discussed in the next section. The electric field is closely associated with the height of scale of the major ion and since it is the source of the upward acceleration, it is not surprising that the minor ion has this scale height. In this case, the concentration of the ion decreases with altitude.

The general solution is a combination of these two extreme solutions. We see that the H^+ concentration may increase or decrease with altitude, depending on the condition of flux that is imposed. It is what occurs at low or mid latitudes, where closed flux tubes prevents the ion to flow upward and then, the first solution prevails with $\phi_2^o = 0$. However, at high latitudes where magnetic flux tubes are stretched or opened, the conditions are such that an upward flux is possible and the second solution prevails, with a flux value close to the limiting flux $\phi_2^{\lim}$.

This situation remains valid as long as the H^+ ion remains a minor ion. Thus, if the ion diffuses according to solution N_3, the ion remains a minor species up to a very high altitude. In the polar ionosphere, H^+ may remain a minor species up to 6000 km. On the other hand when the solution N_4 dominates, since the concentration of H^+ increases while the one of O^+ decreases, there is a point where both concentrations are equal. Above this point, H^+ becomes the major ion and the conditions change; the calculations are no longer valid. At low and mid latitudes, this crossing point may occur around 700 km.

Above this point, the major ion is H^+, which then constrains the plasma scale height while O^+ becomes a minor ion, with a mass 16 times more important than that of the major ion. One can again make calculations with the new data and then one finds that the height of scale of the ion O^+ changes to approach that which it would have if it were alone in the plasma. Then, the electric field only has a weak influence on its structure: its effective mass is very close to its real mass.

The Polar Wind

Although the polar wind includes a contribution specific of a multi-species plasma, it is similar to what gives rise to the solar wind (see Chap. 3) and its name comes from this analogy. In a magnetized ionosphere, polar wind can only develop at very high latitudes, close to the magnetic poles, in the regions where the magnetic field tubes have a volume that is sufficient to maintain the phenomenon. In the plasmasphere, the polar wind can only be a transient phenomenon, because the pressure balance in the tube is reached very quickly in comparison to what happens at high latitudes. This phenomenon is all the more stable as the field lines are stretched and it reaches a maximum on reconnected field lines (see Chap. 7), because the ions are then ejected in the interplanetary medium. In fact, the term polar wind more generally relates to the process of extraction of the ions from the ionosphere at high latitudes and applies to all the ions. Mechanisms of extraction can vary and involve, for example, interactions between electromagnetic waves and ions (Moore et al., 1999).

Since the polar wind develops along open magnetic field lines, it is interesting to account for the divergence of the magnetic flux tube in which the ions flow. If we denote by A the cross section of a magnetic flux tube (the flux of the magnetic field across A is constant, see Sect. 8.2.2), then we can write the conservation of the ions flux as

$$An_2u_2 = \alpha \Rightarrow \frac{\mathrm{d}n_2}{n_2} + \frac{\mathrm{d}u_2}{u_2} + \frac{\mathrm{d}A}{A} = 0\,, \qquad (8.69)$$

where α is a constant. Then, we assume an equation of state for H^+ temperature of the form

$$n_2^{1-\gamma} T_2 = \beta \Rightarrow \frac{\mathrm{d}T_2}{T_2} = (\gamma - 1)\frac{\mathrm{d}n_2}{n_2}\,, \qquad (8.70)$$

where β is a constant and γ is a positive factor (called the specific heat coefficient): $\gamma = 1$ corresponds to the isothermal case and $\gamma = 4/3$ to the adiabatic case. We introduce the speed of sound $c_2 = \sqrt{\frac{\gamma k_b T_2}{m_2}}$ and the Mach number $M_2 = \frac{u_2}{c_2}$. Combining (8.69) and (8.70), we can derive the relation between the Mach number and the velocity

$$u_2 = M_2 c_2 \Rightarrow \frac{\mathrm{d}u_2}{u_2} = \frac{\mathrm{d}M_2}{M_2} + \frac{1}{2}\frac{\mathrm{d}T_2}{T_2} = \frac{2}{\gamma+1}\left(\frac{\mathrm{d}M_2}{M_2} - \frac{\gamma-1}{2}\frac{\mathrm{d}A}{A}\right). \qquad (8.71)$$

With the same approximation as in the previous section, we can write the momentum equation for ion 2 in the form

$$u_2\frac{\mathrm{d}u_2}{\mathrm{d}z} + \frac{k_b T_2}{m_2}\frac{1}{n_2 T_2}\frac{\mathrm{d}}{\mathrm{d}z}(n_2 T_2) + g - \frac{n_2}{m_2} e E_{//} = -\nu_{21} u_2\,, \qquad (8.72)$$

where we have kept the most important terms. Combining (8.69) to (8.72), we obtain the following equation for the Mach number:

$$\frac{2}{\gamma+1}\frac{M_2^2 - 1}{M_2}\frac{\mathrm{d}M_2}{\mathrm{d}z} = \underbrace{-\frac{m_2'}{m_2}\frac{g}{c_2^2}}_{S_1} \underbrace{- \nu_{21}\frac{M_2}{c_2}}_{S_2} + \underbrace{\left(\frac{\gamma-1}{\gamma+1}M_2^2 + \frac{2}{\gamma+1}\right)\frac{1}{A}\frac{\mathrm{d}A}{\mathrm{d}z}}_{S_3} \qquad (8.73)$$

Equation (8.73) determines the evolution of the Mach number along the direction of the flow. This equation is equivalent to the equation of diffusion (8.64), except that the ion temperature is not constant and the contribution of the mean kinetic energy has been added as well as the geometry of the flux tube. It is better suited for high speed flows, which is a characteristic of the polar wind. Indeed, if a subsonic (or supersonic) flow is defined as being a flow of Mach number $M_2 < 1$ (or $M_2 > 1$), (8.73) defines the condition for a transition from subsonic to supersonic.

At low altitudes, close to the region of production of H^+ ions, the flux is low and $M_2 < 1$. In the previous section, we presented the limiting flux solution, which corresponds to an upward motion of H^+ ions. This vertical acceleration is due to the polarization field, which is then balanced by friction of the H^+ ions on O^+ ions: S_1 and S_2, on the right hand side of (8.73), compensate each other (this corresponds to the diffusion process discussed in the previous section). In this region, the contribution of the divergence of the cross section S_3 is negligible because we are close to the ground. Since the collisions decrease when the altitude increases as the ions move upward, the contribution of the left hand side becomes stronger and the right hand term becomes positive, due to the effective negative mass m_2'. This effect is enhanced by the contribution of the cross section which becomes more important at higher altitudes. In that case, if the M_2 is lower than 1, then the Mach number decreases with altitude and the flow remains subsonic; the main part of the energy is contained in the thermal energy.

If M_2 is greater than 1, the Mach number increases with altitude and the flow is supersonic; the kinetic energy is the most important contribution to the energy. At very high altitudes, the contribution of the cross section S_3 becomes dominant and then contributes to accelerate further the ions, as is the case for the solar wind.

The mechanism described here corresponds to the thermal polar wind and the possible energy gain is a few eV. It contributes to the extraction of ions from the ionosphere and thus to the particle transfer from the atmosphere to the magnetosphere, but it does not really contribute to the energization of the ions. Other mechanisms must be invoked for that.

Other Energization Mechanisms

Moore et al. (1999) have discussed the various sources of particle energization in the ionosphere. The most efficient processes are connected to the presence of the magnetic field lines that control the dynamics of the charged particles. As a matter of fact, although the magnetic field does not contribute to the kinetic energy that can be divided into the kinetic energy parallel to the magnetic field and the energy perpendicular to it, it nevertheless controls the distribution between both components of the energy, through the conservation of the magnetic moment. Some wave-particle interactions are likely to heat the particles in the direction perpendicular to the magnetic field, such as, for instance, broadband low-frequency electric wave fields, which cover frequencies from less than one Hz up to several hundred Hz, thus including the gyrofrequencies of the major ion species at least for altitudes from about 1000 km up to a few Earth radii. The left-hand polarized fraction of these waves near the gyrofrequency can heat the ions, which are then likely to transfer part of this perpendicular energy to the parallel direction. The result is a net upward acceleration, because the magnetic field decreases when altitude increases.

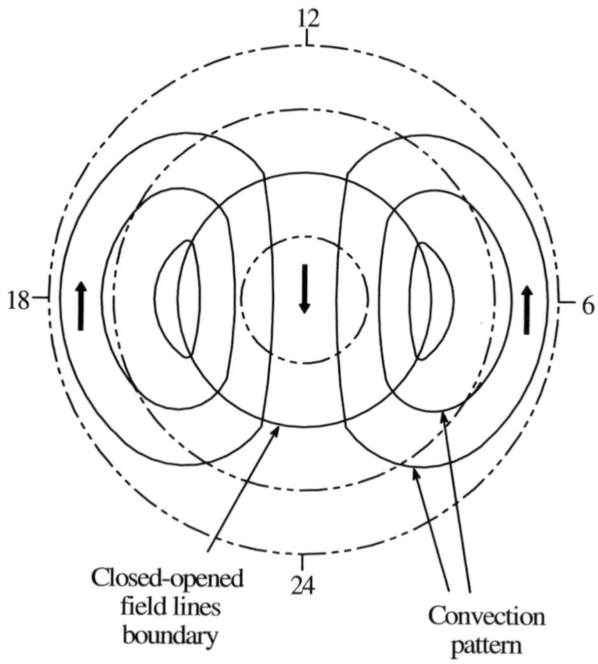

Fig. 8.9. Scheme of the two-cell convection in the high latitude ionosphere. The plasma moves across the magnetic field line along the equipotential of the convection electric field. The resulting motion (*arrows*) is a general antisunward convection in the polar cap ionosphere, from the noon sector towards the midnight sector and a return flow in the auroral regions, both in the dawn and dusk sectors. The closed-opened field line boundary approximately corresponds to the region where the motion changes from antisunward to sunward

8.2.4 Coupling Processes

High Latitude Convection

The solar-wind magnetosphere interaction constrains the magnetospheric plasma to have a convection motion, as described in Chap. 7. The projection of this motion along the conducting magnetic field lines down to the ionosphere results in a two cell convection pattern characterized by a general antisunward convection in the polar cap ionosphere, from the noon sector towards the midnight sector and a return flow in the auroral regions, both in the dawn and dusk sectors (Fig. 8.9).

The convection in the dusk sector is clockwise, while the convection in the dawn sector is counterclockwise. The separation between sunward and antisunward motion corresponds approximately to the boundary between closed and opened field-lines. This, however, is an oversimplified description of the actual distribution of convection in the polar ionosphere whose characteristics strongly depend on the interplanetary magnetic field (IMF) orientation (see Sect. 8.3.2 for a case example).

The purpose of this section is not to discuss in detail the characteristics of the IMF–magnetosphere–ionosphere coupling, but rather to describe the major effects that such an externally driven convection may generate in the ionosphere.

For the sake of simplicity, we suppose in the following that the neutral atmosphere is at rest. In Sect. 8.2.1, we discuss the electrodynamics in the ionosphere and mention that above 150 km the gyrofrequencies are much larger than the collision frequencies and we can neglect the collision frequencies. As a consequence, ions and electrons undergo the same motion that implies having no perpendicular current above 150 km (σ_P and σ_H are negligible and set to 0) and the contribution of the friction can be neglected in (8.30) and (8.31). Furthermore in these equations, the contribution of pressure gradients and gravity force are very small at the ionospheric level on the motion perpendicular to the magnetic field (at least at high latitude) and thus can be neglected. Doing so, we come up with a single equation describing the motion of the plasma in the ground reference frame

$$\boldsymbol{E}_\perp = -\boldsymbol{u}_{\text{conv}} \times \boldsymbol{B}\,, \tag{8.74}$$

where $\boldsymbol{u}_{\text{conv}}$ is the plasma convection velocity (same velocity for ions and electrons) and $\boldsymbol{E}_\perp$ is the convection electric field. This equation expresses that the electric field vanishes in the frame of the particles; this corresponds to the basic principle of the ideal MHD. The convection electric field derives from the convection potential ϕ (i.e. $\boldsymbol{E}_\perp = -\nabla\phi$), which allows a better description of the plasma convection. As a matter of fact, the solution of (8.74) is given by

$$\boldsymbol{u}_{\text{conv}} = \frac{\boldsymbol{E}_\perp \times \boldsymbol{B}}{B^2} = \frac{-\nabla\phi \times \boldsymbol{B}}{B^2}, \tag{8.75}$$

which means that the plasma moves along equipotential lines, perpendicularly to $\boldsymbol{E}_\perp$ and $\boldsymbol{B}$. Thus, the equipotential of ϕ are the streamlines of the convection and hence these equipotential define the convection pattern (see Fig. 8.9).

Frictional Effects

$\boldsymbol{u}_{\text{conv}}$ is the transport velocity of the ions constrained by the convection. It is perpendicular to the magnetic filed and is almost horizontal at high latitudes. If we replace $\boldsymbol{u}_\text{i}$ by $\boldsymbol{u}_{\text{conv}}$ in (8.56), we can express the frictional heating of the ions on the neutrals, implied by the convection (also referred as Joule heating). The variation δT_i of the ion temperature resulting from this heating is then

$$\delta T_\text{i} \simeq \frac{m_\text{n}}{3k_\text{b}} \left(\boldsymbol{u}_{\text{conv}} - \boldsymbol{u}_n\right)^2 . \tag{8.76}$$

Typically, in the terrestrial atmosphere, the temperature increase resulting from a convection electric field $\boldsymbol{E}_\perp$ can be estimated by

$$\delta T_\text{i} = \frac{m_\text{n}}{3k_\text{b}} \left(\frac{E_\perp}{B}\right)^2 = 1.6 \times 10^{-2} m_\text{n} \boldsymbol{E}_\perp{}^2 , \tag{8.77}$$

where m_n is in amu and E_{perp} is in mV m^{-1}.

A convection electric field of 50 mV m^{-1} induces a temperature increase of 650 K, while a 100 mV m^{-1} electric field, which frequently occurs in the auroral zones, induces a temperature increase of 2600 K. Equation (8.77) provides a convenient way of estimating the value of the effective convection electric field $E_\perp$ from the measurement of the departure δT_i of T_i from T_n (St.-Maurice et al., 1999).

The most important effect of the friction has already been mentioned in Sect. 8.1.3. The ion composition can be significantly altered if a strong convection electric field is applied, because the chemical reaction rates increases rapidly with the ion temperature. The effect of the convection becomes sensitive for electric fields higher than 40 mV m^{-1}.

Neutral Dynamics Effects

When a polar cap convection potential is imposed on the ionosphere, the ions reach a steady state in a time scale of $1/\nu_{\text{in}}$ (see 8.32), which is of the order of a few seconds or less below 150 km. The dynamical coupling with the neutral atmosphere (8.27) drives the neutral atmosphere with a similar convection pattern, but with time constants in the ratio n_n/n_i larger than the ionosphere time constant (i.e. of the order of 20 min or more). Reversely, if the external momentum source vanishes, the inertia of the neutral atmosphere keeps it convecting, driving a fossil convection of the ionosphere and consequently drives a remanent potential in the ionosphere–magnetosphere system. (Peymirat et al., 2002) have shown that this fossil convection can occasionally contribute up to 10% of the magnetosphere convection.

Uplifting Effects

Another more subtle but important effect of magnetospheric convection results from simple geometric considerations. Because the convection drift is perpendicular to the magnetic field line and because these field lines are not vertical, except at the magnetic pole, the convective transport towards the pole (or away from the pole) in the polar cap induces a net upwards (or downwards) transport of the whole ionosphere layer, with significant consequences on the layer chemical and diffusive equilibrium. When, for example, the ionospheric F_2 layer is upshifted by this process, the ion production is kept constant (for a given solar illumination), while the recombination is significantly reduced due to the rarefied neutral atmosphere. It follows that, despite the associated decompression of the plasma, the ion chemical lifetime is larger and the ion concentration in the F_2 region increases drastically, creating patches of enhanced plasma densities. A typical case study of this effect can be found in (Blelly et al., 2005).

The Equatorial Ionosphere

Equatorial Electrodynamics

While in the high latitude region, the electrodynamics is controlled by the interaction between the solar wind and the magnetosphere, the electrodynamics in the equatorial region is driven by the thermospheric winds, which are at the origin of the equatorial electrojet, and strongly structured by the magnetic configuration in the equatorial region.

As a matter of fact, in this region the magnetic field is horizontally oriented from south to north and the field aligned direction is then in the meridional plane. As mentioned in Sect. 8.2.1, above about 100 km the conductivity parallel to the magnetic field is so large that the magnetic field lines are almost equipotential and thus only the zonal direction (east-west direction) and the vertical direction contribute to the electrodynamics. Below 100 km, the conductivities are so small that this lower ionosphere region does not contribute significantly to the overall dynamics.

The thermospheric winds are oriented in the eastward direction during the day and impose an eastward electric field E_{P}. In response to this electric field, a vertical Hall current develops, mainly due to the electrons; this current is downward during the day. Then, a vertical charge separation appears in the ionosphere and a vertical electric field E_{H} builds up, which constrains the ions to have a vertical motion. Thus, it acts to reduce the vertical Hall current and eventually compensates it through the contribution due to the non-diagonal terms in the conductivity tensor (8.39). In the end, the electric field is such that no Hall current is present. Averaging on a magnetic field line (by integrating over the entire length of field line, characterized by the curvilinear abscissa s), we get

$$E_{\mathrm{P}} \int \sigma_P \, \mathrm{d}s = E_{\mathrm{H}} \int \sigma_H \, \mathrm{d}s \,. \quad (8.78)$$

The polarization electric field E_{H} must be important in the lower ionosphere, because $\int \sigma_P \, \mathrm{d}s$ is much lower than $\int \sigma_H \, \mathrm{d}s$ there.

In return, this electric field is likely to generate a strong eastward Hall current, known as the equatorial electrojet, which contributes to the enhancement of the current J_P in the zonal direction. This is a characterization of the couplings between the atmosphere and the magnetosphere. However, since the Pedersen conductivity decreases rapidly with altitude, E_{H} similarly decreases. As a result, the polarization field E_{H} is strong only over a reduced latitude range and consequently, the equatorial electrojet has a limited latitudinal extension.

Equatorial Fountain

Above 120 km, ions and electrons are trapped by the magnetic field. Similar to what occurs at high latitude, the Pedersen electric field (due to the neutral wind) is responsible for a convection drift of the plasma, which occurs in the vertical direction: with an eastward electric field, the ionosphere is uplifted. At upper altitudes, this vertical transport is replaced by a diffusion along the magnetic field lines as a result of the gravity force. The overall motion of the ionospheric plasma is then a fountain-like motion in the equatorial region: this is the equatorial fountain. As a consequence of this motion, the plasma accumulates at latitudes about 10–20 degrees north and south of the magnetic equator resulting in density enhancements sometimes referred to as the Appleton anomaly.

8.3 Observations and Modeling

Many experimental techniques allow probing of the ionosphere. We can mention *in situ* measurements with satellites, such as mass spectrometer or Langmuir probes and *remote sensing* techniques, such as the ionosonde (historically the first tool to probe the ionosphere) or its modern version, the dynasonde. (Kohl et al., 1996) provides a good oversight of these techniques that are commonly used for the ionosphere. In this chapter, we focus on two specific ground-based radar techniques, which are well suited for illustration purposes.

8.3.1 Incoherent Scatter

The incoherent scatter technique was initiated by the ideas of Gordon (1958) for exploring near space with radars. The operating frequency is chosen as being sufficiently greater than the plasma frequency and the gyrofrequency, so that, in a first approximation, the wave can be considered propagating as in vacuum. The principle is that electromagnetic waves are scattered by free electrons if they are present in the medium (the mechanism is known as *Thomson scattering*). An electromagnetic wave of angular frequency ω and wave vector κ produces at a distance $\boldsymbol{r}$ of the transmitter an electric field $\boldsymbol{E} = \boldsymbol{E_o}\, e^{i\varphi}$, with the phase $\varphi = (\kappa \cdot \boldsymbol{r} - \omega t)$; in the ionosphere this wave forces the free electrons to oscillate and thus to radiate at the same frequency. If the ionosphere were purely homogeneous, the various phases radiated by the electrons would randomly cancel each other producing a null resultant for the radiated power. Heterogeneities in the distribution of electrons in space mean that a net return can be expected. The phase of the scattered signal may vary randomly with time, but the signal itself exists; this character is at the origin of the denomination *incoherent scattering*. The scattered signal is very weak, as the interaction cross section is of the size of the electron cross section. Also the signal has to be sampled in a wide band $\Delta\nu = \frac{1}{\lambda}\sqrt{\frac{2k_b T_e}{m_e}}$ around the transmitted frequency, due to the Doppler broadening induced by the thermal motion of the electrons (k_b is the Boltzmann constant, m_e the mass of the electron, T_e the temperature of the ionospheric electrons, and $\lambda = 1/\kappa$ is the transmitted wavelength). For a transmitted wave at 1 GHz ($\lambda \simeq 0.3$ m) and an electron temperature $T_e = 1000$ K the bandwidth would be $\Delta\nu \simeq 580$ kHz. The first attempt to perform an incoherent scatter experiment was made by Bowles (1958); the received return power appeared to be close to the expected order of magnitude, but the observed bandwidth was much narrower than predicted by Gordon (1958).

It was understood later that this simple approach would only hold if the electrons could be actually considered as *free*. This would be true for wavelengths significantly smaller than the Debye length (in m) $\lambda_D = 69\sqrt{\frac{T_e}{n_e}}$, where n_e is the electron concentration (in m^{-3}), with typical densities of $n_e = 10^{11}\,m^{-3}$, and the same temperature as above, $\lambda_D \simeq 0.7$ cm. For decimetric or metric radar wavelengths, the electrons cannot be considered as *free* and the collective behavior of the plasma must be considered instead. In this case, the electrons still scatter the transmitted wave, but the signature of the scattering has to be regarded as arising from the density fluctuations associated with longitudinal natural and collective plasma wave structures, such as the ion-acoustic waves, or also the plasma waves around the plasma frequency. The so called *ion-line* is associated with the ion-acoustic waves, in which most of the scattering power concentrates. The scattering occurs with positive (negative) Doppler-shifts from the transmitted frequency corresponding to waves traveling towards (or away from) the radar. Now the signal is concentrated in a much narrower frequency band corresponding to the plasma thermal speed, $\Delta\nu = \frac{1}{\lambda}\sqrt{\frac{2k_b(T_e+T_i)}{m_i}}$. ($m_i$ is the mean mass of the ions and T_i their temperature). For an oxygen-dominant ion with a temperature $T_i = 1000$ K and $T_e = 1000$ K as above, one obtains $\Delta\nu \simeq 4.8$ kHz, although the total scattered power is somewhat less, roughly divided by $1 + T_e/T_i$ (Bauer, 1975).

Curves representing incoherent scatter spectra for various electron-to-ion temperature ratios T_e/T_i are shown in Fig. 8.10; the symmetric two humps, or shoulders, are roughly indicative of the plasma thermal speed $v_{th} = \sqrt{\frac{2k(T_e+T_i)}{m_i}}$. The depth of the central depression is an indicator of the T_e/T_i ratio. As shown in Fig. 8.10, the depression between the two shoulders increases with increasing T_e/T_i ratio. Generally speaking, the shape of the ionic part of the spectrum depends on the electron density, the ion temperature, the electron temperature, and the ion composition. Moreover, if the volume of plasma undergoes a bulk motion of velocity vector $\boldsymbol{u}$ with respect to the radar frame of reference, the total signal undergoes a Doppler shift $\delta\nu = \kappa \cdot \boldsymbol{u}$, which can be a measure of the longitudinal component, along the vector κ, of the ionospheric plasma drifts; with the same conditions as above, a drift of $u_d \simeq \pm 1000\,m\,s^{-1}$ along κ would cause a displacement of the spectrum center from 0 to ±1 in Fig. 8.10. For further reading on incoherent scattering theory and applications, see the cited reviews.

8.3.2 Coherent Radars

Radars probing the ionosphere in the early fifties showed backscattered signals at HF frequencies higher than the ionosphere cutoff frequencies. Initial observations made at high latitudes showed correlations between backscattered signals and geomagnetic activity (Bowles, 1955). These observations also showed the strong aspect-angle dependence of the radar direction, the return signal coming quite exclusively when the radar beam was almost perpendicular to the magnetic

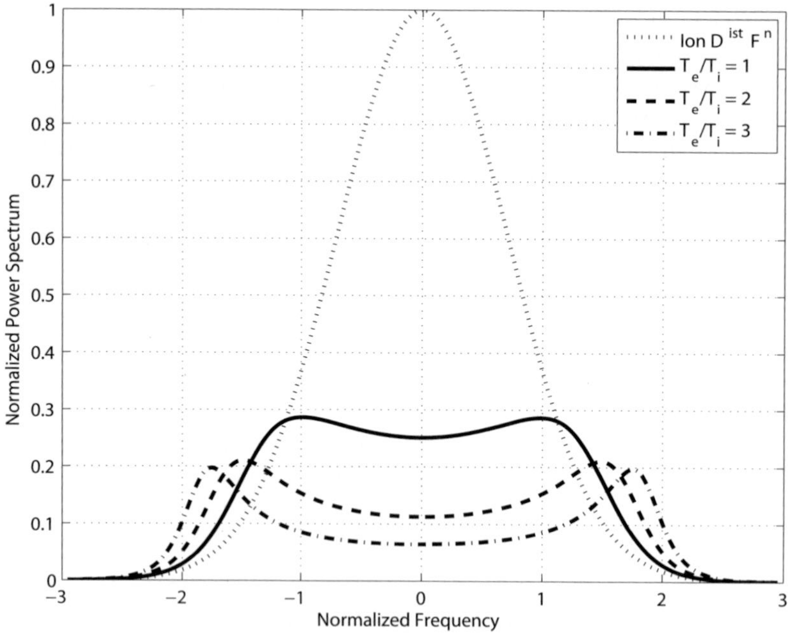

Fig. 8.10. Incoherent scatter spectrum (after Bauer, 1975). The frequency is scaled by the ion thermal Doppler shift $\frac{1}{\lambda}\sqrt{\frac{2kT_i}{m_i}}$. *Complete line*: $T_i = 1000$ K and $T_e/T_i = 1$. *Dashed line*: $T_e/T_i = 2$. *Dashed-dotted line*: $T_e/T_i = 3$. The *dotted line* represents the normalized distribution function of the ions

field line; this was interpreted by Booker (1956) in terms of Bragg scattering of the radio waves by field-aligned electron density irregularities, hence after called *coherent scattering*. The aspect angle dependence condition was easily met in the equatorial regions where this technique was extensively used to study the equatorial electrojet (see, e.g.: Fejer, 1996). At high latitudes, on the contrary, with the dip angle of the magnetic field lines close to 90°, the aspect angle condition limits coherent backscatter to the E region for VHF and UHF radar systems, for which the signal propagates through the ionosphere in straight lines. However, on HF frequencies the ionosphere is a refractive medium and bending of the path of propagation makes it possible to satisfy the aspect angle condition at F region altitudes (see, e.g.: Greenwald, 1996). After initial studies of the equatorial electrojet with this HF radar technique, Hanuise et al. (1981) transported their radar to Scandinavia and initiated regular HF radar observation in the auroral zones; the technique experienced spectacular development after combined incoherent scatter and HF radar observations showed that the drift of the F region electron irregularities measured by HF radars and the ion convection drifts observed by IS radars were measurements of the same quantity, both being induced by the ionosphere $\boldsymbol{E} \times \boldsymbol{B}$ convection (Villain et al., 1985). The versatility of HF radars and their ability to measure ionosphere convection drifts over large areas inspired radar scientists to construct chains of radars all around the auroral zones, with the purpose of studying ionosphere convection on global scales.

Thus the SuperDARN project was born (Greenwald et al., 1995), an unprecedent tool for observing the magnetosphere–ionosphere coupling mechanisms, with almost identical radars running in the 10–20 MHz portion of the HF band; radars are paired, whenever possible, in order to give two-directional components of the convection velocities; one chain of nine radars is operational in the northern hemisphere, covering almost all longitudes between eastern Scandinavia, Greenland and North-America. A second chain of six radars is also operational in the southern hemisphere. The auto-correlation function of the return signals are used to calculate the back-scattered power, spectral width, whose signatures are indicative of the plasma instabilities and of the specific regions of the auroral zones, and the Doppler velocity of the plasma density irregularities used for ionosphere convection studies. True velocities can be inferred from paired radars, but most of the time the ionosphere convection patterns are evaluated by using the combined observations from all the radar sites and fitted to give a global estimate of the potential structure over the polar cap (Ruohoniemi and Baker, 1998). For a complete review of the HF radar capabilities for ionospheric plasma studies and magnetospheric applications, see the review by Greenwald (1996).

Figure 8.11 shows an example of the measurement obtained with the chain of radars. The velocity vectors reconstructed from the pairs of radars are plotted with arrows. The length of each arrow is proportional to the amplitude of the convection velocity and the direction indicates the direction of the flow. Based on these reconstructions, it is possible to estimate the convection potential all over the polar region (from, e.g. (8.74)). Some of the equipotential lines are drawn on the figure. The "+" mark indicates the position of the maximum of potential, while the "×" mark indicates the position of the minimum of potential. For this peculiar example, the potential drop estimated from the measurements is 71 kV. We can clearly see that the convection is much more complex than the one given in Fig. 8.9.

8.3.3 Modeling

Any modeling must take into account the kinetic/fluid duality in the representation of the ionospheric plasma. As far as the dynamics of the thermal ionospheric plasma is concerned, a fluid approach is well suited. Some models use kinetic transport coefficients (Chapman and Cowling, 1970) to describe the transport phenomena, but they are limited to the region where the collisions controls the transport. Some other models use multi-moment transport equations (Schunk, 1977) to describe the plasma dynamics, which are only limited by the distortion of the particle distribution function. The fluid approach is well suited to the electrodynamics coupling in the magnetospheric system. However, it does not allow for a correct representation of the physical processes leading to the creation of the ionosphere (see Sect. 8.1). In that case, only a kinetic approach allows proper modeling of the different aspects that we have discussed in this chapter. They are as follows:

1. Solar EUV photo-production;
2. electron precipitation, which occurs mostly at high latitudes;
3. ion precipitation (mainly protons), also mostly occurring at high latitudes.

Schunk (1996) gives a detailed description of the ionospheric models used to study the ionosphere dynamics and its coupling with the atmosphere or the magnetosphere; in particular those combining both kinetic and fluid representation. However, the complexity of the atmosphere–ionosphere–magnetosphere system is such that not all the processes involved in the couplings are correctly understood and some approximations are required in the models to account for the known processes. Recent physical modeling is based on global circulation models of the ionosphere–thermosphere system coupled with time-varying auroral energy inputs based on empirical models or observations (e.g. Fuller-Rowell et al. (1996)). However, these time-varying studies were mostly successful in dealing with mid-latitude observations of the ionosphere/atmosphere, which are mainly controlled by the diurnal/seasonal solar EUV source, with magnetospheric inputs as global perturbations. At mid-latitudes, these codes were mostly successful in correctly quantifying the storm time *F* region ionospheric response. However, to our knowledge, no modeling case study was able to accurately described the three-dimensional and time-dependent distribution of the ionospheric plasma during quiet and disturbed conditions in the auroral and polar ionospheres. Observed and modeled mid-latitude plasma density comparisons are good, while matching the structure in the polar cap and auroral zone is poor, since models use statistical patterns based on the estimated hemispheric power as high-latitude precipitation and electric field inputs. Indeed, at high latitudes the coupling mechanisms with the magnetosphere and the interplanetary medium are of the same order, or even preponderant, as compared to the solar EUV control; the physical modeling must take into account not only the time-dependent transport of the ionospheric plasma along a flux tube, but also the transport of the flux-tube itself by the ionospheric convection (Blelly et al., 2005).

8.4 Conclusion

This chapter presented a brief introduction to the physics of the Earth's ionosphere, which could be easily extended to the ionosphere of other planets. The complexity of the mechanisms controlling the ionosphere, embedded in the magnetospheric system, does not allow an exhaustive presentation of all the

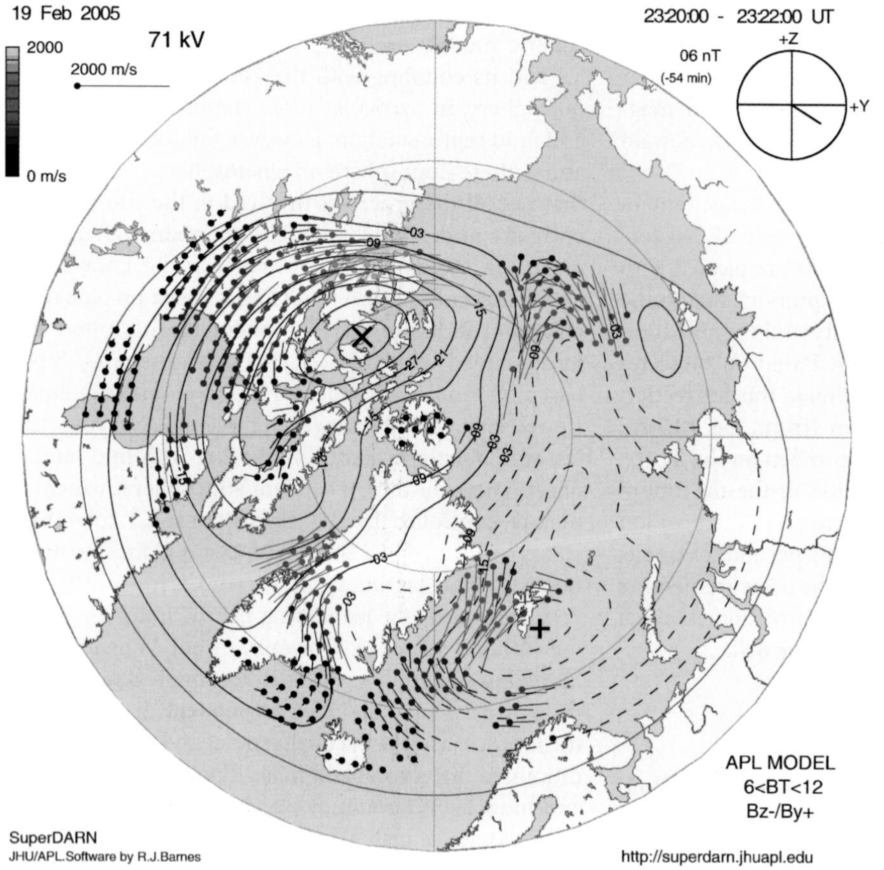

Fig. 8.11. Potential convection map inferred from combined observation from the northern SuperDARN radars. The velocity vectors reconstructed from the pairs of radars are plotted together with the equipotential lines of the convection potential derived from the measurement and a model (Ruohoniemi and Baker, 1998)

processes driving the dynamics of the ionosphere. We focussed on some important aspects of the creation and energization (including bulk flow kinetic energy and thermal energy) of the populations present in the thermal plasma. In particular, we have put forward some key parameters that are important for a correct description of the couplings that take place between the ionosphere and the atmosphere or the magnetosphere. Especially from a theoretical point of view, we have presented how these parameters impact on the ionospheric dynamics and we have shown how these parameters can be inferred from the measurements of the basic parameters of the ionospheric plasma, using these theoretical developments.

Not all the physical processes involved in these couplings are well understood. For instance, we mention some phenomena that can provide an important amount of energy to the thermal plasma such as the instabilities or the wave-particle interactions. They are all based on the capacity of the plasma to interact with the electromagnetic fields $\boldsymbol{E}$ and $\boldsymbol{B}$, because of the presence of charged particles. Such interactions can only be handled using a kinetic approach, which provides information on the particles in the velocity space. They induce distortion of the distribution functions, controlled by the background magnetic field and then lead to anisotropies on the macroscopic parameters such as the pressure. In the end, they may impact the transport, but in a marginal way. A detailed description of such mechanisms was beyond the scope of this chapter and we preferred to concentrate on a fluid representation of the ionospheric plasma, because such an approach is well suited to describing the engine of the thermal motion within the global system, in particular the electrodynamics. Nevertheless, moderate anisotropies can be accounted for with the fluid representation, allowing their use for modeling the high latitude ionosphere, where they are a common feature.

To conclude, in this chapter we have tried to show that the ionosphere is a key region within the magnetospheric system and hence plays an important role in Sun–Earth connections. It strongly contributes to the transfer of particles and energy between the atmosphere and the magnetosphere. The fact that this region is composed of charged particles, that it is a multi-component medium and lays in the domain where the static magnetic field is strong, are major reasons for the complexity of the processes that control the dynamics of the ionosphere. Thus, if the main features of the ionospheric physics are now quite well understood, it is a domain where science is still active. As an illustration, we mention the large effort being made in the solar-terrestrial community to develop experimental means and numerical tools with the goal of achieving a better understanding of the behavior of such a complex plasma lying in this natural laboratory that is the ionosphere.

Acknowledgement. The authors thank EISCAT Director and staff for providing the EISCAT data used in this chapter. EISCAT is an international association supported by the Research Councils of Finland (SA), France (CNRS), the Federal Republic of Germany (MPG), Japan (NIPR), Norway (NFR), Sweden (VR) and the United Kingdom (PPARC).

References

Appleton, E.V., M.A.F. Barnett, Local reflection of wireless waves from upper atmosphere, Nature, 115, 333, 1925.

Bauer, P., Theory of waves incoherently scattered, Philos. Trans. Roy. Soc. London, 280, 167–191, 1975.

Bauer, P., P. Waldteufel, D. Alcaydé, Diurnal variations of atomic oxygen density and temperature determined from incoherent scatter measurements in the ionospheric F region, J. Geophys. Res., 75, 4825–4832, 1970.

Blelly, P.-L., and D. Alcaydé, Electron heat flow in the auroral ionosphere inferred from EISCAT-VHF observations, J. Geophys. Res., 99, 13, 181–13,188, 1994.

Blelly, P.L., and R.W. Schunk, A comparative study of the time-dependent standard 8-, 13- and 16-moment transport formulations of the polar wind, Ann. Geophys., 11, 443–469, 1993.

Blelly, P.-L., C. Lathuillère, B. Emery, J. Lilensten, J. Fontanari, D. Alcaydé, An extended TRANSCAR model including ionospheric convection: simulation of EISCAT observations using inputs from AMIE, Ann. Geophys., 23, 1–13 – SRef–ID: 1432–0576/ag/2005–23–1, 2005.

Booker, H.G., A theory of scattering of nonisotropic irregularities with application to radar reflection from the aurora, J. Atmos. Terr. Phys., 8, 204–221, 1956.

Bowles, K.L., Some recent experiments with VHF radio echoes from aurora and their possible significance in the theory of magnetic storms and auroras, Cornell Univ. Scool Elec. Eng., Tech. Rept. 22, 1955.

Bowles, K.L., Observations of vertical incidence scatter from the ionosphere at 41 Mc/sec, Phys. Rev. Lett., 1, 454–455, 1958.

Breit, G., and M.A. Tuve, A radio method of estimating the hieght of the conducting layer, Nature, 116, 357, 1925.

Buneman, O., Excitation of field aligned sound waves by electron streams, Phys. Rev. Lett., 10, 285–287, 1963.

Chapman, S., and T.G. Cowling, The Mathematical Theory Of Non-uniform Gases, Cambridge University Press, Cambridge, UK, 1970.

Farley, D.T., A plasma instability resulting in field-aligned irregularities in the ionosphere, J. Geophys. Res., 68, 6083–+, 1963.

Fejer, B.G., Natural ionospheric plasma waves, In Modern Ionospheric Science, H. Kohl, R. Rüster, K. Schlegel, eds., Chapt.8, pp. 216–273, 1996.

Fuller-Rowell, T.J., M.V. Codrescu, H. Risbeth, R.J. Moffett, S. Quegan, On the seasonal response of the thermosphere and ionosphere to geomagnetic storms, J. Geophys. Res., 101, 2343–2354, 1996.

Galand, M., R.G. Roble, D. Lummerzheim, Ionization by energetic protons in thermosphere-ionosphere electrodynamics general circulation Model, J. Geophys. Res., 104, 27,973–+, 1999.

Gordon, W.E., Incoherent scattering of radio waves by free electrons with applications to space exploration by radars, Proc. IRE, 46, pp 1824–1829, 1958.

Greenwald, R.A., The role of coherent radars in ionospheric and magnetospheric research, In Modern Ionospheric Science, H. Kohl, R. Rüster, K. Schlegel, eds., Chapt.13, pp 391–414, 1996.

Greenwald, R.A., K.B. Baker, J.R. Dudeney, M. Pinnock, T.B. Jones, E.C. Thomas, J.-P. Villain, J.-C. Cerisier, C. Senior, C. Hanuise, R.D. Hunsucker, G. Sofko, J. Koehler, E. Nielsen, R. Pellinen, A.D.M. Walker, N. Sato, H. Yamagishi, Darn/Superdarn: A global view of the dynamics of high-latitude convection, Space Science Reviews, 71, 761–796, 1995.

Hanuise, C., J.P. Villain, M. Crochet, Spectral studies of F-region irregularities in the auroral zone, Geophys. Res. Lett., 8, 1083–1086, 1981.

Hartree, D., The propagation of electromagnetic waves in a refractive medium in a magnetic field, Proc. Cam. Phil. Soc., 27, 143–162, 1931.

Kohl, H., R. Rüster, K. Schlegel (Eds.), Modern Ionospheric Science, European Geophysical Union, 1996.

Lathuillère, C., P.L. Blelly, J. Lilensten, P. Gaimard, Storm effects on the ion composition, Adv. Space Res., 20(9), 1699–1708, 1997.

Lilensten, J., and P.L. Blelly, The TEC and F_2 parameters as tracers of the ionosphere and thermosphere, J. Atmos. Solar Terr. Phys., 64, 775–793 – doi:10.1016/S1364–6826(02)00,079–2, 2002.

Moore, T.E., R. Lundin, D. Alcayde, M. André, S.B. Ganguli, M. Temerin, A. Yau, Chapter 2-source processes in the high-latitude ionosphere, Space Science Reviews, 88, 7–84, 1999.

Oran, E.S., and D.J. Strickland, Photoelectron flux in the Earth's ionosphere, Planet. Space Sci., 26, 1161–1177, 1978.

Pavlov, A.V., New electron energy transfer rates for vibrational excitation of N_2, Ann. Geophys., 16, 176–182, 1998a.

Pavlov, A.V., New electron energy transfer and cooling rates by excitation of O_2, Ann. Geophys., 16, 1007–1013, 1998b.

Pavlov, A.V., and K.A. Berrington, Cooling rate of thermal electrons by electron impact excitation of fine structure levels of atomic oxygen, Ann. Geophys., 17, 919–924, 1999.

Peymirat, C., A.D. Richmond, R.G. Roble, Neutral wind influence on the electrodynamic coupling between the ionosphere and the magnetosphere, J. Geophys. Res., 107-(A1), SMP–2(1–12) doi:10.1029/2001JA900,106, 2002.

Rees, M.H., Physics and chemistry of the upper atmosphere, Cambridge and New York, Cambridge University Press, 297 p., 1989.

Richards, P.G., J.A. Fennelly, D.G. Torr, EUVAC: A solar EUV flux model for aeronomic calculations, J. Geophys. Res., 99, 8981–8992, 1994.

Richards, P. G., T. N. Woods, and W. K. Peterson, HEUVAC: A new high resolution solar EUV proxy model, *Advances in Space Research*, *37*, 315–322, , 2006.

Rodriguez, J.V., and U.S. Inan, Electron density changes in the nighttime D region due to heating by very-low-frequency transmitters, Geophys. Res. Lett., 21, 93–96, 1994.

Ruohoniemi, J.M., and K.B. Baker, Large-scale imaging of high-latitude convection with super dual auroral radar network HF radar observations, J. Geophys. Res., 103, 20,797–20,811, 1998.

Schunk, R. (Ed.), Solar Terrestrial Energy Program (STEP): Handbook of Ionospheric Models, 1996.

Schunk, R.W., Mathematical structure of transport equations for multispecies flows, Rev. Geophys. Space Phys., 15, 429–445, 1977.

Schunk, R.W., and A.F. Nagy, Electron temperatures in the F region of the ionosphere – Theory and observations, Rev. Geophys. Space Phys., 16, 355–399, 1978.

Schunk, R.W., and A.F. Nagy, Ionospheres, Cambridge University Press, 2000.

Sheehan, C.H., and J.-P. St.-Maurice, Dissociative recombination of N_2^+, O_2^+, and NO^+: Rate coefficients for ground state and vibrationally excited ions, J. Geophys. Res., *109*, 3302–+, doi:10.1029/2003JA010132 2004.

St.-Maurice, J.-P., and R. Laher, Are observed broadband plasma wave amplitudes large enough to explain the enhanced electron temperatures of the high-latitude E region?, J. Geophys. Res., 90, 2843–2850, 1985.

St.-Maurice, J.-P., and R.W. Schunk, Use of generalized orthogonal polynomial solutions of Boltzmann's equation in certain aeronomy problems – Auroral ion velocity distributions, J. Geophys. Res., 81, 2145–2154, 1976.

St.-Maurice, J.-P., and D.G. Torr, Nonthermal rate coefficients in the ionosphere – The reactions of O^+ with N_2, O_2, and NO, J. Geophys. Res., 83, 969–977, 1978.

St.-Maurice, J.-P., C. Cussenot, W. Kofman, On the usefulness of E region electron temperatures and lower F region ion temperatures for the extraction of thermospheric parameters: a case study, Annales Geophysicae, 17, 1182–1198, 1999.

Swartz, W.E., and J.S. Nisbet, Revised calculations of the f region ambient electron heating by photoelectrons, J. Geophys. Res., 77, 6259–6261, 1972.

Turunen, E., H. Matveinen, J. Tolvanen, H. Ranta, D-region Ino Chemistry Model, Solar Terrestrial Energy Program (STEP): Handbook of Ionospheric Models, R.W. Schunk, ed., pp. 1–25, 1996.

Villain, J.P., G. Caudal, C. Hanuise, SAFARI-EISCAT comparison between the velocity of F-region small scale irregularities and the ion drift, J. Geophys. Res., 90, 8433–8444, 1985.

Watson-Watt, R., Weather and wireless, Quart. J. Roy. Meteor. Soc., 55, 273, 1929.

9 Thermosphere

Susumu Kato

The thermosphere consists of neutral- composition as O, O_2 and N_2 which absorb solar radiation for heating, ionization, excitation and dissociation. The composition density widely varies in distribution accompanying the solar activity.

Kinematic viscosity is so large in the thermosphere that turbulence mixing of the atmosphere stops above about 100 km, and the atmosphere tends to be in diffusive equilibrium. Precise mechanism of the cessation of the turbulence mixing remains open to question. Other important dynamics of the thermosphere is of tides and gravity waves (GW) which have been studied for long time. Tides are being studied now with sophisticated GCM (General Circulation Model) simulation reproducing the real complex situation. GW are related to TID (traveling ionospheric disturbance). Certain peculiar behaviors of TID, as showing geomagnetic conjugacy, have recently been discovered, requiring novel understanding of TID which may imply inconsistency with simple manifestation of GW as understood so far. Thermosphere global winds are little known observationally but GCM simulation now presents a plausible model. Intense vertical motion as highly fluctuating and with large amplitudes is sometimes observed in the polar thermosphere, showing non-hydrostatic motion, requiring new approaches. Observations have given remarkable successes in recent thermosphere studies. Besides with incoherent scatter radars, observation with Imager and Fabry-Perot interfermeter at many ground-stations can elucidate precise dynamic behaviors of the mesopause and lower thermosphere; observation with satellites can give global snapshots of these regions as done by UARS et al.; and innovative is observation with GPS.

Contents

Susumu Kato, Thermosphere.
In: Y. Kamide/A. Chian, Handbook of the Solar-Terrestrial Environment. pp. 221–245 (2007)
DOI: 10.1007/11367758_9

9.1 Outline of the Thermosphere

The thermosphere is a region of the Earth's atmosphere. Below it are the mesosphere, the stratosphere and the troposphere, as illustrated in Fig. 9.1. The thermosphere extends, say, from 90 km up to 350–700 km. Above it is another region, called the exosphere, as the uppermost atmosphere, which behaves as a non-continuum in many ways. The thermosphere receives energy and momentum from below and above and very often behaves in a severe and intense manner.

The thermosphere, as other regions of the atmosphere, is characterized by temperature, $T(z)$, increasing with height z (Fig. 9.1). The temperature increase is due to the increasing solar radiation as soft X-ray, UV and EUV; the radiative energy is absorbed mainly by O, O_2, N_2 through heating, ionization, dissociation and excitation. Moreover, radiative cooling occurs due to CO_2 below 120 km and by nitric oxide NO above 120 km, emitting infrared radiation. There are other processes, such as thermal conduction and dynamic heating due to the atmospheric waves that come from below and dissipate in the thermosphere. Furthermore, magnetospheric heating in the form of Joule heating and precipitating particles occurs in the polar region.

Roble et al. (1987) showed the global mean structure of the thermosphere with these heat sources and sinks except for atmospheric waves. They solved the energy conservation and major constituent equations by using a self-consistent global average model. The calculated overall global energy budget under the solar minimum and maximum conditions is given in Table 9.1, assuming geomagnetically quiet periods. They showed that the calculated exospheric temperature is too low if maintained only by radiative processes, and magnetospheric heat sources are necessary to bring the calculated global mean structure into agreement with an empirical model (MSIS-83 model; Hedin, 1983) for both minimum and maximum solar conditions.

As shown in Table 9.1, the infrared radiative cooling by nitric oxide is more effective at the solar maximum than at the solar minimum on energetics in the thermosphere. In addition, it is well known that the radiative cooling by nitric oxide also depends on the auroral activity. Maeda et al. (1989) showed that the radiative cooling by nitric oxide is crucial for preventing the thermospheric temperature from being extremely enhanced during a geomagnetic storm period. Table 9.1 also implies that the thermospheric temperature should fluctuate fairly largely with the solar activity around the mean state as illustrated in Fig. 9.1. The temperature fluctuation varies the density and composition structure of the atmosphere. Figure 9.2 shows the vertical distributions of the three main thermospheric constituents (N_2, O_2, O) in the low solar activity and high solar activity obtained from the NRLMSISE-00 empirical model of the atmosphere (Picone et al., 2002). In this model, the spring equinox, mid-latitude noon, and geomagnetically quiet conditions are assumed. As will be discussed in Sect. 9.2, below about 100 km hydrostatic equilibrium with well-mixed composition occurs and the molecular species are dominant. In this region, the profiles are almost same through solar activities. In the upper thermosphere where the diffusive equilibrium oc-

Table 9.1. Global energy budget (1987). ($\times 10^{11}$ W) [Roble et al., 1987]

	Solar Min.	Solar Max.
Input Solar Energy Absorbed		
EUV	2.1	6.33
*S-R continuum	12.0	15.20
Neutral gas heating		
S-R continuum ($\lambda < 175$ nm)	4.17	5.59
S-R bands ($175 < \lambda < 206$ nm)	0.98	0.96
Neutral-neutral chemistry	0.60	1.99
Ion-neutral chemistry	0.29	0.80
Electron-ion collisions	0.14	0.57
$O(^1D)$ quenching	0.17	0.33
O recombination	0.29	0.38
O_3 absorption	0.29	0.32
**Aurora (particle precipitaion)	0.01	0.01
Joule	0.70	0.71
Total heating	7.64	11.66
IR cooling		
CO_2	5.87	5.95
NO	0.41	3.05
$O(^3P)$	0.23	0.36
Total IR cooling	6.51	9.36
MSIS-83 exospheric temperature (K)	737	1255

*S-R indicates Schumann–Runge

**The energy flux of the auroral particles is assumed to be 0.025 ergs/(cm^2, s) for both conditions. The electric field for the solar minimum and maximum conditions are assumed to be 5.7 and 3.6 mV/m, respectively

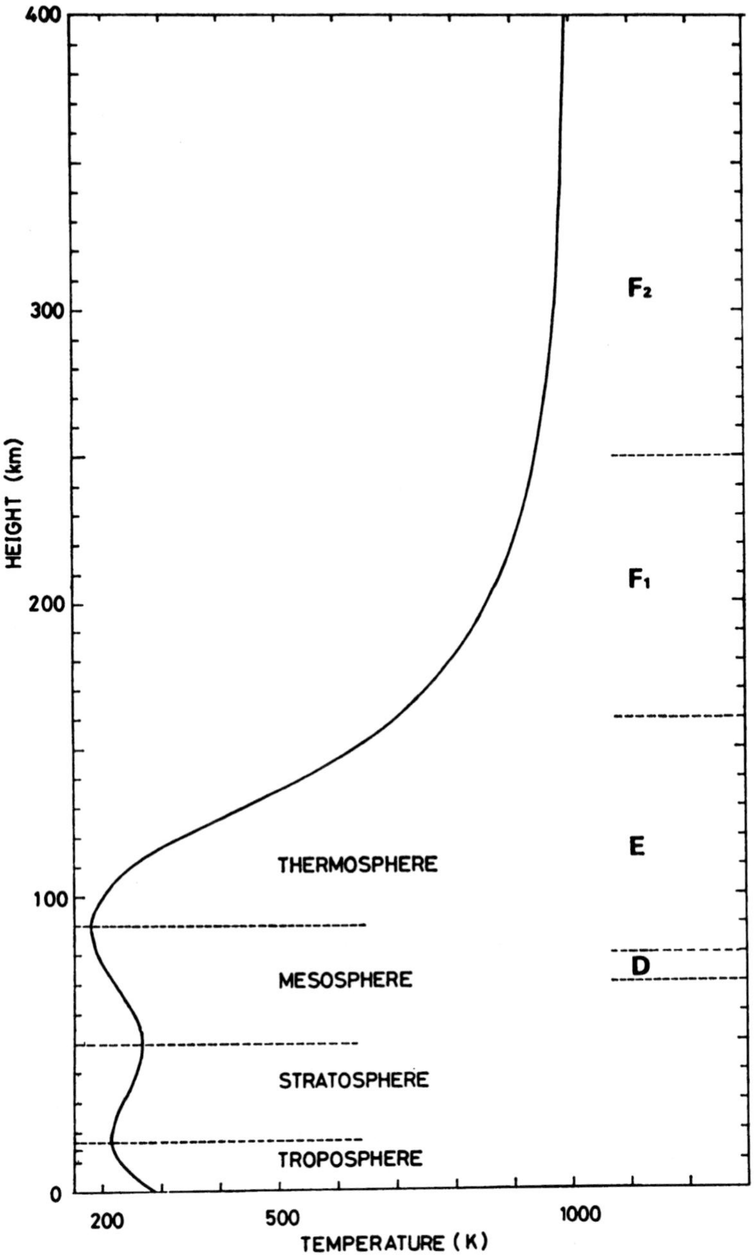

Fig. 9.1. Temperature distribution. The ionosphere consists of four regions, which are called, from below, D, E, F_1, and F_2 layers. F_1 and F_2 layers in combination are called F layer

curs, the profiles are strongly dependent on the solar activity. The number density of atomic oxygen at 400 km is 4.069×10^{13} and 4.963×10^{14} m^{-3} in the solar minimum and the solar maximum, respectively. Also in association with geomagnetic storms, changes in the thermospheric density and composition have been observed globally.

However, note that unlike the lower atmosphere, observation of the global thermosphere is still fairly scarce. Thus, it is significantly important to simulate, under various possible conditions, global behaviors of the thermosphere. In particular, this is the case with the simulation by GCM, now considering the atmosphere from

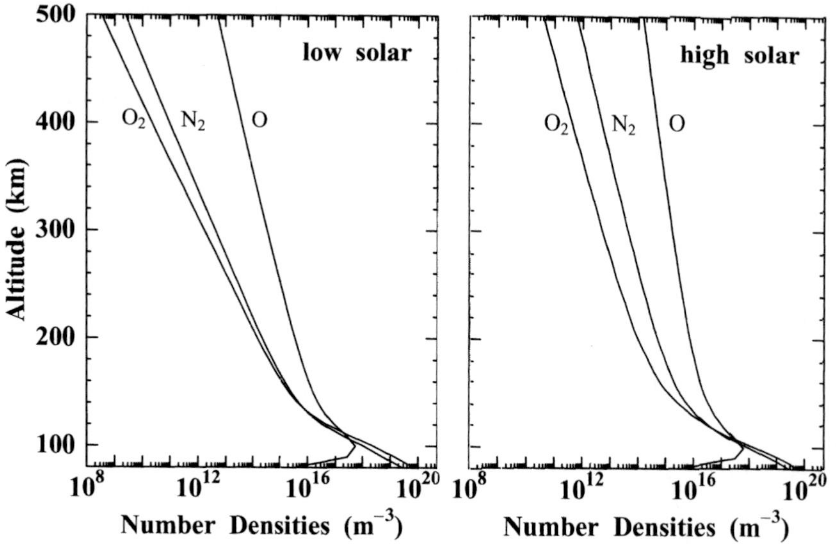

Fig. 9.2. Vertical distributions of three thermospheric main constituents (N_2, O_2, O) in the low solar activity (*left panel*) and high solar activity (*right panel*) obtained from the NRLMSISE-00 empirical model of the atmosphere (Picone et al., 2002). The spring equinox, mid latitude noon, and geomagnetically quiet conditions are assumed. A peculiar behavior of the O profile below 100 km is due to the O_2 dissociation beginning around 100 km [Courtesy of H. Fujiwara]]

the ground up to a height of 500 km (Miyoshi and Fujiwara, 2003, 2004). Certain simulation results will be introduced in Sect. 9.3.

Over almost the same height range as the thermosphere, there is another region called the ionosphere, since the region is ionized or of a plasma state. The ionosphere was discovered in the 1920s to reflect and scatter radio waves, thereby to be used for telecommunication over long distances. Ions and electrons in the ionosphere show interesting behaviors in the presence of electric and magnetic fields, static as well as induced, and in colliding with neutral particles (e.g. Kelly, 1989). Ions closely follow neutral particles in the lower thermosphere, but at altitudes higher than 150 km they tend to be controlled electromagnetically as electrons that are strongly controlled at all heights above 90 km. Thus, there is a peculiar region where charged particles are in peculiar motion between 90 and 150 km, called the dynamo region where ions tend to almost co-move with neutral particles, whilst electrons tend to be free from the neutral particle control. Such a peculiar situation causes the electric current (density) J to be driven and a static electric field to be produced due to polarization in the presence of non-uniform neutrals as well as charged particle distribution. Such quasi-static electric fields are produced to satisfy (Kato, 1980d)

$$\mathrm{div} J = 0 . \quad (9.1)$$

Although any semantic distinction between the thermosphere and ionosphere may be insignificant, in this chapter we adopt a distinction as follows: The thermosphere refers to the neutral particle component, whilst the ionosphere refers to the charged particle component. Here the thermospheric behaviors, controlled by the neutral particle component, will be our main concern, except in Sect. 9.5 where some consideration will be given to the polar thermosphere dynamics, which is controlled by the plasma component. Note that the number density of the neutral particle monotonously decreases from $10^{20}\,/\mathrm{m}^3$ at 80 km to about $10^{15}\,/\mathrm{m}^3$ at 300 km and to 10^{13}–$10^{14}\,/\mathrm{m}^3$ at 500 km, whilst the electron number density, although it varies fairy widely with time, location, solar activity, etc., increases roughly from $10^9\,/\mathrm{m}^3$ at 80 km to $10^{12}\,/\mathrm{m}^3$ at 300 km and then decreases to $10^{11}\,/\mathrm{m}^3$ at 500 km. Clearly in the thermosphere the neutral particles are far more abundant than the charged particles.

9.2 Basic Thermosphere Dynamics

9.2.1 Turbulence and Gravity Waves

The thermosphere, except for the lower part, is in diffusive equilibrium (see Fig. 9.2) and each composition of the atmosphere is in hydrostatic equilibrium with its own scale height, which is inversely proportional to the corresponding molecular or atomic weight as

$$\rho_i(z) \propto \exp\left\{-\int \mathrm{d}z/H_i(z)\right\}/H_i(z) \qquad (9.2)$$

where ρ_i and H_i are, respectively, the mass density and the scale height of each composition specified by subscript i; $H_i = (k_B T)/(m_i g)$ with gravity g, particle mass m_i and the Boltzmann constant k_B. Besides being ionized, the diffusive equilibrium is a clear distinction of the thermosphere from the lower atmosphere consisting of the mesosphere, stratosphere and troposphere, which all are well mixed with one common scale height H with the mean air particle mass m. These lower atmospheres are well mixed due to turbulence, which is almost constantly existent, preventing each composition from tending to diffusive separation.

However, the turbulence mixing tends to cease beyond certain height, i.e. the turbopause beyond which the diffusive equilibrium prevails; the turbopause is found to be at 90–110 km as understood from Fig. 9.2. The turbopause was observed to be at 106 km by Nakamura et al. (1972) with an artificial cloud experiment releasing sodium gas from a rocket to produce illuminating gas trails which changed their form clearly from zigzag to smooth around that height. Cessation of the mixing is understood to be caused by suppression of the turbulence of scales smaller than a certain size.

Batchelor in his famous work (1953) successfully established the isotropic turbulence theory, obtaining the turbulence spectrum in assuming $\mathrm{div}\, V = 0$, where V is the turbulence velocity and considered a stochastic variable. The theory is based on the Navier–Stokes equation as

$$\partial V/\partial t + (V \cdot \nabla)V = -(1/\rho)\nabla p + \mu\nabla^2 V\,. \qquad (9.3)$$

where p is the pressure and μ the kinematic viscosity. Note that $\mu = 40\,\mathrm{m^2/s}$ at a height of 100 km and increases with height, inversely proportional to the atmosphere density (standard atmosphere model; e.g., Kato, 1980e). Batchelor's theory gives κ_d a wavenumber, showing, although approximately, the boundary between the inertial and viscous subrange of turbulence. It is expected that the cessation of turbulent mixing beyond the turbopause would be due to the minimum scale of turbulence corresponding to κ_d which reaches a certain critical value with decreasing of the atmosphere density with height at the turbopause. Batchelor' theory gives

$$\kappa_d = (\varepsilon/\mu^3)^{1/4}\,, \qquad (9.4)$$

where ε is dissipation rate of the total turbulence energy which is supplied externally and removed through viscosity in keeping an equilibrium as (Batchelor, pp. 46–114, 1953)

$$\varepsilon = -(1/2)\,\mathrm{d}V_0^2/\mathrm{d}t > 0\,, \qquad (9.5)$$

where V_0 is the root-mean-square of turbulent velocity V. For isotropic turbulence $V_{0x} = V_{0y} = V_{0z}$ where V_{0x}, V_{0y} and V_{0z} are the root-mean-square of Cartesian coordinate (x,y,z) component of V. Note that $(1/2)\,V_0^2 = (3/2)\,V_{0x}^2$ in (9.5).

In application of the turbulence theory to atmospheric observations, particularly with atmospheric radars, not the velocity turbulence but the *density turbulence* plays an important role. In spite of $\mathrm{div}\, V = 0$ which is assumed in Bachelor' theory, the hydrodynamic equation of continuity gives density perturbation in the presence of vertical gradient of the atmosphere density as

$$\partial\rho/\partial t = -V_z(\mathrm{d}\rho/\mathrm{d}z)\,, \qquad (9.6)$$

where, in the hydrostatic atmosphere, the vertical density gradient is considered to be mainly the ambient density gradient depending only on z. Thus, it is understood with (9.6) that the atmosphere has the density perturbation, which follows the velocity turbulence, obtaining the same density turbulence spectrum as the velocity turbulence spectrum.

There are many works about experimental investigations on atmosphere turbulence characteristics with radars that receive the radio wave echoes from the turbulence (e.g., Hardy et al., 1966; Rastogi and Bowhill, 1976; Lehmacher, 2006). The atmosphere-radar observation is based on tracking of the density turbulence eddies that result in perturbation of the radio refractive index, scattering the radar pulse containing information on the turbulence structure as κ_d. Radio refractive index Λ (Balsely and Gage, 1980) is given as

$$\Lambda = 1 + 3.73 \times 10^{-1}(e/T^2) + 7.76 \times 10^{-5}(P/T) - (Ne/2Nc)\,, \qquad (9.7)$$

where e is the partial pressure of water vapor in mb, P the pressure same to p expressed in mb, Ne the electron number density per m^3, Nc the critical plasma density given as $Nc = 1.24 \times 10^{-2}\, f^2/m^3$ with f as the radio wave frequency per second. $\delta\Lambda$, perturbed Λ, generates the scattering of radar pulses.

As understood from (9.6), vertical gradient of each term containing some atmosphere parameters in the right side of (9.7) produces perturbation due to the vertical velocity of turbulence, obtaining the spectrum same as the velocity turbulence spectrum (Lehmacher et al., 2006); the second term gradient is mainly due to e in the troposphere, the third term gradient is due to ρ in the stratosphere and the last term gradient is due to Ne in the mesosphere and the thermosphere. However, only one particular component of the turbulence spectrum for each term in (9.7), corresponding to half of the radar wavelength, contributes to the radar pulse scattering, returning echoes to the radar (Villars, Weisskopf, 1954). Whilst the formation of perturbation for e and ρ in the troposphere and stratosphere is simple as known from (9.6), the situation in the mesosphere and lower thermosphere is a little complex because, as shown in (9.7), electrons are concerned in the last term. First, note that ions co-move with neutral particles through collisions up to the lower thermosphere. However, electrons are strongly controlled by the geomagnetic field Bo above 60 km and so unable to co-move with neutral particles. Next, it must be understood that any difference in motion between ions and electrons along (gradNe) = (gradNi) ≠ 0 with $Ne = Ni$, where Ni is the ion density, would immediately set up an intense electric polarization field causing electrons and ions to co-move along the gradient of Ne. However, orthogonally both to the polarization field and Bo, Hall drifts take place, driving electrons and ions differently. Note that the drifts have no divergence (Kato, 1980c) and the drifts are orthogonal to the gradient of Ne. Thus, besides the original perturbation along (gradNe), no additional electron density perturbation is produced. Note that if the electron density perturbation originally produced by turbulence tends to move independently from the turbulent eddies, the additional electron density perturbation disappears because of recombination, which is fairly quick (Kelley, p. 407, 1989).

In order to obtain the radar echo scattered by turbulence, the turbulence spectrum component with its wavenumber equal to twice the radar wavenumber or half of the radar wavelength must be in the turbulence

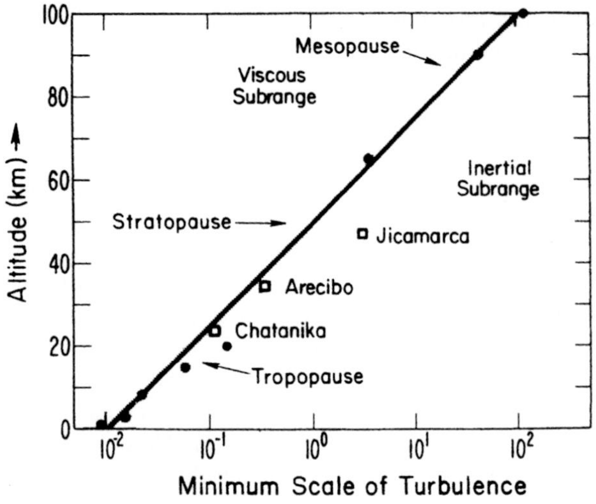

Fig. 9.3. Minimum scale of turbulence $\lambda_{min} = 5.92/\kappa_d$. The diagonal straight line is shown to give an excellent fit to many radar observations. The *solid circles* and the *open squares* denote the maximum height of observed atmospheric echoes for radars operating at different frequencies. The open squares stand for well known radars with their locations. [after Balsely and Gage, 1980; Birkhäuser Verlag]

inertial-subrange bounded by κ_d which decreases with height. As understood by (9.4). Such a favorable radar wavelength at some height approaches κ_d with increasing height, i.e. in increasing the minimum size of turbulence. Finally, at some higher altitude such a spectral turbulence component with a wavenumber equal to twice the radar wavenumber leaves the inertial subrange, entering the viscous subrange, resulting in no radar echo because of insufficient turbulence spectrum intensity in the viscous subrange.

Thus, in radar observation, the radar wave-length decides the minimum size of turbulence through the maximum height for getting the radar echo. The situation is illustrated in Fig. 9.3 which shows the minimum scale of turbulence λ_{min} ($= 5.921/\kappa_d \sim 2\pi/\kappa_d$) based on many radar experiments receiving echoes from turbulent scatterers i.e. $\delta\Lambda$. Note that the correct dependency of λ_{min} on κ_d as shown in Fig. 9.3, has been experimentally well proven.

Since the turbulence eddies move with background winds, the maximum of the echo Doppler spectrum gives the wind velocity projected along the radar beam. The Doppler spectrum spreading is due to the turbulence random motion. As shown in Fig. 9.3, the maximum height able to obtain echoes increases with increasing radar wave-length or with decreasing radar frequency. At 100 km, λ_{min} was found to be about 100 m and can be extrapolated to 1000 m at 120 km.

In (9.4) κ_d depends on the turbulence dissipation rate ε, which is inferred to be about 1 W/kg with λ_{min} = 100 m and $\mu = 40\ m^2/s$.

Further $\varepsilon = \kappa_d^4 \mu^3$ can be roughly estimated to be between a height of 0–100 km with κ_d or λ_{min} from Fig. 9.3 and with μ for the standard atmosphere model (e.g. Kato, 1980e). Then, we know that ε must increase with height; ε at 100 km amounts to a few thousand times of ε near the ground, whilst μ at 100 km is about a million times that near the ground mainly because μ is inversely proportional to the atmosphere density. Such a weaker rate of increase of ε compared to μ with height may have some unknown mechanism for the atmospheric turbulence occurrence.

Note that for the existence of turbulence the Reynolds number, R, is understood in hydrodynamics to be larger than a few thousands, but at 100 km in the lower thermosphere $R = (LV/\mu) = 100–300$ with the minimum turbulence scale for L = 100 m as observed, V = 40–100 m/s, seemingly acceptable, and $\mu = 40\ m^2/s$. This is an intriguing result regarding the turbulence behavior in the upper atmosphere and suggests an inappropriate application of the Reynolds number criterion for turbulence in the upper atmosphere.

Another interesting observation made by Fujiwara et al. (2004) with the European Incoherent Scatter (EISCAT, hereafter) radar shows turbulence at a height of 100–120 km under intense electromagnetic influence. They also observed intense wind shear (Fig. 9.4). The Richardson number of wind shear Ri is given as

$$Ri = N^2/(\partial V_H/\partial z)^2 , \qquad (9.8)$$

where V_H is the horizontal wind velocity and N is the Brunt–Väisälä frequency, as will be explained below. They found Ri to be fluctuating around 0.25 or a slightly larger value; 0.25 implies the upper limit for producing an instability of the flow tending to turbulence. Also on the basis of their observed turbulent velocity, they obtained ε, averaged on many measurements, to be about 1 W/kg at 100 km height, although fluctuating under auroral activities between 0.1–5 W/kg, which amounts to several times the UV heating at 100 km with the density of $4.5 \times 10^{-7}\ kg/m^3$ (e.g. Tohmatsu, p. 485, 1990). The averaged ε is in a good agreement with the radar experiment just prescribed. Note that ε, estimated by the atmosphere-radar experiment shown in Fig. 9.3, depends on the minimum scale of turbulence equal to half of the radar wavelength for obtaining echoes, whilst ε by the EISCAT experiment by Fujiwara et al. (2004) depends on the observed wind shear and eddy diffusion coefficients, which can change significantly with different methods of estimation (Fukao et al., 1994). A good agreement of Σ between the two different observations suggests that the turbolence structure would be similar anywhere.

Although no information is available either about the turbopause height at the EISCAT location or the minimum scale of turbulence there on the basis of the radar observation as shown in Fig. 9.3, the question arises as to why the minimum scale of turbulence λ_{min} is about 100 m for the cessation of turbulent mixing. This implies that the turbulence mixing is possible only with those turbulences whose sale is smaller than 100 m. However, this may not imply non-existence of larger turbulence eddies. It should be remarked that Batchelor's theory of isotropic turbulence considers no gravity in (9.3) and the theory is inappropriate for answering the present question about mixing with large eddies that may be insignificant in lower atmosphere dynamics. The gravity force, if considered in (9.3), would become more effective with an increase of the turbulence size because the non-linear term $(V \cdot \nabla)\ V$ becomes smaller with an increase of the turbulence size. This may imply that large turbulence eddies are not isotropic and the vertical velocity (along gravity) of turbulence eddies larger than certain threshold size may actually be much smaller than the horizontal velocity; a situation that is true for gravity waves as will be seen later (see (9.10)). Since the vertical velocity of turbulence plays the main role in mixing, the gravity effect upon the turbulence structure may be significant for the cessation of the turbulent mixing at about a height of 100 km. The threshold size of the turbulence eddy for the mixing effect would be about 100 m, a subject that can be studied by some simulation.

As for the occurrence of atmospheric turbulence, came an important finding in the 1980s, first theoretically and later observationally. Readers are referred to

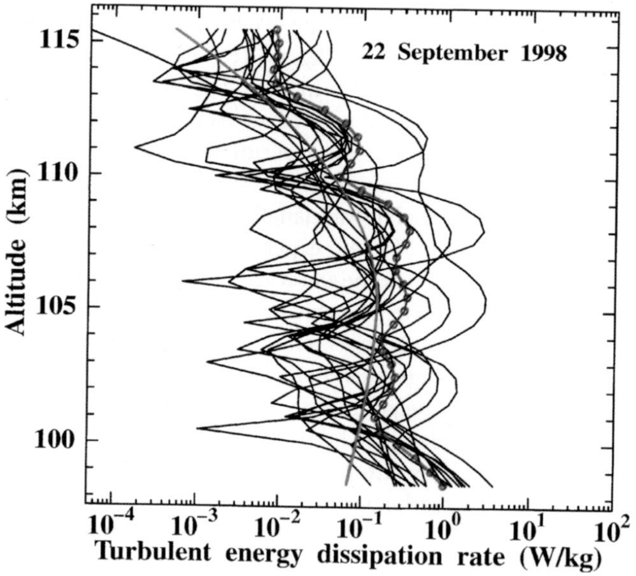

Fig. 9.4. Turbulence energy dissipation based on EISCAT observation. The *black lines* are based on wind shear observation, the *blue line* is the mean value and the *red line* is based on a model (see text) [after Fujiwara et al., 2004; The American Geophysical Union]

the review on the subject by Kato (2005). An important role played by gravity waves (GW, hereafter) is for the mesosphere general circulation. Those GW that are important for this role have periods are as long as a few ten of hours but that are shorter than the inertial oscillation period. Such GW are called inertial GW; the inertial oscillation period is $1/(2\Omega \sin \phi)$, where Ω is the Earth's angular velocity and ϕ the latitude, decreasing from infinity at the equator to 12 h at the pole.

Avoiding any detail about inertial GW playing an important role in the upper mesosphere, it is necessary here to understand that the inertial GW are excited in the troposphere, traveling upwards, being amplified because of the exponentially decreasing ambient atmosphere density, finally reaching the saturation amplitude, breaking into turbulence and releasing wave momentum to the background winds; an idea that explains why turbulence exists in free atmosphere above the planetary boundary layer near the ground where the friction between the ground and atmosphere in motion excites turbulence. Note that the basic physics of saturation, breaking and momentum release is in the second non-linear term in (9.3); after taking the equation of continuity into account, this non-linear term is reduced to *divergence of the Reynolds stress tensor* on the right side of the equation as (Kato, 2005)

$$\nabla(\rho V V)\,, \tag{9.9}$$

where (VV) is a tensor with the GW velocity V; (ρVV) is the Reynolds stress tensor or wave momentum flux, giving an acceleration to the horizontal background winds. Since the vertical wavenumber of GW is much larger than the horizontal one (later understood through (9.10)), along the x direction

$$\rho \partial W_x/\partial t = -\partial\left(\rho V_x V_z\right)/\partial z\,, \tag{9.9'}$$

where W_x is wind along the x-direction. Equation (9.9') is based on the Boussinesq approximation (e.g. Holton, 1992). This shows that the background winds are accelerated through non-linear interaction with GW. Observation of the upper mesosphere with MST (mesosphere, stratosphere and troposphere) radars have successfully proven an acceleration in (9.9') for the background winds upon breaking, as theoretically required for contributing to the general circulation of the mesosphere (e.g. Vincent and Reid, 1983).

It is very likely that some GW may survive the breaking in the mesosphere and reach the lower thermosphere, breaking into turbulence, releasing momentum and heat. In the process of breaking, eddies with increasingly larger wavenumbers than the original GW wavenumber are borne, thereby making $(V \cdot \nabla)V$, in (9.3) so large as to neglect the gravity force even if g is considered in (9.3). Finally, Batchelor's theory of isotropic turbulence can be valid. If GW can propagate up to the lower thermosphere and supply ε for maintaining isotropic turbulence, a smaller Reynolds number as mentioned above may maintain turbulence at a height of 100 km, where μ is fairly large. The situation may be somewhat different at high latitudes (see Fig. 9.4) where some shear instabilities may occur, but the process of breaking into turbulence may be similar.

As understood from (9.3), viscous damping is estimated with the first term, i.e. the inertial tem on the left-hand side of (9.3) in comparison with the second term, the viscous force term, on the right-hand side of (9.3); we realize the viscous damping time to be of the order of (L^2/μ), which amounts to a few days for $L = 10^6$ m and $\mu = 2 \times 10^6$ m^2/s at 300 km height. A more exact treatment has been given by Hines (Hines et al. 1974b). The viscous damping should not be very serious with increasing scales and frequencies as with tides actually observed up to 300 km and with TID (traveling ionosphperic disturbances) observed up to higher altitudes (later shown in Fig. 9.6).

9.2.2 Waves and Winds

Pioneering Work and Historical Developments

The first step forward in the study of thermosphere dynamics is tidal dynamo theory by Stewart in 1883 and Schuster in 1889 (e.g. Kato, 1980d); these two works proved that the systematic daily-geomagnetic-variation (Sq) observed on the ground is caused by a global electric current system flowing somewhere in the upper atmosphere; at the time, neither the ionosphere nor the thermosphere was known. They suspected that the atmospheric pressure tides on the ground, once excited near the ground, travel upwards, reaching the upper atmosphere where tidal wind (V) interact with the geomagnetic field B_0 and sets up an electro-motive force ($V \times B_0$) driving electric currents, thereby producing Sq geomagnetic variation as observed on the ground (Chapman and Bartels, 1940). Figure 9.5 shows the Sq current system based on observation. Now we have confirmed this dynamo region to be between a 90–150 km height, where the ionosphere conductivity is maximum. However, this fantastic idea of the tidal wave propagation up to such a high altitude did present a serious problem in the 1950s. An inconsistency was found between tides observed on the ground and those in the upper atmosphere when tidal winds in the upper atmosphere were deduced from Sq on the basis of the tidal dynamo theory in the 1950s (Kato, 1980d); tides on the ground are mainly semi-diurnal, whilst those in the upper atmosphere are mainly diurnal. Solution of the inconsistency paved the way for establishing the classical tidal theory (CTT hereafter; Kato, 1980b); this theory has determined all modes of tides as well as all other global scale atmospheric waves with eigenfunctions, named the Hough function and eigenvalues of the Laplace tidal equation (LTE, hereafter). Diurnal tides consist of both positive and negative modes corresponding to, respectively, positive and negative eigenvalues of LTE; positive modes can vertically propagate but negative ones cannot. Tides responsible for Sq are mainly of the diurnal first negative mode and are excited, not on the ground, but in situ, i.e. in the lower thermosphere. CTT is of linear theory, neglecting $\partial V_z/\partial t$ (the hydrostatic approach), based on a simple atmosphere model and assuming the atmosphere to be shallow, windless and horizontally uniform, but still gives the basic understanding of tides whose real structures are very complex (see Sect. 9.4). Since the establishment of CTT, many works on tides have been carried out, considering realistic atmosphere models as background winds, non-uniform temperature distribution, radiative cooling, thermal conductions, etc. (e.g. Forbes and Hagan, 1988; Hagan and Forbes, 2002). The approach of these works has been numerical and suitable for understanding the physics of how various realistic complications, not considered in CTT, modify the CTT modes. These works, however, have obtained a limited agreement with observation. Tides both in the lower and upper atmosphere vary from day to day because of complex atmospheric conditions coupling with winds and GW in a non-linear fashion. These factors can hardly be accepted with any simple model. However, now it is very promising to develop GCM simulation originating in weather prediction, taking into account all real situations to reproduce reality. We will discuss some developments of this new approach in Sect. 9.3.

GW is another important subject concerning thermospheric as well as ionospheric dynamics. Long before the 1980s when the inertial GW were found to play an important role for the mesosphere general circulation, as mentioned above in Sect. 9.2.1, early in the 1950s GW with much shorter periods than the inertial oscillation period had become popular among ionosphere physi-

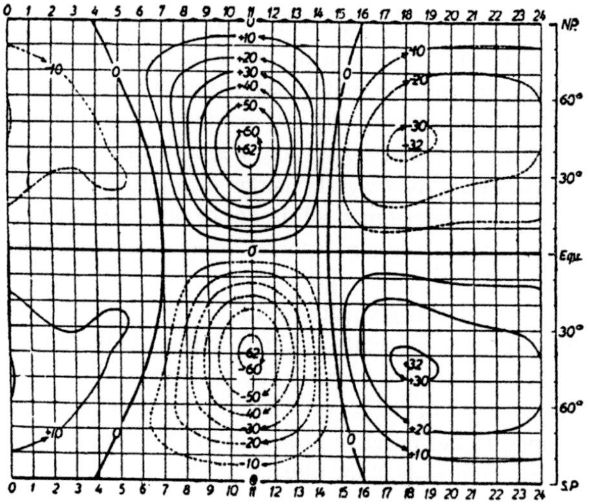

Fig. 9.5. Ionospheric electric-current system in local time (*horizontal axis*) versus latitude (*vertical axis*) for geomagnetic Sq variation in equinoxes in 1902. Between consecutive stream lines, 10000 amperes flow in the direction of the arrows [after Chapman and Bartels, 1940; Oxford]

cists observing TID appearing as wavy fluctuations of the reflection heights of sensing radio waves of particular frequency that corresponds to particular electron density height (see (9.7)). The phase front of the wavy fluctuation descends as illustrated in Fig. 9.6, suggesting TID as a possible GW manifestation (Georges, 1968), as pointed out by Hines in 1960 (Hines et al., 1974a). He successfully showed GW behaviors in the dispersion and propagation characteristics saving the breaking and saturation; one reason would be that the GW he discussed are of shorter periods, usually traveling horizontally faster than the background winds, resulting in no interaction with the winds. His well known dispersion is as follows:

$$m^2 = (N^2/\omega^2 - 1)k^2 - (\Pi^2 - \omega^2)/C^2 \,, \quad (9.10)$$

where k and m are the horizontal and vertical wavenumber, respectively, N = Brunt–Väisälä frequency = $(\gamma - 1)^{(1/2)}$ (g/C) and $2\pi/N$ = 10 min at 200 km (e.g. Kato, 2005), C = sonic speed, Π = acoustic lower cut-off frequency = $\gamma g/(2C)$, γ is the ratio of specific heat capacities between constant pressure and constant volume equal to 1.4 in the lower thermosphere and 1.67 in the upper thermosphere, and ω is the frequency. The dispersion explains the peculiar behaviors of GW in the free atmosphere. Further, Hines (Hines et al., 1974c) suggested that GW travel from the lower atmosphere, reaching the thermosphere where GW heat the region by viscous dissipation. The heating seems to be fairly appreciable, amounting to as much as 10^{-9} W/m^3, a few tens of percent of EUV heating; a similar heating as expected by tidal heating (Lindzen and Blake, 1970). Besides Hines' works, there are many excellent works by Francis in the 1970s (Kato, 1980a) on GW as well as on infrasonic waves regarding the realistic propagation including wave forms, which make an important contribution to TID studies.

It should be remarked that, unlike tides and GW, the thermosphere winds, averaged zonally and in time, have been little understood until recently. Even now, except for the lower part, no real situation of the winds is known observationally (Sutton et al., 2005). A certain simulation for zonal winds will be shown in Sect. 9.3. In equinoctial season, above about 200 km zonal winds are westward due to the Coriolis force acting on the equatorward winds produced by intense heating at high latitudes; below 180 km and above 140 km zonal winds are eastward as in the lower atmosphere. Below this region and above a height of 100 km, however, the winds are strongly affected by drags due to GW arriving from below and breaking as mentioned above, resulting in a westward direction. The GW begins to reverse the mesosphere zonal winds at about 80 km.

Non-hydrostatic Waves (Linear Theory)

With various energy inputs as shown in Sect. 9.1, the thermosphere is a forum for motions with various scales and in violent fluctuation. Sometimes, very large vertical velocities (50–100 m/s or larger), intense temperature and density (a few tens of percent) fluctuation occur as shown later in Sect. 9.5. These observations may stimulate interesting novel studies on non-hydrostatic motion in the thermosphere.

Non-hydrostatic motion $(\mathrm{d}V_z/\mathrm{d}t \neq 0)$ implies that the atmospheric vertical acceleration cannot be neglected in comparison with the Earth's gravity g (amounting to about 10 m/s^2). $\mathrm{d}V_z/\mathrm{d}t$ consists of the first linear term and second non-linear term on the left-hand side of the Navier–Stokes equation (9.3). Evaluation of the order of magnitude gives

$$|\mathrm{d}V_z/\mathrm{d}t| \sim |V_z| \{|\omega| + |m||V_z|\} \,,$$

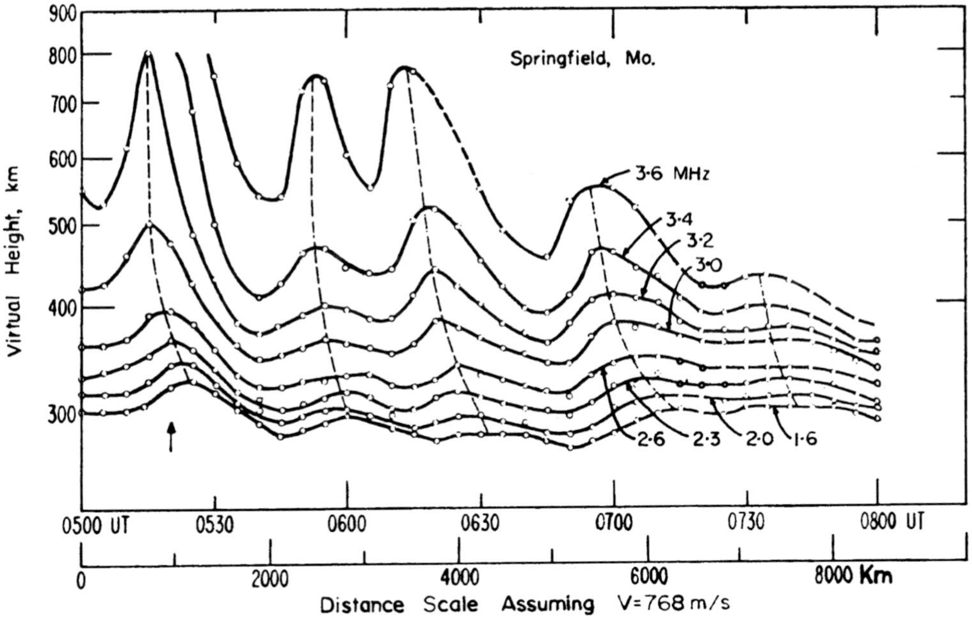

Fig. 9.6. Iso-ionic contours in the virtual height versus times obtained at Springfield, MO, are shown during LSTID for various frequencies of radio signals [after George, 1968. With permission of Elvier]

which shows that the hydrostatic approach, as is usual so far, remains valid if both $|\omega|$ and $|m|$ are small enough even with fairly large V_z. However, recent observation shows that from time to time this is not the case and $|dV_z/dt|$ reaches 10% of g or more as with $|Vz| > 100$ m/s, $|\omega| \sim 10^{-2}$ /s and $|m| \sim 2 \times 10^{-4}$ /m. Then, non-hydrostatic motion really takes place. In spite of such a situation any suitable treatment for non-hydrostatic motion seems unavailable as yet.

However, it seems useful to discuss the ω effect or $\partial V_z/\partial t \neq 0$, because it produces an interesting and simple modification of CTT, which is the basic theory of atmospheric tides as well as other global scale waves. Note that (9.10) by Hines is based on $\partial V_z/\partial t \neq 0$ but assumes no Coriolis force with a plane atmosphere. An appropriate linear theory for understanding the inertial term consideration for global waves was put forward by Kasahara and Quian (2000; hereafter KQT). KQT with $\partial V_z/\partial t \neq 0$ assumes, as in CTT, the atmosphere to be of uniform temperature with no background wind. Since the theory considers no dissipation, it is not applicable to the dissipative thermosphere directly.

However, generally, waves in the thermosphere suffer less viscous effects with higher frequencies and larger scales as understood by (9.3). Accordingly, KQT may be significant, giving an insight into the differences between non-hydrostatic and hydrostatic waves under the real situation of the thermosphere. KQT starts with the following basic equations of motion:

$$\rho_0 \partial u/\partial t + (1/a\cos\phi)\partial p'/\partial\lambda - f\rho_0 v = 0\,, \quad (9.11)$$

$$\rho_0 \partial v/\partial t + (1/a)\partial p'/\partial\phi + f\rho_0 u = 0\,, \quad (9.11')$$

$$\rho_0 \partial w/\partial t + \partial p'/\partial z + \rho' g = 0\,, \quad (9.12)$$

$$\partial \rho'/\partial t - (1/C^2)\partial p'/\partial t - (\rho_0 N^2/g)w = 0\,, \quad (9.13)$$

$$(1/\rho_0 C^2)(\partial p'/\partial t - \rho_0 g w) + \mathrm{div} V = 0\,, \quad (9.14)$$

where $V = (u, v, w)$ = small perturbed velocity, ρ_0 = static density, ρ' = perturbed pressure and density, respectively, λ = longitude, f = Coriolis factor = $2\Omega \sin\phi$, a = the Earth's radius.

The above basic equations are, as in CTT, reduced to two equations that describe the horizontal and vertical structure of the waves. Assume harmonic waves along longitudes as $\exp\{-\mathrm{i}(\omega t + s\lambda)\}$, where s is the longitudinal wavenumber. It is remarkable that *also in the non-hydrostatic case, the latitudinal structure is given by LTE, identical with the structure in CTT, giving the same eigenvalues or the equivalent height h and eigenfunction or the Hough function*, both specified by three indexes as $\Theta(\phi : \omega, s, l)$, where ω is the frequency usually normalized by Ω and I is an integer as the latitudinal structure index.

The equation describing the vertical structure in the non-hydrostatic case, corresponding to the Wilkes equation in the hydrostatic case (Kato, 1980b) is

$$\mathrm{d}^2\eta/\,\mathrm{d}z^2 + (\chi - \Gamma^2)\eta = 0\,, \tag{9.15}$$

where $\chi = \{1/(gh) - 1/(C^2)\}\,(N^2 - \omega^2)\,,$ (9.16)

and $\Gamma = (1-2\kappa)/2H,\ \kappa = (\gamma-1)/\gamma\,.$

η is the vertical structure function dependent only on z defined as w' multiplied by $\rho_0^{(1/2)}$, namely,

$$\eta(z) = \left(w'\rho_0^{(1/2)}\right)\,, \tag{9.17}$$

where $w \propto w'(z)$.

Equation (9.16) shows that in the non-hydrostatic case, the vertical structure is distinct from that in the hydrostatic case, since (9.16) contains ω, which is absent in the hydrostatic case; the non-hydrostatic case replaces N^2 with $(N^2 - \omega^2)$. *This demonstrates clearly, as expected, that for $\omega \ll N$, the vertical structure of non-hydrostatic (linear) waves tends to that of hydrostatic waves.* The KQT originally attempts to find free modes of the non-hydrostatic waves. The free modes satisfy the upper and lower boundary conditions as $\eta = 0$ and without heat sources. The free mode wave amplitudes can be infinitely large with any heating, no matter how small, i.e. resonance. We know that in the hydrostatic case there is only one free mode with one h in (9.16), which for an unbounded isothermal atmosphere is

$$h = \gamma H = 10.5\,\mathrm{km} \text{ with } H = 7.5\,\mathrm{km} \text{ or } T = 256\mathrm{K}\,, \tag{9.18}$$

since originally the lower atmosphere is modeled in the KQT (Fig. 9.7a). However, this is not the case for non-hydrostatic waves because the vertical structure equation, i.e. (9.15) includes ω, implying the possible existence of many h's for free modes with different ω's as indexed with integer k.

In order to satisfy the upper and lower boundary conditions as $\eta = 0$ (9.15) requires a certain relation between h and ω as, although inexplicitly expressed,

$$F(h_\mathrm{k}, \omega) = 0\,. \tag{9.19}$$

Also, as in CTT, LTE gives a function G, which is an algebraic equation in the form of an infinite continued fraction (Kato, 1980b) to decide h with the vertical index k for specified ω, s and l as

$$G(\omega, s, l, h_\mathrm{k}) = 0\,. \tag{9.20}$$

Details of the calculation to solve the simultaneous (9.19) and (9.20) are not shown here but the calculated result is illustrated in Fig. 9.7a, b and c. The KQT free modes have the following main points:

1. Differently from hydrostatic waves, non-hydrostatic waves have free modes with various h's (Fig. 9.7c), consisting of three different kinds of internal mode and one kind of external mode (Fig. 9.7a). Of these three internal modes the first is of positive modes or the global GW mode (G, Fig. 9.7a), traveling both eastwards and westwards. The second is of negative modes or the rotational mode (Fig. 9.7b), traveling only westwards. Note that negative modes in low frequencies has $h_\mathrm{k} > O$. These two modes are the same in the structure as CTT modes, but both kinds exist only for $\omega < N$. The third is of the internal acoustic wave mode for frequencies higher than $\Pi = 1/(9\,\mathrm{min})$ at 300 km (denoted as A in Fig. 9.7a). The fourth is of the external mode (E in Fig. 9.7a). With increasing s, the dispersion is asymptotic to Hines' dispersion, which is understood in Fig. 9.7a. There exists a gap region for the frequency between N and Π where no internal mode exists.

2. All four above modes are discrete and specified by h_k, ω and two integers k, l as $(\omega, s, h_\mathrm{k}, l)$. For acoustic wave modes, the Hough function Θ approximates the associated Legendre function P_1^s (sin ϕ). This is readily understood because of the acoustic wave mode with frequencies so high as to neglect the Earth's rotation (Kato, 1980b). It is also found (Fig. 9.7c) that h_k of the acoustic wave mode is

$$h_\mathrm{k} > \gamma H\,, \tag{9.21}$$

whilst for the other two kinds of internal modes

$$h_\mathrm{k} < \gamma H\,. \tag{9.22}$$

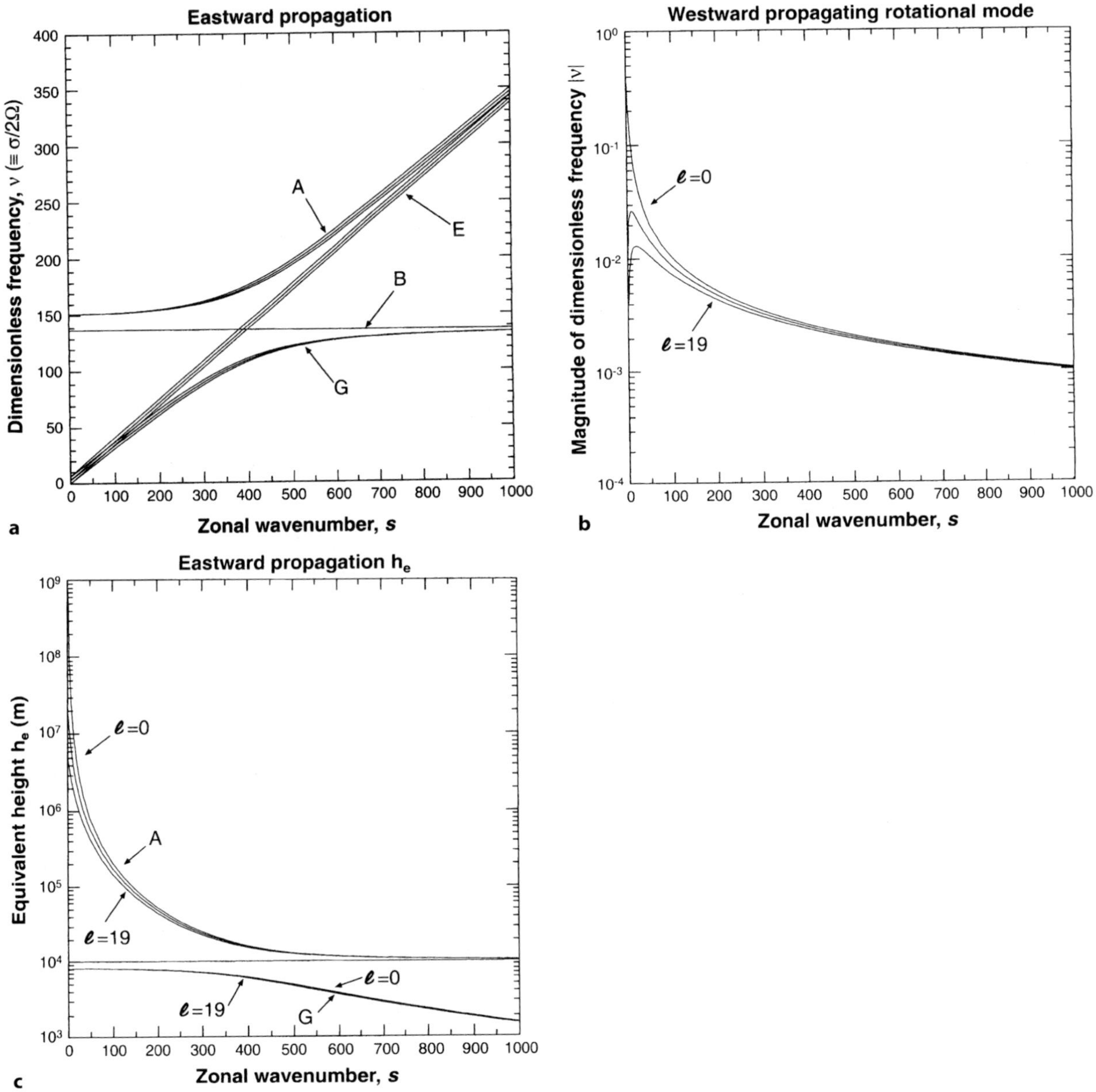

Fig. 9.7a–c. Non-hydrostatic waves of free modes in unbounded atmosphere with a vanishing vertical wavenumber. (**a**) Eastward traveling waves; "A" denotes acoustic modes, "G" GW modes and "E" external modes. These modes are able to travel both eastwards and westwards. The horizontal thin straight line shows the Brunt–Väisälä frequency; in the figure σ denotes the frequency instead ω in text. (**b**) Westward traveling waves of the rotational modes; the waves are able to travel only westwards; l denotes latitudinal index for the Hough function; this mode has only periods longer than 12 hrs. (**c**) Equivalent heights h_e (for h_k in the text) of free modes. Only equivalent heights for eastward traveling modes are shown. The straight horizontal line shows $h = \gamma H$ in the hydrostatic case (see text) [after Kasahara and Qian, 2000; American Meteorological Society]

Although the real thermosphere is dissipative and different from the atmosphere modeled in KQT, the difference between the hydrostatic and non-hydrostatic cases discussed above is significant for understanding the real situation in the thermosphere. Note that, besides heat sources, the amplification of waves in the existence of free modes would be worth considering.

It is expected that various quickly fluctuating disturbances on the solar surface, such as flares and spots in the the solar radiation, as well as solar winds arriving at Earth may produce a highly fluctuating intense and global heating of the upper atmosphere. This consideration presents novel and interesting problems for atmospheric dynamics in the thermosphere. Some observed vertical velocity in the auroral region seems to be so strong, as will be discussed in Sect. 9.5, that a simple consideration of only thermally forced excitation is unrealistic and we may expect some contribution from the free modes. Since the velocity divergence by waves generally increases with frequencies, acoustic waves are more effective than GW in directly producing the density perturbation that may accompany the formation of ionospheric Ne irregularities, as are almost constantly observed; the formation may be through certain plasma instabilities ignited by the acoustic waves. Concluding Sect. 9.2.2, it seems informative that, besides dissipation, viscous forces may generate viscous waves (Hines et al., 1974b) but their actual significance in thermosphere dynamics remains unknown as yet.

9.3 GCM Simulation

The thermosphere is full of unknowns as yet, mainly because of insufficient observations for understanding clearly the global situation. Under the circumstances, it seems helpful or even indispensable to develop excellent GCM simulation systems, which were originally developed for weather forecasting, intending to reproduce reality in its full complications. The simulation results would be helpful to plan effective observation and to encourage a gaining of insight into real complex processes. The first GCM presented in the world was a thermospheric general circulation model developed in the UK (Fuller-Rowell and Rees, 1980). This model has now been updated to be coupled with the mesosphere, ionosphere and plasmasphere (e.g. Fuller-Rowell et al., 1996; Millward et al., 1996; Harris et al., 2002). Another model was called TGCM (thermosphere GCM; Dickinson et al., 1981) and was developed to TIGCM (thermosphere and ionosphere GCM; Roble et al., 1988) and eventually to TIME-GCM (thermosphere, ionosphere, mesosphere electrodynamics GCM; Roble and Ridley, 1994); all models were developed at NCAR in Boulder, Colorado, USA. Another model for the thermosphere-ionosphere-protonosphere system research was developed in Russia (Nagmaladze et al. 1988). Whilst these models locate the lower boundary in the lower atmosphere, there is a different GCM that puts the lower boundary on the ground. This was developed at Kyusyu University in Japan. It was originally called MACMKU (middle atmosphere circulation model at Kyusyu University) and has now been developed for including the thermosphere up to 500 km (hereafter, TCMKU). Details of this system can be found in the references (e.g., Miyoshi and Fujiwara, 2003). It is clear that disturbances such as GW and tides, which are generated near the ground and propagate to the thermosphere, would be correctly taken into account only in TCMKU. However, recently the model at NCAR has somehow been connected to a GCM for lower atmosphere studies (Nozawa et al., 2001).

The models, however, assume hydrostatic motions such as $dV/dt = 0$, which cannot treat severe vertical motions as mentioned in Sect. 9.2.2. When severe vertical fluctuation become serious subjects, the present GCMs need to be improved with regards to resolving power, time aspects, as well as height, latitude and longitude aspects, besides considering the vertical acceleration term.

Nevertheless, many interesting simulation results, mainly about heating and winds, have been obtained with GCMs based on the hydrostatic approach. Just a few of these results will be shown below.

A big difference occurs in the radiative heating between the upper and lower thermosphere at 300 km and 120 km, as shown in Fig. 9.8a and b (Dickinson et al., 1981); in the upper region the heating in daytime varies slightly and suddenly decreases at the arrival time of the terminator to remain so during night, whereas in the lower region the heating varies with the sub-solar point. It is impressive to see such a sharp contrast simply due to the optical depth difference between the two

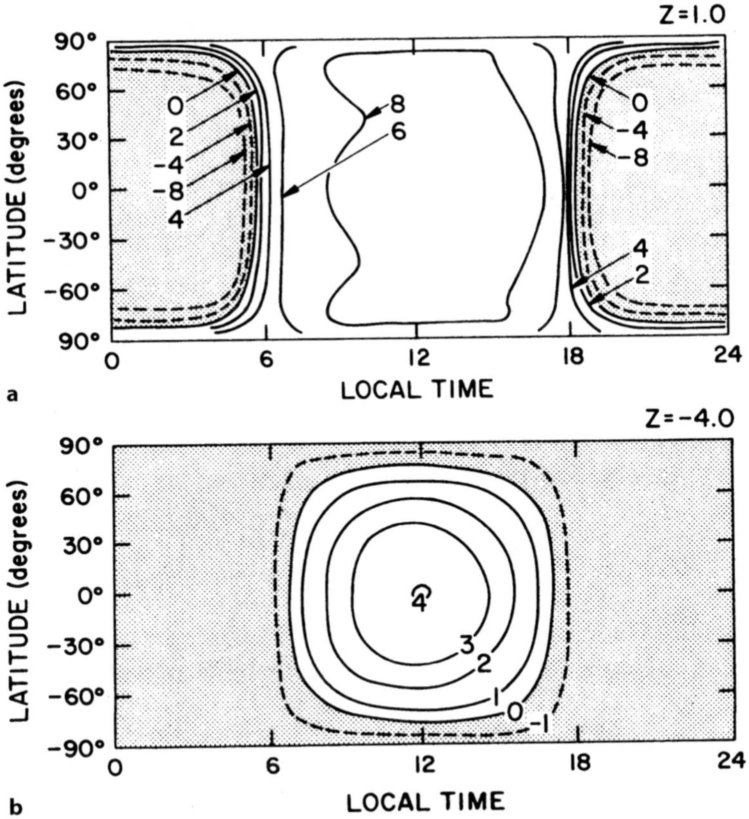

Fig. 9.8a,b. Heating simulation for equinox. The contours give departures from the mean in units (W/kg); (**a**) at 300 km ($Z = 1.0$) and (**b**) at 120 km ($Z- = -4.0$) [after Dickinson et al. 1981; The American Geophysical Union]

heights. A comparison with some suitable observation a such satellite observation would be desirable. As explained in Sect. 9.2.2, winds (in the equinoctial season) vary between the upper and lower thermospheres; the zonally-averaged zonal winds above 200 km are westward and between 100 km and 140–180 km eastward, as in the lower atmosphere. Due to the GW drag (Kato, 2005), winds below 100 km are westward. Figure 9.9a (Miyoshi and Fujiwara, 2003) shows the result; TGCM gives a similar result, which is, however, not shown here.

Diurnal and semi-diurnal tides are excited near the ground, traveling upward, all through the lower atmosphere, constantly changing and then reaching the thermosphere The simulated result (Fig. 9.9b Miyoshi and Fujiwara, 2003) demonstrates the diurnal tides propagating upwards with a vertical wavelength of 25–30 km up to a height of 100 km and zonal wind amplitudes amounting to 50 m/s; the diurnal tides of negative modes that are excited in situ are dominant above 200 km; although not shown altogether, the semidiurnal tide zonal wind reaches maximum at 120–200 km heights.

It would be interesting to test whether the simulation result supports the traditional idea of the Sq tidal dynamo theory, which now is well established (Kato, 1980d). What is to be tested is whether the tidal dynamo electro-motive force, $\int [\sigma](V \times Bo)\,dz$, where $[\sigma]$ is the ionosphere electric conductivity tensor for driving Sq ionospheric electric currents, is mainly of the diurnal tide first negative mode $\Theta\,(1, 1, -2)$ of CTT as mentioned above (Kato, 1980b). The electromotive force is excited in the dynamo region between 90–150 km where the daytime effective ionospheric conductivity is predominant. Note that the propagating tidal modes may be ineffective due to changing phase and $[\sigma]$ in magnitude along z in the integration.

Miyahara and Ooishi (1997) attempted to simulate the Sq current, which varies from day to day (Chapman and Bartels, 1940) because of traveling tides varying

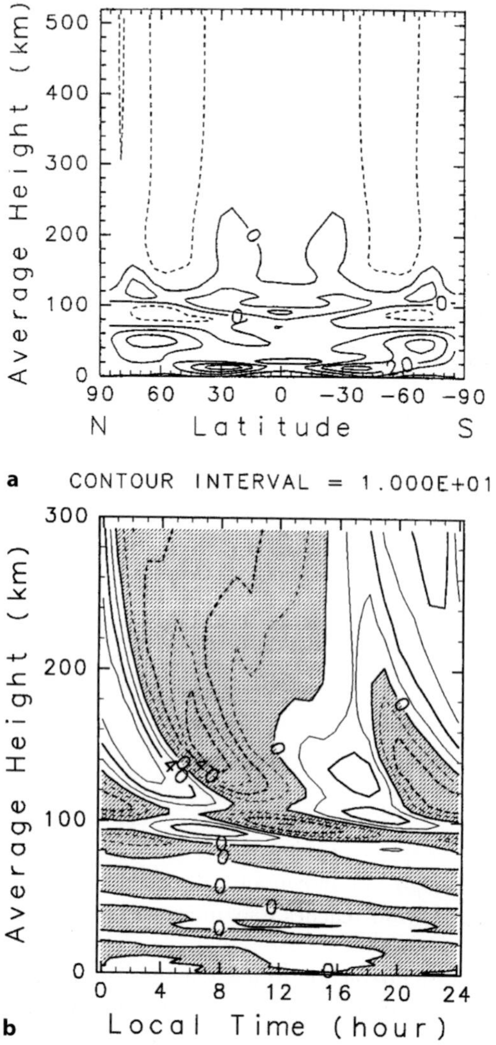

Fig. 9.9a,b. Wind simulation in equinox. (**a**) Mean zonal winds: Contour interval is 10 m/s and *solid* and *dotted lines* are eastward and westward, respectively. (**b**) Diurnal tides; zonal tidal winds as departures from the mean zonal winds at East Long. 0° and South Lat. 24.9°; contour interval is 20 m/s [after Miyoshi and Fujiwara, 2003; The American Geophysical Union]

from day to day through the lower atmosphere. Their GCM simulation was based on an earlier version of TCMKU with the upper boundary at 150 km and the diurnal negative mode was specified, not generated inside the GCM. Hence their simulation was unsuitable for this test on the dynamo theory.

GCM for non-hydrostatic motion will soon find a new development. The first step has recently been taken with a two-dimensional approach that considers $V\ (u, v.w)$ with only y (meridional) and z (height) dependence under the Coriolis force effect (Shinagawa et al., 2003) as discussed in Sect. 9.5. Note that the KQT introduced in Sect. 9.3 may give a basis for non-hydrostatic GCM simulations.

9.4 Observation

Observation of the thermospheric neutral particle behavior has been limited as yet and has mainly been conducted in an indirect way through plasma observation with radio waves. However, recently, through optical observation we were able to obtain some direct information on the neutral particle motion, although this is still not enough. Among other observation methods, the incoherent scatter radar (ISR, hereafter) is still most powerful among ground-based observations of the thermosphere in the sense that it supplies various kinds of information on ionospheric plasma under control of abundant neutral particles at all heights of the thermosphere. An understanding of the physics of the control makes it possible to derive information on the neutral particle behavior through ionosphere observation.

Observation with ISR is based on the Doppler spectrum of radar echoes from the ionosphere plasma caused by the refractive index fluctuation $\delta\Lambda$ due to δNe with the scale equal to half the radar wavelength (see (9.7)); the radar scatterer is not turbulence but plasma waves. The incoherent scattering, or IS, is not a correct expression for describing phenomena of the scattering by the radar, although it was once believed to be incoherent until the radar echo was actually received at the end of the 1950s. It had been expected that the echo returning to the radar would just be a summation of incoherent echoes scattered by each electron moving at random independently from ions in the volume illuminated by the radar beam. However, in the presence of plasma, ion acoustic waves to some extent produce a coherent scattering of each electron. The Thomson scatter is correct in expressing the ISR scattering. The spectrum shows a central shift and the peculiar shape of double peaks, all of which are due to ionosphere plasma characteristics in the presence of winds, the electrostatic field and the geomagnetic field *Bo*, etc. The interested reader is referred to the famous review paper by Evans (1969).

The neutral particle velocity component along *Bo*, $V_{//}$, is obtained from the observed plasma velocity, which is not controlled by either the electric field or the magnetic field. As a result, the neutral and charged particles almost co-move along *Bo* (up to about 120 km). Another useful relation for the observation is that the neutral particle temperature is very close to the ion temperature in the lower thermosphere up to 130 km. Further sophisticated approaches are possible for obtaining other information on the neutral particles from plasma observation on the basis of plasma physics (Kato, 1980c; Kelley, pp. 425–423, 1989). However, a shortcoming of the observations is that there are only five ISR systems available to date, i.e. at Arecibo (Puerto Rico), Millstone Hill (USA), Jicamarka (Peru), Sondrestromfjord (Greenland) and EISCAT in northern Europe. The ISRs are all located in the western hemisphere. The MU radar in Japan can receive IS echoes supplying some data of a low resolution on height and time (Oliver et al., 1988) because of the lower sensitivity given by the product between the average output power and the antenna aperture area. The MU radar sensitivity is 5×10^8 W m^2, whilst other powerful ones are ten times more or larger (Balsely and Gage, 1980).

ISR observation of thermospheric tidal winds has been carried out since the 1970s with the Millstone Hill ISR (Kato, 1980c; recent work by Gonchaenko and Salah, 1998) and the Arecibo radar. Note that some significant comparisons of the observation with GCM simulations as prescribed in Sect. 9.3 is now available. EISCAT observation of the auroral *E* region winds, including the mean winds and tides, has been compared with a GCM (TIME GCM) simulation (Nozawa et al., 2001).

As for TID, LSTID (large-scale TID), which was observed on particular days with the Milestone and Arecibo ISR, was once successfully simulated by assuming a GW excited by Joule heating at the north pole by Roble et al. in 1978 (Kato, 1980a). However, as for MSTID (medium-scale TID), such an agreement between the observation and simulation has not been attempted.

Imager and Fabry–Perot interferometer (FPI) observation (Shepherd, 2002) has recently been done extensively and proven to be a new information source (e.g., Price et al.,1995). Imager and FPI observations depend on receiving particular airglow emission by OH with 843.0 nm maximizing at 86 km, OI with 557.7 nm at 95 km (110 km in the aurora region) and OI with 630.0 nm at 250 km. The first two airglows are useful for finding GW of short periods (shorter than a few hours) in the mesopause and the lower thermosphere. The 630.0 nm airglow only originates in the upper thermosphere and would be useful for understanding the physics of TID in relation to GW (Garcia et al., 2000). Note that the 630.0 nm airglow at night is now known to be well correlated with the *F* region Pedersen conductivity integrated along *Bo* (Makela and Kelley, 2003) although the basic physics of the correlation is not yet well-known.

Another unknown physics aspect was a surprising finding obtained by Otsuka et al. (2004). This is the real existence of the conjugate property of MSTID (Fig. 9.10), which was simultaneously observed, with all sky imagers of 630.0 nm airglow at Sata in southern Japan and Darwin in northern Australia, both locations being geomagnetically conjugate to each other. It is remarkable to have found, at both locations, almost identical imager observation results, consisting of dark and bright bands of the airglow. The conjugate property implies an action of electric field that, in fluctuation with a scale of more than several tens of km, is to be transmitted along the geomagnetic field-line without serious attenuation to the conjugate point, as observationally confirmed (Saito et al., 1995), thus contributing to producing an identical image. This may suggest that one of the two TIDs observed at conjugate points is no simple manifestation of GW but an electron density irregularity produced by an electric field due to the TID caused by GW at the conjugate point. *No* such conjugate property should occur during daytime because of the short-circuited electric field due to the high conductivity in the *E* layer in daytime. A comprehensive observation at conjugate points is desirable.

Satellite observation is most effective for obtaining global snapshots for the study of the thermosphere. One of the most sophisticated atmosphere observations was carried out with the Upper Atmosphere Research Satellite (UARS) which was launched in September 1991 and continued to be in operation for more than ten years. It supplied a tremendous amount of data with sophis-

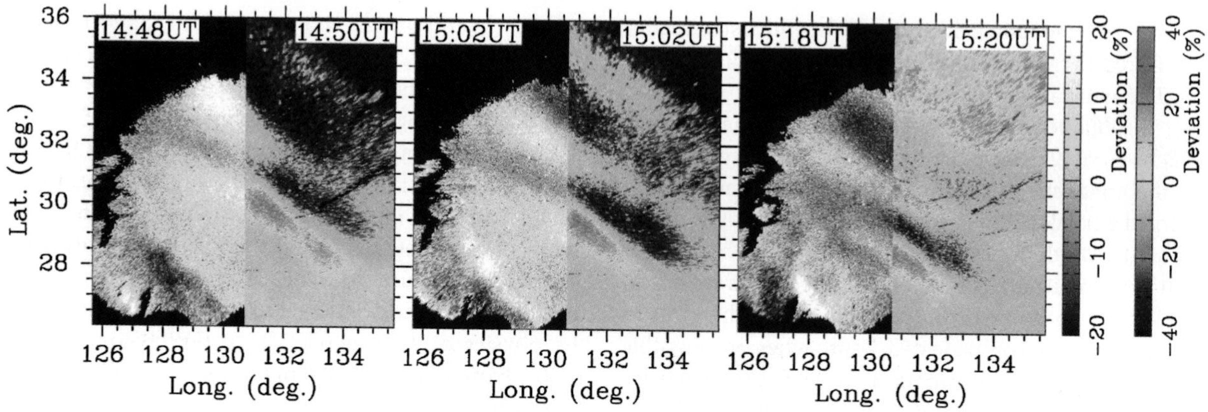

Fig. 9.10. Two dimensional maps of 630.0 nm airglow intensity at Sata (*left*) and Darwin (*right*) at three consecutive times. The observed all-sky images are converted to geographical coordinates assuming that the emission layer is at 250 km height. Color levels show percentage departure from the one-hour average background of the airglow. The Darwin image is mapped along the geomagnetic field line from the southern to northern hemisphere and superimposed onto the right hand side of each corresponding image. The wavy airglow structures caused by MSTID are smoothly connected between the two images in conjugate locations [after Otsuka et al., 2004; The American Geophysical Union]

ticated optical techniques with respect to composition, temperature and even winds of the lower thermosphere (Emmert et al., 2004). After the UARS mission, the Thermosphere, Ionosphere, Mesosphere, Energetics and Dynamics (TIMED) spacecraft has observed the mesosphere and lower thermosphere/ionosphere (MLTI) region up to the present time (December 2005) since its launching on 7 December, 2001. TIMED investigates thermal balance, dynamics, and chemistry of the MLTI region, which is sensitive to influences from the sun above and the atmospheric layers below (e.g. DeMajistre et al., 2004; Woods et al., 2005).

As mentioned in Sect. 9.1, NO, nitric oxide, is important for energetics in the thermosphere. Recent TIMED observations have also revealed the behavior of nitric oxide and its cooling effects in the lower thermosphere. During solar storm events, Mlynczak et al. (2003) observed large infrared radiance enhancements at 5.3 $-\mu$m emitted from nitric oxide in the polar thermosphere with the sounding of the atmosphere using broadband emission radiometry (SABER) experiment on the TIMED satellite. The 5.3 $-\mu$m emission by nitric oxide is indicative of the conversion of solar energy to infrared radiation within the atmosphere, and the emission has the role of a "natural thermostat" by which heat and energy are efficiently lost from the thermosphere to space and to the lower atmosphere. In addition, the nitric oxide density in the lower thermosphere has been measured from the polar-orbiting SNOE satellite as a function of latitude, longitude, and altitude, simultaneously with the solar soft X-ray. Barth and Bailey (2004) showed an excellent correlation of the model calculations of nitric oxide by using the daily variable solar soft X-ray data and the SNOE observations at 110 km between 30° S and 30° N, which supports the hypothesis that the solar soft X-rays are the source of the variability of nitric oxide in the thermosphere at low latitudes (see Table 9.1).

A highlight of satellite observations of thermosphere dynamics so far may be the global tidal structures observed with UARS and, more recently, with TIMED. The high resolution Doppler imager on board UARS is a Fabry–Perot interferometer used for observing the Doppler shift of O_2 atmospheric band emission, which provides wind velocity from the stratosphere to the lower thermosphere (in 65 – 115 km in the daytime and only at 95 km altitude in the nighttime (Hays et al., 1993). Temperatures were determined from two rotational lines (Ortland et al., 1998). This observation has made it possible to clarify the height structure of the migrating diurnal first mode Θ (1,1,1) (Kato, 1980b). The amplitude showed semiannual variation with equinoctial maxima, with a peak amplitude observed to be 70 m/s at 95–100 km altitude, which, as understood through (9.9'), could accelerate zonal winds up to 5–25 m/s in the lower thermosphere (Lieberman and Hays, 1994). On the other hand, the wind imaging interferometer (WINDII) of UARS is a Michelson interferometer (Shepherd, 2002) that can observe mesosphere and lower thermosphere winds (85–110 km) during the day and at night, using the 557.7 nm emission. This was helpful for revealing the structure of diurnal tides with latitudes, longitudes and heights up to the lower thermosphere. Latitudinal structure and seasonal variation of diurnal and semi-diurnal tides were reported by McLandress et al. (1996). They discussed difference of seasonal variation of diurnal tides between subtropical regions and mid-latitude and also the significance of the antisymmetric semi-diurnal mode Θ (2, 2, 3) in the equatorial thermosphere.

Observational data of diurnal tides at 95 km by UARS with HRDI and WINDII was Fourier-analysed at various longitudes, deriving seasonal and latitudinal variation of migrating diurnal tides with $\omega = \Omega$ and $s = 1$, and non-migrating diurnal tides with $\omega = \Omega$ but with three different wavenumbers as $s = 0$ (standing), $s = 2$ (westward propagating), $s = -3$ (eastward propagating). These complex non-migrating qtides were found to have maximum amplitude of 10 m/s (Fig. 9.11a; Forbes et al., 2003). Although the amplitude of the non-migrating tides of these wave numbers was not as strong as that of migrating diurnal tides with $s = 1$, superposition of the migrating and non-migrating tides may cause significant variability in longitude in the mesosphere and lower thermosphere region at the mid and low latitudes as understood from the diurnal eastward tidal winds in Fig. 9.11b. Northward winds of the non-migrating tides (Fig. 9.11a) at 95 km in August are larger in the southern hemisphere than those in the northern hemisphere. However, the superposition of larger migrating tides (Fig. 9.11b) particularly in northward winds seems to produce almost symmetrical northward winds about the equator (top panel in Fig. 9.11b).

Broadband emission radiometry (SABER) of the TIMED satellite observes the temperature between 20 and 120 km altitudes during the day and at night (Mlynzcak, 1997) and the TIMED Doppler interferometer (TIDI) observes Doppler wind between 70 and 115 km during the day and between 80 and 105 km at night with four telescopes (Wu et al., 2006). TIMED observes tides from the south pole to the north pole which due to the high inclination orbit, is advantageous over UARS whose orbit inclination is lower, restricting observation to only mid and low latitudes.

Although the satellite observations mentioned above are powerful in clarifying tidal characteristics in the lower thermosphere, it should be noted that long time windows such as 30–60 days are required to derive tides without ambiguity and, therefore, short time variations (e.g. Nakamura et al., 1997) cannot be observed from a single satellite. Coordination with the ground-based observations such as MF and meteor radars is very important for clarifying such precise behaviors of tides.

The basic limit in the time resolution for single satellite observation has recently been removed with GPS (global positioning system), which is based on 24 satellites simultaneously in orbit at heights of about 20,000 km. They constantly send microwave signals that can be received at any location on the Earth; at least four satellites can be seen to identify the receiving location three-dimensionally. The received signals of GPS can give TEC (total electron content) overhead at many locations, an observation that contributes significantly to observing MSTID in good time and space resolution (e.g. Saito et al., 2001).

9.5 Dynamics of the Polar Thermosphere

Since dynamics of the polar thermosphere is distinct from dynamics of the middle and lower latitude thermosphere, it would be worth devoting a separate section to a discussion about the region. Although dynamics of the polar thermosphere, controlled by plasma, is not of our main concern in this chapter, a certain basic understanding of the plasma control on the polar thermosphere dynamics would be important.

Magnetospheric energy inputs to the ionosphere-thermosphere are known to significantly affect the dynamics of energetics and the composition in the polar thermosphere. At high latitudes a strong ionospheric convection is driven by the magnetospheric convection associated with magnetic storms, leading to a strong thermospheric convection through ion drag. Acceleration of the neutral particle flow by the ion drag force is proportional to the ion density times the ion velocity accelerated by magnetospheric electric field. Note that

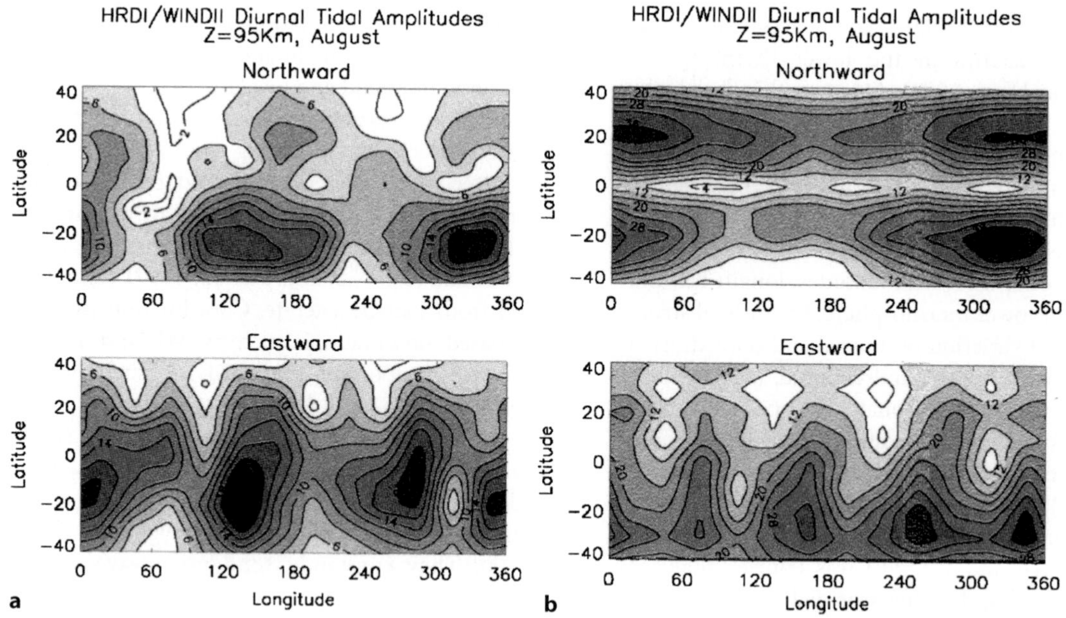

Fig. 9.11. (**a**) Latitude versus longitude contours of the amplitude (m/s) of northward (*top*) and eastward (*bottom*) non-migrating diurnal tides at 95 km during August, reconstructed from those with $s = 0$, 2 and -3 components extracted from UARS/HRDI and WINDII measurements. (**b**) Same as (**a**) except for adding migrating diurnal tides with $s = 1$ to those non-migrating in (**a**) [after Forbes et al., 2003; The American Geophysical Union]

the ion density increase due to precipitation must be taken into consideration. Moreover, physical and chemical coupling processes between ions and neutral particles play a very important role in determining the structure of the polar thermosphere and ionosphere. However, a serious difficulty in this study is the lack of direct measurements of their processes. Such measurements are essential for a correct understanding of the Joule heating rate and the ion drag force, which depend on collisional energy and momentum exchange, and chemical reactions between plasma and neutral particles. Under the circumstances, for simulation, a certain parameterization of the ion-neutral coupling is useful and practical based on satellite observation.

The dynamics explorer (DE) satellites have enabled us to understand experimentally high latitude ion-neutral coupling processes. Due to the low altitude polar orbit and comprehensive instrumentation of DE-2 spacecraft, comprehensive measurements of the thermosphere and the ionosphere have been made successfully (Hoffman et al., 1981). The comprehensive set of measurements have provided direct information in detail about dynamical, energetic, and chemical coupling processes between the thermospheric neutral atmosphere and the ionospheric plasma. Typical data obtained by DE-2 is illustrated in Fig. 9.12a, where in the top panel the ion and neutral particle velocity vectors are shown with respect to the satellite trajectory in the horizontal plane for orbit 1174 as a function of time and latitude. Simultaneously, in the second panel, measured electron, ion, and neutral temperatures are given and denoted, respectively, as $T_{e,i,n}$. In the third panel, the atomic oxygen density $n(O)$, the molecular oxygen $n(N_2)$ and the electron density ne are shown. The bottom panel shows the calculated ion-neutral time constant τ_{ni} to give the frictional and collisional heat exchange between the neutral particles and plasma (Killeen et al., 1984).

After the success of DE observations, the effects of magnetospheric convection and auroral heating on thermospheric motions have been widely investigated using various optical and radar techniques, as well as theoretical models (e.g. Fujiwara et al., 1996). Global-scale numerical models, however, do not always provide re-

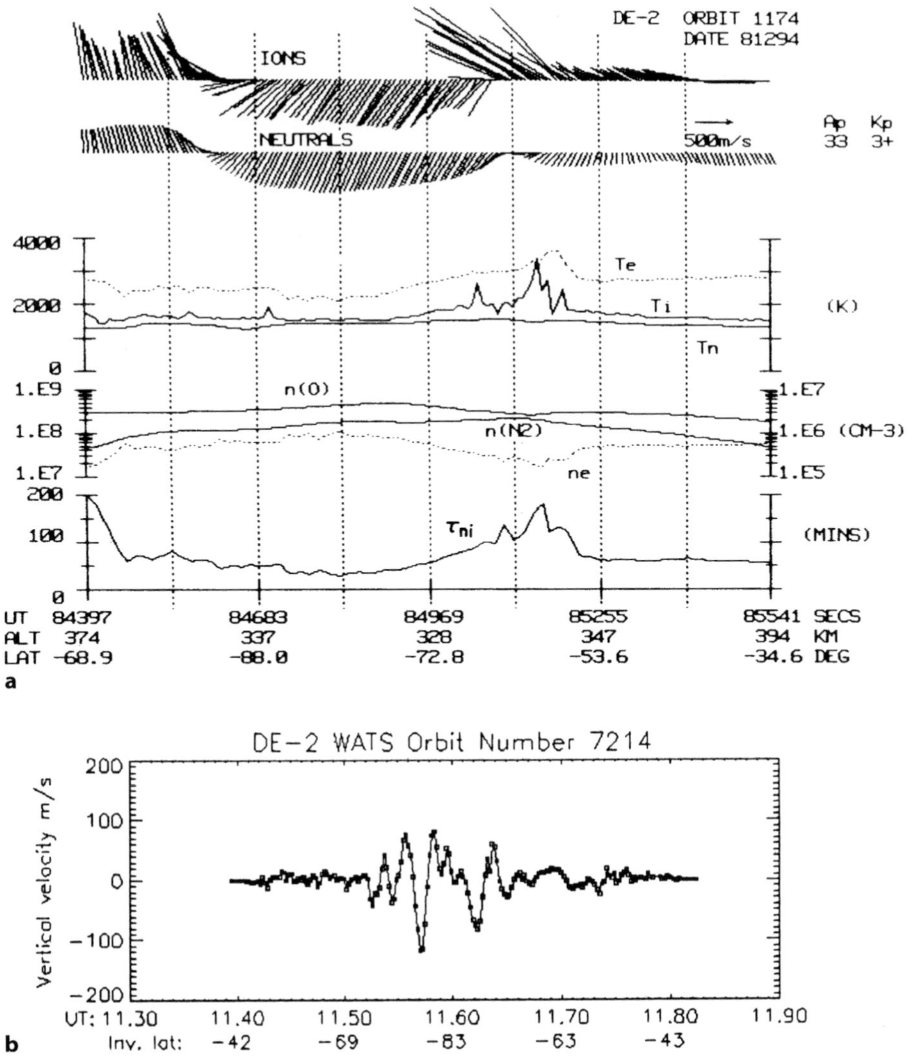

Fig. 9.12. (**a**) Geophysical observables measured along the track of Dynamics Explorer 2 during orbit 1174. The *top panel* shows horizontal velocities of the ion drift and the neutral wind with respect to the DE-2 trajectory (*horizontal axis* in the diagram) versus time, altitude, and latitude of the spacecraft. The *middle panel* shows the electron, ion, and neutral temperatures, respectively, $T_{e,i,n}$ (K) measured along the trajectory. The *bottom panel* shows the $n(O)$, $n(N_2)/cm^3$ (*left scale*) and the electron density ne/cm^3 (*right scale*) and time constant τ_{ni} in minutes [Killeen et al., 1984; The American Geophysical Union]. (**b**) Vertical velocity time series from the wind and temperature spectrometer (WATS) of Dynamics Explorer 2, orbit 7214, day 328, 1982 [Innis and Conde, 2001; The American Geophysical Union]

alistic small-scale structures of the electric field and the plasma density associated with fine structures of the auroral arc. The time-dependent thermospheric response to the energy input over small horizontal scale during highly auroral active periods has been simulated with sophisticated two-dimensional and three-dimensional high-resolution local models (e.g., Fuller-Rowell, 1985; Walterscheid and Lyons, 1992; Sun et al., 1995).

One of the major discoveries of the DE-2 spacecraft was large vertical winds in the thermosphere. Vertical wind speeds in the high latitude thermosphere occasionally reach or exceed 100 m/s (Spencer et al., 1982; Rees et al., 1984; Conde and Dyson, 1995; Innis et al. 1996, 1999, 2001; Innis and Conde, 2001). Figure 9.12(b) shows vertical velocity in time series from WATS, orbit 7214, day 328, 1982 with UT, universal time, and invariant latitudes (Kelley, p. 274, 1989). Note that the vertical velocity observed is highly fluctuating with the frequency spectrum spreading up to the infrasonic range. However, during such large vertical velocity variation in the polar region beyond latitudes 70°, there is little variation in lower latitudes. Similar large vertical winds in the polar thermosphere have also been observed with FPI (Rees et al., 1984; Eastes et al., 1992; Sica et al., 1986; Price et al., 1995; Conde and Dyson, 1995; Johnson et al., 1995; Smith, 2000; Ishii et al., 2001). Large upward and downward winds appear in the polar-side region of the auroral oval boundary, extending over several degrees in latitude and several hours in local time (Innis et al., 1999).

The observed vertical velocities are significantly stronger than the predicted theoretical models (Shinagawa et al., 2003). Such strong vertical velocities demonstrate non-hydrostatic motion in the thermosphere, resulting in significant departures from diffusive equilibrium for atomic oxygen density, molecular nitrogen and oxygen density, a situation that is very different from that shown in Fig. 9.2.

A strong expansion in the thermosphere also drives enhanced equatorward winds, which transport the composition to lower latitudes. As a result of the disturbances, so-called "negative ionospheric storms" implying electron density depletions in the *F* layer at high latitudes takes place. This is due to a decreased concentration ratio $n(\mathrm{O})/n(\mathrm{N_2})$, which decides the *F* layer electron density (Prölss, 1987).

When the heating is impulsive, equatorward "surges" or traveling atmospheric disturbances (TAD) are generated (Richmond and Matsushita, 1975, Hocke and Schlegel, 1996). TAD often take the form of LSTID, which can penetrate to lower latitudes and are seen as sequential rises in heights of the maximum F_2 layer electron density.

9.6 Concluding Remarks

Here, we would like to emphasize some of the important points made in this chapter as follows:

Firstly (see Fig. 9.2), the turbopause is known to be around 100 km height where the minimum scale of turbulence was found, with radar experiments, to be roughly 100 m. An important question now would be as to why and how the turbopause is set up at this height and with this minimum scale of turbulence for the cessation of turbulence.

Secondly (see Fig. 9.12b), intense and highly fluctuating vertical motion is observed at high latitudes and at 200–300 km altitudes in the thermosphere. Under the circumstances, in addition to hydrostatic motion, non-hydrostatic motion must be considered in thermosphere GCM. Note that the existence of various non-hydrostatic waves of free modes or resonance as shown in the KQT may be significant for the excitation, although KQT considers no viscosity.

Thirdly (see Fig. 9.10), TID is a very important ionosphere-thermosphere phenomena and the geomagnetic conjugacy for the occurrence recently found presents the novel problem as to whether TID are simple manifestations of GW in the ionosphere-thermosphere region; a question that is inconsistent with our traditional understanding.

Acknowledgement. The author of this chapter gratefully acknowledges contributions by Drs. H. Shinagawa, T. Nakamura, H. Fujiwara and A. Saito. Dr. Shinagawa' s contribution concerns dynamics of the polar thermosphere. Dr. Nakamura's contribution concerns tidal observation with satellites and Dr. Fujiwara's variation of thermosphere composition and GCM simulation. Their contribution to the revision of the mansucript is also very much appreciated. Moreover, Dr. Saito's discussion about TID observation was invaluable in preparing the work. The author thanks Profs. S. Fukao and T. Ogawa who worked on editing the revised manuscript. The very extensive and friendly assistance of all these people in completing the present work is precious to the author who had a sudden health trouble in summer in 2005. Thanks are also due to Dr. A. Richmond for reviewing the original manuscript.

References

Balsely, B.B., Gage, K.S., The MST radar techniques: potential for middle atmosphere studies, Pure and Applied Geophysics (PAGEOPH), 118 452–493, 1980

Barth, C.A., Bailey, S.M., Comparison of a thermospheric photochemical model with student nitric oxide explorer (SNOE) observations of nitric oxide, J. Geophys. Res., 109(A3), A03304, doi: 10.1029/2003JA010227, 2004

Batchelor, G.K., The Theory of Homogeneous Turbulence, pp. 46 and 114–132, Cambridge Monographs on Mechanics and Applied Mathematics, Cambridge Univ. Press 1953

Chapman, S., Bartels, J., Geomagnetism, Vol. I, pp. 214–243, Oxford At The Clarendon Press, 1940

Conde, M., Dyson, P.L., Thermospheric vertical winds above Mauson, Antarctica, J. Atmos. Terr. Phys., 57, 589-596, 1995

DeMajistre, R., Paxton, L.J., Morrison, D., Yee, J.-H., Goncharenko, L.P., Christensen, B., Retrievals of nighttime electron density from thermosphere ionosphere mesosphere energetics and dynamics (TIMED) mission global ultraviolet imager (GUVI) measurements, J. Geophys. Res., 109, A05305, doi:10.1029/2003JA010296, 2004

Dickinson, R.E., C Ridley, E., Roble, R.G., A three-dimensional general circulation model of the thermosphere, J. Geophys. Res., 86, 1499–1512, 1981

Eastes, R.W., Killeen, T.L., Wu, Q., Winningham, J.D., Hoegy, W.R., Wharton, L.E., Carignan, G.R., An experimental investigation of thermospheric structure near an auroral arc, J. Geophys. Res., 97, 10, 539–549, 1992

Emmert, J.T., Fejer, B.G., Shepherd, G.G., Solheim, B.H., Average nighttime F region disturbance neutral winds measured by UARS WIND II Initial results, Geophys. Res. Lett., 31, L22807, doi:10.1029/2004GL021611, 2004

Evans, J.V., Theory and practice of ionospheric study by Thomson scatter radar, Proc. The IEEE, 57(4), 496–530, 1969

Forbes, J.M., Hagen, M.E., Diurnal propagating tide in the presence mean winds and dissipation: A numerical investigation, Planet. Space Sci., 36, 579–590, 1988

Forbes, J.M., Zhang, X., Talaat, E.R., Ward, W., Nonmigrating diurnal tides in the thermosphere, J. Geophys. Res., 108, 1033, doi:10.1029/2002JA009262, 2003

Fujiwara, H., Maeda, S., Fukunishi, H., Fuller-Rowell, T.J., Evans, D.S., Global variations of thermospheric winds and temperatures caused by substorm energy injection, J. Geophys. Res., 101, 225–239, 1996

Fujiwara, H., Maeda, S., Suzuki, M., Nozawa, S., Fukunishi, H., Estimates of electromagnetic and turbulent energy dissipation rate under the existence of strong wind shears in the polar lower thermosphere from the European incoherent scatter (EISCAT) Svalbard radar observations, J. Geophys. Res., 109, A0736, doi:10. 1029/2003JA010046, 2004

Fukao, S., Yamanaka, M.D., Ao, N., Hocking, W.K., Sato, T., Yamamoto, M., Nakamura, T., Tsuda, T., Kato, S., Seasonal variability of vertical eddy diffusivity in the middle atmosphere 1. Three year observations by the middle and upper atmosphere radar, J. Geophys. Res., 99, 18, 973–18987 1994

Fuller-Rowell, T.J., Rees, D., A three-dimensional time-dependent global model of the thermosphere, J. Atmos. Sci., 37, 2545–2567, 1980

Fuller-Rowell, T.J., A two-dimensional, high-resolution, nested-grid model of the thermosphere 2. Response of the thermosphere to narrow and broad electrodynamic features, J. Geophys. Res., 90, 6567–6586, 1985

Fuller-Rowell, T.J., Codrescu, M.V., Risbeth, H., Moffett, R.J., Quegan, S., On the seasonal response of the thermosphere and ionosphere to geomagnetic storms, J. Geophy. Res., 101(A2), 2343–2354, 1996

Garcia, F.J., Kelley, M.C., Makela, J.J., Huang, C.-S., Airglow observation of mesoscale low-velocity traveling ionospheric disturbances at midlatitudes, J. Geophys. Res., 105, 18, 407–415, 2000

Goncharenko, L.P., Salah, J.E., Climatology and variability of semi-diurnal tides in the lower thermosphere over Milstone Hill, J. Geophys. Res., 103, 20, 715–20726, 1998

Georges, T.M., HF Doppler studies of traveling ionospheric disturbances, J. Atmos. Terr. Phys.30, 735–746, 1968.

Hagan, M.E., Forbes, J.M., Migrating and non-migrating diurnal tides in the middle and upper atmosphere excited by tropospheric latent heat release, J. Geophys. Res., 107, 4754, doi:10.1029/2001JD001236, 2002

Hardy, K.R., Atlas, D., Glober, K.M., Multiwavelength back scattering from the clear atmosphere, J. Geopyhys. Res. 71, 1537–1552, 1966

Harris, M.J., Arnold, N.F., Aylward, A.D., A study into the effect of the diurnal tide on the structure of the background mesosphere and thermosphere using the new coupled middle atmosphere and thermosphere (CMAT) general circulation model, Ann. Geophys., 20, 225–235, 2002

Hays, P.B., Abreu, V.J., Dobbs, M.E., Gell, D.A., Grassl, H.J., Skinner, W.R., The high-resolution Doppler imager on the upper atmosphere research satellite, J. Geophys., Res., 98, 10, 713–723, 1993

Hedin, A.E., A revised thermospheric model based on mass spectrometer and incoherent scatter data: MSIS-83, J. Geophys. Res. 88, 10170–10188, 1983

Hines, C.O. et al. The upper atmosphere in motion, (a) pp. 248–328, (b) pp. 363–420, (c) pp. 741–757, Geophysical Monograph Series (Spilhaus, A.F., Jr., Managing Editor), Amer. Geophys. Union, Washington D.C., 1974

Hocke, K., Schlegel, K., A review of atmospheric gravity waves and travelling ionospheric disturbances, 1982–1995, Ann. Geophys., 14, 917–940, 1996

Hoffman, R.A., Hogan, G.D., Maehl, R.C., Dynamics Explorer spacecraft and ground operations systems, Space Sci. Instrum., 5, 349–367, 1981

Holton, J.R., An Introduction to Dynamic Meteorology, 3rd Edition, pp.118–119, Academic Press, 1992, San Diego, CA

Innis, J.L., Greet, P.A., Dyson, P.L., Fabry–Perot spectrometer observations of the auroral oval/polar cap boundary above Mawson, Antarctica, J. Atmos. Terr. Phys., 58, 1973–1988, 1996

Innis, J.L., Greet, P.A., Murphy, D.J., Conde, M.G., Dyson, P.L., A large vertical wind in the thermosphere at the auroral oval/polar cap boundary seen simultaneously from Mawson and Davis, Antarctica, J. Atmos. Solar-Terr. Phys., 61, 1047–1058, 1999

Innis, J.L., Conde, M., Thermospheric vertical wind activity maps derived from Dynamics Explorer-2 WATS observations, Geophys. Res. Lett., 28, 3847–3850, 2001

Innis, J.L., Greet, P.A., Dyson, P.L., Thermospheric gravity waves in the southern polar cap from 5 years of photometric observations at Davis, Antarctica, J. Geophys. Res., 106, 15489–15499, 2001

Ishii, M., Conde, M., Smith, R.W., Krynicki, M., Sagawa, E., Watari, S., Vertical wind observations with two Fabry–Perot interferometers at Poker Flat, Alaska, J. Geophys. Res., 106, 10, 537–551, 2001

Johnson, F.S., Hanson, W.B., Hodges, R.R., Coley, W.R., Cargnan, G.R., Spencer, N.W., Gravity waves near 300 km over the polar caps, J. Geophys. Res., 100, 23, 993–24, 002, 1995

Kasahara, A., Quian, J.-H., Normal modes of a global non-hydrostatic atmospheric model, Mon. Weath. Rev. 128, 3357–3375, 2000

Kato, S., Dynamics of the upper atmosphere, (a) pp. 35–56, (b) pp. 58–97 (c) pp. 115–140 (d) pp.141–188 (e) pp. 213–215, Center Acad. Pub. Japan, D. Reidel Pub. Co Dordrecht, 1980

Kato, S., Middle atmosphere research and radar observation, Proceeding of the Japanese Academy, Ser. B, Vol. 8, 306–320, 2005

Kelley M.C., The Earth's Ionosphere, Plasma Physics and Electrodynamics, Academic Press, San Diego, CA, 1989

Killeen, T.L., Hays, P.B., Carignan, G.R., Heelis, R.A., Hanson, W.B., Spencer, N.W., Brace, L.H., Ion-neutral coupling in the high-latitude F region: Evaluation of ion heating terms from Dynamics Explorer 2, J. Geophys. Res., 89, 7495–7508, 1984

Lehmacher, G.A., Croskey, C.L., Michell, J.D., Friedrich, M., Lueben, F.-J., Rapp, M., Kudeki, E., Fritts, D.C., Intense turbulence observed above a mesospheric temperature inversion at equatorial latitude, Geopys. Res. Lett., 33, 8, L08808, DOI 10. 1029/2005GL024345, 2006

Lieberman, R.S., Hays, P.B., An estimation of the momentum deposition in the lower thermosphere by the observed diurnal tide, J. Atmos. Sci., 51, 3094–3105, 1994

Lindzen, R.S., Blake, D., Mean heating of the thermosphere by tides, J. Geophys. Res. 75, 6868–6871, 1970

Maeda, S., Fuller-Rowell, T.J., Evans, D.S., Zonally averaged dynamical and compositional response of the thermosphere to auroral activity during September 18–24, 1984, J. Geophys. Res., 94(A12), 16,869–16,883, 1989.

Makela, J.J., Kelley, M.C., Using the 630.0-nm nightglow emission as a surrogate for the ionospheric Pedersen conductivity, J. Geophys. Res., 108, 1253, doi:10.1029/2003JA009894, 2003

McLandress, C., Shepherd, G.G., Solheim, B.H., Satellite observations of thermospheric tides: Results from wind imaging interferometer on UARS, J. Geophys. Res., 101, 4093–4114, 1996

Millward, G.H., Rishbeth, H., Fuller-Rowell, T.J., Aylward, A.D., Quegan, S., Moffett, R.J., Ionospheric F2 layer seasonal and semiannual variations, J. Geophys. Res., 101(A3), 5149–5156, 1996

Miyahara, S., Ooishi, M., Variation Sq induced by atmospheric tides simulated by a middle general circulation model, J. Geomag. Geoelectr., 49, 77–87, 1997

Miyoshi, Y., Fujiwara, H., Day-to-day variations of migrating diurnal tide simulated by a general circulation model, Geophys. Res. Lett. 30, 1789, doi:10.1029/2003GL017695, 2003

Miyoshi, Y., Fujiwara, H., Day-to-day variations of migrating semi-diurnal tide simulated by a general circulation model, Advance Polar Upper Atmos. Research, 18, 87–95, 2004

Mlynczak, M.G., Energetics of the mesosphere and lower thermosphere and the SABER experiment, Adv. Space Res., 20, 1177–1183, 1997

Mlynczak, M., Martin-Torres, F.J., Russell, J., Beaumont, K., Jacobson, S., Kozyra, J., Lopez-Puertas, M., Funke, B., Mertens, C., Gordley, L., Picard, R., Winick, J., Wintersteiner, P., Paxton, L., The natural thermostat of nitric oxide emission at 5.3 μ m in the thermosphere observed during the solar storms of April 2002, Geophys. Res. Lett., 30, doi: 10.1029/2003GL017693, 2003

Nakamura, J., Kimura, H., Matsuoka, T., Aso, T., Kato, S., Wind shear and electron density measurements by rocket, Space Res, XII, 1329–1333, 1972

Nakamura, T., Fritts, D.C., Isler, J.R., Tsuda, T., Vincent, R.A., Reid, I.M., Short period fluctuations of the diurnal tide observed with low-latitude MF and meteor radars during CADRE: Evidence for gravity wave/tidal interactions, J. Geophys. Res., 102, 26225–26239, 1997

Namgaladze, A.A., Korenkov, Yu.N., Klemenko, V.V., Karpov, I.V., Bessarab, F.S., Surotkin, V.A., Gulushchenko, T.A., Naumova, N.M., Global Model of the thermosphere-ionosphere-protonosphere, PAGEOPH, 127, 219–254, 1988

Nozawa, S., Liu, H.-L., Richmond, A.D., Roble, R., Comparison of the auroral E region neutral winds derived with the European incoherent scatter radar and predicted by the National Center for Atmosphere Research Thermosphere-Mesosphere-Electrodynamics general circulation model, J. Geophys. Res., 106, 24, 691–700, 2001

Oliver, W.L., Fukao, S., Sato, T., Tsuda, T., Kato, S., Kimura, I., Ito, A., Saryo, T., Araki, T., Ionospheric incoherent scatter measurements with the MU radar: Observation of F-region electrodynamics, J. Geomag. Geoelectr., 40, 963–985, 1988

Ortland, D.A., Hays, P.B., Skinner, W.R., Yee, J.-H., Remote sensing of mesospheric temperature and 02(1sigma) band volume emission rates with the High Resolution Doppler Imager, J. Geophys. Res., 103, 1821–1835, 1998.

Otsuka,Y., Shiokawa, K., Ogawa, T., Wilkinson, P., Geomagnetic conjugate observation of medium-scale traveling ionospheric disturbances at midlatitude using all-sky airglow imagers, Geophys. Res. Lett., 31, L15803, doi: 10.1029/2004GL020262, 2004

Picone, J.M., Hedin, A.E., Drob, D.P., Aikin, A.C., NRLMSISE-00 empirical model of the atmosphere: Statistical comparisons and scientific issues, J. Geophys. Res., 107(A12), 1468, doi:10.1029/2002JA009430, 2002

Price, G.D., Smith, R.W., Hernandez, G., Simultaneous measurements of large vertical winds in the upper and lower thermosphere, J. Atmos. Terr. Phys. 57, 631–643, 1995

Prölss, G.W., Storm-induced changes in the thermospheric composition at middle latitudes, Planet. Space Sci. 35, 807–811, 1987

Rastogi, P.K., Bowhill, S.A., Scattering of radio waves from the mesosphere–2, Evidence for intermittent mesospheric turbulence, J. Atmos. Terr. Phys., 38, 449–462, 1976

Rees, D., Smith, R.W., Charleton, P.J., McCormac, F.G., Lloyd, N., Steen, A., The generation of vertical thermospheric winds and gravity waves at auroral latitudes I. Observations of vertical winds, Planet. Space Sci., 32, 667–684, 1984

Richmond, A.D., Matsushita, S., Thermospheric response to a magnetic substorm, J. Geophys. Res., 80, 2839–2850, 1975

Roble, R.G., Ridley, E.C., Dickinson, R.E., On the global mean structure of the thermosphere, J. Geophys. Res., 92(A8), 8745–8758, 1987

Roble, R.G., Ridley, E.C., Richmond, A.D., Dickinson, R.E., A coupled thermosphere/ionosphere general circulation model, Geophys. Res. Lett., 15, 1325–1328, 1988

Roble, R.G., Ridley, E.C., A thermosphere-ionosphere-mesosphere-electrodynamics general circulation model (time-GCM): Equinox solar cycle minimum simulations (30–500 km), Geophys. Res. Lett., 21, 417–420, 1994

Saito, A., Iyemori, T., Sugiura, M., Maynard, N.C., Aggson, T.L., Brace, L.H., Takeda, M., Yamamoto, M., Conjugate occurrence of the electric field fluctuations in the nighttime midlatitude ionosphere, J. Geophys. Res., 100, 21, 439–451, 1995

Saito, A., Nishimura, M., Yamamoto, M., Fukao, S., Kubota, M., Shiokawa, K., Otsuka, Y., Tsugawa, T., Ishi, M., Sakanoi, T., Miyazaki, S., Traveling ionospheric disturbances detected in the FRONT campaign, Geophys. Res. Lett., 28, 682–692, 2001

Shepherd, G.G., Spectral imaging of the atmosphere, pp. 102–128, 191–196, Academic Press, San Diego, CA, 2002

Shinagawa, H., Oyama, S., Nozawa, S., Buchett, S.C., Fujii, R., Ishii, M., Thermospheric and ionospheridynamics in the auroral Region, Adv. Space Res. 31, 951–956, 2003

Sica, R.J., Rees, M.H., Romick, J., Hernandez, G., Roble, R.G., Auroral zone thermospheric dynamics 1. Averages, J. Geophys. Res., 91, 3231–3244, 1986

Smith, R.W., The global-scale effect of small-scale thermospheric disturbances, J. Atmos.- Solar-Terr. Phys., 62, 1623–1628, 2000

Spencer, N.W., Wharton, L.E., Carignan, G.R., Maurer, J.C., Thermosphere zonal winds, vertical motions and temperature as measured from Dynamics Explorer, Geophys. Res. Lett., 9, 953–956, 1982

Sun, Z.P., Turco, R.P., Walterscheid, R.L., Venkateswaran, S.V., Jones, P.W., Thermospheric response to morningside diffuse aurora: High-resolution three-dimensional simulations, J. Geophys. Res., 100, 23, 779–793, 1995

Sutton, E.K., Forbes, J.M., Nerem, R.S., Global thermospheric neutral density and Wind response to severe 2003 geomagnetic storms from CHAMP accelerometer data, J. Geophys. Res., 110, A09S40, 1–10, doi:10.1029/2004JA010985, 2005

Tohmatsu, T. Compendium of aeronomy, pp. 457–501, Ter. Sci. Pub. Co., Tokyo, Kluwer Acad. Pub., Dordrecht, 1990

Villars, F., Weisskopf, V.F., The scattering of electromagnetic waves by turbulent atmospheric fluctuations, Phys. Rev., 94, 232–240, 1954

Vincent, R.A., Reid, I.M., HF Doppler measurements of mesosphere gravity wave momentum flux, J. Atmos. Sci., 40, 1321–1333, 1983

Walterscheid, R.L., Lyons, L.R., The neutral circulation in the vicinity of a stable auroral arc, J. Geophys. Res., 97, 19, 489–499, 1992

Woods, T.N., Eparvier, F.G., Bailey, S.M., Chamberlin, P.C., Lean, J., Rottman, G.J., Solomon, S.C., Tobiska, W.K., Woodraska, D.L., Solar EUV experiment (SEE): Mission overview and first result, Geophys. Res. 110, A01312, doi: 10,1029/2004JA010765, 2005

Wu, Q., Killeen, T.L., Ortland, D.A., Solomon, S.C., Gablehouse, R.D., Johnson, R.M., Skinner, W.R., Niciejewski, R.J., Franke, S.J., TIMED Doppler interferometer (TIDI) observations of migrating diurnal and semidiurnal tides, J. Atmos. Solar-Terr. Space, in print, 2006.

Part 3

Space Plasmas

10 Space Plasmas

Chanchal Uberoi

The theoretical understanding of space plasmas which are basically tenuous, magnetized, and have large characteristic dimensions, depends on the study of the interplay between the particle and fluid behaviors. The motion of the individual charged particle in the electric and magnetic fields and the dynamics of the MHD fluid plasmas are discussed with an emphasis on the study of currents arising due to drifting of particles, magnetization and collisions and the interaction of the magnetic fields with plasmas.

The kinetic description of plasma is necessary to understand the wave-particle interactions giving rise to many new phenomena exclusive to plasmas.

Plasma waves are observed in space and this is a much studied topic. The basics of electromagnetic, electrostatic and magnetohydrodynamic waves are discussed. The kinetic theory of waves is given mainly to show the results arising due to resonant wave-particle interactions such as Landau-damping.

The interface between two plasma regions can exist in an equilibrium state in space. The study of surface waves and their instability plays a very important role in many dynamical processes in the magnetosphere. The resonant absorption of Alfvén surface waves along the diffuse boundaries in a non-dissipative system is a low frequency phenomenon similar to collisionless Landau damping of high frequency plasma waves.

Two-stream instability involves high frequency plasma oscillations and an important concept of negative energy waves is associated with the excitation of this instability.

Tearing mode instability is of interest in understanding of the magnetic reconnection processes in space plasmas.

Contents

Chanchal Uberoi, Space Plasmas.
In: Y. Kamide/A. Chian, Handbook of the Solar-Terrestrial Environment. pp. 249–278 (2007)
DOI: 10.1007/11367758_10

10.1 Characteristic Properties of Plasmas

The name plasma was first used by the Nobel laureate, Irving Langmuir, in 1928 (Langmuir, 1928) to describe the glow in the positive column region of a discharge tube containing ions and electrons in about equal number so that the resultant space charge was small. The man-made plasmas formed in the gas discharge can be seen in mercury vapor rectifiers, in electric arcs and in neon and fluorescent lamps. Other examples of man-made plasmas are those seen in explosions and strong shock waves and in fire flames. These studies of man-made plasmas find various applications in the laboratory and industrial processes.

Earlier in 1920 the Indian physicist Meghnath Saha (Saha, 1920) in his studies related to the nature of matter in the interior of the stars gave the well known Saha's ionization equation for the relation between the temperature and the degree of ionization. Using this relation it was understood that plasma is a normal state of matter at temperature of 10,000 ° K or more. Equilibrium between gas and plasma states is a function of temperature similar to liquid-gas. Therefore, it is appropriately called the *fourth state of matter*. It is the most common state of matter in nature. The Sun and the other stars can be considered as lumps of hot plasma. The matter in the solar corona, interstellar space and nearer home, all the matter surrounding our atmosphere at a distance of about 90 km from the surface of the Earth and above is in a plasma state. Nature's plasma covers a broad range of temperature and densities. Figure 10.1 shows typical parameters that are characteristic of some of the plasmas in nature. Interestingly, auroras and lightning are the oldest studied effects of the plasma environment of the Earth.

Basic Definition The existence of plasma, an assembly of charged particles, as a stable and natural state of matter, can be explained by the important fact that in a plasma the interactions of particles is mainly by the long-range Coulomb forces, unlike the case of neutral gases where molecules interact with the short-range forces only. For instance, the electric field due to a point charge decreases only as the cube of the linear dimension, so the dominant interactions in plasma are such that a charged particle interacts with many or all other particles simultaneously. Thus, it is possible for the electric (and magnetic) fields of an assembly of charged particles to act together in a coherent way giving rise to strong cooperative plasma behavior. The dominance of collective particle interactions over the close binary collisions allows an ionized gas to exist as plasma.

The basic definition of plasma therefore reads as follows: *Plasma is a collection of charged particles that is sufficiently dense so that space charge effects can result in strongly coherent behavior.*

Quantitative Criterion The presence of collective effects constitutes the primary plasma criterion. A quantitative measure of this criterion may be obtained from a determination of the distance to which the electric field of an individual charged particle extends before it is effectively shielded by the oppositely charged particles in the neighborhood. Such a calculation was first performed by Debye for an electrolyte. Assuming a large number of neighboring particles so that the electric field can be taken as a continuous function of distance, the shielding distance deduced by Debye is

$$\lambda_{\mathrm{D}} = (\mathrm{K}T/4\pi nq^2)^{1/2}\,, \tag{10.1}$$

where $\mathrm{K} = 1.38 \times 10^{-16}$ ergs/degree Kelvin is the Boltzmann constant with T as the effective temperature in degrees Kelvin (K) of the particles of charge q and density n per unit volume. Although the precise applicability of Debye's results in an ionized gas is open to question, the Debye shielding distance does provide a measure of the distances over which the influence of an individual charged particle is dominant. Beyond λ_{D}, the electric field, and hence the influence, of the individual particle is nil, and collective effects dominate. The quantitative criterion for the existence of a plasma is that the linear dimension L of the system should be larger than a Debye length, $L \gg \lambda_{\mathrm{D}}$. If $q = \mathrm{e}$, the electron charge, in (10.1) we obtain

$$\lambda_{\mathrm{D}} = 740 \left(\frac{\mathrm{K}T_{\mathrm{e}}}{n}\right)^{1/2} \mathrm{cm}, \tag{10.2}$$

where $\mathrm{K}T_{\mathrm{e}}$ expresses the energy of random motion in electron volts and is given by

$$\mathrm{K}T_{\mathrm{e}} = 1 \text{ at } T = 11{,}600\,\mathrm{K}\,.$$

Using (10.2), λ_{D} can be calculated for different plasmas characterized by the parameters n_{e} and $\mathrm{K}T_{\mathrm{e}}$. Note that λ_{D} is independent of the particle mass. Also no assumption is made about the absence of neutral particles; the term plasma is independent of the degree of ionization.

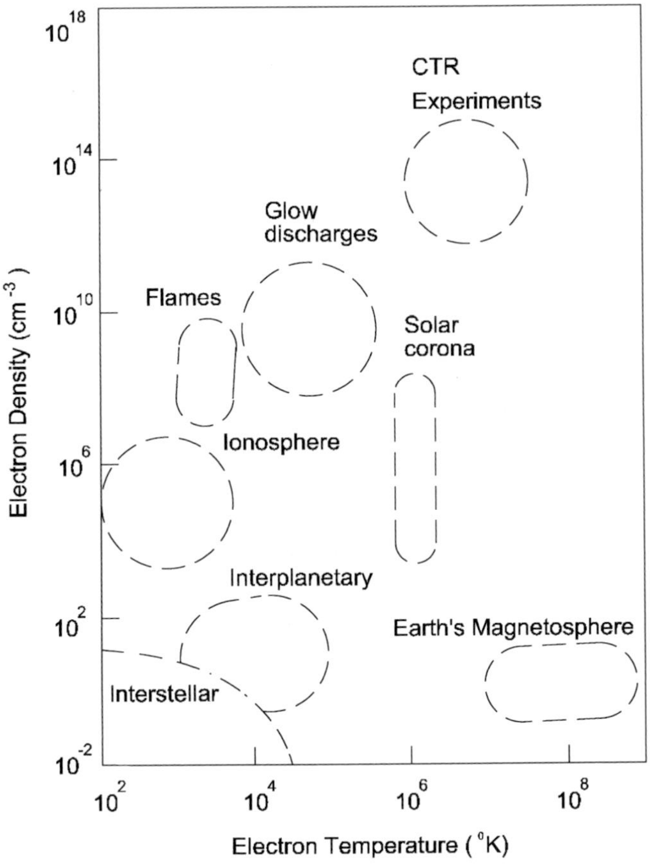

Fig. 10.1. Plasma parameters for a variety of natural plasmas in terms of electron density and temperature. For comparison, laboratory plasmas used for controlled thermonuclear reaction (CTR) experiments are also shown

Plasma Parameter The Debye shielding concept is statistically valid only if there are enough particles within a sphere of radius λ_D, the Debye sphere. The number of particles N_D in a Debye sphere are

$$N_D \equiv \frac{4}{3}\pi n \lambda_D^3 = 1380 T^{3/2}/n^{1/2} . \qquad (10.3)$$

Therefore, in addition to $\lambda_D \ll L$, the collective behavior requires $N_D \gg 1$. This condition is also given in terms of the plasma parameter $g = 1/N_D$ as $g \ll 1$.

As $g \sim n^{1/2}/T^{3/2}$ and since the collision frequency ν of binary collisions decreases with decreasing density n and with increasing temperature T, a smaller g corresponds to less collisions and in the limit $g \to 0$ the plasma becomes collisionless. For space plasmas very commonly density is low and the temperature is high, so the plasma can be treated in a collisionless limit.

Plasma Neutrality The definition $L > \lambda_D$ also means that the potential that is set up when an electron with an average energy of KT_e moves away from an ion decreases very rapidly at distance $r > \lambda_D$ and leaves the bulk of plasma free of large electric potentials. This means that a plasma has the tendency to remain neutral. If n_e is the average electron density and n_i denotes the density of positive ions with charges q_i per particle, we have

$$|n_e - \sum_i q_i n_i| \ll n_e , \qquad (10.4)$$

which is the quasi-neutrality condition and is sometimes used as a more limited definition of the term plasma.

Plasma Oscillations The plasma electron oscillations arise as a consequence of the property of the plasma to try to remain neutral. If the electrons in a plasma are displaced from a uniform background of ions, electric fields will be built up in such a direction so as to restore the neutrality of the plasma by pulling the electrons back to their original position. Because of their inertia, the electrons will overshoot and oscillate around their equilibrium positions with a characteristic frequency namely the radian plasma frequency

$$\omega_{\mathrm{p}} = \left(\frac{4\pi n_{\mathrm{e}}^2}{m_{\mathrm{e}}} \right)^{1/2} . \tag{10.5}$$

The quantity ω_{p} is the characteristic oscillation rate of electrostatic disturbances in plasmas. The numerical value f_{p}, the frequency in cycles per sec can be approximated as:

$$f_{\mathrm{p}} = \frac{\omega_{\mathrm{p}}}{2\pi} \approx 9000\sqrt{n_{\mathrm{e}}}\ \mathrm{Hz}. \tag{10.6}$$

where n_{e} is expressed in cm^{-3}. Collisions between electrons and ions will tend to damp these collective oscillations. Therefore for collective behavior to exist $\omega_{\mathrm{p}}\tau > 1$, where τ is the collision time.

Langmuir discussed plasma oscillations for the first time in 1928 (Langmuir, 1928; Langmuir and Tonks, 1929). In fact, he found that the characteristic behavior of the jelly-like movement of the group of charged particles was similar to that of blood plasma and this led him to use the name "plasma". Plasma frequency is also sometimes called Langmuir frequency.

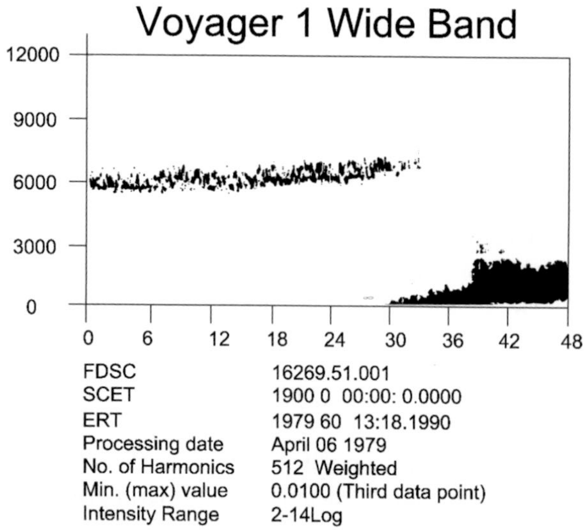

Fig. 10.2. Jovian plasma oscillations: Frequency-time diagram of the electron plasma oscillations detected in Jupiter's magnetosphere by the spacecraft *Voyager*. The plasma probe picked up the signals on 1 March 1979, which when plotted on the frequency-time graph, showed constant frequency oscillations. A frequency of 6000 Hz was calculated to be the plasma frequency in the vicinity of the spacecraft. After the 33 seconds mark, plasma turbulence features are seen

Figure 10.2 shows the frequency-time diagram of electron plasma oscillations detected by Voyager 1 as it crossed Jupiter's bow shock on 1 March 1979.

Low frequency ion oscillations will be discussed in the section on plasma waves.

Summary of Conditions For a plasma to exist we should have $\lambda_{\mathrm{D}} \ll L$, $N_{\mathrm{D}} \gg 1$ and $\omega_{\mathrm{p}}\tau > 1$.

10.2 Particles in Space Plasmas

The interaction of plasma with electromagnetic fields prevalent in space is very important for understanding space plasmas. In order to deal with plasma dynamics in the presence of magnetic and electric fields, it is necessary and important to first understand the behavior of a single charged particle in these fields. This is not only required for an understanding of various space plasma processes, but also for studying the various current systems, and the trapping and acceleration of charged particles.

10.2.1 Motion of Charged Particles in a Uniform Magnetic Field

The Lorentz equation of motion for a general particle of mass m and charge q in electric field $\boldsymbol{E}$ and magnetic field $\boldsymbol{B}$ is given as

$$m\frac{\mathrm{d}\boldsymbol{v}}{\mathrm{d}t} = q\left(\boldsymbol{E} + \frac{1}{c}\boldsymbol{v} \times \boldsymbol{B}\right) \tag{10.7}$$

when $\boldsymbol{E} = 0$ and $\boldsymbol{B} = (0, 0, B)$ is a uniform field. Equation (10.7) gives

$$v_z = \mathrm{const}\frac{\mathrm{d}^2 v_x}{\mathrm{d}t^2} = -\omega_c^2 v_x\,, \quad \frac{\mathrm{d}^2 v_y}{\mathrm{d}t^2} = -\omega_c^2 v_y\,, \tag{10.8}$$

with $\omega_c = qB/mc$ the *Larmor or gyrofrequency*. This shows that a particle with initial velocity $(\boldsymbol{v}_\parallel, \boldsymbol{v}_\perp)$, along and transverse to the magnetic field direction will have constant motion along the magnetic field and a circular trajectory about a point (x_0, y_0) determined by the initial conditions, with a radius r_{L}, the *Larmor radius*

$$r_{\mathrm{L}} = v_\perp/\omega_c = \frac{m v_\perp c}{qB} . \tag{10.9}$$

Thus the trajectory of a charged particle in the presence of a uniform magnetic field is a helix with its axis parallel to $\boldsymbol{B}$. The point $(x_0, y_0, z_0, +v_\parallel t)$ describes the locus of the center of the circle and is called a *guiding center*.

The *pitch angle* of the helical motion is the angle between $\boldsymbol{v}$ and $\boldsymbol{B}$. It is defined as

$$\alpha = \tan^{-1} v_\perp / v_\parallel \,. \tag{10.10}$$

Note that $v_\parallel = v \cos\alpha$, $v_\perp = v \sin\alpha$, where v, $v_\parallel$ and $v_\perp$ are magnitudes of these vectors.

The plasma in a magnetic field, therefore, has a new characteristic length scale r_L and the time scale ω_c. For most plasmas in space $r_\mathrm{L} \ll L$ and also $\tau \ll \omega_c$. In this case, the guiding center theory, also known as the first order orbit theory, provides a valid picture of the behavior of particles. Although the study of the charged particles in the Earth's dipole magnetic field dates back to 1895 when the Norwegian physicist Kristian Birkeland performed geophysical experiments in the laboratory to understand auroras. The guiding center approximation was first used by the Nobel laureate Hannes Alfvén in 1950 [for details, see (Alfvén and Fälthammar, 1963)]. This method approximates the particle orbits in inhomogeneous magnetic fields to first order by a perturbation technique yielding information on particle orbits in the average sense. Another aspect of this theory is that it is very helpful in describing adiabatic particle motions in complex electromagnetic fields providing important adiabatic invariants of motion. The invariants have been very useful in defining trapped and precipitated particles in radiation belts [see, e.g. (Schulz and Lanzerotti, 1974)].

10.2.2 Particle Drifts

We now consider the motion of a charged particle moving in a magnetic field $\boldsymbol{B}$ but subject to various perturbations, such as the presence of an electric field, a small spatial inhomogeneity or a slow change in time, of $\boldsymbol{B}$. In such cases, the motion can be described approximately as gyration around the guiding center. The motion of the guiding center transverse to $\boldsymbol{B}$ is called a drift of the particle.

Electric Drift: $\boldsymbol{E} \times \boldsymbol{B}$ Drift Consider $\boldsymbol{E}$ to lie in the $x - z$ plane so that $E_y = 0$. In this case, (10.7) with same magnetic field direction will give a straightforward acceleration along $\boldsymbol{B}$

$$v_z = q\frac{E_x}{m}t + v_{z0}\,, \tag{10.11}$$

and the equations for v_x and v_y are

$$\frac{\mathrm{d}^2 v_x}{\mathrm{d}t^2} = -\omega_c^2 v_x\,, \quad \frac{\mathrm{d}^2 v_y}{\mathrm{d}t^2} = -\omega_c^2\left(\frac{cE_x}{B} + v_y\right). \tag{10.12}$$

The equation in v_y can be written as

$$\frac{\mathrm{d}^2}{\mathrm{d}t^2}\left(v_y + \frac{cE_x}{B}\right) = -\omega_c^2\left(v_y + \frac{cE_x}{B}\right). \tag{10.13}$$

Equation (10.13) gives Larmor motion plus a superimposed drift v_E of the guiding center in the $-y$ direction (for $E_x > 0$) (Fig. 10.3).

The general formula for electric drift can be written as

$$\boldsymbol{v}_\mathrm{E} = \frac{c\boldsymbol{E} \times \boldsymbol{B}}{B^2} = \frac{c}{q}\frac{\boldsymbol{F} \times \boldsymbol{B}}{B^2}\,, \tag{10.14}$$

where $\boldsymbol{F}$ = the electric force $q\boldsymbol{E}$. Substitution of various other forces for $\boldsymbol{F}$ gives different types of drift velocities. The two cases in which it is easy to find the drift velocity using (10.14) are given below.

Gravity Field Drift The force per unit charge, which is $q\boldsymbol{E}$ in the previous case, becomes $m\boldsymbol{g}_\perp$, the gravitational field component perpendicular to $\boldsymbol{B}$. The drift velocity $\boldsymbol{v}_z$ from (10.14) is

$$\boldsymbol{v}_g = \frac{cm\boldsymbol{g}_\perp}{qB} = \frac{cm}{q}\frac{\boldsymbol{g} \times \boldsymbol{B}}{B^2}\,. \tag{10.15}$$

This drift is charge-dependent and positive and negative charge particles drift in the opposite direction. Hence the gravity field in the presence of the magnetic field gives rise to a guiding center current.

Polarization Drift When $\boldsymbol{E}$ is time-dependent, the drift becomes [for details, the reader is referred to, e.g. (Parks, 1991; Hasegawa and Sato, 1989)]

$$\boldsymbol{v}_p = -\frac{c^2 m}{qB^2}\frac{\mathrm{d}\boldsymbol{E}}{\mathrm{d}t}\,. \tag{10.16}$$

Grad B or ∇B Drift When the magnetic flux density varies in the plane perpendicular to the direction of the magnetic field, the radius of gyration of the particle as given in (10.9) will change over the orbit. As in the previous cases, a drift called grad B or ∇B drift will result.

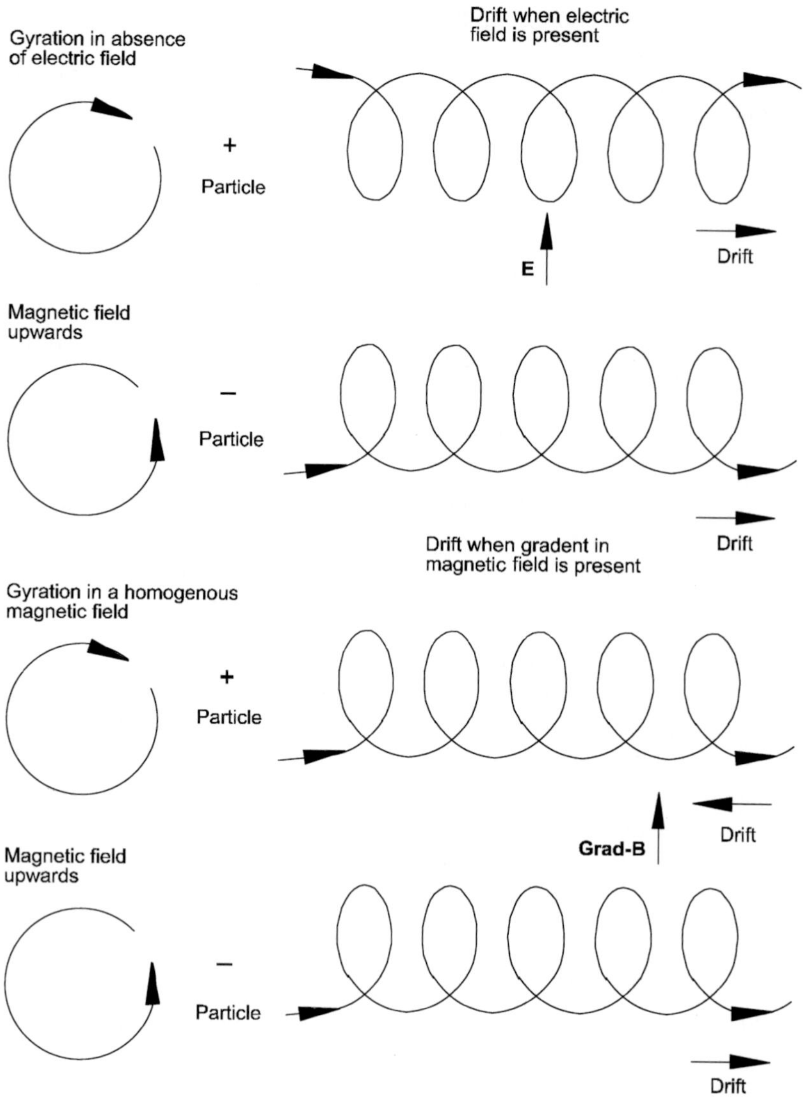

Fig. 10.3. Drift due to the presence of an electric field and the gradient in the magnetic field ($\boldsymbol{E}\times\boldsymbol{B}$ and grad-B drift)

However, in contrast to earlier cases, this drift cannot be found in a simple way but the first-order orbit theory as developed by Alfvén (Alfvén and Fälthammar, 1963) must be used. Here we only present the results and the interested reader is referred to, e.g. Parks (1991) and Hasegawa and Sato (1989) for details

$$\boldsymbol{v}_B = v_\perp r_\mathrm{L}\frac{\nabla_\perp B}{2B}\,,$$

where $\nabla_\perp B$ is the gradient of the scalar, B, in the plane perpendicular to $\boldsymbol{B}$ (Fig. 10.3). Expressing r_L is terms of $v_\perp$

$$\boldsymbol{v}_B = \frac{1}{2}\frac{mcv_\perp^2}{qB^3}(\boldsymbol{B}\times\nabla B)\,. \tag{10.17}$$

Curvature Drift Similarly a drift arises if a particle moves with a velocity $v_\parallel$ along a line of force that is curved with a radius of curvature R. The force encounter by the particle is the centrifugal force $\boldsymbol{F}_\mathrm{c} = mv_\parallel^2\boldsymbol{R}/R^2$, which gives the drift velocity

$$\boldsymbol{v}_\mathrm{R} = \frac{cmv_\parallel^2}{qB^2}\frac{\boldsymbol{R}\times\boldsymbol{B}}{R^2} = \frac{cmv_\parallel^2}{qB^4}\left[\boldsymbol{B}\times(\boldsymbol{B}\cdot\nabla)\boldsymbol{B}\right] \tag{10.18}$$

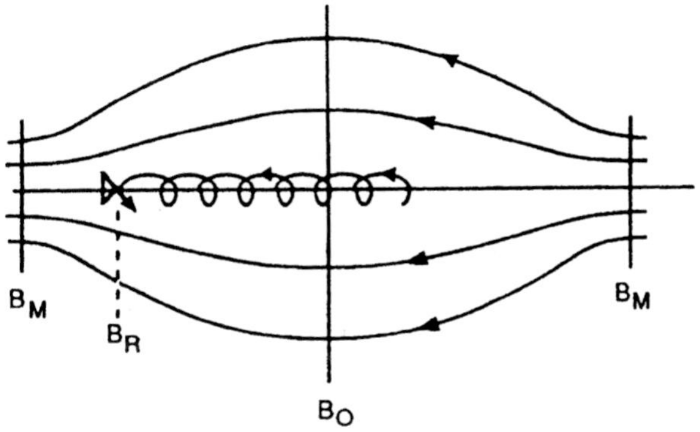

Fig. 10.4. Magnetic mirror geometry. B_0 is the equatorial field and B_M is the field at the throat. All particles mirror in the region where $B < B_M$ are trapped. All particles mirroring where $B > B_M$ will escape the magnetic device

on writing the curvature R in terms of unit vector along the magnetic field as $\boldsymbol{R}/R^2 = -(1/B^2)(\boldsymbol{B} \cdot \nabla)\boldsymbol{B}$.

10.2.3 Magnetic Mirrors

The non-uniform configuration of the magnetic field such that grad B is parallel to $\boldsymbol{B}$ gives the magnetic mirror effect. The *magnetic moment* μ, an important quantity associated with the Larmor motion of a particle, is defined as $\mu = IA$, where I is the current in the loop with area A. For a charged particle $I = q\omega_c/2\pi$ and $A = \pi r_L^2$,

$$\mu = \frac{1}{2}\frac{mcv_\perp^2}{B} . \qquad (10.19)$$

Without formal proof [for proof the reader is referred to, e.g. (Parks, 1991; Spitzer, 1962; Hasegawa and Sato, 1989)], it is important to note that when a particle moves into regions of stronger or weaker $\boldsymbol{B}$, the variation being gradual, its Larmor radius changes, but μ remains invariant of motion. It is called the adiabatic invariant.

The invariance of μ is the basis for one of the primary schemes for plasma confinement, the magnetic mirror (Fig. 10.4). As a particle moves from a weak-field region to a strong field region in the course of its thermal motion, it sees an increasing B and, therefore, its $v_\perp$ must increase in order to keep μ constant. Since its total energy must remain constant, $v_\parallel$ must necessarily decrease. If B is high enough in the "throat" of the mirror, $v_\parallel$ eventually becomes zero and the particle is reflected back to the weak field region due to the magnetic force. The non-uniform field of a simple pair of coils forms two magnetic mirrors between which plasmas can be trapped. This effect occurs on both ions and electrons.

The trapping is not perfect, however. For instance, a particle with $v_\perp = 0$ will have no magnetic moment and will not feel any force along B. A particle with small $v_\perp/v_\parallel$ at the mid-plane ($B = B_0$) will also escape if the maximum field B_m is not large enough. For given B_0 and B_m, which particles will escape? A particle with $v_\perp = v_{\perp 0}$ and $v_\parallel = v_{\parallel 0}$ at the mid-plane will have $v_\parallel = 0$ at its turning point. Let the field be B_m there. Then, the invariance of μ yields

$$(1/2)mcv_{\perp 0}^2/B_0 = (1/2)mcv_\perp^2/B_m , \qquad (10.20)$$

$$v_{\perp 0} = v \sin\alpha, v_{\parallel 0} = v \cos\alpha . \qquad (10.21)$$

Combining the above equations, we find that

$$\frac{\sin^2\alpha_0}{\sin^2\alpha} = \frac{B_0}{B_m} . \qquad (10.22)$$

For $v_\perp = 0$, that is, $\alpha = 0$, $v_\parallel$ only exists and therefore the particle escapes. When $\alpha = \pi/2$, $v_\perp$ exists, but $v_\parallel = 0$. In this case,

$$\sin^2\alpha_0 = B_0/B_m = 1/R_m , \qquad (10.23)$$

where R_m is the mirror ratio (Fig. 10.4). This equation defines the boundary of a region in the velocity space in the shape of a cone, called the loss cone. For $\alpha > \alpha_0$, particles are confined. Particles lying within the loss cone $0 < \alpha < \alpha_0$ are not confined. Consequently, a mirror-confined plasma is never isotropic. Note that the loss cone is independent of q or m. Without collisions, both ions and electrons are equally well confined. When collisions occur, particles are lost when they change their pitch angle in a collision and are scattered into the loss cone. Generally, electrons are lost more easily because they have a higher collision frequency. [For loss cones for the planetary magnetosphere, see p. 114 in (Parks, 1991)].

Adiabatic Invariants We have already discussed one adiabatic invariant, the magnetic moment μ, for a trapped particle oscillating between two magnetic mirrors; the velocity parallel to the magnetic field gives rise to the longitudinal adiabatic invariant. If we denote

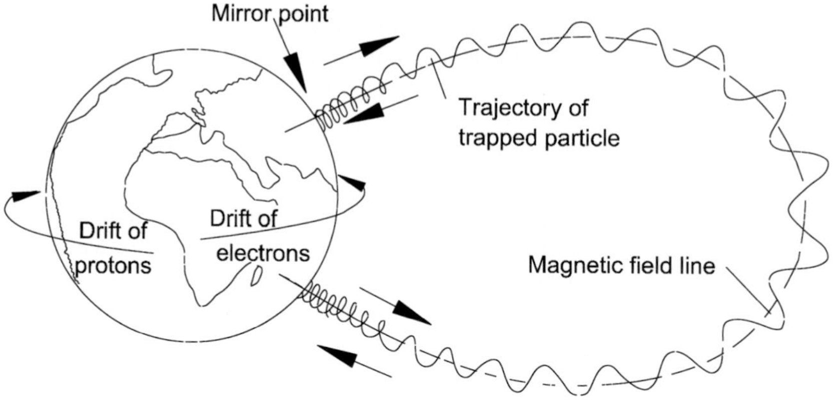

Fig. 10.5. General behavior of charged particles in the Earth's magnetic field. Particles are trapped by the Earth's magnetic field and undergo three types of motion. There is a circling around a magnetic field line (cyclotron motion), a bouncing back and forth between the mirror points in either hemisphere, and a drifting around the planet

by $\mathrm{d}s$ the distance interval along the magnetic field, the invariance is the integral of $v_\parallel\,\mathrm{d}s$ over one period of oscillation back and forth between the mirrors. This invariant remains constant as the distance between mirrors changes slowly. The second invariant is usually represented by

$$J = \int_{s_1}^{s_2} v_\parallel\,\mathrm{d}s = \text{const}\,, \tag{10.24}$$

if B varies slowly compared to periods of motion between the turning points.

μ being constant, the first invariant also implies that the total magnetic flux enclosed by the orbit of motion remains constant.

The third adiabatic invariant arises by carrying this concept to the drift motion. The guiding center drift motion conserves the total magnetic flux within its drift path. The third invariant is conserved as long as the perturbation time scale is longer than the drift times of the particle [For details, see (Northrop, 1963)].

10.2.4 Motion in a Dipole Magnetic Field

The geometry of the Earth's Dipole magnetic field is such that it forms a natural mirror machine trapping the particles spiraling back and forth around the lines of force from one hemisphere to the other. This fact was realized before Van Allen's discovery of the intense radiation on Explorer I and Explorer II and the source of the radiation was soon recognized to be geomagnetically trapped particles. In addition, due to the radial gradient in the Earth's magnetic field, the particles drift around the globe – the positively charged particles from east to west and those with negative charge from west to east due to the grad B drift.

The motion of a charged particle in a dipole field is usually broken down into three components with a certain approximation made by the theorists. The particle rotates rapidly around a field line, it bounces back and forth along a line between its two mirror points and slowly drifts around the Earth (Fig. 10.5). The speeds of these three motions are so different that they can be separated.

10.2.5 Currents

Drift or Guiding Center Currents The guiding center drift velocities (other than $\boldsymbol{E} \times \boldsymbol{B}$ drift) obtained above can produce currents when summed over species in a plasma. There is another important current due to the gyromotion of the individual particles, which should be included in the above list of currents. This is the magnetization current and it is important to understand the relation between the fluid and particle description (see Sect. 10.4.2).

Consider a set of loop currents produced by the Larmor motion of individual particles in a locally homogeneous plasma; the net current appears only at the edge of the plasma because the currents between two neighboring orbits cancel inside the plasma. Hence, the net current penetrated by a line element dI produced by N sets of current loops in an area A, each carrying the current I, is given by

$$I_\mathrm{N} = \oint NI\boldsymbol{A} \cdot \mathrm{d}\boldsymbol{l} = \int \nabla \times \boldsymbol{M} \cdot \mathrm{d}\boldsymbol{S}\,,$$

where $\boldsymbol{M} = NI\boldsymbol{A}$ is the magnetization.

The magnetization current density $\boldsymbol{J}_\mathrm{m}$ is, therefore, given as

$$\boldsymbol{J}_\mathrm{m} = \nabla \times \boldsymbol{M} . \qquad (10.25)$$

In terms of the magnetic moment $\mu = \mu \boldsymbol{b}$, $\boldsymbol{b}$ is the unit vector in the direction of the magnetic field, $\boldsymbol{M} = -n\mu$, where n is the particle density. Writing $w_\parallel = \frac{1}{2} m v_\parallel^2$, $w_\perp = \frac{1}{2} m v_\perp^2$, the guiding center currents can be summarized as follows:

$$\begin{aligned}
J_\mathrm{R} &= \frac{c2nw_\parallel}{B^4}[\boldsymbol{B} \times (\boldsymbol{B} \cdot \nabla)\boldsymbol{B}] : && \text{curvature current,}\\
J_\mathrm{B} &= cnw_\perp \frac{(\boldsymbol{B} \times \nabla \boldsymbol{B})}{B^3} : && \nabla B \text{ current,}\\
J_\mathrm{P} &= \frac{c^2 m_\mathrm{i} n}{B^2}\frac{\mathrm{d}\boldsymbol{E}}{\mathrm{d}t} : && \text{polarization current,}\\
J_\mathrm{F} &= nc\frac{\boldsymbol{F} \times \boldsymbol{B}}{B^2} : && \boldsymbol{F} \times \boldsymbol{B} \text{ current,}\\
J_\mathrm{M} &= -\nabla \times \left[\frac{ncw_\perp}{B^2}\boldsymbol{B}\right] : && \text{magnetization current.}
\end{aligned} \qquad (10.26)$$

However, since all these currents are perpendicular to the magnetic field direction, an important question is whether there are currents parallel to $\boldsymbol{B}$.

Field Aligned Currents The field aligned currents have a very interesting history [see, e.g. (Uberoi, 2000; Potemra, 1984)]. The existence of these currents was first suggested in 1908 by Birkeland in his studies of the dynamics of the Earth's aurora. His idea met with much opposition and the controversy about the existence of field aligned currents continued through the 1960s until satellite data was provided in 1966 to confirm their existence.

Field aligned currents $J_\parallel$ do not contribute to the electromagnetic stress because

$$\boldsymbol{J} \times \boldsymbol{B} = 0 .$$

Therefore, these currents are associated with a "force-free" magnetic configuration. These currents play a very important role in the coupling of magnetosphere and ionosphere and are a source of visual auroras.

Theoretically, the momentum equation cannot be used to obtain $J_\parallel$, because they do not appear explicitly in this equation. Instead they are obtained by using Maxwell's equations [see, e.g. (Parks, 1991; Hasegawa and Sato, 1989)].

10.3 Mathematical Equations for Plasmas

The individual particle interaction gives the insight into the behavior of plasma. However for the understanding of plasma dynamics on the whole it is necessary to describe the plasma from statistical concepts as used in the kinetic theory of gases (Chapman and Cowling, 1953). The motion of a particle of mass m is defined by its position $\boldsymbol{r}$ and its velocity $\boldsymbol{v}$. Each particle can, therefore, be represented by a point $(\boldsymbol{r}, \boldsymbol{v})$ in space called the "phase" space. This space is a six-dimensional space with coordinates (x, y, z, v_x, v_y, v_z). The probability density of points in the $(\boldsymbol{r}, \boldsymbol{v})$ space at time t is proportional to the *distribution function* $f(\boldsymbol{r}, \boldsymbol{v}, t)$. $f(\boldsymbol{r}, \boldsymbol{v}, t)\,\mathrm{d}\boldsymbol{r}\,\mathrm{d}\boldsymbol{v}$ represents the expected number of particles at time t in $(\boldsymbol{r}, \boldsymbol{v})$ space with coordinates $\boldsymbol{r}$ and $\boldsymbol{r} + \mathrm{d}\boldsymbol{r}$ and velocity $\boldsymbol{v}$ and $\boldsymbol{v} + \mathrm{d}\boldsymbol{v}$.

Macroscopic Variables The various macroscopic variables for plasma can be calculated by using the distribution function as follows:

(i) The density $n(\boldsymbol{r}, t)$. This is obtained as

$$n(\boldsymbol{r}, t) = \int_{-\infty}^{\infty} f(\boldsymbol{r}, \boldsymbol{v}, t)\,\mathrm{d}\boldsymbol{v} . \qquad (10.27)$$

The normalized function $\hat{f}(\boldsymbol{r}, \boldsymbol{v}, t)$ is given as

$$\hat{f}(\boldsymbol{r}, \boldsymbol{v}, t) n(\boldsymbol{r}, t) = f(\boldsymbol{r}, v, t) . \qquad (10.28)$$

(ii) The average velocity of the particles

$$\boldsymbol{v}(\boldsymbol{r}, t) = \frac{1}{n(\boldsymbol{r}, t)} \int_{-\infty}^{\infty} f(\boldsymbol{r}, \boldsymbol{v}, t)\boldsymbol{v}\,\mathrm{d}\boldsymbol{v} . \qquad (10.29)$$

(iii) The average random kinetic energy

$$E_\mathrm{av} = \frac{1}{n(\boldsymbol{r}, t)} \int_{-\infty}^{\infty} \frac{1}{2} m(v_x^2 + v_y^2 + v_z^2) f(\boldsymbol{v}, \boldsymbol{r}, t)\,\mathrm{d}\boldsymbol{v} . \qquad (10.30)$$

(iv) The pressure tensor

$$\boldsymbol{P}(\boldsymbol{r}, t) = \frac{n(\boldsymbol{r}, t)m}{n(\boldsymbol{r}, t)} \int_{-\infty}^{\infty} (\boldsymbol{v} - \boldsymbol{V})(\boldsymbol{v} - \boldsymbol{V}) f(\boldsymbol{r}, \boldsymbol{v}, t)\,\mathrm{d}\boldsymbol{v} , \qquad (10.31)$$

where $\boldsymbol{V}$ is the mean velocity (Chapman and Cowling, 1953), which yields

$$P_{xx} = nm \int_{-\infty}^{\infty} v_x^2 \hat{f}(\boldsymbol{r}, \boldsymbol{v}, t)\,\mathrm{d}\boldsymbol{v} ,$$

$$P_{xy} = \int_{-\infty}^{\infty} v_x v_y \hat{f}(\boldsymbol{r}, \boldsymbol{v}, t)\,\mathrm{d}\boldsymbol{v} .$$

Similarly, the other components of the tensor $\boldsymbol{P}(\boldsymbol{r}, t)$, namely

$$\boldsymbol{P}(\boldsymbol{r}, t) = \begin{bmatrix} P_{xx} & P_{xy} & P_{xz} \\ P_{yx} & P_{yy} & P_{yz} \\ P_{zx} & P_{zy} & P_{zz} \end{bmatrix} \tag{10.32}$$

can be defined.

The averaged parameters, referred to as moments of the distribution function, provide information on the macroscopic behavior of a system of particles. These parameters become variables in the large-scale fluid description of plasma in space.

The Maxwellian Distribution Function This is a very important distribution function given as

$$f_{\mathrm{m}}(v) = n \left(\frac{m}{2\pi KT} \right)^{3/2} \exp\left(\frac{-v^2}{v_{\mathrm{th}}^2} \right), \tag{10.33}$$

where

$$v = (v_x^2 + v_y^2 + v_z^2)^{1/2}, v_{\mathrm{th}} = (2KT/m)^{1/2}. \tag{10.34}$$

The average random kinetic energy for this distribution can be obtained by substituting $f_{\mathrm{m}}(v)$ in (10.30) and then integrating by parts

$$E_{\mathrm{av}} = \frac{1}{2} KT \quad \text{for the one-dimensional case,}$$
$$E_{\mathrm{av}} = \frac{3}{2} KT \quad \text{for the three-dimensional case,} \tag{10.35}$$

where K is the Boltzmann constant and T is the absolute temperature.

Thus, for a plasma in thermodynamics equilibrium at temperature T, the average kinetic energy of the particles equals $\frac{1}{2}KT$ per degree of freedom.

Since the Maxwellian distribution is isotropic, the pressure tensor becomes a scalar quantity

$$P_{xx} = P_{yy} = P_{zz} = nKT. \tag{10.36}$$

With other cross-diagonal components being zero we can write $P = nKT$ for the Maxwellian distribution.

The Boltzmann Equation The Boltzmann equation, which was first given by Boltzmann in 1872 in his famous paper, is now generally accepted and is considered to be the central part of statistical mechanics. For details about this equation refer to (Unlenbeck, 1973).

Here we just write the equation that the distribution function $f(\boldsymbol{r}, \boldsymbol{v}, t)$ should satisfy in the phase space $(\boldsymbol{r}, \boldsymbol{v})$

$$\frac{\partial f}{\partial t} + \boldsymbol{v} \cdot \frac{\partial f}{\partial \boldsymbol{r}} + \frac{\boldsymbol{F}}{m} \cdot \frac{\partial f}{\partial \boldsymbol{v}} = \left(\frac{\partial f}{\partial t} \right)_{\mathrm{c}}, \tag{10.37}$$

where $\boldsymbol{F}$ is an external force and $(\partial f/\partial t)_{\mathrm{c}}$ is the change in f due to collisions. As f is a function of several variables, the right side of (10.37) can be shown to be the convective derivative of six-dimensional space $(\boldsymbol{r}, \boldsymbol{v})$. By noting that $m\,\mathrm{d}\boldsymbol{v}/\mathrm{d}t = \boldsymbol{F}$, from Newton's law (10.37) can also be written as

$$\frac{Df}{Dt} = \left(\frac{\partial f}{\partial t} \right)_{\mathrm{c}}. \tag{10.38}$$

Equation (10.37) is generally valid for conservative systems, $\boldsymbol{F}$ may include a magnetic component $(q/c)\boldsymbol{v} \times \boldsymbol{B}$. In absence of collision this equation reduces to Liouville's theorem, which states that for a conservative system f is constant along a dynamical trajectory.

The Vlasov Equation In 1945, Vlasov (Vlasov, 1945) modified the Boltzmann equation for neutral gases to describe the motion of the assembly of charged particles in the phase-space when Coulomb interactions are important but closed binary collisions can be neglected.

In a plasma the collisions between the particles can be divided into two categories as follows: (i) The short-range impulse forces acting on a particle when it makes a close collision with another particle involving heavy momentum exchange. (ii) The long-range Coulomb forces exerted by other particles simultaneously on a particle. The first type of force can be taken into account fairly accurately by considering the actual mechanism of collisions. The second type of force can be replaced by a space average force that neglects the point character of the charges and the short-range collisions. In this smearing process, the plasma frequency and the Debye length remain fixed, while the mean free path λ_{m}, the average distance traveled before collision with another particle, becomes infinite. The Vlasov equation considers the smeared out continuous electric field, but neglects the short-range collisions. In the Vlasov equation, therefore, each species "r" in a plasma is described by the collisionless Boltzmann equation

$$\frac{\partial f_{\mathrm{r}}}{\partial t} + \boldsymbol{v} \cdot \frac{\partial f_{\mathrm{r}}}{\partial \mathrm{r}} + \frac{q_{\mathrm{r}}}{m_{\mathrm{r}}} \left[\boldsymbol{E} + \frac{1}{c} \boldsymbol{v} \times \boldsymbol{B} \right] \cdot \frac{\partial f_{\mathrm{r}}}{\partial \boldsymbol{v}} = 0, \tag{10.39}$$

where E and B are self-consistently determined by Maxwell's equations written in Gaussian units

$$\begin{aligned}\nabla \times \boldsymbol{E} &= -\frac{1}{c}\frac{\partial \boldsymbol{B}}{\partial t}, \\ \nabla \times \boldsymbol{H} &= \frac{4\pi}{c}\boldsymbol{J} + \frac{1}{c}\frac{\partial \boldsymbol{D}}{\partial t}, \\ \nabla \cdot \boldsymbol{B} &= 0,\ \nabla \cdot \boldsymbol{D} = 4\pi\sigma_e,\end{aligned} \tag{10.40}$$

with the change and current density given as

$$\sigma_e = \sum_r q_r \int_{-\infty}^{\infty} f_r \, d\boldsymbol{v} \quad \text{and} \quad \boldsymbol{J} = \sum_r q_r \int_{-\infty}^{\infty} \boldsymbol{v}_r f_r \, d\boldsymbol{v}.$$

The two new vector quantities D and H in (10.40) are, by traditional usage, the electric displacement and magnetic field intensity (B is then called the magnetic induction). The relationships between D and E and H and B are given by the constitutive equations $B = \mu H$, and $D = \varepsilon E$, with the magnetic permeability μ and the dielectric constant ε. For plasmas $\mu = \mu_0$, the permeability of vacuum and in Gaussian units ($\mu_0 = 1$ is used here), giving $B = H$. The dielectric constant ε_0, for vacuum is unity in Gaussian units. However, to understand the interaction of plasmas with electromagnetic waves, the dielectric constant of plasmas is derived in various ways (see Sect. 10.5).

Equations (10.39) and (10.40) give the kinetic description of a collisionless plasma. However, when collisions are to be considered, various collision models are used. The two simple models that are used more frequently are the Bhatnagar–Gross–Krook (BGK) model (Bhatnagar et al., 1954) and the Fokker–Planck equation (Chandrasekhar, 1943).

Conservation Laws To understand plasma processes involving, say, wave particle interaction or trapping of particles, it is necessary to solve the Boltzmann–Vlasov equation. However, the solution of this equation are complicated in most cases. It is fortunate that many results observed in nature and in the laboratory can be described by the simplest fluid description of plasma. The conservation laws or the equations for the macroscopic variables are obtained from the Vlasov equation, by taking moments of the equation. This involves multiplying (10.39) by the various powers of the velocity vector v and then integrating over the velocity space. Depending on the power of v, the relationship between the macroscopic variables given by (10.27)–(10.32) are obtained.

For obtaining the equation for the conservation of particle density $n(r,t)$, we take the zeroth order moment by integrating the Vlasov equation over the entire velocity space.

The Continuity Equation

$$\frac{\partial n(\boldsymbol{r},t)}{\partial t} + \nabla \cdot [n(\boldsymbol{r},t)\boldsymbol{v}(\boldsymbol{r},t)] = 0. \tag{10.41}$$

The first-order moment is obtained by multiply (10.39) by v and integrating over the velocity space. This gives the equation for conservation of momentum.

The Momentum Equation

$$mn\left(\frac{\partial \boldsymbol{v}}{\partial t} + \boldsymbol{v} \cdot \nabla \boldsymbol{v}\right) = nq\left(\boldsymbol{E} + \frac{1}{c}\boldsymbol{v} \times \boldsymbol{B}\right) - \nabla \cdot \boldsymbol{P}, \tag{10.42}$$

where P is the stress or pressure tensor.

The second-order moment will give the conservation of energy. The higher moments introduce new unknowns, not needed in many class of problems for understanding the macroscopic behavior of plasmas. From (10.41) and (10.42) we note that the conservation equations do not form a closed set, as the higher moments introduce new variables. Hence, to obtain the number of equations equal to number of unknowns, some meaningful physical assumptions should be made before the equations can be solved.

10.4 Plasma as an MHD Fluid

Magnetohydrodynamics (MHD) is a branch of continuum mechanics that deals with the motion of electrically conducting material in the presence of electromagnetic fields. The individual particle identity is ignored and only the motion of a group or an ensemble of particles is considered.

The important element of the MHD theory is that it incorporates the effects that arise from the motion of an electrically conducting fluid across magnetic fields. It is well known that when a conductor is moved across a magnetic field, an electromotive force appears in the conductor. Currents driven by this force will then flow in the conductor. Two processes occur. Firstly, the magnetic fields associated with these currents will modify the original magnetic field that created them. Secondly, the fluid motion is modified as it experiences the mechanical force of electromagnetic origin, the local force for conducting fluid. This interaction, also referred to as Maxwell's coupling between the motion, currents and magnetic fields, characterizes the general behavior of MHD fluids.

The MHD equations for plasma are derived by assuming that the plasma is in thermal equilibrium and so it retains Maxwellian distribution. Hence, the equations are applicable for processes in which the temporal changes are slower than the ion cyclotron frequency and the spatial variations less than the ion Larmor radius. Further, if the process under consideration occurs during a time period shorter than a collision time, the assumption of Maxwellian distribution is invalid. In most space plasmas, the inter-particle collision time is very long. For example, the mean free path of a solar wind particle is approximately one AU. Therefore, the applicability of MHD to space plasmas is not questionable. Moreover, although the early development of MHD was closely tied to the discovery of sunspots in 1908, more recently, in situ measurements of space plasmas in our solar system by spacecraft borne experiments have given direct evidence that many classes of observed large-scale electrodynamic phenomena can be understood by using the MHD description of plasmas. In this section, we first develop MHD equations and then discuss some important concepts of MHD theory.

10.4.1 MHD Equations

Plasma consists of electrons, species of ions and also neutrals. Considering a plasma with electrons and only one species of ions, we derive the MHD one-fluid equations for plasma. This method can be easily extended for multi-species plasmas or plasmas with neutrals.

The Continuity Equation The MHD equation of continuity is as obtained by adding the equations of continuity for electrons and ions given in (10.41)

$$\frac{\partial \rho_{\mathrm{m}}}{\partial t} + \operatorname{div}(\rho_{\mathrm{m}} \boldsymbol{u}) = 0\,, \tag{10.43}$$

where $\rho_{\mathrm{m}} = (n_{\mathrm{i}} m_{\mathrm{i}} + n_{\mathrm{e}} m_{\mathrm{e}})$ and the fluid velocity $\boldsymbol{u}$ is given as

$$\boldsymbol{u} = \frac{n_{\mathrm{i}} m_{\mathrm{i}} \boldsymbol{v}_{\mathrm{i}} + n_{\mathrm{e}} m_{\mathrm{e}} \boldsymbol{v}_{\mathrm{e}}}{n_{\mathrm{i}} m_{\mathrm{i}} + n_{\mathrm{e}} m_{\mathrm{e}}}\,. \tag{10.44}$$

The Momentum Equation Equation (10.42) gives the following equations of motion for electrons and ions (because MHD distribution is Maxwellian, the pressure tensor becomes a scalar quantity p):

$$m_{\mathrm{i}} n_{\mathrm{i}} \frac{\mathrm{d}\boldsymbol{v}_{\mathrm{i}}}{\mathrm{d}t} = q_{\mathrm{i}} n_{\mathrm{i}} \left(\boldsymbol{E} + \frac{1}{c} \boldsymbol{v}_{\mathrm{i}} \times \boldsymbol{B} \right) - \nabla p_{\mathrm{i}}\,, \tag{10.45}$$

$$m_{\mathrm{e}} n_{\mathrm{e}} \frac{\mathrm{d}\boldsymbol{v}_{\mathrm{e}}}{\mathrm{d}t} = q_{\mathrm{e}} n_{\mathrm{e}} (\boldsymbol{E} + \frac{1}{c} \boldsymbol{v}_{\mathrm{e}} \times \boldsymbol{B}) - \nabla p_{\mathrm{e}}\,. \tag{10.46}$$

As $n_{\mathrm{i}} = n_{\mathrm{e}}$ and $q_{\mathrm{i}} = q_{\mathrm{e}} = |e|$, adding (10.46) and (10.47), taking $p_{\mathrm{e}} + p_{\mathrm{i}} = p$ as the total pressure, we obtain

$$\rho_{\mathrm{m}} \frac{\mathrm{d}\boldsymbol{u}}{\mathrm{d}t} = \frac{1}{c} \boldsymbol{J} \times \boldsymbol{B} - \nabla p\,, \tag{10.47}$$

where

$$\boldsymbol{J} = e(n_{\mathrm{i}} \boldsymbol{v}_{\mathrm{i}} - n_{\mathrm{e}} \boldsymbol{v}_{\mathrm{e}})\,. \tag{10.48}$$

Equation (10.47) indicates that the acceleration of the center of gravity of the plasma as one fluid is given by the Lorentz force $1/c\boldsymbol{J} \times \boldsymbol{B}$ and ∇p and the electric field is not responsible for the acceleration. This is because plasma does not have an average electric charge. The role of the electric field can be seen from the following equation, which is the generalized Ohm's law.

Ohm's Law Taking the difference of the (10.45) and (10.46) after multiplying by m_{e} and m_{i}, respectively, and ignoring the terms with $(m_{\mathrm{e}}/m_{\mathrm{i}})$, we obtain

$$\boldsymbol{E} + \frac{1}{c} \boldsymbol{u} \times \boldsymbol{B} = -\frac{m_{\mathrm{e}}}{e} \frac{\mathrm{d}\boldsymbol{v}_{\mathrm{e}}}{\mathrm{d}t} - \frac{\nabla p_{\mathrm{e}}}{en} + \frac{\boldsymbol{J} \times \boldsymbol{B}}{cen}\,. \tag{10.49}$$

Further, if the short-range electron-ion collisions effects were retained, the loss of electron momentum would have contributed to the resistivity $\eta = (\nu m_{\mathrm{e}}/e^2 n)$ [see Sect. 10.4.4]. Then (10.49) would be modified to

$$\boldsymbol{E} + \frac{1}{c} \boldsymbol{u} \times \boldsymbol{B} = \frac{-m_{\mathrm{e}}}{e} \frac{\mathrm{d}\boldsymbol{v}_{\mathrm{e}}}{\mathrm{d}t} - \frac{\nabla p_{\mathrm{e}}}{en} + \frac{\boldsymbol{J} \times \boldsymbol{B}}{cen} + \eta \boldsymbol{J}\,. \tag{10.50}$$

This equation is called the generalized Ohm's law. The first term on the right side represents the electron inertia term introducing the scale length of the electromagnetic skin depth c/ω_{pe}. The second term, which represents the electron pressure gradient, introduces the scale of the ion Larmor radius at the electron temperature $\rho_s = \sqrt{KT_{\mathrm{e}}/m_{\mathrm{i}}}/\omega_{\mathrm{ci}}$ (see (10.123)), and the third term, which is referred to as the Hall term, represents the finite frequency effect $\omega/\omega_{\mathrm{ci}}$, all of which are regarded as small parameters in the MHD scale. In the ideal MHD case, (10.50) assumes the form of Ohm's law

$$\boldsymbol{E} + \frac{1}{c}\boldsymbol{u} \times \boldsymbol{B} = 0\,. \qquad (10.51)$$

Summarizing, the set of one-fluid equations when a plasma is assumed to behave as a conductor are as follows:

Modified Maxwell Equations

$$\nabla \times \boldsymbol{E} = -\frac{1}{c}\frac{\partial \boldsymbol{B}}{\partial t}\,, \qquad (10.52)$$

$$\nabla \cdot \boldsymbol{B} = 0\,, \qquad (10.53)$$

$$\nabla \times \boldsymbol{B} = \frac{4\pi}{c}\boldsymbol{J}\,,$$

$$\nabla \cdot \boldsymbol{D} = 0\,.$$

In the MHD theory, the displacement current and free charges are neglected as the phenomena studied are low frequency and accumulation of charged particles does not occur as the system is good electric conductor.

Continuity Equation

$$\frac{\partial \rho}{\partial t} + \nabla \cdot \rho \boldsymbol{v} = 0\,. \qquad (10.54)$$

The Equation of Motion

$$\rho\frac{\mathrm{d}\boldsymbol{v}}{\mathrm{d}t} = \frac{1}{c}\boldsymbol{J} \times \boldsymbol{B} - \nabla p\,. \qquad (10.55)$$

The force $1/c\boldsymbol{J} \times \boldsymbol{B}$ arises as coupling between fluid motion and magnetic field.

Ohm's Law

$$\eta \boldsymbol{J} = \left(\boldsymbol{E} + \frac{1}{c}\boldsymbol{v} \times \boldsymbol{B}\right)\,. \qquad (10.56)$$

To have a closed set of equations with the number of equations and the number of variables being the same, we have the following equation:

The Equation of State

$$\frac{\mathrm{d}}{\mathrm{d}t}(p/\rho^{\gamma}) = 0, \text{ the adiabatic relation.} \qquad (10.57)$$

Here γ is the ratio of specific heats. The appearance of the adiabatic relation is the natural consequence of the assumption of Maxwellian distribution for MHD plasmas.

10.4.2 Motion of the Magnetic Field

We shall now discuss some important features of the motion of the magnetic field in plasmas by using the MHD equations. We shall not discuss the important process of magnetic reconnection since this is discussed in another chapter on the same topic.

Diffusion of the Magnetic Field in Plasmas Using (10.52) and (10.56), we obtain

$$\frac{\partial \boldsymbol{B}}{\partial t} = \frac{c^2\eta}{4\pi}\nabla^2\boldsymbol{B} + \nabla \times (\boldsymbol{v} \times \boldsymbol{B})\,. \qquad (10.58)$$

When $\boldsymbol{v} = 0$

$$\frac{\partial \boldsymbol{B}}{\partial t} = -\lambda\nabla^2\boldsymbol{B}, \quad \text{where} \quad \lambda = \frac{c^2\eta}{4\pi}\,, \qquad (10.59)$$

which has the form of a diffusion equation. The quantity λ can be called the magnetic diffusivity. Equation (10.59) indicates that the field leaks through the plasma from point to point resulting in decay of the field. The dimensional argument indicates a time of decay of the order $L^2\lambda^{-1} = t_D$, where L is a characteristic spatial scale length of $\boldsymbol{B}$. The diffusion time is proportional to L^2 and $\sigma \equiv 1/\eta$. Since the conductivity is very large for MHD fluids, $L^2\sigma$ is very large for space plasmas, making the diffusion time for magnetic fields very long.

Frozen-in Field Concept Now consider a different limiting case. Suppose the plasma is in motion, but has negligible resistivity. Then the induction equation (10.58) gives

$$\frac{\partial \boldsymbol{B}}{\partial t} = \nabla \times (\boldsymbol{v} \times \boldsymbol{B})\,. \qquad (10.60)$$

This equation is identical in form to the vorticity equation of an ordinary homogeneous inviscid fluid. An important theorem derived from this equation states that:

"Fluid elements that lie on a vortex line continue to lie on the same vortex line".

Extending this theorem to conducting fluids, Alfvén in 1942 deduced that

"Fluid elements that lie on a magnetic fluid line remain on the same field line".

This formed the basis for the frozen-in-field concept in magnetohydrodynamics.

The Magnetic Reynolds Number When neither term on the right side of (10.58) is negligible, the time variation of $\boldsymbol{B}$ is the sum of both the parts on the right. Thus the lines of force tend to be carried about with the moving material and at the same time they leak through it. The transport effect dominates the leak if $LV \gg \eta$, where V is the characteristic velocity. By analogy with the Reynolds number in hydrodynamics, $R = LV/v_c$, where v_c is the viscosity, a magnetic Reynold's number R_m is defined as

$$R_m = LV/\eta\,. \tag{10.61}$$

When $R_m \gg 1$, the transport dominates the leak. This condition is very easily satisfied for cosmic plasmas as L and are large σ.

The MHD Generator For a conducting fluid in the limiting case when conductivity is infinite

$$\frac{\partial \boldsymbol{B}}{\partial t} = \nabla \times (\boldsymbol{v} \times \boldsymbol{B})\,, \tag{10.62}$$

which in comparison with Maxwell's equation

$$\frac{\partial \boldsymbol{B}}{\partial t} = -c\nabla \times \boldsymbol{E}\,, \tag{10.63}$$

shows that the electric field in an infinitely conducting medium is given by

$$c\boldsymbol{E}_\perp = -\boldsymbol{v} \times \boldsymbol{B}\,. \tag{10.64}$$

Taking the cross product of this equation with $\boldsymbol{B}$ gives

$$c\boldsymbol{E} \times \boldsymbol{B} = B^2 \boldsymbol{v}\,, \tag{10.65}$$

which in turn yields (similarly to (10.14))

$$\boldsymbol{v}_\perp = \frac{c\boldsymbol{E} \times \boldsymbol{B}}{B^2}\,. \tag{10.66}$$

In an ideal conducting fluid, therefore, there is a relationship between $\boldsymbol{E}$ and $\boldsymbol{v}$ across $\boldsymbol{B}$ such that the existence of motion implies the existence of an electric field or vice versa. In the direction parallel to $\boldsymbol{B}$, charged particles move very freely. This means that the magnetic field acts like a perfect electrical conductor, transmitting perpendicular electric fields and voltages across vast distances with no change in the potential in the direction parallel to $\boldsymbol{B}$. Thus, any flowing magnetized plasma can act as a source of voltage if there is a component of $\boldsymbol{v}$ perpendicular to $\boldsymbol{B}$.

Alfvén Waves The possibility of such waves was first established by the Nobel laureate H. Alfvén in 1942 (Alfvén, 1942). The name given by Alfvén was "electromagnetic-hydrodynamic" waves.

The force $1/c\boldsymbol{J} \times \boldsymbol{B}$ can be interpreted in terms of Maxwell's stresses

$$\frac{1}{c}\boldsymbol{J} \times \boldsymbol{B} = -\mathrm{grad}(B^2/8\pi) + \mathrm{div}(\boldsymbol{BB}/4\pi)\,, \tag{10.67}$$

where the last term denotes the divergence of a dyad. The first term represents the hydrostatic pressure $B^2/8\pi$ and the second gives the tension $B^2/4\pi$.

From the equation of motion it is seen that for an incompressible fluid the hydrostatic pressure can be balanced by the pressure of the fluid, so that only the tension $B^2/4\pi$ remains effective. Analogy with the theory of stretched strings suggests that this tension may lead to the possibility of transverse waves along the lines of force, with a velocity $\boldsymbol{v}_A$ given by

$$v_A^2 = B^2/4\pi\rho\,, \tag{10.68}$$

where ρ is the mass density of the fluid. Since the magnetic lines are frozen in the fluid, the mass density can be taken as ρ, the fluid density.

Alfvén waves play a very important role in the physics of various phenomena in space. We shall discuss these again in Sect. 10.5.

Diamagnetic Current We shall see how currents are produced in MHD fluids by taking the simple case of when fluid is not flowing. The equation of motion in this case gives

$$\nabla p = \frac{1}{c}\boldsymbol{J} \times \boldsymbol{B}\,. \tag{10.69}$$

Equation (10.69) is general, it holds for the boundary layer or for any volume element in the plasma. Taking the cross product with $\boldsymbol{B}$, we obtain

$$\boldsymbol{J}_\perp = \frac{1}{c}\frac{\boldsymbol{B} \times \nabla p}{B^2}\,, \tag{10.70}$$

the current arising from the coupling of the pressure gradient (usually found at the boundary) and the local magnetic field $\boldsymbol{B}$. Interestingly, it exists even if the fluid is not flowing. This is a diamagnetic current as it always flows in such a direction as to reduce the magnetic field intensity in the fluid. This means that the gyromotion plays an important role. From the particle point of view this current can be arrived at by adding the magnetization current density and the gradient drift current density given in (10.26).

Equation (10.70) is based on the assumption that the fluid pressure is scalar. In the case of anisotropic pressure

$$cJ_\perp = \frac{B}{B^2} \times \nabla_\perp p_\perp + \frac{(p_\parallel - p_\perp)}{B^2}[B \times (b \cdot \nabla)b] , \quad (10.71)$$

where $B = B\hat{b}$. For derivation of (10.71), see p. 259 in (Parks, 1991).

10.4.3 Hydromagnetic Equilibrium

Another concept arising from (10.69) is the "confinement" of plasma by the magnetic field. Writing the Lorentz force in term of Maxwell stresses gives

$$\nabla p = \frac{(B \cdot \nabla)B}{4\pi} - \nabla \left(\frac{B^2}{8\pi} \right) . \quad (10.72)$$

If the magnetic field does not vary along B, the straight and parallel field lines, the first term on the right side in (10.72) vanishes, giving

$$\nabla \left(p + \frac{B^2}{8\pi} \right) = 0 ,$$

$$p + \frac{B^2}{8\pi} = \text{const.} \quad (10.73)$$

Equation (10.73) states that the pressure due to the field acting in a perpendicular direction to the lines of force must be equal and opposite to the intrinsic pressure in the plasma.

Writing (10.73) in the form

$$p_1 + \frac{B_1^2}{8\pi} = p_2 + \frac{B_2^2}{8\pi} \quad (10.74)$$

shows that a plasma at a pressure p_1 may be confined by a magnetic field to a particular region even though it is surrounded by regions at a lower pressure p_2. When $B_2^2 > B_1^2$ confinement is possible even when $p_2 = 0$. Inequality $B_1^2 < B_2^2$ shows that a plasma confined by a magnetic field behaves diamagnetically.

Plasma Beta The plasma beta (β) parameter is defined as

$$\beta = \frac{\text{particle pressure}}{\text{magnetic field pressure}} = p/(B^2/8\pi) . \quad (10.75)$$

β measures the relative importance of the particle and magnetic field pressures. A plasma is referred to as a low-beta plasma when $\beta \ll 1$ and a high-beta plasma when $\beta \approx 1$. In space both high and low beta plasmas are encountered.

10.4.4 Transport Coefficients: Electrical Conductivity

Interparticle collisions in a plasma allow fluid quantities such as the density, currents, momentum and energy to be transported across the ambient magnetic field. There are a large number of transport coefficients because of anisotropies produced by the magnetic field. We shall only look at the electrical conductivity, which is a very important transport coefficient for the generation of currents in plasmas including collisions. For details of other transport coefficients see, e.g. (Braginskii, 1965; Parks, 1991; Uberoi 1988; Hasegawa and Sato, 1989).

Consider a simple plasma model consisting of an electron fluid moving under the action of an applied electric field E relative to the ions that are at rest. Assume the cold plasma, then equation of motion including collisions is

$$mn\frac{dv}{dt} = -enE - mn\nu v , \quad (10.76)$$

where ν represents the collision frequency. Collisions occur between the electrons and ions and this can involve significant momentum and energy transfers. Assume plasma to be in steady state and v uniform, then

$$v = -eE/m\nu$$

and, therefore,

$$J = -nev = \frac{ne^2}{m\nu}E \equiv \sigma E . \quad (10.77)$$

Note that σ, the electrical conductivity, depends on $1/\nu$. The numerical value of the resistivity $\eta = 1/\sigma$ can be calculated from the values of collision frequencies. Details of accurate computations can be found in, e.g. (Spitzer

1962; Longmire 1973). For example, for a plasma with a temperature of 100 eV,

$$\eta = 5 \times 10^{-15}\ \text{Ohm-cm}.$$

This value is similar to that of stainless steel.

Equation (10.77) is the familiar Ohm's law and states that the current density and the total electric field are linearly related if the conductivity is a constant. $\boldsymbol{E}$ is then parallel to $\boldsymbol{J}$. This occurs only if the plasma is isotropic. However, the magnetized plasma is anisotropic and, therefore, the currents in general do not flow parallel to the direction of the electric field. The conductivity in this case is not a constant but a tensor and Ohm's law can be written as $\boldsymbol{J} = \sigma \cdot \boldsymbol{E}$, where σ is the conductivity tensor [for the derivation the reader is referred to, e.g. (Parks, 1991; Cramer, 2001; Longmire, 1973; Kelly, 1989). We give the conductivity tensor of a plasma with electrons and one species of ions, in the presence of the magnetic field $\boldsymbol{B} = (0, 0, B)$ and $\boldsymbol{E} = (E_x, 0, E_z)$

$$\sigma = \begin{pmatrix} \sigma_1 & -\sigma_2 & 0 \\ \sigma_2 & \sigma_1 & 0 \\ 0 & 0 & \sigma_0 \end{pmatrix}, \tag{10.78}$$

where

$$\sigma_1 = \frac{nq}{B} c \left(\frac{\omega_{\text{ci}} \nu}{\nu^2 + \omega_{\text{ci}}^2} - \frac{\omega_{\text{ce}} \nu}{\nu^2 + \omega_{\text{ce}}^2} \right)$$

$$\sigma_2 = \frac{nq}{B} c \left(\frac{-\omega_{\text{ci}}^2}{\nu^2 + \omega_{\text{ce}}^2} + \frac{\omega_{\text{ce}}}{\nu^2 + \omega_{\text{ce}}^2} \right)$$

$$\sigma_0 = \frac{nq}{B} c \left(\frac{\omega_{\text{ci}}}{\nu} - \frac{\omega_{\text{ce}}}{\nu} \right).$$

The currents J_x and J_y are in the direction of E_x and J_z is in the E_z direction. Note that for $\omega_{\text{ci}} \gg \nu$ and $\omega_{\text{ce}} \gg \nu$, $\sigma_1 = |\sigma_2| = \sigma_0$ is the ordinary conductivity discussed above.

σ_1 conducts the currents perpendicular to $\boldsymbol{B}$ and in the direction of $\boldsymbol{E}$. This is called Pedersen conductivity and the associated current is the Pedersen current.

σ_2 is the Hall conductivity and the associated Hall-current is perpendicular both to $\boldsymbol{E}$ and $\boldsymbol{B}$ as already seen in the generalized Ohm's law.

σ_0 is the parallel or ordinary conductivity.

10.5 Waves in Space Plasmas

A wide variety of waves have been discussed and observed in plasmas in space. Here we shall only consider the small amplitude or linear waves, since the non-linear processes are discussed in another chapter.

The basic method of studying small amplitude plane waves is to first linearize the equations under consideration and find the solution as a Fourier mode, such that any perturbed quantity will be represented as

$$\boldsymbol{A}(\boldsymbol{r}, t) = \boldsymbol{A}_0 e^{i(\boldsymbol{k}\cdot\boldsymbol{r} - \omega t)}. \tag{10.79}$$

Here A_0 is the amplitude of the wave, it is constant as we are considering presently homogeneous medium. ω and $\boldsymbol{k}$ are the angular frequency and the wave vector that gives the direction of wave propagation. Wave number $k = |\boldsymbol{k}|$ is related to wavelength λ as $k = 2\pi/\lambda$. $\omega(\boldsymbol{k})$ in radians is related to wave frequency f in cycles per second and period T through $\omega = 2\pi f = 2\pi/T$.

The nature of the small amplitude wave can be characterized by the wave vector dependence on the frequency $\omega(\boldsymbol{k})$, which is called the *dispersion relation*. The phase velocity is defined as $v_{\text{ph}} = \omega/k$ or in vector terms as $\omega/k\,\hat{k}$, and the group velocity of the wave as

$$\boldsymbol{v}_{\text{g}} = \frac{\partial \omega}{\partial \boldsymbol{k}}. \tag{10.80}$$

10.5.1 Electromagnetic Waves

For understanding the features of electromagnetic waves, plasma is treated as a dielectric medium or as a "cold" plasma, the thermal effects are neglected. The cold plasma model surprisingly gives many interesting features of electromagnetic waves in plasma to explain the observation of these waves in space.

Dielectric Response of Plasma Consider a high frequency oscillating electric field applied to cold plasma. The equation of motion of electron fluid yields (ions form the stationary background)

$$m_{\text{e}} \frac{\mathrm{d}\boldsymbol{v}_{\text{e}}}{\mathrm{d}t} = -e\boldsymbol{E}, \quad \text{with} \quad \boldsymbol{E} \equiv \boldsymbol{E} e^{-i\omega t}. \tag{10.81}$$

This gives

$$\boldsymbol{v}_{\text{e}} = (-ie/m_{\text{e}}\omega)\boldsymbol{E}. \tag{10.82}$$

To express the plasma current density $\boldsymbol{J}$ in terms of $\boldsymbol{E}$, we note that the electric displacement $\boldsymbol{D}$ includes the vacuum displacement and the current arising from the charge displacements in plasma. Therefore,

$$\frac{\partial \boldsymbol{D}}{\partial t} = \frac{\mathrm{d}\boldsymbol{E}}{\mathrm{d}t} + 4\pi \boldsymbol{J}\,. \tag{10.83}$$

As

$$\boldsymbol{J} = \sum n\boldsymbol{v}q = -en\boldsymbol{v}_{\mathrm{e}}\,, \tag{10.84}$$

using (10.82), (10.83) becomes

$$\boldsymbol{D} = \left(1 - \frac{4\pi n e^2}{m_{\mathrm{e}}\omega^2}\right)\boldsymbol{E}\,, \tag{10.85}$$

which gives the dielectric constant of the cold and unmagnetized plasma as

$$\varepsilon(\omega) = \left(1 - \frac{\omega_{\mathrm{p}}^2}{\omega^2}\right). \tag{10.86}$$

Introducing the refractive index vector $\boldsymbol{n} = c(\boldsymbol{k}/\omega)$ with magnitude $n = |\boldsymbol{n}|$, we have

$$n = [\varepsilon(\omega)]^{1/2}\,. \tag{10.87}$$

Equations (10.86) and (10.87) show that plasma is a dispersive media and an important fact is that the refractive index of plasma in the high frequency range is less than one. The refractive indices of waves in the visible range for substances in solid, liquid and gaseous states are greater than one, hence plasma is unusual in this optical property. An interesting consequence of this is that if a convex lens of a plasma were used for refracting light, it would be a divergent lens and not convergent like a lens of glass. It is now realized that this effect plays a very important role in the operation of gas lasers.

Dispersion of Electromagnetic Waves Consider the necessary Maxwell equations

$$\nabla \times \boldsymbol{E} = -\frac{1}{c}\frac{\partial \boldsymbol{B}}{\partial t}\,, \tag{10.88}$$

$$\nabla \times \boldsymbol{B} = \frac{1}{c}\frac{\partial \boldsymbol{D}}{\partial t} + \frac{4\pi}{c}\boldsymbol{J}_{\mathrm{ext}}\,, \tag{10.89}$$

where $\boldsymbol{J}_{\mathrm{ext}}$ is the current of the external sources.

The induction $\boldsymbol{D}$ is given by the relation

$$\frac{\partial \boldsymbol{D}}{\partial t} = \frac{\partial \boldsymbol{E}}{\partial t} + 4\pi \boldsymbol{J}_{\mathrm{ind}}\,, \tag{10.90}$$

where $\boldsymbol{J}_{\mathrm{ind}}$ is the current density produced by fields $\boldsymbol{E}$ and $\boldsymbol{B}$.

The currents $\boldsymbol{J}_{\mathrm{ext}}$ and $\boldsymbol{J}_{\mathrm{ind}}$ arise from charge displacements, the first from the motion of free charges and the second from the bound charges. The ions and electrons comprising the plasma are the equivalents of the bound charges. The separation between the free and bound charges is therefore unnecessary so (10.90) is the same as (10.83) and (10.89) then gives

$$\nabla \times \boldsymbol{B} = \frac{1}{c}\frac{\partial \boldsymbol{E}}{\partial t} + \frac{4\pi}{c}\boldsymbol{J} = \frac{1}{c}\frac{\partial \boldsymbol{D}}{\partial t}\,. \tag{10.91}$$

Fourier analysis of (10.88) and (10.91) in space and time and combining the two gives on using (10.38)

$$\boldsymbol{k} \times (\boldsymbol{k} \times \boldsymbol{E}) + \frac{\omega^2}{c^2}\varepsilon(\omega)\cdot \boldsymbol{E} = 0\,. \tag{10.92}$$

As

$$\boldsymbol{k} \times (\boldsymbol{k} \times \boldsymbol{E}) = \nabla(\nabla \cdot \boldsymbol{E}) - \nabla^2 \boldsymbol{E}\,, \tag{10.93}$$

we see that Eq. (10.92) gives the wave equation for the transverse electromagnetic wave propagating in vacuum with velocity c, as $\nabla \cdot \boldsymbol{E} = 0$ and $\varepsilon(\omega) = 1$ in vacuum.

For plasma with $\varepsilon(\omega)$ given by (10.86) the dispersion relation for transverse waves can be written as from (10.92) as

$$\omega^2 = c^2k^2 + \omega_{\mathrm{p}}^2\,. \tag{10.94}$$

Note that v_{ph} for a light wave in a plasma is greater than the velocity of light. However,

$$v_{\mathrm{g}} = \frac{\partial \omega}{\partial k} = \frac{c^2}{v_{\mathrm{ph}}} < c\,.$$

At $\omega = \omega_{\mathrm{p}}$, $k = 0$, which means that electromagnetic wave has a cut-off frequency at $\omega = \omega_{\mathrm{p}}$ in a plasma. For $\omega < \omega_{\mathrm{p}}$, k becomes imaginary and the wave is damped with a characteristic length $1/|k|$ (Fig. 10.6).

The above feature of electromagnetic waves propagating in a plasma has many applications, for example in short wave radio communications and for establishing communication with space vehicles. The cut-off frequency is also called the critical frequency. If we take the

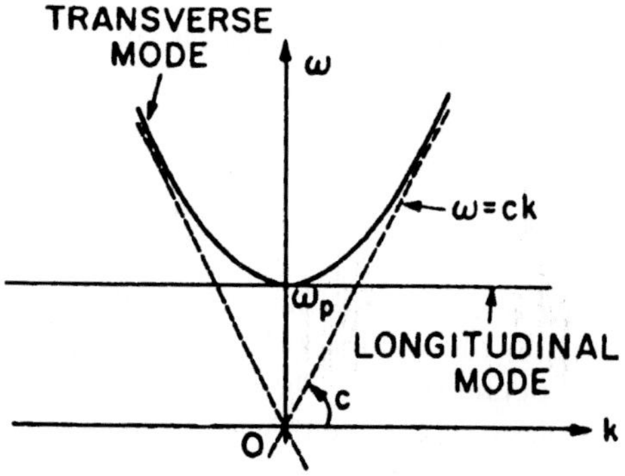

Fig. 10.6. Dispersion relation of electromagnetic waves in a cold, unmagnetized plasma

maximum density of ionosphere plasma to be 10^4 /cm^3 the critical or cut-off frequency is of the order of 10^7 /s. Hence, the frequencies used for radio communication should be less than this. This characteristic property of electromagnetic waves in plasma also causes a total communication blackout during reentry of the space vehicle into the Earth's atmosphere. For a comprehensive treatment of radio waves in ionospheric plasma, see (Kelly, 1989; Budden, 1985).

The other type of wave is a longitudinal wave whose electric field is polarized in the direction of $\boldsymbol{k}$. For this $\nabla \times \boldsymbol{E} = 0$, so (10.92) gives the dispersion relation

$$\omega^2 = \omega_{\mathrm{p}}^2 , \tag{10.95}$$

this mode represents the longitudinal plasma oscillations.

10.5.2 Dielectric Constant for Magnetized Plasma

In the presence of the external magnetic field the dielectric constant assumes the tensor form as the magnetic field makes the plasma medium anisotropic in response to the electromagnetic wave interacting with the plasma.

To find the dielectric tensor write the Lorentz equation of motion for all the particle species in the plasma

$$m_q \frac{\mathrm{d}\boldsymbol{v}_q}{\mathrm{d}t} = q\left(\boldsymbol{E} + \frac{1}{c}\boldsymbol{v}_q \times \boldsymbol{B}\right). \tag{10.96}$$

Solve (10.96) for electrons and ions for small perturbations, assuming the quantities to vary as $\exp\left(i\boldsymbol{k}\cdot\boldsymbol{r} - i\omega t\right)$.

Taking the initial magnetic field to be uniform and in the z-direction only, $\boldsymbol{B} = (0, 0, B_0)$, (10.96) gives the velocity components in terms of $\boldsymbol{E}$, which are then fed into (10.84) for the current density. The resulting expression for $\boldsymbol{J}$ in terms of components of $\boldsymbol{E}$ when used in (10.83) gives (Stix, 1962)

$$\varepsilon \cdot \boldsymbol{E} = \begin{pmatrix} S & -iD & 0 \\ iD & S & 0 \\ 0 & 0 & P \end{pmatrix} \begin{pmatrix} E_x \\ E_y \\ E_z \end{pmatrix}, \tag{10.97}$$

where $S = \frac{1}{2}(R + L)$, $D = \frac{1}{2}(R - L)$, with

$$R, L = 1 - \sum_q \frac{\omega_{pq}^2}{\omega^2}\left(\frac{\omega}{\omega \pm q\omega_{cq}}\right) \tag{10.98}$$

and

$$P = 1 - \sum_q \frac{\omega_{pq}^2}{\omega^2} . \tag{10.99}$$

The matrix on the right side of (10.97) defines the components of the dielectric tensor. Substituting these into (10.92) we obtain the following equation for $n = ck/\omega$, giving the dispersion relation for the electromagnetic waves in the magnetized plasma:

$$\begin{aligned} &An^4 - Bn^2 + C = 0 , \\ &A = S\sin^2\theta + P\cos^2\theta \\ &B = RL\sin^2\theta + PS(1 + \cos^2\theta) \\ &C = PRL . \end{aligned} \tag{10.100}$$

Here θ is the angle between the direction of the magnetic field and the wave propagation direction; n is assumed to be in the x-z plane.

The quadratic nature of the dispersion relation (10.100) shows that for any angle θ there are two modes of wave propagation. In general it also shows that $C = 0$ gives $n^2 = 0$, which is the cut-off frequency at which the wave is reflected and $A = 0$ gives $n^2 \to \infty$, which is the resonant frequency near which the wave is absorbed. The cut-off and resonance frequencies are important for waves in inhomogeneous plasmas.

For simplicity and to understand features of wave propagation in magnetized cold plasma we shall consider two directions of wave propagation $\theta = 0$ and $\theta = \pi/2$.

For $\theta = 0$ (10.100) gives three cases $P = 0$, $n^2 = R$, $n^2 = L$. $P = 0$ involves only E_z as seen from the matrix in (10.97) and gives longitudinal plasma oscillations $\omega^2 = \omega_p^2$.

$n^2 = R$, $n^2 = L$: There are two modes of wave propagation. One is the right circularly polarized mode with $iE_x/E_y = 1$ and the second is the left circularly polarized mode with $iE_x/E_y = -1$.

For the R mode $\omega = \omega_{ce}$ is the resonant frequency and for the L-mode it is $\omega = \omega_{ci}$.

The cut-off frequencies on neglecting ion motion can be easily calculated from (10.100) as

$$\omega_{cR,L} = \frac{1}{2}\left[\pm\omega_{ce} + \left(\omega_{ce}^2 + 4\omega_{pe}^2\right)^{1/2}\right]. \quad (10.101)$$

In Figure 10.2 the R and L wave dispersion curves are plotted. The lower branch for higher frequencies near ω_{ce} in the R-wave represents the electron-cyclotron wave and in the L-wave near ω_{ci}, the ion-cyclotron wave. For very low frequencies both these waves become Alfvén waves and at intermediate frequencies the R-mode becomes an important mode of propagation in space, the Whistler mode (Fig. 10.7). An important feature for two-component plasma is that phase velocity of these modes is never equal in the propagating band, whereas for a multi-species plasma the phase velocities can become equal (Uberoi, 1973).

We shall now give some details about Alfvén and Whistler waves propagating along the direction of $\boldsymbol{B}$ in two-component plasma.

Whistler Waves At intermediate frequencies $\omega_{ci} \ll \omega \ll \omega_{ce}$ the R-mode dispersion relation $n^2 = R$ gives

$$\omega = k^2 c^2 \omega_{ce}/\omega_{pe}^2 . \quad (10.102)$$

An interesting result comes from the calculation of the group velocity, which becomes

$$v_g = \frac{\partial\omega}{\partial k} = \frac{2c(\omega\omega_{ce})^{1/2}}{\omega_{pe}} \quad \alpha\sqrt{\omega}. \quad (10.103)$$

Equation (10.103) shows that the higher frequency component of the wave energy arrives earlier, as an audio signal the wave therefore sounds like a descending whistling tone and hence called a Whistler. [For details, see the discussions on Whistler related phenomena (Helliwell, 1965; Storey, 1953)].

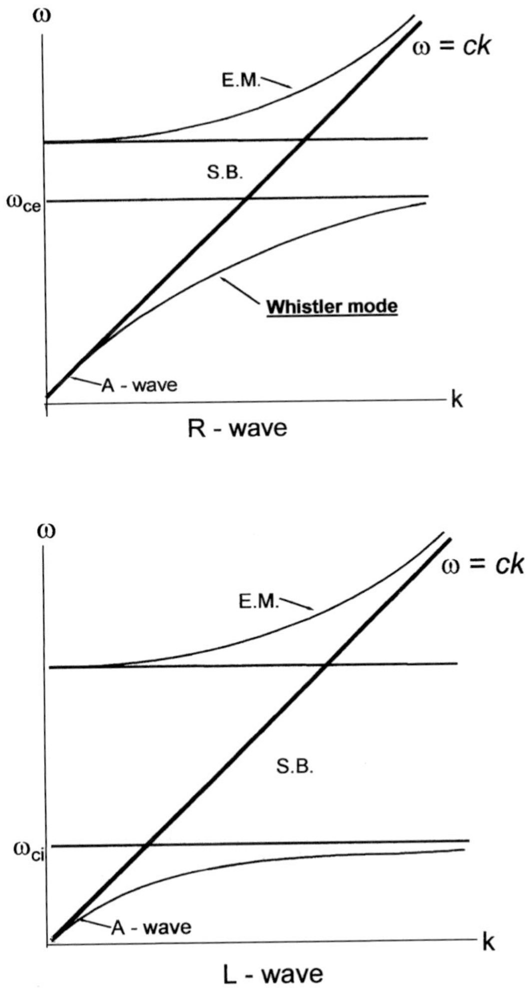

Fig. 10.7. Dispersion relation of R and L electromagnetic waves in magnetized plasma

Alfvén Waves At the very low frequency $\omega \ll \omega_{ci}$, the motion of ions cannot be neglected. In this case, we can show that

$$n_R^2 = n_L^2 = 1 + \frac{4\pi\rho c^2}{B_0^2}, \quad (10.104)$$

where $\rho = n_0 m_i$. This gives

$$\frac{c^2k^2}{\omega^2} = 1 + \frac{c^2}{v_A^2}, \quad (10.105)$$

As $v_A^2 \ll c^2$, this gives the Alfvén wave with velocity $v_A = B_0/(4\pi\rho)^{1/2}$. This is a very important result, because the Alfvén wave, which is an ideal MHD mode, as we shall discuss later, appears as a fundamental mode of wave propagation in plasmas involving low frequency ion oscillations in the presence of a magnetic field. Since (10.105) gives the dielectric constant for a low frequency wave, the Alfvén wave can be treated as an electromagnetic wave, which is modified by the high dielectric constant of the medium. Properties of this wave have been widely discussed [see, e.g. (Cramer, 2001; Hasegawa and Uberoi, 1989)].

For $\theta = \frac{\pi}{2}$, (10.100) has roots

$$n^2 = RL/S \tag{10.106}$$

and

$$n^2 = P \tag{10.107}$$

giving extraordinary and ordinary modes of propagation. In the case of (10.106) the corresponding wave vector is in the y-direction, the presence of both E_x and E_y components show that the wave is mixed transverse and longitudinal. This mode has two resonances ω_{UH} and ω_{LH}, called the upper and lower hybrid frequencies respectively,

$$\omega_{\mathrm{UH}}^2 = \omega_{\mathrm{pe}}^2 + \omega_{\mathrm{ce}}^2 \text{ and } \omega_{\mathrm{LH}}^2 = \omega_{\mathrm{pi}}^2 \frac{\omega_{\mathrm{ce}}^2}{\omega_{\mathrm{ce}}^2 + \omega_{\mathrm{pe}}^2} . \tag{10.108}$$

The cut-off frequencies are same as in the case of the R and L modes, $R = 0$ and $L = 0$.

For $\omega \ll \omega_{\mathrm{ci}}$, the dispersion relation gives $\omega \simeq k v_{\mathrm{A}}$, the wave is the compressional Alfvén mode arising due to compression of the magnetic lines of force. This is an important mode and will be discussed in the section on MHD waves.

The dispersion relation for ordinary wave is

$$\omega^2 = k^2 c^2 + \omega_{\mathrm{pe}}^2 . \tag{10.109}$$

The structure of this wave is identical to the electromagnetic wave in an unmagnetized plasma. The electric field of the ordinary wave is polarized in the direction of the external magnetic field, hence the induced motions of plasma particles are not influenced by the magnetic field.

10.5.3 Electrostatic Waves

Electron Plasma Waves The electrostatic or longitudinal plasma oscillations obtained by using the cold plasma model show various new features when thermal effects are taken into account. In this case, the equation of motion will contain the pressure gradient force. The equations for the electron fluid are

$$\frac{\partial n_{\mathrm{e}}}{\partial t} + \nabla \cdot (n_{\mathrm{e}} \boldsymbol{v}_{\mathrm{e}}) = 0 \tag{10.110}$$

$$m_{\mathrm{e}} n_{\mathrm{e}} \left[\frac{\partial \boldsymbol{v}_{\mathrm{e}}}{\partial t} + (\boldsymbol{v}_{\mathrm{e}} \cdot \nabla) \boldsymbol{v}_{\mathrm{e}} \right] = -\nabla p_{\mathrm{e}} - e n_{\mathrm{e}} \boldsymbol{E} . \tag{10.111}$$

The interaction of the ion and electron fluids is included in the Poisson equation

$$\nabla \cdot \boldsymbol{E} = 4\pi e (n_{\mathrm{i}} - n_{\mathrm{e}}) . \tag{10.112}$$

Equations (10.110)–(10.112) therefore determine the electron fluid motion. If ions form the stationary background, then $n_{\mathrm{i}} = n_0$, where n_0 is the equilibrium density.

To complete the set of equations so that number of equations and unknown variables are the same, the equation of state is taken with

$$p_{\mathrm{e}} = n_{\mathrm{e}} K T_{\mathrm{e}} \tag{10.113}$$

considering adiabatic compression $T_{\mathrm{e}} \alpha n_{\mathrm{e}}^{\gamma-1}$, $\nabla p_{\mathrm{e}}/p_{\mathrm{e}} = \gamma \nabla n_{\mathrm{e}}/n_{\mathrm{e}}$ or

$$\nabla p_{\mathrm{e}} = \gamma K T_{\mathrm{e}} \nabla n_{\mathrm{e}} . \tag{10.114}$$

On linearizing (10.110)–(10.112) and using (10.114) the dispersion relation for electron plasma waves is given as

$$\omega^2 = \omega_{\mathrm{p}}^2 + \frac{3}{2} k^2 v_{\mathrm{th}}^2 , \quad v_{\mathrm{th}}^2 = (2KT_{\mathrm{e}}/m_{\mathrm{e}}) . \tag{10.115}$$

As $\lambda_D^2 = 1/2(v_{\mathrm{th}}^2/\omega_{\mathrm{pe}}^2) = (KT_{\mathrm{e}}/4\pi n_{\mathrm{e}} e^2)$, (10.115) is sometimes written as

$$\omega^2 = \omega_{\mathrm{p}}^2 (1 + 3k^2 \lambda_D^2) . \tag{10.116}$$

The group velocity $v_{\mathrm{g}} = \frac{3}{2} v_{\mathrm{th}}^2 / v_{\mathrm{ph}}$.

Ion-acoustic Waves These waves are analogous to sound waves in neutral fluids. In the latter, sound waves propagate due to collisions. In plasma when low-frequency ion motion is taken into account the ion-acoustic or ion-sound waves propagate through the medium of electric fields.

For electrostatic oscillations, the two-fluid (10.110)–(10.112) with (10.114) are used for electrons and a similar set for ions.

For low frequency waves, m_e can be neglected. As the electrons move very fast relative to slow ion waves, these have time to equalize the temperature everywhere. Therefore, electrons are isothermal and $\gamma_e = 1$.

Putting $m_e \to 0$ and $\boldsymbol{E} = -\text{grad}\phi$ in (10.111) and using the initial condition $\phi = 0$ when $n_e = n_0$, we obtain

$$\phi = (\mathrm{K}T_e/e)\ln\left(\frac{n_e}{n_0}\right). \qquad (10.117)$$

Consider now the ion motions and linearize the equations for ions. With the assumption that $n_e \sim n_i$, we obtain the wave equation

$$m_i\frac{\partial^2 n_e}{\partial t^2} = (\mathrm{K}T_e + \gamma_i \mathrm{K}T_i)\frac{\partial^2 n_e}{\partial x^2}. \qquad (10.118)$$

For ion waves or ion-acoustic waves propagating with velocity

$$\frac{\omega}{k} = \left(\frac{\mathrm{K}T_e + \gamma_i \mathrm{K}T_i}{m_i}\right)^{1/2} \equiv v_s\,. \qquad (10.119)$$

Equation (10.119) shows that ion-acoustic waves do not show any dispersion and travel with constant phase velocity. The inertia is provided by the ions and the thermal motion by electrons. These waves can be supported even when $T_i \ll T_e$ or $T_i \approx 0$.

Dispersion of Ion-acoustic Waves Equation (10.118) was arrived at by the assumption that $n_e = n_i$. If this assumption is removed, following the same procedure as above the dispersion relation for ion-acoustic waves is given as

$$\frac{\omega}{k} = \left(\frac{\mathrm{K}T_e}{m_i}\frac{1}{1+k^2\lambda_D^2} + \frac{\gamma_i \mathrm{K}T_i}{m_i}\right)^{1/2}. \qquad (10.120)$$

For simplicity, if we assume $T_i = 0$, we obtain

$$\omega \approx k v_s\left(1 + \frac{1}{2}k^2\lambda_D^2\right) \quad \text{for } k\lambda_D \ll 1. \qquad (10.121)$$

The second term proportional to k^3 is the dispersive term arising when plasma approximation is not made. This term plays an important role in the study of ion-acoustic solitons.

10.5.4 Magnetohydrodynamic Waves

The unubiquitous presence of Alfvén or, in general, magnetohydrodynamic waves in space has now been well established by in situ observations and they find applications in the understanding of many plasma phenomena in space. We have already seen that these waves arise due to tension $B^2/4\pi$ along the lines of force in Sect. 10.4 and also exist in plasma as an electromagnetic wave at low frequencies. We now derive the dispersion relation of Alfvén waves using the MHD equation and then study the effect of compressibility on these waves in order to understand other important modes of MHD wave propagation.

Dispersion Relation For an incompressible fluid and the assumption of infinite conductivity, the linearized set of MHD equations for uniform plasma are

$$\nabla\cdot\boldsymbol{v} = 0, \quad \rho\frac{\partial \boldsymbol{v}}{\partial t} = -\text{grad}\,p + \frac{1}{4\pi}(\nabla\times\boldsymbol{B})\times\boldsymbol{B}_0 \qquad (10.122)$$

and

$$\frac{\partial \boldsymbol{B}}{\partial t} = \nabla\times(\boldsymbol{v}\times\boldsymbol{B}_0)\,. \qquad (10.123)$$

Consider $\boldsymbol{B}_0 = (0, 0, B_0)$ and a plane polarized wave in the $x - z$ plane along the magnetic field direction and with perturbation perpendicular to the $\boldsymbol{B}_0$ direction. Thus

$$\boldsymbol{B} = (0, B_y, 0), \quad \boldsymbol{v} = (0, v_y, 0)\,. \qquad (10.124)$$

Substituting (10.124) in (10.122) and (10.123), we obtain

$$\frac{\partial^2 B_y}{\partial t^2} = \frac{B_0^2}{4\pi\rho_0}\frac{\partial^2 B_y}{\partial z^2} \qquad (10.125)$$

and

$$v_y = -B_y/(4\pi\rho_0)^{1/2} = -v_A(B_y/B_0)\,. \qquad (10.126)$$

Equation (10.125) gives the dispersion relation for Alfvén waves

$$\frac{\omega^2}{k^2} = \frac{B_0^2}{4\pi\rho_0} \qquad \frac{\omega}{k} = \pm v_A\,. \qquad (10.127)$$

The Alfvén wave is transverse in nature, as can be seen from the fact that the particle velocities are perpendicular to the direction of the wave propagation. The phase velocity of this wave is several orders less than the velocity of light. This property of the Alfvén wave is used in the coupling of transverse electromagnetic waves to sound waves both in fluids and solids.

As the lines of force lie in the y–z plane $\mathrm{d}y/\mathrm{d}z = B_y/B_0$, so (10.126) shows that the velocity of the lines of force is the same as fluid velocity. This is the frozen in the fluid concept discussed in Sect. 4.5.

Compressibility Effects When compressibility is taken into account, the dispersion relation assumes the form

$$\left[\frac{\omega^4}{k^4} - \frac{\omega^2}{k^2}(v_S^2 + v_A^2) + v_S^2 v_A^2 \cos^2\theta\right] \left(\frac{\omega^2}{k^2} - \frac{\omega^2}{v_A^2}\cos^2\theta\right) = 0\,, \qquad (10.128)$$

where $v_S = (\gamma p_0/\rho_0)^{1/2}$ is the collisional sound speed and θ is the angle between the equilibrium magnetic field direction and the wave propagation direction.

Equation (10.128) gives three modes of propagation. The first two are given by the upper and lower signs of the equation

$$\frac{\omega^2}{k^2} = \frac{1}{2}(v_S^2 + v_A^2) \pm \frac{1}{2}[(v_S^2 + v_A^2)^2 - 4v_S^2 v_A^2 \cos^2\theta]^{1/2} \qquad (10.129)$$

and the third by

$$\frac{\omega^2}{k^2} = v_A^2 \cos^2\theta\,. \qquad (10.130)$$

They are termed fast, slow (magnetosonic waves) and intermediate (Alfvén waves) modes, respectively.

When $\theta = 0$, we have Alfvén and sound waves and when $\theta = \pi/2$, only the fast mode persists. It has the same velocity as the Alfvén speed when the magnetic pressure is large compared with the particle pressure and represents the compressional Alfvén mode. This mode arises due to compression of the magnetic lines of force.

10.5.5 Kinetic Theory of Plasma Waves

So far we have considered the dielectric and fluid description of plasmas, which show the similarities plasmas have with dielectric and fluid media. However, the kinetic description of plasmas is very interesting because their novel physical properties are brought out mainly due to an important feature that is unique to them: the wave-particle interaction. It is this feature that gives the most important result of plasma physics: the collisionless damping or Landau damping of plasma waves. This damping is due to particles resonating with the wave. The particles carry away more energy from the wave, which is then damped. The reverse phenomenon of particles giving energy to the wave can also exist, and we call this plasma wave instability.

What are the kinetic effects on electrostatic plasma oscillations? For this we take $f(x, v, t)$ as the one-dimensional distribution function for electrons. The ions are considered to form a stationary background with charge density en_0.

The two equations governing the electron plasma are the Boltzmann–Vlasov equation

$$\frac{\partial f}{\partial t} + v\frac{\partial f}{\partial x} - \frac{eE}{m}\frac{\partial f}{\partial v} = 0 \qquad (10.131)$$

and the Poisson equation (for the average electric field)

$$\frac{\partial E}{\partial x} = 4\pi e\left(n_0 - \int_{-\infty}^{\infty} f\,\mathrm{d}v\right). \qquad (10.132)$$

Consider the small-amplitude oscillations. For linearization of (10.131) and (10.132), assume that $f = f_0 + f_1$ and $E \equiv E_1$. Retaining only the terms of order f_1 and E, we obtain

$$\frac{\partial f_1}{\partial t} + v\frac{\partial f_1}{\partial x} - \frac{eE}{m_e}\frac{\partial f_0}{\partial v} = 0\,, \qquad (10.133)$$

$$\frac{\partial E}{\partial x} = -4ne\int_{-\infty}^{\infty} f_1\,\mathrm{d}v\,. \qquad (10.134)$$

Assume that f_1 and E are proportional to $\exp[i(kx - \omega t)]$. Equation (10.133) can be solved for f_1 in terms of E, i.e.

$$f_1 = \frac{1}{i(kv - \omega)}\frac{eE}{m_e}\frac{\partial f_0}{\partial v}\,. \qquad (10.135)$$

Substituting this equation in the relation (10.134), we obtain

$$\frac{k^2}{\omega_p^2} - \int_{-\infty}^{\infty}\frac{G(v)\,\mathrm{d}v}{v - \omega/k} = 0\,, \qquad (10.136)$$

where

$$\omega_p^2 = \frac{4\pi n_0 e^2}{m_e}\,, \quad \text{and} \quad G(v) = \frac{1}{n_0}\frac{\partial f_0}{\partial v}\,. \qquad (10.137)$$

Equation (10.136) is the dispersion relation for the electrostatic plasma waves since it gives the frequency ω as a function of k for any given velocity distribution $f_0(v)$. In this equation, the integral on the left-hand side is an improper integral with singularity at $v = \omega/k$. This singularity was first recognized by Landau in 1946 (Landau, 1946). He was the first to treat this integral properly, arriving at the most fundamental result in plasma physics, now known as *Landau damping*.

The plasma waves are damped in the absence of any dissipative effect. This phenomenon of wave damping without any energy dissipation by collisions is perhaps the most remarkable result of plasma physics research. It is not just a mathematical result, but a real effect, now well demonstrated in the laboratory.

Landau's initial value problem of treating (10.133) and (10.134) rigorously requires a rather complicated mathematical analysis of the contour integration in the complex v-plane. Without giving any details [which can be found, e.g. in (Uberoi, 1988; Chen, 1974; Jackson, 1960; Dawson, 1961)], we point out that (10.136) admits complex values for ω. The real part gives the plasma waves dispersion relations given in (10.94) and the imaginary part for frequencies close to ω_p^2, i.e. $\omega^2 \approx \omega_p^2$ is given by

$$Im\,\omega = \frac{\pi}{2}\left(\frac{\omega_p}{k}\right)^2 \omega_p G\left(\frac{\omega}{k}\right), \qquad (10.138)$$

where

$$G\left(\frac{\omega}{k}\right) = \frac{1}{n_0}\frac{\partial f_0}{\partial v}\bigg|_{v=\omega/k}. \qquad (10.139)$$

Therefore, when $\partial f_0/\partial v$ is negative, damping occurs; when the slope of the distribution function is positive, there is a possibility of a growing solution (Figs. 10.8 and 10.9). The second case will be discussed in the section on plasma instabilities.

The Bernstein Wave The Vlasov theory in magnetized plasma yields some modes that are not given by the fluid theory. Without giving any details, we would like to mention the Bernstein mode (Bernstein, 1960), which is of considerable interest in space plasma. This mode propagates perpendicular to $\boldsymbol{B}_0$ and is almost purely electrostatic, that is, $\boldsymbol{k}$ is nearly parallel to $\boldsymbol{E}$. In the limit $\boldsymbol{B}_0 \to 0$, it gives plasma oscillations at high frequencies and degenerates to ion-acoustic oscillations at low frequencies. The resonance occurs at a harmonic of the cyclotron frequency, $n\omega_{ce}$. The features of wave propagation in narrow bands thus point out considerable complexity in waves in a Vlasov plasma [for details see, e.g. (Hasegawa and Sato, 1989; Bekefi, 1966)].

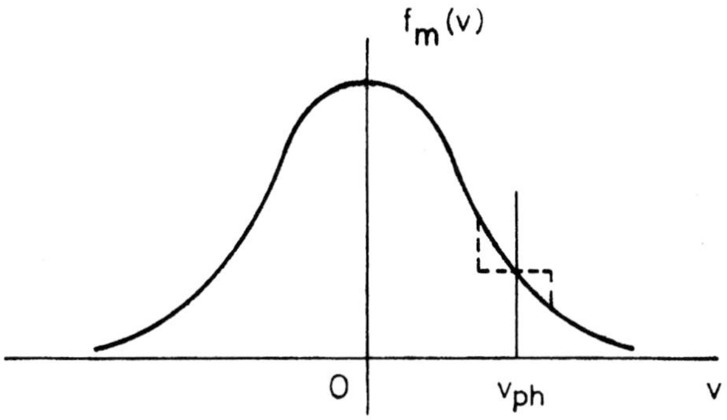

Fig. 10.8. Distortion of a Maxwellian distribution in the region $v \approx v_{ph}$ caused by Landau damping

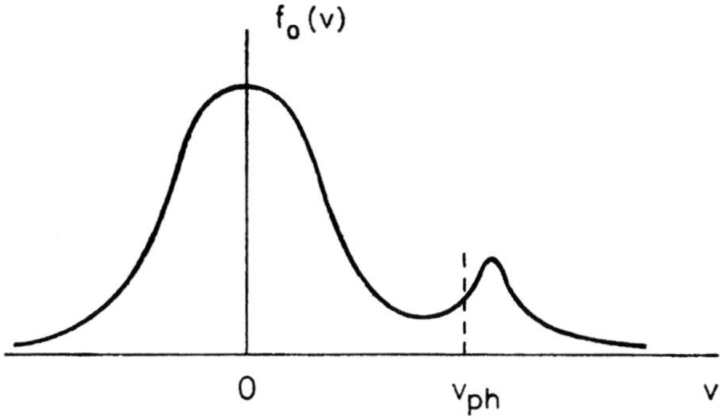

Fig. 10.9. A double-humped distribution and the region where instabilities will develop

10.5.6 Surface Waves

A new set of waves can arise due to discontinuity in the plasma variables across an interface between two plasma media, along which these waves propagate, but decrease exponentially in amplitude with distance from the interface. These waves are carried by a source at the discontinuous boundary and differ from the volume plasma waves. We shall first discuss the electrostatic surface wave in the cold plasma model and then the Alfvén surface waves in the magnetized plasma. The latter is very important in space plasma.

Electrostatic Surface Waves Consider a plasma interface $x = 0$ across which the plasma has a sharp density gradient.

$$n(x) = n_{01} \qquad x \geq 0\,,$$
$$n(x) = n_{02} \qquad x < 0\,. \qquad (10.140)$$

The amplitude of the perturbation will depend on x and the wave propagation is in the y–z plane. The continuity equation, the equation of motion and taking $\boldsymbol{E} = -grad\phi$, and the Poisson equation for electrons (with ions being stationary) give the equation for ϕ as

$$\nabla \cdot (\varepsilon \nabla \phi) = 0\,, \qquad (10.141)$$

which is to be satisfied in both regions 1 and 2.

Since ε is constant in medium 1 and medium 2, (10.141) reduces to the Laplace equation

$$\nabla^2 \phi = 0\,, \qquad (10.142)$$

which is to be solved with the boundary conditions: (i) the continuity of ϕ

$$\phi_1 = \phi_2 \quad \text{at} \quad x = 0 \qquad (10.143)$$

and (ii) the continuity of the normal component of D

$$\varepsilon_1 \frac{\partial \phi_1}{\partial x} = \varepsilon_2 \frac{\partial \phi_2}{\partial x} \quad \text{at} \quad x = 0\,. \qquad (10.144)$$

Seeking the solution of (10.142) only in the y-direction (for simplicity) and choosing the one that decays away from the boundary, we obtain

$$\phi_{1,2} = A_{1,2} e^{-k|x|}\,. \qquad (10.145)$$

Using the boundary conditions the dispersion relation is

$$\varepsilon_1 + \varepsilon_2 = 0 \quad \text{or} \qquad (10.146)$$
$$\omega^2 = (\omega_{p1}^2 + \omega_{p2}^2)/2\,. \qquad (10.147)$$

If $n_2 = 0$, (10.147) gives the well known (Krall and Trivelpiece, 1973) surface wave frequency along the plasma vacuum interface

$$\omega = \omega_{\rm p}/\sqrt{2}\,, \qquad (10.148)$$

which is smaller than the bulk plasma frequency. The surface wave is carried along the surface by the induced charge at the boundary surface.

Alfvén Surface Waves With the sharp density gradient as given in (10.140) also consider the discontinuity in the magnetic field taken along the z-direction

$$\boldsymbol{B}_0 = B_{01}\hat{z} \qquad (x \geq 0)\,,$$
$$\boldsymbol{B}_0 = B_{02}\hat{z} \qquad (x < 0)\,. \qquad (10.149)$$

Considering the MHD model for plasma we look for surface waves. The basic ideal MHD equations after linearization are

$$\left[\frac{\partial^2}{\partial t^2} - \frac{(\boldsymbol{B}_0 \cdot \nabla)^2}{4\pi\rho_0}\right]\boldsymbol{v} = -\frac{1}{\rho_0}\frac{\partial}{\partial t}\nabla\tilde{p} \qquad (10.150)$$
$$\nabla \cdot \boldsymbol{v} = 0\,, \qquad (10.151)$$

where $\tilde{p} = [p + [\boldsymbol{B}_0 \cdot \boldsymbol{B})/4\pi]$ is the total pressure and $\rho_0 = m_{\rm i} n_0$. Eliminating $\tilde{p}$ from (10.150) and (10.151) gives

$$\nabla^2 v_x = 0\,.$$

Therefore, we again obtain the Laplace equation governing the surface perturbations with the boundary conditions v_x and $\tilde{p}$ continuous. From (10.150) $\tilde{p}$ can be written as

$$\tilde{p} = \frac{1}{\omega k^2}\left(\omega^2\rho_0 - \frac{B_0^2 k_\parallel^2}{4\pi}\right)\frac{\partial v_x}{\partial x} \equiv \varepsilon \frac{\partial v_x}{\partial x}\,, \qquad (10.152)$$

where $k_\parallel = k_z$, $k_\perp = k_y$ and $k^2 = k_\parallel^2 + k_\perp^2$, so the boundary conditions at $x = 0$ become

$$v_{x1} = v_{x2}\,,$$
$$\varepsilon_1 \frac{\partial v_{x1}}{\partial x} = \varepsilon_2 \frac{\partial v_{x2}}{\partial x}\,.$$

Following the same method as above, the dispersion relation is

$$\rho_{01}\left(\frac{\omega^2}{k_\parallel^2} - v_{\rm A1}^2\right) + \rho_{02}\left(\frac{\omega^2}{k_\parallel^2} - v_{\rm A2}^2\right) = 0\,. \qquad (10.153)$$

In the case of $B_{01} = B_{02}$ and $\rho_{02} = 0$, the frequency of surface waves for plasma vacuum interface becomes

$$\omega_{\rm As} = \sqrt{2} v_{\rm A}\, k_\parallel\,. \qquad (10.154)$$

Note that the surface wave frequency in this case is higher than the Alfvén wave frequency, which is in contrast to the electrostatic wave where surface frequency is less than the bulk frequency.

The multi-spacecraft ISEE mission has given detailed observations of Alfvén surface waves with periods of about 2 min or more along the magnetopause (Song et al., 1988; Uberoi, 2003).

Resonant Absorption of Alfvén Surface Waves Instead of the sharp discontinuity (10.140) if we have a smooth density transition in the region $-a < x < a$, then instead of the Laplace equation we obtain the wave equation for Alfvén waves as

$$\frac{\mathrm{d}}{\mathrm{d}x}\left[\varepsilon(x)\frac{\mathrm{d}v_x}{\mathrm{d}x}\right] - k^2\varepsilon(x)v_x = 0\,, \qquad (10.155)$$

where $\varepsilon = [\rho_0(x)\omega^2 - k_\parallel^2 B_0^2/4\pi]$ and $k^2 = k_\parallel^2 + k_\perp^2$. We note that (10.155) has a logarithmic singularity at $\omega = k_\parallel v_A(x)$, i.e. at the point $x = x_c$, where the phase velocity of the wave meets the Alfvén wave velocity. The rigorous analysis of this equation (Uberoi, 1972; Stix, 1992)] shows that the surface waves, which are now coupled to the bulk Alfvén wave due to inhomogeneity, as seen from (10.155), are resonantly absorbed near the critical point $x = x_c$. The damping of surface waves due to resonant absorption in a collisionless plasma is similar to the Landau damping of high frequency electrostatic waves. The phenomena of resonant absorption of Alfvén waves has various applications in space plasma processes, for example, in the study of micropulsations (Chen and Hasegawa, 1974), in the understanding of the structure of auroras, heating of plasmas (Hasegawa, 1976) and recently in the understanding of magnetic reconnection at the magnetopause (Uberoi, 1994; Uberoi et al., 1996).

10.6 Equilibria and Their Stability

When all the macroscopic quantities characterizing a plasma are constant in time, the plasma is said to be in equilibrium. In the laboratory and in nature, particular interest attaches to a "confined" plasma, which is held in a steady state within a finite region, surrounded by a magnetic field. Many possible equilibrium configurations and the extent to which these are stable have been studied in the laboratory because of the importance of the confined plasma system in the understanding of the release of the thermonuclear fusion power. Many of these model space configurations, like for example the magnetopause, which is an interface between the solar wind plasma and the magnetospheric plasma or the tail region of the magnetosphere, are important for space plasma processes. However, as space is concerned with large-scale phenomena, we limit the discussions to MHD instabilities and can treat most of the confined regions as plane systems.

The plasma equilibrium state will be stable if all types of infinitesimal perturbations lead to damped oscillations about this equilibrium state, and unstable if one or more types of perturbations grow exponentially. For this, a perturbed plasma MHD system can be studied in terms of waves that are generated by a disturbance. Small disturbances can be approximated by a traveling plane wave solution of the form $\exp[i(\boldsymbol{k}\cdot\boldsymbol{r} - \omega t)]$. In Sect. 10.5.6 the dispersion relation obtained was examined only for real frequencies. For the stability analysis, we examine this relation for complex frequencies as these are associated with unstable wave modes. The frequency of a wave can be generally written as $\omega = \omega_R + i\omega_I$, where ω_R and ω_I are real. Then from the expression $\exp -i\omega t$, we note that ω_I determines whether a wave grows ($\omega_I > 0$), or decays ($\omega_I < 0$) in time. This method of stability analysis is known as normal mode analysis.

The stability or instability of any physical system is determined by its kinetic and potential energy as a function of the system parameters. A very familiar example is a ball on a hill versus a ball in a well. Both are in equilibrium, but one is unstable and the other one stable. The other method sometimes used to find out about the stability of a system is the energy principle method. The change in potential energy ΔV is examined for small displacements about the equilibrium. If ΔV is less than zero, the system is stable.

10.6.1 Interface Instabilities

We shall first derive a general dispersion relation for small disturbances propagating along the plasma interface $x = 0$ between the two plasma systems moving with velocity $\boldsymbol{V}_1$ for $x < 0$ and $\boldsymbol{V}_2$ for $x > 0$. The plane boundary is taken along the $y-z$-direction and magnetic field in the z-direction. We consider an external force $\boldsymbol{F} = F\hat{\boldsymbol{x}}$, a non-magnetic constant force, acting perpendicular to the boundary (for example, the gravitational force or curvature force). Considering the wave that propagates along this surface as $A(x)\exp[i(\boldsymbol{k}\cdot\boldsymbol{r} - \omega t)]$, the linearized MHD equations of plasma will give wave equations in both media, as in the case of Alfvén surface waves. The boundary conditions for the perturbed quantities, the continuity of total pressure, the continuity of the normal component of velocity and the continuity of the normal component of the perturbed magnetic field across the interface $x = 0$ gives the dispersion relation

$$\begin{aligned} &\rho_{01}[(\omega - \boldsymbol{k}\cdot\boldsymbol{V}_1)^2 - (\boldsymbol{k}\cdot\boldsymbol{v}_{A1})^2] \\ &+ \rho_{02}[(\omega - \boldsymbol{k}\cdot\boldsymbol{V}_2)^2 - (\boldsymbol{k}\cdot\boldsymbol{v}_{A2})^2] \\ &= -kF(\rho_{01} - \rho_{02}) . \end{aligned} \tag{10.156}$$

Here $k = (k_y^2 + k_z^2)^{1/2}$. Note that when $V_1 = V_2 = 0$ and $F = 0$, (10.156) gives the dispersion relation for the Alfvén surface waves.

Rayleigh–Taylor Instability This instability in plasma has an analogy in hydrodynamic instability, which arises when a heavier density fluid is balanced above a lighter density fluid in a gravity field. The motion caused by the dense fluid penetrating into the underlying lighter fluid drives the Rayleigh–Taylor (R–T) instability (Chandrasekhar, 1961). From (10.156), when $V_1 = V_2 = 0$, $F = g$ and the magnetic field is absent

$$\omega^2 = \frac{k(\rho_{01} - \rho_{02})}{\rho_{01} + \rho_{02}} g . \tag{10.157}$$

When $\rho_{01} > \rho_{02}$ the system is stable and perturbations are surface gravity waves along the interface. However, if $\rho_{01} < \rho_{02}$, the equilibrium of heavy density fluid above the lighter fluid is unstable as ω is imaginary, giving growing perturbations.

Now consider a plasma-vacuum interface, plasma being supported by the magnetic field, considering the wave propagation such that $\boldsymbol{k} = k\hat{\boldsymbol{y}}$ or waves perpendicular to the magnetic field direction (10.156) gives

$$\omega^2 = -kg ,$$

or ω is imaginary giving growing perturbation making the system unstable with A rate of growth $\sqrt{kg}$, which is same as in the case of R–T instability for fluids. If the fluid has a smooth interface so that the density ρ_0 changes smoothly from medium 1 to medium 2, this dispersion relation takes the form (see, e.g. Parks, 1991)

$$\omega^2 = +\frac{1}{\rho_0}\frac{d\rho_0}{dx} g . \tag{10.158}$$

The relation (10.158) is obtained for short wavelengths. If $\boldsymbol{g}$ and $\nabla\rho_0$ are in the same direction $\omega^2 > 0$, the system is stable. If these are in the opposite direction, $\omega^2 < 0$ and the system is unstable.

This instability for a plasma-vacuum interface can be understood by using the particle orbit theory following the arguments given by Rosenbluth and Longmire (1957). Considering the same geometry as above note that as the $\boldsymbol{g} \times \boldsymbol{B}$ drift is mass-dependent, the ions will drift faster than the electrons. The drift $\boldsymbol{u}_0 = -g/\omega_{ci}\hat{\boldsymbol{y}}$ of ions over the rippled surface (Fig. 10.10) will cause the charges to build up. This charge separation produces an electric field $\boldsymbol{E}_1$ and since the charges change sign between the minimum and the maximum of the ripples $\boldsymbol{E}_1 \times \boldsymbol{B}$ is in the x-direction at the minimum and in the $-x$-direction at the peaks. The amplitude of the ripple will thus grow larger and the boundary becomes unstable. The dispersion relation for R–T instability for the plasma vacuum case can be obtained by taking ion drift velocity and neglecting the electron mass in two-fluid equations for plasma. The electrostatic perturbations then show that for $\omega \ll \omega_{ci}$, the growth rate of R–T plasma instability is exactly the same as R–T fluid instability. Thus the charge separation is able to over-

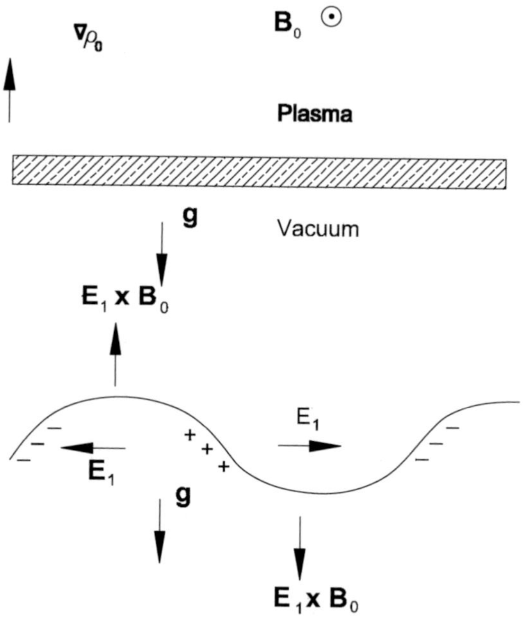

Fig. 10.10. Charge separation in instability of a plasma-vacuum interface in a gravitational field

come exactly the restraining influence of the magnetic field.

Flute Instability The essential mechanism of the Taylor instability is the charge separation produced by the gravitational force $\boldsymbol{g}$. Any force perpendicular to $\boldsymbol{B}$ that is independent of the sign of the charge will cause such a charge separation. Consider, therefore, a plasma configuration with a curved boundary, more appropriate to model space configurations, such as the magnetopause. In this case, the guiding centers will be subject to curvature and gradient drifts. Using this, g has to be replaced by

$$g \to \frac{\boldsymbol{R}}{R^2}\left(v_\parallel^2 + \frac{1}{2}v_\perp^2\right), \qquad (10.159)$$

where $\boldsymbol{R}$ is the radius of curvature of the magnetic line of force, to obtain the instability criterion. If $\boldsymbol{R}$ is in the same direction as the density gradient, the configuration is stable, while if it is in the opposite direction, the configuration is unstable. This also implies that if the magnetic field curves away from the plasma, the electric field produced by the charge-dependent drift will damp the oscillation and restore the equilibrium. On the other hand, if the magnetic field points into the plasma, the electric field produced by the drift will make the perturbations grow and the boundary is unstable. Ripples at the boundary resemble "flutes", and these instabilities are also referred to as flute instabilities.

Kelvin–Helmholtz Instability In the presence of velocity shear a plasma–plasma interface in the absence of any external force can support surface waves that under certain conditions can grow and make the interface unstable. This is called the Kelvin–Helmholtz instability. This instability is very important in solar wind–planetary interactions.

Equation (10.156) for $\boldsymbol{F} = 0$ gives the dispersion relation for K–H instability as

$$\rho_{01}\left[(\omega - \boldsymbol{k}\cdot\boldsymbol{V}_1)^2 - (\boldsymbol{k}\cdot\boldsymbol{v}_{\mathrm{A1}})^2\right] + \rho_{02}\left[(\omega - \boldsymbol{k}\cdot\boldsymbol{V}_2)^2 - (\boldsymbol{k}\cdot\boldsymbol{v}_{\mathrm{A2}})^2\right] = 0\,. \qquad (10.160)$$

Equation (10.160) can be shown to have complex roots [see (Chandrasekhar, 1961)], corresponding to the unstable modes provided the flow shear $\Delta V = |V_1 - V_2|$ is such that

$$\rho_{01}\rho_{02}(\Delta\boldsymbol{V}\cdot\boldsymbol{k})^2 > (\rho_{01} + \rho_{02}) \left[\rho_{01}(\boldsymbol{k}\cdot\boldsymbol{v}_{\mathrm{A1}})^2 + \rho_{02}(\boldsymbol{k}\cdot\boldsymbol{v}_{\mathrm{A2}})^2\right]. \qquad (10.161)$$

A critical velocity shear or a shear threshold is required to produce the K–H instability. When the magnetic field is zero, the K–H instability occurs for all wave numbers for any value of ΔV. The condition (10.161) arises because the tension in the magnetic field will resist any force acting on it to stretch it. Since $\boldsymbol{B}_0$ and $\boldsymbol{V}$ are restricted to be in the x–y plane, this instability is likely to occur in the dawn-dusk flanks of the magnetopause. An important point to note is that the excited mode is the Alfvén surface mode, which becomes unstable for a critical value of the velocity shear.

Firehose Instability The dispersion relation for the Alfvén wave propagating along the magnetic field modified in the presence of anisotropic pressure (Hasegawa, 1971) is

$$\frac{\omega^2}{k^2} = v_{\mathrm{A}}^2\left[1 - \frac{(p_\parallel - p_\perp)}{B^2/4\pi}\right] \equiv v_{\mathrm{A}}^2\left[1 - \frac{1}{2}(\beta_\parallel - \beta_\perp)\right]. \qquad (10.162)$$

This relation shows that for low-β plasma ($p_\parallel$ or $p_\perp \ll B^2/4\pi$) or for nearly isotropic plasmas ($p_\parallel \sim p_\perp$) the phase velocity reduces to the Alfvén speed v_A. The speed is larger than v_{A} if $\beta_\parallel < \beta_\perp$. When $\beta_\parallel - \beta_\perp > 2$, ω becomes purely imaginary and we get instability. Since the physical mechanism of this instability is similar to that which generates oscillations in the water hose when the water pressure exceeds a critical value, this is called "fire-hose instability".

10.6.2 Two-stream Instability

Two-stream instability occurs as an electrostatic instability when electrons and ions are drifting with a relative velocity. Consider the two-fluid equation for ions and electrons with electrons moving with velocity $\boldsymbol{u}_0$ against the stationary ions, for the cold plasma case. The electric field is given by the Poisson equation. The linear dispersion equation for the waves propagating along the direction of $\boldsymbol{u}_0$ with $\boldsymbol{u}_0 = (0, 0, u_0)$, so that the wave number

k is in the z-direction, is as follows:

$$1 = \frac{\omega_{pi}^2}{\omega^2} + \frac{\omega_{pe}^2}{(\omega - ku_0)^2} . \quad (10.163)$$

This equation is a fourth-order algebraic equation in ω. For a given real k its roots can be either real or complex. If complex, these will occur in complex conjugate pairs, so one of the roots will give an unstable wave. The analysis of this equation shows that for a given u_0, the system is unstable for $k < k_c$ or for long wavelength oscillations. The maximum growth rate can be calculated to be

$$Im\omega \approx \omega_p \left(\frac{m_e}{m_i} \right)^{1/3} . \quad (10.164)$$

We shall now briefly discuss the mechanism for two-stream instability.

Writing the dispersion relation as

$$(1 + \varepsilon_s + \varepsilon_p) = 0 , \quad (10.165)$$

where

$$\varepsilon_p = -\frac{\omega_{pi}^2}{\omega^2} \quad \text{and} \quad \varepsilon_s = -\frac{\omega_{pe}^2}{(\omega - ku_0)^2} , \quad (10.166)$$

ε_s is the dielectric constant for the drifting part of plasma and ε_p for the stationary plasma. The electric field energy W of a wave propagating in a lossless dielectric medium can in general be expressed as (Landau and Lifshitz, 1960)

$$W = \frac{\partial(\omega\varepsilon)}{\partial\omega} \frac{\langle E^2 \rangle}{2} , \quad (10.167)$$

where $\langle E^2 \rangle$ is the time average of the square of the electric-field amplitude. Using (10.166) we see

$$W_s = \frac{\omega + ku_0}{(\omega - ku_0)^3} \omega_{pe}^2 \quad (10.168)$$

and

$$W_p = \frac{\partial(\omega\varepsilon_p)}{\partial\omega} = \frac{\omega_{pi}^2}{\omega^2} . \quad (10.169)$$

Thus we see that W_p is always positive, whereas W_s can be negative if $\omega < ku_0$. Hence this mode carries a negative energy and is called a negative energy wave.

The two-stream instability for cold plasma can, therefore, be interpreted as caused by the coupling between the negative energy wave in the drifted electron stream and the positive energy wave in the plasma.

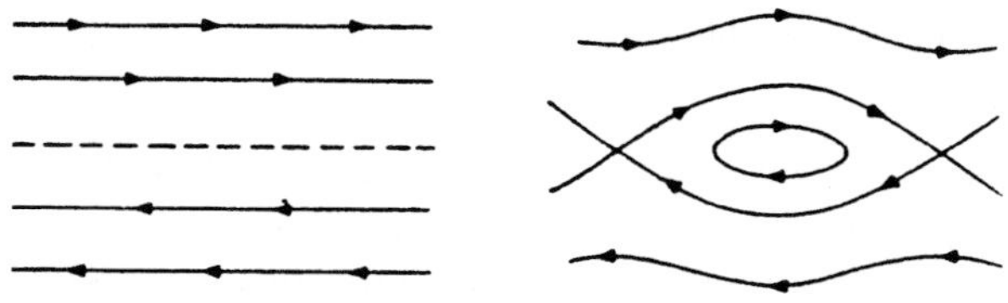

Fig. 10.11. The change of magnetic topology due to excitation of the tearing mode instability

Thermal Effects In the case of thermal effects becoming important, this instability is studied by using kinetic equations and it is excited by negative dissipation produced by the wave-particle interaction. The necessary condition of instability is

$$\frac{\omega \partial f_0}{\partial v}\bigg|_{v=\omega/k} > 0 . \quad (10.170)$$

This means that at $v = \omega/k = v_{ph}$, the unperturbed velocity distribution function has a positive gradient (Fig. 10.9). When considering the low-frequencies the ion dynamics becomes important and it is found that instability persists when $u_0 > v_s$, the ion acoustic speed. For further discussion about thermal effects, see e.g. (Uberoi, 1988; Hasegawa 1971).

10.6.3 Tearing Mode Instability

This is the very important mode of the instability of a neutral sheet first pointed out by Dungey (1958). It requires dissipation like finite resistivity in the excitation. An elaborate analysis of this instability was given by Furth et al. (1963). As the name suggests, the tearing mode breaks or tears up the current sheet into a number of smaller segments or magnetic islands (Fig. 10.11), thus changing the magnetic topology of the system. The instability occurs when an increase in island size leads to a state of lower magnetic energy. For a comprehensive account of this instability, the reader is referred to, e.g. Hasegawa (1971). Tearing mode instability is very important in thex study of magnetic reconnection processes in space plasmas. Here we would like to mention that the excitation of this instability by the long-wavelength Alfvén surface waves and its relationship to Alfvén resonance theory was pointed out recently by Uberoi (2003).

10.7 Conclusion

We have only considered plasmas consisting of electrons and single species of ions. No attempt has been made to give the new features introduced by considering neutrals in partially ionized plasmas and multi-ion species in multi-component plasmas that are known to be present in the near-Earth space environment. However, the basic equations for electron-ion plasmas with can be used with some modifications to study these plasmas. The other topic that has been omitted is radiation processes in plasmas, as this subject has a wide scope and cannot be covered in these limited pages. A comprehensive study of this topic is available in Bekefi (1966).

References

Alfvén, H., Existence of electromagnetic hydrodynamic waves, *Nature*, **150**, 504, 1942.

Alfvén, H., Fälthammar, C.G., Cosmical Electrodynamics, Fundamental Principles, 2nd ed., Oxford University Press, 1963.

Bekefi, G. Radiation Processes in Plasmas, Wiley, New York, 1966.

Bernstein, I.B., Plasma oscillations perpendicular to a constant magnetic field, *Phys. Fluids*, **3**, 489, 1960.

Bhatnagar, P.L., Gross, E.P., Krook, M., A model for collision processes in gases, *Phys. Rev.* **94**, 511, 1954.

Braginskii, S.I., Review of Plasma Physics, Vol. 1, ed. Lentovich, L., Consultant Bureau, New York, 1965.

Budden, K.G., The Propagation of Radio Waves, Cambridge University Press, Cambridge, 1985.

Chandrasekhar, S., Stochastic problems in physics and astronomy, *Rev. Mod. Phys.* **15**, 1, 1943.

Chandrasekhar, S., Hydrodynamic and Hydromagnetic Stability, Clarendon Press, Oxford, 1961.

Chapman, S., Cowling, T.G., The Mathematical Theory of Non-Uniform Gases, Cambridge University Press, Cambridge, UK, 1953.

Chen, F.F., Introduction to Plasma Physics, Plenum Press, New York, 1974.

Chen, L., Hasegawa, A., A theory of long period magnetic pulsations, 1, Steady state excitation of field line resonance, *J. Geophys. Res.* **79**, 1024, 1974.

Cramer, N.F., The Physics of Alfvén Waves, Wiley-VCH, Berlin, Germany, 2001.

Dawson, J., On Landau damping, *Phys. Fluids* **4**, 809, 1961.

Dungey, J.W., Cosmic Electrodynamics, Cambridge University Press, New York, 1958.

Furth, H.P., Killeen, J., Rosenbluth, M.N., Finite resistivity instabilities of a sheet pinch, *Phys. Fluids*, **6**, 459, 1963.

Hasegawa, A., Plasma instabilities in the magnetosphere, *Rev. Geophys. and Sp. Phys.* **9**, 703, 1971.

Hasegawa, A., Particle acceleration by MHD surface wave and formation of aurora, *J. Geophys. Res.* **81**, 5083, 1976.

Hasegawa, A., Sato, T., Space Plasma Physics: 1. Stationary Processes, Springer, Berlin Heidelberg New York, 1989.

Hasegawa, A., Uberoi, C., The Alfvén Wave, Technical Information Center, US Department of Energy, Oak Ridge, Tennessee, 1989.

Helliwell, R.A., Whistlers and Related Ionospheric Phenomena, Stanford University Press, Stanford, CA, 1965.

Jackson, J.D., Longitudinal plasma oscillations, *J. Nucl. Energy, Part C, Plasma Physics*, **1**, 171, 1960.

Kelly, M.C., The Earth's Ionosphere: Plasma Physics and Electrodynamics, Academic Press, IMG, New York, 1989.

Krall, N.A., Trivelpiece, A.W., Principles of Plasma Physics, McGraw Hill, New York, 1973.

Langmuir, I., Oscillation in ionized gases, *Proc. Nat. Acad. Sciences*, XII, 274, 1928.

Langmuir, I., Tonks, L., Oscillations in ionized gases, *Phys. Rev.* **33**, 195, 1929.

Longmire, C.L., Elementary Plasma Physics, Interscience Publishers, New York, 1973.

Landau, L.D., On the Vibrations of the Electronic Plasma, J. Phys. (USSR) **10**, 25, 1946.

Landau, L.D., Lifshitz, E.M., Electrodynamics of Continuous Media, p. 253, Pergamon, Reading, Mass., 1960.

Northrop, T.G., The Adiabatic Motion of Charged Particles, Interscience Pub., New York, 1963.

Parks, G.K., Physics of Space Plasmas: An Introduction, Addision-Wesley, Reading, Mass, 1991.

Potemra, T., ed., Magnetospheric Currents, Geophysics Monograph, **28**, AGU, Washington, DC, 1984.

Rosenbluth, M.N., Longmire, C.L., Stability of plasmas confined by magnetic fields, *Annals of Physics*, **1**, 120, 1957.

Saha, M.N., Ionization in the solar chromosphere, *Phil. Mag.* **40**, 472, 1920.

Schulz, M., Lanzerotti, L.J., Particle diffusion in the radiation belts, Springer, Berlin Heidelberg New York, N.Y., 1974.

Song, P., Elphic, R.C., Russell, C.T., ISEE I and II Observations of the oscillating magnetopause, *Geophys. Res. Lett.* **15**, 744, 1988.

Spitzer, L. Physics of Fully Ionized Gases, Interscience Pub., New York, 1962.

Stix, T.H., The Theory of Plasma Waves, McGraw Hill, New York, 1962.

Stix, T.H., Waves in Plasmas, American Institute of Physics, New York, 1992.

Storey, L.R.O., An investigation of whistling atmospherics, Phil. Trans. Roy. Soc. (London) **246**, 1136, 1953.

Uberoi, C., Alfvén Waves in inhomogeneous magnetic fields, *Phys. Fluids*, **15**, 1673, 1972.

Uberoi, C., Crossover frequencies in multicomponent plasma, *Phys. Fluids*, **16**, 704, 1973.

Uberoi, C., Introduction to Unmagnetized Plasmas, Prentice Hall, New Delhi, India, 1988.

Uberoi, C., Resonant Absorption of Alfvén Waves near a neutral point, *Plasma Physics*, **52**, 215, 1994.

Uberoi, C., Earth's Proximal Space: Plasma Electrodynamics and the Solar System, University Press, India, 2000.

Uberoi, C., Some observational evidence of Alfvén surface waves induced magnetic reconnection, Sp. Science Reviews, vol. 107, p. 197 Proceedings of the WISER Workshop, Chief ed. Chian, A.C.-L., Kluwer Academic Publishers, 2003.

Uberoi, C., Lanzerotti, L.J., Wolfe, A., Surface Waves and Magnetic Reconnection at a Magnetopause, *J. Geophys. Res.* **101**, 24, 979, 1996.

Uhlenbeck, G.E., The validity and the limitations of the Boltzmann Equation, in The Boltzmann Equation, eds., Cohen, E.G.C. and Thirring, W., Springer, Berlin Heidelberg New York, 1973.

Vlasov, A.A., Theory of Vibrational Properties of an Electron Gas and its Applications, *J. Phys. (USSR)*, **9**, 25, 1945.

11 Magnetic Reconnection

Atsuhiro Nishida

Magnetic reconnection is the key mechanism that controls input as well as output of energy and momentum in the magnetosphere. The input occurs on the magnetopause, and the output is initiated in the magnetotail. This article presents basic features of the reconnection in the magnetosphere, namely, (1) where the reconnection line is formed, (2) how the reconnection operates, and (3) what the main consequences of the reconnection are, for the magnetopause and magnetotail reconnections, respectively. The process leading to onsets of reconnection in the near-Earth region of the magnetotail is also addressed.

Contents

Atsuhiro Nishida, Magnetic Reconnection.
In: Y. Kamide/A. Chian, Handbook of the Solar-Terrestrial Environment. pp. 279–310 (2007)
DOI: 10.1007/11367758_11

11.1 Introduction

A characteristic feature of the plasma that distinguishes it from non-ionized gaseous bodies is its strong coupling with the magnetic field. In collisionless plasmas having essentially infinite electric conductivity this coupling is expressed by the frozen-in theorem [Alfvén, 1958] where the magnetic field lines are constrained to move with the plasma. This theorem implies that, if the electrical conductivity were practically infinite everywhere, plasmas of solar wind and of the magnetosphere cannot mix, so that the interaction between the solar wind and the geomagnetic field would be limited to the confinement of the latter in a cavity carved in the former. This obviously contradicts multitudes of observations (constituting a research discipline called solar terrestrial relation) of the influences exerted by the solar wind on the magnetosphere's interior.

Because of this difficulty Alfvén himself suggested that the frozen-in theorem is an ideal which is not realized in the real space environment. In his theory of magnetic storms he adopted a contrasting viewpoint that the interplanetary magnetic field can be superposed on the geomagnetic field and the solar plasma has an unobstructed access into the geomagnetic field. This was based on his belief the space plasma is rife with processes that make the electric conductivity low and the frozen-in theorem inapplicable even if inter-particle collisions are extremely infrequent. It is clear, however, that Alfvén's bold suggestion is untenable because the magnetohydrodynamics built on the frozen-in theorem has been quite successful in explaining multitudes of observed features. A typical example is the magnetic pulsations that have been shown to be explainable as magnetohydrodynamic resonant oscillations of geomagnetic field lines.

In fact, the frozen-in theorem does not have to be broken everywhere in order that the solar wind imparts its energy to the magnetosphere and produces structures and disturbances in its interior. It suffices that the electric conductivity is reduced in a limited region on the boundary surface, because the magnetic field lines of the solar wind and magnetosphere become connected there. These connected field lines, called open field lines, are dragged anti-sunward with the solar wind and sweep the entire boundary surface downstream of this magnetic diffusion region. In this manner the coupling between the solar wind and the magnetosphere's interior can be accomplished if the conductivity is reduced only in a limited region on the magnetopause even if the magnetohydrodynamics is applicable everywhere else. The process is called magnetic reconnection, and the line at which field lines make contact and reconnect is the reconnection line, which is often called X-line because the connected field lines take a shape of the letter X. Magnetic reconnection on the magnetopause generates global convection in the magnetosphere and produces the magnetotail on the nightside, as suggested originally by Dungey [1961].

The magnetotail is constituted primarily by the open field lines which are dragged tailward with the solar wind. These open field lines, in turn, are reconnected in the magnetotail; Open field lines rooted in northern and southern polar caps meet in the distant tail and are reconnected. Of the field lines resulting from this reconnection the earthward ones are closed field lines, that is, have both ends in the ionosphere. The closed field lines thus produced move back toward the dayside magnetopause and complete the global convection. The reconnected field lines on the anti-sunward side of the X-line have both ends in the solar wind and flow away from the magnetotail. In addition to this reconnection in the distant tail, the tail reconnection occurs also in the near-Earth region of the tail where field lines are closed, and this process is associated with the global disturbance called substorm.

Magnetic reconnection converts the magnetic energy into the plasma energy. Energy is dissipated by non-magnetohydrodynamics processes that operate in the magnetic diffusion region. Particles are accelerated further on the reconnected field lines since these field lines have strong curvatures and are stressed. (Acceleration and heating occur also in such regions as the thin current sheet.) Ions and electrons accelerated by reconnection on the magnetopause stream down to the ionosphere at the polar cusp. Reconnection in the tail fills the plasma sheet with hot particles and produces anti-earthward ejection of the accelerated plasma. In the magnetospheric dynamics the magnetic reconnection is the key process that not only supplies plasma, momentum and energy into the magnetosphere but also activates the magnetosphere by energizing particles and generating currents inside the magnetosphere.

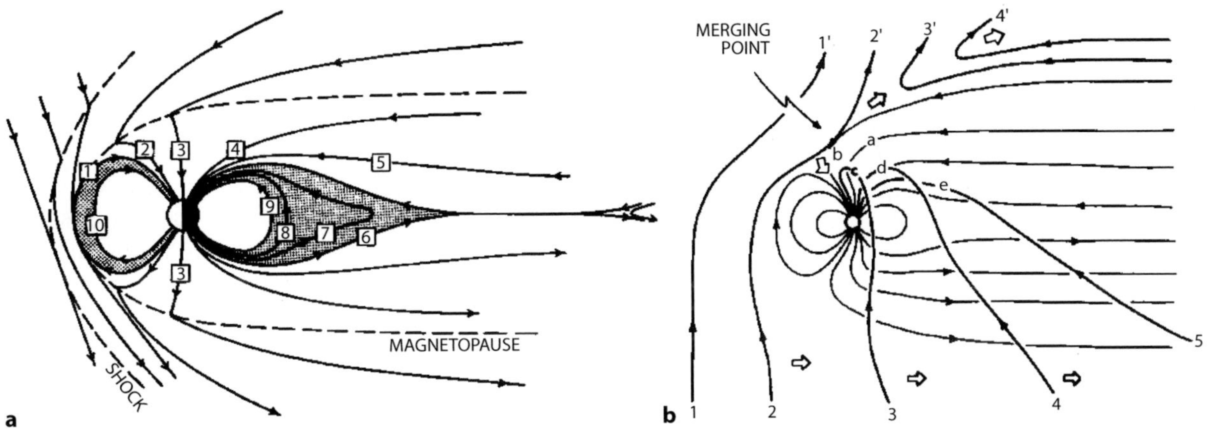

Fig. 11.1a,b. Models of the open magnetosphere for (**a**) the southward IMF [Levy et al., 1964] and (**b**) the northward IMFcondition [Maezawa, 1976]. Convection is illustrated by noting successive positions of a field line after its reconnection on the magnetopause

11.2 Reconnection on the Magnetopause

11.2.1 Formation and Topology of Open Field Lines

In the basic configuration of the open magnetosphere illustrated in Fig. 11.1a, reconnection with the interplanetary magnetic field (IMF) occurs with the closed field lines of the geomagnetic field and drives convection throughout the magnetosphere. In this illustration it is assumed that IMF has the southward component. The closed field line labeled 1 is reconnected with IMF and the open field line moves anti-sunward taking the positions 2, 3, 4, 5 successively. These open field lines constitude the lobes of the magnetotail. Reconnection of the open field lines in the tail forms the closed field line 6, and this moves sunward taking positions 7, 8, 9, and 10 and eventually reaches the position 1 where it is reconnected again with IMF [Levy et al., 1964]. Thus reconnection generates a large scale convection of magnetospheric plasma and associated field lines during which the topology of the field liens changes between the closed and the open. When IMF has the northward component, reconnection occurs on the surface of the tail as illustrated in Fig. 11.1b. The open field line b is reconnected with the IMF 2-2' and makes an open field line b-2. This field line is dragged anti-sunward taking the positions c-3, d-4, e-5, and eventually return to b where it is connected with IMF again. In this latter type of convection the field lines involved keep the open topology all the time. This reconnection also produces the field lines 3' and 4' which are not connected with the geomagnetic field and flow away with the solar wind [Maezawa, 1976]. For convenience we shall call the reconnection of IMF with the dayside closed field lines as "frontside reconnection" and its reconnection with the lobe field lines as "lobe reconnection."

In an idealized situation where IMF is due southward, the reconnection line would be a magnetic neutral line that is formed in the equatorial section of the magnetopause. In actual cases IMF has the east or west component which more often exceeds the north-south component. The equatorial neutral line of the frontside reconnection for the due southward IMF could be generalized to the reconnection line under more general IMF conditions by either of the following two ways. The first is to consider a reconnection line along which projections of IMF and GMF (geomagnetic field) have the same value, so that the components perpendicular to the line have opposite signs though different intensities. The reconnection line of this type, called the component reconnection, is illustrated for the case of the frontside reconnection by a dashed line in Fig. 11.2 for the IMF with (a) the southward polarity and (b) the northward polarity, respectively [Nishida and Maezawa, 1971]. In this figure the dayside magnetopause is modeled by a surface of a doughnut. While the above condition specifies the attitude but not the position of the reconnection line, the line is drawn to pass the dayside nose of the magnetosphere which is the stagnation point of the solar wind flow. This guarantees that a pair of field

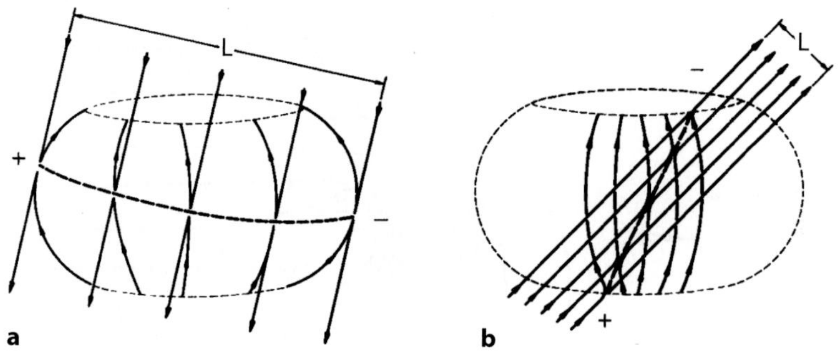

Fig. 11.2a,b. The component reconnection for (**a**) the southward and (**b**) the northward IMF. Reconnection occurs at the reconnection line (*dashed curve*) on the magnetopause which is modeled by the surface of a doughnut. L is the width of the IMF field lines that are reconnected, and plus and minus signs show the potential difference along the reconnection line [Nishida and Maezawa, 1971]

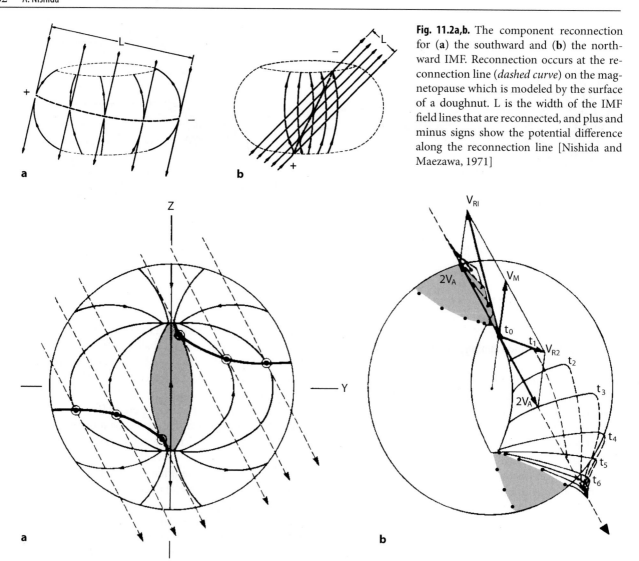

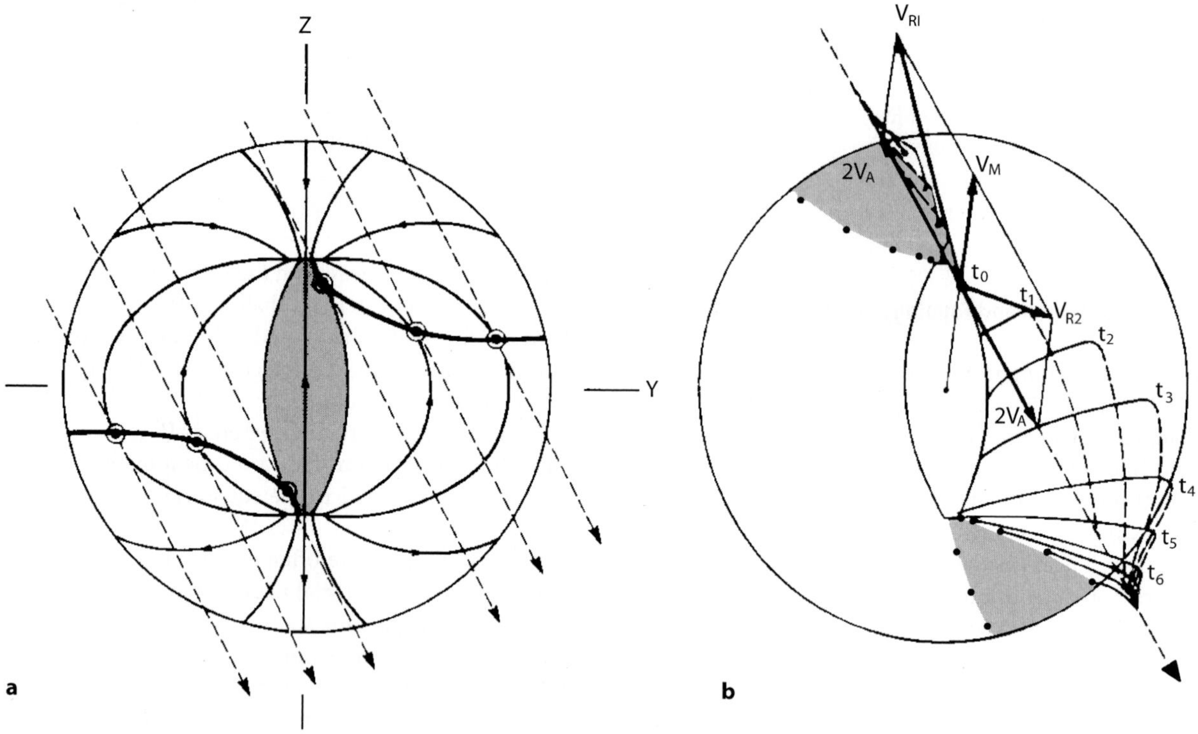

Fig. 11.3a,b. The anti-parallel reconnection in the schematic view of the dayside magnetopause. (**a**) Reconnection of IMF (*dashed lines*) with GMF (*thin solid lines*) takes place at the locus (*thick solid line*) of points where these fields are locally anti-parallel. (**b**) A field line reconnected near the northern hemisphere cusp at time t_0 makes a pair of open field lines that are dragged poleward in both north and south directions [Crooker, 1979]

lines that is produced by the reconnection is separated smoothly away from the reconnection line. The second is to consider the line which traces the points where IMF and GMF are exactly anti-parallel [Crooker, 1979]. As illustrated in Fig. 11.3a, the reconnection of this type, called anti-parallel reconnection, operates at a pair of reconnection lines (illustrated by thick curves) off the equator, and characteristically there is a lune-shaped gap (lightly shaded) around the noon meridian where field directions do not become anti-parallel except when IMF is due southward. Figure 11.3b shows successive positions of a field line which has been reconnected near the northern cusp at t_0 as they are dragged anti-sunward by the solar wind through the time t_1 to t_6.

Searching for clues to determine whether the component or the anti-parallel reconnection is taking place,

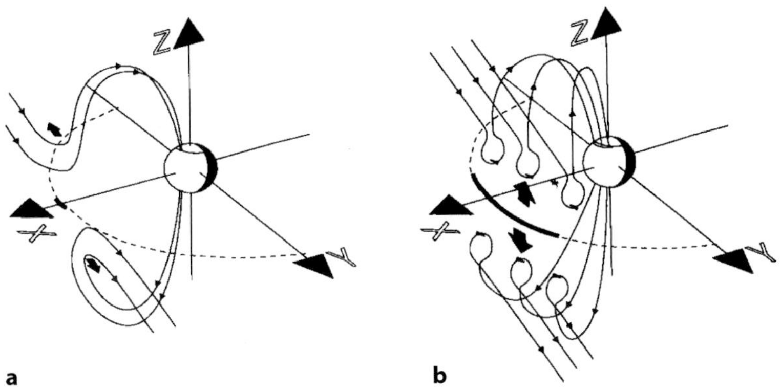

Fig. 11.4a,b. Configurations of reconnected field lines when the reconnection occurs (**a**) as a pulse at a relatively short X-line, and (**b**) as a two-dimensionally symmetric pulse due to time variation of the reconnection rate [Lockwood and Hopgood, 1998]

Petrinec et al. [2003] have found a case where a gap was observed in a longitudinally extended belt of ionospheric emissions at the cusp around noon as predicted by the anti-parallel reconnection model. On the other hand, Kim et al.[2002] have observed frequent reversals in the beam direction along the magnetopause during an interval of large and relatively quiet IMF B_y that suggest the component reconnection in low latitudes.

In comparing the component reconnection and the anti-parallel reconnection with observations, it has to be noted that the anti-parallelism is not the sole condition that determines the reconnection rate. The force acting to separate the reconnected field lines would be stronger where the field lines are antiparallel, but the field strengths that also influence this force tend to be weak at large shear angles and offset the effect of the anti-parallelism [Phan et al., 1996]. In addition, plasma and field lines in the magnetosheath are in motion, so that the force that acts at the reconnection line should be strong enough to overcome this background flow in order that the reconnection can operate continuously at a given position. Otherwise the reconnection line would be washed away. It is also necessary that the open field lines that are produced at different positions on the reconnection line do not move against each other. Otherwise the open field lines become tangled. All these dynamical effects influence the efficiency and smooth progress of reconnection at any point.

Reconnection on the magnetopause does not necessarily operate continuously or over an extended reconnection line. Signatures of spatially and/or temporarily limited reconnection have been identified as Flux Transfer Events (FTEs) [Russell and Elphic, 1979]. Figure 11.4a is a schematic of the structure of FTE. The dashed line shows the equatorial section of the magnetopause where a short reconnection line shown by a heavy solid line is formed as a pulse. The newly opened field tubes move away in the direction of the dark arrow. Figure 11.4b is a schematic for a similar case where the reconnection line is fixed but the rate changes with time so that a bulge is formed in the open field lines [Lockwood and Hapgood, 1998]. The temporal and/or spatial variations of the reconnection line would primarily be due to variations in IMF, but they can be produced also from the dynamics intrinsic to the magnetopause layer.

11.2.2 Structure of the Magnetopause in Terms of Magnetohydrodynamics

Figure 11.5 illustrates the section of the magnetopause in the direction perpendicular to the reconnection line. The separatrices, that is, the field lines that pass the reconnection line, are labeled S1 and S2. The plasma flow (dashed lines) is directed from the magnetosheath toward the magnetosphere and gains speed as it crosses the magnetopause (MP). The magnetic field (solid lines) changes direction across the MP and has a component normal to MP. The tangential component $\boldsymbol{E}_\mathrm{t}$ of the electric field, which accompanies the flow perpendicular to the magnetic field, should be constant across MP in the steady state. In terms of magnetohydrodynamics, these properties identify the magnetopause with a rotational discontinuity or, in other words, an Alfvén wave (intermediate wave).

Several properties have to be seen if MP is a rotational discontinuity. The first item for the test is the presence of the normal component B_n. Although many cases

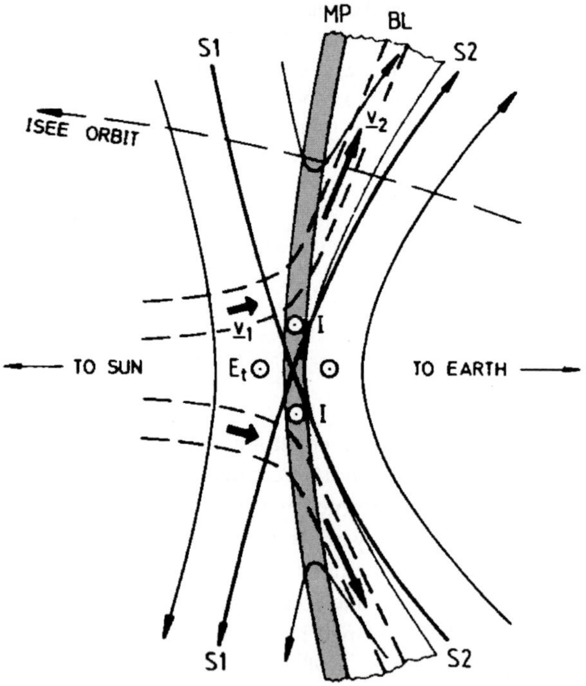

Fig. 11.5. Section of the reconnection region [Sonnerup, 1979]

have been studied and positive results have been obtained, reliable measurements of B_n are often difficult because B_n is usually much smaller than the total field B and could be obscured by local and temporal variabilities in the MP structure. It is preferable instead to use a tangential balance conditions. From continuity of $\boldsymbol{E}_t$ and conservation of mass, namely,

$$\boldsymbol{E}_t = -[\boldsymbol{v}_1 \times \boldsymbol{B}_1]_t = -[\boldsymbol{v}_2 \times \boldsymbol{B}_2]_t \qquad (11.1)$$

$$G \equiv \rho_1 v_{n1} = \rho_2 v_{n2} \qquad (11.2)$$

we obtain

$$B_n(\boldsymbol{v}_{2t} - \boldsymbol{v}_{1t}) = G(\boldsymbol{B}_{2t}/\rho_2 - \boldsymbol{B}_{1t}/\rho_1) \qquad (11.3)$$

The conservation of the tangential momentum is expressible as

$$G(\boldsymbol{v}_{2t} - \boldsymbol{v}_{1t}) = (B_n/\mu_0)[\boldsymbol{B}_{2t}(1-\alpha_2) - \boldsymbol{B}_{1t}(1-\alpha_1)] \qquad (11.4)$$

where the pressure anisotropy factor α is defined by

$$\alpha = (p_{\parallel} - p_{\perp})\mu_0/\boldsymbol{B}^2$$

For a rotational discontinuity, $\boldsymbol{B}_{2t}$ and $\boldsymbol{B}_{1t}$ are not collinear. Hence elimination of $(\boldsymbol{v}_{2t}-\boldsymbol{v}_{1t})$ from equations (11.3) and (11.4) yields

$$v_n = \pm B_n[(1-\alpha)/(\mu_0\rho)]^{1/2} \qquad (11.5)$$

that is, the normal flow speed into or out of the discontinuity is equal to the Alfvén speed based on the normal field component. Using equations (11.2) and (11.5), we obtain

$$\rho_1(1-\alpha_1) = \rho_2(1-\alpha_2) \qquad (11.6)$$

Using the above relation, we obtain from (11.3) and (11.4)

$$(\boldsymbol{v}_{2t} - \boldsymbol{v}_{1t}) = \pm[\rho_1(1-\alpha_1)/\mu_0]^{1/2}(\boldsymbol{B}_{2t}/\rho_2 - \boldsymbol{B}_{1t}/\rho_1) \qquad (11.7)$$

When the direction of the normal is taken positive sunward so that v_n of the magnetosheath flow toward MP has a negative sign, positive and negative signs of (11.5) and (11.7) correspond to $B_n < 0$ and $B_n > 0$, respectively [Sonnerup et al., 1981].

Oppositely directed plasma flows whose velocities satisfy equation (11.7) were observed on a fortuitous occasion when two spacecraft crossed the dawn and flank magnetopause simultaneously on northern and the southern sides of the X-line. Figure 11.6a shows trajectories of these spacecraft, Geotail and Equator-S, in x-y (left panel) and x-z (right panel) projections. They were separated by ~4 R_E in the north-south and ~3 R_E in the east-west direction. The IMF polarity observed by another spacecraft, WIND, was persistently southward. Figure 11.6b shows the flow velocity vectors $(\boldsymbol{v}_{t2}-\boldsymbol{v}_{t1})$ observed $(\Delta \boldsymbol{V}_{obs})$ at two MP crossings by Equator-S (left) and by Geotail (right), and compares them with those predicted $(\Delta \boldsymbol{V}_{th})$ by equation (11.7). The tangential directions are expressed by L and M coordinates in the (L, M, N) system in which N points

Fig. 11.6. (**a**) Positions of Geotail (*red*) and Equator-S (*green*) satellites at the times of two magnetopause crossings expressed in x-y (*left*) and x-z (*right*) coordinates. (**b**) Velocity changes across the magnetopause at Equator-S (*left*) and Geotail (*right*) that were observed (*red*) or predicted theoretically (*blue*) [Phan et al., 2001]

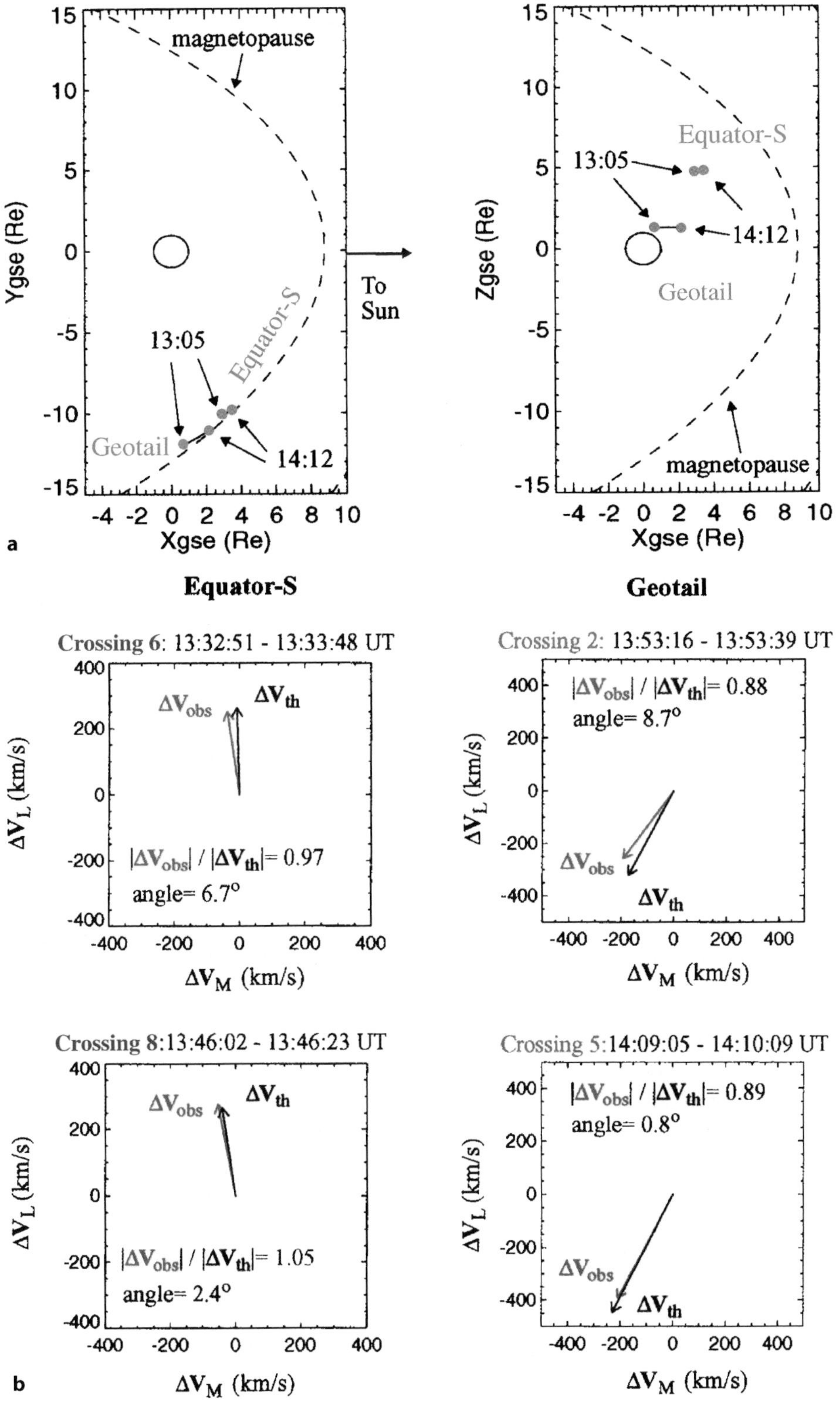
magnetopause
Equator-S
Geotail
13:05
14:12
To Sun
Ygse (Re)
Zgse (Re)
Xgse (Re)
a
Equator-S
Geotail
Crossing 6: 13:32:51 - 13:33:48 UT
ΔVobs
ΔVth
|ΔVobs| / |ΔVth|= 0.97
angle= 6.7°
Crossing 2: 13:53:16 - 13:53:39 UT
|ΔVobs| / |ΔVth|= 0.88
angle= 8.7°
Crossing 8:13:46:02 - 13:46:23 UT
|ΔVobs| / |ΔVth|= 1.05
angle= 2.4°
Crossing 5:14:09:05 - 14:10:09 UT
|ΔVobs| / |ΔVth|= 0.89
angle= 0.8°
ΔVL (km/s)
ΔVM (km/s)
b

outward along the magnetopause normal with L oriented approximately due north and M due west. It can be seen that flows were directed southward at Geotail which was not very far from the equatorial plane but northward at Equator-S which was at the northern middle latitudes, and the observed $\Delta \boldsymbol{V}_{\text{obs}}$ agreed excellently with the predicted $\Delta \boldsymbol{V}_{\text{th}}$. The reconnection at this time was active for more than one hour of the multiple magnetopause crossings since the jets were encountered over the period. The length of the reconnection line was at least 3 R_{E}, and, combined with additional data from the ground-based SuperDARN observations and spacecraft observations made in the subsolar region, this observation suggests that an extended reconnection line was formed along the equatorial magnetopause [Phan et al., 2001].

In order to derive the velocity distribution function of the plasma jets it is useful to refer to a frame that moves with the crossing point of a given field line across the magnetopause. The velocity $\boldsymbol{V}_{\text{f}}$ of this frame, that is, de Hoffman-Teller frame, is given by

$$\boldsymbol{V}_{\text{f}} = (\boldsymbol{E}_{\text{t}} \times \boldsymbol{B}_{\text{n}})/\boldsymbol{B}_{\text{n}}^2$$

In this frame, which is the local rest frame of the field lines, the tangential electric field transforms locally to zero, so that particles do not change their kinetic energy in crossing the magnetopause from one side of the other: the magnetosheath plasma simply streams with constant speed along the field lines across the magnetopause. Figure 11.7 schematically illustrates the effect on the distribution functions of ions when the magnetopause is the rotational discontinuity. Two-dimensional ($\boldsymbol{v}_{\parallel}$, $\boldsymbol{v}_{\perp}$) distributions are at the bottom and their cuts along the $\boldsymbol{v}_{\parallel}$ axis are on the top. The left side panel is for inside the magnetosphere and the right side one is for just outside in the magnetosheath. Source populations are supposed to consist of magnetosheath plasma, cold magnetospheric ions of the ionospheric origin, and energetic ions in the magnetosphere (sometimes referred to as ring current ions). The ions with speeds lower than $v_{\parallel} = V_{\text{f}}$ (labeled 1) in the right panel are the original magnetosheath population that has not encountered the discontinuity. Ions with speeds higher than this (labeled 2) are reflected (right panel) or transmitted (left panel) particles. Since the speeds are unchanged in the de Hoffman-Teller frame at the reflection or transmission these distributions are at the mirror image of the original population with respect to V_{f}; for illustration the densities are assumed to be divided equally between the reflected and transmitted. The cold magnetospheric ions (labeled 3) become the populations 4 and the energetic ions of the magnetosphere (labeled 5) become the populations 6 after reflection or transmission. The distribution functions of the reflected and transmitted magnetosheath ions have a characteristic D-shape because of the cutoff at $v_{\parallel} = V_{\text{f}}$ [Cowley, 1982]. V_{f} is essentially the Alfvén speed at the rotational discontinuity.

In addition to the rotational discontinuity that stands in the earthward flow of the magnetosheath plasma, there is another discontinuity of the same kind that stands in the flow of the magnetospheric plasma toward the magnetopause. Both of these waves are launched from the reconnection line and their wavefronts extend from there. The velocity of de Hoffman-Teller frame for the latter (interior) wavefront is considerably faster than that for the former (exterior) since the higher field strength and lower density make the Alfvén velocity higher. Representative values are 190 km/s for the exterior wave and 600 km/s for the interior wave. Reflection at the interior wave accelerates the cold ions of the magnetospheric origin to higher energies than the ion acceleration at the exterior wave [Lockwood et al., 1996]. The slow mode waves also stand between the pair of the rotational discontinuities since magnetic fields have different strengths across the magnetopause.

Figure 11.8a is an example of the ion distribution functions observed when satellite traversed the magnetosheath side of the boundary region (referred to as MSBL). The positions of measurements are indicated in the magnetic field B_z data in the lower left; spacecraft was on the magnetosheath side when B_z was negative and on the magnetosphere side when it was positive. The left panel is a typical anisotropic magnetosheath He^{2+} distribution with very low flow velocity. The middle panel shows the incident and reflected magnetosheath ions which were observed together in the boundary layer. As seen in the lower panel these distributions were separated by twice the Alfvén speed in the magnetopause layer. The reflected ions are heated. The right panel shows the distribution function of the cold He^{+} ions transmitted from the magnetosphere. The flow speed of these transmitted ions along the magnetic field is somewhat smaller than, but comparable to that of the reflected He^{2+}

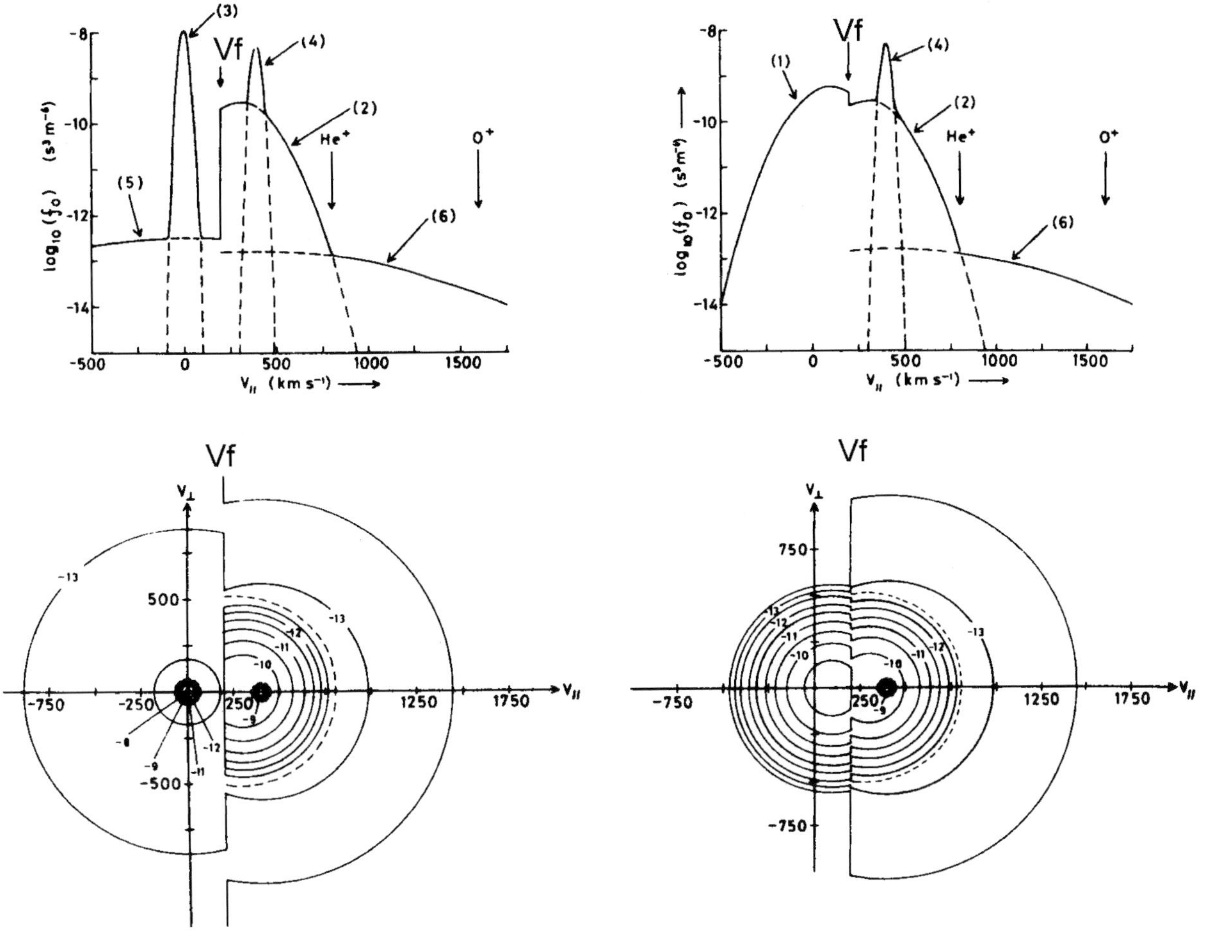

a magnetosphere **b** magnetosheath

Fig. 11.7a,b. Effect of the acceleration by the Alfvén wave on the velocity distribution functions of ions. *Top*: one-dimensional distribution with respect to $v_{||}$, and *bottom*: two-dimensional distribution in the $v_{||}$–$v_{\perp}$ plane [Cowley, 1982]. Distribution on the magnetosphere side of the magnetopause is on the left and that on the magnetosheath side is on the right. V_f represents the de Hoffman-Teller velocity

ions in the middle panel. The distribution functions of both reflected and transmitted ions have the typical D-shape.

Figure 11.8b are the observations on the magnetosphere side of the boundary region (LLBL). (Here the distribution functions are shown in a plane which is approximately tangent to the subsolar magnetopause viewed from the sun with the x direction approximately perpendicular to the ecliptic plane and y approximately in the ecliptic plane toward dusk.) The left panel is the distribution observed in the magnetosphere well away from the boundary region and shows highly anisotropic He^+ distribution in the outer magnetosphere. The middle panel shows that these He^+ ions had picked up a substantial flow velocity in the direction perpendicular to the magnetic field, and in addition to them there were hotter ions flowing in the antiparallel to ***B*** direction at approximately twice the local Alfvén speed. This was the reflected He^+ distribution which had become hotter than the incident one. The right panel

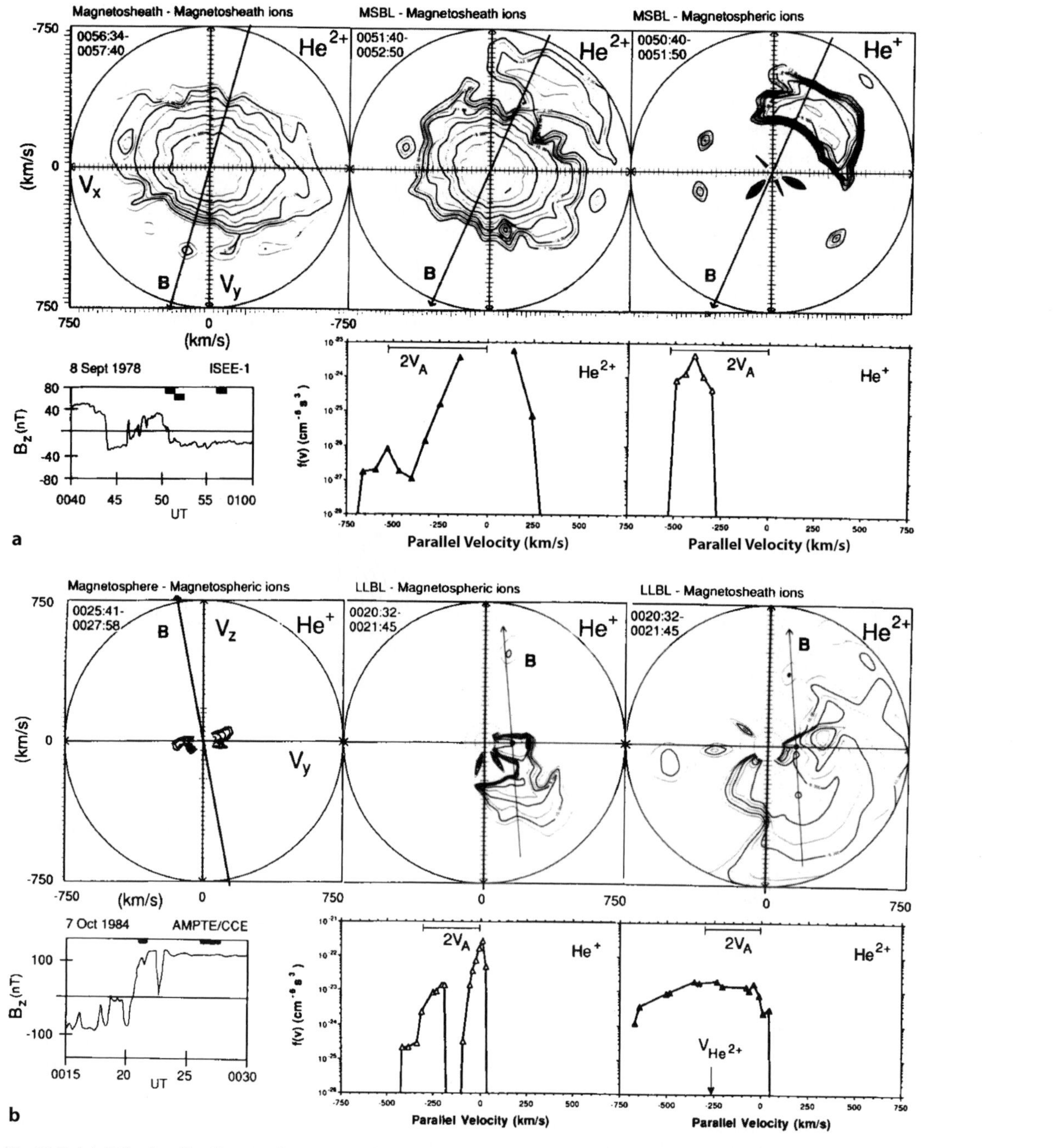

Fig. 11.8. (**a**) Velocity distribution functions of He ions in MSBL, that is, the magnetosheath-side region of the boundary where $B_z < 0$. In the left and middle panels the He^{2+} distribution centered near $\boldsymbol{v} = 0$ is the incident magnetosheath distribution, and the D-shaped distribution in the middle panel represents the reflected magnetosheath ions that flow faster along the magnetic field. Concurrent with this reflected distribution is the transmitted He^{+} distribution in the third panel. (**b**) Velocity distribution of He^{+} ions in LLBL, that is, the magnetosphere-side region of the boundary where $B_z > 0$. The incident cold magnetospheric ions of the left panel become the faster ions in the middle panel upon reflection, and the right panel shows the concurrent distribution of transmitted He^{2+} [Fuselier, 1995]

is the distributions of He^{2+} ions transmitted from the magnetosheath; they originally had near zero velocity in the magnetosheath but gained significant energy as they crossed the magnetopause [Fuselier, 1995].

Distribution functions actually observed tend to be more structured than the idealized ones of Fig. 11.7 because of such effects as time-variations, scattering and heating, and in some cases the D-shaped distribution function seems to be produced by the effect of the time of flight, suggesting that de Hoffman-Teller velocity V_f was lower than the low-speed cutoff velocity due to the time of flight.

The term "Low-Latitude Boundary Layer (LLBL)" has been used to designate a domain of the magnetosphere where plasma of the magnetosheath origin is found to coexist with the magnetospheric plasma. The frontside reconnection generates such a domain on geomagnetic field lines (labeled MP in Fig. 11.5) and offers explanation to a good part of the dayside LLBL observations. "Magnetosheath Boundary Layer (MSBL)" is its counterpart on the magnetosheath side. However, pronounced mixing of these plasmas observed under the northward IMF conditions in the magnetotail suggests that other mechanisms are also operative [Fujimoto et al., 1998]. One of the suggested mechanisms is a two-step reconnection process on the magnetopause, where some of the field lines which have been opened become tangled and re-reconnected to form closed field lines again. The closed field lines produced in this manner would retain the magnetosheath plasma which steamed in when they were open [Nishida, 1989]. Another is the plasma mixing due to non-linear development of the Kelvin-Helmholtz instability at the flanks of the magnetosphere. In support of the latter idea, it has been found that the plasmas from the two sources coexist in the rolled-up vortices at the magnetopause which have the field and particle properties expected from the non-linear development of this instability [Hasegawa et al., 2004].

11.2.3 Direct Consequences of the Magnetopause Reconnection

Because it couples the magnetosphere with the solar wind, reconnection at the magnetopause leads to a variety of phenomena in the magnetosphere. They can be divided into those which result directly and those which occur after the input energy has been stored in the magnetotail. We shall describe the direct consequences in this section.

The particles that are accelerated at the rotational discontinuities at the magnetopause precipitate to lower altitudes. The time taken for a 600 km/s ion, which is in the high energy range of the magnetosheath ions accelerated by reflection at the exterior wave, to travel a distance of 15 R_E is 150 s. The flow is field-aligned, but at the same time the convection makes the ion to drift poleward with a representative speed of 1 km/s, so that when it reaches the low-altitude observing site it is about 1.5° poleward in latitude from the position where the field line was rooted when the acceleration occurred. Since the extent of the poleward shift is larger for slower ions, the ions are dispersed in latitude as they travel down to low altitudes. This effect of the "time of flight" is seen in the form of energy dispersion of precipitating ions when spacecraft traverses the cusp along a latitudinal track. The magnetospheric cold ions accelerated at the interior wave also precipitate, and the ions having different origins and histories of acceleration produce a multiple-layered structure in the ion precipitation in the cusp region [Lockwood et al., 1996]. Precipitation of >900 eV protons from the cusp increases ionization at altitudes around 150 km, while the precipitation of electrons at a temperature of 50 eV produce dominant effects at F2 peak altitudes of around 300 km [Millward et al., 1999].

The rate and site of the magnetopause reconnection are controlled by strength and direction of IMF and so are the profile and speed of the convection that is generated in the magnetosphere and projected to the ionosphere. Since the convection generates Hall current in the ionosphere where electrons can follow the convection but ions cannot due to collisions with the thick neutral atmosphere, geomagnetic observations on the ground can be used to monitor the state of the convection. The geomagnetic counterpart of the magnetospheric convection has been identified as the DP2 variations that are characterized by a twin vortex pattern of the current system. They track the variations of southward B_z of IMF with a short delay of several minutes, demonstrating a rapid, global response of the convection to IMF [Nishida, 1968]. More recent observations using a much larger number of ground magnetometers and the AMIE technique for deriving the global electric field have confirmed this earlier result [Ridley

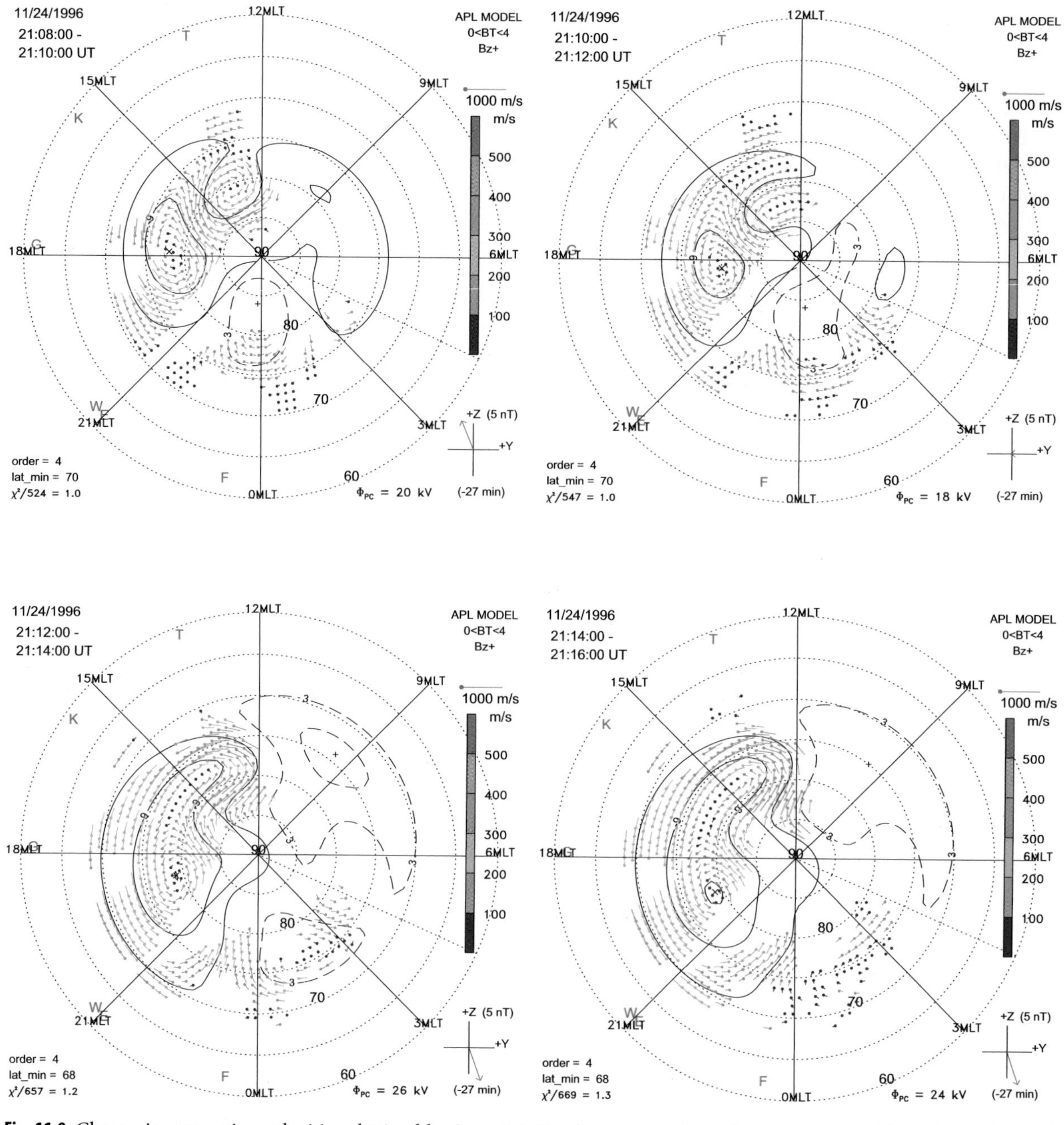

Fig. 11.9. Change in convection velocities obtained by SuperDARN observations through the time of northward IMF (*top-left panel*) to the time of southward IMF (*bottom panels*). IMF directions are indicated on the lower right of each diagram [Ruohonimei and Greenwald, 1998]

et al., 1998]. The convection under the northward IMF has also been identified originally from analysis of the geomagnetic field [Maezawa, 1976]. The effect of the IMF B_y has also been detected in the geomagnetic data in the form of intensifications of the zonal current in the cusp region (Svalgaard-Mansurov effect) which later have been found to represent asymmetries of the twin vortex pattern.

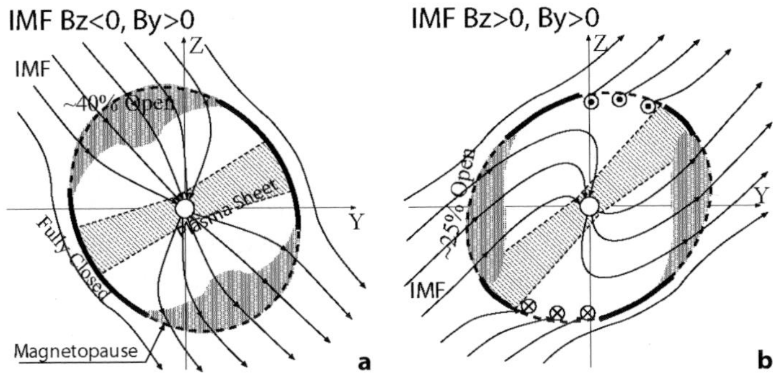

Fig. 11.10a,b. Sections of the distant magnetotail under (**a**) southward IMF and (**b**) northward IMF [Hasegawa et al., 2002]

IMF-dependent modulation of the ionospheric drift is being monitored more directly by SuperDARN network of cohenrent HF radars. Figure 11.9 is an example of the global convection map obtained from analysis of SuperDARN velocity data across an interval of the IMF polarity change from northward to southward. In the top-left panel taken when IMF $B_z > 0$ and $B_y < 0$ there was an anticlockwise convection vortex (called lobe cell) in the postnoon sector in agreement with the convection pattern of Fig. 11.1b. In the bottom-left and bottom-right panels taken after B_z had become negative a large clockwise vortex (called merging cell) covered the entire afternoon sector, as expected for the convection pattern of Fig. 11.1a. This response to IMF began at almost the same time from noon to midnight [Ruohoniemi and Greenwald, 1998].

In the convection pattern of panel (a) for the northward B_z, the dusk sector is dominated by a clockwise vortex which does not belong to the lobe cell. This could be a merging cell produced by the frontside reconnection that could occur in parallel with the lobe reconnection, or a "viscous cell" that is associated with the formation of the LLBL by a mechanism other than the reconnection such as the Kelvin-Helmholz instability. It has been suggested that under the northward IMF a cell of another kind, called "exchange cell", is also formed as the northward IMF is reconnected with the closed field lines on the dayside high-latitude magnetopause [Tanaka, 1999].

Since the magnetotail is formed with field lines that have been carried anti-sunward following the frontside reconnection, its structure keeps a mark of the IMF influence. Figure 11.10 is sections of the distant tail which summarize 1800 crossings in the range of 100 to 210 R_E from the Earth. Panel (a) is for the southward IMF and panel (b) is for the northward IMF. Open field lines thread the magnetopause in the densely shaded sectors where they have the character of the rotational discontinuity. These field lines are present under both polarities of the IMF B_z, but the sector where they cross the magnetopause is distinctly different; it is the polar sector when IMF is southward but the dawn and dusk sectors when IMF is northward. The lightly shaded sectors of the magnetopause have the character of the tangential discontinuity and bound the plasma sheet where the field lines are closed. The plasma sheet is strongly inclined to the equatorial plane when IMF is northward due to the magnetic tension exerted by the IMF along the open field lines. Under the northward IMF conditions there are also field lines (indicated by dots or crosses in panel b) which extend to the solar wind from inside the high-latitude lobe. These are the field lines of the type 3' and 4' in Fig. 11.1b which have undergone the lobe reconnection with IMF [Hasegawa et al., 2002].

The magnetic flux content in the tail lobe is determined by the balance between the frontside reconnection and the reconnection inside the tail. When IMF turns northward the lobe flux content decreases, but it seems to be rather rare that the lobe disappears completely and the entire tail is occupied by the closed field lines. If $B_z \sim |B_y|$ the open field lines are still produced by the frontside reconnection. If $B_z > |B_y|$ for about 4 hours the open field lines disappear, but the southward IMF of a duration of about 6–7 min is enough to produce open field lines. Hence even relatively short fluctuations of B_z to negative or near zero values can keep the open field lines during long intervals where average values of B_z are large and positive, according to the analysis that have used the particle precipitation in the polar cap to estimate the size of the region filled by the closed/open field lines [Newell et al., 1997].

11.3 Reconnection Inside the Magnetotail

In this Chapter we discuss magnetic reconnection in the magnetotail. The magnetotail consists of the lobe and the plasma sheet. Lobe is the bundles of open field lines that occupy the higher latitude part of the magnetotail. The plasma sheet occupies the low latitude part around the midplane. When the magnetic effect of the electric current flowing in the plasma sheet is of the main concern it is called the current sheet. The midplane of the magnetotail is called neutral sheet although the magnetic field intensity does not usually vanish there. Magnetic reconnection takes place at the X-line (reconnection line) that is formed in the neutral sheet. A much broader region around the X-line where the energy conversion takes place is called reconnection region in this Chapter.

11.3.1 Sites of the Magnetotail Reconnection

The open field lines generated at the magnetopause are transported to the magnetotail and reconnected in the distant tail beyond $\sim 100\,R_E$ [Slavin et al., 1985]. In addition to this, the tail reconnection occurs also in the near-Earth region in the range of 15 to 25 R_E where it operates firstly on closed field lines of the plasma sheet and produces tailward ejection of plasmoids, namely, the plasma clouds containing a loop of magnetic field lines [Hones, 1979]. In both distant-tail and near-Earth reconnections, the reconnection rate is time-varying and is reflected in the activities of the magnetosphere.

The near-Earth reconnection has been studied extensively by using a wealth of data obtained by the Geotail satellite whose trajectory has been designed to explore the plasma sheet at distances of 10 to 50 R_E from the Earth. Figure 11.11 shows the positions of Geotail at times of onsets of Pi2 pulsations that signify the occurrence of substorms. Different symbols are used to designate the events where Geotail observed (1) fast earthward flows ($v_x \geq 300$ km/s) with positive B_z and (2) fast tailward flows ($v_x \leq -300$ km/s) with negative B_z within 4 min of the Pi2 onsets, and (3) no fast plasma flows or magnetic disturbances. The data are limited to observations made in the plasma sheet by using the condition that the plasma β was greater than 1 for more than 2 min during 15 min before the Pi2 onset. It is seen that the earthward flow events are observed mostly earthward of $-20\,R_E$, while the tailward flow events are seen mostly tailward of this distance. This suggests that the near-Earth reconnection tends to take place around x of $-20\,R_E$ at the onsets of substorms [Asano et al., 2004b], and the distribution of the disturbance events is centered on the duskside at y of $5\,R_E$ in GSM coordinates [Machida et al., 1999]. (In the Geocentric Solar Magnetospheric Coordinates, x is in the direction of the sun, and z is in the plane containing the Earth's dipole field.)

This interpretation is confirmed by simultaneous observations by two spacecraft which happened to be located on opposite sides of the reconnection region (Fig. 11.12). At the time of this event Interball-Tail was at $[-11.5, -1.9, -2.5]\ R_E$ in GSM coordinates and Geotail was at $[-27.8, -6.4, -1.5]\ R_E$. Interball-Tail (Panel b) which was on the earthward side of the reconnection line detected the earthward flow accompanied by the northward field, and Geotail (Panel a) which was on the tailward side observed signatures of plasmoid, that is, tailward turning of the flow at the first of the vertical lines and the southward turning of the field at the second line. The tailward-flowing northward field observed at Geotail between these lines can be interpreted to represent the closed field lines that had existed between the satellite and the reconnection site. In close association with the intensification of the tailward flow at Geotail at 1128 and the start of the earthward flow at Interball-Tail at 1126, the meridian scanning of data at Poker Flat (Panel c) show that an auroral intensification with a small poleward expansion took place at 1128-1129 [Petrukovich et al., 1998].

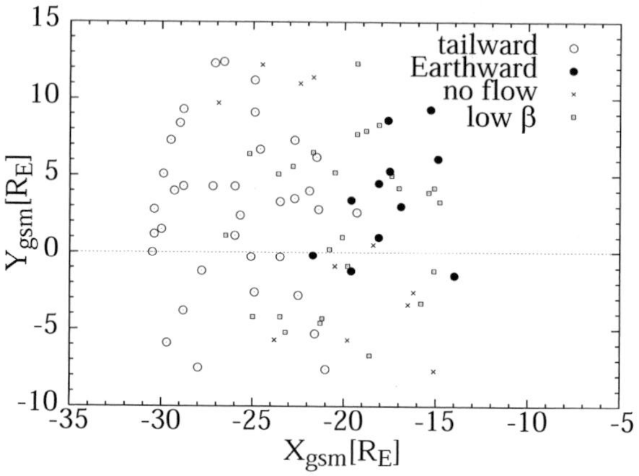

Fig. 11.11. Positions of Geotail when tailward flow (○) or earthward flow (●) was observed at substorm onsets plotted in the GSM x-y plane [Asano et al., 2004b]

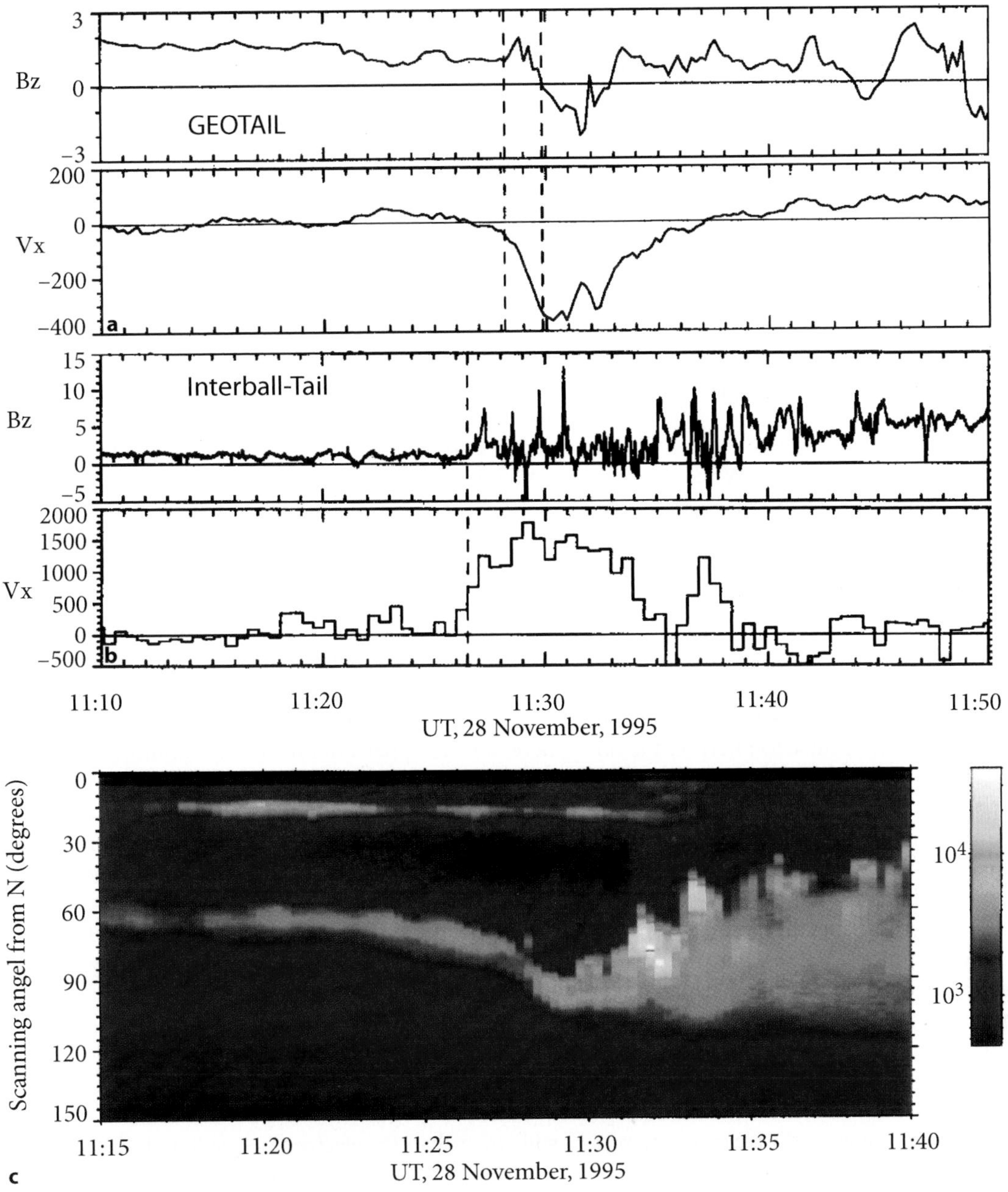

Fig. 11.12a–c. Field and plasma observations at two locations in the magnetotail at the time of auroral intensification. (**a**) Geotail observation of plasmoid signatures and (**b**) Interball-Tail observation of earthward flow, and (**c**) aurora observation at Gillam [Petrukovich et al., 1998]

A comprehensive survey of plasmoids using the Geotail data has shown that azimuthal extents of plasmoids tend to increase as they propagate downtail. In the near-tail region ($X_{GSM'} \geq -50\,R_E$) plasmoids tend to be observed in the premidnight sector ($|Y_{GSM'} - 3\,R_E| \leq 10\,R_E$), but they are observed widely ($|Y_{GSM'}| \leq 20\,R_E$) at greater distances ($X_{GSM'} < -50\,R_E$). (GSM' means the modified GSM coordinates where aberration of the tail axis has been taken into account). The average speed of plasmoids increases from 400 km/s to 700 km/s but their average ion temperature decreases from 4.5 keV to 2 keV between these distance ranges. Typical dimensions are 10 (length) × 40 (width) × 10 (thickness) R_E^3 in the middle and distant tail. Inside plasmoids the thermal energy flux exceeds the bulk flow energy flux and the Poynting flux, and the energy flux released tailward by plasmoids is estimated to be typically about 10^{15} J [Ieda et al., 1998].

11.3.2 Structure of the Reconnection Region

The canonical model of the reconnection in collisionless plasma has been that of Petschek. While in earlier models the site of conversion of the magnetic energy was limited to the magnetic diffusion region surrounding the X-line, Petschek recognized that there exists another way of reducing the magnetic field and increasing the plasma energy, namely, a slow-mode shock wave emanating from the X-line. The energy conversion at the shock front overwhelms the dissipation in the very close vicinity of the X-line and augments the reconnection rate to the levels that are applicable to interpret the observations. The maximum of the reconnection rate that is measured by the speed of the magnetic flux entering the reconnection region is slightly smaller than 0.1 times the Alfvén speed just outside (that is, upstream of) the diffusion region [Vasyliunas, 1975].

The Petschek model describes the plasma in terms of the magnetohydrodynamics (MHD), where the particle motions are averaged over temporal and spatial scales of their Larmor motion. Non-MHD behaviors are allowed for but are thought to be limited to the region of weak magnetic field around the X-line and to the slow shocks. It has been revealed, however, that kinetic properties of ions and electrons govern the plasma dynamics over a much broader space extending outside these specific regions. Observations have shown such non-MHD features as anisotropic and multi-spectral velocity distribution functions of both ions and electrons, Hall current due to decoupling of ion and electron motions, and plasma waves resulting from non-Maxwellian distributions of ions and electrons. Processes that depend on kinetic properties of plasma particles play more influential roles than have originally been envisaged. Simulations have been used extensively to understand the nature of these kinetic properties.

A parameter that is critically important in the subsequent discussion is the thickness of the plasma sheet. A measure of this parameter is the thickness of the current sheet that can be derived by dividing the field strength in the tail lobe by the electric current density that is obtained from the measurements of ion and electron fluxes. The current sheet thicknesses for the earthward flow events and the tailward flow events of Fig. 11.11 are plotted in panels (a) and (b) of Fig. 11.13 for a 50-min interval around the onset times of Pi2 pulsations. Thinning of the current sheet is observed for both types of events, and the observed thickness is reduced to several times the ion inertia length $\lambda = c/\omega_{pi}$ which is 720 km for density of 0.1 /cm^3 as well as the proton Larmor radius which is 671 km for 5 keV ions in a 15 nT field. (Since the characteristic speed of the ion flow in the reconnection region is given by the Alfvén speed, the ion inertia length is comparable to the ion Larmor radius.) For the earthward flow events the thinning starts during the growth phase preceding the onset, while for the tailward flow events it starts at the onset [Asano et al., 2004b]. The values given here are the upper bounds to the minimum thickness because measurements are made at varying distances of the X-line, and in individual events the thickness as small as 500 km has been observed [Asano et al., 2004a]. Panel (c) of Fig. 11.13 shows changes of the current density relative to the Pi2 onset time, and it can be seen that the increase in the current density begins about 10 min before the onset for the earthward flow events and shortly after the onset for the tailward flow events.

When the plasma sheet thickness becomes comparable to the ion inertial length, ions are no longer convected with magnetic field lines. Since the thickness is still larger than the electron inertia length the electrons are tied to magnetic field lines, and the Hall current results from the decoupling of ions and electrons. The Hall current is directed away from the neutral sheet since electrons are convected toward the neutral sheet together with the field lines. This current is channeled to field-aligned currents and forms a current loop in the boundary region of the plasma sheet, and produces magnetic perturbation fields in the cross-tail direction [Sonnerup, 1979].

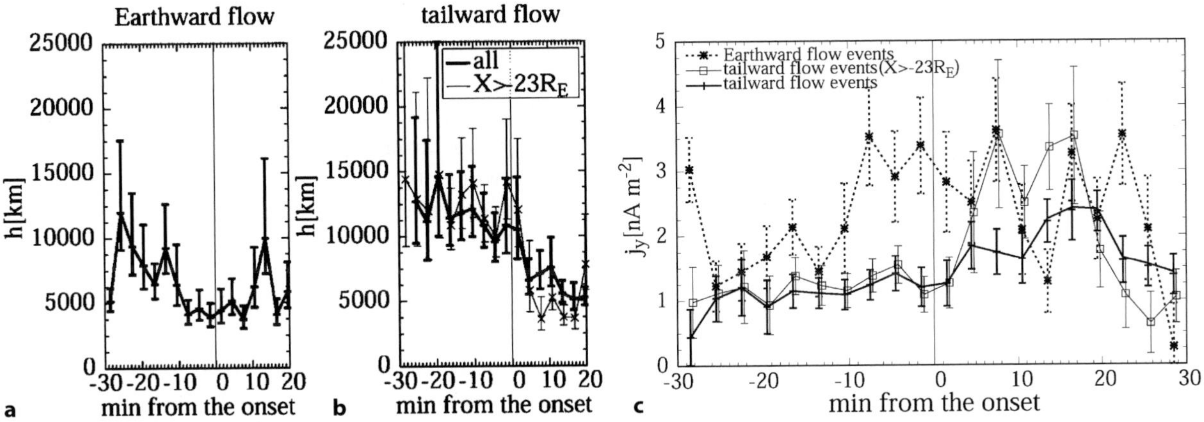

Fig. 11.13a–c. Variations of the current sheet thickness relative to the time of the substorm onset when (**a**) the flow is earthward or (**b**) tailward, and (**c**) the variation of the current density for both types of events [Asano et al., 2004b]

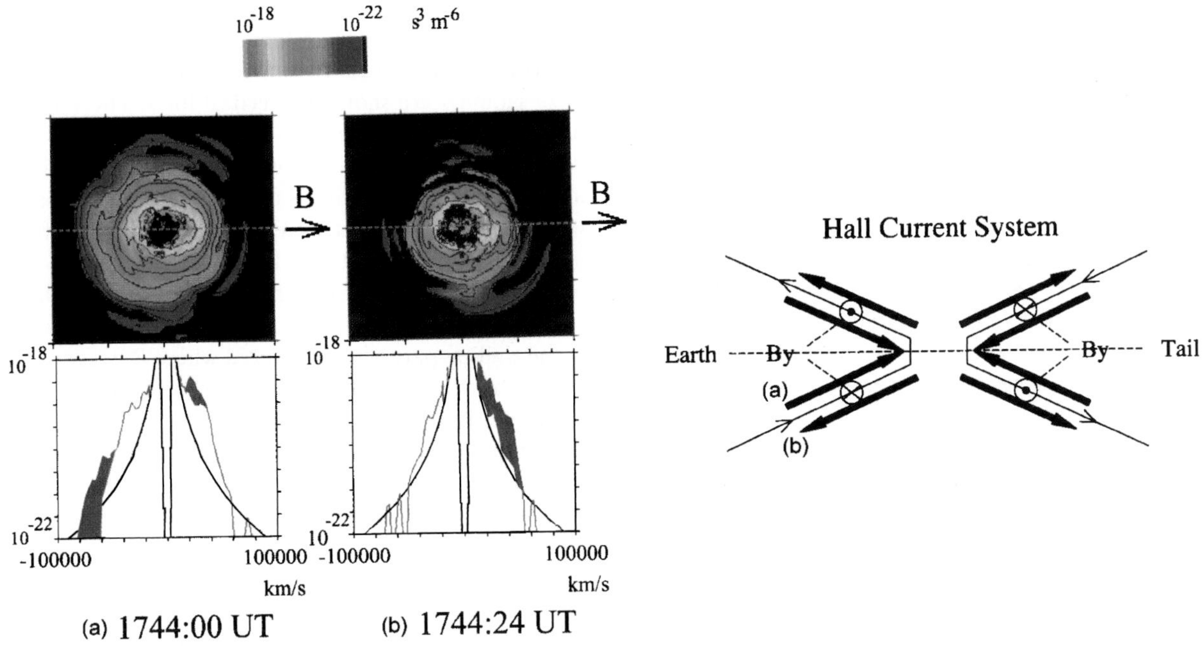

Fig. 11.14. Velocity distribution function of electrons carrying the Hall current in the regions of (a) current toward the X-line and (b) away from it as indicated in the *right panel* [Nagai et al., 2003]

Field-aligned flows of electrons constituting this Hall current system have been observed in the magnetotail. The left panel of Fig. 11.14 shows velocity distribution functions of electrons that have been observed at the points on the earthward side of the X-line which are indicated by (a) and (b) in the right panel. Top panels are two-dimensional cuts in the plane that includes the magnetic field vector and the ion bulk flow vector. The x axis is the magnetic field direction with positive tailward. Bottom panels are the cuts of the above at the red dotted lines and show phase space densities versus $v_{\parallel}$. It is seen that tailward flowing low-energy (< 5 keV) electrons are enhanced at (b), while at (a) higher-energy (> 5 keV) electrons flow earthward. Thus the observations not only have identified the electron flow constituting the Hall current system, but also have revealed that electrons are accelerated substantially in the reconnection region and the energized electrons are ejected away from the reconnection region along magnetic field lines [Nagai et. al., 2003].

The out-of-the-plane magnetic field B_y produced by the Hall current has the quadrapole structure as illustrated in the right panel of Fig. 11.14. This magnetic structure can be interpreted as standing whistler mode waves. The waves propagate away from the neutral sheet along the background field B_z, but they are standing since their phase velocity and the inflow velocity of the plasma are equal in magnitude (which is about 0.1 times the Alfvén speed in the upstream region) but opposite in direction. Waves are non-linear since B_x and B_y are much lager than the background field B_z [Anzner and Scholer, 2001].

The Hall electric field plays a key role in producing the ion outflow from the magnetic diffusion region where the ions are not magnetized. In this region the Hall electric field has the E_x component directed away from the X-line, and as ions travel a distance of the ion inertia length E_x accelerates them outward to V_A regardless of the size of the system. The inflow speed becomes about $0.1V_A$. The quadratic dispersion property of the whistler wave (higher phase speed at smaller spatial scales) is the key to this process. Taking a lobe density of $0.5\,/\mathrm{cm}^3$, $B \sim 15$ nT, we obtain $V_A \sim 1500$ km/s so that the inflow speed is ~150 km/s and a significant fraction of the lobe flux can be reconnected in 5 min [Shay et al., 1999].

The z component E_z of the Hall electric field in the thin plasma sheet is the ambipolar electric field directed toward the neutral sheet. In the off-neutral sheet region where B_x is not zero electrons drift dawnward/duskward under this component of the electric field, and results in a bifurcated layer of the duskward cross-tail current [Asano et al., 2004a and b].

Acceleration occurs also in the thin plasma sheet outside the magnetic diffusion region where ions exert non-adiabatic motions due to strong non-uniformity of the magnetic field. It has been shown that such motions occur when the κ parameter defined by $\kappa = (R_{\min}/\rho_{\max})^{1/2}$ is less than 1, that is, the minimum curvature $R_{\min}$ of field line is smaller than the Lamor radius $\rho_{\max}$ under the field B_z across the neutral sheet [Büchner and Zelenyi, 1989]. The motion for $\kappa < 0.4$ is called the Speiser orbit which is composed of a faster multi-bounce motion in the z direction and a slower half-gyration around B_z in the x-y plane. Ions gain energy while they move during this half gyration in the direction of the electric field E_y associated with the convection. The motion for $\kappa > 1$ is adiabatic and ions simply pass through the plasma sheet along magnetic field lines. In the intermediate range of κ the ion motions are stochastic due to deterministic chaos.

Examples of ion velocity distributions observed in the plasmoid are shown in Panels a though f of Fig. 11.15. The top panel of this Figure is records of the magnetic field obtained on this occasion when Geotail was at about 95 R_E from the Earth. The northward component B_z of the magnetic field (red curve) showed the characteristic positive-then-negative excursion of a plasmoid. The times when the distribution functions were sampled are shown by vertical lines. The velocity distributions are cuts in the B (magnetic field) – C (convection velocity) plane. The outstanding feature of these ion distribution functions is the presence of two beams which are seen to counterstream in the direction of the magnetic field relative to the framework of the convecting plasma. Counterstreaming beams are particularly clear in the bottom three panels which were obtained in the trailing part of the plasmoid where B_z was southward. The velocity difference between these beams was 1000–1500 km/s, which was higher than the convection velocity and the local Alfvén speed. Sometimes more than two peaks were seen, and non-gyrotropic distributions were also observed [Mukai et al., 1998]. These phase-space structures of accelerated ions can be interpreted to be formed because the ions in plasmoids are mixtures of populations which had entered the plasma sheet at various different positions and had been accelerated under spatially and temporally varying values of the κ parameter. This subject will be discussed later.

Hybrid simulations have been performed to understand the structure of the reconnection region in the collisionless plasma, and it has been seen that the magnetic reconnection can proceed rapidly without invoking the slow-mode shock. Figure 11.16 is a result of a hybrid simulation where ions are represented by macroparticles while electrons are assumed to be massless fluid. Distance is scaled by the ion inertia length λ which cor-

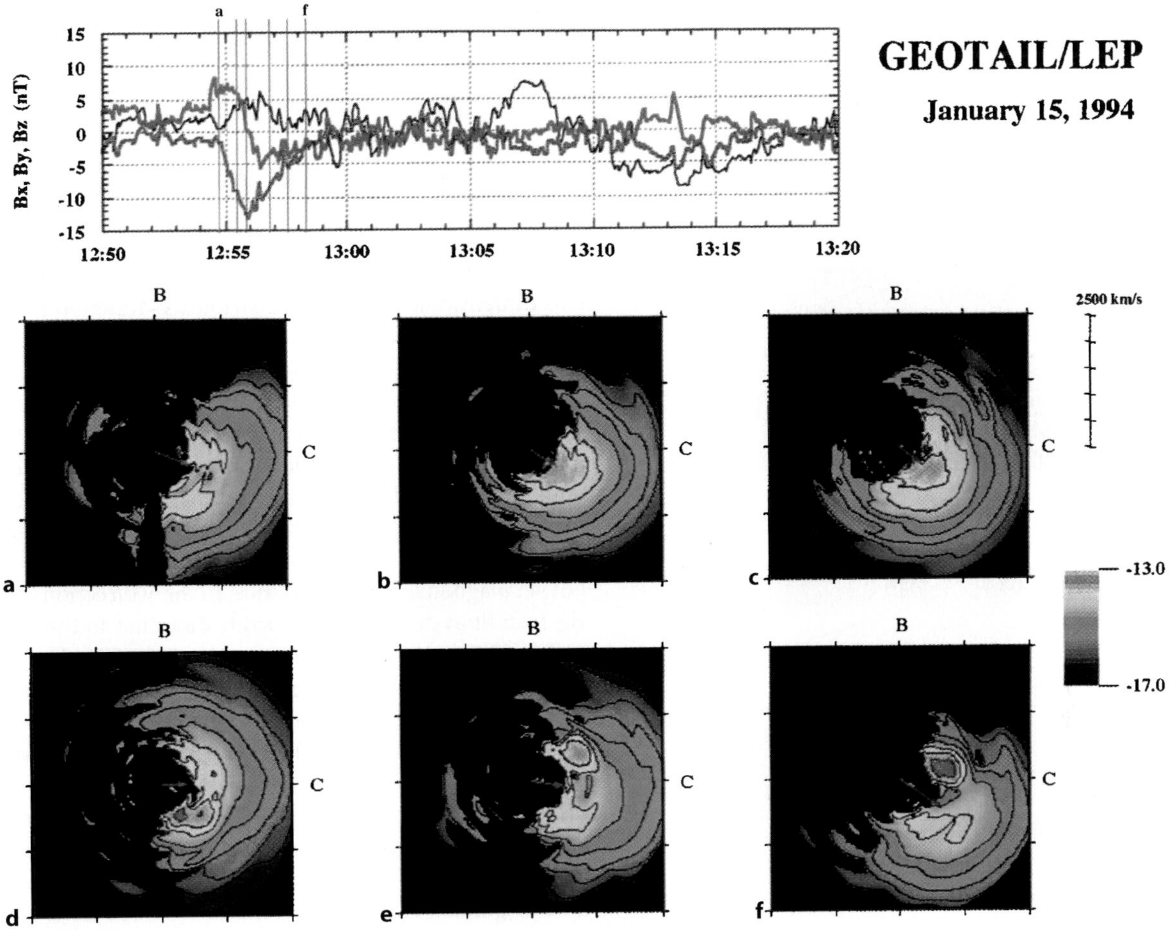

Fig. 11.15a–f. Velocity distribution function of ions observed in a plasmoid at the times a through f indicated in the top panel [Mukai et al., 1998]

responds to $1/6\,R_{\mathrm{E}}$ for a reasonable choice of parameters. The initial half-thickness of the current layer is chosen to be 2.4λ. Time is scaled by the inverse of the ion gyrofrequency $\Omega^{-1} = 1.5$ s. Scale of the simulation box is $240\lambda \times 60\lambda$ but only its central portion is presented in the Figure. (Note that in this diagram the reconnection plane is x-y, while it is x-z elsewhere in this Chapter.) Thick lines in each panel represent the separatrix. The magnetic field lines (top panel) have no sharp bends that characterize the slow-mode shock. The current density J_z (second panel) which flows perpendicular to the plane of the initial magnetic field configuration is bifurcated in the close neighborhood of the X-line but at further distances it constitutes a single layer in the middle of the plasma sheet. Nevertheless, the streamlines (third panel) show that flows turn away from the direction of the X-line at the edge of the plasma sheet at any distance, suggesting that they are governed by the pressure. The fourth panel shows the z component of the vorticity. The cross-tail current is carried mainly by electrons to about 40λ from the X-line but beyond 60λ the ions are the main carrier. These ions are performing the Speiser-type motion in the curved magnetic field lines. Since the current sheet in the field reversal region produces sharp bend in the magnetic field and makes the κ parameter small, the Speiser type ion orbit can be realized self-consistently. Slow shocks seem to be produced further downstream from the X-

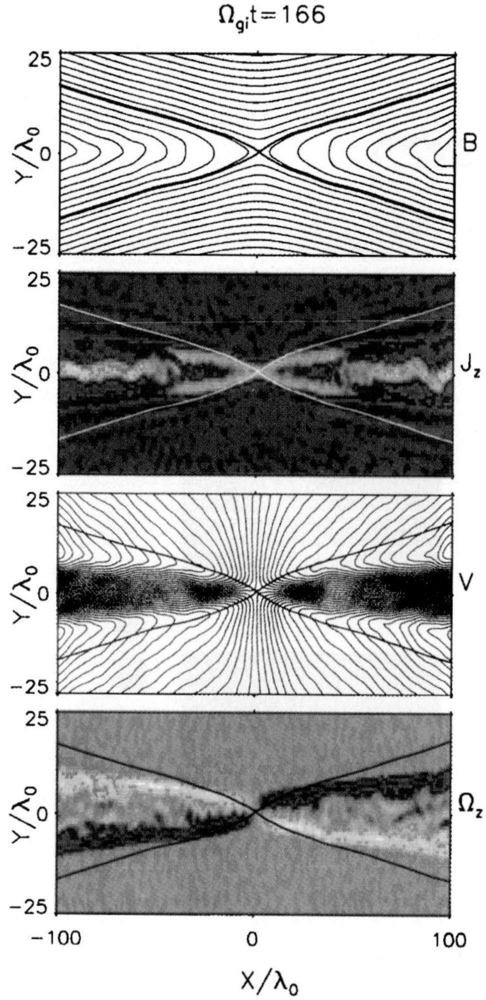

Fig. 11.16. Hybrid simulation of reconnection. From the *top*, magnetic field lines, current in the direction perpendicular to the sheet, streamlines, and vorticity [Lottermoser, Scholer, and Matthews, 1998]

line where the Maxwellian distribution functions have been re-established through the wave-particle interactions [Lottermoser et al., 1998].

Observationally, the slow-mode shocks have been clearly identified at boundaries of the plasma sheet in a number of cases, but there are numerous other instances where such identification was not successful [Saito et al., 1995]. It could be that the slow mode waves do not develop into shocks that satisfy the Rankine-Hugoniot relation of MHD because ions and electrons have non-Maxwellian distribution functions over an extended region and, moreover, the system is not one-dimensional since there are energy fluxes carried by ion/electron flows along field lines.

In order to understand the origin of the non-Maxwellian distribution functions of ions in more detail, trajectories of sample ions have been traced in the simulation box. Figure 11.17 is an example of the results of the simulation. The panel (a) at the top is the two dimensional cuts of the distribution function near the neutral sheet at the distance of about 25λ tailward from the X-line where κ is less than 1. A cut by the V_y-V_x plane at $V_z = 0$ is to the left and a cut the by the V_z-V_x plane at V_y given by the red dashed line in the former is to the right. The red and black crosses indicate the ion bulk speed and electron fluid velocity, respectively. In the x direction the electron fluid drifts away from the reconnection region slightly faster than the ions which are not yet magnetized at this distance. In the y direction the electron fluid drifts in the opposite direction to the ion bulk velocity and the resultant current supplies the y-directional neutral sheet current. The ions are on a half circle in the V_y-V_x plot, and its center is at $(V_y, V_x) \sim (0, -V_A)$ and its radius is $\sim V_A$ where V_A is the upstream lobe Alfvén speed. The positions of sample ions in these distribution functions are shown in the panel (b), and back-tracing of these ions mapped on the x-z and x-y planes are shown in the panel (c). Dark blue ions that have originated from the adjacent lobe are not yet accelerated much and are located near the origin of the velocity planes. Red and yellow ions that have come from the vicinity of the reconnection region have experienced the Speiser type orbit; they have bounced multiple times across the neutral sheet and have gained energy as they exert half gyrations in the $+y$ direction. Green and blue ions show intermediate behaviors between them [Nakamura et al., 1998]. (Before the ions enter the plasma sheet it is seen in the bottom panel that they move in the $-y$ direction, but this is because the magnetic field lines they follow have y component due to the Hall effect.)

Observations by Interball/Auroral and Cluster spacecraft have shown that beams of ions with energies up to $\geq$ 14 keV are produced over a wide range of closed field lines earthward of the reconnection region. These are sporadic and recurrent field-aligned beams with time scales of 2 to 3 min and characterized by decreasing energy with time. The energy dispersion

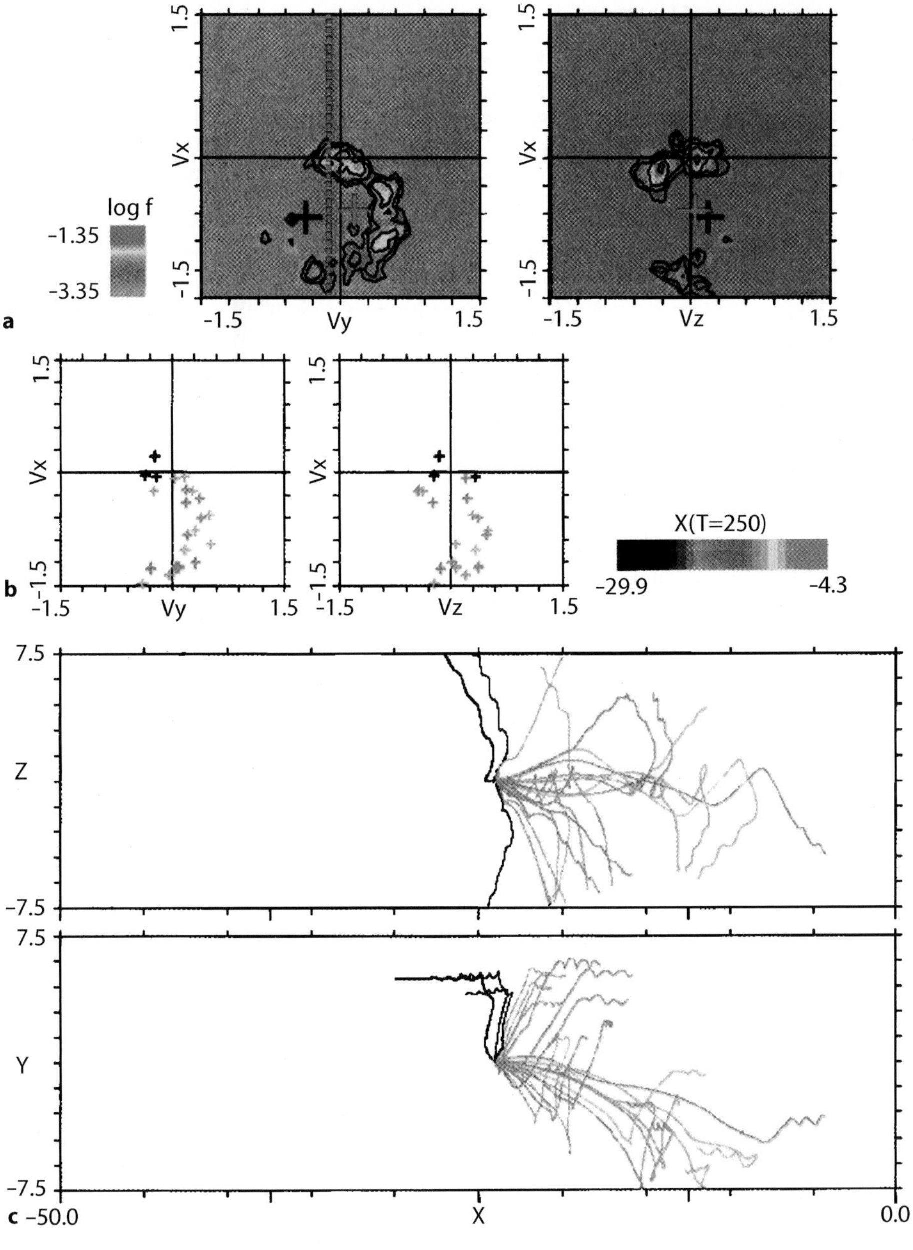

Fig. 11.17a–c. Causative mechanism of the proton spectrum in the downstream region of the X-line is studied by backtracing the motions of ions. (**a**) Spectrum in (V_y, V_x) and (V_z, V_x) planes at the given point, (**b**) positions of representative protons in the spectrum, and (**c**) trajectories of these protons at earlier times in (x, z) and (x, y) planes [Nakamura et al., 1998]

has been interpreted to be the consequence of the time of flight effect so that the beams of this kind have been called TDIS (Time-of-flight Dispersed Ion Structures) [Sauvaud and Kovrazhkin, 2004]. TDIS is correlated with intensifications of the westward ionospheric current and auroral activations at the poleward edge of the bulge period. It has been suggested that TDIS originates from localized and transient reconnections that occur earthward of the near-Earth reconnection line which has the global character. In support of this view Cluster observed a case where TDIS was accompanied by the bipolar electric field that can be related to the Hall effect [Nakamura et al., 2004].

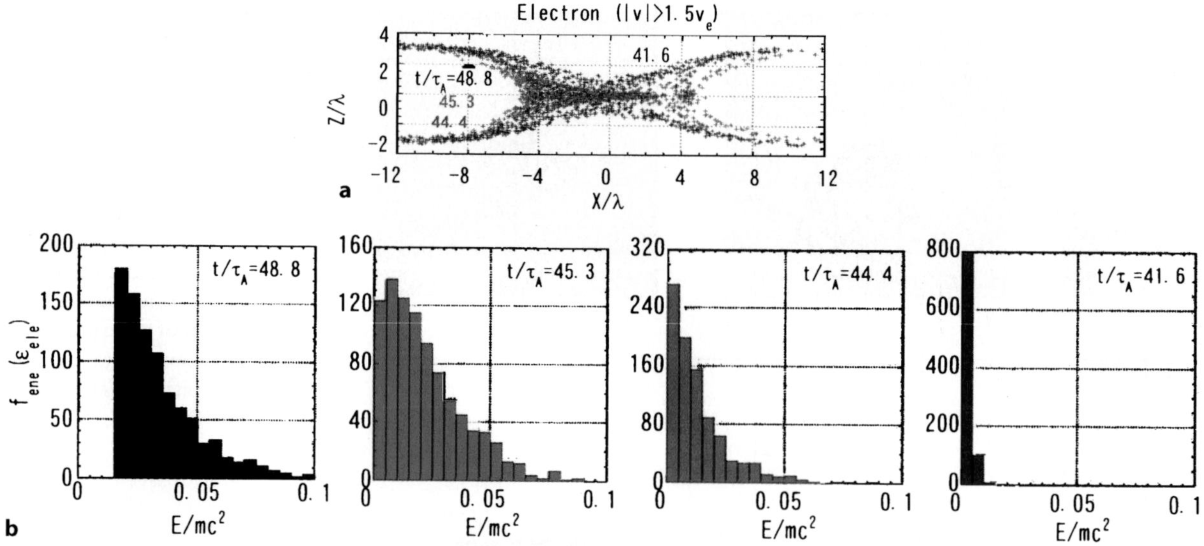

Fig. 11.18a,b. Causative mechanism of the electron spectrum is studied by backtracing the motions of energetic electrons that constitute a *black dot* in the *top panel*. Positions (*top panel*) and spectra (*bottom panel*) at three earlier times are shown with different colors [Hoshino et al., 2001]

Alternatively, it has been suggested that the plasma is heated when the flux tubes produced by the near-Earth reconnection encounter the strong pressure gradient of the closed flux tubes on the earthward side and are decelerated [Sergeev et al., 2000].

Mechanisms for producing suprathermal electrons having energies higher than 20 keV out of the thermal electrons with temperature of 2–3 keV have been studied by using a two-dimensional particle-in-cell simulation where the ion to electron mass ratio is assumed to be 64. Ion inertia length is 1.15 λ' and electron inertia length is 0.14λ', where λ' is initial thickness of the plasma sheet. Figure 11.18 shows how the electrons develop suprathermal tail in their energy spectrum. The electrons sampled are those that constitute the black dot in the upper panel and have energies higher than 1.5 times the initial electron thermal energy at the time 48.8, and their positions and spectra at earlier times are plotted in the upper and lower panels, respectively. Different colors are used to designate electrons at different times. It is seen that at the earliest time 41.6 where blue color is used all of the electrons were in the lobe and had energies less than 0.015 m_ec^2. At the time 44.4 where green is used most of them were in the neutral sheet near the X-line and were being accelerated by performing the Speiser-type motion under the dawn-to-dusk electric field in the positive y direction. The spectrum has extended to $> 0.05\ m_ec^2$. At the time 45.3 where red is used the accelerated electrons were bouncing along the O-type field lines and accelerated further since gradient and curvature drifts in this region carry electrons in the direction of the electric force. The extent of the acceleration is dependent on how many times electrons enter the neutral sheet where acceleration processes operate, and hence the pitch angle scattering by the waves excited by the electrons themselves plays an important role in determining the energy spectrum [Hoshino et al., 2001].

In the magnetotail even more energetic electrons having energies as high as >200 keV have also been detected at the distance of about 30 R_E. They often have sharp-rise and slower-decay forms with time scales of tens of minutes, and occur in intervals of northward as well as southward B_z of the local magnetic field [Baker and Stone, 1977]. The energies of these electrons are much higher than what are derivable from the electric potential difference across the tail, and the mechanism of their acceleration remains as an important subject for investigation.

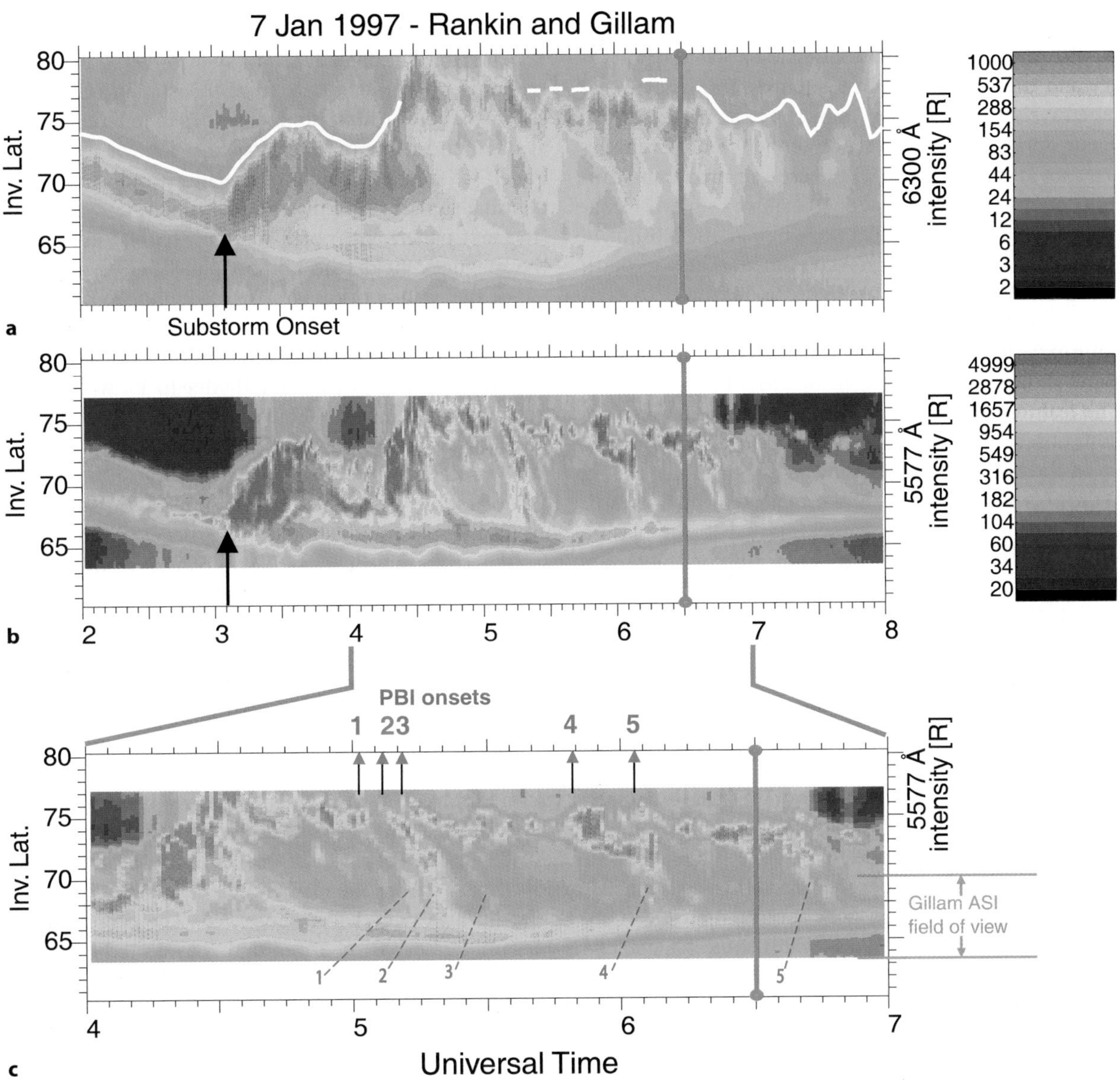

Fig. 11.19a–c. Poleward Boundary Intensifications observed in aurora. (**a**) Identification of the poleward boundary of the auroral oval (*white dashed line*) by 6300 Å emission, (**b**) intensification observed by 5577 Å emission, and (**c**) blow up of part of (**b**) [Zesta et al., 2000]

11.3.3 Consequences of the Magnetotail Reconnection

The reconnection in the magnetotail opens the gate of the energy flow that affects the entire magnetosphere. Some of the resulting disturbances are observable from the ground.

Occurrence of the reconnection in the distant tail activates the aurora at the poleward boundary of the nightside auroral oval which corresponds to the separatrix between the closed field lines of the plasma sheet and the open field lines of the tail lobe. The auroral activity of this type has been called PBI (Poleward Boundary Intensifications). PBIs are intense, transient auroral disturbances that initiate along the poleward boundary and then typically move equatorward. They occur repetitively, so that many individual disturbances can occur during time intervals of about 1 hour. Examples are shown in Fig. 11.19. These are the meridian scanning photometer observations of aurora at two Canadian stations – Rankin Inlet (at geomagnetic latitude of 73.3°) and Gillam (67.1°) – combined. The top panel shows 6300 Å emissions due to precipitations of low energy (<1 keV) electrons which are used to identify the boundary (white curves) of the auroral oval. The middle panel shows 5577 Å emissions due to > 1 keV electrons. The features of our present interest are the intensifications observed after 0440 UT. As can be seen in the blow up in the bottom panel, several intensifications started at the poleward boundary and propagated equatorward, suggesting that hot plasma was injected to the plasma sheet. This interpretation was supported by the observation (not shown) that each of these onsets (labeled 1 though 5) were followed several minutes later by earthward flow bursts in the plasma sheet (at the times labeled by 1' through 5') at the distance of about 30 R_E from the Earth. (The footprint of Geotail satellite which made these flow observations was located within 2° in latitude and 1 hour of local time northwest of Gillam.) Velocity distribution functions at the times of these flows showed signatures of acceleration in the thin current region that would be formed in the region where magnetic field lines are reconnected. PBIs are longitudinally localized, and there can simultaneously be several intensifications along the oval. They are seen as north-south aligned auroral forms in all-sky camera images of aurora [Zesta et al., 2000].

The reconnection in the near-Earth tail is associated very closely with ground signatures of magnetospheric substorms which involve the auroral breakup due to electron precipitation, DP1-type magnetic disturbance due to ionosphere-magnetosphere current flow, and Pi2 magnetic pulsations. This is shown in Fig. 11.20 where some of the representative data obtained for the December 31, 1995 substorm are compared with the Geotail data obtained at $(-30.3, 0.8, -3.7)$ R_E in GSM. Panel (a) is the magnetogram data from 6 Canadian stations which are at invariant latitudes of 61° (PIN) to 74° (RAN) and at about 23 MLT at the time of the event. There was a negative bay in the H component (belonging to the DP1 category) which started at 0558 at GIL. Panel (b) is the meridian scanning photometer data from this station (GIL) and shows that the auroral luminosity began to be enhanced at 0556 and a major intensification followed at 0559. Panel (c) is the Geotail observation and shows the defining signatures of plasmoids, namely, tailward flow (the first panel) and north-to-south bipolar change in the magnetic field (the bottom panel). In the second panel the velocity $\boldsymbol{V}_x$ is separated into components $\boldsymbol{V}_\perp$ and $\boldsymbol{V}_\parallel$ which are perpendicular and parallel to the magnetic field, respectively, and it is seen that the tailward flow was predominantly due to $\boldsymbol{V}_\perp$ with which the magnetic flux was convected. The onset time of the tailward convection at 0554 (indicated by a vertical line) preceded the onset signatures of substorm in aurora and geomagnetic perturbation. In this substorm event the plasmoid should have been produced by reconnection in the near-Earth region earthward of $x = -30\,R_E$ prior to the onset of the substorm [Ohtani et al., 1999].

Figure 11.21 is another example of the plasmoid-substorm relation for a case where global images of aurora are available. The observation of a plasmoid by Geotail is in the left panel and the observation of UV aurora by the Polar satellite is in the right panel. (The aurora image is in the Lyman-Birge-Hopfield band and each shot takes 36 s or 18 s.) In the Geotail data obtained at (-28, 8, -19) R_E the north-to-south turning in B_z (the fourth panel from the top) and the fast tailward flow in V_x (the fifth panel) were observed. The center of the plasmoid where B_z changes sign is indicated by a vertical line. In the Polar data the auroral brightening and poleward expansion started around the middle of the sequence at 0356:12 which is labeled in red letters. This occurred at the projection of Geotail along magnetic field lines which is indicated by a small circle in that plot. In the left panel the onset time of this auroral activation is indicated by a red bar that encompasses the time interval during which the image was taken. It is seen that the auroral brightening and the plasmoid ejection are closely linked together [Ieda et at., 2001].

A complementary evidence for the intimate relation between the auroral activity and the magnetotail dynamics has been obtained from the analysis of the plasma flow velocity when the aurora is quiet. In this study Ieda et al. [2003] used the LBHL images obtained

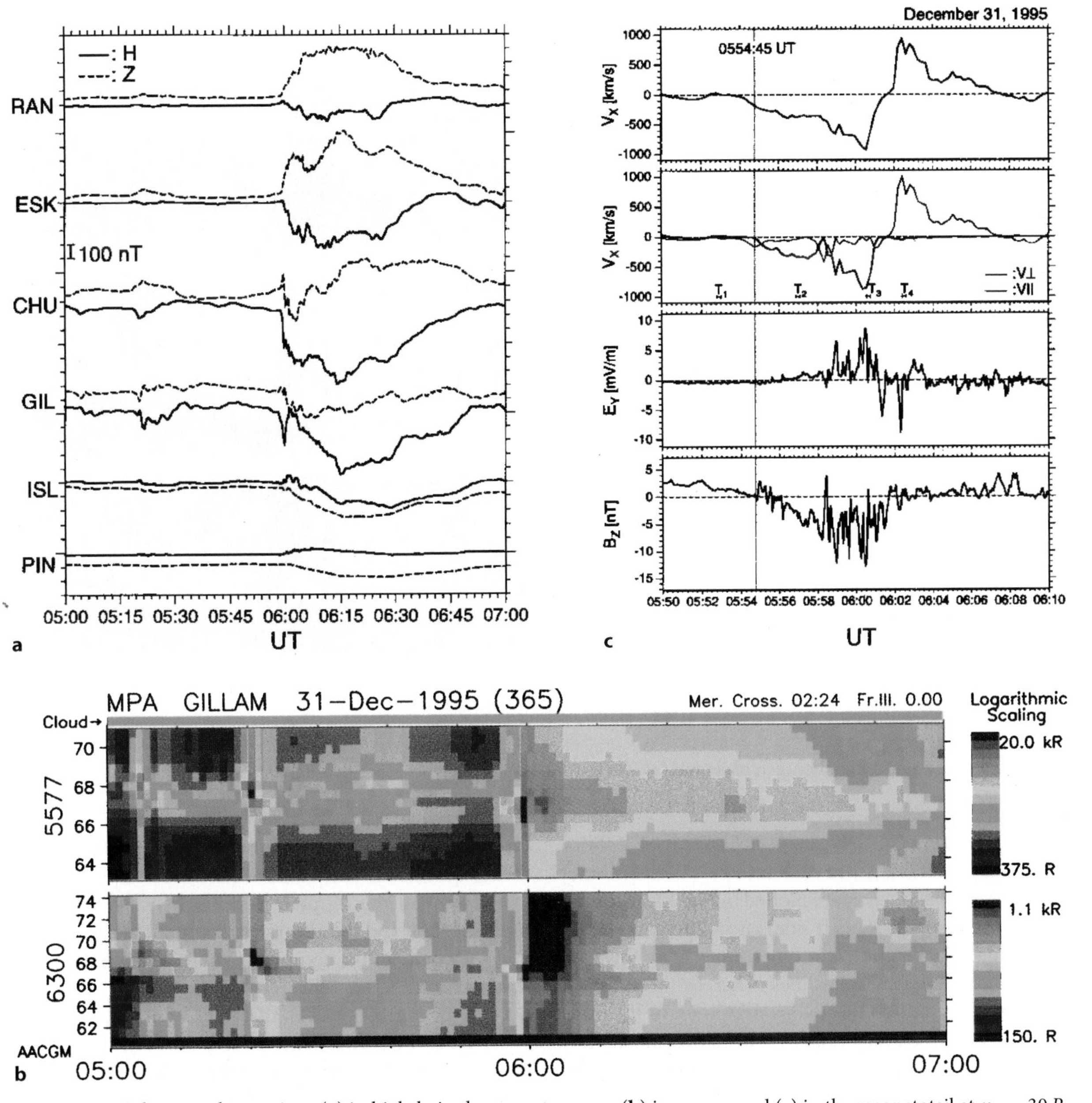

Fig. 11.20a–c. Substorm observations (**a**) in high-latitude magnetograms, (**b**) in aurora, and (**c**) in the magnetotail at $x = -30\,R_E$ [Ohtani et al., 1999]

by the UVI experiment on board Polar and averaged the photon counts over each grid (0.5° in the magnetic latitude MLAT and 0.25 hours in the magnetic local time MLT) in the target area of 60°–75° MLAT and 21–03 MLT. Images were labeled quiet when no more than one grid point had counts above a threshold of 10 photons $cm^{-2}\,s^{-1} \sim 300$ Rayleigh. Twenty percent of images were identified as quiet when Geotail was at $-8 > x > -31\,R_E$ and $|y| < 10\,R_E$ in the tail. Then the intervals when all the 36-s exposures of LBHL images remained quiet for at

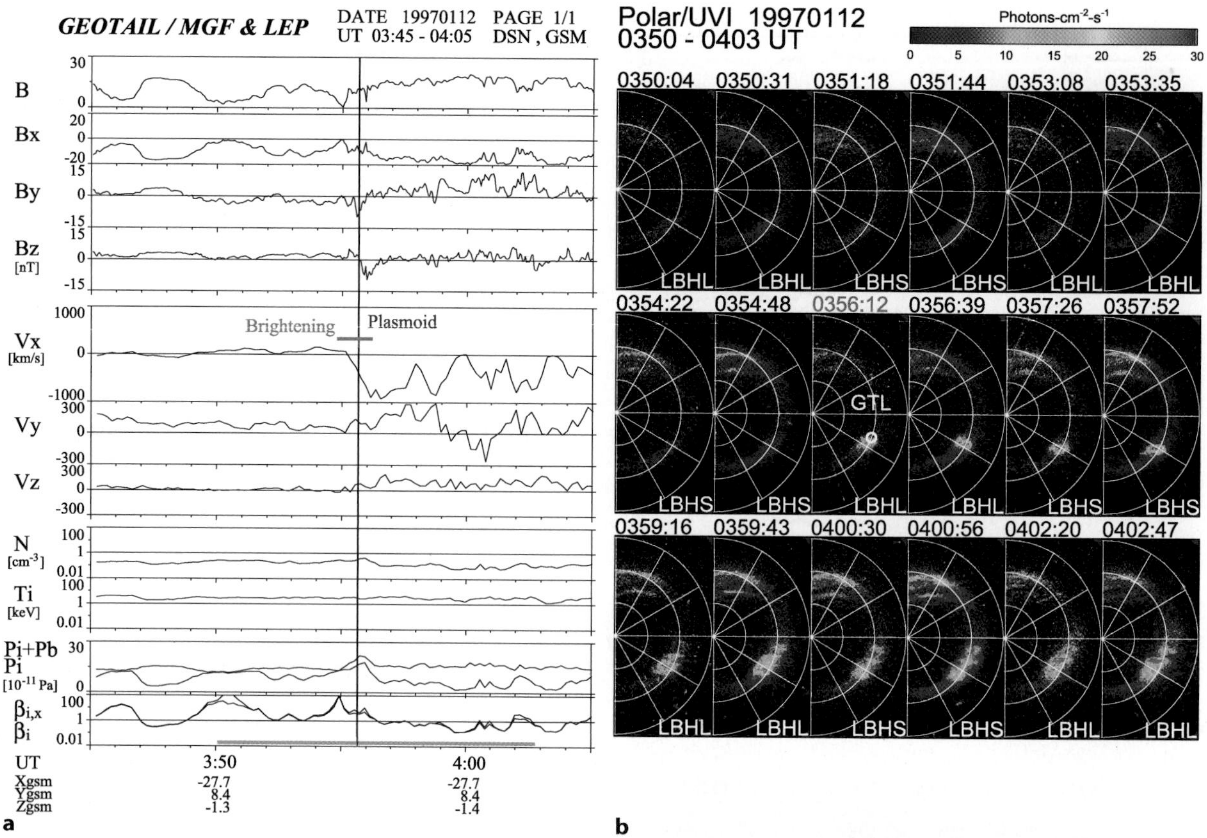

Fig. 11.21a,b. Close association between plasmoid (**a**) and auroral brightening observed by satellite (**b**) [Ieda et al., 2001]

least 30 min were chosen. The occurrence probabilities of V_x, $V_{\text{perp},x}$ and $V_{\text{para},x}$ in such intervals in the plasma sheet (defined by $\beta \geq 0.1$) are plotted in Fig. 11.22 for the ranges $-8 > x > -20\,R_E$ (left side) and $-20 > x > -31\,R_E$ (right side). Probabilities for quiet times are shown by squares connected by solid lines, whereas those for all the data are by dots connected by thin lines. In the middle panels it is seen that $V_{\text{perp},x}$ is very slow (mostly less than 100 km/s) in quiet times. The distributions of the quiet-time $V_{\text{para},x}$ in the bottom panel show the same feature although their wings are slightly broader than that of $V_{\text{perp},x}$. Since it has been shown that fast tailward flows (faster than 300 km/s) tend to be associated with the southward polarity of the magnetic field and can be taken as a product of the near-Earth reconnection while slower tailward flows are associated more frequently with northward polarity and seem to represent the inward/outward fluctuating motion of closed field lines, the above can be interpreted to mean that the aurora is quiet when this reconnection is not operating.

In view of the intimate relation between the plasmoid ejection and the auroral activation it seems certain that they belong to a common dynamical phenomenon that is controlled by the reconnection. This does not necessarily mean, however, that the reconnection region centered at the X-line is the direct source of the energy that supports all of them. Once the reconnection is initiated, the energies can flow from upstream to downstream across the entire length of the reconnected field lines and the energy flux originating from the reconnection region is only a portion of the entire energy supply. This can be seen in Fig. 11.23 where the energy fluxes at three representative locations are plotted against the time from the substorm onset. Panel (a) is z component (positive equatorward) of the energy flux in the lobe and PSBL (plasma sheet boundary layer), (b) is the x compo-

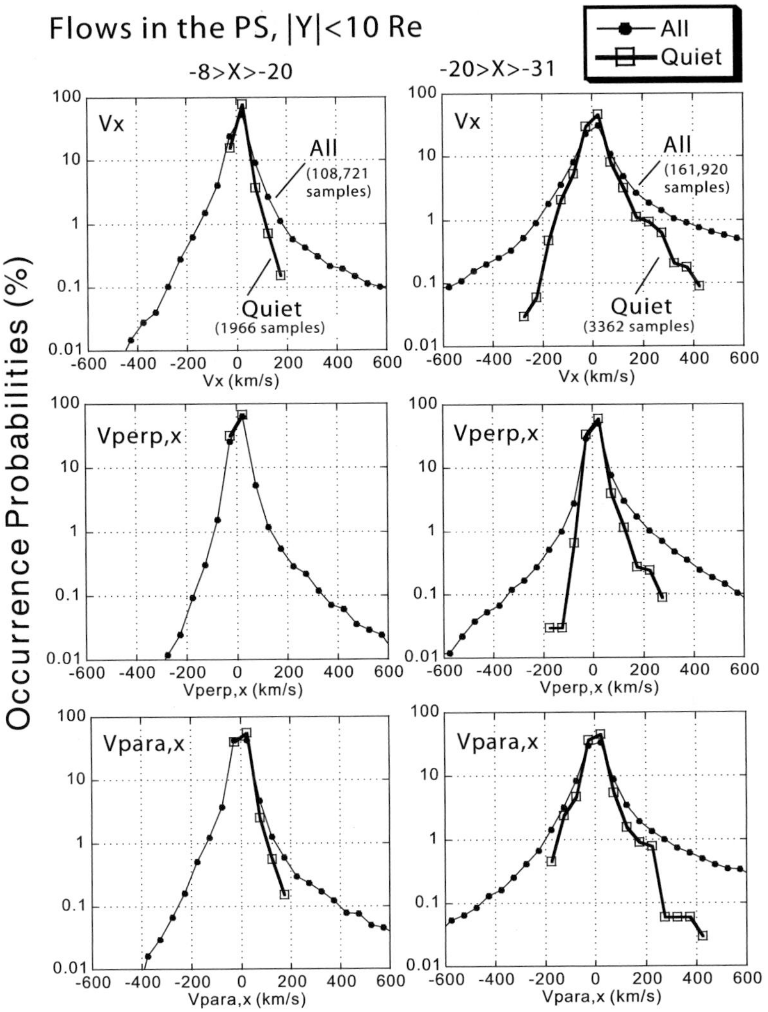

Fig. 11.22. Occurrence probability of flow velocities during quiet times (*thick lines*) as compared with all observations (*thin lines*) at two distance ranges in the magnetotail. Top, middle and bottom panel shows distributions of V_x, $V_{\mathrm{perp},x}$ and $V_{\mathrm{para},x}$, respectively [Ieda et al., 2003]

nent of the flux in the plasma sheet, and (c) is its y component. The left, middle and right panels correspond to three distance ranges, respectively. Poynting (solid lines), kinetic (dotted lines) and thermal (dashed lines) energy fluxes are plotted with different symbols. Time 0 is the substorm onset time determined by the UVI observation on board Polar. In the "dipolarization" region ($-7 \geq x \geq -13\,R_{\mathrm{E}}$) it is the Poynting flux toward the plasma sheet that shows the largest and the most distinctive increase around time 0. The input of this Poynting flux is larger at closer distances from the Earth because V_z does not vary very much with x (not shown) but the magnetic field is stronger closer to the Earth. In the "plasmoid" region ($-23 \geq x \geq -31\,R_{\mathrm{E}}$) it is the tailward thermal energy flux that shows the largest increase after time 0. But an enhancement in the earthward energy flux around time 0 is not evident at any distance range [Miyashita et al., 2003]. The Poynting energy entering the dipolarization region of the plasma sheet would be consumed primarily to let plasma flow equatorward and earthward against pressure, but it could also be used to generate the auroral activity; it seems possible that the auroral activity is supported not so much by the energy that comes directly from the direction of the reconnection region but by the Poynting flux that enters the plasma sheet at much shorter distances from the Earth.

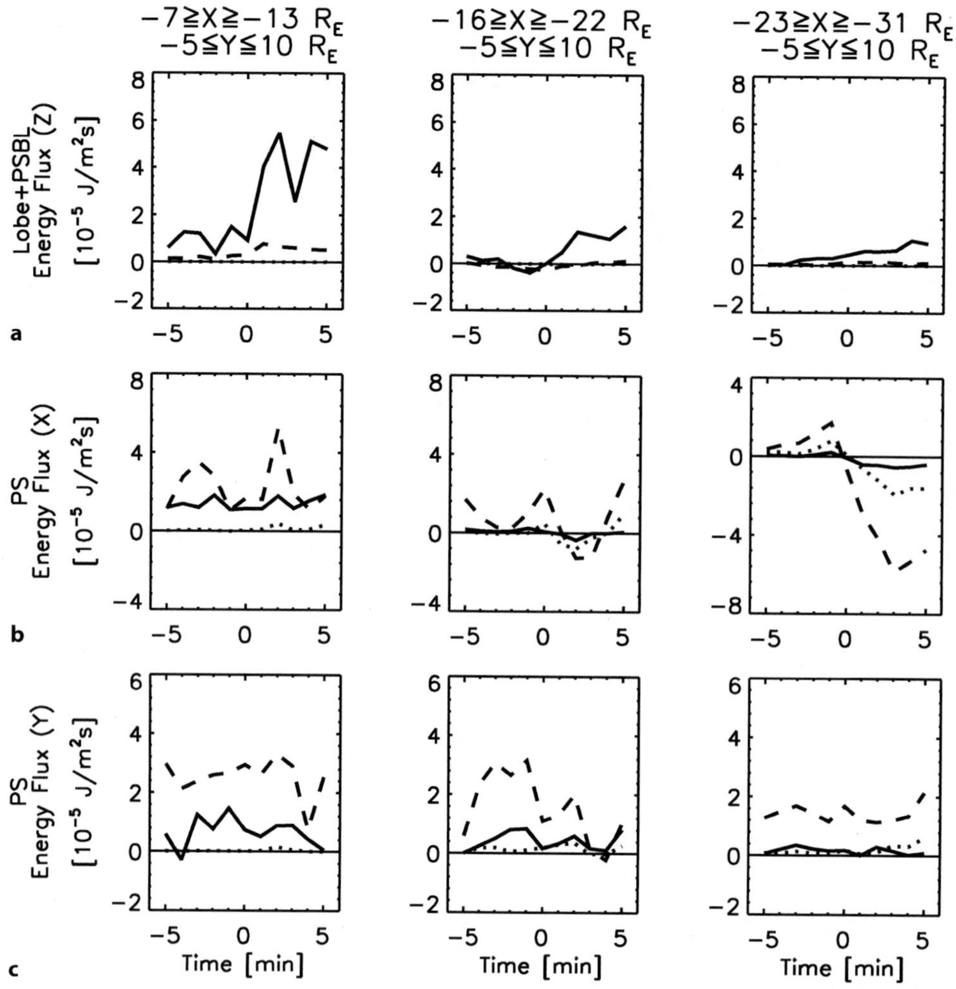

Fig. 11.23a–c. Variation of energy fluxes with substorm time at three distance ranges (*right, middle* and *left panels*). From the top, (**a**) Z component (positive equatorward) of energy fluxes in the lobe and the plasma sheet boundary layer, (**b**) X component of the energy flux in the plasma sheet, and (**c**) Y component of the energy flux in the plasma sheet [Miyashita et al., 2003] The *solid, dotted* and *dashed lines* indicate Poynting, kinetic energy and thermal energy fluxes, respectively

11.3.4 Cause of the Near-Earth Reconnection

While the distant tail reconnection line is the essential counterpart of the frontside reconnection and must occur in order to recover closed field lines from the open field lines, the near-Earth reconnection which begins in the domain of the closed field lines is not an obvious entity. The reason for its occurrence need be clarified.

In terms of the electric current, the magnetic reconnection means that the cross-tail current is locally weakened so much so that B_z is made negative for some distance beyond the region of the current reduction. The weakening of the cross-tail current would occur when the thickness of the current layer is reduced and/or the flow of the current-carrying particles is impeded by instabilities arising from the enhanced current density.

Internal processes that could critically change the cross-tail current structure have been studied extensively and numerous microscopic plasma instabilities have been proposed and tested as candidates [e.g., Lui, 2004]. However, the global development of the reconnection requires more than just local mechanisms that make the cross-tail current weaker. It is necessary that the magnetotail is in the metastable state and the system can be brought into a state of lower energy by the occurrence of reconnection. Otherwise the dynamics that is initiated by the local reconnection would not develop into a truly global feature which the near-Earth reconnection has been observed to represent. The observation of a case where the X-line is convected earthward along with the background plasma [Ohtani et al., 2004] suggests that reconnection in the near-Earth magnetotail caused by a local process does not necessarily develop into the full-fledged global near-Earth reconnection process that produces, among other things, tailward ejection of the plasmoid.

We should also note that the current densities plotted in panel (c) of Fig. 11.13 show enhancements associated with substorms but do not clearly demonstrate a drastic decrease that could be identified with the "current disruption" at the time 0. Thus the above data do not render support to the idea that the magnetic reconnection is caused by the current disruption, although it could still be argued that the disruption was not detected as it took place in a very limited region.

Several simulations have been performed to study how the near-Earth reconnection develops. Tanaka [2000] used a resistive code where the magnetotail becomes more diffusive as it goes further downtail. Following the southward turning of IMF at the frontside magnetopause the plasma sheet thins because the supply of the closed flux from the distant tail cannot balance the enhanced convection which is initiated on the dayside. The configuration change begins abruptly with a change of pressure distribution in the near-Earth plasma sheet. He has suggested that the primary cause of the onset is the dipolarization that results from the breakdown of dynamic stress balance in the plasma sheet, and interpreted this as a manifestation of a self-organized critical phenomenon in a nonlinear system, where the change of the plasma sheet from the thinned to the dipolarized state represents a state (phase) transition of the system. Formation of the near-Earth reconnection line is a part of the new system and it occurs simultaneously with the dipolarization.

Birn et al. [2004] have looked into the effect of the boundary deformations of the magnetotail on its internal structure. They have found that equilibrium configurations that satisfy magnetic flux, entropy and topology conservation cease to exist when the boundary deformations exceeds critical limits. Equilibrium can be lost regardless of the dissipation mechanism, and catastrophe points can be reached for relatively modest perturbations of the boundary. Most important property that influences the critical limit is the overall tail flaring that corresponds to the decrease of the lobe field with distance downtail. This is probably an important direction of research that need to be expanded to include the state of the convective motion.

Large fraction of substorms is triggered by changes in solar-wind or IMF conditions. The effect of the northward turning of IMF is most frequently seen and accounts for almost a half of the substorm onsets, whereas the onsets triggered by solar wind pressure pulse is much less frequent. Figure 11.24 (a) and (b) shows the result of superposed epoch analysis of B_z and B_x components of the lobe field in the tail at x of ~ 15 to $\sim 35\,R_{\mathrm{E}}$, when the AL index in (c) signifies the onset of substorm. The time is relative to the Pi2 onset time and averages for the IMF triggered and non-triggered cases are plotted separately [Hsu and McPherron, 2004]. In both cases, strength of the magnetic field increases and flaring angle increases during $t < 0$, and the field becomes weaker and the flaring disappears sharply at $t > 0$. These are the typical features of the substorm growth phase and expansion phase, respectively, in the tail lobe of the near-Earth region which are associated with the occurrence of the reconnection at the same range of x. It is noteworthy that essentially the same features are observed for both triggered and non-triggered substorms. This suggests that the magnetosphere has been in the metastable state and building up the free energy when it is triggered by the IMF change to recover to the stable state by releasing the extra energy by the magnetic reconnection.

It has to be said that the causative mechanism for the near-Earth reconnection is yet to be fully identified. The mechanism should involve both the kinetic process that enhances the reconnection rate locally and the macroscopic process that brings the overall configuration from a metastable to a stable state. It is essential to consider both aspects together. We have to build a theory that can explain why the near-Earth reconnection occurs in the region of closed field lines at the distance of about 20 R_{E}.

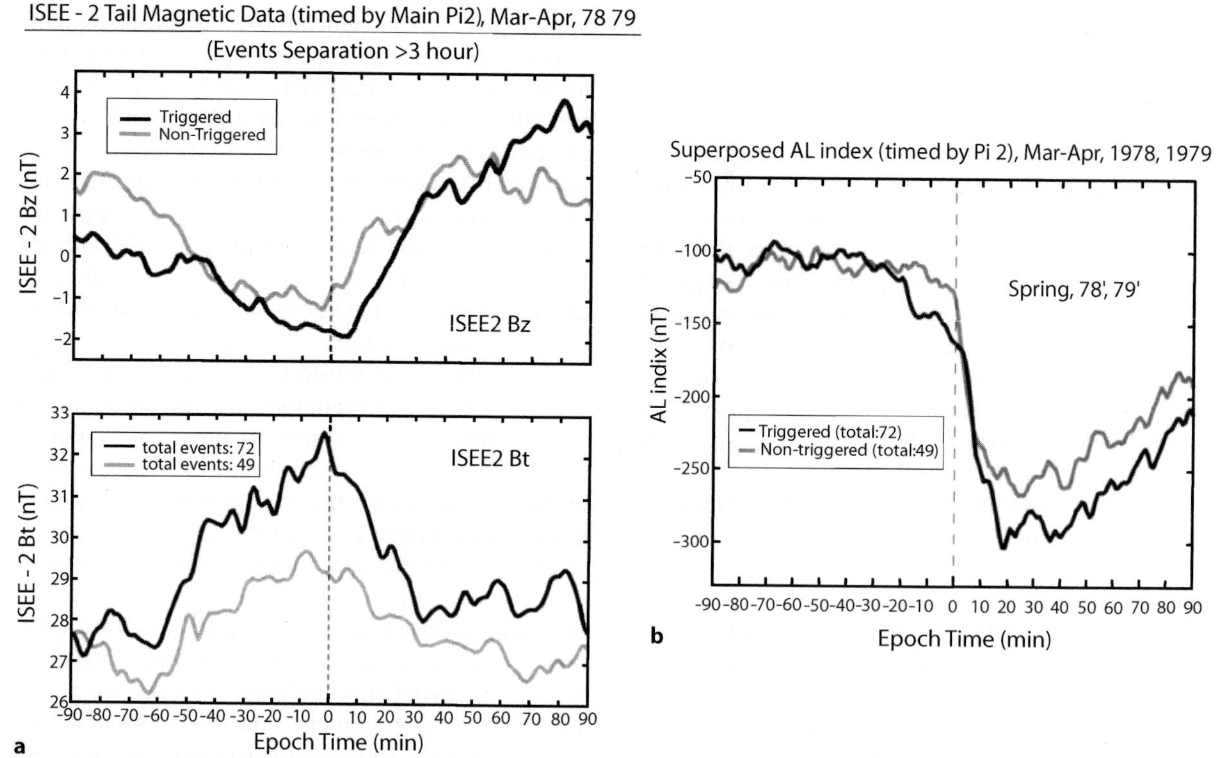

Fig. 11.24. (**a**) Average changes in the magnetic field in the tail lobe and (**b**) in the AL index. Substorms triggered by an IMF perturbation and those which are not are compared [Hsu and McPherron, 2004]

References

Alfvén, H., On the theory of magnetic storms and aurorae, Tellus, 10, 104, 1958

Arzner, K., Scholer, M., Kinetic structure of the post plasmoid plasma sheet during magnetotail reconnection, J. Geophys. Res., 106, 3827, 2001

Asano, Y. et al., Current sheet structure around the near-Earth neutral line observed by Geotail, J. Geophys. Res., 109, A02212, 2004a

Asano, Y., et al., Statistical study of thin current sheet evolution around substorm onset, J. Geophys. Res., 109, A05213, 2004b

Baker, D.N., Stone, E.C., Observations of energetic electrons ($E \geq 200$ keV) in the Earth's magnetotail: Plasma sheet and fireball observations, J. Geophys. Res., 82, 1532, 1977

Birn, J., Dorelli, J.C., Hesse, M., Schindler, K., Thin current sheets and loss of equilibrium: Three-dimensional theory and simulation, J. Geophys. Res., 109, A02215, 2004

Büchner, J., Zelenyi, L.M., Regular and chaotic charged particle motion in the magnetotaillike field reversals, 1. Basic theory of trapped motion, J. Geophys. Res., 94, 11, 821, 1989

Crooker, N.U., Dayside merging and cusp geometry, J. Geophys. Res., 84, 951, 1979

Cowley, S.W.H., The causes of convection in the Earth's magnetosphere: A review of developments during the IMS, Rev. Geophys. Space Phys., 20, 531, 1982

Dungey, J.W., Interplanetary magnetic field and the auroral zones, Phys. Rev. Lett., 6, 47, 1961

Fujimoto, M., Terasawa, T., Mukai, T., The low-latitude boundary layer in the tail-flanks, in *New Perspectives on the Earth's magnetotail*, Nishida, A., Baker, D.N., Cowley, S.W.H., Eds., Geophys. Mono. 105, Amer. Geophys. Union, pp. 33, 1998

Fuselier, S.A., Kinetic aspects of reconnection at the magnetopause, in *Physics of the Magnetopause*, Song, P., Sonnerup, B.U.Ö., Thomsen, M.F., Eds., Geophys. Mono. 90, Amer. Geophys. Union, pp. 181, 1995

Hasegawa, H., Maezawa, K., Mukai, T., Saito, Y., Plasma entry across the distant tail magnetopause, 1. Global properties and IMF dependence, J. Geophys. Res., 107, 10.1029/2001.JA900139, 2002

Hasegawa, H., et al., Transport of solar wind into Earth's magnetosphere through rolled-up Kelvin-Helmholtz vortices, Nature, 430, 755, 2004

Hones, E.W., Jr., Transient phenomena in the magnetotail and their relation to substorms, Space Sci. Rev., 23, 393, 1979

Hoshino, M., Mukai, T., Terasawa, T., Shinohara, I., Suprathermal electron acceleration in magnetic reconnection, J. Geophys. Res., 106, 25, 979, 2001

Hsu, T.-S., McPherron, R.L., Average characteristics of triggered and non-triggered substorms, J. Geophys. Res., 109, A07208, 2004

Ieda, A., et al., Statistical analysis of the plasmoid evolution with Geotail observations, J. Geophys. Res., 103, 4453, 1998

Ieda, A., et al., Plasmoid ejection and auroral brightenings, J. Geophys. Res., 106, 3845, 2001

Ieda, A., et al., Quiet time magnetotail plasma flow: Coordinated Polar ultraviolet images and Geotail observations, J. Geophys. Res., 108, 1345, 2003

Kim, K.-H., et al., Evidence for component merging near the subsolar magnetopause: Geotail observation, Geophys. Res. Lett., 29, 1080, 2002

Levy, R.H., Petschek, H.E., Siscoe, G.L., Aerodynamic aspects of the magnetospheric flow, AIAA J., 2, 2065, 1964

Lockwood, M., Cowley, S.W.H., Onsager, T.G., Ion acceleration at both the interior and exterior Alfvén waves associated with the magnetopause reconnection site: Signatures in cusp precipitation, J. Geophys. Res., 101, 21, 501, 1996

Lockwood, M., Hapgood, M.A., On the cause of a magnetic flux transfer event, J. Geophys. Res., 103, 26, 453, 1998

Lottermoser, R.-F., Scholer, M., Matthews, A.P., Ion kinetic effects in magnetic reconnection: Hybrid simulations, J. Geophys. Res., 103, 4547, 1998

Lui, A.T.Y., Potential plasma instabilities for substorm expansion onsets, Space Sci., Rev., 113, 127, 2004

Machida, S., et al., Geotail observations of flow velocity and north-south magnetic field variations in the near and mid-distant tail associated with substorm onsets, Geophys. Res. Lett., 26, 635, 1999

Maezawa, K., Magnetospheric convection induced by the positive and negative z components of the interplanetary magnetic field: Quantitative analysis using polar cap magnetic records, J. Geophys. Res., 81, 2289, 1976

Millward, G.H., Moffett, R.J., Balmforth, M.F., Roger, A.S., Modeling the ionospheric effects of ion and electron precipitation in the cusp, J. Geophys. Res., 104, 24, 603, 1999

Miyashita, Y., et al., Evolution of the magnetotail associated with substorm auroral breakups, J. Geophys. Res., 108, A9, 1353, 2003

Mukai, T., Yamamoto, T., Machida, S., Dynamics and kinetic properties of plasmoids and flux ropes: Geotail observations, in *New Perspectives on the Earth's magnetotail*, Nishida, A., Baker, D.N., Cowley, S.W.H., Eds., Geophys. Mono. 105, Amer. Geophys. Union, pp. 117, 1998

Nagai, T., et al., Structure of the Hall current system in the vicinity of the magnetic reconnection site, J. Geophys. Res., 108, 1357, 2003

Nakamura, M.S., Fujimoto, M., Maezawa, K., Ion dynamics and resultant velocity space distributions in the course of magnetotail reconnection, J. Geophys. Res., 103, 4531, 1998

Nakamura, R., et al., Flow shear near the boundary of the plasma sheet observed by Cluster and Geotail, J. Geophys. Res., 109, A05204, 2004

Newell, P.T., Xu, D., Meng, C.-I., Kivelson, M.G., Dynamical polar cap: A unifying approach, J. Geophys. Res., 102, 127, 1997

Nishida, A., Coherence of geomagnetic DP2 fluctuations with interplanetary magnetic variations, J. Geophys. Res., 73, 5549, 1968

Nishida, A., Can random reconnection on the magnetopause produce the low latitude boundary layer? Geophys. Res. Lett., 16, 227, 1989

Nishida, A., Maezawa, K., Two basic modes of interaction between the solar wind and the magnetosphere, J. Geophys. Res., 76, 2254, 1971

Ohtani, S., et al., Substorm onset timing: The December 31, 1995, event, J. Geophys. Res., 104, 22, 713, 1999

Ohtani, S., et al., Tail dynamics during the growth phase of the 24 November 1996, substorm event: Near-Earth reconnection confined in the plasma sheet, J. Geophys. Res., 109, A05211, 2004

Petrinec, S.M., Fuselier, S.A., On continuous versus discontinuous neutral lines at the dayside magnetopause for southward interplanetary magnetic field, Geophys. Res. Lett., 30, 1519, 2003

Petrukovich, A.A., et al., Two spacecraft observations of a reconnection pulse during an auroral breakup, J. Geophys. Res., 103, 47, 1998

Phan, T.-D., Paschmann, G., Sonnerup, B.U.Ö., Low-latitude dayside magnetopause and boundary layer for high magnetic shear, 2. Occurrence of magnetic reconnection, J. Geophys. Res., 101, 7817, 1996

Phan, T.D., et al., Evidence for an extended reconnection line at the dayside magnetopause, Earth Planets Space, 53, 619, 2001

Ridley, A.J, Lu, G., Clauer, C.R., Papitashivili, V.O., A statistical study of the ionospheric convection response to changing interplanetary magnetic field conditions using the assimilative mapping of ionospheric electrodynamics technique, J. Geophys. Res., 103, 4023, 1998

Ruohoniemi, J.M., Greenwald, R.A., The response of high-latitude convection to a sudden southward IMF turning, Geophys. Res. Lett., 25, 2913, 1998

Russell, C.T., Elphic, R.C., ISEE observations of flux transfer events at the dayside magnetopause, Geophys. Res. Lett, 6, 33-36, 1979

Saito, Y., et al., Slow-mode shocks in the magnetotail, J. Geophys. Res., 100, 23, 567, 1995

Sauvaud, J.-A., Kovrazhkin, R.A., Two types of energy-dispersed ion structures at the plasma sheet boundary, J. Geophys. Res., 109, A12213, 2004

Sergeev, V.A., et al., Plasma sheet ion injections into the auroral bulge: Correlative study of spacecraft and ground observations, J. Geophys. Res., 105, 18, 465, 2000

Shay, M.A., Drake, J.F., Rogers, B.N., Denton, R.E., The scaling of collisionless, magnetic reconnection for large systems, Geophys. Res. Let., 26, 2163, 1999

Slavin, J.A., et al., An ISEE 3 study of average and subsstorm conditions in the distant magnetotail, J. Geophys. Res., 90, 10, 873, 1985

Sonnerup, B.U.Ö., Magnetic field reconnection, in *Solar System Plasma Physics*, vol. 3, edited by Lanzerotti, L.T., Kennel, C.F., Parker, E.N., pp.45. North-Holland, New York, 1979

Sonnerup, B.U.Ö, et al., Evidence for magnetic field reconnection at the Earth's magnetopause, J. Geophys. Res., 86, 10, 049, 1981

Tanaka, T., Configuration of the magnetosphere-ionosphere convection system under northward IMF conditions with nonzero IMF B_y, J. Geophys. Res., 104, 14, 683, 1999

Tanaka, T., The state transition model of the substorm onset, J. Geophys. Res., 105, 21, 081, 2000

Vasyliunas, V.M., Theoretical models of magnetic field line merging, 1., Rev. Geophys. Space Phys., 13, 303, 1975

Zesta, E., Lyons, L.R., Donovan, E., The auroral signature of earthward flow bursts observed in the magnetotail, Geophys. Res. Let., 27, 3241, 2000

12 Nonlinear Processes in Space Plasmas

Lennart Stenflo and Padma Kant Shukla

We present here a comprehensive review of some of the main nonlinear effects involving wave–wave and wave–particle interactions in space plasmas. Attention is focused on three-wave decay interactions, modulational instabilities, wave localization and the formation of structures caused by ponderomotive forces, differential electron Joule heating, and self-wave interactions of high- and low-frequency electromagnetic waves. We present nonlinear dispersion relations and their analysis, as well as the dynamics of nonlinearly interacting modes with the background plasma. The relevance of our investigation to space plasmas is discussed.

Contents

Lennart Stenflo and Padma Kant Shukla, Nonlinear Processes in Space Plasmas.
In: Y. Kamide/A. Chian, Handbook of the Solar-Terrestrial Environment. pp. 311–329 (2007)
DOI: 10.1007/11367758_12

12.1 Introduction

Nonlinear collective processes in space and laboratory plasmas are of significant interest, e.g. for example [Sagdeev and Galeev, 1969; Karpman, 1971; Kadomtsev and Karpman, 1971; Hasegawa, 1974; Shukla and Dawson, 1984; Horton and Hasegawa, 1994; Shukla, 1999; Shukla and Stenflo, 1995; Chian *et al.*, 1995; Chian, 1997; Stenflo and Shukla, 2000; Shukla, 2004], in understanding the role of large amplitude waves and fields, which are either spontaneously produced or launched by external sources. Such waves can cause a number of nonlinear effects including particle trapping which gives rise to nonlinear frequency shifts and trapped particle instabilities, parametric instabilities exciting low-frequency modes and sidebands, etc. Wave–particle and wave–wave interactions are, in fact, dominant processes in space plasmas, as they determine the dynamics of the plasma systems. In wave–particle interactions, energy from a large amplitude wave is transferred to particles or vice versa, if the phase speeds of the waves are close to the thermal speeds of the plasma particles. Here, we refer to Landau and inverse Landau damping processes. On the other hand, in wave–wave interactions, one can encounter nonlinear excitation of low-frequency waves by the interaction of two high-frequency waves. Such a scenario is quite common in the ionospheric modification experiments with high power radio waves/radar beams, as well as in laser produced plasmas and in magnetically confined fusion plasmas where intense electromagnetic waves are used for heating purposes. Furthermore, in the Earth's auroral zone and in the magnetosphere, as well as in the solar corona one can have finite amplitude electromagnetic ion-cyclotron and Alfvén waves which are subjected to parametric instabilities on account of the ponderomotive force and the Joule heating caused by wave-electron interactions.

In this chapter, we describe the essential physics and the mathematical background of some of the main nonlinear effects associated with high- and low-frequency electromagnetic waves which are relevant to space plasmas. In Sect. 12.2 we discuss stimulated scattering of high-frequency electromagnetic (HF-EM) waves and show how a large amplitude HF-EM can excite sidebands and different types of low-frequency waves in plasmas. A general formulation for resonant three-wave interactions is outlined in Sect. 12.3. Parametric instabilities of magnetic field-aligned Alfvén waves are considered in Sect. 12.4. Excitation of zonal flows by kinetic Alfvén waves is discussed in Sect. 12.5. In Sect. 12.6 we focus on obliquely propagating shear Alfvén waves (SAWs) and present the ponderomotive force induced magnetic field-aligned density perturbations. The latter are incorporated in the description of the amplitude modulated SAWs in Sect. 12.7. Section 12.8 contains a novel mechanism for the electron heating due to wave-electron interactions. Self-interactions between SAWs and drift-Alfvén-Shukla-Varma modes are considered in Sects. 12.9 and 12.10, respectively, where we point out the possibility of self-organization in the form of various types of vortices and structures. Section 12.11 suggests potential applications and new directions.

12.2 Stimulated Scattering of Electromagnetic Waves

As a first simple example, following Stenflo [1990], we consider an electromagnetic wave that propagates in a uniform unmagnetized plasma. We denote the electric field amplitude of this pump wave by $\boldsymbol{E}_0$. The wave frequency ω_0 and wave vector $\boldsymbol{k}_0$ are here related by means of the dispersion relation

$$\omega_0^2 = \omega_{\rm pe}^2 + k_0^2 c^2 \,, \tag{12.1}$$

where $\omega_{\rm pe} = (4\pi n_0 q_{\rm e}^2/m_{\rm e})^{1/2}$ is the electron plasma frequency, n_0 is the equilibrium density, $q_{\rm e}$ is the electron charge, $m_{\rm e}$ the electron mass, and c the speed of light in vacuum.

The electromagnetic pump wave interacts with electrostatic low-frequency fluctuations $(\omega, \boldsymbol{k})$ in the plasma, and produces thus sideband waves $(\omega_+, \boldsymbol{k}_+)$ and $(\omega_-, \boldsymbol{k}_-)$, where $\omega_\pm = \omega \pm \omega_0$ and $\boldsymbol{k}_\pm = \boldsymbol{k} \pm \boldsymbol{k}_0$. If ω is much less than ω_0, we can obtain the electromagnetic sideband wave amplitudes $\boldsymbol{E}_\pm$ from the wave equation

$$\left(\omega_\pm^2 - \omega_{\rm pe}^2\right)\boldsymbol{E}_\pm + c^2 \boldsymbol{k}_\pm \times (\boldsymbol{k}_\pm \times \boldsymbol{E}_\pm) \approx \omega_{\rm pe}^2 (n_1/n_0) \boldsymbol{E}_{0\pm} \,, \tag{12.2a}$$

where n_1 represents the density perturbation due to the low-frequency wave, and where we have denoted $\boldsymbol{E}_{0+} \equiv \boldsymbol{E}_0$ and $\boldsymbol{E}_{0-} \equiv \boldsymbol{E}_0^*$.

The pump wave also interacts with the sideband waves to produce a low-frequency ponderomotive force. If the electron equation of motion, where this force appears, is combined with the continuity equations, the ion equation of momentum, and Poisson's equation, one obtains

$$(1+\chi_e+\chi_i)n_1 = -\frac{k^2}{4\pi m_e \omega_0^2}\chi_e(1+\chi_i)(\boldsymbol{E}_{0+}\cdot\boldsymbol{E}_- + \boldsymbol{E}_{0-}\cdot\boldsymbol{E}_+)\,, \tag{12.2b}$$

where $\chi_e = \chi_e(\omega, \boldsymbol{k})$ and $\chi_i = \chi_i(\omega, \boldsymbol{k})$ are the susceptibilities for the electrons and ions, respectively. They can be written as $\chi_j = (4\pi q_j^2/k^2 m_j) \int \mathrm{d}\boldsymbol{v}\boldsymbol{k} \cdot (\partial F_{0j}/\partial \mathbf{v})/(\omega - \boldsymbol{k}\cdot\boldsymbol{v})$, where F_{oj} represents the velocity distribution function. Assuming in the derivation of Eq. (12.3) that ω_0 is much larger than the ω_{pe}, and inserting $\boldsymbol{E}_\pm$ from Eq. (12.2a) into Eq. (12.2b), one now obtains the nonlinear dispersion relation

$$\frac{1}{\chi_e} + \frac{1}{1+\chi_i} = \frac{k^2|\boldsymbol{k}_+ \times \boldsymbol{v}_0|^2}{k_+^2\left(k_+^2c^2 - \omega_+^2 + \omega_{pe}^2 - \mathrm{i}\omega_+\gamma_+\right)} + \frac{k^2|\boldsymbol{k}_- \times \boldsymbol{v}_0|^2}{k_-^2\left(k_-^2c^2 - \omega_-^2 + \omega_{pe}^2 - \mathrm{i}\omega_-\gamma_-\right)}\,, \tag{12.3}$$

where $\boldsymbol{v}_0 = q_e\boldsymbol{E}_0/m_e\omega_0$, and where, in addition, sideband damping terms with $\gamma_\pm$ of the order of $\nu_e\omega_{pe}^2/\omega_0^2$, where ν_e is the electron collision frequency, have been included. As ν_e here is supposed to be small, we still denote our plasma as essentially collisionless.

In order to investigate stimulated Brillouin scattering in the upper part of the ionosphere, where collisions are of comparatively little interest, one can now consider the low-frequency regime $kv_{ti} < \omega \ll kv_{te}$, where $v_{tj}[= (T_j/m_j)^{1/2}]$ is the thermal speed of particle species j. Supposing, for simplicity, that the electron temperature T_e is larger than the ion temperature T_i, and that the particle velocity distribution functions are Maxwellians, we therefore write the susceptibilities in the form

$$\chi_e \approx \left[1 + \mathrm{i}(\pi/2)^{1/2}\omega/kv_{te}\right]/k^2\lambda_{De}^2\,, \tag{12.4a}$$

and

$$\chi_i \approx -\frac{\omega_{pi}^2}{\omega^2} + \mathrm{i}(\pi/2)^{1/2}\frac{\omega}{\omega_{pi}k^3\lambda_{Di}^3}\exp\left(-\omega^2/2k^2v_{ti}^2\right)\,, \tag{12.4b}$$

where $\lambda_{De} = v_{te}/\omega_{pe}$ and $\lambda_{Di} = v_{ti}/\omega_{pi}$ are the electron and ion Debye radii, respectively, and ω_{pi} is the ion plasma frequency. Inserting numerical values relevant to the upper part of the ionosphere, it then turns out that the threshold value for stimulated Brillouin scattering in fact can be exceeded in ionospheric experiments, for example by employing the VHF transmitter in Tromsø in Norway. Other investigations using the Jicamarca radar in Peru support this conclusion.

The experimental verification of stimulated Brillouin scattering in the ionosphere increased the interest in further extensions of theory and observations. Thus, it was predicted that it should be possible to observe stimulated electromagnetic emissions (SEEs) by means of some other facilities, for example the Max-Planck-Institute heating facility near Tromsø. The theory was consequently extended in order to identify some of the features in SEEs with decay instabilities [Larsson *et al.*, 1976; Stenflo, 2004]. In the ionospheric experiments here we note that ω_0, which is larger or comparable to ω_{pe}, at heights 150 – 400 km is much larger than the electron gyrofrequency ω_{ce}. Generalizing the calculations leading to Eq. (12.3) to a magnetized and slightly nonuniform plasma one thus obtains a nonlinear dispersion relation which is similar to Eq. (12.3), except that the electrostatic susceptibility is now replaced by [Stenflo and Shukla, 2000]

$$\chi_j(\omega, \boldsymbol{k}) = \frac{\omega_{pj}^2}{k^2v_{tj}^2}\left\{1 - \sum_{n=-\infty}^{\infty} I_n(\mathrm{b}_j)\exp(-b_j) \times \left[\omega - \omega_{j*}\left(1 - \frac{n\omega}{b_j\omega_{cj}}\right)\right]G_{nj}(v_z)\right\}\,, \tag{12.5}$$

where I_n is the modified Bessel function of order n, $b_j = k_\perp^2 v_{tj}^2/\omega_{cj}^2$, $\omega_{cj} = q_jB_0/m_jc$, $\omega_{j*} = \boldsymbol{k}\cdot\boldsymbol{v}_{Dj}$ is the drift wave frequency, $\boldsymbol{v}_{Dj} = (cT_j/q_jB_0n_0)\hat{\boldsymbol{z}}\times\nabla n_0$ is the diamagnetic drift velocity in the density gradient, and $G_{nj}(v_z) = \int_{-\infty}^{\infty} F_{0zj}(v_z)\,\mathrm{d}v_z/(\omega - k_zv_z - n\omega_{cj})n_{0j}$. Here, F_{0zj} is the velocity distribution function in the magnetic field direction.

As a simple example, we here call attention to the excitation of electrostatic electron oscillations, neglecting the thermal motion of the particles. We can then approximate χ_i by zero and χ_e by

$$\chi_e \approx -\frac{k_\perp^2\omega_{pe}^2}{k^2\left(\omega^2 - \omega_{ce}^2\right)} - \frac{k_z^2\omega_{pe}^2}{k^2\omega^2}\,, \tag{12.6}$$

where index z and $\perp$ denote directions with respect to the external magnetic field $B_0\hat{\boldsymbol{z}}$. As $\omega_{pe} \gg \omega_{ce}$ at the ionospheric heights, which we have in mind here, the solutions of the dispersion relation $1+\chi_e \approx 0$ are

$$\omega_1 \approx \pm\omega_{pe}\left(1+\omega_{ce}^2 \sin^2\alpha/2\omega_{pe}^2\right), \qquad (12.7a)$$

and

$$\omega_2 \approx \pm\omega_{ce}\cos\alpha\left(1-\omega_{ce}^2 \sin^2\alpha/2\omega_{pe}^2\right), \qquad (12.7b)$$

where α is the angle between $\boldsymbol{k}$ and $\hat{\boldsymbol{z}}$.

Many papers have considered the scattering by the ω_1-modes (stimulated Raman scattering). However, in the upper ionosphere it is in addition possible to observe the scattering off the ω_2-modes (resonance line scattering).

In the lower part of the ionosphere, the collisions between the charged particles and the neutrals play an important role. In order to describe such effects, we adopt for each species a continuity equation, a momentum equation, and an energy equation. This is complemented by the Maxwell equations. Assuming again that ω_0 is much larger than ω_{ce}, we obtain [Stenflo, 1990]

$$\frac{1}{\chi_e}+\frac{1}{1+\chi_i}=\left(1+\frac{4\mathrm{i}\nu_e}{3\Omega}\right)\left[\frac{k^2|\boldsymbol{k}_+\times\boldsymbol{v}_0|^2}{k_+^2\left(k_+^2c^2-\omega_+^2+\omega_{pe}^2-\mathrm{i}\omega_+\gamma_+\right)}+\frac{k^2|\boldsymbol{k}_-\times\boldsymbol{v}_0|^2}{k_-^2\left(k_-^2c^2-\omega_-^2+\omega_{pe}^2-i\omega_-\gamma_-\right)}\right], \qquad (12.8)$$

where $\Omega=\omega-\boldsymbol{k}\cdot\boldsymbol{v}_{de}+\mathrm{i}\tau^{-1}+\mathrm{i}\boldsymbol{k}\cdot\mathbf{H}\cdot\boldsymbol{k}$, $\boldsymbol{v}_{de}$ is the equilibrium electron drift velocity, τ is the electron energy relaxation time, and $\mathbf{H}$ is the heat diffusion tensor. The generalized susceptibilities are

$$\chi=\left[\frac{k^2v_t^2}{\omega_p^2}[1+(\gamma-1)\omega_d/\Omega]-\frac{k^2\omega_d(\omega_d+\mathrm{i}\nu)[1-\omega_c^2/(\omega_d+\mathrm{i}\nu)^2]}{\omega_p^2A}\right]^{-1}, \qquad (12.9)$$

where γ is the ratio of specific heats, ν is the particle collision frequency, $A\approx k^2+\omega_c\boldsymbol{k}\cdot(\hat{\boldsymbol{z}}\times\nabla n_0)/n_0(\omega_d+\mathrm{i}\nu)-k_z^2\omega_c^2/(\omega_d+\mathrm{i}\nu)^2$, and $\omega_d=\omega-\boldsymbol{k}\cdot\boldsymbol{v}_d$. Here we have assumed that $|\nabla n_0/n_0|\ll|\boldsymbol{k}|$.

The dispersion relation (12.8), with (12.9), which governs the stimulated scattering of a large amplitude pump wave in the lower part of the Earth's ionosphere has been generalized to include also the electrostatic sidebands as well as the effect of the Earth's magnetic field on the pump wave and sidebands, e.g. [Larsson *et al.*, 1976; Stenflo, 2004]. Its general form is, of course, then more complex than (12.8). It should be stressed that the nonlinear heating term, represented by the ν_e/Ω term in Eq. (12.8), is often much more important than the usual ponderomotive force term (i.e. $\nu_e \gg \Omega$).

In order to illustrate the use of Eq. (12.8) with (12.9), by means of a simple but important example, we now consider the scattering of electromagnetic waves by low-frequency collisional gradient drift modes, neglecting, for convenience, the thermal and density gradient terms in (12.9). When $\omega \ll \nu_i$ and $\boldsymbol{v}_{di}=0$, where ν_i is the ion-neutral collision frequency and $\boldsymbol{v}_{di}$ is the equilibrium ion drift velocity, we can then replace χ_i by

$$\chi_i \approx -\frac{\omega_{pi}^2}{\omega(\omega+\mathrm{i}\nu_i)} \approx \mathrm{i}\frac{\omega_{pi}^2}{\omega\nu_i}. \qquad (12.10a)$$

Furthermore, assuming that $k_z=0$ and $|\omega-\boldsymbol{k}\cdot\boldsymbol{v}_{de}| \ll \nu_e \ll \omega_{ce}$, we approximate (12.9) by

$$\chi_e \approx \mathrm{i}\frac{\omega_{pe}^2\nu_e}{(\omega-\boldsymbol{k}\cdot\boldsymbol{v}_{de})\omega_{ce}^2}. \qquad (12.10b)$$

By means of Eqs. (12.10a) and (12.10b) we then solve the dispersion relation $1+\chi_e+\chi_i=0$, finding the well known gradient drift modes in their simplest form, i.e.

$$\omega \approx \frac{\boldsymbol{k}\cdot\boldsymbol{v}_{de}}{1+\psi}, \qquad (12.11a)$$

where

$$\psi=\frac{\nu_e\nu_i}{|\omega_{ce}\omega_{ci}|}. \qquad (12.11b)$$

We can now use Eq. (12.8) to consider the threshold values and the growth rates for three and four wave parametric instabilities involving gradient drift modes. After some algebra, it turns out that the threshold values for such processes indeed can be exceeded in several ionospheric heating experiments.

Next, we present the general nonlinear dispersion relation for the parametric interactions of high-frequency electromagnetic waves that are affected by the presence of an external magnetic field. Here the dispersion relation is of the form [Stenflo, 1999; Stenflo and Shukla, 2000]

$$\frac{1}{\chi_e} + \frac{1}{1+\chi_i} = -\frac{k^2c^2}{\omega_0^2}\left\{ \frac{d_+}{(1-A_+^2)\,D_+} \times\left[\left(1+\frac{4i\nu_e}{3\Omega}\right)\boldsymbol{K}_+ - \frac{i\omega_{ce}}{\omega_0}\frac{k_z}{k_{se}^2}(\boldsymbol{K}_+\times\boldsymbol{K}^*)\right]\cdot\boldsymbol{v}_0^* \times\left[\boldsymbol{K}_+^* + \frac{i\omega_{ce}}{\omega_0}\frac{k_z}{k_{se}^2}(\boldsymbol{K}_+^*\times\boldsymbol{K})\right]\cdot\boldsymbol{v}_0 + \frac{d_-}{(1-A_-^2)\,D_-} \times\left[\left(1+\frac{4i\nu_e}{3\Omega}\right)\boldsymbol{K}_- + \frac{i\omega_{ce}}{\omega_0}\frac{k_z}{k_{se}^2}(\boldsymbol{K}_-\times\boldsymbol{K}^*)\right]\cdot\boldsymbol{v}_0 \times\left[\boldsymbol{K}_-^* - \frac{i\omega_{ce}}{\omega_0}\frac{k_z}{k_{se}^2}(\boldsymbol{K}_-^*\times\boldsymbol{K})\right]\cdot\boldsymbol{v}_0^*\right\}, \tag{12.12a}$$

where $d_\pm = k_\pm^2c^2 - \omega_\pm^2 + \omega_{pe}^2$. We have here assumed that a pump wave $(\omega_0, \boldsymbol{k}_0)$ with $\omega_0 \gg \omega$ propagates through the plasma, exciting sidebands $(\omega_\pm, \boldsymbol{k}_\pm)$. The amplitude of the induced electron velocity in the pump field has been denoted by $\boldsymbol{v}_0$.

The dielectric function $D_\pm$ in the denominator on the right-hand side of (12.12a) stands for $D(\omega_\pm, \boldsymbol{k}_\pm)$, where D is here defined as

$$D(\omega,\boldsymbol{k}) \equiv \left(1-\frac{\omega_{pe}^2}{\omega^2}-\frac{k^2v_t^2}{\omega^2}\right)(k^2c^2-\omega^2+\omega_{pe}^2)^2 - \frac{\omega_{ce}^2}{\omega^2}(k^2c^2-\omega^2) \times\left[(k^2c^2-\omega^2)\left(1-\frac{\omega_{pe}^2}{\omega^2}-\frac{k_z^2v_t^2}{\omega^2}\right)+\frac{k_\perp^2c^2\omega_{pe}^2}{\omega^2}\right]. \tag{12.12b}$$

The wave vector $\boldsymbol{k}$ has been written here as $\boldsymbol{k} = k_z\hat{\boldsymbol{z}} + \boldsymbol{k}_\perp$, and v_t^2 is the square of the electron thermal speed times the ratio of specific heats. Furthermore, we have also introduced the notations $\boldsymbol{v}_{0+} = \boldsymbol{v}_0$, $\boldsymbol{v}_{0-} = \boldsymbol{v}_0^*$, $\boldsymbol{K} = \boldsymbol{k} - i(\omega_{ce}/\omega_\nu)\hat{\boldsymbol{z}}\times\boldsymbol{k} - (\omega_{ce}^2/\omega_\nu^2)k_z\hat{\boldsymbol{z}}$, $\boldsymbol{K}_\pm = \boldsymbol{k}_\pm - k_{z\pm}A_\pm^2\hat{\boldsymbol{z}} - iA_\pm\hat{\boldsymbol{z}}\times\boldsymbol{k}_\pm$, $A_\pm = (\omega_{ce}/\omega_\pm)(k_\pm^2c^2 - \omega_\pm^2)/(k_\pm^2c^2-\omega_\pm^2+\omega_{pe}^2)$, $k_{se}^2 = \boldsymbol{k}\cdot\boldsymbol{K}$, and $\omega_\nu = \omega + i\nu_e$.

By means of Eq. (12.12a) it is now straightforward to generalized previous results for scattering and modulational instabilities in a collisional magnetoplasma. Equation (12.12a) also provides the possibility to include wave interactions in a strongly magnetized plasma; e.g. those encountered in plasma filled backward wave oscillators.

Equation (12.12a) was derived for an electron–ion plasma where $m_e \ll m_i$. The left-hand side of (12.12a) is therefore not symmetric when indices e and i are interchanged. In order to illustrate its original symmetric form we therefore consider the special case where all the waves propagate in the z-direction in a collisionless plasma containing several particle species with arbitrary mass ratios, and where the pump as well as the sidebands are transverse. In that case, one can derive the *exact* dispersion relation, valid for any magnitude of the pump wave. For a non-relativistic plasma, we have [Stenflo and Shukla, 2000]

$$\frac{\epsilon}{1+c_0\epsilon/c_+c_-} \equiv -|\boldsymbol{E}_{\perp 0}|^2\left(\frac{c_+^2}{N_+}+\frac{c_-^2}{N_-}\right), \tag{12.13}$$

where

$$\epsilon = 1+\sum\chi,$$

$$\chi = -\frac{\omega_p^2}{\omega^2-k^2v_T^2},$$

$$v_T^2 = v_t^2 + \frac{2q^2k_0^2\omega^2\omega_c|\boldsymbol{E}_{\perp 0}|^2}{m^2k^2\omega_0^2(\omega_0+\omega_c)[\omega^2-(\omega_0+\omega_c)^2]},$$

$$c_0 = \sum\frac{q^2}{m^2}\frac{\chi}{(\omega_0+\omega_c)^2}\left(k-\frac{k_0\omega_c\omega}{\omega_0(\omega_0-\omega+\omega_c)}\right)\times\left(k-\frac{k_0\omega_c\omega}{\omega_0(\omega_0+\omega+\omega_c)}\right),$$

$$c_\pm = \mp\sum\frac{q}{m}\frac{\chi}{(\omega_0+\omega_c)}\left(k-\frac{k_0\omega_c\omega}{\omega_0(\omega_0\mp\omega+\omega_c)}\right),$$

and

$$N_\pm = (k\mp k_0)^2c^2-(\omega\mp\omega_0)^2+\sum\frac{\omega_p^2(\omega_0\mp\omega)}{\omega_0\mp\omega+\omega_c} - |\boldsymbol{E}_{\perp 0}|^2\left\{\sum\frac{q^2}{m^2}\frac{\chi}{(\omega_0+\omega_c)^2}\times\left[k-\frac{k_0\omega_c\omega}{\omega_0(\omega_0\mp\omega+\omega_c)}\right]^2+\frac{c_\pm c_0}{c_\mp}\right\}.$$

The pump electric field amplitude is here denoted by $\boldsymbol{E}_{\perp 0}$, and $\sum$ means summation over all particle species σ. For notational simplicity, we have omitted the index σ

on all symbols. We note that (12.13) agrees with (12.12a) when $m_e \ll m_i$ and $\omega \ll \omega_0$. For other plasmas, e. g. an electron–positron plasma with $m_e = m_p$, Eq. (12.13) is still useful, whereas (12.12a) has to be improved.

The dispersion relation (12.12a) does not cover kinetic effects, which for example appear when the wave frequency is close to a multiple of ω_{ce}. We then have to use Eq. (A1) of Stenflo [1999]. The explicit version of that full dispersion relation is naturally very complex and we will thus not repeat its description here. For illustrative purposes, we just present its simplest version, namely the case where all the waves are longitudinal and propagate in the z-direction in a Vlasov plasma with the velocity distribution function F_0. We then have

$$\tilde{\epsilon}(\omega,k) = \left\{ \frac{\left[\sum \frac{q}{m}\frac{\omega_p^2}{n_0}\int \mathrm{d}v_z \Omega_+^{-3}(\partial F_0/\partial v_z)\right]^2}{\epsilon(\omega_0+\omega, k_0+k)} + \frac{\left[\sum \frac{q}{m}\frac{\omega_p^2}{n_0}\int \mathrm{d}v_z \Omega_-^{-3}(\partial F_0/\partial v_z)\right]^2}{\epsilon(\omega_0-\omega, k_0-k)} \right\} |E_0|^2 , \tag{12.14}$$

where

$$\epsilon(\omega,k) = 1 + \sum \chi(\omega,k) \equiv 1 + \sum \frac{\omega_p^2}{kn_0}\int \mathrm{d}v_z \frac{\partial F_0/\partial v_z}{\omega - kv_z} ,$$

$$\tilde{\epsilon} = \frac{\epsilon}{1 - \dfrac{\left[\sum (q^2/m^2)\chi\right]\epsilon}{\left[\sum (q/m)\chi\right]^2}} ,$$

and

$$\Omega_\pm^3 = (\omega - kv_z)(\omega_0 - k_0 v_z)[\omega_0 \pm \omega - (k_0 \pm k)v_z] .$$

12.3 Resonant Three-Wave Interactions in Plasmas

When two small amplitude waves $(\omega_1, \boldsymbol{k}_1)$ and $(\omega_2, \boldsymbol{k}_2)$ interact, they create perturbations at the sum and different frequencies, for example at the frequency $\omega_1 + \omega_2$ and consequently at the wave vector $\boldsymbol{k}_1 + \boldsymbol{k}_2$. This nonlinear response is generally very small unless it happens that $\omega_1 + \omega_2$ and $\boldsymbol{k}_1 + \boldsymbol{k}_2$ coincide with the frequency and wave vector of another natural mode. As this process is basic for the understanding of more complicated nonlinear plasma phenomena, it has received much attention. The theoretical ideas have been verified in many laboratory experiments. It has in this connection been shown that a rich variety of wave coupling processes play an important role in the physics of laser plasma interactions as well as in space plasmas.

Different techniques, such as coupled mode theory, have been employed to calculate the coupling coefficient for three wave interaction processes. Here, following Stenflo [1994], we will present some basic results of previous works. In order to focus our interest on the simplest possible situation, we then consider the *resonant* interaction between three waves in a plasma. Thus, we assume that the matching conditions

$$\omega_3 = \omega_1 + \omega_2 , \tag{12.15a}$$

and

$$\boldsymbol{k}_3 = \boldsymbol{k}_1 + \boldsymbol{k}_2 \tag{12.15b}$$

are satisfied. It is well known that even a small frequency mismatch can significantly decrease the efficiency of the three wave interaction process. For simplicity, we shall however omit any frequency mismatch here. Considering the development of the magnitudes $E_j (j = 1,2,3)$ of the wave electric field amplitudes $\boldsymbol{E}_j = E_j \hat{\boldsymbol{e}}_j$, where $\hat{\boldsymbol{e}}_j$ is the polarization vector of unit length, one thus derives the three coupled bilinear equations [Stenflo, 1994]

$$\frac{\mathrm{d}E_1^*}{\mathrm{d}t} = c_1 E_2 E_3^* , \tag{12.16a}$$

$$\frac{\mathrm{d}E_2^*}{\mathrm{d}t} = c_2 E_1 E_3^* , \tag{12.16b}$$

$$\frac{\mathrm{d}E_3}{\mathrm{d}t} = c_3 E_1 E_2 , \tag{12.16c}$$

where the asterisk stands for complex conjugate, c_j are coupling coefficients which we shall present below, $\mathrm{d}E_j/\mathrm{d}t = (\partial_t + \boldsymbol{v}_{gj} \cdot \nabla + \gamma_j)E_j$, $\boldsymbol{v}_{gj}$ is the group velocity of wave j, and γ_j accounts for the linear damping rate. Introducing the phases φ_j and $-\theta_j$ of E_j and c_j from $E_j = |E_j|\exp(\mathrm{i}\varphi_j)$ and $c_j = |c_j|\exp(-\mathrm{i}\theta_j)$, defining $\varphi = \varphi_3 - \varphi_1 - \varphi_2$, and normalizing the amplitudes $|E_j|$ to u_j, we rewrite Eqs. (12.16a)–(12.16c) in terms of four equations for the four real quantities u_1, u_2, u_3 and φ, i.e.

$$\frac{\partial u_1}{\partial t} + \gamma_1 u_1 = u_2 u_3 \cos(\varphi + \theta_1) , \tag{12.17a}$$

$$\frac{\partial u_2}{\partial t} + \gamma_2 u_2 = u_1 u_3 \cos(\varphi + \theta_2) , \tag{12.17b}$$

$$\frac{\partial u_3}{\partial t} + \gamma_3 u_3 = u_1 u_2 \cos(\varphi + \theta_3) , \tag{12.17c}$$

and

$$\frac{\partial\varphi}{\partial t} = -\frac{u_2u_3}{u_1}\sin(\varphi+\theta_1) - \frac{u_1u_3}{u_2}\sin(\varphi+\theta_2) - \frac{u_1u_2}{u_3}\sin(\varphi+\theta_3)\,, \qquad (12.17\text{d})$$

where, for simplicity, we have neglected the group velocity term.

In the most well known case with $\theta_1 = \theta_2 = 0$ and $\theta_3 = \pi$, it is possible to express the oscillating solutions of (12.17a)–(12.17d) in terms of elliptic function. Another well known case of interest is obtained when $\theta_1 = \theta_2 = \theta_3 = 0$. This occurs if one of the waves has negative energy as we can see below. A simple solution of (12.17a)–(12.17d), with $\gamma_j = 0$, is then $\varphi = 0$ together with

$$u_j = \frac{1}{t_\infty - t}\,, \qquad (12.18)$$

where t_∞ is an integration constant. We notice that this solution "explodes" at the time t_∞. Such explosive instabilities can also occur for the case where θ_1 and θ_2 are close to zero and θ_3 is close to π if the phases of the coupling coefficients define complex vectors which all point in the same half-plane. In addition, a new equilibrium state, characterized by comparatively large wave amplitudes, can exist if the phase φ is constant and satisfies the relation [Stenflo, 1994]

$$\sum_j \gamma_j \tan(\varphi+\theta_j) = 0\,. \qquad (12.19)$$

We next have to present explicit expressions for the coupling coefficients for different situations of interest. It is then instructive to start with the simplest case, namely the well known resonant interactions between three electrostatic waves propagating in the same direction (the x-direction) in a collisionless unmagnetized plasma. Each wave satisfies thus the dispersion relation

$$\epsilon(\omega,k) \equiv 1 + \frac{4\pi}{k}\sum\frac{q^2}{m}\int \mathrm{d}v_x \frac{\partial F_0/\partial v_x}{\omega - kv_x} = 0\,, \quad (12.20)$$

where $\sum$ stands for summation over the different species with the unperturbed velocity distribution functions F_0.

Starting from the Vlasov equation and keeping only the resonant nonlinear terms one then obtains to second order

$$\frac{\mathrm{d}E_1^*}{\mathrm{d}t} = \frac{c_l E_2 E_3^*}{\partial\epsilon(\omega_1,k_1)/\partial\omega_1}\,, \qquad (12.21\text{a})$$

$$\frac{\mathrm{d}E_2^*}{\mathrm{d}t} = \frac{c_l E_1 E_3^*}{\partial\epsilon(\omega_2,k_2)/\partial\omega_2}\,, \qquad (12.21\text{b})$$

and

$$\frac{\mathrm{d}E_3}{\mathrm{d}t} = -\frac{c_l E_1 E_2}{\partial\epsilon(\omega_3,k_3)/\partial\omega_3}\,, \qquad (12.21\text{c})$$

where the coupling coefficient is

$$c_l = 4\pi\sum\frac{q^3}{m^2}\int \mathrm{d}v_x \frac{\partial F_0/\partial v_x}{(\omega_1 - k_1v_x)(\omega_2 - k_2v_x)(\omega_3 - k_3v_x)}\,. \qquad (12.21\text{d})$$

We note that the same coupling coefficient c_l appears in (12.21a)–(12.21c). This is a manifestation of the energy conservation properties in the three-wave interaction process.

The coupling coefficients for interaction processes in magnetized plasmas are of course much more complex than the comparatively very simple expressions for unmagnetized plasmas. In order to limit the algebra, we shall next focus our interest on the resonant interaction between three waves propagating in arbitrary directions in an one-component, cold, collisionless, uniform magnetoplasma. Starting from the fluid equations for the electrons, combining them with the Maxwell equations, and regarding the ions as a stationary background, we then first linearize the equations to deduce the well known linear dispersion relation

$$S(\omega,k) \equiv \left(1 - \frac{\omega_\mathrm{p}^2}{\omega^2}\right)\left(k^2c^2 - \omega^2 + \omega_\mathrm{p}^2\right) - \frac{k_\perp^2 c^2\omega_\mathrm{p}^2\omega_c}{\omega^3}\frac{a}{1-a^2} = 0\,, \qquad (12.22\text{a})$$

where $k_\perp^2 \equiv k^2 - k_z^2$, and

$$a = \frac{\omega_c\left(k^2c^2 - \omega^2\right)}{\omega\left(k^2c^2 - \omega^2 + \omega_\mathrm{p}^2\right)}\,. \qquad (12.22\text{b})$$

Next, keeping only the resonant nonlinear terms in our equations, we obtain to second order [Stenflo, 1994]

$$\frac{\mathrm{d}E_{1l}^*}{\mathrm{d}t} = \frac{\left(k_1^2c^2 - \omega_1^2 + \omega_\mathrm{p}^2\right)^2}{k_1^2}\frac{C}{\partial S(\omega_1,k_1)/\partial\omega_1}E_{2l}E_{3l}^*\,, \qquad (12.23\text{a})$$

$$\frac{\mathrm{d}E_{2l}^*}{\mathrm{d}t} = \frac{\left(k_2^2c^2 - \omega_2^2 + \omega_\mathrm{p}^2\right)^2}{k_2^2} \frac{C}{\partial S(\omega_2, k_2)/\partial \omega_2} E_{1l}E_{3l}^* , \tag{12.23b}$$

and

$$\frac{\mathrm{d}E_{3l}}{\mathrm{d}t} = -\frac{\left(k_3^2c^2 - \omega_3^2 + \omega_\mathrm{p}^2\right)^2}{k_3^2} \frac{C}{\partial S(\omega_3, k_3)/\partial \omega_3} E_{1l}E_{2l} , \tag{12.23c}$$

where the coupling coefficient is

$$\begin{aligned} C = {} & \frac{qc^4\omega_\mathrm{p}^2 k_1k_2k_3}{m\omega_1\omega_2\omega_3\left(1-a_1^2\right)\left(1-a_2^2\right)\left(1-a_3^2\right)} \\ & \times \left[\frac{\boldsymbol{k}_1\cdot\boldsymbol{K}_1}{\omega_1}\boldsymbol{K}_2\cdot\boldsymbol{K}_3 + \frac{\boldsymbol{k}_2\cdot\boldsymbol{K}_2}{\omega_2}\boldsymbol{K}_1\cdot\boldsymbol{K}_3 + \frac{\boldsymbol{k}_3\cdot\boldsymbol{K}_3}{\omega_3}\boldsymbol{K}_1\cdot\boldsymbol{K}_2 \right. \\ & \left. - \frac{\mathrm{i}\omega_c}{\omega_3}\left(\frac{k_{2z}}{\omega_2} - \frac{k_{1z}}{\omega_1}\right)\boldsymbol{K}_3\cdot\left(\boldsymbol{K}_1\times\boldsymbol{K}_2\right)\right] \\ & \times \left(k_1^2c^2 - \omega_1^2 + \omega_\mathrm{p}^2\right)^{-1}\left(k_2^2c^2 - \omega_2^2 + \omega_\mathrm{p}^2\right)^{-1} \\ & \times \left(k_3^2c^2 - \omega_3^2 + \omega_\mathrm{p}^2\right)^{-1} , \end{aligned} \tag{12.23d}$$

and where $\boldsymbol{K}_{1,2} = \boldsymbol{K}(\omega_{1,2}, \boldsymbol{k}_{1,2})$, $\boldsymbol{K}_3 = \boldsymbol{K}(-\omega_3, -\boldsymbol{k}_3)$ and $\boldsymbol{K}(\omega, \boldsymbol{k}) = \boldsymbol{k} - a^2k_z\hat{\boldsymbol{z}} + \mathrm{i}a(k_y\hat{\boldsymbol{x}} - k_x\hat{\boldsymbol{y}})$. The fields E_{jl} in Eq. (12.23) are here related to the electric field amplitude $\boldsymbol{E}_j$ according to $E_{jl} \equiv \boldsymbol{k}_j\cdot\boldsymbol{E}_j/k_j$, where $k_j = |\boldsymbol{k}_j|$. Also in this second example, we note that the coupling coefficient C in Eq. (12.23a) reappears in Eqs. (12.23b) and (12.23c).

As our third example, we consider the interaction between three magnetohydrodynamic (MHD) waves. The ideal MHD equations describe two kinds of waves, namely Alfvén waves which satisfy the dispersion relation

$$D_\mathrm{A}(\omega, \boldsymbol{k}) \equiv \omega^2 - k_z^2c_\mathrm{A}^2 = 0 , \tag{12.24a}$$

and magnetosonic waves which satisfy

$$D_\mathrm{m}(\omega, \boldsymbol{k}) \equiv \omega^4 - \omega^2k^2\left(c_\mathrm{A}^2 + c_\mathrm{s}^2\right) + k_z^2k^2c_\mathrm{A}^2c_\mathrm{s}^2 = 0 . \tag{12.24b}$$

We have here introduced the Alfvén speed c_A.

The coupling coefficients for MHD wave propagation can be derived in a straightforward way from the MHD equations. As one particular example, we here present the equations governing the interaction between two Alfvén waves [satisfying (12.24a)], propagating in different directions, i.e. ω_1/k_{1z} and ω_3/k_{3z} have different signs, and characterized by the magnitudes B_1 and B_3 of their oscillating magnetic fields, and one magnetosonic wave [satisfying (12.24b)] and characterized by its density perturbation amplitude n_2. The result is

$$\frac{\mathrm{d}B_1^*}{\mathrm{d}t} = i\omega_1 C_\mathrm{AMA}\frac{n_2}{n_0}B_3^* , \tag{12.25a}$$

$$\frac{\mathrm{d}n_2^*}{\mathrm{d}t} = -\frac{8in_0\omega_1\omega_3k_2^2c_\mathrm{A}^2C_\mathrm{AMA}}{\partial D_\mathrm{m}(\omega_2, \boldsymbol{k}_2)/\partial\omega_2}\frac{B_1B_3^*}{B_0^2} , \tag{12.25b}$$

and

$$\frac{\mathrm{d}B_3}{\mathrm{d}t} = -i\omega_3 C_\mathrm{AMA}\frac{n_2}{n_0}B_1 , \tag{12.25c}$$

where the coupling coefficient is

$$C_\mathrm{AMA} = \frac{1}{k_{1\perp}k_{3\perp}}\left[\boldsymbol{k}_{1\perp}\cdot\boldsymbol{k}_{3\perp} - \frac{\omega_2^2\left(\boldsymbol{k}_1\times\boldsymbol{k}_3\right)_z^2}{k_{1z}k_{3z}k_2^2c_\mathrm{A}^2}\right] . \tag{12.25d}$$

Finally, it should be pointed out that the coupling coefficients have also been derived for a general hot magnetized uniform plasma. The explicit expressions for the coupling coefficients have been presented in Stenflo [1994].

12.4 Parametric Instabilities of Magnetic Field-Aligned Alfvén Waves

Let us first discuss the stability of an oscillating magnetic field whose amplitude here is allowed to be arbitrarily large [Lashmore-Davies and Stenflo, 1979]. This is in contrast to most of the theories of parametric excitation where the results are usually restricted to small amplitude pump waves due to the use of a perturbation analysis, or to the difficulty of dealing with the higher harmonics of the pump. In order to perform the analysis for the pump wave whose amplitude is unrestricted it is necessary to consider a rather simple model, namely a uniform, unbounded plasma with a uniform constant magnetic field. The pump wave describes a circularly polarized magnetic field in the direction of the constant magnetic field.

In addition to the large amplitude nature of the pump we do not require the frequency and wavenumber of the pump wave to be connected by some definite relation, i.e. we do not insist that the pump wave is a natural mode of the plasma. For this to be possible, we assume that the pump wave is generated by an external current. It is this current which determines the frequency and wavenumber of the pump wave. This extra freedom regarding the frequency of the pump allows us to investigate a range of frequencies not previously explored. This is of relevance to the theory of ideal MHD instabilities.

Let us now formulate the problem we have just described. To do this, we use an MHD model given by the equations

$$\frac{\partial \rho}{\partial t} + \nabla \cdot (\rho \boldsymbol{v}) = 0 , \qquad (12.26)$$

$$\rho \frac{\partial \boldsymbol{v}}{\partial t} + \rho(\boldsymbol{v} \cdot \nabla)\boldsymbol{v} = -c_s^2 \nabla \rho + \frac{1}{4\pi}(\nabla \times \boldsymbol{B}) \times \boldsymbol{B} - c^{-1}\boldsymbol{J}_{\text{ext}} \times \boldsymbol{B} , \qquad (12.27)$$

and

$$\frac{\partial \boldsymbol{B}}{\partial t} = \nabla \times (\boldsymbol{v} \times \boldsymbol{B}) , \qquad (12.28)$$

where we consider a compressible plasma with an isothermal equation of state. Here ρ, $\boldsymbol{v}$ and $\boldsymbol{B}$ represent the mass density, velocity, and magnetic field, respectively. We have also assumed the presence of an external current source $\boldsymbol{J}_{\text{ext}}$ in order to balance the curl of the external magnetic field, which here is distributed throughout the entire plasma.

A space and time dependent solution of (12.26)–(12.28) is now taken to consist of a helical (or circularly polarized) magnetic field superimposed on the constant magnetic field B_{z0} which is directed along the z-axis. The total magnetic field in the unperturbed state is therefore

$$\boldsymbol{B}_0(z,t) = \hat{\boldsymbol{z}} B_{z0} - B_{\perp 0} \times \left[\hat{\boldsymbol{x}} \sin(\omega_0 t - k_0 z) - \hat{\boldsymbol{y}} \cos(\omega_0 t - k_0 z)\right] . \qquad (12.29)$$

The helical magnetic field results from an external current source

$$\boldsymbol{J}_{\text{ext}} = -J_0 \left[\hat{\boldsymbol{x}} \sin(\omega_0 t - k_0 z) - \hat{\boldsymbol{y}} \cos(\omega_0 t - k_0 z)\right] . \qquad (12.30)$$

The relationship between the amplitude of the helical magnetic field and that of the external current must be

$$J_0 = \frac{k_0 c}{4\pi} B_{\perp 0} \left(1 - \frac{\omega_0^2}{k_0^2 c_A^2}\right) , \qquad (12.31)$$

where $c_A = B_{z0}/\sqrt{4\pi\rho_0}$ is the Alfvén speed. The self-consistency of our space and time dependent solution is completed by the following expression for the plasma quiver velocity

$$\boldsymbol{v}_0 = -v_{\perp 0} \left[\hat{\boldsymbol{x}} \sin(\omega_0 t - k_0 z) - \hat{\boldsymbol{y}} \cos(\omega_0 t - k_0 z)\right] , \qquad (12.32)$$

where $v_{\perp 0}$ is related to $B_{\perp 0}$ by

$$v_{\perp 0} = \frac{\omega_0 B_{\perp 0}}{k_0 B_{z0}} . \qquad (12.33)$$

The helical solution just described does not perturb the plasma density and we therefore note that ρ_0 is constant. We stress that for this particular example the helical fields are *exact* solutions of the nonlinear MHD equations (12.26)–(12.28). The amplitudes of these fields can therefore be as large as we choose.

Next, we consider small perturbations of this helical state. Suppose that there is a density perturbation ρ_1 which varies as $\exp(ikz - i\omega t)$. This density perturbation can beat with the oscillating equilibrium fields $B_{\pm 0} \equiv B_{x0} \pm iB_{y0}$ to generate sideband perturbations which we describe in terms of the perturbed magnetic field variable $B_{\pm 1} = B_{x1} \pm iB_{y1}$. The sideband perturbations are then given by [Lashmore-Davies and Stenflo, 1979]

$$B_{\pm 1} = \alpha_{\pm} \rho_1 B_{\pm 0} , \qquad (12.34)$$

where

$$\alpha_{\pm} = \pm \left(\frac{k \mp k_0}{\rho_0 k_0}\right) \omega_0^2 \left[1 - \frac{k_0 \omega}{k \omega_0} \pm \frac{\omega(\omega \mp \omega_0) k_0}{\omega_0^2 k}\right] \times \left[(\omega \mp \omega_0)^2 - (k \mp k_0)^2 c_A^2\right]^{-1} . \qquad (12.35)$$

Note that the sideband perturbations $B_{\pm 1}$ vary as $\exp[i(k \mp k_0)z - i(\omega \mp \omega_0)t]$.

These perturbations can now beat with the unperturbed helical fields to reinforce the density perturbation. Thus, we have

$$\left(\omega^2 - k^2 c_s^2\right) \rho_1 = \frac{k}{8\pi} \left[\left(k - k_0 + \frac{\omega_0^2}{k_0 c_A^2}\right) B_{-0} B_{+1} + \left(k + k_0 - \frac{\omega_0^2}{k_0 c_A^2}\right) B_{+0} B_{-1}\right] . \qquad (12.36)$$

Substituting (12.34) into Eq. (12.36) we obtain the nonlinear dispersion relation [Lashmore-Davies and Stenflo, 1979]

$$\omega^2 - k^2 c_s^2 = \frac{k}{8\pi}\left[\left(k - k_0 + \frac{\omega_0^2}{k_0 c_A^2}\right)\alpha_+ + \left(k + k_0 - \frac{\omega_0^2}{k_0 c_A^2}\right)\alpha_-\right] B_{\perp 0}^2 . \tag{12.37}$$

Within the limitations of the model, Eq. (12.37) is a very general dispersion relation which is valid for all values of $B_{\perp 0}$ and any combination of ω_0 and k_0. The perturbed frequency ω is also unrestricted (apart from the limitations of the MHD model).

Let us now consider the special case when $\omega_0 = 0$. The equilibrium is then represented by a helical magnetic field. For this case, the dispersion relation (12.37) reduces to the simpler form

$$\omega^2 = k^2 c_s^2 + \frac{\omega^2 B_{\perp 0}^2}{8\pi\rho_0} \times \left[\frac{(k+k_0)^2}{\left[\omega^2 - (k+k_0)^2 c_A^2\right]} + \frac{(k-k_0)^2}{\left[\omega^2 - (k-k_0)^2 c_A^2\right]}\right] . \tag{12.38}$$

This is a cubic equation in ω^2, and thus the solutions are somewhat tedious. However, for a cold plasma ($c_s = 0$) Eq. (12.38) is easily solved and we then find that the equilibrium is always stable no matter how large is the value of $B_{\perp 0}/B_{z0}$. For $k = \pm k_0$ the plasma is marginally stable.

It should be pointed out that Eq. (12.37) has been generalized to large amplitude electromagnetic waves with arbitrary frequencies that are beyond the MHD limit [Stenflo and Shukla, 2000]. It is of interest to note that Eqs. (12.13) and (12.37) have also been generalized to cover kinetic effects [Stenflo, 1981]. As a particular example, we can then for example consider the stimulated Compton scattering of Alfv waves off plasma ions [Shukla and Dawson, 1984]. In that case, we replace the left-hand side of Eq. (12.37) by

$$-\omega_{pi}^2\left(\chi_e^{-1} + \chi_i^{-1}\right) , \tag{12.39}$$

where $\chi_e = (k^2\lambda_{De}^2)^{-1}$ for $\omega \ll kv_{te}$, and $\chi_i = (k^2\lambda_{Di}^2)^{-1}\left[1 + \xi Z(\xi)\right]$, with $\xi = \omega/\sqrt{2}kv_{ti}$ and Z being the well known plasma dispersion function. The stimulated ion Compton scattering theory has been applied to the propagation of cosmic rays through the interstellar medium [Shukla and Dawson, 1984]. It then turns out that the scattering can significantly reduce the magnetic field strength of the hydromagnetic turbulence, indicating that cosmic rays can propagate more freely in the interstellar medium.

12.5 Kinetic Alfvén Waves Driven Zonal Flows

Here we discuss excitation of electrostatic convective cells [Okuda and Dawson, 1973; Shukla *et al.*, 1984] (CCs)/zonal flows (ZFs) by dispersive kinetic Alfvén waves (DKAWs) [Stefant, 1970; Hasegawa and Chen, 1976; Hasegawa and Uberoi, 1982; Shukla and Stenflo, 2000a] in a uniform magnetoplasma. The DKAWs are low-frequency (in comparison with the ion gyrofrequency ω_{ci}) electromagnetic waves, which have dispersion due to the ion polarization and ion gyroradius effects. The DKAW frequency is [Stefant, 1970; Hasegawa and Chen, 1976; Hasegawa and Uberoi, 1982] $\omega = k_z c_A(1 + k_\perp^2\rho^2)^{1/2}$, where k_z is the parallel component of the wavevector $\boldsymbol{k} = \hat{\boldsymbol{z}}k_z + \boldsymbol{k}_\perp$, $\rho = (\rho_s^2 + 3\rho_i^2/2)^{1/2}$ is the effective ion gyroradius, $\rho_s = c_s/\omega_{ci}$ is the ion sound gyroradius and $\rho_i = v_{ti}/\omega_{ci}$ is the ion thermal gyroradius. The electromagnetic fields are $\boldsymbol{E} = -\nabla\phi - c^{-1}\partial A_z/\partial t$ and $\boldsymbol{B}_\perp = \nabla A_z \times \hat{\boldsymbol{z}}$, where ϕ is the scalar potential and A_z is the parallel (to $\hat{\boldsymbol{z}}$) component of the vector potential. The DKAWs are accompanied by a finite density perturbation $n_1 = (n_0 c/B_0\omega_{ci})\nabla_\perp^2\phi$. Thus, they are an admixture of electrostatic and electromagnetic fields. Since the parallel phase speed (ω/k_z) of the DKAWs is much smaller than the electron thermal speed v_{te}, they appear in a plasma with intermediate plasma β, viz. $m_e/m_i \ll \beta = 4\pi n_0(T_e + T_i)/B_0^2 \ll 1$, values. In view of the low-β approximation, the compressional magnetic field perturbation can thus be neglected in the DKAW dynamics. Finite amplitude DKAWs are of significant interest in space [Stasiewicz *et al.*, 2000; Pokhotelov *et al.*, 2004] and laboratory environments [Gekelman, 1999; Vincena *et al.*, 2004], as they produce interesting nonlinear effects [Sagdeev *et al.*, 1978; Yu *et al.*, 1981; Shukla *et al.*, 1984; Shukla and Stenflo, 1999a; Shukla and Stenflo, 1999b; Shukla, Stenflo and Bingham, 1999; Shukla and Stenflo, 2000b,c; Drozdenko and Morales, 2001; Wu and Cho, 2004; Shukla, 2005].

Following Shukla [2005] we now show how DKAWs can excite convective cells/zonal flows (ZFs). Let us consider a uniform electron–ion plasma in an external magnetic field $\hat{\boldsymbol{z}}B_0$. The electron and ion fluid velocities in the presence of nonlinearly interacting low-frequency ($\ll \omega_{ci}$) DKAWs and ZFs are, respectively,

$$\boldsymbol{v}_{e\perp} \approx \frac{c}{B_0}\hat{\boldsymbol{z}} \times \nabla\phi - \frac{cT_e}{eB_0 n_0}\hat{\boldsymbol{z}} \times \nabla n_1 , \tag{12.40}$$

and

$$\boldsymbol{v}_{\mathrm{i}\perp} \approx \frac{c}{B_0}\hat{\boldsymbol{z}} \times \nabla\phi - \frac{c}{B_0\omega_{\mathrm{ci}}}\left[\left(\frac{\partial}{\partial t} + \nu_{\mathrm{in}} + \mu_{\mathrm{i}}\nabla_\perp^2\right)\nabla_\perp\phi + \frac{c}{B_0}(\hat{\boldsymbol{z}} \times \nabla\psi \cdot \nabla)\nabla_\perp\phi + \frac{c}{B_0}(\hat{\boldsymbol{z}} \times \nabla\phi \cdot \nabla)\nabla_\perp\psi\right], \tag{12.41}$$

where $e(= -q_{\mathrm{e}})$ is the magnitude of the electron charge, ν_{in} is the ion–neutral collision frequency, $\mu_{\mathrm{i}} = 0.3\nu_{\mathrm{ii}}\rho_{\mathrm{i}}^2$ is the coefficient of the ion gyroviscosity, ν_{ii} is the ion–ion collision frequency, ψ is the potential of the ZFs, and $T_{\mathrm{e}} \gg T_{\mathrm{i}}$ has been assumed.

The appropriate electron and ion velocities involved in ZFs in the presence of the DKAWs are, respectively,

$$\boldsymbol{u}_{\mathrm{e}\perp} \approx \frac{c}{B_0}\hat{\boldsymbol{z}} \times \nabla\psi + \frac{\langle \nu_{ez}\boldsymbol{B}_\perp\rangle}{B_0}, \tag{12.42}$$

and

$$\boldsymbol{u}_{\mathrm{i}\perp} \approx \frac{c}{B_0}\hat{\boldsymbol{z}} \times \nabla\psi - \frac{c}{B_0\omega_{\mathrm{ci}}}\left[\left(\frac{\partial}{\partial t} + \nu_{\mathrm{in}} + \mu_{\mathrm{i}}\nabla_\perp^2\right)\nabla_\perp\psi + \frac{c}{B_0}\langle(\hat{\boldsymbol{z}} \times \nabla\phi \cdot \nabla)\nabla_\perp\phi\rangle\right], \tag{12.43}$$

where the parallel component of the electron fluid velocity in the DKAW fields is

$$\nu_{ez} \simeq \frac{c}{4\pi e n_0}\nabla_\perp^2 A_z, \tag{12.44}$$

which is obtained from the parallel component of Ampère's law, with $\boldsymbol{B}_\perp = \nabla A_z \times \hat{\boldsymbol{z}}$. In (12.44) we have neglected the parallel ion motion, as we are isolating the ion acoustic waves in our intermediate plasma. The last terms in the right-hand side of (12.42) and (12.43) are the Reynolds stresses of the DKAWS, which reinforce the two-dimensional ZFs.

Substituting (12.40) and (12.41) into $\nabla\cdot\boldsymbol{J} = 0$, where $\boldsymbol{J} = e n_0(\boldsymbol{v}_{\mathrm{i}\perp} - \boldsymbol{v}_{\mathrm{e}\perp} - \nu_{ez}\hat{\boldsymbol{z}})$, we have

$$\frac{\partial}{\partial t}\nabla_\perp^2\phi + \frac{c_{\mathrm{A}}^2}{c}\frac{\partial}{\partial z}\nabla_\perp^2 A_z + \frac{c}{B_0}(\hat{\boldsymbol{z}} \times \nabla\psi \cdot \nabla)\nabla_\perp^2\phi + \frac{c}{B_0}(\hat{\boldsymbol{z}} \times \nabla\phi \cdot \nabla)\nabla_\perp^2\psi = 0, \tag{12.45}$$

where we have assumed that $|\partial\phi/\partial t| \gg (\nu_{\mathrm{in}} + \mu_{\mathrm{i}}\nabla_\perp^2)\phi$. The quasi-neutrality condition $n_{\mathrm{e}1} = n_{\mathrm{i}1} = n_1$ holds in the present dense plasma with $\omega_{\mathrm{pi}} \gg \omega_{\mathrm{ci}}$.

From the parallel component of the inertialess electron equation of motion, we obtain

$$\frac{\partial}{\partial t}A_z + c\frac{\partial}{\partial z}\left(\phi - \frac{T_{\mathrm{e}}n_1}{e n_0}\right) + \frac{c}{B_0}\hat{\boldsymbol{z}} \times \nabla\psi \cdot \nabla A_z = 0. \tag{12.46}$$

On the other hand, the ion continuity equation, together with (12.41), yields

$$\left(\frac{\partial}{\partial t} + \frac{c}{B_0}\hat{\boldsymbol{z}} \times \nabla\psi \cdot \nabla\right)\left(n_1 - \frac{c n_0}{B_0\omega_{\mathrm{ci}}}\nabla_\perp^2\phi\right) = 0. \tag{12.47}$$

The equation for two-dimensional zonal flows is obtained by inserting (12.42) and (12.43) into the electron and ion continuity equations, respectively, and substituting them into Poisson's equation. We obtain

$$\left(\frac{\partial}{\partial t} + \nu_{\mathrm{in}} + 0.3\nu_{\mathrm{ii}}\rho_{\mathrm{i}}^2\nabla_\perp^2\right)\nabla_\perp^2\psi + \frac{c}{B_0}\langle(\hat{\boldsymbol{z}} \times \nabla\phi \cdot \nabla)\nabla_\perp^2\phi\rangle - \frac{c_{\mathrm{A}}^2}{cB_0}\langle(\hat{\boldsymbol{z}} \times \nabla A_z \cdot \nabla)\nabla_\perp^2 A_z\rangle = 0, \tag{12.48}$$

where to lowest order, we use

$$\frac{\partial A_z}{\partial z} + \frac{c}{c_{\mathrm{A}}^2}\frac{\partial\phi}{\partial t} = 0 \tag{12.49}$$

into the last term of Eq. (12.48) to eliminate A_z in terms of ϕ. Equations (12.45)–(12.49) form a closed system of equations for studying the excitation of ZFs by finite amplitude DKAWs.

Next, we derive a dispersion relation for the modulational instability of a constant amplitude DAW pump against zonal flow perturbations. For this purpose, we decompose the high-frequency potentials into those of the pump and the two sidebands, viz.

$$\phi = \phi_{0+}\exp(-\mathrm{i}\omega_0 t + \mathrm{i}\boldsymbol{k}_0\cdot\boldsymbol{r}) + \phi_{0-}\exp(\mathrm{i}\omega_0 t - \mathrm{i}\boldsymbol{k}_0\cdot\boldsymbol{r}) + \sum_{+,-}\phi_\pm\exp(-i\omega_\pm t + \mathrm{i}\boldsymbol{k}_\pm\cdot\boldsymbol{r}), \tag{12.50}$$

and

$$A_z = A_{z0+}\exp(-\mathrm{i}\omega_0 t + \mathrm{i}\boldsymbol{k}_0\cdot\boldsymbol{r}) + A_{z0-}\exp(\mathrm{i}\omega_0 t - \mathrm{i}\boldsymbol{k}_0\cdot\boldsymbol{r}) + \sum_{+,-}A_{z\pm}\exp(-\mathrm{i}\omega_\pm t + \mathrm{i}\boldsymbol{k}_\pm\cdot\boldsymbol{r}), \tag{12.51}$$

where $\omega_\pm = \Omega\pm\omega_0$ and $\boldsymbol{k}_\pm = \boldsymbol{K}\pm\boldsymbol{k}_0$ are the frequency and wavevector of the upper and lower DAW sidebands. The subscripts $0\pm$ and $\pm$ represent the pump and sidebands, respectively.

Assuming further that $\psi = \varphi\exp(-\mathrm{i}\Omega t + \mathrm{i}\boldsymbol{K}\cdot\boldsymbol{r})$, we insert (12.50) and (12.51) into Eqs. (12.45)–(12.47) and Fourier transform them and combine the resultant equations to obtain

$$D_\pm\phi_\pm = \pm\frac{\mathrm{i}c}{B_0}\hat{\boldsymbol{z}}\times\boldsymbol{k}_0\cdot\boldsymbol{K}\left(\omega_0 + \omega_\pm\frac{k_{0\perp}^2 - K_\perp^2}{k_{\perp\pm}^2}\right)\varphi\phi_{0\pm}\,, \qquad (12.52)$$

where $D_\pm = \omega_\pm^2 - k_{z0}^2c_A^2(1 + k_{\perp\pm}^2\rho_s^2) \approx \pm 2\omega_0(\Omega - \boldsymbol{K}_\perp\cdot\boldsymbol{V}_{g\perp} \mp \delta)$, with $\omega_0 = k_{z0}c_A(1 + k_{\perp 0}^2\rho_s^2)^{1/2}$, $\boldsymbol{V}_{g\perp} = \boldsymbol{k}_{0\perp}\rho_s^2k_{z0}^2c_A^2/\omega_0$, and $\delta = k_{z0}^2c_A^2K_\perp^2\rho_s^2/2\omega_0$.

On the other hand, inserting (12.50) and (12.51) into Eq. (12.48) and Fourier transforming the resultant equation, we have

$$(\Omega + \mathrm{i}\Gamma_z)\varphi = \mathrm{i}\frac{2c}{B_0}\frac{\hat{\boldsymbol{z}}\times\boldsymbol{k}_0\cdot\boldsymbol{K}}{K_\perp^2}\left(1 - \frac{\omega_0^2}{k_{z0}^2c_A^2}\right) \times\left(\mathcal{K}_-^2\phi_{0+}\phi_- - \mathcal{K}_+^2\phi_{0-}\phi_+\right)\,, \qquad (12.53)$$

where $\Gamma_z = \nu_{in} + 0.3\nu_{ii}K_\perp^2\rho_i^2$ and $\mathcal{K}_\pm^2 = k_{\perp\pm}^2 - k_0^2 \equiv K_\perp^2 \pm 2\boldsymbol{k}_{0\perp}\cdot\boldsymbol{K}_\perp$. Equation (12.53) reveals that the coupling constant on the right-hand side remains finite only if $\omega_0 \neq k_{0z}c_A$. Thus, dispersion of Alfvén waves is required for the parametric coupling between CCs/ZFs and the DKAWs to remain intact.

Eliminating $\phi_\pm$ from Eq. (12.53) by using (12.52) we finally obtain the nonlinear dispersion relation

$$\Omega + i\Gamma_z = \frac{2c^2\omega_0|\phi_0|^2}{B_0^2}\frac{|\hat{\boldsymbol{z}}\times\boldsymbol{k}_0\cdot\boldsymbol{K}|^2}{K_\perp^2}k_{0\perp}^2\rho_s^2\sum_{+,-}\frac{\mathcal{K}_\pm^2\kappa_\pm^2}{k_{\perp\pm}^2D_\pm}\,, \qquad (12.54)$$

where $\kappa_\pm^2 = K_\perp^2 \pm \boldsymbol{k}_{0\perp}\cdot\boldsymbol{K}_\perp$. We see from Eq. (12.54) that the coupling constant in the right-hand side is proportional to $k_{0\perp}^2\rho_s^2$, which is a feature of the kinetic Alfvén wave dispersion. For long wavelength ZFs with $|\boldsymbol{K}_\perp| \ll |\boldsymbol{k}_{0\perp}|$, Eq. (12.54) reduces to

$$(\Omega + \mathrm{i}\Gamma_z)\left[(\Omega - \boldsymbol{K}_\perp\cdot\boldsymbol{V}_{g\perp})^2 - \delta^2\right] = -2K_\perp^2c^2\delta\frac{|E_{0\perp}|^2}{B_0^2}k_{0\perp}^2\rho_s^2\left|\left(\hat{\boldsymbol{K}}\cdot\hat{\boldsymbol{k}}_{0\perp}\right)\left(\hat{\boldsymbol{z}}\times\hat{\boldsymbol{k}}_0\cdot\hat{\boldsymbol{K}}\right)\right|^2\,, \qquad (12.55)$$

where $|E_{0\perp}|^2 = k_{0\perp}^2|\phi_0|^2$ and $\hat{\boldsymbol{K}}_\perp$ and $\hat{\boldsymbol{k}}_{0\perp}$ are the unit vectors.

We can analyze Eq. (12.55) in two limiting cases. First, we let $\Omega = \boldsymbol{K}_\perp\cdot\boldsymbol{V}_{g\perp} + \mathrm{i}\gamma_m$ in Eq. (12.55) and obtain for $\gamma_m, \Gamma_z \ll |\boldsymbol{K}_\perp\cdot\boldsymbol{V}_{g\perp}|$, the growth rate

$$\gamma_m = \left[\frac{2K_\perp^2c^2k_{0\perp}^2\rho_s^2\delta}{|\boldsymbol{K}_\perp\cdot\boldsymbol{V}_{g\perp}|}\frac{|E_{0\perp}|^2}{B_0^2} \times\left|\left(\hat{\boldsymbol{K}}\cdot\hat{\boldsymbol{k}}_{0\perp}\right)\left(\hat{\boldsymbol{z}}\times\hat{\boldsymbol{k}}_0\cdot\hat{\boldsymbol{K}}\right)\right|^2 - \delta^2\right]^{1/2}. \qquad (12.56)$$

The expression (12.56) shows that a modulational instability sets in if

$$|E_{0\perp}|^2 > \frac{B_0^2\delta|\boldsymbol{K}_\perp\cdot\boldsymbol{V}_{g\perp}|}{2K_\perp^2c^2k_{0\perp}^2\rho_s^2\left|\left(\hat{\boldsymbol{K}}\cdot\hat{\boldsymbol{k}}_{0\perp}\right)\left(\hat{\boldsymbol{z}}\times\hat{\boldsymbol{k}}_0\cdot\hat{\boldsymbol{K}}\right)\right|^2}\,. \qquad (12.57)$$

Second for $\Omega \gg \Gamma_z, \boldsymbol{K}_\perp\cdot\boldsymbol{V}_{g\perp}, \delta$, we have from Eq. (12.55)

$$\Omega^3 \simeq -2K_\perp^2c^2\delta\frac{|E_{0\perp}|^2}{B_0^2}k_{0\perp}^2\rho_s^2\,, \qquad (12.58)$$

which admits a reactive instability whose maximum growth rate is

$$\gamma_r \simeq \sqrt{6}(K_\perp c)^{2/3}\delta^{1/3}(k_{0\perp}\rho_s)^{2/3}\left(\frac{|E_{0\perp}|}{B_0}\right)^{2/3}. \qquad (12.59)$$

We observe from (12.59) that the increment is proportional to the two-third power of $k_{0\perp}\rho_s$ and the DKAW pump electric field strength $|E_{0\perp}|$. For typical laboratory Argon plasmas [Gekelman *et al.*, 2000] with $n_0 = 2\times10^{12}\,\mathrm{cm}^{-3}$, $B_0 = 1.5\,\mathrm{kG}$, $T_e = 10T_i = 10\,\mathrm{eV}$, we have $\beta m_i/m_e \approx 30$, $c_A = 10^8\,\mathrm{cm/s}$, and $\rho_s = 0.25\,\mathrm{cm}$. Taking $k_{0\perp}\rho_s = 0.1$, $\omega_0 \sim 10^5\,\mathrm{s}^{-1}$, $K_\perp/k_{0\perp} \sim 0.1$, and $|E_{0\perp}| \sim 10^{-4}B_0$, we find that $\delta \sim 10\,\mathrm{s}^{-1}$, $|\boldsymbol{K}_\perp\cdot\boldsymbol{V}_g| \sim 100\,\mathrm{s}^{-1}$, and $\gamma_r \sim 10^3\,\mathrm{s}^{-1}$. Thus, the reactive instability can produce ZFs within a millisecond at the expense of the kinetic Alfvén wave energy.

12.6 Ponderomotive Forces and Plasma Density Modifications

Recent observations by the FREJA and FAST spacecrafts [Stasiewicz *et al.*, 2000] and the large plasma device (LAPD) experiments [Gekelman *et al.*, 2000] reveal signatures of nonlinear structures consisting of localized dispersive shear Alfvén wave (DSAW) electric fields and very narrow magnetic-field-aligned density perturbations. Accordingly, in the following, we have to discuss the ponderomotive force of the DSAWs and the associated plasma density modification. Following [Shukla *et al.*, 2004], the quasi-stationary plasma slow response in the DSWA fields is given by

$$m_e\left(\langle \boldsymbol{v}_{e\perp}\cdot\nabla v_{ez}\rangle+\frac{1}{2}\partial_z\langle v_{ez}^2\rangle\right)+\frac{e}{c}\langle(\boldsymbol{v}_{e\perp}\times\boldsymbol{B}_\perp)_z\rangle$$
$$=-eE_z^s-T_e\partial_z ln\, n_e^s\,, \tag{12.60}$$

$$m_i\left(\langle \boldsymbol{v}_{i\perp}\cdot\nabla v_{iz}\rangle+\frac{1}{2}\partial_z\langle v_{iz}^2\rangle\right)-\frac{e}{c}\langle(\boldsymbol{v}_{i\perp}\times\boldsymbol{B}_\perp)_z\rangle$$
$$=eE_z^s-T_i\partial_z ln\, n_i^s\,, \tag{12.61}$$

where the superscript s denotes the quantities associated with the plasma slow motion. The angular bracket denotes averaging over the DSAW period.

We first consider the DSAWs with $k_z v_{te}\ll\omega\ll\omega_{ce}$, so that the appropriate fluid velocities are

$$\boldsymbol{v}_{e\perp}\approx\frac{c}{B_0}\boldsymbol{E}_\perp\times\hat{\boldsymbol{z}}\,,\quad v_{ez}=-\mathrm{i}\frac{eE_z}{m_e\omega}\,, \tag{12.62}$$

and

$$\boldsymbol{v}_{i\perp}\approx\frac{e\omega_{ci}}{m_i(\omega_{ci}^2-\omega^2)}\boldsymbol{E}_\perp\times\hat{\boldsymbol{z}}-\mathrm{i}\frac{c\omega\omega_{ci}\boldsymbol{E}_\perp}{B_0(\omega_{ci}^2-\omega^2)}\,. \tag{12.63}$$

The perpendicular component of the current density is

$$\boldsymbol{J}_\perp^*=-\frac{n_0ec}{B_0}\frac{\omega^2}{(\omega_{ci}^2-\omega^2)}\boldsymbol{E}_\perp\times\hat{\boldsymbol{z}}+\mathrm{i}\frac{n_0ec\omega\omega_{ci}\boldsymbol{E}_\perp}{B_0(\omega_{ci}^2-\omega^2)}\,, \tag{12.64}$$

where the asterisk denotes the complex conjugate. Accordingly, we have a balance between the ponderomotive force and the pressure gradient

$$\frac{i}{n_0\omega}\langle\boldsymbol{J}_\perp^*\cdot\nabla E_z\rangle-\frac{1}{n_0c}\langle(\boldsymbol{J}_\perp^*\times\boldsymbol{B}_\perp)\rangle+\frac{e^2}{4m_e\omega^2}\frac{\partial}{\partial z}|E_z|^2$$
$$=-(T_e+T_i)\frac{\partial}{\partial z}\ln n_e^s\,. \tag{12.65}$$

By using the wave magnetic field

$$\boldsymbol{B}_\perp=-\mathrm{i}\frac{c}{\omega}(\nabla\times\boldsymbol{E})_\perp \tag{12.66}$$

as well as the electric field relationship

$$\frac{\omega^2\omega_{ci}}{(\omega_{ci}^2-\omega^2)\,\omega_{ce}}\nabla\cdot\boldsymbol{E}_\perp=\frac{\partial E_z}{\partial z}\,, \tag{12.67}$$

and the modified inertial Alfvén wave (IAW) dispersion relation

$$\frac{\omega^2(1+k_\perp^2\lambda_e^2)}{k_z^2\lambda_i^2}=\omega_{ci}^2-\omega^2\,, \tag{12.68}$$

where $\lambda_e=c/\omega_{pe}$ and $\lambda_i=c/\omega_{pi}$, we obtain the quasineutral density response

$$n_e=n_0\exp\left[\frac{|B_y|^2}{16\pi n_0(T_e+T_i)}\right]. \tag{12.69}$$

On the other hand, for the modified kinetic Alfvén waves (KAWs) we take $v_{ez}=\mathrm{i}(e\omega/k_z^2T_e)E_z$ and obtain

$$n_e^s=n_0\exp\left[\frac{e^2|\boldsymbol{E}_\perp|^2}{4m_i(T_e+T_i)(\omega_{ci}^2-\omega^2)}\right.$$
$$\left.-\frac{e^2|E_z|^2}{4m_e(T_e+T_i)}\frac{\omega^2}{k_z^4v_{te}^4}\right]. \tag{12.70}$$

Since for the modified KAWs we have

$$E_z=\frac{k_zk_\perp c_s^2|\boldsymbol{E}_\perp|}{\omega_{ci}^2-\omega^2} \tag{12.71}$$

and

$$\omega^2=\frac{k_z^2c_A^2}{1+k_z^2\lambda_i^2}(1+k_\perp^2\rho_s^2)\,, \tag{12.72}$$

the quasi-stationary electron response for the KAWs turns out to be

$$n_e^s=n_0\exp\left[\frac{\alpha e^2|\boldsymbol{E}_\perp|^2}{4m_i(T_e+T_i)(\omega_{ci}^2-\omega^2)}\right], \tag{12.73}$$

where $\alpha=1-\omega^2k_\perp^2m_e/(\omega_{ci}^2-\omega^2)k_z^2m_i$.

The expressions (12.69) and (12.73) reveal that the ponderomotive force of the DSAWs produces magnetic field-aligned electron density compressions.

12.7 Modulated Circularly Polarized Dispersive Alfvén Waves

Here, for illustrative purposes, following Shukla *et al.* [2004], we consider the amplitude modulation of circularly polarized dispersive Alfvén waves (CPDAWs) along the external magnetic field direction. The cold plasma dispersion relation for CPDAWs ($\omega\ll\omega_{ce}$, $\boldsymbol{k}\parallel\hat{\boldsymbol{z}}$) is

$$\frac{k^2c^2}{\omega^2} \approx \mp \frac{\omega_{\rm pe}^2}{\omega\omega_{ce}} - \frac{\omega_{\rm pi}^2}{\omega(\omega \pm \omega_{ci})}, \quad (12.74)$$

which for $\omega \sim kc_A \ll \omega_{ci}$ reduces to

$$\omega \approx kc_A\left(1 \mp \frac{kc_A}{2\omega_{ci}}\right). \quad (12.75)$$

Supposing that the nonlinear interaction between CPDAWs and the plasma slow response produces an envelope of waves which varies slowly, we introduce the eikonal representation

$$\omega \to \omega_0 + i\frac{\partial}{\partial t} - {\rm i}c_{A0}\frac{\partial}{\partial z}, \quad k \to k_0 - {\rm i}\frac{\partial}{\partial z}, \quad (12.76)$$

in (12.75) and operate on the wave electric field $\boldsymbol{E} = E_\perp(\hat{\boldsymbol{x}} \pm {\rm i}\hat{\boldsymbol{y}})$. Assuming that $\partial E_\perp/\partial t \ll \omega_0 E_\perp \equiv \omega_0(E_x \mp {\rm i}E_y)/\sqrt{2}$, we then obtain the derivative nonlinear Schrödinger equation (DNLSE)

$$\left(\frac{\partial}{\partial t}+c_{A0}\frac{\partial}{\partial z}\right)E_\perp - \frac{c_{A0}}{2n_0}\frac{\partial}{\partial z}(n_1E_\perp) \pm \frac{ic_{A0}^2}{2\omega_{\rm ci}}\frac{\partial^2 E_\perp}{\partial z^2} = 0. \quad (12.77)$$

The slow plasma response assumes inertialess electrons

$$0 = -eE^s - \frac{T_{\rm e}}{n_0}\frac{\partial n_1}{\partial z} \mp \frac{\omega_{\rm pe}^2}{4\pi n_0\omega\omega_{ce}}\frac{\partial |E_\perp|^2}{\partial z}, \quad (12.78)$$

and inertial ions

$$m_{\rm i}\frac{\partial v_{\rm i}^s}{\partial t} = eE^s - \frac{\gamma_{\rm i}T_{\rm i}}{n_0}\frac{\partial n_1}{\partial z} - \frac{\omega_{\rm pi}^2}{4\pi n_0\omega(\omega \pm \omega_{\rm ci})}\frac{\partial |E_\perp|^2}{\partial z}. \quad (12.79)$$

Eliminating the ambipolar electric field E^s from (12.78) and (12.79) and using the ion continuity equation we thus obtain the equation for the driven ion sound waves

$$\left(\frac{\partial^2}{\partial t^2} - c_{\rm s}^2\frac{\partial^2}{\partial z^2}\right)n_1 = -\frac{n_0c^2}{B_0^2}\frac{\partial^2|E_\perp|^2}{\partial z^2}. \quad (12.80)$$

The quasi-stationary response is then

$$\frac{n_1}{n_0} = \frac{m_{\rm i}c^2}{(T_{\rm e}+T_{\rm i})}\frac{|E_\perp|^2}{B_0^2}. \quad (12.81)$$

Subsequently, Eq. (12.77) takes the form

$$\left(\frac{\partial}{\partial t} + c_{A0}\frac{\partial}{\partial z}\right)E_\perp - \frac{m_{\rm i}c_{A0}c^2}{2(T_{\rm e}+T_{\rm i})B_0^2}\frac{\partial}{\partial z}\left(|E_\perp|^2E_\perp\right) \pm \frac{{\rm i}c_{A0}^2}{2\omega_{\rm ci}}\frac{\partial^2 E_\perp}{\partial z^2} = 0 \quad (12.82)$$

The DNLSE (12.83) admits a localized DAW electric field envelope accompanied by a background plasma density compression.

12.8 Electron Joule Heating

The dispersive Alfvén waves (DAWs) can produce electron Joule heating due to the wave-electron interaction [Shukla, Bingham, McKenzie and Axford, 1999]. To understand the electron heating, we compute the parallel electron current density J_z for both the IAWs and KAWs. For the IAWs ($k_z v_{\rm Te} \ll \omega \ll \omega_{ce}$), we have

$$J_z = E_z\frac{{\rm i}\omega_{\rm pe}^2}{4\pi\omega}\left[1 - 2{\rm i}\sqrt{\pi}\frac{\omega^3}{k_z^3v_{\rm te}^3}\exp\left(-\frac{\omega^2}{k_z^2v_{\rm te}^2}\right)\right]. \quad (12.83)$$

On the other hand, for the KAWs with $v_{\rm ti} \ll \omega/k_z \ll v_{\rm te}$ we have

$$J_z = -\frac{{\rm i}\omega E_z}{4\pi k_z^2\lambda_{\rm De}^2}\left\{1 + {\rm i}\sqrt{\pi}\frac{\omega}{k_zv_{\rm ti}}\sqrt{\frac{T_{\rm i}}{T_{\rm e}}} \times \left[\sqrt{\frac{m_{\rm e}}{m_{\rm i}}} + \left(\frac{T_{\rm e}}{T_{\rm i}}\right)^{3/2}\exp\left(-\frac{\omega^2}{k_z^2v_{\rm ti}^2}\right)\right]\right\}. \quad (12.84)$$

The rate of change of the electron temperature follows from

$$n_0\frac{{\rm d}T_{\rm e}}{{\rm d}t} = J_zE_z, \quad (12.85)$$

which for the KAWs can be written as

$$n_0\frac{{\rm d}T_{\rm e}}{{\rm d}t} = \sqrt{\frac{\pi}{2}}\frac{\omega^2}{k_zv_{\rm te}}\frac{|E_z|^2}{4\pi k_z^2\lambda_{\rm De}^2}. \quad (12.86)$$

The expression (12.86), with $\omega = k_zc_A(1+k_\perp^2\rho^2)^{1/2}/(1+k_z^2\lambda_{\rm i}^2)^{1/2}$, has been used by Shukla, Bingham, McKenzie and Axford [1999] to explain the solar coronal heating by dispersive electromagnetic ion-cyclotron-Alfvén waves. In the unperturbed solar coronal state, one typically has $n_0 \sim 5 \times 10^9$ cm^{-3}, $T_{\rm e} \sim 6$ million K, and

$B_0 = 100$ G. Thus, we have $\omega_{pe} = 4 \times 10^6$ rad s^{-1}, $\omega_{ci} = 10^8$ rad s^{-1}, $v_{te} = 10^9$ cm s^{-1}, $c_s = 2.2 \times 10^7$ cm s^{-1}, and $8\pi n_0 T_e / B_0^2 = 0.01$. The electron sound gyroradius and the ion skin depth turn out to be 22 cm and 3 m, respectively. Assuming that the perpendicular (parallel) wavelength of the high-frequency dispersive Alfvén waves is one meter (one cm), we have $k_\perp \rho_s \approx 1$ and $k_z \lambda_i = 300$. It then follows from (12.86) that within 5 – 10 s the electron temperature will rise to a value of 60 million K when the parallel electric field of the dispersive Alfvén waves is of the order of ten V cm^{-1}. Thus, the wave-electron interaction is capable of producing the desired heating of the solar corona.

12.9 Self-Interaction Between DSAWs

In this section, we consider self-interactions between low-frequency (in comparison with ω_{ci}), long wavelength (in comparison with the ion gyroradius) DSAWs. For our purposes, the appropriate fluid velocities are then

$$\boldsymbol{v}_{e\perp} \approx \frac{c}{B_0}\hat{\boldsymbol{z}} \times \nabla\phi - \frac{c}{eB_0 n_e}\hat{\boldsymbol{z}} \times \nabla(n_e T_e) + v_{ez}\frac{\nabla_\perp A_z \times \hat{\boldsymbol{z}}}{B_0}, \tag{12.87}$$

$$\boldsymbol{v}_{i\perp} \approx \frac{c}{B_0}\hat{\boldsymbol{z}} \times \boldsymbol{\nabla}\phi - \frac{c}{B_0\omega_{ci}}\left(\partial_t + \frac{c}{B_0}\hat{\boldsymbol{z}} \times \nabla\phi \cdot \nabla\right)\boldsymbol{\nabla}_\perp\phi, \tag{12.88}$$

where v_{ez} is given by Eq. (12.44).

Substituting (12.87) and (12.88) into $\nabla \cdot \boldsymbol{J} = 0$, where $\boldsymbol{J}$ is the plasma current density, and using (12.44) we then obtain

$$\frac{\mathrm{d}}{\mathrm{d}t}\nabla_\perp^2\phi + \frac{c_A^2}{c}\frac{\mathrm{d}}{\mathrm{d}z}\nabla_\perp^2 A_z = 0 \tag{12.89}$$

where $\mathrm{d}/\mathrm{d}t = (\partial/\partial t) + (c/B_0)\hat{\boldsymbol{z}} \times \nabla\phi \cdot \nabla$ and $\mathrm{d}/\mathrm{d}z = (\partial/\partial z) - B_0^{-1}\hat{\boldsymbol{z}} \times \nabla A_z \cdot \nabla$. We have here assumed that $(\omega_{pe}^2/|\omega_{ce}|)|\hat{\boldsymbol{z}} \times \nabla\phi \cdot \nabla| \gg c\partial_z \nabla_\perp^2 A_z$.

For the IAWs, we neglect the parallel electron pressure gradient in the parallel electron momentum equation and find

$$\frac{\mathrm{d}}{\mathrm{d}t}\left(1 - \lambda_e^2\nabla_\perp^2\right)A_z + c\partial_z\phi = 0\,. \tag{12.90}$$

On the other hand, for the KAWs the parallel electron inertia is negligible in comparison with the electron pressure gradient. Thus, we have

$$\frac{\partial A_z}{\partial t} + c\frac{\mathrm{d}}{\mathrm{d}z}\left(\phi - \frac{T_e n_1}{en_0}\right) = 0\,, \tag{12.91}$$

where the electron continuity equation

$$\frac{\mathrm{d}n_1}{\mathrm{d}t} + \frac{c}{4\pi e}\frac{\mathrm{d}}{\mathrm{d}z}\nabla_\perp^2 A_z = 0 \tag{12.92}$$

determines n_1. Equations (12.89)–(12.92) are complex nonlinear partial differential equations containing vector nonlinearities. They admit dual cascades and provide the possibility of self-organization [Hasegawa, 1985] of DSAWs in the form of different types of vortical structures [Petviashvili and Pokhotelov, 1992; Shukla *et al.*, 1995; Jovanovic *et al.*, 1998] (e.g. dipolar and tripolar vortices and a vortex street). Equations (12.90)–(12.92) are thus useful for studying collisionless tearing modes and current filaments in plasmas with sheared magnetic fields [Del Sarto *et al.*, 2003; Pegoraro *et al.*, 2004].

12.10 Nonlinear Drift-Alfvén–Shukla–Varma Modes

In this section, we consider the nonlinear propagation of low-frequency (in comparison with the ion gyrofrequency), drift-Alfvén–Shukla–Varma modes in a nonuniform magnetoplasma containing immobile charged dust impurities, a situation quite common in cosmic environments [Shukla and Mamun, 2002]. It will be shown that the presence of charged dust grains provides the possibility of linear as well as nonlinear couplings between the drift-Alfvén and the Shukla–Varma mode [Shukla and Varma, 1993, Shukla and Eliasson, 2005]. Furthermore, the latter modifies the theory of drift-Alfvén vortices in that the dust density gradient causes a complete localization of the electromagnetic drift-Alfvén vortex, in addition to introducing a bound on the vortex speed.

The equilibrium plasma state now satisfies the quasi-neutrality condition, i.e.

$$n_{i0} = n_{e0} + Z_d n_{d0}\,, \tag{12.93}$$

where n_{j0} is the unperturbed number density of the particle species j (j equals e for the electrons, i for the ions, and d for the negatively charged dust grains) and Z_d is the number of charges residing on the dust grain surface.

In the electromagnetic fields, the electron and ion fluid velocities are given by Eq. (12.87) and

$$\boldsymbol{v}_{i\perp} \approx \frac{c}{B_0}\hat{\boldsymbol{z}} \times \nabla\phi + \frac{cT_i}{eB_0 n_i}\hat{\boldsymbol{z}} \times \nabla n_i - \frac{c}{B_0\omega_{ci}}\left(\partial_t + \boldsymbol{v}_{i\perp}\cdot\nabla\right)\nabla_\perp\phi\,, \quad (12.94)$$

respectively. The parallel component of the electron fluid velocity is given by Eq. (12.44) with $n_0 \to n_{e0}$.

Following Pokhotelov *et al.* [1999], and substituting Eqs. (12.87) into the electron continuity equation, letting $n_j = n_{j0}(x) + n_{j1}$, where $n_{j1} (\ll n_{j0})$ is the particle number density perturbation, and using (12.44) we obtain

$$\frac{dn_{e1}}{dt} - \frac{c}{B_0}\hat{\boldsymbol{z}} \times \nabla n_{e0}\cdot\nabla\phi + \frac{c}{4\pi e}\frac{d}{dz}\nabla_\perp^2 A_z = 0\,. \quad (12.95)$$

On the other hand, substitution of the ion fluid velocity (12.94) into the ion continuity equation yields

$$\frac{dn_{i1}}{dt} - \frac{c}{B_0}\hat{\boldsymbol{z}} \times \nabla n_{i0}\cdot\nabla\phi - \frac{cn_{i0}}{B_0\omega_{ci}}\left(d_t + \boldsymbol{u}_{i*}\cdot\nabla\right)\nabla_\perp^2\phi - \frac{c^2 T_i}{eB_0^2\omega_{ci}}\nabla_\perp\cdot\left[(\hat{\boldsymbol{z}}\times\nabla n_{i1})\cdot\nabla\nabla_\perp\phi\right] = 0\,, \quad (12.96)$$

where $\boldsymbol{u}_{i*} = (cT_i/eB_0 n_{i0})\hat{\boldsymbol{z}} \times \nabla n_{i0}$ is the unperturbed ion diamagnetic drift.

Subtracting (12.96) from (12.95) and assuming $n_{i1} = n_{e1}$, we obtain the modified ion vorticity equation

$$\left(\frac{d}{dt} + u_{i*}\partial_y\right)\nabla_\perp^2\phi + \frac{c_A^2}{c}d_z\nabla_\perp^2 A_z + \omega_{ci}\delta_d\kappa_d\partial_y\phi + \frac{cT_i}{eB_0 n_{i0}}\nabla_\perp\cdot\left[(\hat{\boldsymbol{z}}\times\nabla n_{e1})\cdot\nabla\nabla_\perp\phi\right] = 0\,, \quad (12.97)$$

where $u_{i*} = (cT_i/eB_0 n_{i0})\partial n_{i0}/\partial x$, $c_A = B_0/(4\pi n_{i0} m_i)^{1/2}$ is the Alfvén velocity, $\delta_d = Z_d n_{d0}/n_{i0}$, and $\kappa_d = \partial ln(Z_d n_{d0}(x))/\partial x$. The term $\omega_{ci}\delta_d\kappa_d\partial_y\phi$ is associated with the Shukla–Varma mode [Shukla and Varma, 1993].

By using (12.87) and (12.44), the parallel component of the electron momentum equation can be written as

$$\left(\partial_t + u_{e*}\partial_y\right)A_z - \lambda_e^2 d_t\nabla_\perp^2 A_z + c\frac{d}{dz}\left(\phi - \frac{T_e}{en_{e0}}n_{e1}\right) = 0\,, \quad (12.98)$$

where $u_{e*} = -(cT_e/eB_0 n_{e0})\partial n_{e0}(x)/\partial x$ is the unperturbed electron diamagnetic drift.

Equations (12.95), (12.97) and (12.98) are the desired nonlinear equations for the coupled drift-Alfvén–Shukla–Varma modes in a nonuniform dusty magnetoplasma [Pokhotelov *et al.*, 1999].

Let us now consider stationary solutions of the nonlinear equations (12.95), (12.97) and (12.98), assuming that all the field variables depend on x and $\eta = y+\alpha z-ut$, where u is the translation speed of the vortex along the y-axis, and α the angle between the wave front normal and the (x, y) plane. Two cases are considered. First, in the stationary η-frame, (12.98) for $\lambda_e^2|\nabla_\perp^2| \ll 1$, can be written as

$$\hat{D}_A\left(\phi - \frac{T_e}{en_{e0}}n_{e1} - \frac{u-u_{e*}}{\alpha c}A_z\right) = 0\,, \quad (12.99)$$

where $\hat{D}_A = \partial_\eta + (1/\alpha B_0)[(\partial_\eta A_z)\partial_x - (\partial_x A_z)\partial_\eta]$. A solution of (12.99) is

$$n_{e1} = \frac{n_{e0}e}{T_e}\phi - \frac{n_{e0}e(u-u_{e*})}{\alpha c T_e}A_z\,. \quad (12.100)$$

Writing (12.95) in the stationary frame, and making use of (12.100) it can be put in the form

$$\hat{D}_A\left(\lambda_{De}^2\nabla_\perp^2 A + \frac{u(u-u_{e*})}{\alpha^2 c^2}A_z - \frac{u-u_{e*}}{\alpha c}\phi\right) = 0\,. \quad (12.101)$$

A solution of (12.101) is

$$\lambda_{De}^2\nabla_\perp^2 A_z + \frac{u(u-u_{e*})}{\alpha^2 c^2}A_z - \frac{u-u_{e*}}{\alpha c}\phi = 0\,. \quad (12.102)$$

The modified ion vorticity equation (12.97) for cold ions can be expressed as

$$\hat{D}_\phi\left[\nabla_\perp^2\phi + \frac{u_{c*}}{u}\phi\right] = 0\,, \quad (12.103)$$

where $\hat{D}_\phi = \partial_\eta - (c/uB_0)[(\partial_x\phi)\partial_\eta - (\partial_\eta\phi)\partial_x]$, and $u_{c*} = -c_s\delta_d\kappa_d/\rho_s$. Here $c_s = (n_{i0}/n_{e0})^{1/2}(T_e/m_i)^{1/2}$ and ρ_s are the ion sound and sound gyroradius in dusty plasmas.

Combining (12.102) and (12.103) we obtain

$$\hat{D}_\phi\left(\nabla_\perp^2\phi + \frac{p}{\rho_s^2}\phi + \frac{u-u_{e*}}{\alpha c\rho_s^2}A_z\right) = 0\,, \quad (12.104)$$

where $p = (c_s/u)\left[\delta_d\kappa_d\rho_s + (u_{e*}-u)/c_s\right]$. A typical solution of (12.104) is

$$\nabla_\perp^2\phi + \frac{p}{\rho_s^2}\phi + \frac{u-u_{e*}}{\alpha c\rho_s^2}A_z = C_1\left(\phi - \frac{uB_0}{c}x\right)\,, \quad (12.105)$$

where C_1 is an integration constant.

Eliminating A_z from (12.102) and (12.105), we obtain a fourth order inhomogeneous differential equation

$$\nabla^4\phi+F_1\nabla_\perp^2\phi+F_2\phi+C_1\frac{u^2(u-u_{e*})B_0}{\alpha^2c^3\lambda_{De}^2}x=0\,, \quad (12.106)$$

where $F_1=(p/\rho_s^2)-C_1+u(u-u_{e*})/\alpha^2c^2\lambda_{De}^2$ and $F_2=(u-u_{e*})^2/\alpha^2c^2\lambda_{De}^2\rho_s^2+(p-C_1\rho_s^2)u(u-u_{e*})/\alpha^2c^2\times\lambda_{De}^2\rho_s^2$. We note that in the absence of charged dust we have $\delta_d\kappa_d=0$ and $F_2=0$. Accordingly, the outer solution, where $C_1=0$, of Eq. (12.106) has a long tail for $(u-u_{e*})(\alpha^2c_A^2-u^2)>0$. On the other hand, inclusion of a small fraction of dust grains would make F_2 finite in the outer region. Here, we have the possibility of well behaved solutions. In fact, (12.106) admits spatially bounded dipolar vortex solutions. In the outer region ($r>R$), where R is the vortex radius, we set $C_1=0$ and write the solution of (12.106) as [Liu and Horton, 1986; Shukla *et al.*, 1986]

$$\phi=[Q_1K_1(s_1r)+Q_2K_1(s_2r)]\cos\theta\,, \quad (12.107)$$

where Q_1 and Q_2 are constants, and $s_{1,2}^2=-[-\alpha_1\pm(\alpha_1^2-4\alpha_2)^{1/2}]/2$ for $\alpha_1<0$ and $\alpha_1^2>4\alpha_2>0$. Here, $\alpha_1=(p/\rho_s^2)+u(u-u_{e*})/\alpha^2c^2\lambda_{De}^2$ and $\alpha_2=[(u-u_{e*})^2+u(u-u_{e*})p]/\alpha^2c^2\lambda_{De}^2\rho_s^2$. In the inner region ($r<R$), the solution reads [Yu *et al.*, 1986]

$$\phi=\left[Q_3J_1(s_3r)+Q_4I_1(s_4r)-\frac{C_1}{\lambda_{De}^2}\frac{u^2(u-u_{e*})B_0}{\alpha^2c^3F_2}r\right]\times\cos\theta\,, \quad (12.108)$$

where Q_3 and Q_4 are constants. We have defined $s_{3,4}=[(F_1^2-4F_2)^{1/2}\pm F_1]/2$ for $F_2<0$. Thus, the presence of charged dust grains is responsible for the complete localization of the vortex solutions both in the outer as well as in the inner regions of the vortex core.

Second, we present the double vortex solution of Eqs. (12.95)–(12.98) in the cold plasma approximation. Thus, we set $T_j=0$ and write (12.97) and (12.98) in the stationary frame as

$$\hat{D}_\phi\left(\nabla_\perp^2\phi-\frac{\omega_{ci}\delta_d\kappa_d}{u}\phi\right)-\frac{c_A^2\alpha}{uc}\hat{D}_A\nabla_\perp^2A_z=0\,, \quad (12.109)$$

and

$$\hat{D}_\phi\left[\left(1-\lambda_e^2\nabla_\perp^2\right)A_z-\frac{\alpha c}{u}\phi\right]=0\,. \quad (12.110)$$

It is easy to verify that (12.110) is satisfied by

$$\left(1-\lambda_e^2\nabla_\perp^2\right)A_z-\frac{\alpha c}{u}\phi=0\,. \quad (12.111)$$

By using (12.109) one can eliminate $\nabla_\perp^2A_z$ from (12.111), yielding

$$\hat{D}_\phi\left[\nabla_\perp^2\phi-\frac{\omega_{ci}\delta_d\kappa_d}{u}\phi+\frac{\alpha^2c_A^2}{u^2\lambda_e^2}\phi-\frac{\alpha c_A^2}{uc\lambda_e^2}A_z\right]=0\,. \quad (12.112)$$

A typical solution of (12.112) is

$$\nabla_\perp^2\phi+\beta_1\phi-\beta_2A_z=C_2\left(\phi-\frac{uB_0}{c}x\right)\,, \quad (12.113)$$

where $\beta_1=\left(\alpha^2c_A^2/u^2\lambda_e^2\right)-\omega_{ci}\delta_d\kappa_d/u$, $\beta_2=\alpha c_A^2/uc\lambda_e^2$, and C_2 is an integration constant.

Eliminating A_z from (12.111) and (12.113) we obtain

$$\nabla_\perp^4\phi+F_1\nabla_\perp^2\phi+F_2\phi-\frac{C_2uB_0}{\lambda_e^2c}x=0\,, \quad (12.114)$$

where $F_1=\lambda_e^{-2}\left(\alpha^2c_A^2/u^2-1\right)-(\omega_{ci}\delta_d\kappa_d/u)-C_2$ and $F_2=\lambda_e^{-2}\left(C_2+\omega_{ci}\delta_d\kappa_d/u\right)$.

Equation (12.114) is similar to (12.106) and its bounded solutions [similar to (12.107) and (12.108)] exist, provided that $u^2+\lambda_e^2\omega_{ci}\delta_d\kappa_du>\alpha^2c_A^2$ and $\kappa_d>0$. In the absence of the dust, we have $F_2=0$ in the outer region ($C_2=0$), and the outer solution of the dust-free case has a long tail (decaying as $1/r$).

12.11 Summary and Conclusions

We have focused on the present understanding of some of the most significant nonlinear effects in plasmas. We have thus described the underlying physics of the nonlinear mode couplings between high and low-frequency electromagnetic waves in unmagnetized and magnetized plasmas, and the interplay between the ponderomotive force and the Joule heating nonlinearities on various parametric processes that occur in collisional and collisionless plasmas. Furthermore, we have developed a general theory for three-wave decay interactions and presented a specific example for the magnetic field-aligned Alfvén-sound wave couplings. Considering multi-dimensional wave–wave interactions, we have investigated the generation of zonal flows by the Reynolds stresses of the dispersive Alfvén waves. The ponderomotive force of the latter creates density compressions in the plasma. We have also considered the amplitude modulation of the magnetic field-aligned dispersive Alfvén waves, leading to the trapping of localized Alfvén wave energy in self-created density perturbations. The self-interactions between large amplitude inertial and kinetic Alfvén waves as well as drift-Alfvén–Shukla–Varma modes are governed by a system of differential equations in which the Jacobian nonlinearities play a very important role with regard to dual cascading and self-organization in the form of different types of vortices. In conclusion, we stress that wave-wave and wave-particle interactions and their nonlinear evolution are of great importance in understanding the salient properties of enhanced plasma lines and fluctuation spectra, enhanced zonal flows, nonthermal plasma particle heating, and coherent nonlinear structures in the Earth's ionosphere and in the auroral zone, as well as in the solar corona, cosmic environments and in inertial confinement fusion schemes. Hopefully, forthcoming observations by means of EISCAT, CLUSTER, SOHO, CASSINI, and LAPD will reveal the signatures of the wave-wave couplings and associated coherent structures, as discussed in this chapter.

This work was partially supported by the Deutsche Forschungsgemeinschaft through the Sonderforschungsbereich 591 "Universelles Verhalten Gleichgewichtsferner Plasmen: Heizung, Transport und Strukturbildung", as well as by the Swedish Research Council.

References

Chian, A.C.-L., A.S. de Assis, C.A. de Azevedo, P.K. Shukla, and L. Stenflo, *Alfvén Waves in Cosmic and Laboratory Plasmas, Physica Scripta*, **T60**, 1, 1995.

Chian, A.C.-L., Nonlinear wave–wave interactions in astrophysical and space plasmas, *Astrophys. Space Sci.*, **242**, 249, 1997.

Del Sarto, D., F. Califano, and F. Pegoraro, Secondary instabilities and vortex formation in collisionless-fluid magnetic reconnection, *Phys. Rev. Lett.*, **91**, 235001, 2003.

Drozdenko, T., and G. Morales, Nonlinear effects resulting from the interaction of a large-scale Alfvén wave with a density filament, *Phys. Plasmas*, **8**, 3265, 2001.

Gekelman, W., Review of laboratory experiments on Alfvén waves and their relationship to space observations, *J. Geophys. Res.*, **104**, 14417, 1999.

Gekelman, W., S. Vincena, N. Palmer, P. Pribyal, D. Leneman, C. Mitchell, and J. Maggs, Experimental measurements of the propagation of large-amplitude shear Alfvén waves, *Plasma Phys. Control. Fusion*, **42**, B15, 2000.

Hasegawa, A., Decay of a plasmon into two electromagnetic waves, *Phys. Rev. Lett.*, **32**, 817, 1974.

Hasegawa, A., Self-organization processes in continuous media, *Adv. Phys.*, **34**, 1, 1985.

Hasegawa, A. and L. Chen, Kinetic processes in plasma heating by resonant mode conversion of Alfvén wave, *Phys. Fluids*, **19**, 1924, 1976.

Hasegawa, A., and C. Uberoi, *The Alfvén Waves* (National Technical Information Service, Springfield, VA), 1982.

Horton, W., and A. Hasegawa, 1994, Quasi-two-dimensional dynamics of plasmas and fluids, *Chaos*, **4**, 227.

Jovanovic, D., F. Pegoraro, and J.J. Rasmussen, Tripolar shear Alfvén vortex structures, *J. Plasma Phys.*, **60**, 383, 1998.

Kadomtsev, B.B. and V.I. Karpman, Nonlinear waves, *Sov. Phys.-Usp.*, **14**, 40, 1971.

Karpman, V.I., High-frequency electromagnetic fields in plasma with negative dielectric constant, *Plasma Phys.* **13**, 477, 1971.

Larsson, J., L. Stenflo, and R. Tegeback, Enhanced fluctuations in a magnetized plasma due to the presence of an electromagnetic wave, *J. Plasma Phys.*, **16**, 37, 1976.

Lashmore-Davies, C.N., and L. Stenflo L., On the MHD stability of a helical magnetic field of arbitrary amplitude, *Plasma Phys.*, **21**, 735, 1979.

Liu, J., and W. Horton, The intrinsic electromagnetic solitary vortices in magnetized plasma, *J. Plasma Phys.*, **36**, 1, 1986.

Okuda, H., and J.M. Dawson, Theory and numerical simulation on plasma diffusion across a magnetic field, *Phys. Fluids*, **16**, 408, 1973.

Pegoraro, F., D. Borgogno, F. Califano, D. Del Sarto, E. Echkina, D. Grasso, T. Lieseikina, and F. Porcelli, Developments in the theory of collisionless reconnection in magnetic configurations with a strong guide field, *Nonlinear Proc. Geophys.*, **11**, 567, 2004.

Petviashvili, V.I., and O.A. Pokhotelov, *Solitary Waves in Plasmas and in the Atmosphere* (Gordon and Breach, London), 1992.

Pokhotelov, O.A., O.G. Onishchenko, P.K. Shukla, and L. Stenflo, Drift-Alfvén vortices in dusty plasmas, *J. Geophys. Res.*, **104**, 19797, 1999.

Pokhotelov, O.A., O.G. Onishchenko, R.Z. Sagdeev, M.A. Balikhin, and L. Stenflo, Parametric interaction of kinetic Alfvén waves with convective cells, *J. Geophys. Res.*, **109**, A03305, 2004.

Sagdeev, R.Z., and A.A. Galeev, *Nonlinear Plasma Theory* (Benjamin, New York), 1969.

Sagdeev, R.Z., V.D. Shapiro, and V.I. Shevchenko, Excitation of convective cells by Alfvén waves, *Sov. Phys. JETP*, **27**, 340, 1978.

Shukla, P.K., *Nonlinear Plasma Science, Physica Scripta*, **T82**, 1, 1999.

Shukla, P.K., *Nonlinear Physics in Action, Physica Scripta*, **T113**, 1, 2004.

Shukla, P.K., Excitation of zonal flows by kinetic Alfvén waves, *Phys. Plasmas* **12**, 012310, 2005.

Shukla, P.K., R. Bingham, J.F. McKenzie, and W.I. Axford, Solar coronal heating by high-frequency dispersive Alfvén waves, *Solar Phys.*, **186**, 61, 1999.

Shukla, P.K., G.T. Birk, and R. Bingham, Vortex streets driven by sheared flow and applications to black aurora, *Geophys. Res. Lett.*, **22**, 671, 1995.

Shukla, P.K., and J.M. Dawson, Stimulated Compton scattering of hydromagnetic waves in the interstellar medium, *Astrophys. J.*, **276**, L49, 1984.

Shukla, P.K., and B. Eliasson, Low-frequency compressional electromagnetic waves in a nonuniform dusty magnetoplasma, *Phys. Lett. A*, **337**, 419, 2005.

Shukla, P.K., and A.A. Mamun, *Introduction to Dusty Plasma Physics* (Institute of Physics, Bristol), 2002.

Shukla, P.K., and L. Stenflo, Nonlinear Alfvén waves, *Physica Scripta*, **T60**, 32, 1995.

Shukla, P.K., and L. Stenflo, Nonlinear phenomena involving dispersive Alfvén waves, in *Nonlinear MHD Waves and Turbulence*, Eds. T Passot and P L Sulem (Springer, Berlin), pp 1–30, 1999a.

Shukla, P.K., and L. Stenflo, Plasma density cavitation due to inertial Alfvén wave heating, *Phys. Plasmas*, **6**, 4120, 1999b.

Shukla, P.K. and L. Stenflo, Generalized dispersive Alfvén waves, *J. Plasma Physics*, **64**, 125, 2000a.

Shukla, P.K., and L. Stenflo, Generation of localized density perturbations by shear Alfvén waves *Phys. Plasmas*, **7**, 2738, 2000b.

Shukla, P.K., and L. Stenflo, Comment on ion Larmour radius effect on rf ponderomotive forces and induced poloidal flow in Tokamak plasmas, *Phys. Rev. Lett.*, **85**, 2408, 2000c.

Shukla, P.K., L. Stenflo, and R. Bingham, Nonlinear propagation of inertial Alfvén waves in auroral plasmas, *Phys. Plasmas*, **6**, 1677, 1999.

Shukla, P.K., L. Stenflo, R. Bingham, and B. Eliasson, Nonlinear effects associated with dispersive Alfvén waves in plasmas, *Plasma Phys. Control. Fusion*, **46**, B349, 2004.

Shukla, P.K., and R.K. Varma, Convective cells in nonuniform dusty plasmas, Phys. Fluids B, **5**, 236, 1993.

Shukla, P.K., M.Y. Yu, H.U. Rahman, and K.H. Spatschek, Nonlinear convective motion in plasmas, *Phys. Rep.*, **105**, 227, 1984.

Shukla, P.K., M.Y. Yu, and L. Stenflo, Electromagnetic drift vortices, *Phys. Rev. A*, **34**, 3478, 1986.

Stasiewicz, K., P. Bellan, C. Chaston, C. Kletzing, R. Lysak, J. Maggs, O. Pokhotelov, C. Seyler, P. Shukla, L. Stenflo, A. Streltsov, and J.E. Wahlund, Small scale Alfvénic structure in the aurora, *Space Sci. Rev.*, **92**, 423, 2000.

Stefant, R.Z., Alfvén wave damping from finite gyroradius coupling to the ion acoustic mode, *Phys. Fluids*, **13**, 440, 1970.

Stenflo, L., Self-consistent Vlasov description of a magnetized plasma in a large amplitude circularly polarized wave, *Physica Scripta*, **23**, 779, 1981.

Stenflo, L., Stimulated scattering of large amplitude waves in the ionosphere, *Physica Scripta*, **T30**, 166, 1990.

Stenflo, L., Resonant three-wave interactions in plasmas, *Physica Scripta*, **T50**, 15, 1994.

Stenflo, L., Theory of stimulated scattering of large amplitude waves, *J. Plasma Phys.*, **61**, 129, 1999.

Stenflo, L., Comments on stimulated electromagnetic emissions in the ionospheric plasma, *Physica Scripta*, **T107**, 262, 2004.

Stenflo, L., and P.K. Shukla, Theory of stimulated scattering of large-amplitude waves, *J. Plasma Phys.*, **64**, 353, 2000.

Vincena, S., W. Gekelman, and J. Maggs, Shear Alfvén wave perpendicular propagation from the kinetic to the inertial regime, *Phys. Rev. Lett.*, **93**, 105003, 2004.

Wu, D.J., and J.K. Cho, J, Recent progress in nonlinear kinetic Alfvén waves, *Nonlinear Process. Geophys.*, **11**, 631, 2004.

Yu, M.Y., P.K. Shukla, and K.H. Spatschek, Alfvén wave excitation of the magnetostatic mode, *Phys. Fluids*, **24**, 1799, 1981.

Yu, M.Y., P.K. Shukla, and L. Stenflo, Alfvén vortices in a strongly magnetized electron-positron plasma, *Astrophys. J. Lett.* **309**, L63, 1986.

Part 4

Processes in the Solar-Terrestrial Environment

13 The Aurora

Bengt Hultqvist

The aurora has been studied for many hundred years but the detailed physical mechanisms giving rise to its various characteristics are to a large extent still poorly known. The physics of the aurora is basically plasma physics but many branches of physics are involved, such as interaction of energetic particles with matter and electrodynamics. Great progress in the physical understanding has been achieved in the last couple of decades by means of specially designed satellites placed in suitable orbits and provided with high data rates. This chapter gives a brief review of what is known about the auroral phenomena. In general, references are given only to the early papers for each specific subject and no completeness has been possible with the limited space available. A rough idea of what the chapter deals with can be obtained from the table of content.

Contents

Bengt Hultqvist, The Aurora.
In: Y. Kamide/A. Chian, Handbook of the Solar-Terrestrial Environment. pp. 333–354 (2007)
DOI: 10.1007/11367758_13

13.1 Introduction

The visual manifestation of the interaction between the hot magnetospheric plasma and Earth's upper atmosphere in the form of aurora is the most spectacular effect of the physical processes involved and it has played an important role in the development of the understanding of Earth's environment outside the lower atmosphere. For a very long time only observations by the naked eye were at the disposal of those pioneers, who tried to understand the light phenomena in the sky. Figure 13.1 shows an example of a beautiful auroral display photographed with a film with a similar spectral sensitivity as the human eye.

Several very old descriptions of aurora have been found in ancient Chinese and European written documents. Already in the early part of the eighteenth century the French scientist DeMairan discovered through extensive statistical investigations on auroral frequency of occurrence in Northern Scandinavia its connection with the long-term temporal variations in the occurrence of sunspots and he was aware of the connection between the almost total disappearance of sunspots and of aurorae during the Maunder Minimum ('little ice age') in the 17th century. In spite of DeMarian's observational results, it was not until the end of the 19th century that some physical ideas about the Sun-Earth connections were presented. Becquerel proposed that aurora is caused by the same kind of energetic particles as he had discovered (radioactive alpha particles) coming from the Sun. Goldstein launched a similar hypothesis, but his particles were electrons.

The Norwegian Birkeland started his famous series of terrella experiment in the last years of the 19th century. These experiments made it clear that the geomagnetic field plays an important role for the localisation of the aurora on Earth. A connection between the occurrence of aurora and geomagnetic disturbances had been found already in 1741 by Celcius and Hiorter. Birkeland imitated Earth by means of a small sphere containing a magnetic coil, which was covered with a substance that sends out light when it is hit by accelerated electrons in a vacuum chamber. Schuster pointed out in 1911 that a beam containing particles all charged in the same way would spread strongly because of mutual repulsion and Lindemann suggested a few years later that the particle beam from the Sun is electrically neutral, i.e. a plasma beam. After that short history review we go to the observations of the aurora and their interpretations.

13.2 Geographical Distributions

The auroral zone. That the probability for occurrence of aurora increases with increasing latitude was known in Europe by the old Greeks and probably earlier, but that the occurrence probability does not increase all the way to the pole seems to have been known first by Muncke in the 1830s. Fritz (1881) published a book on the aurora, where he collected practically all recorded observations of aurora from several hundred years before Christ and identified the northern hemisphere 'auroral zone' for the first time. The central curve of the auroral zone connects those points on Earth where the probability of seeing aurora is the largest. All observations were made in the night. Whereas for the first auroral zone Fritz used all observations of aurora wherever it was in the sky, one has later used only observations in the zenith as the basis for the auroral zone. If the width of the zone is defined as the distance between the latitudes where the observation probability is 50% of that at the center curve, one finds it to be some 12 latitude degrees, i.e. about 1300 km. The auroral zone has the shape of an oval roughly centred on the geomagnetic pole at a distance of about 23°. The deviation from a circle has been found to be fully due to the deviations of the geomagnetic field from a dipole field (Hultqvist, 1959).

The auroral oval. The northern and southern auroral zones are statistically defined from visual observations of aurora during the night and do not describe the instantaneous distribution of aurora over the globe. The data collected during the International Geophysical Year 1957–58 (IGY) made it possible for Feldstein and co-workers (Feldstein 1960; Feldstein and Solomatina, 1961; Feldstein and Starkov, 1967) to determine the auroral occurrence probability also on the dayside of Earth and identify the full auroral oval. In the middle of the day the aurora occurs at about 80° magnetic latitude as compared with some ten degrees further equatorwards at midnight. On the night-side of Earth the auroral oval mostly coincides roughly with the auroral zone geographically.

As shown in Fig. 13.2, the size of the oval depends on the disturbance level of the geomagnetic field (in-

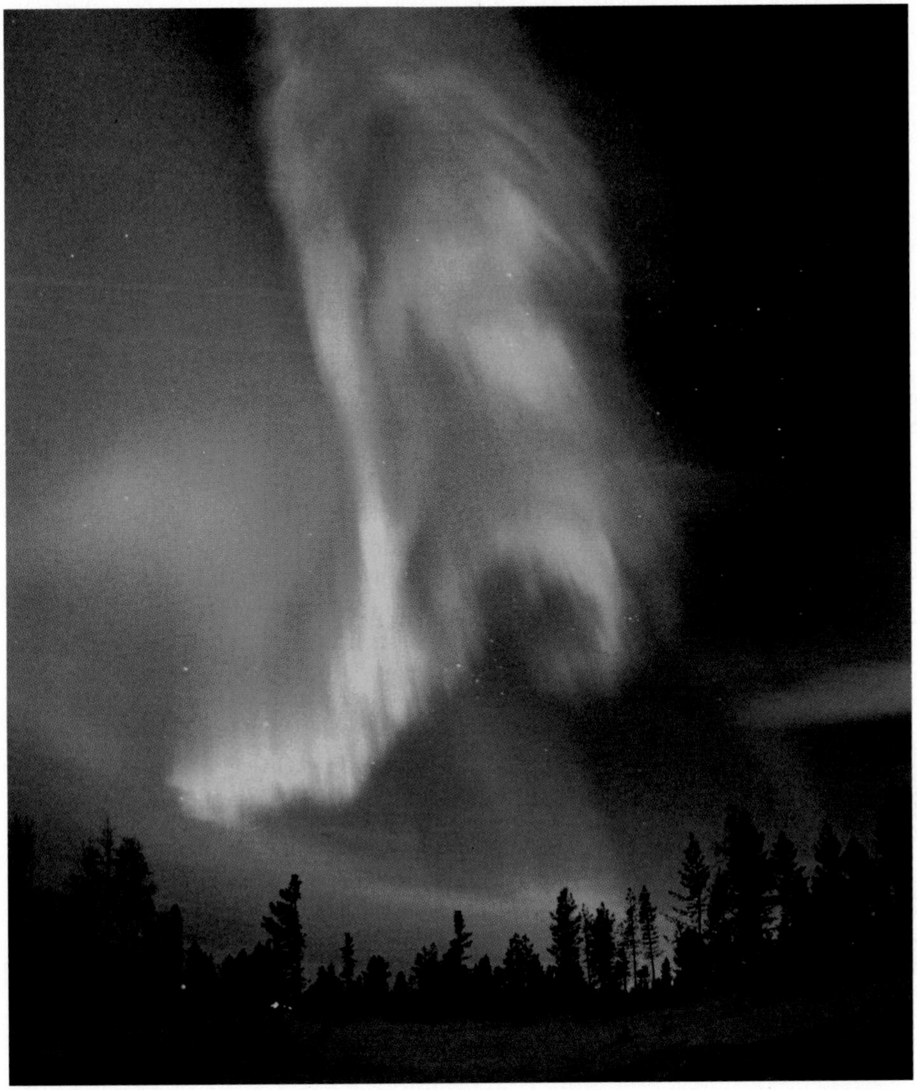

Fig. 13.1. An example of a dramatic auroral display in the break-up region of an auroral substorm. What is not seen in the picture is the rapid variations in time and space that characterize auroral break-ups (courtesy of Torbjörn Lövgren, Swedish Institute of Space Physics)

dex Q, which indicates the magnetic activity at high latitudes - value between 0 and 11 - with a time resolution of 15 minutes). In the very strongest magnetospheric storms, of which there usually are a few in each solar 11-year cycle, the auroral oval widens more than shown in Fig. 13.2, to the extent that it reaches the European continent and disappears from northern Scandinavia.

Polar cap aurora. Aurora does not occur only in the auroral ovals but also in the polar cap, polarward of the oval. This is illustrated in Fig. 13.3, which contains a mass-plot of discrete auroral arcs, observed in the period 1963–1974 in Greenland for the AE interval 41 - 50 nT. As may be seen in the figure, the arcs in the central polar cap are roughly aligned in the noon-midnight direction. They are therefore referred to as 'Sun-aligned arcs', 'high-latitude auroras', 'polar cap arcs', 'theta aurora' or 'transpolar arcs'. The term 'polar cap aurora' implies that the aurora is on open magnetic field lines, but it is not known if that really is the case.

As shown in Fig. 13.3, there are more Sun-aligned arcs on the morning side of the polar cap than on the evening side, and the morning-side arcs generally move

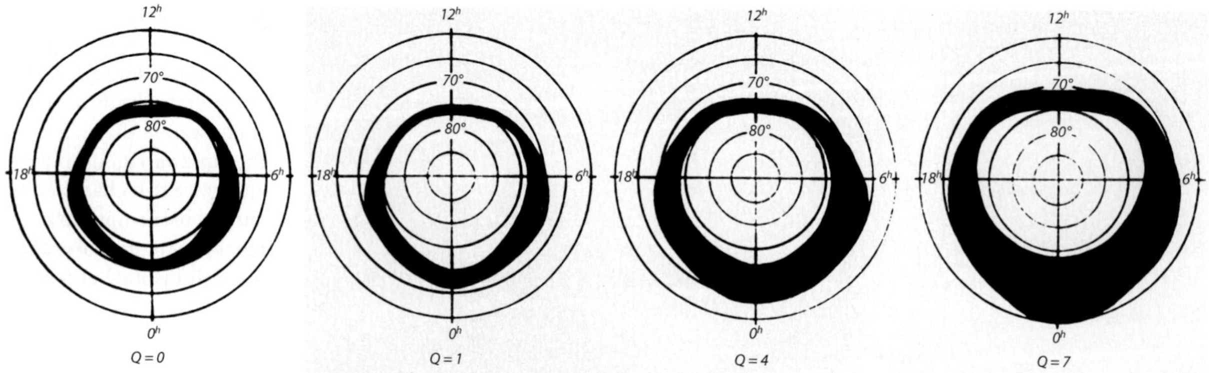

Fig. 13.2. The Feldstein auroral oval in magnetic local time and corrected geomagnetic coordinates and its variation with magnetospheric disturbance level in the northern hemisphere (after Feldstein and Starkov, 1967)

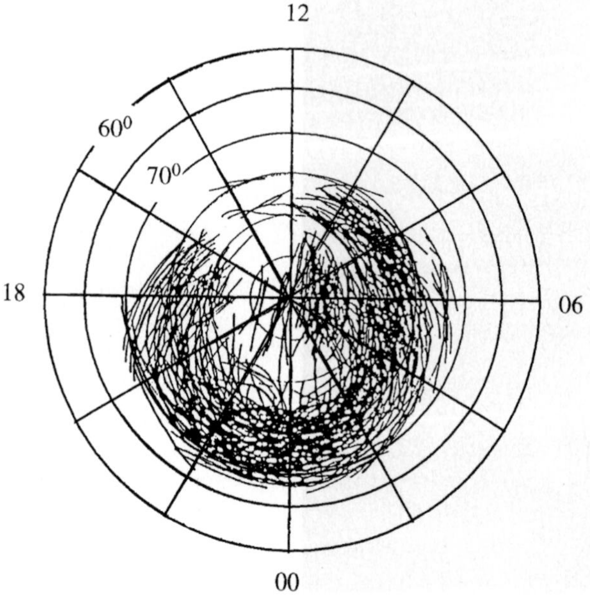

Fig. 13.3. Mass plot of discrete auroral arcs observed in the period 1963–1974 in Greenland, in corrected geomagnetic coordinates for fairly low magnetospheric disturbance level (AE = 41 – 50 nT; after Lassen and Danielssen, 1989)

pole-ward until they disappear at the centre of the polar cap. Each arc corresponds to a local shear convection flow on the polar cap, which suggests that local converging electric fields, similar to those in the night-side auroral oval, accelerate electrons downward (Shiokawa et al., 1996).

The polar-cap aurora occurs when the magnetosphere is quiet, i.e. when the interplanetary magnetic field (IMF) is northward. As soon as a substorm starts, the polar-cap aurora disappears (see e.g. Hultqvist, 1974, Fig. 25). Why and how this happens is not clear, and there are many other open questions about polar-cap auroras, such as

- where are the source regions of the arcs?
- what causes the converging electric fields (the V-shaped electric potential surfaces)?
- why do the arcs move polarward?

Dayside aurora. The dayside aurora has become possible to be observed from ground even at noon by means of observatories established on Spitzbergen in the Svalbard archipelago, which are located close to the northern geomagnetic cusp region near noon and provide unique possibilities for optical observations of dayside aurora. But it can also be recorded in the UV spectral range from satellites (see Fig. 13.4).

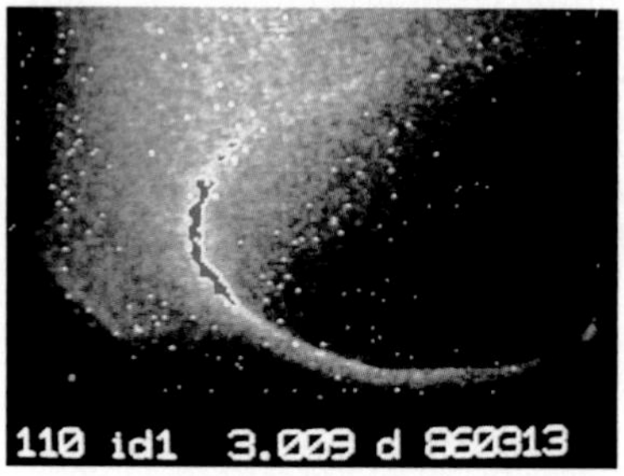

Fig. 13.4. Strong dayside aurora observed in the UV by means of the camera on board the Swedish Viking satellite on 13 March 1986

There are generally no strong auroral emissions near local noon, and the noon region of the auroral oval is therefore generally termed the 'midday auroral gap'. There exist, however, always some low-level emissions (generally red) also there. More intense, structured green emissions mostly occur at higher latitudes at earlier and later local times. These emissions are produced by electrons accelerated by magnetic-field-aligned electric fields (Sandholt et al., 1993). In Fig. 13.5 there are two latitudinally separated cusp region auroral forms (types 1 and 2 in the figure). Type 1 aurora is associated with electron precipitation from the boundary region between the cusp and the low-latitude boundary layer (LLBL), whereas type 2 comes from the cusp/mantle region. Type 3 aurora is diffuse and observed equatorward of type 1. It is generally green and is caused by electrons drifting in from the night-side (plasma sheet). Types 4 and 5 aurorae form broad latitudinal regions of multiple arcs with both red and green emissions. The direction of motion of the poleward-moving auroral forms (PMAFs) depends on the sign of the B_y component of the IMF, as indicated in Fig. 13.5.

The characteristic energies of the aurora-generating electrons on the dayside are generally lower (100 – 500 eV) than on the nightside (several keV). The peak energy flux has been found to be of the order 10 mW/m^2 in the cusp region (Sandholt et al., 1989). According to Newell and Meng (1992) the aurora-generating electron precipitation into the cusp region (100 km by 1500 km) amounts to 1–4% of the precipitation in the total hemisphere.

Stable auroral red (SAR) arcs. In strongly disturbed conditions so called SAR arcs are observed at quite low L-values (2.7 – 4). They are caused by the increased fluxes of ring current ions within the plasmasphere, where the cold plasma amplifies precipitation of ions into the upper atmosphere because it amplifies the ion-cyclotron instability, which causes scattering of ions into the loss cone. The amplified waves heat the electrons in the ionosphere so much that they can excite the 630 nm oxygen triplet. Cold plasma densities of the order of 10^3 cm^{-3}, seem, however, to be needed for the instability to work at keV energies, and the unstable region at these energies is therefore located well inside the plasmapause during quiet and moderately disturbed conditions. It does not vary its location or extension significantly except during magnetic storms. The process appears to work more or less continuously at a low level and to be strongly intensified during storms (Hultqvist et al., 1976).

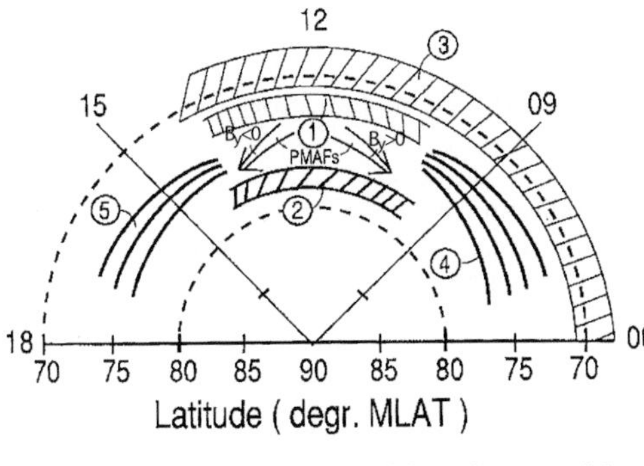

Fig. 13.5. Schematic overview of dayside auroral forms provided by P.E Sandholt (after Fig. 3.10 in Hultqvist et al., eds., 1999)

13.3 Spectrum, Optical Intensity and Power

Emission spectrum. The greenish colour that characterizes most aurorae in the nightside auroral oval is due to the so called 'auroral' line with a wavelength of 557.7 nm. This wavelength is close to that where the human eye is most sensitive; that is the reason for the dominance of the green line in the visual impression. The first wavelength determination was made by Ångström in the late 1800s. The emission is due to a forbidden transition with a lifetime of 0.9 s in atomic oxygen and it is difficult to produce in the laboratory. Only when advanced vacuum spectrographs became available in the first half of the 20th century, the auroral line was produced and investigated in the laboratory together with the so called coronal lines of the Sun.

The red colour of very strong auroral events (so called 'high red aurora' or 'red aurora of type A'), which generally reach lower latitudes than the green aurora, is also due to emissions from atomic oxygen (spectral triplet 630.0, 636.4, and 639.1 nm). They are also forbidden with a very long lifetime of 110 s. The reason for the red emissions to be strongest at great altitudes (>200 km) is that collisions deactivate the excited oxygen atoms before they radiate at lower altitudes. The same kind of effect for the 557.7 nm line limits the extension downward in altitude to about 100 km, but the decrease downward of the green line is also due to the fast decrease of the number density of atomic oxygen below 100 km. Aurora at altitudes below 100 km is generally red ('red aurora of type B'). This colour originates mainly in permitted emissions from nitrogen molecules. The auroral spectrum contains a large number of atomic lines and molecular bands in addition to those mentioned above. The reader is referred to Omholt (1971) or Vallance Jones (1974) for more information about the optical spectrum of the aurora.

Energetic electrons precipitating into the upper atmosphere is the dominating cause of aurora but sometimes the Balmer lines of the hydrogen atom (Hα at 656.3 nm and Hβ at 486.1 nm) are seen in the spectra. They were discovered by Vegard in 1939. Meinel could show a decade later, through analysis of their line shape, that protons enter the upper atmosphere with original velocities of at least 4000 km/s. Larger displacements of the Balmer lines than corresponding to this velocity is not seen in the spectra because the protons at higher velocities do not attach an electron and form a hydrogen atom. A proton with an initial energy above 90 keV on average attaches an electron and loses it again some 700 times during the braking in the atmosphere and emits about 10 Hβ photons.

The hydrogen emissions are generally hardly visible for the naked eye. They tend to be displaced equatorward relative to the ordinary aurora in the evening and poleward in the morning.

Optical intensity and power. The aurora is one of the most intense light phenomena in the sky. This is demonstrated in Table 13.1, where some characteristic flux values from various radiation sources are shown.

Table 13.1. Comparison of intensities of various radiation sources in the sky

Source Power	mW/m^2
The Sun	1.4×10^6
The full moon	3000×10^{-3}
Strong aurora	1000×10^{-3}
Total star light	1.8×10^{-3}
The airglow	16×10^{-3}
Lyman-α	10×10^{-3}
UV sources in night sky (123 – 135 nm)	0.1×10^{-3}
Cosmic radiation	4×10^{-3}

The value for aurora in the table is for a strong one. The intensity varies over a very wide range. Whereas the very strongest aurorae, which occur very rarely, can be even stronger than that in the table, the majority are much weaker and the whole scale down to the level of the airglow occurs.

The aurora is often grouped into four classes with regard to intensity: IBC I-IV. IBC is the acronym for 'International Brightness Coefficient'. The four classes are defined in terms of the number of photons of the 557.6 nm auroral line as specified in Table 13.2.

An emission rate of 10^6 photons per second in a column with a cross section of 1 cm^2 and going to infinity in the direction of view is named 1 Rayleigh (1R).This unit is not limited to the wavelength 557.7 nm as the IBC classes are.

The amounts of energy involved in the auroral phenomena are quite considerable, globally of the order of 10^{11} W on average. For the very strongest global auroral events, lasting an hour, this corresponds to the energy released in the explosion of a large hydrogen bomb. Practically all energy supply to the aurora is in the form of precipitation of energetic electrons and ions. Of the supplied particle energy only 0.2–1% is transformed into visible auroral light. Most of it is used for ionising the upper atmosphere. About 50 ionizations (each using on average 35 eV) occur for each emission of a photon with the wavelength 391.4 nm (3.2 eV energy). Expressed in a different way, a particle energy flux of about 3 mW/m^2 (i.e. 2×10^{12} eV/cm^2 s, or for an average electron energy of 2 keV, 10^9 electrons per cm^2 s) is required to cause the 10^9 photons per cm^2 (column) and second in an aurora of intensity 1 kR in the 391.4 nm band. A major part of the supplied particle energy finally ends up as heat in the upper atmosphere. The lower limit of particle energy

Table 13.2. International Brightness Classes (IBC) in terms of kiloRayleighs

IBC	Number of photons per cm^2 (column) and sec.	Number of kiloRayleigh, kR
I	10^9	1
II	10^{10}	10
III	10^{11}	100
IV	10^{12}	1000

flux required to produce aurora visible to a well-adapted eye is of the order of 1 mW/m^2.

13.4 Auroral Forms and Structuring

The aurora has an enormous richness of structures in space and time. The spatial scales go from global to sizes of the order of 100 m. The temporal scales also vary widely from hours – or from the solar cycle period of 11 years if you want – to fractions of seconds in blinking, shimmering or twisting auroral forms with detailed small structures running fast along the large-scale forms. It is certainly this richness in shapes and temporal variations that contributes most to peoples' fascination in the aurora, and they are also the most important characteristics that are not at all understood yet.

Auroral forms. Auroral forms may contain ray structure or they may not. To those containing ray structure belong isolated rays, draperies and corona, and those without are called homogeneous arcs, homogenous bands and diffuse and pulsating surfaces. Corona is seen only at high latitudes, where the magnetic field lines are almost vertical and the rays of the corona seem to come from the same point in the direction of the magnetic field lines.

A number of parallel homogeneous arcs recorded by an all-sky camera and the corresponding measurements of the energetic electrons causing the arcs on board the FAST satellite are shown in Fig. 13.6.

An arc and a band, folded at one end, can be seen in Fig. 13.7. A strongly rayed drapery can be seen in Figs. 13.8 and 13.9 contains a high red aurora with green aurora at lower altitudes. A form of aurora not identified from ground is shown in Fig. 13.10. It is a string of lightspots at such a distance that only one of the spots at a time can be seen from ground. It therefore took a satellite camera to see the whole structure. The intense spots are separated by less than one hour in local time on the night-side and by about an hour and a half on the day-side. Each spot varies in intensity with time but is fairly fixed in location. The spots are most likely folds or surges, unresolved by the image, and presumably produced by velocity shear, which gives rise to Kelvin–Helmholtz instability and increased electron precipitation (Lui et al., 1987).

Altitude distribution. The lower limit of the aurora is generally located at about 100 km altitude. A very large number of altitude measurements were made in Norway by Störmer, Vegard and Harang in the first half of the 1900s. The lower border altitude depends on the intensity of the aurora. Whereas the lower border was found by Harang (1944) to be located at 114 km for weak aurorae, the corresponding value was 95 km for very strong aurorae. The difference between different auroral forms with regard to the lower altitude limit is demonstrated in Table 13.3. The lowest altitude reported of the lower limit is 65 km.

Whereas arcs and bands have an extension in altitude of a few tens of km, rays may have an altitude extension of several hundred km. There are observations of aurorae, which reach above the shadow of Earth and thus are located in the sunlit part of Earth's upper atmosphere. They have in rare cases been found to reach 1100 km altitude.

The intensity of aurora varies with altitude in a regular way as shown in Fig. 13.11. It is the average variation of intensity with altitude for a number of different auroral forms that are given there. All forms except rays have the peak intensity less than 10 km above the lower limit. The curve for rays is completely different from the others. Why this is so is still a mystery.

Motions and periodic variations. Entire auroral forms frequently move in roughly east-west direction as noted first by Tromholt before 1885. Störmer was the first to make systematic investigations of drift motions in aurora. The velocities measured in visual aurora are

Table 13.3. Average altitude of the lower limit of the aurora. The altitude is given in km and is followed by the number of measurements within parenthesis

Auroral form	Southern Norway Störmer (1953)	Störmer (1946, 1948)	Northern Norway Vegard and Krogness (1920)
Homogeneous arcs	104 (41)	107 (120)	109 (355)
Arcs containing rays	101 (32)		107 (888)
Homogeneous bands	94 (10)		
Pulsating arcs	88 (4)	103 (430)	
Pulsating surfaces	93 (19)		106 (160)
Diffuse surfaces	88 (46)	94 (158)	
Rays	127 (52)		113 (61)
Bands with ray-structure	103 (126)	86 (405)	110 (409)
Draperies	104 (64)		

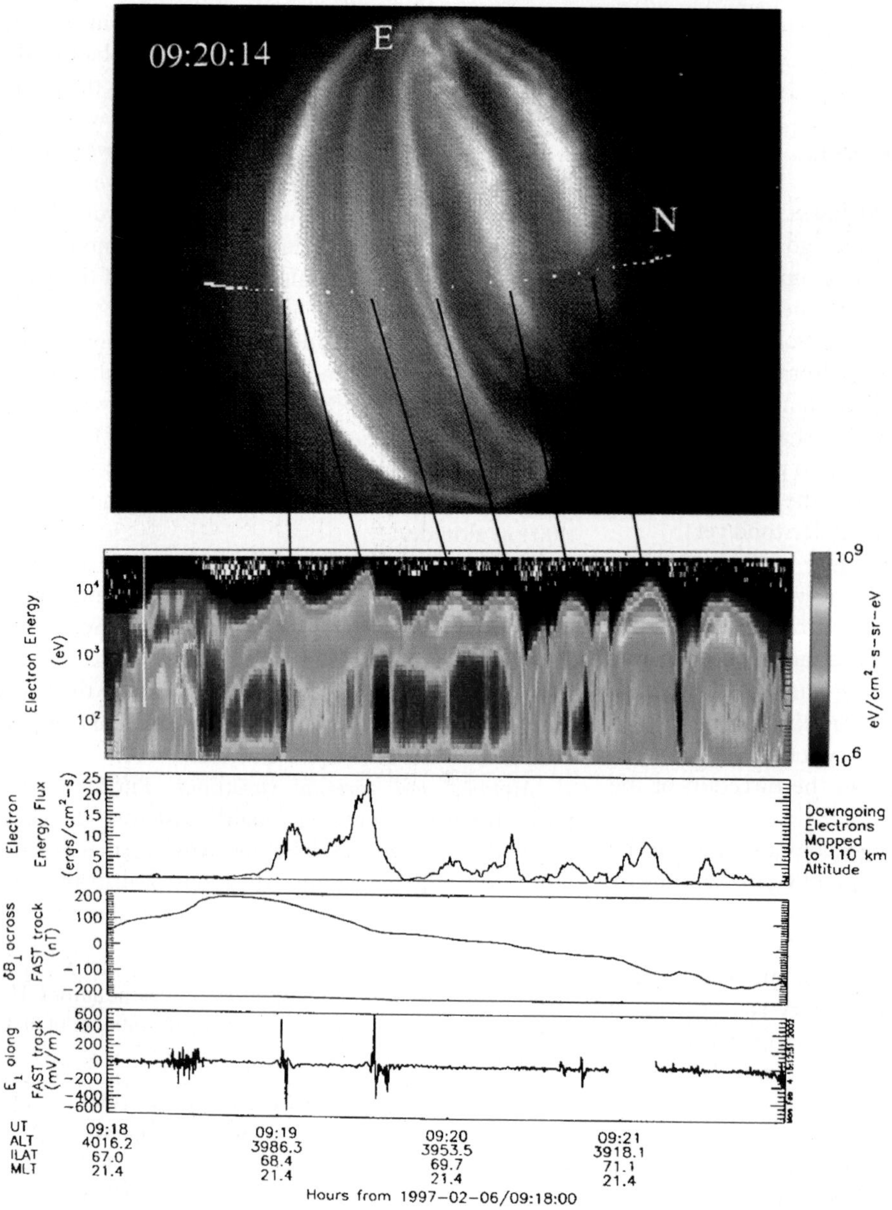

Fig. 13.6. Several parallel arcs recorded by an all-sky camera on an aircraft when the FAST satellite passed (location mapped to 110 km indicated by the *white dots*). The *second panel* shows the electron energy spectrum measured by FAST. The *third panel* gives the precipitated energy flux on a linear scale mapped to 110 km altitude. The *two lowest panels* show associated disturbances in the magnetic and electric fields. The satellite passage had a duration of 4 minutes and there were some variations in the arcs in that period (after Stenbeck-Nielsen et al., 1998)

generally higher than the drift velocities observed in the ionosphere. Speeds above 120 km/s for rays in bands have been observed (Davis and Hicks, 1964), which is higher than the speed of sound. It is therefore clear that the drift motions seen in auroral forms generally is the motion of the electron beams generating the aurora rather than the motion of the ionospheric plasma.

A daily variation has been observed in these drift motions of auroral structures. Most aurorae move westward well before midnight and eastward on the morning side with overlap of the west and east directions in a rather wide interval around magnetic midnight. This pattern of motion agrees with the average pattern of convection of the plasma, which is caused by the interaction of the solar wind and the magnetosphere. (However, the fast and often irregular motions in dynamic aurora are not related to the plasma flow and are often much faster.) In addition to drift motions there are sometimes varia-

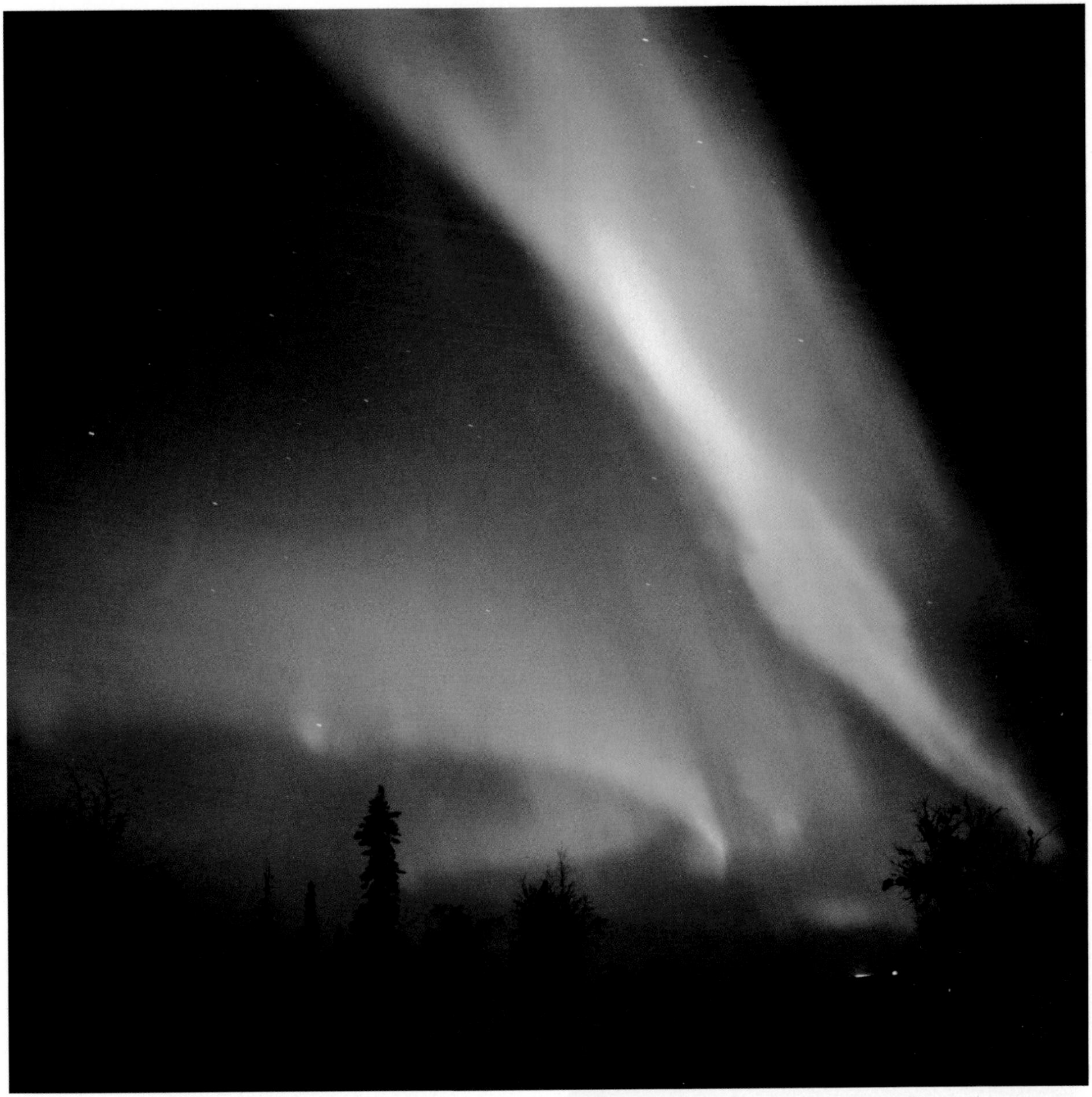

Fig. 13.7. A folded auroral band with some ray structure (courtesy Torbjörn Lövgren, Swedish Institute of Space Physics)

tions in arcs and bands, which look like wave motions. Another type of variation is the so-called pulsating aurora, in which e.g. a quiet arc disappears and returns in the same region periodically with a period of a few seconds.

A lot of investigations have been devoted to the daily variations of the occurrence of aurora. It is, however, quite a complex question, which depends on geographical position, probably also on season and phase of the solar cycle and possibly on the type of aurora. The only fairly clear tendency in the older studies from ground on the night-side seems to be a maximum in the probability of auroral occurrence around magnetic midnight. But the satellite observations, on the other hand, indi-

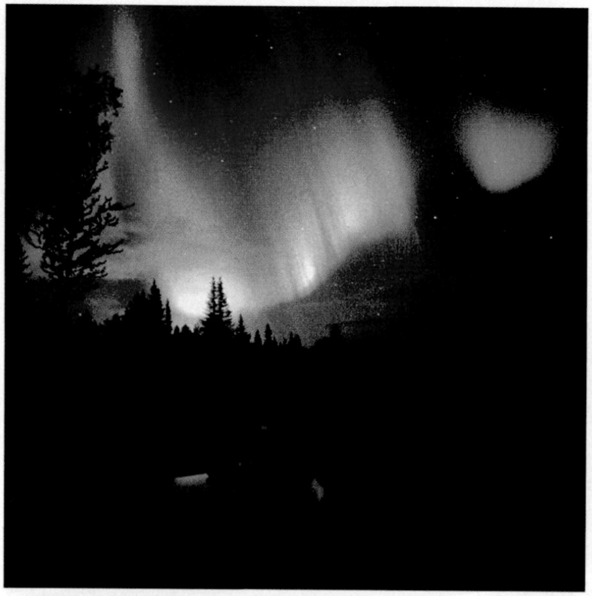

Fig. 13.8. Strongly rayed auroral drapery (courtesy Torbjörn Lövgren, Swedish Institute of Space Physics)

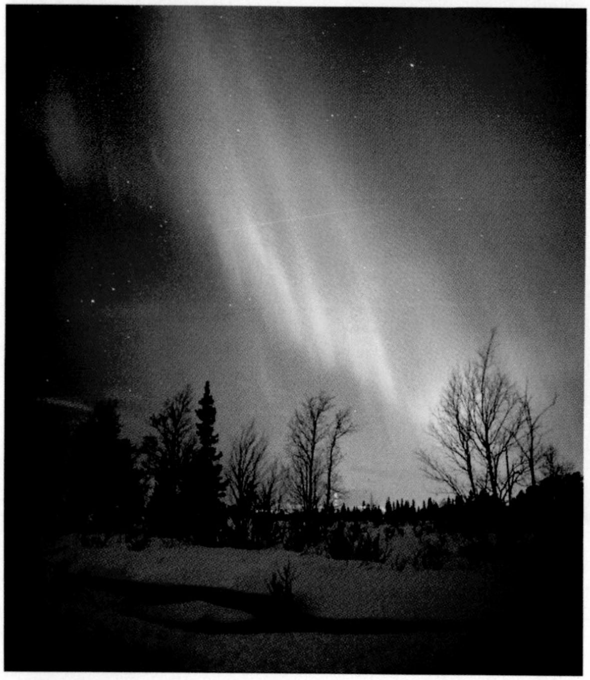

Fig. 13.9. High red aurora with green emissions at the bottom (courtesy Torbjörn Lövgren, Swedish Institute of Space Physics)

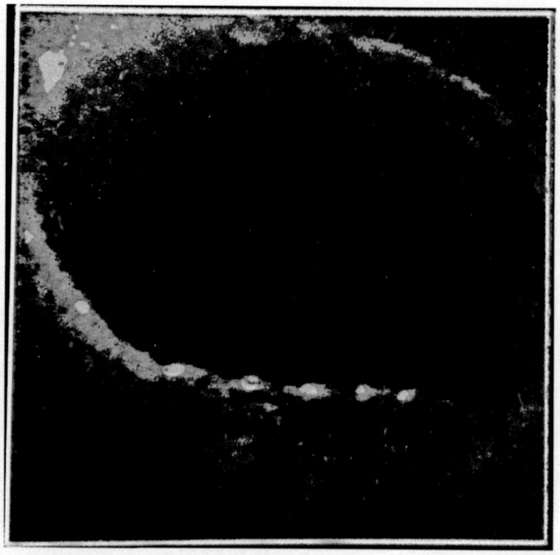

Fig. 13.10. Pearl-string aurora along the night-side auroral oval observed by means of Viking (courtesy J.S. Murphree)

cate that the electron and ion precipitation is more permanently present in the region around the geomagnetic cusp on the dayside, although with less energy flux, than on the night-side (see e.g. Hulqvist and Lundin, 1987). So the question of daily variation has to be very well specified before trying to answer it.

13.5 Auroral Substorms and Storms

Auroral substorms are a part of magnetospheric substorms, which show up in practically all plasma and field variables in the magnetosphere. The term substorm was suggested by Chapman (see Akasofu, 1968, the Foreword). His student, S.-I. Akasofu (1964) succeeded in describing the auroral substorm on the basis of the very large number of pictures which were taken from many all-sky cameras – each covering a circle of the sky with a diameter of the order of 500 km – during the IGY. The figure of Akasofu (1964), which still holds in many respects, summarizes the different phases of a substorm (see Fig. 13.12).

The first indication of an auroral substorm according to Akasofu is a brightening of a quiet arc near the low-latitude limit of a quiet aurora, or a sudden formation of an arc if there was none (B in Fig. 13.12). In most

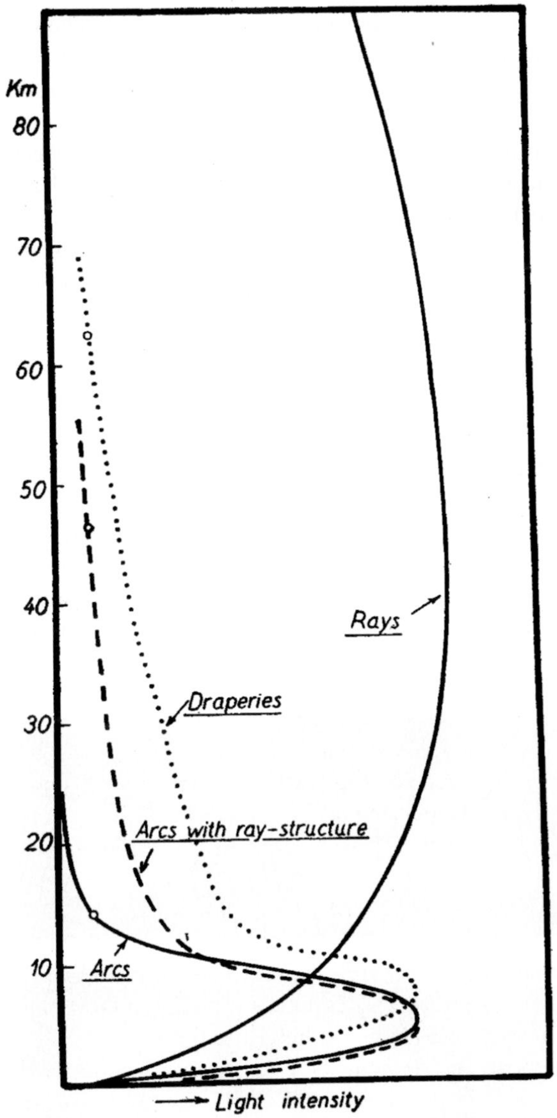

Fig. 13.11. The mean variation of light intensity in the vertical direction for different auroral forms (after Harang, 1951)

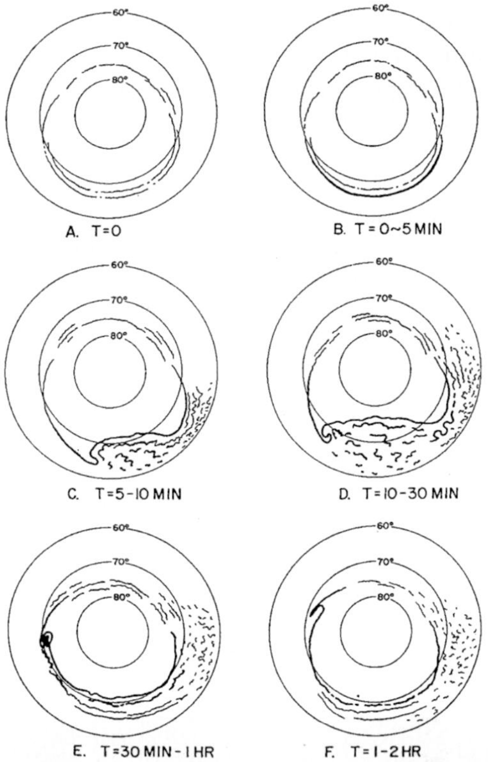

Fig. 13.12. Schematic diagram to show the development of auroral substorms (after Akasofu, 1968)

cases this is followed by a rapid pole-ward expansion of the auroral region with new arcs forming at higher latitudes, resulting in an auroral bulge in the midnight sector (C in the figure). The auroral 'break-up', with very dynamic aurora of high intensity, rapid motions and variations and frequently many different colours, occurs within the auroral bulge, (see e.g. Fig. 13.1), which expands in all directions (D in the figure). In the evening sector a large-scale fold appears, which expands west-ward mainly by new folds forming further west and the old ones disappearing. This process is often called 'the west-ward-travelling surge' (WTS). In the morning-side arcs disintegrate into 'patches' and move eastward with typical speeds of the order of a km per second. When the expanding bulge has attained its highest latitude and starts to contract, the recovery phase of the auroral substorm begins (E in Fig. 13.12). The WTS may continue to expand west-ward considerable distances during the recovery phase of the substorm but it degenerates eventually into irregular bands. East-ward moving patches remain in the morning sector until the end of the recovery phase (F in Fig. 13.12). At the end, the general situation will be similar to that just before the onset of the substorm.

Later research in the space era has shown that substorms have a well-defined initial "growth phase" before the expansive phase. One has also found that the expansion of the auroral region is generally not associated with the individual auroral forms moving but with new forms appearing. For a more complete discussion of the auroral substorm and its relations to other aspects of the magnetospheric substorm the reader is referred to the special chapter on substorms in this book.

The auroral manifestation of the magnetospheric substorm is the most spectacular one. But the substorm shows up in practically all aspects of the magnetosphere (as mentioned above). The magnetosphere maintains its energy and mass content within certain limits primarily by means of the kind of macroscopic relaxation instability (i.e. an instability affecting more or less the whole system, which returns it to a lower level of total energy) that the magnetospheric substorm is.

The storm-time aurorae extend to lower latitudes (sometimes much lower) and withdraws from the normal auroral latitudes. They are frequently so called high red aurora of type A but may also contain much green emissions as well as red emissions of type B (lower border). They are often very intense, variable and dramatic and extend over many hours of several nights. Whether storm-time aurora can be described as the superposition of substorms, or contains something in addition, is still not clear.

Aurorae are effects of a complex interaction between the magnetosphere and the ionosphere, which has been studied intensively in recent years mainly by means of satellite measurements. We shall here not go into this field but only indicate, by means of Fig. 13.13, that the auroral processes, involving plasma sources and acceleration, affect large parts of the magnetosphere.

13.6 Auroral Electrodynamics and Energetic Particle Precipitation

As indicated in Fig. 13.13, the aurora is an element in a complex electrodynamical system that constitutes the magnetosphere. In the last decade or two much research has been devoted to the electrodynamics

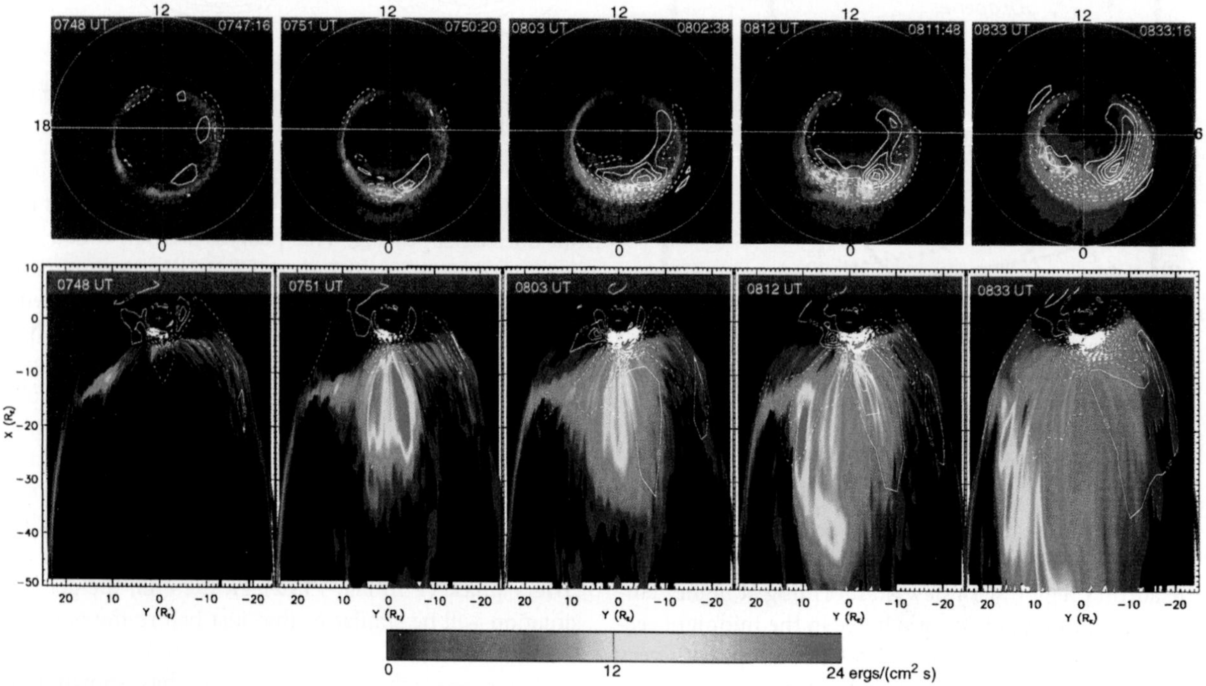

Fig. 13.13. *Top*: Distributions of field-aligned currents and auroral images over the northern hemispshere during a substorm expansion phase (on 9 January 1997). *Bottom*: Corresponding equatorial maps of field-aligned currents and auroral emissions. *Solid contours* represent currents that flow into the ionosphere and *dashed contours* currents flowing out of the ionosphere. The contour interval is 0.3 $\mu A/m^2$ for the ionospheric maps and 0.1 nA/m^2 for the equatorial maps (after Lu et al., 2000)

of the magnetosphere. In this section some results directly related to the aurora will be briefly summarized.

Connections with current systems. Detailed investigations have demonstrated that the primary auroral electrons, precipitated into the upper atmosphere from the magnetosphere, carry the upward-directed magnetic field-aligned current observed to disturb the geomagnetic field. The downward directed return currents are found outside of the auroral form and they are carried by upward-moving electrons (see Paschmann et al., 2003, Sect. 4.2).

Magnetic field-aligned electric fields. The dispute about the existence of an electric field component directed along the magnetic field lines, originally proposed by Alfvén, has been definitely settled by observations in space. Figure 13.14 shows a schematic of Earth's auroral magnetosphere-ionosphere coupling circuit. The region with a parallel electric field is generally found in the altitude range 5000 – 10,000 km. It accelerates electrons downward into the atmosphere. The reason for the acceleration is generally thought to be low density of current carriers in the altitude range mentioned, which, in combination with the downward converging magnetic field that causes the majority of the electrons to mirror and move outwards before they reach the ionosphere, causes current continuity problems. A part of the voltage available in the circuit is therefore applied to the low-density region, thereby accelerating the electrons downward (and ions upward) to the extent that the current can be carried through that low-density region.

The detailed nature and location of the generator is not known, but it is believed to be somewhere in the direct-interaction-region between the solar wind and the magnetosphere.

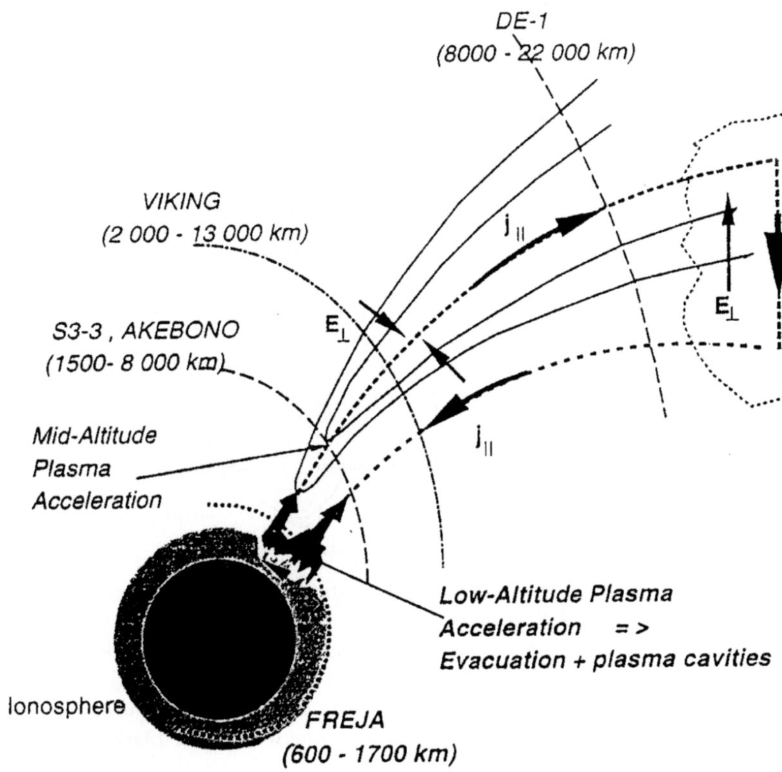

Fig. 13.14. Schematic of Earth's auroral magnetosphere-ionosphere coupling circuit, indicating the dynamo region, the ionosphere and aurora region, and the region in between, where many controlling mechanisms are located. The magnetic field lines are equipotentials at high altitudes but not in the acceleration region (after Lundin et al., 1995)

The kind of V-shaped potential surfaces indicated in Fig. 13.14 are associated with opposite horizontal plasma convection on the two sides of the centre line. In situations where shear convection is produced, e.g. by the direct interaction with the solar-wind, the shear motion may produce potential distributions of the shape shown in Fig. 13.14 and electrons may be accelerated into the ionosphere producing aurora. Similar effects are expected at the magnetopause where, in combination with Kelvin–Helmholtz instability, the pearl-string aurora, shown in Fig. 13.10, may be formed.

A result of satellite investigations in the last decade is that an $E_{||}$ is involved also in closing the return current outside of the auroral form, in this case pointing downwards. This was indicated first by Viking and confirmed by the Freja spacecraft, from the measurements of which the potential distribution shown in Fig. 13.15 was derived (Marklund et al., 1998). The region with a downward $E_{||}$ appears to be associated with dark regions (black aurora) within aurora reported earlier from ground based observations (Roywik and Davis, 1977; Trondsen and Cogger, 1977; Kimball and Hallinan, 1998a,b). In Fig. 13.15 an auroral picture that fits very well to the potential schematic below it can be seen. The black filament in the middle is very clear.

Diffuse aurora. The generation of diffuse aurora does not involve acceleration in parallel electric fields. Energetic electrons from the plasma sheet in drift orbits around Earth are pitch-angle scattered into the loss cone and stopped in the atmosphere, producing aurora. Although this mechanism sounds simple, it is associated with one of the unsolved problems of auroral physics: what causes the pitch-angle scattering?

In the 1970s it was believed that the cause of the scattering were strong electrostatic waves observed by means of the OGO 5 satellite (Kennel et al., 1970). With newer wave instruments the amplitude decreased and careful analysis by means of GEOS-1 and -2 demonstrated that the wave amplitudes were too low to cause the pitch-angle scattering that gives rise to diffuse aurora (Belmont et al., 1983).

Electrodynamic models of auroral forms. Much research about the relationship between auroral forms and electric fields, currents and ionospheric conductivity patterns in and around the forms has been conducted in the last two decades, using ground-based

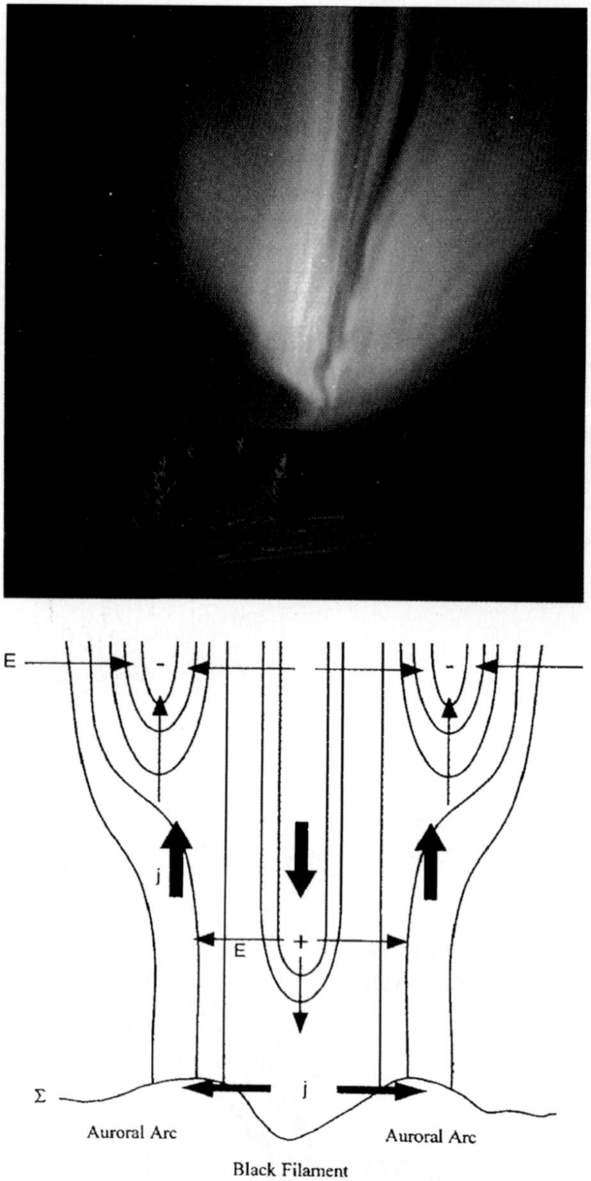

Fig. 13.15. Photograph of two parallel auroral arcs, separated by a dark region referred to as black aurora (courtesy Torbjörn Lövgren, Swedish Institute of Space Phyics). The *bottom part* shows the electric field and current configuration for a double auroral arc system, with the inverse configuration characteristic of black aurora in between (after Marklund et al., 1998)

measurement techniques (all-sky cameras, magnetometer and radar networks, and incoherent scatter radars) in combination with in-situ measurements with spacecraft-born instruments. For a complete review of the results obtained, the reader is referred to Chap. 6 in Paschmann et al., eds., (2003). Here only a brief summary of some aspects closely related to a few auroral forms will be given.

Most studies have been made of stable, discrete auroral arcs. Depending on (1) the magnitude and direction of the background convection electric field (of magnetospheric origin), (2) the relationships between the ionospheric conductivities inside and outside of the arc, and (3) the magnetic field-aligned currents (FAC) associated with the arc, quite a variety of variations of the electric field when passing from one side of the arc, through the arc to the other side are expected (Marklund, 1984) and many of them have been observed (Marklund et al., 1982, 1983). The potential difference at great altitudes above the auroral forms is generally much larger than in the ionosphere, as indicated already by the schematic in Fig. 13.14.

The simple models have been compared with one exceptionally complete set of measurements, when the FAST satellite passed through the magnetic field-aligned Tromsø beam of the EISCAT incoherent scatter radar system at the time when a quiet auroral arc slowly drifted through the antenna beam (around 19 UT on February 19, 1998).

A summary of the observational results is shown in the schematic in Fig. 13.16. The field-aligned currents are shown at the top (auroral primary electrons to the left and the return-current electrons to the right). The conductivity is enhanced within the arc and the electric field is depleted. The electric field has its maximum value just outside the arc. The optical features depend strongly on the electron flux at the upper end of the energy spectrum, and as the conductivity and electric field are modified by the precipitating electrons, there is a feedback on the magnetospheric source plasma. The auroral arc is thus a fundamental element of the interaction between the magnetosphere and the ionosphere and the ionosphere is an active – not only passive – part of the electromagnetic system of the magnetosphere.

The westward travelling surge (WTS) is the poleward boundary on the evening side of the auroral substorm bulge (see C in Fig. 13.12). Extensive synoptic investigations of the electrodynamic characteristics of WTSs have been carried out and are summarized in Sect. 6.2 of Paschmann et al., eds., (2003). A model has been designed by Opgenoorth et al. (1983) by deter-

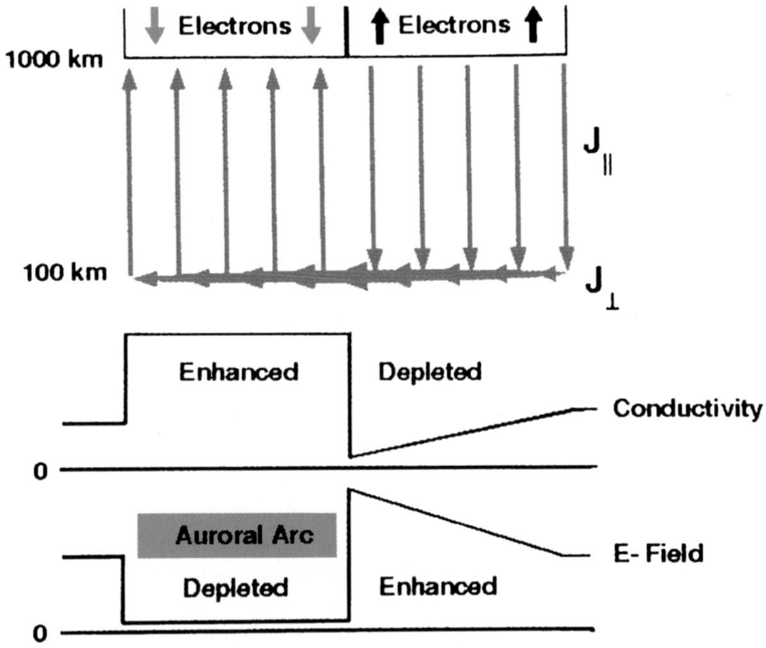

Fig. 13.16. Schematic of the geometry and relative intensity of magnetospheric field-aligned currents, ionospheric closure currents, height-integrated conductivity and corresponding horizontal electric field (after Paschmann et al., eds., 2003, p. 269)

mining distributions of the ionospheric conductances in such a way that the three-dimensional current system resulting from them and the given electric field produces a ground magnetic field effect that agrees with the observations by the magnetometer network as closely as possible. One conclusion of the model is that the westward electrojet flows along the surge until it terminates at the westward end of the surge and diverges back into the magnetosphere. The FAC in the nose of the WTS is quite intense and the primary auroral electrons reach higher energies than elsewhere (up to 100 keV).

Similar investigations to those of arcs and WTS discussed above for some additional auroral forms ('omega bands' and 'streamers') can also be found in Chapt. 6 of Paschmann et al. (2003) but they are not dealt with here.

13.7 Correlations of Aurora with Various Solar and Geophysical Phenomena

The major auroral storms, like magnetic storms and related major disturbances of other magnetospheric variables are connected with strong disturbances on the Sun, in connection with which plasma clouds are emitted from the Sun (see chapters on the Sun). If these clouds reach Earth (after a couple of days and nights) they interact with the magnetosphere in passing it and give rise to all the storm phenomena mentioned (see e.g. Chap. 15 on 'Substorms'). A correlation is therefore expected to exist between auroral occurrence and many other magnetospheric phenomena. We will not discuss all of these connections but only a few of them, which are of major importance scientifically or even practically. But first some relationships with solar phenomena will be summarized.

As mentioned in the introduction, the French scientist de Mairan, working in northern Scandinavia, discovered already in the early part of the 18th century a correlation between the solar activity manifested in sunspot number, and the occurrence frequency of aurora. He seems also to have been aware of the so called 'little ice age' in the 17th century (the Maunder minimum) when the occurrence of sunspots more or less disappeared and also the aurora. The average temperature on Earth's surface fell significantly in that period of about 70 years (to the extent that the Swedish king could bring a whole army over the ice on a sea strait between Danish islands and make a surprise attack on the Danish army from behind). A good statistical correlation between sunspot number and auroral occurrence frequency was demonstrated for the period 1761–1877 by Tromholt (1902).

There is a clear tendency for aurora to reappear after the Sun has made one rotation around its axis (period 27 days), but that tendency is weaker than for magnetic storms and it has not been found to be significant over more than one solar rotation.

The latitude of the auroral region was found to depend on the solar cycle phase already by Tromholt in 1882; the region reaching somewhat lower latitudes at solar maximum than at solar minimum.

Dependence on the direction of the interplanetary magnetic field (IMF). The auroral substorms, like the total magnetospheric substorms, depend on the direction of the IMF in such a way that southward-directed IMF lines favours the occurrence of substorms and northward-directed IMF disfavours substorm occurrence ('half-wave rectifier'). These mechanisms are not dealt with here, but the reader is referred to the chapter on substorms.

Dependence on daylight. Newell et al. (1996) have found that the occurrence probability of intense electron events in the records of the DMSP satellites, defined as having energy flux greater than 5 mW/m^2, was different for magnetic field lines for which the foot is sunlit and those for which it is on the nightside. The occurrence probability was about a factor of three higher in dark conditions than in sunlight.

The beams of accelerated electrons that cause the discrete auroral forms thus occur mainly in darkness: the winter hemisphere is favoured over the summer hemisphere and night is favoured over day. Newell and colleagues have also shown that discrete aurorae rarely occur in presence of diffuse aurora, which is strong enough to provide sufficient conductivity to support the electric currents between the magnetosphere and the ionosphere. Their observations thus clearly indicate that the ionospheric conductivity is a key factor controlling the occurrence of discrete aurorae.

Correlation with magnetic disturbances. The first relationship between aurora and another geophysical phenomenon was that with disturbances in the geomagnetic field, observed in 1741 by Celsius and Hiorter, when there was an aurora over Uppsala. They observed that their compass needle changed direction slightly. By correspondence with their colleague Graham in London, they could conclude that the magnetic disturbance was not only a local one close to the aurora but covered a sizeable part of Earth's surface. The substorm variations in the geomagnetic field and in the aurora have been closely studied as shown in the previous section on the electrodynamics of the aurora. For instance, the westward travelling surge of aurora, which carries the westward jet current, also defines the region where the so called 'negative bay' in the horizontal component of the magnetic field is seen. To the west of the nose of the WTS the disturbance in the magnetic horizontal component is positive.

Auroral absorption of radio waves in the short wavelength (3 – 30 m) region, was a more or less serious problem for radio communication around the globe before satellites took over the long-distance communication and 'short-wave' radio waves were then the only means for global distance communication. Such auroral absorption events affected only great circles which crossed or touched the auroral regions on their way from transmitter to receiver, but for such routes the communication deteriorated for days when there was a strong storm. The absorption of the radio waves is caused by the most energetic part of the primary auroral particles, which penetrates deep into the atmosphere (D-layer) where collisions are more frequent than at greater altitudes. The collisions transfer the ordered motion of the ionospheric electrons caused by the radio waves into stochastic motions (heat).

The structured ionisation in the upper atmosphere produced by the auroral particles *scatter radio waves* in a wide range of HF and VHF frequencies and that also affects communication qualities.

Ionospheric emissions in the VLF and ELF bands are also produced by the auroral particles. A large number of observations of *sound in connection with aurora* have been reported. Some of these reports have come from experienced aurora researchers. It is claimed that sound is heard in a few percent of active aurorae especially in years near solar maximum. It is mostly claimed that the sound has frequencies close to the upper limit of the frequency range that the human ear can record. However, the gas density at the altitudes where aurora occurs is normally so low that propagation of sound down to Earth's surface cannot take place. The aurora has to reach at least as low as 65 km above Earth's surface for propagation to be possible in the audible frequency range. But even if propagation sometimes were possible, the variations in light and sound emissions could not at all be recorded simultaneously at Earth's surface, as has been reported, because of the very low propagation speed of the sound waves compared to that of light. Delays of the sound by tens of minutes would be expected. A hypothesis that has been put forward is that the acoustic phenomena may be due to electric fields, which are generated in connection with strong aurorae and may at the surface of Earth give rise to some sort of point discharge that makes some sound.

For compressional atmospheric waves with periods above 1/2 s, so-called infrasonic waves, propagation from 100 km to Earth's surface is possible and such waves have also been observed in connection with aurora.

13.8 Aurora as a Source of Plasma

Before the first ion composition measurements of magnetospheric plasma were made, there was a general belief among space physicists that the ionosphere is negligible as a plasma source for the magnetosphere and that all plasma in the magnetosphere is of solar wind origin. The view was based on the fact that most processes transporting ionospheric plasma into the magnetosphere were unknown at that time. When the first ion mass spectrometer was launched on a small US military low-orbiting satellite by the Lockheed group in the late 1960s they found that the keV ions, which precipitated into the atmosphere from the magnetosphere, contained an appreciable fraction of O^+ ions, which can originate only in the ionosphere. The results were so surprising that they launched another ion mass spectrometer to have their original results confirmed before they published the discovery (Shelley et al., 1972).

S3-3, another small US satellite, was launched in 1976 into a polar orbit with an apogee at 8000 km at high latitudes and it observed field-aligned ion beams with a large fraction of O^+ ions coming out of the ionosphere (Shelley et al., 1976). When the first ion mass spectrometer was sent into the central magnetosphere somewhat later, it was concluded that the ionosphere

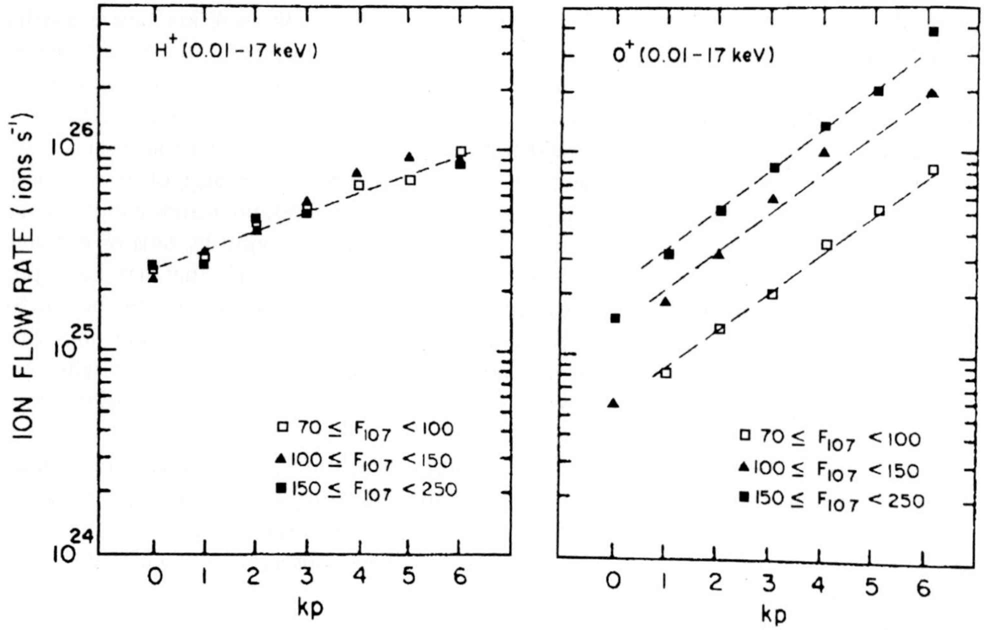

Fig. 13.17. Outflow rates of H^+ and O^+ ions of 0.01 – 17 keV energy observed on DE-1 at 16,000 – 24,000 km altitude, integrated over all MLT and latitudes above 56° in both hemispheres as function of K_p index, for different phases of the solar cycle ($F_{10.7}$) (after Yau et al., 1988)

is a plasma source of similar importance as the solar wind within about 8 R_E geocentric distance (Geiss et al., 1978).

Measurements of ion outflow from the high-latitude ionosphere into the magnetosphere, are summarized in Fig. 13.17 (after Yau et al., 1988). The ion outflow rate for different phases of the solar cycle ($F_{10.7}$ values) and for different disturbance levels (K_p) are summarized in the figure for H^+ and O^+ ions. The H^+ and O^+ flow rates are of similar orders of magnitude. As can be seen, the flow rate of H^+ is virtually independent of the solar cycle phase but increases by a factor of three from undisturbed to quite disturbed conditions in the magnetosphere. The O^+ flow, on the contrary, varies with both solar cycle phase and magnetic activity level. The O^+ flow is larger than the H^+ flow at all activity levels when the solar cycle phase is close to the peak. On the other hand, the H^+ flow rate is greater than the O^+ values at all K_p values for the low range of F values. Fig. 13.17 also demonstrates that the ion outflow rate is above 10^{26} ions/sec for $K_p > 3$ and $F_{10.7} > 150$. These global flow rates of ions from the ionosphere (and electrons) are an order of magnitude larger than the inflow of charged particles associated with auroral phenomena, although the energies are different. This has to do with the fact that charged particles from the magnetosphere have to be closely aligned with the magnetic field lines in order not to be mirrored back into the magnetosphere before reaching the levels of the ionosphere where collisions prohibit them from moving out again. The ionospheric particles, which by various mechanisms achieve a velocity component upward along the magnetic field lines, are all able to go into the magnetosphere, irrespective of pitch-angle, provided their parallel velocity component is enough to overcome the gravitational force.

The outflow from the auroral regions is generally somewhat higher than the outflow of the low-energy polar wind. The bulk flow rates of thermal and suprathermal ions (polar wind and upwelling ions) are, however, of similar magnitude as the outflow rates of more energetic ions at greater altitudes (UFIs), except for high K_p values. The bulk thermal and suprathermal flux from the upper ionosphere may thus be enough as source of energised fluxes at higher altitudes, except at high K_p values. Situations with high K_p values are generally transient, and the bulk flow rates may suffice for time-limited high K_p situations, especially as ionospheric density depletions (plasma cavities) produced by the transport dynamics along magnetic field lines may, at the lower edge of the cavity, give rise to appreciable fluxes of O^+ ions which are normally bound gravitationally (Singh et al., 1989).

The source of free energy that powers the outflow of ionospheric plasma is a combination of solar UV radiation and solar wind kinetic electromechanical energy, but how solar wind kinetic energy is transferred, eventually driving the outflow of ionospheric plasma, is yet to be understood in most details. The heating/acceleration mechanisms that cause the outflow of heavy ions are known to include topside frictional heating, transverse heating by low-frequency waves at greater altitudes (which increase the magnetic moment of the ions and thereby drive them outwards) and parallel acceleration and heating, where parallel electric fields are enhanced by strong auroral parallel currents. For more details about the mechanisms the reader is referred to Chap. 2 in Hultqvist et al., eds. (1999) and Paschmann et al., eds. (2003).

13.9 The Aurora as a Universal Phenomenon

Whereas aurora on Earth has been part of the human environment as long as people have been living at high latitudes, aurora on other planets have been observed only in the last few decades by means of space-borne instrumentation and the most powerful ground-based telescopes. The first investigations of aurora on planets other than Earth were made by means of spectrometers and cameras on Pioneers 10 and 11 and the two Voyager spacecraft when they passed Jupiter in the 1970s. Figure 13.18 shows a Jovian aurora observed by the Hubble Space Telescope (Clarke, 2000).

The Voyager spacecraft continued to Saturn and recorded aurora there in 1980. Uranus and Neptune also show auroral emissions as discovered by means of Voyager 2 in 1986 and 1989, respectively. In the solar system all planets with an internal magnetic field and an atmosphere exhibit auroral phenomena. They differ in many respects on the various planets but they

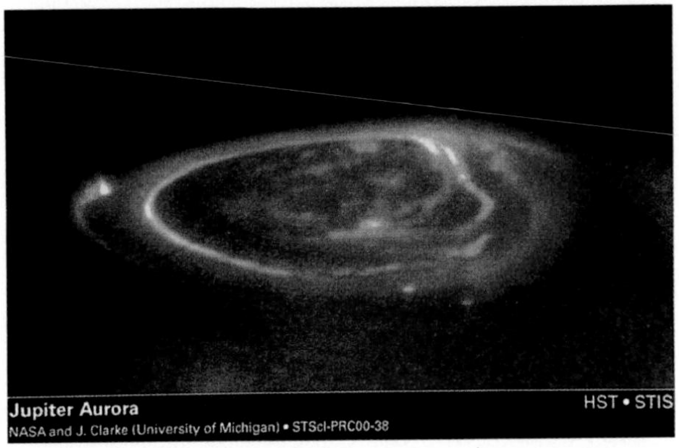

Fig. 13.18. Aurora on Jupiter observed by the Hubble Space Telescope in UV (Clarke, 2000). The *left-most bright feature* is the 'spot' associated with the moon Io (NASA Space Telescope Science Institute release)

are all similar in that the electromagnetic emissions are generated by the precipitation of energetic particles into the upper atmosphere of the planets. A few characteristics of aurora on some planets in the solar system will be presented below. The reader is referred to Chap. 9 in Paschmann et al., eds. (2003) for more details.

Jovian aurora is different from aurora on Earth in many ways. On Jupiter hydrogen emissions dominate the aurora instead of oxygen emissions. Spectra from the H_2 continuum have been identified on Jupiter for the first time in the history of astronomy. They are generated by precipitating electrons. EUV emissions are generated in the altitude range 250 – 1300 km with the bulk of the integrated intensity occurring below 700 km altitude. Primary electron energies appear to be in the several tens of keV range and the total power input from precipitating particles has been estimated at the order of 10^{14} W. Contrary to on Earth the Jovian auroral belt co-rotates with the planet, mainly because of the strong magnetic field on Jupiter.

The moon Io plays a major role for the Jovian auroral morphology. Volcanoes on Io emit large amounts of sulphur and oxygen ions together with electrons. The magnetic field lines on Jupiter, rotating with the planet, pass Io with a speed of 57 km/s and provide a total power of order 10^{12} W available for accelerating charged particles and running an electric current system along the magnetic field lines and through the Jovian ionosphere. The emissions show up as a bright spot at the low-latitude edge of the auroral region. These emissions are not due to the energetic S^+ and O^+ ion from Io, because those ions are thought to deliver their energy in the Jovian atmosphere at low altitudes, where CH_4 prohibit any emissions to escape into space. The main auroral oval is at an L value of about 30 and it is quite intense (100 kR to 2 MR) and very narrow (widths as low as 80 km have been reported).

Aurora on Saturn has been found to extend between latitudes 78° and 81.5°. The optical emissions are mainly from molecular hydrogen excited by electron collisions. They are more variable in intensity than on Jupiter. The total northern auroral emissions in UV has been estimated at 4×10^{10} W.

Aurora on Uranus was not seen by Voyager 2 until a few days before closest approach. It has similar spectral characteristics as on Jupiter and Saturn but the morphology is much more complex because Uranus' axis of rotation is only 8° from the orbital plane and the magnetic dipole axis makes an angle of 60° with the rotation axes. Voyager 2 recorded a brightness of 1.5 kR for HLy-α and 9 kR for H_2 Lyman and Werner bands, which requires electron precipitation on the order of 10^{11} W. Temporal variations indicative of substorm-like phenomena have been observed.

Aurora on Neptune is substantially weaker than on the earlier mentioned planets. Weak aurora-like emissions in H Ly-α were seen on the night-side of the planet when Voyager 2 passed. A weak aurora is consistent with low fluxes of soft electrons and ions observed over the polar region. The qualitative reaction between particle and aurora remains to be worked out. For additional information see Broadfoot et al. (1989).

Auroral acceleration in astrophysical objects. The plasma processes generating auroral emissions on planets in the solar system are quite likely to occur in various kinds of astrophysical objects. Some conditions that have to be fulfilled for auroral mechanisms to occur is the existence of a strong current concentration and filamentation. An outline of methods for analysing the potential or even magnitude of acceleration of particles in distant astrophysical systems, of the same general kinds as in the aurora, can be found in Haerendel (2001).

13.10 Concluding Remarks

The above presentation is a compressed description of what we know of the aurora. More details about the ground-based measurements on aurora and references concerning the pre-space-era work can be found in the books by Omholt (1971) and Vallance Jones (1974). For a fairly complete review of knowledge achieved in the decades of the space era the reader is referred to a recently published book by Paschmann et al., eds. (2003).

Although there has been strong progress in the understanding of the complex physical processes involved in aurora in the space era, and especially in the last couple of decades with the availability of high time resolution instruments on spacecraft in the right kind of orbits, there are still many important auroral features that are not understood. Most important of these are the very large ranges of scales for the structuring in space and time, with widths of some forms as low as 100 m sometimes. The generator mechanisms and the location of the generators for the electric fields associated with aurora are poorly understood. Even the height distribution of some auroral emissions and their variations between forms is not understood and the basic generation mechanism for diffuse aurora, i.e. what waves are responsible for the pitch-angle scattering, is unknown. The understanding of the role of the ionosphere in controlling the aurora is also far from complete. Many of these gaps in our knowledge and understanding are of basic importance not only in connection with aurora but for plasma physics in general.

References

Akasofu, S.-I.: 1964, 'The development of the auroral substorm'. *Planet. Space Sci.* **12**, 273

Akasofu, S.-I.: 1968, *Polar and Magnetospheric Substorms.* Dordrecht: D. Reidel

Belmont, G., D. Fontaine and P. Cann: 1983, 'Are equatorial electron cyclotron waves responsible for diffuse auroral precipitation?' *J. Geophys. Res.* **88**, 9163

Broadfoot, A.L. et al.: 1989, 'Ultraviolet spectrometer observations of Neptune and Triton'. *Science* **246**, 1459

Clarke, J.T.: 2000, 'Satellite footprints seen in Jupiter aurora', Space Telescope Science Institute, Press Release STScI-PRC00-38, Baltimore, MD

Davis, T,N., and G.T. Hicks: 1964, 'Television cinematography of auroras and preliminary measurements of auroral velocities'. *J. Geophys. Res.* **69**, 1931

Feldstein, Ya.I.: 1960, 'Geographical distribution of the auroras and azimuths of the arcs'. *Investigations of the Aurorae* No 4, 61 (in Russian)

Feldstein, Ya.I., and E.K. Solomatina. 1961, in *Geomagn. and Aeron,* (US translation), **1**, 475

Feldstein, Y.I. and G.V. Starkov: 1967, 'Dynamics of auroral belt and polar geomagnetic disturbances'. *Planet. Space Sci.* **18**, 501

Fritz, H.: 1881, *Das Polarlicht.* Brockhaus, Leipzig

Geiss, J., H. Balsiger, P. Eberhardt, H.P. Walker, L. Weber, and D.T. Young: 1978, 'Dynamics of magnetospheric ion composition as observed by the the GEOS mass spectro-meter', *Space Sci. Rev.* **22**, 537

Haerendel, G: 2001, 'Auroral acceleration in astrophysical plasmas'. *Physics of Plasmas* **8**, 2365

Harang, L: 1951, *The Aurorae*, London: Chapman & Hall, Ltd.

Hultqvist, B.: 1959, 'Auroral isochasms'. *Nature* **183**, 1478

Hultqvist, B: 1974, 'Rocket and satellite observations of energetic particle precipitation in relation to optical aurora'. *Annales de Geophys.* **30**, 223

Hultqvist, B., W. Riedler, and H. Borg: 1976, 'Ring current protons in the upper atmosphere within the plasmasphere'. *Planet. Space Sci.* **24**, 783

Hultqvist, B. and R. Lundin: 1987, 'Some Viking results related to dayside magnetosphere-ionosphere interactions', *Annales Geophys.* **5A**, 503

Hultqvist, B., M. Øieroset, G. Paschmann, and R. Treumann (eds.): *Magnetospheric Plasma Souces and Losses*, Vol. 6 of *Space Science Series of ISSI.* Dordrecht: Kluwer Academic Publishers

Kennel, C., F. Scarf, R. Fredricks, J. McGehee, and F. Coroniti: 1970, 'VLF electric field observations in the magnetosphere'. *J. Geophys. Res.* **75**, 6136

Lassen, K., and C. Danielsson: 1989, 'Distribution of auroral arcs during quiet geomagnetic conditions'. *J. Geophys. Res.* **94**, 2587.

Lu, G., M. Brittnacher, G. Parks, and D. Lummerzheim: 2000, *J. Geophys. Res.* **105**, 18483

Lui, T.Y., D. Venkatesan, G. Rostoker, J.S. Murphree, C.D. Anger, L.L. Cogger and T.A. Potemra: 1987, 'Dayside auroral intensifications during an auroral substorm'. *Geophys. Res. Lett.* **14**, 415.

Lundin, R., G. Haerendel, and S. Grahn: 1994, 'The Freja Science mission'. *Space Science Rev.* **70**, 405

Marklund,G., I. Sandahl and H. Opgenoorth: 1982, 'A study of the dynamics of a discrete auroral arc'. *Planet. Space Sci.* **30**, 179

Marklund, G., W. Baumjohann and I. Sandahl; 1983, 'Rocket and ground-based study of an auroral breakup event'. *Planet. Space Sci.* **31**, 207

Marklund, G.: 1984, 'Auroral arc classification scheme based on the observed arc-associated electric field pattern'. *Planet. Space Sci.* **32**, 193

Marklund, G. et al.: 1998, 'Observations of the electric fields and their relationship to black aurora'. *J. Geophys. Res.* **103**, 4125

Mauk, B.H., B.J. Anderson, and R.M. Thorne: 2004, 'Magnetosphere-Ionosphere Coupling at Earth, Jupiter and beyond'. In *Atmospheres in the Solar System: Comparative Aeronomy,* Geophysical Monograph 130, AGU

Newell, P.T., and C.-I. Meng: 1992, 'Mapping the dayside ionosphere to the magnetosphere according to particle precipitation characteristics', *Geophys. Res. Lett.* **19**, 609

Newell, P., C.-I. Meng, and K. Lyons: 1996: 'Suppression of discrete aurora by sunlight'. *Nature* **381**, 766

Omholt, A.: 1971, *The Optical Aurora.* Heidelberg: Springer Verlag

Opgenoorth, H., R. Pellinen, W. Baumjohann, E. Nielsen, G. Marklund, and L. Eliasson: 1983: 'Three dimensional current flow and particle precipitation in a westward travelling surge (observed during the Barium-Geos rocket experiment)', *J. Geophys. Res.* **88**, 3183

Paschmann, G., S. Haaland, and R. Treumann (eds.): 2003, *Auroral Plasma Physics*, Dordrecht: Kluwer Academic Publishers, Vol. 15 of Space Science Series of ISSI

Royrvik, O., and T. Davis: 1977, 'Pulsating aurora: Local and global morphology'. *J. Geophys. Res.* **82**, 4720

Sandholt, P.E., B. Jacobsen, B. Lybeck, A., Egeland, P.F. Bythrow, and D.A. Hardy: 1989, 'Electrodynamics of the polar cusp ionosphere: A case study'. *J. Geophys. Res.* **94**, 6713

Sandhiolt, E.E., J. Moen, A. Rudland, D. Opsvik, W.F. Denig, and T. Hansen: 1993, 'Auroral event sequences at the dayside polar cap boundary for positive and negative interplanetary magnetic field By', *J. Geophys. Res.* **98**, 7737

Sandholt, P.E., C.J. Farrugia, M. Øieroset, P. Stauning, and S.W.H. Cowley: 1998, 'The dayside aurora and its regulation by the interplanetary magnetic field, in Moen, J., A. Egeland

and M. Lockwood (eds.): *Polar Cap Boundary Phenomena*, NATO ASI Series, 189

Shelley, E.G., R.G. Johnson, and R.D. Sharp: 1972, 'Satellite observations of energetic heavy ions during a geomagnetic storm'. *J. Geophys. Res.* **77**, 6104

Shelley, E.G., R.D. Sharp, and R.G. Johnson: 1976, Satellite observations of an ionospheric acceleration mechanism'. *Geophys. Res. Lett.* **3**, 654

Shiokawa, K. et al.: 1996, 'Quasi-periodic poleward motions of sun-aligned auroral arcs in the high-latitude morning sector: A case study'. *J. Geophys. Res.* **101**, 19789

Singh, N., K.S. Hwang, D.G. Torr, and P. Riehands: 1989, 'Temporal features of the outflow of heavy ionospheric ions in response to a high-altitude plasma cavity'. *Geophys. Res. Lett.* **16**, 29

Stenbeck-Nielsen, H., T. Hallinan, D. Osborne, J. Kimball, C. Chaston, J. McFadden, G. Delory, M. Temerin, and C. Carlson: 1998, 'Aircraft observations conjugate to FAST'. *Geophys.Res. Lett.* **25**, 2073

Tromholt, S.: 1902, 'Katalog der in Norwegian bis June 1878 beobachteten Nordlichter', Kristiania (Oslo)

Trondsen, T. and L. Cogger: 1997, 'High-resolution television observations of black aurora'. *J. Geophys. Res.* **102**, 363

Vallance Jones, A: 1974, *Aurora*. Dordrecht: D. Reidel Publishing Company

Yau, A.W., W.K. Peterson, and E.G. Shelley: 1988, 'Quantitative parametrization of energetic Ionospheric ion outflow', in T.E. Moore and J.H. Waite Jr. (eds.): *Modeling Magnetospheric Plasma*, Geophys. Monograph 44, 211, AGU

14 Geomagnetic Storms

Yohsuke Kamide and Yuri P. Maltsev

The present solar-terrestrial research has its roots in studies of geomagnetic storms, well before the discovery of the solar wind and various plasma regions in the Earth's magnetosphere. Geomagnetic storms are phenomena which originate in the solar corona and occur in the entire Sun-Earth system, including the Earth's upper atmosphere. A geomagnetic storm is defined by the existence of the main phase during which the horizontal component of the Earth's magnetic field decreases drastically. This depression, strongest at low latitudes, is thought to be caused by the enhanced ring current in the inner magnetosphere. Its growth and decay is typically monitored by the *Dst* index. During a geomagnetic storm, intense substorms take place successively at high latitudes. Beginning with a brief description of observational characteristics of geomagnetic storms, this chapter discusses how solar wind energy is deposited into and is dissipated in the constituent elements that are critical to magnetospheric and ionospheric processes during geomagnetic storms. Although most of the *Dst* variance during magnetic storms can be reproduced by changes in the electric field in the solar wind and the residuals are uncorrelated with substorms, recent satellite observations of the ring current constituents show the importance of ionospheric ions. This implies that ionospheric ions, which are associated with intense substorms, are accelerated upward and contribute to the energy density of the storm-time ring current. It is thus important to identify the role of substorm occurrence in the enhancement of magnetospheric convection driven by solar wind electric field. In evaluating the contribution of various current systems in the *Dst* decreases during geomagnetic storms, we contend that not only the ring current but also the tail current is important.

Contents

Yohsuke Kamide and Yuri P. Maltsev, Geomagnetic Storms.
In: Y. Kamide/A. Chian, Handbook of the Solar-Terrestrial Environment. pp. 355–374 (2007)
DOI: 10.1007/11367758_14

14.1 Introduction

Studies of the solar-terrestrial environment or space physics have their roots in inspections of geomagnetic storms in the 19th century. The discipline of this research was called as Geomagnetism, which aimed at locating and estimating the intensities of electric currents that generate world geomagnetic disturbances during geomagnetic storms. Since the end of the 19th century, it became clear that great magnetic perturbations observed on the Earth's surface are caused by currents flowing in the inner magnetosphere (according to the present-day terminology) as well as in the region of auroras. It was in the 1950–60s when the existence of the solar wind and the magnetosphere was predicted and subsequently discovered by means of satellite measurements: see Sect. 14.4. The average configuration of the magnetosphere was modeled: see Sect. 14.7. A number of plasma regions within the magnetosphere were discovered and the role of each plasma region in geomagnetic storm processes was identified.

It was Alexander von Humboldt (1769–1859) who first used "magnetisches Ungewitter" (magnetic thunderstorms) to describe the variability of geomagnetic needles, which was associated with the occurrence of auroras in the polar sky. Humboldt thought that worldwide magnetic disturbances on the ground and auroras in the polar sky are two manifestations of the same phenomenon, that is geomagnetic storms. He maintained a lifelong interest in geomagnetic disturbances, establishing a number of magnetic stations around the world, notably in Britain, Russia, and in lands then under British and Russian rule, e.g., at Bombay, Toronto, and Sitka. It was found by Humboldt that the significant decrease in the daily mean value of the horizontal component is one of the chracteristic features of the storm-time disturbance. Scientists during the First Polar Year (1882–1883) defined "geomagnetic storms" as intense, irregular variabilities of geomagnetic field which occur as a consequence of solar disturbances. S. Chapman thought that worldwide geomagnetic disturbances during geomagnetic storms are a result of electric currents encircling the Earth, called the ring current. It was then proposed that charged particles originated from the Sun would drift around Earth, causing the decrease in the geomagnetic horizontal intensity at the Earth's surface.

The final goal of the study of the solar-terrestrial environment would be to be able to predict the chain of processes that occur in the entire Sun-Earth system (Gopalswamy et al., 2006). Research of geomagnetic storms is the result of a convergence of multi-disciplinary sciences which developed from several traditional fields of research, such as solar physics, geomagnetism, auroral physics, and aeronomy. Geomagnetic storms are multi-faceted phenomena that originate at the solar corona and occur in the solar wind, the magnetosphere, the ionosphere, and the thermosphere. What in the solar wind causes geomagnetic storms? How is the ring current energized during geomagnetic storms? How do changes in the magnetosphere-ionosphere system affect the Earth's upper atmosphere throughout the chain of storm processes? These are some of the major questions relating to geomagnetic storms. Because geomagnetic storms take place in a wide range of plasma regions, they must be understood as a chain of processes from the Sun to Earth. It may thus be more appropriate to use the terminology space storms instead of geomagnetic storms: see Daglis (2001).

14.2 What is a Geomagnetic Storm?

The characteristic signature of a geomagnetic storm, or simply a magnetic storm, is a depression in the *H* component of the magnetic field lasting normally over one to several days. This depression is caused by the ring current flowing westward in the magnetosphere, and can be monitored by the *Dst* index. The *Dst* index is calculated by

$$Dst = \frac{1}{N} \sum_{n=1}^{N} \frac{H - H_q}{\cos \phi} \tag{14.1}$$

where H is the H component of the magnetic field disturbance at a given station, H_q is the same component over the quietest days, N is the total number of the stations, and ϕ is the station latitude. The $\cos \phi$ factor is to normalize magnetic disturbances at various latitudes to the values at the equator.

It is generally accepted that storms are time periods with a minimum of *Dst* less than −50 nT. According to this criterion, 20 – 50 storm events occur annually, depending on solar activity. It is now commonly assumed that the magnitude of magnetic storms can be defined

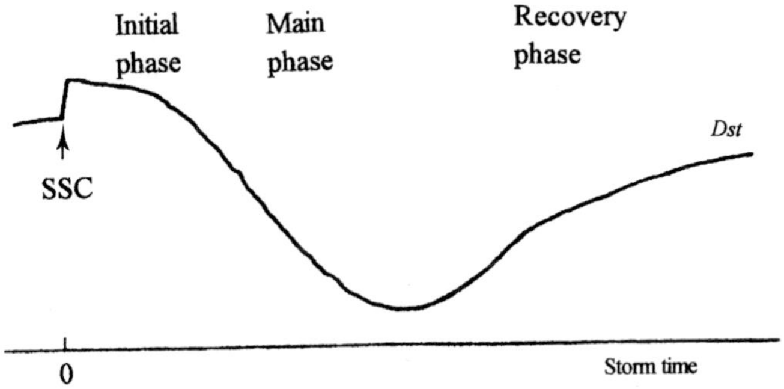

Fig. 14.1. Schematic illustration of the *Dst* variation for a typical geomagnetic storm. For the definition of the initial, main, and recovery phases, see text

by the minimum *Dst* value (Gonzalez et al., 1994). In the early years of research, a general picture of a typical magnetic storm emerged which had the features shown in Fig. 14.1. This picture of the storm involved a sudden positive increase in the *H* component (storm sudden commencement or SSC) followed by a period of arbitrary length in which the elevated field does not change significantly: the initial phase. This phase is followed by the development of a depressed *H* component transpiring over a period from one to a few hours: the main phase. The storm concludes by a slow recovery over hours to tens of hours: the recovery phase.

The SSC is understood as being the effect of a compression of the front side of the magnetosphere by enhanced solar wind pressure. The depression of the magnetic field during the main phase was explained by Singer as the effect of a ring current carried primarily by energetic ions. It decays due primarily to charge-exchange, Coulomb interaction and wave-particle interaction processes associated with neutral particles in the volume of space occupied by the ring current particles.

It was found that the direction of the north-south component of the interplanetary magnetic field (IMF) regulates the growth and decay of the ring current. In fact it became possible to reproduce successfully the growth of the ring current, i.e., variations of *Dst*, using, as input, only the component of the interplanetary electric field in the ecliptic plane normal to the Sun-Earth line: see Burton et al. (1975). It was further recognized that the initial phase simply represented a period of time after the onset of the SSC during which the IMF was oriented northward, i.e., little energy was entering the magnetosphere regardless of the speed and number density of particles in the solar wind. More importantly, it was suggested that an SSC is not a necessary condition for a storm to occur and hence the initial phase is not an essential feature to be a geomagnetic storm (Joselyn and Tsurutani, 1990). In fact, the only basic element of a storm is the significant development of a ring current, that is the existence of the main phase.

The major question which then arises regarding the nature of magnetic storms involves the physical processes leading to the growth of the ring current. This question was apparently answered by Akasofu and Chapman (1961) who noted that during the main phase, there was violent auroral electrojet activity in the midnight sector with the amplitudes of the disturbances there far exceeding the magnetic perturbation associated with the ring current itself. These disturbances at auroral latitudes, i.e., polar substorms, were in some way thought to be responsible for the growth of the ring current. It was later shown that the storm time ring current was carried by energetic ions with energies typically in excess of several tens of keV (see Williams, 1987). The question of how ring current particles attain their energies and whether substorms play an important role in that process are still open. In other words, the relationship between substorms and storms is currently poorly understood, and therefore basic questions remain unanswered regarding the hypothesis of whether a magnetic storm is simply a superposition of intense substorms (Akasofu, 1968).

In recent times it has become clear that during geomagnetic storms, significant amounts of atomic oxygen are transferred from the auroral ionosphere into the plasma sheet and ultimately form a significant component of ring current population. Since charge exchange processes affect oxygen ions and protons differently, the observed decay of the ring current can reflect the different behavior of the components of the ring current due to the two ion species.

The description of storm manifestations can be found in the reviews by, for example, Feldstein (1992), Gonzalez et al. (1994), Kamide et al. (1998), Daglis et al. (2003), and Maltsev (2004) and proceedings of the conferences, such as Tsurutani et al. (1997; 2006) and Sharma et al. (2004).

14.3 Ring Current as a Dominant Signature of Geomagnetic Storms

The *Dst* decrease, which is the main manifestation of geomagnetic storms, is caused by the ring current encircling Earth. The growth of the ring current begins with the so-called injection of particles into the inner magnetosphere from the magnetotail. The ring current is carried primarily by energetic (10 – 200 keV) ions in $L \sim 2-7$. The concept of trapping charged particles in the magnetic field by, e.g., Störmer and Singer, was understood well before the discovery of trapped radiation by Van Allen. See textbooks for charged-particle motions under the influence of magnetic and electric fields in the magnetosphere.

The principal property of a geomagnetic storm is the creation of an enhanced ring current, producing a decrease in the horizontal component of the geomagnetic field on the Earth's surface. The strength of this perturbation on the Earth's surface is approximately given by the so-called Dessler–Parker–Sckopke relationship (Dessler and Parker, 1959; Sckopke, 1966):

$$\Delta B/B_0 = 2E/3E_{\mathrm{m}} \qquad (14.2)$$

where ΔB is the field decrease at the center of the Earth caused by the ring current, B_0 (~0.3 gauss) is the average equatorial surface field, E is the total energy of the ring current particles, and E_{m} ($= 8 \times 10^{24}$ ergs) is the total magnetic energy of the geomagnetic field outside Earth. According to the above relationship, the *Dst* value is, in a first approximation, linearly proportional to the total energy of the ring current particles. This is the reason why the *Dst* index is being used practically as a measure of the magnitude of geomagnetic storms. Vasyliunas (2006) showed that the Dessler–Parker–Schopke formula can be generalized to include the effect of ionospheric currents.

Parker (1957) established a hydromagnetic formalism, relating the magnetospheric currents to particle pressures both parallel and perpendicular to the magnetic field. The total current $\boldsymbol{j}$ summing over motions of individual particles is: $\boldsymbol{j} = \boldsymbol{j}_{\mathrm{D}} + \boldsymbol{j}_{\mathrm{c}}$, where $\boldsymbol{j}_{\mathrm{D}}$ is the drift current caused by the magnetic field gradient and field line curvature, and $\boldsymbol{j}_{\mathrm{c}}$ is the current driven by gyration effects within the particle distribution.

Following the discovery by IMP-1 that the Earth's magnetic field is consistently confined and distorted by the solar wind, various plasma regions in the magnetosphere, including the ring current and the plasma sheet, were identified. As energetic particles are injected into the inner magnetosphere on the night side, they are influenced by forces due to curvature and gradient of the Earth's magnetic field. Because of these forces, protons drift westward from nightside toward dusk and electrons drift eastward from nightside toward dawn, comprising the net effects as a ring current encircling Earth westward. A geomagnetic storm is nothing but an enhancement of this ring current.

The ring current is not symmetric in local time. Conducting extensive particle measurements in the magnetosphere, Frank (1967) was the first who discovered the asymmetric nature of the ring current. Measurements of the differential energy spectrums of protons and electrons over the energy range extending from 200 eV to 50 keV were used. The total energy of particles of this energy range was found to be sufficient to account for the depression of the geomagnetic field in terms of the *Dst* index. Figure 14.2 shows intensities of protons as functions of L during the different phases of a magnetic storm. It is evident that a severe increase in proton intensities over $3 < L < 5.5$ is apparent in the main phase observations, with a maximum located near $L = 3.5$, and that by the recovery phase, this distribution has substantially decreased in intensities with a peak positioned at $L = 4.5$.

It should be noted that *Dst* includes the magnetic effects not only of the symmetric ring current but of other currents, such as ionospheric, field-aligned, and tail currents (e.g., Campbell, 1996). Using a numerical model-

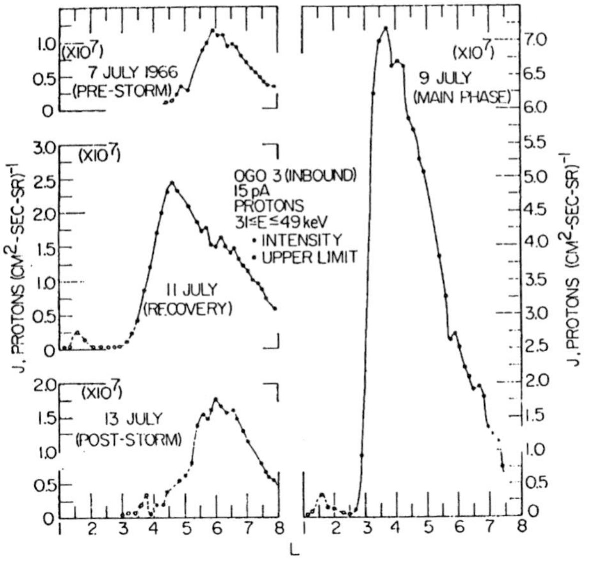

Fig. 14.2. Directional intensities of protons ($31 < E < 49$ keV) as functions of L during the pre-storm, main phase, recovery phase, and post-storm periods of an intense geomagnetic storm. After Frank (1967)

ing of various current systems, Alexeev et al. (1996) have demonstrated that the ground effect of the tail current during the main phase of geomagnetic storms can be of the same order as the ring current. Dremukhina et al. (1999) have applied this model to major geomagnetic storms, showing that during the main phase, the contribution of the tail current to *Dst* is roughly equal to that of the symmetric ring current, although the ring current becomes dominant during the recovery phase.

The role of ionospheric particles in the development of the ring current during geomagnetic storms became evident after the CHEM (charge energy mass) spectrometer onboard the AMPTE/CCE mission, which was the first experiment to investigate the near-Earth magnetotail with multi-species ion measurements extending in the higher-energy (> 20 keV) range. The great storm of February 1986 was studied by Hamilton et al. (1988), showing that 60–80% of the ring current density near the maximum of the storm was of ionospheric origin. The next opportunity for multi-species measurements in the inner magnetosphere was provided by the MICS (Magnetospheric Ion Composition Spectrometer) experiment onboard the CRRES mission. Daglis (1997) studied the importance of the ionospheric ion component in the ring current during several storms observed by CRRES, showing that during the main phase of great storms, the abundance of ionospheric-origin ions (O^+ in particular) in the inner magnetosphere is extraordinarily high. The outstanding feature was the concurrent development of *Dst* and of the O^+ contribution to the total particle energy density: see Fig. 14.3 (Daglis et al., 1999).

The increased relative abundance of ionospheric O^+ ions in the ring current during storms, besides influencing the ring current enhancement, influences the decay rate of the ring current, since the charge-exchange lifetime of O^+ is considerably shorter than the H^+ lifetime for ring current energies. This implies that O^+-dominated ring current will decay faster. Such a fast initial ring current decay, associated with a large O^+ component during the storm main phase, has indeed been observed: although "decay" of the ring current can also occur as a result of the escape of energetic ions through the dayside magnetopause. This O^+ dominance coupled with the observations that a significant fraction of H^+ is also ionospheric in origin, suggests that the cause of the ring current during great storms lies in the upward acceleration of ionospheric ions along field lines associated with the occurrence of intense substorms. Note, however, that the relative importance of ionospheric-origin and solar wind-origin ions varies considerably from storm to storm.

14.4 Solar Wind Causes of Geomagnetic Storms

The evidence is overwhelming that solar wind dawn-to-dusk electric fields drive intense geomagnetic storms: see McPherron (1997). These electric fields are caused by a combination of solar wind velocity and southward IMF. Of these two parameters, the southward field is probably more important because of its far greater variability. At least, two primary mechanisms are known to be the main sources of enhanced dawn-to-dusk electric fields of substantial duration in the interplanetary medium. They are interplanetary coronal mass ejections (ICME) and corotating interaction regions (CIRs): see Fig. 14.4a and b, respectively (Tsurutani and Gonzalez, 1997). CMEs are impulsive solar/coronal ejecta that occur near the maximum sunspot phase of the solar cycle. Most geoeffective ICMEs are magnetic clouds, a subset

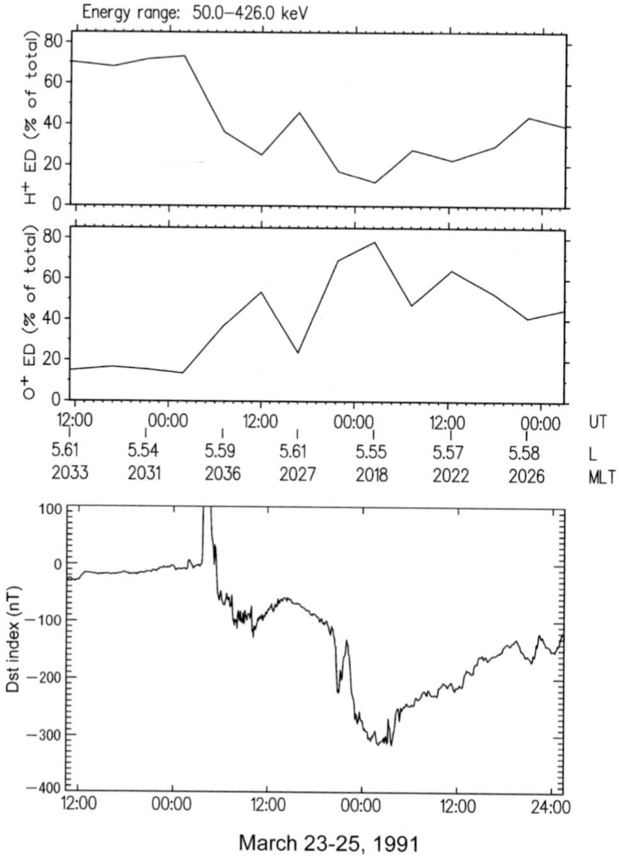

Fig. 14.3. Time profile of the contribution of H^+ and O^+ to the total energy density of energetic ion population in the outer ring current during the great geomagnetic storm of March 1991 (*top two panels*) and the *Dst* index (*bottom panel*). Adapted from Daglis, I.A., G. Kasotakis, E.T. Sarris, Y. Kamide, S. Livi, and B. Wilken, Variations of the ion composition during a large magnetic storm and their consequences, *Phys. Chem. Earth, 24*, 229–232, 1999

of ejecta characterized by large north-south components of the IMF. During the declining phase of the solar cycle, on the other hand, coronal holes tend to dominate, expanding from the polar regions to equatorial regions: they emit fast plasma continuously, resulting in CIRs.

CMEs are distinct particle and field structures with field orientations which are most numerous near solar maximum, causing most major geomagnetic storms at that phase of the solar cycle and possibly at other phases as well (e.g., Gosling et al., 1991). The ejections that are most effective in creating magnetic storms are known to be the ones that are fast, with speeds exceeding the ambient wind speed by the magnetosonic wave speed, so that a fast forward shock is formed. As a fast plasma and field structure propagates from the Sun through interplanetary space, it sweeps up and compresses the slower plasma and field ahead, creating a "sheath" between the shock and the interplanetary manifestation of the ejecta.

Figure 14.4a illustrates the schematic of the regions of possible intense IMF for such solar ejecta, in which two types of possible satellite crossings, T1 and T2, are shown. The sheath fields leading the fast ejecta often contain substantial north-south field components, possibly due to compression and draping of the ambient IMF over the ICMEs. Both the remnant ejecta fields and plasma and those of the sheath can be geoeffective, depending on the field orientations. Roughly 80% of fast ejecta do not cause major storms, however, because of the lack of large southward field components persisting for three hours or longer.

During the declining phase of the solar cycle, another type of solar/coronal event dominates. During this phase, the coronal holes have expanded from polar locations and extend into, and sometimes across, the equatorial regions. Fast (750 – 800 km/s), tenuous plasma is continuously emitted from these solar regions. Because these regions are long-lived and evolve

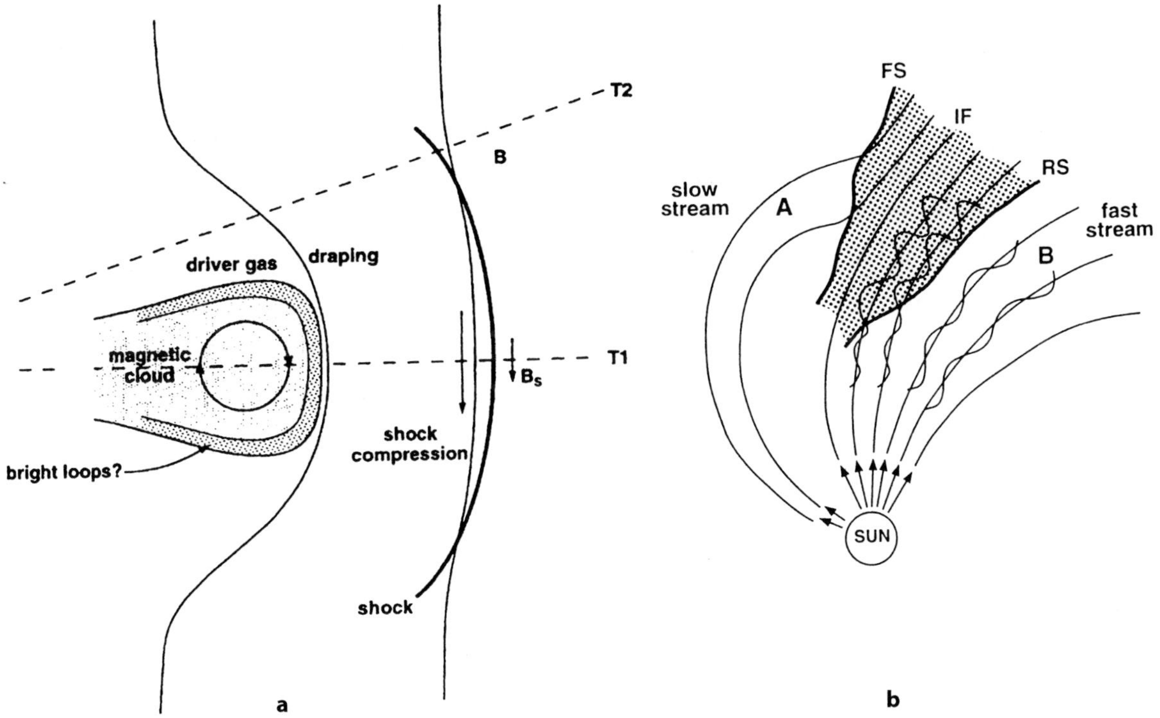

Fig. 14.4. (**a**) Schematic of a magnetic cloud region of intense IMF seen typically during solar maximum. T1 and T2 are two types of satellite crossings in the interplanetary structure. (**b**) Schematic of the formation of corotating interaction regions (CIRs) during the descending phase of the solar cycle. The interaction between a high-speed stream (B) and a slow-speed stream (A) are shown together with the CIR (shaded). The forward shock (FS), interface surface (IF), and reverse shock (RS) are also indicated. After Kamide et al. (1998)

relatively slowly, they appear to "corotate" with the Sun. If a coronal hole is near the ecliptic plane, the Earth's magnetosphere will be "bathed" in this stream once per solar rotation. Typically, a heliospheric neutral sheet/plasma sheet lies ahead of the fast stream in interplanetary space. The characteristics of the plasma sheet wind include low speed (~350 km/s) and high density (tens of particles/cm^3). The interaction of the fast stream with the slow stream ahead creates a CIR. CIRs are bounded on the leading and trailing edges by forward- and reverse-propagating compressional waves, respectively. Figure 14.4b shows schematically the formation of CIRs in which magnetic field fluctuations are present in the high speed stream proper.

In addition to ICMEs and CIRs, there are "modulators" that do not directly drive magnetic storms without an ICME or CIR, but increase/decrease the geoeffectiveness of the ICMEs and CIRs effects through the different phases of individual geomagnetic storms, the sunspot cycle, and seasons. A 27-day modulation in geomagnetic activity has been noted since the 19th century. This periodicity, attributed to solar regions called "M-regions" by Bartels, was later discovered to arise from high-speed solar wind streams originating in coronal holes. While ICMEs often contain sustained southward fields accompanied by fast wind speeds, the high-speed wind from coronal holes generally has relatively low field magnitude and a radial orientation not conducive to production of steady and substantial north-south fields. However, the interaction of this fast wind with slower, denser streamer wind, forming a CIR, produces geoeffective field compressions and deflections. Geomagnetic storms associated with CIR-like plasma signatures rarely have minimum $Dst < -100$ nT, and generally lack the sudden commencements often occurring for ICME-driven storms.

The coronal hole wind often holds continuous Alfvénic activity, consisting of large-amplitude quasi-periodic fluctuations in the IMF orientation, in-phase with similar fluctuations in the flow direction, with periods from tens of minutes to a few hours. In the interplanetary region following CIRs, the southward field components caused by these waves can cause magnetic reconnection, small injections of plasma into the magnetosphere, and prolonged recovery phases of the storms. Events of this type are known as "high-intensity, long-duration, continuous *AE* activity" (HILDCAA) events (Tsurutani and Gonzalez, 1997). Although the average B_Z component in HILDCAAs is zero, the half-wave induced reconnection in the magnetospheric response results in a continuous occurrence of substorms or other disturbances in the magnetosphere.

14.5 Magnetospheric Geometry During Geomagnetic Storms

The magnetic configuration in the magnetosphere changes significantly during geomagnetic storms, with the magnetic field lines at the nightside stretching, magnetic flux in the tail increasing, and dayside polar cusps shifting to lower latitudes.

14.5.1 Auroral Electrojets

The auroral electrojets are known to shift equatorward drastically during geomagnetic storms. During the main phase of intense storms, the westward electrojet can cover the latitudes from 50° to 80° on the night side, its total intensity reaching often as intense as 10 MA. The eastward electrojet flows in the dusk sector at latitudes lower than those of the westward electrojet. With *Dst* varying from 0 to −400 nT, the minimum latitude appeared to lower down from 67° to 52° (*L* changed from 7 to 2.7), the rate of the equatorward shift subsiding with storm intensity increasing.

Feldstein et al. (1997) suggested the following approximation for the corrected geomagnetic latitude Λ of the westward electrojet center under *Dst*:

$$\Lambda = 65.2 + 0.035\,Dst\,. \tag{14.3}$$

Note that the *AE* indices underestimate the intensity of the auroral electrojets during intense geomagnetic storms because of the equatorward shift of the auroral electrojets from the standard *AE* observatories.

14.5.2 Auroral Oval

The latitude of the particle precipitation regions, e.g. of the auroral oval, also decreases as the storm develops. The equatorward edge of the auroral oval can be approximated by a circle centered at midnight at the latitude of 87° under quiet conditions and 85° under moderately and strongly disturbed conditions (Starkov and Feldstein, 1967). The radius of the circle exhibits a very good correlation with *Dst*. Meng (1984) studied the behavior of the equatorward edge of the auroral oval in the course of intense storms, revealing that the latitude of this edge follows approximately the *Dst* variation.

The points in Fig. 14.5, taken from Starkov (1993), indicate the observed latitudes of the equatorward edge of the auroral oval in the dusk-to-midnight sector, the dashed line being the least-squares approximation:

$$\Lambda = 74.9° - 8.6\,\log_{10}|Dst|\,. \tag{14.4}$$

The poleward boundary of the auroral oval also moves to lower latitudes with storm intensifying, although the correlation with *Dst* is lower than for the equatorward boundary. There is a class of auroras associated with energetic proton precipitation, which also shifts equatorward.

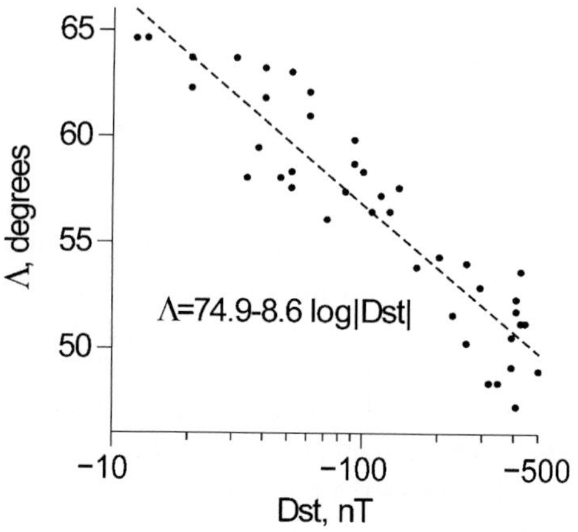

Fig. 14.5. Latitude of the equatorward edge of the auroral oval in the dusk-to-midnight sector versus storm intensity. The *points* indicate observations, and the *dashed line* is the least-squares approximation fit. After Starkov (1993)

14.5.3 Standoff Distance

Up to distances of ~$40R_E$ downtail, the magnetopause can be well approximated by an ellipsoid. The standoff distance r_s (the distance to the subsolar point) is controlled mostly by the solar wind dynamic pressure p. Under weakly disturbed conditions this distance can be taken as (Roelof and Sibeck, 1993)

$$r_s = r_{sa}(p_a/p)^{1/6} \tag{14.5}$$

where $r_{sa} \approx 10R_E$ is the average standoff distance, and $p_a \approx 2.3$ nPa is the average solar wind dynamic pressure. With the growth of geomagnetic activity, r_s most commonly decreases due to an increase in p and other factors. Dayside magnetospheric erosion, defined as inward magnetopause motion under invariable solar wind dynamic pressure, is a phenomenon closely connected with a storm time decrease of the cusp latitude. A decrease in r_s under p being constant begins once the IMF turns southward.

A minimum standoff distance $r_s = 4.7R_E$ was reached during the strongest storm of the 20th century on March 13, 1989, when *Dst* was as low as −600 nT. During a storm on February 8–9, 1986 the dayside magnetopause was detected at $5.2R_E$ (Hamilton et al., 1988), *Dst* being −257 nT.

Rufenach et al. (1989) studied 64 magnetopause crossings by a geosynchronous satellite ($r = 6.6R_E$). Although the exact location of the subsolar point was not known because MLT of the crossing sites varied from 06 to 18 MLT, it can be inferred that the average r_s was smaller than $6.6R_E$, considering that the subsolar point is nearer to Earth than any other magnetopause point. The average *Dst* for these 64 crossings was −108 nT, and the average solar wind dynamic pressure $\langle p \rangle \approx 15$ nPa.

14.5.4 Stable Trapping Boundary

The main energy content of the ring current is provided by 10 to 200 keV protons. The ring current is formed by trapped particles. Under stationary conditions, the drift trajectory of the particles with 90° pitch-angle is described by the equation $q\Phi + \mu B = \text{constant}$, where q is the charge of a particle, Φ is the electric potential, $\mu = w_\perp/B$ is the magnetic moment (the first adiabatic invariant), and $w_\perp$ is the energy of a particle perpendicular to the magnetic field B. For energetic particles, which are weakly influenced by the electric field, the drift trajectories in the equatorial plane coincide nearly with the contours $B = \text{constant}$. It should be pointed out, however, that in the ring current region the electric field is reduced due to the shielding effect of the field-aligned currents. For energetic particles with 90° pitch-angle, the stable trapping boundary is the contour $B = B_s$, B_s being the magnetic field at the subsolar point. This boundary is somewhat different for particles with other pitch-angles, but we can consider that the majority of the trapped particles are inside the contour $B = B_s$.

During intense magnetic storms this contour approaches Earth. On the nightside the contour $B = B_s$ is displaced from $7R_E$ under quiet conditions up to $4R_E$ under $Dst = -400$ nT. A decrease in the stable trapping area leads to contraction of the radiation belts. Under quiet conditions maximum flux of high-energy ($E >$ 1 MeV) electrons is observed at $L \approx 5$. With storm intensifying, however, the high-energy electron belt is approaching Earth.

14.6 Storm-Time Magnetic Fields and Electric Fields in the Magnetosphere

14.6.1 Spatial Distribution of the Electric Currents

The overall pattern of large-scale electric currents includes the magnetopause currents and the currents inside the magnetosphere, both transverse and aligned to the magnetic field. The magnetopause currents shield magnetic fields associated with the sources inside Earth. There are two regions of the large-scale field-aligned currents (FACs): the Region 1 FACs, flowing at the polar cap boundary, and the Region 2 FACs at the equatorward edge of the auroral oval.

The currents that are transverse to the magnetic field, are confined to the near-equatorial layer, $-3 < z < 3R_E$. Maltsev and Ostapenko (2004) used the database of Fairfield et al. (1994) to restore the spatial distribution of the currents under various levels of *Dst*. The results are shown in Fig. 14.6. One can see that the longitudinal distribution of the westward electric current is quite asymmetric in the day-night direction, the nightside current intensity being several times greater than that at the dayside. The dawn-dusk asymmetry is rather weak. The radial distribution is not

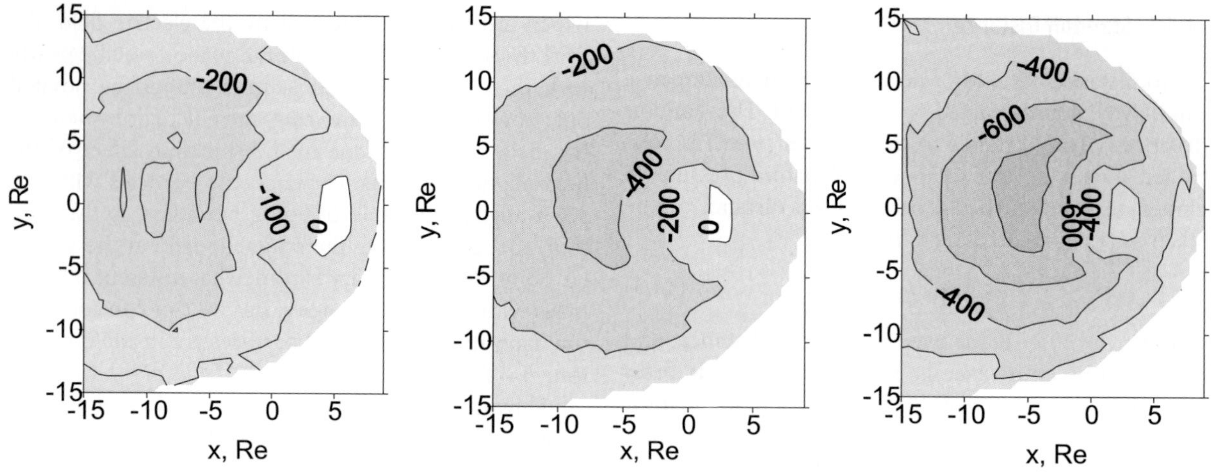

$Dst = 10$ nT $\quad$ $Dst = -20$ nT $\quad$ $Dst = -70$ nT

Fig. 14.6. Isocontours of the eastward current flowing in the layer $-3 < z_{SM} < 3R_E$. The currents are expressed in kA$/R_E$. Negative contour levels indicate that currents flow westward. After Maltsev and Ostapenko (2004)

very sensitive to either geomagnetic activity or solar wind conditions, with the maximum current located at $5-7R_E$. The following approximations were obtained for the total westward current flowing between radial distances of 4 and $9R_E$ at the nightside and dayside, respectively

$$I_{\text{night}}(\text{MA}) = 1.75 - 0.041\,Dst\,, \tag{14.6}$$

$$I_{\text{noon}}(\text{MA}) = 0.22 - 0.013\,Dst\,. \tag{14.7}$$

Note that the model of Tsyganenko (2002) predicts nearly a consistent response of these currents to *Dst*: $\mathrm{d}I_{\text{night}}/\mathrm{d}Dst = -0.047$, $\mathrm{d}I_{\text{noon}}/\mathrm{d}Dst = -0.020$.

14.6.2 Contribution of Different Current Systems to *Dst*

There are basically five types of magnetospheric currents, as schematically shown in the left panel of Fig. 14.7: the magnetopause currents shielding the magnetic field of the Earth's dipole; symmetric ring current; magnetotail current system including the cross-tail current and closure currents on the magnetopause; Region 1 FACs closed in the ionosphere and on the magnetosphere flanks or in the solar wind; and Region 2 FACs closed through the partial ring current. Each of the five basic current systems is closed. One can see from the right column that, except for the Region 1 FACs, all current systems contribute to *Dst*, i.e.

$$Dst = k(\delta B^{\text{mp}} + \delta B^{\text{RC}} + \delta B^{\text{ct}} + \delta B^{\text{pr}}) \tag{14.8}$$

where the four quantities on the right-hand side designate the variations of magnetic disturbances at the Earth's center produced by the divergence-free part of the magnetopause currents, symmetric ring current, cross-tail current along with the closure currents on the magnetopause, and partial ring current, respectively; and k (≈ 1.3) is related to the induction currents within Earth.

The magnetic field of the shielding current on the magnetopause can be found from the Mead (1964) model: $B_z^{\text{mp}}(0) \approx 0.3B_s$, where B_s is the magnetic field at the subsolar point on the magnetopause. The field B_s can be obtained from the pressure balance condition as $B_s = \sqrt{2\mu_0 p_{sw}}$, in which μ_0 is the magnetic permeability of vacuum, and p_{sw} is the solar wind pressure at the stagnation point that is nearly equal to the dynamic pressure of the solar wind protons.

Note that the noon-side current (14.7) presents the purely symmetric part of the ring current, without a contribution of either partial ring current or cross-tail current. A symmetric ring current of 1 MA magnitude flowing at distance of $6R_E$ produces the disturbance $B_z^{\text{rc}}(0) \approx -16$ nT at the Earth's center. Multiplying (14.7) by -16, we have

$$B_z^{\text{rc}}(0) = -3.5 + 0.21\,Dst\,. \tag{14.9}$$

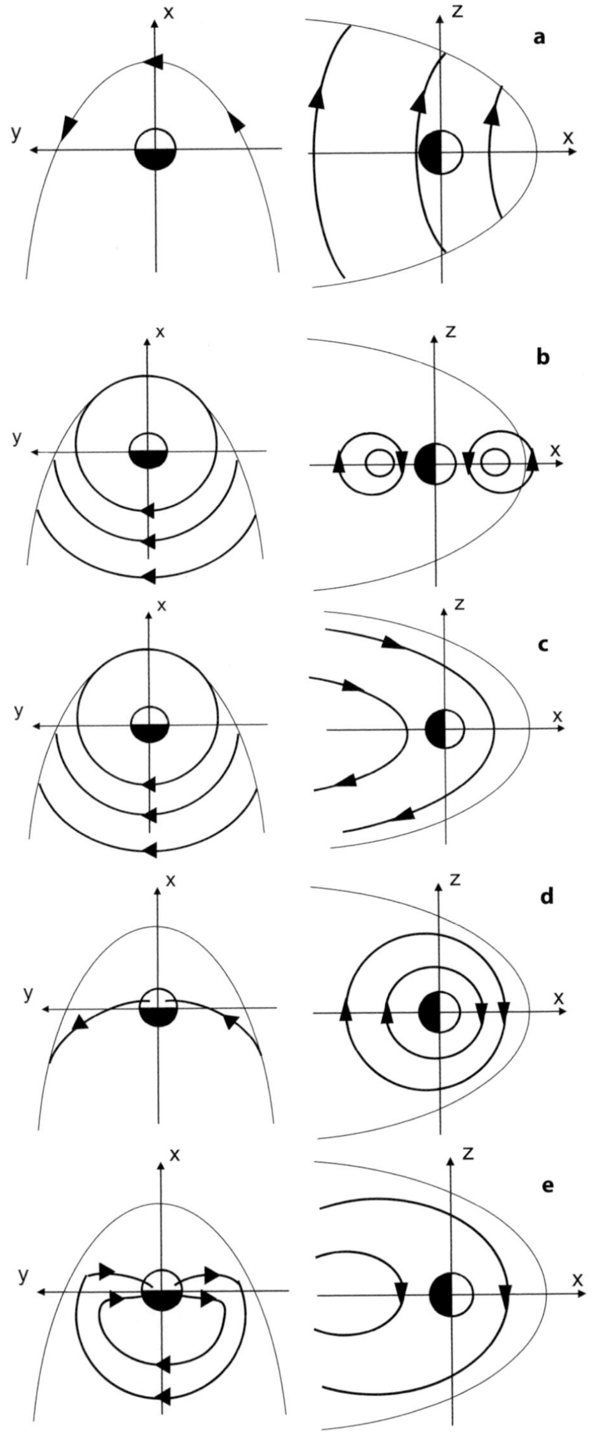

Fig. 14.7a–e. Electric currents (*the left column*, equatorial plane) and the corresponding magnetic fields (*the right column*, noon-midnight meridian plane) of the five basic current systems. From *top* to *bottom*: (**a**) the divergence-free part of the magnetopause currents, (**b**) symmetric ring current, (**c**) cross-tail current (the closure currents on the magnetopause are not shown), (**d**) Region 1 FACs, and (**e**) Region 2 FACs with the partial ring current

With the induction currents ($k \approx 1.3$) included, we obtain the ground magnetic effect to be

$$kB_z^{rc}(0) = -4.6 + 0.27\,Dst \qquad (14.10)$$

indicating that the contribution of the symmetric ring current to *Dst* turns out to be 27%.

Subtracting (14.7) from (14.6) we can obtain the sum of the partial ring current (PRC) and near-Earth cross-tail current (NCT)

$$I^{PRC} + I^{NCT} = 1.53 - 0.028\,Dst\,. \qquad (14.11)$$

Assuming that this combined current is confined within 18 to 06 local time sector in the nightside, at distance of 6 R_E, each MA of the current produces a disturbance of −8 nT in the Earth's center or $-8 \times 1.3\,(= k) = 10.4$ nT on the Earth's surface at low latitudes. The corresponding relationship of the disturbance to *Dst* becomes the form

$$k\left(B_z^{PRC} + B_z^{NCT}\right) = -16 + 0.29\,Dst\,. \qquad (14.12)$$

Consequently, the total contribution of the partial ring and near-Earth cross-tail currents to *Dst* is about 29%. Using the paraboloid magnetospheric model, Alexeev et al. (1996) and Dremukhina et al. (1999) predicted about 50% contribution of the magnetotail current to *Dst*.

The contributions of various currents to *Dst* can be evaluated not only from magnetic data but also from energetic particle observations. In particular, the symmetric ring current effect can be estimated from the well-known Dessler–Parker–Sckopke formula (Dessler and Parker, 1959; Sckopke, 1966)

$$B_z^{rc}(0) = -\frac{\mu_0}{4\pi}\frac{2E_{RC}}{M_E} \qquad (14.13)$$

in which E_{RC} is the total energy of the particles trapped in the dipolar magnetic field, and M_E is the Earth's magnetic moment. This is equivalent to Eq. (14.1). The total energy is given by

$$E_{RC} = \int_V w \, dV$$

where w is the plasma energy density, and dV is the elementary volume.

The energy density is related to pressure as

$$w = p_\perp + \frac{1}{2} p_\parallel$$

with $p_\perp$ and $p_\parallel$ being the perpendicular and parallel (to the magnetic field) pressure, respectively. Under quiet conditions, these quantities are maximum near $L = 3$ ($p_\perp \approx 20$ nPa, $p_\parallel \approx 7$ nPa) both in daytime and nighttime (Lui and Hamilton, 1992). At $L = 9$ the pressure turns out to be nearly isotropic ($p_\perp \approx p_\parallel \approx 1$ nPa).

A statistical test of relation (14.13) was performed by Greenspan and Hamilton (2000). During 80 storm events, AMPTE/CCE registered the energy content E_{RC} at distances from $L = 2$ to $L = 7$. Figure 14.8 shows the result of the *Dst* dependence on E_{RC}. The left panel corresponds to the 40 storms during which the satellite was located on the nightside, while the right panel is referred to the 40 storms with the measurements performed on the dayside. One can see the lack of any correlation between the dayside energy content and *Dst*, but at the nightside there is a high correlation. We should keep in mind that it is solely the symmetric ring current that flows on the dayside, while on the nightside, there are also cross-tail and partial ring currents. The right panel of Fig. 14.8 suggests that the symmetric ring current is practically not related to *Dst* and can hardly be considered as a principal cause of the geomagnetic depression.

The second statistical test of (14.13), performed by Turner et al. (2001), yields somewhat different results. The authors restore the total energy content E_{RC} in four MLT sectors from particle observations by the POLAR satellite. Their statistics included more than 1000 satellite passes through the ring current region along the polar orbit at altitudes from ~1 to ~8R_E. The time period covered was two and a half years. For moderately disturbed conditions ($Dst > -50$ nT), Turner et al. (2001) revealed nearly symmetric ring current, which provided 75% contribution to *Dst*. For storms with $Dst = -100$ nT, the ratio between the dayside and nightside currents is reduced, so that the symmetric ring current contribution to *Dst* dropped to 40%.

Presumably, the discrepancy between the results of the two extensive statistical studies can be attributed to the different types of satellite orbits. The AMPTE/CCE orbit was near the equatorial plane, so that all the trapped particles at a given L-shell were included. POLAR, whose orbit is limited in the meridian plane, however, could observe only part of the trapped particles.

14.6.3 Storm-Time Electric Fields in the Magnetosphere

The spatial distribution of the electric potential consists primarily of two cells, with maximum/minimum located at the dawn/dusk polar cap boundary. On the basis of Dynamics Explorer-2 (DE2) observations, Weimer (1995) constructed the ionospheric potential distributions for various IMF clock angles in the GSM system: see Fig. 14.9. The necessary condition for the initiation of the storm main phase is a prolonged southward IMF of large values, so that the electric potential distribution in the bottom row of Fig. 14.9 is typical for storm conditions.

The potential drop U between the centers of the dawn and dusk cells grows with the southward IMF increasing. In the study of Doyle and Burke (1983) the following empirical relation was obtained:

$$U(\text{kV}) = 55.3 + 14E_y(\text{mV/m}) \qquad (14.14)$$

where $E_y = -VB_z$ is the duskward component of the interplanetary electric field, and V is the solar wind speed.

According to Boyle et al. (1997), however, the approximation that fits best has the form

$$U = 10^{-4}V^2 + 11.7B\sin^3(\theta/2) \qquad (14.15)$$

in which B is the IMF modulus, V is the solar wind speed in km/s, and $\theta = \arccos(B_z/B)_{GSM}$.

During moderate storms the potential drop across the polar cap varies in a certain range, with the average value rarely exceeding 150 kV and only rarely reaching 200 kV. At the main phase of a storm on July 8–9, 1991 ($Dst_{min} = -190$ nT), the value of U was greater than 200 kV, according to DMSP observations in the inner magnetosphere (Wilson et al., 2001). Magnetospheric

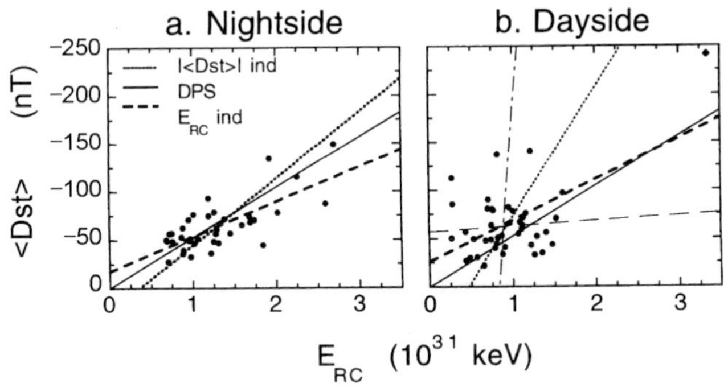

Fig. 14.8. The *Dst* index versus the energy content (*dots*) according to AMPTE observations: (*Left*) at the nightside, (*Right*) at the dayside. The *solid lines* present the dependence expected from the Dessler–Parker–Sckopke formula with the amplifying coefficient $k = 1.3$ due to the induction currents within Earth. After Greenspan and Hamilton (2000)

observations show at times a significant enhancement of the electric field during superstorms. A very strong relative increase (nearly 60 times) in the electric field was observed by CRRES during a storm on March 24, 1991 (Wygant et al., 1998).

14.7 Discussion on Open Issues of Geomagnetic Storm Dynamics

14.7.1 Influence of the Solar Wind Parameters on *Dst*

In examining storm time behavior of *Dst*, we recognize that this index is typically presented as a sum of the rapidly and slowly varying components as

$$Dst = DCF + Dst^* \tag{14.16}$$

where *DCF* includes the rapidly varying (on a time scale of a few minutes) component, which is controlled by the solar wind dynamic pressure; and Dst^* is the component varying on the time scale of more than several hours. It is generally accepted that

$$DCF = a\sqrt{p_{\rm sw}^{\rm dyn}} \tag{14.17}$$

where a is a coefficient and $p_{\rm sw}^{\rm dyn}$ is the dynamic pressure of the solar wind protons. In the study of O'Brien and McPherron (2000) using observations over the period of 30 years, it is shown that the value of a amounts to $7.26\,{\rm nT}/({\rm nPa})^{1/2}$. During storms occurring under southward IMF, the value of a decreases by a factor of 2.

The behavior of the slowly varying component of *Dst* is commonly described by the following differential equation:

$$\frac{{\rm d}Dst^*}{{\rm d}t} = Q - \frac{Dst^*}{\tau} \tag{14.18}$$

where τ is the relaxation time, and Q is the coupling function (i.e., the rate of energy injections) resulting from solar wind-magnetosphere coupling. Burton et al. (1975) found that Q is linearly proportional to the duskward component of the solar wind electric field. Later a number of studies were performed aiming to relate the coupling function to other solar wind parameters. Table 14.1 summarizes the principal features of Q and τ for the main phase ($\tau_{\rm mp}$) and recovery phase ($\tau_{\rm rp}$) of storms.

From Table 14.1 one can see that there is a deal of controversy with regard to the functional forms of Q and τ. An attempt was made to minimize this misleading effect by Maltsev and Rezhenov (2003). For this purpose one can treat the relationship between ${\rm d}Dst^*/{\rm d}t$ and a certain solar wind parameter with all the others varying in a narrow range. More than 100,000 hourly values of the solar wind observations from the OMNI database were processed. It was found that (1) τ is governed by the electric field component $E_{\rm y}^{\rm r} = -VB_{\rm s}$, not by Dst^*, (2) a dependence of ${\rm d}Dst^*/{\rm d}t$ on the solar wind speed persists both for southward and northward IMF, (3) no noticeable correlations with the solar wind density or horizontal IMF component were revealed, and (4) the dependence on the ε parameter under the electric field $E_{\rm y}^{\rm r} = -VB_{\rm s}$ being fixed appeared to be weaker than that on the electric field $E_{\rm y}^{\rm r}$ under constant ε. The results can be presented as $Q = 1.05 - E_{\rm y}^{\rm r}(4.00 + Dst^*/47.2) - V/243$ and $\tau = 15.4$ hrs. Most of the results presented in Table 14.1 suggest nearly linear coupling of *Dst* with the solar wind parameters and strongly non-linear decay, dependent on both *Dst* and Q. However, the technique for calculating Q and τ does not seem to yield a unique solution, and we may have to enforce non-linear coupling and linear decay.

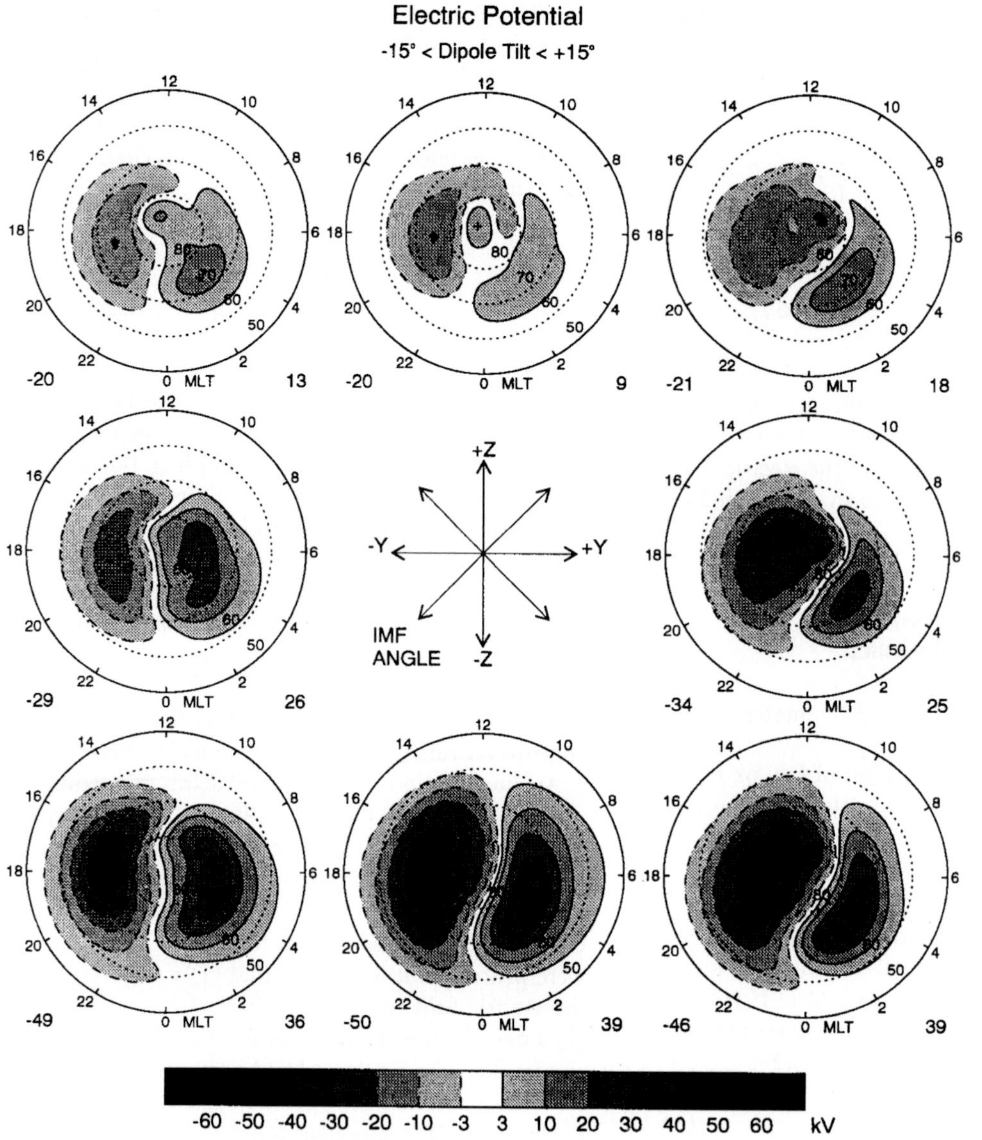

Fig. 14.9. The potential distribution in the high-latitude ionosphere as a function of the IMF clock angle. After Weimer (1995)

14.7.2 Influence of the Substorm Expansion Phase on *Dst*

A substorm is a magnetospheric process of energy accumulation in the magnetotail as a result of solar wind-magnetosphere coupling, followed by a subsequent explosive release of energy onto the polar ionosphere and into the inner magnetosphere (Rostoker et al., 1980). A typical substorm lasts from 1 to 3 hours. The stage of the energy accumulation, i.e., the growth phase, is accompanied by stretching of magnetic field lines tailward. The expansion phase is signified by dipolarization of the magnetospheric magnetic field, and by a sudden brightening of the auroral arcs and their rapid expansion poleward and westward and intensifications of the auroral electrojet in the ionosphere. Field-aligned currents increase with the onset of substorm expansion phase. They supply energized O^+ ions of ionospheric origin to the ring current, so that O^+ ions often become a dominant

Table 14.1. Coupling function Q, which relates the *Dst* index to solar wind parameters, and relaxation times of the storm-associated currents τ_{mp} and τ_{rp} for the main and recovery phases, respectively, as summarized by different studies

No	Reference	Q	τ_{mp}	τ_{rp}
1	Pudovkin et al. (1988)	$Q = -3.5 + 4.3V(0.56 - B_z)\,10^{-3}$ for $V(0.56 - B_z) \geq 1.5 \times 10^3$	$\tau = 3.0 + 9.8\,e^{-Q/4.5}$	$\tau = 6.6 + 0.07\,\lvert Dst^*\rvert$
		$Q = 3$ nT/hr for $0 \leq V(0.56 - B_z) \leq 1.5 \times 10^3$		
		$Q = 0$ for $0 \leq V(0.56 - B_z) < 0$		
2	Feldstein (1992)	$Q = 8.2 \times 10^{-3} V\,(B_z - 0.67\sigma) - 14.1 \times 10^{-3}\,(V - 300) + 9.4$ for $V(B_z - 0.67\sigma) < -1146$	Weak and moderate storms ($Dst^* > -160$ nT): $\tau = 1.6 + 13\,e^{0.08Q}$	$\tau = 5.4 + 10\,e^{0.025\,Dst_0}$
		$Q = -14.1 \times 10^{-3}\,(V - 300)$ for $V(B_z - 0.67\sigma) > -1146$	Strong storms ($Dst^* < -160$ nT): $\tau = 2.4 + 13\,e^{0.07Q}$	$\tau = 10 + 1.84\,e^{0.07\,Dst_0}$
3	Gonzalez et al. (1994)	$Q \propto -\varepsilon$	$\tau = 4$ hr for $Dst \geq -50$ nT; $\tau = 0.5$ hr for $-50 > Dst \geq -120$ nT; $\tau = 0.25$ hr for $Dst < -120$ nT	
4	Valdivia et al. (1996)	$Q = -3.1E_y^r$	$\tau = 12.5/(1 - 0.0012\,Dst^*)$	
5	O'Brien and McPherron (2000)	$Q = -4.4\,(E_y - 0.5)$ for $VB_s > 0.49$ mV/m	$\tau = 2.40 \exp\left[9.74/\left(4.69 + E_y^r\right)\right]$	
		$Q = 0$ for $VB_s < 0.49$ mV/m		
6	Maltsev and Rezhenov (2003)	$Q = 1.05 - 4.00E_y^r - V/243$	$\tau = 15.4/(1 + 0.326E_y^r)$	

Q is expressed in nT/hr; τ is in hrs, B_z and B_s as well as the IMF variability σ are in nT; the electric field component $E_y = -VB_z$ is in mV/m, the solar wind velocity V is in km/s; the quantity $E_y^r = -VB_s$ is the duskward electric field component; B_s is the southward IMF component in the GSM coordinates, with adopting $B_s = B_z$ under $B_z < 0$ and $B_s = 0$ under $B_z > 0$; and $\varepsilon = VB^2 \sin^4(\theta/2)\, l_o^2$ where B is the IMF modulus, $\theta = \arctan(B_y/B_z)$, and $l_o \approx 7R_E$ is the effective transverse size of the magnetosphere

ring current component in substorm periods (Daglis et al., 1997).

Nearly all storms at the main phase are accompanied by intense substorms. Among rare storm events not accompanied by substorms are those occurred, for example, on November 24, 1981 (Yahnin et al., 1994) and on July 15, 1997 (Zhou et al., 2003).

A statistical study of storm-substorm relationships was presented by Iyemori and Rao (1996). Using the superimposed epoch analysis of more than 100 substorms, they built average *SYM-H* index that is a one-minute resolution analog of *Dst*, having taken the substorm expansion onset for a zero moment. The result is shown separately for substorms at the storm main phase and those at the recovery phase, but for both phases, *SYM-H* (or *Dst*) was found to show a slight weakening after the expansion onset of substorms.

Huang et al. (2004) examined a number of samples of magnetic field dipolarization associated with substorm onsets in the nightside magnetosphere, showing that all of them lead to a noticeable increase in *Dst* by 20 – 40 nT, i.e. to storm subsiding.

These observations have led a number of researchers to the conclusion that there is no direct relation between substorms and storms, although both phenomena develop under the same condition, namely, the southward IMF. The average substorm initiates under IMF $B_z \approx -2$ nT, while the main phase of the average storm proceeds under IMF $B_z \approx -5$ nT. This implies that an average storm is accompanied by substorms with the intensity higher than the average intensity. Although the ring current must be strengthened by energized ion injections from the magnetotail and ionosphere, this strengthening can be overwhelmed by the competing effect of magnetotail current reduction at substorm expansion onsets. This can explain why the resulting geomagnetic depression, as measured by *Dst*, weakens after substorm explosive onset.

14.7.3 What Causes Storm-Time Equatorward Shift of the Auroral Oval?

It is generally accepted that the equatorward edge of the auroral oval at the nightside is coincident with the outer boundary of the trapping region of energetic electrons. In Fig. 14.10, the region of the trapped energetic electrons is shown to be confined to the inner magnetosphere, while the auroral oval is associated with more distant magnetospheric domains. We should then explain why the boundary between the inner and outer magnetospheric regions moves earthward during magnetic storms. Within the framework in Fig. 14.10, such a reconfiguration of the magnetosphere is caused by the large-scale currents in the magnetosphere.

It is clear from Fig. 14.7 that the FACs produce mostly a longitudinally asymmetric distortion of the trapping region boundary, without changing significantly its radial size. In the model of Alexeev et al. (2000) the Region 1 FAC, whose intensity is 5 MA, displaces the auroral oval sunward by 8° on the dayside and by 3° at night. The shielding currents on the magnetopause are capable of decreasing the extent of the inner magnetospheric region but cannot change

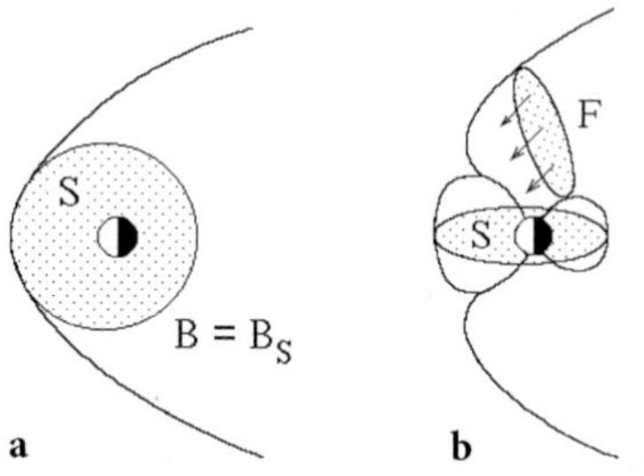

Fig. 14.10a,b. Magnetospheric model used in the calculations of storm time geomagnetic depression: (**a**) equatorial plane, (**b**) view from the duskside. S is the equatorial cross-section of the inner magnetosphere confined by the contour $B = B_s$, B_s being the magnetic field in the subsolar point on the magnetopause. F is the magnetic flux in the outer magnetosphere. The outer magnetosphere contains both the closed magnetic field lines of the plasma sheet and open field lines of the magnetotail lobes

noticeably its ionospheric projection. Thus a significant decrease in the auroral oval latitude can only be achieved by redistribution of the magnetic flux in the magnetosphere, specifically by its enhancement in the outer magnetosphere.

The first attempt to interpret such a magnetic flux redistribution in terms of the ring current (Siscoe, 1979a) indicated that such a redistribution could be accomplished only if the ring current flew beyond the stable trapping region. In other words, it is the cross-tail current that should be considered as an ultimate cause of this effect. In Siscoe (1979b) the contribution of the ring current, defined as a current flowing within the stable trapping region, to the observed storm time enhancement of the magnetic flux in the polar cap was estimated as only 25%. Schulz (1997) suggested that every 100 nT of the magnetic disturbance in the Earth's center associated with the ring current (or 130 nT with the induction currents inside the Earth included) can displace the auroral oval toward the equator by 2°. From Fig. 14.5, however, it is seen that in reality this displacement is, at least, two times greater.

The reason why the ring current can only slightly affect the magnetic flux redistribution between the inner and outer magnetosphere is that the area of the magnetic depression region, associated with the ring current, is rather small, compared to that of the cross-tail current.

The total effect of all magnetospheric currents can be expressed as follows (Maltsev et al., 1996):

$$B_z^{\mathrm{ext}}(0) = \frac{2}{3}\sqrt{2\mu_0 p_{\mathrm{sw}}} + B_z^{\mathrm{rc}}(0) - \frac{F}{3S} \qquad (14.19)$$

where p_{sw} is the solar wind dynamic pressure in the stagnation point on the magnetopause, and $B_z^{\mathrm{rc}}(0)$ is the magnetic field of the ring current; and the physical meaning of the magnetic flux F and cross-section S is illustrated in Fig. 14.10.

The high-latitude magnetic flux can be written as

$$F = \pi R_{\mathrm{E}}^2 (2B_{\mathrm{e}}) \sin^2\theta_{\mathrm{a}} \qquad (14.20)$$

where $B_{\mathrm{e}} = 31{,}000$ nT is the dipole magnetic field at the Earth's equator, θ_{a} is the colatitude of the equatorward edge of the auroral oval averaged over longitude. The area of the equatorial cross-section S is equal to $S = \pi r_{\mathrm{s}}^2$, where r_{s} is the radius of the contour $B = B_{\mathrm{s}}$. In the case of dipolar magnetic field, $r_{\mathrm{s}} = R_{\mathrm{E}}/\sin^2\theta_{\mathrm{a}}$.

For strong storms, however, this expression transfers to $r_s = (3/2)R_E/\sin^2\theta_a$ (Maltsev et al., 1996). As a result, we have for storm-time conditions

$$S = \frac{9}{4}\frac{\pi R_E^2}{\sin^4\theta_a} . \quad (14.21)$$

The equatorward boundary of the auroral oval can be approximated by a circle centered at latitude of 85° at midnight during disturbed periods, i.e.,

$$\Lambda = 85^\circ - \theta_a . \quad (14.22)$$

By combining (14.19) through (14.22), one can obtain the relation of Λ to $Dst = kB_z^{ext}(0)$. The solid lines in Fig. 14.11 show the dependence Λ (Dst) for two levels of the solar wind dynamic pressure: $p_{sw} = 0$ and 4 nPa. The dashed curve shows the observed dependence from empirical formula Eq. (14.3). It is seen that the observed dependence is close to the calculated one, indicating that the cross-tail current, or the magnetic flux in the tail, is as important as the ring current. A major role of the magnetotail currents in the storm time decrease of the auroral oval latitude has also been pointed out by other observations.

14.7.4 Why Does the IMF Southward Component Affect *Dst*?

A theoretical framework for the coupling function Q has been developed by Arykov and Maltsev (1996) who concentrated on the magnetotail current effect on Dst, by including the cross-tail current, closure currents on the magnetopause, and the partial ring current. During storms, the third term $(-F/3S)$ on the right-hand side of (14.19) dominates. Having differentiated (14.19) with respect to t and keeping in mind that, according to (14.20) and (14.21), S is proportional to F^{-2}, we have

$$\frac{dB_z^{ext}(0)}{dt} = -\frac{1}{S}\frac{dF}{dt} . \quad (14.23)$$

It is well known that the magnetic flux in the tail grows when the IMF is southward. According to (14.23), this results in strengthening the geomagnetic depression $kB_z^{ext}(0)$. The high latitude magnetic flux satisfies the Maxwell equation

$$\frac{dF}{dt} = C \quad (14.24)$$

where C is the electric field circulation along the contour $B = B_s$, which separates the inner and outer magnetospheric regions: see Fig. 14.10. The circulation C can be divided into the dayside and nightside portions:

$$C = C_{day} + C_{night} . \quad (14.25)$$

For the dayside portion we adopt

$$C_{day} = U \quad (14.26)$$

where U is the convection-associated potential drop between the dawn and dusk boundaries of the inner magnetosphere. The nightside portion C_{night} can be presented as

$$C_{night} = -\frac{F - F_0}{\tau_F} \quad (14.27)$$

where F_0 is the undisturbed quantity of the flux, τ_F is the relaxation time. With the induction currents inside the Earth included, the geomagnetic disturbance H is

$$H = kB_z^{ext}(0) \quad (14.28)$$

where $k \approx 1.3$. The potential U is equal to

$$U = \chi U_{PC} \quad (14.29)$$

where U_{PC} is the convection potential drop between the dawn and dusk flanks of the whole magnetosphere, which map onto the polar cap boundary, and χ is a certain coefficient ($\chi < 1$). After substituting (14.24)–(14.29) into (14.23), we have

$$\frac{dH}{dt} = Q - 3\frac{H}{\tau_F} - k\frac{F_0}{S\tau_F} \quad (14.30)$$

where

$$Q = -k\chi\frac{U_{PC}}{S} \quad (14.31)$$

is the coupling function, which relates the magnetotail current to the solar wind conditions.

Numerous studies have explored a statistical relation between the U_{PC} and solar wind parameters: see Table 14.1. To compare this theoretical elaboration with observations, we can use empirical formula (14.14) by Doyle and Burke (1983). Substituting (14.14) into (14.31), with $k = 1.3$, $\chi = 0.5$, and $S = 7.5 \times 10^{15}$ m^2 (corresponding to the circle with a radius of $7.7R_E$), we obtain $Q(\text{nT/hr}) = -4.4E_y(\text{mV/m}) - 16$. One can see that the thus calculated coupling function is

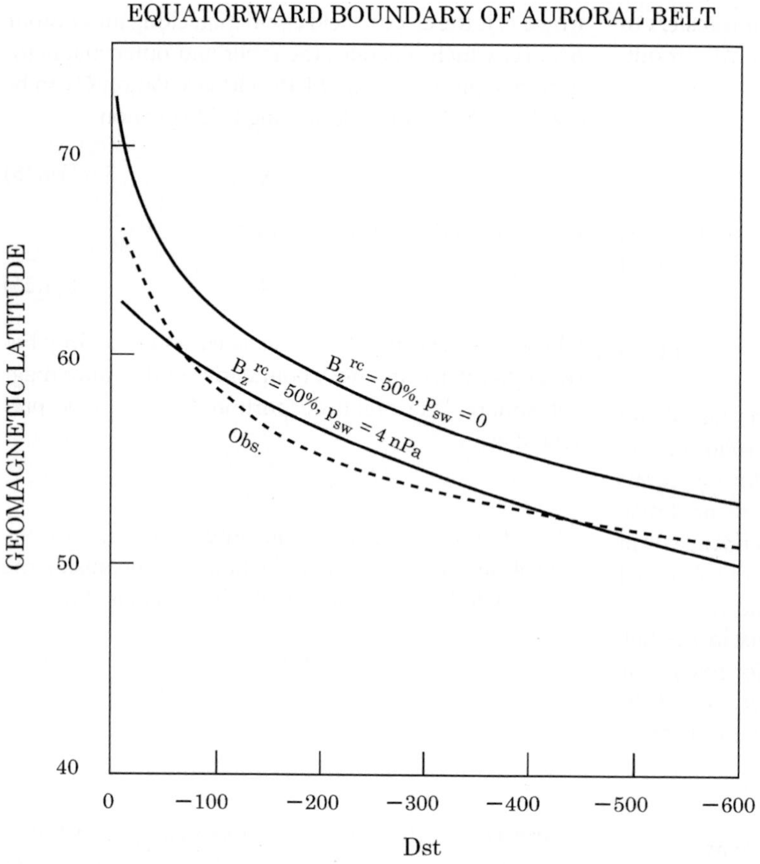

Fig. 14.11. Latitude of the equatorward edge of the auroral oval Λ versus magnetic storm intensity. The *solid lines* indicate the calculations, and the *dashed line* refers to the observations by Feldstein et al. (1997). After Maltsev et al. (1996)

close to what O'Brien and McPherron (2000) found empirically. Thus, it is quite possible that the relation of geomagnetic depression to solar wind parameters be almost entirely interpreted in terms of the magnetotail current system.

From comparing (14.18) with (14.30), we can see that $\tau = \tau_F/3$. That is, the decay of the storm time geomagnetic depression is related to relaxation of the magnetic flux in the tail. A theoretical framework for this process has yet to be developed.

14.8 Summary

Space storms, or geospace storms, are multi-faceted phenomena that occur in the entire solar-terrestrial environment. In this sense, geomagnetic storms are only one of the manifestations of space storms. Beginning with a brief description of the average character of geomagnetic storms along with some historical account, this chapter describes magnetospheric-wide dynamical changes during geomagnetic storms, which are caused directly and indirectly by the solar wind. Several unsolved questions are also discussed.

Acknowledgement. The authors would like to thank Y.I. Feldstein and I.A. Daglis for their constructive comments on an earlier version of the manuscript. This work was supported in part by "A Cooperative Research Program Between Japanese and US Space Weather Centers of Excellence" (grants AOARD-02-4013 and 00-40229) of the Asian Office of Aerospace Research and Development (AOARD), and in part by the Grant-in-Aid for Creative Scientific Research "The Basic Study of Space Weather Prediction" (grant 17GS0208, PI K. Shibata) of the Ministry of Education, Culture,

Sports, Science and Technology (MEXT) of Japan. This work was also supported in part by the Russian Basic Research Foundation (grants 03-05-65379 and 03-05-20003-BNTS) and by the Division of Physical Sciences of the Russian Academy of Sciences (program DPS-18).

References

Akasofu, S.-I., *Polar and Magnetospheric Substorms*, D. Reidel Publ. Co., Dordrecht, Holland, 1968

Alexeev, I.I., E.S. Belenkaya, and C.R. Clauer, A model of region 1 field-aligned currents dependent on ionospheric conductivity and solar wind parameters, *J. Geophys. Res.*, *105*, 21,119–21,127, 2000

Alexeev, I.I., E.S. Belenkaya, V.V. Kalegaev, Ya.I. Feldstein, and A. Grafe, Magnetic storms and magnetotail currents, *J. Geophys. Res.*, *101*, 7737–7747, 1996

Arykov, A.A., and Yu.P. Maltsev, Direct-driven mechanism for geomagnetic storms, *Geophys. Res. Lett.*, *23*, 1689–1692, 1996

Boyle, C.B., P.H. Reiff, and M.R. Hairston, Empirical polar cap potentials, *J. Geophys. Res.*, *102*, 111–125, 1997

Burton, R.K.., R.L. McPherron, and C.T. Russell, An empirical relationship between interplanetary conditions and *Dst*, *J. Geophys. Res.*, *80*, 4204–4214, 1975

Campbell, W.H., Geomagnetic storms, the *Dst* ring-current myth, and lognormal distribution, *J. Atmos. Terr. Phys.*, *58*, 1171–1187, 1996

Daglis, I.A., The role of magnetosphere-ionosphere coupling in magnetic storm dynamics, in *Magnetic Storms*, *Geophys. Monogr. Ser.*, vol. 98, edited by B.T. Tsurutani, W.D. Gonzalez, Y. Kamide, and J.K. Arballo, pp. 107–116, American Geophysical Union, Washington, DC, 1997

Daglis, I.A., Space storms, ring current and space-atmosphere coupling, in *Space Storms and Space Weather Hazards*, edited by I.A. Daglis, NATO Sci. Series, pp. 1–42, Kluwer Academic Pub., 2001

Daglis, I.A., R.M. Thorne, W. Baumjohann, and S. Orsini, The terrestrial ring current: Origin, formation, and decay, *Rev. Geophys.*, *37*, 407–438, 1999

Daglis, I.A., J.U. Kozyra, Y. Kamide, D. Vassiliadis, A.S. Sharma, M.W. Liemohn, W.D. Gonzalez, B.T. Tsurutani, and G. Lu, Intense space storms: Critical issues and open disputes, *J. Geophys. Res.*, *108*, 1208, DOI 10.1029/2002JA009722, 2003

Dessler, A.J., and E.N. Parker, Hydromagnetic theory of geomagnetic storms, *J. Geophys. Res.*, *64*, 2239–2252, 1959

Doyle, M.A., and W.I. Burke, S3-2 measurements of the polar cap potential, *J. Geophys. Res.*, *88*, 9125–9133, 1983

Dremukhina, L.A., Y.I. Feldstein, I.I. Alexeev, V.V. Kalegaev, and M.E. Greenspan, Structure of the magnetospheric magnetic field during magnetic storms, *J. Geophys. Res.*, *104*, 28,351–28,360, 1999

Fairfield, D.H., N.A. Tsyganenko, A.V. Usmanov, and M.V. Malkov, A large magnetosphere magnetic field database, *J. Geophys. Res.*, *99*, 11,319-11,326, 1994

Feldstein, Y.I., Modelling of the magnetic field of magnetospheric ring current as a function of interplanetary parameters, *Space Sci. Rev.*, *59*, 83–165, 1992

Feldstein, Y.I., A. Grafe, L.I. Gromova, and V.A. Popov, Auroral electrojets during geomagnetic storms, *J. Geophys. Res.*, *102*, No A7, 14,223–14,235, 1997

Frank, L.A., Direct detection of asymmetric increase of extraterrestrial "ring current" proton intensities in the outer radiation zone, *J. Geophys. Res.*, *72*, 3753–3767, 1967

Gonzalez, W.D., J.A. Joselyn, Y. Kamide, H.W. Kroehl, G. Rostoker, B.T. Tsurutani, and V.M. Vasyliunas, What is a geomagnetic storm? *J. Geophys. Res.*, *99*, 5771–5792, 1994

Gopalswamy, N., R. Mewaldt, and J. Torsti, editors, Solar Eruptions and Energetic Particles, *Geophys. Monogr. Ser.*, vol. 165, AGU, Washington, D.C., 386 p., 2006

Gosling, J.T., D.J. McComas, J.L. Phillips, and S.J. Bame, Geomagnetic activity associated with Earth passage of interplanetary shock disturbances and coronal mass ejections, *J. Geophys. Res.*, *96*, 7831–7841, 1991

Greenspan, M.E., and D.C. Hamilton, A test of the Dessler-Parker-Sckopke relation during magnetic storms, *J. Geophys. Res.*, *105*, 5419–5430, 2000

Hamilton, D.C., G. Gloeckler, F.M. Ipavich, W. Studemann, B. Wilken, and G. Kremser, Ring current development during the great geomagnetic storm of February 1986, *J. Geophys. Res.*, *93*, 14,343-14,355, 1988

Huang, C.-S., J.C. Foster, L.P. Goncharenko, G.D. Reeves, J.L. Chau, K. Yumoto, and K. Kitamura, Variations of low-latitude geomagnetic fields and *Dst* index caused by magnetospheric substorms, *J. Geophys. Res.*, *109*, doi:10.1029/2003JA010334, 2004

Iyemori, T., and D.R.K. Rao, Decay of the *Dst* field of geomagnetic disturbance after substorm onset and its implication to storm-substorm relation, *Ann. Geophys.*, *14*, 608–618, 1996

Joselyn, J.A., and B.T. Tsurutani, geomagnetic sudden impulses and storm sudden commencements, *Eos Trans., Amer. Geophys. Union*, *71*, 1808, 1990

Kamide, Y., W. Baumjohann, I.A. Daglis, W.D. Gonzalez, M. Grande, J.A. Joselyn, R.L. McPherron, J.L. Phillips, E.G.D. Reeves, G. Rostoker, A.S. Sharma, H.J. Singer, B.T. Tsurutani, and V.M. Vasyliunas, Current understanding of magnetic storms: Storm-substorm relationships, *J. Geophys. Res.*, *103*, 17,705-17,728, 1998

Lui, A.T. Y., and D.C. Hamilton, Radial profiles of quiet time magnetospheric parameters, *J. Geophys. Res.*, *97*, 19,325-19,342, 1992

Maltsev, Y.P., Points of controversy in the study of magnetic storms, *Space Sci. Rev.*, *110*, 227–277, 2004

Maltsev, Y.P., and A.A. Ostapenko, Azimuthally asymmetric ring current as a function of *Dst* and solar wind conditions. *Ann. Geophys.*, *22*, 2989–2996, 2004

Maltsev, Y.P., and B.V. Rezhenov, Relation of *Dst* index to solar wind parameters, *Int. J. Geomag. Aeron.*, *4*, 1–9, 2003

Maltsev, Y.P., A.A. Arykov, E.G. Belova, B.B. Gvozdevsky, and V.V. Safargaleev, Magnetic flux redistribution in the storm time magnetosphere, *J. Geophys. Res.*, *101*, 7697–7704, 1996

McPherron, R.L., The role of substorms in the generation of magnetic storms, in *Magnetic Storms*, edited by B.T. Tsurutani, W.D. Gonzalez, Y. Kamide, and J.K. Arballo, pp. 131–148, Geophys. Monograph 98, American Geophysical Union, Washington, D.C., 1997

Mead, G.D., Deformation of the geomagnetic field by the solar wind, *J. Geophys. Res.*, *69*, 1181–1195, 1964

Meng, C.-I., Dynamic variation of the auroral oval during intense magnetic storms, *J. Geophys. Res.*, *89*, 227–235, 1984

O'Brien, T.P., and R.L. McPherron, An empirical phase space analysis of ring current dynamics: Solar wind control of injection and decay, *J. Geophys. Res.*, *105*, 7707–7719, 2000

Parker, E.N., Newtonian development of the dynamical properties of ionized gases at low density, *Phys. Rev.*, *107*, 924, 1957

Roelof, E.C., and D.G. Sibeck, Magnetopause shape as a bivariative function of interplanetary magnetic field Bz and solar wind dynamic pressure, *J. Geophys. Res.*, *98*, 21,421–21,450, 1993

Rostoker, G., S.-I. Akasofu, J. Foster, R.A. Greenwald, Y. Kamide, K. Kawasaki, A.T.Y. Lui, R.L. McPherron, and C.T. Russell, Magnetospheric substorms – Definition and signatures, *J. Geophys. Res.*, *85*, 1663–1668, 1980

Rufenach, C.L., R.F. Martin, Jr., and H.H. Sauer, A study of geosynchronous magnetopause crossings, *J. Geophys. Res.*, *94*, 15,125–15,134, 1989

Schulz, M., Direct influence of ring current on auroral-oval diameter, *J. Geophys. Res.*, *102*, 14,149–14,154, 1997

Sckopke, N., A general relation between the energy of trapped particles and the disturbance field near the earth, *J. Geophys. Res.*, *71*, 3125–3130, 1966

Sharma, A.S., Y. Kamide and G.S. Lakhina, editors, *Disturbances in Geospace: The Storm-Substorm Relationship*, 268 pp, Geophys. Monogr. Ser., vol. 142, AGU, Washington, D.C., 2004

Siscoe, G.L., A quasi-self-consistent axially symmetric model for the growth of a ring current through earthward motion from a prestorm configuration, *Planet. Space Sci.*, *27*, 285–295, 1979a

Siscoe, G.L., A *Dst* contribution to the equatorward shift of the aurora, *Planet. Space Sci.*, *27*, 997–1000, 1979b

Starkov, G.V., Planetary morphology of auroras. In monograph "*Magnetosphere-Ionosphere Physics. Brief Handbook*", ed. by Yu.P. Maltsev, "Nauka" Publ., St-Petersburg, 85–90, 1993

Starkov, G.V., and Y.I. Feldstein, Variations of auroral oval boundaries, *Geomagnetism and Aeronomy*, *7*, 62–71, 1967

Tsurutani, B.T., and W.D. Gonzalez, The interplanetary causes of magnetic storms: A Review, in *Magnetic Storms*, *Geophys. Monogr. Ser.*, vol. 98, edited by B.T. Tsurutani, W.D. Gonzalez, Y. Kamide, and J.K. Arballo, pp. 77–90, AGU, Washington, D.C., 1997

Tsurutani, B.T., R. McPherron, W.D. Gonzales, G. Lu, J.H.A. Sobral, and N. Gopalswamy, editors, *Recurrent Magnetic Storms, Geophys. Monogr. Ser.*, vol. 167, AGU, Washington, D.C., 340 p., 2006

Tsurutani, B.T., W.D. Gonzalez, Y. Kamide, and J.K. Arballo, editors, *Magnetic Storms, Geophys. Monogr. Ser.*, vol. 98, AGU, Washington, D.C., 266 p., 1997

Tsyganenko, N.A., A model of the near magnetosphere with a dawn-dusk asymmetry, 2, Parameterization and fitting to observations, *J. Geophys. Res.*, *107*, doi:10.1029/2001JA000220, 2002

Turner, N.E., D.N. Baker, T.I. Pulkkinen, J.L. Roeder, J.F. Fennell, and V.K. Jordanova, Energy content in the storm time ring current, *J. Geophys. Res.*, *106*, 19,149–19,156, 2001

Vasyliunas, V.M., Ionospheric and boundary contributions to the Dessler–Parker–Schopke formula for *Dst*, *Ann. Geophys.*, *24*, 1058–1097, 2006

Weimer, D.R., Models of high-latitude electric potential derived with a least error fit of spherical harmonic coefficients, *J. Geophys. Res.*, *100*, 19,595–19,607, 1995

Williams, D.J., Ring current and radiation belts, *Rev. Geophys.*, *25*, 570–578, 1987

Wilson, G.R., W.J. Burke, N.C. Maynard, C.Y. Huang, and H.J. Singer, Global electrodynamics observed during the initial and main phases of the July 1991 magnetic storm, *J. Geophys. Res.*, *106*, 24,517–24,539, 2001

Wygant, J., D. Rowland, H.J. Singer, M. Temerin, F. Mozer, and M.K. Hudson, Experimental evidence on the role of the large spatial scale electric field in creating the ring current, *J. Geophys. Res.*, *103*, 29,527–29,544, 1998

Yahnin, A.G., M.V. Malkov, V.A. Sergeev, R.G. Pellinen, A. Fulamo, S. Vennerström, E. Friis-Christensen, K. Lassen, C. Danielsen, G. Craven, and C. Deehr, Features of steady magnetospheric convection, *J. Geophys. Res.*, *99*, 4039–4051, 1994

Zhou, X.-Y., B.T. Tsurutani, G. Reeves, G. Rostoker, W. Sun, J.M. Ruohoniemi, Y. Kamide, A.T.Y. Lui, G.K. Parks, W.D. Gonzalez, and J.K. Arballo, Ring current intensification and convection-driven negative bays: Multisatellite studies, *J. Geophys. Res.*, *108*, doi:10.1029/2003JA009881, 2003.

15 Substorms

Gordon Rostoker

A magnetospheric substorm is a term ascribed to magnetospheric and ionospheric activity that takes place during an interval of time that begins with enhanced energy input from the solar wind into the magnetosphere and ends with a decrease of that energy input to its original value. The study of substorms breaks down into three major thrusts. First of all, there are the observations which have been well documented since the latter part of the 19th century. Secondly, there are the studies of the physical processes whereby energy from the solar wind enters the magnetosphere and is dissipated within geospace, that have been vigorously pursued since the early 1960's. As part of this effort, several different frameworks have been proposed, each based in the physics of magnetized space plasmas, that purport to reflect the way in which substorm activity develops. Finally, since the early 1980's, the evolution of high performance computing has led to the ability of researchers to simulate the magnetosphere-ionosphere system and to probe its response to changing solar wind conditions. In this chapter, we shall look at substorms in terms of the first two thrusts mentioned above. We shall begin by outlining the historical development of the substorm concept, which is essential for understanding how that concept developed as increasingly better data became available. Such an approach is necessary so older papers that contributed to the development of the context can be more critically assessed.

Contents

Gordon Rostoker, Substorms.
In: Y. Kamide/A. Chian, Handbook of the Solar-Terrestrial Environment. pp. 375–395 (2007)
DOI: 10.1007/11367758_15

15.1 Observations

15.1.1 Historical Perspective

The term substorm was introduced by Akasofu (1964) to describe a repeatable sequence of auroral disturbances that he identified using allsky cameras concentrated in the Alaska sector supplemented by pictures from widely spaced allsky cameras in other parts of the world. This sequence was termed an *auroral substorm* and is shown in Fig. 15.1 as a series of polar plots in which the evolution of bright auroral forms is detailed. The evolution is broken down into an expansion/expansive phase and a recovery phase. The onset of the *expansion phase* (EP) is marked by brightening of an auroral arc in the midnight sector. It was not unusual for there to be several parallel arcs across the midnight sector, and the so-called breakup arc was said to be typically the most equatorward of these arcs. The auroral substorm then developed through the poleward expansion of the region of activated auroras, that region featuring a sharp western edge known as the westward traveling surge. The disturbed region actually expanded both westward and eastward during the EP (Panels A–D). The EP was then followed by an interval termed the *recovery phase* during which the auroras returned to their pre-expansive phase condition (Panels E–F), the entire substorm lasting over an interval of 1 – 2 h.

While the term substorm was only coined in the early 1960's, the disturbance itself had been studied for many decades before the term came into existence. Perhaps the most comprehensive studies carried out well before the so-called "space age" began, were those by Birkeland (1908) who investigated auroral and magnetic disturbances in the high arctic during the Norwegian Aurora Polaris Expedition of 1902–1903. The "polar elementary storms" identified and defined by Birkeland were actually the magnetic disturbances that accompanied Akasofu's auroral substorm and were subsequently renamed polar magnetic substorms by Akasofu. Birkeland's 1908 publication contained a remarkably comprehensive treatment of polar elementary storms, including the proposal that the magnetic variations were caused by a three dimensional current system shown in Fig. 15.2. Over the following decades, the substorm phenomenon was studied using several different observational techniques, but most researchers did not relate the measurements using their specific techniques to the measurements of the same phenomena by others using different techniques. Akasofu (1968) brought together all the observations using the different techniques under the umbrella of the *magnetospheric substorm*. The various observational techniques and what they measured were summarized by Akasofu (1968) and synthesized to create the overall substorm concept.

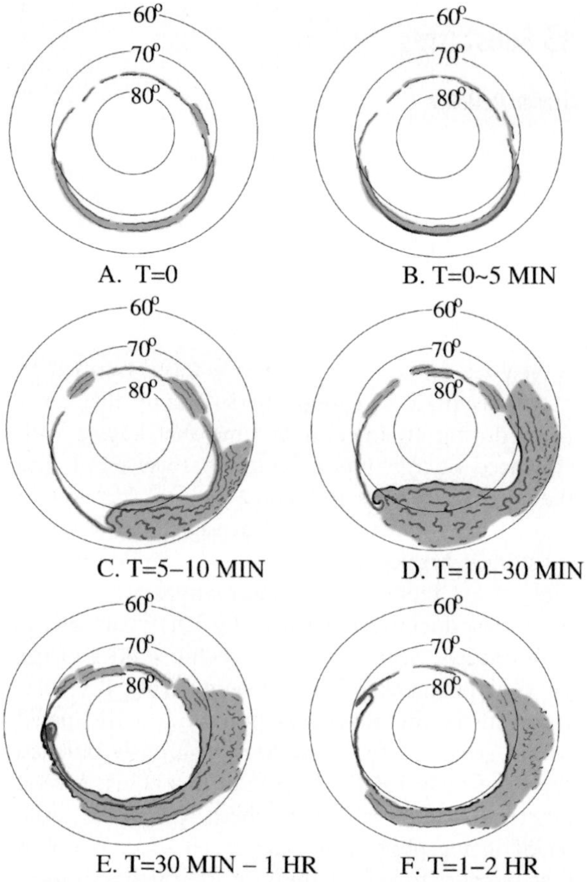

Fig. 15.1. The development of an auroral substorm (modified after Akasofu, 1964). This sequence describes the evolution of the expansive phase, and was based on local Alaskan allsky camera observations and limited allsky camera data from other locations around the world. This cartoon representation was built on observations of relatively bright discrete aurora, as the allsky cameras of the day could not detect diffuse aurora (represented by *green shading* in the figure)

At the time these definitions were proposed, the measuring techniques were relatively crude compared to those employed nowadays. Auroras were monitored primarily by allsky cameras that were only capable of

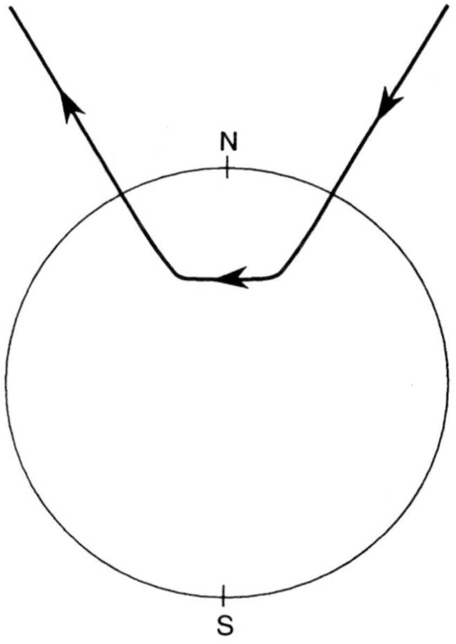

Fig. 15.2. The three dimensional current system proposed by Birkeland (1908) to describe the magnetic disturbance associated with a polar elementary storm

detecting bright auroras with a nominal sample time of one minute. Nowadays, digital imagers using charge coupled devices (CCDs) can easily detect the diffuse aurora that permit the entire latitudinal width of the auroral oval to be assessed. Chart recordings of the magnetic field at irregularly spaced locations have been replaced by digital logging from carefully distributed arrays of stations (cf. Rostoker et al., 1995). X-rays are rarely monitored on balloon borne platforms, as their use was originally to provide information about precipitating energetic electrons, and these are now measured directly aboard polar orbiting satellites. Thus, the technique oriented definitions of the substorm are now more of historical interest, and the substorm is now viewed more as a morphology for which a physical explanation is sought.

15.1.2 Observational Basis for the Evolution of the Substorm Concept: Pre-Satellite Era

While the polar elementary storm defined by Birkeland represented the first named feature of what we now call a substorm, the studies of the phenomenon over the following century did not follow a well defined path and, in many cases, provided conflicting pictures for the researchers who were looking for physical explanations. The observations on which the study of substorms developed over the first half of the 20th century were primarily the visible aurora and the magnetic field fluctuations that accompanied changes in auroral luminosity. As the magnetic field measurements were more readily acquired than auroral observations on a global scale, they form the basis for most of the discussions that led to the modern view of the substorm. In the balance of this section, the development of the phenomenology of the substorm based on ground magnetometer measurements will be detailed.

After the pioneering work of Birkeland, for the next four decades there were few studies that impacted the area of what would later be considered as substorm research. Those studies that were done tended to use hourly averages or three hour averages of the magnetic field perturbations at individual sites. The results were usually portrayed as equivalent current systems in which it was assumed that the observed horizontal component of the perturbation magnetic field was due to an overhead current sheet of infinite extent. Accordingly, at each station the horizontal magnetic field disturbance vector was rotated through 90° and closed current loops were constructed based on the equivalent current vectors so obtained. One of the early studies of this nature was by Silsbee and Vestine (1942) using three hour magnetic field average values from a sparse global array of ground stations. Their equivalent current system featured a westward electrojet in the morning sector and a weaker eastward electrojet in the afternoon sector of the high latitude. A more comprehensive study involving hourly averaged data was carried out by Harang (1946) using an approximately north–south line of magnetometers through Scandinavia which covered 24 hours of local time as the earth rotated. In contrast to the Silsbee and Vestine, Harang studied the horizontal and vertical disturbance vectors from which he was able to infer more information about the electrojet structures. Harang portrayed his data in the form of contour plots of the magnetic field perturbations, one of which is shown in Fig. 15.3. From this plot, it was inferred that the high latitude currents were dominated by an eastward electrojet in the evening sector and a westward electrojet in

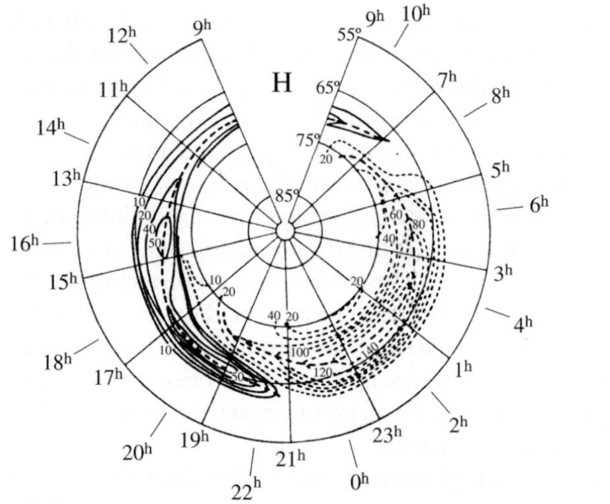

Fig. 15.3. Contour plot of the H-component disturbance at high latitudes based on hourly averaged values. Times are given in UT (*inner numbers*) and magnetic local time (*outer numbers*). This plot is consistent with the presence of an eastward electrojet in the evening sector (*solid contour lines*) and a westward electrojet in the morning sector (*dashed contour lines*). It also points to westward current flowing at the poleward edge of the evening sector eastward electrojet (after Harang, 1946)

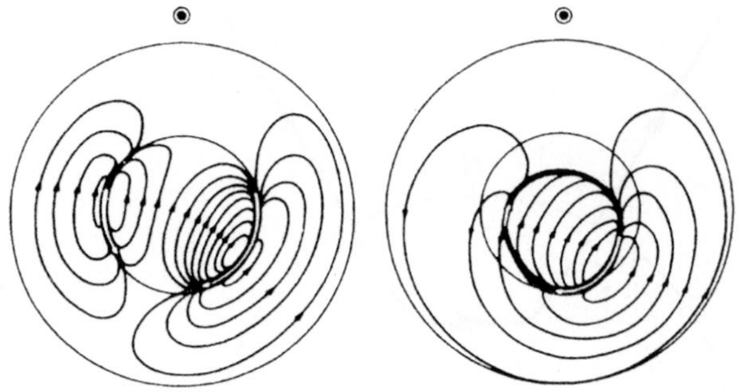

Fig. 15.4. One cell and two cell equivalent current system, each of which have been proposed in the past to describe the substorm disturbance (after Rostoker, 1996). The currents are seen looking down on the pole with the position of the sun being indicated by the *circular symbol at the top* of each panel

the morning sector, similar to the result of Silsbee and Vestine described earlier. It should be noted that the equivalent current system that would have been inferred from Birkeland's measurements involved only a westward electrojet in the ionosphere, and therefore was inconsistent with the results of both Silsbee and Vestine and of Harang.

A major advance came with the work of Fukushima (1953) who constructed equivalent current systems for geomagnetic bays using instantaneous values of the magnetic field perturbations from a global array of observatories during bay disturbances. (Prior to the appearance of the term substorm, the magnetic disturbance associated with a substorm was known as a bay because its appearance on a magnetogram looked like a bay on the coastline of a continent.) Fukushima discovered that, at times during periods of strong activity, the equivalent current pattern resembled that of Birkeland (one cell) and at times that of Silsbee and Vestine and Harang (two cell), these patterns being shown in Fig. 15.4. Later studies did nothing but stress this apparent inconsistency. After Akasofu (1964) introduced the substorm terminology, Sugiura and Heppner (1965) argued that the substorm disturbance was best represented by the two cell system while Akasofu et al. (1966) maintained that a one cell system provided the best description. Rostoker (1969) resolved the controversy by suggesting that both types of equivalent current patterns were present during substorms. Figure 15.5 shows a magnetogram from the high latitude station of Tromsø. This magnetogram features four clear bay disturbances between 1830 and 2230 UT, each with a time scale of about one hour and characterized by a one cell equivalent current system. However, a longer time scale disturbance (indicated by the dashed line) is evident in the H-component and it is this disturbance which is characterized by the two cell equivalent current system.

15.1.3 Observational Basis for the Evolution of the Substorm Concept: Satellite Era

In the late 1960's, there was an important development that amended the Akasofu picture of the substorm. This was the identification of a period of time, in advance of EP onset, during which it was suggested that energy was being stored in the near–earth magnetotail. McPherron (1970) used ground magnetometer data to identify this period of time, which was termed the *growth phase* of a substorm. Around the same time, early observations of the geostationary orbit environment in the midnight sector indicated that, during growth phase episodes the magnetic field was distorted from

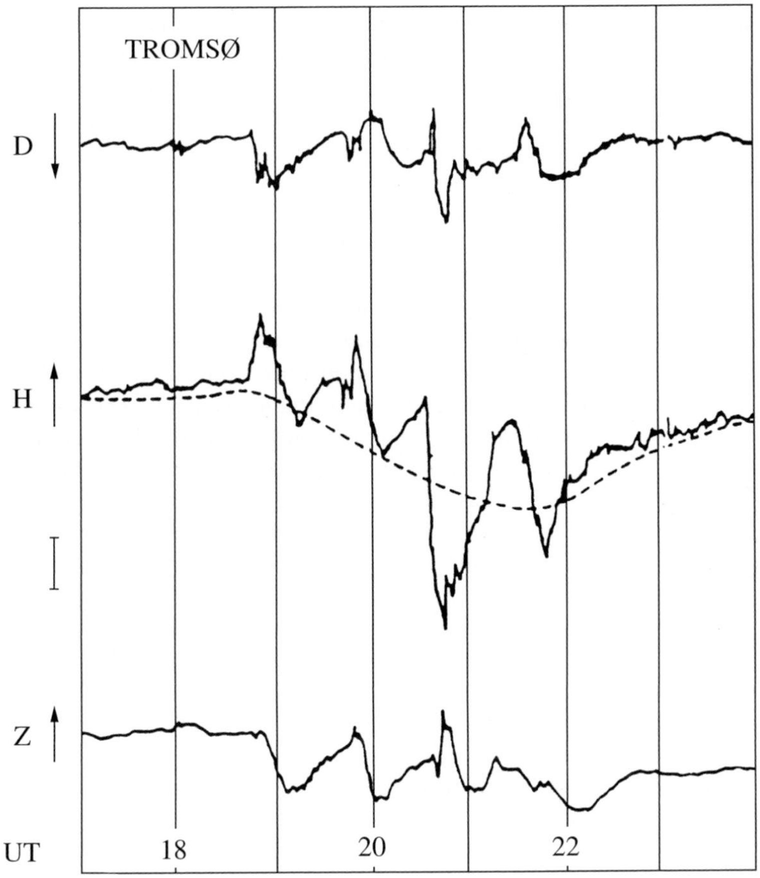

Fig. 15.5. Tromsø magnetogram showing four polar magnetic substorm disturbances superposed on a long period disturbance seen best as a negative perturbation in the H-component (after Rostoker, 1969). It is now understood that the polar magnetic substorms are best described by a one cell equivalent current system, while the longer period disturbance is best described by a two cell equivalent current system

its dipolar state to become more tail like. In contrast, during substorm EPs the magnetic field at geostationary orbit suddenly returned to a more dipolar configuration (cf. Cummings et al., 1968), the episode being termed *dipolarization.*

For most of the 1970's, very little was published about the disturbances that produced the two cell equivalent current system during substorm activity, with the exception of the work by Kamide and Fukushima (1972) that drew attention to the asymmetric ring current and the eastward electrojet in the afternoon sector that was connected to it through field-aligned currents that flowed downward near noon and upward in the evening sector. Attention was focused primarily on the onset and subsequent development of the early stages of the EP which was best represented by the one cell equivalent current system. This emphasis came as a result of suggestions regarding the nature of the real current system associated with the substorm EP. The equivalent three dimensional current system that had been proposed for the substorm EP, shown in Fig. 15.6a, involved an eastward current in the equatorial plane. At around the same time, Akasofu and Chapman (1972) and McPherron et al. (1973) proposed that the three dimensional current flow was initiated by a diversion of crosstail current into the high latitude ionosphere (cf. Fig. 15.6b). The eastward equivalent current in the equatorial plane could then be understood as a reduction in the westward crosstail current. This real three dimensional current system produced a magnetic perturbation that was consistent with a one cell ionospheric equivalent current system and in more modern times, has been termed the *substorm current wedge.*

Towards the end of the 1970's, Perreault and Akasofu (1978) developed an expression quantifying the flow of

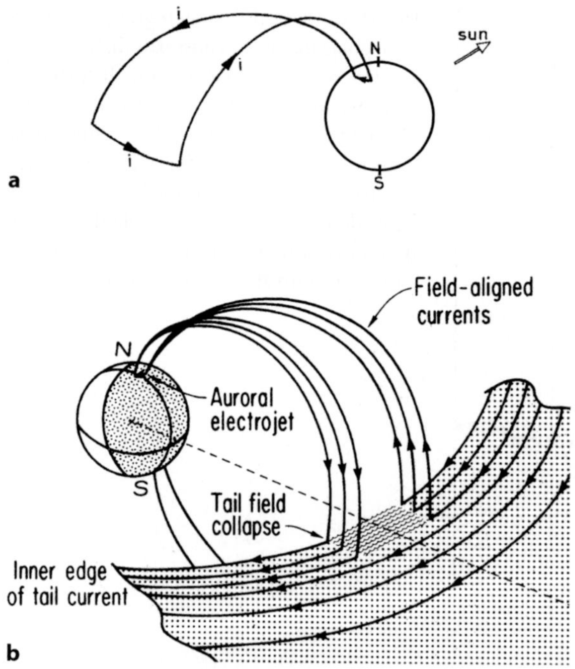

Fig. 15.6. (**a**) Equivalent three-dimensional equivalent current system for a polar magnetic substorm (after Bonnevier et al., 1970). (**b**) Tail current disruption leading to a real three dimensional current system proposed for a polar magnetic substorm (after McPherron et al., 1973)

energy into the magnetosphere as a function of the solar wind parameters, viz.

$$\varepsilon = l_o^2 v B^2 \sin^4(\theta/2) \tag{15.1}$$

where v is the solar wind speed, B the magnitude of the interplanetary magnetic field (IMF), θ the polar angle of the component of the IMF normal to the sun-earth line measured from the northward geomagnetic axis, and l_o(~7 Re) a constant with the dimension of distance. They quantified the dissipation of energy within the magnetosphere by the parameter

$$U_T = U_A + U_J + U_R \tag{15.2}$$

where U_A is the rate of energy dissipation associated with the collision of precipitating energetic particles in the ionosphere, U_J is the rate of energy dissipation associated with Joule heating in the ionosphere and U_R is the rate of injection of energy into the ring current and tail current. The parameters U_A and U_J were evaluated using the auroral electrojet index AE (cf. Davis and Sugiura, 1966) while U_R was a complicated function of the ring current index D_{st}. Akasofu (1979) summarized the response of the magnetosphere (U_T) to the input of energy from the solar wind ε in the graphical form shown in Fig. 15.7. This response had two characteristic behaviors. The first of these was the *directly driven process* in which a portion of the energy entering the magnetosphere was dissipated within the magnetosphere-ionosphere system with a time lag of a few minutes (related to the propagation time from the outer regions of the magnetosphere to the ionosphere). The second, called the *loading–unloading* process, involved the storage of the remaining energy input in the magnetic field and plasma of the magnetotail with the loading on a time scale of the increase in the directly driven activity and the unloading occurring suddenly (the initiation being on the time scale of less than a minute). In retrospect, the growth of directly driven activity paralleled the growth phase of the substorm, while the onset of unloading marked the start of the substorm EP. The directly driven activity was characterized by the two cell equivalent current system, while the unloading led to real current flow that was characterized by the one cell equivalent current system.

The work of Perreault and Akasofu refocused the substorm community on the fact that two different types of activity co-existed during episodes of substorm activity, ending a period of fixation on the current wedge as the only signature of a substorm. While Rostoker (1969) had pointed out that the two cell (directly driven) and one cell (unloading) patterns co-existed, and Pytte et al. (1978) had presented disturbances called convection bays in which the two cell system was clearly dominant, the emphasis on the importance of directly driven activity in the substorm process was an important development. In the following years it became apparent that directly driven activity was, in terms of energy dissipation, a significant if not the dominant component of substorm activity (cf. Clauer et al., 1983; Goertz et al., 1993). The electric current picture of a substorm as solely a three dimensional current loop as envisioned by Birkeland and given a physical explanation through disruption of the crosstail current (cf. Fig. 15.6b) was supplanted by a two component picture as shown in Fig. 15.8 and summarized in detail by Kamide and Kokubun (1996).

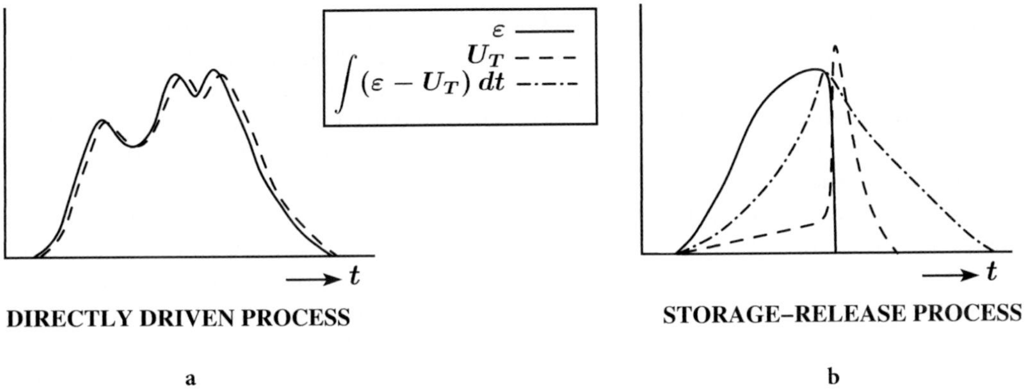

Fig. 15.7. Response of the magnetosphere-ionosphere current systems (U_T) to forcing from the solar wind ε (modified after Akasofu, 1979). Panel (**a**) shows the directly driven system response, which follows closely the changes in ε with a time delay related to the communication time from the magnetospheric boundary layers to the ionosphere. Panel (**b**) includes the triggering of an expansive phase through a reduction of energy flow into the magnetosphere, typically through a northward turning of the IMF. Note that not all expansive phases are necessarily triggered by northward turnings of the IMF

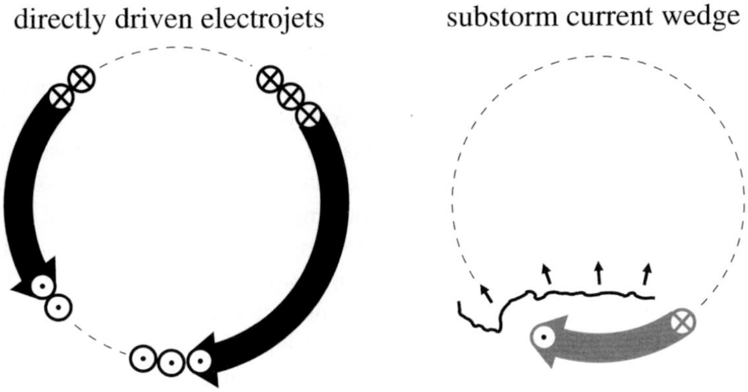

Fig. 15.8. Auroral electrojets associated with the two types of activity during substorms, viz. directly driven (*left panel*) and release of stored magnetotail energy (*right panel*)

15.1.4 The Solar Wind as a Driver of Substorm Activity

The concept of magnetic field line reconnection as a means of facilitating energy transfer from the solar wind to the magnetosphere outlined by Dungey (1961) began the process of understanding and predicting the onset of substorm activity. The theoretical concept was validated observationally in the middle 1970's (cf. Fairfield and Cahill, 1966; Rostoker and Fälthammar, 1967) and the explanation of the substorm EP in terms of magnetotail reconnection was advanced by Atkinson (1966) and confirmed observationally by Camidge and Rostoker (1970). The concept involved the storage of energy that entered the magnetosphere from the solar wind, the energy being stored as magnetic field energy in the plasma sheet and tail lobe and as kinetic energy of drift of plasma sheet particles. At some time after the energy began to enter the magnetotail, the substorm EP would facilitate the sudden dissipation of the stored energy through ionospheric heating and the growth of a ring current circling the earth. While increased substorm activity was generally attributed to enhanced energy input from the solar wind associated with a more southward pointing interplanetary magnetic field (IMF), the timing of EP onset vis-à-vis the start of energy input was initially somewhat of a puzzle. It became necessary to distinguish between the start of substorm activity and the onset of the EP (cf. Rostoker et al., 1980) and ultimately it became clear that EP onset was often initiated by a reduction of energy input from the solar wind into the magnetosphere. This reduction in energy input was normally caused by a turning towards the north of the IMF (cf. Caan et al., 1977; Rostoker, 1983) although it could also be achieved by a marked reduction in solar wind speed. The question of whether or not all substorm EP onsets could be attributed to northward turnings of the IMF has been a topic of intense study over the last two decades, with some researchers arguing that almost all EP onsets are caused by northward turnings (e.g. Lyons et al., 1997) and others arguing that many are triggered internally (e.g. Henderson et al., 1996). Most recently, Hsu and McPherron (2004) have suggested that both IMF triggered and internally triggered substorms occur, but the IMF triggered substorms are the stronger of the two types.

15.1.5 Evolution of a Substorm Optically and Magnetically: a Case Study

In this section, an example of substorm development will be presented that will provide a guide to researchers who require a template to which to refer in attempting to understand their observations. We will show the changes in several relevant parameters that are typically presented to characterize substorm disturbances. The event presented will highlight the importance of recognizing that there are two distinct regions of auroral activation during substorm activity, namely the region near the equatorward edge of the oval and the region near the poleward edge in the evening sector. Very similar auroral and magnetic disturbances are detected during substorm activity in both regions of disturbances, even though these region of the ionosphere map to quite different regions of the magnetotail and may involve different source mechanisms (cf. Rostoker, 2002). This feature of the nighttime auroral region has been termed the double oval by Elphinstone et al. (1995), and the implications of this structure have only begun to be studied systematically in the last decade.

Our case study will employ data from an interval of time on December 7, 1999. Figure 15.9 shows the solar wind parameters during the interval of interest. All times should be shifted by ~37 min to reflect the propagation time from the ACE satellite to the front of the magnetosphere. Figure 15.10 shows data from meridian scanning photometers while Fig. 15.11 shows ground based magnetograms from key ground stations along the meridian line where the photometer is located which is close to the region of substorm development. Finally, Fig. 15.12 shows data from two geostationary satellites on either side of the meridian at which EP development was centered, and from one satellite further back in the tail close to the region which is expected to be disturbed by the substorm.

The interplanetary conditions associated with the substorm activity (Fig. 15.9) are typical in the sense that the solar wind speed is relatively constant (at ~610 km/s) with an equally stable number density (at ~2/cm^3). This speaks to the fact that the kinetic energy flux of the solar wind is not correlated with the probability of substorm EP occurrence. The IMF shows many fluctuations over a broad frequency range, including several sharp northward and southward turnings. The EP onset that occurred at ~0432 UT could readily be attributed to the sharp northward turning of the IMF at ~0355 UT (Fig. 15.9) taking into account the ~37 min delay for the solar wind detected at ACE to impact the earth's magnetopause. However, typical of the timing issues often noted in correlating substorm EP onsets with IMF northward turnings, the ~0544 UT EP onset is not precisely correlated with either of the possible northward turnings detected by ACE. Using the same ~37 min delay, the northward turning at ~0448 UT would be expected to impact the magnetopause at ~0525 UT while the northward turning at ~0514 UT would be expected to impact the magnetopause at ~0551 UT. There are, however, uncertainties associated with propagating structures from the L1 Lagrangian point to earth orbit which are estimated by Blanchard et al. (2000) to be of the order of ~10 min. As well, there is good evidence that the structure of the IMF is complex to the extent that the field configuration detected at satellites such ACE as is not always detected just in front of the magnetopause. Finally, other sharp northward turnings (e.g. ~0528 UT in Fig. 15.9) may have no associated EP, so simply having a northward turning of the IMF does not guarantee an EP onset. While it is thought that the magnetosphere must be preconditioned in terms of stored energy in the magnetotail before a northward turning of the IMF can trigger a substorm EP, at this time it is not possible to predict that any particular northward turning trigger a particular event. From the photometer data shown in Fig. 15.10, the substorm EP onset that occurred at ~0432 UT conforms to what is typically observed in an

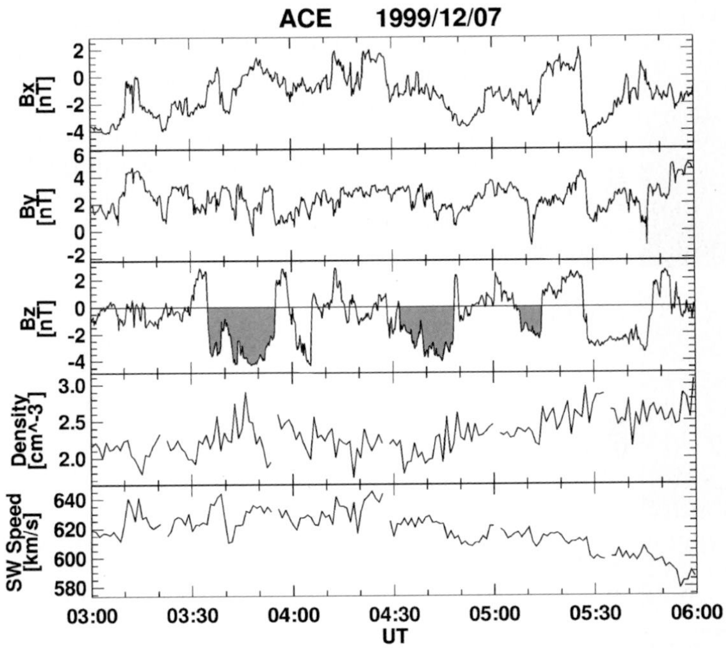

Fig. 15.9. Solar wind parameters for the substorm event of December 7, 1999. (Courtesy the ACE SWEPAM instrument team and the ACE Science Center) The interplanetary magnetic field is expected to impact the front of the magnetosphere ~37 min after it was detected by ACE

isolated substorm event. It begins with an intensification near the equatorward edge of the auroral region, at the poleward edge of the λ4861 hydrogen emission region. Since the hydrogen emissions are attributed to the precipitation of energetic protons with energies of tens of keV, and these energies are characteristic of the proton population near the inner edge of the plasma sheet, this indicates that onset takes place in quasi-dipolar geometry probably earthward of ~12 Re (Samson et al., 1992). It is apparent that the λ6300 red line emissions extend to poleward of 74° at the time the EP begins; from the work of Blanchard et al. (1994) this indicates that all field lines threading the earth's surface equatorward of ~74° are closed and therefore that the development of this EP takes place predominately on closed field lines. The EP occurs in three steps, the second and third steps involving impulsive activations occurring progressively further poleward following the pattern outlined by Kisabeth and Rostoker (1974). The disturbed region reaches its highest latitude by ~0510 UT, after which the activity begins to die down. The equatorward drift of the high latitude auroras after this time strongly suggests that energy continues to be fed into the magnetosphere from the solar wind and the IMF data shown in Fig. 15.9 is consistent with this view.

In the ~15 min leading up to the second EP onset detected after ~0540 UT, the double oval aspect of the evening sector auroral oval is clearly evident. Activations can occur both on the equatorward branch of the double oval (i.e. EP onset and pseudobreakups) and on the poleward branch (i.e. poleward boundary intensifications or PBIs). The initial intensification at ~0544 UT occurs at the poleward edge of the λ4861 hydrogen emission region and for this event the poleward expansion is very rapid. Later, the activations penetrate into the region of open field lines which appears to have expanded equatorward to ~70° by the time the ~0544 UT intensification takes place. This suggests that the magnetotail lobe area (which maps to the open field line region of the polar cap) has expanded since the ~0432 UT EP, and its magnetic field energy is subsequently made available to power the substorm activity. By ~0700 UT the PBI's have died away and the oval became quiet, settling at a latitude higher than that before the EP activity had commenced at ~544 UT.

The magnetic activity is shown in magnetogram format in Fig. 15.11 as perturbations in the north–south (X) component of the disturbance field at stations along the meridian on which the photometer was located. The weak nature of the disturbance associated with the ~0432 UT onset is due to the fact that the center of

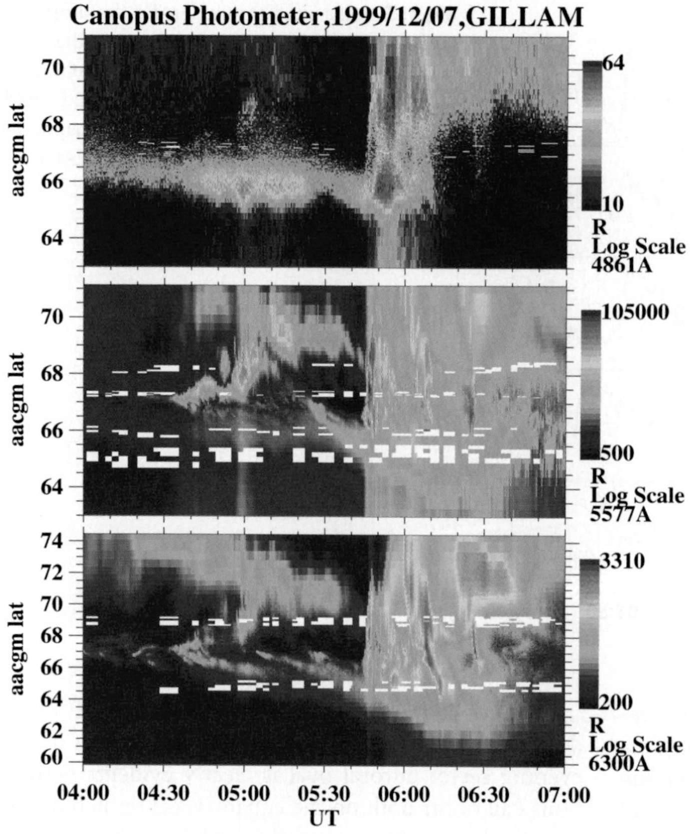

Fig. 15.10. Meridian scanning photometer data from the site of Gillam (67.4° aacgm) located on the meridian through Fort Churchill for the substorm event of December 7, 1999 (courtesy the Canadian Space Agency). After the substorm EP onset at ~0432 UT, a double oval is clearly apparent. The second substorm EP at ~0543 UT is initiated on the equatorward branch of the double oval

activity lies far to the east. After the ~0544 UT brightening, the region of auroral activations moves to the edge of the polar cap in the matter of a few minutes through a series of steps, each one indicated by the vertical lines in Fig. 15.10. The rapid poleward movement of the high latitude edge of the electrojet mirrors the poleward expansion seen in the auroral luminosity as presented in Fig. 15.10. It should be noted that the disturbances seen during this event at the poleward edge of oval should be thought of as PBIs, and they are at least as large as the perturbations associated with expansive phase development in the closed field line region.

The disturbance in the midnight sector of the near-Earth magnetosphere shown in Fig. 15.12 gives an indication of the size of the disturbed region and how the timing of the magnetospheric perturbations relates to those seen in the auroral ionosphere for the event of December 7, 1999. This behaviour is to be contrasted to that expected during a "classical" substorm event shown in Fig. 15.13, in which EP onset is preceded by a brief period of intense growth followed by sudden dipolarization that is turbulent in character within the disturbed region. For the ~0432 UT EP onset, GOES 8 is located ~2 h before local magnetic midnight and GOES 10 is located near the dusk meridian. GOES 10 sees no detectable signature of this EP. GOES 8 sees a weak dipolarization, delayed by ~20 min from onset, that coincides with the second auroral intensification (cf. Fig. 15.10) in the event. This is typical behaviour for instances in which the geostationary spacecraft are outside the azimuthal region of EP onset.

In the minutes leading up to the ~0544 UT intensification, the magnetic field at GOES 8 becomes more tail like (i.e. the signature of growth), while there is no detectable signature at GOES 10 far to the west. GOES 8 does not detect a classical EP dipolarization at ~0544 UT, despite the fact that it is located at approximately local magnetic midnight very close to the meridian on which the intensification took place.

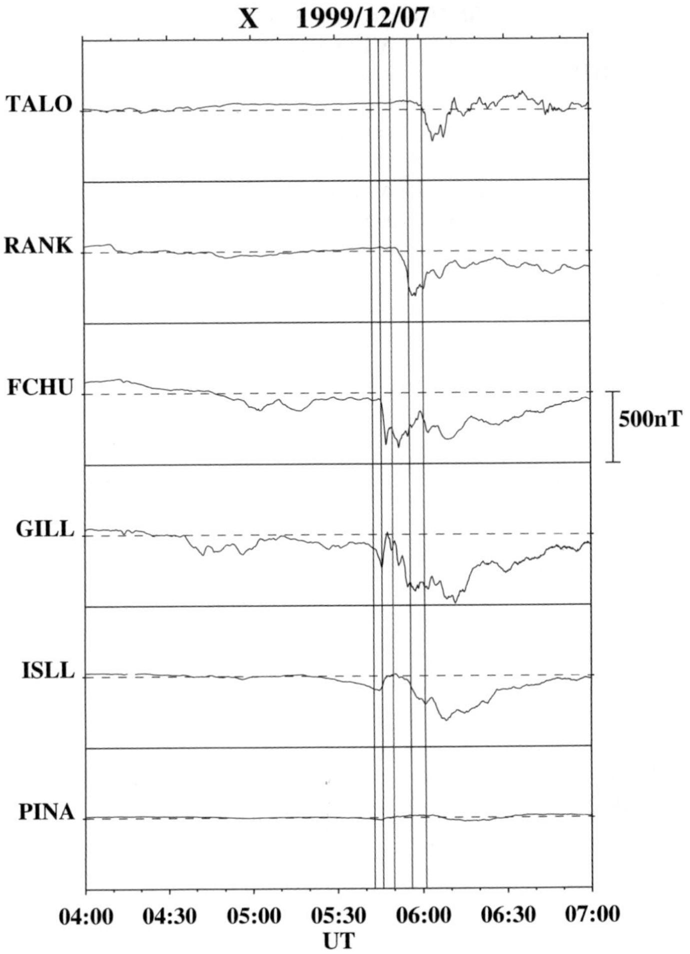

Fig. 15.11. Magnetograms from stations along the Churchill line of magnetometers stretching from PINA (61.1° aacgm) to TALO (79.4° aacgm) for the substorm event of December 7, 1999. The *first vertical line* at ~0544 UT marks the initial substorm intensification, while the *following four vertical lines* mark clearly defined important intensifications. The *red vertical line* at 0550 UT indicates the time that the current wedge begins to grow and, as such, marks the onset of the substorm EP. (Data courtesy the Canadian Space Agency)

Instead, it continues to see growth until the time of the second major optical intensification at ~0550 UT (cf. Fig. 15.10) at which time dipolarization commences. Further study of the event reveals that the substorm current wedge signature at low latitudes does not appear until the ~0550 UT intensification, in approximate agreement with the time of dipolarization detected by GOES 8 and confirmed by the brightening of the λ4861 hydrogen emissions detected by the Gillam photometer (cf. Fig. 15.10). Therefore, despite the obvious auroral brightening seen in the photometer data at ~0544 UT, the actual onset of the EP is at ~0550 UT. This event demonstrates the care that must be exercised in the timing of substorm EP onsets that are later correlated with changes in the magnetotail particles and fields.

Some important information on this event is provided by the Interball satellite, located ~16 Re behind the earth slightly on the dusk side of the noon-midnight meridian plane. This satellite detected the start of dipolarization right at the time of the initial auroral brightening at ~0544 UT, this behaviour continuing until ~0600 UT. At that time, the magnetic field becomes turbulent and dipolarization ceases shortly thereafter. The Interball data can be interpreted as the collapse of the crosstail current, starting close to the earth and expanding rapidly tailward in a manner reported in the past by Jacquey et al. (1991) and by Ohtani et al. (1992a). The simultaneous decrease in the Bx and By components of the magnetic field together with increasing Bz, and the subsequent onset of turbulent behavior is understood as the result of plasma sheet

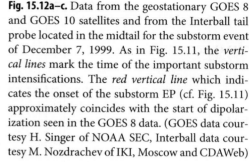

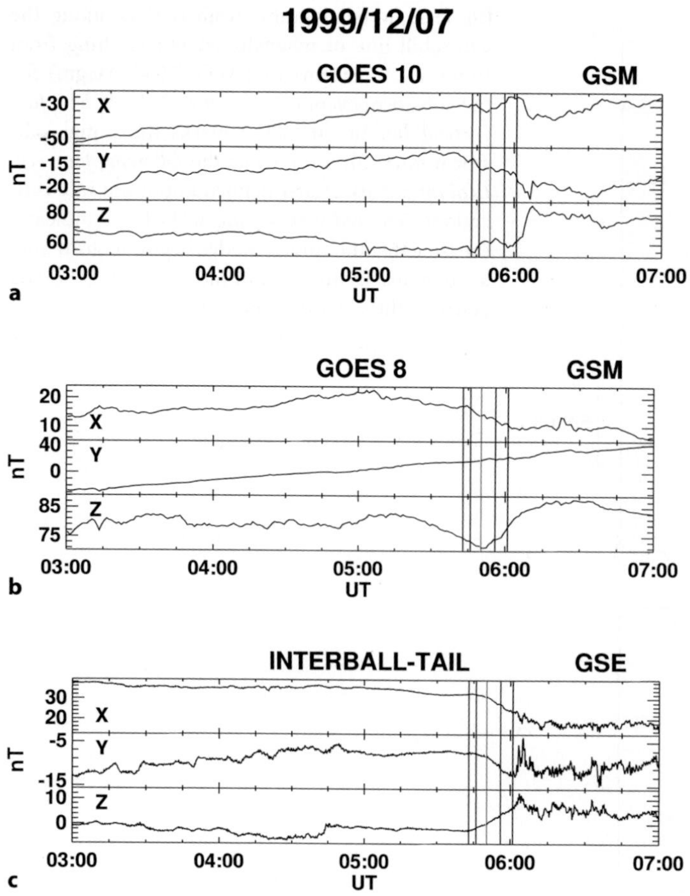

Fig. 15.12a–c. Data from the geostationary GOES 8 and GOES 10 satellites and from the Interball tail probe located in the midtail for the substorm event of December 7, 1999. As in Fig. 15.11, the *vertical lines* mark the time of the important substorm intensifications. The *red vertical line* which indicates the onset of the substorm EP (cf. Fig. 15.11) approximately coincides with the start of dipolarization seen in the GOES 8 data. (GOES data courtesy H. Singer of NOAA SEC, Interball data courtesy M. Nozdrachev of IKI, Moscow and CDAWeb)

thickening, with Interball initially in the north tail lobe and moving through the plasma sheet boundary layer into the central plasma sheet ~0600 UT.

Finally, it should be noted that the substorm EP onset is accompanied by the appearance of energetic particles at geostationary orbit as first reported by Deforest and McIlwain (1971). It was claimed that these particles were injected from further back in the magnetotail through the action of an electric field [McIlwain, 1974]. By tracking the trajectories of the drifting energetic particles, an injection boundary was defined as an initial condition. This injection boundary was claimed to have a double-spiraled configuration by Mauk and Meng (1983). Using data from geostationary satellites at different longitudes, Reeves et al. (1991) have shown that the longitudinal extent of the injected particles can be defined thus quantifying the spatial extent of the onset region. Particle and field data from geostationary spacecraft clearly play a critical role in understanding the development of magnetospheric substorms.

15.2 Physical Frameworks for Understanding Substorms

15.2.1 The Near-Earth Neutral Line (NENL) Framework

The suggestion by Dungey (1961) that merging of the IMF with frontside terrestrial magnetic field lines and subsequent reconnection in the tail provided the mechanism whereby solar wind plasma could enter the magnetosphere and established convection as a primary means of transport of mass and energy throughout the system. Initially it was believed that the magnetotail

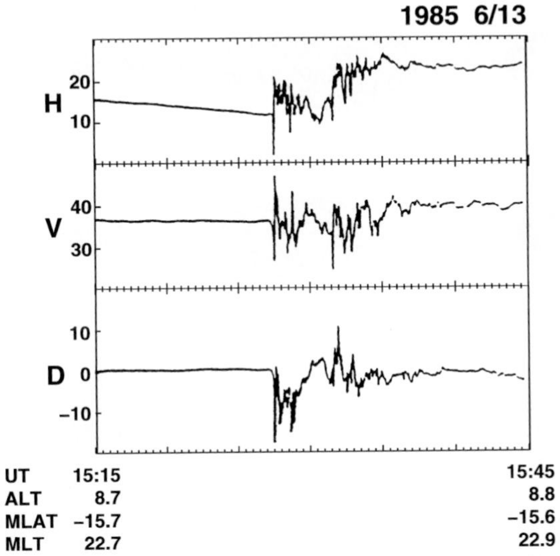

Fig. 15.13. Classical substorm expansive phase disturbance at geostationary orbit for a case when the satellite lies in the in region where the disturbance is initiated (after Ohtani et al., 1992b). The dipolarization (positive H) associated with the substorm EP is preceded by a short lived negative perturbation in the H-component attributed to a sudden intensification of the crosstail current near the inner edge of the plasma sheet

stretched anti-sunward to beyond 100 Re behind the earth, and Dungey (1965) thought it could extend to at least 1000 Re. It was understood that there would be a neutral line somewhere in the distant tail and earthward of that line the plasma flow would be towards the earth. However, early theoretical considerations (cf. Siscoe and Cummings, 1969) suggested that a neutral line might form closer to the earth during times of geomagnetic bay (i.e. substorm) activity and they placed that neutral line at closer to ~10 Re behind the earth. Early observations by Camidge and Rostoker (1970) placed that substorm related neutral line beyond 21 Re behind the earth, and over some years there was considerable argument about the positioning of that substorm related neutral line. That issue was settled by Baumjohann et al. (1989) who used AMPTE IRM data to demonstrate that any substorm related neutral line had to be located tailward of the apogee of that satellite (which was ~19 Re). More recently, Geotail satellite measurements have shown that neutral line to lie between ~20 – 30 Re behind the earth at times of onset of substorm EP activity (cf. Nagai et al., 1998).

Although several researchers were contemplating the possibility of the formation of a near-earth neutral line behind the earth in association with substorm EP onset, the concept was most clearly enunciated by E.W. Hones (cf. Hones, 1976) and formalized in the cartoon presentation shown in Fig. 15.14. In this concept, a southward turning of the IMF causes a thinning of the plasma sheet which leads to the formation of a new neutral line close to the earth. Initially, closed plasma sheet field lines reconnect setting up a flow pattern involving earthward and tailward flow away from the neutral line. Ultimately, open field lines begin to reconnect (cf. panel 6 of Fig. 15.14), at which time a closed field line structure termed a plasmoid moves down the tail transporting mass and energy back into the solar wind. The earthward flow provides the energy responsible for the substorm in this scenario.

At the time of its development, it was believed that the near-earth neutral line mapped to the region of the auroral breakup in the ionosphere. This associated the open-closed field line boundary with the region of the auroral breakup that signals substorm EP onset. The most comprehensive review of the NENL framework was prepared by Baker et al. (1996) at a time when there was some concern about the mapping of the auroral breakup region into the magnetotail. As discussed earlier in this chapter, Samson et al. (1992) had provided strong evidence that the substorm EP onset region was located well within the closed field line region. Shortly after the review by Baker et al., the NENL model was renovated in such a way as to maintain the original concept while acknowledging that EP onset occurred deep within the closed field line region. It was proposed by Shiokawa et al. (1997) that reconnection at the near-earth neutral line initiated bursty bulk flows (BBFs) of the type reported and studied by Angelopoulos et al. (1992, 1994). These flows transported mass and energy earthward rapidly, and the deceleration of the BBFs at the interface between the dipolar and tail magnetic field regions led to the formation of the substorm current wedge and the auroral breakup that signals EP onset. This braking of plasma flows has been modeled using an MHD code by Birn et al. (1999) to demonstrate that the electric currents associated with substorm current wedge can be generated.

There is still ongoing research into the NENL model concentrating on the order of events leading up to and

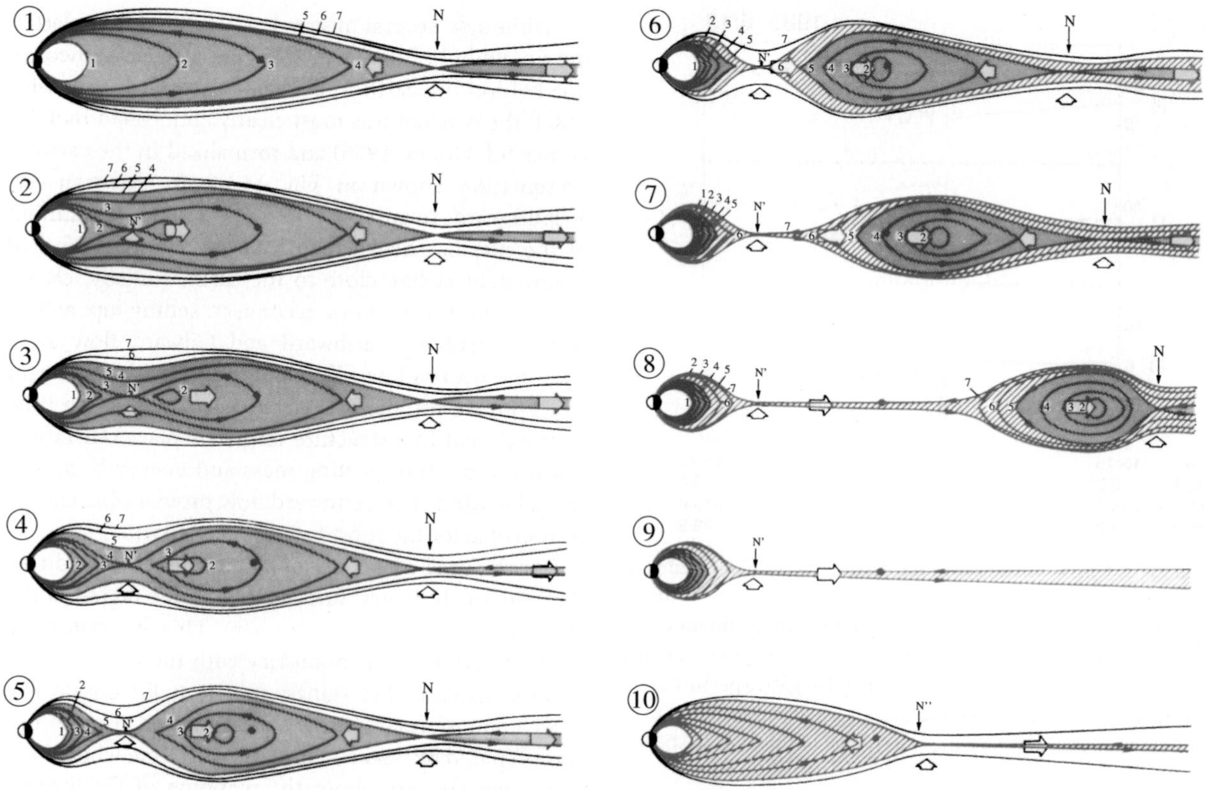

Fig. 15.14. Cartoon of the life cycle of a magnetospheric substorm viewed as a projection in the noon-midnight meridian plane for the near-earth neutral line model (after Hones, 1984). Initially, the reconnection at the near-earth neutral line involves closed field lines. When open field lines begin to be reconnected at the near-earth neutral line, a closed field line structure called a plasmoid begins to move downtail and after its passage the magnetotail recovers to its pre-substorm configuration through tailward movement of the neutral line. The hatching in *panels* 5 – 10 indicates plasma populating previously open field lines that have reconnected at the near-earth neutral line

during the development of the EP. The primary question that needs to be answered is whether the onset of reconnection at the near-Earth neutral line precedes or follows EP onset closer to earth. Because of the complex nature of the changing plasma and field configuration in the tail, multiple satellites appropriately located will be required to answer this important question.

15.2.2 Near-Earth Current Disruption Framework

It is generally acknowledged that reconnection of magnetic field lines in the earth's magnetotail leads to earthward convective plasma flow, and the energy from this flow ultimately powers substorm EP activity. It is, however, a matter of dispute as to when, with respect to substorm EP onset, this reconnection rate increases suddenly to provide the required energy to the volume of space close to the inner edge of the tail current sheet where the onset is believe to take place. The NENL framework considers the onset or sudden increase of reconnection as the start of energy transport to the near-Earth onset region through BBF's. Within minutes, the braking of the convective plasma flow in the BBF near the inner edge of the plasma sheet provides the energy required to initiate EP onset. This process involves conversion of the kinetic energy ofthe BBF's to electromagnetic energy associated with inertial currents that couple the magnetosphere to the auroral ionosphere (Shiokawa et al., 1997). The formation of the substorm current wedge is further encouraged by pressure gradients near the inner edge of the plasma sheet that result from the pileup of the magnetic field transported by the BBF's (Birn et al., 1999).

In the near-Earth current disruption framework, it is believed that the energy required to power the initial stages of EP activity is provided during the growth phase of the substorm, a process that can occur over many tens of minutes and is marked by stretching of the tail magnetic field close to the earth. This stretching is the magnetic field signature of the development of an intense radially confined region of crosstail current near the inner edge of the tail current sheet. Kauffman (1987) investigated the magnitude of the current density required to distort the magnetic field at geostationary orbit so as to produce the amount of stretching observed, and found it to be ~300 mA/m, approximately an order of magnitude larger than the current density associated with the normal crosstail current sheet outside of ~12 Re. In the early 1990s, evidence arose that the crosstail current density near geostationary orbit increased suddenly just prior to substorm onset (see Fig. 15.13). A picture then emerged that did not call on a sudden increase in tail reconnection to trigger EP onset, but rather called on some plasma instability in the onset region to initiate EP activity.

Several possible instabilities have been introduced to account for the explosive onset of the substorm EP. Lui et al. (1991) proposed that a cross-field current instability could be responsible for the observed current disruption and the chaotic magnetic field behavior that accompanied the dipolarization of the magnetic field and Lui et al. (1992) provided strong observational support for this concept. It is interesting to note that Lui et al. (1993) showed that fast plasma flows, both in the earthward and tailward directions, could be produced from tail current disruption. This suggests that fast plasma flows are not necessarily indicators of magnetic reconnection.

Another mechanism that has been proposed by Voronkov et al. (1997) for explaining onset of the substorm EP in the near-Earth midnight sector is a hybrid vortex instability which appears because of the coupling of Kelvin–Helmholtz (KH) and Rayleigh–Taylor instabilities. This hybrid vortex mode grows faster than a KH mode, extracts ambient potential energy, and leads to the development of vortex cells that have a larger spatial extent than a simple KH vortex. It can be understood as a shear ballooning mode, and has found observational support in combined satellite and ground based observations together with modeling studies (cf. Voronkov et al., 1999, 2000).

There are several other plasma instabilities that have been identified that might lead to the disruption of the near-Earth crosstail current, most of which require the thinning of a pre-existing current sheet. These instabilities are summarized by Lui (2004) and remain as serious candidates for explaining how substorm EP onset can be initiated deep in the closed field region near the inner edge of the plasma sheet.

The essence of the current disruption framework is then that, after energy is stored in the near-Earth plasma sheet over the growth phase period of the substorm, the onset is triggered through the development of a plasma instability that disrupts the crosstail current flow in the volume of space to which the EP auroral arcs map. The disturbance then spreads tailward where, at a later time, it may cause the tail reconnection rate to increase making more energy available for the continuation of the substorm activity.

15.2.3 Boundary Layer Dynamics Model

One of the dominant features of the development of a substorm is the progressively westward appearance after EP onset of auroral surge forms that sometimes can be seen across large portions of the evening sector. These forms can stretch out as far as the dusk meridian, and their structure is highly suggestive of some kind of instability that creates long wavelength disturbances (Fig. 15.15). The boundary layer dynamics (BLD) framework was proposed by Rostoker and Eastman (1987), and its ability to explain quasi-periodic surge structures along the evening sector auroral oval was invoked by Rostoker (1987) and by Kidd and Rostoker (1991).

Figure 15.16 is a cartoon that contains the essence of the BLD framework, showing the projection of plasma convective flows on the plane of the neutral sheet. Rather than showing the electric field, this figure identifies the regions of space charge that are responsible for the electric field configuration. The space charge distribution is calculated from the relationships shown at the bottom of the figure. Field-aligned currents connect the

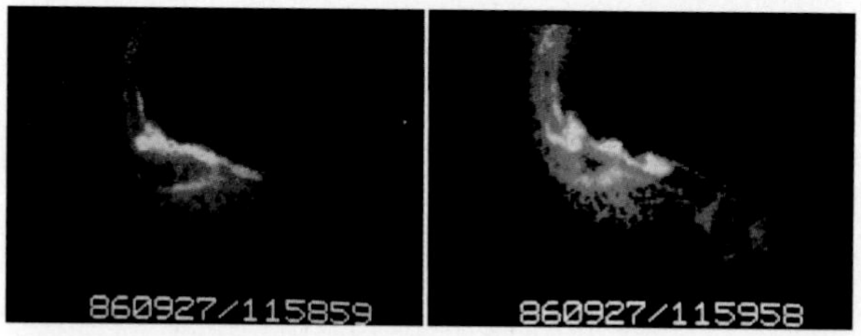

Fig. 15.15. Images taken 59 s apart on September 27, 1986 by the CCD imager aboard the Viking satellite showing surge development at the high latitude edge of the evening sector auroral oval near dusk. The rapid development of the surge forms (growth time ~1 minute and scale size ~500 km) and the clear spatial periodicity of the surge structures is highly suggestive of the action of a Kelvin–Helmholtz instability

regions of space charge to the ionosphere, and are a consequence of the magnetosphere-ionosphere systems attempting to minimize any space charge. In the absence of substorm EP activity, the figure above represents the directly driven activity associated with relatively steady convection in the magnetosphere. In the BLD framework, the substorm growth phase involves the buildup of space charge close to the earth which acts to shield the inner magnetosphere from the magnetotail electric field associated with earthward convection. The term $\boldsymbol{J} \cdot \boldsymbol{v}$ in the formalism shown at the bottom of Fig. 15.16 is associated with the shielding space charge, and can be significant when applied to the intense radially localized crosstail current buildup near the inner edge of the plasma sheet identified by Ohtani et al. [1992b]. The substorm EP onset occurs near the inner edge of the plasma sheet in the midnight sector, much as in the Current Disruption Framework described earlier and for the same reasons. It triggers a collapse of the crosstail current that moves slowly back in the magnetotail (cf. Fig. 15.12). However, information about the sudden disruption also travels rapidly back into the tail at fast mode Alfven speeds of a few hundred km/s, initiating or enhancing pre-existing reconnection somewhere in the ~20 – 50 Re range behind the earth a few minutes after EP onset. This enhanced reconnection causes increased earthward convection which, in turn, causes the growth of Kelvin–Helmholtz (KH) wave activity at the interface between the low latitude boundary layer and the central plasma sheet as shown in Fig. 15.17. The growth rate is expected to be of the order of ~2 min (cf. Rostoker, 1987) and the regions of wave crests map into the ionosphere where they are ultimately responsible for the formation of auroral surge forms. This process involves the development of a parallel electric field along magnetic field lines on which the KH instability develops (cf. Thompson, 1983) and the magnetic field disturbance associated with resultant field-aligned currents causes the characteristic windup of the surge activity in the ionosphere at the foot of the field lines involved. The auroral surges thus created should be associated with PBI's discussed earlier in this chapter. A comprehensive description of the BLD framework for substorms is presented in Rostoker (1996).

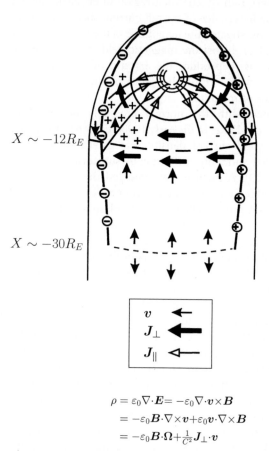

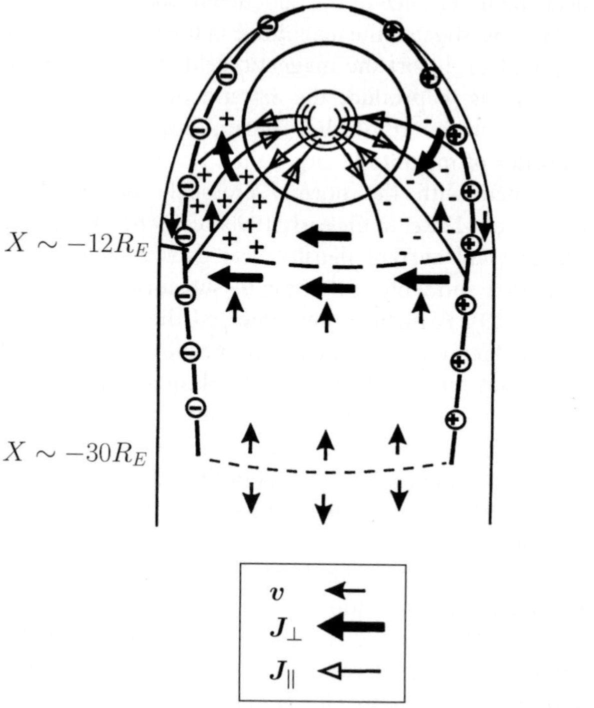

$$\rho = \varepsilon_0 \nabla \cdot \boldsymbol{E} = -\varepsilon_0 \nabla \cdot \boldsymbol{v} \times \boldsymbol{B}$$
$$= -\varepsilon_0 \boldsymbol{B} \cdot \nabla \times \boldsymbol{v} + \varepsilon_0 \boldsymbol{v} \cdot \nabla \times \boldsymbol{B}$$
$$= -\varepsilon_0 \boldsymbol{B} \cdot \boldsymbol{\Omega} + \frac{1}{C^2} \boldsymbol{J}_\perp \cdot \boldsymbol{v}$$

Fig. 15.16. Cartoon showing, as a projection in the equatorial plane, the space charge distribution, plasma convective flow, and perpendicular current flow in the magnetotail as suggested in the boundary layer dynamics model of a substorm. Also shown are the related field-aligned currents and the development of a Maxwell equation explaining the placement of the regions of space charge in the figure

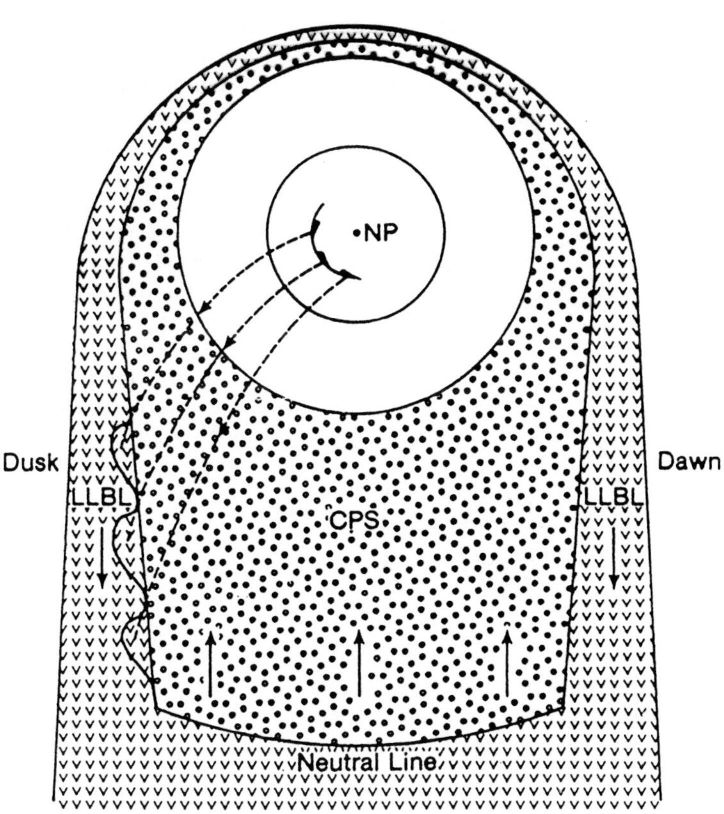

Fig. 15.17. Surge development along the interface between the plasma sheet and the low latitude boundary layer and the mapping of the disturbances to the ionosphere (after Kidd and Rostoker, 1991). Waves along the LLBL/PS interface are a consequence of a Kelvin–Helmholtz instability. The establishment of the field-aligned currents linking the plane of the neutral sheet to the high latitude ionosphere through the development of a parallel electric field is explained in Thompson (1983)

15.3 Final Comments

In this chapter, the development of the substorm concept has been traced from its origins in an effort to put the past literature into perspective. The acquisition of new data from increasingly sophisticated instrumentation both on the ground and in space has, from time to time, forced an evaluation of the frameworks developed to understand the substorm phenomenon. This makes it important to evaluate the older literature critically, as conclusions reached on the basis of older and less comprehensive data may now be invalid.

As an example of this reassessment, observations of the auroral oval from high altitude satellites have allowed the complete substorm cycle to be observed. The original Akasofu (1964) scheme shown in Fig. 15.1 of this chapter can now be viewed as applicable only to the life cycle of the substorm EP, but the entire substorm cycle including growth is better viewed in the way shown in Fig. 15.18. From this perspective, growth is marked by equatorward motion of the auroras, while recovery actually describes the EP in the Akasofu scheme, in which the region of active auroras expands to its most poleward extension. The recovery phase of the Akasofu substorm (i.e. equatorward drift of auroral arcs) can then be viewed as an indication of the buildup of stored energy in the near-Earth plasma sheet leading up to the next EP.

As is evident from the frameworks that have been outlined in Sect. 15.2 of this chapter, our understanding of the physics of the substorm process has not yet crystallized, in large measure due to the inadequacy of the data sets with which researchers must work. With the development of multipoint measurements in space using flotillas of satellites (e.g. the Cluster concept) some of the non-uniqueness that presently plagues interpretation of the existing data sets may be ameliorated. Perhaps the most important inadequacy in the study of the substorm phenomenon is the inability to map from the ionosphere to the plane of the neutral sheet in the magnetotail. The work of Donovan (1993) has demonstrated that magnetic

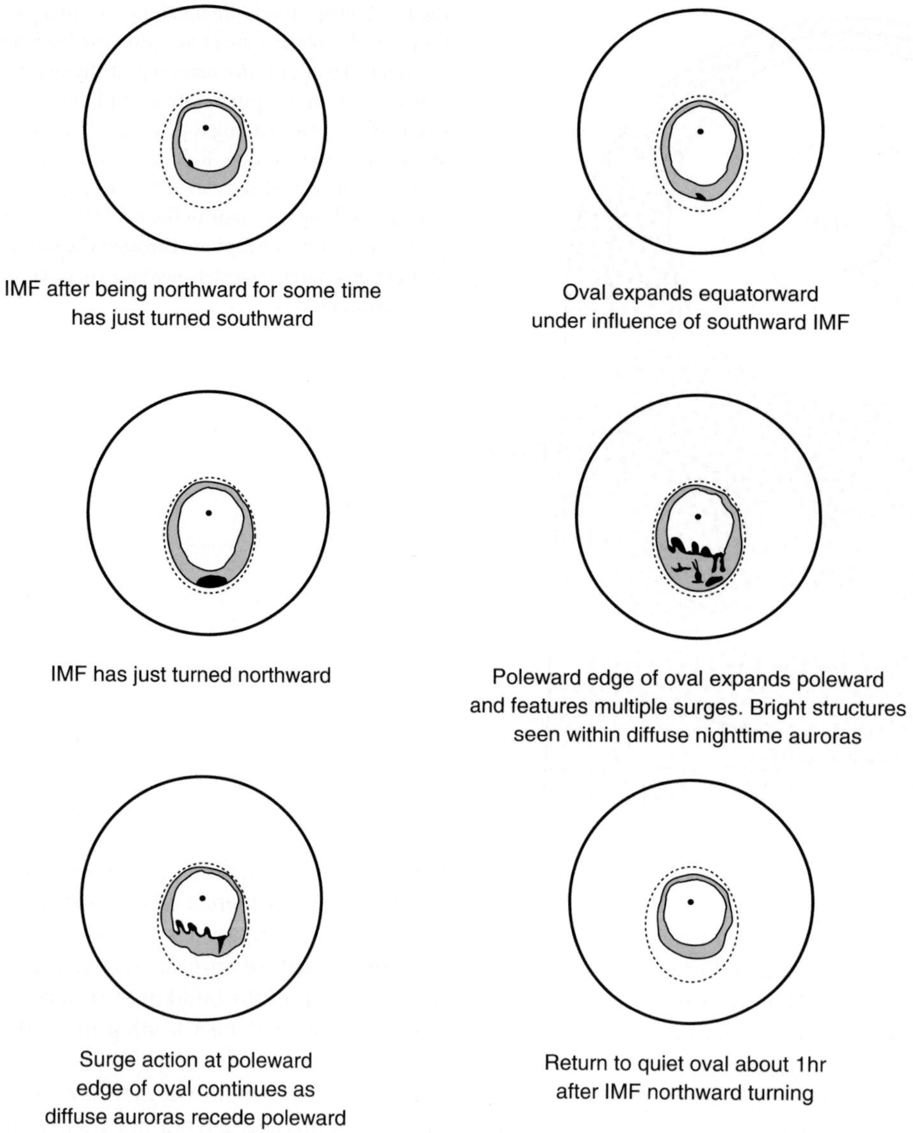
IMF after being northward for some time
has just turned southward
Oval expands equatorward
under influence of southward IMF
IMF has just turned northward
Poleward edge of oval expands poleward
and features multiple surges. Bright structures
seen within diffuse nighttime auroras
Surge action at poleward
edge of oval continues as
diffuse auroras recede poleward
Return to quiet oval about 1 hr
after IMF northward turning

Fig. 15.18. A global perspective of the development of a magnetospheric substorm as seen in the behavior of the auroral oval over the entire life cycle of the disturbance. The pattern of activity described in the Akasofu (1964) definition of the auroral substorm is applicable to the disturbance across the midnight sector that is initiated when the IMF turns towards the north

field lines are skewed significantly towards the flanks through the effects of the Birkeland current sheets that are associated with directly driven activity. The greatest challenge facing the substorm community is to be able to map from the ionosphere to the magnetospheric equatorial plane. If this matter is dealt with successfully, it will then be possible to accurately relate the multipoint measurements in the ionosphere (e.g. SuperDARN, IMAGE, CANOPUS/CARISMA to name a few coordinated networks of instrumentation) to individual observations in the magnetosphere made by instruments aboard orbiting satellites. This will then allow efforts to synthesize a comprehensive model framework for substorms of the type attempted by Lui (1991) to ultimately achieve success.

Acknowledgement. This work was supported by Science and Engineering Research Canada.

References

Akasofu, S.-I., The development of the auroral substorm, *Planet. Space Sci.*, **12**, 273, 1964

Akasofu, S.-I., *Polar and Magnetospheric Substorms*, D. Reidel, Norwell, Mass., 1968

Akasofu, S.-I., What is a magnetospheric substorm?, in *Dynamics of the Magnetosphere*, ed. by S.-I. Akasofu, p. 447, D. Reidel, Dordrecht: Holland, 1979

Akasofu, S.-I., and S. Chapman, *Solar-Terrestrial Physics*, Oxford Univ. Press, 1972

Akasofu, S.-I., S. Chapman and C.-I. Meng, The polar electrojet, *J. Atmos. Terr. Phys.*, **30**, 227, 1966

Angelopoulos, V., W. Baumjohann, C.F. Kennel, F.V. Coroniti, M.G. Kivelson, R. Pellet, R.J. Walker, H. Luhr, and G. Paschmann, Bursty bulk flows in the inner central plasma sheet, *J. Geophys. Res.*, **97**, 4027, 1992

Angelopoulos, V., C.F. Kennel, F.V. Coroniti, R. Pellat, M.G. Kivelson, R.J. Walker, C.T. Russell, W. Baumjohann, W.C. Feldman, and J.T. Gosling, Statistical characteristics of bursty bulk flow events, *J. Geophys. Res.*, **99**, 21,257, 1994

Atkinson, G., A theory of polar substorms, *J. Geophys. Res.*, **71**, 5157, 1966

Baker, D.N., T.I. Pulkkinen, V. Angelopoulos, W. Baumjohann, and R.L. McPherron, Neutral line model of substorms: Past results and present view, *J. Geophys. Res.*, **101**, 12,975, 1996

Baumjohann, W., G. Paschmann, and C.A. Cattell, Average plasma properties in the central plasma sheet, *J. Geophys. Res.*, **94**, 6597, 1989

Birkeland, K., The Norwegian Aurora Polaris Expedition 1902-1903, vol. 1, 1st sect., Aschhoug and Co., Oslo, 1908

Birn, J., M. Hesse, G. Haerendel, W. Baumjohann, and K. Shiokawa, Flow braking and the substorm current wedge, *J. Geophys. Res.*, **104**, 19,895, 1999

Blanchard, G.T., L.R. Lyons and J. Spann, Predictions of substorms following northward turnings of the interplanetary magnetic field, *J. Geophys. Res.*, **105**, 375, 2000

Blanchard, G.T., L.R. Lyons, J.C. Samson, and F.J. Rich, Locating the polar cap boundary from observations of 6300 Å auroral emission, *J. Geophys. Res.*, 100, 7855, 1994

Bonnevier, B., R. Boström and G. Rostoker, A three-dimensional model current system for polar magnetic substorms, *J. Geophys. Res.*, **75**, 107, 1970

Caan, M.N., R.L. McPherron and C.T. Russell, Characteristics of the association between the interplanetary magnetic field and substorms, *J. Geophys. Res.*, **82**, 4837, 1977

Camidge, F.P., and G. Rostoker, Magnetic field perturbations in the magnetotail associated with polar magnetic substorms, *Can. J. Phys.*, **48**, 2002, 1970

Clauer, C.R., R.L. McPherron and C. Searls, Solar wind control of the low-latitude asymmetric magnetic disturbance field, *J. Geophys. Res.*, **88**, 2123, 1983

Cummings, W.D., J.N. Barfield and P.J. Coleman, Jr., Magnetospheric substorms observed at the synchronous orbit, *J. Geophys. Res.*, **73**, 6687, 1968

Davis, T.N. and M. Sugiura, Auroral electrojet activity index AE and its universal time variations, *J. Geophys. Res.*, **71**, 785, 1966

Deforest, S.E. and C.E. McIlwain, Plasma clouds in the magnetosphere, *J. Geophys. Res.*, **76**, 3587, 1971

Donovan, E.F., Modeling the magnetic effects of field-aligned currents, *J. Geophys. Res.*, **98** , 13, 529, 1993

Dungey, J.W., Interplanetary magnetic field and the auroral zones, *Phys. Rev. Lett.*, **6**, 47, 1961

Dungey, J.W., The length of the magnetospheric tail, *J. Geophys. Res.*, **70**, 1753, 1965

Elphinstone, R. D., J.S. Murphree, D.J. Hearn, L.L. Cogger, I. Sandahl, P.T. Newell, D.M. Klumpar, S. Ohtani, J.A. Sauvaud, T.A. Potemra, K. Mursula, A. Wright, M. Shapshak, The double oval UV auroral distribution. 1. Implications for the mapping of auroral arcs, *J. Geophys. Res.*, **100**, 12075, 1995

Fairfield, D.H., and L.J. Cahill, Jr., Transition region magnetic field and polar magnetic disturbances, *J. Geophys. Res.*, **71**, 155, 1966

Fukushima, N., Polar magnetic storms and geomagnetic bays, *J. Fac. Sci. Tokyo Univ.*, **8**, 293, 1953

Goertz, C.K., L.-H. Shan and R.A. Smith, Prediction of geomagnetic activity, *J. Geophys. Res.*, **98**, 7673, 1993

Harang, L., The mean field of disturbance of polar geomagnetic storms, *Terr. Mag. Atmos. Electr.*, **51**, 353, 1946

Henderson, M.G., G.D. Reeves, R.D. Belian, and J.S. Murphree, Observations of magnetospheric substorms occurring with no apparent solar wind/IMF trigger, *J. Geophys. Res.*, **101**, 10,773, 1996

Hones, E.,W., Jr., The magnetotail: Its generation and dissipation, in *Physics of Solar-Planetary Environments. Proceedings of the International Symposium on Solar-Terrestrial Physics*, ed. by D.J. Williams, p. 558, American Geophysical Union, Washington, 1976

Hones, E.W., Jr., Plasma sheet behaviour during substorms, in *Magnetic Reconnection in Space and Laboratory Plasmas*, ed by E.W. Hones, Jr., p. 178, American Geophysical Union Monograph 30, Washington, 1984

Hsu, T., and R.L. McPherron, Average characteristics of triggered and nontriggered substorms, *J. Geophys. Res.*, **109**, A07208, doi:10.1029/2003JA009933, 2004

Jacquey, C., J.A. Sauvaud, and J. Dandouras, Location and propagation of the magnetotail disruption region during substorm expansion: Analysis and simulation of an ISEE multi-onset event, *Geophys. Res. Lett.*, **18**, 389, 1991

Kamide, Y., and N. Fukushima, Positive geomagnetic bays in evening high-latitudes and their possible connection with partial ring current, *Rep. Ionos. Space Res. Japan*, 26, 79, 1972

Kamide, Y., and S. Kokubun, Two-component auroral electrojet: Importance for substorm studies, *J. Geophys. Res.*, **101**, 13,027, 1996

Kaufmann, R.L. Substorm currents: Growth phase and onset, *J. Geophys. Res.*, **92**, 7471, 1987

Kisabeth, J.L. and G. Rostoker, The expansive phase of polar magnetic substorms, I. Development of the auroral electrojets and the auroral arc configuration during a substorm, *J. Geophys. Res.*, **79**, 972, 1974

Kidd, S.R. and G. Rostoker, Distribution of auroral surges in the evening sector, *J. Geophys. Res.*, **96**, 5697, 1991

Lui, A.T.Y., A synthesis of magnetospheric substorm models, *J. Geophys. Res.*, **96**, 1849, 1991

Lui, A.T.Y., Potential plasma instabilities for substorm expansion onsets, *Space Sci. Rev.*, **113**, 127, 2004

Lui, A.T.Y., C.-L. Chang, A. Mankovsky, H.-K. Wong and D. Winske, A cross-field current instability for substorm expansions, *J. Geophys. Res.*, **96**, 11,389, 1991

Lui, A.T.Y., R.E. Lopez, B.J. Anderson, K. Takahashi, L.J. Zanetti, R.W. McEntire, T.A. Potemra, D.M. Klumpar, E.M. Greene and R. Strangeway, Current disruptions in the near-Earth neutral sheet region, *J. Geophys. Res.*, **97**, 1461, 1992

Lui, A.T.Y., P.H. Yoon and C.-L. Chang, Quasi-linear analysis of ion Weibel instability, *J. Geophys. Res.*, **98**, 153, 1993

Lyons, L.R., G.T. Blanchard, J.C. Samson, R.P. Lepping, T. Yamamoto, and T. Moretto, Coordinated observations demonstrating external substorm triggering, *J. Geophys. Res.*, **102** 27,039, 1997

Mauk, B.H., and C.-I. Meng, Characterization of geostationary particle signatures based on the 'injection boundary' model, *J. Geophys. Res.*, **86**, 3055, 1983

McIlwain, C.E., Substorm injection boundaries, in *Magnetospheric Physics*, ed. by B.M. McCormac, p. 143, D. Reidel, Dordrecht, Netherlands, 1974

Perreault, P., and S.-I. Akasofu, A study of geomagnetic storms, *Geophys. J. R. Astron. Soc.*, **54**, 547, 1978

McPherron, R.L., Growth phase of magnetospheric substorms, *J. Geophys. Res.*, **75**, 5592, 1970

McPherron, R.L., C.T. Russell and M.P. Aubry, Satellite studies of magnetospheric substorms on August 15, 1968, 9, Phenomenological model for substorms, *J. Geophys. Res.*, **78**, 3131, 1973

Nagai, T., M. Fujimoto, Y. Saito, S. Machida, T. Terasawa, R. Nakamura, T. Yamamoto, T. Mukai, A. Nishida, and S. Kokubun, Structure and dynamics of magnetic reconnection for substorm onsets with Geotail observations, *J. Geophys. Res.*, **103**, 4419, 1998

Ohtani, S., S. Kokubun and C.T. Russell, Radial expansion of the tail current disruption during substorms, *J. Geophys. Res.*, **97**, 3129, 1992a

Ohtani, S., K. Takahashi, L.J. Zanetti, T.A. Potemra, R.W. McEntire, and T. Iijima, Initial signatures of magnetic field and energetic particle fluxes at tail reconfiguration: explosive growth phase, *J. Geophys. Res.*, **97**, 19,311, 1992b

Pytte, T., R.L. McPherron, E.W. Hones, Jr., and H.I. West, Jr., Multiple-satellite studies of magnetospheric substorms: Distinction between polar magnetic substorms and convection-driven negative bays, *J. Geophys. Res.*, **83**, 663, 1978

Reeves, G.D., R.D. Belian, and T.A. Fritz, Numerical tracing of energetic particle drifts in a model magnetosphere, *J. Geophys. Res.*, **96**, 13,977, 1991

Rostoker, G., Classification of polar magnetic disturbances, *J. Geophys. Res.*, **74** , 5161, 1969

Rostoker, G., Triggering of expansive phase intensifications of magnetospheric substorms by northward turnings of the interplanetary magnetic field, *J. Geophys. Res.*, **88**, 6981, 1983

Rostoker, G., The Kelvin–Helmholtz instability and its role in the generation of electric currents associated with Ps 6 and westward traveling surges, in *Magnetotail Physics*, ed. by A.T.Y. Lui, p. 169, Johns Hopkins University Press, Baltimore, 1987

Rostoker, G., Phenomenology and physics of magnetospheric susbtorms, *J. Geophys. Res.*, **101**, 12,955, 1996

Rostoker, G., Identification of substorm expansive phase onsets, *J. Geophys. Res.*, **107(A7)**, 1137, doi:10.1029/2001JA003504, 2002

Rostoker, G., and C.-G. Fälthammar, Relationship between changes in the interplanetary magnetic field and variations in the magnetic field at the earth's surface, *J. Geophys. Res.*, **72**, 5853, 1967

Rostoker, G., S.-I. Akasofu, J. Foster, R.A. Greenwald, Y. Kamide, K. Kawasaki, A.T.Y. Lui, R.L. McPherron and C.T. Russell, Magnetospheric substorms – Definition and signatures, *J. Geophys. Res.*, **85**, 1663, 1980

Rostoker, G., J.C. Samson, F. Creutzberg, T.J. Hughes, D.R. McDiarmid, A.G. MacNamara, A.V. Jones, D.D. Wallis and L.L. Cogger, CANOPUS – a ground-based instrument array for remote sensing the high latitude ionosphere during the ISTP/GGS program, *Space Sci. Rev.*, **71**, 743, 1995

Samson, J.C., L.R. Lyons, P.T. Newell, F. Creutzberg and B. Xu, Proton aurora and substorm intensifications, *Geophys. Res. Lett.*, **19**, 2167, 1992

Shiokawa, K., W. Baumjohann and G. Haerendel, Braking of high-speed flows in the near-Earth tail, *Geophys. Res. Lett.*, **24**, 1179, 1997

Silsbee, H.C. and E.H. Vestine, Geomagnetic bays, their occurrence frequency and current systems, *Terr. Mag. Atmos. Electr.*, **47**, 195, 1942

Siscoe, G.L. and W.D. Cummings, On the cause of geomagnetic bays, *Planet. Space Sci.*, **17**, 1795, 1969

Sugiura, M., and J.P. Heppner, The earth's magnetic field, in *Introduction to Space Science*, ed. by W.N. Hess, p. 5, Gordon and Breach, New York, 1965

Thompson, W.B., Parallel electric fields and shear instabilities, *J. Geophys. Res.*, **88**, 4805, 1983

Voronkov, I., R. Rankin, P. Frycz, V.T. Tikhonchuk and J.C. Samson, Coupling of shear flow and pressure gradient instabilities, *J. Geophys. Res.*, **101**, 9639, 1997

Voronkov, I., E. Friedrich, and J.C. Samson, Dynamics of the substorm growth phase as observed using CANOPUS and SuperDARN instruments, *J. Geophys. Res.* **104**, 28,491, 1999

Voronkov, I., E.F. Donovan, B.J. Jackel, and J.C. Samson, Large-scale vortex dynamics in the evening and midnight auroral zone, *J. Geophys. Res.*, **105**, 18,505, 2000

16 Ultra Low Frequency Waves in the Magnetosphere

Umberto Villante

Geomagnetic pulsations are the ground manifestation of ultra low frequency hydromagnetic waves propagating in the magnetosphere. Frequencies typically range between $f \approx 1$ mHz and $f \approx 10$ Hz; ground amplitudes range from less than 0.1 nT to tens or hundreds of nT and generally increase with latitude up to auroral/cusp regions. The distinct periodicity of most events suggests an interpretation in terms of standing waves reflecting between ionospheres of opposite hemispheres and hydromagnetic resonance is the basic process to interpret most aspects of geomagnetic pulsations.

The Kelvin–Helmholtz instability at the magnetopause is considered an important energy source for continuous low frequency events ($f \approx 1-10$ mHz); an additional contribution might come from cavity/waveguide modes of the magnetosphere. "*Upstream waves*" generated by particles reflected from the bow shock along interplanetary magnetic field lines are important exogenic sources for pulsations in the mid-frequency band ($f \approx 10-100$ mHz). High frequency pulsations ($f \approx 0.1-10$ Hz) are traveling waves related to ion-cyclotron instabilities occurring within the magnetosphere. Irregular pulsations represent transient signals associated with dramatic changes of the state of the magnetosphere, related to substorm manifestations.

The identification of field line resonance processes represents an important tool for several aspects of magnetospheric diagnostics: a quantitative determination of the set of field line eigenfrequencies can be used to model the plasma distribution along the magnetospheric field lines from equatorial to high latitudes, to monitor temporal variations of the magnetospheric plasma concentration and to highlight interesting aspects of plasmasphere/ionosphere coupling.

Contents

Umberto Villante, Ultra Low Frequency Waves in the Magnetosphere.
In: Y. Kamide/A. Chian, Handbook of the Solar-Terrestrial Environment. pp. 397–422 (2007)
DOI: 10.1007/11367758_16

16.1 Introduction

Geomagnetic (or more simply "*magnetic*") pulsations are the ground manifestation of ultra low frequency (ULF) hydromagnetic waves propagating in the magnetosphere. Originally termed "*micropulsations*", they were first identified by Celsius (who compared compass measurements in Uppsala with auroral fluctuations, 1741), by Nervander, 1840's and by Stewart, 1859. More than fifty years ago, Dungey (1967) argued that micropulsations could be interpreted in terms of Alfvén waves excited on geomagnetic field lines. In fact, their distinct periodicity led Dungey to suggest an interpretation of magnetic pulsations in terms of standing waves reflecting between ionospheres of opposite hemispheres.

The origin of magnetic pulsations is in the interplanetary medium, in the magnetosphere and, possibly, on the Sun itself. However, waves detected on the ground are not the same waves that enter the magnetosphere: indeed, wave energy is transformed by several processes, and ground signals are electromagnetic waves radiated from currents induced in the ionosphere by the impinging hydromagnetic waves. The properties of ground pulsations also depend on the conductivity of the Earth underneath the observer.

Magnetic pulsations typically have frequencies between $f \approx 1$ mHz and $f \approx 10$ Hz, with highest frequencies being determined by the hydrogen gyrofrequency in the magnetosphere ($f > 1$ Hz) and lowest frequencies corresponding to propagation times across the magnetosphere (Fig. 16.1). Ground amplitudes range from less than 0.1 nT (at the highest frequencies) to tens or hundreds of nT and generally increase with latitude up to auroral/cusp regions.

As for other areas of geophysics and space physics, the International Geophysical Year (1957–58) stimulated a great impetus for research on magnetic pulsations. By the early 1970's, well over 5000 papers had been published on this topic. Since then, pulsations have also been used as an important tool in magnetospheric dynamics ("*geomagnetic storms*" and "*substorms*"), for determining magnetospheric plasma density and for diagnostics of important processes such as "*magnetic reconnection*". Geomagnetic pulsations also represent the source field for electromagnetic induction studies of the Earth's crust, mantle and oceans. ULF

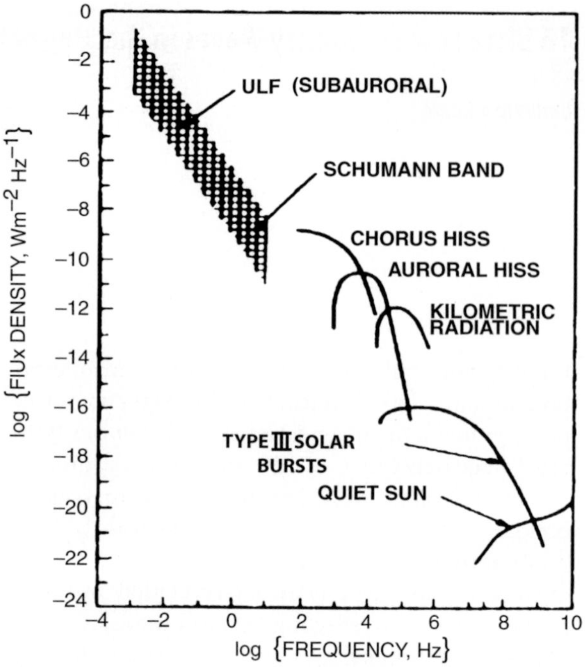

Fig. 16.1. The spectrum of natural signals. A power spectrum representing the natural situation on Earth. ULF waves correspond to the lowest frequency band. [Lanzerotti et al., 1990]

geomagnetic signals may occasionally be emitted in association with earthquake occurrence.

The International Association of Geomagnetism and Aeronomy (IAGA), classified geomagnetic pulsations into two classes, continuous, Pc, and irregular, Pi. Their further separation in period subclasses does not reflect any definite physical difference; rather, from a physical point of view, it would be more reasonable to divide Pc into three distinct frequency bands: low frequency ($f \approx 1 - 10$ mHz, see Fig. 16.8a for a typical example; these waves have wavelengths comparable to the dimensions of the magnetosphere), mid-frequency ($f \approx 10 - 100$ mHz, Fig. 16.6d), and high frequency pulsations ($f \approx 0.1 - 10$ Hz, Fig. 16.10a). However, pulsations sharing the same frequency band often present different characteristics, reflecting their different origin. Table 16.1 summarizes the classification scheme and major energy sources (adapted from Samson, 1991).

The occurrence of pulsations and their characteristics depend on the conditions of the solar wind (SW) and on the state of the magnetosphere. Changes

Table 16.1. The IAGA classification scheme (1964)[1] [Samson, 1991]

	T(s)	Frequency	Sources
Pc1	0.2 – 5	High:	Ion-cyclotron instability
Pc2	5 – 10	0.1 – 10 Hz	in magnetosphere.
Pc3	10 – 45	Mid: 10 – 100 mHz	Proton-cyclotron instability in the SW; Kelvin–Helmholtz instability.
Pc4	45 – 150	Low:	Kelvin–Helmholtz instability;
		1 – 10 mHz	Drift-mirror instability;
Pc5	150 – 600		Bounce resonance.
Pi1	1 – 40		Field aligned current driven instabilities.
Pi2	40 – 150		Abrupt changes in convection in the magnetotail; Flux transfer events.

in the orientation of the interplanetary magnetic field (IMF) can have dramatic effects on the characteristics of waves seen on the Earth. The morphological and physical properties of pulsations also depend on the geomagnetic latitude and longitude (or local time)[2].

As for other geomagnetic studies, different regions of interest are usually considered in terms of latitude, or L parameter: the equatorial region ($L < 1.5$, or $\Lambda < \approx 35°$), the low latitude region ($1.5 \leq L \leq 3$; or $\approx 35° < \Lambda < \approx 55°$), the middle latitude region ($3 \leq L \leq 5$; or $\approx 55° < \Lambda < \approx 63°$, i.e. up to the expected position of the plasmapause[3]); the high latitude region (from $L \approx 5$ out to the last closed field line); the polar cap, with open field lines which extend into the tail, or connect to IMF lines. High latitudes encompass several important zones such as the auroral oval and the cusp[4]. The study of pulsations in Antarctica is very interesting as Antarctica extends up to latitudes (corresponding to oceans in the northern hemisphere) where local field lines penetrate extreme magnetospheric regions where several generation mechanisms are active (Arnoldy et al., 1988). At those latitudes Antarctica also allows geomagnetic measurements in a wide longitudinal range.

The classification scheme in terms of L is also adopted for magnetospheric studies; in this case L identifies the line of force which maps to a given ground latitude: $L < 4$ and $L > 6$ usually identify the "*inner*" and "*outer*" magnetosphere, respectively. However, since the magnetosphere is a highly dynamic system, the state of the magnetosphere also partially determines the magnetic projection of different regions. Major factors which control magnetospheric dynamics are the SW dynamic pressure (ρV_{sw}^2, ρ and V_{sw} being the SW mass density and flow velocity) and the rate of transport

[1] In 1973 IAGA added two new classes (Pc6 > 600 s, and Pi3 > 150 s) to the classification scheme. Pi3 include fluctuations associated with storm sudden commencements (Psc5 and Psc6), and substorms (Pip, 100 – 400 s; Ps6, 5 – 40 min; [Saito, 1978]). For more detailed discussions on sources and theoretical aspects, see [Dungey, 1967; Hughes, 1983; Southwood and Hughes, 1983; Samson, 1991].

[2] The Z-axis of the geomagnetic coordinate system is parallel to the magnetic dipole axis, the geographic coordinates of which are ≈ 11.02° (colatitude) and ≈ −70.70° (east longitude). The geomagnetic latitude (λ) is measured from the geomagnetic equator and is positive northward; the geomagnetic longitude (ϕ) is measured from the meridian that contains the south geographic pole and is positive eastward. The relationship between geomagnetic and geographic longitude is such that the geomagnetic longitude is ≈ 70° greater than the geographic longitude, except near the poles. The magnetic local time (MLT) is defined as the geomagnetic longitude of the observer minus the geomagnetic longitude of the Sun expressed in hours plus 12 hr. L is the magnetic shell parameter which identifies the geocentric equatorial distance of a field line, measured in Earth radii (R_E). A related parameter is the invariant latitude $\Lambda = cos^{-1}(1/L)^{1/2}$, which is the latitude where a line of force intersects the Earth's surface. For example, for a line that extends up to 10 R_E in the equatorial plane, Λ is ≈ 71.6°. At high latitudes where field lines are open or non dipolar, L values become meaningless.

[3] The position of the plasmapause (the outer boundary of the plasma population corotating with the Earth) depends on local time (and other factors such as geomagnetic activity) and varies roughly between $L \approx 4$ (in the dawn sector) and $L \approx 6$ (in the dusk sector). Typically the electron density drops off by two orders of magnitude across the plasmapause.

[4] The auroral oval ($\lambda \approx 67°$–$75°$), more extended equatorward on the nightside, maps on closed field lines into the plasmasheet. The cusp is a funnel-shaped region separating closed field lines extending sunward from those extending tailward. The dayside cusp is highly confined in latitude; however, its position is dependent on IMF conditions ($\Lambda \approx 70°$–$85°$), moving equatorward during southward IMF orientation. Through the cusp, the magnetosheath plasma has direct access to the ionosphere.

of the southward magnetic flux ($B_s V_{sw}$, B_s being the southward IMF component).

Dungey (1967) also introduced the concept of hydromagnetic resonance, a basic process for interpreting most aspects of magnetic pulsations, and identified the Kelvin–Helmholtz instability (KHI) at the Chapman–Ferraro layer (i.e. the magnetopause) as an important energy source for low frequency events. In addition, "*upstream waves*" generated by particles reflected from the bow shock along IMF lines are considered important exogenic sources for daytime pulsations in the mid-frequency band. High frequency Pc are thought to be generated by ion-cyclotron instabilities occurring within the magnetosphere. Pi2 pulsations (Fig. 16.11a) represent transient signals associated with dramatic changes of the state of the magnetosphere; they can only be extensively treated in the context of substorm manifestation.

Theoretical and experimental aspects of geomagnetic pulsations have been discussed in several books and review papers (Dungey, 1967; Saito, 1969, 1978; Jacobs, 1970; Lanzerotti and Southwood, 1979; Rostoker, 1979; Hughes, 1983; Russell and Hoppe, 1983; Southwood and Hughes, 1983; Odera, 1986; Arnoldy et al., 1988; Samson, 1991; Takahashi, 1991, 1998; Allan and Poulter, 1992; Anderson, 1994; Fazakerley and Russell, 1994; Hughes, 1994; Le and Russel, 1994; Engebretson, 1995; Kivelson, 1995; Villante and Vellante, 1997; Kangas et al., 1998; Olson, 1999; McPherron, 2002); pulsations in the magnetosphere have been reviewed by Anderson (1994); hydromagnetic waves upstream of the bow shock have been discussed in a special issue of the *Journal of Geophysical Research* (June, 1991), by Russell and Hoppe (1983), and by Le and Russell (1994); magnetosheath waves have been examined by Fazakerley and Russell (1994). For more detailed references about aspects discussed in the next paragraphs, the reader is referred to these reviews.

Following the classical approach to the physics of magnetic pulsations, we will summarize basic elements of hydromagnetic waves and several aspects of the major processes related to the sources of these oscillations.

16.2 Linear Theory of Hydromagnetic Waves

In a plasma imbedded in a magnetic field, B, hydromagnetic (or "*magnetohydrodynamic*", MHD) waves arise at low frequencies (i.e. lower than both the plasma frequency, $\omega_{ps} = (n_s e^2/\varepsilon_0 m_s)^{1/2}$, and the ion gyrofrequency, $\Omega_i = eB/m_i$, where n_s and m_s are the number density and the mass of particles) as a combined effect of mechanical and electromagnetic forces.

16.2.1 The Uniform Field

Following Dungey's approach (1967) we assume a uniform fluid, with a density ρ, in a uniform magnetic field $\boldsymbol{B}_0$, and consider small amplitude disturbances in the electric field, velocity, current density, and magnetic field ($\boldsymbol{e}$, $\boldsymbol{u}$, $\boldsymbol{j}$, and $\boldsymbol{b}$) that vary like $e^{i(\boldsymbol{k}\cdot\boldsymbol{r}-\omega t)}$. In these conditions, basic hydromagnetic equations are the fluid momentum equation, which in a "*cold plasma*" (i.e. when the thermal pressure can be neglected with respect to the magnetic pressure) simplifies to

$$\rho\frac{\partial \boldsymbol{u}}{\partial t} = \boldsymbol{j}\times\boldsymbol{B}_0 \rightarrow -i\omega\rho\boldsymbol{u} = \boldsymbol{j}\times\boldsymbol{B}_0 \tag{16.1}$$

the hydromagnetic form of Ohm's law

$$\boldsymbol{e} = -\boldsymbol{u}\times\boldsymbol{B}_0 \tag{16.2}$$

and Maxwell's equations

$$\nabla\times\boldsymbol{e} = -\frac{\partial \boldsymbol{b}}{\partial t} \rightarrow \omega\boldsymbol{b} = \boldsymbol{k}\times\boldsymbol{e} \tag{16.3}$$

$$\nabla\times\boldsymbol{b} = \mu_0\boldsymbol{j} \rightarrow i\boldsymbol{k}\times\boldsymbol{b} = \mu_0\boldsymbol{j} \tag{16.4}$$

in which displacement currents have been neglected with respect to the conduction current. Equation (16.1) shows that $\boldsymbol{j}$ is perpendicular to $\boldsymbol{u}$; $\boldsymbol{e}$, $\boldsymbol{u}$, and $\boldsymbol{B}_0$ are mutually perpendicular (Eqs. (16.1) and (16.2)), and the same is true for $\boldsymbol{j}$, $\boldsymbol{k}$, and $\boldsymbol{b}$ (Eqs. (16.3) and (16.4)). As a consequence ($\boldsymbol{j}$, $\boldsymbol{e}$, $\boldsymbol{k}$), ($\boldsymbol{u}$, $\boldsymbol{b}$, $\boldsymbol{k}$) and ($\boldsymbol{u}$, $\boldsymbol{b}$, $\boldsymbol{B}_0$) must be coplanar. These conditions present two cases requiring that either $\boldsymbol{j}$ and $\boldsymbol{e}$, or $\boldsymbol{u}$ and $\boldsymbol{b}$, are parallel to $\boldsymbol{k}\times\boldsymbol{B}_0$.

In the first case (Fig. 16.2a), previous equations provide a dispersion relation such as

$$\omega = \pm k V_a \tag{16.5}$$

where

$$V_a = B_0/(\mu_0\rho)^{1/2} \tag{16.6}$$

is the phase velocity ("*Alfvén velocity*") which is independent of the direction of propagation. The corresponding wave mode is identified as "*fast*" mode. It has group velocity parallel to $\boldsymbol{k}$, propagates isotropically, and energy can be transported in any direction. The "*fast*" mode has $\boldsymbol{j} \cdot \boldsymbol{B}_0 = 0$, so carries no current along magnetic field lines. In general, $\boldsymbol{b}$ has a component parallel to $\boldsymbol{B}_0$; so, this mode, which is analogous to an ordinary sound wave in a fluid, can trasmit pressure variations. When this mode propagates perpendicular to $\boldsymbol{B}_0$, it is seen as compression and rarefaction of both the magnetic field and the plasma density.

In the second case (Fig. 16.2b), considering that $\boldsymbol{u}$ and $\boldsymbol{b}$ are both perpendicular to $\boldsymbol{B}_0$, we obtain

$$\omega = \pm k V_a \cos \vartheta \qquad (16.7)$$

where ϑ is the angle between $\boldsymbol{k}$ and $\boldsymbol{B}_0$. This is a transverse mode ("*shear Alfvén mode*") which only bends the field lines. $\boldsymbol{b}$ is perpendicular to $\boldsymbol{B}_0$ and the magnetic field magnitude is constant (to a linear approximation), even in the presence of a wave: the magnetic pressure is constant and the wave is noncompressive. The group velocity is parallel to $\boldsymbol{B}_0$ and the energy is guided along the ambient magnetic field. In general, geomagnetic pulsations ultimately originate in the magnetosphere either as fast or Alfvén modes, or a combination of these two modes[5].

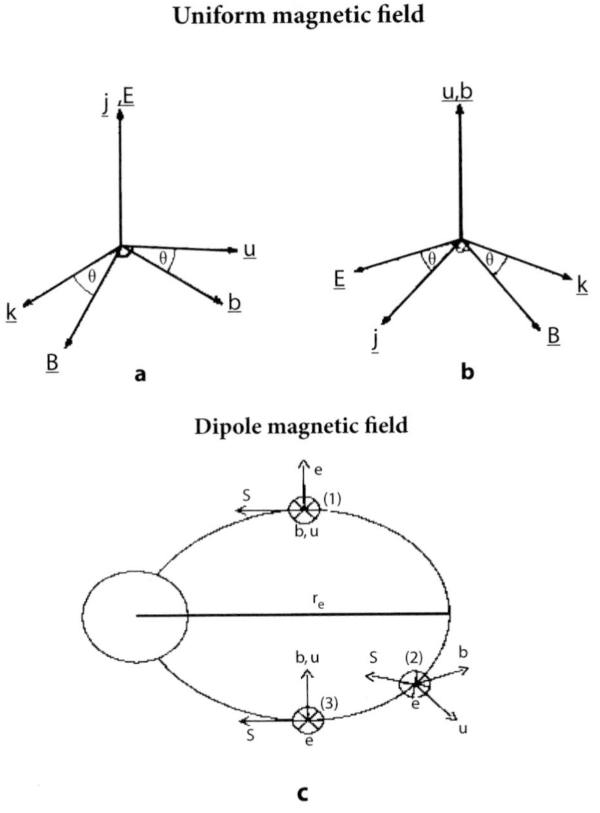

Fig. 16.2a–c. The characteristics of hydromagnetic wave modes. *Top panel.* The relative orientation of the vector fields: (**a**) the fast mode; (**b**) the shear mode. [Dungey, 1967]. *Bottom panel.* (**c**) The different wave modes in a dipole field. $\boldsymbol{S}$ is the Poynting vector: it is parallel to $\boldsymbol{B}_0$ for the toroidal mode (*1*) and for the guided poloidal mode (*3*); it is across $\boldsymbol{B}_0$ for the cavity mode (2). [Vellante, 1993a]

16.2.2 The Dipole Field

In the inner magnetosphere $V_a \approx 1000$ km/s, and pulsation periods of ≈ 100 s or longer are commonly detected. This leads to estimated wavelengths of $\approx 15 - 20$ R_E, i.e. comparable with the size of the entire magnetosphere. It suggests that the homogeneous plasma approximation is not appropriate for magnetospheric pulsations. If the field is not uniform, previous equations require additional terms related to the spatial derivatives of $\boldsymbol{B}_0$. These additional terms couple the Alfvén and fast mode and in general it is impossible to separate two distinct wave modes. Interesting simplifications arise when the axial symmetry of $\boldsymbol{B}_0$ (as for a dipole field) is considered. In cylindrical coordinates (r, ϕ, z) an axial symmetric magnetic field has $B_{0\phi} = 0$ and $\dfrac{\partial B_{0r}}{\partial \phi} = \dfrac{\partial B_{0z}}{\partial \phi} = 0$. In this case, assuming a longitudinal variation such as $e^{i(m\phi - \omega t)}$ for perturbed quantities, it is easy to derive the following equations (Dungey, 1967; Hughes, 1983)[6]:

$$\left[\frac{\omega^2}{V_a^2} + \frac{1}{r^2 B_0} \nabla_{\|} \left(r^2 B_0 \nabla_{\|}\right)\right] \left(\frac{u_\phi}{r}\right) = \frac{m\omega}{r^2} \frac{b_{\|}}{B_0} \qquad (16.8)$$

[5] It is worth noting that in the absence of "*cold plasma*" conditions, two additional compressional modes arise: the "*fast magnetoacoustic wave*", with a phase velocity greater than V_a and C_s ($C_s = (\gamma p/\rho)^{1/2}$ being the sound speed), and the "*slow magnetoacoustic wave*", with a phase velocity less than V_a and C_s. If $\vartheta = 0°$, phase velocities are V_a and C_s, so that the motion is a superposition of a pure Alfvén wave and a pure sound wave, both traveling along $\boldsymbol{B}_0$. If $\vartheta = 90°$, the phase velocity is $(V_a^2 + C_s^2)^{1/2}$, and waves propagate perpendicularly to $\boldsymbol{B}_0$. A similar situation may occur in the magnetosphere, at geocentric distances of $\approx 4 - 5$ R_E, where plasma and magnetic pressure may be comparable (Lanzerotti and Southwood, 1979).

[6] Interesting aspects of these oscillations emerge in the dipole coordinate system ((Radoski, 1972), and papers referenced).

$$\left[\frac{\omega^2}{V_a^2} + r^2 B_0 \nabla_{\|} \left(\frac{\nabla_{\|}}{r^2 B_0}\right)\right](re_\phi) = -\mathrm{i}\omega B_0 r \nabla_\perp \frac{b_{\|}}{B_0} \quad (16.9)$$

$$\mathrm{i}m\frac{u_\phi}{r} + \frac{1}{rB_0}\nabla_\perp re_\phi = \mathrm{i}\omega\frac{b_{\|}}{B_0} \quad (16.10)$$

where $\nabla_{\|} = \hat{\boldsymbol{e}}_{B_0} \cdot \nabla$, and $\nabla_\perp = (\hat{\boldsymbol{e}}_{B_0} \times \nabla)_\phi$, $\hat{\boldsymbol{e}}_{B_0}$ being the unit vector along $\boldsymbol{B}_0$.

This system has eigenperiods corresponding to eigenfunctions which must satisfy certain boundary conditions. Previous equations can be decoupled in two limits. If the wave is axisymmetric ($m = 0$, which means that the signals are in phase around an entire circumference) the right hand side of (16.8) vanishes. The left hand side takes the form of a one-dimensional wave equation with the only spatial derivative along $\boldsymbol{B}_0$. It corresponds to a transverse mode ("*toroidal mode*" with the characteristics of an Alfvén mode, Fig. 16.2c). It can be interpreted as a torsional oscillation of the lines of force, or magnetic shell, which oscillate azimuthally, independently of all others. The magnetic and velocity perturbations ($\boldsymbol{b}$, $\boldsymbol{u}$) are azimuthal and the electric field ($\boldsymbol{e}$) is normal to the field lines. Such a wave is guided by the magnetic lines of force. In the absence of ionospheric effects, this mode would correspond to oscillations of the east–west (D) geomagnetic field component. In the same limit (16.9) represents a mode ("*poloidal mode*", with the characteristics of a fast mode), in which $\boldsymbol{e}$ is azimuthal, $\boldsymbol{u}$ and $\boldsymbol{b}$ are in the meridian plane and the whole cavity oscillates coherently ("*cavity mode*"). On the Earth's surface, in the absence of ionospheric effects, this "*field-aligned*" mode would affect the vertical (Z) and north–south (H) components.

The other limiting case corresponds to a disturbance confined to a narrow range of longitude ($m \to \infty$, which means that adjacent field lines are highly decoupled and perform independent azimuthal oscillations). In this case we obtain (16.9) a transverse wave mode which propagates along $\boldsymbol{B}_0$ ("*guided poloidal mode*"); $\boldsymbol{e}$ is in the azimuthal direction, while $\boldsymbol{b}$ and $\boldsymbol{u}$, perpendicular to the field line, lie in the plane of $\boldsymbol{B}_0$.

16.3 Sources of Geomagnetic Pulsations

16.3.1 Upstream Waves

A major source for geomagnetic pulsations in the mid-frequency range is considered the penetration into the magnetosphere of waves generated in the upstreaming SW ("*foreshock region*") by protons reflected by the bow shock along IMF lines. Briefly, such waves are generated when the following resonant condition is matched

$$|\omega - \boldsymbol{k} \cdot \boldsymbol{V}_\mathrm{p}| = \Omega_\mathrm{p} \quad (16.11)$$

where ω is the wave frequency, $\boldsymbol{V}_\mathrm{p}$, the velocity of reflected protons (both in the SW frame of reference), and Ω_p their gyrofrequency. In the spacecraft frame of reference the wave frequency would be:

$$\omega_\mathrm{s} = \omega + \boldsymbol{k} \cdot \boldsymbol{V}_\mathrm{sw} \quad (16.12)$$

Considering that both V_p ($\approx 600 - 1200$ km/s) and V_SW ($\approx 300 - 800$ km/s) are usually much larger than V_a ($\approx 40 - 100$ km/s in the interplanetary medium), previous equations provide (16.5)

$$\omega_\mathrm{s} \approx \Omega_\mathrm{p} \frac{V_\mathrm{sw}}{V_\mathrm{p}} \frac{\cos \vartheta_1}{\cos \vartheta_2} \quad (16.13)$$

where ϑ_1 is the angle between $\boldsymbol{k}$ and $\boldsymbol{V}_\mathrm{sw}$ and ϑ_2 is the angle between $\boldsymbol{k}$ and $\boldsymbol{V}_\mathrm{p}$. Equation (16.13) reveals that wave frequency is dependent on IMF magnitude, SW speed, and the bulk speed of backstreaming ions. Assuming typical values for V_sw, and for the orientation and magnitude of V_p, previous equation gives the numerical relation (see also Takahashi et al., 1984; Le and Russell, 1996):

$$f\,(\mathrm{mHz}) \approx (6 \pm 1)\,B\,(\mathrm{nT}) \quad (16.14)$$

which, for typical values of IMF magnitude ($\approx 3 - 10$ nT), leads to the prediction of upstream waves mostly in the Pc3 regime.

Since these waves propagate slowly, they are convected downstream towards the bow shock and, under certain conditions, can enter the magnetosphere. Within the magnetosphere, these oscillations may couple to field lines or propagate directly through magnetospheric cavity resulting in the detection of ground pulsations.

The orientation of the IMF (and the "*cone angle*", θ_XB, between the IMF and the Earth-Sun line) is also an important parameter influencing the structure of the bow shock[7], the location of the foreshock region

[7] When the angle between the IMF and the shock normal is large (*quasi-perpendicular shock*), the field is almost paral-

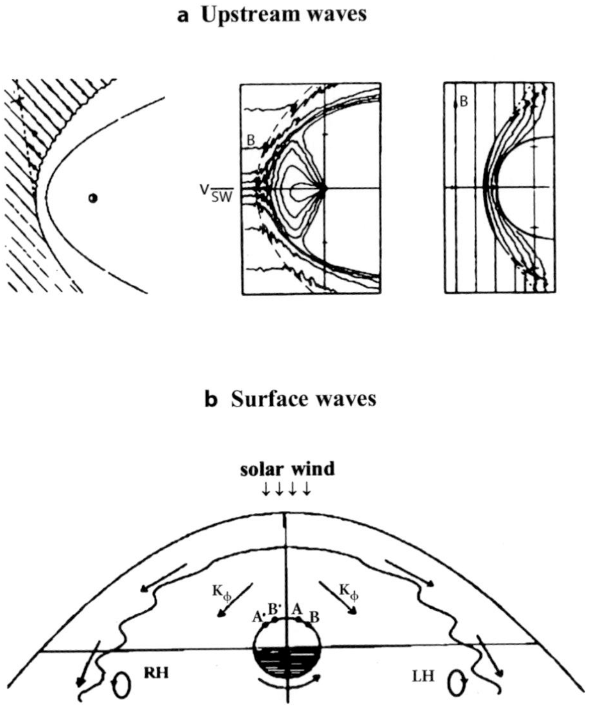

Fig. 16.3a,b. The generation and penetration of pulsations. (**a**) The location of the foreshock region in the equatorial plane: *Left panel*. Upstream waves are generated by protons along the spiral IMF lines. *Central panel*. For a radial IMF, waves are generated in a foreshock region symmetric around the subsolar point. *Right panel*. For a perpendicular IMF, waves are generated in narrow regions close to the bow shock flanks. [Greenstadt et al., 1981; Formisano, 1984]. (**b**) A schematic representation of the generation and propagation of surface waves on the magnetopause. K_ϕ is the azimuthal component of the wave vector. A and B identify two ground stations. Since waves propagate tailward, they would be observed to propagate from station A to station B when stations are located on the morning side, and from B′ to A′ when they are located on the afternoon side. The polarization of surface waves driven by the KHI changes sense near local noon

lel to the shock boundary and the shock appears as a sharp discontinuity. When it is small (*quasi-parallel shock*), the field is almost perpendicular to the shock boundary and the transition becomes turbulent. Hence a quasi-parallel shock is also a possible source of ULF waves. Under spiral IMF orientation, a quasi-parallel and a quasi-perpendicular bow shock structure are predicted on the dawn and dusk side of the magnetosphere, respectively (Fig. 16.3a, *top panel*).

and wave transmission through the magnetosphere (Fig. 16.3a). Under nominal conditions, the spiral IMF predicts a foreshock region (and a possible wave penetration) on the morning side of the magnetosphere; convected downstream through magnetosheath, these waves, in general, would not reach the magnetopause easily. For a radial IMF, a wide and symmetric foreshock region is predicted around the subsolar point, and waves are mostly convected toward the magnetosphere. For a perpendicular IMF, upstream waves would only be generated in narrow regions close to the bow shock flanks and swept away by the SW flow.

16.3.2 Kelvin–Helmholtz Instability

Surface waves at the magnetopause boundary layer in the low- and mid-frequency range are expected to arise as a consequence of the relative motion of the SW and magnetospheric plasmas. These waves are amplified when V_{sw} exceeds a critical value determined by:

$$(\boldsymbol{k}\cdot\boldsymbol{V}_{\mathrm{sw}})^2 > \left(\frac{1}{n_{\mathrm{sw}}}+\frac{1}{n_{\mathrm{M}}}\right) \times\left[n_{\mathrm{sw}}\left(\boldsymbol{k}\cdot\boldsymbol{V}_{\mathrm{a}}\right)^2_{\mathrm{sw}}+n_{\mathrm{M}}\left(\boldsymbol{k}\cdot\boldsymbol{V}_{\mathrm{a}}\right)^2_{\mathrm{M}}\right] \quad (16.15)$$

in which n is the number density and subscripts SW and M identify the SW and the magnetosphere, respectively. KHI waves have a phase velocity in the same sense as the wind which drives them (i.e. westward in the morning and eastward in the afternoon; Fig. 16.3b): as a consequence, a phase inversion across the noon meridian should appear in experimental observations. Beneath the boundary, a fluid element will have an approximately elliptical motion (with the rotation being in the opposite sense along the dawn and the dusk flank) similar to that of the fluid motion associated with a wave on a free surface in a pure gravitational field. As the field lines are frozen into the plasma, they will also rotate generating elliptical polarized waves which propagate to the Earth via the field lines.

16.3.3 Ion-Cyclotron Instability

Consider the case of a wave with frequency ω and phase velocity V_{a}, and a proton with a velocity $V_{\parallel}$ (i.e. parallel to the magnetic field) traveling in opposite directions along the magnetic field. In the proton frame of reference the wave has a Doppler shifted frequency $\omega' = \omega(1 + V_{\|}/V_a)$. Provided the proton has $V_\perp \neq 0$, it will be gyrating around B in a left-handed sense. When ω' and Ω_i are the same, the wave electric field and the proton velocity are in resonance. If the wave electric field is antiparallel to the ion velocity, the particle will be slowed down and the wave will gain energy. In fact, the distribution of the "*pitch angle*", the angle between the particle velocity and the magnetic field, peaks at large angles within the magnetosphere. This means that particles have more perpendicular than parallel energy: as a net effect waves gain energy and transverse, left-handed, circularly polarized waves are expected to arise at $f \approx 0.1-5$ Hz (McPherron, 2002).

16.4 Effects of the Ionosphere and Field Line Eigenperiods

The simplest approach to evaluating the effects of the ionosphere is to consider it a perfect conductor. So the electric field vanishes and the energy of transverse waves guided along the magnetic field is confined to the region of space between ionospheres of opposite hemispheres. This approach allows the evaluation of the expected wave periods numerically integrating the corresponding wave equations. Alternatively, approximate values of the eigenperiods T_n for standing oscillations can be obtained using the WKB approximation, which is valid when the wavelength is short compared to the scale variation of B_0 and ρ; in particular, for the toroidal mode

$$T_n = \frac{2}{n}\int_l \frac{ds}{V_a} \qquad n = 1, 2, 3 \dots \qquad (16.16)$$

where integration is carried out along field lines anchored on opposite ionospheres. Assuming that the geomagnetic is a centred dipole, (16.16) becomes

$$T_n = \frac{16\pi R_E^4}{nM\mu_0^{1/2}\cos^8\lambda_0}\int_0^{\lambda_0} \rho^{1/2}\cos^7\lambda\, d\lambda \qquad (16.17)$$

where M is the dipole moment, and λ_0 is the geomagnetic latitude of the foot of the line of force of a given shell.

Figure 16.4a shows the classical representation of the expected L-dependence of the fundamental period of the uncoupled toroidal and guided poloidal mode as well as the WKB solution for a simple radial dependence of the magnetospheric plasma density. Clearly, the L-variation corresponds to a latitudinal variation at ground magnetometric arrays. Figure 16.4a also predicts a continuous spectrum of T_n eigenvalues, with a general tendency for increasing periods with increasing latitude (between $\approx 10-170$ s, with the exception of a narrow region which corresponds to the ρ cutoff at the plasmapause). Obviously, given the ionosphere height, Eq. (16.17) cannot be applied below $L \approx 1.1-1.2$; a similar conclusion holds for high latitudes, where the dipole approximation becomes poor. Eigenfrequencies of the field lines for more realistic field geometries and more sophisticated density distributions have been evaluated by several authors: in general, the ρ-dependence makes T_n estimates dependent on local time, magnetospheric activity, the solar cycle, etc.

In addition, a significant portion of the field line lies within the ionosphere at low latitudes: as a consequence, mass loading due to ionospheric heavy ions lowers the expected eigenfrequencies (whose harmonic values are not integer multiples of the fundamental mode) and provides maximum values of the eigenfrequency at $\lambda \approx 30°$ ($L \approx 1.3$; Fig. 16.4b). Figure 16.4c shows the structure of the wave perturbation along the magnetic field line for the two lowest harmonics.

If the ionosphere were a perfect conductive medium, waves would be confined within the magnetosphere and no signal would be detected on the ground. In reality, the incoming wave drives horizontal current sheets in the ionosphere because of its finite and anisotropic conductivity and the input signal is not completely shielded ([Hughes, 1994] and papers referenced). In fact, the ionosphere smears the rapid variations at ground level and features with scale length smaller than ≈ 100 km are strongly damped. In addition, the Pedersen current along $\boldsymbol{e}$ generates a magnetic field opposite to the wave field, while the Hall current along $\boldsymbol{e} \times \boldsymbol{B}_0$ generates a magnetic field perpendicular to the wave field. So, the net effect consists in a 90° rotation of the original signal (Fig. 16.4d), which is left-handed in the northern hemisphere and right-handed in the southern hemisphere, when looking downward[8].

[8] In addition, the excitation of pulsations by rapid changes of the ionospheric conductivity induced by solar flare X-ray

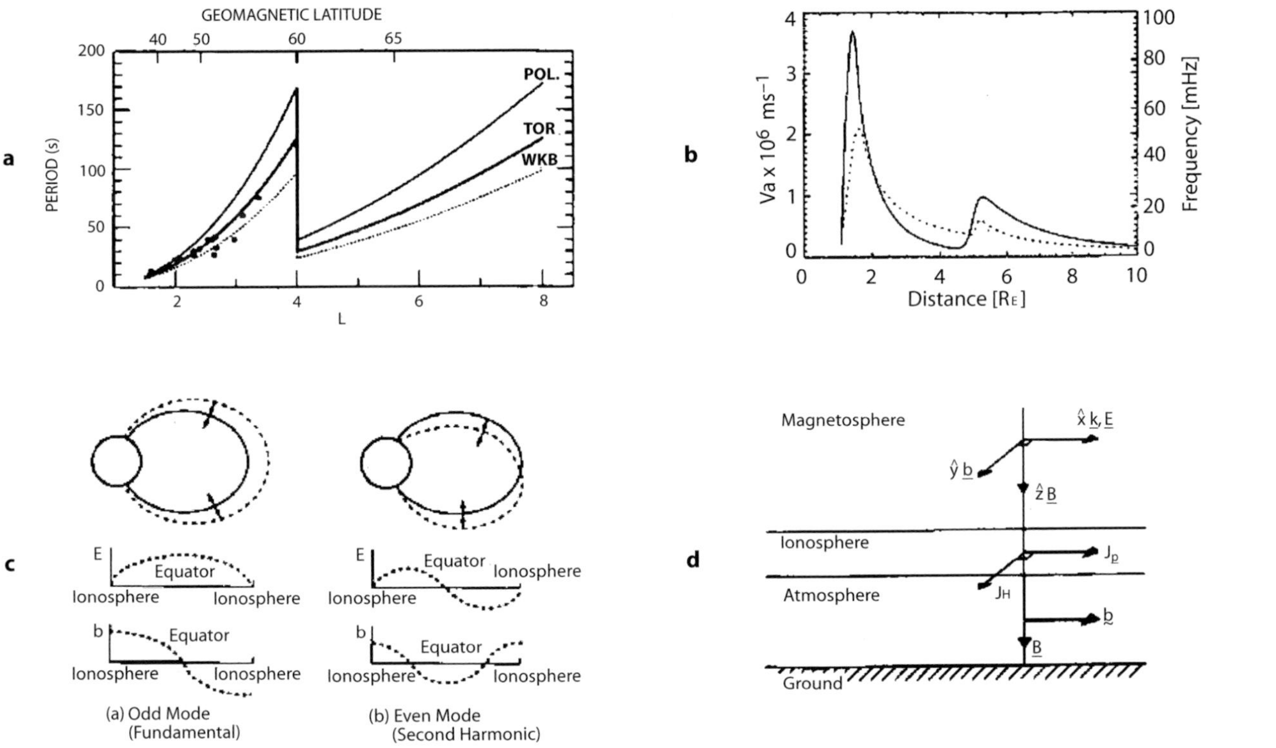

Fig. 16.4a–d. The effect of the ionosphere. (**a**) The L-variation of the fundamental period of the toroidal mode, guided poloidal mode and WKB solution for a dipole field, assuming a r^{-3} dependence for the plasma density in the plasmasphere (located at $\approx 4\,R_E$), and a r^{-4} dependence beyond the plasmapause. Experimental points represent dominant pulsation periods observed at different latitudes. [Villante and Vellante, 1997]. (**b**) The behaviour of the FLR frequency (*dotted line*) and Alfvén velocity (*solid line*); in this case the plasmapause is located at $\approx 5\,R_E$. [Waters et al., 2000]. (**c**) A schematic representation of the field displacements (*dashed lines*) in a fundamental and second harmonic mode of FLR and the corresponding perturbation of the electric and magnetic fields. (**d**) A schematic representation of the ionospheric effects on an incident Alfvén wave. [Hughes, 1983]

16.5 Field Line Resonance

Although questioned by some authors, the field line resonance (FLR) is the principal mechanism and energy source for ground pulsations. Southwood (1974) and Chen and Hasegawa (1974) examined a simple model in which the magnetospheric field line resembles a damped harmonic oscillator in the presence of a driving force. FLR occurs when the frequency of the incoming (driving) wave is comparable to the field line eigenfrequency. If the incoming wave is monochromatic at f^*, the coupling will be strongest at the closed field line for which f^* is a resonant frequency (i.e. it matches the frequency of the wave that can stand on that field line). A schematic illustration of this process is shown in Fig. 16.5a (*left panel*) where KHI waves are represented as wiggly lines moving away from the local noon and line thickness represents wave amplitude. As can be seen, amplitude progressively decreases inward, but peaks locally at resonant L-shells (Kivelson, 1995). In the magnetosphere, the fast and Alfvén mode can interact to generate azimuthal standing oscillations. In this case the amplitude of the wave azimuthal component is enhanced; however, as a consequence of the ionospheric rotation, the occurrence of such resonance processes should lead to an enhancement of the H wave amplitude in ground ob-

and EUV fluxes represents an interesting example of generation mechanisms not associated with SW and/or magnetospheric processes.

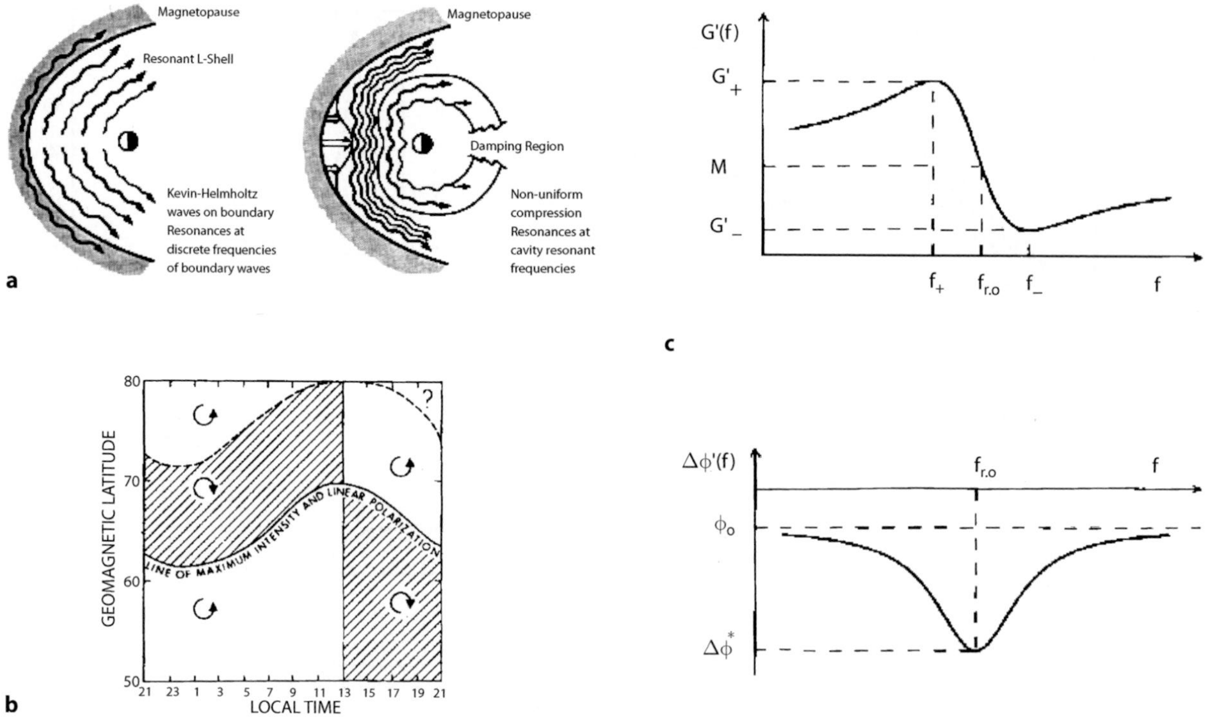

Fig. 16.5a–c. Theoretical aspects of the resonance processes. (**a**) A schematic representation of wave perturbation through the dayside magnetosphere produced by the KHI at the magnetopause (FLR, *left panel*), and by a compression of the magnetopause nose (cavity mode, *right panel*). [Kivelson, 1995]. (**b**) The diurnal and latitudinal variation of the polarization pattern in the northern hemisphere for pulsations with $f \approx 5$ mHz; for higher frequencies the entire pattern shifts equatorward. [Samson et al., 1971]. (**c**) The amplitude ratio $G(f)$ and the phase difference $\Delta\phi$ between H signals for a pair of stations separated in latitude. $f_{r,0}$ is the resonant frequency at the middle point between stations. [Vellante et al., 2002]

servations. For a broadband driving source such as upstream waves, a continuum of field lines may resonate, provided the eigenfrequencies are within the frequency of the source; in addition, multiple harmonics can be generated on a single shell.

The important aspects of the polarization pattern for FLR were examined considering a field line eigenfrequency that varies across the field lines. Near resonance coupling between modes creates a narrow amplitude maximum. As Southwood (1974) argued, the phase of the radial and azimuthal component are expected to change in a way that the wave polarization changes sense at both the resonant field line and at the amplitude minimum between the resonant field line and the magnetopause. Waves propagating westward (eastward) are expected to have a right-handed (left-handed) polarization poleward of the resonant field line, and left-handed (right-handed) equatorward of it. In addition, as noted earlier, a polarization switch around local noon is expected for KHI waves. This leads to the prediction of a polarization pattern which is consistent with the one obtained at northern auroral latitudes by Samson et al. (1971) (Fig. 16.5b).

Some aspects of FLR can be easily understood in terms of forced, damped harmonic transverse oscillations of field lines anchored on opposite hemispheres, driven by incoming fast mode waves. In this simplified scheme the field line oscillation can be described as:

$$\ddot{b}_\phi + 2\gamma \dot{b}_\phi + \omega_r^2 b_\phi = \omega_r^2 b_z c\,(\sin \omega_d t) \quad (16.18)$$

where b_ϕ is the amplitude of the transverse resonance wave, b_z is the amplitude of the driving wave at frequency ω_d, ω_r is the resonant frequency, c is a coupling factor between the incoming wave and the field line, and γ is a damping factor (Menk et al., 1994). Assuming a broad band source and a low damping, in addition to transient fluctuations at $\approx \omega_r$ which decay exponentially, it is possible to obtain a steady state solution with amplitude and phase given by

$$A(\omega) = \frac{\omega_r^2 b_z c}{\left[\left(\omega_r^2 - \omega_d^2\right)^2 + 4\gamma^2 \omega_d^2\right]^{1/2}} \quad (16.19)$$

$$\phi(\omega) = \tan^{-1}\left[\frac{-2\gamma\omega_d}{\omega_r^2 - \omega_d^2}\right] \quad (16.20)$$

Clearly, at $\omega_r = \omega_d$, the amplitude assumes maximum values and the phase reverses. If data from two stations closely spaced in latitude are available, a comparison of signal amplitudes ("*gradient method*") or phases ("*cross-phase method*"), can be used for determining the resonant frequency. A theoretical expression for the meridional structure of the complex amplitude of the H component in a restricted region around the resonant point is given by:

$$H(x, f) = \frac{\varepsilon H_r(f)}{\varepsilon + i\left[x - x_r(f)\right]} \quad (16.21)$$

where x is the meridional coordinate, $x_r(f)$ is the resonant point, ε is the resonance width, and $H_r(f)$ is the amplitude at the resonant point (Fig. 16.5c; Vellante et al., 2002 and papers referenced).

16.6 Cavity Resonance

As for toroidal modes, expected periods may also be evaluated for poloidal "*cavity*" modes propagating through the magnetosphere and reflected by its boundaries. The modern concept of magnetospheric cavity mode was introduced by Kivelson et al. (1984), who proposed a simple box geometry with perfectly reflecting boundaries, in which the magnetosphere rings as a whole at its own eigenfrequencies. At the same time, Allan et al. (1986) suggested that impulsive stimuli at the magnetopause can set up compressional cavity resonances which drive FLR within the magnetosphere. This model was further developed for more realistic conditions. Obviously, the cavity mode eigenperiods are determined by the cavity dimensions. The presence of several boundaries reflecting signals (the magnetopause, the plasmapause, the ionosphere, etc.) concurs to determine a discrete spectrum of expected frequencies. Harrold and Samson (1992), who considered the bow shock as an outer boundary, proposed discrete frequencies at $f \approx 1.3, 1.9, 2.5, 3.4$ and 4.2 mHz. Magnetospheric dimensions, on the other hand, are continuously changing due to the continuous variations of the SW and IMF parameters. In addition, given its long tail, the magnetosphere would be better represented as an open-ended waveguide, in which compressional modes propagate antisunward and energy resonates radially between an outer boundary (such as the bow shock or the magnetopause) and an inner turning point (Samson et al., 1992). All these aspects make estimates (and identification) of magnetospheric cavity/waveguide modes uncertain. Figure 16.5a (*right panel*) shows a schematic representation for possible cavity resonances driven by impulsive variations of SW pressure. It is worth noting that as a consequence of coupling with FLR, the global mode has the characteristics of an Alfvén mode near the resonant field line, and is similar to a compressional mode away from the resonant field line. Multiharmonic cavity modes and toroidal resonances may also be excited when SW pressure pulses impinge the magnetopause.

16.7 Low Frequency Pulsations

During the 1960's, hydromagnetic fluctuations in the distant magnetic field were observed by several spacecraft. Explorer 12 provided a direct observation of magnetic pulsations, both compressional and transverse, with an approximate period of 2 – 3 min[9]. Estimates of the field line eigenperiods confirmed their interpretation in terms of standing waves. Large amplitude compressional fluctuations ($\delta B/B \approx 0.2-0.5$), with periods of $\approx 5-10$ min, were interpreted in terms of slow modes in which particle pressure and magnetic pressure were in anti-phase. Similar fluctuations,

[9] The magnetospheric field exhibits magnetic fluctuations in radial, azimuthal, or parallel (i.e. along B) directions. Depending on which component is dominant, the wave is usually termed "*poloidal*", "*toroidal*", or "*compressional*". More generically, "*transverse*" waves may include both azimuthal and radial perturbations.

with periods of ≈ 10 min or longer, were observed at $L > 8$ (indicating a source near the magnetopause) with highest occurrence rates in the morning and afternoon sectors: they were polarized in the meridian plane with comparable compressional and transversal components. At geosynchronous orbit ($L \approx 6.6$), waves were observed to propagate away from noon on both sides of the magnetosphere, a feature consistent with KHI predictions.

In the meanwhile, evidence from ground observations was also very important: indeed, the same wave trains were observed simultaneously at the foot of both lines of force (as expected for guided modes), while conjugate observations were matched cycle for cycle (as expected for standing waves). More in general, the agreement between the expected and the dominant periods observed at different latitudes during daytime intervals (Fig. 16.4a), as well as the correspondence between conjugate observations, were considered outstanding arguments to interpret Pc3/5 pulsations in terms of oscillations of lines of force which are rooted at magnetically conjugate points. Simultaneous observations at high latitude conjugate stations and at a satellite revealed symmetric motions of the field line at northern and southern end, and indicated the equator as the nodal plane of an odd mode standing wave, with a period close to the expected for the fundamental mode at those latitudes. Nowadays, conjugate phenomena are observed from low to high latitudes in a wide frequency range.

However, despite such basic conclusions, the theoretical elements summarized in the previous paragraphs should be only considered as rough indicators in comparing theory and observations (Rostoker, 1979). Indeed, pulsations can rarely be interpreted in terms of pure toroidal or poloidal modes, the two being invariably coupled: ground elliptical polarization, on the other hand, is clearly indicative of coupling of toroidal and poloidal modes.

At ground stations, the maximum intensity (up to hundreds of nT) of low frequency pulsations tends to follow the approximate position of the auroral oval (Samson, 1991). Between $\lambda \approx -69°$ and $\lambda \approx -86°$ the low frequency power, shows a non monotonic behaviour, with a power minimum at $\lambda \approx -80°$, followed by a further increase; moreover, pulsation activity in the auroral zone and in the polar cap appears decoupled. In the Pc4 range the maximum wave energy may occasionally occur inside the plasmapause.

A pronounced morning/afternoon asymmetry (with higher power level in the morning) has been reported at auroral and cusp latitudes, together with a secondary enhancement near local midnight. Nighttime enhancement, correlated with substorms, is mainly due to more irregular pulsations sharing the same frequency band. In this sense such enhancement is not indicative of significant nighttime Pc5 activity. Conversely, in the polar cap, the fluctuation power only maximizes around magnetic local noon, when stations approach the dayside cusp. This feature suggests a minor influence of substorm related events deep in the polar cap.

Pc5 pulsations, due to their frequency, are usually observed from auroral to cusp latitudes. Their occurrence and intensity are correlated to SW speed, in particular in the dawn sector. The dependence of the pulsation power on SW speed was stronger at auroral latitudes than at near cusp latitudes and a threshold value ($\approx 450-500$ km/s) has been proposed above which SW control of the pulsation power is dominant. Comprehensive auroral surveys revealed statistical evidence for antisunward propagation and reversal of the polarization sense in latitude and around local noon (Fig. 16.5b). In the meanwhile, increased activity was found during times of high SW speeds. In general, local morning fluctuations were attributed to KHI at the dawn magnetopause (which might be less stable than the dusk flank), while the less frequent afternoon events were attributed to corotating SW pressure pulses impinging the post-noon magnetopause. At lower latitudes ($\lambda \approx 50°–60°$), morning pulsations revealed clockwise polarization and westward propagation suggesting that these waves are mid-latitude signatures of SW driven FLR occurring at higher latitudes; conversely, the polarization characteristics of afternoon events were interpreted in terms of ground signatures of compressional cavity modes. Satellite studies also showed that transverse azimuthally polarized waves (correlating to ground observations) predominate in the morning sector, while compressional events (poorly correlated with ground observations) predominate in the afternoon sector (Anderson, 1994).

Low frequency events are more rarely observed at low latitudes, where they are interpreted as signatures of global compressional modes or large scale-cavity resonances. They would be difficult to explain in terms of KHI waves because of the damping rate of the surface wave mode in the radial direction. Their amplitude (and occurrence) considerably decreases with decreasing latitude; it sharply enhances at the dip equator (i.e. where the Z component vanishes) because of the anomalous ionospheric conductivity.

Sudden impulses (SI)[10] and storm sudden commencements (SSC) are often accompanied at middle and high latitudes by long period, often irregular, pulsations which have been interpreted in terms of FLR; however, this interpretation is in conflict with the large angles of the polarization axes with respect to the H orientation identified in recent analysis. The occurrence of damped Pc4 pulsations that accompanied magnetic disturbances has been reported at $L \approx 4$, exterior to the plasmasphere boundary. An extended investigation of the long period wave response to magnetic storms from equatorial to cap latitudes revealed that narrow band Pc5 activity typically occurs during the recovery phase in the dawn-noon sector at each site; during the main phase the wave activity is broadband and the strongest power is observed above $f \approx 2$ mHz in the auroral zone, and below $f \approx 2$ mHz at middle and low latitudes (Posch et al., 2003). Storm time magnetospheric pulsations might be generated by ring current particles as a result of internal plasma instabilities.

On rare occasions (few times per year), highly monochromatic, amplitude modulated signals appear in the Pc4 range (termed "*giant*" or "P_g" pulsations, with a peak to peak amplitude of ≈10 – 40 nT at ground level and few nT in space). First discussed by Birkeland, these pulsations mostly occur during geomagnetically quiet conditions, between midnight and noon, within a few degrees of the auroral oval (Rostoker, 1979). Giant pulsations are believed to result from plasma instabilities within the magnetosphere.

[10] The SI (SSC) itself at ground level consists of a composite superposition of several wavelike signals ultimately driven by impinging SW discontinuities and related to magnetopause and ionospheric currents. The different waveforms in different sectors can be interpreted in terms of the relative importance of different contributions.

16.8 Mid-Frequency Pulsations

Mid-frequency pulsations (with amplitudes ranging between fractions of nT and several nT) are a common daytime feature of ground observations; they typically reach maximum amplitudes at the position of the dayside cusp ($\lambda \approx 70°$) and appear to decline rapidly as the point of observations moves to higher latitudes. For many years, the polar cap has been considered to be characterized by very low activity in the mid-frequency range. However, more recently, several investigations have shown that this might not be the case (Chugunova et al., 2003). Villante et al. (2002) found peaks of correlation between the Pc4 power and SW speed at dawn and dusk at Terra Nova Bay (CGM λ[11] = −80°): this suggests a major role of KHI at these frequencies.

Although several aspects of the penetration of upstream wave activity through the magnetopause still needs further investigation, it is clear that a significant fraction of the external wave energy enters the magnetosphere. Attempts to measure the transfer function of the magnetopause had little success. However, the occurrence of Pc3 waves at $L \approx 6.6$ meets highly favourable conditions in the morning sector when SW velocity is high and ground observations reveal a close relationship between the amplitude of daytime pulsations and SW velocity. As for lower frequency pulsations, a similar, but somewhat weaker, dependence on SW speed has also been determined for the energy and occurrence of mid-frequency pulsations (Odera, 1986). This feature is also consistent with the general tendency of the pulsations activity to recur with the same period of the corotating high velocity SW streams. Figure 16.6a shows the close correlation between the energy of mid-frequency pulsations and SW velocity detected at low latitudes ($L \approx 1.6$, Yedidia et al., 1991).

As mentioned earlier, the dependence of the pulsation activity on SW speed suggests a major role of KHI

[11] The corrected geomagnetic (CGM) coordinate system has proven to be an excellent tool in organizing geophysical phenomena controlled by the Earth's magnetic field. For a point in space CGM coordinates are evaluated by tracing the field line of the International Geomagnetic Field Reference (IGRF) through the specific point to the dipole geomagnetic equator then returning to the same altitude along the dipole field and assigning the obtained dipole coordinates as CGM coordinates to the starting point.

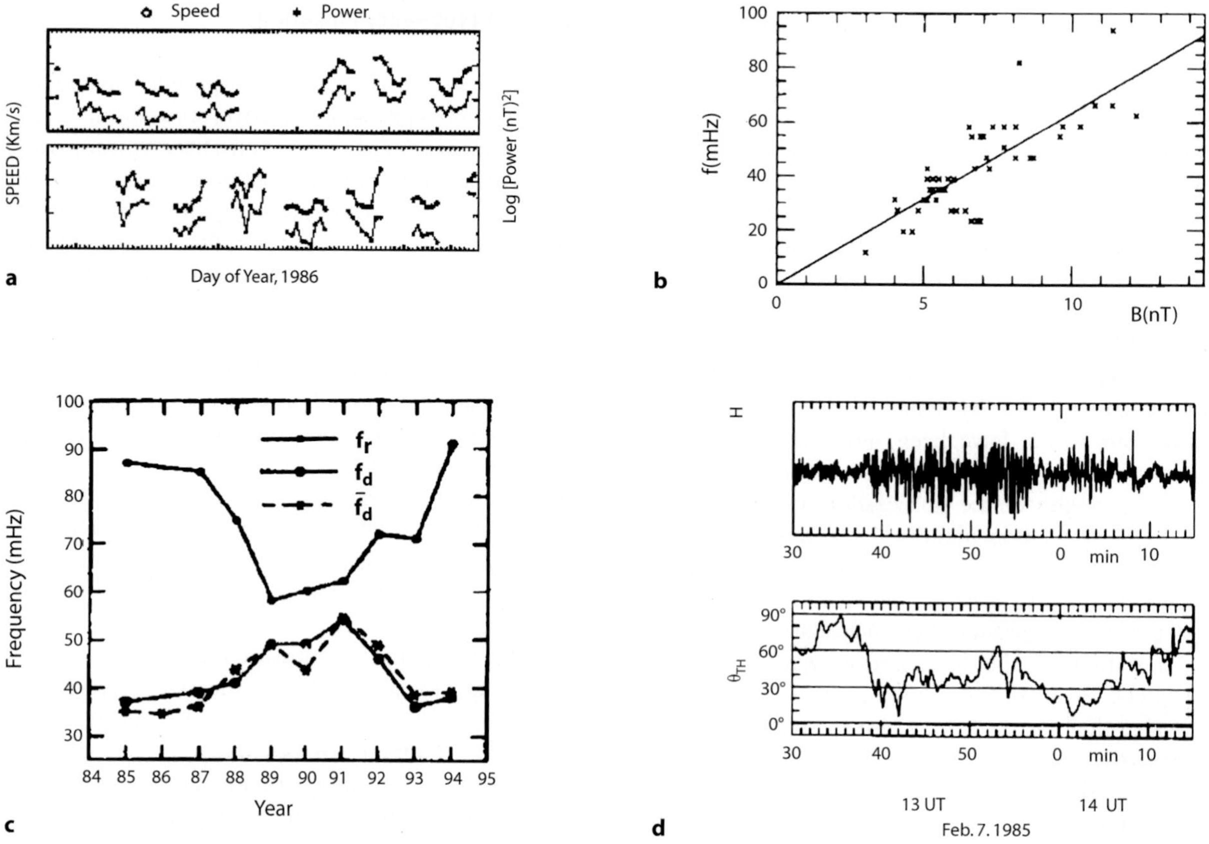

Fig. 16.6a–d. The relationships with SW and IMF parameters at $L \approx 1.6$ (**a**) A comparison between the daily average power in the Pc3 band and SW velocity. [Yedidia et al., 1991]. (**b**) The relationship between the frequency of pulsations and IMF magnitude. [Villante et al., 1992]. (**c**) The solar cycle variation of the frequency of the "*resonant*" mode, f_r, of the "*upstream*" mode, f_d, and its predicted value, f_u (which is proportional to IMF strength). [Vellante et al., 1993b]. (**d**) A comparison between the onset of the pulsation activity (Pc3) and the cone angle. [Vellante et al., 1996]

in the generation of surface waves and/or in the amplification of already existing waves, as they are convected and transmitted through the high latitude magnetopause. However, since increasing power with increasing SW speed is found for most classes of pulsations and in a wide latitudinal range, caution should be adopted before considering this correlation as definitive prove of a given source (Anderson, 1994): indeed, a dependence on SW speed might also reflect an effect of magnetospheric compression which makes the wave source closer to observational points as well as a more efficient generation and/or transmission process of upstream waves.

Upstream waves as a source of ground pulsations have been confirmed by several investigations. An important element is the relationship (16.14) between IMF strength and the frequency of ground pulsations, with an average coefficient well within the limits of theoretical predictions (5.9 ± 0.4, Fig. 16.6b). The relationship between IMF strength and the frequency of waves in the foreshock region was the same as for ground pulsations. Le and Russell (1996) and Takahashi et al. (1984) noted the additional influence of the "*cone angle*" θ_{XB} on wave frequency, both in the foreshock region and in the magnetosphere. Within the magnetosphere both standing Alfvén waves and compressional waves have been identified, and the frequency of the compressional fluctuations has also been related to IMF strength. Engebretson et al. (1986) reported harmonically structured pulsations in the outer dayside magnetosphere in association with upstream waves and with periods governed by local resonance conditions; in addition, almost monochromatic compressional fluctuations with periods identical to the those detected in the SW were occasionally observed. In a similar scenario, two kinds of pulsations would be reasonably expected at ground stations (Villante and Vellante, 1997): a wide band, irregular wave form with a period dependent on IMF strength (from upstream waves propagating through the magnetosphere) and a more regular wave form with latitude dependent period (from standing oscillations along local field lines). In fact, an analysis conducted at $L \approx 1.6$ in 1985 showed that dayside events tend to occur predominantly in two separate period ranges ($T_1 = 25 \pm 5$ s and $T_2 = 13 \pm 2$ s) and these observations were considered consistent with an upstream source spectrum peaked at T_1, together with a coupling resonant mechanism at T_2, roughly the fundamental period of the local field line (Vellante et al., 1989). When extended to a longer interval (1985–1994, Fig. 16.6c), the same analysis revealed a clear solar cycle variation of both dominant periods, consistent with different values of IMF strength (T_1) and plasmaspheric density along the field line (T_2) in different phases: obviously, the two periods may intermingle through the solar cycle due to the different SW and magnetospheric conditions.

In agreement with model predictions (Fig. 16.3a), IMF orientation also plays a significant role, in that ground pulsations occur more frequently when θ_{XB} is small. Figure 16.6d shows an example of the relationship between the pulsation onset and favourable θ_{XB} values at $L \approx 1.6$. Studies on the transmission of upstream waves into the magnetosphere revealed that small θ_{XB} provide quasi-parallel shock conditions at the subsolar bow shock and allow the convection of the turbulent magnetosheath plasma toward the nose of the magnetosphere. More in general, combined large V_{sw} and small θ_{XB} values were found to enhance significantly the pulsation occurrence in the dayside magnetosphere.

A careful statistical analysis of the ground polarization pattern at low latitudes revealed a polarization reversal occurring 1 – 2 hours before noon, i.e. consistent with an upstream wave penetration on the morning flanks of the magnetosphere during spiral IMF conditions (Fig. 16.3a). During intervals related to radial IMF orientation, the polarization reversal was found to occur closer to local noon, as expected for a more symmetric wave penetration around the subsolar point (Villante et al., 2003). Nevertheless, a greater pulsation occurrence in the morning sector of the magnetosphere with respect to the afternoon sector was also observed for IMF orientation far from the spiral (Takahashi et al., 1984).

In general, the relationship between upstream waves and ground pulsations is better at low than at high latitudes: their frequency, on the other hand, is such that they preferentially excite FLRs at low and middle latitudes. Nevertheless, Villante et al. (2002) found a higher correlation between the pulsation power and SW speed in the morning, an explicit θ_{XB} control and a linear relationship between frequency and IMF magnitude in the Pc3 range, at Terra Nova Bay. These features suggest that the role of upstream waves might also be significant at high latitudes. The occurrence of almost monochromatic Pc3-4 events in high latitude regions is interpreted either in terms of higher harmonics of local FLRs, or in terms of fast modes propagating earthwards in the equatorial plane and refracted and diffracted by the changing refractive index of the plasma environment. In addition to the conventional approach which assumes direct transmission of upstream waves from the subsolar magnetopause, Engebretson et al. (1991) suggested an "*ionospheric transistor*" model in which wave transmission may also occur as an indirect process involving modulation of the dayside Birkeland current[12]. Observations at South Pole (CGM $\lambda = -74°$) provided quantitative proof of pulsations driven by modulated electron precipitation near the magnetospheric boundary and indicated the cusp entry as an important source for pulsation energy (Olson and Fraser, 1994). At cusp latitudes, the burst-like Pc3/4 signals were highly localized, a result which is consistent with the modulation of precipitating electron beams. At high latitudes, Chugunova et al.

[12] The Birkeland (or field-aligned, FAC) currents, parallel or antiparallel to B, are current systems which, at high latitudes, link the SW-magnetosphere system to the ionosphere.

(2003) identified a peak of activity in the morning sector which might indicate an additional propagation path via the magnetotail lobe.

At middle and low latitudes, most aspects of the mid-frequency pulsations are interpreted in terms of resonant phenomena related to the penetration of upstream fluctuations. Below $L \approx 3$, the occurrence of Pc3 waves was considered consistent with compressional wave modes coupling to shear Alfvén resonances. Mid-frequency pulsations are also detected in the equatorial region, where they typically show a strong polarization along H. Here, in the absence of FLR, these observations are interpreted either in terms of compressional waves propagating in the equatorial plane, or in terms of waves propagating into the high latitude ionosphere, generating large current oscillations which cause the Pc3/4 observations in the equatorial region.

16.9 FLR and Magnetospheric Diagnostics

In addition to what has been previously discussed, very interesting results were obtained by radar measurements (Walker, 1980): indeed, in agreement with FLR theory, they showed (Fig. 16.7) that the wave electric field changed in phase by $\approx 180°$ over about 1° of latitude and that this corresponded to the half width of the wave amplitude maximum (Hughes, 1994).

Several aspects of magnetospheric research have also been important in gaining a better understanding of low- and mid-frequency pulsations and their interpretation in terms of resonance phenomena. Spatially limited polarized pulsations consistent with a second harmonic standing wave were identified; resonant processes were observed to provide a significant effect on the azimuthal component of the magnetospheric field, and clear evidence for series of harmonic structures and for simultaneous resonant oscillations of a continuum of field lines was detected at $L \approx 2-9$ (Fig. 16.8a,b; Takahashi and McPherron, 1982; Engebretson et al., 1986).

A vast amount of literature has been published on ground signatures of FLR at middle and high latitudes: in particular, the peak of the H component occurs at a latitude which is frequency dependent, and is typically accompanied by rapid phase variation. In fact, f_r decreased from $f \approx 15$ mHz to $f \approx 1$ mHz between $\lambda \approx 60°–78°$, (Samson et al. 1971; Samson and Rostoker, 1972). Waters et al. (1995) found a FLR at $f \approx 3$ mHz at $\lambda \approx 71°$, decreasing to $f \approx 7$ mHz at $\lambda \approx 65°$. On the other hand, given the variable length of the field line with the local time and the different plasma characteristics in different magnetospheric regions, for a given frequency, the resonance latitude has a local time dependence and typically shows an arch structure through the day.

As noted earlier, important results on polarization ($f \approx 5$ mHz) were obtained by Samson et al. (1971). They determined a complex pattern (Fig. 16.5b), in which two or more polarization reversals, depending on latitude ($\lambda \approx 60°–80°$), were observed through the day, with polarization changes occurring approximately at noon and across the line of maximum amplitude (where a linear polarization was detected). As mentioned previously, these results were interpreted in terms of surface waves which excite a FLR deep into the magnetosphere on the field line whose eigenfrequency matches the wave frequency and also create a narrow wave amplitude maximum.

At higher latitudes (Terra Nova Bay; $f \approx 1-4$ mHz) several reversals of the polarization pattern were

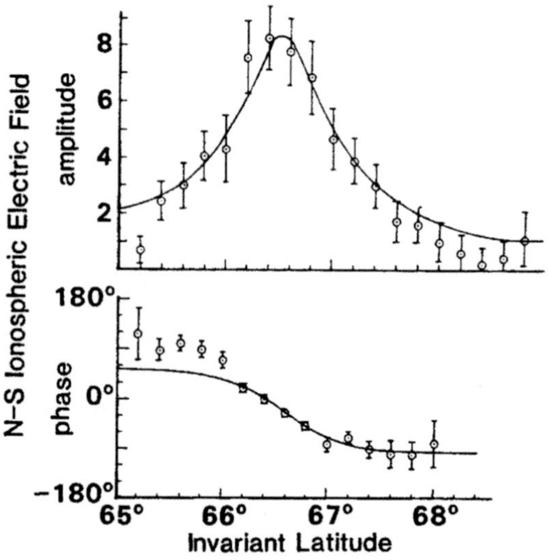

Fig. 16.7. Radar observations of resonance. Latitudinal profiles of the amplitude and phase of the oscillating electric field with a period of ≈ 250 s (STARE radar observations). The *solid line* is a model calculation. The narrow peak amplitude and the phase change by $\approx 180°$ are both features predicted for FLR. [Walker, 1980]

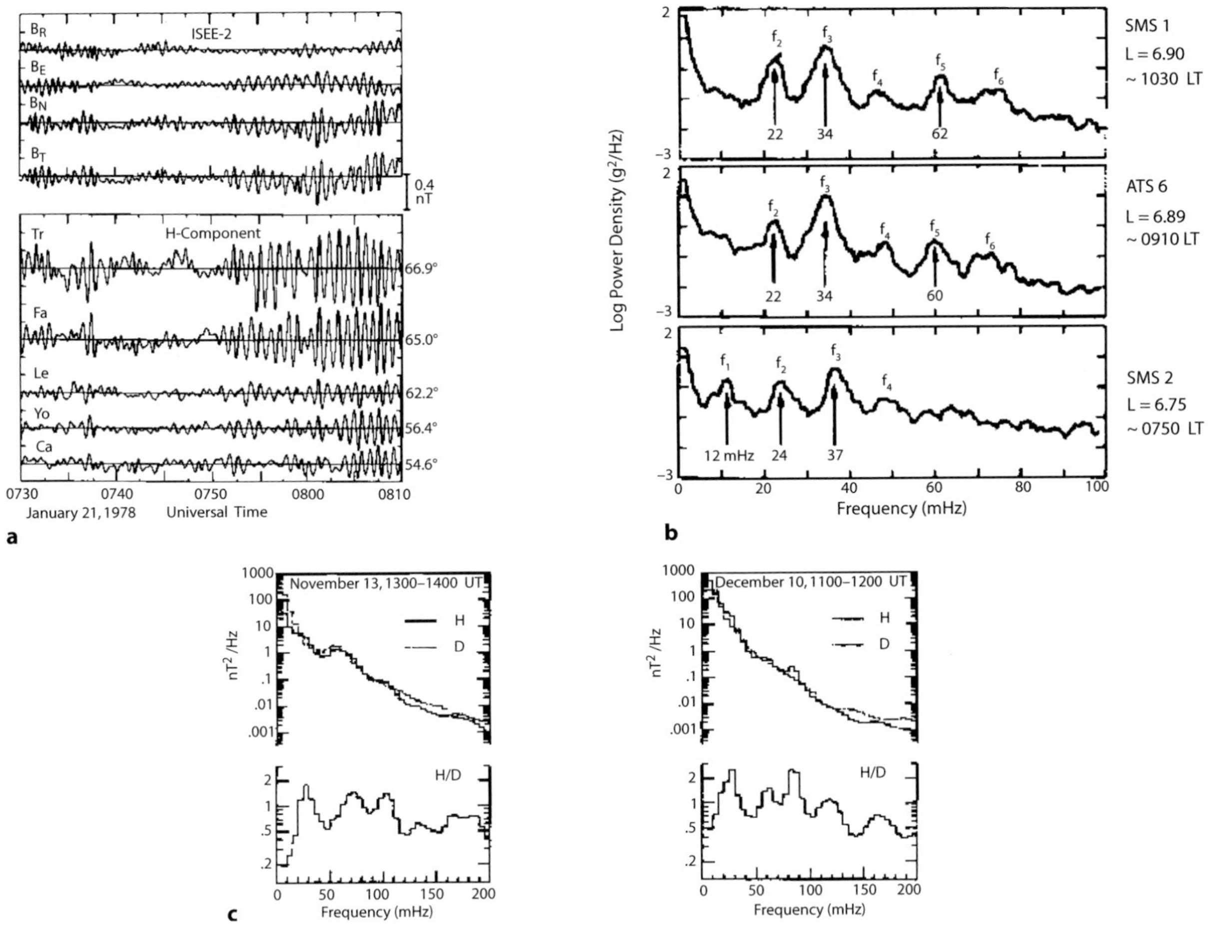

Fig. 16.8a–c. Experimental aspects of the Field Line Resonance. (**a**) An example of Pc5 pulsations simultaneously observed in the magnetosphere and on the ground. [Kivelson, 1995]. (**b**) Spectra of the azimuthal component observed by three satellites. Each spectra observed a series of harmonic peaks. [Takahashi and McPherron, 1984]. (**c**) H and D power spectral densities and the corresponding ratio for different time intervals at $L \approx 2.3$. The peaks in the ratio identify the resonant frequency and its harmonics. [Vellante et al., 1993b]

identified through the day, suggesting resonance effects of lower latitude field lines. It is clear (Fig. 16.5b), that a polar cap station may cross the higher latitude line of polarization reversal at different local times through the day. On the other hand, resonant oscillations of closed field lines have been commonly observed at somewhat lower latitudes (South Pole) during closed magnetospheric conditions, when the cusp is expected to be located poleward with respect to the station (Francia et al., 2005).

In general, resonant effects are hardly identified from a single station because spectral properties tend to reflect the source spectrum. Baransky et al. (1990) proposed the peaks in the ratio between the H and D spectra to indicate resonant frequency, f_r. With such a technique, up to five harmonics were identified at $L \approx 2.3$ (Fig. 16.8c). The introduction of the "*gradient*" and "*cross-phase*" methods between nearby stations allowed a significant improvement in the f_r identification and permitted the identification of a continuous f_r

variation even in ground measurements. For example, by means of the cross-phase technique at $L \approx 1.3-2$, f_r was found to increase with decreasing latitude up to $L \approx 1.6$ and then to decrease at lower latitudes, and this feature was considered consistent with the predicted effects of the mass loading of heavy ions (Fig. 16.4b). Vellante et al. (2002, and papers referenced) adopted different spectral techniques to determine the resonance characteristics ($L \approx 1.6-1.8$) during the main phase of a magnetic storm; f_r estimates were found to be significantly higher than expected, suggesting unusual conditions of the ionosphere-plasmasphere system during this particular event. In addition, a high frequency resolution analysis led authors to suggest the possible occurrence of FLR driven by cavity/waveguide modes.

These arguments suggest that the clear identification of FLRs is an important tool for several aspects of magnetospheric diagnostics. A quantitative determination of the set of field line eigenfrequencies can be used to model the plasma distribution along the field lines from equatorial to high latitudes, to monitor temporal variations of the magnetospheric plasma concentration, to highlight aspects of the plasmasphere also in comparison with other methods (Menk et al., 1999; Waters et al., 1996; Takahashi and McPherron, 1982). In fact, resonant frequencies are dependent on V_a (16.6), a function of the magnetic field and plasma density along the field line: since the magnetic field is well known, the density can be determined by the observed resonant frequencies. Over past few years, the eigenfrequency method has been improved in several aspects. Using this technique a latitudinal magnetometric chain, with a typical pair spacing of 75 – 200 km, is capable of monitoring the radial distribution of the mass density from the last closed field lines through the plasmapause into the low latitude magnetosphere (Fraser, 2003); moreover, as the chain rotates, it allows to obtain a map of the plasma mass density from dawn to dusk. The results of a case event provided excellent agreement between simultaneous ground and spacecraft observations, particularly in the noon sector (Waters et al., 1996). Interesting results have also been obtained by Menk et al. (1999) who derived the mass density profile of the dayside plasmapause ($3.3 < L < 15$) with spatial and temporal resolution of $\approx 0.15-0.4\,R_E$ and $\approx 20-60$ min. The possibility of determining f_r by ground measurements is particularly useful at low latitudes, which are difficult to monitor with spacecraft because rapid satellite motion causes spectral broadening and phase shear. Vellante et al. (2004) concluded that the difference between ground and spacecraft f_r estimates was consistent with the fast satellite motion through the resonant region at $L \approx 1.6-1.8$. They also provided an unprecedented direct confirmation of the 90° rotation of the polarization ellipse through the ionosphere. The boundary between closed and open field lines is generally identified by means of particle measurements. However, since FLR does not occur on open field lines, it is possible to identify such boundary as the latitude of the last field line where FLR is detected (Fraser, 2003). Lanzerotti et al. (1999) suggested that the demarcation in latitude between the appearance or not of specific spectral tones may indicate the location of the dayside magnetopause. Mathiè et al. (1999) identified FLRs on closed field lines at $\lambda \approx 73.5°$, under quiet geomagnetic conditions; they also suggested that during perturbed conditions the closed/open boundary might be located at $\lambda \approx 71°-72°$.

16.10 Cavity/Waveguide Modes

From an experimental point of view the evidence for cavity/waveguide modes is still sparse. Within the magnetosphere, search for cavity modes has mostly concentrated on looking for compressional waves with L-independent frequencies or for enhancements in the Alfvén continuum at expected cavity mode eigenfrequencies. Rickard and Wright (1995) found some correspondence between spacecraft measurements and simulations of the magnetic field signals expected along the spacecraft trajectory for a waveguide mode. Mann et al. (1999) investigated multisatellite and ground observations of a tailward propagating compressional wave and interpreted the experimental observations in terms of a magnetospheric waveguide mode. More recently, Waters et al. (2002) proposed a set of criteria for improving the identification of cavity modes in spacecraft data which, in addition to amplitude characteristics, include tests for signal phase information.

Several investigations at ground auroral latitudes have reported evidence for long period waves at the "*discrete*" frequencies $f \approx 1.3$, 1.9, 2.6 and 3.4 mHz

(Samson et al., 1992). As noted earlier, such low frequencies, also known as cavity mode frequencies (CMS), would involve the bow shock as an outer boundary (Harrold and Samson, 1992). Evidence of similar signals has also been reported at low latitudes. In some cases the same oscillation modes were observed simultaneously at low and Antarctic latitudes as well as in the magnetosphere. These results were tentatively considered consistent with features expected for global compressional modes or large scale cavity/waveguide resonances. In this context, the observed variability of the "*discrete*" frequencies might be interpreted considering that CMS frequencies represent a set of the most frequently occurring eigenfrequencies; however they are subject to some variability, due to the changing nature of the waveguides. Samson et al. (1995) suggested that cavity/waveguide modes of the plasmasphere were responsible for low latitude Pc3 as well ($L \approx 1.4-2.2$). In fact, standing Alfvén waves might be excited on local field lines by coupling to the waveguide modes; as ground magnetometers respond to ionospheric currents over a range of latitude, the measured power spectral density might present a multiharmonic fine envelope structure centered at the frequencies of the harmonics of the local standing wave (Takahashi, 1991). These arguments were further developed by Waters et al. (2000), who proposed a model able to reproduce several features of the low latitude power spectra, indicating that the interaction between waveguide and FLR modes might be important in understanding several aspects of low latitude observations (Menk et al., 1999).

In this context it is important to mention that several cases have been presented in which fluctuations in SW density and in magnetospheric field were highly correlated and often matched some of the CMS frequencies (Fig. 16.9a). Kepko and Spence (2002) argued that for those events the discrete frequencies were an inherent property of the SW and were not related to possible cavity or waveguide modes. They also speculated a possible solar source related to solar p-modes in the mHz range. More in general, Francia et al. (1999) identified a dramatic correlation between continuous variations of the H component (on time scale of several minutes) and variations of the square root of the SW dynamic pressure (Fig. 16.9b), suggesting that ground measurements closely respond to rapid, small amplitude variations of the magnetopause current. In conclusions, before a definitive interpretation of the "*discrete*" frequencies modes in terms of cavity/waveguide modes can be made, further investigation is required of spacecraft and ground observations.

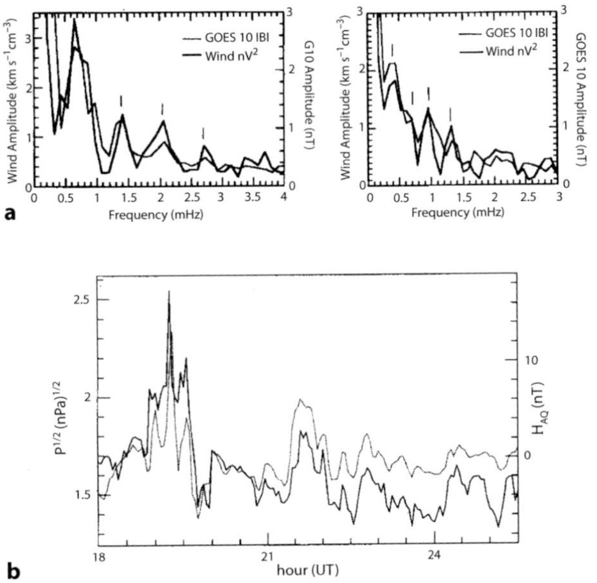

Fig. 16.9a,b. Aspects of pulsations at "*discrete*" frequencies. (**a**) Two examples of the correlation between the power spectra of SW dynamic pressure and those of the magnetospheric field at geostationary orbit. [Kepko et al., 2002]. (**b**) An example of the strong correlation between continuous variations of SW dynamic pressure and continuous variations of the H component at low latitudes. [Francia et al., 1999]

16.11 High Frequency Pulsations

Unlike lower frequency standing pulsations, high frequency pulsations are traveling waves which often show spectacular amplitude and frequency modulation. This characteristic is represented by the term "*pearl necklace*" as a reference to a common class of quasi-periodic sequences of pulsations which appear as structured wave packets and represent the most common Pc1 manifestation at low and middle latitudes. Such "*structured*" pulsations (which typically appear at conjugate points as a repetitive burst of waves, with modulated envelopes, Fig. 16.10a) typically range in frequency from 0.2 to 5 Hz, with repetition periods from 100 to 300 s; there is often correlation between the wave period and the repetition period with a proportionality factor of ≈ 100. Pc1 pulsations with a less regular behaviour (typically identified as "*unstructured*" pulsations) are the dominant manifestation at higher latitudes ($L \geq 5$). An important class is represented by the intervals of pulsations of diminishing period ("*IPDP*") which show a typical frequency rise from $f \approx 0.1$ to $f \approx 1$ Hz in ≈ 20 min; the "*hydromagnetic chorus*" is a mixture of structured and unstructured pulsations between $f \approx 0.2-0.5$ Hz. Actually, the high frequency range encompasses a large number of ground pulsations with different characteristics; several subtypes of high frequency pulsations, based on their spectral structures, have been classified and each subtype has a preferential local time occurrence (Fig. 16.10b; Saito, 1969; Fukunishi et al., 1981). High frequency pulsations have maximum amplitude ($\approx 0.1-1$ nT) in the auroral zone and much smaller amplitudes at equatorial latitudes.

The typical tendency of high frequency pulsations to recur on consecutive days, approximately at the same hours, or to disappear for several days or weeks has been emphasized in several investigations. In addition, they appear more frequently during winter months, a feature which is considered consistent with a more efficient ionospheric attenuation during the summer. Their longer term occurrence appears anticorrelated, at least at high latitudes, with the sunspot number and several mechanisms (related to plasmapause position, the presence of heavy ions in the magnetospheric plasma and ionospheric waveguides) have been proposed to interpret this inverse relationship.

Short period fluctuations with different characteristics appear in each phase of geomagnetic storms (Kangas et al., 1998). For example, IPDP show strong association with SI/SSC, in particular in the noon sector and also tend to occur during the main phase, mostly in the afternoon-evening sector; structured pulsations tend to occur during the recovery phase. IPDP are also connected to substorm activity.

Pc1/2 pulsations, on the other hand, are a common feature of the magnetosphere. Since early observations, magnetospheric events have mostly been related to unstructured pulsations (with few exceptions), and few structured events have been detected beyond geostationary orbit. A statistical analysis of satellite

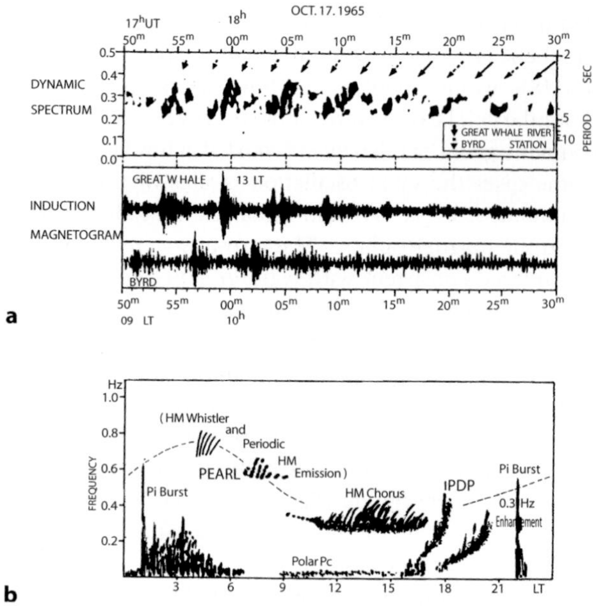

Fig. 16.10a,b. Classical aspects of the high frequency pulsations. (**a**) A classical representation of the alternate appearance of pearl events in conjugate hemispheres and the corresponding dynamic spectra. [Saito, 1969]. (**b**) The local time dependence of the occurrence of various types of high frequency pulsations at auroral latitudes. [Kokubun, 1970]

data revealed a main peak of occurrence near the magnetic equator beyond $L \approx 7$ and a lower maximum at the plasmapause. Engebretson et al. (2002) identified a close correspondence between Pc1/2 events in the outer dayside magnetosphere and high latitude observations; they suggested that most events were associated with significant compressions of the magnetosphere. However, space/ground comparisons are not straightforward in that the field line guidance stops at the ionosphere, and ground measurements are influenced by the ionospheric waveguide.

The origin of high frequency pulsations is mostly based on the occurrence of electromagnetic ion-cyclotron instability (EMIC) in the magnetosphere. In this sense, the observed predominance of left-handed polarization as well as the gap of spectral power close to the helium gyrofrequency are important experimental aspects consistent with theoretical predictions. In agreement with theory, high frequency pulsations appear to be generated in the equatorial plane of geomagnetic shells between $L \approx 3-6$ (Anderson et al., 1996); they propagate to Earth along geomagnetic field lines. Structured pulsations might originate on field lines located near or inside the plasmapause (Fraser et al., 1984). It was suggested, over forty years ago, that these events were related to wave packets guided along field lines, bouncing from hemisphere to hemisphere with losses being compensated by wave growth at the equator. Satellite observations do not explicitly support the bouncing wave packet model; however, alternative generation mechanisms still need firm observational support. Unstructured pulsations and Pc1 bursts (occasionally extending to Pc2 frequencies) are also observed during daytime intervals on field lines related to cusp, suggesting an origin related to plasma instabilities near the dayside magnetopause.

16.12 Irregular Pulsations

As previously noted, Pi2 events occur as transient and damped signals associated with dramatic changes of the state of the magnetosphere which occur at the substorm expansive phase (McPherron, 1979; Baumjohann and Glassmeier, 1984). Such irregular events are observed during nighttime intervals, from high to low latitudes. Their maximum amplitude is detected at auroral latitudes, close to the region of the substorm enhanced westward ionospheric electrojet, while a secondary maximum is detected around the plasmapause. Southwood and Stuart (1979) interpreted these waves as a transient response to sudden changes in the magnetosphere, acting to communicate and balance stress between magnetosphere and ionosphere (Allan and Poulter, 1992).

Mid-latitude events are typically very monochromatic, while high latitude events have much more complicated power spectra. Below $L \approx 2$, Pi2 spectra contain up to four harmonics and a similar multiharmonic structure has also been observed in magnetospheric events. Despite these differences, Pi2 tend to have dominant frequencies independent of latitude. At low latitudes, Pi2 are characterized by a clear initial phase (Fig. 16.11a) and for this reason they are often used to identify substorm onset time, although several aspects suggest caution (for example, the delay time between pulsation onset and the first auroral brightening).

It is generally accepted that the original source of Pi2 pulsations is the energy and momentum impulsively released as the magnetic field of the near Earth tail suddenly changes from an elongated configuration to a dipolar configuration at substorm onset. Several observational aspects are interpreted in terms of the substorm current wedge model (SCW, Fig. 16.11b) proposed by McPherron (1979) (for the formation of SCW and its association Pi2 pulsations see also Baumjohann and Glassmeier, 1984). In this model the tail current is interrupted at the substorm onset, current flows along field lines into the ionosphere, couples to the westward electrojet, and returns to the equatorial plane via an upward FAC. Basically, the most important signature for interpreting Pi2 in terms of SCW oscillations is the observed variation of the orientation of the polarization axis with longitude (a feature which also extends to low latitudes). Figure 16.11c shows the predicted gradual rotation from northeast (west of the event source longitude), to north/south (close to the centre of the current wedge), and to northwest (east of the event source in the northern hemisphere (Lester et al., 1989; Li et al., 1998; and papers referenced). The period of Pi2 pulsations is related to the fundamental eigenperiod of the toroidal mode along the field line where the auroral breakup starts (Fig. 16.11d). As for toroidal Pc5/4, which roughly share the same frequency band, this suggests interpreting higher latitude Pi2 waveforms in terms of standing waves reflected between conjugate ionospheres: in this case, the rapid damping would be a consequence of the much lower ionospheric conductivity in the nighttime hours. The different waveforms as well as the observation of dayside events suggest additional Pi2 sources, such as plasmapause surface waves and cavity resonances of the inner magnetosphere at middle and low latitudes. For example, Yeoman et al. (1990) proposed the superposition of waves from the auroral current system with plasmaspheric cavity resonances. Takahashi et al. (1995) conducted a statistical analysis of Pi2 pulsations and found that Pi2 were detected in the nightside and primarily at $L < 5$; they also suggested cavity mode resonances excited in the inner magnetosphere ($2 < L < 5$), bounded below by the ionosphere and at high altitude by an

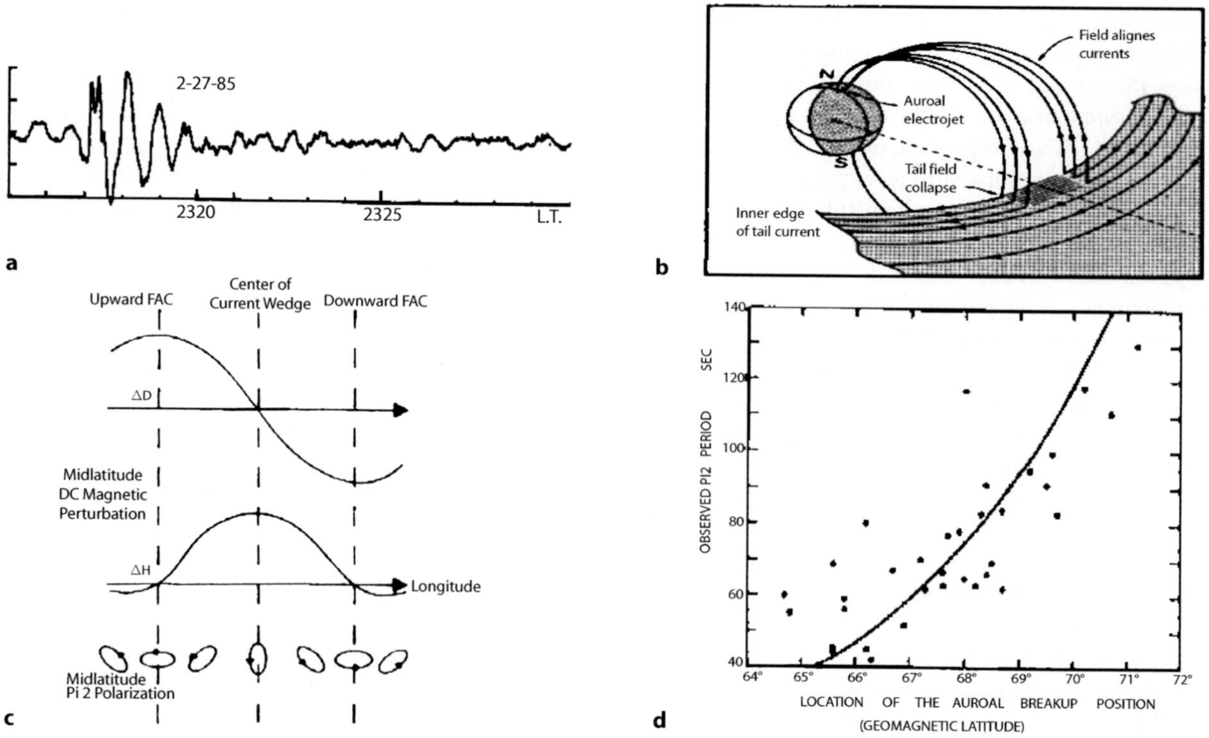

Fig. 16.11a–d. Aspects of the Pi2 pulsations. (**a**) A low latitude Pi2. [Villante et al., 1990]. (**b**) A representation of the SCW. [McPherron et al. 1973]. (**c**) The variation of the H and D component due to the substorm manifestation at middle latitudes and the Pi2 polarization pattern. [Lester et al., 1984]. (**d**) The period of Pi2 event vs. the fundamental period of the toroidal mode at the latitude of the auroral breakup. [Kuwashima and Saito, 1981]

Alfvén velocity gradient. Olson (1999), who reviewed theoretical and experimental aspects, proposed a global scenario in which the Pi2 signal encompasses a class of pulsations generated by the same event: the onset of FAC associated with the current disruption in the near Earth plasmasheet and the impulsive response of the inner magnetosphere to compressional waves generated at the substorm occurrence or intensification (Fig. 16.12). In this scheme, oscillations in the SCW currents produce high and middle latitude Pi2 signals, while compressional waves, traveling inward, stimulate FLR and surface waves at the plasmapause which can be observed near the plasmapause footprint. In addition, at low latitudes, other resonant and global modes of the inner magnetosphere can be observed. Numerical calculations of cavity quencies gave results consistent with the low latitude multiharmonic observations.

Higher frequency components of Pi2 waves in the ionospheric cavity may produce Pi1 pulsations, which represent an additional typical manifestation of auroral and subauroral latitudes; they occur after substorm onset and are correlated with pulsating aurora. A comparison of the results from high latitude stations revealed that the significant Pi1 activity associated with substorms detected at $\lambda \approx -74°$ becomes weak at $\lambda \approx -80°$. This feature is consistent with the lack of power enhancement during nighttime hours deep in the polar caps.

Broad-band bursts of PiB pulsations (Pi1+Pi2, from several Hz to ≈20 mHz) are observed at FAC onset. Typical burst duration is ≈1 – 3 min and 3 – 5 PiB impulses occur in ≈3 – 10 min (Kangas et al., 1998). PiB are also used as high time resolution monitors of substorm development.

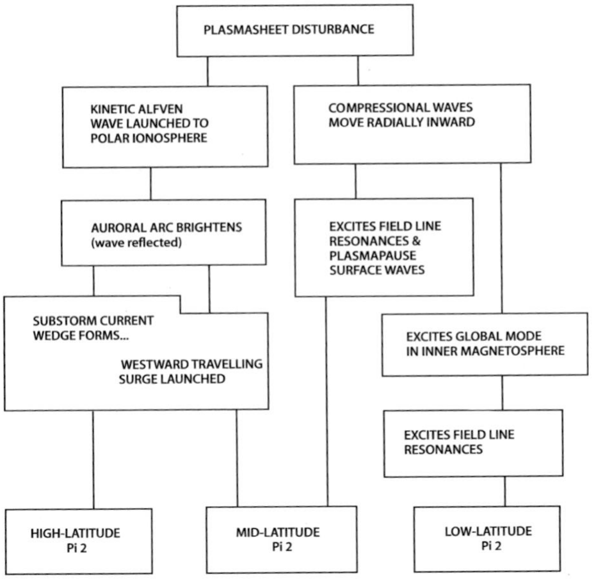

Fig. 16.12. The diagram outlining the flow of energy that manifests itself as Pi2 oscillations in the magnetosphere and at ground level. [Olson, 1999]

PiC pulsations appear in a narrowband of frequencies which is continuous in time, sometimes up to several hours. They are generally seen in the morning with periods of tens of seconds (at the lower frequency end of Pi1) in correlation with auroral luminosity variations. They have been modeled in terms of a current system of patch of enhanced conductivity in the ionosphere.

Large amplitude, damped fluctuations (denoted "*magnetic impulse events*", MIE, or "*traveling current vortices*", TCV) are also observed from $\lambda \approx 70°$ to $\lambda \approx 80°$. Most of these transients events consist in single-cycle pulsations, with periods between ≈100–600 s, and are possibly associated with flux transfer events (FTE, patchy reconnection events between IMF and magnetospheric lines across the dayside magnetopause) and FAC, or SW pressure pulses.

16.13 Concluding Remarks

As is clear from the arguments above, ULF waves, endemic within the magnetosphere, are involved in major manifestations of the magnetospheric dynamics and play a significant role in the energy transfer from the SW to the magnetosphere. In the author's opinion (and experience) some interesting arguments for study in the near future, among others, are:

- the penetration mechanism of external waves into the magnetosphere (via the magnetopause nose, flanks and magnetotail lobes), the role of IMF strength and direction, the wave propagation inside the magnetosphere, the characteristics of the wave energy transport, the role of different instability processes;
- the definite identification of cavity/waveguide modes, the possible correspondence of "*discrete*" frequencies modes with simultaneous compressional and/or Alfvenic fluctuations at the same frequencies in the near Earth SW, and their possible association with solar oscillations in the mHz range;
- additional sources of Pi2 manifestations at middle and low latitudes (plasmapause surface waves, cavity resonances of the inner magnetosphere, etc.);
- the correlation between pulsations and auroral manifestations, with a special emphasis on pulsating aurora;
- the role of ULF waves in the energization and transport of radiation belt particles;
- the improvement of experimental methods for magnetospheric diagnostics, particularly important at low latitude where spacecraft measurements are generally not available.

Acknowledgement. Author is grateful to drs. L. J. Lanzerotti (New Jersey Inst. Tech.), P. Francia, M. Vellante (University of L' Aquila) for useful comments.

References

Review papers

Allan, W. and E.M. Poulter, ULF waves: their relationship to the structure of the magnetosphere, Rep. Progr. Phys. 55, 533, 1992.

Anderson, B.J., An overview of spacecraft observations of 10 s to 600 s period magnetic pulsations in the Earth's magnetosphere, in Solar Wind Sources of Magnetospheric Ultra-Low Frequency Waves, ed. by M.J. Engebretson, K. Takahashi, M. Scholer, AGU Geophys. Mon., 81, 25, 1994.

Arnoldy, R.L., L.J. Cahill Jr., M.J. Engebretson, L.J. Lanzerotti, and A. Wolfe, Review of hydromagnetic wave studies in the Antarctic, Rev. Geophys., 26, 181, 1988.
Dungey, J.W., Hydromagnetic waves, in Physics of Geomagnetic Phenomena, ed. by S. Matsushita and W.H. Campbell, Academic Press, New York, 1967.
Engebretson, M.J., Catching the wave: ULF research in the U.S. since 1991, Rev. Geophys., 33, 693, 1995.
Fazakerley, A.N., and C.T. Russell, Theory and observation of magnetosheath waves, in Solar Wind Sources of Magnetospheric Ultra-Low Frequency Waves, ed. by M.J. Engebretson, K. Takahashi, M. Scholer, AGU Geophys. Mon., 81, 147, 1994.
Hughes, W.J., Hydromagnetic waves in the magnetosphere, in Solar-Terrestrial Physics, ed. by R.L. Carovillano and J.M. Forbes, Reidel, 1983.
Hughes, W.J., Magnetospheric ULF waves: a tutorial with historical perspective, in Solar Wind Sources of Magnetospheric Ultra-Low Frequency Waves, ed. by M.J. Engebretson, K. Takahashi, M. Scholer, AGU Geophys. Mon., 81, 1, 1994.
Jacobs, J.A., Geomagnetic micropulsations, in Physics and Chemistry in Space, Vol. 1, ed. by J.G. Roeder, Springer-Verlag, 1970.
Kangas, J., A. Guglielmi, and O. Pokhotelov, Morphology and physics of short-period magnetic pulsations: A review, Space Sci. Rev., 83, 435, 1998.
Kivelson, M.G., Pulsations and magnetohydrodinamic waves, in Introduction to Space Physics, ed. by M.G. Kivelson and C.T. Russell, Cambridge Univ. Press, 1995.
Lanzerotti, L.J., and D.J. Southwood, Hydromagnetic waves, in Solar System Plasma Physics, Vol. III, ed. by C.F. Kennel, L.J. Lanzerotti, and E.N. Parker, North-Holland, 1979.
Le, G., and C.T. Russell, The morphology of ULF waves in the Earth's foreshock, in Solar Wind Sources of Magnetospheric Ultra-Low Frequency Waves, ed. by M.J. Engebretson, K. Takahashi, M. Scholer, AGU Geophys. Mon., 81, 87, 1994.
McPherron, R.L., Magnetic pulsations: their sources and relation to solar wind and geomagnetic activity, Workshop on Electromagnetic Induction in the Earth, Santa Fè, 2002.
McPherron, R.L., C.T. Russell, and M.P. Aubry, Satellite studies of magnetospheric substorms on August 15, 1968, 9, Phenomenological model for substorms, J. Geophys. Res., 78, 3131, 1973.
Odera, T.J., Solar wind controlled pulsations: a review, Rev. Geophys., 24, 55, 1986.
Olson, J.V., Pi 2 pulsations and substorm onsets: a review, J. Geophys. Res., 104, 17499, 1999.
Rostoker, G., Geomagnetic micropulsations, in Fundamentals of Cosmic Physics, Vol. 4, Gordon and Breach, 1979.
Russell, C.T., and M.M. Hoppe, Upstream waves and particles, Spa. Sci. Rev., 31, 155, 1983.
Saito, T., Geomagnetic pulsations, Space Sci. Rev., 10, 319, 1969.
Saito, T., Long period irregular magnetic pulsations, Space Sci. Rev., 21, 427, 1978.
Samson, J.C., Geomagnetic pulsations and plasma waves in the earth's magnetosphere, in Geomagnetism, Vol. 4, ed. by J.A. Jacobs, Academic Press, 1991.
Southwood, D.J., and W.J. Hughes, Theory of hydromagnetic waves in the magnetosphere, Space Sci. Rev., 35, 301, 1983.
Takahashi K., ULF waves in the magnetosphere, Rev. Geophys. suppl., 1066, 1991.
Takahashi K., ULF waves: 1997 IAGA division 3 reporter review, Ann. Geophys., 16, 787, 1998.
Villante, U., and M. Vellante, Experimental aspects of low latitude ground pulsations, Intern. School Space Science, Course on Solar System Plasma Physics, SIF, Conference Proceedings, 56, 189, 1997.

Regular papers

Allan, W., S.P. White, and E.M. Poulter, Impulse-excited hydromagnetic cavity and field line resonances in the magnetosphere, Planet. Space Sci., 34, 371, 1986.
Anderson, B.J., R.E. Denton, G. Ho, D.C. Hamilton, S.A. Fuselier, and R.J. Strangeway, Observational test of local proton cyclotron instability in the Earth's magnetosphere, J. Geophys. Res., 101, 21527, 1996.
Baransky, L.N., S.P. Belokris, Y.E. Borovkov, and C.A. Green, Two simple methods for the determination of the resonance frequencies of magnetic field lines, Planet. Space Sci., 38, 1573, 1990.
Baumjohann, W., and K.-H. Glassmeier, The transient response mechanism and Pi2 pulsations at substorm onset. – Review and outlook, Planet. Space Sci., 32, 1361, 1984.
Chen, L., and A. Hasegawa, Theory of long-period magnetic pulsations, 1, Steady state of excitation of field line resonance, J. Geophys. Res, 79, 1024, 1974.
Chugunova, O.M., V.A. Pilipenko, and M.J. Engebretson, Statistical features of Pc3–Pc4 pulsations at very high latitudes, in Physics of Auroral Phenomena, Apatity, 103, 2003.
Engebretson, M.J., L.J.Jr. Cahill, R.L. Arnoldy, B.J. Anderson, T.J. Rosenberg, D.L. Carpenter, U.S. Inan, and R.H. Eather, The role of the ionosphere in coupling ULF wave power into the dayside magnetosphere, J. Geophys. Res., 96, 1527, 1991.
Engebretson, M.J., W.K. Peterson, J.L. Posch, M.R. Klatt, B.J. Anderson, C.T. Russell, H.J. Singer, R.L. Arnoldy, and H. Fukunishi, Observations of two types of Pc 1-2 pulsations in the outer dayside magnetosphere, J. Geophys. Res., 107 (A12), 1451, doi: 10.1029/2001JA000198, 2002.
Engebretson, M.J., L.J. Zanetti, T.A. Potemra, and M.H. Acuna, Harmonically structured ULF pulsations observed by the

AMPTE CCE magnetic field experiment, Geophys. Res. Lett, 13, 905, 1986.

Formisano V., The Earth's bow shock fine structure, in Correlated Interplanetary and Magnetospheric Observations, ed by, D.E. Page, Reidel, 1984.

Francia, P., L.J. Lanzerotti, U. Villante, S. Lepidi, and D. Di Memmo, A statistical analysis of low frequency magnetic pulsations at South Pole, J. Geophys. Res.110 (A2), A02205, doi:10.1029/2004JA010680, 2005.

Francia, P., S. Lepidi, U. Villante, P. Di Giuseppe, and A.J. Lazarus, Geomagnetic response at low latitude to continuous solar wind pressure variations during northward interplanetary magnetic field, J. Geophys. Res., 104, 19923, 1999.

Fraser, B.J., Recent developments in magnetospheric diagnostics using ULF waves, Space Sci. Rev., 107, 149, 2003.

Fraser, B.J., W.J. Kemp, and D.J. Webster, Pc 1 pulsation source regions and their relationship to the plasmapause, in Achievements of the IMS, ESA publication SP – 217, 609, 1984.

Fukunishi, H., T. Toya, K. Koike, M. Kuwashima, and M. Kawamura, Classifications of hydromagnetic emissions based on frequency-time spectra, J. Geophys. Res., 86, 9029, 1981.

Greenstadt, E.W., R.L. McPherron, and K. Takahashi, Solar wind control of daytime mid-period geomagnetic pulsations, in ULF pulsations in the magnetosphere, ed. by D.J. Southwood, Reidel, 1981.

Harrold, B.J., and J.C. Samson, Standing ULF modes of the magnetosphere: A theory, Geophys. Res. Lett., 19, 1811, 1992.

Kepko, L., H.E. Spence, and H.J. Singer, ULF waves in the solar wind as direct drivers of magnetospheric pulsations, Geophys. Res. Lett., 29(8), 1197, doi: 10.1029/2001IGLO14405, 2002.

Kivelson, M.G., J. Etcheto, and J.G. Trotignon, Global compression oscillations of the terrestrial magnetosphere: The evidence and a model, J. Geophys. Res., 89, 9851, 1984.

Kokubun, S., Rep. Ionos. Space Research. Jap., 24, 24, 1970.

Kuwashima, M., and T. Saito, Spectral characteristics of magnetic Pi 2 pulsations in the auroral region and lower latitudes, J. Geophys. Res., 86, 4686, 1981.

Lanzerotti, L.J., C.G. MacLennan, and A.C. Fraser-Smith, Background magnetic spectra: $\approx 10^{-5}$ to $\approx 10^{5}$ Hz, Geophys. Res. Lett., 17,1593, 1990.

Lanzerotti, L.J., A. Shono, H. Fukunishi, and C.G. MacLennan, Long period hydromagnetic waves at very high geomagnetic latitudes, J. Geophys. Res., 104, 28423, 1999.

Le, G., and C.T. Russell, Solar wind control of upstream wave frequency, J. Geophys. Res., 101, 2571, 1996.

Lester, M., W.J. Hughes, and H.J. Singer, Longitudinal structure in Pi 2 pulsations and the substorm current wedge, J. Geophys. Res., 89, 5489, 1984.

Lester, M., H.J. Singer, D.P. Smits, and W.J. Hughes, Pi 2 pulsations and the substorm current wedge: Low-latitude polarization, J. Geophys. Res., 94, 17133, 1989.

Li, Y., B.J. Fraser, F.W. Menk, D.J. Webster, and K. Yumoto, Properties and sources of low and very low latitude Pi 2 pulsations, J. Geophys. Res., 103, 2343, 1998.

Mann, I.R., A.N. Wright, K.J. Mills, and V.M. Nakariakov, Excitation of magnetospheric waveguide modes by magnetosheath flows, J. Geophys. Res., 104, 333, 1999.

Mathie, R.A., F.W. Menk, I.R. Mann, and D. Orr, Discrete field line resonances and the Alfvén continuum in the outer magnetosphere, Geophys. Res. Lett., 26, 659, 1999.

McPherron, R.L., Magnetospheric substorms, Rev. Geophys. and Space Phys., 17, 657, 1979.

Menk, F.W., B.G. Fraser, C.L. Waters, C.W.S. Ziesolleck, Q. Feng, S.H. Lee, and P.W. McNabb, Ground measurements of low latitude magnetosphere field line resonances, Solar Wind Sources of Magnetospheric Ultra-Low Frequency Waves, ed. by M.J. Engebretson, K. Takahashi, and M. Scholer, AGU Monogr., 81, 299, 1994.

Menk, F.W., D. Orr, M.A. Clilverd, A.J. Smith, C.L. Waters, D.K. Milling, and B.J. Fraser, Monitoring spatial and temporal variations in the dayside plasmasphere using geomagnetic field line resonances, J. Geophys. Res., 104, 19955, 1999.

Olson, J.V., and B.J. Fraser, Pc 3 pulsations in the cups, in Solar Wind Sources of Magnetospheric Ultra-Low-Frequency Waves, Geophysical Monograph 81, ed. by M.J. Engebretson, K. Takahashi, and M. Scholer: AGU Mon., 81, 325, 1994.

Posch, L.J., M.J. Engebretson, V.A. Pilipenko, W.J. Hughes, C.T. Russell, and L.J. Lanzerotti, Characteristics of long-period ULF response to magnetic storms, J. Geophys. Res., 108 (A1), 1029, doi: 101029/2002JA 009386, 2003.

Radoski, H.R., The effects of asymmetry on toroidal hydromagnetic waves in a dipole field, Planet. Spa. Sci., 20, 1015, 1972.

Rickard, G.J., and A.N. Wright, ULF pulsations in a magnetospheric waveguide: comparison of real and simulated satellite data, J. Geophys. Res., 100, 3531, 1995.

Samson J.C., B.G. Harrold, J.M. Ruohoniemi, R.A. Greenwald, and A.D.M. Walker, Field line resonances associated with MHD waveguides in the magnetosphere, Geophys. Res. Lett., 19, 441, 1992.

Samson, J.C., J.A. Jacobs, and G. Rostoker, Latitude-dependent characteristics of long-period geomagnetic micropulsations, J. Geophys. Res., 76, 3675, 1971.

Samson, J.C., and G. Rostoker, Latitude-dependent characteristics of high latitude field line resonances, J. Geophys. Res., 77, 6133, 1972.

Samson, J.C., C.L. Waters, F.W. Menk, and B.J. Fraser, Fine structure in the spectra of low latitude field line resonances, Geophys. Res. Lett., 22, 2111, 1995.

Southwood, D.J., Some features of field line resonances in the magnetosphere, Planet. Space Sci., 22, 483, 1974.

Southwood, D.J., and W.F. Stuart, Pulsations at the substorm onset, in Dynamics of the magnetosphere, ed. by S.-I. Akasofu, Reidel, 341, 1979.

Takahashi K., and R.L. McPherron, Harmonic structure of Pc 3–4 pulsations, J. Geophys. Res., 87, 1504, 1982.

Takahashi, K., R.L. McPherron, and T. Terasawa, Dependence of the spectrum of Pc3–4 pulsations on the interplanetary magnetic field, J. Geophys. Res., 89, 2770, 1984.

Takahashi, K., S. Ohtani, and B.J. Anderson, Statistical analysis of Pi 2 pulsations observed by the AMPTE CCE spacecraft in the inner magnetosphere, J. Geophys. Res., 100, 21929, 1995.

Vellante, M., Pulsazioni geomagnetiche, Annali di Geofisica, 36, supp. 5-6, 79, 1993.

Vellante, M, M. De Lauretis, M. Foester, S. Lepidi, B. Zieger, U. Villante, V.A. Pilipenko, and B. Zolesi, Pulsation event study of August 16, 1993: geomagnetic field line resonances at low latitudes, J. Geophys. Res., 107, 10.1029/2001JA900123, 2002.

Vellante., M., H. Lhur, T.L. Zhang, V. Wesztergom, U. Villante, M. De Lauretis, A. Piancatelli, M. Rother, K. Schwingenschuh, W. Koren, and W. Magnes, Ground/satellite signatures of field line resonance: a test of theoretical predictions, J. Geophys. Res., 107, 10.1029/2004JA010126, 2004.

Vellante, M., U. Villante, R. Core, A. Best, D. Lenners, and V.A. Pilipenko, Simultaneous geomagnetic pulsation observations at two latitudes: resonant mode characteristics, Ann. Geophys, 11, 734, 1993.

Vellante, M., U. Villante, M. De Lauretis, and P. Cerulli-Irelli, An analysis of micropulsation events at a low-latitude station during 1985, Planet. Space Sci, 37, 767, 1989.

Vellante M., U. Villante, M. De Lauretis, G. Barchi, Solar cycle variation of the dominant frequencies of Pc3 geomagnetic pulsations at L = 1.6, Geophys. Res. Letters, 23, 12, 1505, 1996.

Villante U., M. Vellante, M. De Lauretis, P. Cerulli-Irelli, R. Orfei, Micropulsation measurements at low latitudes, Proceedings of the IV Cosmic Physic National Conference, Il Nuovo Cimento, 13C, 93, 1990.

Villante, U., P. Francia, M. Vellante, and P. Di Giuseppe, Some aspects of the low latitude geomagnetic response under different solar wind conditions, Space Sci. Rev., 107, 207, 2003.

Villante U., S. Lepidi, M. Vellante, A.J. Lazarus, and R.P. Lepping, Pc3 activity at low geomagnetic latitudes: a comparison with solar wind observations, Planet. Spa. Sci.,40, 1399, 1992.

Villante, U., M. Vellante, and G. De Santis, An analysis of Pc 3 and Pc 4 pulsations at Terra Nova Bay (Antarctica), Annales Geophys., 18, 1412, 2002.

Walker, A.D. M., Modelling of Pc5 pulsation structure in the magnetosphere, Planet. Space Sci., 28, 213, 1980.

Waters, C.L., B.G. Harrold, F.W. Menk, J.C. Samson, and B.J. Fraser, ULF waveguide mode waves at low latitudes, 2. A model, J. Geophys. Res., 105, 7763, 2000.

Waters, C.L., J.C. Samson, E. and F. Donovan, The temporal variation of the frequency of high latitude field line resonances, J. Geophys. Res., 100, 7987, 1995.

Waters, C.L., J.C. Samson, and E.F. Donovan, Variation of the plasmatrough density derived from magnetospheric field line resonances, J. Geophys. Res., 101, 24737, 1996.

Waters, C.L., K. Takahashi, D.-L. Lee, and B.J. Anderson, Detection of ultralow-frequency cavity modes using spacecraft data, J. Geophys. Res., 107 (A10), 1284, doi: 10.1029/2001JA000224, 2002.

Yedidia B.A., M. Vellante, U. Villante, and A.J. Lazarus, A study of the relationship between micropulsations and solar wind properties, J. Geophys. Res., 96, 3465, 1991.

Yeoman, T.K., D.K. Milling, and D. Orr, Pi 2 polarization patterns on the U.K. SAMNET, Planet. Space Sci., 38, 589, 1990.

17 Space Weather

Louis J. Lanzerotti

Ever since the development of the electrical telegraph in the mid-nineteenth century, the effects of Earth's space environment on technologies have posed challenges to designers and operators of many technical systems. The possible systems that can be impacted by the space environment have grown over the last century and one-half from ground based communications and electrical grid technologies, to space-based systems that include communications, national security, precision location determination, and human space flight. The need for ever more detailed understanding of the space environment and its response to solar-produced disturbances continues to grow in order to be able to predict and mitigate detrimental operations and disruptions by space-originating processes. This chapter outlines some history of the effects of space processes on technologies, and discusses the wide range of contemporary technologies whose designs and operations are influenced by the basic fact that Earth's space environment is not benign. The aspects of the space environment that can affect human technologies wherever they may be deployed – from Earth's surface to the outer reaches of the solar system – are identified as space weather.

Contents

Louis J. Lanzerotti, Space Weather.
In: Y. Kamide/A. Chian, Handbook of the Solar-Terrestrial Environment. pp. 423–443 (2007)
DOI: 10.1007/11367758_17

17.1 Introduction

The discovery of the intense fluxes of trapped charged particles (electrons and ions) around Earth by Van Allen (Van Allen et al., 1958) and additional measurements of the radiation by Vernov and Chudakov (1960) confirmed, if anyone had bothered to ask before that time, that the space environment around Earth was certainly not benign. The Earth's near-Earth space was apparently filled with radiation of sufficient intensity and energy to cause significant problems for materials and for electronics that might be launched into it. The space radiation would also prove to be detrimental under some conditions to human space flight. And so, because of radiation, the low-orbit Telstar© 1 (launched July 10, 1962; *Bell System Technical Journal*, 1963), the first commercial telecommunications spacecraft, suffered anomalies in one of its two command lines within a couple of months of its launch. And within five months both command lines had failed. While clever engineering by Bell Laboratories personnel resurrected the satellite for more than a month in early 1963, by the end of February of that year it had gone silent for good, a victim of the solar-terrestrial environment (e.g., Reid, 1963).

Thus, it was clear that the satellites that had been proposed by Arthur Clark (1945) and by John Pierce (1954) for telecommunications use prior to the onset of the space age (generally attributed to the successful launch of Sputnik 1 in 1957) would now have to be designed to withstand the Earth's radiation environment. This meant that the semiconductor electronic parts (which were the obvious choice for even the earliest spacecraft and instrument designs) would have to be carefully evaluated and qualified for flight. In addition, the space radiation environment would have to be carefully mapped, and the time dependencies of the environment be well understood if adequate designs were to be implemented to ensure the success of the missions.

Early in its life, the U.S. National Aeronautics and Space Administration (NASA) initiated programs for satellite communications. This began with a contract to the Hughes Aircraft Corporation for geosynchronous (GEO) Syncom satellites (the first launched in February 1963) and a low orbit communications program under the name Relay (the first of which was launched in December 1962). NASA also initiated an Applications Technology Satellite (ATS) program (ultimately six satellites were launched into various orbits; two of these were unsuccessful due to launch vehicle failures) to investigate and test technologies and concepts in a number of space applications. In addition to communications, this included practical objectives such as navigation, meteorology, and health delivery.

ATS-1 was launched into a geosynchronous orbit in December 1966. Important elements of the payload were three separate experiments containing charged particle detectors of various configurations that were designed to characterize the geosynchronous space environment. The three sectors – commercial (AT&T Bell Laboratories), military (Aerospace Corporation), and academic (University of Minnesota) – involved in this applications spacecraft demonstrated the wide-ranging interest and importance of the conditions in the space environment surrounding Earth. The experiments all provided exciting data on such topics as the trapped radiation at the GEO orbit, the large changes in the radiation intensities with geomagnetic activity, and the relatively ready access of solar-produced particles to GEO.

Thus, the discovery by Van Allen, motivated largely by intellectual curiosity about cosmic rays and the aurora (Van Allen, 1983), led in the subsequent decades – and continuing to today – not only to intense scientific investigations of the radiation phenomena, but also to much more engineering-related work that has been devoted specifically to mitigating the effects of the radiation on any technologies that are placed into it. Today, aspects of the space environment that can affect human-constructed technologies wherever they might exist – from Earth's surface to the outer reaches of the solar system – are called 'space weather'. Of course, most interest to date has centered on the effects of the Sun and of the near-Earth space environment on technologies on Earth's surface and in the near-Earth space. The several spacecraft that have ventured into the space environments of the outer planets over the last nearly forty years have had to be designed to survive the enhanced radiation that is found in these planet's radiation belts. As humans venture back to the Moon, and perhaps on to Mars in future decades, the interplanetary radiation environment and its modifications by solar activity will become of more significance for space weather considerations.

That the space environment was not likely to be totally benign to technologies should not have been a surprise to those who may have considered the question.

The Austrian physicist Victor Hess had demonstrated at the beginning of the 20th century that cosmic rays originated outside the Earth's atmosphere. Many authors, including Birkeland (1908) and others (see Chapman and Bartels (1941), Cliver (1994), and Siscoe (2005) for considerable historical perspective) had long discussed the possibility that charged particles, likely from the Sun, played a key role in producing the aurora and geomagnetic activity at Earth.

But not only scientific curiosity drove studies of geomagnetic activity and its causes during the last half of the 19th and the first part of the 20th century. Interest in this natural phenomena was also importantly motivated by the fact that 'modern' technologies such as the telegraph and early wireless communications were found to be disrupted during times of geomagnetic disturbance.

Other chapters of these volumes are devoted to the details of the physics and the geophysics of processes in the solar and the terrestrial environments. This chapter concentrates on those technologies that can be affected by the Sun and by the space environment of Earth. The chapter does this through providing an historical development of the subject from the mid-eighteenth century to the present. Broad outlines of the underlying physical processes are noted as necessary, with the expectation that details of the operative geophysics will be obtained by the reader from the other chapters.

17.2 Early Technologies: Telegraph and Wireless

It was a surprise to find strange, spontaneous currents flowing on the lines of the early telegraphs that were rapidly being installed in many locales in Europe and the eastern U.S. following the invention of the first commercially successful system by Samuel F. B. Morse. W.H. Barlow, an employee of the Midland Railway Company in England, established a measurement program to investigate these strange currents on the telegraph line that was installed along several of the rail lines to provide communications between stations. Barlow (1849) wrote, "The observations described ... were undertaken in consequence of certain spontaneous deflections having been noticed in the needles of the electric telegraph on the Midland Railway, the erection of which was carried out under my superintendence as the Company's engineer."

The hourly means of Barlow's data for the Derby to Birmingham route, shown in Fig. 17.1, illustrate the galvanometer fluctuations at Derby for a two week data interval in May 1847, during the peak of the 9th sunspot cycle. In addition to some hour-to-hour variations, there are distinct diurnal variations as well: large right-hand swings of the galvanometer during local day and left-hand swings during local night. The systematic daily change in the galvanometer readings, while not explicitly recognized by Barlow in his paper, is likely the first measurement of the diurnal component of the geomagnetically-induced Earth currents (often referred to in subsequent literature of the late 19th and early 20th centuries as "telluric currents"). Such diurnal variations have long been recognized to be produced by solar-induced effects on Earth's dayside ionosphere.

Barlow, in discussing his measurements, further noted that "... in every case which has come under my observation, the telegraph needles have been deflected whenever aurora has been visible". This observation of a possible connection between a natural phenomenon and a new technology was dramatically confirmed in early September 1859 during the peak of the 10th sunspot cycle. At that time, under intense aurora activity, the arcing and sparking of keys and armatures were reported from a wide range of telegraph stations, including the eastern U.S., England, Scandinavia, Belgium, France, Switzerland, Prussia, Wurttemberg, Austria, and Tuscany. In Christiania, Norway, "...

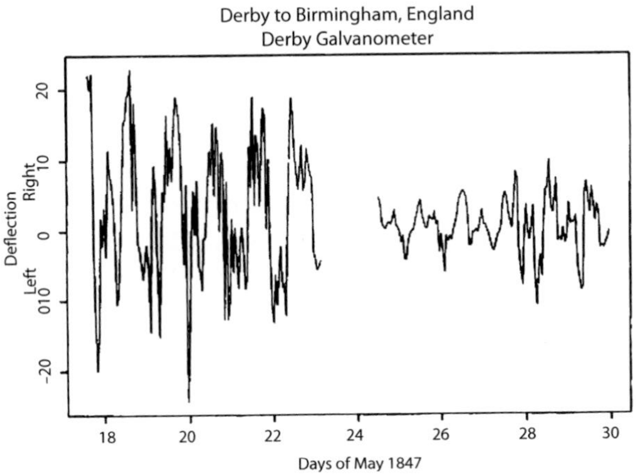

Fig. 17.1. Galvanometer readings on telegraph line along Derby to Birmingham railroad route for two weeks in May 1847 (derived from Barlow, 1849)

sparks and uninterrupted discharges were from time to time observed. Pieces of paper were set on fire ...." (Am. J. Sci., 1860). As Prescott (1860) reported, on the telegraph line from Boston to Portland (Maine), on "Friday, September 2d, 1859" the operators "continued to use the line (without batteries) for about two hours, when, the aurora having subsided, the batteries were resumed."

The aurora and large geomagnetic disturbances (Stewart, 1861) followed by about fifteen hours the dramatic observation by Carrington (1861) from his observatory at Redhill of the first solar flare: "The observation of this very splendid (sunspot) group on September 1st has had some notoriety. ... I ... witnessed a singular outbreak of light which lasted about 5 minutes, and moved sensibly over the contour of the spot ...." The association of the solar event and subsequent aurora and the disturbances on the new telegraph technology was noted by many. However, association does not prove causality. And so, controversy and uncertainty existed for decades afterwards as to the reality of effects by the Sun on Earth (e.g., Siscoe, 2005). Much of the controversy (described from a somewhat different perspective in Soon and Yaskell (2003)) that swirled among the scientific community in the later two decades of the 18th century was centered around the eminent British researcher William Thomson (Lord Kelvin), who showed in a number of publications (e.g., Kelvin, 1892) that the Sun could not emit sufficient energy to cause the observed geomagnetic disturbances (and by inference, any effects on the telegraph lines).

The solar event of 1859 (e.g., recent discussion by Cliver and Svalgaard (2005)) was followed by several decades of attention by telegraph engineers and operators to the effects on their systems of Earth electrical currents. The invention of intercontinental wireless communications, with the long wavelength radio transmissions from Poldhu Station, Cornwall, to St. John's, Newfoundland, by Marconi in December 1901 eliminated Earth currents as a source of disturbances on any communications that were sent through the atmosphere.

Marconi's achievement (for which he was awarded the Nobel Prize in Physics in 1909) was only possible because of the existence of the reflecting ionosphere (definitively identified only some two decades later by Briet and Tuve (1925) and by Appleton and Barnett (1925)) about 100 km above Earth's surface. Physical changes in this reflecting layer (hypothesized by Oliver Heaviside in 1902) were critical to the success of reliable telegraphic (and later voice) communications by this new technology.

Indeed, there were large frustrations in the use of wireless for trans-ocean (as well as continental) telegraph communications in the first couple of decades of the 20th century. Because of the very low frequencies then being used, reliable communications were only possible at night, when the solar UV (and X-rays) were not ionizing the upper atmosphere (although these Sun-induced physical effects were not specifically identified or understood at this time). And the communications were found to be significantly less disturbed by static during winter conditions than during summer, when atmospheric lightning could cause havoc.

The change and the advance in technology from cable telegraph to wireless did not obviate the effects of Sun-originating disturbances on communications. The same electrical currents that could produce "spontaneous" electrical currents within Earth and that would flow in the telegraph lines could also affect the reception and fidelity of transmitted wireless signals. Marconi (1928) commented on this when he wrote that "... the times of bad fading (of radio signals) practically always coincide with the appearance of large sun-spots and intense aurora-boreali usually accompanied by magnetic storms...." These are "...the same periods when cables and land lines experience difficulties or are thrown out of action."

Wireless public voice telephone service was established across the Atlantic in 1927 (at a cost of about $75 for a three minute call – a huge sum at that time). The technical literature of the early wireless era showed clearly that solar-originating disturbances could seriously affect wireless communications. Engineers pursued various strategies to mitigate the effects. One of these is illustrated in Fig. 17.2, where it is demonstrated that the lower frequency transmissions to England were not disrupted during the time of the magnetic disturbance, in contrast to the higher frequency (Anderson, 1929). Of course, the lower frequency had a lower bit rate, so transmissions at this lower frequency were impacted, but by not as much as was the higher frequency signal with its huge dB losses over many days.

The public awareness of the effects illustrated in Fig. 17.2 are demonstrated by a headline that appeared

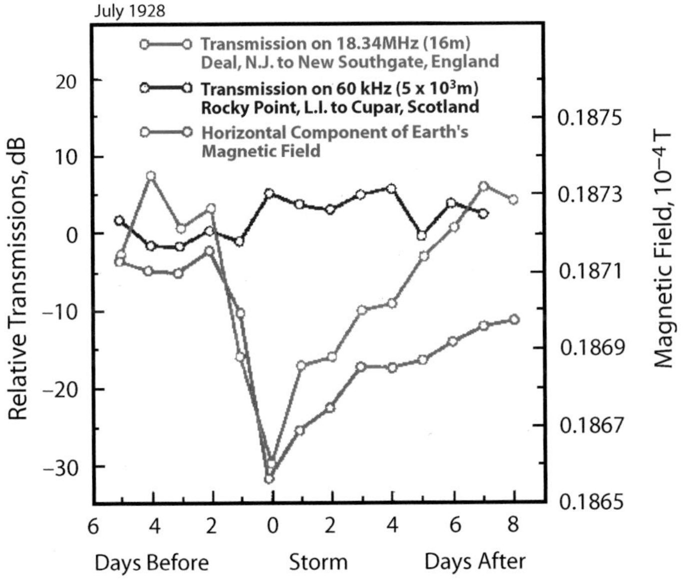

Fig. 17.2. Trans-Atlantic wireless propagation on two frequencies during a magnetic storm in July 1928 (adapted from Anderson, 1929)

over a front page article in *The New York Times* (Sunday, January 23, 1938): "Violent magnetic storm disrupts short-wave radio communication", with a sub-heading that elaborated "Transoceanic services transfer phone and other traffic to long wave lengths as sunspot disturbance strikes". The engineering work-around that shifted the cross-Atlantic wireless traffic from short to long wave lengths prevented the complete disruption of communications. This magnetic storm was one of the twenty-five largest in terms of the geomagnetic Dst index during the 70-year interval 1932–2002 (Cliver and Svalgaard, 2005).

17.3 Growth in Electrical Technologies

The first significant effects of solar-terrestrial processes on technologies other than communications occurred during the 17th solar cycle. In March 1940 a large geomagnetic storm caused ten electrical power transformer banks to trip in the region of the Ontario Hydro commission, and numerous related problems occurred in electrical power systems in the northeast and northern United States (Davidson, 1940). Voltages as high as 2 V/km were recorded on a telephone line (Fig. 17.3) between Minneapolis, Minnesota, and Fargo, North Dakota (Germaine, 1940).

Of considerable importance during this 17th solar cycle (February 1942) was the discovery of radio frequency bursts from a celestial source – the Sun. Intense solar radio emissions were discovered to be interfering with the radar that were being used in the United Kingdom to provide warnings of German aircraft during the Second World War (Hey, 1946). It was during these same solar events that the first measurements were made of solar particle events at ground level – a major surprise – although again the report was not published until after the war had ended (Forbush, 1946). The Sun could obviously be a very prodigious source of high energy particles and of high intensity radio waves, in addition to its being a source of geomagnetic storms.

Figure 17.4 illustrates, on a plot of sunspot numbers for the last 20 years the dates of some significant effects of solar-terrestrial processes on various technologies. Many of them have the same underlying physical cause (e.g., Lanzerotti, 1983; Lanzerotti and Gregori, 1986; Boteler, 1998): under disturbed geomagnetic conditions largely driven by increased solar wind activity, greatly increased (over steady-stare conditions) electrical current systems in the magnetosphere and the ionosphere produce large variations in the time rate of change of the geomagnetic field at Earth's surface. These time-varying fields in turn induce voltage potential differences across large areas of Earth's surface. These po-

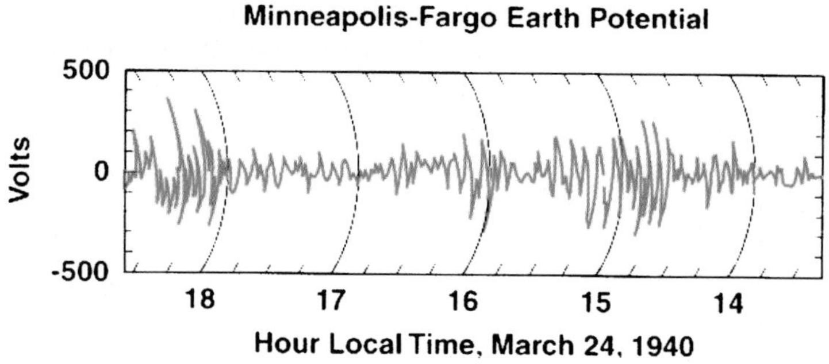

Fig. 17.3. Earth potential measured by an analog strip chart recorder on the Minneapolis, Minnesota, to Fargo, North Dakota, telephone cable during large March 1940 geomagnetic storm. Adapted from Germaine, 1940

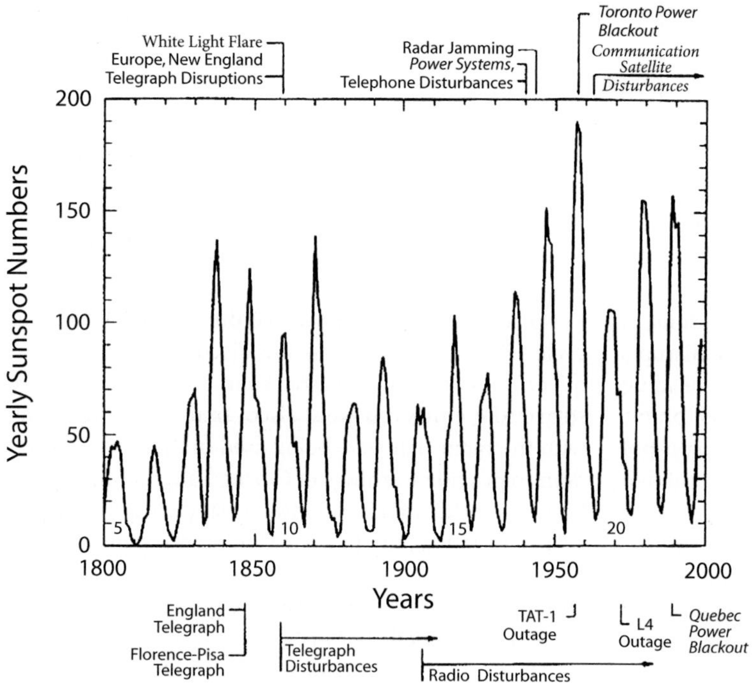

Fig. 17.4. Yearly sunspot numbers with times indicated of selected major impacts of the solar-terrestrial environment on largely ground-based technical systems. The *numbers above the horizontal axis* are the conventional numbers of the sunspot cycles

tentials (in effect, temporary 'batteries') cause currents to flow within the Earth along the least resistive paths – which can be long power or telecommunications cables that use the Earth as ground returns.

The effect of these induced currents on technical systems depends upon the systems in which they are flowing. In the case of long telecommunications lines, Earth potentials can produce overruns of the compensating voltage swings that are designed into the system power supplies (e.g., Anderson et al., 1974). Several modes of system degradation or failure can occur in power grids (Albertson et al., 1973, 1974; Pirjola et al., 2005; Boteler et al., 1989; Kappenman et al., 1981; Kappenman, 2003, 2004). Such anomalous Earth currents can also degrade and cause disruptions to measuring systems that are installed on pipelines to monitor corrosion (e.g., Pirjola and Lehinen, 1985; Campbell, 1986; Viljanen, 1989).

After the start of the space age, but before communications satellites, telephone transmissions on the first trans-Atlantic voice cable (TAT-1 from Newfoundland to Scotland) were disrupted in February 1958 by a very large geomagnetic storm (Winckler et al., 1959). During this same event power circuits tripped in the Toronto, Canada, area, plunging the region into "… temporary darkness broken only by the strange light of the aurora overhead" (Brooks, 1959). A large magnetic storm in August 1972 disrupted a Chicago to west coast telecommunications line in the Chicago to Iowa link (Anderson et al., 1974; Boteler and van Beek, 1999). The entire province of Quebec suffered a power outage for nearly a day as transformers failed during the geomagnetic storm of March 1989 (Czech et al., 1992). At the same time, the first trans-Atlantic fiber optic cable (TAT-8) was rendered nearly inoperative by the large potential difference that was established between the terminals on the coasts of New Jersey and England (Medford et al., 1989).

17.4 The Space Age and Space Weather

As discussed above, the space age did not obviate the need for attention to the effects of Sun and near-Earth space processes on ground-based technologies that are critical for modern-day life. Indeed, while many communications moved into space with the advent of the space age, the power grid (for example) remains firmly anchored to the ground. What the space age did demonstrate, conclusively, was that the placing into space of ever-advancing technologies – for both civilian as well as national defense purposes – meant that ever more sophisticated understanding of the space environment is now required to ensure reliable operations of the systems (e.g., Lanzerotti et al., 1997, 1999; Song et al., 2001; Lanzerotti 2001a,b; Daglis, 2001, 2004; Scherer et al., 2005). The operations of both ground-based and space-based systems have often encountered unanticipated surprises because of solar-terrestrial effects (e.g., Barbieri et al., 2004; Webb and Allen, 2004).

In addition, the increasing diversity of technical systems that can be affected by space weather processes is accompanied by continual changes in the dominance of one technology over another for specific uses. This is especially true for communications, civil as well as military. The last transatlantic coaxial copper cable (TAT-7, laid in 1983 from New Jersey to France) carried fewer than 4000 simultaneous voice messages, while TAT-8 (the first fiber cable – two pairs of single mode fibers, laid in 1988 from New Jersey to France with a branching point to England) could carry nearly 40,000. In 1988 satellites were the dominant carriers of transoceanic messages and data; only about two percent of this traffic was by undersea cable. By the year 2000, the wide bandwidth provided by the vast fiber networks that had been deployed meant that more than 80% of the voice and data traffic was now via ocean cable (Mandell, 2000). (This did not necessarily result in profitability for either cable or satellite companies; indeed, several companies entered into bankruptcy in the early 21st century due to poor capital investments and technical decisions.) Much satellite traffic has converted to multipoint streaming media for broadcasting (including satellite radio), a considerable amount of which is now directed to home users. Thus, space weather processes can affect some systems, and their end users, differently than others. While a communications cable might be impacted by a geomagnetic storm, transmissions from satellites might be relatively unaffected.

Many contemporary technologies that must include considerations of the Earth's space environment in their design considerations and/or operations are listed in Table 17.1 (adapted from Lanzerotti et al., 1999). Figure 17.5 (Lanzerotti, 2001a) schematically illustrates these effects. The systems in the table are grouped into broad categories that have similar physical origins in the solar-terrestrial system.

17.4.1 Ionosphere and Earth Currents

The basic chain of events behind the production of earth potentials and their effects on technical systems that consist of long conductors was outlined in the previous section. While the effects of these earth currents on systems have been extensively studied since the time of Barlow during the 9th solar cycle, there remain some key outstanding issues that continue to prevent total mitigation strategies, especially for power grids (e.g., Kappenman and Radasky, 2005). The most important of these is that the time variations and the spatial dependencies of the space electrical currents (in both the ionosphere and the magnetosphere) remain poorly understood or predictable from one geomagnetic storm to the next (there are numerous analogies to the predictability of cyclones and hurricanes and their tracks (Siscoe, 2005)). This is of especial importance since the potentials that are induced across Earth's surface are very much dependent upon the conductivity structure of the Earth that underlies the affected ionosphere. Similar electrical current variations in the space environment can produce very different Earth potential drops depending upon the nature and orientation of the underground Earth conduc-

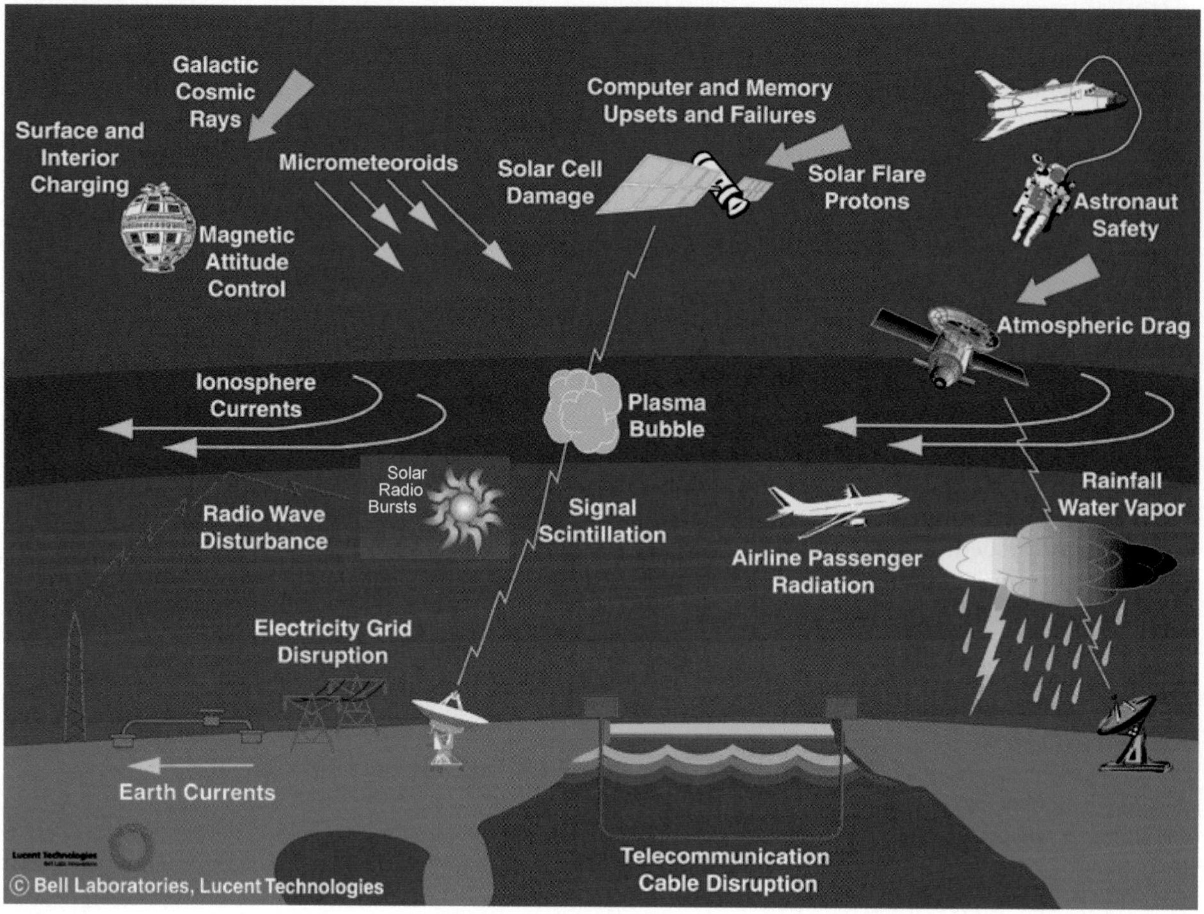

Fig. 17.5. Solar-terrestrial effects on ground- and space-based technologies

tivity structures in relationship to the variable overhead electrical currents.

Modeling of these effects on systems is becoming quite advanced in many cases (e.g., Lanzerotti et al., 2001c; Kappenman, 2003). However, the use of the modeled results for "predictions" for practical purposes is difficult at present, even when accurate knowledge of the interplanetary conditions close to Earth is at hand. The transfer function from the interplanetary medium to Earth's surface, and taking into account (as needs to be done) solid Earth effects, is poorly known. Further, this area of research involves a close interplay between space plasma geophysics and solid Earth geophysics (e.g., Lanzerotti and Gregori, 1986; Gilbert, 2005), and is thus one that is not often addressed collaboratively by two very distinct research communities (except by the limited group of researchers who pursue electromagnetic investigations of the solid Earth). The educational backgrounds and the terminologies of the research groups are often very different.

17.4.2 Ionosphere and Wireless

The ionosphere is both a facilitator and an intruder in numerous communications applications. Military, police and fire emergency agencies, and commercial enterprises in many nations (as well as a vast base of amateur radio operators – "hams" – across many nations) rely on wireless links that make intensive use of frequencies from kHz to hundreds of MHz that require

Table 17.1. Impacts of solar-terrestrial processes on technologies

- Ionosphere variations
  - Induction of electrical currents in Earth
    - Power distribution systems
    - Long communications cables
    - Pipelines
  - Interference with geophysical prospecting
    - Resource exploration
    - Archeological studies
  - Source for geophysical prospecting
  - Wireless signal reflection, propagation, attenuation
    - Commercial radio and TV
    - Radio direction finding
    - RF monitoring systems
  - Satellite signal interference, scintillation
    - Commercial telecom and broadcasts
    - GPS systems
- Magnetic Field Variations
  - Attitude control of spacecraft
  - Radio direction finding
  - Navigation (compass)
- Solar Radio Bursts
  - Interference with radar
  - Excess noise in wireless communications systems
- Particle Radiation
  - Solar cell damage
  - Semiconductor device damage and failure
  - Faulty operation of semiconductor devices
  - Spacecraft charging: surface and interior
  - Astronaut safety
  - Aircraft crew and passenger safety
- Micrometeoroids and Artificial Space Debris
  - Solar cell damage
  - Damage to mirrors, surfaces, materials, complete vehicles
  - Spacecraft attitude control
- Atmosphere
  - Low altitude spacecraft drag
  - Attenuation and scattering of wireless signals

the ionosphere as a reflector. Changes in the reflections (that is, changes in the conductivity of the ionosphere) that is produced by solar activity, from solar UV and X-ray emissions as well as by magnetic storms, can significantly alter the propagation paths (e.g., Eccles et al., 2005).

Point-to-point high frequency (HF) wireless communications can be seriously affected by the Sun's interactions with Earth's space environment. Modern-day users of such systems are well familiar with many anecdotes of solar-produced effects and disruptions. As one example, near the peak of the 21st solar cycle the *Los Angeles Times* reported that a distress signal from a downed commuter plane in 1979 was received by an Orange County, California, fire department – which responded only to discover that the signal had originated from the accident site in West Virginia. An Associated Press release on October 30, 2003, reported that "airplanes flying north of the 57th parallel (had) experienced some disruptions in (HF) radio communications … due to the geomagnetic storm from solar flares", in the declining phase of the 23rd solar cycle.

At frequencies of around a GHz and higher, the production of "bubbles" and other irregularities in the ionosphere densities can serve as a prime source of "scintillations" in signals transmitted from satellite to ground. These variable disturbances in the signals can cause detrimental effects in surveillance, communications, and navigation systems. Engineers at the COMSAT Corporation first discovered these effects after the initial deployment of the INTELSAT satellite network at geosynchronous orbit (GEO) (Taur, 1973).

Such ionosphere disturbances are the cause of major problems in the employment of single frequency signals from the Global Positioning System (GPS) for precise positional location on Earth. The prospective European Galileo system will have similar ionosphere issues to deal with. Evolution to a dual (or more) frequency positioning system may well eliminate many of the more severe problems. Nevertheless, much research is on-going in identifying and understanding some of the underlying physical processes (Kintner et al., 2005). A major U.S. Air Force satellite experiment (the C/NOFS satellite mission) that is directed toward deeper understandings of ionosphere disturbances and their effects on trans-ionosphere signals is in progress (e.g., de la Beaujardiere et al., 2005; Kelley et al., 2005).

17.4.3 Solar Radio Noise

Solar radio noise and solar radio bursts were discovered six decades ago by Southworth (1945) and by Hey (1946) during early research on radar at the time of the Second World War. Solar radio bursts produced unexpected (and unrecognized at first) jamming of this new technology that was under rapid development and deployment for war-time use (Hey, 1973). Extensive post-war research established that solar radio emissions can exhibit a wide range of spectral shapes and intensity levels (e.g., Kundu, 1965; Castelli et al., 1973; Guidice and Castelli, 1975; Barron et al., 1985). Solar radio continues to be an active and productive area of research directed toward understanding solar activity and solar-terrestrial phenomena (e.g., Bastian et al., 1998; Gary and Keller, 2004).

Solar radio noise and bursts are still of considerable relevance for the interference that they can produce in radar systems and in ground terminals when such technologies find themselves oriented toward the sun. The rapid growth in the wireless business over the last decade produced the need to evaluate the possibility that solar noise might affect this communications system, largely through disturbances to cell site base stations (e.g., Lanzerotti et al., 2002; Gary et al., 2004).

Shown in Fig. 17.6 are cumulative probability distributions (for solar minimum and solar maximum intervals) of solar radio bursts per day above 2 GHz that have been analyzed from a forty-year compilation of bursts by the NOAA National Geophysical Data Center (Nita et al., 2002). The exponents λ of power law fits to the histogram distributions are shown; the solid and the dashed lines are fits to the actual (histogram) distributions and to the geographically-corrected (because of the non-uniform distribution of observing sites around the world) distributions, respectively. The roll-over of the distribution at the lowest flux density is believed to be a result of decreased instrument sensitivities at the very lowest levels; the fall-off at the highest flux levels may indicate energy limits on solar processes. Using such distributions, and taking into account the time interval over which the data were acquired, the probability of a burst affecting a specific receiver can be estimated. Consistent with the conclusions of Bala et al. (2002), bursts with amplitudes $>10^3$ solar flux units (sfu) at $f \sim 1$ to 2 GHz could cause potential problems in a wireless cell site on average of once every three to four days during solar maximum, and perhaps once every twenty days or less during solar minimum.

17.4.4 Space Radiation Effects

Human Space Flight

Five decades after Van Allen's discovery, space radiation still places severe constraints on many aspects of robotic and human space flight, both within the Earth's magnetosphere as well as outside. Just as facilities on Earth must be designed to withstand expected extremes in atmospheric weather in the regions in which the facilities are located, so too must "facilities" that are placed into space be designed for the space environments that they will encounter during their expected lifetimes.

The implications of the space radiation environment for human space flight are obviously quite serious, since the safety and survivability of humans are involved. Thus, the radiation environment for the low altitude-orbiting space station is monitored, and predictions are made for the possible occurrence of solar events that might be expected to produce radiation at certain locations in the station's orbit inclined at 57°. Radiation levels outside the relative shielding of the magnetosphere have taken on a larger importance following the enunciation of the United States' Space Exploration Vision – back to the Moon and on to Mars – and discussions of the new missions, robotic and human, that will be expected in association with it.

In the earlier human flight era, it is likely that astronauts enroute to the Moon or on its surface during the solar event of August 1972 – when the AT&T L4 continental cable was disrupted (Fig. 17.4) – would have suffered serious radiation effects, even a potentially lethal radiation dose (Wilson et al., 1999; Parsons and Townsend, 2000; National Research Council, 2000; Townsend et al., 2002). Fortunately, the last two Apollo flights, Apollo 16 and Apollo 17, bracketed this event by several months.

Wilson et al. (1999) used a compiled set of integral solar proton event fluence spectra for six very large events (Fig. 17.6) in order to discuss the effects of such events on astronauts, and to determine shielding needs for exploration. Figure 17.6 shows clearly that the August 1972 event had the hardest spectra with the highest fluences of all of the events, with large fluences at proton energies above 50 MeV that could readily penetrate spacecraft shielding, as well as space suits should astronauts be on the lunar or Martian surface. Public awareness of the implications for astronauts of an event such

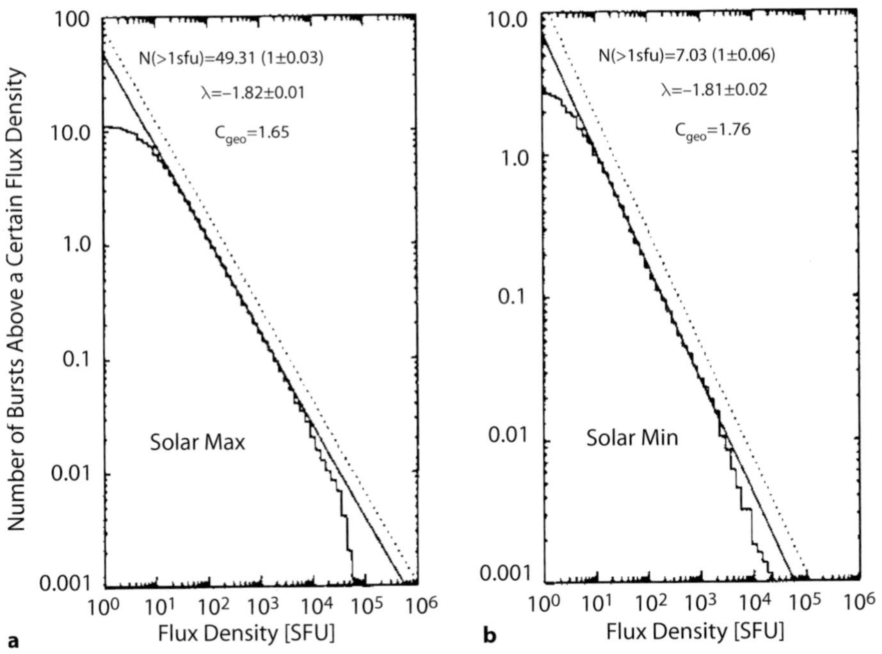

Fig. 17.6a,b. Comparison of cumulative probability histograms of the number of solar radio noise bursts above given flux densities at solar maximum and solar minimum. From Nita et al., 2002

as the one in August '72 was enhanced by the 1982 novel *Space* by James Michener.

Of growing importance for potential radiation exposure of humans has been an increase in the monitoring of radiation at aircraft altitudes (e.g., Lewis et al., 2001; Taylor et al., 2002; Stassinopoulos et al., 2003; Spurný et al., 2004; Getley et al., 2005). In fact, under one monitoring program, the first measurement at aircraft altitudes of a "ground level" solar particle event was made in 2003 during a Quantas airline flight from Los Angeles to New York (Getley, 2004). A special "users" meeting is now held for airline interests during the yearly Space Weather Week that is organized by the NOAA Space Environment Center in Colorado (Murtagh et al., 2004). Airline concerns about radiation levels at flight altitudes have increased with the advent of cross-polar routes from locations such as the east coast of the United States to Asian countries. The mapping of energetic solar events and galactic cosmic rays to aircraft altitudes and high latitude routes is accomplished with sophisticated software codes (e.g., Smart et al., 2000; Smart and Shea, 1997, 2005) that often also use information on concurrent disturbances to the background geomagnetic field to determine regions of the airspace of most risk to an event.

Robotic Space Flight

Since radiation effects on the electronics and materials used in human space missions will be similar to those in purely robotics missions, the two can be discussed together. At the outset of the space age, and the advent of communications satellites, "... the basic processes occurring in the solar cells and transistors by the electrons and protons (that were encountered in space were) poorly understood" (McCormac, 1966). This situation is much improved today because of the huge advances that have been made in the semiconductor electronics industry in understanding electronic materials, including their modifications by photons and charged particles. Nevertheless, the study of space radiation and its effects on electronics systems remains a major research area (e.g., Shea and Smart, 1998; Koons et al., 1999; Baker et al.; 2004, Li et al., 2005). A textbook discussion of the space environment and implications for satellite design is contained in Tribble (1995). Some 200 or so in-use communications satellites now occupy prime orbital space at geosynchronous altitude, and many additional spacecraft, often national security related, are in this vicinity or in highly elliptical orbits with various inclinations. Accurate and timely information is required on the spatial and temporal dependencies of the radiation environment for design and operational purposes.

The low energy plasma population of the magnetosphere (few eV to few keV energy particles) is important as it can produce electrical charging of spacecraft surface materials (used principally for temperature control) that encase a satellite (DeForest, 1972; Garrett, 1981; Grard et al., 1983). This plasma population can also modify surface materials and solar arrays through sputtering and by chemical reactions, the latter being of particular importance for materials that are on low altitude orbit satellites that can be bombarded by the high fluxes of oxygen atoms.

If good electrical connections are not established between the various surface materials on a spacecraft, and between the materials and the solar arrays, differential charging (i.e., large potential differences) on the several surfaces can produce lightning-like discharges between the materials. These discharges produce both electromagnetic interference and outright damage to components and subsystems (e.g., Vampola, 1987; Gussenhoven and Mullen, 1983). A special U. S. Air Force satellite program, the SCATHA program, was conducted in the early 1980's (launched in January 1979) to specifically investigate mitigation methods for spacecraft charging (McPherson and Schober, 1976).

The levels of charging that might be expected depends sensitively upon the location of a spacecraft in its orbit and on the plasma state of the magnetosphere. At GEO, the location of most commercial communications spacecraft, the plasma state can vary by large amounts under differing geomagnetic conditions. Under conditions of enhanced geomagnetic activity the cross-magnetosphere electric field will convect earthward the plasma sheet in the Earth's magnetotail. When this occurs, on-board anomalies from surface charging effects can increase; this increased charging tends to be most prevalent in the local midnight to dawn sector of the GEO orbit (Mizera, 1983).

An example of a study of surface charging on commercial spacecraft surfaces is shown in Fig. 17.7. Two surface-mounted charge plate sensors were flown on the former AT&T Telstar 4 GEO satellite to monitor the surface charging effects, and this figure shows the statistical distributions of charging on one of the sensors in January 1997 (Lanzerotti et al., 1998). The solid line in each panel corresponds to the charging statistics for the entire month, while the dashed lines omit data from a large magnetic storm event on January 10th. Charging voltages as large as −800 V were recorded on the charge plate sensor during the magnetic storm, an event during which a permanent failure of the Telstar 401 satellite occurred (although the failure was not attributed specifically to the space conditions).

The intensities of higher energy particles in the magnetosphere (MeV energy electrons to tens of MeV energy protons) can change by orders of magnitude over the course of minutes, hours, and days. These intensity changes occur through a variety of processes, including plasma energization processes within the magnetosphere and ready access of solar particles to GEO and the outer magnetosphere. In general, it is prohibitively expensive (in comparing, for example, the tradeoffs between adding additional spacecraft shielding mass versus additional orbit control gas or additional revenue-producing transponders) to provide sufficient shielding of all interior spacecraft subsystems in order to minimize radiation problems.

The range of a 100 MeV proton in aluminum (a typical spacecraft material) is ~40 mm. The range of a 3 MeV electron is ~6 mm. These particles can therefore penetrate deeply into the interior regions of a satellite. In addition to producing transient upsets in signal and control electronics, such high energy particles can also cause electrical charges to build up in interior insulating materials such as those that are used in coax cables. If the charge buildup is sufficiently large, these interior materials will eventually suffer electrical breakdowns. Electromagnetic interference and damage to electronics will occur.

Examples of the types of spacecraft anomalies from a period of intense solar flares, coronal mass ejections, and geomagnetic activity in October–November 2003 is shown in Table 17.2 (adapted from Barbieri and Mahmot (2004)). The compilers note that, with the exception of the orbit change of the TRMM mission, all of the impacts shown were caused by "solar energetic particles … or similarly accelerated particles in geospace." The purely communications satellites included in the Table, the NASA Tracking and Data Relay Satellite System (TDRSS), suffered electronic errors during the interval of the solar-origin events.

The significant uncertainties in placing, and retaining, a spacecraft in a revenue-returning orbital loca-

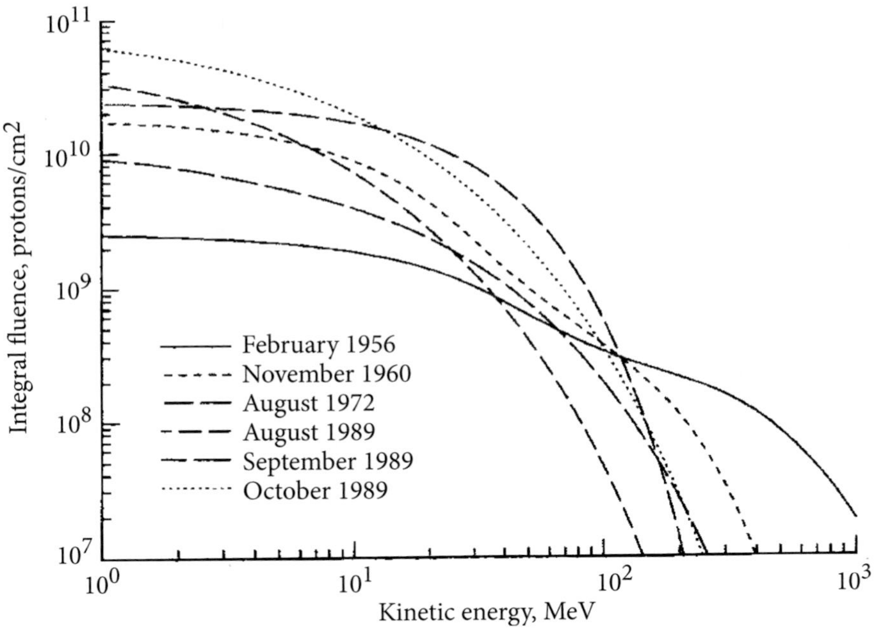

Fig. 17.7. Solar proton integral fluence spectra measured at 1 A.U. in selected solar events occurring over the last five decades. From Wilson et al., 1999

Table 17.2. Summary of space weather impacts on selected spacecraft in October–November 2003 (adapted from Barbieri and Mahmot, 2004)

		SPACE WEATHER IMPACT				
Spacecraft Mission	Change in Operation Status	Electronic Errors	Noisy House-keeping Data	Solar Array Degradation	Change in Orbit Dynamics	High Levels Accumulated Radiation
Aqua	None	X				
Chandra	Instrument safed					X
CHIPS	Control loss	X				
Cluster	None			X		
Genesis	Auto safed	X				
GOES 9,10	None		X			
ICESat	None	X				
INTEGRAL	Command safe					
Landsat 7	Instrument safed					
RHESSI	Abs. time seq. stop	X				
SOHO	Instrument safed			X		
Stardust	Auto safed	X				
TDRSS	None	X				
TRMM	Added delta V				X	
WIND	None			X		

tion has led to a large business in risk insurance and re-insurance for one or more stages in a satellite's history (government satellites are self-insured). The loss of a spacecraft, or one or more communications transponders, from adverse space weather conditions is only one of many contingencies that can be insured against. It is frequently not possible to determine if an anomaly is space weather-related or from some other cause, i.e., from one or more manufacturing deficiencies to 'other'.

The consulting firm Futron Corporation (2002) notes that in the four years prior to its study, satellite insurance rates increased by 129% while the number of major on-orbit anomalies rose by 146%. In some years the space insurance industry is profitable, and in some years there are serious losses in net revenue after paying claims (e.g., Todd, 2000). For example, Todd (2000) states that in 1998 there were claims totaling more than $1.17 billion after salvage, an amount just less than about twice that received in premiums. These numbers can be quite volatile from year to year.

17.4.5 Magnetic Field Variations

The designs of those GEO spacecraft that use the Earth's magnetic field for attitude control must take into account the high likelihood that the satellite will find itself outside the dayside magnetosphere on those occasions when there is very high solar wind pressure. Enhanced solar wind velocities and densities, such as those that can occur in a coronal mass ejection event, can easily distort the dayside magnetopause and push it inside the GEO orbit. The highly spatial- and time-varying magnetic fields that occur at the boundary and outside the magnetosphere can seriously disrupt a magnetically-stabilized satellite if appropriate precautions have not been incorporated in the design of the on-board control systems. The magnetic field outside the magnetosphere will have a polarity that is predominantly opposite to that in which the satellite is normally situated so that a complete "flip" of the orientation could occur when the magnetopause is crossed.

17.4.6 Micrometeoroids (and Space Debris)

Spacecraft, robotic and manned, can be seriously disoriented, damaged, or left inoperable by the impacts of solid objects in space, such as micrometeoroids. The debris that is left in orbit from space launches and from satellites that fragment or break up for whatever the reason (e.g., Beech et al., 1995; 1997; McBride, 1997) is another source of solid debris in space, although under the strict definition of space weather as consisting of only natural phenomena and processes, only micrometeoroids would qualify to be included for discussion. The U.S. Air Force systematically tracks thousands of space debris items that are circling Earth, most of which are in low altitude orbits. Topics on space debris are documented and updated regularly, including a quarterly newsletter, on the NASA space debris home site at the Johnson Space Center in Houston, Texas: http://www.orbitaldebris.jsc.nasa.gov/.

There was considerable concern around the turn of the millennium about the possibility that, because of the location of the Earth with respect to the Leonid meteoroid stream, the stream at that time could pose a hazard to operating spacecraft (e.g., Yeomans et al., 1996; McBride and McDonnell, 1999). Few detrimental effects were observed, a portion of the good news perhaps arising from the safeing procedures that were followed for many spacecraft, including the temporary changing of the orientations of solar panels.

Plotted in Fig. 17.8 is the altitude versus estimated diameter for space objects detected by the Haystack Observatory (Massachusetts) during October 2002 through September 2003 (Stansbery et al., 2005). The concentration of objects above about 700 km range is striking. The lifetimes in orbit of debris (and of micrometeoroids) are significantly influenced by the residual atmosphere that the objects encounter at their orbital altitudes.

The effect of the atmosphere for sweeping objects out of orbit is considerably more effective at altitudes below about 600 km. And the density of this residual atmosphere is strongly dependent upon the solar cycle; the density is considerable higher at a given altitude during solar maximum than during solar minimum.

There is little information on the micrometeoroid flux at GEO. Personal anecdotal experience indicates that such objects that strike a GEO communications spacecraft often result in a need for automatic or manual intervention for the re-orientation of the satellite. Most of the perturbations are small such that transponder lock is not necessarily lost and that there is no damage to the satellite. But micrometeoroid effects do occur not infrequently and need to be accounted for in operational procedures and manuals.

17.4.7 Atmosphere: Low Altitude Spacecraft Drag

The same atmospheric drag effect that operates on low altitude space debris and micrometeoroids, as discussed

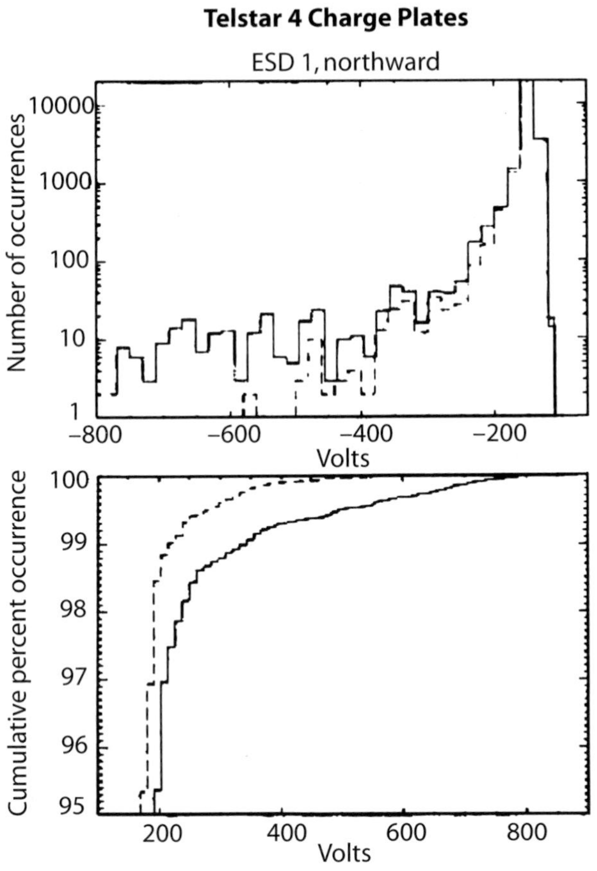

Fig. 17.8. Statistical distribution of surface charging recorded on the northward-facing charge plate sensor on the Telstar 4 spacecraft during the month of January 1997 (*solid line*) and for the same month with data from January 10th (the date of a large magnetic storm; *dashed line*) removed. The *upper panel* records (in approximately 25 volt bins) the number of voltage occurrences in each voltage bin; the *lower panel* plots the cumulative percent voltage occurrence above 95% in order to illustrate the extreme events seen by the communications spacecraft

above, also can significantly influence the orbits of low altitude spacecraft. The ultraviolet emissions for the Sun change by more than a factor of two at wavelengths ~170 nm during a solar cycle (Hunten et al., 1991), significantly larger that the order 0.1% changes in the visible spectrum. This heating by the increased UV emissions causes the atmosphere to expand sufficiently to raise the "top" of the atmosphere by several hundred km during solar maximum. These greater densities result in increased drag on all objects in low orbits – including space stations and the Hubble Space Telescope. In 2001 (solar maximum conditions), just before it was de-orbited, the MIR space station was decreasing in altitude by about 1.5 km/day because of increased atmosphere density and the resultant increase in drag on the station. The International Space Station decreases in altitude between 100 km and 150 km per year, depending upon phase of the solar cycle, and thus needs regular reboosting in its orbit.

Skylab, the first U.S. space station in the 1970s, was lost in July 1979 due to the effects of atmospheric drag in the solar maximum period of the 21st solar cycle. The space shuttle was not flight-ready in time to carry out its planned boost of Skylab to a higher orbit. Telecommunications satellites that fly in low earth orbit use some of their orbit control fuel to maintain orbit altitude during the buildup to, and during, solar maximum conditions (e.g., Picholtz, 1996).

17.4.8 Atmosphere Water Vapor

At frequencies in the Ka band that are planned for high bandwidth space-to-ground applications (as well as for point-to-point communications between ground terminals), water vapor in the neutral atmosphere is the most significant natural phenomena that can seriously affect the signals (Gordon and Morgan, 1993). It would appear that, in general, the space environment can reasonably be ignored when designing around the limitations that are imposed by rain and water vapor in the atmosphere.

A caveat to this claim would arise if it were definitely to be proven that there are effects of magnetosphere and ionosphere processes (and thus the interplanetary medium) on terrestrial weather and climate (e.g., Rycroft et al., 2000). For example, it is well recognized that even at GHz frequencies the ionized channels caused by lightning strokes, and possibly even the charge separation in clouds, can reflect radar signals. Lightning and cloud charging phenomena may produce as yet unrecognized noise sources for low-level wireless signals. Thus, if it were to be learned that ionosphere electrical fields influenced the production of weather disturbances in the troposphere, the space environment could be claimed to affect even those wireless signals that might be disturbed by lightning. Much further research is required in this area of speculation. Advances are perhaps coming with the intense research

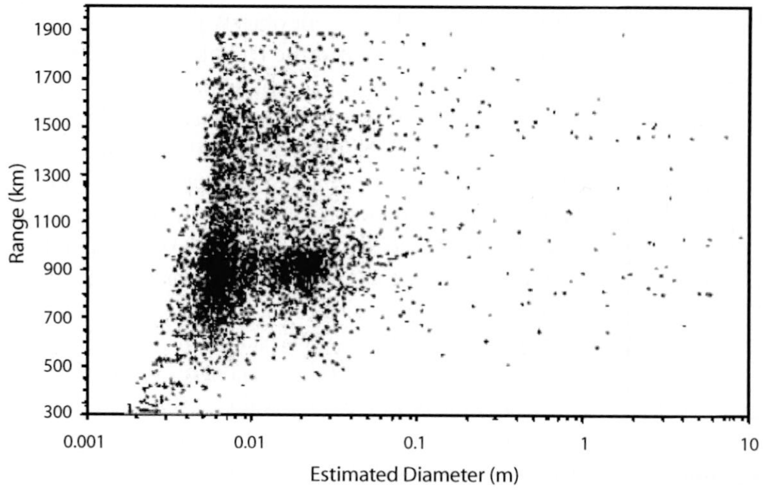

Fig. 17.9. Estimated diameter for orbital objects detected by the Haystack Observatory during October 2002–September 2003 as a function of altitude (range). From Stansbery et al., 2005

efforts directed toward understanding the blue jets and red sprites that are found above certain thundercloud systems (e.g., Sentman and Wescott, 1995; Fukunishi et al., 1996; Winckler et al., 1996).

17.5 Policy Issues

A growth in the number and types of policy issues associated with space weather has accompanied the growth of the interest in, and the importance of, the subject to all sectors of the global economy – commercial, public, and defense. Until quite recently, space weather engineering problems were usually addressed by the sector impacted. Commercial enterprises such as the electrical power industry and the communications industry that experienced problems from solar-terrestrial processes generally employed resources from their internal engineering and research staffs to address the industry-specific questions. The government, both civil and military, tended to do similarly, using in-house laboratories. Of course, communication existed as necessary between the civil and governmental sectors, with the communications often being facilitated by much governmental (and some industrial) technical results published in the research literature. Just as in the past for atmospheric weather services (Siscoe, 2005), the growing need for more predictive and operational capabilities for space weather has attracted entrepreneurial private firms to bid for the provision of services.

A chapter in the 2002 Decadal Survey of Solar and Space Physics, *The Sun to the Earth – And Beyond* (National Research Council, 2002), was devoted to the effects of solar-terrestrial processes on technology and society. In addition to a brief overview of the impacts of space weather, the chapter addresses a number of policy issues from the vantage point of the United States, and makes recommendations related to them. Many of these policy concerns are broader than just one country, and therefore warrant serious consideration as space weather continues to grow in importance internationally.

These more global issues include the following:

1. Monitoring the solar-terrestrial environment. "Effective monitoring of the space environment requires identification of those research instruments and observations that are needed to provide the basis for modeling interactions of the solar-terrestrial environment with technical systems and for making sound technical design decisions."
2. Transition from research to operations. "An important task facing the space weather community during the coming decade will be to establish, maintain and evolve mechanisms for the efficient transfer of new models of the solar-terrestrial environment into the user community. This will involve such issues as establishing verification and validation programs for models and system-impact products, and for prioritizing operational needs."

3. Data acquisition and availability. "Developing successful space weather mitigation strategies involves the ability to predict space weather effects on specific technological systems as well as to predict space weather in general; it also requires knowledge of extreme space weather conditions. Among other issues involved in this is that serious consideration should be given to establishing a centralized database of extreme space weather conditions that covers as many relevant space weather parameters as possible."
4. The public and private sectors in space weather applications. "Both the government and private industry (in the U.S.) are involved in acquiring, assessing, and disseminating information and models related to the solar-terrestrial environment in the context of its relevance for technological systems. Therefore, it is appropriate to determine the appropriate roles for each sector in space-weather-related activities." The policy issues that arise from this can be quite difficult and are occasionally contentious. There are numerous analogies with the government and private sectors in meteorology and weather forecasting, as Fisher (2004a,b) has discussed. Many recent discussions of private and government interests have involved the definition of value-added services and which sector of the economy should be most relied upon to provide them.

For space weather research and forecasting to be most useful at the most economical costs, it will be necessary for the above policy issues, as well as those that are now unrecognized but will undoubtedly arise, to be continued to be discussed by all involved parties. The research community often does not consider such matters in its day-to-day activities. Nevertheless, policy related to the applications of solar-terrestrial research can have a strong determinant in deciding research directions and funding.

17.6 Summary

Over the last more than a century and one-half, the array of technologies that are imbedded within space-affected environments have vastly increased. The set of underlying physical phenomena that can affect these technologies is limited in number, even if perfect understanding of the physics of the phenomena remains fragmentary. The increasing sophistication of technologies, and how they relate to the environments in which they are imbedded, means that ever more detailed understanding of the physical phenomena is needed.

At the same time, most present-day technologies that are affected by space phenomena are the underpinnings of dynamic and economically important businesses. This is certainly the case for the communications and the electric power industry. These technologies, and the businesses that control them, can not wait for optimum scientific knowledge to be acquired before new technical embodiments are created, implemented, and marketed. If companies were to seek perfectionist understanding of nature as exhibited in space weather phenomena before marketing new services or equipment, the companies would in all likelihood be left behind in the marketplace. A balance is needed between deeper understanding of physical phenomena and "engineering" solutions to crises that that can arise in space weather monitoring and forecasting. The research community must be able to understand and operate creatively and with great adaptability within this dynamic environment.

Acknowledgement. Portions of this chapter are based upon past reviews and observations of space weather research and engineering as reported in several papers (Lanzerotti, 1983; Lanzerotti et al., 1999; Lanzerotti, 2001a,b). My knowledge and views of the subject have been shaped in essential ways over the last four decades by my industrial research experiences, and by numerous colleagues in industry and in academia especially C.G. Maclennan, D.J. Thomson, G. Siscoe, J.H. Allen, D. Baker, J.B. Blake, G. Paulikas, A. Vampola, H.C. Koons, and L. J. Zanetti.

References

Albertson, V.D., J.M. Thorson, R.E. Clayton, and R.E. Tripathy, Solar induced currents in power systems: cause and effects, *IEEE Trans. Power App. & Sys.*, **PAS-92**, 471, 1973

Albertson, V.D., J.M. Thorson, and S.A. Miske, The effects of geomagnetic storms on electrical power systems, *IEEE Trans. Power App. & Sys.*, **PAS-93**, 1031, 1974 *American Journal of Science*, **29**, 386, 1860

Anderson, C.N., Notes on the effects of solar disturbances on transatlantic radio transmissions, *Proc. I.R.E.*, **17**, 1528, 1929

Anderson, C.W. III, L.J. Lanzerotti, and C.G. Maclennan, Outage of the L4 system and the geomagnetic storm of 4 August 1972, *The Bell Sys. Tech. J.*, **53**, 1817, 1974

Appleton, E.V., and M.A.F. Barnett, Local reflection of wireless waves from the upper atmosphere, *Nature*, **115**, 333, 1925

D.N. Baker, S.G. Kanekal and J.B. Blake, Characterizing the Earth's outer Van Allen zone using a radiation belt content index, *Space Weather*, Vol. 2, No. 2, S02003 10.1029/2003SW000026, 2004

Bala, B., L.J. Lanzerotti, D.E. Gary, and D.J. Thomson, Noise in wireless systems produced by solar radio bursts, *Radio Sci.*, **37**, doi:10.1029/2001RS002481, 2002

Barlow, W.H., On spontaneous electrical currents observed in the wires of the electric telegraph, *Phil. Trans. R. Soc.*, **61**, 1849

L.P. Barbieri and R.E. Mahmot, October–November 2003's space weather and operations lessons learned, *Space Weather*, Vol. 2, No. 9, S0900210.1029/2004SW000064, 2004

Barron, W.R., E.W. Cliver, J.P. Cronin, and D.A. Guidice, Solar radio emission, in *Handbook of Geophysics and the Space Environment*, ed. A.S. Jura, Chap. 11, AFGL, USAF, 1985

Bastian, T.S., A.O. Benz, and D.E. Gary, Radio emission from solar flares, *Ann. Rev. Astron. Astrophys.*, **36**, 131, 1998

Beech, M., P. Brown, and J. Jones, The potential danger to satellites from meteor storm activity, *Q. J. R. Astr. Soc.*, **36**, 127, 1995

Beech, M., P. Brown, J. Jones, and A.R. Webster, The danger to satellites from meteor storms, *Adv. Space Res.*, **20**, 1509, 1997.

Bell System Technical Journal, Special Telstar Issue, **42**, Parts 1, 2, 3, July 1963

Birkeland, K., *Norwegian Aurora Polaris Expedition 1902–1903*, Vol. 1, On the causes of magnetic storms and the origin of terrestrial magnetism, Christiania, H. Aschehoug & Co., 1908

Boteler, D.H., R.M. Shier, T. Watanabe, and R.E. Horita, Effects of geomagnetically induced currents in the B.C. Hydro 500 kV system, *IEEE Trans. Power Delivery*, **4**, 818, 1989

Boteler, D.H., R.J. Pirjola, and H. Nevanlinna, The effects of geomagnetic disturbances on electrical systems at the Earth's surface, *Adv. Space Res.*, **22**, 17, 1998

Boteler, D.H. and G.J. van Beek, August 4, 1972 revisited: A new look at the geomagnetic disturbance that caused the L4 cable system outage, *Geophys. Res. Lett.*, **26**, 577, 1999

Breit, M.A. and M.A. Tuve, A test of the existence of the conducting layer, *Nature*, **116**, 357, 1925

Brooks, J. A reporter at large: The subtle storm, *New Yorker*, February 19, 1959

Campbell, W.H., An interpretation of induced electrical currents in long pipelines caused by natural geomagnetic sources of the upper atmosphere, *Surveys Geophys.*, **8**, 239, 1986

Carrington, R.C., *Observations of the Spots on the Sun from November 9, 1853, to March 24, 1861*, Williams and Norgate, London, 167, 1861

Chapman, S., and J. Bartels, *Geomagnetism*, Oxford U. Press, 1941

Castelli, J.P., J. Aarons, D.A. Guidice, and R.M. Straka, The solar radio patrol network of the USAF and its application, *Proc. IEEE*, **61**, 1307, 1973

Clark, Arthur C., Extra-Terrestrial Relays – Can Rocket Stations Give World-Wide Radio Coverage? *Wireless World*, 305, October 1945

Cliver, E.W., Solar activity and geomagnetic storms, *Eos*, Trans. AGU, **75**, 569, 1994

Cliver, E.W., and L. Svalgaard, The 1859 solar–terrestrial disturbance and the current limits of extreme space weather activity, *Solar Phys.*, **224**, 407–422, 2005

Czech, P., S. Chano, H. Huynh, and A. Dutil, The Hydro-Quebec system blackout of 13 March 1989: System response to geomagnetic disturbance, *Proc. EPRI Conf. Geomagnetically Induced Currents,* **EPRI TR-100450**, Burlingame, CA, 19, 1992

Daglis, I.A., ed., *Space Storms and Space Weather Hazards*, Proceedings of Nato Advanced Study Institute, Springer, 2001

Daglis, I.A., ed., *Effects of Space Weather on Technological Infrastructure*, Kluwer, 2004

Davidson, W.F., The magnetic storm of March 2, 1940 – Effects in the power system, *Edison Electric Inst. Bulletin*, 365, 1940

DeForest, S.E., Spacecraft charging at synchronous orbit, *J. Geophys. Res.*, **77**, pp. 651–659, 1972

De la Beaujardiere, O., J.M. Retterer, M. Kelley, D. Hunton, and L. Jeong, Forecasting Low-Latitude Ionospheric Disturbances, *Space Weather*, 2005

Eccles, J.V., R.D. Hunsucker, D. Rice, J.J. Sojka, Space weather effects on midlatitude HF propagation paths: Observations and a data-driven D region model, *Space Weather*, Vol. 3, No. 1, S0100210.1029/2004SW000094, 2005

Fisher, G., Lessons From the U.S. Meteorological Public-Private Sector Services Partnership, *Space Weather*, **2**, No. 3, S0300510.1029/2003SW000061, 2004a

Fisher, G., Challenges Facing the U.S. Space Weather Public-Private Sector Partnership *Space Weather*, **2**, S04003, 10.1029/2003SW000060, 2004b

Forbush, S.E., Three unusual cosmic-ray increases possibly due to charged particles from the sun, *Phys. Rev.*, **70**, 771–772, 1946

Fukunishi, H., Y. Takahashi, M. Kubota, K. Sakanoi, U.S. Inan, and W.A. Lyons, Elves: lightning-induced transient luminous events in the lower ionosphere. *Geophs. Res. Lett.*, **23**, 2157–60, 1996

Futron Corp., Satellite insurance rates on the rise – market correction or overreaction? Bethesda, MD, 2002

Garrett, H.B., The charging of spacecraft surfaces, *Rev. Geophys.*, **19**, 577, 1981

Gary, D.E., and C.U. Keller, eds., *Solar and Space Weather Radiophysics*, Springer, Heidelberg, 2004

Gary, D.E., L.J. Lanzerotti, G.M. Nita, and D.J. Thomson, Effects of solar radio bursts on wireless systems, in *Effects of Space Weather on Technology Infrastructure*, ed. I.A. Daglis, Kluwer, 203–213, 2004

Germaine, L.W., The magnetic storm of March 24, 1940 – effects in the communication system, *Edison Electric Inst. Bulletin*, July, 1940

Getley, I.L., Observation of solar particle event on board a commercial flight from Los Angeles to New York on 29 October 2003, *Space Weather*, Vol. 2, No. 5, S05002 10.1029/2003SW000058, 2004

Getley, I.L., M.L. Duldig, D.F. Smart and M.A. Shea, Radiation dose along North American transcontinental flight paths during quiescent and disturbed geomagnetic conditions, *Space Weather*, Vol. 3, No. 1, S0100410.1029/2004SW000110, 2005

Grard, R., K. Knott, and A. Pedersen, Spacecraft charging effects, *Space Sci. Rev.*, **34**, pp. 289–304, 1983

Gilbert, J.L., Modeling the effect of the ocean-land interface on induced electric fields during geomagnetic storms, *Space Weather*, **3**, S04A03, doi:10.1029/2004SW000120, 2005

Gordon, G.D., and W.L. Morgan, *Principals of Communications Satellites*, John Wiley, New York, 178–192, 1993

Guidice, D.A., and J.P. Castelli, Spectral characteristics of microwave bursts, in *Proc. NASA Symp. High Energy Phenomena on the Sun*, Goddard Space Flight Center, Greenbelt, MD, 1972

Gussenhoven, M.S., and E.G. Mullen, Geosynchronous environment for severe charging, *J. Spacecraft Rockets*, **20**, 26, 1983

Hey, J.S., Solar radiations in the 4–6 metre radio wavelength band, *Nature*, **158**, 234, 1946

Hey, J.S., *The Evolution of Radio Astronomy*, Neale Watson Academic Pub., Inc., New York, 1973

Hunten, D.M., J.-C. Gerard, and L.M. Francois, The atmosphere's response to solar irradiation, in *The Sun in Time*, ed. C.P. Sonnet, M.S. Giampapa, and M.S. Matthews, Univ. Arizona Press, Tucson, 463, 1991

Kappenman, J.G., V.D. Albertson, and N. Mohan, Current transformer and relay performance in the presence of geomagnetically-inducded currents, *IEEE Trans. Power App. & Sys.*, **PAS-100**, 1078, 1981

Kappenman, J.G., The Evolving Vulnerability of Electric Power Grids, *Space Weather*, Vol. 2, S01004 10.1029/2003SW000028, 2004

Kappenman, J.G., Storm sudden commencement events and the associated geomagnetically induced current risks to ground-based systems at low-latitude and midlatitude locations, *Space Weather*, **1**, 1016 10.1029/2003SW000009, 2003

Kappenman, J.G., and W.A. Radasky, Too Important to Fail, *Space Weather*, **3**, S05001 10.1029/2005SW000152, 2005

Kelley, M., J. Makela, and O. de la Beaujardiere, Convective Ionospheric Storms: A Major Space Weather Problem, *Space Weather*, **4**, 503C05 10.1029/2005SW000144, 2006

Kelvin, Lord W.T., Proc. R. Soc. London A, **52**, 302, 1892

Kintner, P.M., B.M. Ledvina and E.R. de Paula, An amplitude scintillation test pattern standard for evaluating GPS receiver performance, *Space Weather*, Vol. **3**, No. 3, S03002 10.1029/2003SW000025, 2005

Koons, H.C., J.E. Mazur, R.S. Selesnick, J.B. Blake, J.F. Fennel, J.L. Roeder, and P.C. Anderson, *The Impact of the Space Environment on Space Systems*, Engineering and Technology Group, The Aerospace Corp., Report TR-99(1670), El Segundo, CA, 1999

Kundu, M.R., *Solar Radio Astronomy*, Interscience, New York, 1965

Lanzerotti, L.J., Geomagnetic induction effects in ground-based systems, *Space Sci. Rev.*, **34**, 347, 1983

Lanzerotti, L.J., and G.P. Gregori, Telluric currents: The natural environment and interactions with man-made systems, in *The Earth's Electrical Environment*, National Academies Press, Washington, D. C., 1986

Lanzerotti, L.J., D.J. Thomson, and C.G. Maclennan, Wireless at high altitudes – environmental effects on space-based assets, *Bell Labs Tech. J.*, **2**, 5, 1997

Lanzerotti, L.J., C. Breglia, D.W. Maurer, and C.G. Maclennan, Studies of spacecraft charging on a geosynchronous telecommunications satellite, *Adv. Space Res.*, **22**, 79, 1998

Lanzerotti, L.J., C.G. Maclennan, and D.J. Thomson, Engineering issues in space weather, in *Modern Radio Science*, ed. M.A. Stuchly, Oxford, 25, 1999

Lanzerotti, L.J., Space weather effects on technologies, in *Space Weather*, ed. P. Song, H. Singer, and G. Siscoe, American Geophysical Union, Washington, 11, 2001a

Lanzerotti, L.J., Space weather effects on communications, in *Space Storms and Space Weather Hazards*, Proc. NATO Advanced Study Institute on Space Storms and Space Weather hazards, ed. I.A. Daglis, Kluwer, 313, 2001b

Lanzerotti L.J., L.V. Medford, C.G. Maclennan, J.S. Kraus, J.A. Kappenman, and W. Radasky, Trans-Atlantic geopotentials during the July 2000 solar event and geomagnetic storm, *Solar Physics*, **204**, 351, 2001c

Lanzerotti, L.J., D.E. Gary, D.J. Thomson, and C.G. Maclennan, Solar radio burst event (6 April 2001) and noise in wireless communications systems, *Bell Labs Tech. J.*, **7**, 159, 2002

Lewis, B.J., M.J. McCall, A.R. Green, L.G.I. Bennett, M. Pierre, U.J. Schrewe, K. O'Brien, and E. Felsberger, Aircrew exposure from cosmic radiation on commercial airline routes, *Radiat. Prot. Dosimetry*, **93**, Append. 1, 293–314, 2001

Li, X., D.N. Baker, M. Temerin, G. Reeves, R. Friedel and C. Shen, Energetic electrons, 50 keV to 6 MeV, at geosynchronous orbit: Their responses to solar wind variations, *Space Weather*, Vol. **3**, No. 4, S0400110.1029/2004SW000105, 2005

Mandell., M., 120,000 leagues under the sea, *IEEE Spectrum*, **50**, April 2000

Marconi, G., Radio communication, *Proc. IRE*, **16**, 40, 1928

McBride, N., The importance of the annual meteoroid streams to spacecraft and their detectors, *Adv. Space Res.*, **20**, 1513, 1997

McBride, N., and J.A.M. McDonnell, Meteoroid impacts on spacecraft: sporadics, streams, and the 1999 Leonids, *Planet. Space Sci.*, **47**, 1005, 1999

McCormac, B.M., Summary, in *Radiation Trapped in the Earth's Magnetic Field*, ed. B.M. McCormac, D. Reidel Pub. Co., Dordrecht, Holland, 887, 1966

McPherson, D.A., and W.R. Schober, Spacecraft charging at high altitudes: The SCATHA satellite program, *Progress in Astronautics and Aeronautics*, Vol. **47**, pp. 15–30, MIT Press, 1976

Medford, L.V., L.J. Lanzerotti, J.S. Kraus, and C.G. Maclennan, Trans-Atlantic Earth potential variations during the March 1989 magnetic storm, *Geophys. Res. Lett.*, **16**, 1145, 1989

Michener, J., *Space*, Random House, 1982

Mizera, P.F., A summary of spacecraft charging results, *J. Spacecraft Rockets*, **20**, 438, 1983

Murtagh, W., L. Combs, and J. Kunches, A Workshop for the Aviation Community, *Space Weather*, **2**, S06004 10.1029/2004SW000081, 2004

National Research Council, *The Sun to the Earth – And Beyond*, National Academies Press, Washington, 2002

Nita, G.M., D.E. Gary, L.J. Lanzerotti, and D.J. Thomson, Peak flux distribution of solar radio bursts, *Astrophys. J.*, **570**, 423, 2002

Parsons, J.L., and L.W. Townsend, Interplanetary crew dose for the August 1972 solar particle event, *Radiation Research*, **153**, 729–733, 2000

Picholtz, R.L., Communications by means of low Earth orbiting satellites, in *Modern Radio Science 1996*, ed. J. Hamlin, Oxford U. Press, 133, 1996

Pierce, John R., Orbital radio relays, *Jet Propulsion*, p. 153, April 1955

Pirjola, R. and M. Lehinen, Currents produced in the Finnish 400 kV power transmission grid and in the Finnish natural gas pipeline by geomagnetically induced electric fields, *Annales Geophysicae*, **3**, 485, 1985

Pirjola, R., K. Kauristie, H. Lappalainen, A. Viljanen, and A. Pulkkinen, Space weather risk, *Space Weather*, **3**, S02A0210.1029/2004SW000112, 2005

Prescott, G., *Theory and Practice of the Electric Telegraph*, IV ed., Tichnor and Fields, Boston, 1860

Reid, E.J., How can we repair an orbiting satellite?, in *Satellite Communications Physics*, ed. R.M. Foster, Bell Telephone Laboratories, **78**, 1963

Rycroft, M.J., S. Israelsson, and C. Price. The global atmospheric electric circuit, solar activity, and climate change. *Journal of Atmospheric and Solar-Terrestrial Physics*, **62**, 1563–76, 2000

Scherer, K., H. Fichtner, B. Heber, and U. Mall, eds., *Space Weather, The Physics Behind a Slogan,* Lecture Notes in Physics, Vol. **656**, Springer, 2005

Sentman, D.D., and E.M. Wescott. Red sprites and blue jets: thunderstorm-excited optical emissions in the stratosphere mesosphere and ionosphere. *Physics of Plasmas*, 2(6):pt.2, 2514–22, 1995

Shea, M.A. and D.F. Smart, Space weather: The effects on operations in space, *Adv. Space Res.*, **22**, 29, 1998

Siscoe, G., Space weather forecasting historically through the lens of meteorology, in *Space Weather – Physics and Effects*, ed. V. Bothmer and I.A. Daglis, Springer-Praxis, p. 5, 2007.

Smart, D.F., M.A. Shea, and E.O. Flückiger, Magnetospheric Models and Trajectory Computations, *Space Sci. Rev.*, **93**, 305–333, 2000

Smart, D.F., and M.A. Shea, World Grid of Cosmic Ray Vertical Cutoff Rigidities for Epoch 1990.0, *25th International Cosmic Ray Conference*, Contributed Papers, **2**, 401–404, 1997

Smart, D.F., and M.A. Shea, A review of geomagnetic cutoff rigidities for earth-orbiting spacecraft, *Adv. Space Res.*, **36**, 2012, 2005

Song, P., H.J. Singer, and G.L. Siscoe, ed., *Space Weather*, Geophysical Monograph 125, American Geophysical Union, Washington, 2001

Soon, W.W.-H., and S.H. Yaskell, *The Maunder Minimum and the Variable Sun-Earth Connection*, World Publishing, Singapore, 2003

Southworth, G.C., Microwave radiation from the Sun, *J. Franklin Inst.*, **239**, 285–297, 1945

Spurný, F., K. Kudela, and T. Dachev, Airplane radiation dose decrease during a strong Forbush decrease, *Space Weather*, **2**, S05001, doi:10.1029/2004SW000074, 2004

Stansbery, E., J. Foster, and C. Stokely, Haystack orbital debris radar measurements update, *The Orbital Debris Quarterly News*, **9**, 3, http://www.orbitaldebris.jsc.nasa.gov/newsletter/pdfs/ODQNv9i1.pdf, January 2005

Stassinopoulos, E.G., C.A. Stauffer and G.J. Brucker, A systematic global mapping of the radiation field at aviation altitudes, *Space Weather*, **1**, S01005 10.1029/2003SW000011, 2003

Stewart, B., On the great geomagnetic disturbance which extended from August 28 to September 7, 1959, as recorded by photography at the Kew Observatory, *Phil. Trans. Roy. Soc. London*, **151**, 423, 1861

Taur, R.R., Ionospheric scintillations at 4 and 6 GHz, *COMSAT Tech. Rev.*, **3**, 145, 1973

Taylor, G.C., R.D. Bentley, T.J. Conroy, R. Hunter, J.B.L. Jones, A. Pond, and D.J. Thomas, The evaluation and use of a portable TEPC system for measuring in-flight exposure to cosmic radiation, *Radiat. Prot. Dosimetry*, **99**, 435–438, 2002

Todd, D., Letter to *Space News*, pg. 12, March 6, 2000

Townsend, L.W., J.W. Wilson, J.L. Shinn, and S.B. Curtis, Human Exposure to Large Solar Particle Events in Space, *Adv. Space Res.*, **12**, 339–348, 2002

Tribble, A.C., *The Space Environment, Implications for Spacecraft Design*, Princeton U. Press, Princeton, N.J., 1995

Vampola, A., The aerospace environment at high altitudes and its implciaitons for spacecraft charging and communications, *J. Electrost.*, **20**, 21, 1987

Van Allen, J.A., Ludwig, G.H., Ray, E.C., and Mcllwain, C.E., Observation of high intensity radiation by satellites 1958 α and β, *Jet Propulsion* **28**, 588, 1958

Van Allen, J.A., *Origins of Magnetospheric Physics*, Smithsonian Institution, Washington, 1983

Vernov, S.N., and A.E. Chudakov, Terrestrial corpuscular radiation and cosmic rays, *Adv. Space Res.*, **1**, 751, 1960

Viljanen, A., Geomagnetically induced currents in the Finnish natural gas pipeline, *Geophysica*, **25**, 135, 1989

Webb, D.F., and J.H. Allen, Spacecraft and ground anomalies related to the October–November 2003 solar activity, *Space Weather*, **2**, S03008, doi:10.1029/2004SW000075, 2004

Wilson, J.W., F.A. Cucinotta, J.L. Shinn, L.C. Simonsen, R.R. Dubey, W.R. Jordan, T.D. Jones, C.K. Chang, and M.Y. Kim, Shielding From Solar Particle Events in Deep Space, *Radiat. Meas.*, **30**, 361–392, 1999

Winckler, J.R., L. Peterson, R. Hoffman, and R. Arnoldy, Auroral X-rays, cosmic rays, and related phenomena during the storm of Feb. 10-11, 1958, *J. Geophys. Res.*, **64**, 597, 1959

Winckler, J.R., W.A. Lyons, T.E. Nelson, and R.J. Nemzek. New high-resolution ground-based studies of sprites. *J. Geophy. Res*,**101**,6997-7004, 1996

Yeomans, D.K., K.K. Yau, and R.R. Weisman, The impending appearance of comet Tempel-Tuttle and the Leonid meteors, *Icarus*, **124**, 407, 1996

18 Effects of the Solar Cycle on the Earth's Atmosphere

Karin Labitzke

Until recently it was generally doubted that the solar variability in the "11-year sunspot cycle" (SSC), as measured by satellites, has a significant influence on weather and climate variations. But several studies, both empirical and modelling, have in recent years pointed to probable and certain influences. For instance, Labitzke suggested in 1982 that the sun influences the intensity of the north polar vortex (i.e., the Arctic Oscillation (AO)) in the stratosphere in winter, and that the Quasi-Biennial Oscillation (QBO) is needed to identify the solar signal. At present there is no agreement about the mechanism or mechanisms through which the solar variability effect is transmitted to the atmosphere. But there is general agreement that the direct influence of the changes in the UV part of the solar spectrum (6 to 8% between solar maxima and minima) leads to more ozone and warming in the upper stratosphere (around 50 km) in solar maxima. This leads to changes in the vertical gradients and thus in the wind systems, which in turn lead to changes in the vertical propagation of the planetary waves that drive the global circulation. Therefore, the relatively weak, direct radiative forcing of the solar cycle in the stratosphere can lead to a large indirect dynamical response in the lower atmosphere.

Contents

Karin Labitzke, Effects of the Solar Cycle on the Earth's Atmosphere.
In: Y. Kamide/A. Chian, Handbook of the Solar-Terrestrial Environment. pp. 445–466 (2007)
DOI: 10.1007/11367758_18

18.1 Introduction

Nearly all the earth's energy derives from the sun, and it is therefore natural to look for links between variations in the sun's irradiance and changes in the atmosphere and oceans. One of the first attempts to measure the total solar radiation (the "solar constant") for this purpose was made by C.G. Abbot (1913) and, despite many difficulties, he succeeded in obtaining a mean value of the solar constant for the period 1902–1912. Abbot et al. (1913) showed that there is also a change in the radiation from maximum to minimum in the "11-year sunspot cycle" (SSC) in the sense that less radiation was emitted in solar minimum than in maximum of the oscillation, despite the greater spottedness of the sun in the maximum.

In this chapter we are dealing with the 11-year SSC and its influence on the Earth's atmosphere. For studies on longer time scales the reader is referred to the works of Beer et al. (2000), Cubasch and Voss (2000), Reid (2000) and Langematz et al. (2005), among others.

In 1978 the first satellite observations of total solar radiation began; qualitatively the satellite observations confirm Abbot's result that the values are higher in the solar maxima, but the variation from maxima to minima within the 11-year solar cycle is very small (0.1% difference between the extremes) in the satellite data (Fröhlich, 2000). The satellite observations of the total solar irradiance included the variability of the ultraviolet radiation; the variability of this quantity is considerably larger than that of the total solar radiation: 6 to 8% in those wavelengths in the ultraviolet (200 to 300 nm) that are important in the production of ozone and middle atmosphere heating (Chandra and McPeters, 1994; Haigh, 1994; Hood, 2003, 2004; Lean et al., 1997). This is the most likely mechanism through which the changes in the sun's radiation can influence the atmosphere; see discussion in Sect. 18.7.

Until recently it was generally doubted that the solar variability in the SSC, as measured by satellites, has a significant influence on weather and climate variations (see, e.g., Pittock's review (1983) and Hoyt and Schatten, 1997). But several studies, both empirical and modeling, have in recent years pointed to probable and certain influences. For instance, Labitzke (1982) suggested that the sun influences the intensity of the north polar vortex (i.e., the Arctic Oscillation (AO)) in the stratosphere in winter, and that the Quasi-Biennial Oscillation (QBO, see below) is needed to identify the solar signal (e.g., Labitzke, 1987; Labitzke and van Loon, 1988, 2000; Salby and Callaghan, 2000; Ruzmaikin and Feynman, 2002). Kodera (2004), and van Loon et al. (2004) show strong connections between the SSC and important characteristics of the tropical oceans and the lower atmosphere over them.

At present there is no agreement about the mechanism or mechanisms through which the solar variability effect is transmitted to the atmosphere. The correlations between cosmic rays and clouds (e.g., Svensmark and Friis-Christensen (1987), Udelhofen and Cess (2001), and Kristjánsson et al. (2004) are still a matter of debate, see De Jager and Usoskin (2006).

In Sect. 18.2 we explain which data and methods were used to analyze the solar variability signal in the observations and in Sect. 18.3 we describe the variability in the stratosphere and troposphere against which the influence of solar variability must be measured. Then follows, in Sect. 18.4, a summary of our diagnostic studies of the solar variability effect during the past 50 years, supplemented by the results of similar studies by others. Section 18.5 gives a short overview of studies of the solar variability signal in the troposphere, and in Sect. 18.6 the importance of the QBO throughout the year is discussed.

In Sect. 18.7, we present proposed mechanisms and some modeling experiments and a Summary is given in Sect. 18.8.

18.2 Data and Methods

In addition to the FU-Berlin analyses of the Northern Hemisphere stratosphere, which start in 1956 and terminate in 2001 (Labitzke and Collaborators, 2002), the **global** re-analyses by NCEP/NCAR (Kalnay et al., 1996) are used, mainly for the period 1968–2004 (except for Figs. 18.5, 18.6, and 18.12 where the data start in 1958). The re-analyses are less reliable for earlier periods, mainly because of the lack of radiosonde stations over the Southern Hemisphere, the lack of high reaching balloons in the early years and the scarce satellite information before 1979. However, we note that the inclusion of the early data nevertheless yields similar results (Labitzke et al., 2006).

The monthly mean values of the 10.7 cm solar flux are used as a proxy for variations through the SSC. The flux values are expressed in solar flux units: 1 s.f.u. = 10^{-22} W m^{-2} Hz^{-1}. This is an objectively measured radio wave intensity, highly and positively correlated with the 11-year SSC and particularly with the UV part of the solar spectrum (Hood, 2003).

For the *range* of the SSC, the mean difference of the 10.7 cm solar flux between solar minima (about 70 units) and solar maxima (about 200 units) is used, i.e., 130 units. Any linear correlation can be represented also by a regression line with $y = a + bx$, where x in this case is the 10.7 cm solar flux and b is the slope. This slope is used here, multiplied by 130, in order to get the differences between solar minima and maxima (Labitzke, 2003).

It is difficult to determine the statistical significance of the correlations because we have often less than four solar cycles and the number of degrees of freedom are therefore limited. However, using the same data Ruzmaikin and Feynman (2002) as well as Salby and Callaghan (2004) found a high statistical significance of their results, and similar to ours; (see also Labitzke et al., 2006).

The QBO is an oscillation in the atmosphere which is best observed in the stratospheric winds above the equator, where the zonal wind direction changes between east and west (see below). The period of the QBO varies in space and time, with an average value near 28 months at all levels; see reviews by Naujokat (1986) and Baldwin et al. (2001). Because the QBO modulates the solar variability signal, it is necessary to stratify the data into years for which the equatorial QBO in the lower stratosphere (at about 45 hPa, i.e. about 22 km altitude), (e.g., Holton and Tan, 1980) was in its westerly or easterly phase (QBO data set in Labitzke and Collaborators, 2002).

18.3 Variability in the Stratosphere

The arctic stratosphere reaches its highest variability in winter. Figure 18.1 gives an example of the variability of the stratosphere during the northern winter (February) and summer (July). It is remarkable that in the lower and middle stratosphere the standard deviations in the arctic winter are three to four times larger than those in the antarctic winter; this is due to the fact that the Major Mid-Winter Warmings which create the large variability of the Arctic, do usually not penetrate to the lower stratosphere over the Antarctic. But the variability is large in the upper stratosphere over the Antarctic where so-called Minor Mid-Winter Warmings occur frequently, (Labitzke and van Loon, 1972).

When the antarctic westerly vortex breaks down in spring (September–November) the middle stratosphere varies so much from one spring to another that the standard deviation at the South Pole in October (Labitzke and van Loon, 1999, their Fig. 2.11) approaches that at the North Pole in January and February. In summer the variability is low in both hemispheres, below 1 K. A relative maximum of variability is observed on the equator due to the QBO.

In Fig. 18.2, 30-hPa temperatures at the North Pole in **February** since 1956 are plotted; this shows how the large standard deviation over the Arctic comes about. Out of the 50 years, 7 years are about one standard deviation below the average and 7 years are about one standard deviation above. In most of these warm years Major Mid-Winter Warmings occurred (see Labitzke and van Loon, 1999), and these warmings are associated with a breakdown of the cold westerly vortex. It is of great interest to note that the overall **temperature trend** is almost zero – (see, e.g., discussions about temperature trends in the Arctic by Pawson and Naujokat (1999) and Labitzke and Kunze (2005)) – and that the correlation with the solar cycle is small.

The state of the arctic westerly vortex in northern winter is influenced by several factors (van Loon and Labitzke, 1993):

- The **QBO**, Fig. 18.3, consists of downward propagating west and east winds in the stratosphere with an average period of about 28 month; this pattern is centered on the equator. A historical review and the present explanation of the QBO can be found in Labitzke and van Loon (1999). The QBO modulates the arctic and also the antarctic polar vortex, (Labitzke, 2004a) but this modulation changes sign depending on the phase in the solar cycle, see Sect. 18.4.1 and 18.4.2.
- Another quantity whose effect is felt in the stratosphere is the **Southern Oscillation (SO)**. The SO is defined as a "see-saw" in atmospheric mass (evidenced by sea-level pressure) between the Pacific Ocean and the Australian-Indian region, (see, e.g., Labitzke and van Loon, 1999). Its influence is global and reaches into the stratosphere. The anomalies

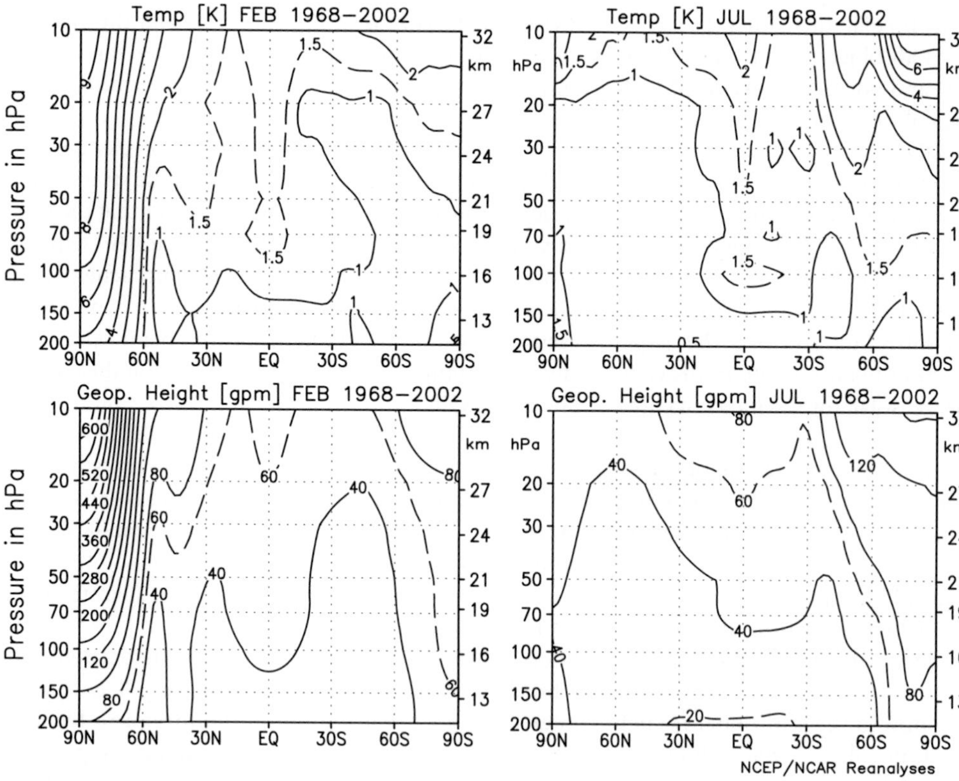

Fig. 18.1. Vertical meridional sections of the standard deviations for February (*left*) and July (*right*) of the zonal mean monthly mean temperatures (K), *upper panels*, and of the zonal mean monthly mean geopotential heights (geopot. m), *lower panels*, for the period 1968–2002 (NCEP/NCAR re-analyses)

in the lower stratosphere associated with extremes of the SO are described in van Loon and Labitzke (1987), where they are discussed in terms of other influences such as the QBO and volcanic eruptions. In the warm extremes of the SO the stratospheric temperatures and heights at arctic latitudes are most of the time well above normal (about 1 standard deviation), and conversely in the cold extremes.

- The stratosphere is also influenced by different types of waves which penetrate under certain conditions from the troposphere to the stratosphere. These are mainly the very large-scale planetary waves (with horizontal wavelengths between 5 and 10 thousand km). Further, different types of the so-called gravity waves reach the stratosphere where they deposit their momentum. The heights which they reach, depend on the vertical profiles of the zonal winds.
- Tides are very important for the dynamics in the upper stratosphere and mesosphere. They develop in these regions mainly through thermal forcing of the rotating earth's atmosphere by the sun.
- Yet another influence on the stratosphere is **solar variability** which until recently received little attention. The influence of the 11-year SSC will be discussed in the next section.

18.4 Influences of the 11-Year Sunspot Cycle on the Stratosphere

18.4.1 The Stratosphere During the Northern Winter

Based on results published in 1982, Labitzke (1987) found that a signal of the 11-year SSC emerged when the arctic stratospheric temperatures and geopotential

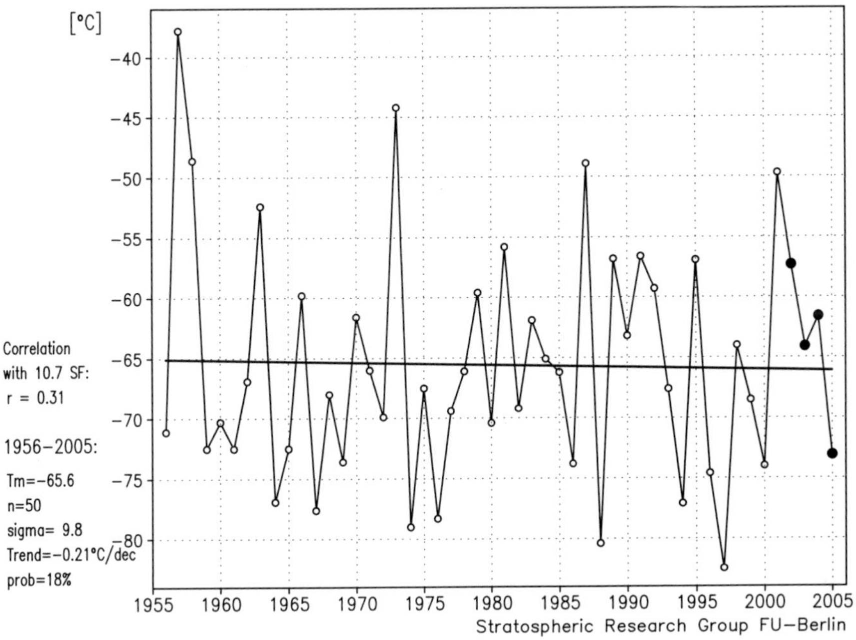

Fig. 18.2. Time series of monthly mean 30-hPa temperatures (°C) at the North Pole in **February**, 1956–2005. Both, a linear trend and the correlation with the 11-year solar cycle have been computed. (Data: Meteorological Institute, Free University Berlin until 2001 (Labitzke and Collaborators, 2002); then ECMWF.)

heights were grouped into two categories determined by the direction of the equatorial wind in the stratosphere (QBO), Fig. 18.3. The reality and significance of using this approach have been confirmed by Naito and Hirota (1997), Salby and Callaghan (2000, 2004, 2006) and Ruzmaikin and Feynman (2002).

An example of this approach is given in a scatter diagram (Fig. 18.4) for the 30-hPa heights over the North Pole in February when the modulation of the solar signal by the QBO is at its maximum. The correlations between the 30-hPa heights and the solar cycle are shown, with the winters in the east phase of the QBO in the left part of the figure, and the winters in the west phase of the QBO in the right part. The abscissa indicates the SSC. The correlations are clearly very different in the two groups, with negative correlations over the Arctic in the east phase of the QBO and large positive correlations there in the west phase. (The correlation for all years is 0.1, not shown.) The numbers in the scatter diagrams are the years of the individual Februarys. The extremes of the SO are marked, as well as the three eruptions of tropical volcanoes. Obviously, the volcanoes do not influence the correlations, but appear to be connected with Warm Events (WE) of the SO, without leading to a warm stratosphere (Labitzke and van Loon, 1989). As mentioned above, there is a tendency for the WE (El Niños) of the SO to be connected with a warm polar stratosphere (greater heights) and a weak polar vortex, and conversely so in the cold extremes of the SO (van Loon and Labitzke, 1987).

While the data shown in Fig. 18.4 are only for the atmosphere above the North Pole, Fig. 18.5 shows on the left hand side the correlations for the whole Northern Hemisphere, with the winters in the east phase of the QBO in the upper part of the figure, and the winters in the west phase of the QBO in the lower part. Again, the pattern of correlations is clearly very different in the two groups, with negative correlations over the Arctic in the east phase and large positive correlations there in the west phase. Outside of the Arctic the correlations are positive and strong in the east phase, but very weak in the west phase.

The respective height differences between solar maxima and minima are given on the right hand side of Fig. 18.5. In the east phase of the QBO, because the stratosphere is colder, the heights tend to be below normal over the Arctic in solar maxima (about one standard deviation, see Fig. 18.1); they are above normal towards the equator. In the west phase, the arctic heights tend to be well above normal (about two standard deviations) in solar maxima; there are only very small anomalies outside the Arctic.

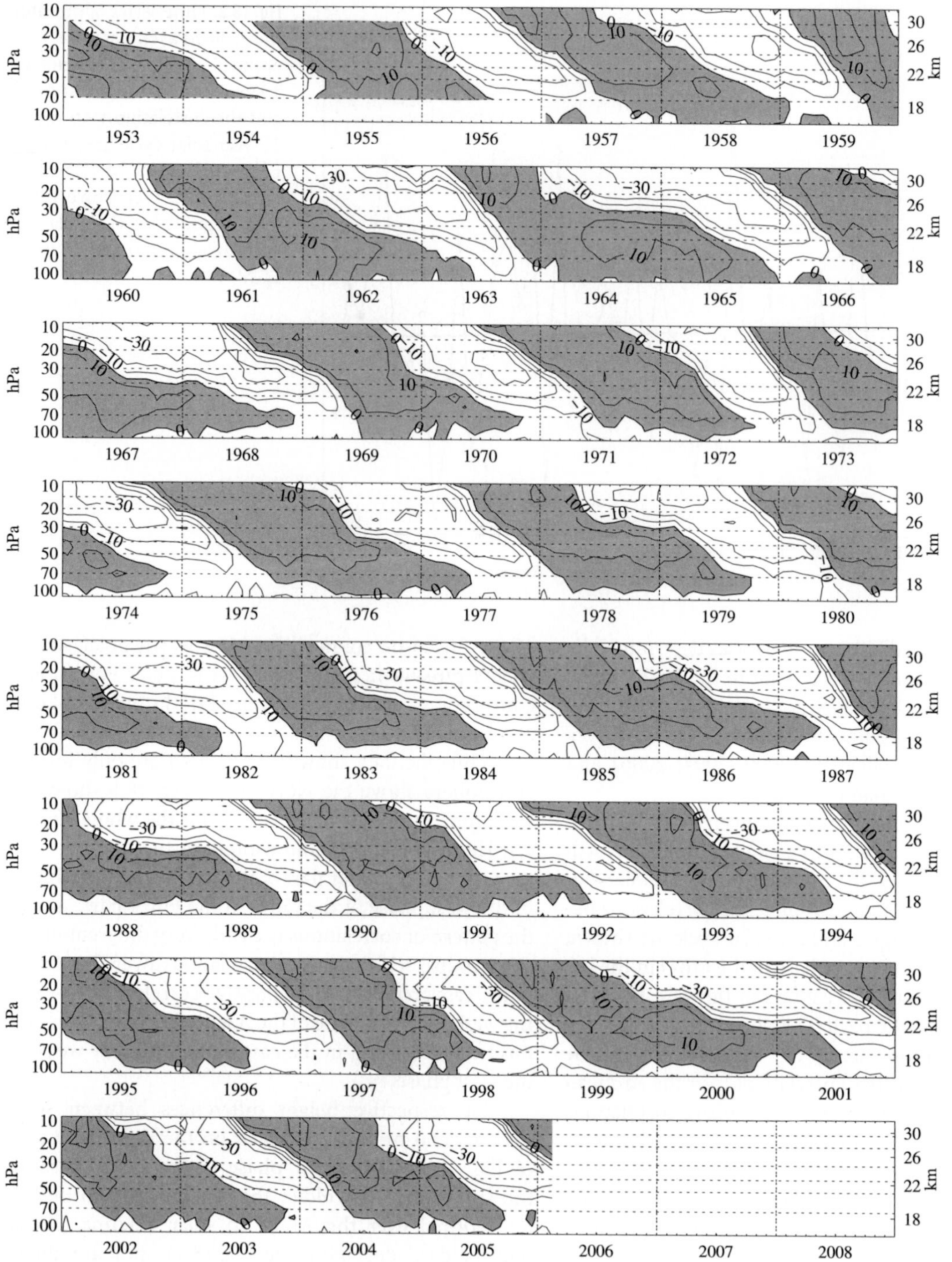

Fig. 18.3. Time-height section of monthly mean zonal winds (m/s) at equatorial stations: Canton Island, 3°S/172°W (Jan 1953–Aug 1967), Gan/Maledive Islands, 1°S/73°E (Sep 1967–Dec 1975) and Singapore, 1°N/104°E (since Jan 1976). Isopleths are at 10 m/s intervals; winds from the west are *shaded* (updated from Naujokat, 1986)

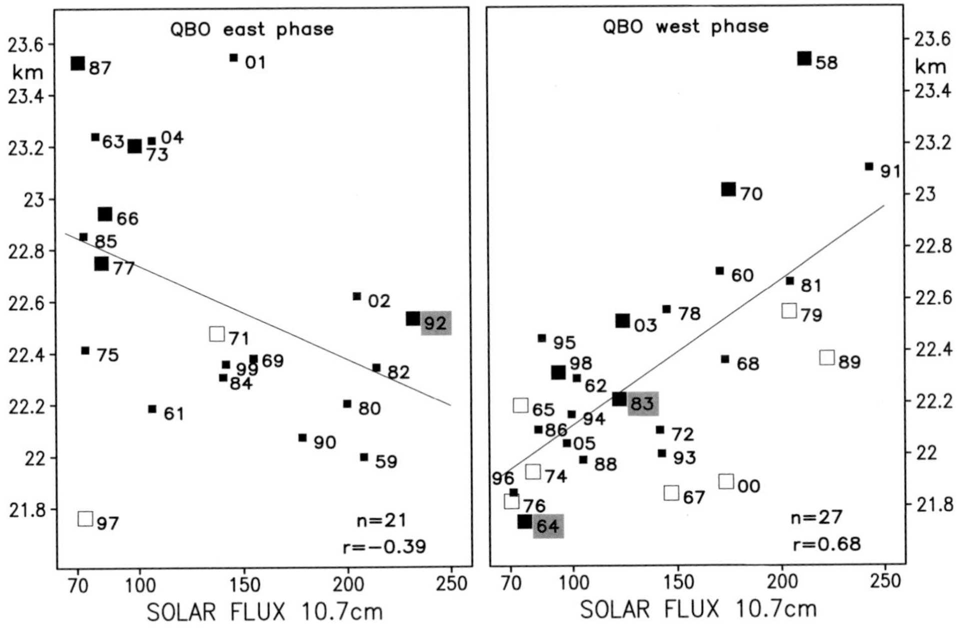

Fig. 18.4. Scatter diagrams of the monthly mean 30-hPa geopotential heights (km) in **February** at the North Pole plotted against the 10.7 cm solar flux. *Left*: years in the east phase of the QBO ($n = 21$); *right*: years in the west phase ($n = 27$). The *numbers* indicate the respective years, *shaded* are 3 Februarys after large volcanic eruptions. *Large filled symbols* indicate Warm Events, *large open symbols* indicate Cold Events, respectively, in the Southern Oscillation; r = correlation coefficient. Data: FU-Berlin 1958–2001; until 2005 ECMWF (van Loon and Labitzke (1994), updated)

Figure 18.6 shows a vertical meridional section of correlations between the solar 10.7 cm flux and zonally averaged temperatures, as well as the corresponding temperature differences between solar maxima and minima. When all years are used in February, the correlations and the corresponding temperature differences (top left and right, respectively) are small. But, in the east phase of the QBO, the correlations of the zonally averaged temperatures with the solar data are positive from 60°N to the South Pole in the summer hemisphere, and negative north of 60°N, in the winter hemisphere. On the right hand side in the middle panel are the zonally averaged temperature differences between solar maxima and minima in the east phase of the QBO which correspond to the correlations on the left side; the shading is the same as that in the correlations where it denotes correlations above 0.4.

In the west phase of the QBO (Fig. 18.6, bottom), the correlations with the solar 10.7 cm flux are highly positive over the Arctic (+0.7) and near zero or weakly negative elsewhere. The large positive correlations are associated with the frequent Major Mid-Winter Warmings which occur when the QBO is in the west phase at solar maxima (e.g., van Loon and Labitzke, 2000). The arctic temperatures and heights in the stratosphere are then determined by strong subsidence. Outside the Arctic the lower latitudes are expected to warm at solar maximum but, because of the subsidence and warming in the Arctic, the warming to the south is dynamically counterbalanced by a rising motion, and cooling, well into the southern (summer) hemisphere.

The height and temperature changes shown in Figs. 18.5 and 18.6 indicate that the solar cycle influences the "Mean Meridional Circulation (MMC)", also called the "Brewer–Dobson Circulation (BDC)". Forced by planetary waves the MMC regulates wintertime polar temperatures through downwelling and adiabatic warming (Kodera and Kuroda, 2002; Kuroda and Kodera, 2002; Hood and Soukharev, 2003; Labitzke, 2003; Hood, 2004; Salby and Callaghan, 2004).

During the **west phase** of the QBO the MMC is **intensified** during solar maxima (and *vice versa*

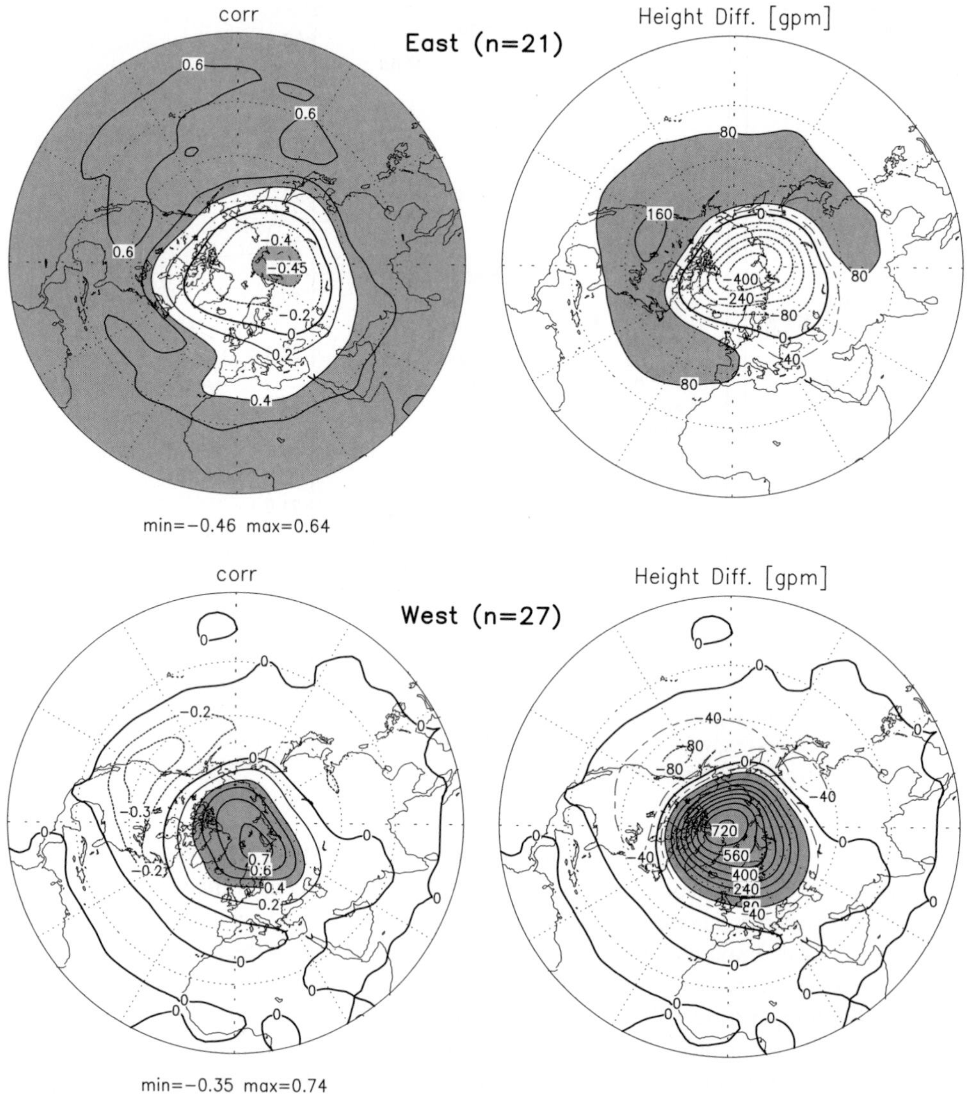

Fig. 18.5. QBO effects in the Northern Hemisphere. *Left*: correlations between the 10.7 cm solar flux (representing the 11-year solar cycle) and detrended 30-hPa heights in **February**, *shaded* for emphasis where the correlations exceed 0.4. In the *upper panel*: data for years in the east phase of the QBO are presented. In the *lower panel*, data for years in the west phase of the QBO are given. *Right*: respectively, height differences (geopot. m) between solar maxima and minima. (NCEP/NCAR re-analyses, period: 1958–2005); (Labitzke (2002, Fig. 7), updated)

during solar minima), with large positive anomalies over the Arctic (intensified downwelling and warming), and concurrent weak anomalies (anomalous upwelling/adiabatic cooling) over the tropics and subtropics, as shown in the lower maps in Fig. 18.5 and the lowest panels in Fig. 18.6. During the **east phase** the MMC is **weakened** in solar maxima, with reduced downwelling (anomalous upwelling/cooling) and negative anomalies over the Arctic in solar maxima, and concurrent anomalous downwelling with positive anomalies over the tropics and subtropics.

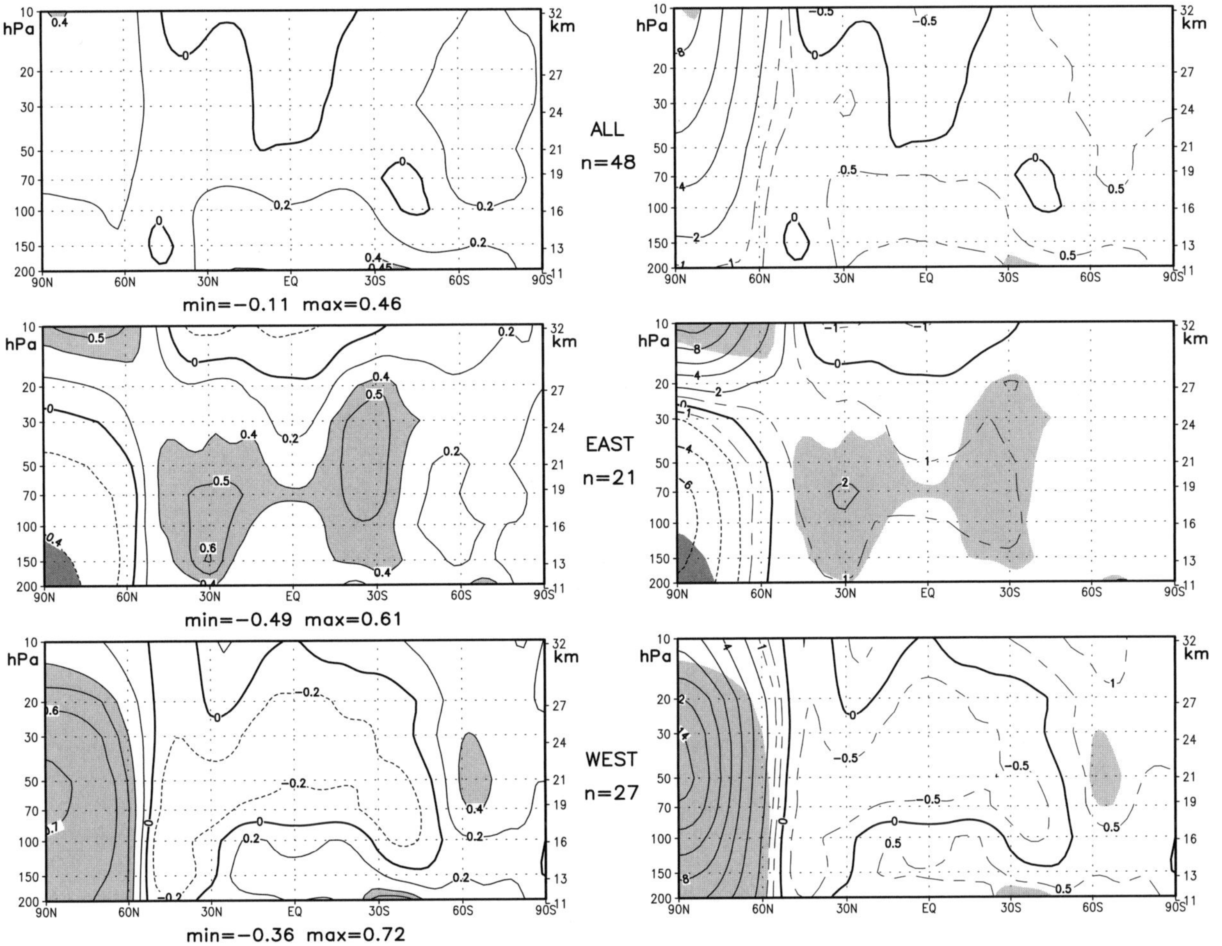

Fig. 18.6. Vertical meridional sections between 200 and 10 hPa (about 11 and 32 km) of, on the *left*, the correlations between the detrended zonally averaged temperatures for **February** and the 10.7 cm solar flux (*shaded* for emphasis where the correlations are larger than 0.4), and, on the *right*, the respective temperature differences (K) between solar maxima and minima, *shaded* where the corresponding correlations on the *left hand side* are above 0.4. The *upper panels* show all years, the *middle panels* only years in the east phase of the QBO, and the *lower panels* only years in the west phase of the QBO (NCEP/NCAR re-analyses, 1958–2005). (Labitzke (2002), updated)

18.4.2 The Stratosphere During the Northern Summer

The interseasonal shift between hemispheres of the solar variability – stratosphere relationship is evident in Fig. 18.7: the curves on the left hand side in this figure show the correlations between the 30-hPa zonally averaged temperature in May–August, the four months centered on the northern summer solstice (dashed line). The biggest correlations, above 0.4, lie between 10°N to 50°N, and a secondary peak is found at 15°S. This picture reverses in the four months from November–February, which are centered on the southern solstice (solid line), when the largest correlations are found between 15°S and 65°S, and a secondary peak is found at 15°N. The temperature differences between solar maxima and minima at about 24 km altitude are given on the right hand side of Fig. 18.7. They are almost everywhere positive, with the largest differences (more than 1 K) over the summer hemispheres.

The correlations on the left in Fig. 18.8 show for July the vertical distribution of the solar relationship from the upper troposphere to the middle stratosphere.

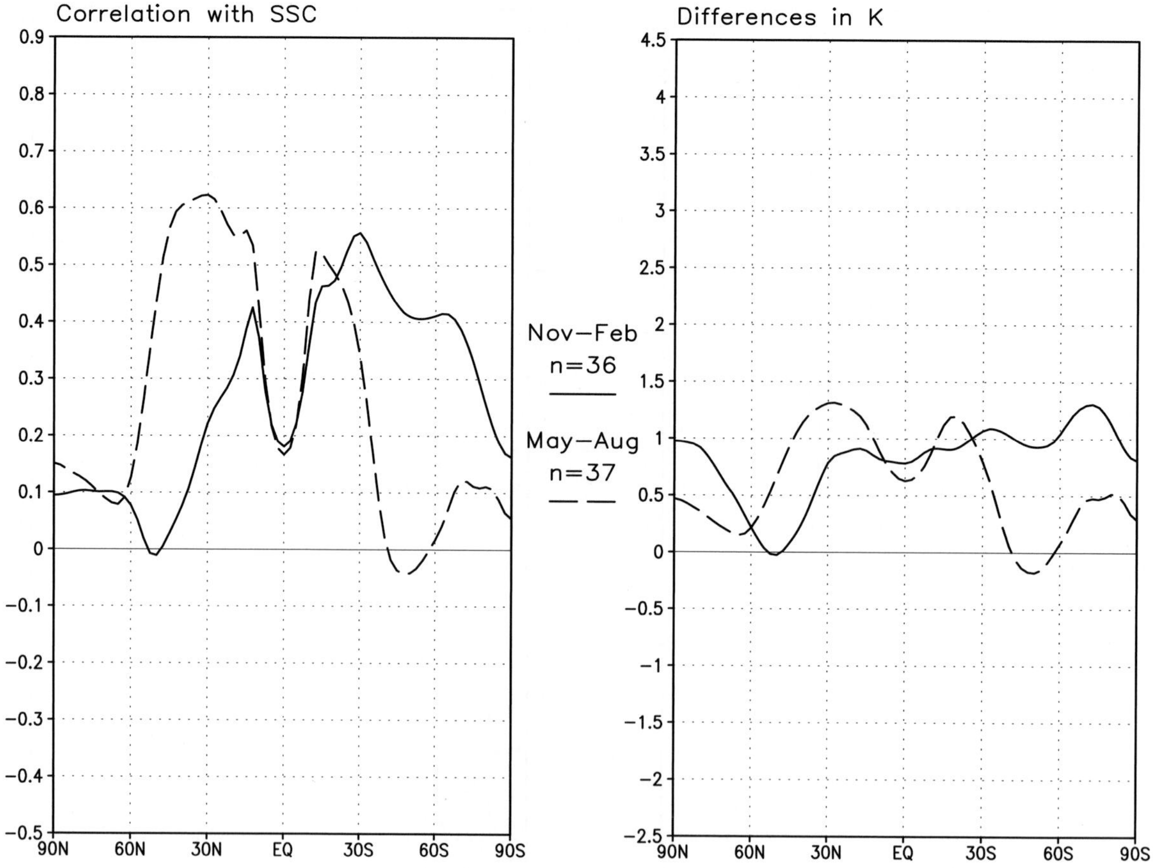

Fig. 18.7. *Left*: correlations between the 10.7 cm solar flux and the detrended zonally averaged 30-hPa temperatures in May–June–July–August (*dashed line*) and November–December–January–February (*solid line*)(van Loon and Labitzke (1999, Fig. 9), updated.) *Right*: The respective temperature differences (K) between solar maxima and solar minima (NCEP/NCAR re-analyses, 1968–2004)

Again, the data are grouped in all years and according to the east and west phases of the QBO; the corresponding temperature differences are on the right in the diagram. The results for the east phase (middle panels) are most striking: two centers with correlations above 0.8 are found over the subtropics between 20 and 30 hPa; further down, the double maximum in the east phase changes into one maximum, centered on the equatorial tropopause (Labitzke 2003, Fig. 18.4). The temperature differences (right hand side) between solar maxima and minima are large, more than two standard deviations in some regions. This warming, i.e. positive anomalies, can only be explained by downwelling over the subtropics and tropics (Kodera and Kuroda, 2002; Shepherd, 2002) which – in other words – means a weakening of the BDC for solar maxima/east phase of the QBO, as discussed above for the northern winter.

The solar variability signal is much weaker in the west phase years. It hints at an intensification of the Hadley Circulation (HC) over the Northern Hemisphere, with stronger rising motions over the equator (warming due to latent heat release) and some anomalous heating (downwelling) over the subtropics of the northern summer hemisphere.

Spatially, the interseasonal movement in Fig. 18.7 is illustrated in Fig. 18.9 for the northern summer (July). At the top of the figure and for the period 1968–2004 the subtropical to mid-latitude peak dominates at all longitudes in the Northern Hemisphere, and the secondary peak in the Southern Hemisphere spans that

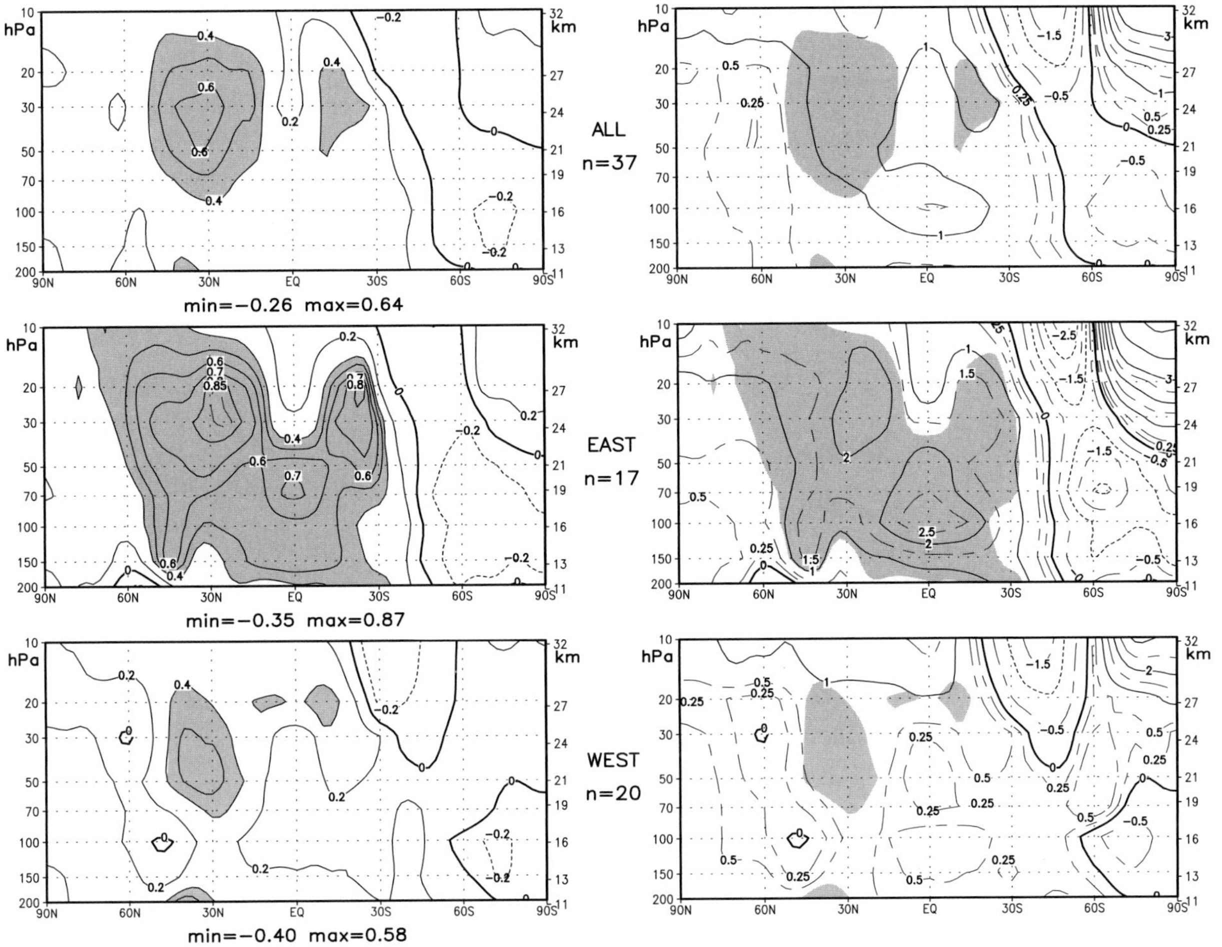

Fig. 18.8. Vertical meridional sections between 200 and 10 hPa (11 to 32 km) of, on the *left*, the correlations between the detrended zonally averaged temperatures for **July** and the 10.7 cm solar flux (*shaded* for emphasis where the correlations are larger than 0.4) and, on the *right*, the respective temperature differences (K) between solar maxima and minima, *shaded* where the corresponding correlations on the *left hand side* are above 0.4. The *upper panels* show all years, the *middle panels* only years in the east phase of the QBO, and the *lower panels* only years in the west phase of the QBO (NCEP/NCAR re-analyses, 1968–2004). (Labitzke (2003, Fig. 4), updated)

hemisphere. However, when the data are divided into the east and west phases of the QBO, the picture is different. Originally we made this division according to the phase of the QBO only in the winter data. However, it turns out that it is also a valid approach for the rest of the year (Labitzke, 2003, 2004b, 2005).

In the east phase of the QBO (Fig. 18.9 middle) both the major and the minor peaks are accentuated, whereas in the west phase of the QBO the solar variability relationship is weaker. In other words, *the correlations for all years in the top panel are dominated by the QBO/east years.*

This is further emphasized with the time series shown in Fig. 18.10. The point above the Gulf of Mexico was chosen for the high correlation in the east years, with $r = 0.87$ at 25°N/90°W, while r equals only +0.36 in the west years. In the east phase the temperature difference between solar maxima and minima is 2.6 K, which is more than 2 standard deviations, as shown in Fig. 18.1.

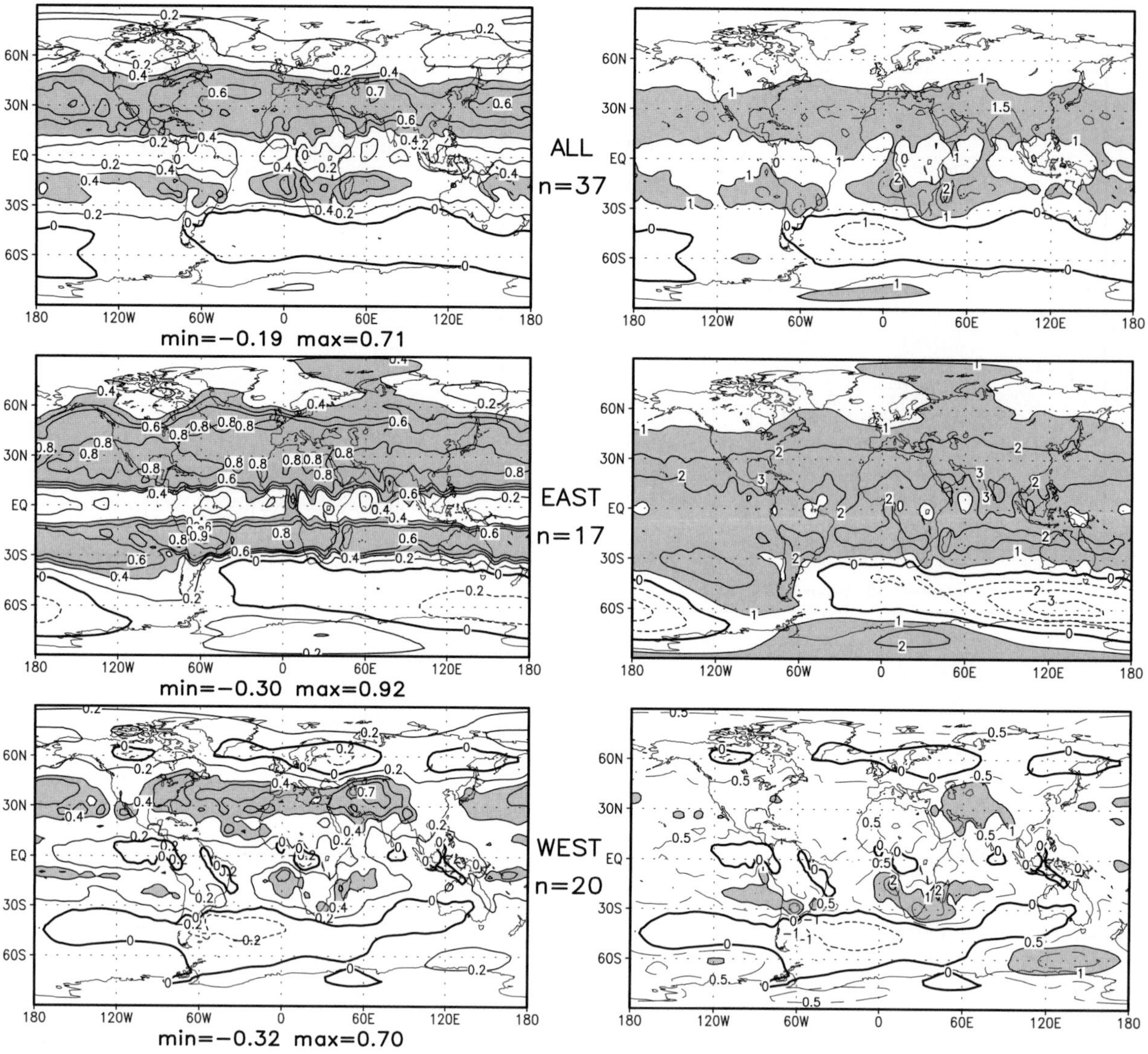

Fig. 18.9. *Left*: correlations between the 10.7 cm solar flux and the detrended 30-hPa temperatures in **July**, *shaded* for emphasis where the correlations are above 0.4. *Right*: the respective temperature differences (K) between solar maxima and minima, *shaded* where the differences are above 1 K. The *upper panels* show all years, the *middle panels* only years in the east phase of the QBO, and the *lower panels* only years in the west phase of the QBO (NCEP/NCAR re-analyses, 1968–2004). (Labitzke (2003, Fig. 1), updated)

Figure 18.11 shows the 30-hPa height differences (between solar maxima and minima) over the subtropics and tropics in July: again, the east phase dominates the solar signal (top of figure). In addition, the anomalous zonal (west-east) wind in the equatorial belt is affected by the solar variability on the decadal scale. At the top of the figure an anomalously high value is centered over the equator. It means that an anomalous anticyclonic circulation is centered on the equator in the solar maximum east years, connected with anomalous winds from the west. Therefore, during solar maxima in QBO/east years the low-latitude east

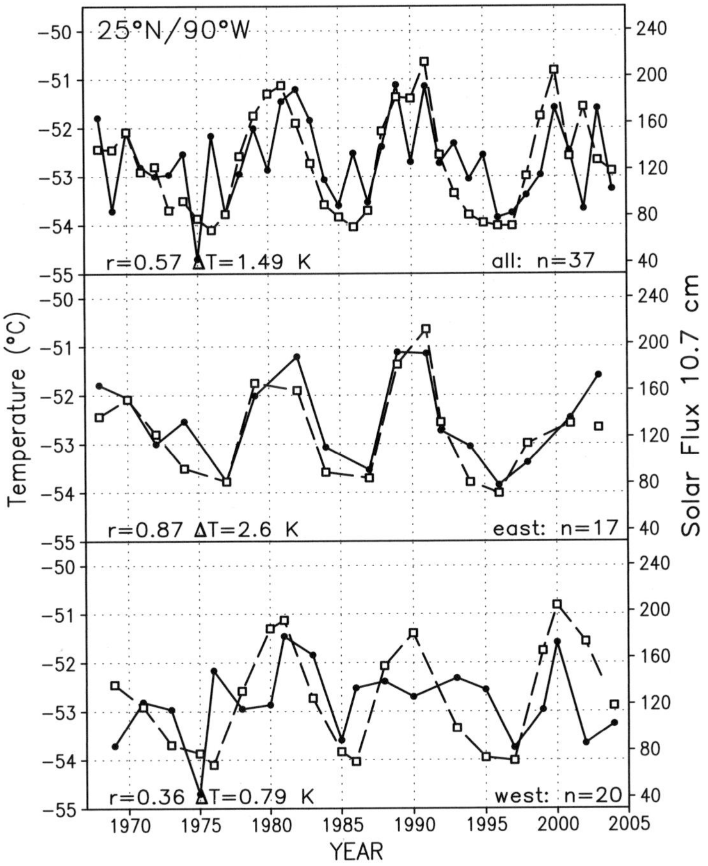

Fig. 18.10. Time series of monthly mean values of the 10.7 cm solar flux (*dashed lines*) and detrended 30-hPa temperatures in **July** at the gridpoint 25°N/90°W. The *upper panel* shows all years ($n = 37$), the *middle panel* only years in the east phase of the QBO ($n = 17$), and the *lower panel* only years in the west phase of the QBO ($n = 20$). r = correlation coefficients; ΔT = temperature difference (K) between solar maxima and minima (NCEP/NCAR re-analyses, 1968–2004)

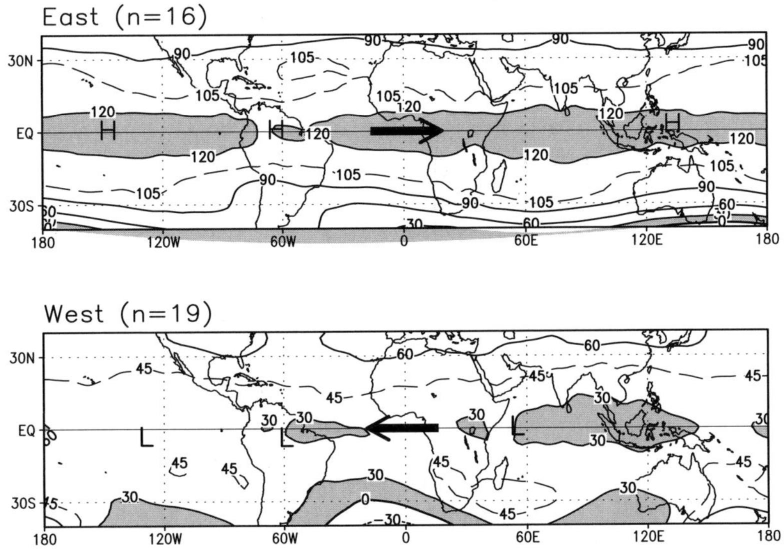

Fig. 18.11. Maps of the 30-hPa height differences (geopot. m) in **July** between solar maxima and minima. The *upper map* is for years in the east phase of the QBO, and the *lower map* for years in the west phase of the QBO. *Arrows* indicate the direction of the anomalous winds. (NCEP/NCAR re-analyses, 1968–2002). (Labitzke (2003), adopted from Fig. 5)

wind is weakened, and conversely in the solar minimum years.

In the west phase of the QBO (bottom of Fig. 18.11), the geopotential heights are lowest on the equator in the solar maxima and the anomalous winds are from the east, around the anomalous low on the equator, and conversely in the solar minima and west years. *The QBO thus not only modulates the solar signal on the decadal scale, but is itself modulated by the solar variability* (Salby and Callaghan, 2000; Soukharev and Hood, 2001; Labitzke, 2003).

18.5 The Solar Signal in the Troposphere

There are several indications that solar variability forcing affects the troposphere too. For instance, Labitzke and van Loon (1992, 1995) and van Loon and Labitzke (1994) noted that radiosonde stations in the tropics and subtropics of the Northern Hemisphere showed a marked difference in the vertical distribution of temperature between maxima and minima in the solar decadal oscillation, the temperatures being higher in the maxima in the *troposphere* and stratosphere, and lower or little changed in the tropopause region. The gridpoint data from the NCEP/NCAR analyses agree well with the radiosondes used by van Loon and Labitzke, even though they extend their analysis by two more solar periods. And van Loon and Shea (1999, 2000) demonstrated that three-year running, area-weighted means of the zonally averaged temperature of the entire Northern Hemisphere – in the layer between 700 hPa and 200 hPa in July–August – followed the decadal solar oscillation, with higher temperatures in the solar maxima than in the minima. The temperature of the nearly 9 km thick layer correlated with the solar oscillation: $r = 0.65$ for July–August (Fig. 18.12) and $r = 0.57$ for the 10-month average March to December (not shown).

Furthermore, van Loon and Labitzke (1998, their Figs. 10 and 11) demonstrated that the first empirical orthogonal function (EOF 1) in the 30-hPa temperatures and heights follows the interannual course of the solar 10.7 cm flux, and accounts for over 70% of the interannual variance in the summer of both hemispheres. This eigenvector at 30 hPa is well correlated with the temperatures in the troposphere.

Gleisner and Thejll (2003), Coughlin and Tung (2004), Kodera (2004), and van Loon et al. (2004, 2007) found similar positive anomalies in the troposphere of the tropics and subtropics, associated with an increase in the solar irradiance. van Loon et al. (2004) stress that differences in rainfall, vertical motion, and outgoing long wave radiation point to differences between solar maxima and minima in the tropical Walker Cell

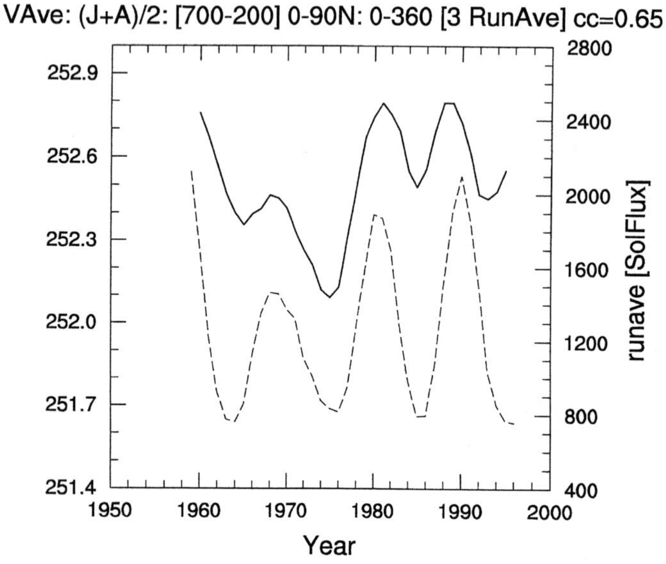

Fig. 18.12. *Solid line*: three year running means of the temperature (K) in the tropospheric layer between 700 hPa and 200 hPa, (i.e. 2 – 11 km), averaged over the Northern Hemisphere in **July–August** and area weighted. *Dashed line*: the 10.7 cm solar flux, used as an index of the solar variability. (van Loon and Shea, 2000)

(vertical, west to east oriented) and in the Hadley Cell (vertical, south to north oriented).

It should be noted, however, that the natural variability of the troposphere is so large that it is difficult to obtain stable results with too small data sets. Therefore, in this Section the data were, on the whole, not divided according to the phase of the QBO. A complete understanding of the mechanisms which transfer the direct solar variability signal from the upper to the lower stratosphere and to the troposphere is still missing (Matthes et al., 2006). The amplitude of the solar variability signal will be established only after data during more solar cycles have become available.

18.6 The QBO–Solar–Relationship Throughout the Year

For an easier comparison of the differences between solar maxima and minima in the west and east phase of the QBO, respectively, Fig. 18.13 gives the zonal mean 30-hPa height differences separately for two-month means. The height of a pressure level is the result of the integrated temperatures **below** the respective pressure level, and in this presentation the integral of the temperature differences is below 30-hPa, (see also Fig. 18.15).

Again, **during the east phase of the QBO**, the size of the solar variability signal is impressively large over a wide range of latitudes throughout the year. And the differences are positive except for high latitudes during winter when they turn to negative values, while the west phase signal becomes strongly positive; see discussion above in Sect. 18.4.1. The maximum of the height differences is situated directly over the equator most of the time, with lower values poleward, and therefore the anomalous winds connected with this structure are from the west. This implies, for the QBO/east situation, that the winds are weaker during solar maxima, as discussed above for July with Fig. 18.11.

This structure, with the largest height differences between solar maxima and minima directly over the equator, exists from March through October, that is for 8 months. Only during the northern winter does a very weak anomalous height gradient exist over the tropics. Therefore *one finds an influence of the solar cycle on the tropical QBO for the whole year*, as discussed by Soukharev and Hood (2001) and Labitzke (2003), (cf. discussion of Fig. 18.14).

During the west phase of the QBO, most of the year the solar signal is much weaker than during the east phase. During the northern spring and summer (April through August) a clear signal exists from the northern subtropics to the Arctic, with a secondary maximum over the Southern Hemisphere. The positive height differences over the Arctic in late winter are again connected with the warmer arctic stratosphere in late winter during the west phase of the QBO in solar maxima. At the same time the solar variability signal during the southern summer (i.e. northern winter) is much reduced compared with the results obtained during northern summer. This is due to the dynamical interactions between high and low latitudes, as discussed above.

Over the equatorial region the height differences show a structure completely different from the other phase of the QBO. A minimum of the height differences is found over the equator for most of the year, implying anomalous winds from the east, that is a weaker QBO-west wind in solar maximum.

Figure 18.14 summarizes the differences discussed above for Fig. 18.13. As it is practically impossible to derive an annual mean for the QBO data, the data shown in Fig. 18.14 display a *constructed annual mean* of the differences between solar maxima and minima, **where the data given in Fig. 18.13 are linearly averaged**, and similarly so for the 30-hPa temperatures. The main features are clearly identified. In the middle stratosphere the signal of the 11-year SSC (differences between solar maxima and minima) is much stronger during the east phase of the QBO, from 60°N to 50°S. This is the case for the 30-hPa temperatures and heights. The height differences (Fig. 18.14, bottom) have a clear maximum over the equator during the east phase, and a clear minimum during the west phase. These differences are connected with the weaker QBO winds (in both phases) during solar maxima, as described above.

The summarized solar variability signal in the 30-hPa temperatures shows two maxima in the subtropics in both phases of the QBO. However, there are large differences between the two phases of the QBO in the tropics, with almost no signal directly over the equator in the west phase and a strong signal (almost 1 K) in the east phase (Fig. 18.14, top).

Figure 18.15 shows from the ground till 32 km the vertical meridional section of the *constructed annual means* for all years (upper panels) and for the two different phases of the QBO, with the differences between

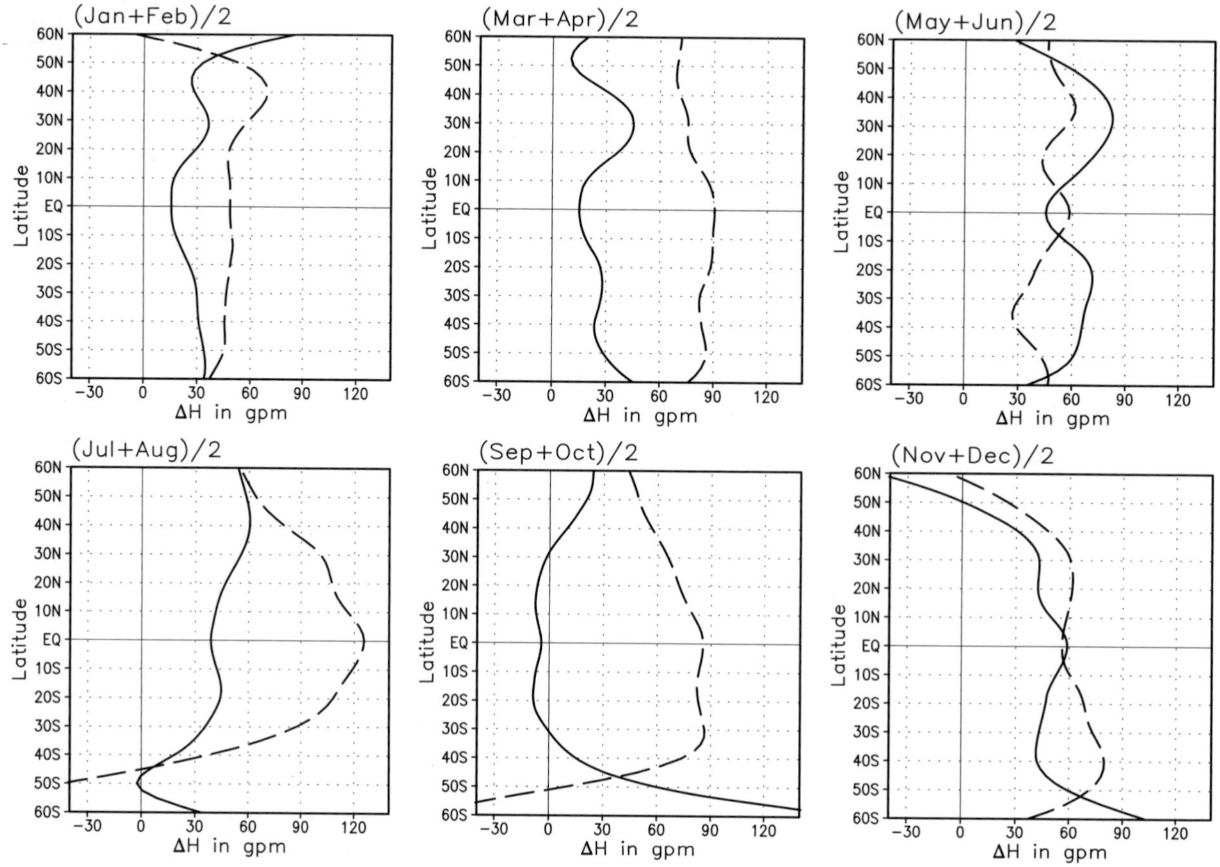

Fig. 18.13. Meridional profiles (60°N to 60°S) of the detrended zonal mean 30-hPa height differences (geopt. m) between solar maxima and minima (2-month means); *dashed* for the east phase of the QBO, and *solid* for the west phase (NCEP/NCAR reanalyses, 1968–2002, Labitzke, 2004b)

solar maxima and minima of the temperatures (K) (left) and of the geopotential heights (gpm) (right). *The temperature differences for all years (upper left) are practically positive from the ground up to* 32 km, *i.e., as high as the data are available. Both the troposphere and the stratosphere are warmer during solar maxima than during solar minima* (Labitzke, 2005).

In the middle and lower panels, the solar variability signal is convincingly summarized for the two different phases of the QBO, and all the details discussed before for the respective winters and summers stand out clearly. They can be summarized as follows:

(1) Temperature differences:

a) Over the latitude range 30°N to 30°S, from the upper troposphere to 10 hPa the solar variability signal over the tropics and subtropics is much larger in the east phase. This must be explained by anomalous downwelling (or reduced upwelling, i.e. less adiabatic cooling), which indicates a weakened BDC, during solar maxima/QBO-east phase.

b) Over both polar regions the solar signal is dominated by the conditions during winter (Sect. 18.4): it is positive in the west phase, reflecting the warmer polar stratosphere during solar maxima/QBO-west phase in winter, (i.e. an enhanced BDC with enhanced downwelling during solar maxima), see discussion above. And it is negative in the east phase which indicates a weakened BDC during solar max-

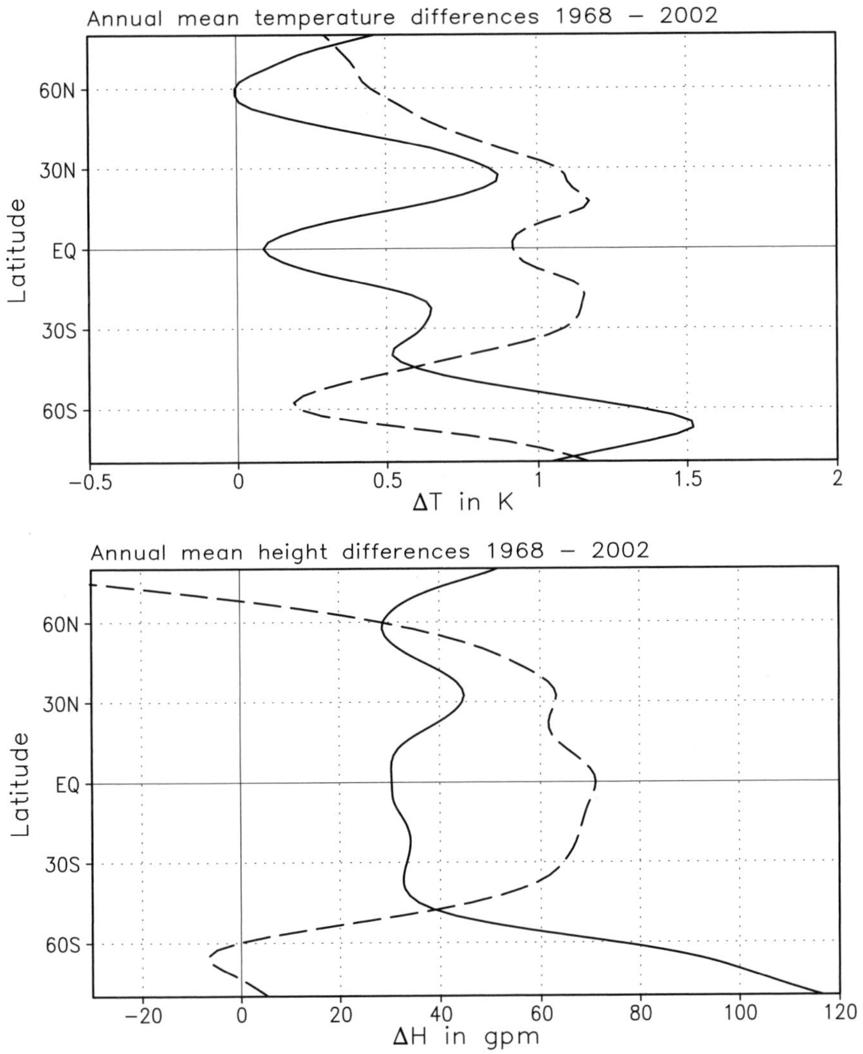

Fig. 18.14. *Top*: meridional profiles (80° N to 80° S) of the *Constructed Annual Mean* 30-hPa temperature differences (K). *Bottom*: the same, but for the 30-hPa height differences (geopot. m). These are both arranged for the QBO west phase (*solid lines*) and QBO east phase (*dashed lines*), respectively. (NCEP/NCAR re-analyses, 1968–2002, Labitzke, 2004b)

ima/east phase, i.e. less downwelling over the poles in winter.

c) The warmer polar stratosphere during solar maxima/QBO-west phase in winter is connected with dynamically enhanced upwelling over the tropics, explaining the smaller solar signal there in the annual mean.

(2) Height differences:

a) The height differences relate to the integral of the temperature differences below (hydrostatic relationship). Accordingly, in the QBO/east phase they are larger than in the QBO/west phase in the stratosphere over the tropics and subtropics and weaker (negative) over the polar regions.

b) There is a maximum of the height differences over the equator in the QBO/east phase and a minimum in the QBO/west phase, reflecting the influence of the solar cycle on the QBO, see discussion above for Fig. 18.11.

18.7 Models and Mechanisms

Based on the observations, the results presented above demonstrate conclusively the existence of a signal of the

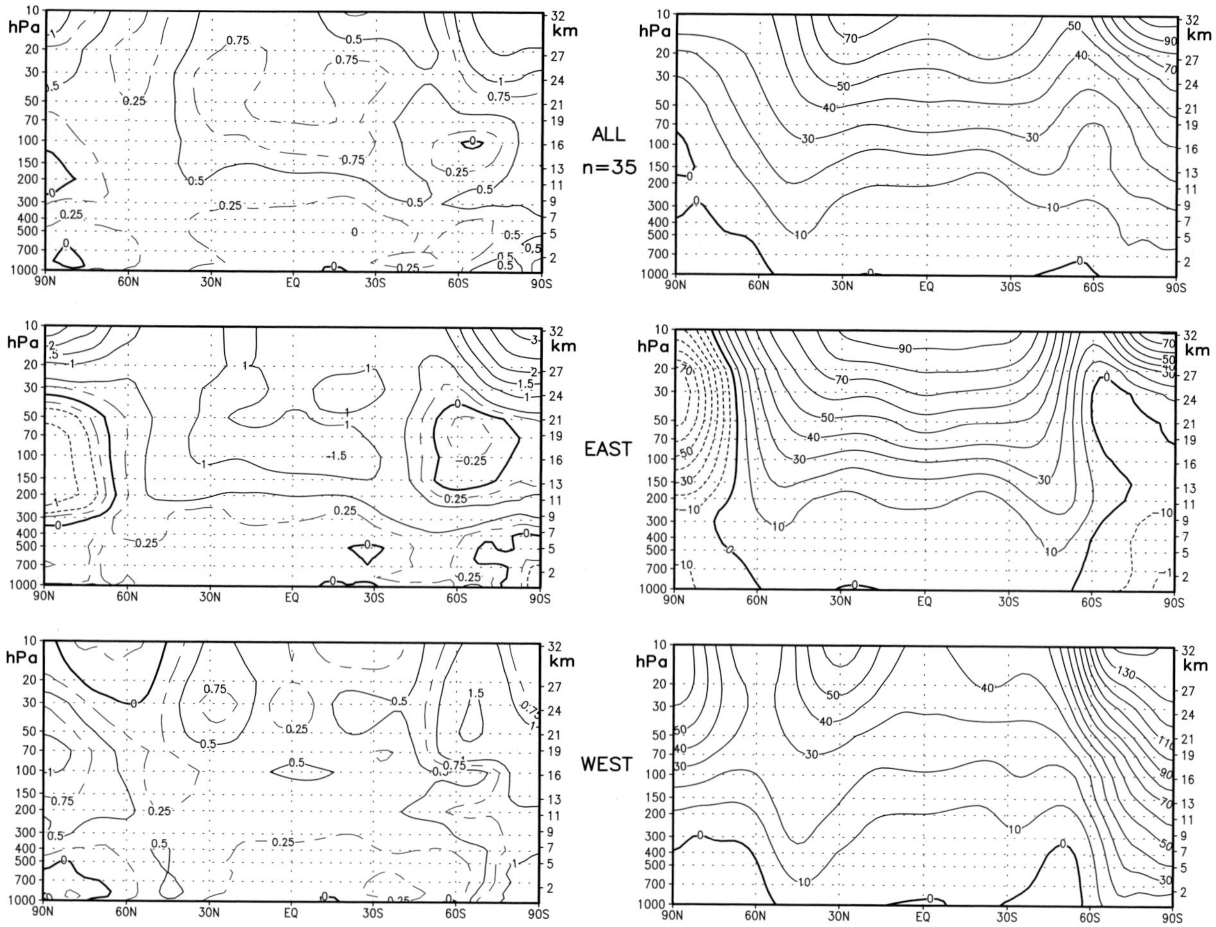

Fig. 18.15. Vertical meridional sections of the *Constructed Annual Means*. *Left*: of the temperature differences (K) between solar maxima and minima. *Right*: of the height differences (geopot. m). The *upper panels* show all years, the *middle panels* only years in the east phase of the QBO, and the *lower panels* only years in the west phase. (NCEP/NCAR re-analyses, 1968–2002, Labitzke, 2005)

11-year SSC in the stratospheric and tropospheric temperatures and heights. There have been many model studies with General Circulation Models (GCMs) to investigate the impact of changes in the "solar constant", but the change from solar maxima to minima within the SSC is only about 0.1% and the influence on the atmosphere (in the models) is very small.

Kodera et al. (1991) and Rind and Balachandran (1995) were the first to use GCMs with a better resolution of the stratosphere to study the effects of increases in solar UV flux. Later, Haigh (1996, 1999) and Shindell et al. (1999) carried out computer experiments where they imposed in the General Circulation Models (GCMs) realistic changes in the UV part of the solar spectrum and estimates of the resulting ozone changes.

There is general agreement that the *direct* influence of the changes in the UV part of the spectrum (6 to 8% between solar maxima and minima) leads to more ozone and warming in the upper stratosphere (around 50 km) in solar maxima (Haigh, 1994; Hood et al., 1993; Hood, 2004). This leads to changes in the thermal gradients and thus in the wind systems, which in turn lead to changes in the vertical propagation of the planetary waves that drive the global circulation. Therefore, the relatively weak, *direct* radiative forcing of the solar cycle in the stratosphere can lead to a large *indirect* dynamical response in the lower atmosphere through a modulation of the polar night jet (PNJ) as well as through a change in the Brewer–Dobson Circulation (BDC) (Kodera and Kuroda, 2002).

Some of the model results were found to be of similar structure to those seen in the analysis of data in the stratosphere by, e.g., van Loon and Labitzke (2000). However, so far the sizes of the changes were much smaller than observed, especially during summer. This is probably due to the fact that the GCMs do not produce a QBO and that these models are not coupled to the oceans, so that the most important natural forcings have not yet been included in the modelling.

Recently, Matthes et al. (2004), using the "Freie Universität Berlin–Climate Middle Atmosphere Model (FUB–CMAM)", introduced in addition to the realistic spectral UV changes and ozone changes a relaxation towards observed equatorial wind profiles throughout the stratosphere, representing the east and west phases of the QBO, as well as the Semiannual Oscillation (SAO) in the upper stratosphere. The importance of the SAO in the upper stratosphere has been stressed by Gray et al. (2001a,b). During the arctic winter a realistic poleward–downward propagation of the polar night jet anomalies, significantly weaker planetary wave activity and a weaker mean meridional circulation under solar maximum conditions are reproduced in the FUB–CMAM. This confirms the solar variability signal observed in the upper stratosphere, by, e.g., Kodera and Yamazaki (1990) and Kuroda and Kodera (2002). The observed interaction between the sun and the QBO is captured and stratospheric warmings occur preferentially in the west phase of the QBO, during solar maxima (cf. Sect. 18.4). And the solar signal from the upper stratosphere influences tropospheric circulation patterns in the model as suggested from observations.

It should be pointed out that other GCM studies have so far failed to produce such a good correspondence with the observed magnitude and temporal evolution of the zonal wind anomalies in northern winters.

18.8 Summary

It is now widely accepted by the scientific community that there exists a strong signal of the 11-year solar cycle in the Earth's atmosphere throughout the year – but it can be identified better if the data are stratified according to the phase of the QBO. If the unstratified data are used, no clear signal emerges. Stratification of the data leads to a reduction of the number of years in each group. But the results shown here indicate clearly that we cannot identify the full size of the solar signal if we are using the undivided data, and therefore we cannot understand the mechanisms involved.

The results given here support earlier work suggesting that the mean meridional circulation systems, particularly the BDC, are affected by the variable sun. The strong warming of the lower stratosphere over the tropics and subtropics during the maxima of the solar cycle **in the east phase** can probably be explained with anomalous downwelling (adiabatic warming) which works against, and weakens, the BDC. This appears to be the case during most of the year for the years in the east phase of the QBO.

During the QBO/west phase the solar variability signal results in winter/spring in warmer/weaker polar vortices over both polar regions during maxima of the SSC. Thus the sign of the anomalies is reversed between high and low latitudes during this time of year. Over the tropics and subtropics the solar variability signal in the **QBO/west phase** is most of the time weaker than in the east phase, but there are indications of a strengthening of the HC.

The 30-hPa height differences between solar maxima and minima given in Figs. 18.13 and 18.14 show clearly, for most months and in the *constructed annual mean*, a maximum over the equator during the east phase of the QBO and a minimum during the west phase. This reflects a weaker QBO during solar maxima in both phases.

Recent simulations of the middle atmosphere, using a General Circulation Model (GCM) and introducing the changes in solar UV radiation, ozone and profiles of the winds over the equator, simulating the east and west phase of the QBO, respectively, result in a realistic simulation of the variability of the arctic polar vortex in northern winter (Matthes et al., 2004). The simulated signal over the tropics is, however, smaller than the results shown here.

A complete understanding of the mechanisms which transfer the direct solar variability signal from the upper to the lower stratosphere and into the troposphere is still missing. The amplitude of the solar variability signal will only be established after the data accumulated during more solar cycles have become available.

Acknowledgement. I thank the members of the Stratospheric Research Group, FUB for professional support and Dipl. Met. Markus Kunze for doing the computations and graphics. Special thanks go to Harry van Loon for long discussions and to Michael Rycroft for reviewing the manuscript. The 10.7 cm solar flux data are from the World Data Center A, Boulder, Colorado.

References

Abbot, C. G., F. E. Fowle, and L. B. Aldrich, Annals of the Astrophysical Observatory of the Smithsonian Institution, Vol. III, Washington, 1913

Baldwin, M. P. and coauthors, The quasi-biennial oscillation. *Rev. Geophys.*, **39**, 179–229, 2001

Beer, J., W. Mende, and R. Stellmacher, The role of the Sun in climate forcing. *Quart. Sci. Rev.*, **19**, 403–415, 2000

Chandra, S., and R. D. McPeters, The solar cycle variation of ozone in the stratosphere inferred from Nimbus 7 and NOAA 11 satellites. *J. Geophys. Res.*, **99**, 20665–20671, 1994

Coughlin, K. T., and K. K. Tung, 11-year solar cycle in the stratosphere extracted by the empirical mode decomposition method. *Advances in Space Research*, **34**, 323–329, 2004

Cubasch, U., and R. Voss, The influence of total solar irradiance on climate. *Space Sci. Rev.*, **94**, 185–198, 2000

De Jager, C., and I. Usoskin, On possible drivers of Sun-induced climate changes. *J. Atmos. Sol.-Terr. Phys.*, **68**, 2053–2060, 2006

Fröhlich, C., Observations of irradiance variations. *Space Sci. Rev.*, **94**, 15–24, 2000

Gleisner, H., and P. Thejll, Patterns of tropospheric response to solar variability. *Geophys. Res. Lett.*, **30**, (13), 1711, doi: 10.1029/2003GL17129, 2003

Gray, L., E. F. Drysdale, T. J. Dunkerton, and B. Lawrence, Model studies of the interannual variability of the northern hemisphere stratospheric winter circulation: The role of the Quasi-Biennial Oscillation. *Q. J. Roy. Met. Soc.*, **127**, 1413–1432, 2001a

Gray, L., S. J. Phipps, T. J. Dunkerton, M. P. Baldwin, E. F. Drysdale, and M. R. Allen, A data study of the influence of the equatorial upper stratosphere on northern hemisphere stratospheric sudden warmings. *Q. J. Roy. Met. Soc.*, **127**, 1985–2003, 2001b

Haigh, J. D., The role of stratospheric ozone in modulating the solar radiative forcing of climate. *Nature*, **370**, 544–546, 1994

Haigh, J. D., The impact of solar variability on climate. *Science*, **272**, 981–984, 1996

Haigh, J. D., A GCM study of climate change in response to the 11-year solar cycle. *Q. J. Roy. Met. Soc.*, **125**, 871–892, 1999

Holton, J., and H. Tan, The influence of the equatorial Quasi-Biennial Oscillation on the global circulation at 50 mb. *J. Atmos. Sci.*, **37**, 2200–2208, 1980

Hood, L. L., Thermal response of the tropical tropopause region to solar ultraviolet variations. *Geophys. Res. Lett.*, **30**, No. 23, 2215, doi: 10.1029/2003 GL018364, 2003

Hood, L. L., Effects of solar UV variability on the stratosphere. – In: Solar variability and its effect on the Earth's atmosphere and climate system, *AGU Monograph Series*, Eds. J. Pap et al., American Geophysical Union, Washington D.C., 283–304, 2004

Hood, L. L., and B. Soukharev, Quasi-decadal variability of the tropical lower stratosphere: the role of extratropical wave forcing. *J. Atmos. Sci.*, **60**, 2389–2403, 2003

Hood, L. L., J. L. Jirikowic, and J. P. McCormack, Quasi-decadal variability of the stratosphere: influence of long-term solar ultraviolet variations. *J. Atmos. Sci.*, **50**, 3941–3958, 1993

Hoyt, D. V., and K. H. Schatten, The role of the Sun in climate change. Oxford University Press, New York, 279 pp, 1997

Kalnay, E., R. Kanamitsu, R. Kistler, W. Collins, D. Deaven, L. Gandin, M. Iredell, S. Saha, G. White, Y. Zhu, M. Chelliah, W. Ebisuzaki, W. Higgins, J. Janowiak, K. C. Mo, C. Ropelewski, J. Wang, R. Reynolds, R. Jenne, and J. Joseph, The NCEP/NCAR 40-year re-analysis project. *Bull. Am. Meteor. Soc.*, **77**, 437–471, 1996

Kodera, K., Solar influence on the Indian Ocean Monsoon through dynamical processes. *Geophys. Res. Lett.*,**31**, L24209, doi: 10.1029/2004GL 020928, 2004

Kodera, K., and Y. Kuroda, Dynamical response to the solar cycle. *J. Geophys. Res.*, **107**, (D24), 4749, doi:10.1029/2002JD002224, 2002

Kodera, K., and K. Yamazaki, Long-term variation of upper stratospheric circulation in the northern hemisphere in December. *J. Met. Soc. Japan*, **68**, 101–105, 1990

Kodera, K., M. Chiba, and K. Shibata, A general circulation model study of the solar and QBO modulation of the stratospheric circulation during northern hemisphere winter. *Geophys. Res. Lett.*, **18**, 1209–1212, 1991

Kristjánsson, J. E., J. Kristiansen, and E. Kaas, Solar activity, cosmic rays, clouds and climate – an update. *Advances in Space Res.*, **34**,407–415, 2004

Kuroda, Y., and K. Kodera, Effect of solar activity on the polar-night jet oscillation in the northern and southern hemisphere winter. *J. Met. Soc. Japan*, **80**, 973–984, 2002

Labitzke, K., On the interannual variability of the middle stratosphere during the northern winters. *J. Met. Soc. Japan*, **60**, 124–139, 1982

Labitzke, K., Sunspots, the QBO, and the stratospheric temperature in the north polar region. *Geophys. Res. Lett.*, **14**, 535–537, 1987

Labitzke, K., The solar signal of the 11-year sunspot cycle in the stratosphere: Differences between the northern and southern summers. *J. Met. Soc. Japan*, **80**, 963–971, 2002

Labitzke, K., The global signal of the 11-year solar cycle in the atmosphere: When do we need the QBO? *Meteorolog. Z.*, **12**, 209–216, 2003

Labitzke, K., On the signal of the 11-year sunspot cycle in the stratosphere over the Antarctic and its modulation by the Quasi-Biennial Oscillation (QBO). *Meteorolog. Z.*, **13**, 263–270, 2004a

Labitzke, K., On the signal of the 11-year sunspot cycle in the stratosphere and its modulation by the Quasi-Biennial Oscillation (QBO). *J. Atmos. Sol.-Terr. Phys.*, **66**, 1151–1157, 2004b

Labitzke, K., On the solar cycle–QBO relationship: a summary. *J. Atmos. Sol.-Terr. Phys.*, **67/1-2**, 45–54, 2005

Labitzke, K. and Collaborators, 2002. The Berlin Stratospheric Data Series; CD from Meteorological Institute, Free University Berlin

Labitzke, K., and M. Kunze, Stratospheric temperatures over the Arctic: Comparison of three data sets. *Meteorolog. Z.*, **14**, 65–74, 2005

Labitzke, K., and H. van Loon, The stratosphere in the Southern Hemisphere, Chapter 7, 113–138, in: Meteorology of the Southern Hemisphere, *Met. Monogr.*, **13**, No. 35 (C. W. Newton, Ed.), 1972

Labitzke, K., and H. van Loon, Associations between the 11-year solar cycle, the QBO and the atmosphere. Part I: The troposphere and stratosphere in the northern hemisphere winter. *J. Atmos. Terr. Phys.*, **50**, 197–206, 1988

Labitzke, K., and H. van Loon, The Southern Oscillation. Part IX: The influence of volcanic eruptions on the Southern Oscillation in the stratosphere. *J. Clim.*, **2**, 1223–1226, 1989

Labitzke, K., and H. van Loon, Association between the 11-year solar cycle and the atmosphere. Part V: Summer. *J. Clim.*, **5**, 240–251, 1992

Labitzke, K., and H. van Loon, Connection between the troposphere and stratosphere on a decadal scale. *Tellus*, **47 A**, 275–286, 1995

Labitzke, K., and H. van Loon, The Stratosphere (Phenomena, History, and Relevance), 179 pp. Springer Verlag, Berlin Heidelberg New York, 1999

Labitzke, K., and H. van Loon, The QBO effect on the global stratosphere in northern winter. *J. Atmos. Sol.-Terr. Phys.*, **62**, 621–628, 2000

Labitzke, K., M. Kunze, and S. Brönnigmann, Sunspots, the QBO and the stratosphere in the north polar region – 20 years later. *Meteorolog. Z.*, **15**, 335–363, 2006

Langematz, U., A. Clausnitzer, K. Matthes, and M. Kunze, The climate during the Maunder Minimum, simulated with the Freie Universität Berlin climate middle atmosphere model (FUBCMAM). *J. Atmos. Sol.-Terr. Phys.*, **67/1-2**, 55–59, 2005

Lean, J. L., G. J. Rottman, H. L. Kyle, T. N. Woods, J. R. Hickey, and L. C. Puga, Detection and parameterisation of variations in solar mid - and near-ultraviolet radiation (200 – 400 nm). *J. Geophys. Res.*, **102**, 29939–29956, 1997

Matthes, K., U. Langematz, L.L. Gray, K. Kodera, and K. Labitzke, Improved 11-year solar signal in the FUB-CMAM. *J. Geophys. Res.*, **109**, doi: 10.1029/ 2003 JD 004012, 2004

Matthes, K., Y. Kuroda, K. Kodera, and U. Langematz, Transfer of the solar signal from the stratosphere to the troposphere: Northern winter. *J. Geophys. Res.*, **111**, D06108, doi: 10.1019/2005JD 006283, 2006

Naito, Y., and I. Hirota, Interannual variability of the northern winter stratospheric circulation related to the QBO and the solar cycle. *J. Met. Soc. Japan*, **75**, 925–937, 1997

Naujokat, B., An update of the observed Quasi-Biennial Oscillation of the stratospheric winds over the tropics. *J. Atmos. Sci.*, **43**, 1873–1877, 1986

Pawson, S., and B. Naujokat, The cold winters of the middle 1990s in the northern lower stratosphere. *J. Geophys. Res.*, **104**, 14,209–14,222, 1999

Pittock, A. B., Solar variability, weather and climate: An update. *Q. J. Roy. Met. Soc.*, **109**, 23–55, 1983

Reid, G. C., Solar variability and the Earth's climate: Introduction and overview. *Space Sci. Rev.*, **94**, 1–11, 2000

Rind, D., and N. K. Balachandran, Modelling the effects of UV variability and the QBO on the troposphere-stratosphere systems. Part II: The troposphere. *J. Clim.*, **8**, 2080–2095, 1995

Ruzmaikin, A., and J. Feynman, Solar influence on a major mode of atmospheric variability. *J. Geophys. Res.*, **107** (14), doi: 10.1029/2001JD001239, 2002

Salby, M., and P. Callaghan, Connection between the solar cycle and the QBO: The missing link. *J. Clim.*, **13**, 2652–2662, 2000

Salby, M., and P. Callaghan, Evidence of the solar cycle in the general circulation of the stratosphere. *J. Clim.*, **17**, 34–46, 2004

Salby, M., and P. Callaghan, Relationship of the quasi-biennial oscillation to the stratospheric signature of the solar cycle. *J. Geophys. Res.*, **111**, D06110, doi: 10.1029/2005JD006012, 2006

Shepherd, T. G., Issues in stratosphere–troposphere coupling. *J. Met. Soc. Japan*, **80**, 769–792, 2002

Shindell, D., D. Rind, N. K. Balachandran, J. Lean, and J. Lonergan, Solar cycle variability, ozone and climate. *Science*, **284**, 305–308, 1999

Soukharev, B., and L. L. Hood, Possible solar modulation of the equatorial quasi-biennial oscillation: Additional statistical evidence. *J. Geophys. Res.*, **106**, 14,855–14,868, 2001

Svensmark, H., and E. Friis-Christensen, Variation of cosmic ray flux and global cloud coverage–a missing link in solar-climate relationships. *J. Atmos. Sol.-Terr. Phys.*, **59**, 1225, 1997

Udelhofen, P. M., and R. D. Cess, Cloud cover variations over the United States: An influence of cosmic rays or solar variability? *Geophys. Res. Lett.*, **28**, 2617–2620, 2001

van Loon, H., and K. Labitzke, The Southern Oscillation. Part V: The anomalies in the lower stratosphere of the Northern Hemisphere in winter and a comparison with the Quasi-Biennial Oscillation. *Monthly Weather Rev.*, **115**, 357–369, 1987

van Loon, H., and K. Labitzke, Interannual variations in the stratosphere of the Northern Hemisphere: A description of some probable influences. Interactions Between Global Climate Subsystems, The Legacy of Hann, *Geophys. Monograph* 75, IUGG **15**, 111–122, 1993

van Loon, H., and K. Labitzke, The 10–12 year atmospheric oscillation. Review article in: *Meteorolog. Z.*, N.F., **3**, 259–266, 1994

van Loon, H., and K. Labitzke, The global range of the stratospheric decadal wave. Part I: Its association with the sunspot cycle in summer and in the annual mean, and with the troposphere. *J. Clim.*, **11**, 1529–1537, 1998

van Loon, H., and K. Labitzke, The signal of the 11-year solar cycle in the global stratosphere. *J. Atmos. Sol.-Terr. Phys.*, **61**, 53–61, 1999

van Loon, H., and K. Labitzke, The influence of the 11-year solar cycle on the stratosphere below 30km: A review. *Space Sci. Rev.*, **94**, 259–278, 2000

van Loon, H., and D. J. Shea, A probable signal of the 11-year solar cycle in the troposphere of the Northern Hemisphere. *Geophys. Res. Lett.*, **26**, 2893–2896, 1999

van Loon, H., and D. J. Shea, The global 11–year signal in July–August. Geophys. Res. Lett., **27**, 2965–2968, 2000

van Loon, H., G. E. Meehl, and J. M. Arblaster, A decadal solar effect in the tropics in July–August. *J. Atmos. Sol.-Terr. Phys.*, **66**, 1767–1778, 2004

van Loon, H., G. A. Meehl, and D. Shea, Coupled air sea response to solar forcing in the Pacific region during northern winter, *J. Geophys. Res.*, **112**, D02108, doi: 10.1029/2006JD007378, 2007

Part 5

Planets and Comets in the Solar System

19 Planetary Magnetospheres

Margaret Galland Kivelson

The study of planetary magnetospheres began almost a half century ago with the launch of Sputnik and Explorer 1, the first artificial satellites of the Earth. The exploration of other magnetospheres started not long after. Our understanding of our own space environment has grown ever deeper with the passing years as flotillas of spacecraft have gradually acquired measurements whose interpretation provides a good (although as yet imperfect) understanding of Earth's environment in space. Our exploration of the magnetospheres of other planets has also progressed brilliantly but the high cost of planetary probes inevitably implies that we understand less about remote magnetospheres than about our own. Fortunately even limited data are of immense value in advancing the study of comparative magnetospheres because they reveal how magnetospheric processes respond to changes of scale, of rotation rate and of solar wind structure in the vicinity of the planet. This article addresses the topic of planetary magnetospheres by contrasting their properties with those familiar at Earth. The differences are related to key dimensionless parameters of the plasma flowing onto the different bodies of the solar system and to key properties of the central bodies such as the strength and symmetry of the magnetic field at the planet's surface, the size and rotation period of the planet, the nature of its plasma sources and the conductivity of its surface layers.

Contents

Margaret Galland Kivelson, Planetary Magnetospheres.
In: Y. Kamide/A. Chian, Handbook of the Solar-Terrestrial Environment. pp. 469–492 (2007)
DOI: 10.1007/11367758_19

19.1 Introduction

Laboratory scientists have the luxury of being able to probe their samples repeatedly under controlled conditions over a range of underlying parameters such as density and pressure. Magnetospheric scientists have little control over the conditions of their investigations. To be sure, the responses of the terrestrial magnetosphere to changes in solar wind dynamic pressure and magnetic field orientation have been extensively analyzed, but the variations are uncontrolled, narrowly bounded and some important internal parameters of the system do not change. Fortunately some other planetary magnetospheres exist and some of their properties differ significantly from those applicable to the terrestrial magnetosphere. This chapter emphasizes the physical parameters that control the outcome of the interaction of a flowing plasma with a magnetized body and describes some of the interesting ways in which the magnetospheres of other magnetized bodies in the solar system differ from the one with which we are familiar. Armed with such information, we can speculate on how Earth's magnetosphere itself may have changed over geological time.

19.2 Parameters that Control Magnetospheric Configuration and Dynamics

A magnetosphere forms when a plasma flows onto a magnetized body such as a planet or a moon. Critical to the form of the interaction are the properties of the plasma, some of which are effectively expressed in terms of dimensionless ratios including the ratio of the Alfvén speed and the sound speed to the speed of the plasma measured in the rest frame of the planet and the ratio of the plasma pressure to the magnetic pressure. At Earth the changing orientation of the interplanetary magnetic field contributes significantly to temporal variations, implying that orientation is a control parameter but the importance of this element of solar wind control varies from one planet to another. Other parameters that govern the interaction are intrinsic to the planet: its radius and rotation rate, the strength and symmetry of the magnetic field at its surface, the conductivity of its surface and upper atmosphere, its neutral exosphere and the location and composition of any moons and rings. Finally, the scale of the interaction region is determined by the dimensionless parameter that relates the energy density of the incident solar wind to the energy density in the magnetic field and the magnetospheric plasma near the boundary.

One must consider how to restrict the subject of this chapter, recognizing that comets and planets or moons lacking permanent internal magnetic fields also perturb the solar wind, creating regions of disturbed flow that have much in common with the magnetospheres of magnetized planets. The reader is referred to Chap. 20 (The Solar-Comet Wind Interaction) for a discussion of the cometary interaction. That discussion reveals that an unmagnetized body, like a magnetosphere, greatly modifies plasma properties in the space surrounding it and that the external field drapes around the body extending the interaction region downstream in the antisolar direction. Analogous interaction regions form around the unmagnetized moons of Jupiter (Io, Europa, and Callisto), the largest moon of Saturn (Titan). None of these cases will be discussed in this chapter, which instead focuses on the true magnetospheres of the solar system: Mercury, Earth, Jupiter, Saturn, Neptune, Uranus and Jupiter's moon Ganymede. To this list some would like to add Mars, which lacks a planet-wide field but does have regions where the magnetic field is sufficiently intense to prevent the solar wind from flowing onto some parts of its surface. The localized fields create structures that resemble solar arcades. Table 19.1 gives some of the key parameters for the magnetospheres that are discussed in this chapter. Extensive tables of properties of the bodies discussed in this chapter can be found in Kivelson and Bagenal (2005).

19.2.1 Properties of the Flowing Plasma

A magnetosphere responds to various forms of pressure in the plasma that confines it. In a magnetized plasma, the total pressure P, exerted in the direction of the flow, is given by

$$P = \rho u^2 + p + B^2/2\mu_o \qquad (19.1)$$

where the terms represent the dynamic pressure, the thermal pressure and the magnetic pressure expressed in terms of the density, ρ, flow velocity, $\boldsymbol{u}$, thermal pressure, p, and magnetic field, $\boldsymbol{B}$. The thermal pressure has been assumed isotropic. In steady state, at the boundary of the magnetosphere the external pressure balances the internal pressure. The form of the magnetosphere is

Table 19.1. Properties of planet and of plasma flowing onto its magnetosphere

	Radius (km)	Surface equatorial field (nT)	Dipole tilt and sense	Sidereal rotation period	Density of of external plasma**	Dynamic pressure of external plasma (nPa)*	Magnetic field of external plasma (nT)**
Mercury	2440	140 to 400	~10°*	59 days	~50/cm^{-3}	15	20
Earth	6373	31,000	+ 10.8 °	23.9 h	8/cm^{-3}	2	8
Mars	3390	<10	–	24.6 h	3.5/cm^{-3}	1	3.5
Jupiter	71,398	428,000	– 9.6°	9.8 h	0.3/cm^{-3}	0.1	1
Saturn	60,330	22,000	0.0°	10.7 h	0.1/cm^{-3}	0.03	0.5
Uranus	25,559	23,000	−59°	15.5 h	0.02/cm^{-3}	0.005	0.3
Neptune	24,764	14,000	−47°	15.8 h	0.008/cm^{-3}	0.002	0.2
Ganymede	2634	720	4°	7.2 days	100 AMU/cm^{3}	1	100

* The dipole tilt of Mercury is not well determined. This value from Slavin (2004)
** The properties of the solar wind vary greatly; hence the values are approximate

Table 19.2. Parameters relevant to the structure and dynamics of planetary magnetospheres*

	(a) $(B^2_{\text{surf}}/2\mu_o)/\rho_{\text{ext}}u^2_{\text{ext}}$ (b) $(B^2_{\text{surf}}/B^2_{\text{ext}})$	Upstream magnetosonic Mach number**	Distance to magnetopause (planetary radii or noted)	$0.1\, v_{\text{sw}}/v_{\text{rot}}$ near nose of magnetopause
Mercury	(a) ~1	6	1.5	3×10^4
Earth	(a) 4×10^5	7	10	9
Mars	(a) <0.04	8	–	–
Jupiter	(a) 7×10^8	10	70	0.04
Saturn	(a) 7×10^7	12	20	0.2
Uranus	(a) 4×10^7	13	18	0.7
Neptune	(a) 4×10^7	15	24	0.6
Ganymede	(b) 50	0.4	1.6	∞

* The properties of the solar wind vary greatly; hence the values are approximate average values
** The values of the magnetosonic Mach numbers are from a model tabulated by Slavin et al. (1985), rounded to integer values

dictated by the dominant term in the external pressure. When dynamic pressure dominates as in the solar wind, the magnetosphere is bullet-shaped and extended along the direction of external plasma flow as in Fig. 19.1, left, which represents the magnetosphere of Mercury and compares it with Earth's magnetosphere. When the magnetic pressure dominates the ambient plasma, as in the vicinity of Jupiter's magnetized moon Ganymede, the magnetosphere formed by interaction with the incident flowing plasma is rod- or cylinder-shaped and roughly aligned with the external magnetic field as illustrated in Fig. 19.1, right. The form of the magnetosphere is thus seen to depend on the ratios of the differing forms of pressure in the surrounding plasma. Even when the dynamic pressure dominates, as in the solar wind, its contribution to magnetospheric confinement is a function of the angle between the flow and the local surface normal. Dynamic pressure controls the sunward-facing boundaries of planetary magnetospheres, including Earth's, whereas thermal and magnetic pressure confine the magnetosphere on the distant flanks where the flow direction is roughly antiparallel to the normal.

Dimensionless parameters that express the relative importance of the three terms in (19.1) are the Alfvén Mach number u/v_A where $v_A = B/(\mu_o\rho)^{1/2}$ is the Alfvén speed whose square is the ratio of the energy density in the flow to the magnetic energy density, the sonic Mach number whose square is the ratio of the dynamic

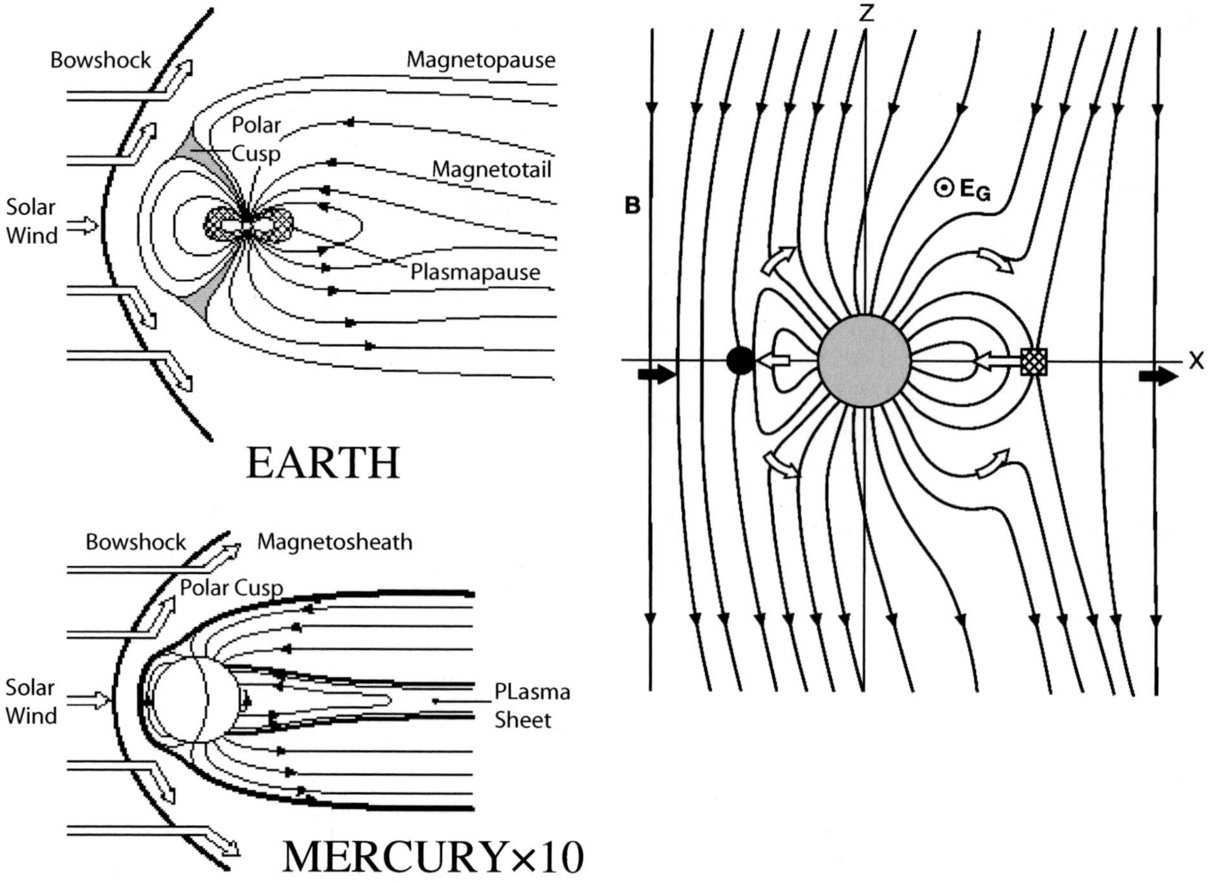

Fig. 19.1. *Left*: Schematic of Mercury's magnetosphere (*below*) compared with Earth's magnetosphere (*above*) (Kivelson and Bagenal, 1998). *Right*: Schematic of a cut through the plane of the flow and the upstream field through the center of Ganymede's magnetosphere. In all schematics, the plasma flow is from the *left* and is represented by *broad arrows*

to the thermal pressure (to within a factor of order 1) and the plasma $\beta[= p/(B^2/2\mu_o)]$ which is the ratio of the thermal to the magnetic pressure. For normal and even for most extreme conditions in the solar wind, the dynamic pressure dominates, so both Mach numbers are >1; a shock forms upstream of all the planets and the magnetospheres are bullet-shaped. Ganymede, embedded in the flowing plasma of Jupiter's magnetosphere, is the exception. In its environment (see Table 19.2), the Alfvén Mach number is normally <1; no upstream shock forms and the magnetic pressure dictates the structure of the magnetosphere.

External and internal plasmas interact not only through hydromagnetic forces but also through reconnection of magnetic fields, a process efficient in accelerating particles and increasing the stress on the system. Therefore, it is not only the magnitude of the magnetic field but also its direction that is relevant to the dynamics of a magnetosphere. At Earth, reconnection with the solar wind is fundamental to geomagnetic disturbances. It has been securely established that the rate of energy input into the magnetosphere is controlled by $\boldsymbol{u}_{sw} \times \boldsymbol{B}_{sw}$ (where $\boldsymbol{u}_{sw}$ and $\boldsymbol{B}_{sw}$ are the flow velocity and the magnetic field of the solar wind and the negative of the cross product is the electric field). Maximum power for a fixed solar wind speed, density and field magnitude arises where $\boldsymbol{B}_{sw}$ is antiparallel to Earth's equatorial field, a configuration that favors reconnection on the low latitude dayside magnetopause. The significance of the field orientation in the upstream plasma is discussed for other bodies in later sections.

19.2.2 Properties of the Planet or Moon

It is not only the external plasma conditions that control the configuration of a magnetosphere and its dynamics. Various properties of the central planet or moon such as the planet's rotation rate, the strength of its magnetic field as characterized by its surface magnitude and the orientation of the dipole moment relative to the spin axis are also critical. The planetary radius that establishes the spatial scale of the interaction region varies by one and a half orders of magnitude between Mercury and Jupiter. The ratio of the time for the solar wind to flow from the nose of the magnetosphere to the terminator plane to the period of planetary rotation gives a measure of the relative importance of rotation. For Jupiter this ratio is roughly a third of the planetary rotation period and rotation dominates much of magnetospheric dynamics. At Earth, where the ratio is 1/540, rotation is far less important. For Ganymede and Mercury, the effects of planetary rotation are negligible as will be discussed below. Magnetospheric dynamics differ greatly in the two limits.

Widely separated regions within a magnetosphere are strongly coupled by field-aligned currents and this implies that the electrical conductivity of the central body is a key parameter in constraining the dynamics of the system. Currents may close in an ionosphere or through the surface/interior of the body.

Finally it is interesting to recognize that some magnetospheres are significantly affected by plasma introduced within their boundaries either from an ionospheric source or when neutrals that escape from the exosphere or from rings and moons gravitationally bound to the planet are later ionized. Ionization transfers mass to the plasma. Charge exchange extracts momentum from it. The giant planet magnetospheres, especially those of Jupiter and Saturn, owe many of their unique properties to the presence of such plasma sources.

19.2.3 Dimensionless Ratios Controlling Size and Dynamics

In steady state, the total pressure given in (19.1) must be the same on the two sides of the magnetopause, the boundary of the magnetosphere. For planets other than Jupiter and Saturn, the internal pressure near the boundary is dominated by the magnetic pressure. The scale of the magnetosphere is thus controlled by

$$(B_{\text{surf}}^2/2\mu_o)/P_{\text{ext}} \tag{19.2}$$

where P_{ext} is the total pressure of the external plasma. For planetary magnetospheres, the relevant ratio is that of the magnetic pressure of the internal magnetic field at the surface of the body, to the dynamic pressure of the external plasma expressed in terms of the solar wind density, ρ_{sw}, and the flow speed. Thus the critical ratio, S_{M}, is

$$S_M = (B_{\text{surf}}^2/2\mu_o)/\rho_{\text{sw}}u_{\text{sw}}^2 \tag{19.3}$$

where B_{surf} is the dipole field magnitude at the surface of the planet. When S_{M} is of order 1, the magnetosphere cannot extend far above the surface in the sunward direction. When $S_{\text{M}} \gg 1$, the standoff distance can be tens of planetary radii. It follows from Table 19.2 that the magnetosphere of Mercury cannot extend far above the surface, that the global field of Mars cannot stand off the solar wind, and that all of the other magnetized planets have magnetospheres that extend to large distances above the surface of the body. Ganymede's nose distance is determined by the ratio of magnetic pressures and is found to lie about 1 R_{G} (Ganymede radius = 2634 km) above the surface.

As described in Sect. 19.2.1, the dynamics of the terrestrial magnetosphere are controlled to a considerable extent by reconnection with the magnetic field of the solar wind. Conditions for reconnection with the solar wind are at least intermittently satisfied at all the magnetized planets. However, the relative importance of reconnection and internally driven rotation in driving the dynamics of the system varies from planet to planet. Rotation is particularly important at Jupiter where the large spatial scale and the short rotation period imply that, in the outer magnetosphere, centrifugal stresses dominate those imposed by reconnection. A comparison of the speed of plasma corotating with the planet just inside the magnetopause (u_{corot}) with the maximum convective speed that is imposed on the plasma of the outer magnetosphere by the cross magnetosphere electric field arising from dayside reconnection (u_{reconn}) confirms this statement. (The convective speed refers to flow perpendicular to the magnetic field.) Assuming a reconnection efficiency, α, u_{reconn} can be estimated from the electric field of the solar wind, $|E_{\text{sw}}| = u_{\text{sw}}B_{\text{sw}}$ as $u_{\text{reconn}} = \alpha u_{\text{sw}}B_{\text{sw}}/B_{\text{msf}}$ with B_{msf} the magnetic field of the outer magnetosphere. The field strength typically increases by roughly a factor of 5 between the solar wind and the dayside outer magnetosphere. Accordingly, reconnection-imposed flow just inside the dayside magnetopause is $0.2\alpha u_{sw}$ and for a characteristic solar wind speed of 400 km/s a flow speed of $\sim 80\alpha$ km/s can be attributed to reconnection. Estimates of α are of order 0.1 (Kennel and Coroniti, 1977) or as high as 0.18 (Slavin and Holzer, 1978). For comparison with the effects of rotation, one must correct for the fact that the plasma of the outer magnetosphere does not corotate with the planet. Corotation requires coupling to the ionosphere through field-aligned currents linking to the ionosphere. At large distances, observations show that corotation is not fully imposed. An efficiency factor β can be introduced to account for the fraction of the corotation speed that is actually observed near the magnetopause. At Jupiter, β is ~ 0.3 to 0.5. Then the rotation speed is $\beta r_M \Omega_p$ where r_M is the magnetopause nose distance and Ω_p is the angular frequency of planetary rotation. Using $\alpha = 0.2$ as an approximate upper limit, the dimensionless parameter $\sim 0.1 u_{sw}/r_M \Omega_p$ must be larger than 1 for reconnection to dominate internal rotation. The ratio is >1 for Earth and Mercury whose magnetospheres are dominated by reconnection, $\ll 1$ for Jupiter, which is dominated by rotation, and somewhat smaller than 1 for Saturn, implying that both rotation and reconnection are important.

19.3 A Tour of Planetary Magnetospheres

It is convenient to tour the planetary magnetospheres in groups. The first group, the mini-magnetospheres, includes Mercury and Ganymede. In both cases rotational effects are either negligible or absent and the properties of their inner boundaries differ greatly from those familiar at Earth. They differ in some ways from one another because they form in very different plasma environments. The giant magnetospheres of the rapid rotators, Jupiter and Saturn, form a second group distinguished by important effects of planetary rotation and the significant contribution of internal plasma sources such as moons and rings. The third group, Uranus, Neptune and the heliosphere, contains magnetospheres that do not readily fall into either of the first two categories. A few remarks on Mars conclude the tour of planetary magnetospheres. Selected properties of the central bodies and of the plasma within which they are embedded are given in Table 19.1. Dimensionless parameters relevant to the discussion are given in Table 19.2.

19.3.1 Mini-Magnetospheres

In this chapter, the designation mini-magnetosphere applies to magnetospheres for which the shortest distance to the magnetopause is less than or of the order of one planetary radius above the surface, a requirement that singles out the magnetospheres of Mercury and Ganymede. An excellent review of Mercury's magnetosphere is provided by Slavin (2004). For background on Ganymede's magnetosphere, see Kivelson et al. (2004). These magnetospheres are so small that radiation belts, familiar from studies of Earth, cannot form. They rotate so slowly that the concept of a plasmasphere must be abandoned. In both systems, volatiles from which are formed pickup ions may be important to consider. Length and time scales differ greatly from those familiar from the study of Earth. One can argue that the distant neutral line in Mercury's magnetotail will form beyond 30 planetary radii downtail, a distance covered by the solar wind in ~ 3 min. Contrast this with Earth where it takes the solar wind about 1 hour to flow to the downtail distant neutral line. Data support the view that time scales are governed by these characteristic values.

Simple parallels to Earth do not apply at Ganymede where a low beta plasma flowing at sub-Alfvénic speed confines the magnetosphere. Unique to this magnetosphere are the absence of an upstream shock, the unusual configuration that links it to Jupiter's ionospheres and the quasi-steady form of the external magnetic field that leads to a steady form of reconnection. Much of what we know about Ganymede comes from Galileo's flybys, but simulations now underway are revealing interesting aspects of the unmeasured portions of the system.

Structure and Dynamics

The magnetospheres of Mercury and Ganymede share many characteristics with Earth's magnetosphere. As can be seen in the schematic of Fig. 19.1, a distinct boundary, the magnetopause, separates the internal and external plasmas and in both cases, the polar cusp permits direct penetration of external plasma. The internal fields are dominated by dipolar fields with northward equatorial field orientation. In the presence of an external magnetic field oriented southward, reconnection appears to link internal and external fields.

At Earth stochastic variations of the magnitude and orientation of the external magnetic field control much of the internal dynamics of the system such as storms and substorms. When the interplanetary magnetic field remains southward oriented, magnetic flux is added to the magnetosphere. Substorms return the newly added magnetic flux to the solar wind.

Little is known about Mercury's magnetosphere because measurements are available only from two brief flybys by the Mariner 10 spacecraft that were within the magnetosphere for only about 30 min. Figure 19.1 shows the dayside magnetopause standing above the surface but it is likely that at times of extremely high solar wind dynamic pressure, the dayside magnetopause moves down to the surface.

During the pass through Mercury's magnetosphere shown at the top of Fig. 19.2, the interplanetary magnetic field seems to have remained northward oriented. The smooth variation of the magnetic field magnitude reflects changes linked to the change of distance from the planet. During the pass shown at the bottom of the figure, the interplanetary field was initially northward-oriented but it rotated southward during the pass. On this pass, several substorms were observed. One can estimate the rate of transport of magnetic flux in the magnetotail toward the neutral sheet, assuming that 10% of the solar wind electric field is imposed within the magnetosphere and that the lobe field is comparable with the solar wind field at large downstream distances. For Earth these assumptions imply that it requires ~50 min for a flux tube to flow across a lobe of width ~20 R_E and that during this time the solar wind flows ~200 R_E downstream, reaching the typical distance of the distant neutral line in the tail. For southward oriented IMF, terrestrial substorms recur on average every ~3 h or roughly 3 times the estimated transport time. For Mercury, the same analysis implies a transport rate of 1 min per R_M (R_M is a Mercury radius = 2439 km) across the tail. In the 3 min required for a flux tube in the magnetotail to move north-south across a lobe of width ~3 R_M (see Fig. 19.1), the solar wind would flow ~30 R_M downstream, a plausible location for a neutral line. If substorms at Mercury recur at intervals of a few minutes or roughly the estimated transport time, then the occurrence of multiple substorms during a brief Mercury encounter is plausible. It is uncertain whether the substorm at Mercury includes a phase during which flux is stored in the magnetotail as in the growth phase of terrestrial substorms or if the magnetosphere responds to changes in the solar wind without delay.

Although one must await data from the upcoming MESSENGER (arrival at Mercury on March 18, 2011) and Bepi-Columbo (to be launched in 2012) missions to document the properties of its magnetosphere, it is amusing to anticipate that because of Mercury's considerable orbital eccentricity, some features of the magnetosphere are likely to vary at the 88 day orbital period. With aphelion at 0.47 AU and perihelion at 0.36 AU, the average solar wind Mach number should vary by a factor of 1.3 and the field magnitude and plasma density by a factor of 1.7 every Mercury year. Consequently it is likely that average properties of the bow shock (shock strength, standoff distance) and of "hermeamagnetic" activity may be slowly modulated.

The plasma and field properties of the Jovian plasma flowing onto Ganymede, also vary periodically with a 172 h period because of Jupiter's dipole tilt. (Short period fluctuations are also present but only at amplitudes of order 10% of the background levels.) The external field changes little in magnitude but slowly rocks radially through an angle of ±50° always having a southward orientation.

Despite the field configuration consistently favorable to reconnection, there have been no reports of activity at Ganymede analogous to terrestrial substorms. One must consider whether the dwell in the magnetosphere has been sufficiently long for Galileo to have observed substorm activity during the 6 flybys. Again we must estimate the expected interval between substorms. At the ~150 km/s flow speed of the Jovian plasma, it takes ~3 min for the external flow to carry plasma across the ~10 R_G (R_G, Ganymede radius = 2634 km) width of the magnetosphere to the downstream neutral line (see Fig. 19.1). Analogy with Earth suggests that some small multiple of this number provides a reasonable estimate of substorm recurrence time. If substorms are similar to those observed at Earth, one would expect them to occur every 10 or so minutes. Galileo's multiple passes through Ganymede's magnetosphere provided more than 2.5 h of data within the magnetopause. The fact that substorms were not identified during the Galileo

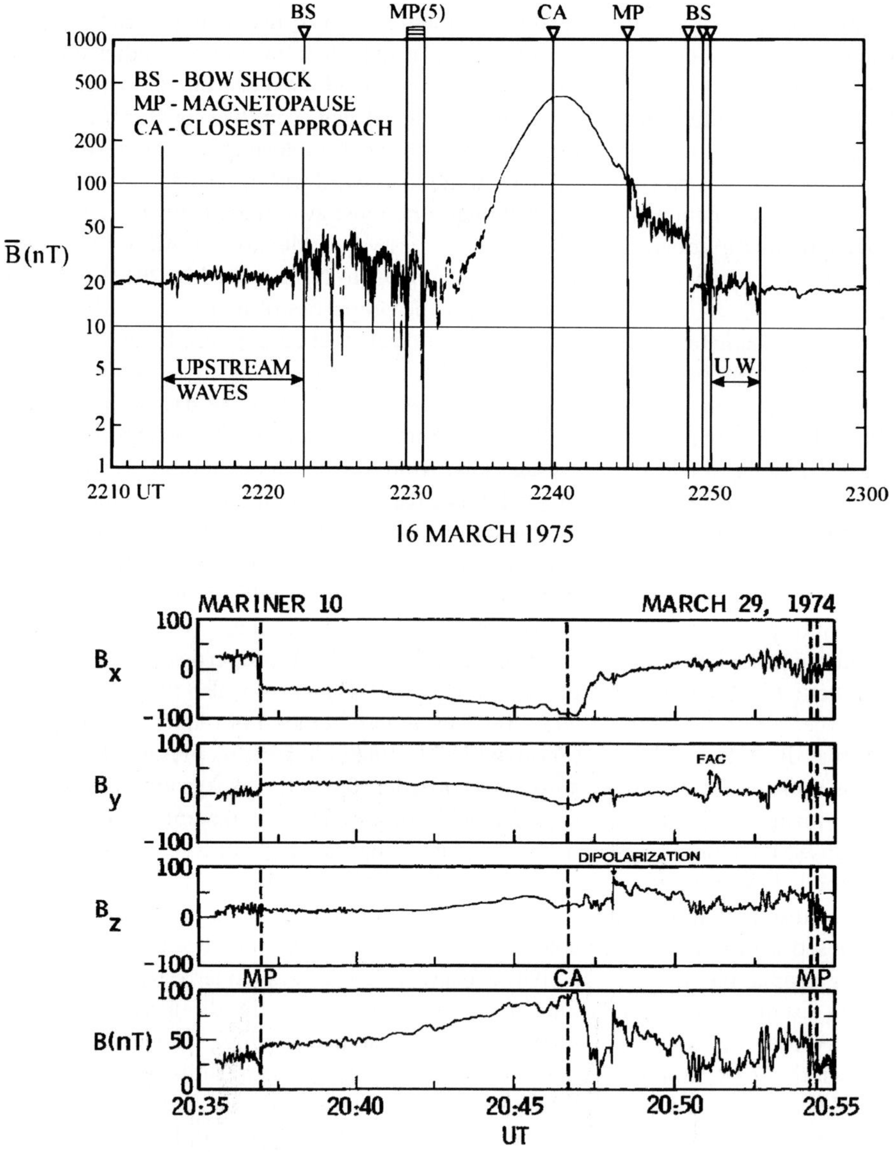

Fig. 19.2. *Above*: The magnitude of the magnetic field observed by Mariner 10 during the third fly-by of Mercury on March 16, 1975 (Grard and Balogh, 2001). *Below*: Mariner 10 magnetic field observations made during the first flyby on March 29, 1974 (Slavin, 2004)

flybys suggests either that substorms do not occur or that they have characteristics quite different from those observed at Earth. The cycling of magnetic flux from the external Jovian plasma through Ganymede's magnetosphere appears not to function through unsteady internal reconnection. Reconnection and subsequent transport may be a relatively steady state process, similar to what at Earth would be termed steady magnetospheric convection. If this is the case, the data from Ganymede gives insight into a particular type of process that occurs at Earth. However, more complete documentation is needed to support this interpretation.

In the magnetospheres of Mercury and Ganymede, rotation is absent or irrelevant. Ganymede rotates about its axis once every 7.15 Earth days, but this is also the period of its orbital motion, so the direction of the external plasma flow changes in phase with the rotation. Thus, relative to the principal axis of the magneto-

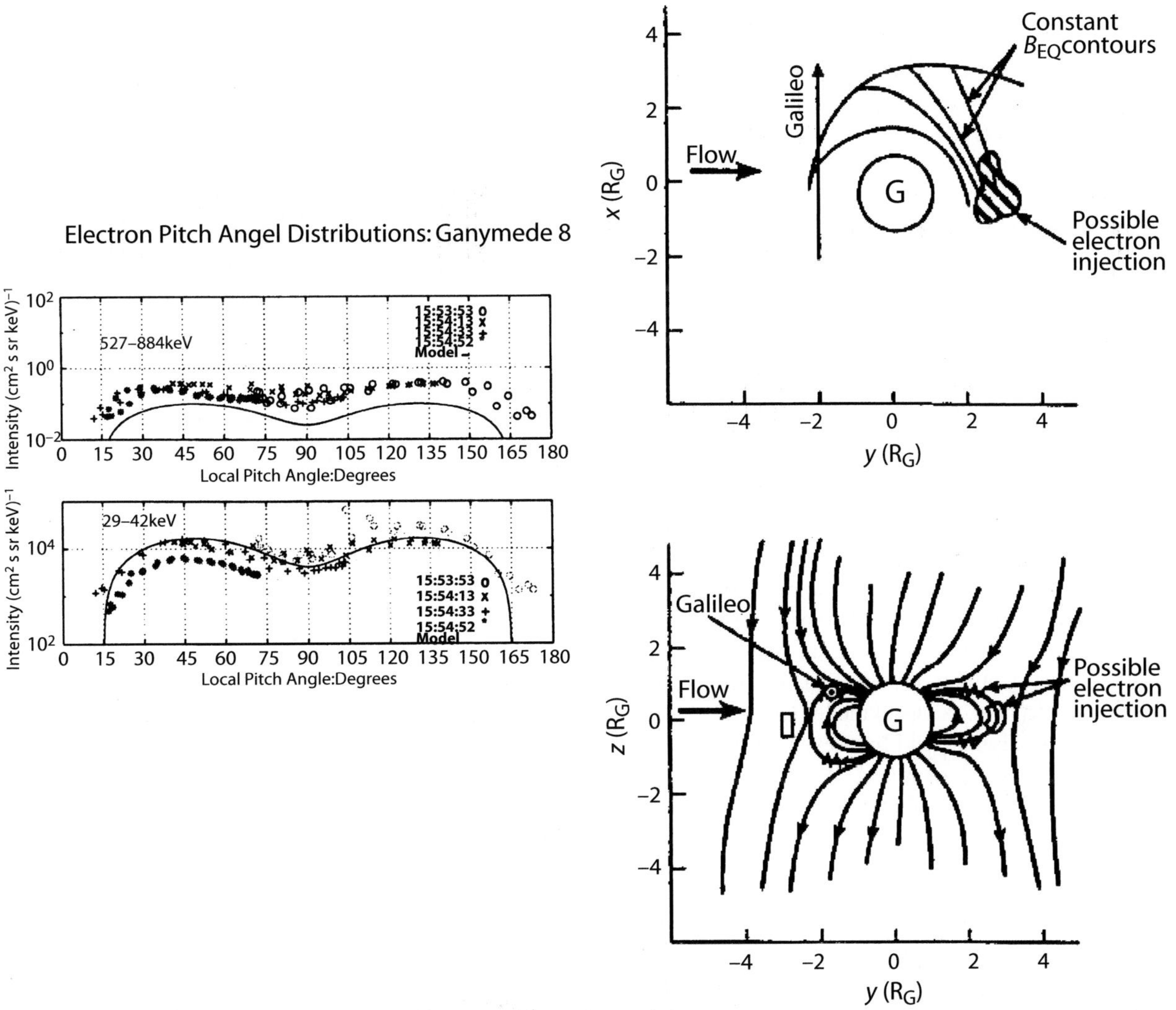

Fig. 19.3. Energetic electrons in the upstream magnetosphere of Ganymede (Eviatar et al., 2000). *Left*: Pitch angle distributions at two different energies, both showing the butterfly distribution produced by drift shell splitting. *Right*: Schematics in the equatorial plane showing nominal electron drift paths (*above*) and a cut through the field and the flow showing proposed particle injection regions and regions in which electrons are detected (*below*)

sphere, aligned with the direction of upstream flow, Ganymede does not rotate at all. Mercury's rotation period is 59 Earth days, but relative to the planet-Sun line, the principal axis of the magnetosphere, the rotation period (P_M) is 176 days. At Earth, rotation is dominant inside the plasmapause, a boundary between relatively high density plasma ($\gtrsim 100\,\text{ions/cm}^3$) with a predominantly ionospheric source and low density magnetospheric plasma. The characteristic distance of the plasmapause from the center of rotation is determined by the location where the corotation speed equals the speed of convection imposed by the solar wind. This distance, L_{pp}, expressed in units of planetary radii is given by

$$L_{pp} = (20\pi B_{surf} R_P / v_{sw} B_{sw} P_M)^{1/2} \tag{19.4}$$

Here R_p is the planetary radius and it is assumed that reconnection with the solar wind occurs at 10% efficiency. For Earth, the critical distance is $L_{pp} \sim 6$ whereas at Mercury $L_{pp} \sim 0.02$ and at Ganymede $L_{pp} = 0$, i.e., there can be no plasmasphere for either system because the nominal plasmapause location lies deep within the planet.

Energetic Particles in Mini-Magnetospheres

Despite the small scales of the two magnetospheres that we are considering, both have significant populations of energetic particles (tens to hundreds of keV per ion). The mechanisms through which particles are accelerated to such high energies are not yet fully established, but the loss processes are quite well identified and only the existence of efficient acceleration can account for the fluxes that are observed.

Two sources of energetic particles, the magnetopause and the neutral sheet in the magnetotail, are probable, both providing acceleration through reconnection. At Mercury, the increase of energetic electron fluxes by more than four orders of magnitude occurs at the times identified as substorms by the rotation of the magnetic field. At Ganymede, energetic electrons are found on dipolar field lines even without evidence of substorm-like behavior (Fig. 19.3).

For both magnetospheres, the drift paths of energetic particles are controlled by the convection electric field and gradient-curvature drift. Any low energy particles whose source is in the center of the magnetotail have a high probability of being absorbed by the central body (planet or moon) and little chance of drifting around it to the upstream side. Characteristic drift paths for energetic electrons are indicated for Ganymede in Fig. 19.3. In a small magnetosphere, losses also occur through pitch angle scattering into the large loss cones. Even on the outermost closed drift paths, the loss cone becomes bigger than 30° at some portion of a nominal circular drift path as illustrated for Mercury in Fig. 19.4. Assuming strong diffusion, 13% of the particles on this drift path are lost each drift period and the fractional loss per drift period increases rapidly as the radial distance decreases. For equatorial particles of charge q and perpendicular energy $W_\perp$ at distance LR_M from the dipole center, the drift period is $qB_o 2\pi R_M^2/3W_\perp L$ or $85/W_\perp(\text{keV})L$ (in minutes) for Mercury. A 50 keV particle at $L = 1.5$ has a drift period of ~1.1 min, implying that energetic particles injected into the magnetosphere are likely to remain for only a few minutes. Drift periods at Ganymede are roughly twice as long at a given L.

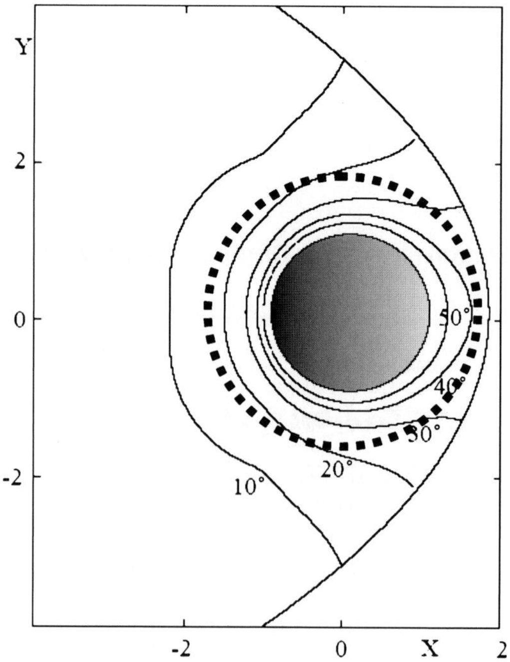

Fig. 19.4. Contours of constant loss cone in Mercury's magnetosphere and a nominal circular drift orbit of a particle in the outer magnetosphere. Modified from Grard and Laakso (2005)

There are no measurements from the day side of Mercury's magnetosphere, but passes on closed field lines upstream of Ganymede show that the flux of energetic electrons increases with distance from Ganymede and falls off in an energy dependent manner as seen in Fig. 19.5. Electrons accelerated by reconnection downstream of Ganymede drift around on the Jupiter-ward side only if their energy is sufficiently high for gradient-curvature drift to dominate. However, they are on open drift trajectories as illustrated in Fig. 19.3, so some radial diffusion is needed to bring them onto closed drift trajectories. The fact that significant fluxes of electrons are found on the dayside magnetosphere despite strong loss mechanisms and that the fluxes are rather symmetric about the central magnetospheric plane containing the magnetic field and the flow suggests that either radial diffusion is strong or that several different injection mechanisms must be acting. The processes that account for the energetic electron populations of the mini-magnetospheres are not yet well understood.

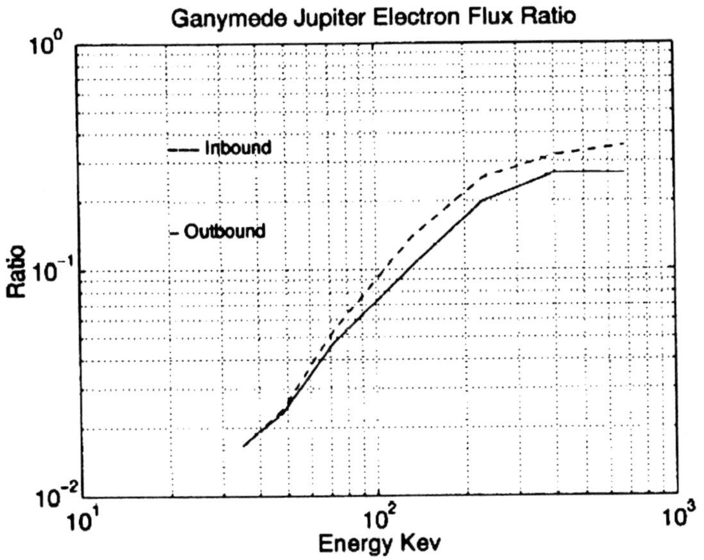

Fig. 19.5. Ratio of electron intensities observed on Ganymede 8 at closest approach on closed magnetic field lines on the upstream side of the magnetosphere to the intensity of ambient Jovian magnetospheric electrons beyond Ganymede's magnetopause as a function of energy. (Eviatar et al., 2000)

Waves in Mini-Magnetospheres

Standing waves have been identified on closed flux tubes. The wave periods are of the order of an ion gyroperiod, so the waves differ from field line resonances typical at Earth. The waves observed at Mercury were quite monochromatic whereas at Ganymede, the waves displayed harmonic structure. The largest amplitude waves in these magnetospheres, also in the ion gyroperiod range, are on the magnetopause boundary and are most probably Kelvin–Helmholtz waves. An example from a pass through Ganymede's magnetosphere is shown in Fig. 19.6. In an analysis of surface waves at Mercury, K.-H. Glassmeier et al. (2003) have pointed out that the applicable gyrotropic theory requires the introduction of a non-diagonal dielectric tensor. The same situation must apply at Ganymede where it should be possible to investigate how the properties of surface waves change as the angle between the internal and external fields change. There is still much to be learned about waves in Ganymede's magnetosphere in anticipation of future analyses of the wave properties of Mercury's magnetosphere.

The Closure of Magnetospheric Currents

Despite the small scale of the mini-magnetospheres, the gyroperiods and gyroradii of the thermal plasma ions are sufficiently small that a magnetohydrodynamic (MHD) description is appropriate for interpreting most of their properties. In this limit, currents are divergenceless, so where the gradient of the perpendicular current is non-vanishing, a non-vanishing parallel or field-aligned current must arise. (Here perpendicular and parallel are directions relative to the magnetic field.) A field-aligned current was identified in the first Mercury flyby in conjunction with substorm activity (see Fig. 19.2). Computer simulations reveal Chapman–Ferraro currents on Ganymede's magnetopause and parallel currents flowing along the field towards and away from Jupiter's ionosphere. The existence of an aurora provides additional reason to believe that such currents are present. A puzzle then arises. How do these currents close? Atmospheres for both bodies are probably time-varying and patchy, so the existence of a gravitationally bound ionosphere is unlikely. Ions produced from the clouds of newly ionized neutrals that are sputtered off the surfaces, referred to as pickup ions, can carry current across the field and may be implicated in current closure and some attention has been paid to the possibility of current closure through the surfaces. However, neither the conducting paths through which field-aligned currents close nor the effect on the dynamics of the system of the current closure paths are fully understood. There is much more to be learned about how magnetospheres work by studying these two small systems.

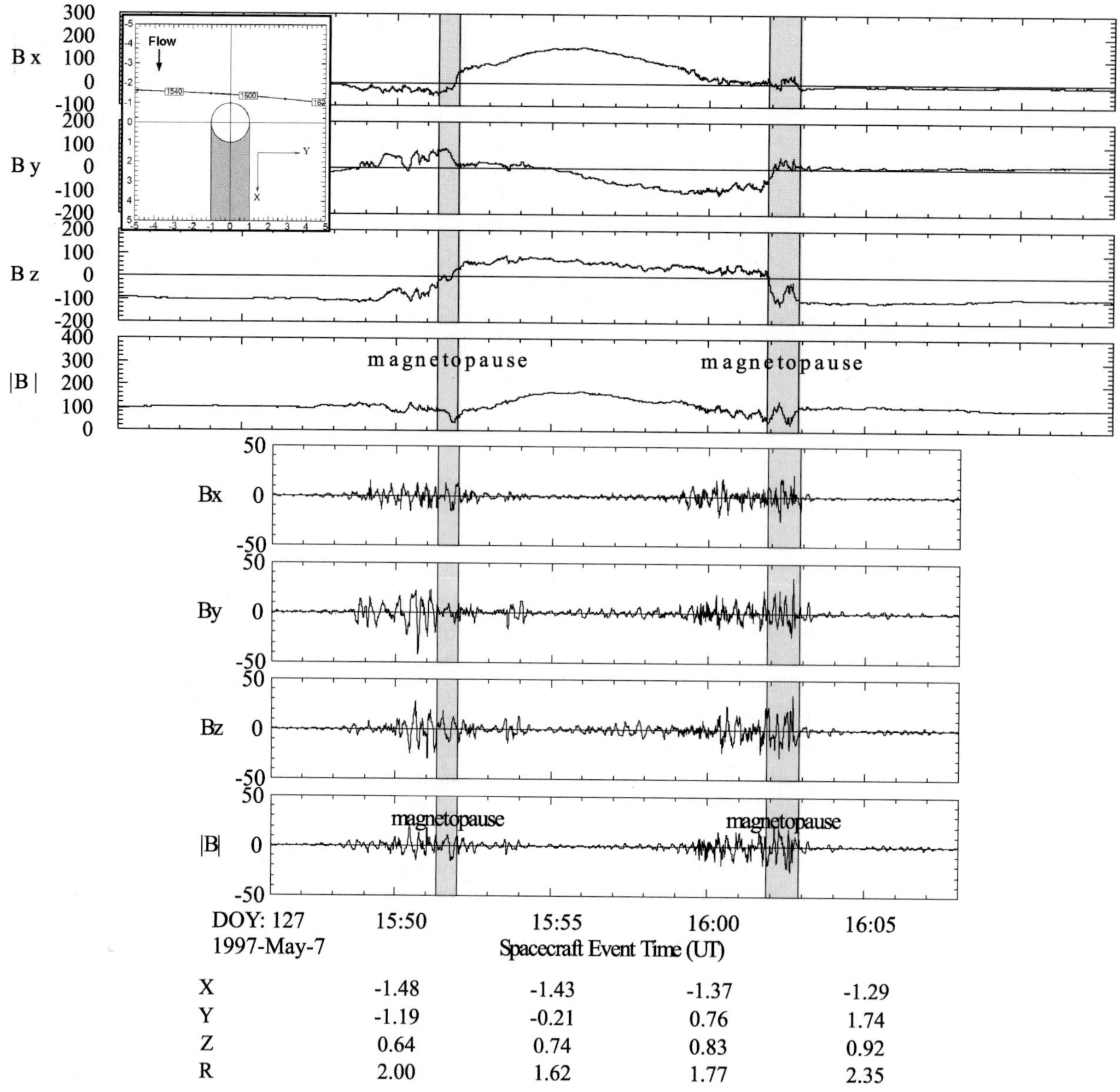

Fig. 19.6. Magnetic field data from a pass on May 7, 1997 in the upstream region of Ganymede's magnetosphere. *Above*: The magnetic field magnitude and components in a coordinate system defined by the flow (x-direction) and the background field direction (in the x–z plane). *Below*: detrended data revealing large amplitude waves near the magnetopause crossings. *Inset* shows the projection of the trajectory on the equatorial plane of the magnetosphere with the wake region shaded

19.3.2 Giant Magnetospheres of Rapidly Rotating Planets

The giant magnetospheres of the solar system are those of Jupiter and Saturn. Descriptions of Jupiter require a vocabulary rich in superlatives (see Bagenal et al., 2004). In scale, Jupiter dwarfs the other planets of the solar system. It has the largest mass, spins fastest around its axis, and has the largest magnetic moment. It seems natural that it should also have the largest magnetosphere. It will become clear that additional unique features relate to dominant role of rotational acceleration, the relatively low momentum density of the solar wind at Jupiter's orbit, and the importance of the four large Galilean moons as plasma sources.

Jupiter – the First Discoveries

Decimetric radiation. For Jupiter, the existence of a magnetic field was inferred in the 1950s from the properties of radio emissions at decimetric wavelengths (tenths of centimeters wavelength or GHz frequency). The radiation is polarized roughly transverse to Jupiter's spin axis in a plane that rocks up and down by about ±10° every Jupiter rotation. The emissions are explained as synchrotron radiation from energetic electrons gyrating near the equatorial plane of a dipolar field. The observed rocking of the plane of polarization was used to infer (correctly) that Jupiter's dipole moment is tilted by about 10° from the spin axis. However, the decimetric radiation gives no information on the magnitude of the field.

Decametric radiation and Io control. The missing information regarding field magnitude was provided from analysis of the very intense decametric emissions (tens of meters wavelength or ~ tens of MHz frequency) modulated at roughly the spin period. The periodicity of this radiation corresponds to the rotation period of the internal magnetic field and, in the absence of a solid surface, is used to define the rotation rate of Jupiter.

Decametric radiation at Jupiter is emitted at the gyrofrequency (f_g) of electrons moving in near circular orbits perpendicular to the magnetic field. Here

$$f_g = qB/2\pi m \tag{19.5}$$

where q is the charge, B is the field magnitude, and m is the particle mass, so, by measuring the frequency, one determines the field magnitude in the source region. A cutoff at the high frequency end corresponds to emission from the region where the magnetic field reaches its largest value, just above the atmosphere of Jupiter. The observed cutoff implies a surface field of ~ 0.001 T. Direct spacecraft measurements revealed that Jupiter's dipole field intensity is 0.0004 T at the equator and several times larger near the pole, providing confirmation of the early estimates. With such a large field (more than ten times the maximum dipole field strength at Earth's surface), there was no doubt that Jupiter would have a magnetosphere; the low solar wind density expected at Jupiter's orbit suggested that its boundary, the magnetopause, would be very distant from Jupiter's cloud tops. Estimates placed the subsolar point at a distance near 50 R_J ($R_J \equiv$ Jupiter radius = 74,000 km) and in situ measurements show that this estimate gives a rough lower bound to the magnetopause location.

Decametric emissions revealed yet another aspect of Jupiter's magnetosphere before the first spacecraft measurements became available. The intensity of the radiation is controlled by the orbital location of the closest large moon, Io, relative to the Earth-Jupiter line, providing the first hint that Io is important to phenomena occurring in Jupiter's magnetosphere.

Neutral clouds of sodium and sulfur ions. In 1973, ground-based observations uncovered yet another surprise. Again the discovery was related to Io, which was found to move around its orbit enshrouded in a cloud of neutral sodium. Sodium turns out to be a marker of the many different neutral species that are liberated from Io into Jupiter's magnetosphere and following the detection of the sodium cloud, ionized sulfur was also observed remotely near Io's orbit.

Spacecraft exploration. Shortly after the discovery of the sodium cloud, Pioneer 10 became the first spacecraft to probe the magnetosphere of Jupiter. Within a few years, direct spacecraft measurements (by Pioneers 10 and 11 and Voyager 1 and 2) confirmed the basic interpretation of the remote observations and uncovered new information. The magnetosphere is even bigger than initial estimates had suggested. Its size can change rapidly. A torus of heavy ions and electrons stretches outward from the orbit of Io. The shape of Jupiter's magnetosphere is flatter than Earth's. Intense fluxes of energetic charged particles fill much of the interior. Exploration continued with an encounter by Ulysses as it swung around Jupiter on its way to a pass over the pole of the sun. As the century drew to a close, Galileo became the first spacecraft to go into orbit around one of the giant planets. Not only did this orbiting spacecraft explore regions never previously encountered, but also it remained within the magnetosphere long enough to reveal the variability of the system over months to years and to investigate the dynamical processes that contribute to the transport of mass and momentum within the magnetosphere.

Structure of Jupiter's Magnetosphere

The particles that populate Earth's magnetosphere come either from the ionosphere or from the solar wind. At Jupiter, such sources are also present, but their contributions are small compared with the sources introduced by ionization of neutrals from the Galilean moons. A useful estimate is that Io injects one ton of plasma per second into its environment. Other moons are weaker sources of neutrals and of the ions formed from them, but their production rates are not negligible.

The heavy ion plasma introduced in the vicinity of the moons controls much of the magnetospheric structure as can be understood by considering the forces that act upon the plasma. The physics of the system is largely described in MHD terms, i.e., in terms of a theory that combines the laws of fluid motion with those of electromagnetic theory. Let us focus on two useful equations:

$$\rho\left(\frac{\partial \boldsymbol{u}}{\partial t} + \boldsymbol{u}\cdot\nabla\boldsymbol{u}\right) = -\nabla p + \boldsymbol{j}\times\boldsymbol{B} + \text{inertial forces} \tag{19.6}$$

$$\boldsymbol{j}/\sigma = \boldsymbol{E} + \boldsymbol{u}\times\boldsymbol{B} \text{ and if } \sigma\to\infty, \quad \boldsymbol{E} + \boldsymbol{u}\times\boldsymbol{B} = 0 \tag{19.7}$$

(in SI units) where (19.6) shows how forces accelerate the plasma and (19.7) is Ohm's law for an electrically conducting fluid in motion. Here ρ is the mass density, $\boldsymbol{u}$ is the fluid flow velocity, p is the thermal pressure, $\boldsymbol{j}$ is the electrical current density, $\boldsymbol{B}$ is the magnetic intensity, $\boldsymbol{E}$ is the electric field and σ is the electrical conductivity. In most magnetospheric applications, one may assume that the plasma conductivity is infinite and adopt the second form of (19.7). In this limit, there is a direct correspondence between the flow and the electric field.

The plasma of the magnetosphere is linked by the magnetic field to Jupiter's ionosphere (period ~10 h). Over much of the magnetosphere, field-aligned currents link the magnetospheric plasma with the ionospheric plasma, closing through the equatorial plasma to exert a $\boldsymbol{j}\times\boldsymbol{B}$ force directed in the sense of Jupiter's rotational motion. The plasma is said to corotate if its angular velocity is that of Jupiter; in the inner magnetosphere the flow is close to corotational. In the middle magnetosphere, corotation is not fully imposed, in which case one talks of corotation lag.

In the rotating system, inertial forces are the inward force of gravity and the outward centrifugal force of the corotating plasma. Beyond a few R_J the gravitational force is negligible but the outward centrifugal force becomes increasingly important with radial distance. The bulk plasma rotating within the magnetosphere experiences centrifugal acceleration, $r\omega^2$ with r the distance from Jupiter's spin axis, and ω the angular velocity of the plasma about Jupiter's spin axis, typically somewhat less than Jupiter's angular velocity.

In the equilibrium system, the centrifugal force of rotating heavy ion plasma and the pressure gradient force of energetic particles are both directed outwards. They are balanced by an inward $\boldsymbol{j}\times\boldsymbol{B}$ force exerted by a disk of azimuthal current. The effect is seen as a stretching of the field lines near the equator in the dayside region between ~15 R_J and ~50 R_J in Fig. 19.7, a schematic representation of a noon-midnight cut through the magnetosphere. The stretched field lines curve sharply as they cross the equator where they exert a curvature force great enough to contain the plasma.

Although many other factors are important in distinguishing the Jovian magnetosphere from others, it is the fact that centrifugal forces are comparable in importance to the other forces through much of the magnetosphere that is critical. In turn, the importance of these inertial forces can be attributed to the rapid rotation of Jupiter, its large size, and the massive amount of plasma introduced into the magnetosphere by the Galilean moons.

Magnetic configuration. Magnetospheres are often described by working inward from the solar wind, but here we shall proceed outward. This approach is natural in a system dominated by the internal sources of momentum that we have described. We start by completing an overview of the magnetic configuration. The internal planetary magnetic field imposes the structure in the inner magnetosphere, the region within ~15 R_J of Jupiter as indicated in Fig. 19.7. At the surface in the northern polar regions, the tilted dipole field points outward from the planet (opposite to Earth's field). Near the equator, the field is oriented southward. Close to the planet, the dipolar field is modified by contributions from higher order multipole moments. Their effect decreases rapidly with distance.

On the day side beyond the orbit of Io (between ~15 R_J and ~50 R_J where heavy ion plasma modifies the magnetic structure as described above), lies a disk-like plasma sheet at all local times in a region referred to as the middle magnetosphere. Beyond ~50 R_J in the outer

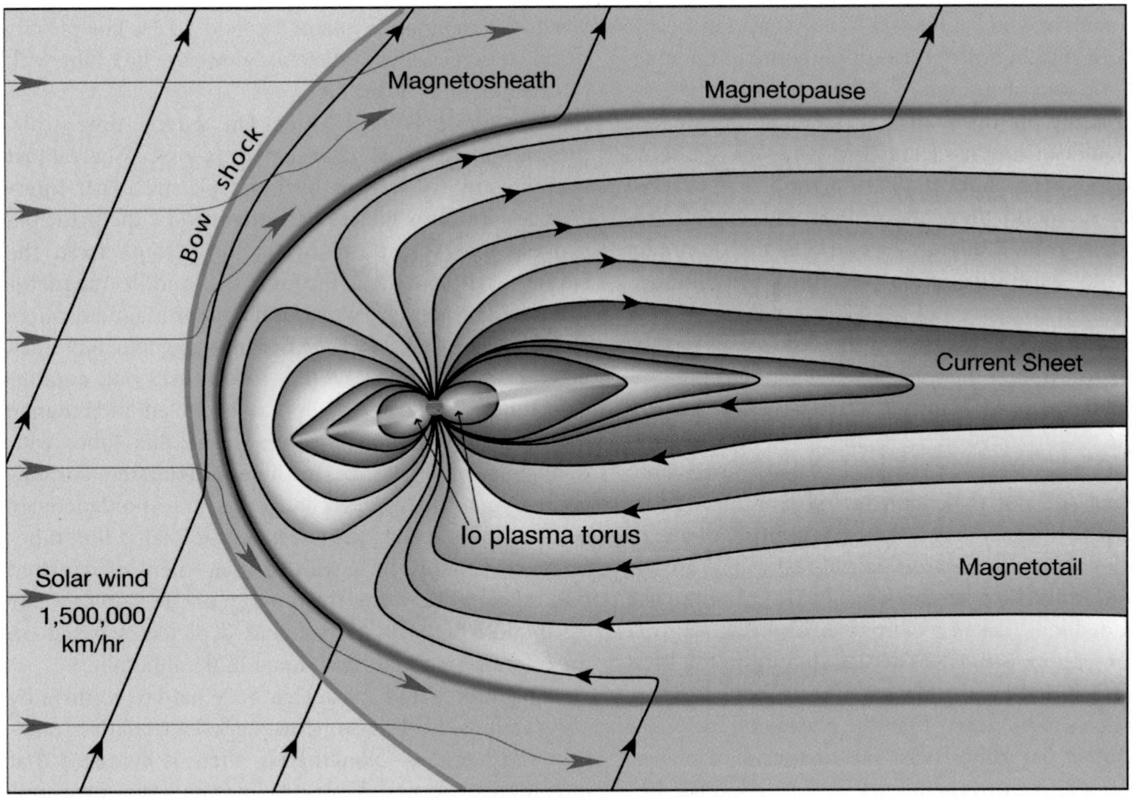

Fig. 19.7. Schematic of the Jovian magnetosphere showing the distended field lines of both day and night side magnetospheres that require an azimuthal current (gold) to flow in the minimum field region. (Courtesy of F. Bagenal, 2004.)

magnetosphere, the field lines no longer stretch radially away from the planet. On the day side of the planet, their orientation is on average roughly dipolar, with southward orientation dominating near the equatorial plane. The field in this region is very disordered and fluctuations of large amplitude are typical. On the night side, the disk-like structure persists to much larger distances. The field structure is similar to that of Earth's magnetotail, although the data are still inadequate to specify the nightside configuration fully.

The magnetopause location is extremely variable. The observed distances to the subsolar region of the magnetopause range from less than 50 R_J to more than 100 R_J, a set of distances that can be consistently understood in terms of pressure balance arguments. The heavy ion plasma spinning around the planet at a fraction of the rate of planetary rotation reduces the gradient of total pressure in Jupiter's dayside magnetosphere relative to that of a vacuum dipole field. Changes of solar wind dynamic pressure that produce a displacement of some fraction of the distance to the nose of the magnetopause at Earth produce a much larger fractional displacement of the magnetopause at Jupiter.

Beyond the magnetopause is found the shocked solar wind plasma of the magnetosheath, bounded, as at Earth, by a bow shock that stands sunward of the magnetosphere in the solar wind. The bow shock slows the solar wind and diverts its flow from the antisolar direction. The standoff distance between the magnetopause and the bow shock is smaller than predicted by simple scaling of the standoff distance observed at Earth. The reduction is, however, consistent with expectations for a magnetosphere somewhat flattened in the north-south direction relative to the roughly circular transverse cross section of Earth's magnetosphere. That distortion of the magnetospheric shape is consistent with the radially extended structure of the magnetic field through much of the magnetosphere.

Plasma sources and characteristics. As at Earth, the ionosphere and the solar wind supply some of the magnetospheric ions at Jupiter. Ions from these sources are predominantly protons. The sources of the heavy ions are the Galilean moons. Clouds of neutrals, sputtered off the atmospheres and surfaces by impacts of charged particles, surround these moons. As was already apparent from ground-based observations, the dominant source is Io. Several processes including photoionization, impact ionization, and charge exchange ionize the neutrals, and thereby create a heavy ion plasma in the equatorial portion of the inner magnetosphere.

Consider a neutral initially at rest with respect to one of the moons. Beyond $\sim 2\,R_{\mathrm{J}}$, the speed of corotation exceeds the Keplerian speed and plasma flows onto the trailing sides of the moons at relative speeds that increase with distance from Jupiter. Ionization of a neutral produces an ion–electron pair at rest in the moon's frame and embedded within the flowing plasma. The newly added ions (called pickup ions) and electrons extract momentum as they are accelerated up to the local flow speed. This slows the plasma. If the plasma near the equator flows more slowly than the plasma off the equator, magnetic flux tubes twist out of meridian planes. The twist, referred to as bendback or corotation lag, implies $\partial B_{\varphi}/\partial\theta < 0$. This inequality implies the presence of an outward-directed radial current density (j_r) and an associated Lorentz force that accelerates the slowed flow. The current circuit closes through field-aligned currents that couple the equatorial plasma to Jupiter's ionosphere and extract momentum from its rotation. This type of field distortion and the associated coupling between the equatorial regions and Jupiter's ionosphere develops wherever the plasma is not fully corotating.

Through much of the magnetosphere, the density is dominated by the low energy (~ 100 eV ions) plasma but the pressure is dominated by energetic ions with energies above ~ 10 keV. From Io's orbit out to $\sim 50\,R_{\mathrm{J}}$, the low energy plasma is confined to a disk of $\sim 1\,R_{\mathrm{J}}$ north-south thickness. The confinement is another manifestation of the importance of the centrifugal force. The field-aligned component of the latter force is directed towards the centrifugal equator, the point on the field line that lies farthest from the spin axis. More thorough analysis shows that within about $20\,R_{\mathrm{J}}$, the plasma density is highest at a position between the centrifugal equator and the magnetic equator and that the peak density shifts towards the magnetic equator beyond $20\,R_{\mathrm{J}}$. The plasma density decreases with distance along the flux tube with a scale height of order $1\,R_{\mathrm{J}}$.

Plasma transport and losses. On a long time scale, plasma sources must balance plasma losses. Several loss mechanisms for magnetospheric plasma exist. Interchange motion is generally thought to be the principal process that transports heavy ion plasma from the source at Io outward through the middle magnetosphere. (The loss of plasma in the outer magnetosphere is discussed in Sect. 19.3.2.) Because magnetic flux must be conserved, when one flux tube moves out, another flux tube moves in to replace it. In an interchange motion, the exchange involves entire flux tubes with their associated plasma. In the approximately corotating plasma torus, interchange occurs spontaneously because the outward displacement of loaded flux tubes accompanied by the inward displacement of depleted flux tubes reduces the free energy of the system. (The distinction between loaded and depleted is based on the total plasma mass contained in the flux tube.)

The interchange model has been hard to confirm by observations, leading some to suggest alternative transport mechanisms. Nonetheless, there is evidence that interchange occurs. Voyager's plasma wave measurements found signatures consistent with intermingled low and high-density flux tubes in the middle magnetosphere. The Galileo Orbiter provided compelling evidence that adjacent flux tubes can have very different plasma content. Small flux tubes (probably of order 1000 km across) with low density plasma at high pressure were detected in a background of higher density, lower pressure plasma just beyond Io's orbit. It is not yet clear whether the distribution of interchanging flux tubes is ordered relative to Jupiter's surface, Io's location, or local time or occurs randomly. The shape of the equatorial cross sections of interchanging flux tubes remains uncertain. Proposed forms include irregular "blobs" and radial fingers of outward and inward moving flux.

Evidence for interchange has also been found at Saturn, where, as at Jupiter, centrifugal stress is important. At Earth, the outer plasmasphere can become unstable to interchange if it extends beyond geostationary orbit. The theoretical arguments were first expounded in the 1980s and good evidence of small scale interchanging flux tubes was found in the Cassini earth-flyby data two decades later.

The properties of the heavy ion plasma at Jupiter are also affected by the process of charge exchange in which a neutral particle becomes ionized and loses an electron to an ion of the plasma. The newly formed ion is accelerated by the convection electric field to the flow velocity of the background plasma and acquires thermal speed equal to that flow speed. The process does not change the charge density of the plasma but, depending on the thermal energy of the original ion, the process may cool or heat the plasma. The newly formed neutral atom retains the velocity of the original ion, which is close to corotation velocity. Lacking a charge, it is unaffected by the magnetic field and escapes from the system on an almost linear path. Neutral matter is thereby distributed in an extended disk surrounding Jupiter. The neutral sodium halo, spread over distances of order 500 R_J near Jupiter, has been observed from Earth.

Twist and warp of the equatorial current sheet. In the schematic of Fig. 19.7, the distended field lines, stretched radially outward by the torus of heavy ions, appear to lie in meridian planes and to be symmetric about a current sheet in the magnetic equatorial plane which is the center of the plasma torus. The actual current sheet surface is warped, and field lines twist out of meridian planes. One might think that the forces associated with the twisted field configuration would ultimately accelerate the plasma to corotational speed and that the twist would disappear. Yet bendback persists. One reason is that pickup is not confined to the immediate vicinity of the moons but occurs throughout the plasma torus, extracting momentum from and slowing the flow, predominantly in the near-equatorial regions. Another reason is that the interchange of flux tubes or other radial diffusion processes transport the bulk plasma radially outward. If plasma moving outward conserves angular momentum, it begins to lag corotation. Again the lag is largest near the equator, causing the field to bend back.

Contributing to the warping of the current sheet is an additional effect related to the north-south motion of Jupiter's magnetic equator relative to points on the rotational equator. As Jupiter spins, the magnetic equator at its surface rocks up and down; associated field perturbations are carried outward from the surface of Jupiter by Alfvén waves whose finite propagation speed introduces a distance-dependent lag to the response. The current sheet appears wavy in meridian planes and the surface appears warped.

Energetic particles: sources, transport and losses. It is customary to regard particles in the >10 keV range as energetic particles. At Jupiter, the energetic ions include protons and a large fraction of heavy ions; energies extend into the tens of MeV per nucleon range. Typically the energetic particle flux decreases with energy according to a power law. Within the magnetosphere, even for protons the energies are far higher than expected for direct acceleration of solar wind particles. Pickup of heavy ions from neutrals does not produce energetic particles.

How then are energetic ions produced? The mechanisms responsible for accelerating ions to energies at which they are observed are not fully established even in the case of Earth's radiation belts. It is known that particles gain energy as they move spatially inward along a gradient of magnetic field magnitude because, in the particle's frame, the magnetic field is increasing in time. If the motion is slow, they conserve the quantity

$$\mu = W_\perp / B \text{ where } W_\perp = \frac{1}{2} m v_\perp^2 \qquad (19.8)$$

μ is referred to as the first adiabatic invariant. Here $W_\perp$ is the kinetic energy associated with motion perpendicular to the magnetic field and $v_\perp$ is the magnitude of the perpendicular velocity of a particle. Inward displacement into an increasingly strong magnetic field increases a particle's energy, but even displacement from the magnetopause to the inner magnetosphere can explain only the low energy end of the energetic particle spectrum. The heavy ions pose an even more serious dilemma because their source is in the high field region and adiabatic outward displacement will cause them to lose, not gain, energy.

Some explanations of the acceleration process at Jupiter have been proposed. Two involve recycling. In order to understand how recycling works, one needs to consider how charged particles move in a magnetic field. Projected into planes perpendicular to the local field, the particles gyrate with the perpendicular velocity $\boldsymbol{v}_\perp$ around a magnetic field line as the gyration center slowly drifts. The radius of the circular orbit centered at the gyration center is referred to as the gyroradius, ρ_g, where

$$\rho_g = |\boldsymbol{v}_\perp| / 2\pi f_g \,. \qquad (19.9)$$

Along the magnetic field, particles move with a parallel velocity $\boldsymbol{v}_\parallel = \boldsymbol{v} - \boldsymbol{v}_\perp$ where $\boldsymbol{v}$ is the total velocity. A stationary magnetic field does not affect the total energy of the particle and hence $|v|$ cannot change but the ratio of $|v_\perp|$ to $|v_{||}|$ and correspondingly the pitch angle $\alpha = \tan^{-1}(|v_\perp|/|v_{||}|)$ can change. As a particle moves off the equator, the field increases in magnitude and $|v_\perp|$ increases to satisfy (19.8). Necessarily $|v_{||}|$ decreases. When $|v_{||}| = 0$, $|v_\perp|$ is the total speed. At this point, the particle starts back to the equator. The location where the reversal of $|v_{||}|$ occurs on a field line, the place where its bounce motion takes it farthest from the equator, is called the mirror point of the particle motion. One recycling model supposes that ions move in from the magnetopause, gaining energy as they move into the stronger field. Having gained energy on their inward path, ions near their mirror points are scattered across field lines by interaction with waves. Recalling that the field lines of a dipole field come close together as they approach the pole, one sees that even short scattering distances across field lines close to the ionosphere can displace an ion onto a field line that returns to the equator far from the initial field line. If the scattering process is sufficiently fast, (19.8) does not apply, and the ions arrive at the equatorial point of the new field line with some of the energy that they acquired on their inward pass. They gain additional energy on their next inward displacement and repeat the scattering process. Several repetitions of such a cycle can, in principle, accelerate particles to the high energies observed.

The model described above was designed principally to account for the acceleration of ions from the solar wind, ions whose source region is the outer magnetosphere. However, molecular ions from Jupiter's ionosphere and heavy ions from the satellites account for roughly half of the energetic ion population. For heavy ions, a different recycling model has been developed. Here one considers the fate of a neutral atom produced by charge exchange. As described previously, such neutrals move away from Jupiter at high speed. There is a small but finite probability that the neutral will be re-ionized before leaving the magnetosphere. If so, $|v_\perp|$ of the pickup ion will correspond approximately to the local rotation speed of the plasma, and in the outer magnetosphere the ion energy can be several keV. Again (19.8) may be used to argue that if this new heavy ion moves closer to Jupiter, it will gain energy in proportion to the increase of B. It is not clear that this process can account fully for the observed particle energy spectra, but it does partially account for the presence of energetic heavy ions. At Earth, some models for energetic particle acceleration invoke electromagnetic wave interactions that can scatter particles in energy. Such processes may also contribute to the acceleration of particles in other magnetospheres.

Transport of energetic particles is similar in some ways to transport of low energy particles. Through the inner magnetosphere and middle magnetosphere, energetic particles typically move azimuthally around Jupiter as does the corotating low energy plasma. The azimuthal velocity of energetic particles differs slightly from that of low energy particles because energetic particles also experience a non-negligible magnetic field gradient drift. Ions drift faster than corotation and electrons drift more slowly. In addition, energetic particles participate in flux tube interchange described previously with the energetic particle flux highest on the low density, inward-moving flux tubes. In the outer magnetosphere, there is a strong local time element in transport. Independent of energy, particle flow down the tail is an effective loss process.

In the discussion of particle acceleration, radial transport was invoked. Radial transport arises partly through stochastic fluctuations. If inward transport and outward transport are equally probable, the effect of random motion is to spread the distribution away from its peak value. Protons introduced into the outer magnetosphere from the solar wind are carried inward by radial diffusion while flux tubes plentifully loaded with low energy heavy ions are transported radially outward from Io's orbit in the inner magnetosphere. Interchange motion, a driven motion that is not stochastic, dominates diffusion outside of the orbit of Io, but this is not the case inside the orbit of Io. Thus, transport inward from Io's orbit proceeds slowly, driven by fluctuations imposed by winds at the feet of the flux tubes in Jupiter's ionosphere.

Energetic ions moving inward can be lost as a result of pitch angle scattering. In this process, interaction with plasma waves can decrease (towards 0°) or increase (towards 180°) a particle's pitch angle. Particles whose velocities are nearly aligned with the magnetic field do not mirror before they enter the atmosphere of Jupiter where they interact with neutrals. Within the atmosphere, particles are either neutralized or lose energy through collisions. Close to Jupiter, loss occurs as particles in the near-equatorial regions collide with the neutral exosphere of Jupiter.

Dynamics of Jupiter's Magnetosphere

Flow bursts in the tail provide a loss mechanism for the Jovian plasma. Arguments in Sect. 19.2 suggest that processes driven by reconnection with the solar wind magnetic field are likely to be unimportant at Jupiter relative to the processes driven by centrifugal stresses. Unlike Earth's magnetosphere in which tail reconnection returns magnetic flux to the solar wind and accelerates solar wind plasma earthward, Jupiter's magnetotail must provide a channel for release of Iogenic plasma with little return of magnetic flux to the solar wind. In considering how plasma containment breaks down in the magnetotail, one needs to recognize that the equatorial portions of the outermost flux tubes move out substantially as the plasma rotates from noon through dusk and into the night sector. The rate of rotation is comparable with the bounce times of particles of energy less than 1 keV, and particles moving outward as they circulate into the magnetotail gain energy from the centrifugal pseudo-potential. It can be shown that this effect results in anisotropy with $p_{\|}$ becoming larger than $p_{\perp}$. Pressure gradient and Lorentz forces can counter the centrifugal forces acting on the rotating plasma in the inner and middle magnetosphere. In the middle magnetosphere, the inward force exerted by magnetic field curvature constrains the plasma of the plasma disk. Farther out, where $p_{\|} - p_{\perp} - B^2/\mu_o > 0$, the plasma-field configuration becomes unstable to the firehose instability. In the magnetotail, it seems likely that the instability becomes explosive and bubbles of plasma surrounded by magnetic field blow off down the tail. The bubbles of plasma are thought to stream down the tail as indicated schematically in Fig. 19.8. High speed outflow in the post-midnight magnetotail was first observed in Voyager energetic particles and subsequently found on multiple passes of Galileo. The outflow is analogous to that found in Earth's magnetotail where bubbles of plasma (plasmoids or flux ropes), confined by wound-up magnetic fields, form during substorms and are returned to the solar wind following acceleration down the tail, but it seems probable that Jupiter's dynamics are driven by the internal instability discussed here and that the flows are not linked to the solar wind magnetic field orientation as they are at Earth.

At Jupiter, newly injected energetic particles have been observed in the inner magnetosphere arriving with a clear energy-dependent dispersion. The dispersion

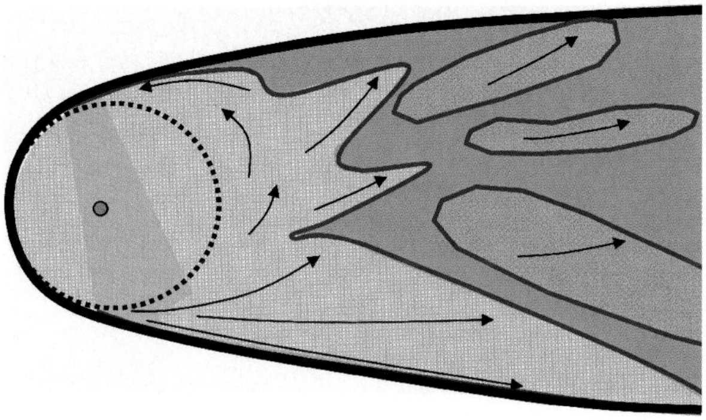

Fig. 19.8. Schematic of flow in Jupiter's equatorial magnetotail (Kivelson and Southwood, 2005)

is consistent with drift from a localized source remote from the spacecraft, the energy-dependence of the drift velocity accounting for the dispersion of arrival times. There seems not to be a preferred local time for the source location and the process is not well understood.

Energetic electrons (> 100 keV) lost from Jupiter's magnetosphere can be identified in the solar wind where measurements show that the high energy electron flux decreases with distance from Jupiter and its amplitude is modulated at the ten-hour periodicity of Jupiter's rotation. The spectral index (ratio of flux in adjacent energy channels) of MeV electrons varies with a 10 h period even at distances of order 1 AU from Jupiter.

Jupiter's Aurora

Jupiter's aurora provides visible evidence of the dynamics of the magnetosphere. As seen in Fig. 19.9, the form of the aurora differs markedly from that found at Earth where the most intense emissions are intermittent and are localized on the night side at latitudes just equatorward of the open-closed field line boundaries. At Jupiter, strong emissions are seen at latitudes substantially lower than the open-closed field line boundary in a region that forms a distorted oval about the pole. Emissions are intense at all local times in this region referred to as the main oval and do not change dramatically with universal time. Magnetic mapping indicates that the main oval is produced in regions linked to the middle magnetosphere, a region in which significant field-aligned currents arise by mechanisms described earlier in this paper. It is widely accepted that field-aligned electric fields arise where the currents link

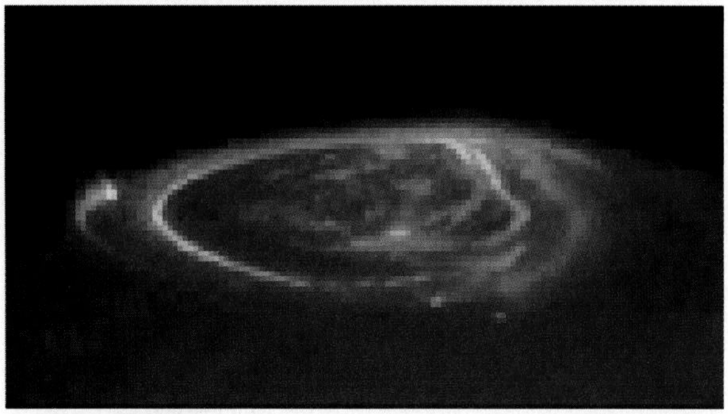

Fig. 19.9. Jupiter's northern aurora imaged by the Hubble Space Telescope (Clarke et al., 1998). The *ring-like bright feature* is referred to as the main oval and *maps* to the plasma disk in the middle magnetosphere. The *isolated bright spots at lower latitudes map* magnetically to the positions of the moons Io, Ganymede and Europa (*left* to *right*). The region of low emission poleward of the main oval maps to the morning sector of the magnetosphere

to Jupiter's ionosphere and that electrons accelerated by such fields excite the observed radiation. Emissions at higher latitudes are time-variable and tend to concentrate in the dusk sector of the polar ionosphere. They are most likely driven by currents that develop to maintain rotational motion as plasma moves outward between noon and dusk and on to the night side of the planet.

Of particular interest are the auroral glows present at the locations where the magnetic field links the Galilean moons to the ionospheres both north and south. These localized bright spots result from field-aligned currents that flow from the conducting bodies through the Jovian plasma that surrounds them. The field aligned current linking Io with the ionosphere is the source of the decametric emissions previously described. The ionospheric footprint of Io extends into a long trail of emission along the locus of magnetic field lines linked to Io's orbit, suggesting that restoring corotation in plasma that has slowed near Io requires a significant fraction of a Jovian rotation period.

Saturn's Magnetosphere

Saturn is similar to Jupiter in many significant ways. Its radius is 84% that of Jupiter and its rotation rate differs little. Moons and rings provide major sources of heavy ion plasma, which are spun up to near corotation by coupling to the ionosphere. The rapid rotation stretches flux tubes radially and a plasma disk is often observed on the day side as well as the night side of the magnetosphere. Saturn's magnetosphere has been explored by Pioneer 11, Voyagers 1 and 2 and is at present being investigated more completely by the Cassini orbiter. Initial reports were published in *Nature* (**433**, 17, Feb 2005) and *Science* (**307**, 125, Feb 2005) and new results have appeared regularly since that time.

Saturn is unique in the extent of the ring system (Jupiter has only a very tenuous ring) and the number of reasonably large moons within its magnetosphere. Both rings and moons are plasma sources. The near-equatorial plasma density rises abruptly just at the outer edge of the A ring and then decreases with distance from Saturn. An extended neutral component is reported to be comparable in density to the plasma in much of the magnetosphere. Until Cassini reached Saturn, it was thought that Titan, the largest moon and one with a dense atmosphere, was a dominant source of magnetospheric ions. However, the surprising discovery of a water plume (Dougherty et al. 2006) at the tiny moon, Enceladus, has led to the recognition that the smaller moons and the rings, which provide ions derived from water ice, actually dominate the plasma sources.

Saturn's surface magnetic field is substantially smaller than Jupiter's and correspondingly its magnetosphere is substantially smaller, its sunward extent being comparable with the distance between Jupiter and the inner edge of the Jovian plasma sheet (see Fig. 19.7). This means that rotational acceleration must be considered but does not dominate at Saturn as it does in the middle and outer parts of Jupiter's magnetosphere. It is not yet established whether there are substorms at Saturn or if there is some rotation-driven mechanism for losing plasma or if multiple processes contribute to plasma losses.

Saturn's moons are not only sources but also sinks for energetic particles. Some of the plasma that flows towards a moon encounters its atmosphere or its surface and is removed from the flow. The remainder of the plasma diverts around the moon and closes in its wake, much as water in a stream parts to flow around a rock. The interaction of energetic ions with a moon may differ greatly from that of a fluid because it depends on pitch angle, bounce phase, and thermal energy. At Saturn, with the moons in the magnetic equator, particles with 90° equatorial pitch angle are strongly absorbed, but energetic particles with pitch angles near 0° or 180° move long distances along the flux tube in the time required to flow across the moon's diameter. Depending on bounce phase, they may or may not encounter the moon as their projected gyrocenters move across the moon's surface. Because very energetic particles have large gyroradii (see (19.9)), their trajectories near the moon may intersect the moon even when their gyrocenters lie pass far from the moon's surface. Thus, the cross section for loss can greatly exceed a circle with the diameter of the moon.

Energetic electrons have small gyroradii even near the outer moons, but their large $v_{||}$ implies that they bounce many times as the plasma flows across the moon. Thus, a moon's near wake is void of energetic electrons.

The voids in electron fluxes and the minima in ion fluxes just downstream of a moon are referred to as microsignatures of the moon. Detailed analysis of the energy and pitch angle dependent microsignatures is useful for the analysis of the magnetic field and gives insight into the nature of the interaction with the moon. The particle depletion in the immediate wake of a moon fills in through radial diffusion at increasing azimuthal distance from a moon. The variation with downstream distance can be used to infer the radial diffusion rate for the energetic particles. In the steady state, the rates are inferred from the slope of the distributions of particles with fixed adiabatic invariants. The solution relies heavily on knowledge of sources and sinks. In a microsignature, one knows precisely when and where the dropout of flux was produced and how long it has taken the plasma to reach the spacecraft. With this information, diffusion coefficients and rates are more accurately established. The inferred diffusion rates roughly agree with the rates determined in other ways. Where there are discrepancies, one must not immediately assume that the rates inferred from microsignatures are pertinent more generally because the plasma conditions may be atypical in the immediate vicinity of a moon.

Like Jupiter and Earth, Saturn's aurora instructs us on aspects of ionosphere-magnetosphere coupling. Saturn's auroral emissions are localized at rather high latitudes and appear to link to the outer boundary of the magnetosphere where they are likely to reflect acceleration associated with reconnection.

Saturn's magnetic dipole moment is closely aligned with the spin axis, so, in contrast with Jupiter, its plasma sheet does not flap up and down as the planet rotates. Despite the axial symmetry of the magnetic field, variations of the magnetic field and the charged particle fluxes at the planetary spin period of 10.7 h are persistent at all locations in the magnetosphere. The periodicity was first identified in the radio wave spectra. Like Jupiter, Saturn emits radio waves modulated at approximately the planetary spin period but, because the radiation is emitted near the electron cyclotron frequency at relatively low altitude and Saturn's dipole field at the surface is weaker than Jupiter's, the modulated emissions are in the kilometric band rather than the decametric band. The periodic modulation is somewhat puzzling because in the absence of dipole tilt there is no clear explanation for the varying intensity; however, high order magnetic multipoles that cannot be ruled out by observations may introduce azimuthal asymmetry at low altitudes. Another puzzle is that the observed period changes over time, having increased by about 6 min between 1980–81 when it was identified by Voyager and 2004–2005 when it was again measured by Cassini. Variable periods were found by Ulysses in the intervening years and there is some recent evidence that the period is increasing systematically with universal time (personal communication: A. Lecacheux, 2006). A close correlation is found at Saturn between rotation-averages of the intensity of kilometric radiation and the integrated auroral input power. Kilometric radiation from Earth's auroral ionosphere (known as auroral kilometric radiation or AKR) is also known to correlate with auroral activity.

Variations of the magnetospheric magnetic field at the period of planetary rotation were discovered in the Voyager data by Espinosa and Dougherty (2000). They interpreted the spin period modulation as a signature of radial transport mechanically imposed at an azimuthally localized region close to the planet. Although signs of periodicity were subtle in the flyby data of the first spacecraft to encounter Saturn, spin period modulation of particles and fields properties is dramatically evident in the data acquired by Cassini on its orbital tour. Localized enhancements of energetic particle flux appear periodically on the night side of the planet and rotate around the planet. The magnetic field amplitude and orientation varies with a 10.7 h period. The relative phases of radial and azimuthal field components follow the pattern that would be imposed by a two cell convective flow pattern that rotates with the planet inside of ~15 R_S (R_S is the radius of Saturn = 60,278 km) and expands or contracts radially beyond that distance.

19.3.3 Unclassified Magnetospheres

The magnetospheres referred to as unclassified are those of Uranus and Neptune. Bagenal (1992) gives insightful descriptions of their unusual properties. These magnetospheres are comparable in scale to Saturn's. Radio emissions, detected as Voyager approached Uranus in 1981, provided the first suggestion that Uranus did have a magnetosphere.

The special character of these systems is linked to the large angles between the planetary dipole axis and the spin axis (see Table 19.1) as well as the presence of strong higher order multipoles of the internal magnetic field. In one rotation period, their magnetospheric configurations vary markedly as the angle between the planetary field and the solar wind velocity changes. The magnetospheres that arise in this case are highly asymmetric and vary greatly in structure at the period of planetary rotation (see Fig. 19.10). The plasma density remains low because of the unstable structure of the magnetosphere.

The configuration of Uranus' magnetosphere also changes in important ways as the planet moves around the sun because the planet spins around an axis that lies only 8° out of the orbital plane. At the time of Voyager's flyby in 1981, the spin axis was nearly aligned with the solar wind flow. In this unique alignment, the flow imposed by planetary rotation is nearly orthogonal to the flow imposed by magnetic reconnection with the solar wind. A magnetotail develops, with two lobes separated by a current-carrying region of high plasma density much as at Earth, but at Uranus the structure rotates around the Uranus–Sun line at the period of planetary rotation. As illustrated in Fig. 19.10a, changing orientation propagates antisunward producing a twisted tail that can be clearly identified in simulations (Toth et al., 2004).

In the 25 years since the Voyager flybys, the planet has moved far along its orbit (84-year period) and the spin axis is now closely aligned with the direction of planetary motion; the solar wind flow is not far from orthogonal to the spin axis as is the case at Earth. The magnetospheric configuration must be much more Earth-like. However, even in the present configuration, the large tilt of the dipole moment should continue to impose significant variability on the structure of the magnetosphere.

The spin axis of Neptune is tilted by only 29.6° to its orbital plane but the dipole axis is tilted by −47° and this configuration produces a magnetospheric structure that varies dramatically within each spin period. As illustrated in Fig. 19.10b, twists in the tail and circular tail current sheets appear in simulations (Zieger et al., 2004) and make it clear that a quasi-steady configuration capable of populating the magnetosphere with plasma is never attained. The simulations of Neptune's changing magnetospheric configuration are of particular interest because of their bearing on our understanding of possible magnetospheres that may have developed during intervals of dipolar reversals at Earth.

Primary plasma sources are moons at Neptune (analogous to Jupiter and Saturn) and the planet's ionosphere at Uranus. Energetic particles, probably accelerated through reconnection in the magnetotail, are observed in both systems, but because of the unusual magnetospheric geometry of these two systems, the energy density in such particles remains small and they do not seem to contribute a significant ring current.

19.3.4 Mars: a Special Case

Extensive exploration of Mars has, in recent years, provided insight into the plasma and field environment of this interesting planetary system (Nagy et al., 2004). Mars lacks a planetary dipole moment sufficient to form a magnetosphere, but localized crustal magnetic anomalies, widespread in the southern hemisphere, are so strongly magnetized that they must form magnetic bubbles capable of holding the solar wind off at altitudes of several hundred kilometers over regions of similar scale. The effects of these strongly magnetized regions on the ionosphere are interesting to contemplate. The magnetic bubbles must arch above the surface in forms similar to solar arcades; reports of encounters with the ionosphere at exceptionally high altitude above the regions of intense magnetic field are consistent with this expectation. Reconnection with the magnetic field of the solar wind should produce open field lines in the vicinity of the arches on the day side of the planet. On these field lines precipitating particles are likely

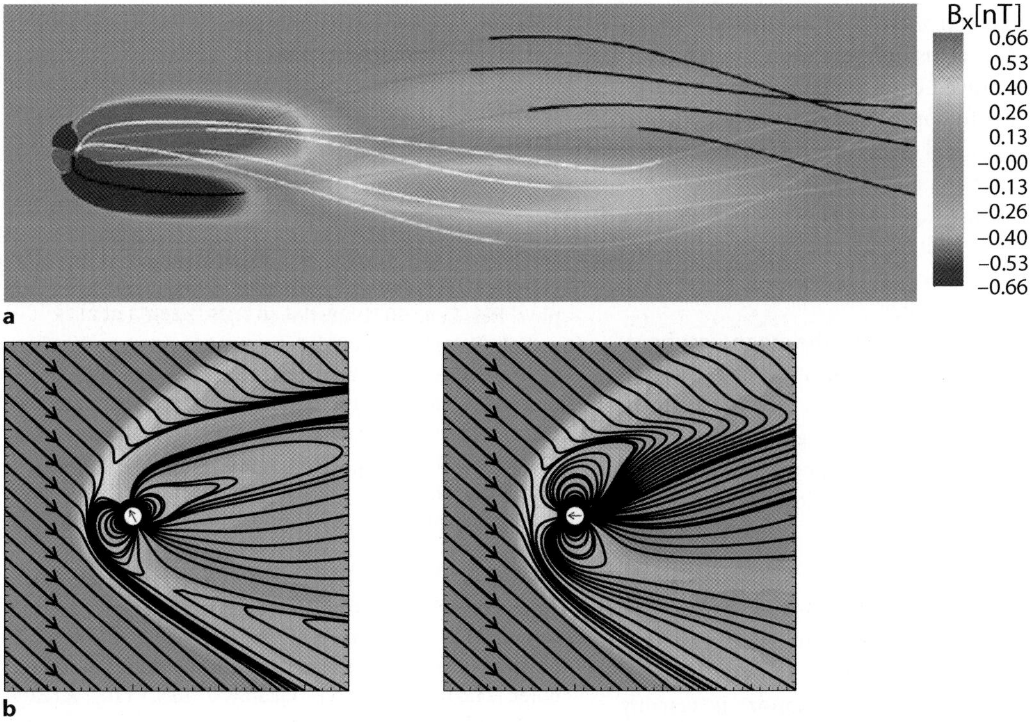

Fig. 19.10. (**a**) From a simulation by Toth et al. (2004). Uranus' magnetosphere with spin axis oriented as it was at the time of the Voyager flyby showing the twisted magnetic configuration of the tail. Magnetic field lines connected to the northern (*black lines*) and southern (*white lines*) poles. Colors represent the x-component of the field. The *color scale* is saturated near the planet. (**b**) From a simulation by Zieger et al. (2004). Diurnal variation of the magnetic field configuration and pressure in an equatorial dipolar paleomagnetosphere for dipole axis at 30° to the normal to the ecliptic plane (*left*) and at 90° (*right*). The configurations are close to those those relevant to Neptune's magnetosphere at different times during a planetary rotation period

to heat the ionosphere. It has been proposed that on the night side, magnetic shielding of the ionosphere within the closed magnetic bubbles may limit access of ionizing electrons and thus imply reduced ionospheric densities above the crustal anomalies. Although some magnetospheric phenomena occur in the regions of anomalously intense magnetic field, their limited spatial extent precludes the development of most magnetospheric phenomena, so the suggestive description of these regions as "mini-magnetospheres" is, in the view of this author, not appropriate.

19.4 Summary: some Lessons for Earth

The magnetospheres of the solar system come in many forms and sizes. By exploring the different magnetospheres we begin to appreciate that Earth's magnetosphere may have been very different in past epochs. When the magnetic dipole reverses, its magnitude may decrease; one can conceive of times when Earth's magnetosphere resembled Mercury's, with the magnetopause lying close to the surface and can think of how this would have affected Earth. For example, at such times, energetic particles would have had ready access to the surface and radiation belts would have been evanescent. Atmospheric escape could have been enhanced. The same situation would have arisen if a magnetosphere had formed in the earliest days of solar system evolution when a T-Tauri solar wind was far more powerful than today's solar wind.

It is quite likely that during magnetic reversals the dipole moment merely rotated, possibly producing a magnetosphere that resembled the highly unstable the magnetospheres of Uranus and Neptune.

Was Earth's magnetosphere ever dominated by rotation as is Jupiter's? It seems unlikely, even though planetary rotation has slowed over the eons. But by studying Jupiter, we learn to appreciate the role of centrifugal stresses and are primed to identify their subtle effects in Earth's magnetosphere. For example, beyond geostationary orbit, the centrifugal radial stresses dominate gravitational stresses and there is some evidence that bits of the plasmasphere can be lost through a process analogous to interchange at Jupiter.

Finally, one must recognize that the magnetospheres we have encountered may be duplicated elsewhere in the galaxy in the vicinity of other stars. One must expect radio emissions, modulated fluxes of escaping particles and planetary auroras in these distant systems, possibly providing new tools for investigation of extrasolar planets.

Acknowledgement. Many of the ideas in the paper matured through lengthy discussions with David J. Southwood, Krishan Khurana and Raymond J. Walker and their continuing interest and support is warmly acknowledged. Work on Ganymede benefited from interactions with Jon Linker and Xianzhe Jia who provided computer-based images of their simulations. The author is grateful to reviewer, Fran Bagenal, for suggesting numerous improvements to the original version of this paper. This work was supported in part by NSF's Division of Atmospheric Sciences under grant NSF ATM 02-05958.

References

Bagenal, F. (1992) Giant planet magnetospheres, Ann. Rev. Earth Planet. Sci. 20, 289

Bagenal, F., Dowling, T., McKinnon, W., editors (2004) Jupiter: The Planet, Satellites and Magnetosphere, Cambridge Univ. Press

Clarke, J.T., Ballester, G.E., Trauger, J., Ajello, J., Pryor, W., Tobiska, K., Connerney, J.E.P., Gladstone, G.R., Waite, J.J.H., Jaffel, L.B., Gerard, J.-C. (1998) Hubble Space Telescope imaging of Jupiter's UV aurora during the Galileo orbiter mission, J. Geophys. Res., 103, 20,217–36

Dougherty, M.K., Khurana, K.K., Neubauer, F.M., Russell, C.T., Saur, J., Leisner, J.S., Burton, M.E. (2006) Identification of a dynamic atmosphere at Enceladus with the Cassini Magnetometer, Science, 311, 1406

Espinosa, S.A., Dougherty, M.K. (2000) Periodic perturbations in Saturn's magnetic field, Geophys. Res. Lett., 27, 2785–88

Eviatar, A., Williams, D.J., Paranicas, C., McEntire, R.E., Mauk, B.M., Kivelson, M.G. (2000) Trapped energetic electrons in the magnetosphere of Ganymede, J. Geophys. Res., 105, 5547

Glassmeier, K.-H., Mager, P.N., Klimushkin, D.Y. (2003) Concerning ULF pulsations in Mercury's magnetosphere, Geophys. Res. Lett., 30, 1928, doi:10.1029/2003GL017175

Grard, R., Balogh, A. (2001) Returns to Mercury: science and mission objectives, Planet. Space Sci., 49, 1395–1407

Grard, R., Laakso, H. (2005) The plasma environment around Mercury, http://solarsystem.estec.esa.nl/~hlaakso/pubs/Grard_Mercury.pdf, consulted in 2005

Holzer, R.E., Slavin, J.A. (1978) Magnetic flux transfer associated with expansions and contractions of the dayside magnetosphere, J. Geophys. Res., 83, 3831–9

Kennel, C.F., Coroniti, F.V. (1977) Jupiter's magnetosphere, Ann. Rev, Astron. Astrophys. 15, 389–436

Kivelson, M.G., Southwood, D.J. (2005) Dynamical consequences of two modes of centrifugal instability in Jupiter's outer magnetosphere, J. Geophys. Res, 110, A12209, doi:10.1029/2005JA011176

Kivelson, M.G., Bagenal, F. (1998) Planetary magnetospheres, The Encyclopedia of the Solar System, Weissman, P., McFadden, L.-A., Johnson, T., Eds.-in-Chief, Academic Press, pp. 477

Kivelson, M.G., Bagenal, F., Kurth, W.S., Neubauer, F.M., Paranicas, C., Saur, J. (2004) Chapter 21 – Magnetospheric interactions with satellites, in Jupiter: The Planet, Satellites and Magnetosphere, edited by Bagenal, F., Dowling, T., McKinnon, W., Cambridge Univ. Press, 513

Nagy, A.F. et al. (2004) The plasma environment of Mars, Space Sci. Rev., 111, 33

Slavin, J. (2004) Mercury's magnetosphere, Adv. Space Res., 33, 1859

Slavin, J.A., Smith, E.J., Sibeck, D.G., Baker, D.N., Zwickl, R.D., Akasofu, S.-I. (1985) An ISEE 3 study of average and substorm conditions in the distant magnetotail, J. Geophys. Res., 90, 10,875–95

Toth, G., Kovacs, D., Hansen, K.C., Gombosi, T.I. (2004) Three-dimensional MHD simulations of the magnetosphere of Uranus, J. Geophys. Res., 109, A11210, doi:10.1029/2004JA010406

Zieger, B., Vogt, J., Glassmeier, K.-H., Gombosi, T.I. (2004) Magnetohydrodynamic simulation of an equatorial dipolar paleomagnetosphere, J. Geophys. Res., 109, A07205, doi:10.1029/2004JA010434

20 The Solar-Comet Interactions

D. Asoka Mendis

The central theme of this review is the role of the plasma tails of comets as free natural probes of the solar wind. It was the behavior of the plasma tails of comets that provided the earliest indication of the continuous outflow of corpuscular radiation from the sun, which we now call the solar wind. In this role comets have not been entirely superseded by the advent of artificial space probes since these, with few recent exceptions are confined to the regions close to the ecliptic plane. Long period comets, on the other hand approach the sun at all inclinations and a few of them get closer to it than any artificial probe has, or will, in the near future. So comets provide us information both of the global properties of the solar wind as well as its spatial and temporal variations. Following a brief summary of the origin and nature of the cometary nucleus a detailed account is given of the interaction of a comet approaching the sun with solar EM radiation and the solar wind since this is what is central to all observed cometary phenomena (outgassing, ionization, and the formation of dust and plasma tails. The review is concluded with some speculations on the expected contributions of forthcoming space missions to the furtherance of our understanding of the subject.

Contents

D. Asoka Mendis, The Solar-Comet Interactions.
In: Y. Kamide/A. Chian, Handbook of the Solar-Terrestrial Environment. pp. 493–515 (2007)
DOI: 10.1007/11367758_20

20.1 Introduction

Comets, or more precisely their solid nuclei, are among the smallest members of the solar system with typical mass $\sim 10^{-11}\ M_{\oplus}$. Yet this very small mass has contributed to their importance in two essential ways. Unlike the larger bodies in the solar system they have undergone little metamorphic change due to the effects of gravity, internal heat and weathering. Particularly those so-called "new" long period comets, which may be entering the inner solar system for the first time since their formation, may well represent the most pristine material in the solar system and therefore are highly significant cosmogonically. Incidentally, it has also been proposed that water and organic materials, that are essential for the evolution of life, were transported to the primitive earth by comets during an early episode of rapid bombardment. Also the negligible cometary gravity cannot hold back the sublimating gases, due to solar heating, and the entrained dust, as comets approach the sun along their highly elliptical orbits. Consequently the solar radiation and the solar wind interact with the cometary gas and dust over length scales that are typically 10^5–10^6 times their nuclear dimensions, at around 1 AU. So paradoxically it is this very smallness of the cometary nuclei that lead to such extensive dust and plasma tails, thereby providing their spectacular visual appearance when close to the sun.

The important role of the plasma tails of comets as free natural probes of the interplanetary medium has been realized since the pioneering work of Biermann (1951). Indeed, it was the behavior of the plasma tails of comets that provided the earliest indication of the continuous outflow of corpuscular radiation from the sun, which we now call the solar wind. In this role comets have not been entirely superseded by the advent of artificial space probes, because while the later (with a few recent exceptions, e.g. Ulysses and Galileo) are largely confined to regions close to the ecliptic plane, long period comets approach the sun at all inclinations. Also few of them get closer to the sun than any artificial probe has or will in the near future. So comets can, in principle, provide us with information both of the global properties of the solar wind as well as its spatial and temporal variations (e.g. high-speed streams, magnetic sector boundaries, coronal mass ejections, etc.).

Finally, some studies have suggested similarities between certain auroral, magnetospheric and geomagnetic processes and processes believed to be operating in the ionospheres and plasma tails of comets. If this turns out to be correct then comets will also provide us with a powerful tool to study these terrestrial phenomena. The advantage that comets have over the earth is that in the case of comets we can visually observe the global morphology and development of these phenomena.

Central to all observed cometary phenomena is the interaction of the comet with the solar radiation and the solar wind. It is the heating of the incoming cometary nucleus by solar radiation, which leads to the release of gas and dust. It is the ionization of the neutral cometary molecules by solar UV radiation and also by the magnetized solar wind, via charge exchange, that eventually leads to the plasma tail. Furthermore it is the pressure of the solar EM radiation on the cometary dust that is responsible for the dust tails. So this dual interaction of comets with the sun will provide the underlying theme of this review. However, its main emphasis will be on the continuing role of comets as natural probes of the interplanetary medium. Here I will adopt a historical approach to show how views advanced from early speculation to our present understanding, which have been greatly advanced during the last 20 years due to in-situ spacecraft observations.

I recognize that the general readership of this review are not experts in cometary science. So I will begin with a brief summary of the nature and origin of comets. For those who would like more detailed information on these topics I recommend the following collections of reviews published after the spacecraft encounters with comets Giacobini–Zinner and Halley: (Mendis, 1988; Huebner, 1990; Mason 1990; Newburn et al., 1991).

20.2 Cometary Reservoirs

In Sect. 20.1 it was noted that comets are among the smallest members of the solar system, with typical masses $\sim 10^{-11}\ M_{\oplus}$. However, they are by far the most numerous. Based on detailed dynamical analysis of the orbital evolution of comets, as well as their known orbital distributions, it is now recognized that there is a vast region of comets surrounding the sun, containing

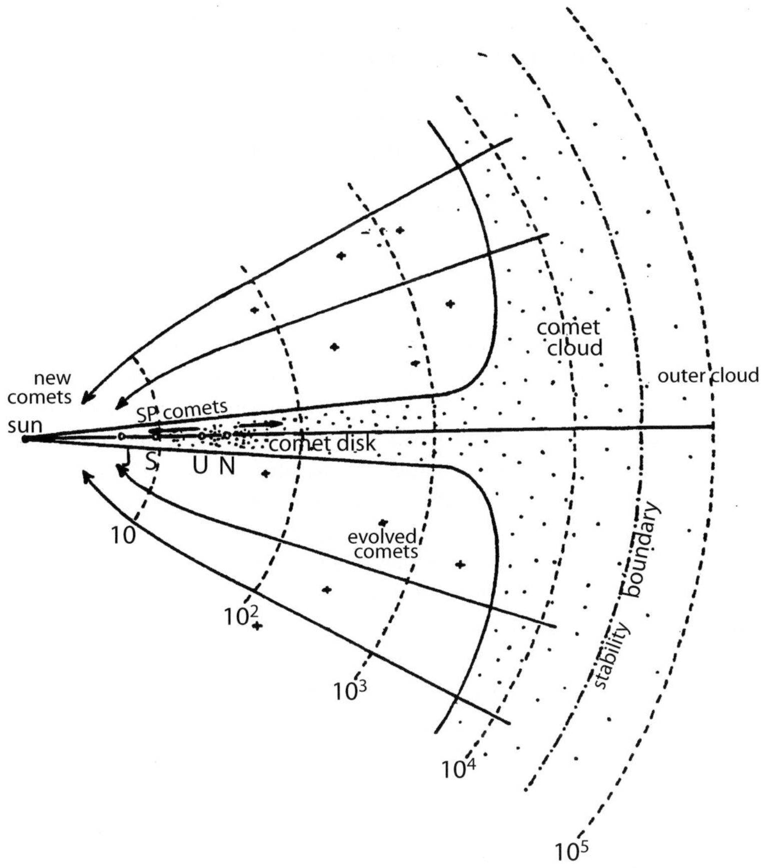

Fig. 20.1. A schematic of the meridional section of the spatial distribution of comets around the sun. The *inner* disc is referred to as the Kuiper belt, while the *thick outer shell* is referred to as the Oort cloud. (From the chapter by Fernandez and Ip, in Newburn et al., 1991)

perhaps as much as 6 trillion comets (see e.g. see chapter by Fernandez and Ip in Newburn et al., 1991 for comprehensive review). The outer part of this cometary reservoir, called the "Oort cloud" after the Dutch astronomer Jan Oort who deduced its existence in 1950, is a spherical shell, containing about a trillion comets, with an inner radius of $\leq 10^4$ AU and an outer radius of $\sim 2 \times 10^5$ AU. This outer radius, which incidentally is about half way to the nearest star, represents the furthest extent of solar gravitational control. Beyond that the tidal force due to the mass distribution in the galactic disc would dominate. The denser inner cometary reservoir, which may contain as many as 5 trillion comets, is believed to be a flattened disc-like distribution with its inner edge at ~ 40 AU which diverges outwards, becoming more spherical at its outer boundary $\sim 10^4$ AU. (See Fig. 20.1.) The inner disc-like region is referred to as the "Kuiper belt" of comets after Gerard P. Kuiper who in 1951 speculated its existence, as the leftover debris from the planetesimal accretion of the outer icy planets.

The next obvious question is how the Oort cloud of comets was formed. It has been recognized for a long time that there was not enough matter for comets to form there in-situ, and that gravitational scattering by the outer planets of a fraction of the planetesimals, that were responsible for their formation by agglomeration, were responsible for the Oort cloud. Wile some planetesimals (comets) were completely ejected from the solar system and other fell into the sun or collided with the inner planets during this process, a fraction was placed in the region of the Oort cloud, which were further scattered into a spherical distribution by the gravitational perturbations of passing stars (as Oort originally proposed) as well as by other gravitational perturbations such as those caused by the occasional passage of giant molecular clouds. While Kuiper believed that the outermost plant, Pluto was mainly responsible for the gravitational scattering from the Kuiper belt, it is now recognized that Uranus and Neptune were the main scatters although the giant planets Jupiter and Saturn could also contribute some fraction.

Comets have been divided into two classes based on their orbital characteristics. Those having periods >200 years (which also approach the sun at all inclinations) are called "long period" comets, whereas those with periods less than 200 years (and which are generally confined to inclinations $\leq 40°$) are called "short period comets. Indeed a good fraction of the latter have periods $\leq$20 years and even smaller inclinations. Earlier it was believed that the source region of both the long and short period comets was the outer Oort cloud, which is most strongly effected by external gravitational perturbations. More recent work, show that while the outer Oort cloud is indeed the source region for the long period comets, [and perhaps a small fraction of the short period comets including the large inclination ones: e.g. P/Halley (T = 76 yrs, i = 162°) and P/Swift-Tuttle (T = 120 yrs, i = 114°)], the source region of the short period comets is the Kuiper belt. This belt is also a continuing source of replenishment of comets for the outer Oort cloud.

While the existence of the outer Oort cloud is widely accepted on the basis of compelling theoretical arguments, no member of it has been observed, in-situ, due to their large distances from earth. On the other hand an increasing number of members of the Kuiper belt have been observed since 1992. While the earliest member discovered by ground-based observations (designated 1992 QB1) was admittedly much larger than any known comet, having a diameter ~320 km, many smaller ones having diameters in the range 10 – 20 km have since been observed using the Hubble space telescope (e.g. see Weissman, 1999, and the references therein).

20.3 The Nature of the Cometary Nucleus

Although not resolved by ground-based observations, the existence of a discrete cohesive nucleus composed of volatile ices and dust has been widely accepted for some time. The basic model proposed by Fred Whipple (1950) described as the "icy conglomerate" model (and popularized as a "dirty snowball") wherein the nucleus is considered an admixture of ices such as H_2O, CO_2, CO, CH_4, NH_3, etc., and nonvolatile (meteoretic) dust was first used by Whipple to explain puzzling dynamical effects on cometary orbits; the so-called non-gravitation recoil effects due to non-isotropic outgassing. Thereafter this model, with some essential modifications became the basis of all subsequent work on the dynamics, physics and chemistry of comets. These essential modifications had to do with nature of the ices, the possible formation of nonvolatile mantles (as was already anticipated by Whipple) due to the incomplete entrainment of the surface dust by the out-flowing gases, as well as the possible chemical differentiation (i.e. layering) of the sub-surface ices according to volatility (with the least volatile H_2O ice closest to the surface, and the more volatile ices further down), due to thermal processing (see Mendis, 1988). Early on, combined observations of a given comet far away from the sun (where it behaved like a non-outgassing asteroidal body) with those near the sun (where it was strongly outgassing) were used to estimate both its radius and its albedo. This led to typical sizes of a few km and albedos in excess of 0.6. (For a detailed review of the work up to the cometary space missions, see Mendis et al., 1985)

We still do not know the nature of cometary nuclei in the Oort cloud and probably will not into the foreseeable future. But due to spacecraft missions to four short period comets (VEGA and Giotto to 1P/Halley; DSP1 to 19P/Borrelly, Stardust to 81P/Wild2, and Deep Impact to Tempel 1) we now have fairly good images of their nuclei (Fig. 20.2).

Several features of these nuclei, some that were totally unexpended by the cometary community, are obvious in these pictures. Comet Wild 2 with dimensions (~5.5 km × 4 km × 3.3 km) is the closest to sphericity. While the larger comet Borrelly with dimensions (~8 km × 3.2 km × 3.2 km) looks like a potato or bowling pin with a distinct neck, the even larger comet Halley (15.3 km × 7.2 km × 7.2 km) looks more like a peanut. Comet Temple 1, which was imaged most recently, before and after it was impacted by the projectile fired from the Deep Impact mission spacecraft, also seems potato shaped, with a size comparable to Borrelly. The non-spherical shapes of the small bodies were not surprising; what was surprising was their darkness. The overall geometric albedo of comets Halley, Borrelly and Wild 2 are respectively estimated as ~(0.02 – 0.04), (0.01 – 0.03), (0.03 ± 0.015). No numerical value for the albedo

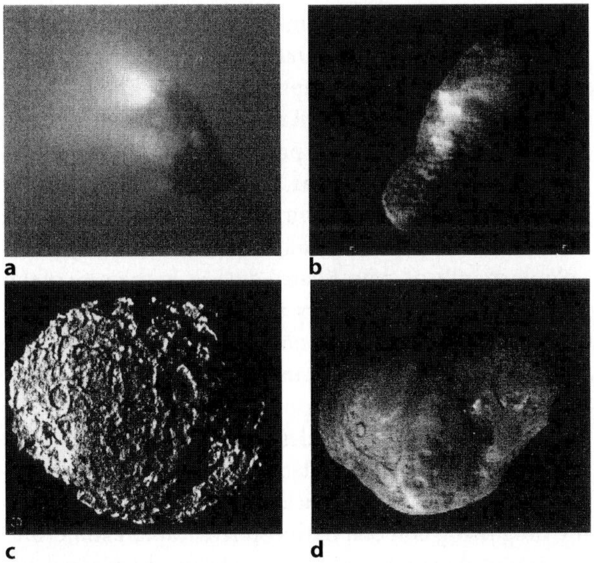

Fig. 20.2a–d. Nuclei of three comets imaged so far from fly-by by spacecraft; (**a**) 1P Halley, (**b**) 19P Borrelly, (**c**) 81P Wild 2, (**d**) Tempel 1. See text for descriptions (Adopted from Weaver, 2004)

of Temple 1 has yet been reported, but it is also believed to be comparable to the other three. These were much smaller than the expectations for even the dirtiest snowballs. Also it is clear that we are seeing not an icy surface or even a pile of dust lying on it. Rather the surfaces seem to be well consolidated dusts with numerous surface features with linear scales varying from a few hundred to a few meters; the maximum resolution being highest for Tempel 1. These features have variously been described as valleys, mountains, mesas, craters, ridges, fractures, etc. All three comets were sufficiently close to the sun when photographed ($d = 0.83$ AU for Halley; $d = 1.36$ AU for Borrelly; $d = 1.8$ AU for Wild 2) and had significant gas (mainly H_2O) production rates ($\sim 7 \times 10^{29}$ mols/s for Halley; $\sim 2 \times 10^{28}$ mols/s for Borrelly; and $\sim 5 \times 10^{27}$ mols/s for Wild 2). While the outflow of H_2O and other gases from the comets are not observed visually, the dust which is entrained by the out-flowing gas acts as a tracer. It is clearly seen in the case of Halley's Comet that the dust is not flowing out isotropically but rather in the form of highly localized sunward jets. While too faint to be seen in the images of the other three comets the dust emission from them are also in the form of jets (collimated beams and fans) mainly in the sunward hemispheres. According to Uve Keller (see his contribution in Huebner, 1990), all the jets observed at Halley occupied $\sim 10\%$ of the surface. So it appears that most, if not all, of the cometary outgassing arises from this small fraction of the surface. The situation is qualitatively similar in the other three comets also. The lack of any ice on the surfaces of all four comets is also supported by the fact that the surface temperatures of three of them estimated by IR measurements are well in excess of 300 K (no measurements are available for Temple 1). Of course from the observed production rates of the volatiles it is clear that H_2O is by far the dominant one. In the case of comet Halley the production rate of H_2O (by number, was $\geq 80\%$ of the total production rate, with CO appearing to be the next most abundant species, with a production rate of $\geq (10-15)\%$ (e.g. see Mendis, 1988). In the case of the other three comets the production rate of minor species are not available, but in each case H_2O is presumably the most abundant volatile species. What is not clear so far is the chemical composition of the outer crust. One important clue comes from the chemical composition of the dust in Halley's Comet. The dust composition analyzer on board the VEGA spacecraft indicated the presence of at least three broad classes of grains. While one class is composed entirely of low-Z elements (mainly CHON), a second class is similar in composition to C1 chondrites but enriched in C, and the third is more similar to the second but less enriched in H. There is also some indirect support for the existence of fragments of the organic polymer $(H_2CO)_n$ as well as several other hydrocarbons (e.g. $C_3^+H_3^+$) from the ion mass spectrometer onboard Giotto. Consequently it is tempting to suppose that the outer mantle is enriched with dark organic material contributing to its low albedo, although it has also been proposed (in the case of Halley) that the mantle is highly porous and that this porosity itself may be largely responsible for the low albedo.

The unexpected observation, in all four comets, that the outgassing, traced by dust jets, is so anisotropic and localized to a small fraction of the surface needs to be addressed. At present this observation is certainly not well understood. We may however speculate that this is associated with the non-uniformity of the inactive (nonvolatile) crust overlying the active subsurface mix of volatile ices (presumably mainly H_2O) and nonvolatile dust. The thickness and porosity of the outer crust could be highly variable and it may also have fissures. In such a case the outgassing would be preferentially from the regions of low crystal thickness, high porosity or fissuring. Such regions of crystal weakness may also be the regions from where dust could be more easily entrained by the out-flowing gasses. It is hoped that the data analysis from the "Deep Impact" mission, which launched a projectile from the spacecraft onto comet Temple 1 on July 4, 2005, which created a huge "football field" sized crater, would provide some useful information about the nature of the cometary nucleus. Very preliminary data in fact seem to indicate a fragile dust layer overlying an H_2O–ice dominated interior.

As discussed above, our present knowledge of the structure of the cometary nucleus is limited to four highly evolved short period comets. The structures of those remaining in the Oort cloud or the Kuiper belt are unknown. As I pointed out in Sect. 20.1, these comets are generally regarded to be the most pristine material in the solar system. They remain there in a deep freeze, and so unlike the periodic comets, have not been processed by solar heating. However it has been pointed out by a number of authors that they too may not be totally "pristine" due to irradiation by cosmic rays. Fred Whipple estimates that an outer layer to a depth of ~1 m would be processed by this radiation in cosmogonic times. In this outer layer, which he refers to as the "outer frosting," he believes that the damage to the crystalline structure of the ice would be total, leading to amorphous ice with considerable free energy. He argues that this would in turn lead to the shedding of this "outer frosting" due to even moderate heating during their first excursion into the inner solar system, and attributes the enhanced activity of "new" comets at relatively large heliocentric distance to this cause. It has also been argued that, when they first condensed at the very low temperatures in the outer solar system, the cometary ices were in an amorphous form and underwent a phase transition to the crystalline form only on being first heated above a critical temperature ~75 K (for a detailed review, see Mendis et al., 1985).

20.4 Interaction with Solar Radiation

Central to essentially all cometary activity is the interaction of its nucleus with solar EM radiation. As the nucleus approaches the sun in its elliptical orbit it is increasingly heated by solar radiation. Eventually volatile ices contained therein evaporate and expand outward carrying along a quantity of dust. If the ices are directly exposed to solar radiation, presumably as in the case for a "new" comet entering the inner solar system for the first time, the ices will sublimate and expand essentially into a vacuum. If the volatile ices are covered by a less volatile mantle of consolidated dust, as in the case of the four short period cometary nuclei observed to date by spacecraft, the heated gases have first either to percolate through a presumably porous mantle or to vent through crevices or fissures on it.

These escaping molecules are then photodissociated and photoionized by the solar UV radiation. While these are the two primary chemical processes taking place in the outgassing cometary atmosphere, rapid exothermic ion-molecule reactions in the collision-dominated inner coma (typically $r \leq 10^4$ km for a medium bright comet such as P/Halley at 1 AU) continuously reshuffle the chemical species there. Consequently chemical modeling of the cometary atmosphere/ionosphere have had to deal with extensive chemical networks sometimes including over 1000 chemical reactions among about 100 "parent" molecules and daughter species (e.g., see Mendis et al., 1985; Huebner, 1990, for detailed reviews). The most important of these ion-molecule reactions in an H_2O-dominated cometary atmosphere is:

$$H_2O^+ + H_2O \rightarrow H_3O^+ + OH + 1.1\,\text{eV}.$$

Indeed all atmospheric/ionospheric models of comets since then have predicted that this stable hydronium ion H_3O^+ would be the dominant ion in the inner atmosphere/ionosphere, unless the percentage of NH_3 or Na was substantial. A typical ionospheric profile (in this case for the initial nuclear chemical abundance ratio of $H_2O : CO_2 : CO : N_2 = 1.0 : 0.15 : 0.15 : 0.1$) is shown in Fig. 20.3. Notice the increasing dominance of H_3O^+ with decreasing nuclear distance. The spectrum of H_3O^+ is unknown. Consequently one of the major discoveries of the ion mass spectrometers on board the NASA/ICE spacecraft to comet P/Giacobini–Zinner, and Giotto and VEGA spacecraft to comet P/Halley was the detection of this ion ($(\frac{m}{q}) = 19$) together with the other water group ions ($(\frac{m}{q}) = 18$, 17 and 16 for H_2O^+, OH^+ and O^+ respectively).

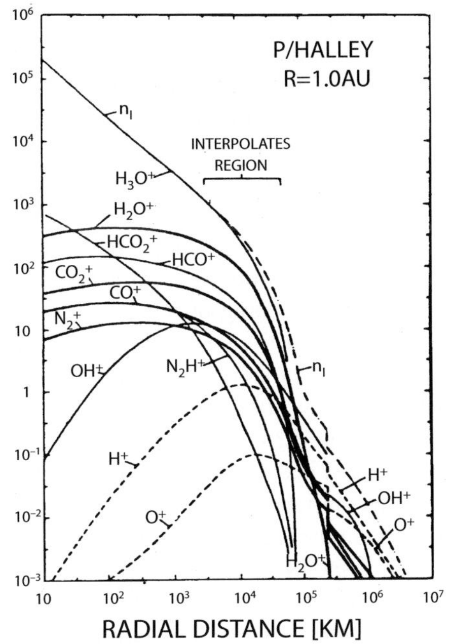

Fig. 20.3. Calculated radial profiles of the ion composition of comet Halley at a heliocentric distance of 1 AU assuming a nuclear chemical abundance ratio of $[H_2O] : [CO_2] : [CO] : [N_2] = 1.0 : 0.15 : 0.15 : 0.1$. (From Ip, 1980)

Spectroscopic observations of comets go back a long way to the early days of astronomical spectroscopy. The dominant violet bands of CN, seen since in the coma of every comet, was first observed in the comet 1881b. Advances in spectroscopic techniques together with the expansion of the spectral range both longward to (IR and radio) and shortward to (UV and X-ray) have resulted in the detection in many more species (both neutral and ionized) since then. A further extension was the use of (neutral and ion) mass spectroscopy to detect chemical species in-situ at comets P/Halley and P/Giacobini–Zinner. In this case, since what is measured is the (molecular) mass-to-charge ratio $(\frac{m}{q})$, uncertainty is involved due to aliasing or overlapping. For instance, $(\frac{m}{q}) = 18$ could represent both H_2O^+ or NH_4^+ and compelling theoretical arguments and calculations are needed to support one candidate over another. In particular there is little doubt that the dominant $(\frac{m}{q})$ peak at 19 is due to the hydronium ion H_3O^+. A composite list of all identifications to date is given in Table 20.1. In each case (atoms, molecules and ions), the species listed above the horizontal lines were detected by EM emissions, while those listed below these lines were deduced from neutral and ion mass spectroscopy. In the latter case the possible aliasing is indicated. In all cases the square brackets indicate a certain degree of uncertainty.

Besides the ionic species listed in Table 20.1, there are certainly others with higher $(\frac{m}{q})$ values. The PICCA ion analyzer onboard Giotto measured several broad peaks at high $(\frac{m}{q})$ values. Observing five distinct peaks, taken to be centered at 45, 61, 75, 91 and 105 and noting that they show alternate differences of 16 and 14, Walter Huebner identified them as dissociation fragments of a linear chain of polymerized formaldehyde $(\ldots -CH_2-O-CH_2-O\ldots)$ (see Huebner, 1990). Since H_2CO has also been observed spectroscopically and arguments have been advanced for $(H_2CO)_n$ to be a component of the "dust," this proposal is plausible. However it was subsequently pointed out by Mitchell et al. (1989) that these observed peaks are rather broad, $\Delta(\frac{m}{q}) \geq 3$, and that these observed peaks could equally well be at 45, 61, 75, 90 and 105. These together with one at 31

Table 20.1. List of Chemical Species Identified in Comets

Atoms	Molecules	Ions
H	C_2, $^{12}C^{13}C$	C^+, Ca^+
O	CH	CO^+, CH^+
C	CN, ^{13}CN	CN^+, OH^+
S	CO, CS	N_2^+, SH^+
Na	NH, OH	H_2O^+, CO_2^+
K	S_2	H_2S^+, NH_4^+
Ca	[SO], [NO], [SH]	————————
V	C_2, NH_2	H^+, C^+
Mn	H_2O, HCN, HCO	O^+ (CH_4^+, NH_2^+)
Fe	$[H_2S]$	N^+ (CH_2^+)
CO	NH_3, H_2CO	Na^+, Fe^+
Ni	CH_3CN	CH^+, CS^+
————————	$[CH_3OH]$	O_2^+, S_2^+
	————————	OH^+ (NH_3^+, CH_5^+)
	O (CH_4, NH_4)	CS^+, SO^+
	OH (NH_3)	CO_2^+, CS_2^+
	H_2O	H_2O^+ (NH_4^+)
	CO (N_2, C_2H_4)	H_3S^+, C_3H^+
	CO_2	H_3O^+
	H_2CO	$H_5O_2^+$, $C_3H_3^+$
		HCS^+, H_3CS^+

show variation by steps of 14, 16, 14, 15 and 15 which could correspond to the dissociative loss of the molecular fragments CH_2, O and NH from a large organic molecule. Consequently Mitchell et al. (1989) identify these peaks with a different set of molecular ions, which they believe are themselves fragments of molecules that constitute the observed CHON dust particles. So it is possible that either one or both sets (or even others) could contribute the aforementioned peaks.

In this connection, it is now recognized that these dust particles, as well as others, could represent a distributed source for several observed molecules and molecular fragments. Earlier it was generally believed the nucleus alone was responsible for all the observed species; the stable molecules outgassing directly from it, with unstable fragments (e.g. CN, CH, etc.) being photodissociation fragments of them.

While the cometary atmospheric species appear to originate in asymmetric jets near the nucleus, the spatial distribution of the neutrals is quite spherical at larger nuclear distances, and this is as expected due to lateral pressure gradients within these jets which will lead to rapid spherical divergence. With regard to their spatial distribution, most neutral species are confined to a region with a radial dimension $\leq 10^5$ km, in the case of a typical medium bright comet like P/Halley at around 1 AU from the sun. Ultraviolet observations of several bright comets however show huge Ly-α halos surrounding them with linear dimensions $\sim 10^7$ km. The Ly-α comes from resonance scattering of solar UV radiation on atomic hydrogen, which is clearly a photodissociation product of the dominant H_2O. The reason why the halo is so extensive is due the fact that the average excess energy of the photodissociation of H_2O ($\sim$1.9 eV) is largely carried by the lighter fragment, H, which can acquire a speed in excess of 10 km s^{-1}. Also the lifetime of H against loss by photoionization and charge exchange at 1 AU $\geq 10^6$ s.

Unlike the neutral species the cometary ions observed from the ground are distributed along a long tail, (often $10^7 - 10^8$ km long) pointing almost directly away from the sun. The reason for this will be discussed in the next section. Typically the strongest emissions are due to CO^+. This is not because it is the most abundant species there (e.g., O^+, OH^+, H^+ are all likely to be more abundant) it is because this ion has very strong resonance bands at optical wavelengths.

As stated earlier, Table 20.1 is a composite list; only a fraction of the species there are observed in a given comet. For instance regarding the metallic atoms, while Na is generally observed at medium distances from the sum, all other metallic species are observed only in sungrazing comets and probably come from the vaporization of dust.

As discussed earlier, it is the gas drag on the dust that is responsible for the dust outflow from the comet. At the same time, the reverse drag of the dust on the gas causes partial choking of the gas outflow and enables it to be transonic. The dynamical effect of the dust is essentially that of a Laval nozzle; it enables the gas to start subsonically at the surface, then smoothly traverse the sonic point and become supersonic beyond. The dust plays another important role in the cometary atmosphere, partaking in the transfer of visual and near-IR radiation which is responsible for heating and sublimating the cometary nucleus. Detailed models of the dynamics and thermodynamics of the dusty cometary atmosphere have been considered by several authors and will not be further discussed here. For detailed reviews see (e.g., see Mendis et al., 1985; and the chapter by Crifo in Newburn et al., 1991). The final point I wish to make here is the direct effect of solar radiation on the dust. As the dust expands outwards it soon decouples from the gas, due to the spatial dilution of the latter ($n_g \propto \frac{1}{r^2}$), and attain terminal speeds which vary inversely as the square root of the grain size (e.g., see Mends et al., 1985). Once that happens it is decelerated and pushed back into the tail by solar radiation pressure, to form the observed dust tails. This will be further discussed in the next section, but it is noteworthy that the suggestion, by Sevente Arrhenius in 1990, that solar radiation pressure was responsible for the acceleration of the dust in the comet tail was the first proposed application of this force in the astronomical context although its existence was discovered theoretically in 1873 by James Clarke Maxwell.

A major goal of cometary atmospheric analysis is to determine the composition of the nuclear ices. Prior to the spacecraft missions to comets several lines of argument strongly suggested that H_2O ice was indeed the dominant one, at least in the upper layers, of all observed comets. The in-situ and remote sensing observations of comet Halley also supports this view. Indeed, at Halley, during the fly-by, $\geq 80\%$ of the total production rate of molecules was due to H_2O, with CO being the next most abundant volatile species, with a production rate ≥ (10 – 15)% of the total production rate [on the basis of the International Ultraviolet Explorer (IUE) observations]. The compositions of the minor parent molecular species are less certain, but arguments have been made on the basis of the ion mass spectroscopic observations that the production rates of NH_3 and CH_4 are each of the order of 1% (e.g. see Mendis, 1988 for a detailed review). As a result of chemical differentiation of the nucleus, that we discussed earlier, it is not possible to infer the mole fractions of the respective ices (H_2O, CO, NH_3, and CH_4) in the volatile mix of the nucleus from this data. Perhaps a knowledge of the heliocentric variation of the production rates of the various parent-molecules which may be obtained during the Rosetta spacecraft encounter with comet 67P/Churyumov-Gerasimenko in the next decade may provide a key to solving this important problem.

20.5 The Interaction with the Solar Wind

Increasingly sophisticated quantitative modeling supporting increasingly detailed knowledge of the cometary plasma environment, particularly following its first in-situ observations by spacecraft in the mid nineteen eighties, have led to a major advance in our understanding of the nature of the comet-solar wind interaction, within the last two decades. The real beginning of the modern era of comet-solar wind interaction studies however goes all the way back to the nineteen fifties.

In 1951 Ludwig Biermann used the existing observations of the orientations of the plasma tails of comets to infer the existence of the solar wind, which will be discussed in the following section on the use of comets as probes of the solar wind. Following this discovery, Biermann directed his attention to the important question of the coupling between the solar wind and the cometary plasma tail by studying the acceleration of several inhomogenities (e.g., "knots" and "condensations") observed down the plasma tails of comets.

There is more than sufficient momentum flux in the solar wind to explain the observed acceleration in the cometary plasma tail, provided there is an efficient mode of coupling between the two plasmas. In 1953 Biermann suggested that the coupling was due to long-range Coulomb collisions between the two groups of ions, leading to cometary ion accelerations, $a \sim \frac{e^2 nw}{\sigma m_c}$ where m_c is cometary ion mass, σ is the electrical conductivity (esu), and n and w are the solar wind number density and speed, respectively. Taking $w = 1000\,\text{km/s}$, $n = 10^3\,\text{cm}^{-3}$, $\sigma = 5 \times 10^{12}$ esu, and $m_c = 28$ amu, Biermann obtained $a \sim 100\,\text{cm/s}^2$. While this value of a is consistent with the accelerations inferred also from the kinematics of inhomogenities ("knots" and "condensations") observed down plasma tails, the high values assumed for w and n were not considered too high at the time.

In an important paper Alfvén (1957) criticized Biermann's mechanism for the production of large accelerations in cometary plasma tails, in particular noting that the high solar wind densities were inconsistent with inferences from coronal white light measurements. Biermann (1951) had already noted that the solar wind plasma would probably carry a magnetic field. Alfvén (1975) developed this idea qualitatively to produce his "magneto-hydrodynamic model" for the interaction of the solar wind with the cometary plasma (see Fig. 20.4). Briefly the idea is that as the solar wind, with its "frozen-in" magnetic field flows into and past the comet, this magnetic field gets "hung-up" in the cometary plasma (ionosphere) and is dragged into the tail, as shown. While Alfvén notes that this picture is strikingly similar to the "folding umbrella" morphology

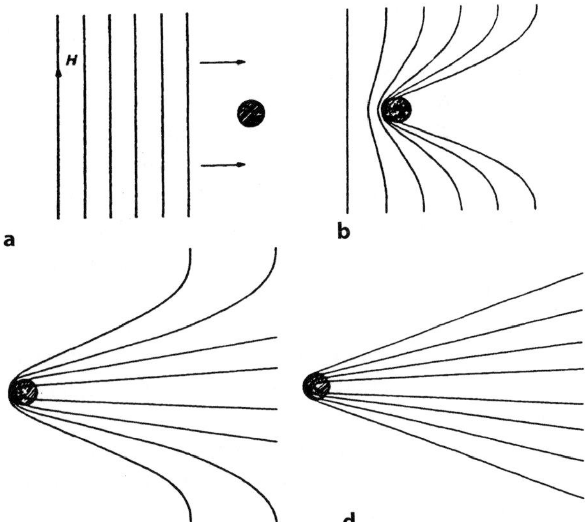

Fig. 20.4a–d. Schematic representation of the "piling-up" of the interplanetary magnetic field convected by the solar wind against the cometary ionosphere. (From Alfvén, 1957)

of plasma rays and streamers in cometary tails, he also notes that observed wavy patterns moving at high velocities may be due to the propagation of hydromagnetic waves down the tail. While Alfvén did not develop this phenomenological model quantitatively he made the important point that the plasma tail of the comet must be regarded as an integral part of the comet, fastened to the head by the magnetic field which channel the tail plasma. In other words the cometary plasma tail is a true "windsock" as opposed to Biermann's (1951) view which may essentially be described as a "smoke-trail."

While Alfvén (1957) identified a central intermediary of the cometary solar wind interaction the second crucial one was identified ten years later by Biermann et al. (1967). This is mass loading of the inflowing solar wind with heavy cometary ions produced either by photoionization or charge exchange, which makes the solar wind interaction with comets so different from its interaction with either strongly magnetized planets (e.g., Earth, Jupiter, Saturn) or with essentially unmagnetized planets with dense atmospheres (e.g., Venus). Noting that the solar wind flow was both supersonic and super-alfemic, and solving the steady-state 1-D hydrodynamic equations along the sun-comet axis, with further approximations, Biermann et al. recognized that the solution led to an unrealistic self-reversal of the flow unless $\hat{x} = \frac{\rho w}{\rho_\infty w_\infty} \leq \frac{\gamma^2}{(\gamma^2-1)}$, where ρ and w are the contaminated solar wind mass density and speed, while γ is the ratio of the specific heats. Taking $\gamma = 2$ as an indirect concession to the existence of the magnetic field they notice that this critical value of $\hat{x} = 4/3$, implying that this corresponded to only a few percent contamination of the solar wind with the heavy cometary ions (e.g., CO^+, CO_2^+, N_2^+, H_2O^+, etc). They also recognized that the implications of this was that a shock will form upstream of the comet to divert the solar wind around the comet before this critical value was reached. Already in 1964, Ian Axford had already pointed out that, since the expanding cometary ionosphere would act as an obstacle to the solar wind, a bow shock should form typically at a distance of $\sim (10^4 - 10^5)\, R_n$ upstream of the cometary nucleus (R_n being the cometary radius). Biermann et al. (1967) arbitrarily assumed that the Mach number, M, of this shock would be ≈ 10 (as in the case of the earth), but it was subsequently shown by Max Wallis in 1973 that this shock would be much weaker ($M \approx 2$). This is because, not only is the inflowing solar wind gradually slowed down by the mass loading, more importantly, it is also heated because the newly assimilated cometary ions have a thermal speed comparable to the local solar wind speed. Also Wallis showed that this shock would occur considerably closer to the cometary nucleus than assumed by Biermann et al. (1967). While subsequent 2-D and 3-D MHD models by several authors have validated Wallis' 1973 contention based on a simple 1-D hydrodynamic model (e.g. see Mendis et al., 1985), a semi-kinetic two-component (solar wind protons and cometary ions) model developed by Wallis and Ong in 1975 and subsequently extended by others provides an explicit kinetic description of cometary ion assimilation process. For a concise review of this model see Ip and Axford (1990). Here we show the solution of this model in the case $M = 2$ and $B = 0$ (Fig. 20.5), appropriate to comets Giacobini–Zinner and Halley. What is most notable here is the rapid increase of thermal pressure as the bow shock is approached; the pressure has increased ~30 times while the flow speed has decreased only ~25%. An unrealistic feature of this model (in common with the previous ones) is the rapid decrease of the flow speed, approaching zero, at cometocentric distance $\sim 10^4$ km for comet Giacobini–Zinner and 2×10^5 km for P/Halley. This unrealistic singularity in the flow velocity arises from the limitation of cometary ion acceleration by solar wind plasma alone. Inclusion of the pressure gradient of the magnetic field ($B \neq 0$) as well as cooling effects associated with charge exchange between the hot pick-up ions and the cool neutrals in the coma, allow ion flow to continue beyond these points.

The kinetics of the pick-up process also leads to an understanding of the momentum coupling between the solar wind and cometary plasma. The nature of the process depends on the orientation of the interplanetary magnetic field (IMF) to the solar wind flow direction. In the special case when IMF is normal to the flow the pick-up of cometary ions are entirely due to macroscopic fields: the IMF and the associated motional electric field. The newly created cometary ions gyrate around the local magnetic field with a gyro-speed of w_{sw}, while their guiding centers move with the magnetic field while conserving their magnetic moment at the point of origin. When the solar wind flows obliquely to the IMF the coupling between the solar wind and cometary ions is dominated by microscopic electric and magnetic fields generated

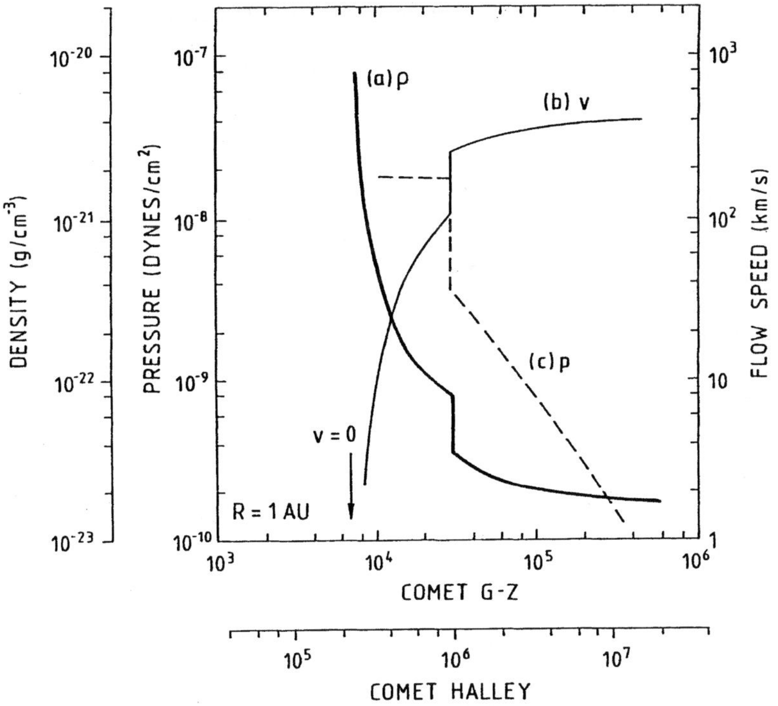

Fig. 20.5. Radial variations of (a) the number density, (b) the axial flow velocity, and (c) the thermal pressure, from a one-dimensional analytical model of the cometary accretion flow. Values are scaled to Comets Giacobini–Zinner and Halley at a heliocentric distance of 1 AU. (From Ip and Axford, 1990)

by various plasma instabilities. The newly formed ions move along the local magnetic field line with a speed of $w_{\|} = w_{sw} \cos\theta$, while gyrating around this field line with a speed of $w_{\perp} = w_{sw} \sin\theta$, (where θ = angle between the IMF and the solar wind direction). The beaming of this gyrotropic ring distribution along the magnetic field leads to various plasma instabilities, and the resulting waves cause both pitch angle scattering and energy diffusion of the ions. It can be shown (see e.g., see chapter by Flammer in Newburn et al., 1991) that the position of the cometary bow shock depends on, among other parameters, its Mach number and on the ratio of specific heats, γ. Giotto observations at comet Halley showed that while its Mach number, $M \approx 2$, its position was between those corresponding to $\gamma = 2$ (gyrotropic ring) and $\gamma = 5/3$ (isotropic shell), which was also consistent with the observation of the ion velocity distribution corresponding to an incomplete shell.

Regarding the structure of the cometary bow-shock what is definitely know from in-situ observations at comet Halley is that it is much thicker than the terrestrial one. While the thickness of the latter is of the order of a proton gyro-radius ($\approx$100 km) the thickness of the former is of the order of a picked up ion gyro-radius ($\geq 10^4$ km). Numerical studies, particularly in the quasi-perpendicular case, show this to be the case. It has also been pointed out that due to the dominance of the plasma pressure by energetic pick-up ions, despite their being only a small fraction of the solar wind plasma, that cometary bow shocks should have similarities to diffusive cosmic ray shocks [e.g., see Ip and Axford, 1990].

As the sub-sonic mass accreting solar wind flows towards the nucleus it continues to slow down while its magnetic field continues to increase, and is eventually brought to stagnation (along the sun-comet axis). The existence of a tangential discontinuity surface (loosely called the cometary ionopause) was already anticipated in the early work of Biermann et al. (1967), and the basic mechanism responsible for its formation, which is the balance of the electromagnetic $\boldsymbol{j} \times \boldsymbol{B}$ force and the drag of the out-flowing cometary neutrals on the plasma just outside it, was first proposed in 1982 by Ip and Axford, who went on to make an estimate of its linear dimension along the sun-comet axis. In order to do so, they needed to know the strength of the magnetic field just outside the ionopause, which they estimated by assuming that

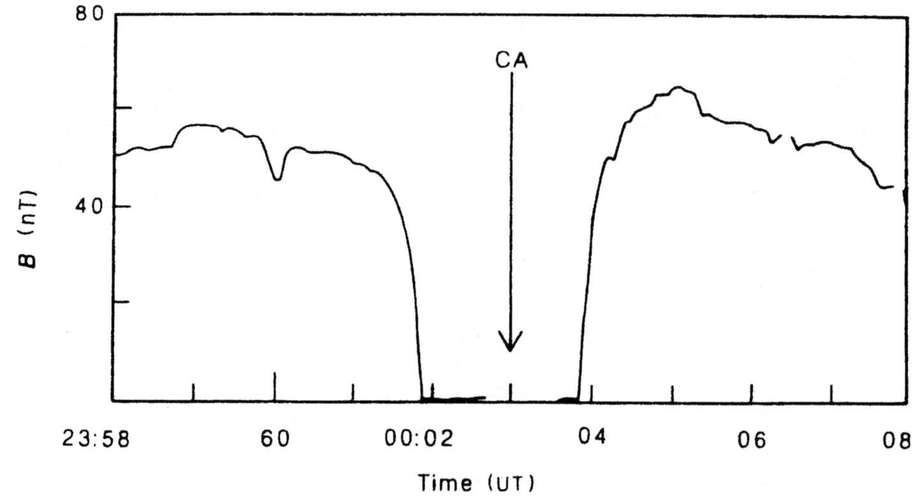

Fig. 20.6. The radial variation of the magnetic field magnitude around closest approach (CA) measured by the magnetometer onboard Giotto. A "magnetic field-free cavity" of linear extent ~ 8500 km is clearly seen. (From Neubauer et al., 1986)

the entire ram pressure of the solar wind was converted to magnetic pressure at the stagnation point.

Subsequently several authors (e.g., Cravens, Ip and Axford, and Haerendel) went on to calculate the radial profile (along the sun-comet axis) of the magnetic field in the "magnetic barrier" region just outside the ionopause, while Wu subsequently calculated the 2-D shape of the ionopause, showing it to have a "tear drop" shape (see Ip and Axford, 1990 for a detailed review).

One of the clearest and most dramatic discoveries of the Giotto mission to Halley's comet was the detection of this ionopause by the magnetometer onboard Giotto (Neubauer et al., 1986). An essentially magnetic field-free cavity, containing purely cometary ions, separated from the inflowing contaminated solar wind ions by a sharp boundary was observed at a distance ~4700 km inbound and ~3800 km outbound (see Fig. 20.6). These encounter distances (by the spacecraft moving at an angle of about 107° to the sun-comet axis) was consistent with the theoretical expectations.

A schematic of the global morphology of the overall comet-solar wind interaction, which summarizes our present knowledge is shown in Fig. 20.7. The possible existence of an "inner-shock" where the supersonically out-flowing cometary ions are decelerated and diverted into the flanks, was proposed by Wallis and Dryer in 1976, and the 2-D structure of the shocked layer between the ionopause and this inner shock was calculated by Houpis and Mendis in 1980 (see Mendis et al., 1985). The shocked layer calculated by the above authors is quite thick ($\sim 10^3$ km).

A 1-D, time-dependent photochemical model of the inner cometary coma developed by Tom Cravens in 1989 showed the existence of a thin (≈ 50 km) layer of enhanced plasma density just inside the cavity boundary. In this layer, which Cravens calls the cavity transition layer, the ions are essentially removed by dissociative recombination; there is no lateral flow. Cravens also describes this layer as "shock like" and identifies its inner boundary as the inner shock. Subsequently Damas and Mendis (1992) extended the earlier 2-D shock layer model of Houpis and Mendis by including the photochemistry. They too find that the shocked layer is now very thin (~100 km along the encounter trajectory of Giotto); the ions being largely removed by dissociative recombination rather than by lateral flow into the flanks. While the spacecraft evidence for the existence of this inner shock layer is not clear cut, the existence of a thin density spike [≤ (100 – 200)km] observed by the Giotto ion mass spectrometer (IMS) for ions of the H_2O group (Goldstein et al., 1989) as well as by the Giotto PICCA analyzer for a number of heavy ions, are at least consistent with it. Perhaps the more detailed observations anticipated during the Rosetta mission to comet Churyumov-Gerasimenko may shed more light on this issue.

As is obvious from Fig. 20.7, there is another global feature called the "cometopause" between the bow shock and the ionopause. This transition region where the solar wind proton density drops relatively fast while the cometary ion density increases towards the nucleus, was observed by the VEGA spacecraft as a rather thick

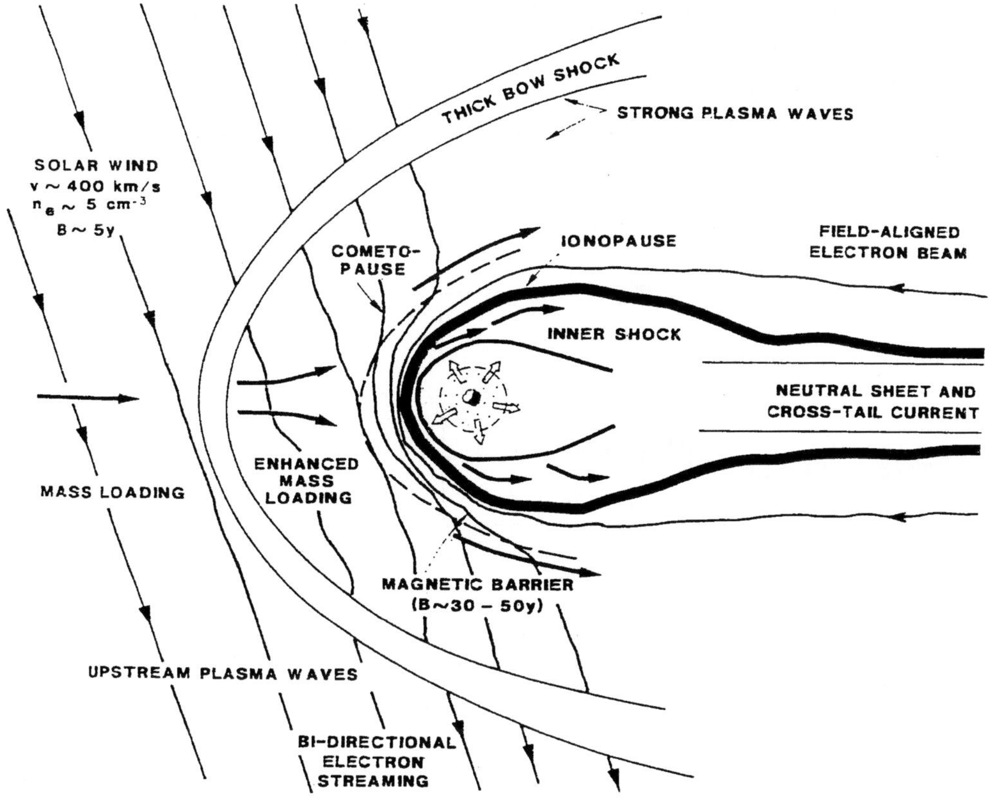

Fig. 20.7. Schematic representation of the global morphology of the solar wind interaction with a well developed cometary atmosphere. (From Mendis, 1988)

region ($\sim 10^4$ km) at a cometocentric distance $\sim 10^5$ km. While a similar transition was observed during the Giotto encounter, it was much more diffuse. Its nature has been discussed by several authors (see Ip and Axford, 1990 for a review) but it seems fair to say that we do not have a good understanding of its nature at the present time.

A basic consequence of the comet-solar wind interaction, also shown in Fig. 20.7, is the draping of the IMF around the comet. This magnetotail model of Alfvén (1957) was spectacularly confirmed by the magnetometer onboard the NASA/ICE spacecraft as it flew through the tail of comet Giacobini–Zinner (Smith et al. 1986). Two magnetic lobes of opposite polarity separated by a current-carrying neutral sheet was clearly observed (see Fig. 20.8); note in particular the dramatic flipping of the component of the magnetic field, B_x, which is along the sun-comet line, as the spacecraft transits the tail axis).

So far the plasma environments of three comets have been observed in-situ by particle and field experiments carried several onboard spacecraft. Comet Giacobini–Zinner, by the ICE spacecraft in 1985, comet Halley by ICE, Sikigake, Suisei, VEGA1 and 2 and Giotto, in 1986 and comet Grigg-Skjellerup by Giotto in 1992. What was seen, in all cases, was that these environments were far from quiescent, being characterized by high level of plasma wave activity and turbulence. These observations also led to the detection of a plethora of wave modes. The nature of the waves, e.g., their growth, level of non-linearity and non-coherence (turbulence) varied from comet to comet as well as from place to place within a given comet. This was not surprising considering the fact that not only were the three encounter geometries very different, but also both the cometary parameters (e.g. the production rate of neutrals) and the solar wind parameters (e.g. the magnitude and orientation of the IMF) were vastly different during these three

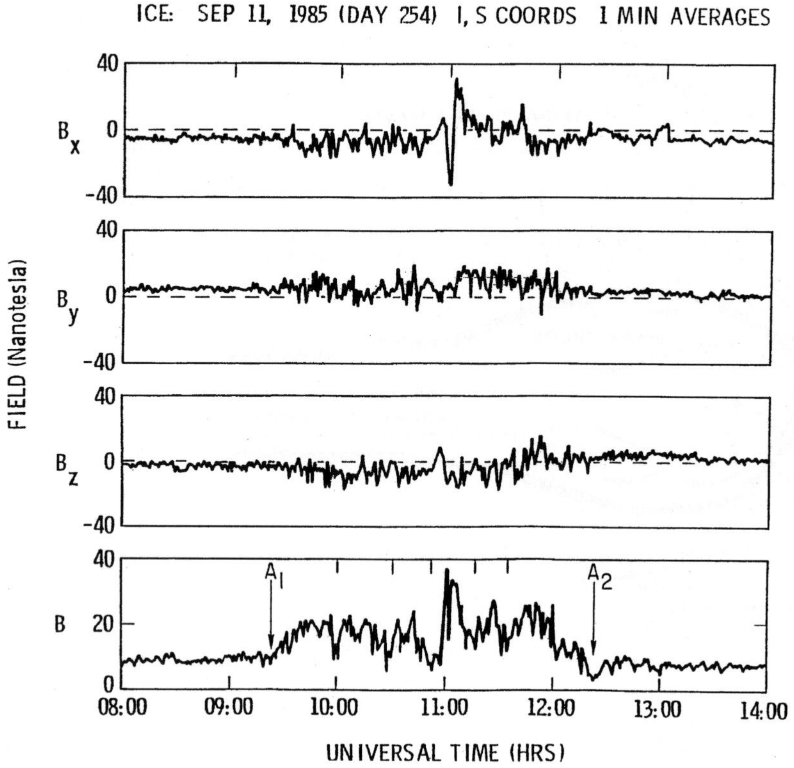

Fig. 20.8. An overview of the magnetic field morphology observed during the NASA/ICE encounter with the tail of Comet Giacobini–Zinner (September 11, 1985). One-minute averages of the three orthogonal components of the magnetic field in *I*, *S* coordinates, as well as the total magnitude, are shown. Here B_x is along the Sun-comet line. (From Smith et al. 1986)

encounters. Consequently not only were different wave modes observed during the three encounters but even the characteristics of the same mode varied from comet to comet. For instance the resonantly excited very low frequency ion cyclotron waves associated with the H_2O group ions ($\nu \sim 10^{-2}$ Hz) were observed at each comet. But their natures were quite different. In the case of comet P/Halley (which had the largest neutral production rate) these waves had the lowest amplitudes but the highest level of non-coherence (turbulence) In the case of comet P/Girgg-Skjellerup (which had the lowest neutral production rate) these waves had the highest amplitudes, with a moderate level of turbulence. In the case of comet P/Giacobini-/Zinner (which had an intermediate production rate of neutrals) these waves had intermediate amplitudes and also a moderate level of turbulence. At the same time they exhibited the highest level of nonlinearity. Theoretical studies have focused on the roles of the variable solar wind and cometary conditions in determining the nature and development of the waves. The most important solar wind parameter appears to be the angle between the IMF and the solar wind flow direction. The most important cometary parameter appears to be the production rate of neutrals, which controls the extent of the region where mass loading of the solar wind with heavy cometary ions is significant.

Plasma waves and turbulence in the mass loaded cometary environment became an important area of investigation following the in-situ observations at comets, leading to sophisticated theoretical models. This is not only because of the central role that wave particle interactions play in the momentum transfer from the solar to the cometary plasma as we discussed earlier, but also because mass loading takes place in many other solar system situations including the interaction of the solar wind with the atmospheres of unmagnetized planets and the interstellar medium, and comets provide an excellent accessible natural laboratory for the study of mass loaded plasmas under a variety of conditions. The future Rosetta mission which will study the temporal variations of the comet-solar wind interaction over a long period of time, as a comet approaches the sun and gradually develops an atmosphere, will undoubtedly increase our understanding of the subject. There are several excellent comprehensive reviews of the in-situ observations as well as the theoretical studies of plasma waves at comets. Here I will mention only the latest of these (i.e. Szegö et al. 2000) since it references all the earlier reviews. Also, for a useful concise summary see Ip and Axford (1990).

The final topic I will consider in this section is the emission of X-rays from comets, which was first detected in comet C/Hyakutake (1996B2) by the Röntgen X-ray satellite, ROSAT (Lisse et al., 1996), and subsequently in many other comets by both ROSAT and, more recently, by CHANDRA. While the first observations came as something of a surprise, the emission of X-rays from comets had already been anticipated by Hudson et al. in 1981 (see Mendis et al., 1985) on the basis of a cometary analog of the terrestrial situation, viz sporadic X-ray bremsstrahlung as energetic (keV) electrons precipitated into the cometary atmosphere during a cometary sub-storm. They searched comet Bradfield (1979) using the orbiting Einstein X-ray observatory and placed an upper limit $\sim 10^{14}$ erg/s for the total X-ray power at the comet, from its non-detection. While the subsequent observations did show some variability in the X-ray power, what was more significant was the existence of an underlying steady component, with a typical power $\sim 10^{15}$ erg/s at the comet. Several mechanisms have been proposed to explain this steady X-ray emission. One of these (e.g., Shapiro et al., 1999) once again assume that X-rays result from the penetration of energetic electrons into the cometary atmosphere leading to a combination of bremesstrahlung and line K-shell radiation from cometary neutrals (e.g., O). These electrons are continuously energized to a few hundred eV, by lower hybrid waves generated by the relative motion between the newly picked-up cometary photoions and the solar wind plasma. Another mechanism, originally proposed by Cravens (1997) and subsequently further investigated by several others, involves charge exchange between high-Z solar wind minor ions and cometary neutrals. A couple of mechanisms involving very small dust particles have also been proposed. While Owns et al. (1998) argue that the observed X-ray emissions is a continuum that can best be fitted with a bremsstrahlung spectrum from a plasma at ~ 0.3 keV temperature, it is fair to say that the observations cannot yet rule out any of the proposed models. It is also possible that more than one of the proposed mechanisms contribute to the observed emission.

20.6 Comets as Probes of the Solar Wind

Ever since the plasma tails of comets were used to infer the existence of the solar wind, as discussed in Sect. 20.5, they have been used as free, natural probes of the solar wind, providing useful information both of its global properties as well as its spatial and temporal variations (e.g., high-speed streams, magnetic sector boundaries, coronal mass ejections, etc.).

Prior to our present understanding of the nature of the plasma tails of comets, there were interesting speculations that go a long way back. Here I will adopt a historical approach which will also enable me to briefly discuss the nature of the second type of cometary tail, the dust tail.

Since many comets are observed visually just before sunrise or soon after sunset, the fact that their spectacular tail points, more or less, directly away from the sun should have been apparent even to the very early observers. The earliest known written record of this fact

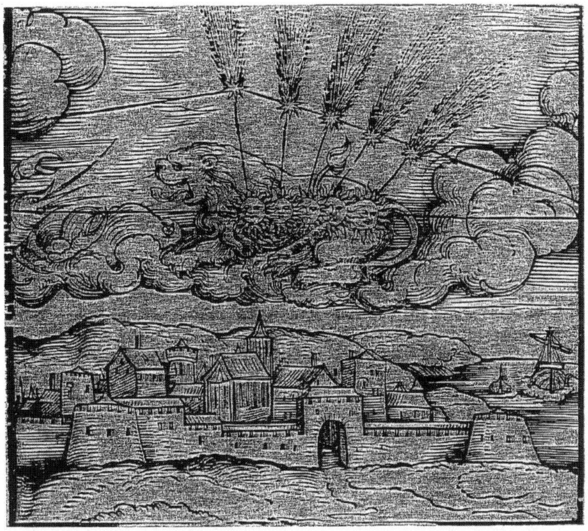

Fig. 20.9. Peter Apian's August 1531 observations of a comet (Halley) in the constellation Leo were used to demonstrate the antisolar nature of cometary tails. Woodcut illustration from Apian's Practica auff dz. 1532 Jar.... (Landshut). (Courtesy of the Crawford Library, Edinburgh, Scotland). (From Yeomens, 1991)

is due to Li Chung-feng in 635 AD in The History of the Chin Dynasty. The first Western record of this fact is almost 900 years later when the German mathematician Peter Apian drew attention to it (see Fig. 20.9). There the orientation of the tail of comet P/Halley as it moved in the constellation of Leo during its 1531 AD apparition, is clearly shown to be in the anti-sunward direction. Incidentally, it is worth noting that reason for this long time lag between Chinese and Western records lay in the dominance, in the West, of the Aristotelian dogma that comets were merely "exhalations from the earth," and not celestial objects worthy of astronomical investigation. Interestingly this view prevailed, for a while, even after the Danish astronomer Tycho Brahe used parallax measurements to show that the great comet of 1577 was at a supra-lunar distance, and therefore was a truly celestial (not terrestrial) phenomenon. (For more detailed discussions of this subject, see Yeomes, 1991)

Even after it was generally accepted that the comet tail points in the anti-sunward direction, the reason for this remained unresolved well into the twentieth century. Central to this was the confusion caused by the fact that there are in fact two major types of tails, type 1 and type 2 (see Fig. 20.10). Following the spectroscopic studies, beginning the middle of last century, we now know that narrow (type 1) tail, which points almost directly away from the sun, is composed of plasma (observed from the ground by resonance scattering of solar radiation by various ions; with its bluish color being due to the strong contribution of the violet bands of CO^+) while the broader, more featureless (type 2) tail (which lags behind the type 1 tail; the cometary orbital motion being to the right in this figure) is composed of dust, observed by scattered sunlight, leading to its more yellow hue. This was of course unknown to the early workers, who mostly thought that both tails were composed of gas.

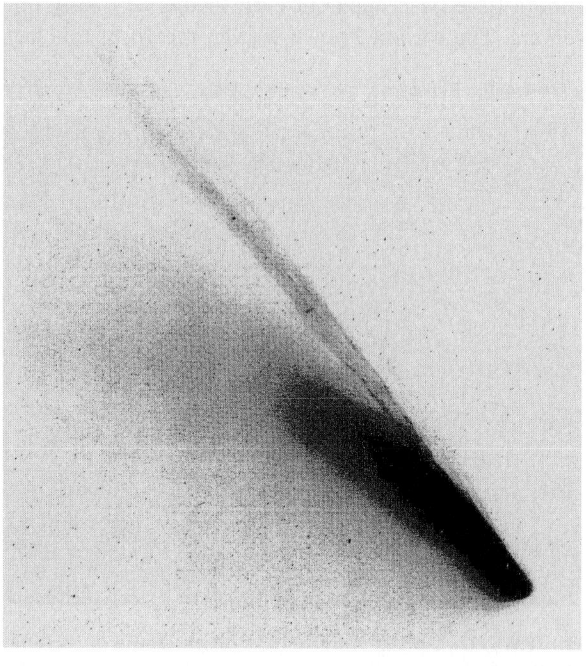

Fig. 20.10. Comet Markos (1957d), showing the straight structured type 1 tail and a broad, homogenous type 2 tail, which lags behind the type 1 tail. Cometary orbital motion is toward the right of the figure. (Courtesy of Mount Wilson and Palomar Observatories). (From Brandt, 1967)

Edmund Halley's recognition, based on their orbital elements, that the comets observed in 1531, 1607 and 1682, were one and the same, moving in a highly elliptical orbit, and his successful prediction that this comet (now fittingly named Halley's Comet) would return again in 1758 had firmly established the validity of Isaac Newton's Law of Universal Gravitation postulated in 1689. It soon became clear that while the motion of center of mass of the comet could be explained by the sunward force of solar gravity, the anti-sunward orientation of comet tails required an outward directed (repulsive) force emanating from the sun. It is remarkable that as early as 1812 the Dutch astronomer, Heinrich Olbers correctly speculated that the cometary tail consisted of "minute particles driven away in the anti-solar direction by a solar repulsive force that is *electrical* in nature."

The first major contribution to the quantitative study of comet tails was due to Friedrich Bessel in 1836. He assumed that, whatever the nature of this repulsive force, it would vary inversely as the square of the distance, just like the gravitational force. He was then able to successfully calculate the observed shape of the type 2 (dust) tails assuming that they were moving under an effective (reduced) solar acceleration of magnitude $\frac{\mu G M_\odot}{r^2}$ which implies that the ratio of this outward radial force, F_r, to the gravitational force F_g is given by $\left|\frac{F_r}{F_g}\right| = 1 - \mu$. Bessel's mechanical theory was extensively used and analyzed by Fedor Bredikhin who showed that all the type 2 tails he analyzed could be explained by assuming that $1 - \mu \sim (0.7 - 2.2)$. He also recognized that if the same model was applied to type 1 tails, typically $(1 - \mu) \geq 100$.

As discussed in Sect. 20.4, the physical nature of this outward force was first recognized by Svante Arrhenius to be the solar radiation (pressure) force. This idea was further developed by Karl Schwarzchild, in 1901, who showed that with

$$1 - \mu = \frac{C}{a\rho_d}, \text{ where } C \sim 6 \times 10^{-5}\, \mathrm{Q_{pr}\, g\, cm^{-2}}$$

(Q_{pr} being the scattering efficiency for radiation pressure), $1 - \mu \sim 1$ (typically) and cannot exceed ~ 20 for any grain size or reasonable density, for known compositions.

Following the spectroscopic discovery that type 1 tails were not composed of dust, but of various ions (e.g., CO^+, CO_2^+, N_2^+, etc.), in the early twentieth century, it was established that radiation pressure on these ions typically gave $1 - \mu \sim 0.1$ (see Mendis et al., 1985). So clearly some other force was required to explain the high accelerations observed in these tails. Even before the nature of the type 1 tails was known, there were speculations that the agency responsible for the observed acceleration were "solar coronal particles." The central observation leading to the understanding of the nature of type 1 tails was due Hoffmeister in 1943 who noticed that the axis of type 1 tails always lagged slightly behind the sun-comet axis and also that the tangent of this lag (or aberration) angle, ε was proportional to the transverse component of the orbital velocity of the comet. Ludwig Biermann (1951) drew the obvious conclusion from this that there was continuous outflow of plasma from the sun and that due to its interaction with outflowing solar plasma, the cometary plasma tail pointed in the direction of this flow as observed by the moving comet, which explains the aberration (see Fig. 20.11). That there was plasma outflow from the sun was inferred earlier by scientists studying geomagnetic variations, but they believed that it was intermittent. Plasma tails of comets are of course visible continuously (at least when their heliocentric distance ≤ 2 AU). So Biermann's (1951) paper is credited with first convincing detection of *continuous* outflow of plasma from the sun, referred to as the Solar Wind, following the classic theoretical paper by Eugene Parker in 1958. As is clear from Fig. 20.11:

$$\tan \varepsilon = \frac{V \sin \gamma - w_\varphi \cos i}{w_r - V \cos \gamma} \tag{20.1}$$

where i is the inclination of the comet's orbital plane to the solar equator (e.g., Brandt, 1967).

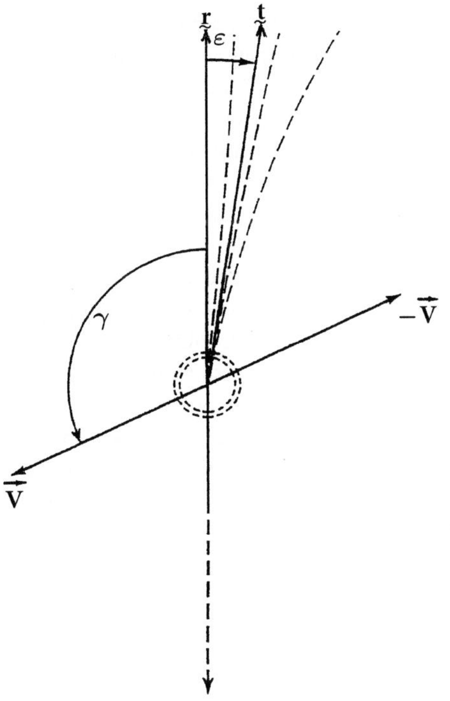

Fig. 20.11. Geometry of the solar wind interaction with the cometary plasma tail. See text. (From Belton and Brandt, 1966)

Biermann (1951) assumed that the solar wind flow was strictly radial and also that the cometary plasma tail lay in the orbital plane of the comet. Then $\tan \varepsilon = \frac{V \sin \gamma}{w_r - V \cos \gamma} \approx \frac{V \sin \gamma}{w_r}$ (assuming that the solar wind speed, $w_r \gg |V \cos \gamma|$, the radial component of the comet's orbital speed). With typical values for the transverse component of the comet's orbital speed $V \sin \gamma \approx 20$ km/s and $\varepsilon \sim 3°$, one obtains $w_r \sim 400$ km/s for the "typical" solar wind speed.

Beginning in the mid nineteen sixties, John C. Brandt together with several co-authors embarked on a systematic study of the global properties of the solar wind, using plasma tail orientation. These studies, which were statistical in nature, and which aimed at deriving average values used a large compilation of tail (both type 1 and type 2) orientations numbering over 1600 (Belton and Brandt, 1966). Examination of (20.1), where it is assumed that the plasma tail and the cometary orbit are coplanar, shows that if w_ϕ is not zero, it will show up as a systematic difference in the mean aberration angles for direct (prograde) and retrograde orbits (because $\cos i$ changes sign) with the value corresponding to the retrograde sample, larger. In 1967 Brandt obtained $\langle\varepsilon\rangle_D \approx 3.7°$ and $\langle\varepsilon\rangle_R \approx 5.5°$ where the suffixes D and R refer to direct and retrograde, with the mean value for the entire sample $\langle\varepsilon\rangle \approx 4.7°$. This gave $w_\varphi \approx 9$ km/s and $w_y \approx 450$ km/s for the entire sample. In a subsequent analysis of the same data where the conditions of coplanarity, as well as the assignment of weights to allow for geometrical circumstances was dropped, yielded $w_y \approx 415$ km/s and $w_\varphi \approx 6$ km/s. Later analyses also obtained estimated the meridional flow speed, w_θ, assuming a theoretical model for its variation with the polar angle θ (e.g., see Mendis et al., 1985, for a detailed review). In 1975 Brandt et al. next used the aberration of plasma tails in an effort to calculate the variation of the radial solar wind speed with heliographic latitude, b, assuming the linear variation:

$$w_r = w_{r0} + \frac{\mathrm{d}w_r}{\mathrm{d}|b|}|b| \qquad (20.2)$$

and obtained $\frac{\mathrm{d}w_r}{\mathrm{d}|b|} = -0.9 \pm 0.7$ km/s deg. This result was at odds with variation of the solar wind with heliographic latitude inferred by Bill Coles et al. in 1980 using radio scintillation observations (see Mendis et al., 1985) which gave:

$$\frac{\mathrm{d}w_r}{\mathrm{d}|b|} \approx (2-3)\,\text{km/s deg.} \quad \text{(during the declining phase of the solar cycle)}$$

$$\approx 0 \quad \text{(at solar maximum)}$$

This latter inference has subsequently been supported by in-situ measurements of the solar wind speed by the Ulysses spacecraft. Near solar minimum the solar wind speed shows a characteristic U-shaped profile with speeds increasing monotonically from ≤ 450 km/s near the equator to ≥ 750 km/s near the poles. Also while the speed fluctuates greatly near the equator it becomes quite uniform at high latitudes. The profile is much flatter during solar maximum, with a very high degree of fluctuation at all latitudes.

The reason for the apparent discordance between comet observations and the other observations discussed above are apparent. On the one hand the comet calculations used a plasma tail aberration sample which spanned 75 years, i.e., almost four solar cycles. On the other hand this sample was highly weighted toward equatorial with only 58 of a total of 700 observations being polar (e.g., see Brandt and Snow, 2000).

Clearly a more direct use of comet tails in this connection is to follow individual, high-inclination comets, measuring their plasma tail aberration at varying heliographic latitudes. This has been done more recently (Brandt and Snow, 2000) for three comets: de Vico ($i = 85.4°$) in 1995, Hyakutake ($i = 124.9°$) in 1996 and Hale-Bopp ($i = 89.4°$) in 1997. Their tail aberrations, which are systematically smaller at higher latitudes implied typical solar wind velocities ~750 km/s at high latitudes, and ~450 km/s in the equatorial region. Also while the plasma tails, in the polar regions, were sharp (as would be expected of a steady solar wind) they appear disturbed in the equatorial region (as would be expected from a highly varying solar wind). Since all these observations correspond to a phase in the solar cycle, near solar minimum (8/17/96), they are entirely consistent with the observations of the Ulysses spacecraft. Brandt and Snow (2000) speculate that the boundary, between the polar region where the comet tail appears undisturbed and the equatorial region where the tail appears highly disturbed, is the maximum poleward extension of warped heliospheric current sheet (HCS). If correct, this is a useful extension of the use of comet plasma tails as natural probes of the interplanetary medium. Interestingly, Brandt and Snow (2000) have reexamined the original data set of Belton and Brandt (1966) and find only three comets there are usable, in this connection; these being the ones with sufficient observations both in the equatorial and polar solar wind regions, with only two of these, Markos (1957d) and Brooks (1911c) providing reliable data. Both these comets exhibit "transregional" behavior like the three more recent ones. Comet Markos goes from having a highly disturbed plasma tail to a relatively undisturbed one around 65° N heliographic latitude, in August 1957, whereas comet Brooks does so around 28° N in October, 1911. Brandt and Snow (2000) argue that these two observations are compatible with the fact that the Markos transition took place near solar maximum, when the HCS has high poleward extensions, whereas the Brooks transition took place around solar minim when the HCS is confined to lower latitudes.

Besides the global properties of the solar wind, the cometary plasma tail can also provide information about its temporal variations. The long wavelength waves which are occasionally observed, propagating down the plasma tails of comets and which were already attributed to MHD waves (Alfvén, 1957) have since been discussed by several authors. In 1973 Ershkovich and Chernikov attributed them to the Kelvin–Helmholtz instability exited by the velocity shear between the solar wind and cometary flows in the tail, when this exceeds a critical value (see Erchkovich, 1980, for a detailed review). A more direct inference of the sudden variation of the solar wind speed is the appearance of a "kink" or bend down the tail, where the aberration angle changes from a larger value in the more distant part to a smaller value in the part closer to the nucleus. An obvious conclusion reached by many observers is that the comet has encountered a fast solar wind stream; the inner region, with the smaller aberration angle, is already immersed in this fast stream, whereas further out the influence of this stream has not yet been felt. These two regions are separated by an intermediate region which has not yet reached equilibrium with the ambient medium and shows up as a kink in the tail. This phenomenon as well as others associated with the time varying fine structure observed in the plasma tails of comets have been discussed by numerous authors. For a detailed review see e.g., Mendis et al. (1985). Here I will limit myself to a discussion of the most dramatic temporal phenomenon observed in the plasma tails of several comets, viz their occasional total separation from the head of the comet. This is also the subject that has received, by far, the most attention in more recent times.

While this phenomenon was already described by Bernard as early as 1920 in connection with what he called the "rejection" of the tail of comet 1919b, its real study of began with a pioneering paper by Niedner and Brandt (1978) wherein they attributed this phenomenon, clearly observed in the tail of comet Kohoutek (1973XII), in January 1974, and which they called a "disconnection event" (DE) to the crossing of an interplanetary magnetic sector boundary by the comet. Their basic mechanism is shown in Fig. 20.12. Here magnetic reconnection occurs at the comet's head as a magnetic field of the opposite polarity is pushed against the old field during the passage of the sector boundary. As the tail, gradually peeled off by this process, drifts away from the comet, a new one containing a magnetic field of the opposite polarity is generated. Despite the unavoidable uncertainty involved in the timing (e.g., the extrapolation of solar wind conditions measured at a different time and place by spacecraft to the location of the comet) these authors, in a continuing series of papers (which now covers over 100 DE's), have made a strong case for the connection between DE's and magnetic section boundary transversals (e.g., see Brandt, 1990 and Brandt and Snow, 2000). While it is fair to say that the above model of Niedner and Brandt (1978) is the leading one at present, it is not the only one. While several authors have proposed that the DE's could be a consequence of increased pressure-induced effects (e.g., the flue instability; the Rayleigh–Taylor instability) during the encounter of high speed solar wind streams (which are often though not always associated with magnetic sector boundaries) and shocks. Others have also proposed that the responsible mechanism is magnetic reconnection in the tail itself rather than in the head. (For a detailed review of these alternative mechanisms, see Brandt, 1990). The coordination of extensive ground-based observations of large scale phenomena in the plasma tail of comet Halley with in-situ spacecraft observations during its fly-by in 1986 provided an excellent opportunity to discriminate between the competing models, but the results could not provide an unambiguous answer. Niedner and Schwingenschu (1987) have argued that ground-based observations of a DE between 8 and 10 March, 1986 was associated with a reversal of the IMF detected by the VEGA 1 and VEGA 2 spacecraft. Saito et al. (1986), on the other hand compared data from the Sakigake spacecraft with observation from the ground and noted that there were no apparent signs of DE's between 11 and 14 March, 1986, although the spacecraft crossed the heliospheric current sheet at least four times during this period. These latter authors, therefore note that more than a mere crossing of such a sheet is necessary for the production of a DE, with the encounter geometry presumably playing a essential role. It is also possible that more than one process could trigger DE's. Clearly this is an area that needs further investigation. There have also been several numerical simulations of this process with different authors reaching different conclusions. Most recently Konz et al.

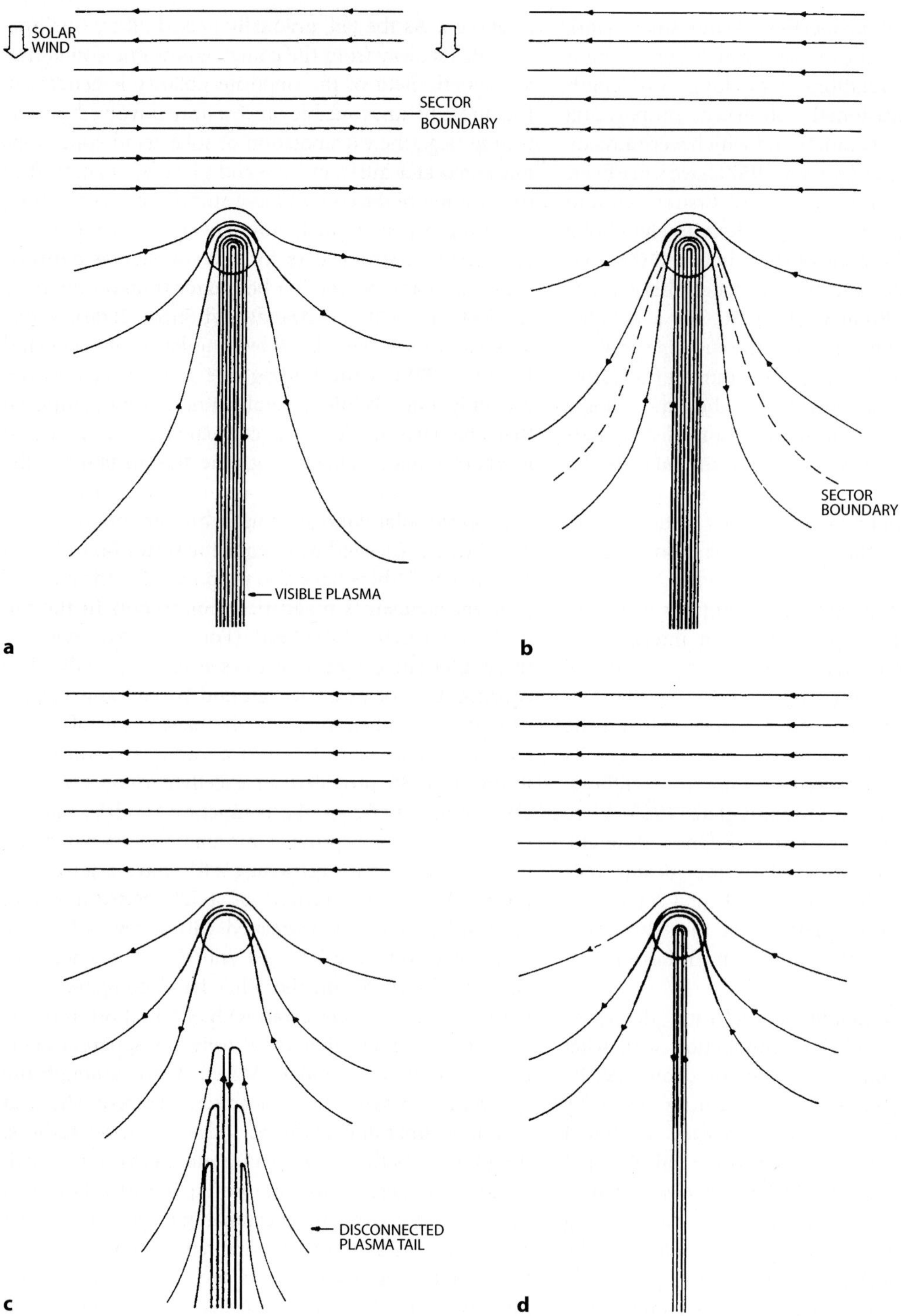

Fig. 20.12a–d. Magnetic sector boundary model of plasma tail disconnection event. (From Niedner and Brandt, 1978)

(2004) have critiqued these earlier simulations. They point out that ideal MHD simulations do not allow for field line reconnection and that it could arise only as an artifact of the chosen algorithm where some numerical diffusion may lead to it. What is needed is a localized violation of the ideal Ohm's law, which is provided in the above simulation by a current driven production of anomalous resistivity. The same conclusion was reached earlier in 1978 by Morrison and Mendis who showed that the generation of anomalous resistivity by current-carrying electrons, which may reach the Alfvén energy $\left(\frac{B_t^2}{4\pi n_e}\right)$, could lead to the tearing of the cometary cross-tail current sheet along current flow lines (see Mendis et al., 1985). While the simulation of Konz et al. (2004) leads to both day-side and tail reconnection, it like the earlier simulations uses normal geometry for the interaction. In view of the suggestion by Saito et al. (1986) noted earlier, future numerical simulations should also attempt to model oblique interaction geometries.

Very recent observations have also shown the effects of the encounter of a fast-moving (interplanetary) coronal mass ejection (ICME) by a comet; in this case comet 153P/Ikeya-Zhang (Jones and Brandt, 2004). The impact produces a highly unusual large scale disturbance down the tail, which is characterized by highly "scalloped" appearance. The reason for this unusual appearance is not clear although the above authors speculate that it may be due to the ICME magnetic field wrapping around preexisting tail density enhancements. Nonlinear development of waves exited by Kelvin–Helmholtz instability is also a possibility. Whatever its cause, if more observations in the future could establish that such a scalloped appearance of the tail is characteristic of ICME-comet interactions, then comets could also be used as natural probes of ICMEs.

My discussion, so far, has been limited to the case when a comet has a well developed plasma tail (typically at heliocentric distances $\leq$3 AU) which arises from the interaction of the solar wind with a well developed cometary atmosphere; the case discussed at length in Sect. 20.5. I will conclude this discussion by considering the possible use of comets as probes of the solar wind at large heliocentric distances. It was shown by Mendis and Flammer in 1984 (see chapter by Flammer in Newburn et al., 1991) that the interaction of the solar wind with a comet approaching the sun had not one mode but multiple modes of interaction. At large distances ($d \geq 5$ AU, for a H_2O-controlled comet) the comet has no significant atmosphere and the solar wind flows directly on the surface. As the comet moves closer and an atmosphere begins to develop. Mass loading of the inflowing solar wind can cause it to develop a weak collisionless bow shock upstream of the comet (for comet Halley this would happen when $d \approx 2.5$ AU). However at this time there is not sufficient momentum in the outflowing cometary atmosphere to stagnate the inflowing solar wind ahead of the nucleus. So the subsonic solar wind flows all the way to the nucleus. Closer to the sun (i.e., when $d \approx 2.2$ AU for comet Halley) the contaminated subsonic solar wind is brought to stagnation ahead of the nucleus by a well developed cometary ionopause. This is probably the time when the beginning of a plasma tail is first observed (Brandt, 1990). Indeed the "turn-on time" for the plasma tail of comet Halley (inbound) was when $d \approx 1.8$ AU, while corresponding "turn-off time" (outbound) was when $d \approx 2.3$ AU, which are in reasonable agreement with the predictions of Flammer (Brandt, 1990).

It had been argued by Mendis et al. in 1981 that the unimpeded flow of the solar wind to the cometary nucleus can cause differential electrostatic charging of the nucleus and thereby lead to the electrostatic levitation and blow-off of fine loose dust on it. They show that this happens mainly on the unlit (night) side which can achieve a large negative electrostatic potential $\propto V_{SW}^2$ due to the buildup of space charge there. The large sporadic brightness fluctuations of comet Halley at heliocentric distances between 8 and 11 AU (inbound), which appear to be entirely associated with dust outbursts, has been attributed to this cause by Flammer et al. (1986), who showed a one-to-one correspondence between these outbursts and the possible interaction of the comet with a high-speed solar wind stream emanating from a coronal hole. The Rosetta mission, which will first intercept its target comet at a distance where the outgassing is expected to be very small, could be used to check the validity of this model. If it turns out to be correct, then the use of comets as natural probes of the solar wind could be extended to much larger distances.

Acknowledgement. The author thanks Prof. W-H. Ip and an anonymous reference for useful comments. He also acknowledges support from the DOE grant: DE-FG02-04ER54804.

References

Alfvén, H. (1957) On the theory of comet tails, Tellus, 9, 92

Belton, M.J.S., Brandt, J.C. (1966) Interplanetary gas. XIII. A category of comet tail orientations, Astrophys. J. Suppl., 13, 125

Biermann, L., Brosowski, B., Schmidt, H.U. (1967) The interaction of the solar wind with a comet, Solar Phys. 1, 254

Brandt, J.C., (1967), Introduction to the solar wind, W.H. Freeman and Co., San Francisco, 107

Biermann, L. (1951) Kometenschweife und Solare Korpuskularstrahlung, Zeit. Astrophys. 29, 274

Brandt, J.C. (1990) The large-scale plasma structure of Halley's comet, 1985-1986, in Comet Halley: Investigations, Results, Interpretation, Vol 1, (Ed. J. Mason), Ellis Harwood, NY, p 33

Brandt, J.C., Snow, M. (2000) Helospheric latitude variation of properties of cometary plasma tails: A test of the Ulysses comet watch paradigm, Icarus, 52

Cravens, T.E. (1997) Comet Hyakutake X-ray source: Change transfer of solar wind heavy ions, Geophys. Res. Lett., 24, 105

Damas, M.C., Mendis, D.A. (1992) A three dimensional axisymmetric photochemical flow model of the cometary "inner" shock layer, Astrophys. J., 396, 704

Ershkovich, A. (1980) Kelvin–Helmholtz instability in type 1 comet tails and associated phenomena, Space Sic. Rev., 25, 3

Flammer, K.R., Jackson, B., Mendis, D.A. (1986) On the brightness variations of comet Halley at large heliocentric distances, Earth Moon and Planets, 35, 203

Goldstein, B.E., Altwegg, K, Balsinger, H., et al. (1989) Obervations of a shock and recombination layer at the contact surface of comet Halley, J. Geophys. Res., 94, 17251

Huebner, W.F. (Ed.) (1990) Physics and Chemistry of Comets, Springer-Verlag, Berlin

Ip, W.-H. (1980) Cometary atmospheres—1. Solar wind modification of the outer ion coma, Astron. Astrophys., 92, 95

Ip, W.H., Axford, W.I. (1990) The plasma, in Physics and Chemistry of Comets, (Ed.) Huebner, W.F., Springer-Verlag, Berlin, p 177

Jones, G.H., Brandt, J.C. (2004) The interaction of comet 153P/Ikeya-Zhang with interplanetary coronal mass ejections: Identification of fast ICME signatures, Geophys. Res. Lett., 31, L20505

Konz, C. Burk, G.T., Lesch, H. (2004) Plasma—neutral gas simulation of reconnection events in cometary tails, Astron. Astrophys., 415, 791

Lisse, C.H., Dennerly, K., Englhauser, J., et al. (1996) Discovery of X-ray and estreme ultraviolet emission from comet C/Hyakutake 1996B2, Science, 205

Mason, J. (Ed.) (1990) Comet Halley: Investigations, Results, Interpretations (Vols 1 and 2), Ellis Harwood, NY

Mendis, D.A. (1988) A postencounter view of comets, Ann. Rev. Astron. Astrophys., 26, 11

Mendis, D.A., Houpis, H.L.F., Marconi, M.L. (1985) The Physics of Comets, Fund. Cosmic Phys., 10, 1

Mitchell, D.L., Lin, R.P., Anderson, K.A., et al. (1989) Complex organic ions in the atmosphere of comet Halley, Adv. Space Res., 9(2), 35

Neubauer, F.M., Glassmeir, K.H., Pohl, M., et al. (1986) First results from the Giotto magnetometer experiment at comet Halley, Nature, 321, 352

Newburn, R.L., Jr., Neugebauer, M., Rahe, J. (Eds,) (1991) Comets in the post-Halley era (Vols 1 and 2), Kluwer Acad. Pub., Dordrecht

Niedner, M.B., Jr. (1982) Interplanetary gas XXVIII. A study of the three-dimensional properties of interplanetary sector boundaries using disconnection events in cometary plasma tails, Astrophys. J. Suppl., 48, 1

Niedner, M.B., Jr., and Brandt, J.C., (1978) Interplanetary gas XXII. Plasma tail disconnection events in comets: Evidence for magnetic field line reconnection at interplanetary sector boundaries? Astrophys. J. 223, 655

Niedner, M.B., Jr., Schwingenschuh, K. (1987) Plasma tail activity at the time of VEGA encounters, Astron. Astrophys., 187, 103

Oort, J. (1950) The structure of the cloud of comets surrounding the solar system and a hypothesis concerning its origin, Bull. Astr. Institute. Neth., 111, 91

Owens, A., Parmar, A.N., Osterbrock, T., et al. (1998) Evidence for dust-related X-ray emission of comet C/1995 01 (Hale-Bopp), Astrophy. J., 493, L47

Saito, K., Saito, T., Aoki, T., Yumoto, K. (1986) Possible models on disturbances of the ion tails of comet Halley during the 1985–1986 apparition, ESA SP-250(3), 155

Shapiro, V.D., Bingham, R., Dawson, J.M., et al. (1999) Energetic electrons produced by lower hybrid waves in the cometary environment and X-ray emission: Bremsstrahlung and K shell radiation, J. Geophys. Res., 104, 2537

Smith, E.J., Tsurutani, B.T., Slavin, J.A., et al. (1986) International Cometary Explorer encounter with comet Giacobini–Zinner: Magnetic field observations, Science, 232, 382

Szegö, K. Glassmier, K.-H., Bingham, R., et al. (2000) Physics of mass loaded plasmas, Space Science Reviews, 94, 429

Weaver, A., (2004) Not a rubble pile?, Science, 304, 1760
Weissman, P.R. (1999) Cometary reservoirs, in The New Solar System, (Eds.) Beatty, J.K., Peterson, C.C., Chaiken, A., Cambridge Univ. Press, 59
Whipple, F.L. (1950) A comet model—1. The acceleration of comet Encke. Astrophys. J. 111, 375
Yeomens, D.K. (1991) Comets: A chronological history of observation, science, myth and folklore, John Wiley, NY

About the Authors

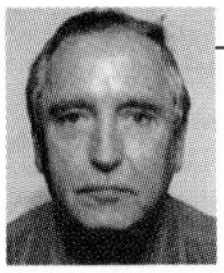

Denis Alcaydé

CESR-CNRS
9 Av. Cl. Roche
31028 Toulouse Cedex 4
France
denis.alcayde@cesr.fr

Chap. 8

Dr Denis Alcaydé spent his career at the Centre d'Etude Spatiale des Rayonnements, a CNRS Laboratory in Toulouse, France, where he defended his PhD thesis in 1975 on the physics of the ionosphere and aeronomy, making use of the St Santin-Nançay incoherent scatter radar facility. The thermospheric temperature data that he analysed for his work were used as input for the MSIS/CIRA neutral atmosphere reference model. He then moved his central interests to the high latitude ionosphere where he focussed particularly on the study of magnetosphere–ionosphere–thermosphere coupling. He took an active part in the development and use of the European EISCAT incoherent scatter radars. With his research group, with Pierre-Louis Blelly in particular, he initiated the development of a fluid-kinetic model of the ionosphere, which is now a reference tool for the analysis of auroral observations. He is CoI of CIS-Cluster and MIMI-Cassini experiments. For more than two decades he was successively a member of the EISCAT Scientific Advisory Committee, of the EISCAT Council, and he was recently appointed to the EISCAT Management Committee. He actively participated in the development of the CDPP (Plasma Physics Data Center) in Toulouse, where EISCAT data are archived together with a large amount of data from space missions. He was the Ionosphere Topical Editor of the EGS magazine Annales Geophysicae from 1995 to 1999 and he was Chief Editor from 1999 to 2005.

Pierre-Louis Blelly

LPCE
3A Avenue de la Recherche Scientifique
45100 Orl'eans, France
pierre-louis.blelly@cnrs-orleans.fr

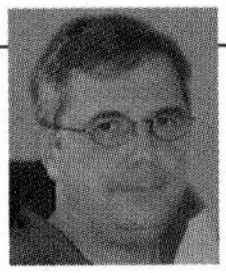

Chap. 8

Dr Pierre-Louis Blelly graduated from two French Grandes Ecoles (graduate engineering schools): l'Ecole Polytechnique in 1987 and l'Ecole Nationale Supérieure de l'Aéronautique et de l'Espace in 1989. He defended his PhD in 1992 on the study and modelling of the polar wind in the high latitude ionosphere, combining EISCAT incoherent radar measurements and numerical simulations. He started his career at the "Centre d'Etudes Spatiales des Rayonnements" (CESR) in Toulouse, France, and he is now the head of the "Laboratoire de Physique et Chimie de l'Environnement" (LPCE) in Orléans (France). He is a specialist in numerical modelling of terrestrial ionosphere and has developed numerous models, the TRANSCAR family models, which are now acknowledged as reference models for the study of the high latitude ionosphere. He gained a large amount of experience in the analysis (and subsequent correction) of various kinds of experimental data (particle spectrometers, optical measurements, incoherent scatter data, etc.) by the use of sophisticated numerical models of the ionosphere, and in their interpretation in the framework of atmosphere–ionosphere–magnetosphere coupling. He has also developed an expertise in the modelling of planetary ionospheres and he contributes to the development of numerical models for Mars and Titan.

Jean-Louis Bougeret

Chap. 6

Laboratoire d'Etudes Spatiales et d'Instrumentation en Astrophysique
LESIA, UMR CNRS 8109, Observatoire de Paris, Universit'es Paris 6 et Paris 7
5, place Jules Janssen, 92195 Meudon Cedex, France
jean-louis.bougeret@obspm.fr

Dr Jean-Louis Bougeret graduated from Ecole Nationale Supérieure de l'Aéronautique in 1968. He then started to work at the Paris Observatory in the field of solar radio astronomy, analyzing data from the first Nançay Radioheliograph and from the Stereo radio experiment on the Soviet Mars-3 interplanetary probe. He defended his PhD in 1978 on the directionality of solar radio storms. He was NAS/NRC Research Associate at Goddard Space Flight Center (1981–1982) and Research Associate at the University of Maryland and GSFC (1983). He pursued his work on interplanetary solar radio emissions and transient solar and interplanetary disturbances. He participated in the development and was PI of the ARTEMIS multichannel radio spectrograph program until 1995, when a new instrument was developed and installed in Greece. He was PI for the Radio and Plasma Wave Investigation (WAVES) on the NASA GGS/WIND spacecraft during the scientific definition and technical development until its launch and thereafter Co-PI. He is PI of the Radio and Plasma Wave Investigation (S/WAVES) on the NASA STEREO spacecraft. His research interests currently include radiophysics, solar physics, long wavelength astrophysics, propagation of radio waves, space plasmas, instrumentation in ground-based and space radio astronomy, data acquisition and processing in real-time. He is presently "Directeur de Recherche au CNRS" and the Director of LESIA (Laboratoire d'Etudes Spatiales et d'Instrumentation en Astrophysique) of the Paris Observatory, a laboratory associated with CNRS, the University Pierre et Marie Curie (UPMC), and the University Paris Diderot.

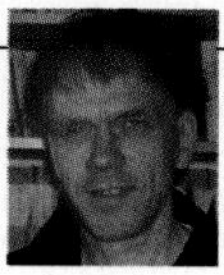

Axel Brandenburg

Chap. 2

NORDITA
Roslagstullsbacken 23
AlbaNova University Center
106 91 Stockholm, Sweden
brandenb@nordita.dk

Professor Axel Brandenburg received his PhD at the University of Helsinki in 1990. After two postdoctorate positions at Nordita in Copenhagen and at the High Altitude Observatory at the National Center for Atmospheric Research in Boulder/Colorado, he returned to Nordita as Assistant Professor for just over a year before he took up a position as full Professor of Applied Mathematics at the University of Newcastle upon Tyne. In 2000 he moved once again to Nordita, this time as full Professor. Axel Brandenburg is working in the field of astrophysical fluid dynamics and has recently also started working on topics of astrobiology. He is particularly interested in the question of magnetic field generation from turbulent motions with applications to the sun and stars, accretion discs, galaxies, and the early universe. His work on accretion disc turbulence was the first to show that magneto-rotational and dynamo instabilities lead to a sustained doubly-positive feedback. Recently, he contributed to clarifying the long-standing question of the suppression of the dynamo effect in generating large scale fields. He is also responsible for the maintenance of the PENCIL CODE, which is a public domain code for astrophysical fluid dynamics that is well suited for large clusters with distributed memory. In 2003 he was involved in magnetohydrodynamics simulations with 1024^3 meshpoints, which were the largest of its kind.

Peter J. Cargill

Space and Atmospheric Physics, The Blackett Laboratory
Imperial College
London SW7 2BW, UK
p.cargill@imperial.ac.uk

Chap. 5

Peter Cargill is a Professor of Physics in the Space and Atmospheric Group in the Department of Physics at Imperial College. He has worked on many areas of solar and space plasma physics in the last 30 years including solar coronal heating, solar flares, particle acceleration, CMEs, interplanetary and magnetospheric turbulence and magnetospheric boundary layers. Achievements of note include pioneering models of plasma flows in the solar corona, the first predictions of the observables associated with nanoflare heating, resolution of the puzzling temperatures seen on long-lived flares, some of the first numerical models of the kinetic processes at magnetospheric boundary layers and models of interplanetary CMEs. In addition he playes a leading role in the recent European Space Agency (ESA) decadal review (Cosmic Vision 2015–2025) through his chairmanship of the Solar System Working Group.

Abraham C.-L. Chian

National Institute
for Space Research (INPE) and
World Institute for Space Environment Research (WISER)
P. O. Box 515, São José dos Campos-SP
12227-010, Brazil
achian@dge.inpe.br

Chap. 1

Professor Abraham C.-L. Chian is a senior scientist at the National Institute for Space Research (INPE) in Brazil, where he is engaged in space plasma research. After graduating from Cornell University with a BSc degree with distinction in Electrical Engineering and Space Science, he obtained a PhD degree in Applied Mathematics and Theoretical Physics from The University of Cambridge and a PhD degree in Economics from The University of Adelaide. He has been active in a broad range of research areas including space plasma physics, plasma astrophysics, nonlinear dynamics, computational science, and economic dynamics. He has done pioneer work applying the chaos theory to study waves, instabilities and turbulence in space plasmas, with many innovative findings in Langmuir turbulence, Alfvén turbulence, drift wave turbulence, and solar and planetary radio emissions. In 2001, Professor Chian founded the World Institute for Space Environment Research (WISER), which operates as a network of centres of excellence dedicated to promoting international cooperation in research and training in multidisciplinary studies of the earth–ocean-space environment, solar-terrestrial relations and the impact of space weather and space climate on the Earth's weather, climate, environment and technology. He has been a guesteditor of various journals including Space Science Reviews, Journal of Atmospheric and Solar-Terrestrial Physics, Advances in Space Research, Nonlinear Processes in Geophysics, and Astrophysics and Space Science. He is a recipient of a Latin American award from the American Geophysical Union, a Marie Curie Senior Research Fellowship from the European Commission, and professorial fellowships from University of Adelaide and Nagoya University.

Louise K. Harra

Mullard Space Science Laboratory
University College London
Holmbury St Mary, Dorking, Surrey, RH5 6NT, UK
lkh@mssl.ucl.ac.uk

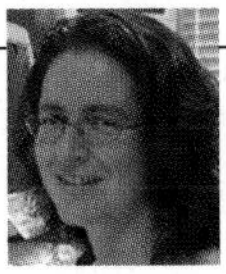

Chap. 5

Professor Louise K. Harra is Deputy Director of the Mullard Space Science Laboratory, UCL and director of research. She was awarded the Philip Leverhulme prize for her research. She is the PI for the EIS instrument on Hinode. She is the UK lead in the imager consortium for Solar Orbiter, and has been promoting the science of this mission through linking up the fields of solar physics and space plasma physics. She is an honorary Professor at the National Astronomical Observatory, Beijing.

Bengt Hultqvist

Swedish Institute of Space Physics
PO Box 812, 98128 Kiruna, Sweden
hultqv@irf.se

Chap. 13

Professor Bengt Hultqvist started his research in space physics at the beginning of the International Geophysical Year in 1957, when he became the director of the new Kiruna Geophysical Observatory in Sweden, an institution that later changed name to Kiruna Geophysical Institute (in 1973) and to the Swedish Institute of Space Physics (in 1987). He was the director of these institutions until his retirement in 1994. In 1995 he became one of the two first directors of the new International Space Science Institute (ISSI) in Bern, Switzerland. Since 2001 he has been Secretary General of the International Association of Geomagnetism and Aeronomy (IAGA). In the early 1960s he initiated the process of having the European Sounding Rocket Range located in Kiruna, he also initiated the Swedish national satellite programme, with Viking as the first satellite, and the European Incoherent Scatter Facility (EISCAT). His research has mainly dealt with the acceleration of charged particles in the magnetosphere and he has authored several books and edited a larger number, some of which are ISSI monographs. He is a member of half a dozen scientific academies, a Fellow of the American Geophysical Union and has received medals from COSPAR, EGU, Academy of Science in Moscow, the King of Sweden, and several Swedish academies and other institutions.

Yohsuke Kamide

Research Institute for Sustainable Humanosphere
Kyoto University
Uji 611-0011, Japan
kamide@rish.kyoto-u.ac.jp

Chaps. 1 and 14

After receiving the MS and PhD degrees from the University of Tokyo, Yohsuke Kamide worked at the University of Alaska, the University of Colorado, the NOAA National Geophysical Data Center, the National Center for Atmospheric Research, the NOAA Space Environment Laboratory, and Kyoto Sangyo University before settling at the Solar-Terrestrial Environment Laboratory of Nagoya University as Professor. He served as its director from 1999 to 2005. After retiring from the Solar-Terrestrial Environment Laboratory, he is currently a part-time research Professor of the Research Institute for Sustainable Humanosphere of Kyoto University. The author of more than 350 scientific papers and 15 books and a participant in a number of joint international research programs, he has made significant contributions to solar-terrestrial physics, including the role of the solar wind in dynamic processes in the magnetosphere, modeling of magnetosphere-ionosphere coupling, and storm-substorm relationships. His pioneering work proposing the so-called KRM method has become an important basis of the algorithm for mapping ionospheric parameters, which is commonly used in the scientific community as a standard tool in specifying the state of the Earth's electrodynamic environment. He has been editor of the Journal of Geophysical Research and Geophysical Research Letters, as well as guest editor and board member of various journals for over twenty years. In recognizing these contributions, the Royal Astronomical Society awarded him the Price Medal. He is also a Fellow of the American Geophysical Union and an Associate of the Royal Astronomical Society.

Susumu Kato

22-15 Fujimidai, Otsu 520-0846, Japan
kato@kurasc.kyoto-u.ac.jp

Chap. 9

Susumu Kato is a Professor Emeritus of Kyoto University. Since 1952 he has been working on the dynamics of the middle and upper atmosphere both along theoretical and experimental lines. He worked on the establishment of the classical tidal theory in the 1960s and the construction of the MU radar near Kyoto, the best facility for observing the middle and upper atmosphere dynamics, in 1984. He has received many honors: He was awarded the Royal Society Appleton Prize in 1987 and the Japanese Academy Prize in 1987. He is a Fellow of the AGU and a Foreign Associate of the U. S. National Academy of Engineering.

Margaret G. Kivelson

Institute of Geophysics and Planetary Physics and Department of Earth and Space Sciences
UCLA, Los Angeles, CA 90095-1567
mkivelson@igpp.ucla.edu

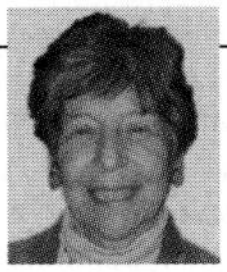

Chap. 19

Margaret G. Kivelson received her AB, AM and PhD degrees in physics from Radcliffe College of Harvard University. After working at the RAND Corporation for a decade she moved to UCLA where she is now Distinguished Professor of Space Physics. Kivelson's research uses theory and data analysis for the study of the magnetized plasmas of the Earth and other bodies of the solar system. Her special interests include the magnetic fields generated internally in the Galilean moons of Jupiter, transport and acceleration of particles in space plasmas, and the properties of ultralow frequency waves in the magnetosphere. The results of her research have been reported in some 300 research papers. She has contributed to the scientific teams for numerous spacecraft missions including the Galileo mission to Jupiter (on which she was Principal Investigator for the Magnetometer), the Cassini mission to Saturn, and the Cluster and Themis multi-spacecraft missions to investigate dynamic processes in the terrestrial magnetosphere. She is co-editor of Introduction to Space Physics, a widely used textbook on space physics. Recipient of a Guggenheim Fellowship, she is an elected fellow of the International Academy of Astronautics, the American Academy of Arts and Sciences, the National Academy of Sciences and the American Philosophical Society. She is a Fellow of the American Geophysical Union, the American Physical Society, and the American Association for the Advancement of Science. She was awarded the Alfvén Medal of the European Geophysical Union and the Fleming Medal of the American Geophysical Union in 2005. Throughout her career, she has actively worked for the advancement of women in science.

Karin Labitzke

Institute for Meteorology
Free University of Berlin, Germany
Carl-Heinrich-Becker-Weg 6–10, 12165 Berlin
karin.labitzke@mail.met.fu-berlin.de

Chap. 18

Professor Karin Labitzke studied meteorology and physics in Germany, in Berlin and Hamburg, and completed her PhD in 1962 in Berlin. Her thesis on stratospheric warmings was based on material of the IGY 1957/58. In 1963 she became Head of the Stratospheric Research Group at the Free University of Berlin where she taught many PhD students and carried out her research. She has made original contributions across a wide range of key issues, particularly connecting the troposphere with the stratosphere and the mesosphere. Her publication in 1987, connecting the 11-year sunspot cycle and quasi-biennial oscillation (QBO) in order to find a significant signal of the solar variability in the stratosphere was highly recognized. Together with Harry van Loon she has written a monograph about the stratosphere (Springer) and more than 20 publications on the Sun-QBOeffect. This work is ongoing.

Karin Labitzke extended her research over several visits to Japan, New Zealand and to the USA, where she was an Affiliate Scientist at the National Center for Atmospheric Research from 1980 to 2001. She was a member of the Advisory Board for the Government on Global Climate Issues. She participates in many international organisations like SCOSTEP, where she represents Germany, and IUGG and COSPAR. She was President of the German Meteorological Society from 1991 to 1995. In recognition of her contributions to meteorology, in 2005 she received the "Alfred-Wegener-Medaille" from the German Meteorological Society. She is married and has one daughter.

Louis J. Lanzerotti

Center for Solar-Terrestrial Research,
New Jersey Institute of Technology, Newark,
New Jersey 07102 USA
ljl@ADM.NJIT.EDU

Chap. 17

Louis J. Lanzerotti received his BS degree in engineering physics from the University of Illinois and AM and PhD degrees in physics from Harvard University. He retired from AT&T and Lucent Technologies Bell Laboratories after 37 years and in 2002 was appointed Distinguished Research Professor of Physics at the New Jersey Institute of Technology, Newark, New Jersey. Lanzerotti's principal research interests have included space plasmas, geophysics, and engineering problems related to the impacts of atmospheric and space processes and the space environment on space and terrestrial technologies. He has served as principal investigator or co-investigator for a number of United States interplanetary and planetary space missions and is currently a principal investigator on the NASA Radiation Belts Storm Probes mission. He has also conducted geophysical research in the Antarctic and the Arctic. He has co-authored one book, co-edited three books, and is an author of more than 500 refereed engineering and science papers. He is founding editor for Space Weather, The International Journal of Research and Applications, published by the American Geophysical Union. He has seven patents issued or filed. Lanzerotti has been elected a member of the National Academy of Engineering and of the International Academy of Astronautics. He is a Fellow of the Institute of Electrical and Electronics Engineers, the American Institute of Aeronautics and Astronautics, the American Geophysical Union, the American Physical Society, and the American Association for the Advancement of Science. He is the recipient of two NASA Distinguished Public Service Medals, the NASA Distinguished Scientific Achievement Medal, the COSPAR William Nordberg Medal, and the Antarctic Service Medal of the United States.

Yuri P. Maltsev* (deceased)

Polar Geophysical Institute, Apatity
Murmansk Region 184200, Russia

Chap. 14

Born on 12 October 1945 in Murmansk, Dr Yuri Maltsev entered the Physics Department of the Leningrad (now St. Petersburg) State University in 1962. In 1967 he graduated from the university and returned to his native country to work at the Polar Geophysical Institute in Apatity, Murmansk region. Going through Laboratory Assistant (1968–1970) and Junior Researcher (1971–1972), he obtained a PhD degree in 1973 for his thesis "The Electric Field and Current System of the Magnetospheric Substorm." He was then Senior Researcher from 1973 to 1990. His doctoral thesis "Disturbances in the Magnetosphere–Ionosphere System" was accepted in 1986. He was Head of the Theoretical Department of the Polar Geophysical Institute from 1990 to 2005 and Professor from 2004 to 2005. He authored more than 200 papers on various topics in the solar-terrestrial environment including solar wind–magnetosphere–ionosphere coupling, geomagnetic pulsations, geomagnetic disturbances, magnetic storms and substorms, magnetospheric current systems, plasma physics, auroral phenomena, and Schumann resonance. He was the founder of the International Annual Seminar "Physics of Auroral Phenomena": see http://www.pgi.kolasc.net.ru/ and a leader of the movement for "Perestroika", for liberal reforms and market economy in Russia. He also published four textbooks, including the "Handbook on Magnetospheric–Ionospheric Physics" (in Russian), Nauka Publishing House, St.-Petersburg (1993). On 3 June 2005 he tragically drowned on a fishing trip.

D. Asoka Mendis

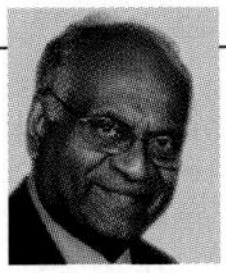

Department of Electrical Engineering
University of California, San Diego
La Jolla, CA 92093, USA
mendis@ece.ucsd.edu

Chap. 20

Professor Asoka Mendis obtained his bachelor degree in mathematics from the University of Ceylon in 1960 and his doctoral degrees in astrophysics (PhD, 1967; DSc, 1978) from the University of Manchester, England. Since 1969 he has pursued his academic career at the University of California, San Diego, where he is presently an emeritus Professor of applied physics, in the Department of Electrical and Computer Engineering. His main areas of research have been solar system physics, with emphasis on comets and planetary rings, and the physics of dusty plasmas. In the area of comets he has made major contributions to the physics of the cometary nucleus and the nature of solar wind interaction with the dusty cometary atmosphere. Moreover, he was an active participant in the early space missions to comets Giacobini-Zinner and Halley. In the area of planetary rings, he pioneered the gravito-electrodynamic study of electrically charged dust in planetary magnetospheres. His influential contribution to the physics of dusty plasmas, with its wide-ranging applications to both space and the laboratory, was recognized by the IEEE journal, Plasma Science, through the publication of a special issue devoted to the subject, to honor him on his 65th birthday in 2001. He is a member of several national and international scientific organizations including AGU, AAAS, IAU, COSPAR, and IAA.

Atsuhiro Nishida

The Graduate University for Advanced Studies (Sokendai)
Shonan Village, Hayama
Kanagawa 240-0193, Japan
hirosoph@dj.mbn.or.jp

Chap. 11

Professor Atsuhiro Nishida has been on the forefront of space research since he received his doctorate from the University of British Columbia in 1962. He spent most of his carrier at the Institute of Space and Astronautical Science (ISAS), Japan, and served as its Director General from 1996 to 2000. He has made original contributions across a wide range of key issues including plasmapause formation, magnetospheric convection, magnetic reconnection both on the magnetopause and in the magnetotail, and the energization of trapped particles. He wrote the monograph "Geomagnetic Diagnosis of the Magnetosphere" (Springer) in 1978 about his early work. In the 1980s he proposed and managed the Geotail satellite program to investigate the dynamics of the magnetotail; this collaborative program between ISAS and NASA played a leading role in the international magnetospheric research in 1990s. In recognition of his contributions he has received awards from the American Geophysical Union, the European Geophysical Society, the Russian Astronautics Association, the Japan Academy, and COSPAR. As of 2007 he is Executive Director of the Graduate University for Advanced Studies (Sokendai), which manages graduate education at about 20 inter-university research institutions in Japan, and President of the Asia Oceania Geosciences Society (http://www5.ocn.ne.jp/~anishida/).

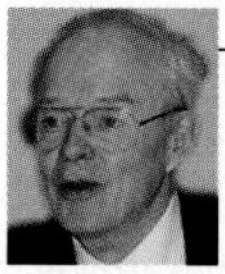

Eugene N. Parker

1323 Evergreen Road,
Homewood, IL 60430, USA
parker@odysseus.uchicago.edu

Chap. 4

Professor Eugene N. Parker studied physics at Michigan State University from 1944 to 1948 and then at Caltech for a PhD in 1951. He was an instructor in the Department of Mathematics at the University of Utah for two years and then Assistant Professor in the Department of Physics. In 1955 he accepted a position as Research Associate in the Cosmic Ray Group at the Institute for Nuclear Studies (now the Enrico Fermi Institute) at the University of Chicago. In 1957 he was appointed Assistant Professor in the Department of Physics and the Institute for Nuclear Studies. He became an Emeritus in 1995. He is widely known for developing the theory of the solar wind and for predicting the spiral shape of the solar magnetic field in the solar system, called the Parker spiral pattern. His work has increased understanding of the solar corona, the solar wind, and the magnetic fields even of Earth. His contributions to the field are well summarized in the citation for the 2003 James Clerk Maxwell Prize of the American Physical Society: "For seminal contributions in plasma astrophysics, including predicting the solar wind, explaining the solar dynamo, formulating the theory of magnetic reconnection, and the instability which predicts the escape of the magnetic fields from the galaxy." His books, especially "Cosmical Magnetic Fields" (Oxford), "Spontaneous Current Sheets in Magnetic Fields: With Applications to Stellar X-rays" (Oxford), and "Conversations on Electric and Magnetic Fields in the Cosmos" (Princeton) have stimulated generations of students, engineers, and researchers. In 1967, he was elected to the National Academy of Sciences. He has received a number of honors, including Hale Prize, John Adam Fleming Medal, William Bowie Medal, Chapman Medal, National Medal of Science, Karl Schwarzschild Medal, Bruce Medal, Gold Medal of the Royal Astronomical Society, and Kyoto Prize for Basic Science.

Monique Pick

LESIA
UMR 8109 (CNRS)
Observatoire de Paris
Observatoire de Paris, 92195 Meudon, France
monique.pick@obspm.fr

Chap. 6

Dr Monique Pick graduated from l'Ecole Supérieure de Physique et Chimie, Paris in 1957. She defended her PhD on the study of solar radio bursts and their association with solar energetic particles in 1961 . She started her career in the Laboratory for Radioastronomy (Observatoire de Paris), which was in charge of the Nançay Radioastronomy Station. She was the project scientist for the Nançay Radio Heliograph until 1985 and then became the head of this laboratory for eight years. She has worked in many areas of solar and interplanetary activity including flares and coronal mass ejections, particle acceleration, the solar-terrestrial relationship. She has been involved in several space missions such as SMM, ULYSSES, ACE, SOHO and STEREO. Dr Pick has been a member of various working groups: Solar System Working Group of ESA, a scientific representative in SCOSTEP, Vice President and President of IAU Commission 10. She has been responsible for various international symposia and workshops.

E.R. Priest

University of St. Andrews, Scotland
eric@mcs.st-andrews.ac.uk

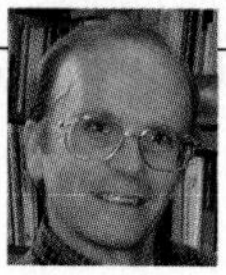

Chap. 3

Eric Priest is the James Gregory Professor of Mathematics at St. Andrews University and head of the Solar Theory Group there. He is a Fellow of the Royal Societies of Edinburgh and of London and is currently Vice-President of the Royal Astronomical Society. He has written a standard monograph on solar magnetohydrodynamics (which he is at present rewriting) and co-authored a book on magnetic reconnection with Terry Forbes. His main research interests are the MHD of the solar atmosphere, especially MHD instabilities and magnetic reconnection, and their roles in heating the Sun's corona and in solar flares.

Gordon Rostoker

Department of Physics
University of Alberta
Edmonton, Alberta, Canada, T6G 2J1
rostoker@space.ualberta.ca

Chap. 15

Gordon Rostoker is a Professor Emeritus of Physics at the University of Alberta. He received his PhD in geophysics (space physics) at the University of British Columbia in 1966. After spending a year an a half as a National Research Council of Canada Postdoctoral Fellow at the Royal Institute of Technology in Stockholm, Sweden, he returned to Canada joining the Faculty at the University of Alberta in 1968. Over the years, Dr Rostoker has served on several advisory boards for the National Research Council of Canada, the Natural Sciences and Engineering Research Council of Canada and the Canadian Space Agency. Over the period 1980–1986 he was Editor of the Canadian Journal of Physics. From 1989–1999, he was Principal Investigator of the Canadian Space Agency's CANOPUS program, which involved a ground based network of remote observatories distributed across western Canada that monitored auroral and magnetic disturbances. On the international scene, Dr Rostoker was Chairman of the Steering Committee of SCOSTEP's Solar-Terrestrial Energy Program from 1987 to 1997. He also served as Vice-President of the International Association for Geomagnetism and Aeronomy from 1995 to 1999. In 1998, he was elected as International Secretary of the American Geophysical Union and served in that capacity until the end of 2002. In 1979, Dr Rostoker was the recipient of the Steacie Prize, awarded each year to a researcher under the age of 40 who has made outstanding contributions to the natural sciences in Canada. In 1995, Dr Rostoker was elected to the Royal Society of Canada.

Michael Schulz

Space Physics Department (ADCS)
Lockheed Martin Advanced Technology Center
Palo Alto, California 94304 (USA)
mike.schulz@lmco.com

Chap. 7

Michael Schulz (BS, Michigan State University, 1964; PhD, Massachusetts Institute of Technology, 1967) is a group leader in the Space Physics Department at Lockheed Martin's Advanced Technology Center (Palo Alto, CA). From 1967 to 1969 he worked at Bell (Telephone) Laboratories (Murray Hill, NJ) and from 1969 to 1993 at The Aerospace Corporation (El Segundo, CA) before joining Lockheed Martin in 1993. He is co-author (with L.J. Lanzerotti) of the book Particle Diffusion in the Radiation Belts (Springer, 1974), a 1977 Fellow of the American Physical Society, winner of the 1982 Trustees' Distinguished Achievement Award (The Aerospace Corporation's highest honor), and a 2003 Fellow of the American Geophysical Union. His numerous contributions to magnetospheric and heliospheric physics include non-spherical source-surface models of planetary and solar magnetic fields, characterization of the heliospheric current sheet, and eigenfunction methods to treat magnetospheric charged-particle diffusion processes, as well as to calculate rigid rotation of coronal and heliospheric magnetic fields on a global scale, despite the Sun's differential (latitude-dependent) rotation rate at the photosphere. He has developed variational techniques to extract empirical values of particle-transport coefficients from observational data, as well as accurate analytical expressions for various characteristic frequencies, adiabatic invariants, and transport coefficients associated with charged-particle motion in model magnetospheres, for use in bounce-averaged guiding-center simulations. He has also developed accurate analytical expressions for eigenfrequencies of oscillating magnetic field lines (geomagnetic pulsations).

Padma Kant Shukla

Department of Physics
Umeå University
SE-90187 Umeå, Sweden
ps@tp4.rub.de

Chap. 12

Padma Kant Shukla received his PhD degree in physics from Banaras Hindu University, Varanasi, India in 1972, and his PhD degree from Umeå University, Umeå, Sweden in 1975. He holds professorships at the physics departments of four universities (Ruhr-University Bochum, Umeå University, Universidade Tecnica de Lisboa, and the University of Strathclyde), and is associated with the Max-Planck Institute for Extraterrestrial Physics and the Centre for Fundamental Physics at the Rutherford Appleton Laboratory. Professor Shukla is Editor of the Journal of Plasma Physics and the New Journal of Physics, Associate Editor of IEEE Transactions on Plasma Science, and an Editorial Board member of Plasma Physics and Controlled Fusion. He directs colleges and workshops on plasma physics at the Abdus Salam ICTP, Trieste. His main research interests include theoretical plasma physics, nonlinear physics, nonlinear structures in plasmas and fluids, and collective processes in space physics and astrophysics. Professor Shukla has been honored with the APS Nicholson Medal for Human Outreach, the Gay Lussak–Humboldt Prize from the French Ministry of Science and Education, and a Doctor Honoris Causa degree from the Russian Academy of Sciences. Professor Shukla is a Foreign Member of the Royal Swedish Academy of Sciences, a Fellow of the American Physical Society and a Fellow of the Institute of Physics (London).

Lennart Stenflo

Umeå University
Department of Physics
SE-90187 Umeå, Sweden
Lennart.Stenflo@physics.umu.se

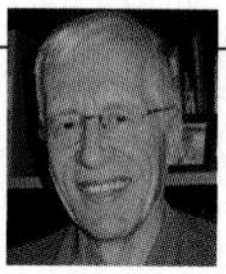

Chap. 12

Lennart Stenflo received his PhD degree in physics from Uppsala University, Sweden in 1968. Since 1971 he has been a Professor at Umeå University, Sweden. His research interests include plasma physics and space physics. In 1999 an international conference "Nonlinear Plasma Science; in honour of Professor Lennart Stenflo" was held in Faro, Portugal (Proceedings edited by P.K. Shukla, in the special issue T 82 (1999) of Physica Scripta). Later he was again honored with a conference in Trieste, Italy (Proceedings in Physica Scripta issue T 113 (2004)). Professor Stenflo is a Member of the Royal Swedish Academy of Sciences and its Nobel Committee for Physics. He is also a Foreign Member of the Russian Academy of Sciences.

Chanchal Uberoi

Indian Institute of Science
Department of Mathematics
Bangalore 560012, India
cuberoi@math.iisc.ernet.in

Chap. 10

Chanchal Uberoi was born in Quetta, Baluchistan. Despite many interruptions in her early education caused by the partition of India in 1947, she managed to obtain her Master's from the Osmania University, Hyderabad, and her Ph D, in 1965, from the Indian Institute of Science, Bangalore, where she came to spend the major part of her career as Professor in the Department of Mathematics and, later, as Dean of Science, the first woman to do so in the near-century history of the Institute. She continues her research as Professor Emeritus. Her important contributions on waves and instabilities in space and solid state plasmas have been published in more than 150 refereed scientific papers. Her pioneering work on the resonant absorption of Alfvén waves has found wide applications in space and laboratory plasmas. She was given the STC 1983 award by the Society for Technical Communications, Tennessee, for the best technical book, "The Alfven Wave", which she co-authored. Her interest in plasma education and popularizing space science can be seen from her books, "The Unmagnetized Plasma", "The Earth's Proximal Space" and a co-edited book published as part of the activity of the ISRO Educational Programme in Space Science and Technology. She has been a Leverhulme Trust Fellow at Cardiff, UK, the first Donald Menzel Fellow, Harvard Observatory, USA, and a Research Associate of the ICTP, Italy. She is a member of many prestigious academic societies, including the Advisory Committee of ICPP. She has played important roles in the cause of Women in Science in India.

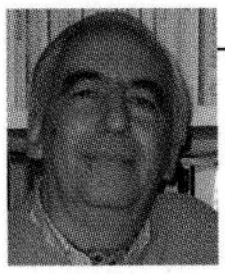

Umberto Villante

Physics Dept. University
Via Vetoio, 67100 L'Aquila, Italy
umberto.villante@aquila.infn.it

Chap. 16

Professor Umberto Villante has been involved in space and magnetospheric physics since he graduated in University of Rome "La Sapienza" (1970). He was a post-doctoral fellow at the Center for Space Research, Massachusetts Institute of Technology (1972–73) and he spent most of his academic career at the University of L'Aquila (Italy) where he served as Director of the Physics Department and Dean of the Science School. He has also served as Director of the Italian Group of Cosmic Physics and a member of the scientific councils of several scientific institutions. He has made original contributions across a wide spectrum of issues including the interplanetary magnetic field and the current sheet, the radial evolution of solar wind waves and discontinuities, the structure of the magnetosphere at long distances downstream from the Earth. He also dedicated his studies to the aspects of ULF waves in the magnetosphere with experimental activity both at low latitudes and in Antarctica. Presently, he is full Professor at the University of L'Aquila, Head of the Solar-Terrestrial Physics Group, Director of the "Area di Ricerca in Astrogeofisica", and a member of the Executive Committee of the World Institute for Space Environment Research. He is also a member of several scientific academies. Professor Villante has been Director of the International School of Space Science of the Consorzio Interuniversitario di Fisica Spaziale since its foundation.

Index

D

E

J

K

L

M

N

T